FUNDAMENTALS OF APPLIED DYNAMICS

James H. Williams, Jr.

School of Engineering
Professor of Teaching Excellence
Charles F. Hopewell Faculty Fellow
Professor of Applied Mechanics

Department of Mechanical Engineering
Massachusetts Institute of Technology
Cambridge, MA 02139, USA

JOHN WILEY & SONS, INC.
New York • Chichester • Toronto • Singapore

Acquisitions Editor Charitey Robey
Editorial Assistant Susanne Dwyer
Marketing Manager Debra Riegert
Senior Production Editor Cathy Ronda
Text Design Laura Nicholls
Manufacturing Manager Mark Cirello
Photo Researcher Elyse Rieder
Illustration Anna Melhorn
Photo Editor Lisa Passmore
Cover Photo Ronald Sheridan/Tony Stone Images, Inc.
Author Photo Michael Quan
Cover Design Lynn Rogan
Photo Credits Page 2: Chris Butler/AstroStock. Page 6: Courtesy
Art Reference Bureau. Page 7: Courtesy New York Public
Library. Page 8: Courtesy the British Museum. Page 10: Courtesy
of the Bildarchiv Foto Marburg. Page 17: Picture Perfect
Page 18: Courtesy New York Public Library. Page 21: Bettmann Archive
Page 26: Courtesy Columbia University Libraries.
Figure 1-6: Courtesy Lick Observatory.

This book was set in 9.5/11.5 ITC Century Light by Publication
Services and printed and bound by Donnelley/Crawfordsville. The
cover was printed by New England Book Company.

Recognizing the importance of preserving what has been written, it
is a policy of John Wiley & Sons, Inc. to have books of enduring value
published in the United States printed on acid-free paper, and we
exert our best efforts to that end.

The paper on this book was manufactured by a mill whose forest
management programs include sustained yield harvesting of its
timberlands. Sustained yield harvesting principles ensure that the
number of trees cut each year does not exceed the amount of new
growth.

Library of Congress Cataloging in Publication Data:
Williams, James. H., Jr.
  Fundamentals of applied dynamics / James H. Williams, Jr.
    p. cm.
  Includes index.
  ISBN 0-471-10937-1 (cloth : alk. paper)
  1. Dynamics. I. Title.
QA845.W68 1996
620.1'04–dc20

Formulation of Equations of Motion
Via Lagrange's Equations
for Holonomic Mechanical Systems

- Generalized Coordinates: $\quad \xi_j \qquad (j = 1, 2, \ldots, n)$

  Admissible Variations: $\quad \delta\xi_j \qquad (j = 1, 2, \ldots, n')$

  If Holonomic, then set $n' = n$ and proceed.

  (If Nonholonomic, seek alternative formulation.)

- Generalized Forces: $\qquad \delta\mathcal{W}^{nc} = \displaystyle\sum_{i=1}^{N} \boldsymbol{f}_i^{nc} \cdot \delta\boldsymbol{R}_i$

  $$= \sum_{j=1}^{n} \Xi_j \delta\xi_j$$

- Lagrangian: $\qquad \mathcal{L} = T^*(\xi_j, \dot{\xi}_j) - V(\xi_j)$

- Equation Derivation: $\quad \dfrac{d}{dt}\left(\dfrac{\partial\mathcal{L}}{\partial\dot{\xi}_j}\right) - \dfrac{\partial\mathcal{L}}{\partial\xi_j} = \Xi_j$

  $$(j = 1, 2, \ldots, n)$$

Hamilton's Principle for Mechanical Systems

$$\text{V.I.} = \int_{t_1}^{t_2}\left[\delta(T^* - V) + \sum_{j=1}^{n}\Xi\delta\xi_j\right]dt$$

Kinetic Coenergy T^* for Rigid Bodies
Undergoing Plane Motion[†]

	Kinetic Coenergy Function
Fixed-Axis Rotation About o	$T^* = \dfrac{1}{2}(I_{zx})_o \boldsymbol{\omega}_z^2$
General Plane Motion	$T^* = \dfrac{1}{2}M\boldsymbol{v}_c \cdot \boldsymbol{v}_c + \dfrac{1}{2}(I_{zx})_c\boldsymbol{\omega}_z^2$

[†]Equations are written on the assumption that motion occurs in the xy plane.

Units and Conversion Factors for Mechanical Quantities
Between U.S. Customary and SI Systems

Fundamental Quantities and Units

Quantity	U.S. Customary Units		SI Units	
	Unit	Symbol	Unit	Symbol
Mass	slug[1]	—	kilogram	kg
Length	foot	ft	meter	m
Time	second	sec	second	s
Force	pound	lb	newton[2]	N

(U.S. Customary: slug, foot, second, pound are Base units. SI: kilogram, meter, second are Base units.)

$^{1}\text{slug} = \dfrac{\text{lb} \cdot \text{sec}^2}{\text{ft}}$

$^{2}\text{N} = \dfrac{\text{kg} \cdot \text{m}}{\text{s}^2}$

Conversion Factors for Fundamental and Derived Quantities

Quantity	U.S. Customary to SI	SI to U.S. Customary
Length	1 in. = 25.40 mm	1 m = 39.37 in.
	1 ft = 0.3048 m	1 m = 3.281 ft
	1 mi = 1.609 km	1 km = 0.6214 mi
Area	1 in.^{2} = 645.2 mm^2	1 m^2 = 1550 in.^{2}
	1 ft^2 = 0.0929 m^2	1 m^2 = 10.76 ft^2
Volume	1 in.^{3} = 16.39(10^3) mm^3	1 mm^3 = 61.02(10^{-6}) in.^{3}
	1 ft^3 = 0.02832 m^3	1 m^3 = 35.31 ft^3
Velocity	1 in./sec = 0.0254 m/s	1 m/s = 39.37 in./sec
	1 ft/sec = 0.3048 m/s	1 m/s = 3.281 ft./sec
	1 mi/hr = 1.609 km/h	1 km/h = 0.6214 mi/hr
Acceleration	1 in./sec^2 = 0.0254 m/s^2	1 m/s^2 = 39.37 in./sec^2
Force	1 lb = 4.448 N	1 N = 0.2248 lb
Mass	1 slug = 14.59 kg	1 kg = 0.06854 slug
Pressure or stress	1 psi = 6.895 kPa	1 kPa = 0.1450 psi
	1 ksi = 6.895 MPa	1 MPa = 145.0 psi
Bending moment or torque	1 ft · lb = 1.356 N · m	1 N · m = 0.7376 ft · lb
Work or energy	1 ft · lb = 1.356 J	1 J = 0.7376 ft · lb
Power	1 ft · lb/sec = 1.356 W	1 W = 0.7376 ft · lb/sec
	1 hp = 745.7 W	1 kW = 1.341 hp

DEDICATION

To all the *ordinary* men and women everywhere
who seek affirmation of self and who in silence,
with dignity and pride, struggle to make a better
life for themselves and their children; I dedicate
this book.

ABOUT THE AUTHOR

James H. Williams, Jr. (Mechanical Designer, 1965—Newport News shipyard Apprentice School; S.B., 1967 and S.M., 1968—Massachusetts Institute of Technology; Ph.D., 1970—Trinity College, Cambridge University) is the School of Engineering Professor of Teaching Excellence, Charles F. Hopewell Faculty Fellow, and Professor of Applied Mechanics in the Mechanical Engineering Department at the Massachusetts Institute of Technology. He has received many awards and published numerous papers and reports in conjunction with his teaching, consulting, and research in design; the mechanical characterization of advanced fiber reinforced composites; wave propagation in large space structures; in-process and post-process quality control; reliability; dynamic fracture; nondestructive evaluation with emphasis on acoustic emission, thermal, and ultrasonic responses of composites; dynamic behavior of structures subjected to seismic excitation; and the development of computerized database systems for composite materials selection. If unavailable at his office, he can likely be found *attempting* to hit a 200-yard three-iron to an elevated green somewhere in the Boston area.

ACKNOWLEDGMENTS

In listing some of the individuals who have so substantially supported the writing of this book, I begin to wonder, just what did I contribute? Perhaps I am simply witnessing another verification that while thunder takes the credit it is lightning that does the work.

Stephen H. Crandall has been my colleague and mentor for the past quarter century. His influence on every aspect of my teaching and research has been both subtle and profound. I am grateful to him as a beacon and a *friend*.

Samson S. Lee, who, as an undergraduate, took my dynamics course and, later as an MIT Lecturer, became my eminent research collaborator, in the late 1970s and early 1980s began to teach dynamics with me. It was during that period that he and I initiated almost every thing that might be characterized as distinctive in this book. *No* assessment of his contributions to this book is likely to be excessive.

I began the formal writing of this book four years ago. As every writer, athlete, or performer of any major enterprise or competition knows, there is a period called *crunch time*. During crunch time—the final year—of this writing project, my former student and colleague as an MIT Lecturer, Hyunjune Yim, was on the scene to collect and reorganize materials, make calculations of new examples, and check calculations of old examples. The care, accuracy, and thoroughness with which he performed every request leave me enormously indebted to him.

In the physical construction of the manuscript, my graduate student Liang-Wu Cai conducted all the computerized graphics and most of the wordprocessing. Further, although most of the end-of-chapter problems and their solutions have been accumulated over the past twenty years, ultimately someone had to organize, compile, and extend the collection; Yim and Cai assisted me in that effort.

My colleague and friend Kenneth R. Manning devoted a month of his 1994 summer reviewing my historical references and *attempting* to rein me in and, in the process, established an insurmountable debt in my ledger. Also, I am deeply grateful to my colleague, Martin Bernal of Cornell University, who reviewed some of my historical writings and encouraged me to undertake a quantitative optics analysis of a longstanding bulwark in the attack on Ancient Egyptian astronomy (Section 1-11).

During the 1994 spring term, mechanical engineering associates Rohan C. Abeyaratne taught from a draft of my notes for this book and offered numerous suggestions, and Jean-Jacques E. Slotine reviewed my brief discussion of state-space stability analysis. In formulating my expressions of gratitude I recollect useful technical conversations with the following faculty colleagues: Roshan L. Aggarwal, Ernesto E. Blanco, Louis D. Braida, Martha L. Gray, Robert J. Hansman, Jr., Hermann A. Haus, James K. Roberge, Dirk J. Struik, Cardinal Warde, all of MIT, and John G. Papastavridis of the Georgia Institute of Technology and Clifford A. Truesdell III of the Johns Hopkins University.

In the delicate refinement of each of them, former advisors and colleagues Frank A. McClintock, through his advocacy of extensive use of appendixes, and Edward W. Parkes and Ernest Rabinowicz, through their embrace of *not cultivating one's garden too tightly*, strongly influenced this book.

In addition to Samson S. Lee, Hyunjune Yim, and Liang-Wu Cai, already mentioned, former MIT Teaching Assistants include Raymond J. Nagem, David M. Wootton, Ismail Kutan, Aitor Galdos, Shiva Ayyadurai, and Hyun Soek Yang. In one manner or another, each of these former students and/or colleagues contributed sample problems and ideas to this book. Several former MIT undergraduates contributed ideas—some used, some not—and among them Cynthia R. Estrella, Theodore T. Kim, Eric L. Scott, David B. Thaller and John S. Zaroulis stand out.

I am enormously pleased with the copyediting by Shelley Flannery and the quality publication efforts expended by the John Wiley & Sons staff as listed on page iii. I am also pleased with the excellent critical reviews by (alphabetically) Professors Anthony K. Amos (Pennsylvania State University), William B. Bickford (Arizona State University), Hossein Haj-Hariri (University of Virginia), Kathleen C. Howell (Purdue University), James M. Longuski (Purdue University), William J. Palm (University of Rhode Island), Bahram Ravani (University of California-Davis), and Slobodan R. Sipcic (Boston University). In one manner or another, the critical reviews by each of these academicians influenced the final manuscript, especially the comments and suggestions by Professors Haj-Hariri and Longuski.

One of the remarkable attributes of MIT is that the senior administration is so accessible to the faculty. At a very important stage of this writing project—the beginning—Charles M. Vest; President of MIT, injected a large dose of moral support into it. Further, in addition to being Professor of Applied Mechanics, I am the School of Engineering Professor of Teaching Excellence and Charles F. Hopewell Faculty Fellow (MacVicar Fellows Program). These honors and their attendant support provide compelling evidence of the patronage and encouragement for this enterprise from Charles M. Vest; Mark S. Wrighton, formerly Provost of MIT, now Chancellor of Washington University; Joel Moses, formerly Dean of Engineering, now Provost; and Nam P. Suh, Head of Mechanical Engineering.

Finally, from the very first word of this book to the very last word, Jette Marianne Hansen was faithfully and passionately with me.

I am filled with so much gratitude for the generous contributions of these individuals that there is no room in *any* part of my life for sorrow or regrets. I have been allowed to sing my song. Indeed, in the George Bernard Shaw sense,[1] my life is tragic in that my reality has exceeded my dreams.

James H. Williams, Jr.
Cambridge, Massachusetts
1995

---

[1]There are only two great tragedies in life: One is to fulfill none of your dreams; the other is to fulfill all of your dreams (with apologies to George Bernard Shaw, *Man and Superman*).

PREFACE

SPLASH! The only copy of the freshly edited version of several chapters of *the* fundamental dynamics textbook I am writing is thrust overboard into the Caribbean Sea, as the peñero nearly capsizes along the Archipielago Los Roques. Believing that there is a Force in the universe that is forever prepared to hurt us if we do not proceed calmly and take all things in stride, I relax. I withdraw. I reflect. I come to understand that there is no such entity as *the* book, only *a* book among all those that are the culminations of countless hours of moods and thoughts, among all those that do not splash into the Sea. Like so many before it, perhaps this loss too, if probed deliberately and deeply enough, will be almost worth the enlightenment.

As the manuscript strikes the surface of the water, almost instantly there is calm. Time stops, sounds cease, breaking waves hang in midair. My vision is telescoped onto the sinking manuscript; deeper and deeper it sinks into the crystal clear Caribbean Sea. I am inspired by the clarity of the water. Believing as I do that students find books on dynamics either (i) too difficult to read or/and (ii) so encumbered by a series of special cases that little scope is developed, my mission becomes clearer. As I edit my next draft, I must seek

- *Clarity* to quicken readers who will pore over my words in their independent efforts to acquire details;
- *Structure* to empower readers to approach with confidence new problems; and
- *Perspective* to embolden readers to acquire *a broad view of the world*, because without it they will forever remain incomplete.

Repetition will be used as a technique in support of each of these goals. Learning is volatile and requires a layering process. After one layer is allowed to set, but before it evaporates, another layer must be applied. And so it builds, layer upon layer, until the principles for a lifetime are established. I have been told that one can become an excellent golfer by developing "muscle memory" through repetition, though I have no personal evidence of this. On the other hand, I believe that superb engineering or scientific skills can be developed only through the process of repetition.

Clarity

Most authors seek clarity. As I use it here, clarity is intended to emphasize that this book is written *for* students. Students, whether taking a formal course or engaging in self-study, should expect to be able to read this book and quickly understand what they read, more so than they may have come to expect of many engineering textbooks. This does not eliminate the teacher but should elevate the level of student-teacher interaction. A textbook should be literally a *manual of instruction,* not simply a compilation of topics, often requiring intense engagement by its readers in order to fulfill their implicit agreement with the author.

No newly encountered concept is incipiently clear. Repetition serves the goal of clarity. Many questions students have asked over the years and my responses to them often drive both the character and the length of my presentation. Examples sometimes precede theory; and although it may not be obvious—that is, until you decide to read comparable material elsewhere—there is a deemphasis on formal mathematics in the body (nine chapters) of the book.

Throughout the book, examples will focus on the analysis of highly idealized, though very useful, models. In this way, emphasis on the principles of dynamics can be maintained. To a very modest extent, end-of-chapter problems require some modeling of physical conditions.

Structure

A good textbook must be more than clear; it must also be structured. Students, as do all analysts, need to feel secure in order to confidently confront and engage *new* problems. A textbook that fails to empower the reader in this manner has itself failed. The structure of a sound regimen is a vital aspect of such confidence. Structure will be found throughout the book, most notably associated with the *discipline* itself, the deliberate *style* of the presentation, and the *format* of the text, all as discussed below.

Discipline: (i) The two principal approaches to analysis in dynamics are drawn. In mechanical problems, these are called the *direct* (or force-momentum) *approach* and the *indirect* (or work-energy) *approach.* (ii) Regardless of the approach used, three fundamental requirements must be satisfied. These points are the focus of Chapter 2 (Eq. (2-1)). (iii) Within the indirect approach, four steps to equation formulation are delineated. This point is emphasized in Chapter 5 (Table 5-1). These disciplinary aspects of structure are associated more with *formulation* of the equations of motion as opposed to the *solution* of the equations of motion.

Style of Presentation: The repetition that serves clarity also serves structure. The unorthodox style of repeating phraseology in deriving equations of motion for different systems is intentional and without apology. In so doing, I seek to emphasize a consistency of thought and technique. I want to reinforce a mode of thought as much as I want to deliver the results; indeed, a major part of the "results" is the *mode of thought* that I seek to inculcate.

Format: (i) The disciplinary structure outlined above is exploited in numerous examples. The principles are comparably few and easy to articulate; the applications are exceedingly broad and rich. The purpose of the examples is to expose the variety of nuances that arise time after time in a broad range of models. The examples are highly structured in a consistent regimen of steps, becoming increasingly difficult as the reader proceeds. (ii) Summary tables, often in the form of flowcharts, are provided throughout the book. The flowcharts that appear primarily in the *solutions*-oriented—as opposed to the *formulation*-oriented—chapters are provided to reinforce the appropriate structure. (iii) The eleven appendixes serve several purposes, including providing (1) mathematical proofs that are not new but not readily accessible in elementary or intermediate textbooks—Appendix A is of this kind; (2) relevant material that is new but extends beyond the goals or level of the body of the book—Appendix B is of this kind; (3) material that simply shifts many detailed and repetitive mathematical manipulations away from the chapters of the book—Appendixes D,

H, I, J, and K are of this kind; and (4) theoretical foundations, sometimes opinionated and unorthodox in their presentation, which are omitted from the body of the book and which represent the underpinnings of the discipline—Appendixes C, E, F, and G are of this kind. Therefore, these appendixes provide underlying theoretical and mathematical support for the body, thus streamlining the main presentation and enhancing its structure. Furthermore, these appendixes will allow the teaching of the subject in a more theoretical format, particularly if the book is used for an introductory (post)graduate course. In an initial exposure to this material, few of the appendixes will be necessary, particularly for the undergraduate who will likely find much of the material to be beyond that which is urgent.

Perspective

The perspective I seek to enhance is an appreciation for the history of dynamics as well as a framework for future learning within this or any discipline. Historical references throughout the book serve this goal. Also, I am reaching out to students who will study dynamics in the future as well as those who will not. In some respects, I want students to view me as a somewhat knowledgeable colleague, but as a colleague nonetheless. I am seeking to generate interest in potential applications, intellectual interest in the underlying concepts, and the ability to read comparable and more advanced literature. Some of the materials—especially several of the appendixes—which are not used substantially in the book serve these goals of perspective for future study. Ultimately, as the book's title suggests, I am trying to probe and expose the underlying *fundamentals* of applied dynamics. There is every attempt not to "sweep under the rug" or circumvent any of the subtleties of the subject. I have seen enough obscurity in textbooks to last a lifetime.

Chapter 1 is my own unorthodox and personalized historical perspective. It is at once as obvious to me as much as it may be debatable to others. It is virtually impossible to write anything on the history of science without incurring the ire of one or more "professionals." The history of any subject is invariably richer than any categorization or sequencing of its features can reveal. Here, I want to raise and refresh the spirit of several of the great contributors to the discipline of dynamics. My purpose is to provoke and to encourage readers who care about such matters to read broadly and thoughtfully and then to formulate their own perspectives. Furthermore, very few, if any, ideas, theories, or inventions can be cited as having sprung exclusively from the mind of a single individual. The names of great individuals as cited here serve as mileposts in the evolution of civilization as much as for purposes of attribution and are not attempts at deification. The concept of the illuminating light bulb in the head of an isolated genius is frequently and substantially a fabrication—a convenience for various mythologies and comic strips, but little else. I recommend that professors who use this book in their teaching of dynamics *not* assign Chapter 1 as required reading since students will read it on their own. . . . Hold it; forget what you just read! I recommend that professors who use this book in their teaching of dynamics assign Chapter 1 as required reading, or perhaps just for skimming. Its contents may not be fully digested in the first reading. Some students may come to regard it like a symphony orchestra in the city, which, though rarely attended, provides comfort simply in the knowledge that it's there; a journey available, but nevertheless a journey declined. Other students may return to it—including its bibliography—as a useful reference, even beyond the purview of dynamics.

Chapter 2 emphasizes two major points: (i) Whether obvious or obscure, the ultimate technical, as opposed to philosophical, goal of all our analyses is *engineering design*; and (ii) the formulation of the equations of motion for mechanical systems necessitates the satisfaction of three fundamental requirements—*kinematical, force-dynamical,* and *constitutive* in character. A third major topic is skirted, namely, the quantitative modeling of engineering systems. An early version of this book dealt with modeling to a much greater degree than here; it was cut due to length restraints. Ultimately, I concluded that the goal of emphasizing the *fundamentals of applied dynamics* would be diluted too much by the inclusion of the topic of quantitative modeling.

I want students to understand that throughout this book the goal of the analyses is *design*. We are clearly investigating dynamics, but we want to maintain a cognizance of design. Here's what I mean.

In most courses in *dynamics*, the results of the analyses are the *motions* of bodies or the *force* accompanying those motions or simply the *equations of motion*. Indeed, these three types of "results" are the respective cores of Chapters 3, 4, and 5; and if students fully appreciate these three aspects of dynamics, they will be well along in their understanding of the scope of the discipline. However, a course in dynamics that terminates there and leaves students with the impression that these are *the* results has not fulfilled its potential. Equations of motion are obviously intermediate results, but frequently, so too are the motions and forces obtained from those equations. To be of practical engineering value, such forces and motions must be related to some design criteria or specifications, including, for example, stresses and strains that should be assessed to be acceptable or unacceptable. In this regard, important design features including the geometry and materials properties of components can be isolated to assess their impact on stresses and strains in individual members. By the term *design,* I intend to emphasize this point. While detailed design goals cannot be explored here, if students who use this book realize that the calculated motions and forces often represent intermediate results leading to a stress analysis, materials selection, or (and) the satisfaction of a manufacturing or operating specification, then my exhortations on this matter will have been adequately rewarded.

Chapter 3 is a presentation of three-dimensional kinematics, *kinematics* being the study of the geometry of motion without regard for forces. Chapter 3 generalizes kinematics in a manner that I believe the reader will find to be unusual in its approach and efficacy as well as highly accessible. Chapter 3 (and its cited appendixes, which I do not recommend during an initial encounter) will richly reward all the careful attention the reader chooses to devote to it.

Chapter 4 formally introduces the *direct approach,* variously called *vectorial mechanics* or the *momentum approach.* With Newton's laws as the foundation, several "kinds" of dynamical problems are defined. Then, following the reemphasis of the three fundamental requirements (kinematical, force-dynamical, and constitutive), several examples reveal the structure and simplicity of many dynamical formulations. One feature of the flexibility of the book is that the reader (or instructor) may spend a week or two—as I generally do—or month(s) on this chapter, or skip it altogether without loss of continuity. Indeed—*and this is a very important point*—colleagues who choose to build a course substantially around momentum principles of the direct approach will find a large number of end-of-chapter problems that can serve as lecture examples, suitable for a broad range of instructional goals.

This textbook is very detailed in its presentation of the work-energy approach (indirect approach). This choice of emphasis is not because students fully understand the momentum principles that they have previously encountered in their physics courses. After all, as I rediscover every time I teach dynamics, a mature understanding of momentum principles is a lifelong pursuit. It's simply that to repeat the principles of dynamics in the same style with many of the same problems as previously encountered is a reliable prescription for boring most students, especially the more brilliant ones, particularly those who regularly watch MTV. My instructional philosophy is to remind students that they have already studied momentum principles, and as I revisit such principles (which they too must repeat along with me) I want to broaden the disciplinary and philosophical context of those familiar ideas to include and compare work-energy principles, both approaches in some semblance of their respective proper historical origins.

On a separate note regarding Chapter 4, in problems where there is a constant force, there will be a constant acceleration; the model of gravity in the vicinity of the Earth's surface is the most frequently cited example of this kind. Such problems are frequently treated exclusively via kinematics, which is not strictly correct, notwithstanding Galileo Galilei. Accelerations imply the presence of forces, requiring some constitutive relation that connects the two. An appreciation of this requirement between forces and accelerations is important in understanding the underlying structure of dynamical formulations, and is thus important in reinforcing the strong analogy between the mechanics of solids and dynamics which I cite.

Chapter 5 formally introduces the *indirect approach*, variously called *lagrangian dynamics* or *variational dynamics*. The presentation is particularly deliberate. The frequently maligned and misunderstood concepts of energy and coenergy are introduced and delineated. Here we adopt Hamilton's principle as our fundamental tenet, and Lagrange's equations as a derivable corollary that is useful for lumped-parameter systems. Table 5-1 summarizes the structure of the formulation of the equations of motion for lumped-parameter systems. The four-step procedure in Table 5-1 is used repeatedly as it will unquestionably become a part of the student's analytical lexicon. With the approach exposed in Table 5-1, only the details of its application remain. Each of the examples that follow Table 5-1 contains one or more nuances that expose the details of the application of the lagrangian formulation. It is important to appreciate each of these subtleties, as indeed these are the fine points that will be repeatedly utilized in the end-of-chapter problems, and ultimately in professional practice.

Chapter 6 expands both the momentum formulation of Chapter 4 and the lagrangian formulation of Chapter 5 to *extended bodies*, but with an emphasis on the lagrangian approach. Prior to the principal business of the lagrangian formulation, a summary of momentum principles for particles and extended bodies is given in Table 6-1. This table and the accompanying analyses lay out clearly (i) the relationship between linear momentum and angular momentum for particles, (ii) the relationship between linear momentum and angular momentum for extended bodies, and (iii) the independence of linear momentum and angular momentum for extended bodies. The concept of a rigid body is defined. Again, as in Chapter 4, colleagues who choose to emphasize momentum principles will find a large number of end-of-chapter problems that can serve as lecture examples, suitable for a broad range of instructional goals. In extending the momentum and lagrangian formulations for particles to the momentum and lagrangian formulations for rigid bodies, the very important *inertia tensor* is derived, as are the *kinetic coenergy function* and the *potential energy function*. Several nuance-revealing examples complete the chapter.

Chapter 7 presents a lagrangian formulation of lumped-parameter electrical and electromechanical devices. In an accompanying appendix (Appendix G), Maxwell's equations are used to derive the required energy functions, as well as Kirchhoff's rules, which are used in the lagrangian formulation of such systems. To my knowledge, equations of motion for operational amplifiers are formulated via this approach for the first time. (I have foregone the introduction of transistors via this approach.) An illuminating summary table (Table 7-2: $\mathcal{LOVE}$) is provided for electrical networks; the structure of the table should figuratively catapult itself toward the reader. (My *philosophical* view is that love is an imitative neural configuration in an individual as when he or she is eating fine chocolate.)

Chapter 8 contains the response analyses of one- and two-degree-of-freedom systems. Concepts of stability are introduced. Although the role of complex analysis is introduced, the chapter can be completed with little encounter of complex variables. For example, in obtaining the response of such systems to harmonic excitation, the particular solutions are obtained via complex variables as an alternative to working exclusively with real variables. Such a parallel presentation is designed to allay some of the discomfort to which students surrender when encountering the eerie little $i = \sqrt{-1}$.

Chapter 9 is a detailed presentation of the formulation of equations of motion for one-dimensional continuum models. In addition to the continuing emphasis on fundamentals and underlying disciplinary structure, the reader will find a renewed display of detail in both the formulation and the solution phases of the analyses. These details—although sometimes complicated and nearly always omitted in textbooks—are exhibited because they too possess a structure I hope will be revealed. Readers who think the presentation is too detailed should simply skip it; readers who want to learn about the formulation, solution, and the disciplinary structure of continuum analyses will find them introduced here.

It has been said variously that *to teach is to choose*. The number of topics that were drafted but not inducted is significant. For example, among these topics are the following: three-dimensional rigid body dynamics, especially Euler angles and related material on gyroscopes; orbital dynamics; nonholonomic systems; variational exposition on transistors; electric motors and generators; dynamic reciprocity; Rayleigh's principle; extended state-space analysis; and Fourier analysis and related spectral concepts. In omitting these topics, I was able to hold to the length of the volume in your hands, and still not compromise my expressed goals.

Several points may be made regarding the administrative aspects of the book: (i) The designations of the measures of the book, in decreasing order, are chapters, sections, subsections, and subdivisions; (ii) concerning units, both the metric system and the British or U.S. customary system are used at will; and (iii) although a comprehensive bibliography is given in the book, in general, references within the text are held to a minimum by the mere expedient of ignoring them.

The book is constructed to appeal to students—undergraduate and (post)graduate—and practicing engineers who seek (i) an accessible comprehensive introduction to applied dynamics and/or (ii) a coherent perspective of the theoretical underpinnings of the discipline, particularly via extensive use of the appendixes. It is the use of the appendixes, the choice of homework problems, and the pace that will distinguish initial and intermediate-level courses.

The prerequisites for a course using the book are quite modest. They include a first course in differential and integral calculus, including vector analysis, and a first course in physics in which calculus is used. In the United States, these are typically taken in the freshman year in most engineering and science curricula. Also, first courses in differential equations and the mechanics of solids are recommended, although such courses may be studied contemporaneously.

The book can be used for a one-quarter introductory course (Chapters 1 through 4 or 5; or Chapters 1 through 4 plus the direct approach in Chapter 6); a one-semester course (Chapters 1 through 6 plus dabs of one or more of the other chapters); or a full-year course (Chapters 1 through 9). Thus, as suggested by the one-quarter course, the book can be used exclusively for the direct approach in a first quarter, the indirect approach in a second quarter, and vibration in a third quarter. Further, with the appropriate omissions, an intermediate-level course might cover essentially the entire book in a semester.

A potential introductory use of the book could entail skimming Chapter 1; covering all of Chapter 2, with emphasis on Examples 2-1 and 2-2; then, in succession, covering Chapter 3 (except Section 3-3) through Example 3-8, Chapter 4 through Example 4-5, and Chapter 5 through Example 5-20. Such a course would provide an *elementary yet broad* treatment of particle dynamics. By extending the coverage modestly into Chapter 6, plane rigid body dynamics could be incorporated into the course. Such a course would accomplish, at least, two important goals: (1) students will have acquired a broad but not overwhelming perspective of the subject and (2) students will have been well prepared to read most of the remainder of the book with little or no assistance.

End-of-chapter problems are cast in three categories. The first category (Category I: Back to Physics), consisting of approximately one-sixth of the problems, is of an elementary nature. In chapters containing kinematics (Chapter 3), momentum principles (Chapters 4 and 6), electrical networks (Chapter 7), and vibration (Chapter 8), students will find these problems comparable to those already encountered in freshman physics. Indeed, these problems have been extracted directly from the superb elementary physics textbook by Halliday, Resnick, and Walker (HRW), also published by John Wiley & Sons, Inc. It is these problems that should reinforce my cited level of entry for this textbook as first-level undergraduate. In HRW, readers will find alternative and often enlightening discussions of many of the topics presented in this textbook. Readers are strongly encouraged to return to HRW or any physics textbook for such enhancing discussions. The second category (Category II: Intermediate), consisting of approximately one-half of the problems, begins at the level of the typical engineering sophomore; such problems are covered in books listed in Section 10-4 of the bibliography. These problems further underscore the undergraduate level of entry for this textbook. The problems in Category II, however, become increasingly difficult and subtle, and it is their solutions that should make clearer both the breadth and the efficacy of the approaches emphasized here. That is—*and this is important*—problems that may be familiar to the reader will be addressed distinctly differently here, in accordance with the structural paradigm I advocate. The third category (Category III: More Difficult), comprising the remainder of the problems, consists of exercises that are generally more difficult than those in Category II.

Clearly, categories II and III—indeed all three categories—could have been mingled; they are in Chapter 7. There is an instructional pedagogy embedded in providing these categories that is somewhat counterbalanced by an intellectual drawback; after all, the "real world" does not present itself to us so neatly categorized. To teach is to choose.

The path through the subject that I cut is both a subtle and personalized one. It is a path that explores both elementary and intermediate-level concepts in dynamics, and which is directed to the uninitiated readers for whom this book may represent their sole professional encounter with the discipline, as well as to readers who seek disciplinary foundations for advanced study. Essentially every mathematical manipulation can be constructed by visual reference only; the remainder may require extremely modest operations. Keep in mind that this is neither dynamics for poets nor a formal treatise; it is a dynamics textbook—a manual of instruction.

I take seriously the statement often attributed to Albert Einstein that "Everything should be made as simple as possible, but no simpler," adding here that *most things should also be repeated*. When asked "How long should a man's legs be?" Abraham Lincoln is reputed to have responded "Long enough to reach the ground." In writing several of the sections that are indeed longer than the reader is likely to find in comparable material in other textbooks, I sought only to make the concepts as simple as possible but long enough—with enough detail—so that students, in completing the material, would feel that their feet were on solid ground. They should have acquired *clarity* of my exposition, a sense of its formal *structure*, and a *perspective* of both its historical development and its potential application.

Learning should be fun. Enjoy.

CONTENTS

BIBLIOGRAPHY   684

APPENDIX A   FINITE ROTATION   688

APPENDIX B   GENERAL KINEMATIC ANALYSIS   694

APPENDIX C   MOMENTUM PRINCIPLES FOR SYSTEMS OF PARTICLES   705

CHAPTER ONE

OUR NICHE IN THE COSMOS

1-1   INTRODUCTION[1]

Dynamics is the most important intellectual discipline in the history of humankind; it is dynamics that has enabled humanity to find our niche in the cosmos.

The history of a subject is most easily appreciated and understood only after one has acquired some understanding of the subject, its terminology, and its methodology; not before. On the other hand, one should not go crabbing without some idea of what a crab is. So, before launching into the quantitative details of our subject, it may serve our broader intellect to consider the origins, the development, and the relevance of dynamics to the cultural issue of who we are as a species. We therefore present a brief and slightly unorthodox retrospective of dynamics. In this history we hope both to support a broader perspective of the origins of the discipline and to encourage an expanded view of its pertinence to our intellectual evolution.

Dynamics is a scientific discipline that encompasses studies of systems undergoing changes of state. The origins of dynamics lay in the earliest quizzical musings of humankind—as people gazed skyward and wondered about the changing patterns of celestial bodies, or peered across a plain and wondered how fast they must run to escape approaching danger or how hard they must strike to fell their food. The quantification of such questions may have been initiated when, securing the abode each night, they wondered "how many" meant that all were present. Who or what caused the Sun to rise, traverse the sky, and set; the rivers to flow; and the rain to fall? Thus, change has been the omnipresent lodger in the grainy substance which constitutes humankind's awareness.

Since 1929, when Edwin Hubble (1889–1953) concluded that the galaxies are all rushing away from one another, we have been compelled to believe that the universe is an evolving rather than static entity. It is a statute of the present scientific canon that time, energy, matter, and space began about fifteen billion years ago in an epochal singularity commonly called the *Big Bang*. This, of course, is only a model that, like *all* models, is limited and subject to change, but it does provide a means for revealing potentially interesting consequences of who we are as a species. Indeed, it is a model that may be judged by our progeny to be a bit of quaint nonsense held by many of those—but by no means all—who resided around the

---

[1]Read the preface if you have not already done so.

Artistic depiction of Big Bang.

twentieth century. According to the big bang hypothesis, the early universe had no carbon or oxygen. So these elements, as well as every iron atom in every blood cell of every human being, having been manufactured in the depths of stars and spewed out by supernovae, physically link us with the heavens. Minuscule changes in the forces of the early universe would have produced a very different cosmos—it would have flown apart or collapsed back onto itself—never to have spawned the consciousness of humankind. Have we been written into nature and was the universe designed to produce us—as the so-called *anthropic principle* suggests; or are we the result of untold numbers of big bangs and many billions of failed experiments? Does humankind have a central role to play or is our species, as *scientific reductionism* would suggest, a product of impersonal interactions of forces, a product simply awaiting either the cold following an unceasing universal expansion or the heat due to an annulling cosmic collapse?

More than any other scientific discipline, dynamics has brought us an appreciation of the depths of these questions and has already delivered some of the answers. Dynamics has caused philosophers to alter their perspectives. Dynamics has caused clerics to alter their messages. It is only in retrospect that we know that dynamics, while extremely rich in its consequences and applications, is based upon only a few principles. Yet the intellectual conquest of those principles has commanded the undertakings of many of history's greatest minds, and much of our success is historically rather recent.

It has been stated variously that to teach is to choose. Rather than endeavoring to write a history that is detailed and complete, we have chosen to write a brief retrospective that is factual and, importantly, once started, inevitable in marching along its path toward our present mindset. There are, perhaps, far too many historical omissions here, but having selected the invention of writing as our entry, we never had to jump-start this retrospective; the record flowed, it was logical, it developed with continuity.

The entire structure of technological and scientific progress can be imagined as an interconnected lattice of advances, where the initiation points of one generation or era are the culminations of all the progress of prior engagements. Western recorded history began with the Ancient Egyptians and the Mesopotamians. To us, this is an unchallenged fact. To conclude that our technological and scientific history began at a later period, after the marvelous contributions of these two great civilizations, is to suggest an enigmatic logic that is truly impossible to defend.

1-2 WHY HISTORY?

The history of a scientific discipline is important in developing a due sense of respect for our predecessors as well as a sense of perspective of our own relationship to the evolving tapestry of civilization. The history of a technical discipline is at once a part of both the global history of humankind and the local history of the individual. No matter how brief or shallow initially, a historical retrospective of one's scientific discipline opens a door that never again can be completely closed. No longer can one be totally comfortable studying a discipline without also desiring a clearer view of and a respect for those whose accomplishments have led the way toward our current foothold. Also, the thoughtful retrospective must acknowledge that no step is more important than the first.

In our brief retrospective, it is worth noting that we shall be concerned only with history—distinguished from very useful archaeological reconstructions and prehistory by the existence of *written* documents, physical scripted records chronicling some degree of human consciousness. Extant physical objects conjoined with such documents are within the purview of our considerations. By this approach, we go back to the invention of writing in Ancient Egypt and Mesopotamia. These societies not only represent the historical beginning of our own civilization; the history of these two long-enduring great coeval societies spans more than half of the recorded histories of all mankind. Ancient Egypt and Mesopotamia can be viewed as nothing short of the cultural bedrock of western civilization.

1-3 IMPORTANCE OF MATHEMATICS IN THE DEVELOPMENT OF MECHANICS[2]

It is a philosophical assumption, even mystical in quality, that the character and functioning of the universe can be penetrated and understood by human beings. The initial uses of mathematics to dispel the apparent arbitrariness of nature were little more than acts of faith. To those who first recognized that mathematics—probably initially through primitive notions of number, magnitude, form, and temporal order—could assist in the understanding of qualitatively diverse objects and phenomena, we owe the initiation of a magnificent succession of intellectual constructs of quantitative models of the universe.

Most branches of mathematics have been invented and developed to study one or more classes of problems. It is conceivable that prehistoric humans created the counting rudiments of arithmetic to keep track of mates and children, or simple possessions. The administration of the affairs of state of Ancient Egypt and Mesopotamia required a substantial knowledge of arithmetic, geometry, and elementary algebra. Plane geometry was needed to compute perimeters and areas of land for planting and tax purposes, and solid geometry originated for finding volumes, perhaps for reservoirs and equitable irrigational water distribution. Trigonometry was invented and developed by builders of major ancient monuments and by astronomers to compute distances, and much later (seventeenth century) calculus was invented to study the motions of bodies.

Furthermore, mathematics cannot be considered in the absence of astronomy. The needs for irrigation, flood control, and navigation, as well as for intellectual inquiry, were

---

[2]*Mechanics*, the science of forces and motions of bodies, is often used to denote the disciplines of *statics* when the body is motionless and *dynamics* when the body is moving, both solids and fluids being encompassed by these delineations. In this book, we shall be concerned primarily with the dynamics of solids.

met by astronomy, which in turn depended on mathematics for its quantification. And, as we shall see, it was astronomy that provided a major focus and conquest for Isaac Newton's unifying dynamics. Without regard for the history of astronomy and mathematics, no unqualified appreciation for the history of dynamics can ensue.

A thoughtful person cannot help but marvel at the fact that the same set of numbers used to count automobiles also can be used to count apples and stars. Why should this be so? It is no less remarkable that the same engineering models enable us to calculate the rustling motion of a leaf in a breeze, the vibration of the wing of a modern passenger airplane, the shaking of a structure during an earthquake, the ride quality of a passenger in an automobile, or the response of a lunar module during landing. And yet, in the absence of other concepts, counting anything tells one very little. It is those additional concepts that define a profession such as engineering or a discipline such as dynamics, or any discipline. It is through such concepts, with mathematics as the linchpin, that engineers have exhibited creativity for predicting a broad range of evolving natural phenomena and for building a technological society.

On the other hand, mathematics owes much of its development to mechanics. In many instances, mechanics and mathematics are inseparable. Indeed, examples from Newton's explanations of calculus, which he called the "theory of fluxions," clearly show that he thought of his derivatives primarily as velocities (though more generally as flowing quantities). Many of the world's greatest mathematicians of the seventeenth (e.g., Newton and Leibniz), eighteenth (e.g., Euler and Lagrange), and nineteenth (e.g., Gauss and Hamilton) centuries developed their mathematics in search of solutions to mechanics problems. More than any other characteristic, it is in being mathematical that the mechanics of the past few centuries has progressed and distinguished itself from the mechanics in all the previous millennia.

1-4   OUR SOURCES FROM ANTIQUITY: GETTING THE MESSAGE FROM THERE TO HERE

In presenting a retrospective of any quantitative subject, it is always a choice of considerable interest as to where to begin. No less a choice was encountered by us. To begin in the seventeenth century with dynamics, as is often done, appears to be somewhat contestable. Why not begin at the beginning—with the invention of writing—and use written documentation and related physical artifacts as our sources?

Estimates of the initial formations of nomadic bands range in the several tens of thousands of years ago. The conventional view is that the first *urban cultures* arose about seven to nine thousand years ago following developments in husbandry and domestication of animals. (A few authors date early *urban* Nile valley cultures to about 18,000 B.C., though this is unconventional.) Within the basins of several great river valleys, the needs for cooperation led to the organization of the labor force, the society, and ultimately the political structures.

Along the Nile, the intertwining societies of Nubia and Ancient (Upper) Egypt had emerged by the fifth millennium B.C. Nubia—an oval region from near Aswan in the north to near Khartoum in the south—is called Cush in biblical, Assyrian, and Egyptian literature. Along the Tigris and Euphrates, peasant farming villages organized themselves to harness the waters flowing into their plain. By the fourth millennium B.C. an irrigation system of dams and canals was being constructed in Mesopotamia (presently Iraq, and extending into Iran). Along the Indus (presently Pakistan), two large cities called Mohenjo-Daro and Harappa and

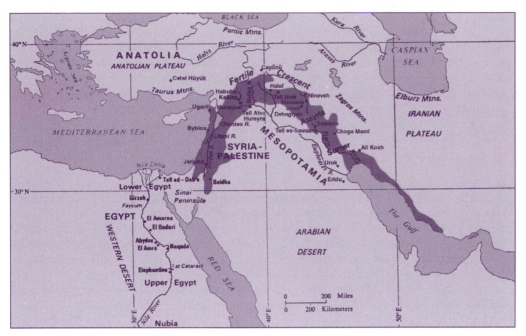

Ancient Western Asia and Northeast Africa showing Egypt and Mesopotamia.

dozens of smaller towns were established by about 3000 B.C.; but by about 1500 B.C. had not continued to thrive. Along the Yellow River in China by about 1700 B.C., invaders from inner Asia had led the building of cities with the emergence of social structures and superb craftsmen; but unification of China lay several centuries ahead. In terms of their influence on the rise of western civilization, the societies of Ancient Egypt and Mesopotamia were seminal; the Indus and Yellow riverine civilizations considerably less so. It was in Ancient Egypt and Mesopotamia that writing was invented.

1-4.1 Invention of Writing

The fourth millennium B.C. was a period of remarkable cultural development, bringing with it the use of writing. As in Ancient Egypt during the fourth millennium B.C., so also in Mesopotamia there was a high order of civilization. Although no one is able to say with certainty when writing was actually invented, the existence of an Ancient Egyptian calendar since around 4230 B.C. suggests that writing in Ancient Egypt was in use by this date. (Advocates of the "long chronology" generally believe that some form of writing in Ancient Egypt preceded the invention of the calendar, whereas advocates of the "short chronology" place the invention of writing in Ancient Egypt throughout the period from 3500 B.C. to 3100 B.C. See Subsection 1-5.1.) Writing in Mesopotamia is generally agreed to have been invented around 3300 B.C.

The invention of Ancient Egyptian hieroglyphics (from the Greek for "sacred carvings"), beginning with the use of pictographs, took many years, perhaps centuries. Hieratic writing, a cursive form of hieroglyphics, was developed much later as a quicker and more convenient method of recording an agreement, conveying a message, or making a calculation. Most Ancient Egyptian documents that have come down to us were originally written in hieratic script; thus, with Egyptian scribes as with present-day handwriters, no two individuals display the same penmanship. So Egyptologists generally transliterate the "cursive" hieratic into the "printed" hieroglyphics first, and then translate the hieroglyphics into a modern language.

1-4.2   Hieroglyphics

In August 1798, in the Egyptian town of Rashîd (which Europeans generally called "Rosetta") near Alexandria, a French soldier unearthed a three-foot long slab of black basalt on which was engraved an inscription in three scripts: hieroglyphics, demotic (New Egyptian), and Greek. The inscription, dated March 27, 196 B.C., was a decree honoring the good deeds of King Ptolemy V Epiphanes (203–181 B.C.). More importantly, it provided the long-sought key that enabled scholars to decipher the mysterious language of Ancient Egyptian hieroglyphics.

The most important contributions to the decipherment of hieroglyphics were made by Thomas Young (1773–1829), an English physician and physicist (of "Young's modulus" esteem), and Jean François Champollion (1790–1832), the French linguist. Their efforts opened the field of Egyptology to linguists, biblical scholars, and historians of the ancient Near East.

Rosetta Stone containing an inscription (ca. 196 B.C.) in three scripts: classical hieroglyphics, cursive demotic, and Greek.

1-4.3   Cuneiform

In Mesopotamia, where clay was abundant, wedge-shaped marks were impressed with a stylus upon soft tablets that were then baked hard in ovens or by the heat of the Sun. This type of writing is known as *cuneiform* (from the Latin word "cuneus" for wedge) because of the shape of the individual impressions. The meaning to be transmitted in cuneiform was determined by the patterns or arrangements of the wedge-shaped impressions.

Until about a century ago, the messages of the cuneiform tables remained muted because the script had not been deciphered. By the mid-nineteenth century, significant progress in the reading of cuneiform was made when it was discovered that the Behistan Cliff (in modern Iran) carried a trilingual account of the accession of Darius I to the throne of Persia in 521 B.C., the inscription being in Persian, Elamite, and Babylonian. Knowledge of Persian consequently supplied a key to the reading of the Elamite and Babylonian. Even after this important discovery, decipherment and analysis of tablets with mathematical content proceeded slowly, and it was not until the second quarter of the twentieth century that awareness of Mesopotamian mathematical contributions became appreciable.

1-4.4   Ancient Egyptian Papyri

Only a small fraction of the Ancient Egyptian documents that have come down to us relate to matters of scientific interest. Nevertheless, there are several Ancient Egyptian papyri

Mesopotamian tablet recording solar eclipse in July 1062 B.C.

("paper" made from the papyrus reed) that somehow have survived the ravages of time over some four millennia. The most extensive document of a mathematical nature is a papyrus roll about 33 centimeters wide and $5\frac{1}{2}$ meters long. It was purchased in 1858 in Egypt by a Scottish antiquary, Henry Rhind; thus, it is generally known as the *Rhind Mathematical Papyrus* (RMP) and, less often, as the *Ahmes Papyrus* in honor of the scribe by whose hand it was copied about 1650 B.C. The scribe Ahmes wrote that the material was copied from writings made about two hundred years earlier. Indeed, there are indications that most of this knowledge had been handed down from Imhotep, the legendary architect and physician to the Pharaoh Djoser (in Greek, Zoser), who, also as history's first great engineer, supervised the building of Djoser's pyramid about 2950 B.C.*

---

*Ancient Egyptian dynastic dates vary widely among the writings of historical scholars and are frequently the source of intense debate. For example, it is common to find dates ranging from 3400 B.C. to 2600 B.C. as the period during which Imhotep flourished. We adopt no definitive position regarding such dates or disputes. For consistency and without prejudgment, we shall generally, though not exclusively, adopt the dates cited by Bernal [1991], which correspond substantially with the chronologies of Breasted [1946] and Mellaart [1979]. On the matter of the invention of the calendar, we follow Breasted. A widely accepted chronology is given by Baines and Malek [1980]. The Baines and Malek date of 2920 B.C. for the beginning of the First Dynasty appears as 3400 B.C. in Bernal's chronology. The two chronologies essentially merge by the beginning of the Twelfth Dynasty at 1991 B.C. for Baines and Malek and 1979 B.C. for Bernal. In general, believing it inessential to do otherwise, we shall round off many of the Ancient Egyptian and Mesopotamian dates to an approximate decade.

The *Rhind Mathematical Papyrus* contains some eighty-seven mathematical problems. These problems are preceded by a table of the division of 2 by the odd numbers from

Rhind Mathematical Papyrus (copied ca. 1650 B.C. from earlier writings): one of the oldest extant technical documents.

3 to 101, the answers being expressed as the sums of unit fractions. (Recall that unit fractions are fractions with unit numerators. Thus, an example of a table entry would be the equivalent of $\frac{2}{5} = \frac{1}{3} + \frac{1}{15}$. We write "the equivalent of" because such notations as the equal sign ($=$) and the plus sign ($+$) had not yet been invented. See Table 1-1 on page 25.) This RMP table is the most extensive and complete of all arithmetical tables found in Ancient Egyptian papyri. It was likely one of the most useful of all the scribes' reference tables, and most certainly it would have been valued in greater light than a modern mathematician would regard a set of logarithmic or trigonometric function tables. That this assessment is not an overstatement is evidenced by the fact that many variations and extensions of the table have been found on much later papyri, wooden, and other tablets, dating variously to as much as 2,000 years after the RMP itself was first written.

No matter how sophisticated and advanced the mathematics of the Greeks, Romans, Arabs, and Byzantines may have been, no one in antiquity was able to devise a more efficient technique for dealing with simple common fractions. Indeed, even during the current century, this table has drawn a diverse and contradictory set of opinions regarding its efficacy and methods.

1-4.5 Mesopotamian Clay Tablets

The Mesopotamian cuneiform clay tablets—many of them dating back some 4,000 years—were far less vulnerable to the ravages of time than were the Ancient Egyptian papyri; hence there is available today a much larger body of evidence about Mesopotamian than about Ancient Egyptian mathematics. From one locality alone, the site of Ancient Nippur (in modern Iraq), approximately 50,000 tablets have been recovered, though only a small fraction of these represents any mathematical or scientific significance. Despite the availability of these clay tablets, it was the Ancient Egyptian hieroglyphic rather than the Mesopotamian cuneiform that was deciphered first in modern times.

1-5 ANCIENT EGYPTIAN ASTRONOMY AND MATHEMATICS

The question of the peopling of Ancient Egypt is a strongly contested one. Osteological measurements; tests on the skin fragments of skeletons; iconographical studies; and literary references and observations by Hellenic scholars, especially Herodotus, who by some western scholars is called "the father of history," are all being pursued in efforts to address this question. This is not an issue addressed in this textbook nor is it relevant to our goals.

The great achievement of prehistoric Egypt was control of the land. The Ancient Egyptians—whoever they were—managed to clear the land for cultivation, drain the swamps, and build dikes against the incursions of flood waters. Gradually, the benefit of using canals for irrigation was learned. Such work required more organized communal efforts and this cooperation led to the growth of local political structures within each district.

An event of major historical significance was the annexation of Lower Egypt into Upper Egypt* by the Upper Egyptian ruler, whom tradition designates as Menes, while most

---

*Because the Nile flowed north, the ancient Egyptians regarded southern Egypt, which was upstream, as Upper Egypt, and they regarded northern Egypt, which was downstream, as Lower Egypt.

Narmer palette (ca. 3400 B.C.) showing Narmer punishing kneeling en-
emy.

archaeological sources call him Narmer (ca. 3400 B.C.). Narmer began the first of the thirty
dynasties, or ruling families, into which the Ancient Egyptian historian Manetho (ca. 280 B.C.)
divided the long line of rulers down to the time of Alexander the Great. Ancient Egyptian
society has been called the most pleasant of all ancient civilizations.

1-5.1   Ancient Egyptian Astronomy

The documentation that has come down to us on Ancient Egyptian astronomy is not at
all comparable to that available on medicine and mathematics. Yet, a major legacy of Ancient Egyptian astronomy is the annual calendar, which is thought to have been instituted
around 4230 B.C.* The division of the day into twenty-four hours is an Ancient Egyptian inven-

---

*This date is somewhat contested, as many historians suggest an origin around 2770 B.C., called
the "short chronology." The "long chronology" of 4230 B.C. is supported by, among other things, the
fact that *annual* dating records were made centuries before 2770 B.C. In several respects, the Bernal
chronology resolves some of the contentions between the advocates of the long and short chronologies;
however, it leaves open the issue of its own incongruence with the Sothic cycle.

tion, although the length of those hours was originally not uniform but depended on the seasons.

The Ancient Egyptian year consisted of 365 days: 12 months, each of 30 days, plus 5 year-end feast days that were dedicated to the gods Osiris, Horus, Seth, Isis, and Nephthys. The year-end days were the gods' birthdays. The Ancient Egyptian calendar has been called the only intelligent calendar that ever existed in human history. The true length of the sidereal year (the returning of the Sun to its same position relative to the stars) is approximately $365\frac{1}{4}$ days, which the Ancient Egyptians knew. This meant that the 365-day civil year slipped backward through the sidereal year, approximately 1 day every 4 years, 30 days or about 1 month every 120 years, and thus 365 days or 1 year, approximately every 1,460 years. (Note that the difference between the years 4230 B.C. and 2770 B.C. is 1,460 years.) Therefore, with respect to a specific date, the occurrence of a season changed very slowly, though not significantly during the lifetime of a particular individual.

When engaging much of the literature on this subject, the reader will encounter the 1,460-year period defined as the *Sothic cycle*. The adjective Sothic (originally from the Ancient Egyptian *Septit* and later changed by the Greeks to *Sothis* for the "dog star") refers to Sirius, when visible, the brightest star in the sky. Sirius is the dominant star in the constellation Canis Major—Great Dog—and thus is called the "dog star." It is widely believed that the annual heliacal rising of Sirius of $365\frac{1}{4}$ days was the basis for the Ancient Egyptian calendar of $365\frac{1}{4}$ days. So, the $365\frac{1}{4}$-day calendar is often called the *Sothic year* calendar, as opposed to the civil year calendar of 365 days. The *Sothic cycle* is simply the 1,460-year period required for the Sothic and civil calendars to synchronize.

The fact that the Ancient Egyptian calendar was an invariable (constant-duration) calendar is of enormous significance. The strictly lunar calendar of the Mesopotamians or any of the Greek calendars, which were invented later and depended not only on the Moon but also on local politics for intercalations, were obviously inferior. (See Subsection 1-6.1.) It is a serious problem to determine the number of days between two given Mesopotamian or Greek new year's days, say twenty years apart. In Ancient Egypt, this interval was simply 20 times 365. In 46 B.C. Julius Caesar introduced the one day per four years (leap year) into the Egyptian calendar, leading to the calendar that regulates our lives today. (The fact that the year is actually 365.242199 days instead of 365.25 days led Pope Gregory XIII in 1582 to proclaim that the years 1700, 1800, and 1900 would not be leap years, as they should have been under the Julian calendar, but the year 2000 will be.)

1-5.2 Ancient Egyptian Mathematics

People of the Stone Age (before ca. 6000 B.C.) had little use for fractions, but with the advent of more advanced cultures during the Bronze Age (ca. 4000–1000 B.C.) the need for the fractional concept and for fractional notations arose. From the *Rhind Mathematical Papyrus* (RMP) it is clear that the Ancient Egyptians proposed and solved problems requiring all four fundamental operations for fractions: addition, subtraction, multiplication, and division. Although many historians take for granted the scribes' ability to add and subtract, there are no clear indications as to how these two operations were performed. (There are subtleties in addition and subtraction that, without careful thought, are easily missed. Counting neither implies nor ensures the ability to add or subtract: For example, counting to 30,000 is quite different from adding a series of numbers equaling 30,000.) On the other hand, one can deduce quite clearly that multiplication was performed by "duplication" or successive doubling. (For example, the multiplication of 14 by 9 would be performed by adding 14 to itself to obtain 28, then adding 28 to itself to obtain 56, and applying duplication again to get 112, which is 8 times 14. Since $9 = 8 + 1$, the answer to 9 times 14 is $112 + 14$ or 126.) For division, the duplication process was reversed as the divisor, instead of the multiplicand, was successively doubled. Furthermore, they were capable of finding the square roots of perfect squares and good approximations to the square roots of imperfect squares.

Many of the practical problems were expressed around examples for finding the required grain to produce a specified amount of bread and beer, two of the most favorite commodities in Ancient Egypt. The Ancient Egyptians were skilled in manipulations involving the fraction $\frac{2}{3}$. To the fraction $\frac{2}{3}$ they assigned a special role in arithmetic processes, so that in finding one-third of a number, they first found two-thirds of the number and subsequently took half of the result! They knew and used the fact that two thirds of the unit fraction $\frac{1}{P}$ is the sum of the two fractions $\frac{1}{2P}$ and $\frac{1}{6P}$. They were also aware that double the unit fraction $\frac{1}{2P}$ is the unit fraction $\frac{1}{P}$. However, it appears as though, apart from the fraction $\frac{2}{3}$, the Ancient Egyptians regarded the general proper rational fraction of the form $\frac{m}{n}$ not as an elementary entity, but as part of an uncompleted process. Whereas today we think of $\frac{3}{5}$ as a single irreducible fraction, Ancient Egyptian scribes thought of it as reducible to the sum of the three unit fractions $\frac{1}{3}$ and $\frac{1}{5}$ and $\frac{1}{15}$.

Many of the calculations in the *Rhind Mathematical Papyrus* appear to be study exercises for students. Whereas many of them are of a practical nature, in some cases Ahmes seems to have had mathematical *recreation* in mind. Thus, Problem 79 simply states "seven houses, 49 cats, 343 mice, 2401 ears of spelt, 16807 hekats." Here Ahmes appears to be dealing with a well-known problem in which in each of seven houses there are seven cats, each of which eats seven mice, each of which would have eaten seven ears of grain, each of which would have produced seven measures of grain. The author of this problem evidently did not seek the practical answer, namely, the number of measures of grain that was saved, but the impractical sum of the numbers of houses, cats, mice, ears of spelt, and measures of grain. This recreational exercise by Ahmes seems to be a forerunner of the familiar nursery rhyme:

> *As I was going to St. Ives,*
> *I met a man with seven wives;*
> *Every wife had seven sacks,*
> *Every sack had seven cats,*
> *Every cat had seven kits,*
> *Kits, cats, sacks, and wives,*
> *How many were going to St. Ives?*

The Ancient Egyptian problems described above are considered to be arithmetic,* but there are others that are best classified as algebraic. These do not concern specific com-

---

*Arithmetic—more broadly number theory—is by no means a trivial subject, having attracted the interests of many famous mathematicians and having occupied a substantial portion of the career of one of the greatest mathematicians of all time, Karl Friedrich Gauss (1777–1855).

modities, such as bread and beer, nor do they require specific numerical calculations. Instead they require the equivalent of solutions of linear equations of the form $x + ax = b$ or $x + ax + bx = c$, where a, b, and c are known constants and x is unknown. Also, analyses of the Berlin and Kahun Papyri indicate the solutions of simultaneous algebraic expressions involving equations of both the first and second degree.

Problem 56 of the *Rhind Mathematical Papyrus* is of special interest in that it contains rudiments of trigonometry and a theory of similar triangles. In the construction of the pyramids it had been essential to maintain a uniform slope for the faces, and it may have been this concern that led the Ancient Egyptians to introduce a concept equivalent to the cotangent of an angle. In modern engineering, it is customary to measure the steepness of a straight line as the ratio of the *rise* to the *run*. In Ancient Egypt it was customary to use the reciprocal of this ratio. This *was* trigonometry; not *trilaterometery*—the study of three-sided polygons (trilaterals)—as it has been pejoratively called.

The Ancient Egyptian rule for finding the area of a circle is one of the outstanding mathematical achievements of antiquity. In Problem 50 of the RMP, Ahmes assumes that the area of a circular field with a diameter of nine units is the same as the area of a square with a side of eight units. If this assumption is compared with the modern formula $A = \pi r^2$, we ascertain an Ancient Egyptian rule for assigning π a value of $\left(\dfrac{4}{3}\right)^4$ or about 3.16, a commendably close approximation. (See Figure 1-1.)

That the number $\left(\dfrac{4}{3}\right)^4$ did indeed play a role comparable to our constant π seems to be confirmed by the Ancient Egyptian rule for the circumference of a circle, according to which the ratio of the area of a circle to its circumference is the same as the ratio of the area of the circumscribed square to its perimeter. This observation represents a geometrical relationship of major *theoretical* mathematical significance. Recognition by the Ancient Egyptians of interrelationships between geometrical figures has often been overlooked, and yet it is this instance in which they came closest in attitude to their more theoretically minded successors. It is now clear that some of the geometric comparisons made in the Nile Valley, such as those on the perimeters and areas of circles and squares, are among the first exact theoretical statements in history concerning geometric figures.

While the *Rhind Mathematical Papyrus* is the most extensive mathematical document from Ancient Egypt, another scientifically important document is the *Moscow Mathematical Papyrus* (MMP). About 8 centimeters wide and $5\dfrac{1}{2}$ meters long, it was written by an unknown scribe about 1890 B.C. Although it is neither as mathematically extensive or as well

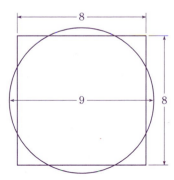

FIGURE 1-1   Ancient Egyptian theory defining equal areas of circle and square, called "squaring the circle," for which $\pi \cong 3.16$.

written as the RMP, the MMP contains twenty-five mostly practical problems, except for two that have special significance.

Problem 10 in the MMP appears to calculate the surface area of a hemispherical dome, thus providing the earliest estimation of a curvilinear surface area. If, in fact, this is the case—and it certainly appears to be so—this result antedates the commonly quoted oldest known computation of a hemispherical surface by about 1,500 years. In problem 14, the scribe correctly computes the volume of the frustum of a pyramid in accordance with the modern formula:

$$\text{Volume} = \frac{h}{3}(a^2 + ab + b^2)$$

where h is the altitude, and a and b are the lengths of the sides of the upper and lower square bases, respectively. Since such a formula is not given, the general basis for the calculation remains unknown. Nevertheless, the high probability that such a result could not have been found empirically clearly indicates that the Ancient Egyptians were familiar with this formula. This too is a remarkable scientific achievement.

It is impressive that the great achievements in government, architecture, astronomy, irrigation, construction, and science by the Ancient Egyptians were accomplished without much of the mathematical structure that today we often accept without thought. The Ancient Egyptians had no zero, no decimal point, and no equals or square-root signs; indeed they had no plus, minus, multiplication, or division signs. Their enormous accomplishments were achieved via a mathematics that was substantially based on two elementary concepts. The first was their knowledge of the twice-times tables. The second was their ability to compute two-thirds of any number, whether fractional or integral. This suggests a simplicity and an efficiency of concepts that we presently consider to be the hallmarks of great science and engineering.

With their mathematics and astronomy, the Ancient Egyptians can be said to have lain the recorded foundations of quantitative science and engineering. While some of the arithmetic and technology may have preceded the Ancient Egyptian civilization, there is no written record of such achievements. Their astronomy, which they used to invent the calendar, was the beginning; and no *first step* in any achievement can be characterized as anything less than significant. And how can the combination of their invention of quantitative geometry and astronomy be taken other than as the commencement of *kinematics,* the study of the geometry of motion?

1-6   MESOPOTAMIAN ASTRONOMY AND MATHEMATICS

As in Ancient Egypt during the first dynasty (ca. 3400–3200 B.C.), at about the same time in the Mesopotamian valley there was a high order of civilization. The Mesopotamians had built homes and temples, and powerful rulers had united the local principates into an empire that completed vast public works, such as a system of canals to irrigate the land and to control flooding. The name "Mesopotamian" includes a series of peoples—Akkadians, Amorites, Assyrians, Chaldeans, Elamites, Hittites, Kassites, Persians, Selucids, and perhaps most notably, the Babylonians and Sumerians—who concurrently or successively occupied the region primarily between the Tigris and Euphrates rivers, a region known as Mesopotamia and currently a part of modern Iraq, extending into Iran. Despite the broad range of peoples who at one time or another settled in Mesopotamia, it was their use of cuneiform script that provided a major bond between them.

1-6.1   Mesopotamian Astronomy

Like Ancient Egyptian astronomy, Mesopotamian astronomy served several purposes. Primarily, it was used to keep a calendar. The year, the month, and the day are astronomical quantities, which had to be obtained to determine planting dates, to celebrate religious holidays, and to cast astrological horoscopes. In Mesopotamia, partly because of the connection of the calendar with religious holidays and ceremonies and partly because the heavenly bodies were believed to be gods, the priests kept the calendar.

From about 1600 B.C., the Mesopotamians compiled astronomical tables of stars and planets, and by 800 B.C. they were able to identify the planets visible to the naked eye. Mercury, Venus, Mars, Jupiter, and Saturn were known! Also, special conjunctions and eclipses of the Sun and Moon were either in their data or readily obtained from them. Astronomers could predict the Moon phases and eclipses to within minutes. Their data indicate that they knew (or could have calculated) the length of the sidereal year (the time for the Sun to return to its same position relative to the stars) to within $4\frac{1}{2}$ minutes. The division of the circle into 360 degrees originated in Mesopotamian astronomy. The Alexandrian astronomer Claudius Ptolemæus (Ptolemy) followed the Mesopotamians in this convention. Indeed, Ptolemy (ca. A.D. 70–147) was in possession of substantial Mesopotamian astronomical records including accounts of eclipses dating to about 747 B.C.

The Mesopotamian calendar was lunar. The month began when the first crescent appeared after the new Moon. A lunar calendar is difficult to maintain, however; although it is convenient to have the month contain an integral number of days, lunar months, reckoned as the time between successive conjunctions of the Sun and Moon (that is, from new Moon to new Moon), vary from 29 to 30 days. Hence a problem arises in deciding which months are to have 29 days and which 30 days. A more important problem is making the lunar calendar agree with the seasons over a period of decades. The resolution of this problem is quite complicated.

The Mesopotamian lunar calendar required extra months intercalated so that 7 such intercalations every 19 years kept the lunar calendar nearly synchronized with the sidereal year. Thus, 235 (12 × 19 plus 7) lunar months were equal to 19 sidereal years. The summer solstice was systematically computed, and the winter solstice and the equinoxes were placed to create equal intervals throughout the year. This calendar was used by the Hebrews and the Greeks as well as by the Romans up to 46 B.C., the year the Julian calendar, based on the Ancient Egyptian calendar, was adopted.

1-6.2   Mesopotamian Mathematics

Although there are numerous clay cuneiform tablets, those pertaining to Mesopotamian mathematics are fewer in number than the Egyptian mathematical papyri. Also, the Mesopotamian mode of writing on clay tablets tended to discourage the compilation of long treatises; and so there is no Mesopotamian record comparable to the *Rhind Mathematical Papyrus,* or even the *Moscow Mathematical Papyrus.* The most important mathematical cuneiform tablets date from the period of Hammurabi's reign as king of Babylon (ca. 1790–1750 B.C.).

It is clear that the Mesopotamians could handle all the fundamental arithmetic operations as could the Ancient Egyptians. They also had the correct rules for computing the areas of plane figures such as triangles, rectangles, and trapezoids. While they achieved greater algebraic breadth than the Ancient Egyptians, the Mesopotamians may have been less concerned about accuracy than the Ancient Egyptians. (This is not globally true, however, since as noted from the Yale Collection [No. 7289], the Mesopotamians showed calculations of $\sqrt{2}$ as 1.41422, which differs from the true value by about 0.000006.) In instances where their interests overlapped, the Ancient Egyptians tended to be more accurate. For example, in astronomical measurements for the length of the sidereal year or in the calculation of the value of π (3.16 for the Ancient Egyptians versus 3 for the Mesopotamians, although there is evidence that the Mesopotamians sometimes used $3\frac{1}{8}$), the Ancient Egyptians excelled.*

On the other hand, the Mesopotamians were not only familiar with linear equations but were also capable of solving quadratic and some cubic equations. Even some formulations in Mesopotamian mathematics called for the solution of two simultaneous equations and solutions akin to compound interest analysis. The so-called Pythagorean theorem, for example, does not appear explicitly in Ancient Egyptian documents although there is evidence of their familiarity with this result, whereas clay tablets from the Old Mesopotamian period (ca. 1800 B.C.) clearly illustrate that the Mesopotamians widely used the Pythagorean theorem.

---

*The calculation indicated here is for a quantity that was the equivalent of π; that is, the ratio of a circle's circumference to its diameter. The notation "π" was not consistently used in mathematics, however, until the eighteenth century.

As early as about 2100 B.C., two important features exhibited by the Mesopotamian number system were the sexagesimal (base 60) and positional notation. Whereas the Ancient Egyptians denoted each higher unit by a new symbol, the Mesopotamians used the same symbol but indicated its value by its position. For example, the number 1 followed by a 3 meant 63 (that is, $1 \times 60 + 3 = 63$) or 2 followed by 6 followed by 3 meant 7563 (that is, $2 \times 60^2 + 6 \times 60 + 3 = 7563$). Such a positional notation accounts for the effectiveness of our current system of writing numbers, in which 655 means $6 \times 10^2 + 5 \times 10 + 5$. The fact that they extended their positional notation to fractions meant that the Mesopotamians had in their command the systematic computational power that decimal notation affords modern society.

Two significant ambiguities existed in the Mesopotamian positional system: They had no decimal point and they had no zero, although they sometimes left a space where a zero was intended. To our knowledge, the decimal point or its equivalent was never clearly addressed by them; presumably the exact interpretation of a number had to be made from the context of its use. It was not until about 300 B.C. that a special sign, consisting of two small oblique wedges, was invented to serve as the zero.

Much of Mesopotamian computation was accomplished using tables that ranged from simple multiplication tables to lists of reciprocals, of square roots and cubic roots, and of exponentials. Finally, the effectiveness of their computations was not limited to their use of tables but involved the development of procedural algorithms, such as a procedure for calculating the $\sqrt{2}$, which is equivalent to what we now commonly denote as a two-term approximation to the binomial series.

1-7   MATHEMATICS OF THE MAYANS, INDIANS, ARABS, AND CHINESE

Several major civilizations will be treated somewhat less than those of Ancient Egypt and Mesopotamia in this review of mathematics and astronomy in the Pre-Christian era. Each of these civilizations is of greater than or about the same antiquity as those of Greece and Rome, although none appears to be as old as those of Ancient Egypt and Mesopotamia. These

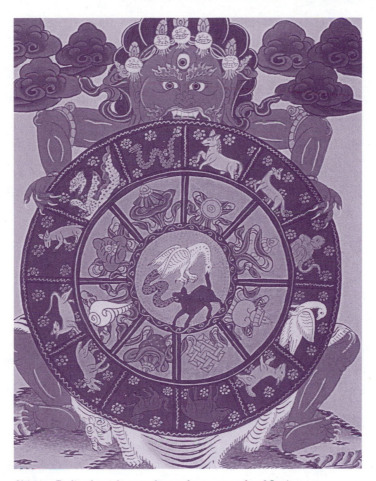

Chinese Zodiac based upon the twelve-year cycle of Jupiter.

include the Mayan, Indian, Arabic, and Chinese societies. The primary reason for their lesser emphasis in this review is that they appear to have had far less influence on the mathematics and astronomy that evolved into western science.

The Mayans, who may have settled as early as 2500 B.C. in Southern Mexico and Central America, developed a calendar and some elementary arithmetic and hieroglyphic writing, but not beginning until about the second century of the Christian era. Rather than accurately correlating the length of the year, the Mayan calendar was used for counting days—past and future—and was sophisticated in its ability to track Venus. Furthermore, the Mayans were sophisticated builders of stone structures as represented, for example, by their temples, courts, and observatories at the famous archeological site called Chichén Itzá.

Significant contributions in algebra were made by the Indians and the Arabs, but they seem to be confined to the post–ancient Greek period, generally after A.D. 400* for the

---

*Records of Indian sects during the approximate period 800 B.C. to A.D. 200 reveal a strong religious, not mathematical aim. The mathematics of this period was elementary, involving planar figures such as squares and rectangles.

Indians and after A.D. 600 for the Arabs. Such contributions appear to have an Alexandrian influence.

The major mathematical records of Chinese mathematics during the Pre-Christian period are the *Chou Pei Suan Ching** (The Arithmetical Classic of the Gnomon and the Circular Paths) and the *Chiu Chang Suan Shu* (Nine Chapters on the Mathematical Art), composed about 250 B.C. Areas for triangles, rectangles, and trapezoids are properly calculated, and the so-called Pythagorean theorem is known. The area of a circle is computed as one-twelfth the square of the circumference—a correct result if the value of π were 3, although no explicit mention of π or its equivalent is cited. Fractions, percentages, square roots, and cube roots are discussed. Simultaneous linear equations in three unknowns are solved. And algebraic word-and-number problems requiring the solutions of quadratic equations are given. From the theory of algebraic equations, it is known that if n is a positive

---

*Estimates for the date of the composition of the *Chou Pei Suan Ching* vary widely, ranging from about 1200 B.C. to about 300 B.C.

Depiction of book burning ordered by Emperor Shih Huang-ti.

integer not divisible by the positive prime p, then $n^{p-1} - 1$ is divisible by p. This fundamental theorem was proved in 1732 and published in 1738 by Leonhard Euler. Apparently, Gottfried Leibniz proved it in 1683 but did not publish it. Well, the Chinese knew this result as early as ca. 400 B.C., at least for the special case of $n = 2$.

Further, the earliest report by anyone of the comet "discovered" by and named for Edmund Halley (1656–1742) in the seventeenth century was actually recorded in China in the *Chun Qiu* (Spring and Autumn Annals), where it was noted that in the autumn of 613 B.C. a comet was observed traveling into the Big Dipper. The Chinese also developed a calendar that was sufficiently intricate to incorporate the approximately twelve-year cycle of Jupiter, which to this day remains important in their astrological system. The twelve positions of Jupiter in the Chinese Zodiac are often identified as the Year of the Rat, the Year of the Buffalo, . . . , the Year of the Pig.

Much of the early progress in Chinese mathematics was hampered in 213 B.C. when the emperor Shih Huang-ti of the Ch'in Dynasty ordered the burning of all books and the burying of all scholars, reducing mathematical interests primarily to those of commerce and maintaining the calendar. Although by 202 B.C. during the Han Dynasty there was a revival of mathematical scholarship, most notably during the first century A.D. in the *Sun-Tsu Suan Ching* (Mathematical Classic of Sun-Tsu), Chinese mathematics appears to have had little, if any, direct influence on western culture.

1-8   THE FIRST GREAT ENGINEERING SOCIETY

At a time when Europe was to remain a backwater of disconnected villages and unlettered tribes for more than another thousand years, the first great engineering culture had been invented in Ancient Egypt. So vast were the engineering achievements of the Ancient Egyptians of four to five millennia ago that they continue to baffle both amateur and professional scientists and engineers.

The focus here is on the engineering aspects of a few major extant structures in Ancient Egypt; not on their art, their daily function, or the style of their architecture, about which much more has been written. No adequate summary of the variety of engineering structures can be reviewed in the limited space allocated here. Even so, this extraordinarily modest summary should reveal some of the radical precedent-setting engineering accomplishments of this ancient society; accomplishments that required innovation, planning, and execution comparable to the greatest engineering projects of all time.

The Ancient Egyptians invented a broad range of construction techniques for stone structures, including quarrying, dressing, foundation preparation, roofing, joining, and arches. The invention of stone columns as an architectural as well as structural element is due to the Ancient Egyptians. At Karnak, Luxor, and Heliopolis monolithic obelisks and built-up columns in excess of 30 meters and weighing 250 tons survive to the present. Although sandstone was more widely used for roofing great monuments, the more-difficult-to-dress granite slabs were also used, such as in the Great Pyramid and the Temple of the Sphinx. Ingenious structural techniques were invented to dovetail columns to the roofing slabs they supported, sometimes with cramps and dowels. (See Figure 1-2.) In order to induce runoff of rainwater, roof slabs were slightly tapered to discharge the water at a desired site, depending on the location of the main gutter.

About twenty miles southwest of Cairo, among a complex of fifteen royal pyramids at Saqqara, is the Step Pyramid of Djoser (in Greek, Zoser). The major significance of the Step

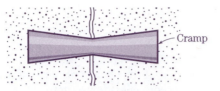

(*a*) Dovetailed wooden cramps up to five feet long were in use.

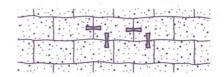

(*b*) Dovetailed cramps uniting stones in the wall.

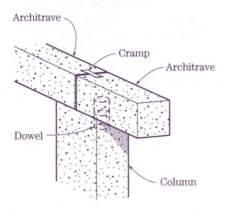

(*c*) Dovetailed cramp and dowel catenate architraves atop column.

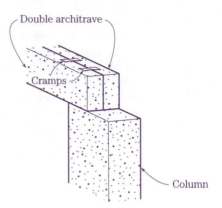

(*d*) Dovetailed cramps integrate double architraves atop column.

FIGURE 1-2 Use of dovetailed cramps and dowels in stone joints of major structures as invented by Ancient Egyptians during third and second millennia B.C. (These are schematics of characteristic features, not precisely detailed.)

Pyramid (ca. 2950 B.C.) is that, as the first pyramid built in Ancient Egypt, it is the world's earliest monumental stone structure comparable to its size. The design of the Step Pyramid is attributed to Imhotep, who was legendary as an author, architect, diplomat, poet, and physician. He is considered to be the inventor of hewn-stone construction, and clearly one of the world's first great engineers.

Ten miles west of Cairo is the Giza plateau, containing a massive complex of pyramids, causeways, temples, and tombs, dominated by three pyramids. The largest of these is the Khufu (in Greek, Cheops) Pyramid, commonly known as the "Great Pyramid," one of the most famous monuments in the world. (See Figure 1-3.) To describe the Great Pyramid of Khufu (ca. 2870 B.C.) briefly is not to do it justice.

A pyramid is least of all a pile of stones rising to an apex. Pyramids such as the Great Pyramid are sophisticated monuments requiring the complete range of engineering surveying and masonry. The preparation and leveling of the thirteen-acre construction site for the Great Pyramid required ingenuity, and was accomplished using the fact that the surface of still water assumes a constant level. (See Figure 1-4.)

The Great Pyramid rises in 201 stepped tiers to the height of a modern forty-story building, originally about 146.3 meters tall. Its square base covers about a dozen football fields or seven midtown blocks in Manhattan (New York City); being about 230.4 meters on a side with 20 centimeters being the difference between the longest and shortest sides. There is a maximum error from right angle between any two sides of about $3\frac{1}{2}$ minutes of arc.

Step Pyramid of Djoser. (ca. 2950 B.C.).

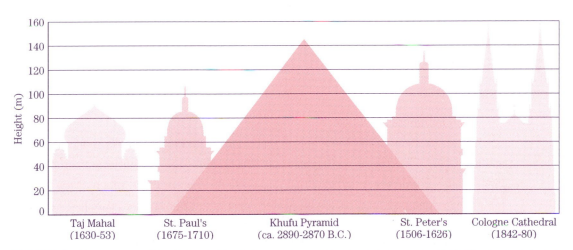

FIGURE 1-3 Comparison of Khufu Pyramid with four of the world's most famous stone structures, showing periods of construction and heights. In terms of solid masonry, the Khufu Pyramid contains more stone than the other four combined. (St. Paul's Cathedral is actually a reconstruction, and construction of the Cathedral of Cologne was completed during the indicated period, having been initiated in 1284.)

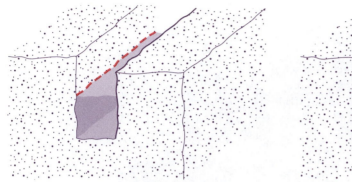

(a) Narrow connecting trenches are cut into entire surface, filled with water, and line is drawn at water level.

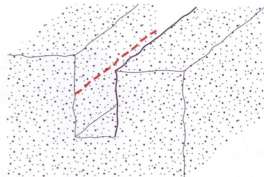

(b) Water is drained, clearly exposing marked line.

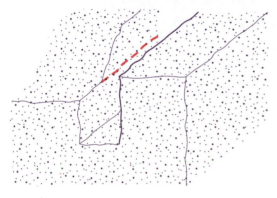

(c) Stone above marked line is removed by chipping and grinding.

(d) Remaining trenches are packed with stone.

FIGURE 1-4 Four-step procedure used for leveling thirteen-acre surface for construction of Khufu Pyramid, where a small segment of a single trench is sketched. The entire thirteen-acre site was crisscrossed with such narrow connecting trenches.

The Great Pyramid contains approximately 2,300,000 stone blocks weighing mostly about $2\frac{1}{2}$ tons, but some up to 70 tons each.* The granite stones comprising the roof of the king's chamber typically weigh about 50 tons each. See Figure 1-5 in which characteristic, not precisely detailed, features are illustrated. Most of the $2\frac{1}{2}$-ton blocks are limestone and are quarried either around the site or across the Nile at Tura. The more massive blocks weighing tens of tons are granite and were quarried nearly *1,000 kilometers* to the south

---

*The volume of brick and mud required to build the ramps for raising the stones into position exceeded the volume of the Pyramid itself; that is, assuming, as we believe, that ramps were used to lift the stone blocks to the required heights.

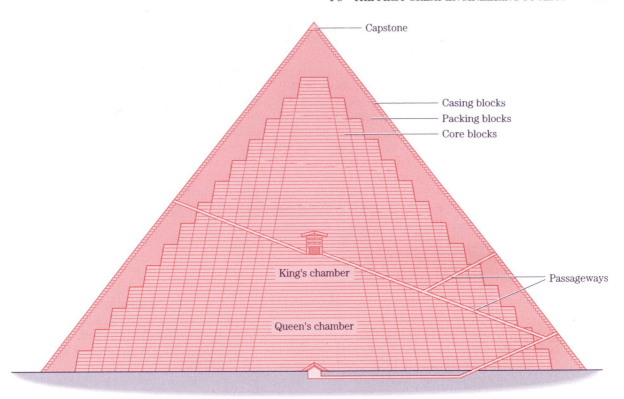

FIGURE 1-5 Typical major pyramid consisting of core, packing, and casing blocks, showing chambers and connecting passageways. Note stress-relieving configuration of roof in king's chamber. (These are schematics of characteristic features, not precisely detailed.)

in Aswan. Over the approximately twenty-year period of its construction, the completion of the Great Pyramid would have required quarrying, roughing out, transporting, finishing, raising, and placing about 315 blocks each day! The precision of the matching joints in the casing stones was within 0.25 mm, so that even today it's difficult to slip a piece of paper or a razor blade between those casing stones that remain.

To plan the project; lay it out; prepare the foundation; organize the quarrying and transportation; maintain the tools and lifting equipment; provide food, water, and medical care for the work force; and ensure that the work force (of about 100,000) was properly trained and employed required the efforts of an engineering genius. The construction of the Great Pyramid using twentieth-century construction and management techniques would pose significant problems.

The question as to whether the Ancient Egyptian society was a great *scientific* society versus a great *technological* society has been the subject of debate. In addition to being an oftimes unenlightened query because the delineation between science and state-of-the-art engineering is at best indistinct, the nature of the question as it is generally framed requires a psychological analysis of people who lived several millennia in the past. (See Section 1-9.) The question of whether the Ancient Egyptian society was a great *engineering* society is easily answered; it was. Many of their immense accomplishments continue to stare at us, astounding and confounding us thousands of years after their invention, design, and construction.

As the Greeks would subsequently introduce into western civilization the first definitive distinction between theory and application, the Ancient Egyptians created the philosophical reality of *transcendental expression*. Having successfully controlled the forces of nature on a grand scale via massive irrigation projects, the Ancient Egyptians were stimulated to transcend their primal needs for survival. By building grand temples and monuments such as the Great Pyramid, through great engineering they produced the first society that chose to express itself aesthetically, transcendentally, and beyond their daily needs in order to establish a union between posterity and themselves. With this knowledge, we can no longer view the prominent edifices of modern structures without reflecting upon the creative aspirations of those who *invented* transcendental expression on a grand scale.

1-9   ADVERSE CRITICISM OF ANCIENT EGYPTIAN AND MESOPOTAMIAN MATHEMATICS

Several historians have adversely criticized the mathematical contributions of the Ancient Egyptians and, to a lesser extent, those of the Mesopotamians. The irony of this negativism is that those same historians generally express a sense of awe for the technical (as opposed to scientific) achievements of these two societies, especially those of the Ancient Egyptians.

The mildest form of these adverse criticisms is the view that all pre-Hellenic[3] mathematics was "purely utilitarian" or "intuitive." More common are criticisms regarding "their lack of scientific attitude of mind" or "their lack of rigor and proof" or the disdaining remark that they "used mathematics but did not study mathematics," all referring to the Ancient Egyptians.

The most uncharitable remarks tend to demean Ancient Egyptian and Mesopotamian mathematics as "scrawlings" and "child's play" or express the opinion that what the Greeks created compared with the Ancient Egyptians and the Mesopotamians was as different "as gold from tin." (Another author has characterized both Ancient Egyptian and Mesopotamian cosmologies as "fancifully childish.") One writer has stated that Ancient Egyptian mathematics "did not contribute positively to the development of mathematical knowledge" and, in fact, its effect has been "rather negative" as a "retarding force upon numerical procedures." Another author has stated that the Ancient Egyptians made no contribution to "mathematical formulation and notation."

While it is true that the Ancient Egyptians achieved major engineering success without such notation as the plus sign (+), the minus sign (−), and the decimal point, the same must be said of most of the societies that historically succeeded them. For as Table 1-1 clearly shows, it took all of western civilization more than 4,000 years from the building of the Great Pyramid to the invention and standardization of the plus sign, the minus sign, and much of the elementary notation we take for granted. Mathematical and scientific concepts—no matter how apparent they are postfactum—are never born fully matured or complete. To criticize the *inventors of written mathematics* because they did not meet some current notions of *scientific* thought is ludicrous.

[3]See Section 1-10.

TABLE 1-1   Elementary Mathematical Notation, Author, and Approximate Date of Publication

Notation		Author	Approximate Date of Publication
=	(Equality)	Robert Recorde (1510–1558)	1557
.	(Decimal Point)	Giovanni A. Magini (1555–1617)	1592
+	(Addition)	François Viéta (1540–1603)	1600[†]
−	(Subtraction)		
>, <	(Inequality)	Thomas Harriot (1560–1621)	1631 (Posthumously)
×	(Multiplication)	William Oughtred (1574–1660)	1631
÷	(Division)	Johann Heinrich Rahn (1622–1676)	1659
·	(Multiplication)	Gottfried W. Leibniz (1646–1716)	1698
π	$(3.14159\cdots)$	William Jones (1675–1749)	1706
≥, ≤	(Inequality-Equality)	Pierre Bougeur (1698–1758)	1730
e	$(2.71828\cdots)$	Leonhard Euler (1707–1783)	1736
$f(x)$	(Function of x)		1740
$\sum$	(Summation)		1755
i	$(\sqrt{-1})$		1794 (Posthumously)

[†]In his 1489 book, Johann Widmann used "+" and "−", but not exclusively to denote addition and subtraction, respectively.

To criticize the founding societies of western civilization exclusively in the context of limited modern disciplinary criteria is to display a lack of depth of appreciation of humankind's culture over the millennia. It is among the worst examples of noncontextual criticism.

Regarding the criticism of pure utility, it is true that the major focus of mathematics was to conduct the commerce of these societies. So what? This is no less true today, or in any age. Certainly, this is the manner in which every engineer or scientist who has ever lived uses mathematics. On the other hand, a distinct link between purpose and practice in all Ancient Egyptian and Mesopotamian mathematics is difficult to establish. For example, what possible practical problem could have led the Mesopotamians to consider equations such as $ax^4 + bx^2 = c$ or $ax^8 + bx^4 = c$, which they recognized as quadratic equations in disguise; that is, quadratics in x^2 and x^4, respectively? Further compilations on the Mesopotamian tablet named Plimpton No. 322 at Columbia University suggest mathematical tabulations having significance in what we today call the "theory of numbers." An examination of the work of great mathematicians in more modern times would clearly show that such greats as Newton and Euler, to name only two, were virtually always grounded in practical applications. Indeed, it was only during the nineteenth century with the advent of so-called "pure mathematics" that one can claim a nonutilitarian proclivity for any but a handful of mathematicians throughout all periods of history. Finally, regarding the criticism of pure utility, we ask "are biologists who have undertaken the purely utilitarian goal of seeking a cure for cancer doing science?"

Adverse criticisms concerning a "lack of proof" can be refuted by considering concluding sentences in problems in the RMP that illustrate Ahmes' "proof" (which by today's standards might be viewed as a "check") that the numerical solution to the problem at hand was correct for the values used. For the Ancient Egyptians such checks appear to have represented both method and proof. Sentences from the RMP such as "Thus findest the area" or "These are the correct and proper proceedings" clearly indicate a mindset directed toward justification, reckoning, or proof. It is worth noting that concepts such as *proof* or *rigor* are not invented fully matured, but evolve as do all human inventions. Even some mathematical formulations given by Augustin L. Cauchy (1789–1857), considered by many to be the "founder of rigor," are lacking by the standards set for current graduate students.

Mesopotamian tablet named Plimpton 322 (ca. 1800 B.C.) confirming use of "Pythagorean theorem" more than a thousand years before Pythagoras lived.

It is generally acknowledged that the Ancient Egyptians and the Mesopotamians did not reason as their beneficiaries, the Greeks, did. It is not to the denigration of the achievements of the Ancient Egyptians and Mesopotamians to acknowledge the great intellectual contribution of the insistence on deductive proof by the Greeks. This transition was simply one of numerous important mathematical watersheds in the progress of humankind. Although it is impossible to draw a single chronological route through the centers of mathematics, at one time or another the world's dominant sites have evolved through Ancient Egypt, Mesopotamia, Greece, the Arabic-Indian axis, Italy, France, Holland, England, Switzerland, Russia, Scandinavia, and Germany. Indeed, had it not been for the religious persecution that drove the Bernoullis from Antwerp, Belgium too would have had its turn. To discard that which preceded Greece gives rise to suggestions that anything that preceded Italy should be disregarded, or France, or England, or everything that preceded last week. . . .

Generally, though not always, if the Ancient Egyptians found a mathematical method or result, they did not seek the *philosophical why* it worked. So, while the Ancient Egyptians and Mesopotamians are sometimes criticized for this mode of science, this is the same mindset adopted by Galilei in the seventeenth century. Today it is broadly recognized that a fundamental distinction between early Greek scientists and modern scientists is the modern discarding of the philosophical why. Whereas the early Greeks were preoccupied with the philosophical why, modern scientists are pragmatic in their focus on quantitatively predicting and controlling the external manifestations of a physical event. (Modern scientists are generally not concerned with the philosophical reason—the *why*—that force equals the product of mass and acceleration, but how to use such a relationship in predicting and controlling physical phenomena. However, even this perspective may be coming unglued as some modern physicists pursue the anthropic principle.) It was Galilei who clearly embodied this pragmatism and it is for this reason that he, in somewhat emulating the Ancient Egyptians and Mesopotamians, is often called the "father of modern science."

To suggest that the arithmetic, geometric, algebraic, and trigonometric mathematical records from civilizations from more than 4,000 years ago are little more than "scrawlings" indicates critical remarks that go beyond scientific integrity and smacks of demagoguery. What possible motives could underlie such harsh and uncharitable criticisms from modern and otherwise respected mathematicians? (The mostly self-taught bicycle designers and manufacturers Wilbur and Orville Wright developed no aerodynamic theory, but does that suggest that their 12-second 7-mile-per-hour erratic flight on December 17, 1903 contributed nothing to the history of powered flight? They conducted hundreds of tests on different airfoil shapes leading to the flight that has been characterized as "the most historic moment in aviation history." With no college education—Wilbur (1867–1912) satisfied the requirements for a high school diploma, but Orville (1871–1948) did not—they are commonly regarded as "the premier aeronautical engineers of history.") While engineering often builds upon scientific understanding, it is more common that initial engineering precedes scientific underpinnings. Indeed, there is an easily recognizable interplay between the evolution of both so-called scientific and so-called engineering understanding, to the extent that one is precariously imprudent to claim a clear distinction between the two.

Further, it has been stated by various historians that no connection between geometry and the physical world preceded Euclid (ca. fourth century B.C.) or, because of their use of geometrical graphics, Giovanni da Casale and Nicole Oresme in the fourteenth century. It would be exceedingly difficult to argue that the Egyptian pyramids were either not geometrical (and not understood to be so) or not physical.

Greek writers during the first millennium B.C. were nearly unanimous in their praise of Ancient Egyptian superiority in mathematics and astronomy. Major Greek scholars of that period (notably Thales, Pythagoras, Herodotus, Aristotle, Euclid,* Archimedes, Eudoxos, and many others) visited and/or learned Egyptian and/or studied in Egypt. Yet several twentieth-century writers treat with skepticism claims by those Greek scholars that there was anything to learn in Ancient Egypt, thus suggesting that these greatest of Greek scholars were foolish and wasting their time in Egypt or were simply ignorant in their attributions to the Ancient Egyptians. It is as if a writer 2,500 years into our future, in extolling twentieth-century American science and engineering, claimed that any reference to Newton or Euler or Lagrange or Maxwell by twentieth-century scientists was mistaken and ignorant in its assignation of credit.

*The question of whether Euclid was Egyptian or Greek is an open one. Where he spent most of his professional life is not questioned; that was in Alexandria.

There can be no question that the Ancient Greeks stood on the shoulders of the Ancient Egyptians and the Mesopotamians. But, they did not simply stagnate in their inheritance. When they inherited or discovered an exact mathematical procedure, the Ancient Greeks sought to establish its universal truth by *a priori* symbolic arguments that could reveal the logic of their analysis. This was a radical imposition on the meaning of knowledge. Thus, as much as any specific intellectual contribution, it is the broad development, indeed a near insistence, of deductive proof that makes the Ancient Greeks important in western scientific history. While such an acknowledgment pays due respect to the Ancient Greeks, it should in no way diminish the scientific foundations established by the Ancient Egyptians and Mesopotamians.

It is easy to be critical and dogmatic regarding the scientific merit of calculations made 4,000 years ago; however, such evaluations are a matter of judgment. Then, as in modern society, the Ancient Egyptians and the Mesopotamians were primarily concerned with daily living. Nevertheless, a fair reading of their literature must acknowledge that even under the obligations of daily immediacy, some of the mathematics from both societies display the earmarks of recreation and perspective. (Of all the calculations made in a single day in the modern world, think of how minuscule is that fraction that is scientific; most of the calculations by even scientists and engineers—from balancing a checkbook to estimating the loss of pressure in a piping system, to calculating the stress in a beam, to totaling a golf scorecard—could hardly be considered as scientific.)

Ancient Egypt and Mesopotamia were two awe-inspiring technological societies, each extending for about a thousand miles, managing economies, collecting taxes, maintaining armies, designing and building irrigation systems, and constructing large granaries. Furthermore, the magnificence, immensity, and grandeur of the architecture of their temples and monuments† establish the Ancient Egyptians as the world's first great engineering society. To conclude that mathematics and science did not contribute to those enormous achievements is to utilize the most strident, if not farcical, criteria of assessment.

†The Mesopotamians extensively used unfired mudbrick, a material that is highly vulnerable to time and weathering.

1-10   EVOLUTION THROUGH THE HELLENIC ERA

Although it is sometimes possible in retrospect to associate a single event, ruler, or war with a change in "age" or period, this is not generally the case. Time and history flow continuously, so such periods and even the historical figures associated with them are simply mechanisms of convenience for organizing historical perspective.

The first Olympic Games were held in 776 B.C. in Olympia, Greece. By this date, Greek writers, most notably Homer, had begun to produce excellent literature. Thus began one of the great periods in Greek history that would subsequently be judged to have ended in 323 B.C. with the death of Alexander the Great, "founder" of Alexandria (332 B.C.) in Egypt. (Actually, the captured town of Rakotis, including its annexed western suburb, was renamed.) The period 776–323 B.C. is sometimes called the Hellenic era, derived from the name Hellenes, which the unlettered invaders from the north who settled along the Mediterranean called themselves. The Hellenes, who brought no scientific, literary, or mathematical tradition with them, traveled to the intellectual centers in Ancient Egypt and Mesopotamia where they encountered well-established institutions.

Virtually nothing is known about Greek mathematics at the beginning of the Hellenic era. Then, around the sixth century B.C., onto the scene came two men, Thales of Miletus (ca. 624–548 B.C.) and Pythagoras of Samos (ca. 580–500 B.C.) who, according to some historians, appeared to have played a role in mathematics that was comparable to Homer's in literature. However, there is reason to question such claims. As far as their personal scholarly contributions are concerned, both Thales and Pythagoras are somewhat enigmatic as no known mathematical compositions from either has survived, nor has it been established that either ever composed such work. Claims for such work by Thales and Pythagoras are based largely on oral tradition as no extant historical documents have ever been found. Thus, as major scientific or technological contributors, they fail to meet the criteria of published documents and/or extant conjoined physical objects that we set forth at the outset.

Despite the approximate nature of the periods of their lifetimes, what's broadly acknowledged about Thales and Pythagoras, however, is that they spent considerable time in Ancient Egypt and Mesopotamia. In particular, Pythagoras spent at least twenty-two years in Egypt studying astronomy, geometry, and the "mysteries," which had an evolutionary influence on Greek philosophy. Legend has it that in 585 B.C. Thales predicted a solar eclipse, thus astonishing his countrymen. Legend, however, does not generally offer that Thales was probably in possession of astronomical tables provided by the Mesopotamian Nebuchadnezzar. The command "know thyself" has been attributed variously to both Thales and Socrates; yet, many Ancient Egyptian temples displayed inscriptions addressed to neophytes, and among them was the common injunction "know thyself." Moreover, an interesting case has been intimated for the Ancient Egyptian influence on the musical and vibrational string analyses often attributed to Pythagoras.[4] And, as we have stated earlier, the Ancient Egyptians appear to have known and the Mesopotamians routinely used the "Pythagorean theorem" more than 1,000 years before Pythagoras was born.

While writing may have existed for about six millennia, it is certain that mathematics, as arithmetic or geometry, has existed much longer. Archaeological artifacts having numerical significance have been found from at least 30,000 years ago. Thus, statements about either the geographical or temporal origins of mathematics are at best hazardous. Nevertheless, it

---

[4]See McClain [1976].

is worth noting that throughout the Hellenic Era, Greek authors are near unanimous in their praise of and assignation to the Ancient Egyptians for priority of the invention of mathematical sciences. In particular, both Herodotus—called by some western scholars the "father of history"—and Aristotle place the origins of applied geometry in Ancient Egypt.

The reestablishing of territorial boundaries following the periodic inundations of the Nile required a developed system of land measurement and land management. Approximately 2,500 years ago Herodotus recorded the Ancient Egyptian use of arithmetic and geometry for establishing land boundaries and taxes, thus originating geometry and bequeathing it to Greece. Therefore, according to the Greeks, the Ancient Egyptian and Mesopotamian societies have had major technical and philosophical influence on western culture, extending well into the Hellenic era.

Just beyond the Hellenic era into the Hellenistic or Alexandrian Age,* two great Greek mathematicians who are not necessarily in the mainstream of this retrospective are briefly mentioned on account of their towering historical significance. These are Euclid, who lived in Alexandria about 300 B.C. and taught at the school known as the Museum, highly regarded in its day; and Archimedes (ca. 287–212 B.C.), who was educated in Alexandria, returned to live in Syracuse, and is considered to have been among the greatest mathematicians of antiquity.

> *The Hellenistic Age (323–30 B.C.) is generally considered to be the period between the death of Alexander the Great and the conquest of Egypt by Rome.

Little is known about Euclid's personal life, but at the Museum he was distinguished for his teaching ability, as opposed to research or administrative activities. Because so little is known about his origin, yet it is certain that he spent most of his life in Egypt, it is reasonable to ask whether Euclid was Egyptian or Greek. For a person so famous, it is surprising that not even his death was recorded. It was his skill of exposition that is considered to be the key to the success of his greatest work, the *Elements,* which was composed about 300 B.C. and has been copied and recopied numerous times. Probably no other mathematical book has had as many editions, as the *Elements* remains one of the most influential books of all time.

Archimedes is called by some western scholars the "father of mathematical physics." Beginning from simple postulates, in his *On the Equilibrium of Planes* he developed the rules for statics involving levers. Although Aristotle had given some comparable results about a century earlier, whereas Aristotle's approach had been nonmathematical, Archimedes' development was similar to Euclid's in the *Elements* in that it was mathematical. In his *On Floating Bodies,* Archimedes, beginning with simple postulates, developed major results on fluid pressures on a variety of shapes and on buoyancy. It is legend that Archimedes made his buoyant-force discovery while considering the problem of determining whether the crown for King Hiero of Syracuse was made of pure gold. Bathing and reflecting on the buoyancy that caused his body to float in the bath, in the excitement of his discovery he jumped from his bath and ran home naked, repeatedly shouting "Eureka" (I have discovered it). Furthermore, Archimedes was not an ivory-tower scholar. The number of ingenious peacetime and wartime devices attributed to him is renowned. However, balance scales, which required some knowledge of the rules of the lever, and "Archimedes' screw" were in use in Ancient Egypt centuries before the Greek scientist was born. Like Pythagoras and others before him, Archimedes is believed to have acquired the ideas for several of his inventions during his student years in Ancient Egypt. In assisting Syracuse in its defense against the Romans during the years 214 to 212 B.C., Archimedes invented catapults and other devices to attack Roman soldiers and ships. In 212 B.C. when Syracuse fell, so did Archimedes as he was slain by a Roman soldier, despite orders from the Roman general Marcellus to spare the life of the great mathematician.

1-11   THE UNIFICATION OF CELESTIAL AND TERRESTRIAL MOTION

For the ancients, any views on motion and dynamics were only a part of a much larger philosophical system that included religion, ethics, and cosmology; a system in which beauty, symmetry, and completeness took precedence over minor technical or scientific discrepancies. Cosmology is a branch of science that is concerned with the origin and evolution of the universe as a whole. In Ancient Egyptian cosmology (ca. 3000 B.C.), the universe was created out of chaotic matter, which itself had always existed, eternal, without a beginning or an end. From this matter evolved all future things: sky, stars, earth, air, fire, plants, animals, and human beings, and thus self-awareness. Ra, the demiurge (of whom Plato spoke) who completed creation, was the first consciousness to emerge from the chaotic primordial matter. Thus, Ra was the first god, having neither mother nor father.

Osiris, who, through the Neolithic period, was the god of agriculture, then died and rose from the dead to save humankind. He was the son of Ra; he was the god of redemption. It was this cosmology that created the concept of the Trinity, which permeated Ancient Egyptian religion. It was this cosmological philosophy that evolved into Greek philosophy, and into which the concepts of motion were enumerated; earlier by Parmenides, Zeno, Melissus, and Democritus; later by Aristotle.

1-11.1   Celestial Motion

The concepts of the motions of celestial bodies that would come naturally to an observer today are substantially the same as those that likely came to the Ancient Egyptians and the Mesopotamians. A summary of such concepts was given in the writings of Aristotle* (384–322 B.C.). In the Aristotelian system all accessible matter is composed of fire, air, earth,

---

*Although many of his concepts in astronomy and mechanics were wrong, Aristotle's ideas merit our review because of the major influence they had on the subsequent development and nondevelopment of both disciplines. Also, it is common among some scholars to question the authorship of various treatises attributed to Aristotle, including the three books *Problems of Mechanics*, *Physics*, and *Treatise on the Heavens*.

---

and water, each having its natural state. Fire and air reside above earth and water; thus, flames and smoke rise, rocks and water fall. Aristotle's view regarding celestial motion may be summarized as follows:

1. The Earth does not move.
2. All heavenly bodies are eternal and are not subject to the terrestrial (earthbound) laws of motion.

3. All celestial motion is circular.

4. The Sun, Moon, planets, and the fixed stars (what we, in fact, today call stars, as opposed to planets) all revolve about the Earth in circular paths on perfect crystal spheres. (See Figures 1-6 and 1-7.)

(As discussed in Subsection 1-11.2, according to Aristotle, the natural state of earthbound or terrestrial objects is rest.)

Today, we know that the apparent motions of the stars, as depicted for the northern hemisphere in Figure 1-6, are almost exclusively due to the motion of the Earth. Most visible stars, including the Sun, are orbiting the center of our galaxy; however, since it takes about two hundred million years to complete a single such orbit, this motion is much less than that represented in Figure 1-6. Since the stars are fixed relative to one another as they move along circular paths, they do appear to be attached to a crystal sphere. Those star trails that are sufficiently close to the celestial pole—the axis about which they appear to rotate—will be seen as complete circles, indicating stars that never dip below the horizon. (The Ancient Egyptians are said to have called these stars "the ones that know no destruction." The planets, which were not a part of this scheme, appeared to wander among this system of "fixed stars.")

The Aristotelian concepts are quite sensible, and in their sensibility foretell the subtle brilliance that would challenge and ultimately overturn them. These views became firmly entrenched and were not supplanted for nearly two millennia. Certainly, there were grumblings all along the way, but it was not until the European Renaissance that sufficiently compelling contrary evidence would supplant them and seal their demise. One of the earliest significant challenges came, within a century after Aristotle, from astronomers who reinterpreted the apparent celestial motions of the planets and the Sun.

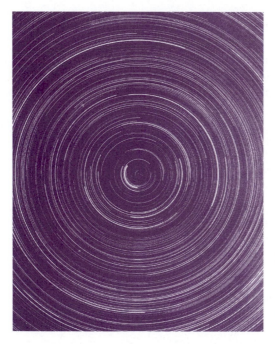

FIGURE 1-6   A time-lapse photograph of the sky makes stars appear to be carried around the celestial pole of heaven as if they were fixed to a hollow rotating sphere. The relatively bright trail near the center represents Polaris (North Star) near the northern pole.

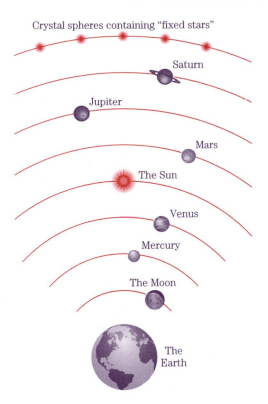

Crystal spheres containing "fixed stars"

Saturn

Jupiter

Mars

The Sun

Venus

Mercury

The Moon

The Earth

FIGURE 1-7 Geocentric scheme of the universe that emerged during the Hellenic period, based upon Ancient Egyptian, Mesopotamian, and Greek astronomy, philosophy, and speculation.

Aristarchus (ca. 310–230 B.C.) suggested a *heliocentric system* in which the Earth and planets rotated about the Sun, and the stars were fixed. This concept, which reputedly had been suggested a century earlier by Philolaos of Crete (a disciple of Pythagoras) and Nicete of Syracuse, offered a logical explanation for the variations in the brightness of the planets and in the retrograde motion of the planets—the apparent reversal of direction of the planets. Both effects, of course, were due to the Earth's orbital motion being superposed onto the orbital motions of each of the observed planets.

If this heliocentric system were true, there should be a semiannual parallax; that is, a change in the apparent location of a near star as the Earth orbited the Sun. (See Figure 1-8). However, due to the lack of precision in the astronomical measurements of that time, this parallax could not be observed. Thus, such a heliocentric system offered little advantage over the Aristotelian *geocentric system,* and hence found few supporters, especially since it contradicted the teachings that had been passed down through Aristotle.

Eratosthenes (ca. 275–195 B.C.) is credited with having made the first fairly accurate measurement of the Earth's diameter, about 200 B.C. At noon on the first day of summer, he observed in Syene, Egypt (near modern Aswan) that sunlight reached the bottom of a vertical well, indicating that Syene was on a direct line from the Sun to the center of the Earth. (See Figure 1-9). At the same date and time in Alexandria, he noticed that the Sun was not directly overhead, but south of the zenith, such that its rays made an angle, with respect to the vertical, of about 7.5°. Since the Sun's rays striking the two cities could be assumed to be parallel, his calculation revealed that the arc along the Earth's surface should subtend an angle of about 7.5°. Since the distance between Alexandria and Syene had been measured,

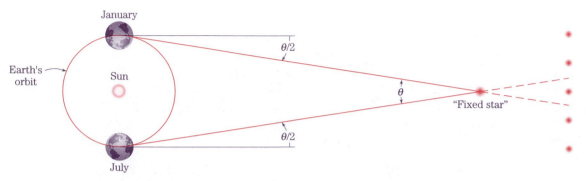

FIGURE 1-8   As Earth moves from January to July, the apparent location of a near "fixed star" appears to shift relative to more distant background stars. Angle θ depends on the diameter of Earth's orbit and the distance between the Sun and the fixed star. This is the parallax effect.*

> *With one eye closed, hold your thumb at arm's length in front of your face, and view a distant object. Now view the object using only the other eye. The apparent shift of your thumb, relative to the distant object, is due to the parallax effect.

the circumference of the Earth was $\dfrac{360°}{7.5°}$ times that distance. As there has been some confusion regarding the actual length of the unit he used,† appraisals of the accuracy of his answer vary from 1 to 20 percent difference from the correct value of about 40,000 km.

> †Eratosthenes used the unit called a "stadium" and there is some doubt regarding which of the various Greek stadia he used.

Hipparchus (ca. 190–125 B.C.) compiled a substantial star catalog consisting of about 850 entries. He divided the stars into six magnitude categories, according to their apparent brightnesses, and specified the magnitude of each star. By comparing his observations with older observations that he had inherited primarily from the Mesopotamians, Hipparchus made the remarkable discovery that the position in the sky of the north pole star had changed during the previous century and a half. He concluded that the direction of the axis about which the celestial sphere appears to rotate actually changes slowly. It has been argued by some authors that this phenomenon, due to what is known today as *precession,* was noted by the Mesopotamians two centuries (ca. 350 B.C.) before Hipparchus announced it.

The true explanation for the apparent motion of the north pole star is that the direction of the Earth's rotational axis changes slowly due to the gravitational influences of the Sun and the Moon, just as a spinning toy top's axis sweeps out a conical path. This motion of the direction of the Earth's axis is called "precession of the equinoxes" and requires about 26,000 years for one cycle. (See Figure 1-10.)

Claudius Ptolemæus (ca. A.D. 70–147), commonly known as Ptolemy, was the last great astronomer of antiquity. His major work was a series of thirteen volumes entitled the *Mathematical Composition,* known more popularly by its Arabic title the *Almagest* (literally, *the greatest*). The portion of the *Almagest* that deals with astronomy is substantially based

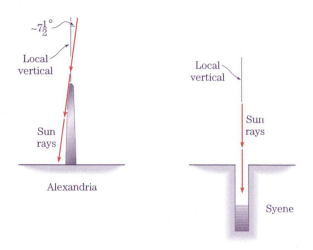

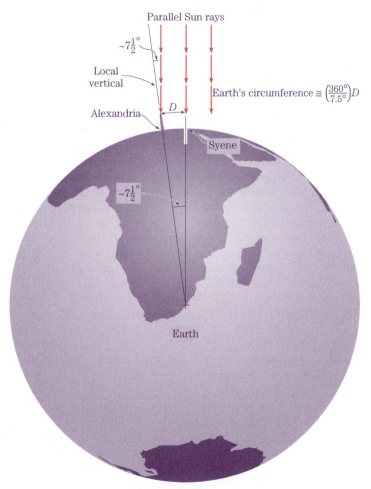

FIGURE 1-9   Eratosthenes' technique for measuring the Earth's circumference.

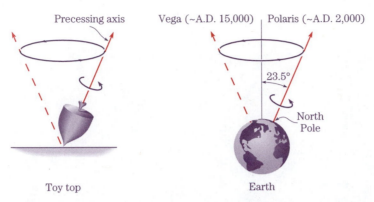

FIGURE 1-10   Precession of Earth's axis is analogous to precession of a spinning toy top. In about A.D. 15,000, Earth's pole star will no longer be Polaris, but Vega.

on the astronomical observations of the Mesopotamians and Hipparchus, including those of Ptolemy himself. Of the 1,022 stars listed by Ptolemy, 850 were passed down by Hipparchus, who himself had inherited star catalogs from the Mesopotamians. No other document from antiquity rivals the *Almagest* for its profound influence on the conceptions of the universe and none, with the exception of Euclid's *Elements,* achieved unchallenged authority for hundreds of years.

The astronomical observations in Ptolemy's possession were sufficiently accurate to indicate that all celestial bodies were not circling the Earth in perfectly Earth-centered orbits. There was a distinction between the "fixed stars"—the real stars that appeared to maintain fixed patterns among themselves (Figure 1-6)—and the "wandering stars" or planets. (See Figure 1-11a.) The word *planet* meant "wanderer" in Greek. The wanderers included the Sun, the Moon, and the then-known true planets, Mercury, Venus, Mars, Jupiter, and Saturn. Thus, these more accurate data caused Ptolemy, a convinced geocentrist, to relax the rigid circular geocentric assumptions. Not only did the planets display retrograde motion or apparent reversals in their paths, but their brightnesses varied and the angular rates at which they appeared to orbit the Earth varied. To maintain a fixed Earth, which was more or less at the center of the universe, Ptolemy adopted and modified a complex system of epicycles where the Earth was slightly off the center of the major orbit, called the *deferent.* (See Figure 1-11b.) This was a colossal and ingenious astronomical system, which for the first time in history enabled the prediction of the locations of the planets with acceptable accuracy. (The correct explanation for the apparent retrograde motion is illustrated in Figure 1-12.) The Ptolemaic geocentric system survived for more than 1,300 years, and even then it did not die quietly.

Nicholas Copernicus (1473–1543), who expressed concern about the complexities of the Ptolemaic system, was the major rejuvenator of the concept of the heliocentric system. Based upon observations of the motions of Mercury and Venus in particular, Copernicus was able to propose a quantitative scheme in which all the known planets were placed in *circular orbits* in their correct relative positions with the periods of revolution about the Sun for each planet specified. He noted that planets closer to the Sun had shorter periods than those farther from the Sun. Regarding the parallax as predicted by Aristarchus, Copernicus could not measure it, so he concluded that the fixed stars were too far away for the parallax to be measured. Interestingly, Copernicus in his *De Revolutionibus Orbium Coelestium* (On

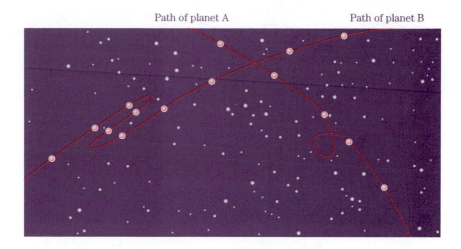

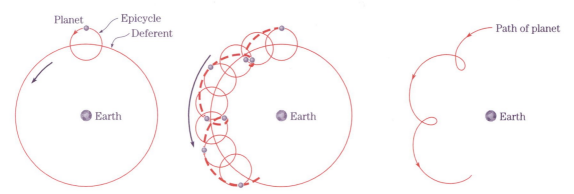

FIGURE 1-11   Geocentric solar system suggested that retrograde motion of planets in (*a*) could be modeled by analyses based on epicycles and deferents shown in (*b*).

the Revolutions of the Celestial Spheres) also expressed the belief that gravity is a "natural quality" given to the Earth by the "divine providence," and suggested that this same natural quality "belongs to the Sun, the Moon and to the wandering lights."

As expected by Copernicus, almost no one welcomed his heliocentric ideas. First of all, in some instances the Copernican system was not as accurate as the Ptolemaic system in predicting the locations of the planets. But even more important, the geocentric system was well-established doctrine in both natural philosophy and religion. In fear of the potential consequences for his recusant views, Copernicus withheld the publication of his ideas until he was near death in 1543 although he had completed his work about 1530. While the heliocentric system offered considerable simplicity, both the geocentric and heliocentric systems provided satisfactory explanations of astronomical observations. Besides, the enormous distances that were necessary to explain the lack of a measurable parallax of the fixed stars were considered to be ludicrous. However, by far the most damning aspect of the heliocentric system was that it required the heretical conclusion that we and our Earth were not at the center of the universe.

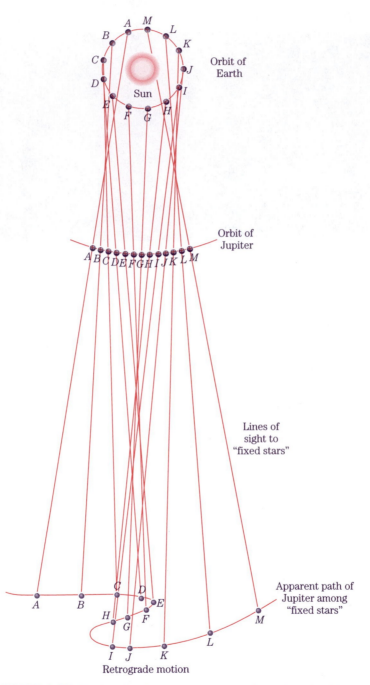

FIGURE 1-12 Explanation for apparent retrograde motion of a planet such as Jupiter.

Before the Danish astronomer Tycho Brahe (1546–1601), the precision of astronomical data was not very high, typically about $\frac{1}{6}^{\circ}$ or 10 minutes of arc. With the patronage of King Frederick II of Denmark, Brahe constructed an observatory, containing state-of-the-art observation equipment, on the island Hven across the Sound from Hamlet's castle at Elsinore. All of Brahe's astronomical data were obtained with instruments using the naked eye; the use of lenses in the telescope had not yet been invented by the Dutch optician Hans Lippershey (1587–1619). The precision of Brahe's angular measurements was approximately two minutes of arc.*

> *The width of the thumb held at arm's length subtends about $1°$ or 60 minutes of arc. Today, radio astronomers are capable of angular precision of the order of 10^{-3} seconds of arc. Such precision corresponds to an arc subtended by a quarter ($0.25) held in Cambridge, Massachusetts when viewed from Cambridge, England.

Brahe rejected Ptolemy's geocentric system because of the complexity of its epicycles. Also, he rejected Copernicus' heliocentric system primarily because of religious predisposition. Under the tension created by the more accurate data that he obtained and his rejection of the geocentric and heliocentric systems, Brahe created his own system. In the "Tychonic system," as it is called, the Moon and the Sun orbited the universe-centered and fixed Earth, but all the (known) five planets orbited the Sun—an amalgamation of two systems.

One of Brahe's goals was to detect and to measure the parallax of the fixed stars, but in this he failed.† In the end, Brahe's most significant contribution was neither in his theorizing nor his new geocentric system but in the very accurate data that he systematically collected over the years and left to his hand-picked successor Johannes Kepler, despite their mutual apprehension of one another.

> †Even with Brahe's increased accuracy, in order to detect the parallax, a fixed star could be no farther than 10^{11} km. Today, we know that the closest star (beyond the Sun) is Alpha Centauri, at a distance of the order of 10^{13} km. In the sixteenth century, the vastness of the known universe was difficult to grasp; today, it still is.

Johannes Kepler (1571–1630) was an indefatigable character who throughout much of his career used his analytical skills to cast horoscopes in order to earn a living. He reportedly said that astronomy would die of hunger if her daughter, astrology, did not earn enough bread for two. A convinced Copernican and once a brief protégé of Tycho Brahe, Kepler took possession of much of Brahe's voluminous and substantially undigested data on the motions of the planets. Kepler attempted to fit Brahe's data to the Copernican system but discovered that the assumption of circular orbits could not be reconciled. There was a nagging error of about 8 minutes of arc even in the best *circular* orbital fit for Mars; pre-Brahe, such an error might have been acceptable, but post-Brahe, it was unacceptable. Confronted with the choice between theory and data, the superb mathematician Kepler, who was among several who came close to inventing the calculus, chose data.

The focus of Kepler's efforts was on the data for Mars that Brahe had accumulated since about 1580. (Mars has the largest orbital eccentricity among those planets for which Kepler had accurate data.) Ultimately, Kepler discarded the assumption of a circular orbit in favor of an *elliptical* path about the Sun, with the Sun at one of the foci. Today, we designate this as the first of Kepler's three laws of planetary motion, which we summarize in modern terminology.

Law I: Law of Ellipses (1609)

The orbit of each planet is an ellipse with the Sun at one of its foci.

Law II: Law of Equal Areas (1609)

A directed line from the Sun to a planet sweeps out equal areas in equal times.

Law III: Harmonic Law (1619)

The square of the period of orbit of each planet is proportional to the cube of its semimajor axis.[*]

---

[*]The dates shown are the dates of publication; Kepler never collected these three laws as indicated here.

Although Kepler was a gifted mathematician, the articulation of these laws was an enormous task that required nearly twenty years for its completion.

Kepler's first and second laws are illustrated in Figure 1-13. First, the Sun is at one focus of the elliptical planetary orbit. Second, if the time between any two successive indicated planetary locations (such as AB or FG or JK) is the same, then the corresponding area swept out between any two successive indicated planetary locations is the same. The third law may be expressed as

$$T^2 = CR^3$$

where T is the time for one orbit around the Sun (that is, one year for Earth) and R is the distance of the semimajor axis of the planetary orbit. The constant C is the same for

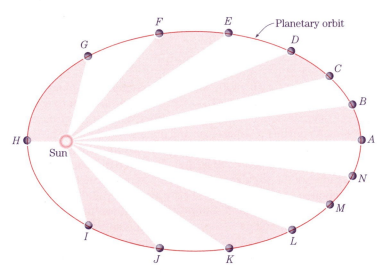

FIGURE 1-13   Illustration of Kepler's first and second laws, where time between any two successive indicated locations is the same, and all shaded and unshaded areas are identical.

all the planets. The third law is plotted in Figure 1-14 for all the planets, normalized by the Earth–Sun distance and the Earth's period, one year. Recall that one astronomical unit (AU) is defined as the length of the semimajor axis of the Earth's orbit; namely, 1 AU is approximately 1.497×10^{11} m ($\sim 93 \times 10^6$ mi).

Alas, the reward was simplicity. Nevertheless, although the motion of the planets was geometrically described, an understanding of the underlying dynamic principles governing their behavior had to wait another fifty years, in the person of Isaac Newton.

Galileo Galilei (1564–1642) understood that as far as pure description was concerned, the geocentric and heliocentric systems were equivalent, the difference being simply a matter of the choice of a reference frame. But the heliocentric system was simpler. Also, because of the writings of Copernicus, the astronomical data of Brahe, the work of Kepler, and his own telescopic observations, especially of Jupiter and its moons, Galilei became convinced that the Earth was not the center of all rotation in the solar system. Many people thought that Galilei had invented the telescope, although, as far as we can discern, he never claimed to have done so. Here, it is interesting to examine Table 1-2, which contains a comparison of the approximate precision of the measurements of some famous astronomers. (It may be noted that regarding the measurements cited for the Khufu Pyramid, we cannot assess to what extent these deviations may be due to 4,900 years of differential settling of its foundation.)

On January 7, 1610 Galilei discovered four of the Jovian moons, and subsequently noted that the moons farthest from Jupiter had the longest periods and that as a group the moons accompanied the planet in its orbit around the Sun. This reinforced the idea of the same possibility for Earth and its moon in relation to the Sun. Galilei's Copernican views brought him into direct conflict with the Vatican. His *Dialogue Concerning the Two Chief World Systems* (four dialogues on the two principal systems of the solar system, those of Copernicus and Ptolemy) resulted in his appearance before the Inquisition. On April 30, 1633 he was obliged to renounce his Copernican beliefs as "absurd, cursed, and detested," and required to remain under technical arrest at his residence in Arcetri, near Florence, for the remaining nine years of his life. Although Galilei endured personal angst upon such a ruling,

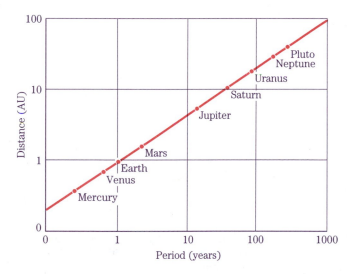

FIGURE 1-14 Plot of Kepler's third law, normalized with respect to Earth–Sun distance and Earth's period.

TABLE 1-2 Approximate Precision of Some Famous Astronomers

Astronomers	Approximate Precision (Seconds of Arc)	Approximate Date
Ancient Egyptians	2[†] 33 182 213	2900 B.C.
Ptolemy	600	100
Brahe (without telescope)	120	1600
Galilei (with telescope)	30	1600
Modern Astronomers	10^{-3}	2000

[†] These are the deviations from right angles of the four corners of the Khufu Pyramid. The data are for the northwest, southwest, northeast, and southeast corners, respectively. The orientations of the four sides of the Khufu Pyramid (in seconds of arc) are as follows:

North   148   S. of W.
East   330   W. of N.
South   117   S. of W.
West   150   W. of N.

Note that these orientations are more a measure of accuracy than precision. Furthermore, we find it significant that what is interpreted as inaccuracies in the orientations of the four sides all possess the same counterclockwise sense (Figure 1-15). This type of systematic error suggests that the presently interpreted misorientation of the Khufu Pyramid may be due substantially to the nutation and precession (Figure 1-11) of the Earth during the past 4,900 years, and/or to continental drift.

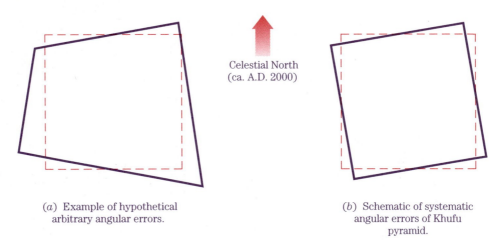

Celestial North (ca. A.D. 2000)

(a) Example of hypothetical arbitrary angular errors.

(b) Schematic of systematic angular errors of Khufu pyramid.

FIGURE 1-15 Illustrations of (a) hypothetical arbitrary angular errors of approximate square and (b) systematic angular errors of Khufu Pyramid; both with respect to present celestial north. (Angular errors are exaggerated for interpretation.)

he no doubt found it preferable to being burned at the stake, a fate that would have likely befallen an equally heretical but less popular personality.

During the 1980s the Vatican continued its quest to rationalize and resolve its dispute with Galilei's heliocentric views. On November 10, 1979, marking the first centenary of the birth of Albert Einstein, Pope John Paul II expressed renewed hope toward this goal, and on July 3, 1981 he constituted a commission to "study the Galileo case more deeply." In an address on May 9, 1983, marking the 350th anniversary of Galilei's *Dialogue Concerning the Two Chief World Systems* (the book was published in 1632), Pope John Paul II stated:

> *"We cast our minds back to an age when there had developed between science and faith grave incomprehension, the result of misunderstandings or errors, which only humble and patient re-examination succeeded in gradually dispelling... the church's experience during the Galileo affair and after it has led to a more mature attitude and to a more accurate grasp of the authority proper to her.... The church herself learns by experience and reflection, and she now understands better the meaning that must be given to freedom of research...."*

And although the commission reported in 1984 that the church had "committed an objective error," into the 1990s the Pope continued to revisit the lessons of the Galilei case, in part, as a paradigm for considering modern scientific issues such as chaos.

In concluding our abbreviated discussion of celestial motion, we shall comment briefly on one hypothesis that purports to explain, without the benefit of astronomy, the Ancient Egyptians' technique for achieving the high levels of precision and accuracy associated with the orientation of the Khufu Pyramid. While we have no interest in evaluating the numerous concepts, suggestions, or concoctions devised to explain the phenomenal precision found in the Khufu Pyramid, one particular proposal[5] merits comment for the following reasons:

1. The proposal denies the Ancient Egyptians' ability to devise any astronomically based procedure to orient the Khufu Pyramid.

2. The proposal is widely cited by modern historians who subscribe to the view that the Ancient Egyptians possessed no significant astronomical facility.

As we shall discuss succinctly, the proposed supposition that astronomy was not used is logically inconsistent, and the merits of the proposed hypothesis for using shadows are technically inadequate.

Without reiterating the details of the proposal here, we simply note that it requires the following steps.

1. The surveyor must be given a preliminary orientation that is "a reasonably accurate estimate of the SN/EW directions."

2. The surveyor then conducts repeated delineations of the edge of a shadow cast by a convenient gnomon about the given directions.

First of all (as we shall see in Chapter 3), it is meaningless to define a direction such as "north" without establishing a reference with respect to which it is to be defined;

---

[5]See Neugebauer [1983].

otherwise *any* direction is north, with perfect accuracy. Second, in the context used here, "north" has no meaningful reference except an astronomical one, assuming the Ancient Egyptians had no compass to direct them toward magnetic north. Apparently sensing the necessity of these two points, the author of the proposal states that the king, "looking at the northern constellations," might give the surveyor the "preliminary orientation." Thus, we are told that Ancient Egyptian surveyors could accurately locate north without the benefit of astronomy, but only after they were directed toward the north via reference to the stars. This is simply illogical. (Furthermore, note that by the time of construction of the Khufu Pyramid, the Ancient Egyptians had centuries of astronomical experience, including the design of their annual calendar. No historian we have read has insinuated either that the annual calendar was not being used by the Ancient Egyptians by the time of Khufu or that the annual calendar had any basis other than an astronomical one.)

Notwithstanding the logical inconsistencies of the proposal, let's turn to considerations that address its quantitative technical difficulties. If Ancient Egyptian surveyors had attempted to use a shadow-based technique, they would have been defeated in their efforts to acquire the precision and accuracy that they bequeathed for us to admire. As readers may observe for themselves, the delineation of the edge of a shadow may appear to be crisp; but not if viewed increasingly closer. As our calculations—which are based on modern optics and are beyond the scope of this textbook—reveal, within an edge region of a few hundred seconds of arc, the combination of a time-varying penumbra's location and width and, during the presence of colored light, Fresnel diffraction fringes—striations of spatially varying intensity—along the fully illuminated edge of the shadow would certainly frustrate attempts to delineate the edge of a shadow with the precision listed in Table 1-2.

Thus, the resolution of a shadow to the required precision would have led the Ancient Egyptians to engage in some sophisticated geometrical optics and, perhaps, fringe averaging. To our knowledge, no one has suggested that the Ancient Egyptians discovered Fresnel diffraction, as they might have if they had used the shadow-based proposal at the required level of precision. Such a feat might impress modern engineers to an extent far greater than the precision of the Khufu Pyramid itself.

In our opinion, centuries prior to the reign of Khufu, the Ancient Egyptians understood the information represented in Figure 1-6 and, with the star trails as a reference, simply chose their center to be *north*. Such was *celestial north* at the time of Khufu; such is *celestial north* today. It's just that simple.

1-11.2   Terrestrial Motion

From Aristotle's writings, *Treatise on the Heavens* and *Physics*, several of his mechanical views regarding terrestrial motion may be summarized as follows:

1.   The natural state of all earthbound bodies is rest (zero velocity).
2.   Objects fall in a straight line.
3.   Heavy objects fall faster than light objects.
4.   If a given weight falls from a certain height in a particular time interval, a weight that is twice as large will fall from the same height in half that time interval.

Further, with some simplification, these views suggest that velocity is proportional to force, although Aristotle never wrote this, perhaps because he recognized that a small force would never move a large object. For nearly 2,000 years,* there had been challenges and rumblings; nevertheless, this was essentially the state of understanding of terrestrial motion into which Galilei was born.

---

*The Ancient Egyptians and Mesopotamians are often criticized for having presided over a relatively static level of scientific development between 2000 B.C. and the Hellenic era.

In his *Dialogue Concerning the Two Chief World Systems,* Galilei constructed a thought experiment in which all external forces and impediments were removed from a body moving along a horizontal plane. All air resistance, imperfections in the ball, and forces were idealized away. Although Galilei understood that such an experiment could not be realized practically, he concluded that the hypothesized conditions would result in perpetual motion. Thus, he concluded the natural or uninfluenced state of motion of a body is *constant velocity;* not zero velocity or rest! This was the seminal idea, the basis for inertia and the key to unleash the development of dynamics. Like most great ideas, post hoc it appears to be very simple; but it really isn't. It is essentially the first of Newton's three laws.

From a letter dated October 16, 1604, it is known that as early as that time Galilei had concluded the correct law for falling bodies: that distance is proportional to the square of time. In his *Discourses and Mathematical Demonstrations Concerning Two New Sciences,* Galilei cleverly discussed his ideas and experiments on projectile motion and falling bodies. He concluded that bodies of different weight, when dropped from the same height at the same time, would reach the ground at the same (or nearly the same) time. He correctly attributed the small differences in the time of fall to differences in air resistance. Thus, he concluded that all bodies, when dropped in a vacuum, would reach the ground at the same time. (As far as we can ascertain, Galilei never dropped objects from the leaning tower of Pisa, as the popular tale suggests.)

1-11.3   Unification

Isaac Newton (1642–1727) was a profound natural philosopher in the most general sense. He made major contributions in mathematics, astronomy, mechanics, heat transfer, and optics. His work in any one of these disciplines would rank him among the best scientists of all time. He was a bold intellect who frequently engendered strong debate among his contemporaries. Most notable were his contention with Robert Hooke (1635–1703)—of "Hooke's law of elasticity"—who criticized Newton's 1672 paper concerning the nature of color as a refraction of white light, and the controversy with Gottfried Leibniz (1646–1716) concerning the independence and priority of the invention of the calculus. Newton's response to Hooke's criticism was to publish nothing further on light during Hooke's lifetime.

Much has been written about the priority of the invention of the calculus. The records indicate that Newton had the calculus first (Newton in 1665–66; Leibniz in 1673–76), but Leibniz published it first (Leibniz in 1684–86; Newton in 1704–36). For fifteen years, Leibniz enjoyed the undisputed honor of having been the inventor of the calculus. In 1699 that position was challenged, resulting in a bitter debate, including accusations of plagiarism against Leibniz. This debate hurt Leibniz personally but it hurt English mathematicians even more; indeed, for the entire eighteenth century. The controversy created an intellectual chasm between English and Continental European mathematicians; and since English mathematicians chose to use Newton's methods and notation, they became ignorant of the brilliant discoveries on Continental Europe by mathematicians using Leibniz's superior notation. Today, it is generally agreed that Leibniz was an independent inventor of the calculus.

Newton's major work in mechanics is entitled *Philosophiæ Naturalis Principia Mathematica* (Mathematical Principles of Natural Philosophy) and is generally simply called *The Principia*. Published in 1687, *The Principia* gives a brilliant distillation of the accumulated dynamics of its time, including Newton's generalizations and extensions beyond terrestrial dynamics to celestial dynamics. Also contained in *The Principia* are analyses of motions of bodies in resisting media, and calculations for the isothermal velocity of sound. Though not always rigorous, logical, or even correct, *The Principia* is considered by many to be the greatest scientific document ever written; it was certainly influential. Newton had accomplished what all theoreticians seek: an underlying principle that unifies an apparently diverse set of natural phenomena.

The Principia contains three books, and at the beginning of Book I, in one of the most widely quoted translations, Newton* states the three laws that he presumes to govern the motions of all objects in the universe.

---

*Newton's studies on terrestrial mechanics were, first and foremost, influenced by the works of Galilei; to a lesser extent by Christiaan Huygens (1629–1695)—admired by both Leibniz and Newton—who expressed his principle of inertia as "Any body in motion tends to move in a straight line with the same velocity as long as it does not meet an obstacle" (1700); and apparently even less, if at all, by René Descartes (1596–1650), who in 1629 wrote a comparable statement to Newton's first law. Newton's studies on celestial mechanics were principally influenced by Kepler's pioneering work.

Law I:

Every body continues in its state of rest, or of uniform motion in a right line, unless it be compelled to change that state by forces impressed upon it.

Law II:

The change of motion is proportional to the motive force impressed; and is made in the direction of the right line in which that force is impressed.

Law III:

To every action there is always opposed an equal reaction; or, the mutual actions of two bodies upon each other are always equal, and directed to contrary parts.

In the Latin in which Newton wrote, these three laws contain only fifty-nine words.

The law of universal gravitation is also enunciated in Book I, where Newton hypothesized that there is a universal attraction between all bodies everywhere in space, and that between two bodies this takes the form of a simple equation

$$f = \frac{km_0 m}{r^2}$$

where f is the attractive gravitational force, k is the gravitational constant, m_0 and m are the masses of the two bodies attracting each other, and r is the distance between the centers of the two bodies. (The celebrated story of the falling apple leading to the concept of gravity is based on the very slender evidence of an anecdote by an eighty-four-year-old reminiscing

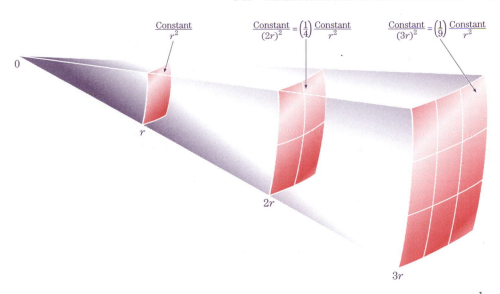

FIGURE 1-16 Illustration suggesting why intensity of many field variables is proportional to $\frac{1}{r^2}$. Intensity at $2r$ is $\frac{1}{4}$ intensity at r since area at $2r$ is 4 times area at r; intensity at $3r$ is $\frac{1}{9}$ intensity at r since area at $3r$ is 9 times area at r.

Newton.) Figure 1-16 suggests the reason for the $\frac{1}{r^2}$ character of the gravitational force. With the three laws of motion and the law of universal gravitation, in 1665–66 (though not published until 1687) Newton mathematically showed in Book 3 that the planetary orbits should indeed be those described by Kepler's laws. Thus, all motion in the universe, terrestrial and celestial, had been unified!

1-12 VARIATIONAL PRINCIPLES IN DYNAMICS[6]

Since the latter half of the seventeenth century, the discipline of dynamics has evolved along two primary lines. One line, which we shall informally call the "direct approach," is often denoted as *vectorial dynamics*. Vectorial dynamics takes as its primary parameters force and momentum. In vectorial dynamics, Newton's laws are considered directly in order to derive the equations of motion for a system.

Newton's contemporary, the natural philosopher Leibniz, advocated an alternative approach to derive the equations of motion for a system. The "indirect approach," as we shall informally call it, is often denoted as *variational dynamics*.* In variational dynamics

---

*The terms *variational dynamics* and *analytical mechanics* are often used interchangeably.

---

[6]In the discussion of this section, we shall use some terminology that is likely to be new to the reader. Nevertheless, we shall plow onward, as all such terminology will be defined in the appropriate technical context in subsequent chapters.

of mechanical systems, the "momentum" of Newton is replaced by a function akin to "kinetic energy"* and the "forces" of Newton are replaced by the "work done by the forces." As

> *We shall see that the function that is generally called *kinetic energy* will be defined differently later in this textbook.

Newton is often considered to be the originator of vectorial dynamics, Leibniz may be considered to be the originator of the indirect approach, although many writers, with justification as indicated later in this section, attribute to John Bernoulli and/or d'Alembert this position. In any event, like vectorial dynamics, variational dynamics represents one of the magnificent chapters in the intellectual history of humankind. A coarse summary of these two dynamics formulation approaches is presented in Figure 1-17.

The earliest extant formulation of the principle of virtual work is given variously in Aristotle's three works *Treatise on the Heavens*, *Problems of Mechanics*, and *Physics*. In these, the law of the lever is derived from an unclear formulation of the *principle of virtual velocities,* ending with the conclusion that forces on opposite sides of the lever balance each other if they are inversely proportional to the velocities. (This can be considered to be a balance of power or work produced by the forces on opposite ends of the lever. The reader should not allow the unfortunate use of the word *virtual* to be confusing; simply think of "virtual motion" as suggesting an "imagined geometrically compatible motion.") The phrase *virtual velocities* survived until the nineteenth century when it was replaced by *virtual displacements.* We shall prefer to use the phrase *admissible variation* for such imagined compatible displacements.

While Galilei's achievements as an experimentalist are well known, his theoretical accomplishments are less known. In considering Aristotle's principle of virtual velocities, Galilei noted that it is not simply the velocity that's important, but the component of the velocity in the direction of the force that produces work.

Gottfried Wilhelm Leibniz (1646–1716) in 1695 proposed to replace the time rate of change of Newton's linear momentum with a quantity he called the *vis viva* (Latin for "living force"), which corresponds to what is commonly called the *kinetic energy,* except the $\frac{1}{2}$ was

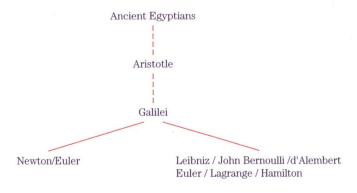

FIGURE 1-17   Brief historical schematic of two major dynamics formulation approaches.

not present. Also, Leibniz replaced the force of Newton with the work done by the force. The concepts of energy and work were consistent with the later concepts of variational dynamics.

In extending these ideas—extensions not attributable to Leibniz—we note that the work of the conservative forces could be replaced by another quantity, the negative of the potential energy. One of the great advantages of this indirect approach is that all forces that do no work, such as forces of constraint on particles, need not be considered—an idea principally due to John Bernoulli. Second, accelerations do not have to be computed, as only velocities are necessary. Third, in general, operations are on scalars, not vectors. Thus, the force–momentum definitions of Newton were clearly better equipped to contribute to vectorial dynamics, whereas the work–energy concepts of Leibniz became the cornerstones of variational dynamics. Again, this discussion should not be interpreted to suggest that the concepts of work and energy were thoroughly stated by Leibniz. Beyond Leibniz, much remained to be done to complete this alternative formulation approach.

Jean I (Anglicized, John or Germanic, Johann) Bernoulli (1667–1748), in a letter written in 1717, considered the generalization of the principle of virtual velocities and so argued that the principle of virtual velocities, today called the *principle of virtual work,* is a general principle of statics. Although this principle had been bandied about for at least two millennia, John Bernoulli is generally acknowledged to have formulated the principle's first precise statement. He concluded that for all admissible variations (compatible infinitesimal displacements), the sum of all the work done by the forces must vanish if the forces are in equilibrium.

Pierre-Louis Moreau de Maupertuis (1698–1759) expressed the hypothesis that in all dynamic processes there is a certain quantity, called the *action,*[*]

---

[*]The "action" which was ultimately defined in different ways by Euler, Lagrange, and Hamilton, not Maupertuis, is a scalar quantity having the dimensions of the product of energy and time (or, equivalently, the product of momentum and displacement). We note that these quantities are yet to be defined in this text.

---

which is minimized. The idea was based on the cosmic and theological idea that a perfect Nature would create only a perfect universe that would not tolerate waste. So, the laws describing the behavior of matter had to reflect the perfection worthy of divine creation and the principle of least action seemed to satisfy this criterion because it showed that Nature was economical. Also, Maupertuis claimed that the principle of least action was the first scientific proof of the wisdom and therefore of the existence of God. After all, even Newton's first law, which states that an unforced object in motion follows a straight line, the shortest path, provided an example of the Creator's economy.

In modern and mechanical terms, the *least action* concept is illustrated in Figure 1-18. Suppose a system, which may be in a gravitational field but which is otherwise not subjected to external loads, starts at point A at time t_1 and arrives at a destination point B at time t_2. Now suppose we consider alternative or "imagined motions" of the system that also start at point A at time t_1 and arrive at point B at time t_2. These *imagined motions* consist of all the combinations of position, velocity, and acceleration that we might envision within our imagination, limited only by their consistency with the constraints on the system. Now, in modern terminology, we define the *action* $\mathcal{A}$ as follows:

$$\mathcal{A} = \int_{t_1}^{t_2} [(\text{Kinetic Coenergy Function}) - (\text{Potential Energy Function})]\, dt$$

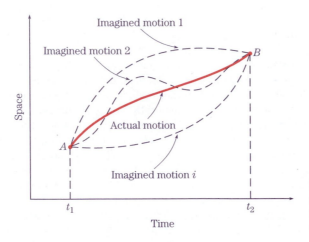

FIGURE 1-18 Sketch of actual and imagined motions of a mechanical system between initiation point A at t_1 and destination point B at t_2.

where the reader may think of the kinetic coenergy function as akin to kinetic energy and the potential energy function as akin to potential energy. Then, if the action $\mathcal{A}$ is computed for all the different motions, actual and imagined, the action associated with the actual motion will be less than the action for any other motion imaginable! *If* it is so, this is clearly an alternative—not necessarily equivalent—articulation of Newton's laws. Also, the *principle of least action*, as it is now known, is a revolutionary perspective of the behavior of the world, with far-ranging implications. Can it be applied to nonmechanical systems, that is, beyond the applicability or the goals of Newton's laws?

The universality of the hypothesis of least action by Maupertuis was in keeping with the philosophical temper of many intellectuals, scientists and nonscientists, of the sixteenth, seventeenth, and eighteenth centuries. These included philosophers and poets as well as mathematicians. Most notable among the mathematicians was Leonhard Euler, considered by some to be the greatest of that century, or any century. Indeed, in times of controversy— and there were several—regarding the priority of the discovery of the principle of least action, the copiously charitable Euler went to the defense of Maupertuis. Maupertuis could not have had a stronger analytical ally, for what he lacked in mathematical abilities Euler possessed in great abundance.

Leonhard Euler (1707–1783) clarified the principle of least action and laid the groundwork for the calculus of variations; the mathematics of variational dynamics. It is, however, Lagrange who is said to have founded the calculus of variations (which, considering Euler's substantial contributions, is only partially true). It was Lagrange who generalized the principle of least action. Furthermore, the application of Leibniz's differential notation rather than Newton's fluxions by Euler and others—most notably Lagrange, Laplace, and the Bernoullis—to the principles given in *The Principia* was among the great intellectual achievements of the eighteenth century. In conjunction with his enormous output, Euler's choice of Leibniz's notation led the way toward our modern notation and techniques in mechanics. Euler was the most prolific mathematician of all time, having averaged about 800 printed pages per year throughout his long life. It is noteworthy that Euler was legendary for his generosity, modesty, and charity throughout his professional career.

Jean LeRond d'Alembert (1717–1783) effectively reduced the problem of dynamics to a problem in statics by the process of adding, to the given externally applied forces, a new force—the force of *inertia*. For a particle of constant mass, this new force is simply the negative of the product of its mass times its acceleration. This new force, the force of inertia, when combined with the externally applied forces produces *dynamic equilibrium,* (i.e., $F - ma = 0$). The apparent triviality of this operation has made d'Alembert's "principle" open to ridicule and distortion. As an equation of statics instead of dynamics, it allows the principle of virtual work, originally derived for statics, to be extended directly to dynamics. Whether such a principle represents a significant milestone in analytical dynamics is still debated; that it influenced the work of Lagrange and Hamilton and therefore the development of dynamics itself is not debatable. The pursuit and development of minimum principles can thus be said to have been strongly influenced by d'Alembert.

As already indicated, it was by no means rare for advocates of minimum principles to attribute their validity to the efficient design of the universe by its Creator. In his *Traité de Dynamique, Discours Préliminaire,* published in 1743, d'Alembert took issue with this kind of metaphysical rhetoric, of which Maupertuis was one among many proponents.

Joseph-Louis Lagrange (1736–1813) adopted the concepts and postulates of Galilei, Huygens, Newton, John Bernoulli, Euler, and d'Alembert, and transformed them into the foundations and approaches of variational dynamics. In his *Méchanique Analytique,* first published in 1788, Lagrange created a new and very powerful tool for deriving the equations of motion. He strongly advocated the superiority of variational principles because of their power to characterize the position of a mechanical system by any set of geometrical variables that was suited to the character of the problem. Thus, without a required reference to physical or geometrical considerations, dynamics problems could be formulated, provided that certain energies of the system could be expressed in analytical forms of those geometrical variables, called *generalized coordinates,* though not so named by Lagrange.

Lagrange saw himself as a mathematician and sought mathematical elegance in his contributions. He saw his book as being primarily a mathematical contribution. According to Lagrange, "No diagram will be found in this work. The methods that I explain in it require neither constructions nor geometrical or mechanical arguments, but only the algebraic operations inherent to a regular and uniform process. Those who love Analysis will, with joy, see mechanics become a new branch of it and will be grateful to me for thus having extended its field."

The Newton-Leibniz controversy over the priority of the invention of the calculus had led to somewhat of an intellectual isolationism within Britain during the eighteenth century. Those educated in and followers of the geometrical methods of Newton were either not aware of or not prepared to exploit the tremendous advances in mechanics made on the Continent. In 1813, a group of young Cantabrigians under the leadership of George Peacock (1791–1858), John Herschel (1792–1871), and Charles Babbage (1792–1871) formed the Analytical Society at Trinity College. By reading papers and preparing translations of European publications, they conducted a famous struggle that was to spearhead the introduction of the Continental notation into the renowned Cambridge mathematical tripos, a portion of the famous system of examinations that had been initiated in 1725. Thus, the turning point in British mathematics had occurred, commencing one of the most prolific centuries in British science and mathematics.

William Rowan Hamilton (1805–1865) extended the formulation of Lagrange by giving the first exact formulation of the principle of least action. By transforming Lagrange's form of d'Alembert's principle via an integration with respect to time, Hamilton derived what is known as *Hamilton's principle.* An unconventional and perhaps controversial interpretation of the generality of Hamilton's principle is suggested here by us. Most authors advocate the view that as a consequence of Hamilton's principle having been derived from Lagrange's form of d'Alembert's principle, both principles are equivalent to Newton's laws, and their scopes are the same. It is our opinion that what Hamilton did in deriving his principle from Lagrange's form of d'Alembert's principle was not a derivation but simply a demonstration of equivalence of the two for mechanical systems. Indeed, in some respects, a higher degree of generality of his principle was understood by Hamilton as he viewed his theory of mechanical dynamics to be applicable also to optics.

1-13   THE INTERNATIONALISM OF DYNAMICS

Having reviewed some of the great achievements in mathematics and mechanics throughout history, we note that the writings of many of these celebrated thinkers were often undeveloped and unclear, leaving much to be done in clarifying their ideas. The systematizing of scientific and engineering concepts is extraordinarily important and is often as difficult as their discovery. Furthermore, in many instances, the attribution of a concept or analysis to a particular individual is as much an indication of the political, sociological, and scientific thought of an era as it is the suggestion of adulation of a persona. The notion of the genius, working largely in isolation, is a sociological and political concoction designed to achieve the goals of a particular nation, institution, or community; a notion that, in modern research, is almost certain to be laden with exaggeration.

The historical roots of dynamics—indeed much more of western civilization—lay in Ancient Egypt, subsequently followed by Mesopotamia. The Ancient Egyptians created a *transcendental expression,*[7] which they rendered, in part, in their temples and massive monuments, establishing themselves as the first great engineering society in history. This was accomplished via, among other achievements, the invention of the annual calendar from their astronomy and the invention of (historical) mathematics via their arithmetic, geometry, algebra, and trigonometry. There is incisive evidence that their immense practical ability with numbers and precision was accompanied by an interest in the intrinsic character of numeration for its own sake. The cornerstone in a significant part of the intellectual edifice of western civilization had been lain. Then, along came the Greeks, who broadly proclaimed their intellectual debt to the Ancient Egyptians but who did not stagnate as they clearly articulated the realm of *abstract theoretical expression*. And the torch was then passed again.

In conclusion, we observe that the development of dynamics has been a distinctly international achievement. Beyond the Ancient Egyptians and the Mesopotamians, by birthplace,[8] Aristotle and Aristarchus were born in Greece, Archimedes in Sicily, Eratosthenes in Cyrene (now Shahhat in modern Libya), Hipparchus in Nicæa (now Iznik in modern Turkey), Euclid (?) and Ptolemy in Egypt, Copernicus in Poland, Brahe in Skåne, Denmark (now in Sweden), Kepler and Leibniz in Germany, Galilei and Lagrange in Italy, Huygens and Daniel Bernoulli in the Netherlands, Newton in England, John and James Bernoulli and Euler in Switzerland, Maupertuis and d'Alembert in France, and Hamilton in Ireland.

---

[7]Notwithstanding all the political considerations, transcendental expression was the primary motivation for Americans going to the Moon, and will play a significant role in wherever else interplanetary travel may take humankind.

[8]While the places in which ideas were initiated and/or flourished are more significant than the origins of the authors of those ideas, birthplaces provide a simpler and more accessible indicator of the internationalism of the roots of dynamics.

1-14   OUR NICHE IN THE COSMOS

We are at the end of our selectively extracted retrospective wherein we have attempted to acknowledge those who contributed most significantly to the conceptual and analytical aspects of our subject. Let's now step back and consider the grand picture, a view implied by the first sentence in this chapter.

Cosmology is a branch of science that is concerned with the study of the past, present, and future of the universe as a whole. What is the totality of the universe; how did it begin; how has it evolved; and where is it going? These are questions that in one form or another have been asked since there have been human beings who have looked skyward and wondered what is the universe and how does it function. And in those musings, all members of our species became unified in taking the first step toward discovering our niche in the cosmos.

We have discussed the mathematics and astronomy of the two great ancient civilizations from which western mathematics, astronomy, and technology evolved. In extending our philosophical perspective, for example, consider the cosmology and religion of Ancient Egypt. They believed that the gods had created the universe out of chaos; that they themselves lived in an immutable and eternal world; that their deities were likened to shepherds tending their human flocks; that the demise and resurrection of their supreme god Osiris hinted at the possibility of immortality for themselves; and that the faithful follower of Osiris could overcome human frailty and even death, and share in his eternal bliss. The cosmology of the Hellenes refined these tenets in which the crystal spheres of the heavens were serene, immutable, and eternal. In this scheme, only here on this lowly Earth were there disorder, death, and decay. But that place, the Earth, was nevertheless at the center of the universe; hence humankind could understandably surmise ourselves to be the underlying purpose of creation. With very modest alterations, this was the western world perspective at the beginning of the Renaissance; the world view into which the modern chapter of dynamics was born.

Then along came Copernicus, Kepler, Galilei, Newton, and our modern astronomers and cosmologists. And by the time they had strutted across the historical stage, all humanity lived on a speck of dust, orbiting an ordinary star, in one among hundreds of billions of galaxies, in an ill-defined place in the universe. Although the ancients had committed us to death and destruction, they had at least placed us at the center of it all; now some of our greatest scientists have taken even that centrality from us.

It is via dynamics that we have discovered the vastness of the universe, the immense emptiness of interstellar space, and our own *physical* insignificance in the cosmos. Consider the following analogies of scale. If the Sun were the size of a dime ($\sim$ 1 centimeter), the Earth would be about two full walking steps away from the Sun, and our solar system would fit within the boundaries of a football field, let's say in Cambridge, Massachusetts. On that scale the nearest star, Alpha Centauri, would be in another city, approximately the distance of New York City from Cambridge. Our galaxy, the Milky Way, would have a diameter that would extend more than twenty times the Earth–Moon distance of about 384,000 kilometers. Now, performing a further reduction of the Milky Way to the size of a typical residential bedroom (thus, making the diameter of the Sun an order of magnitude smaller than the hydrogen atom), the nearest galaxy, Andromeda, would be 700 meters away; and the billions of galaxies in the known universe would be visible every 700 meters in all directions to a distance of Cambridge-to-Washington, D.C. (800 kilometers).

This knowledge has of necessity moved heaven and hell, and affected all similar philosophical concepts. Consider that if at the time of his death in 322 B.C., Aristotle's soul had left his body traveling at the speed of light on its journey toward heaven, today it would not have escaped even our own galaxy. So, where is heaven?

The human experience is not very different from what it was 5,000 years ago in Ancient Egypt. We own land, are taxed, and, as the Ancient Egyptians were, are sometimes helped by our government following natural disasters. Our daily needs and daily concerns have not changed very much. We continue to ponder our philosophical and spiritual role in the cosmos. But in one respect, humanity's perspective of itself has changed forever: We know—where knowing in the scientific sense is always subject to change—our niche in the cosmos. We are insignificantly small, and perhaps therefore *precious*. It is the intellectual discipline called *dynamics* that has enabled us to consider quantitatively these cosmological questions. It is for that reason that dynamics is the most important intellectual subject in the history of humankind. It is through dynamics that we have discovered, at least physically, our niche in the cosmos.

CHAPTER TWO

DESIGN, MODELING, AND FORMULATION OF EQUATIONS OF MOTION

2-1 INTRODUCTION

In this chapter we introduce some general perspectives relating to the design, modeling, and formulation of the equations of motion of systems. A *system* is any combination of interconnected matter and concepts that are combined to accomplish an objective. The range of our considerations in this chapter will be limited to mechanical systems. In this chapter we seek only to emphasize that (1) *design,* sometimes called *synthesis,* is the primary goal of all engineering, (2) *mathematical modeling* is one of several difficult phases of design, (3) quantitative criteria can be used, albeit in a limited sense, in mathematical modeling, and (4) system equation formulation—an important part of mathematical modeling—must satisfy three fundamental requirements. Thus, the primary focus of this chapter has more to do with developing a sense of perspective than with expanding the reader's specific design, modeling, and equation formulation skills.

Throughout the textbook we shall be concerned exclusively with *state-determined system models* of physical systems. Such models are characterized by the following features:

1. Based upon one or more physical principles and a set of analysis requirements, a mathematical formulation of the system model can be derived.

2. At an initial time t_0, a set of system variables can be specified.

3. For all time $t \geq t_0$, all inputs to the system can be specified.

The significance of these general features is that they are the necessary and sufficient conditions to establish the system's behavior for all time. The first feature is primarily concerned with the formulation of the equations of motion; the second and third features relate primarily to the solution of those equations for a specified set of initial conditions and subsequent inputs. An important consequence of these features is that the history of the model (i.e., for all times $t < t_0$) plays no role in the model's dynamics, such as might be due to hysteresis.

Differential equations will be the primary mathematical descriptions resulting from these models of dynamic systems. If the system models are constructed such that time is the only independent variable of the system's coordinates, the dynamics will be represented by ordinary differential equations (ODEs). If the system models are devised such that time and one or more spatial coordinates are independent variables of the system's coordinates, the dynamics will be represented by partial differential equations (PDEs).

We shall focus considerable attention on the formulation of these differential equations. These *equations of motion,* as they are called, can be obtained via two major, distinctly different approaches, which we denote as the *direct approach* and the *indirect approach*. In Section 2-3 we consider an example to amplify the differences between these two distinct approaches—distinct formulations which are often intermingled and confused (see Figure 1-17). In the direct approach, the equations for the individual system elements are combined in accordance with the laws governing their interconnections within the system. The algebraic reduction of these elemental equations and laws leads to one or more equations of motion. In the indirect approach, the equations for the individual system elements are used to define energy and work expressions. The application of a governing principle to these energy and work expressions leads to one or more equations of motion. In Example 2-1, we summarize several contrasting aspects of these two formulation approaches, and in Example 2-2, we reemphasize the importance of ultimately relating the results of all our analyses to the design process.

Historically, dynamics has meant the scientific discipline of mechanics, which studies the spatial motions of interacting bodies subjected to forces. While the foundations of dynamics lay in the mathematical and astronomical writings of the Ancient Egyptians and the Mesopotamians, many authors credit its mathematical (sometimes called *rational*) beginning with Galilei's kinematics and its culmination with the laws of Newton (linear momentum principle) and Euler (angular momentum principle). Furthermore, as we indicated in Chapter 1, the indirect approach to the formulation of the equations of motion for mechanical systems was cultivated by Leibniz, John Bernoulli, d'Alembert, Euler, Lagrange, and Hamilton.

In the context of design, engineers may be concerned with calculating the stresses and strains in machines or components due to imposed motion or forces. It is in this regard that the underlying disciplinary structure of dynamics can be recognized as being analogous to the underlying disciplinary structure of the "mechanics of solids" or "strength of materials." (In general terms, this analogy in structure can be extended to numerous disciplines.) Indeed, until the last quarter of the nineteenth century, engineers found little use for dynamics because the operating speeds of machines were sufficiently low that stresses and strains could be adequately determined via statics formulations of strength of materials. It is this similarity in the underlying structures of statics and dynamics that should be clear in the articulation of conditions (2-1), which will be designated as Eq. (2-1).

As the range of application and consideration of mechanical analyses was expanded to include liquids and gases, and as society's interests in time-varying phenomena broadened to include not only bodies but processes such as electrical, thermal, and chemical activity, the meaning of dynamics has grown to encompass these phenomena. Thus, dynamics has come to mean the study of the time profile of the state of a system or a process. As we shall see, particularly in Chapter 7, the interpretation of the indirect approach can be expanded to incorporate some of these nonmechanical physical phenomena.

2-2   DESIGN AND MODELING

2-2.1   The Design Process

Design is a process of continuous interplay between the goals of accomplishment and the means of accomplishment. The method by which the engineer achieves these desired goals is called the *engineering design process*, the resulting solution being the *engineering design*.

An engineering design may be small or large, simple or complex. It may be as small as the design of a conventional fastener, requiring the effort of a single engineer for a few hours. Or, it may be as complex as the design of a national transportation system, requiring the efforts of thousands of engineers over several years.

Engineering designs may produce new *products* (reinforced plastic automobile parts), new *processes* (new automobile assembly techniques), or new *systems* of combinations of products and processes (new factories to fabricate electric cars or, of course, the electric cars themselves may be the new systems). Furthermore, neither the engineering design nor the route to achieving it is unique. Different designs may be equally objectively good. That is, different designs may satisfy all the quantitative design specifications or requirements to the same extent.

Design approaches may be thought to range from the purely heuristic to the purely axiomatic, although no knowledgeable individual would suggest that modern design practice resides at either of these extremes. Nevertheless, within these bounds there is a broad range of design philosophy, characterizable from apprentice empirical design, in which there is little effort to provide a scientific basis, to engineering design, in which an attempt to introduce scientific principles is undertaken. Despite this broad range of design philosophy, most approaches can be accommodated by the flow diagram represented in Figure 2-1, a diagram that is by no means universally embraced. The series of entries in the diagram is indicative of an overall sense of ordering and flow of the general aspects of the design process. Further, the looping bold arrows along both sides of the aspects of design are intended to emphasize the presence of omnipresent feedback and feedforward of information between *all* aspects throughout the design process. For example, the enormous amount of interaction between the designers, engineers, and the skilled fabricators during the construction of a submarine or an aircraft carrier (even within hours of launch and, indeed, sometimes *after* launch) would surprise most readers. The same could be said for many complex systems.

Finally, it is important to understand that the ultimate purpose of all engineering analysis is design. Although it may not always be apparent to a staff engineer who has been assigned an isolated calculation, to the research engineer who has been charged with the development of a technique, or to the student engineer who is conducting university research, the ultimate goal of the output of each of these individuals is design; the ultimate goal in each instance being a new or improved product or process.

2-2.2 The Modeling Process

Modeling is the reduction of a concept, system, or problem to a simpler representational approximation. The purpose of modeling is to devise a tractable approximation for the purpose of analyzing a real or imagined problem. It is these substitutional representations of imagined or existing problems or systems that are called *models*. Models may assume the form of tables of numbers, graphs, data plots, blueprints, or physical or computerized scaled constructions to assist in the visualization of the relational fit of components in a design; for example, architects often create such scaled constructions. If the behavior of the characteristics of interest can be mapped into a mathematical structure, the resulting mapping is known as a *mathematical model*. Our interests shall be limited to the mathematical modeling of mechanical, electrical, and electromechanical engineering systems.

Mathematical modeling of engineering systems combines analytical reasoning and technical insights that are often disguised as intuition. The analytical reasoning and technical

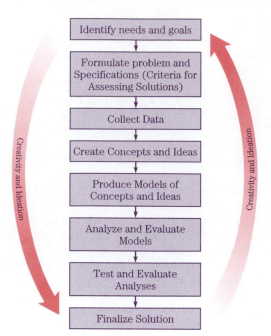

FIGURE 2-1   A perspective of the design process.

insights are based on both scientific principles and aesthetic notions, which are combined to explain features that are observed and to reveal features that are to be observed. Also, mathematical modeling is an abstraction process in response to a specific question, and thus both the results and the limitations of the model should be anticipated in the context of the specific goal.

Good mathematical modeling is that which satisfies the specifications without undue complexity. Good mathematical modeling is a very valuable engineering undertaking; it is also frequently subtle and difficult.

In analyzing mathematical models, perhaps the single most important fact to keep in mind is that *the model being analyzed is not the real system.* The purpose of a model is to enhance the understanding of a real system; and it is to the extent that a model succeeds in this purpose that the analysis of the model may succeed.

2-2.3   Our More Modest Goals

One of the major goals of this text is to derive mathematical expressions, or *equations of motion,* from the symbolized representations of system elements or components such as masses, springs, capacitors, or resistors. These symbolized representations will be the analytical hypotheses about fundamental engineering elements, their relative arrangements, and their interconnections as we attempt to model a dynamic characteristic of interest of a physical system under investigation.

Having selected the derivation of the equations of motion as one of our major goals, we forgo the very important modeling process as a focus of this text. Beyond the derivation of the equations of motion, we explore some of the elementary solution techniques of those equations. It is just such solutions that provide a major portion of the knowledge that will enable the readers to assess the quality of their models, as well as the models and analyses of others. Thus, this text is substantially confined to the "Analyze and Evaluate Models" stage of the design process depicted in Figure 2-1, with modest discussions of the bridges that connect this stage with adjacent stages.

2-3   DIRECT AND INDIRECT APPROACHES FOR FORMULATION OF EQUATIONS OF MOTION

When confronted with an engineering design problem, the engineer must first of all define the problem and the goal of the solution. Shortly thereafter, in the context of this book, some decision must be made regarding whether the required analysis is within the broad discipline of dynamics. (See Example 8-4.) Such a determination may not be easy. Having done so, however, the engineer may note that the dynamic analysis of an engineering system may be categorized in the following steps:

1. Identify the system and its components to be considered.
2. Define and isolate an idealized model of the system and its components, including the inputs and the outputs.
3. Formulate the equations of motion for the idealized model by applying the governing disciplinary principles.
4. Solve the equations of motion for the desired outputs.
5. Check, assess, and interpret the solution for the design goal.
6. If necessary, iterate steps 1 through 5.

The focus of this text will be on steps 3 and 4, with greater emphasis on step 3. Again, steps 3 and 4 relate primarily to the "Analyze and Evaluate Models" stage of the design process, with the other steps relating to adjacent stages in Figure 2-1. Here we consider an example in order to illustrate the general requirements that must be satisfied in the formulation of a mechanical dynamics problem.

The fundamental requirements governing the formulation of the equations of motion for mechanical dynamic systems are analogous to the requirements governing mechanical systems in equilibrium. These may be summarized as follows:

1. Geometric requirements on the motions;
2. Dynamic requirements on the forces; and   (2-1)
3. Constitutive requirements (relations between forces and motions) for all the system elements and fields, including the velocity–momentum relation for the masses.

We dignify this set of requirements with an equation number, indeed, the only equation number in this entire chapter. A similar set of requirements can be listed for electrical and electromechanical systems, as we shall discuss in Chapter 7. The focus of this section is to explore the two fundamental approaches for satisfying these basic requirements: the direct approach and the indirect approach.

In the direct approach, each of Requirements 1, 2, and 3 is expressed as an equation, and algebraic reduction is used to produce one or more equations of motion. In the indirect approach, Requirements 1 and 3 are used to express functions that are constrained such that Requirement 2 is satisfied. We emphasize that irrespective of which approach is used, both approaches must satisfy the three general requirements. We illustrate with an example and then summarize some of the differences between the two approaches. We assume that the analyses presented in the example are substantially familiar to the reader; however, the discussion of the analyses that contrasts the two approaches is likely to be less familiar.

■ **Example 2-1:** Package Transfer System

Figure 2-2 shows a model of a package transfer system where a package or container, released at A, is transferred to a second location at C. Such a concept is used in warehouses where packages, released from storage or inventory at A, are distributed to a dock or customer pickup at C.

In the situation of interest, the package is modeled as a sliding body of negligibly small geometric dimensions. The sliding body of mass 10 kg is released from rest (height of 10 m above horizontal surface BD) along a frictionless circular surface AB of radius 10 m. It then slides along the horizontal surface, having a kinetic coefficient of friction with respect to the body of 0.2, and encounters a spring at point C. The spring constant k is 20,000 N/m and the distance d is 15 m. Note that gravity acts. How much will the spring be compressed?

□ **Solution:**

Direct Approach

The three fundamental requirements for the formulation of the equations of motion for a mechanical dynamic system may be expressed quantitatively as listed below.

1. Geometric requirements on motions: If a body moves along a known path that can be characterized as one-dimensional, the position, with respect to a specified reference, at any time t along the path may be noted as $s(t)$. Then

$$v = \frac{ds}{dt} \quad \text{and} \quad a = \frac{dv}{dt} \tag{a}$$

which gives

$$dt = \frac{ds}{v} = \frac{dv}{a}$$

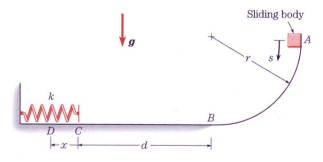

FIGURE 2-2   Body sliding along the surface of the package transfer system.

or

$$v\,dv = a\,ds \tag{b}$$

where v is velocity and a is acceleration.

Integration of Eq. (b) over any interval denoted by its end points 1 and 2 gives

$$\int_{v_1}^{v_2} v\,dv = \int_{s_1}^{s_2} a\,ds$$

or

$$v_2^2 - v_1^2 = 2\int_{s_1}^{s_2} a\,ds \tag{c}$$

where, v_1 and s_1 and v_2 and s_2 are the velocities and positions at end points 1 and 2, respectively. It is understood that in order to use Eq. (c), the acceleration a should be expressed in terms of the position s or be constant.

2. Dynamic requirements on forces: The linear momentum principle is

$$f = \frac{dp}{dt} \tag{d}$$

where f is the net external force, and p is the linear momentum.

3. Constitutive requirements for system elements and gravitational field:

$$(i)\ \text{Momentum–velocity:} \qquad p = mv \tag{e}$$
$$(ii)\ \text{Spring force:} \qquad f_s = ks \tag{f}$$
$$(iii)\ \text{Gravitational force:} \qquad f_g = mg \tag{g}$$
$$(iv)\ \text{Frictional force:} \qquad f_f = \mu_k N \tag{h}$$

where m = mass, f_s = spring force, k = spring constant, g = gravitational constant ($9.81\ \text{m/s}^2$), f_f = frictional force, μ_k = kinetic coefficient of friction, and N = normal force.

As we shall indicate during the next few chapters, many of the variables in Eqs. (a) through (h) are vectors; however, in this essentially one-dimensional problem, it is not necessary to retain this characteristic.

Motion Between Point A and Point B: First we sketch a free-body diagram of the package at an arbitrary position between points A and B, indicated by the angle θ, as sketched in Figure 2-3. During this portion of the motion, there are two forces acting on the body: the gravitational force f_g and the normal force N due to the surface contact. The acceleration along the surface is also shown in Figure 2-3.

Since we are interested in the package's motion along the surface, it is convenient to resolve all forces into components either along or normal to the surface. In the direction along the surface, the total force is

$$f = f_g \cos\theta \tag{i}$$

Substitution of Eq. (g) into Eq. (i), and substitution of the resulting expression and Eq. (e) into Eq. (d) give

$$mg \cos\theta = ma$$

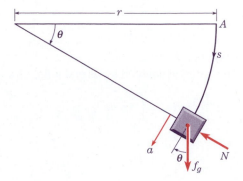

FIGURE 2-3   Free-body diagram of the package as it moves between points A and B.

where Eq. (a) has also been used. Thus,

$$a = g \cos \theta \tag{j}$$

Next, note that

$$s = r\theta \quad \text{and} \quad ds = r\,d\theta \tag{k}$$

Also note that at point A, $v_A = 0$ since m is released from rest. Substitution of Eqs. (j) and (k) into Eq. (c) yields

$$v_B^2 - v_A^2 = 2\int_0^{\frac{\pi}{2}} g \cos \theta (r\,d\theta) = 2gr \tag{l}$$

or

$$v_B = \sqrt{2gr} \tag{m}$$

Although it is not necessary for solving this problem, we can also calculate the force imposed by the surface on the body by analyzing the motion in the direction normal to the surface. In the normal direction there is no acceleration, so the sum of the forces equals zero, leading to

$$N - mg \sin \theta = 0$$

or

$$N = mg \sin \theta \tag{n}$$

Motion Between Point B and Point C: Again, we begin by sketching a free-body diagram for this case as shown in Figure 2-4. During this portion of the motion, there are

FIGURE 2-4   Free-body diagram of the package as it moves between points B and C. (Note that all forces pass through the center of a negligibly small body.)

three forces acting on the body as shown: gravity, the normal surface force, and the frictional force, which is always opposite in direction to the velocity.

Because there is no motion in the direction normal to the surface, we can easily verify that

$$N = mg \tag{o}$$

In the direction along the surface, Eqs. (d), (e) and (h) give

$$-\mu N = ma$$

where Eq. (a) has been used; or

$$a = -\frac{\mu N}{m} = -\mu g \tag{p}$$

where Eq. (o) has been used.

Integration of Eq. (p) between points B and C using Eq. (c) gives

$$v_C^2 - v_B^2 = -2\mu g(s_C - s_B)$$

or, since $(s_C - s_B) = d$,

$$v_C^2 = 2gr - 2\mu gd$$

where Eq. (m) has been used for v_B^2. Thus,

$$v_C = \sqrt{2g(r - \mu d)} = 11.72 \text{ m/s} \tag{q}$$

where, in evaluating Eq. (q) for the discussion below, the positive square root has been selected by physical reasoning; that is, v_C must be positive (or possibly zero) at point C. (Following Eq. (y) below, we shall consider the alternative.)

Motion Beyond Point C (that is, Between Points C and D):   After the package comes into contact with the spring, one additional force, the spring force, acts on the package. Because one end of the spring is fixed at the wall, the direction of the spring force applied to the package is opposite to the displacement of the movable end of the spring, x. The free-body diagram for the package in this case is sketched in Figure 2-5.

Again, the package has no motion in the direction normal to the surface, so the normal force N is due exclusively to gravity and is equal to the body's weight, or

$$N = mg \tag{r}$$

In the direction along the surface, the total force f is

$$f = -f_f - f_s \tag{s}$$

FIGURE 2-5   Free-body diagram of the package as it moves beyond point C. (Note that all forces pass through the center of a negligibly small body.)

where the negative signs are due to the fact that s is positive toward the left. (Refer to Figure 2-2 or 2-3.) According to the constitutive relations in Eqs. (e), (f), (g) and (h), the total force in Eq. (s) becomes

$$f = -\mu mg - k(s - s_C) \tag{t}$$

where s and s_C are the position of the package and the position of the free end of the undeformed spring, respectively. Substitution of Eqs. (e) and (t) into Eq. (d) gives

$$a = -\frac{k}{m}(s - s_C) - \mu g \tag{u}$$

where Eq. (a) has also been used.

Integrating Eq. (u) from point C to point D, using Eq. (c), and noting that

$$x = s_D - s_C$$

and further that at point D, $v_D = 0$, yield

$$-v_C^2 = 2 \int_{s_C}^{s_D} \left[-\frac{k}{m}(s - s_C) - \mu g \right] ds$$

$$= -\frac{k}{m}x^2 - 2\mu gx \tag{v}$$

Recalling the expression for v_C in Eq. (q) enables Eq. (v) to be rewritten as

$$-2g(r - \mu d) = -\frac{k}{m}x^2 - 2\mu gx$$

or

$$kx^2 + 2\mu mgx - 2mg(r - \mu d) = 0 \tag{w}$$

Equation (w) is a quadratic equation in the unknown deflection x, and because it represents the value of x at point D, it is indeed the maximum deflection. So, the value of the maximum spring deflection x_{max} is

$$x_{max} = \frac{-2\mu mg + \sqrt{4\mu^2 m^2 g^2 + 8kmg(r - \mu d)}}{2k}$$

$$= \frac{-\mu mg + \sqrt{\mu^2 m^2 g^2 + 2kmg(r - \mu d)}}{k} \tag{x}$$

Thus,

$$x_{max} = 0.261 \text{ m} \tag{y}$$

where the well-known solution for the quadratic equation has been used to solve Eq. (w), and the numerical values of the parameters have been used to evaluate Eq. (x). The choice of this specific solution from the two possible solutions—the two solutions to Eq. (x) are due to the positive and negative values of the square-root term—is based on the same physical reasoning used in Eq. (q) to evaluate $v_C \geq 0$; that is, x_{max} must be positive (or possibly zero).

Note that in order to proceed beyond Eq. (q), it was necessary that v_C was greater than zero, or that

$$r > \mu d \tag{z}$$

Clearly, the condition in Eq. (z) was satisfied in this problem. If this condition were not satisfied, the body would stop at a point, say E, between points B and C prior to coming into contact with the spring. An expression analogous to Eq. (q) could be used to find the stopping distance corresponding to point E. (For example, in the equation preceding Eq. (q), set the left-hand side to zero and consider d to be the unknown stopping distance.) Finally, x will always be nonnegative (i.e., positive or zero) in Eq. (y) if Eq. (z) is satisfied.

Indirect Approach

Here we consider an alternative method as we solve the problem again. This alternative method, which has work and energy as its fundamental quantities, is designated as the *indirect approach*.

Motion Between Point A and Point B: Referring to Figure 2-2 we note that along the frictionless surface AB, energy is conserved. So, at point B

$$(\text{K.E.S.})_B = (\text{P.E.S.})_A = mgr \tag{aa}$$

where $(\text{K.E.S.})_B$ denotes the kinetic energy state at B and $(\text{P.E.S.})_A$ denotes the potential energy state at A. In writing Eq. (aa), we have used the familiar expression for gravitational potential energy, $mg\text{“}h\text{”} = mgr$, where h is typically interpreted to be the height of a body above a reference, the reference in this case being the horizontal surface.

Motion Between Point B and Point D: Between points B and D, the work-energy expression gives

$$(\text{K.E.S.})_B = \text{Work Done} + (\text{P.E.S.})_D \tag{ab}$$

which expresses the idea that the energy available in the kinetic energy state at B goes into either the work done by the frictional force or the potential energy state of the compressed spring. The evaluation of Eq. (ab) gives

$$mgr = f_f s + \frac{1}{2}kx^2$$

$$= \mu mg(d + x) + \frac{1}{2}kx^2 \tag{ac}$$

where we have used the expressions μmg for the frictional force and $\frac{1}{2}kx^2$ for the stored potential energy state of the spring, as well as Eq. (aa).

Equation (ac) may be rewritten as

$$kx^2 + 2\mu mgx - 2mg(r - \mu d) = 0 \tag{ad}$$

which, of course, is the same as Eq. (w).

One conclusion is striking: the indirect approach appears to lead more quickly to the equation of interest. While this is not always true, it is very often the case, thus representing a distinct advantage of the indirect approach over the direct approach in many problems.

Discussion

To begin, we should hasten to comment that it is expected that the reader has encountered very similar analyses as presented above. Otherwise, there could be no justification for some of the omitted background of our presentation, especially in the case of the indirect approach. The goal here is not to present these two analyses but to contrast them.

First, we emphasize that irrespective of the approach, the three fundamental requirements listed as Eq. (2-1) *must* be satisfied. While this is nearly obvious in the direct approach, it is certainly not so clear in the indirect approach. In the direct approach some part of Requirements 1, 2, and 3 of Eq. (2-1) must be considered at each stage of the analysis. For example, in analyzing the motion between points A and B, we satisfied the geometric requirements via use of Eqs. (a) and (c); we satisfied the dynamic requirements on the forces via use of Eq. (d); and we satisfied the constitutive requirements via use of Eqs. (e) and (g). In the indirect approach, there is a subtle delineation of the forces along the following lines: those forces that do work conservatively (e.g., gravitational and spring forces), resulting in energy conversion or energy storage; those forces that do work nonconservatively (e.g., frictional forces), resulting in energy dissipation; and those forces that do no work as they constrain the motion of the body (e.g., normal surface forces that are orthogonal to the motion), being workless constraint forces. In constructing the kinetic and potential energy state functions, it is necessary to use portions of Requirements 1 and 3 for the conservative forces. In constructing the work expression ("Work Done"), it is necessary to use portions of Requirements 1 and 3 for the nonconservative forces. The relationship between these energy state functions and the work expression is such that Requirement 2 is satisfied. Thus, the satisfaction of the three requirements during analyses via the indirect approach is somewhat subtle; however, because the indirect approach is a major focus of this book, we shall devote considerable effort to exploring the details by which this is achieved.

To conclude this introductory comparison between the direct approach and the indirect approach, we construct the brief summary given in Table 2-1. The term "intermediate forces" in Table 2-1 is used to denote forces that are not of primary interest, such as those between points A and B or between points B and C. The summary comments in Table 2-1 represent a set of contrasts that are of a general character. As such, it is a table worth referencing as we proceed through the book. In particular, as we shall see, velocities are somewhat easier to obtain than accelerations, and scalars are somewhat easier to manipulate than vectors; thus, the indirect approach is often simpler to use than the direct approach. On the other hand, the direct approach more readily provides us with the forces that act on and within the system.

TABLE 2-1   Summary of Some Contrasts Between Direct Approach and Indirect Approach

Direct Approach	Indirect Approach
• Accelerations required	• Velocities required
• Generally vectors required	• Generally scalars required
• Free-body diagrams useful	• Free-body diagrams not useful
• All forces considered	• Workless (or constraint) forces not considered
• All forces handled via same expression	• Conservative and nonconservative forces handled separately
• Intermediate forces more readily available	• Intermediate forces less readily available

■ **Example 2-2:**   So What?

Several analyses and calculations were conducted in Example 2-1 to find the maximum deflection in the spring. Although the primary goal was to illustrate and delineate the direct and indirect approaches, as an engineering exercise the example is somewhat deficient.

Thus, the reader who is interested in engineering is entitled by that interest to ask here and throughout this book "So what?"

☐ **Solution:**   As stated in Subsection 2-2.1, the ultimate engineering purpose of our analyses is design. There are many questions that could be posed about the package transfer system that could be addressed using the analyses conducted in Example 2-1. Many such questions would be related to the maximum spring *deflection* and to the *forces* that act on the package throughout its transit, including its deceleration toward point D. Note that we emphasize deflection(s) as well as forces, since the reader should recall that failure of engineering structures is often initiated by too much deformation, even though the forces in a system may not cause yielding or rupture. Further, in this example the forces, if sufficiently large, could cause damage to the package or the contents of the package. Although we shall defer the calculation of such forces to Chapter 4 and beyond, we briefly consider one of a number of forces encountered by the package in Example 2-1.

In particular, we consider the maximum force applied to the package by the spring in bringing it to a halt. From Eqs. (f), (x), and (y) in Example 2-1, it follows that the maximum force developed in the spring $(f_s)_{max}$ is

$$(f_s)_{max} = kx_{max} = \sqrt{\mu^2 m^2 g^2 + 2kmg(r - \mu d)} - \mu mg = 5220 \text{ N} \tag{a}$$

which is the (magnitude of the) maximum force applied to the package by the spring in bringing it to a halt.

The result in Eq. (a) leads to several likely design issues. For example, is the force sufficiently large to damage the package or its contents or both? Furthermore, from an analytical perspective, will the spring retain its assumed linear characteristics throughout the range of its calculated deformation? If not, appropriate analytical modifications should be considered. Also, it is clear that Eq. (a) provides an expression to guide the design or changes in system parameters for decreasing the maximum force. (Note that Eq. (z) in Example 2-1 is related to this issue as well as to Eq. (a) here.) In summary, we want to emphasize that, whether obvious or obscure, the ultimate goal of all our analyses is the assessment of the performance or reliability of an existing or proposed design. That's the "so what."

CHAPTER THREE

KINEMATICS

3-1   INTRODUCTION

In this chapter, we explore *kinematics: the study of the geometry of motion.* Kinematics is concerned primarily with the time behavior of position, velocity, and acceleration of systems as well as components of systems. It is kinematics by which the first of the three requirements in Eq. (2-1) is satisfied. In kinematics no reference is made to the forces that cause the motion or the forces that are generated as a consequence of the motion; only the description of the motion is of interest. Although kinematics is conducted without reference to forces, ultimately the calculation of the dynamic forces on elements and systems requires knowledge of kinematics. Thus, a thorough knowledge of kinematics is crucial to acquiring a thorough knowledge of system motions and forces, quantities that are often important inputs to the design process.

The invention of kinematics is duly credited to the Ancient Egyptians. The discernment (1) that all the stars are essentially fixed relative to one another, (2) that they appear to rotate about the Earth as if they were points on a rigid inverted crystal bowl or sphere, (3) that the Sun wanders among these "fixed stars," and (4) that it is the return of the Sun to a particular location among these fixed stars each $365\frac{1}{4}$ days that provides the basis for an annual calendar was an intellectual kinematic achievement of the first rank. This is a legacy of the Ancient Egyptians. It is literally the means by which humankind has kept time throughout history and will no doubt do so for the foreseeable future; it is literally the invention of *time* as a quantitative scientific concept. It also fixes the invention of kinematics. The surveying by the Ancient Egyptians in order to orient with great accuracy many of their major structures, as well as the invention of astrology, were based upon kinematic astronomical measurements. Following these successes, the development of the lunar calendar, the casting of horoscopes, and the cataloging of numerous stars and the motions of Mercury, Venus, Mars, Jupiter, and Saturn by the Mesopotamians were great achievements in kinematics. These were major kinematic accomplishments that were integrated into the daily lives of the citizens of these two societies; landmarks of human history in the spirit of Ptolemy, Copernicus, Brahe, and Kepler, all of whom would use and build upon them.

Modern kinematics can be easily traced, at least, to the Mertonians. In the fourteenth century, several scholars at Merton College, Oxford, composed definitions for velocity and acceleration for uniformly accelerated motion. Between 1328 and 1355, the work of Thomas Bradwardine, William Heytesbury, Richard Swineshead, and John of Dumbleton established the kinematic analyses that provided the foundations for Galilei's studies on falling bodies.

Being mathematical, the analyses of the Mertonians provided a formulation of the concept of instantaneous speed, thus foreshadowing the concepts of derivatives. This work was quickly accepted in continental Europe, initially leading most significantly to the analyses of Nicole Oresme (ca. 1323–1382) and Giovanni da Casale (ca. 1325–1375), who discovered how to represent kinematics by geometrical graphs. This reestablished another useful connection between geometry and the physical world, a tradition the Ancient Egyptians had introduced via their monuments several millenia earlier, and which Ptolemy had notably continued in his astronomy, for example, in his use of epicycles.

Galileo Galilei was a pioneer of the modern concept of the testing of a scientific hypothesis based on careful experimental design, observation, and measurement. He discovered and expressed the *law of falling bodies,* as it is often called. (This is not an exclusively kinematic result, although it is frequently presented as such.) His conclusion that in a vacuum all bodies—heavy and light, large and small—fall in the same manner was in part a consequence of meticulous experimentation and brilliant insight. For in reaching such a conclusion, Galilei had to imagine a vacuum that he could not produce and ignore air resistance that he could not eliminate. His deduction that an isolated body moves, and will continue to move, with a constant velocity was a great intellectual achievement. And, to the extent he attempted to dismantle the Aristotelian dichotomy of celestial motion as circular and terrestrial motion as rectilinear, Galilei can be said to have led the progression toward the unification of celestial and terrestrial motion; though mankind's success in this undertaking would have to await Isaac Newton, born in the year of Galilei's death.

The mathematical approach used in our kinematic analyses will be vector analysis. The use and manipulation of quantities that have both magnitude and direction were obviously familiar to the Ancient Egyptians and later the Mesopotamians in their astronomical analyses of position, displacement, and velocity. Further, the Ancient Egyptians' use of the lever in balance scales and construction clearly indicates their familiarity with notions of forces and moments. Yet, those uses of directed quantities were primarily geometrical and numerical, as such use would remain until the rise of postnewtonian mechanics in the eighteenth century. During the eighteenth and nineteenth centuries, the concepts of *vectors* and the mathematical techniques for manipulating them—*vector analysis*—emerged slowly, until by the start of the twentieth century the ideas of Herman Grassman (1809–1877) in Germany and Josiah Willard Gibbs (1839–1903) in the United States, and the notation of Oliver Heaviside (1850–1925) in England, had triumphed over those of Hamilton (who apparently had coined the word *vector* from the Latin and who in 1843 developed a similar but more cumbersome concept known as *quaternions*) in the analysis of vectors.

3-2 POSITION, VELOCITY, AND ACCELERATION

The position, velocity, and acceleration of a body are defined only with respect to a specified reference frame. For example, the velocity of a bus traveling along a highway may be different for observers in different reference frames. A passenger in a car traveling in the same direction as the bus will observe a different velocity of the bus than that observed by a passenger traveling in a car in the opposite direction. A pedestrian will observe still another velocity of the bus. Thus, we must be clear in specifying the reference frame used in each kinematic description.

We begin by considering a rigid rectangular reference frame $OXYZ$ and a point P that moves with respect to the reference frame as indicated in Figure 3-1. Note that $OXYZ$ is a

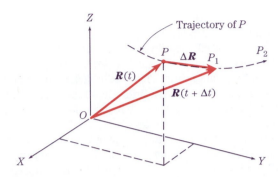

FIGURE 3-1 Displacement of point P with respect to frame $OXYZ$ during interval Δt is $\Delta \boldsymbol{R} = \mathbf{PP}_1$.

right-handed coordinate system; all coordinate systems throughout this book will be right-handed. (Recall that for right-handed coordinate systems, a curling of the fingers on the right hand from the X direction toward the Y direction will cause the thumb to point in the Z direction.) The location of P with respect to any specified reference frame is defined as the *position vector* in that reference frame. Let $\boldsymbol{R}$ denote the position vector OP, where the boldface designates the vector character of the quantity. The vector $\boldsymbol{R}$ has the dimension of length. Typical units are the foot (ft) in the British or U.S. customary system or the foot-pound-second (FPS) system, and the meter (m) in the International System of Units (SI) or the metric system. In general, the vector $\boldsymbol{R}$ will vary with time t; that is, $\boldsymbol{R} = \boldsymbol{R}(t)$. We shall be interested in changes in $\boldsymbol{R}$ measured with respect to the reference frame $OXYZ$.

It is important to note that it may be left open concerning whether the reference frame $OXYZ$ itself is moving with respect to some other reference frame. However, as an intuitive aid to understanding the discussions that follow, it may be of benefit to think of $OXYZ$ as being *fixed in absolute space,* or, as we shall sometimes alternatively write, *fixed to ground.* It is the concept of "fixed in absolute space" that is, in fact, ambiguous—though intuitively beneficial—and which we shall examine in Chapter 4 (Subsection 4-2.4). Suffice it to say that the kinematics with respect to such "fixed" reference frames will dominate our ability to conduct our dynamical analyses; thus, throughout the book we shall reserve $OXYZ$ exclusively for such "fixed" reference frames.

We shall ignore all relativistic effects; so, we adopt the classical premise that there is a universal time t that is independent of the choice of reference frame. The common unit of time is the second ("sec" in the British or U.S. customary system, and "s" in the metric system).

In Figure 3-1, the timewise path or *trajectory* of the point P with respect to the reference frame $OXYZ$ is indicated by the dashed curve PP_1P_2. The position vector $\boldsymbol{R}$ is shown at two times that differ by the time increment Δt. The corresponding *displacement vector* is $\Delta \boldsymbol{R} = \mathbf{PP}_1$. The *velocity $\boldsymbol{v}$* of the point P with respect to the reference frame is the time rate of change of position, namely,

$$\boldsymbol{v} = \lim_{\Delta t \to 0} \frac{\Delta \boldsymbol{R}}{\Delta t} = \frac{d\boldsymbol{R}}{dt} \tag{3-1}$$

The direction of $\boldsymbol{v}$ is the limiting direction of the displacement vector $\Delta \boldsymbol{R}$ as $\Delta t \to 0$; that is, tangent to the trajectory at the point P. The vector $\boldsymbol{v}$ has the dimensions of length per unit time. Common units of velocity are foot per second (ft/sec) and meter per second (m/s). *Speed* is a scalar quantity defined as the magnitude $|\boldsymbol{v}|$ of the velocity vector $\boldsymbol{v}$.

In general, the velocity vector v also varies with time; that is, $v = v(t)$. The time rate of change of v is the *acceleration* vector a of the point P with respect to the reference frame, namely,

$$a = \frac{dv}{dt} = \frac{d^2R}{dt^2} \tag{3-2}$$

The vector a has the dimensions of length per unit time per unit time. Common units of acceleration are foot per second per second (ft/sec^2) and meter per second per second (m/s^2). Accelerations are often quoted in multiples of g, a standardized sea-level acceleration, due to gravity and relative to the rotating Earth at a latitude of 45°. In the U.S. customary system g is 32.17 ft/sec^2, and in the metric system g is 9.81 m/s^2. *Deceleration* is sometimes used (confusingly) for acceleration when the speed is decreasing. In general, we shall refrain from the use of this equivocal term.

Because positions, velocities, and accelerations are vectors, they can be represented by directed line segments, which also suggests that they can be combined or decomposed according to the laws of vector algebra. For example, the total velocity of a point resulting from two individual velocities is represented by a vector that is the vector sum of the vectors representing the individual velocities. However, all entities that can be represented by directed line segments do not combine according to the laws of vector addition. As we shall discuss in Subsection 3-3.4, finite rotations are one such type of entity. Finally, as we shall see repeatedly, kinematics is often the most difficult step in obtaining the equations of motion in dynamics.

■ **Example 3-1:**   Student Rolling in Nonslipping Thin Hoop

A student who enjoys rolling around in thin circular hoops is sketched in Figure 3-2a. (A thin circular hoop is a model of a ringlike body having negligible radial thickness. Here, it is assumed to be undeformable or rigid.) We are interested in studying the kinematics of two points on the student's body when the thin hoop of radius r rolls in a plane without slip along a flat surface, as sketched in Figure 3-2b, where the radial line CP was initially along the Y axis. We want to find the velocity and acceleration of the student's navel (point C), which is located at the center of the hoop, and then we want to find the velocity and acceleration of the crown of the student's head (point P, which is located essentially at the periphery of the thin hoop).

❑ **Solution:**   As indicated in Figure 3-2b, we model the problem as a thin circular hoop that rolls without slip in the XY plane along the X axis. The $OXYZ$ reference frame is considered to be "fixed in space." Note that we are assuming that the student's navel at point C and crown at point P remain fixed relative to the thin hoop. Since the hoop does not slip and we know the initial location of the hoop, the position vector of any point on the hoop can be written in terms of the angle $\theta(t)$, where $\theta(t)$ is defined with respect to any line that is parallel to the Y axis, positive in the clockwise direction. (Given the assumptions and nonslip condition, the fact that we can locate any point in this system via the single variable θ is a slightly subtle issue that we shall explore in detail in Section 5-4). In particular, the displaced hoop is sketched in Figure 3-2b where the radial line CP, which was originally along the Y axis, is at the angle θ with respect to the vertical. Thus, the velocities and accelerations that we are seeking can be expressed in terms of the time history of the angle θ and, as we shall see, the hoop's radius.

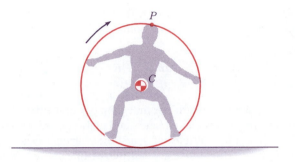

(a) Student in rolling thin hoop.

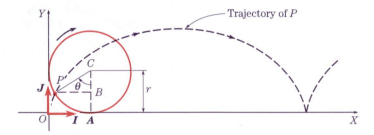

(b) Trajectory of point P of rolling nonslipping hoop is cycloid.

FIGURE 3-2   Student in thin hoop rolling without slip.

The kinematics of the points C and P can be found by finding the position vector $\mathbf{R}(t)$, for each of them, and then successively differentiating to calculate their respective velocity vector $\mathbf{v}(t)$ and acceleration vector $\mathbf{a}(t)$. Except in cases where $\mathbf{v}(t)$ and $\mathbf{a}(t)$ can be written by inspection—which are few—this is the most reliable approach for conducting a successful kinematic analysis. This idea, which may appear to be trivial, is enormously important as it is the foundation of *every* kinematic analysis, whether explicit or implicit. Indeed, the most important sentence in this chapter is as follows:

- Find $\mathbf{R}(t)$ with respect to the reference frame, differentiate to find $\mathbf{v}(t)$, and differentiate again to find $\mathbf{a}(t)$!

First, for point C, its coordinates in the XY reference plane are

$$X_C = OA \quad \text{and} \quad Y_C = AC \tag{a}$$

where X_C and Y_C are the coordinates of C along the X and Y axes, respectively. The no-slip*

---

*The "no-slip" condition is frequently and erroneously called a "rough surface" condition. "Roughness" and "friction" are two substantially different concepts, and therefore should not be used interchangeably.

---

condition requires that the distance OA along the X axis and the circular arc PA are equal in length. Further, since PA is equal to $r\theta$, Eqs. (a) may be rewritten as

$$X_C = r\theta \quad \text{and} \quad Y_C = r \tag{b}$$

In obtaining Eq. (b), we have accomplished the most difficult of the several tasks required for analyzing the motion of point C; calculation of the velocity and acceleration is simply a matter of time differentiation.

So, the position vector $\boldsymbol{R}_C$ extending from the reference frame origin to point C is

$$\boldsymbol{R}_C(t) = r\theta\boldsymbol{I} + r\boldsymbol{J} \tag{c}$$

where $\boldsymbol{I}$ and $\boldsymbol{J}$ are the unit vectors attached to the X and Y axes, respectively. Note that since the X and Y axes are "fixed," the unit vectors $\boldsymbol{I}$ and $\boldsymbol{J}$ do not change in either magnitude or direction, so they should be treated as constants in the differentiations that follow. Thus, applying Eqs. (3-1) and (3-2) to Eq. (c) gives

$$\boldsymbol{v}_C = \frac{d\boldsymbol{R}_C}{dt} = r\frac{d\theta}{dt}\boldsymbol{I} = r\dot{\theta}\boldsymbol{I} \tag{d}$$

$$\boldsymbol{a}_C = \frac{d^2\boldsymbol{R}_C}{dt^2} = r\frac{d^2\theta}{dt^2}\boldsymbol{I} = r\ddot{\theta}\boldsymbol{I} \tag{e}$$

where for representing the time derivatives of $\theta(t)$ we have introduced the notation[1]

$$\dot{\theta} = \frac{d\theta}{dt} \quad \text{and} \quad \ddot{\theta} = \frac{d^2\theta}{dt^2} \tag{f}$$

The results in Eqs. (d) and (e) are generally easy to interpret. The center of the hoop, point C, does not move in the Y direction. The velocity and acceleration of point C are familiar results from elementary physics. The emphasis here should be placed on the method, not the result.

Now, we consider the kinematics of point P. The coordinates of point P in the XY reference plane are

$$X_P = OA - PB$$
$$Y_P = AC - BC \tag{g}$$

where X_P and Y_P are the coordinates of P along the X and Y axes, respectively. Furthermore, as indicated above, $OA = PA$. In terms of the parameters sketched in Figure 3-2, Eqs. (g) may be rewritten as

$$X_P = r\theta - r\sin\theta$$
$$Y_P = r - r\cos\theta \tag{h}$$

Again, in obtaining Eqs. (h), we have accomplished the most difficult of the several tasks required for analyzing the motion of point P; only time differentiation remains. As an interesting aside, we note that Eqs. (h) are the parametric equations of a cycloid. The dashed line in Figure 3-2 is a sketch of Eqs. (h), which is the trajectory of point P.

So, the position vector $\boldsymbol{R}_P$ extending from the origin to point P is

$$\boldsymbol{R}_P = r(\theta - \sin\theta)\boldsymbol{I} + r(1 - \cos\theta)\boldsymbol{J} \tag{i}$$

---

[1]The use of this notation in this problem should cause no confusion; however, in Section 3-5 a slightly specialized meaning for the use of ($\dot{\square}$) as a time derivative of ($\square$) will be adopted. Nevertheless, here as well as in all other instances throughout this book, the notation ($\dot{\square}$) will represent the time derivative of the quantity $\square$, as observed in the reference frame in which $\square$ is defined.

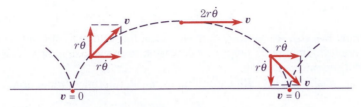

FIGURE 3-3 Instantaneous velocity vector $\boldsymbol{v}$ depends on θ and $\dot{\theta}$ and is tangent to the trajectory, except at the roadbed contact point where it is *zero*.

Differentiation of Eq. (i) according to Eq. (3-1) gives the velocity vector as

$$\boldsymbol{v}_P = r\dot{\theta}[(1 - \cos\theta)\boldsymbol{I} + \sin\theta\boldsymbol{J}] \tag{j}$$

The velocities at five locations along the trajectory are sketched in Figure 3-3.

Specifically, if in Figure 3-3 the leftmost point of contact with the roadbed is for $\theta = 0$, the five points shown along the trajectory correspond to $\theta = 0$, $\dfrac{\pi}{2}$, π, $\dfrac{3\pi}{2}$, and 2π. Perhaps the most interesting observation from Figure 3-3 is that at locations where the point P comes into contact with the roadbed $(0, \pm 2\pi, \cdots)$, its velocity is instantaneously *zero* irrespective of the speed of the rolling hoop! This fact is likely to surprise many readers. As nonintuitive as this result may seem, it should be noted that when point P is in contact with the roadbed (1) its horizontal velocity must be zero because of the no-slip condition and (2) its vertical velocity must be zero because it neither moves into nor lifts off the roadbed. Although we shall not discuss this point further at this time, it is a useful result that we shall use in several analyses, primarily in Chapter 6.

By using Eq. (3-2) and recalling the rules for differentiating a product, the acceleration of point P can be obtained by differentiating Eq. (j) as

$$\frac{d\boldsymbol{v}_P}{dt} = r\ddot{\theta}[(1 - \cos\theta)\boldsymbol{I} + \sin\theta\boldsymbol{J}]$$
$$+ r\dot{\theta}[(-)(-\sin\theta)\dot{\theta}\boldsymbol{I} + (\cos\theta)\dot{\theta}\boldsymbol{J}]$$

or

$$\boldsymbol{a}_P = r\ddot{\theta}[(1 - \cos\theta)\boldsymbol{I} + \sin\theta\boldsymbol{J}]$$
$$+ r\dot{\theta}^2[\sin\theta\boldsymbol{I} + \cos\theta\boldsymbol{J}] \tag{k}$$

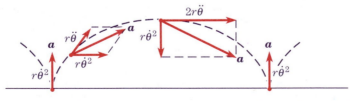

FIGURE 3-4 Instantaneous acceleration vector $\boldsymbol{a}$ depends in general on θ, $\dot{\theta}$ and $\ddot{\theta}$.

The accelerations at four of the same five locations along the trajectory are sketched in Figure 3-4. The acceleration of point P depends on θ, $\dot{\theta}$, and $\ddot{\theta}$. Note that where the point P comes into contact with the roadbed, $(0, \pm 2\pi, \ldots)$, its acceleration is directed away from the roadbed (that is, toward the center of the hoop) and has the magnitude $r\dot{\theta}^2$. Again, the emphasis here should be placed on the method, not the result; although the results in both Figures 3-3 and 3-4 are interesting and useful.

3-3   PLANE KINEMATICS OF RIGID BODIES

Although in principle we can analyze the motion of systems of any degree of complexity, in practice we must seek ways of simplifying an analysis. One extremely powerful simplification is afforded by the concept of a *rigid body*. No real object is rigid, in the sense that it does not deform under load; yet there are many circumstances in which a rigid body can be a useful *model* for a real system. This, of course, is a mathematical idealization. (In Chapter 6, we shall discuss the concept of a rigid body in more detail.) So, in analyses where the deformations of a body are negligible compared with the overall motion of the body, the body is often modeled as a rigid body. While there may be significant information that is suppressed by such a model (most notably, its elastic deformation), the concept of a rigid body is nevertheless a very useful one in dynamics.

Suppose, for example, that we are interested in the motion of an automobile traveling over a bumpy road. It is likely that the deformation of the car body will be small compared with the pitching, heaving, and rolling motions of the car; so, adopting the assumption that the car body is rigid will enormously simplify the analysis without significantly affecting the result obtained for the overall motion of the car. In a subsequent analysis, the assumption of rigidity can be relaxed if it ceases to be useful. For example, once both the motion of the car and the forces applied to the car body by the suspension system have been determined, the assumption that the car is rigid can be relaxed, and a study of the *deformation* of the nonrigid car traversing the bumpy road can be undertaken.

3-3.1   The General Motion of a Rigid Body

To appreciate why the concept of a rigid body is so useful, consider the following option. We could model a nonrigid body as a collection of N points, bound by flexible connectors. (In Chapter 4, we shall endow such points with properties.) In order to produce a useful approximation of the motion of the actual body, N would have to be large. Assuming each point could move throughout space, each point would require three spatial numbers or coordinates to locate it completely. Thus, the system of N points that model the body would require $3N$ such spatial numbers or coordinates to locate it. If, by contrast, we were able to assume that such a body were rigid, we would find that the entire body would require only six spatial numbers or coordinates to locate it.

To see that a rigid body requires only six spatial numbers or coordinates to locate it completely, consider Figure 3-5. We can specify the position of the body relative to an $OXYZ$ reference frame by specifying the location of a rigid triangle ABC fixed in the body as follows. First, specify the three coordinates of point A; the body is now free to rotate only, but about any axis through A. Second, specify the direction of the line AB; this requires two scalar quantities—say two of the three direction cosines of the line AB or two spherical angles.

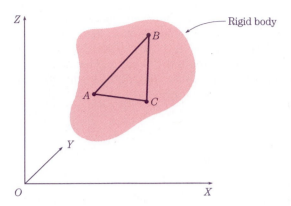

FIGURE 3-5   Specification of location of rigid body.

Now the body can rotate only about the line *AB*. Third, specify the angle of rotation about *AB* to locate the plane containing the line *AC*, using a single coordinate to represent the rotation of the body about *AB*, making six coordinates in all. Other choices of coordinates for locating the body could be made, but any alternative scheme would also require the specification of six independent spatial quantities.

From the preceding discussion, we can see that, in general, the kinematic analysis of rigid bodies includes both linear and angular quantities. In order to establish several important concepts about the angular or rotational motions of rigid bodies, we shall limit most of our discussion here to the plane kinematics of rigid bodies.

3-3.2   Types of Plane Motion of a Rigid Body

A rigid body executes plane motion when all parts of the body move in parallel planes, where a fixed reference plane may be defined as one of such planes. Thus, each point of the body remains at a constant distance from the fixed reference plane during plane motion. This idealization is appropriate for analyzing many rigid body problems in engineering. The plane motion of a rigid body may be divided into several categories, as discussed below.

- **Translation** is any motion in which every line in the body remains parallel to its past and future positions at all times. In translation there is *no rotation of any line in the body at any time.* In *rectilinear translation,* all points in the body move in parallel straight lines (see Figure 3-6*a*). In *curvilinear translation,* all points move along congruent curves (see Figure 3-6*b*). Clearly, rectilinear translation is a special case of curvilinear translation. Note that in each of the two cases of translation, the motion of the body is completely specified by the motion of any point in the body, since all points undergo the same motion. Also, note that all lines such as *AB* in Figures 3-6*a* and 3-6*b* remain parallel to their past and future positions. Thus, in rigid body translation, any kinematics describing the motion of a point in the body also completely specify the motion of the body. This can be quite subtle, as we shall illustrate in Example 3-2.

- **Rotation** about a fixed axis is the angular motion about that axis. If the axis of rotation passes through the body, as does the axis through point *A* in Figure 3-6*c*, points on the axis do not move. If the axis of rotation does not pass through the body, as does the axis through point *O* in Figure 3-6*d*, all points of the body move. In either case, all points in the body move in circular paths about the axis of rotation, and all lines in the body that lie in the plane of the motion rotate through the same angle during the same time. Note that such lines do not have to pass through the axis for this to be true. Thus, in both Figures 3-6*c* and 3-6*d* lines *CD* and *EF* in the plane of the body rotate through the same angle during the same time as does line *AB*.

- **General plane motion** of a rigid body is a combination of translation and rotation (see Figure 3-6*e*). This easily demonstrable fact in plane motion is expressed by Chasles' theorem:* the most general motion of a rigid body is equivalent to a translation of some point (which may be inside or outside of the body) plus a rotation about an axis passing through that point. Figure 3-7 gives a planar interpretation of Chasles' theorem where the motion in Figure 3-7*a* is depicted as the superposition of the motions in Figures 3-7*b*, representing a translation of point *C*, and 3-7*c*, representing a rotation about an axis passing through point *C*. Although our demonstration and discussion have been limited to plane motion, Chasles' theorem is valid for the general motion of any rigid body in three-dimensional space.

---

*This theorem is named for the French geometer Michel Chasles (1793–1880), who stated it in 1830.

3-3.3 Angular Displacement, Angular Velocity, and Angular Acceleration

We continue the discussion of rigid bodies undergoing plane motion. From Example 3-1 forward, we have used the word *rotation* somewhat casually, but we believe without ambiguity. In order to proceed quantitatively, however, we must become more precise in our terminology. A detailed discussion of rotation of rigid bodies is one of the more challenging aspects of kinematics. Our tasks will be simplified greatly, however, by limiting our focus on rigid bodies to planar motion.

Figure 3-8 shows a rigid body moving parallel to the XY plane as it undergoes fixed-axis rotation about OZ, where the reference frame $OXYZ$ is fixed in space. The angular positions of two arbitrary lines in the XY plane, designated as 1–1 and 2–2 and fixed to the body, are defined as θ_1 and θ_2, respectively, measured from the X axis of the reference frame. Note that lines 1–1 and 2–2 do not necessarily extend radially from the origin. The values of θ_1 and θ_2 are generally measured in radians, which are dimensionless. The angle γ is a constant angle between lines 1–1 and 2–2. So, from Figure 3-8,

$$\theta_2 = \theta_1 + \gamma \tag{3-3}$$

During a finite time interval, the body undergoes a finite rotation or *angular displacement* $\Delta\boldsymbol{\theta}$ as it rotates about the axis OZ. The rotation itself can be designated by a directed line segment $\Delta\boldsymbol{\theta}$ whose length represents the magnitude of the rotation and whose sense is determined by the right-hand rule: When the axis OZ is grasped with the right hand so that the fingers curl in the sense of the rotation $\Delta\boldsymbol{\theta}$, the right-hand thumb points in the sense of the directed line segment. Once the axis of rotation has been chosen, the magnitude and the sense of the rotation determine the directed line segment $\Delta\boldsymbol{\theta}$.

If all rotations are always in the same (positive or negative) direction, which is the case for plane motion, then finite rotations can be treated as vectors, as we have chosen to indicate

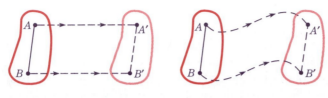

(a) Rectilinear translation.　　(b) Curvilinear translation.

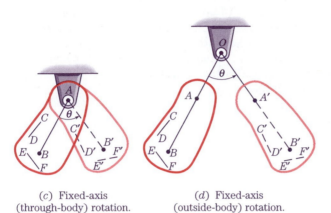

(c) Fixed-axis
(through-body) rotation.

(d) Fixed-axis
(outside-body) rotation.

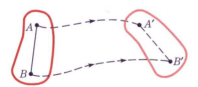

(e) General plane motion.

FIGURE 3-6   Types of rigid-body plane motion; motions go from unprimed to primed configurations.

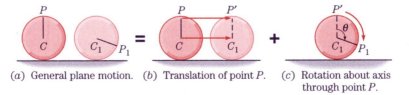

(a) General plane motion.　(b) Translation of point P.　(c) Rotation about axis through point P.

FIGURE 3-7   General plane motion may be considered as superposition of a translation and a rotation.

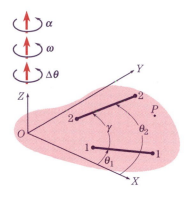

FIGURE 3-8   Rigid body moving parallel to XY plane, undergoing fixed-axis rotation about OZ; where $\Delta\boldsymbol{\theta}$, $\boldsymbol{\omega}$, and $\boldsymbol{\alpha}$ are all drawn in positive sense.

them by our boldfaced vector notation on $\Delta\boldsymbol{\theta}$. However, as we discuss below, for nonplanar rotations, finite rotations are *not* vectors.

From Eq. (3-3), during a finite time interval Δt,

$$\Delta\boldsymbol{\theta}_2 = \Delta\boldsymbol{\theta}_1 \tag{3-4}$$

The first and second time derivatives of Eq. (3-4) give

$$\lim_{\Delta t \to 0} \frac{\Delta\boldsymbol{\theta}_2}{\Delta t} = \lim_{\Delta t \to 0} \frac{\Delta\boldsymbol{\theta}_1}{\Delta t}$$

or

$$\frac{d\boldsymbol{\theta}_2}{dt} = \frac{d\boldsymbol{\theta}_1}{dt} \quad \text{and} \quad \frac{d^2\boldsymbol{\theta}_2}{dt^2} = \frac{d^2\boldsymbol{\theta}_1}{dt^2} \tag{3-5}$$

For the rigid body rotating in a plane, Eqs. (3-5) may be more simply expressed as

$$\boldsymbol{\omega} = \frac{d\boldsymbol{\theta}}{dt} \tag{3-6}$$

and

$$\boldsymbol{\alpha} = \frac{d\boldsymbol{\omega}}{dt} = \frac{d^2\boldsymbol{\theta}}{dt^2} \tag{3-7}$$

where $\boldsymbol{\omega}$ is the *angular velocity* and $\boldsymbol{\alpha}$ is the *angular acceleration*. Common units for $\boldsymbol{\omega}$ and $\boldsymbol{\alpha}$ are radians per second and radians per second per second, respectively. The subscripts on the differential angles in Eqs. (3-6) and (3-7) have been dropped since, as suggested by Eq. (3-4), all such differential angles are the same for a rigid body rotating about a common axis. Of the three angular vectors—angular displacement, angular velocity, and angular acceleration—our discussions and analyses will be centered on angular velocity. The angular velocity $\boldsymbol{\omega}$ can also be represented by a directed line segment. The line of action of $\boldsymbol{\omega}$ is called the *instantaneous axis of rotation of the body with respect to the reference frame.* The length of the directed line segment of $\boldsymbol{\omega}$ is called the *angular speed* and is defined as the magnitude $|\boldsymbol{\omega}|$.

As a group, Eqs. (3-4) through (3-7) emphasize the point made in discussing Figures 3-6c and 3-6d: in plane rotation of a rigid body, all lines fixed on the body in its plane of motion undergo the same angular displacement, the same angular velocity, and the same angular acceleration. This fact leads us to consider the body sketched in Figure 3-8 again in

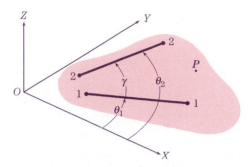

FIGURE 3-9   Rigid body undergoing general plane motion in XY plane.

Figure 3-9, where it is undergoing general plane motion. (Note that OZ does not pass through the body in Figure 3-9.) All the arguments and all the conclusions made in conjunction with Figure 3-8 hold also for Figure 3-9; Figure 3-8 was simply an intermediate vehicle. So, one might ask, "Do the angular displacement, angular velocity, and angular acceleration vectors still go along the OZ axis in Figure 3-9?" The correct answer may be mildly disquieting. First, let's note that the location of $\Delta\boldsymbol{\theta}$, $\boldsymbol{\omega}$, and $\boldsymbol{\alpha}$ along OZ in Figure 3-8 was for the comfort of the reader. Since all line segments in Figure 3-8, as well as in Figure 3-9, undergo the same angular motions, $\Delta\boldsymbol{\theta}$, $\boldsymbol{\omega}$, and $\boldsymbol{\alpha}$ could have been sketched *any place,* as long as they were perpendicular to the XY plane. As vectors—which have a magnitude and a direction but *not* a place—the vectorial representation of these angular motions have no place!

■ **Example 3-2:**   **Rotating and Nonrotating Disks**

A larger disk that rotates about its center at a constant angular velocity $\boldsymbol{\omega}_0$ is shown in four successive views in Figures 3-10 and 3-11. As suggested by the painted dot marker on the larger disk, the combined four views represent a single rotation of 360° of the larger disk.

A hole in the larger disk contains a smaller disk supported on bearings that behave in a special manner. (Actually, the bearings prevent any torque transmission from the larger disk to the smaller disk; but we have no means of knowing this yet.) If the set screw locks the smaller disk to the larger disk, the system behaves as sketched in Figure 3-10; whereas if the set screw is not engaged, the system behaves as sketched in Figure 3-11. The reference frame $OXYZ$ is fixed. What is the angular velocity of the smaller disk with respect to $OXYZ$ in Figure 3-10 and in Figure 3-11?

□ **Solution:**   Because the smaller disk can be considered as a rigid body undergoing plane rotation, the decisive issue is whether any line on (or in) the body undergoes rotation.

In Figure 3-10, the smaller disk is locked into the larger disk, and the "A" painted on the disk undergoes the same angular velocity as the larger disk, $\boldsymbol{\omega}_0$. This can be verified by noting that, with respect to the fixed reference frame, the "A" rotates at $\boldsymbol{\omega}_0$, undergoing 360° of rotation during the four views in Figure 3-10.

In Figure 3-11, the smaller disk moves around the center of the larger disk, but it does *not* rotate as it moves. (Note that it does rotate with respect to the larger disk, but not with respect to $OXYZ$.) This can be verified by noting that, with respect to the fixed reference frame, the "A" does not rotate since its orientation never changes. Thus, in Figure 3-11, the angular velocity of the smaller disk is always zero. *Every* point in the smaller disk moves with respect to $OXYZ$ with the identical velocity!

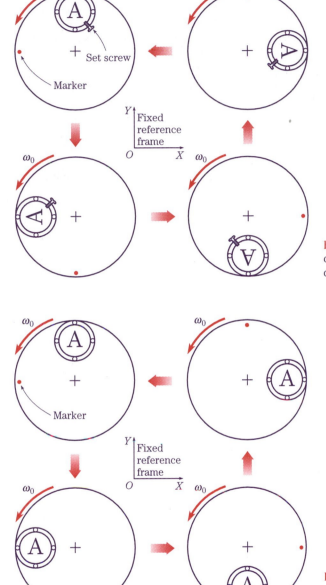

FIGURE 3-10 Four successive views of larger disk rotating at constant $\boldsymbol{\omega}_0$, containing smaller rotating disk.

FIGURE 3-11 Four successive views of larger disk rotating at constant $\boldsymbol{\omega}_0$, containing smaller nonrotating disk.

Conclusions: In Figure 3-10, the smaller disk undergoes the same angular rotation as the larger disk. (This case is analogous to Fig. 3-6*d*.) In Figure 3-11 the smaller disk undergoes only curvilinear translation, as all points within it move in parallel lines and the motion of the entire disk can be completely specified by the motion of any point within the disk. (This case is analogous to Fig. 3-6*b*.) Finally, we shall see in Chapter 6 that the proper calculation of the energy stored by the motion of the smaller disk depends on the kinematics that we have just found. Wrong kinematics guarantees wrong dynamics. (In particular, see Example 3-3, which follows, and its reconsideration in Example 6-15 in Chapter 6.)

■ **Example 3-3:**   **How Many Rotations Does Smaller Cylinder Undergo?**

A smaller cylinder of radius r rolls without slip on a larger cylinder of radius R, as sketched in Figure 3-12. The angle θ locates the center C of the smaller cylinder with respect to a *fixed* vertical reference, and point A, which is on the rolling cylinder, and point D, which is on the stationary cylinder, are coincident when θ is zero, that is, when the rolling cylinder is in position 1. Find the angular displacement of the smaller cylinder as a function of θ.

Before proceeding with the solution, determine whether you can answer the following little riddle. If $R = 3$ m and $r = 1$ m, how many complete rotations does the smaller cylinder undergo as it makes one complete transit around the larger cylinder? Stop reading; answer the riddle.

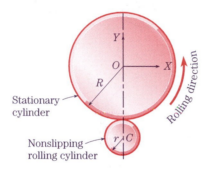

(*a*) Two cylinders shown in original positions.

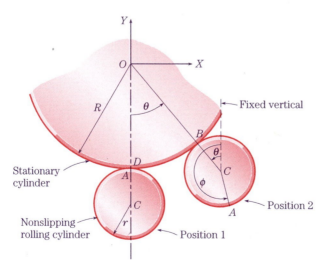

(*b*) Enlarged view showing displaced rolling cylinder.

FIGURE   3-12 Smaller cylinder rolls without slip on larger stationary cylinder, where $OXYZ$ is fixed reference frame.

☐ **Solution:**   The smaller cylinder is shown in its initial position (Position 1) and in an arbitrarily displaced position (Position 2) in Figure 3-12b, where θ is the angle between the *fixed* vertical reference and the line OBC, and ϕ is the angle between the *fixed* vertical reference and CA. (Line CA may be imagined to be a line that is painted onto the smaller cylinder.) Note that by elementary plane geometry, θ is also the angle between the *fixed* vertical reference and CB. The no-slip condition requires that

$$DB = BA \tag{a}$$

or

$$R\theta = r(\phi - \theta) \tag{b}$$

Solving for ϕ gives

$$\phi = \left(\frac{R + r}{r}\right)\theta \tag{c}$$

which represents the answer to the stated problem.

Now, let's address the riddle when $R = 3$ m and $r = 1$ m. From Eq. (c), if $\theta = 2\pi$, $\phi = 8\pi$; so the smaller cylinder makes four complete rotations!

Referring to Figure 3-12a and Eq. (c), give a physical interpretation to this result. (Hint: Equation (c) is $\phi = \left(\dfrac{R}{r} + 1\right)\theta$. Considering the two terms in parentheses, the $\dfrac{R}{r}$ is analogous to Figure 3-2 and the 1 is analogous to Figure 3-10, during which there is no rolling.)

Finally, we note that in this example as well as in our previous (and subsequent) discussions, we have consistently used the word *rotation* to characterize changes in orientation. Words such as "revolution," "orbit," "spin," "turn," and "somersault" are often used incorrectly as synonyms for rotation. Even more importantly, we emphasize that just as linear displacements have been consistently defined with respect to a fixed reference, so too angular displacements must be defined with respect to a fixed reference.

3-3.4   A Cautionary Note about Finite Rotations

The discussion of *angular* displacements, velocities, and accelerations of a rigid body has so far parallelled the discussion of *linear* displacements, velocities, and accelerations of a point. In both cases it is possible to represent displacements, velocities, and accelerations by directed line segments. There is, however, a significant difference between the character of the directed line segments that represent angular displacements and the directed line segments that represent linear displacements. This difference emerges when nonplanar motions of finite bodies are analyzed, or equivalently, when we must consider combining a number of angular displacements. Such considerations are usually explored under the category of *finite rotations*.

If a point P undergoes several linear displacements in succession, the resulting total linear displacement can be represented by a directed line segment that is the vector sum of the directed line segments that represent the individual displacements. However, when a finite body undergoes several finite angular displacements about different axes in succession, the result depends on the order in which the individual angular displacements are taken. So, even though finite angular displacements can be represented by directed line segments, in general, the combination of these directed line segments does not satisfy the laws of vector algebra. The exception to this statement is the case for plane motion, where all the angular displacements are about the same axis. Under planar motion, the algebra of angular displacements does degenerate to vector algebra. A more mathematical presentation of this discussion is given in Appendix A.

The nonvectorial character of finite angular displacements can be illustrated by demonstrating that they do not satisfy the commutative operation of vector addition, which may be written as $A + B = B + A$. Consider the example depicted in Figure 3-13a, where, in response to telemetric signals, the satellite undergoes successive 90° rotations about the Z axis and then about the X axis. The desired orientation is achieved and shown at the far right of Figure 3-13a. If the order of the instructions to rotate were reversed, as sketched in Figure 3-13b, the total angular displacement would be entirely different, resulting in a likely unsatisfactory orientation.

Finally, although finite angular displacements, in general, do not combine according to the laws of vector addition, *infinitesimal* angular displacements do obey the laws of vector addition (see Appendix A). So, it follows from the laws of differentiation that angular velocities also must satisfy the laws of vector addition; that is, if the infinitesimal $d\boldsymbol{\theta}$ satisfies the laws of vector addition, so will $\boldsymbol{\omega}$ according to

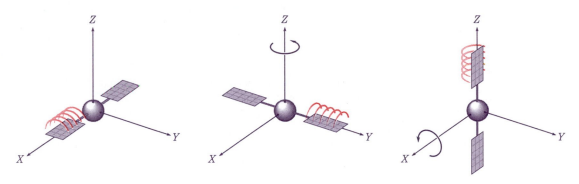

(a) Satellite undergoing successive 90° rotations about Z then about X.

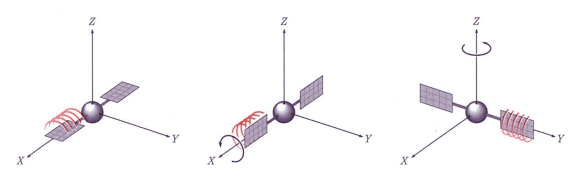

(b) Satellite undergoing successive 90° rotations about X then about Z.

FIGURE 3-13   Finite rotations are not commutative; the order of the maneuvering signals sent to the satellite is important.

$$\boldsymbol{\omega} = \lim_{\Delta t \to 0} \frac{\Delta \boldsymbol{\theta}}{\Delta t} = \frac{d\boldsymbol{\theta}}{dt} \tag{3-8}$$

Thus, if the satellite in Figure 3-13a undergoes $\boldsymbol{\omega}_Z$ and then $\boldsymbol{\omega}_X$, and the satellite in Figure 3-13b undergoes $\boldsymbol{\omega}_X$ and then $\boldsymbol{\omega}_Z$, in both cases the total angular velocity would be the same: $\boldsymbol{\omega} = \boldsymbol{\omega}_X + \boldsymbol{\omega}_Z = \boldsymbol{\omega}_Z + \boldsymbol{\omega}_X$.

3-4   TIME RATE OF CHANGE OF VECTOR IN ROTATING FRAME

In many applications, the velocity and acceleration of a rigid body as well as of a point are most conveniently described by using intermediate rigid frames that may translate and rotate relative to the "fixed" reference frame. For example, the motion of a turbine rotor on a ship with respect to the shore is conveniently described with respect to an intermediate frame that is fixed in the ship, which itself moves relative to the shore. Before discussing kinematic analysis utilizing intermediate frames, we consider the problem of relating the time rate of change of a vector as seen by an observer in a moving frame to that seen by an observer in a second reference frame. (The second reference frame may be moving with respect to a third reference frame. However, for convenience it may be useful to think of the second reference frame as being "fixed.")

In Figure 3-14 the rectangular frame $oxyz$ has angular velocity $\boldsymbol{\omega}$ with respect to the second reference frame $OXYZ$. The origins o and O of the two reference frames are coincident and remain so throughout this discussion. A vector $\boldsymbol{A}$ having a constant magnitude is fixed in $oxyz$ and is carried around by the motion of this frame. From the second reference frame $OXYZ$, the vector $\boldsymbol{A}$ appears to be changing its orientation, which is due to $\boldsymbol{\omega}$. It can be shown that the rate of change of $\boldsymbol{A}$ observed from the second reference frame $OXYZ$ is

$$\frac{d\boldsymbol{A}}{dt} = \boldsymbol{\omega} \times \boldsymbol{A} \tag{3-9}$$

where the symbol "$\times$" denotes the vector cross-product. The derivation of Eq. (3-9) is easier to visualize when $\boldsymbol{A}$ extends from the origin, such as the constant magnitude vectors $\boldsymbol{a}_1$ or $\boldsymbol{a}_2$ in Figure 3-14. So, if we can show that

$$\frac{d\boldsymbol{a}_1}{dt} = \boldsymbol{\omega} \times \boldsymbol{a}_1 \qquad \text{and} \qquad \frac{d\boldsymbol{a}_2}{dt} = \boldsymbol{\omega} \times \boldsymbol{a}_2 \tag{3-10}$$

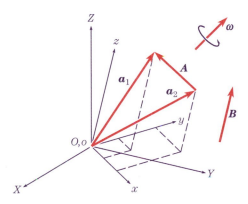

FIGURE 3-14   Vector $\boldsymbol{A}$ is fixed in reference frame $oxyz$, which has angular velocity $\boldsymbol{\omega}$ with respect to reference frame $OXYZ$; vector $\boldsymbol{B}$ is defined in $oxyz$ but is arbitrary.

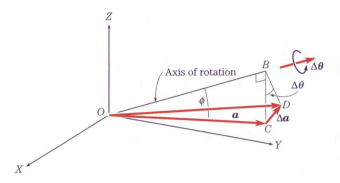

FIGURE 3-15   Incremental change of a during incremental rotation $\Delta\theta$ is Δa.

because $A = a_1 - a_2$, subtraction of the second of Eqs. (3-10) from the first gives Eq. (3-9).

So, consider in Figure 3-15 a vector a of constant magnitude that extends from O and is carried around by the rotating frame, which for simplicity is not shown. During an incremental time Δt, an incremental angular displacement $\Delta\theta$ ($\Delta\theta \approx \omega\Delta t$) of the rotating frame changes the constant magnitude vector a from OC to OD. The magnitude of Δa is

$$|\Delta a| = CD \approx BC\Delta\theta = OC \sin\phi\Delta\theta = |a||\Delta\theta| \sin\phi \qquad (3\text{-}11)$$

Dividing Eq. (3-11) by Δt and taking the limit as $\Delta t \to 0$ give the magnitude of da/dt as

$$\left|\frac{da}{dt}\right| = |a||\omega| \sin\phi \qquad (3\text{-}12)$$

where $\Delta\theta \approx \omega\Delta t$ has been used. Since Δa is nearly orthogonal to the plane containing a and ω, the limiting quantity $\dfrac{da}{dt}$ is orthogonal to this plane. Note also that the vector cross-product $\omega \times a$ is in the same direction as $\dfrac{da}{dt}$ and has the same magnitude as $\dfrac{da}{dt}$ as given in Eq. (3-12). Thus, we may write

$$\frac{da}{dt} = \omega \times a \qquad (3\text{-}13)$$

Finally, by use of Eqs. (3-10), Eq. (3-9) can be verified for a vector having arbitrary orientation, but which is fixed in a rotating frame.

■ **Example 3-4:**   Time Derivative of Constant Magnitude Vector

The vector a of constant magnitude is defined in the $oxyz$ reference frame as sketched in Figure 3-16.

(a)   The $oxyz$ reference frame is rotating at $\omega = \omega_1 = \omega_1 k$ with respect to the "fixed" reference frame $OXYZ$ as indicated. Find the time derivative of a as seen from $OXYZ$.

(b)   Repeat the problem for $\omega = \omega_2 = \omega_2 j$ as the angular velocity of $oxyz$ with respect to $OXYZ$.

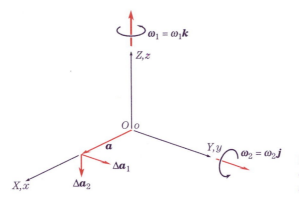

FIGURE 3-16 Time derivative of vector $\boldsymbol{a}$ of constant magnitude, which is defined in a rotating reference frame $oxyz$.

☐ **Solution:** This analysis represents a direct application of Eqs. (3-9) or (3-13). It should be clear that $\dfrac{d\boldsymbol{a}}{dt} = 0$ for all observers in $oxyz$.

(a) First of all, for observers in $OXYZ$ the vector increment $\Delta\boldsymbol{a}_1$ should appeal to the reader's intuition as the change in $\boldsymbol{a}$ during a time increment Δt, due to $\boldsymbol{\omega}_1$. Note that, due to $\boldsymbol{\omega}_1$, $\Delta\boldsymbol{a}_1$ is in the positive Y direction.

For observers in $OXYZ$, use of either of Eqs. (3-9) or (3-13) gives

$$\frac{d\boldsymbol{a}}{dt} = \boldsymbol{\omega} \times \boldsymbol{a} = \omega_1\boldsymbol{k} \times a\boldsymbol{i} = \omega_1 a\boldsymbol{j} \qquad (a)$$

where the vector cross-product mnemonic in Figure 3-17 has been used for convenience. The use of the mnemonic expresses the fact that when traversing the circle in the direction of the arrows, the result is positive; and when traversing opposite to the direction of the arrows, the result is negative. Thus, for example, $\boldsymbol{i} \times \boldsymbol{j} = \boldsymbol{k}$ and $\boldsymbol{k} \times \boldsymbol{i} = \boldsymbol{j}$ or $\boldsymbol{j} \times \boldsymbol{i} = -\boldsymbol{k}$ and $\boldsymbol{i} \times \boldsymbol{k} = -\boldsymbol{j}$.

(b) The vector increment $\Delta\boldsymbol{a}_2$ should appeal to the intuition as the change in $\boldsymbol{a}$ during a time increment Δt, due to $\boldsymbol{\omega}_2$. Note that, due to $\boldsymbol{\omega}_2$, $\Delta\boldsymbol{a}_2$ is in the negative Z direction.

As in part (a),

$$\frac{d\boldsymbol{a}}{dt} = \boldsymbol{\omega} \times \boldsymbol{a} = \omega_2\boldsymbol{j} \times a\boldsymbol{i} = -\omega_2 a\boldsymbol{k}. \qquad (b)$$

Readers are strongly encouraged to make the concept explored in this example a part of their intuition. That is, a vector—characterized by a magnitude and a direction—can be

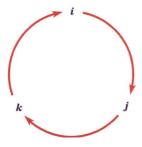

FIGURE 3-17 Vector cross-product mnemonic.

changed by changing its magnitude or its direction (or both simultaneously). But if a vector's magnitude is constant, it can be changed only by changing its direction; and the time rate of change for such a vector is given by Eq. (3-9) or (3-13).

Next, we develop a more general form corresponding to Eqs. (3-9) and (3-13). Let B be an arbitrary vector defined in $oxyz$, having changing magnitude *and* changing direction, possibly simultaneously. The instantaneous component representation of B referred to $oxyz$ is

$$B = B_x i + B_y j + B_z k \tag{3-14}$$

Note that B is any arbitrary vector such as sketched in Figure 3-14. In general, the scalar components B_x, B_y and B_z will vary with time, and, due to $\boldsymbol{\omega}$, the unit vectors attached to $oxyz$ will vary in orientation when viewed from the "fixed" reference frame $OXYZ$. Note that for observers in $oxyz$, B_x, B_y, and B_z will change with time, but the unit vectors i, j, and k will be constant in magnitude and direction. Using the notation $\dfrac{d}{dt}$ to indicate time derivatives as observed from the reference frame $OXYZ$, the derivative of Eq. (3-14) is

$$\begin{aligned}
\frac{dB}{dt} &= \frac{dB_x}{dt} i + \frac{dB_y}{dt} j + \frac{dB_z}{dt} k + B_x \frac{di}{dt} + B_y \frac{dj}{dt} + B_z \frac{dk}{dt} \\
&= \dot{B}_x i + \dot{B}_y j + \dot{B}_z k + B_x \frac{di}{dt} + B_y \frac{dj}{dt} + B_z \frac{dk}{dt}
\end{aligned} \tag{3-15}$$

Note that the time derivatives of the scalar components as indicated by the superior dots are the same for observers in all frames; in this case, for observers in both $oxyz$ and $OXYZ$.

The right-hand side of Eq. (3-15) can be conveniently separated into two parts. The first three terms can be considered to be the rate of change of B as viewed from the $oxyz$ frame, and the last three terms give the contribution due to the rotation of $oxyz$ with respect to $OXYZ$. We choose to denote the first three terms by $\left(\dfrac{\partial B}{\partial t}\right)_{\text{rel}}$, where the subscript "rel" indicates that the time rate of change is with respect to the $oxyz$ frame. The "rel" has its root in the word *relative,* which is used to denote "with respect to the $oxyz$ frame." In the next section, the $oxyz$ frame will be called an intermediate reference frame. The last three terms can be simplified by applying Eqs. (3-9) or (3-13) to evaluate the time derivatives of the unit vectors, which are vectors of fixed magnitude but changing direction. Thus, use of Eqs. (3-9) or (3-13) gives

$$\begin{aligned}
B_x \frac{di}{dt} + B_y \frac{dj}{dt} + B_z \frac{dk}{dt} &= B_x \boldsymbol{\omega} \times i + B_y \boldsymbol{\omega} \times j + B_z \boldsymbol{\omega} \times k \\
&= \boldsymbol{\omega} \times (B_x i + B_y j + B_z k) \\
&= \boldsymbol{\omega} \times B
\end{aligned} \tag{3-16}$$

Thus, Eq. (3-15) can be written as

$$\frac{dB}{dt} = \left(\frac{\partial B}{\partial t}\right)_{\text{rel}} + \boldsymbol{\omega} \times B \tag{3-17}$$

Equation (3-17) relates the time rates of change of an arbitrary vector as viewed from the two reference frames, $oxyz$ and $OXYZ$. The time derivative observed from the "fixed" reference frame $OXYZ$ is the sum of the time derivative observed from the rotating reference frame

$oxyz$ plus the vector cross-product $\boldsymbol{\omega} \times \boldsymbol{B}$. This cross-product term accounts for the angular velocity of the rotating reference frame with respect to the "fixed" reference frame. When the vector $\boldsymbol{B}$ has constant scalar components or is said to be "frozen" in $oxyz$, $\left(\dfrac{\partial \boldsymbol{B}}{\partial t}\right)_{\text{rel}} = 0$; then Eq. (3-17) reduces to Eq. (3-9) or Eq. (3-13). Furthermore, if $\boldsymbol{B}$ is parallel to $\boldsymbol{\omega}$ (that is, parallel to the axis of rotation), the cross-product vanishes and the time rates of change of $\boldsymbol{B}$ as viewed from the two frames are identical. So, when the rate of change of $\boldsymbol{\omega}$ itself is observed from the two reference frames, we note that

$$\frac{d\boldsymbol{\omega}}{dt} = \left(\frac{\partial \boldsymbol{\omega}}{\partial t}\right)_{\text{rel}} \tag{3-18}$$

Finally, because the vector $\boldsymbol{B}$ in Eq. (3-17) is arbitrary, we may define the very useful *operational form*

$$\frac{d}{dt} = \left(\frac{\partial}{\partial t}\right)_{\text{rel}} + \boldsymbol{\omega} \times \tag{3-19}$$

Equation (3-19) provides a means for determining the time derivative with respect to $OXYZ$ of a vector defined in $oxyz$, when $oxyz$ is rotating at $\boldsymbol{\omega}$ with respect to $OXYZ$.

Again, we express the fundamental concept that a vector has both magnitude and direction; thus a vector can be changed by changing either its magnitude or its direction (or both). As we shall illustrate in Example 3-5, the first term on the right-hand side of Eq. (3-17) accounts for the change in scalar components in $oxyz$ of the vector and the second term on the right-hand side of Eq. (3-17) accounts for the direction change of the vector. This is the essence of Eqs. (3-17) and (3-19).

■ **Example 3-5:**   **Time Derivative of Arbitrary Vector**

The shaft sketched in Figure 3-18 is rotating at $\boldsymbol{\omega}_2$ with respect to its bearings, which are mounted on a platform. The platform is rotating at $\boldsymbol{\omega}_1$ with respect to the "fixed" reference frame $OXYZ$. Find the first and second time derivatives of $\boldsymbol{\omega}_2$ with respect to $OXYZ$.

❑ **Solution:**   For convenience, $oxyz$ is defined as a frame that is fixed in the platform. At the instant shown, $oxyz$ is coincident with $OXYZ$ as sketched in Figure 3-18. Note that $\boldsymbol{\omega}_2$ may be considered simply as an arbitrary *vector* that is defined in $oxyz$, which itself is rotating at $\boldsymbol{\omega}_1$ with respect to $OXYZ$.

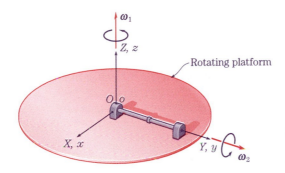

FIGURE 3-18   Shaft rotating at $\boldsymbol{\omega}_2$ with respect to platform which is rotating at $\boldsymbol{\omega}_1$ with respect to "fixed" reference frame $OXYZ$.

A direct application of Eq. (3-19) gives the first time derivative of $\boldsymbol{\omega}_2$, with respect to $OXYZ$, as

$$\frac{d\boldsymbol{\omega}_2}{dt} = \left[\left(\frac{\partial}{\partial t}\right)_{\text{rel}} + \boldsymbol{\omega}_1 \times\right]\boldsymbol{\omega}_2$$

where $\boldsymbol{\omega}_1$ here corresponds to $\boldsymbol{\omega}$ in Eq. (3-19). Thus

$$\frac{d\boldsymbol{\omega}_2}{dt} = \dot{\boldsymbol{\omega}}_2 + \boldsymbol{\omega}_1 \times \boldsymbol{\omega}_2 \tag{a}$$

where $(\dot{\square})$ denotes $\dfrac{\partial}{\partial t}(\square)$.

As suggested earlier, Eq. (a) may be given a simple interpretation. By noting that a vector has magnitude and direction, the first term on the right-hand side of Eq. (a) represents the magnitude change of $\boldsymbol{\omega}_2$, and the second term represents the change in direction of $\boldsymbol{\omega}_2$. Thus, the first term on the right-hand side of Eq. (a) represents the increase (or decrease) of the angular speed of the shaft relative to its bearings, and the second term accounts for the fact that $\boldsymbol{\omega}_1$ is changing the direction of $\boldsymbol{\omega}_2$ toward the negative X direction, at the instant shown.

Another application of Eq. (3-19) on Eq. (a) gives the second time derivative of $\boldsymbol{\omega}_2$, with respect to $OXYZ$, as

$$\begin{aligned}
\frac{d^2\boldsymbol{\omega}_2}{dt^2} &= \left[\left(\frac{\partial}{\partial t}\right)_{\text{rel}} + \boldsymbol{\omega}_1 \times\right]\left[\dot{\boldsymbol{\omega}}_2 + \boldsymbol{\omega}_1 \times \boldsymbol{\omega}_2\right] \\
&= \left(\frac{\partial}{\partial t}\right)_{\text{rel}}[\dot{\boldsymbol{\omega}}_2 + \boldsymbol{\omega}_1 \times \boldsymbol{\omega}_2] + \boldsymbol{\omega}_1 \times [\dot{\boldsymbol{\omega}}_2 + \boldsymbol{\omega}_1 \times \boldsymbol{\omega}_2] \\
&= \ddot{\boldsymbol{\omega}}_2 + \dot{\boldsymbol{\omega}}_1 \times \boldsymbol{\omega}_2 + \boldsymbol{\omega}_1 \times \dot{\boldsymbol{\omega}}_2 + \boldsymbol{\omega}_1 \times \dot{\boldsymbol{\omega}}_2 + \boldsymbol{\omega}_1 \times (\boldsymbol{\omega}_1 \times \boldsymbol{\omega}_2) \\
&= \ddot{\boldsymbol{\omega}}_2 + \dot{\boldsymbol{\omega}}_1 \times \boldsymbol{\omega}_2 + 2\boldsymbol{\omega}_1 \times \dot{\boldsymbol{\omega}}_2 + \boldsymbol{\omega}_1 \times (\boldsymbol{\omega}_1 \times \boldsymbol{\omega}_2)
\end{aligned} \tag{b}$$

The time derivative of the angular acceleration is sometimes called the *angular jerk*. The *jerk* (time derivative of acceleration) and the *angular jerk* are used sometimes by engineers to characterize the ride quality of passenger vehicles. Equation (b) is more difficult to physically interpret than Eq. (a) but, as we shall see in Section 3-5, each term can be associated explicitly with terms that are commonly encountered in kinematics.

3-5   KINEMATIC ANALYSIS UTILIZING INTERMEDIATE FRAMES

We are now prepared to derive two of the most useful kinematic expressions in which rectangular frames in relative motion are used to study the motion of a point. As sketched in Figure 3-19, the "fixed" reference frame is $OXYZ$. The intermediate frame is $oxyz$, with unit vectors $\boldsymbol{i}, \boldsymbol{j}, \boldsymbol{k}$. The frame $oxyz$ translates and rotates with respect to the frame $OXYZ$. The point P is located with respect to $OXYZ$ by the position vector $\boldsymbol{R}$, and it is also located with respect to the intermediate frame $oxyz$ by the *relative position vector* $\boldsymbol{r}$. The goal is to find the velocity and acceleration of P with respect to the "fixed" reference frame $OXYZ$.

The relative position vector may be written as

$$\boldsymbol{r} = x\boldsymbol{i} + y\boldsymbol{j} + z\boldsymbol{k} \tag{3-20}$$

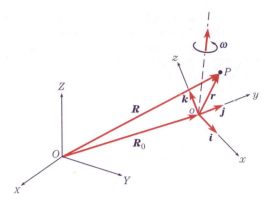

FIGURE 3-19   Intermediate frame $oxyz$ translates and rotates with respect to "fixed" reference frame $OXYZ$.

where x, y, and z are the components along the respective axes. According to Eqs. (3-1) and (3-2), the velocity $\boldsymbol{v}_{\text{rel}}$ and acceleration $\boldsymbol{a}_{\text{rel}}$ of P with respect (or relative) to $oxyz$ are

$$\boldsymbol{v}_{\text{rel}} = \left(\frac{\partial \boldsymbol{r}}{\partial t}\right)_{\text{rel}} = \dot{x}\boldsymbol{i} + \dot{y}\boldsymbol{j} + \dot{z}\boldsymbol{k}$$

$$\boldsymbol{a}_{\text{rel}} = \left(\frac{\partial \boldsymbol{v}_{\text{rel}}}{\partial t}\right)_{\text{rel}} = \left(\frac{\partial^2 \boldsymbol{r}}{\partial t^2}\right)_{\text{rel}} = \ddot{x}\boldsymbol{i} + \ddot{y}\boldsymbol{j} + \ddot{z}\boldsymbol{k} \tag{3-21}$$

From the previous section, the notation $(\partial/\partial t)_{\text{rel}}$ and the superior dots are used to indicate time differentiation observed from $oxyz$; for example, $\dot{x} = \left(\dfrac{\partial x}{\partial t}\right)_{\text{rel}}$ and $\ddot{x} = \left(\dfrac{\partial^2 x}{\partial t^2}\right)_{\text{rel}}$. Referring to Figure 3-19, the position vector $\boldsymbol{R}$ of the point P with respect to the origin of $OXYZ$ is the sum of $\boldsymbol{R}_0$ and $\boldsymbol{r}$. Then the velocity of P with respect to $OXYZ$ is

$$\boldsymbol{v} = \frac{d\boldsymbol{R}}{dt} = \frac{d\boldsymbol{R}_0}{dt} + \frac{d\boldsymbol{r}}{dt} \tag{3-22}$$

Here, we *emphasize* that the left-hand side and middle term expressions in Eq. (3-22) are precisely Eq. (3-1), as it is really Eq. (3-1) that we are evaluating. The right-hand expression in Eq. (3-22) is introduced as a matter of convenience. In considering the operational form in Eq. (3-19) for the right-hand side of Eq. (3-22), we note that the first term $\dfrac{d\boldsymbol{R}_0}{dt}$ is already defined with respect to $OXYZ$, so the operator in Eq. (3-19) does not apply to it; but the second term $\dfrac{d\boldsymbol{r}}{dt}$ is defined with respect to $oxyz$, so the operator in Eq. (3-19) must be appropriately applied to it. So,

$$\boldsymbol{v} = \frac{d\boldsymbol{R}_0}{dt} + \left[\left(\frac{\partial \boldsymbol{r}}{\partial t}\right)_{\text{rel}} + \boldsymbol{\omega} \times \boldsymbol{r}\right]$$

$$\boldsymbol{v} = \frac{d\boldsymbol{R}_0}{dt} + \boldsymbol{v}_{\text{rel}} + (\boldsymbol{\omega} \times \boldsymbol{r}) \tag{3-23}$$

where $\boldsymbol{v}_{\text{rel}}$ is defined by the first of Eqs. (3-21). The first term on the right-hand side of Eq. (3-23) is the velocity of the origin of the intermediate frame; the second term is the velocity of P with respect to the intermediate frame; and the third term is the contribution due to the rotation of the intermediate frame. In general, the velocities v and v_{rel} are different except when the intermediate frame is neither translating nor rotating with respect to the "fixed" reference frame. An exception to this remark occurs when the intermediate frame is rotating only and r is directed along the axis of rotation.

The acceleration of P with respect to the "fixed" reference frame is

$$a = \frac{dv}{dt} = \frac{d}{dt}\left[\frac{dR_0}{dt} + v_{\text{rel}} + (\omega \times r)\right]$$

$$a = \frac{dv}{dt} = \frac{d^2R_0}{dt^2} + \frac{dv_{\text{rel}}}{dt} + \frac{d}{dt}(\omega \times r)$$

$$a = \frac{dv}{dt} = \frac{d^2R_0}{dt^2} + \frac{dv_{\text{rel}}}{dt} + \frac{d\omega}{dt} \times r + \omega \times \frac{dr}{dt} \qquad (3\text{-}24)$$

Here, again, we emphasize that the left-hand side and middle expressions in Eq. (3-24) are precisely Eq. (3-2), as it is really Eq. (3-2) that we are evaluating. Equation (3-23) has been introduced into Eq. (3-24) simply as a matter of convenience. In applying the operational form in Eq. (3-19) to the terms on the right-hand side of Eq. (3-24), we note that the first term $\frac{d^2R_0}{dt^2}$ is already defined with respect to $OXYZ$, so the operator in Eq. (3-19) does not affect it; the second and fourth terms do contain variables defined in $oxyz$, so the operator in Eq. (3-19) must be appropriately applied to them; and Eq. (3-18) (or even Eq. (3-19) can be used to rewrite the third term. Thus, Eq. (3-24) becomes

$$a = \frac{d^2R_0}{dt^2} + \left[\left(\frac{\partial}{\partial t}\right)_{\text{rel}} + \omega \times\right]v_{\text{rel}} + \dot{\omega} \times r + \omega \times \left[\left(\frac{\partial}{\partial t}\right)_{\text{rel}} + \omega \times\right]r$$

$$a = \frac{d^2R_0}{dt^2} + \left(\frac{\partial v_{\text{rel}}}{\partial t}\right)_{\text{rel}} + \omega \times v_{\text{rel}} + \dot{\omega} \times r + \omega \times \left[\left(\frac{\partial r}{\partial t}\right)_{\text{rel}} + \omega \times r\right]$$

$$a = \frac{d^2R_0}{dt^2} + a_{\text{rel}} + \omega \times v_{\text{rel}} + \dot{\omega} \times r + \omega \times v_{\text{rel}} + \omega \times (\omega \times r) \qquad (3\text{-}25)$$

where we have used Eqs. (3-21) for v_{rel} and a_{rel}, and the ordinary rules of differentiation. Note that the third and the fifth terms on the right-hand side of Eq. (3-25) are identical. Thus, the acceleration of point P with respect to the "fixed" reference frame $OXYZ$ can be written as

$$a = \frac{d^2R_0}{dt^2} + a_{\text{rel}} + 2\omega \times v_{\text{rel}} + \dot{\omega} \times r + \omega \times (\omega \times r) \qquad (3\text{-}26)$$

The first term in Eq. (3-26) is the acceleration of the origin of the intermediate frame with respect to $OXYZ$. The second term is the relative acceleration of P with respect to $oxyz$. The third term $2\omega \times v_{\text{rel}}$ is called the *Coriolis acceleration*.* The fourth term $\dot{\omega} \times r$ shall be

---

*The Coriolis acceleration was very elusive, engaging the interests of several of the greatest mechanicians during much of the eighteenth century. Euler almost got it right in 1755. Ultimately, in the study of hydraulic machines, the French engineer Gustave G. Coriolis (1792–1843) in two papers (1831, and more clearly in 1835) described the forces due to this acceleration component that now bears his name.

called the *Euler acceleration,*[2] which gives the contribution due to the angular acceleration $\dot{\boldsymbol{\omega}}$. The last term is called the *centripetal acceleration.*

It is important to note that our choice to consider $OXYZ$ to be "fixed" was simply a matter of intuitive convenience. In fact, Eqs. (3-23) and (3-26) represent the velocity and acceleration of a point described in *any* pair of reference frames that possess the relative correspondence indicated in Figure 3-19. The reference frame $OXYZ$ does not have to be stationary. Examples 3-9 through 3-11 will reinforce this idea; Section 3-6 will extend it.

Accelerations are not easy to visualize. This is true for each of the terms in Eq. (3-26). (Several of the problems at the end of the chapter are designed specifically to help build and reinforce the reader's intuition of these terms.) Of the five terms in the acceleration expression in Eq. (3-26), the Coriolis and centripetal terms are perhaps the most difficult to visualize. The example that follows represents an attempt to assist in developing intuition about these two terms. It merits careful study.

■ **Example 3-6:**   Anatomy of Centripetal and Coriolis Accelerations

A turntable is rotating at a *constant* angular velocity ω_0 about its fixed center C (Figure 3-20). An ant (which we model as a point) is walking along a radius of the turntable at a constant velocity v_0, relative to the turntable. The ant is currently at point 1 and wants to get to the nearby point 2 (although it does not know why it wants to get to point 2). Although ω_0 and v_0 are constants, we seek the acceleration of the bug, relative to a fixed reference frame, which we denote as $OXYZ$.

☐ **Solution:**   Figure 3-21 is a sketch of the motion of the ant as viewed from the fixed reference, where we show the ant at two times, a small time increment Δt apart. When viewed from the fixed reference, the velocity of the ant at point 1 has two components: v_0 in the radial direction and $\omega_0 r$ in the tangential direction. Also, we see that by the time the ant traverses Δr, point 2 is not where it was when the ant wanted to get there, but is at point 2'. Thus, the approximate acceleration of the ant with respect to the fixed reference frame is the difference between its velocity at points 2' and 1, divided by the time increment Δt. In the limit as $\Delta t \to 0$, we shall find the acceleration of the ant, with respect to $OXYZ$.

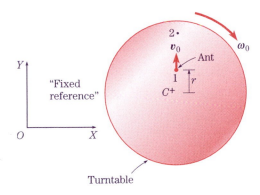

FIGURE 3-20   Ant moving at constant velocity v_0, relative to turntable that is rotating at constant ω_0.

---

[2]The adoption of this name is in accordance with the suggestion of C. Lanczos [1977] and the discussion of his Eq. (45.14).

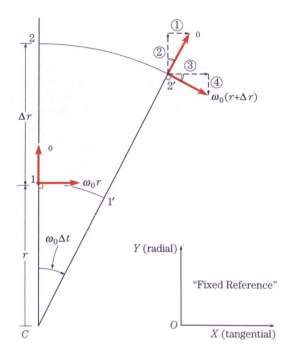

FIGURE 3-21   Velocity components of ant moving at constant velocity v_0, relative to turntable that is rotating at constant ω_0.

The angle between the two radii, corresponding to the time increment Δt, is $\omega_0 \Delta t$. For small Δt, we take

$$\sin \omega_0 \Delta t \approx \omega_0 \Delta t \qquad \text{and} \qquad \cos \omega_0 \Delta t \approx 1 \tag{a}$$

where we have used the first term in each of the Taylor series expansions for sine and cosine functions, thus neglecting terms in Δt of second and higher powers. As the reader may choose to verify, such an approximation has no effect on the final answers since including higher-order terms would simply result, in the limit as $\Delta t \to 0$, in vanishing terms (and wasted paper).

1. Acceleration in radial direction ("fixed" Y direction):   First, the change in radial velocity Δv_{radial} is simply the new radial velocity, which is at point $2'$, minus the old radial velocity, which is at point 1.

$$\Delta v_{\text{radial}} = [\text{New Radial Velocity}] - [\text{Old Radial Velocity}]$$
$$= [\text{Term } ② + \text{Term } ④] - [v_0] \tag{b}$$

where, from Figure 3-21, term ② and term ④ refer to the radial velocity components at point $2'$. Then

$$\Delta v_{\text{radial}} = [v_0 \cos(\omega_0 \Delta t) - \omega_0(r + v_0 \Delta t) \sin(\omega_0 \Delta t)] - [v_0] \tag{c}$$

where $\Delta r = v_0 \Delta t$ has been used in writing Eq. (c). Expanding Eq. (c) gives

$$\Delta v_{\text{radial}} = \left[v_0 - \omega_0^2 r \Delta t - \omega_0^2 v_0 (\Delta t)^2 \right] - [v_0] \tag{d}$$

where Eqs. (a) have been used.

Dividing both sides of Eq. (d) by Δt and taking the limit as Δt goes to zero give

$$a_{\text{radial}} = \lim_{\Delta t \to 0} \frac{\Delta v_{\text{radial}}}{\Delta t} = -\omega_0^2 r \tag{e}$$

This, of course, is the *centripetal acceleration,* due to term ④. The velocity v_0 has no effect on a_{radial}; v_0 could just as well have been zero. At the instant sketched in Figure 3-21, the velocity $\omega_0 r$ is being "pulled" downward in the negative Y (radial) direction. This acceleration always points radially "inward" toward the axis of rotation.

2. Acceleration in tangential direction ("fixed" X direction):   Next, the change in the velocity in the tangential direction $\Delta v_{\text{tangential}}$ is simply the new tangential velocity, which is at point $2'$, minus the old tangential velocity, which is at point 1.

$$\Delta v_{\text{tangential}} = [\text{New Tangential Velocity}] - [\text{Old Tangential Velocity}] \tag{f}$$
$$= [\text{Term} ① + \text{Term} ③] - [\omega_0 r]$$

where term ① and term ③ refer to the tangential velocity components at point $2'$ as sketched in Figure 3-21. Then

$$\Delta v_{\text{tangential}} = [v_0 \sin(\omega_0 \Delta t) + \omega_0 (r + v_0 \Delta t) \cos(\omega_0 \Delta t)] - [\omega_0 r] \tag{g}$$

where again $\Delta r = v_0 \Delta t$ has been used in writing Eq. (g). Expanding Eq. (g) gives

$$\Delta v_{\text{tangential}} = [\omega_0 v_0 \Delta t + \omega_0 r + \omega_0 v_0 \Delta t] - [\omega_0 r] \tag{h}$$

where Eqs. (a) have been used. Dividing both sides of Eq. (h) by Δt and taking the limit as Δt goes to zero give

$$a_{\text{tangential}} = \lim_{\Delta t \to 0} \frac{\Delta v_{\text{tangential}}}{\Delta t} = \omega_0 v_0 + \omega_0 v_0 = 2\omega_0 v_0 \tag{i}$$

Equation (i) is clearly the *Coriolis acceleration.* Note that while the centripetal acceleration is affected by the position of the ant on the turntable, the Coriolis acceleration is independent of the position of the ant.

Equation (h) (in conjunction with Figure 3-21 and Eq. (f)) is instructive. Note that the effect of ω_0 on v_0 (Term ①) is always exactly the same as the effect of v_0 on ω_0 (Term ③); or equivalently, the effect of ω_0 in changing the *orientation* of v_0 is exactly the same as the effect of v_0 in carrying $\omega_0 r$ to a different radius, changing its *magnitude.* It is the sum of these two identical contributions that gives the "2" in the Coriolis acceleration in Eq. (i).

Finally, although we shall not pursue such detailed study, we note that a better understanding of the kinematics of many problems can be enhanced by studying comparably detailed vector sketches.

Several examples will now be presented. Each of the examples has one or more distinctive features that are intended to amplify some aspect of the use of the kinematic results derived in this chapter. The problems are arranged in order of increasing difficulty, and the reader will likely find the sum of these examples rewarding in future endeavors in kinematics.

It can be shown that it is possible to mechanize many kinematics analyses. The reader is referred to Section 3-6, in which expressions for velocity and acceleration are derived when *two* intermediate reference frames are used. Further, Appendix B contains a generalization of this type of kinematic analysis when an *arbitrary* number of intermediate reference

TABLE 3-1 Summary of Major Kinematic Results for Points

	Name	Variable or Equation	Equation Number
Motion of Point (defined in $OXYZ$) with respect to $OXYZ$. (Fig. 3-1)	Position Vector	$R(t)$	
	Velocity Vector	$v = \dfrac{dR}{dt}$	(3-1)
	Acceleration Vector	$a = \dfrac{d^2R}{dt^2}$	(3-2)
Motion of Point (defined in $oxyz$) with respect to $OXYZ$. (Fig. 3-19)	Position Vector	$R(t) = R_0(t) + r(t)$	
	Differential Operator	$\dfrac{d}{dt} = \left(\dfrac{\partial}{\partial t}\right)_{\text{rel}} + \boldsymbol{\omega}\times$	(3-19)
	Velocity Vector	$v = \dfrac{dR_0}{dt} + v_{\text{rel}} + \boldsymbol{\omega} \times r$	(3-23)
	Acceleration Vector	$a = \dfrac{d^2R_0}{dt^2} + a_{\text{rel}} + 2\boldsymbol{\omega} \times v_{\text{rel}}$ $+ \dot{\boldsymbol{\omega}} \times r + \boldsymbol{\omega} \times (\boldsymbol{\omega} \times r)$	(3-26)

frames is used. The generalizations in Appendix B are presented for future study and are not encouraged during an initial encounter. The examples given below are intentionally not forced into such mechanization since the goal here is to encourage the reader to *think* about each step of the analyses in terms of the somewhat elementary results presented thus far in this chapter.

The major kinematic results for calculating the motion of points are summarized in Table 3-1. It is important conceptually to appreciate that Eqs. (3-1) and (3-23) are equivalent expressions, and that, indeed, Eq. (3-23) follows from Eq. (3-1). Similarly, Eqs. (3-2) and (3-26) are equivalent expressions, as Eq. (3-26) follows from Eq. (3-2). Table 3-1 should be reconsidered after the examples that follow have been studied.

■ **Example 3-7:** **Motion of Point in Channel on Rotating Platform**

A platform is rotating at a constant angular velocity $\boldsymbol{\omega}_0$ with respect to the "fixed" reference frame $OXYZ$, as sketched in Figure 3-22. A small work piece labeled point P is moving with constant velocity v_0 along a channel, relative to the channel that is located as shown. We want to find the velocity and acceleration of point P with respect to the "fixed" reference frame.

❒ **Solution:** In the problem specification, we are given convenient reference frames. It should be emphasized that the choice of such reference frames is an important part of the solution. Real objects are not generally encountered with a convenient set of reference axes painted on them. Often, by virtue of symmetry or the definition of the problem itself, some choices of reference frames are to be preferred. For example, the fixed reference frame $OXYZ$ and the intermediate reference frame $oxyz$ attached to the platform are coincident at their origins and remain so throughout the analysis. Also, note that the x axis is parallel to the channel.

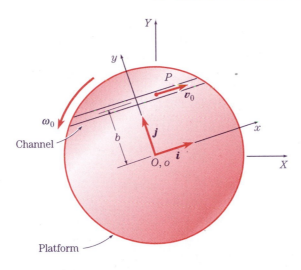

FIGURE 3-22   Point P representing work piece moving at constant velocity v_0 relative to platform that is rotating at constant ω_0.

This solution provides a straightforward application of Eqs. (3-23) and (3-26), but we want to discuss the solution in detail. First, we must decide in which frame we want to write our kinematic variables. We do this by writing the following significant statement:

Motion of P (defined in oxyz) with respect to OXYZ.   By this underlined statement, we emphasize that the kinematic variables will be expressed or defined in $oxyz$, and that the velocity and acceleration will be calculated with respect to $OXYZ$. *All* unambiguous kinematic analyses *must* be accompanied by such a statement! We must be clear regarding both the reference frame in which the motion is *defined* and the reference frame *with respect to which* the motion is calculated. (Read this paragraph at least three times!)

We define the unit vectors in $oxyz$ to be i, j, k and the unit vectors in $OXYZ$ to be I, J, K. Recalling Eqs. (3-23) and (3-26), we write

$$v = \frac{d\boldsymbol{R}_0}{dt} + v_{\text{rel}} + \boldsymbol{\omega} \times \boldsymbol{r} \tag{a}$$

and

$$a = \frac{d^2\boldsymbol{R}_0}{dt^2} + a_{\text{rel}} + 2\boldsymbol{\omega} \times v_{\text{rel}} + \dot{\boldsymbol{\omega}} \times \boldsymbol{r} + \boldsymbol{\omega} \times (\boldsymbol{\omega} \times \boldsymbol{r}) \tag{b}$$

In this problem, we need only apply each of Eqs. (a) and (b) once; as the underlined statement above indicates, we shall calculate the motion with respect to $OXYZ$, which is precisely the answer sought. So, let us identify each of the terms in Eqs. (a) and (b). There is no derivation here; we must simply observe the problem at hand and use our knowledge of vectors to write each term.

$$\boldsymbol{R}_0 = 0 \qquad \frac{d\boldsymbol{R}_0}{dt} = 0 \qquad \frac{d^2\boldsymbol{R}_0}{dt^2} = 0 \tag{c}$$

Equations (c) simply express the fact that the origin of $oxyz$ is coincident with the origin of $OXYZ$, has no velocity with respect to $OXYZ$, and has no acceleration with respect to $OXYZ$. That is, points o and O are coincident and remain coincident throughout the analysis. Further,

$$\boldsymbol{\omega} = \omega_0 \boldsymbol{k} \qquad \dot{\boldsymbol{\omega}} = 0 \tag{d}$$

Equations (d) denote the constant angular velocity, and therefore the zero value angular acceleration. Note that, in accordance with our underlined statement, the i, j, k unit vectors of $oxyz$ are used. By observation,

$$r = xi + bj \tag{e}$$

Note that we always know the y-component of the position of P but that the x-component is a variable. In order to find v_{rel} and a_{rel}, we use Eqs. (3-21). So

$$v_{rel} = \left(\frac{\partial r}{\partial t}\right)_{rel} = \dot{x}i + \dot{b}j = v_0 i$$

$$a_{rel} = \left(\frac{\partial^2 r}{\partial t^2}\right)_{rel} = \ddot{x}i + \ddot{b}j = 0 \tag{f}$$

where we have introduced the fact that $\dot{x} = v_0$, where v_0 is a constant given in the problem statement; so $\ddot{x} = 0$.

Now, we have defined all the terms in Eqs. (a) and (b), so we can find the velocity and acceleration of P as

$$v = 0 + v_0 i + \omega_0 k \times (xi + bj) = (v_0 - \omega_0 b)i + \omega_0 x j \tag{g}$$

and

$$a = 0 + 0 + 2\omega_0 k \times v_0 i + 0 + \omega_0 k \times [\omega_0 k \times (xi + bj)]$$

$$= 2\omega_0 v_0 j - \omega_0^2 xi - \omega_0^2 bj = 2\omega_0^2 v_0 j - \omega_0^2 r \tag{h}$$

which are the velocity and acceleration sought. The two terms on the right-hand side of Eq. (h) are the Coriolis and centripetal accelerations. Also, the centripetal acceleration has been written in a form that reinforces the fact that it always points radially inward toward the axis of rotation.

It is important to appreciate the fact that although the answers in Eqs. (g) and (h) are expressed in terms of $oxyz$ unit vectors, these are nevertheless the velocity and acceleration of point P with respect to the "fixed" reference $OXYZ$. Indeed, that is precisely the meaning of Eqs. (3-23) and (3-26), and furthermore it is in accordance with our underlined statement. Upon an initial encounter, it is often thought that because the motion in Eqs. (g) and (h) is expressed in terms of the i, j, k unit vectors, these answers are somehow not with respect to $OXYZ$. This is not the case, as we shall now explain.

It is possible to express the answers in terms of the $OXYZ$ unit vectors but—if it were necessary, which in general it is not—we would need more information to do so. In particular, for this problem we would need to know the value of an angle, say θ, that would locate $oxyz$ with respect to $OXYZ$ (see Figure 3-23). This might be given in the problem statement or, since ω_0 is constant, it could be calculated as $\theta = \omega_0 t + \theta_0$, given the constant θ_0 at time $t = 0$. Note from Figure 3-23 that

$$I = \cos\theta i - \sin\theta j$$

$$J = \sin\theta i + \cos\theta j \tag{i}$$

$$K = k$$

Thus, by using Eqs. (i), Eqs. (g) and (h) can be expressed in terms of I, J, K. In general, we shall take the forms in Eqs. (g) and (h) to be satisfactory answers; and, indeed, these are generally the preferred forms.

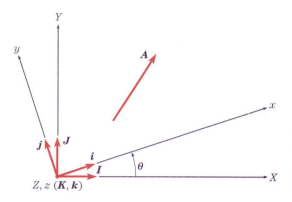

FIGURE 3-23 Relationship between unit vectors in $oxyz$ and unit vectors in $OXYZ$ requires knowledge of θ. An arbitrary vector A is not affected by choice of reference frame.

To emphasize the fact that once a vector has been determined, its magnitude and direction are unaffected by the reference frame in which it is expressed, we show an arbitrary vector A in Figure 3-23. The vector itself—its magnitude and its direction—is totally independent of the choice of reference frame; this is true for all reference frames, irrespective of their directions or locations in space. If we know A in terms of $oxyz$, we know A! We *may* want to express A in terms of some other reference frame, but that desire or the knowledge itself does not change A.

One final point: Now that we have the acceleration in Eq. (h), so what? Well, the significance of Eqs. (g) and (h) is that if we want to apply our mechanics correctly, we shall need such equations. For example—though we are jumping ahead of ourselves—in order to apply Newton's second law to the work piece in Figure 3-22, we shall need exactly the acceleration in Eq. (h). It is the acceleration in Eq. (h)—and no other acceleration—for which force equals mass times acceleration. (*After* reading Example 3-8, immediately take a peek at Example 4-1 in Chapter 4.)

■ **Example 3-8:**   Antenna Deployed by Maneuvering Airplane

An airplane moving with velocity v_1 and acceleration a_1, both with respect to ground (i.e., "fixed" space), is banking at ω_1 and climbing at ω_2, both with respect to "fixed" space, as sketched in Figure 3-24. An antenna A (weight 2 lb) is being deployed from the airplane. The airplane is in a horizontal orientation (with respect to gravity) and the antenna A is at a vertical distance of 10 ft from the centerline of the airplane, moving with velocity v_2 and acceleration a_2, both defined with respect to the airplane. Find the velocity and acceleration of the antenna A relative to "fixed" space, at the instant shown. Assume that v_1 is 200 ft/sec, a_1 is 100 ft/sec^2, v_2 is 6 ft/sec, a_2 is 0.1 ft/sec^2, ω_1 is 3 rad/min, and ω_2 is 2 rad/min. $OXYZ$ is a reference frame that is "fixed" in space, and $oxyz$ is an intermediate frame that is attached to the airplane.

□ **Solution:**   Again, by virtue of Figure 3-24 in the problem specification, we are given a convenient pair of reference frames. The unit vectors in $oxyz$ are i, j, k and in $OXYZ$ are I, J, K. In this problem, we note that $i = I$, $j = J$, $k = K$. This solution also provides a straightforward application of Eqs. (3-23) and (3-26), but with some subtle differences from the previous examples. We proceed as before with our emphatic statement regarding the motion of the point of kinematic interest; we *must* be clear regarding the reference frame in

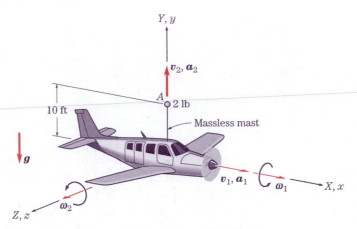

FIGURE 3-24 Sketch of airplane deploying antenna at tip of massless mast.

which the motion is defined and the reference frame with respect to which the motion is calculated.

Motion of A (defined in oxyz) with respect to OXYZ. Recalling Eqs. (3-23) and (3-26), we write

$$v = \frac{d\mathbf{R}_0}{dt} + v_{\text{rel}} + \boldsymbol{\omega} \times r \tag{a}$$

and

$$a = \frac{d^2\mathbf{R}_0}{dt^2} + \mathbf{a}_{\text{rel}} + 2\boldsymbol{\omega} \times v_{\text{rel}} + \dot{\boldsymbol{\omega}} \times r + \boldsymbol{\omega} \times (\boldsymbol{\omega} \times r) \tag{b}$$

Despite all the various motions that appear to be occurring, we need to apply each of Eqs. (a) and (b) only once. From the statement of the problem, we can identify the various terms to be the following:

$$\mathbf{R}_0 = 0 \qquad \frac{d\mathbf{R}_0}{dt} = 200\boldsymbol{i}\,\frac{\text{ft}}{\text{sec}} \qquad \frac{d^2\mathbf{R}_0}{dt^2} = 100\boldsymbol{i}\,\frac{\text{ft}}{\text{sec}^2}$$

$$\mathbf{r} = 10\boldsymbol{j}\text{ft} \qquad v_{\text{rel}} = 6\boldsymbol{j}\,\frac{\text{ft}}{\text{sec}} \qquad \mathbf{a}_{\text{rel}} = 0.1\boldsymbol{j}\,\frac{\text{ft}}{\text{sec}^2} \tag{c}$$

$$\boldsymbol{\omega} = \omega_1\boldsymbol{i} + \omega_2\boldsymbol{k} = \left(\frac{3}{60}\boldsymbol{i} + \frac{2}{60}\boldsymbol{k}\right)\frac{\text{rad}}{\text{sec}} \qquad \dot{\boldsymbol{\omega}} = 0$$

Note that the expression for $\boldsymbol{\omega}$ emphasizes the vectorial character of angular velocity. All the other expressions among Eqs. (c) should be found to be straightforward.

Substitution of Eqs. (c) into Eqs. (a) and (b), and some elementary calculation give

$$v_A = 199.667\boldsymbol{i} + 6.000\boldsymbol{j} + 0.500\boldsymbol{k}\,\frac{\text{ft}}{\text{sec}} \tag{d}$$

and

$$\mathbf{a}_A = 99.600\boldsymbol{i} + 0.064\boldsymbol{j} + 0.600\boldsymbol{k}\,\frac{\text{ft}}{\text{sec}^2} \tag{e}$$

which are the velocity and acceleration of the antenna, which has been modeled as a point.

Was it necessary to define the unit vectors I, J, K in $OXYZ$? Not really. (Since $I = i$, $J = j$ and $K = k$, then I, J, K become superfluous.) Was it necessary to specify gravity, or the antenna's weight? No. (It will be found in Chapter 4 that in situations where the kinematics are completely specified, as in the case here, such quantities as gravity or weight become significant only when we seek forces.) Finally, we emphasize that in this example, as in most of the following examples, the velocity and acceleration that have been found are valid only *at the instant* shown. At a later instant, the parameters in the problem will be different, requiring another calculation. A series of updated kinematics can be obtained either analytically or numerically by using an electronic computer.

We shall use the results of this example in Chapter 4.

■ **Example 3-9:   Bird on Mobile**

On the street level of the Porter Square Ⓣ Station,[3] there is an artistic mobile structure for the pleasure of passersby, which, along with a little bird, inspired this example. The structure, modeled as sketched in Figure 3-25a, consists of a large $+$ and four smaller Y's, one each attached to a tip of the $+$. The $+$ is horizontal at all times and rotates at a constant angular velocity ω_1 (with respect to ground or "fixed" space) about an axis through its center. Also, each of the four Y's remains at all times in a vertical plane and rotates at a constant angular velocity ω_2 (with respect to its $+$ tip) about an axis through its center. At the instant shown, a bird of mass m is on a leg of one of the Y's, which is oriented as indicated in Figure 3-25b. Relative to the Y, the bird is running with a velocity v_0 and an acceleration a_0, at the instant shown. At the same instant, a gust of wind exerts a force F_w on the bird in the X direction.

Find the acceleration of the bird, which may be modeled as a point, at the instant shown. (Clearly, the acceleration sought is with respect to $OXYZ$; this is the only acceleration which is of ultimate interest.)

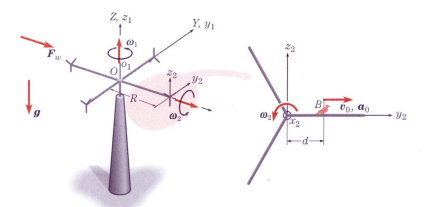

FIGURE 3-25   Sketch of artistic mobile structure with bird running along horizontal leg of Y.

---

[3]A Cambridge, Massachusetts subway station.

❏ **Solution:**   As usual, we are have undertaken the task of defining several convenient reference frames. Again, real bodies do not have convenient reference frames painted on them; the choice of reference frames is an important intellectual part of any analysis. Also, we are aware that the reference frames that have been defined are by no means the only (nor perhaps even the best) reference frames.

In this example, we have multiple rotating frames: $o_1x_1y_1z_1$ rotates at $\boldsymbol{\omega}_1$ with respect to ground; $o_2x_2y_2z_2$ rotates at $\boldsymbol{\omega}_2$ with respect to $o_1x_1y_1z_1$, and $OXYZ$ is attached to ground. These references for $\boldsymbol{\omega}_1$ and $\boldsymbol{\omega}_2$ are given in the problem statement. In general, in order to find the velocity and acceleration of the bird, we could move "outward" from $OXYZ$ to $o_2x_2y_2z_2$, or "inward" from $o_2x_2y_2z_2$ to $OXYZ$. Either approach is valid; either approach can be used. In most instances, however, if the outer rotation (here, $\boldsymbol{\omega}_2$) is defined with respect to the frame undergoing the inner rotation (here, $\boldsymbol{\omega}_1$), it is easier to work "inward." On the other hand, if the outer rotation (here, $\boldsymbol{\omega}_2$) is defined with respect to ground, it is easier to work "outward." In this problem, the former situation holds, so we shall work "inward."

Here, a sequential approach will be used: We shall repeatedly use Eqs. (3-23) and (3-26), with reference to Figure 3-19. In this solution, we shall obtain the answer in two steps. Nevertheless, in each step we *must* be clear regarding the reference frame in which the motion is defined and the reference frame with respect to which the motion is calculated. In particular, (1) using the frame $o_2x_2y_2z_2$ as an intermediate frame, we shall find the motion of the bird with respect to frame $o_1x_1y_1z_1$, which is fixed in the rotating $+$; then (2) using the frame $o_1x_1y_1z_1$ as an intermediate frame, we shall find the motion of the bird with respect to the "fixed" frame $OXYZ$.

1. *Motion of bird (defined in $o_2x_2y_2z_2$) with respect to $o_1x_1y_1z_1$.*   The statement of the bird's motion as written in the underlined statement is important and should leave no doubt about the intended use of the reference frames for the kinematics.

Equation (3-23) is

$$v \equiv v_{(o_1x_1y_1z_1)} = \frac{d\boldsymbol{R}_0}{dt} + \boldsymbol{v}_{\text{rel}} + \boldsymbol{\omega} \times \boldsymbol{r} \tag{a}$$

and Eq. (3-26) is

$$a \equiv a_{(o_1x_1y_1z_1)} = \frac{d^2\boldsymbol{R}_0}{dt^2} + \boldsymbol{a}_{\text{rel}} + 2\boldsymbol{\omega} \times \boldsymbol{v}_{\text{rel}} + \dot{\boldsymbol{\omega}} \times \boldsymbol{r} + \boldsymbol{\omega} \times (\boldsymbol{\omega} \times \boldsymbol{r}) \tag{b}$$

where $\boldsymbol{v}_{(o_1x_1y_1z_1)}$ and $\boldsymbol{a}_{(o_1x_1y_1z_1)}$ are the velocity and acceleration, respectively, of the bird with respect to $o_1x_1y_1z_1$. By recourse to the problem statement, we note that

$$\frac{d\boldsymbol{R}_0}{dt} = \frac{d^2\boldsymbol{R}_0}{dt^2} = 0 \qquad \boldsymbol{v}_{\text{rel}} = v_0\boldsymbol{j} \qquad \boldsymbol{a}_{\text{rel}} = a_0\boldsymbol{j}$$
$$\boldsymbol{\omega} = \omega_2\boldsymbol{i} \qquad \dot{\boldsymbol{\omega}} = 0 \qquad \boldsymbol{r} = d\boldsymbol{j} \tag{c}$$

Substituting Eqs. (c) into Eqs. (a) and (b), performing the indicated vector cross-products, and collecting the terms having common unit vectors give

$$v_{(o_1x_1y_1z_1)} = 0 + v_0\boldsymbol{j} + \omega_2\boldsymbol{i} \times d\boldsymbol{j} = v_0\boldsymbol{j} + d\omega_2\boldsymbol{k}$$
$$a_{(o_1x_1y_1z_1)} = 0 + a_0\boldsymbol{j} + 2\omega_2\boldsymbol{i} \times v_0\boldsymbol{j} + 0 + \omega_2\boldsymbol{i} \times (\omega_2\boldsymbol{i} \times d\boldsymbol{j}) \tag{d}$$
$$= \left(a_0 - d\omega_2^2\right)\boldsymbol{j} + 2v_0\omega_2\boldsymbol{k}$$

2. *Motion of bird (defined in $o_1x_1y_1z_1$) with respect to OXYZ.* Once again, the statement of the bird's motion as written in the underlined statement is important and should leave no doubt about the intended use of the reference frames for the kinematics.

In anticipation of using Eq. (3-26), by recourse to the problem statement and part 1 of the solution above, we note that

$$\frac{d\boldsymbol{R}_0}{dt} = \frac{d^2\boldsymbol{R}_0}{dt^2} = 0 \qquad \boldsymbol{v}_{\text{rel}} = \boldsymbol{v}_{(o_1x_1y_1z_1)} \qquad \boldsymbol{a}_{\text{rel}} = \boldsymbol{a}_{(o_1x_1y_1z_1)}$$

$$\boldsymbol{\omega} = \omega_1\boldsymbol{k} \qquad \dot{\boldsymbol{\omega}} = 0 \qquad \boldsymbol{r} = R\boldsymbol{i} + d\boldsymbol{j} \tag{e}$$

An interesting and important observation here is that

$$\boldsymbol{v}_{\text{rel}} = \boldsymbol{v}_{(o_1x_1y_1z_1)} \qquad \boldsymbol{a}_{\text{rel}} = \boldsymbol{a}_{(o_1x_1y_1z_1)}$$

Realizing we are interested only in the acceleration of the bird with respect to $OXYZ$, we substitute Eqs. (e) into Eq. (3-26), perform the indicated vector cross-products, and collect terms having common unit vectors to obtain

$$\begin{aligned}\boldsymbol{a} \equiv \boldsymbol{a}_{(OXYZ)} &= 0 + \left[(a_0 - d\omega_2^2)\boldsymbol{j} + 2v_0\omega_2\boldsymbol{k}\right] + 2\omega_1\boldsymbol{k} \times (v_0\boldsymbol{j} + d\omega_2\boldsymbol{k}) \\ &\quad + 0 + \omega_1\boldsymbol{k} \times [\omega_1\boldsymbol{k} \times (R\boldsymbol{i} + d\boldsymbol{j})] \\ &= -(2\omega_1v_0 + R\omega_1^2)\boldsymbol{i} + (a_0 - d\omega_1^2 - d\omega_2^2)\boldsymbol{j} + 2v_0\omega_2\boldsymbol{k}\end{aligned} \tag{f}$$

where $\boldsymbol{a}_{(OXYZ)}$ is the acceleration of the bird with respect to $OXYZ$.

We shall return to this example in Chapter 4 where we shall discover that we have come far toward finding the forces on the bird. As noted in Example 3-8, here too we did not need to use the mass m of the bird, the effect of gravity, or the information about the gust of wind. So what; who said that one must use all the information that is provided or that enough information will always be provided?

■ **Example 3-10:**   Robot Manipulating Work Piece

A robot named JT is rolling with respect to the shop floor at a constant speed of 0.5 m/s and carrying a work piece 1 m long, as sketched in Figure 3-26. Each of the links of the robot arm is 0.75 m long, and the second link has an end gripper that holds the work piece which may be considered as rigid. At the instant shown, the link AB is at an angle of 45° with respect to the vertical and the link BD is horizontal. Also, the link AB is rotating at ω_1 (1 rev/3 s), and the link BD is rotating at ω_2 (1 rev/2 s) and $\dot{\omega}_2$ (0.5 rad/s^2), all with respect to the shop floor. At the instant shown, find the velocity and acceleration of the center of the work piece, labeled point C, as sketched in Figure 3-26. (Clearly, the velocity and acceleration sought are with respect to the shop floor, which is assumed to be "fixed to ground.")

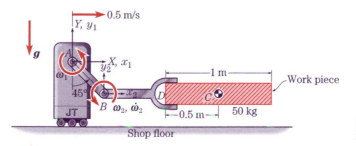

FIGURE 3-26   JT manipulating work piece.

☐ **Solution:**   For the convenient set of reference frames defined in Figure 3-26, we seek the velocity and acceleration of the center of the rigid work piece. Reference frame $Bx_2y_2z_2$ is *fixed* in link BD, reference frame $Ax_1y_1z_1$ is *fixed* in link AB and reference frame $OXYZ$ is fixed with respect to the shop floor (ground).

We shall successively apply Eqs. (3-23) and (3-26), namely,

$$v = \frac{d\mathbf{R}_0}{dt} + \mathbf{v}_{\text{rel}} + \boldsymbol{\omega} \times \mathbf{r} \tag{a}$$

and

$$\mathbf{a} = \frac{d^2\mathbf{R}_0}{dt^2} + \mathbf{a}_{\text{rel}} + 2\boldsymbol{\omega} \times \mathbf{v}_{\text{rel}} + \dot{\boldsymbol{\omega}} \times \mathbf{r} + \boldsymbol{\omega} \times (\boldsymbol{\omega} \times \mathbf{r}) \tag{b}$$

Also note that the orientations of the three frames indicate that the corresponding unit vectors are equal—that is, $\mathbf{i}, \mathbf{j}, \mathbf{k}$ may be considered as the respective unit vectors in $OXYZ$, $Ax_1y_1z_1$ and $Bx_2y_2z_2$, without confusion. This is so because of the identical orientation of the respective axes in all three frames. The consistent alignment of the respective axes in multiple frames is a convenient choice whenever the motion at a specified instant is sought.

Consider the following important observation. The definition of $\boldsymbol{\omega}_2$ (and $\dot{\boldsymbol{\omega}}_2$) here should be contrasted with the definition of $\boldsymbol{\omega}_2$ in Example 3-9. Here $\boldsymbol{\omega}_2$ (and $\dot{\boldsymbol{\omega}}_2$) is defined with respect to $OXYZ$; in Example 3-9, $\boldsymbol{\omega}_2$ was defined with respect to $o_1x_1y_1z_1$. As opposed to the "direction" of solution in Example 3-9, where we worked "inward" toward ground, here we find it is easier to work "outward" from ground. As always, we *must* be clear regarding the frame in which the motion is defined and the reference frame with respect to which the motion is calculated. In particular, (1) using the frame $Ax_1y_1z_1$ as an intermediate frame, we shall find the motion of point B with respect to the "fixed" reference frame $OXYZ$; then (2) using the frame $Bx_2y_2z_2$ as an intermediate frame, we shall find the motion of point C with respect to the "fixed" reference frame $OXYZ$.

1. *Motion of point **B** (defined in $Ax_1y_1z_1$) with respect to $OXYZ$.* From the problem statement and in anticipation of using Eqs. (a) and (b), we may identify the following terms:

$$\mathbf{R}_0 = 0 \qquad \frac{d\mathbf{R}_0}{dt} = 0.5\mathbf{i}\left(\frac{\text{m}}{\text{s}}\right) \qquad \frac{d^2\mathbf{R}_0}{dt^2} = 0$$

$$\mathbf{r} = 0.75\sin 45°\mathbf{i} - 0.75\cos 45°\mathbf{j}\,(\text{m})$$

$$\mathbf{v}_{\text{rel}} = 0 \qquad \mathbf{a}_{\text{rel}} = 0 \tag{c}$$

$$\boldsymbol{\omega} = \boldsymbol{\omega}_1 = \frac{1\text{rev}}{3\text{s}}\mathbf{k} = \frac{2\pi}{3}\mathbf{k}\left(\frac{\text{rad}}{\text{s}}\right) \qquad \dot{\boldsymbol{\omega}} = \dot{\boldsymbol{\omega}}_1 = 0.$$

An interesting observation is that for rigid linkages, as well as for other types of extended undeformable bodies, $\mathbf{v}_{\text{rel}}$ and $\mathbf{a}_{\text{rel}}$, which represent magnitude changes in $\mathbf{r}$, vanish. Also, although it is not needed, we give the value of $\mathbf{R}_0$ to encourage the proper interpretation of the various terms for the frames defined in Figure 3-26 and Figure 3-19. Substituting Eqs. (c) into Eqs. (a) and (b), and conducting some elementary calculations give

$$\mathbf{v}_{B(OXYZ)} = (1.611\mathbf{i} + 1.111\mathbf{j})\left(\frac{\text{m}}{\text{s}}\right)$$

$$\mathbf{a}_{B(OXYZ)} = (-2.326\mathbf{i} + 2.326\mathbf{j})\left(\frac{\text{m}}{\text{s}^2}\right) \tag{d}$$

where the subscripted notation on $\boldsymbol{v}_B$ and $\boldsymbol{a}_B$ is written to emphasize that these results are the velocity and acceleration of point B with respect to $OXYZ$.

2. *Motion of point $\boldsymbol{C}$ (defined in $Bx_2y_2z_2$) with respect to $OXYZ$.* From the problem statement and the results obtained as Eqs. (d), in anticipation of using Eqs. (a) and (b), we may write

$$\boldsymbol{R}_0 = 0.75 \sin 45° \boldsymbol{i} - 0.75 \cos 45° \boldsymbol{j} \, (\text{m})$$

$$\frac{d\boldsymbol{R}_0}{dt} = \boldsymbol{v}_{B(OXYZ)} \qquad \frac{d^2\boldsymbol{R}_0}{dt^2} = \boldsymbol{a}_{B(OXYZ)}$$

$$\boldsymbol{r} = 1.25\boldsymbol{i} \, (\text{m}) \qquad \boldsymbol{v}_{\text{rel}} = 0 \qquad \boldsymbol{a}_{\text{rel}} = 0 \qquad (\text{e})$$

$$\boldsymbol{\omega} = \boldsymbol{\omega}_2 = \frac{1\text{rev}}{2\text{s}} = \pi\boldsymbol{k}\left(\frac{\text{rad}}{\text{s}}\right) \qquad \dot{\boldsymbol{\omega}} = \dot{\boldsymbol{\omega}}_2 = 0.5\boldsymbol{k}\left(\frac{\text{rad}}{\text{s}^2}\right).$$

Again, although it is not needed, we give the value of $\boldsymbol{R}_0$ to encourage comparison of the frames defined in Figure 3-26 with those in Figure 3-19. Substituting Eqs. (e) into Eqs. (a) and (b), and conducting some elementary calculations give

$$\boldsymbol{v}_{C(OXYZ)} = (1.611\boldsymbol{i} + 5.038\boldsymbol{j})\left(\frac{\text{m}}{\text{s}}\right)$$

$$\boldsymbol{a}_{C(OXYZ)} = (-14.66\boldsymbol{i} + 2.951\boldsymbol{j})\left(\frac{\text{m}}{\text{s}^2}\right) \qquad (\text{f})$$

which are the velocity and acceleration sought.

We shall return to these results in Chapter 6, where we consider the dynamics of rigid bodies.

■ **Example 3-11:** Pilot in Airplane on Aircraft Carrier

A jet airplane is about to take off from the deck of a maneuvering aircraft carrier as sketched in Figure 3-27. At the instant shown, the ship is moving at $\boldsymbol{\omega}_1$ and $\boldsymbol{\omega}_2$, as indicated, where *both* angular velocities are defined with respect to "fixed" space or ground. The airplane is moving with velocity $\boldsymbol{v}_0$ and acceleration $\boldsymbol{a}_0$, both defined with respect to the aircraft carrier. The airplane is also climbing with angular velocity $\boldsymbol{\omega}_3$ and angular acceleration $\boldsymbol{\alpha}_3$, both defined with respect to the aircraft carrier. Find the velocity and acceleration of the pilot P, who is modeled as a point, both with respect to "fixed" space.

❑ **Solution:** Again, we note that the difficult task of selecting a useful set of reference frames has been conducted for the reader. Reference frame $o_2x_2y_2z_2$ is fixed in the airplane, reference frame $o_1x_1y_1z_1$ is fixed in the ship, and reference frame $OXYZ$ is fixed to ground. We note that, at the instant shown, $\boldsymbol{i}, \boldsymbol{j}, \boldsymbol{k}$ may be considered as the unit vectors along the respective x, y, z axes in each of the various reference frames.

In this example as in all kinematics examples, the solution technique here is one technique of sequentially using Eqs. (3-23) and (3-26), not *the* technique, as no such single technique exists for these problems. In particular, in this solution, (1) using the frame $o_2x_2y_2z_2$ as an intermediate frame, we shall find the motion of the pilot P with respect to frame $o_1x_1y_1z_1$, which is fixed in the maneuvering aircraft carrier; then (2) as an intermediate step, we shall find the motion of the origin of the frame $o_1x_1y_1z_1$ with respect to the "fixed" reference frame $OXYZ$; and finally (3) using the frame $o_1x_1y_1z_1$ as an intermediate frame, we shall find the motion of P with respect to the "fixed" reference frame $OXYZ$.

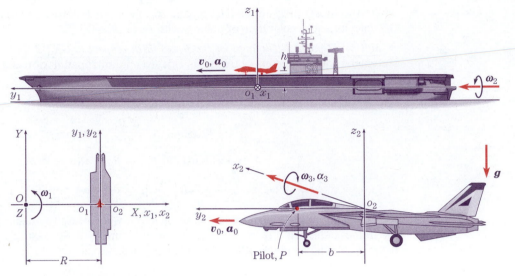

FIGURE 3-27 Sketch of takeoff of airplane from aircraft carrier.

The introduction of step 2 into our solution here makes this analysis slightly different from the previous examples. Nevertheless, after completing this example, the reader should reconsider the significance of having introduced this intermediate step. Further, because $\boldsymbol{\omega}_3$ and $\boldsymbol{\alpha}_3$ are defined with respect to $o_1 x_1 y_1 z_1$, it should be noted that the "direction" of the outlined solution is "inward" toward ground, analogous to Example 3-9.

1. Motion of $\boldsymbol{P}$ (defined in $o_2 x_2 y_2 z_2$) with respect to $o_1 x_1 y_1 z_1$. In preparation for using Eqs. (3-23) and (3-26), by recourse to the problem statement, we identify the following terms:

$$\frac{d\boldsymbol{R}_0}{dt} = v_0\boldsymbol{j} \qquad \frac{d^2\boldsymbol{R}_0}{dt^2} = a_0\boldsymbol{j} \qquad \boldsymbol{r} = b\boldsymbol{j}$$

$$\boldsymbol{v}_{\text{rel}} = 0 \qquad \boldsymbol{a}_{\text{rel}} = 0$$

$$\boldsymbol{\omega} = \omega_3\boldsymbol{i} \qquad \dot{\boldsymbol{\omega}} = \alpha_3\boldsymbol{i}$$

(a)

Then, by use of Eqs. (3-23) and (3-26), the velocity and acceleration can be found as

$$\boldsymbol{v}_{(o_1 x_1 y_1 z_1)} = \frac{d\boldsymbol{R}_0}{dt} + \boldsymbol{v}_{\text{rel}} + \boldsymbol{\omega} \times \boldsymbol{r} = v_0\boldsymbol{j} + 0 + \omega_3\boldsymbol{i} \times b\boldsymbol{j}$$

$$= v_0\boldsymbol{j} + b\omega_3\boldsymbol{k}$$

(b)

and

$$\boldsymbol{a}_{(o_1 x_1 y_1 z_1)} = \frac{d^2\boldsymbol{R}_0}{dt^2} + \boldsymbol{a}_{\text{rel}} + 2\boldsymbol{\omega} \times \boldsymbol{v}_{\text{rel}} + \dot{\boldsymbol{\omega}} \times \boldsymbol{r} + \boldsymbol{\omega} \times (\boldsymbol{\omega} \times \boldsymbol{r})$$

$$= a_0\boldsymbol{j} + 0 + 0 + \alpha_3\boldsymbol{i} \times b\boldsymbol{j} + \omega_3\boldsymbol{i} \times (\omega_3\boldsymbol{i} \times b\boldsymbol{j})$$

$$= \left(a_0 - b\omega_3^2\right)\boldsymbol{j} + b\alpha_3\boldsymbol{k}$$

(c)

where $\boldsymbol{v}_{(o_1 x_1 y_1 z_1)}$ and $\boldsymbol{a}_{(o_1 x_1 y_1 z_1)}$ are the velocity and acceleration, respectively, of the pilot with respect to $o_1 x_1 y_1 z_1$.

2. *Motion of o_1 (defined in $o_1x_1y_1z_1$) with respect to $OXYZ$.* The position of point o_1 with respect to the reference frame $OXYZ$ is

$$\boldsymbol{R}_0 = R\boldsymbol{i} \tag{d}$$

This position vector $\boldsymbol{R}_0$ is carried around at the angular velocity $\boldsymbol{\omega}_1$ according to the problem statement. So, by use of Eq. (3-19), the velocity of point o_1 is

$$\boldsymbol{v}_{o_1} = \frac{d\boldsymbol{R}_0}{dt} = \left[\left(\frac{\partial}{\partial t} \right)_{\text{rel}} + \boldsymbol{\omega}_1 \times \right] \boldsymbol{R}_0$$

$$= \left(\frac{\partial R\boldsymbol{i}}{\partial t} \right)_{\text{rel}} + \omega_1 \boldsymbol{k} \times R\boldsymbol{i} = R\omega_1 \boldsymbol{j} \tag{e}$$

and the acceleration of point o_1 is

$$\boldsymbol{a}_{o_1} = \frac{d^2\boldsymbol{R}_0}{dt^2} = \left[\left(\frac{\partial}{\partial t} \right)_{\text{rel}} + \boldsymbol{\omega}_1 \times \right] \frac{d\boldsymbol{R}_0}{dt}$$

$$= \omega_1 \boldsymbol{k} \times R\omega_1 \boldsymbol{j} = -R\omega_1^2 \boldsymbol{i} \tag{f}$$

Refer to Example 3-5 for the detailed conduct of the operations using Eq. (3-19) in Eqs. (e) and (f); the calculations shown in Eqs. (e) and (f) are simplified reductions of potentially more complex operations. Furthermore, although it would have been slightly inefficient to do so, we could have used Eqs. (3-23) and (3-26) to obtain the results in Eqs. (e) and (f). Finally, as indicated next, the intermediate results obtained as Eqs. (e) and (f) are simply in preparation for using Eqs. (3-23) and (3-26) in step 3.

3. *Motion of $\boldsymbol{P}$ (defined in $o_1x_1y_1z_1$) with respect to $OXYZ$.* In anticipation of the use of Eqs. (3-23) and (3-26), we note that

$$\frac{d\boldsymbol{R}_0}{dt} = R\omega_1 \boldsymbol{j} \qquad \frac{d^2\boldsymbol{R}_0}{dt^2} = -R\omega_1^2 \boldsymbol{i} \tag{g}$$

which come from Eqs. (e) and (f) in step 2 above. Also,

$$\boldsymbol{v}_{\text{rel}} = v_0 \boldsymbol{j} + b\omega_3 \boldsymbol{k} \qquad \boldsymbol{a}_{\text{rel}} = \left(a_0 - b\omega_3^2 \right) \boldsymbol{j} + b\alpha_3 \boldsymbol{k} \tag{h}$$

which come from Eqs. (b) and (c) in step 1 above. And, from Figure 3-27,

$$\boldsymbol{r} = b\boldsymbol{j} + h\boldsymbol{k} \qquad \boldsymbol{\omega} = \omega_2 \boldsymbol{j} + \omega_1 \boldsymbol{k} \qquad \dot{\boldsymbol{\omega}} = 0 \tag{i}$$

Note that $\dot{\boldsymbol{\omega}} = 0$ because both $\boldsymbol{\omega}_1$ and $\boldsymbol{\omega}_2$ are constant and are both defined with respect to $OXYZ$. The use of Eqs. (3-23) and (3-26) leads directly to

$$\boldsymbol{v}_{P(OXYZ)} = \frac{d\boldsymbol{R}_0}{dt} + \boldsymbol{v}_{\text{rel}} + \boldsymbol{\omega} \times \boldsymbol{r}$$

$$= R\omega_1 \boldsymbol{j} + (v_0 \boldsymbol{j} + b\omega_3 \boldsymbol{k}) + (\omega_2 \boldsymbol{j} + \omega_1 \boldsymbol{k}) \times (b\boldsymbol{j} + h\boldsymbol{k})$$

$$= (h\omega_2 - b\omega_1)\boldsymbol{i} + (v_0 + R\omega_1)\boldsymbol{j} + b\omega_3 \boldsymbol{k} \tag{j}$$

and

$$\boldsymbol{a}_{P(OXYZ)} = \frac{d^2\boldsymbol{R}_0}{dt^2} + \boldsymbol{a}_{\text{rel}} + 2\boldsymbol{\omega} \times \boldsymbol{v}_{\text{rel}} + \dot{\boldsymbol{\omega}} \times \boldsymbol{r} + \boldsymbol{\omega} \times (\boldsymbol{\omega} \times \boldsymbol{r})$$

$$= -R\omega_1^2 \boldsymbol{i} + \left[\left(a_0 - b\omega_3^2\right)\boldsymbol{j} + b\alpha_3\boldsymbol{k}\right] + 2\left(\omega_2\boldsymbol{j} + \omega_1\boldsymbol{k}\right) \times \left(v_0\boldsymbol{j} + b\omega_3\boldsymbol{k}\right)$$

$$+ 0 + \left(\omega_2\boldsymbol{j} + \omega_1\boldsymbol{k}\right) \times \left[\left(\omega_2\boldsymbol{j} + \omega_1\boldsymbol{k}\right) \times \left(b\boldsymbol{j} + h\boldsymbol{k}\right)\right]$$

$$= \left(-R\omega_1^2 - 2\omega_1 v_0 + 2b\omega_2\omega_3\right)\boldsymbol{i} + \left(a_0 - b\omega_3^2 - b\omega_1^2 + h\omega_1\omega_2\right)\boldsymbol{j}$$

$$+ \left(b\alpha_3 + b\omega_1\omega_2 - h\omega_2^2\right)\boldsymbol{k} \tag{k}$$

which are the velocity and acceleration sought.

In Chapter 4, we shall return to these results in order to consider the forces on the pilot. Note that both the pilot's weight and gravity are irrelevant in the kinematic analysis (although gravity is sometimes confusingly incorporated into kinematics).

3-6 GENERALIZATIONS OF KINEMATIC EXPRESSIONS

In the examples presented in the previous section, we noted that the choice of reference frames and the sequential use of Eqs. (3-23) and (3-26) in each example resulted in *a* technique, not *the* technique. There is no such thing as *the* technique or method in a kinematic analysis. On the other hand, in Examples 3-9 and 3-10, we suggested that when multiple intermediate reference frames are used, there may be a preferred technique (based on the definitions of $\boldsymbol{\omega}_2$ and $\dot{\boldsymbol{\omega}}_2$ in those examples). Now, we explore the generalization of that idea.

In this section, we investigate the expressions for velocity and acceleration in the case when two intermediate reference frames are used. Figure 3-28 shows a point P that is defined in the intermediate reference frame $o_2x_2y_2z_2$, which itself is defined in another intermediate reference frame $o_1x_1y_1z_1$. Frame $o_1x_1y_1z_1$ is rotating at the angular velocity $\boldsymbol{\omega}_1$ and angular

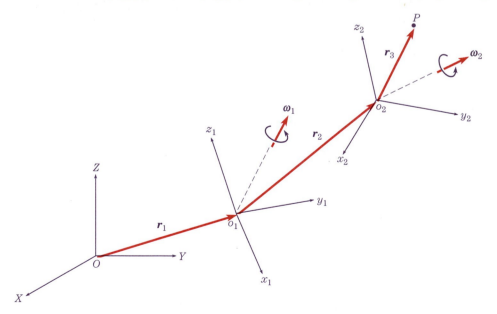

FIGURE 3-28   Point P is defined in intermediate reference frame $o_2x_2y_2z_2$, which is moving with respect to another intermediate reference frame $o_1x_1y_1z_1$.

acceleration $\dot{\boldsymbol{\omega}}_1$, with respect to the "fixed" reference frame $OXYZ$. We consider two cases: the case in which the angular motions of $o_2x_2y_2z_2$, $\boldsymbol{\omega}_2$ and $\dot{\boldsymbol{\omega}}_2$, are defined with respect to frame $o_1x_1y_1z_1$; and the case in which the angular motions of $o_2x_2y_2z_2$, $\boldsymbol{\omega}_2$ and $\dot{\boldsymbol{\omega}}_2$, are defined with respect to frame $OXYZ$.

In order to find the motion of point P with respect to the "fixed" reference frame $OXYZ$, the formulas for a single intermediate reference frame may be successively used. Great care must be taken to identify correctly each term in Eqs. (3-23) and (3-26). These equations represent the motion of a point with respect to the "inner" reference frame ($OXYZ$ in Figure 3-19),[4] where $\boldsymbol{r}$, $\boldsymbol{v}_{\rm rel}$ and $\boldsymbol{a}_{\rm rel}$ are the position, velocity, and acceleration of the point with respect to the "outer" reference frame ($oxyz$ in Figure 3-19), $\boldsymbol{\omega}$ and $\dot{\boldsymbol{\omega}}$ are the angular velocity and angular acceleration of the "outer" frame with respect to the "inner" frame, $\boldsymbol{R}_0$ is the position of the origin of the "outer" reference frame defined with respect to the "inner" reference frame, and the time derivatives of $\boldsymbol{R}_0$ are time rates of change observed from the "inner" frame.

CASE 1: If ω_2 and $\dot{\omega}_2$ Are Defined with Respect to "Fixed" Frame $OXYZ$

In this case, where $\boldsymbol{\omega}_2$ and $\dot{\boldsymbol{\omega}}_2$ are defined with respect to the "fixed" frame $OXYZ$, we successively find the motion of o_2 with respect to $OXYZ$ and then the motion of P with respect to $OXYZ$. In the first step, $OXYZ$ is the inner frame and $o_1x_1y_1z_1$ is the outer frame that corresponds to the intermediate frame of Figure 3-19 and Eqs. (3-23) and (3-26). In the second step, $OXYZ$ is the inner frame again, but $o_2x_2y_2z_2$ is the outer frame that corresponds to the intermediate frame of Figure 3-19 and Eqs. (3-23) and (3-26). Thus, in this manner, we successively apply Eqs. (3-23) and (3-26).

Motion of point o_2 (defined in $o_1x_1y_1z_1$) with respect to $OXYZ$

From Figure 3-28, for the direct use of Eqs. (3-23) and (3-26), we identify the terms

$$\boldsymbol{R}_0 = \boldsymbol{r}_1 \qquad \boldsymbol{r} = \boldsymbol{r}_2 \qquad \boldsymbol{\omega} = \boldsymbol{\omega}_1$$

If we adopt the notation $(\dot{\square})$ to represent the time derivative of the quantity $\square$, as observed in the reference frame in which $\square$ is defined, we may write

$$\frac{d\boldsymbol{R}_0}{dt} = \dot{\boldsymbol{r}}_1 \qquad \frac{d^2\boldsymbol{R}_0}{dt^2} = \ddot{\boldsymbol{r}}_1$$

$$\boldsymbol{v}_{\rm rel} = \dot{\boldsymbol{r}}_2 \qquad \boldsymbol{a}_{\rm rel} = \ddot{\boldsymbol{r}}_2 \qquad \dot{\boldsymbol{\omega}} = \dot{\boldsymbol{\omega}}_1$$

Substitution of these terms into Eqs. (3-23) and (3-26) gives

$$\boldsymbol{v}_{o_2(OXYZ)} = \dot{\boldsymbol{r}}_1 + \dot{\boldsymbol{r}}_2 + \boldsymbol{\omega}_1 \times \boldsymbol{r}_2 \qquad (3\text{-}27)$$

and

$$\boldsymbol{a}_{o_2(OXYZ)} = \ddot{\boldsymbol{r}}_1 + \ddot{\boldsymbol{r}}_2 + 2\boldsymbol{\omega}_1 \times \dot{\boldsymbol{r}}_2$$
$$+ \dot{\boldsymbol{\omega}}_1 \times \boldsymbol{r}_2 + \boldsymbol{\omega}_1 \times (\boldsymbol{\omega}_1 \times \boldsymbol{r}_2) \quad (3\text{-}28)$$

where $\boldsymbol{v}_{o_2(OXYZ)}$ and $\boldsymbol{a}_{o_2(OXYZ)}$ are the velocity and acceleration of point o_2 with respect to $OXYZ$, respectively.

Motion of point P (defined in $o_2x_2y_2z_2$) with respect to $OXYZ$

Having the motion of point o_2 with respect to $OXYZ$, we can consider a vector emanating from O to the point o_2 in Figure 3-28. Then, for direct use of Eqs. (3-23) and (3-26), this is the vector $\boldsymbol{R}_0$, which can be expressed as

$$\boldsymbol{R}_0 = \boldsymbol{r}_1 + \boldsymbol{r}_2$$

By definition, the first and second time derivatives of this vector in $OXYZ$ are the velocity and acceleration of point o_2 with respect to $OXYZ$, respectively. As we have just found in Eqs. (3-27) and (3-28):

$$\frac{d\boldsymbol{R}_0}{dt} = \boldsymbol{v}_{o_2(OXYZ)} \qquad \frac{d^2\boldsymbol{R}_0}{dt^2} = \boldsymbol{a}_{o_2(OXYZ)}$$

Now the outer reference frame is $o_2x_2y_2z_2$. The position vector of P in the outer frame is

$$\boldsymbol{r} = \boldsymbol{r}_3$$

and in turn

$$\boldsymbol{v}_{\rm rel} = \dot{\boldsymbol{r}}_3 \qquad \boldsymbol{a}_{\rm rel} = \ddot{\boldsymbol{r}}_3$$

The angular motion of the outer frame ($o_2x_2y_2z_2$) with respect to the inner frame ($OXYZ$) is

$$\boldsymbol{\omega} = \boldsymbol{\omega}_2 \qquad \dot{\boldsymbol{\omega}} = \dot{\boldsymbol{\omega}}_2$$

Substitution of the terms immediately above into Eqs. (3-23) and (3-26) gives

---

[4]By reexamining the derivations of Eqs. (3-23) and (3-26), the reader should realize that the "inner" reference frame $OXYZ$ was not necessarily fixed.

$$\boldsymbol{v}_{P(OXYZ)} = \boldsymbol{v}_{o_2(OXYZ)} + \dot{\boldsymbol{r}}_3 + \boldsymbol{\omega}_2 \times \boldsymbol{r}_3$$

$$= \dot{\boldsymbol{r}}_1 + \dot{\boldsymbol{r}}_2 + \dot{\boldsymbol{r}}_3 + \boldsymbol{\omega}_1 \times \boldsymbol{r}_2 + \boldsymbol{\omega}_2 \times \boldsymbol{r}_3 \quad (3\text{-}29)$$

and

$$\boldsymbol{a}_{P(OXYZ)} = \boldsymbol{a}_{o_2(OXYZ)} + \ddot{\boldsymbol{r}}_3 + 2\boldsymbol{\omega}_2 \times \dot{\boldsymbol{r}}_3$$

$$+ \dot{\boldsymbol{\omega}}_2 \times \boldsymbol{r}_3 + \boldsymbol{\omega}_2 \times (\boldsymbol{\omega}_2 \times \boldsymbol{r}_3)$$

$$= \ddot{\boldsymbol{r}}_1 + \ddot{\boldsymbol{r}}_2 + \ddot{\boldsymbol{r}}_3 + 2\boldsymbol{\omega}_1 \times \dot{\boldsymbol{r}}_2 + 2\boldsymbol{\omega}_2 \times \dot{\boldsymbol{r}}_3 + \dot{\boldsymbol{\omega}}_1 \times \boldsymbol{r}_2$$

$$+ \dot{\boldsymbol{\omega}}_2 \times \boldsymbol{r}_3 + \boldsymbol{\omega}_1 \times (\boldsymbol{\omega}_1 \times \boldsymbol{r}_2) + \boldsymbol{\omega}_2 \times (\boldsymbol{\omega}_2 \times \boldsymbol{r}_3)$$

$$(3\text{-}30)$$

which are the velocity and acceleration expressions sought for Case 1 as defined above.

In conclusion, it is very useful to note that in this case where $\boldsymbol{\omega}_2$ and $\dot{\boldsymbol{\omega}}_2$ are defined with respect to $OXYZ$, the better strategy is to "move" from $OXYZ$ "outward" to $o_2x_2y_2z_2$. This approach is not necessary (see Problem 3-52), but it is, in general, the easier procedure to conceptualize and to compute. In particular, Example 3-10 involving the robot was of this type, and was analyzed by "moving" from $OXYZ$ to $o_2x_2y_2z_2$. Example 3-10 (and all such problems) could have been solved in a single step via Eqs. (3-29) and (3-30).

CASE 2: If $\boldsymbol{\omega}_2$ and $\dot{\boldsymbol{\omega}}_2$ Are Defined with Respect to Intermediate Frame $o_1x_1y_1z_1$

In this case, where $\boldsymbol{\omega}_2$ and $\dot{\boldsymbol{\omega}}_2$ are defined with respect to the intermediate frame $o_1x_1y_1z_1$, we successively find the motion of P with respect to $o_1x_1y_1z_1$ and then the motion of P with respect to $OXYZ$. In the first step, $o_1x_1y_1z_1$ is treated as the inner frame and $o_2x_2y_2z_2$ is the outer frame that corresponds to the intermediate frame of Figure 3-19 and Eqs. (3-23) and (3-26). In the second step, $OXYZ$ is the inner frame and $o_1x_1y_1z_1$ is the outer frame that corresponds to the intermediate frame of Figure 3-19 and Eqs. (3-23) and (3-26). Thus, in this manner, we successively apply Eqs. (3-23) and (3-26).

Motion of point P (defined in $o_2x_2y_2z_2$) with respect to $o_1x_1y_1z_1$

The outer reference frame here is $o_2x_2y_2z_2$ and the inner reference frame is $o_1x_1y_1z_1$. From Figure 3-28, for the direct use of Eqs. (3-23) and (3-26), we identify the terms

$$\boldsymbol{R}_0 = \boldsymbol{r}_2 \qquad \frac{d\boldsymbol{R}_0}{dt} = \dot{\boldsymbol{r}}_2 \qquad \frac{d^2\boldsymbol{R}_0}{dt^2} = \ddot{\boldsymbol{r}}_2$$

$$\boldsymbol{r} = \boldsymbol{r}_3 \qquad \boldsymbol{v}_{\text{rel}} = \dot{\boldsymbol{r}}_3 \qquad \boldsymbol{a}_{\text{rel}} = \ddot{\boldsymbol{r}}_3$$

$$\boldsymbol{\omega} = \boldsymbol{\omega}_2 \qquad \dot{\boldsymbol{\omega}} = \dot{\boldsymbol{\omega}}_2$$

The motion of P with respect to $o_1x_1y_1z_1$ can be found by substituting the terms immediately above into Eqs. (3-23) and (3-26), giving

$$\boldsymbol{v}_{P(o_1x_1y_1z_1)} = \dot{\boldsymbol{r}}_2 + \dot{\boldsymbol{r}}_3 + \boldsymbol{\omega}_2 \times \boldsymbol{r}_3 \qquad (3\text{-}31)$$

and

$$\boldsymbol{a}_{P(o_1x_1y_1z_1)} =$$
$$\ddot{\boldsymbol{r}}_2 + \ddot{\boldsymbol{r}}_3 + 2\boldsymbol{\omega}_2 \times \dot{\boldsymbol{r}}_3 + \dot{\boldsymbol{\omega}}_2 \times \boldsymbol{r}_3 + \boldsymbol{\omega}_2 \times (\boldsymbol{\omega}_2 \times \boldsymbol{r}_3) \quad (3\text{-}32)$$

where $\boldsymbol{v}_{P(o_1x_1y_1z_1)}$ and $\boldsymbol{a}_{P(o_1x_1y_1z_1)}$ are the velocity and acceleration of P with respect to $o_1x_1y_1z_1$, respectively.

Motion of point P (defined in $o_1x_1y_1z_1$) with respect to OXYZ

Now the inner reference frame is $OXYZ$ and the outer reference frame is $o_1x_1y_1z_1$. From Figure 3-28, the position vector from O to the origin of the outer reference frame is

$$\boldsymbol{R}_0 = \boldsymbol{r}_1$$

and

$$\frac{d\boldsymbol{R}_0}{dt} = \dot{\boldsymbol{r}}_1 \qquad \frac{d^2\boldsymbol{R}_0}{dt^2} = \ddot{\boldsymbol{r}}_1$$

Also,

$$\boldsymbol{\omega} = \boldsymbol{\omega}_1 \qquad \dot{\boldsymbol{\omega}} = \dot{\boldsymbol{\omega}}_1$$

The position vector of P with respect to the reference frame $o_1x_1y_1z_1$ is

$$\boldsymbol{r} = \boldsymbol{r}_2 + \boldsymbol{r}_3$$

and $\boldsymbol{v}_{\text{rel}}$ and $\boldsymbol{a}_{\text{rel}}$ are the velocity and acceleration of P with respect to $o_1x_1y_1z_1$, respectively, which we have just found as Eqs. (3-31) and (3-32). Thus

$$\boldsymbol{v}_{\text{rel}} = \boldsymbol{v}_{P(o_1x_1y_1z_1)} \qquad \boldsymbol{a}_{\text{rel}} = \boldsymbol{a}_{P(o_1x_1y_1z_1)}$$

Application of Eqs. (3-23) and (3-26) once more, using the terms immediately above, gives

$$\boldsymbol{v}_{P(OXYZ)} = \dot{\boldsymbol{r}}_1 + \boldsymbol{v}_{P(o_1x_1y_1z_1)} + \boldsymbol{\omega}_1 \times (\boldsymbol{r}_2 + \boldsymbol{r}_3)$$

$$= \dot{\boldsymbol{r}}_1 + \dot{\boldsymbol{r}}_2 + \dot{\boldsymbol{r}}_3 + \boldsymbol{\omega}_1 \times \boldsymbol{r}_2 + (\boldsymbol{\omega}_1 + \boldsymbol{\omega}_2) \times \boldsymbol{r}_3$$

$$(3\text{-}33)$$

and

$$\begin{aligned}
\boldsymbol{a}_{P(OXYZ)} &= \ddot{\boldsymbol{r}}_1 + \boldsymbol{a}_{P(o_1x_1y_1z_1)} + 2\boldsymbol{\omega}_1 \times \boldsymbol{v}_{P(o_1x_1y_1z_1)} \\
&\quad + \dot{\boldsymbol{\omega}}_1 \times (\boldsymbol{r}_2 + \boldsymbol{r}_3) + \boldsymbol{\omega}_1 \times [\boldsymbol{\omega}_1 \times (\boldsymbol{r}_2 + \boldsymbol{r}_3)] \\
&= \ddot{\boldsymbol{r}}_1 + \ddot{\boldsymbol{r}}_2 + \ddot{\boldsymbol{r}}_3 + 2\boldsymbol{\omega}_1 \times \dot{\boldsymbol{r}}_2 + 2(\boldsymbol{\omega}_1 + \boldsymbol{\omega}_2) \times \dot{\boldsymbol{r}}_3 \\
&\quad + \dot{\boldsymbol{\omega}}_1 \times \boldsymbol{r}_2 + (\dot{\boldsymbol{\omega}}_1 + \dot{\boldsymbol{\omega}}_2) \times \boldsymbol{r}_3 + \boldsymbol{\omega}_2 \times (\boldsymbol{\omega}_2 \times \boldsymbol{r}_3) \\
&\quad + 2\boldsymbol{\omega}_1 \times (\boldsymbol{\omega}_2 \times \boldsymbol{r}_3) + \boldsymbol{\omega}_1 \times [\boldsymbol{\omega}_1 \times (\boldsymbol{r}_2 + \boldsymbol{r}_3)]
\end{aligned}$$

$$(3\text{-}34)$$

which are the velocity and acceleration expressions sought for Case 2 as defined above. Furthermore, it may be of interest to note that Eq. (3-34) is identical to Eq. (B–65) in Appendix B.

In conclusion and in contrast with Case 1, it is very useful to note that in Case 2, where $\boldsymbol{\omega}_2$ and $\dot{\boldsymbol{\omega}}_2$ are defined with respect to $o_1x_1y_1z_1$, the better strategy is to "move" from $o_2x_2y_2z_2$ "inward" to $OXYZ$. This approach is not necessary (see Problem 3-53 at the end of this chapter), but it is, in general, the easier approach to conceptualize and to compute. In particular, Example 3-9 involving the bird on the mobile is of this type, and was analyzed by "moving" from $o_2x_2y_2z_2$ to $OXYZ$. Example 3-9 (and all such problems) could have been solved in a single step via Eqs. (3-33) and (3-34). Example 3-11 could have been solved also in an analogous single step!

Generalization

The two pairs of results—Eqs. (3-29) and (3-30) for Case 1 and Eqs. (3-33) and (3-34) for Case 2—constitute potentially very beneficial kinematic expressions. Their value lies in the fact that they represent complete expressions that may be implemented without further differentiation. The reason for this benefit is that the various terms in these expressions are exactly those terms that are required in a complete problem statement.

Having obtained the results for Cases 1 and 2 above, the question might arise as to whether these results can be generalized to n intermediate reference frames, where n is an arbitrary number. This is precisely what is done in Appendix B, although this subject matter is not recommended for students during their first encounter with this material.

Problems for Chapter 3

Category I: Back to Physics

Problem 3-1 (*HRW36, 14P*)[5]: A high-performance jet plane, practicing radar avoidance maneuvers, is in horizontal flight 35 m above the level ground. Suddenly, the plane encounters terrain that slopes gently upward at $4.3°$, an amount difficult to detect; see Figure P3-1. How much time does the pilot have to make a correction if the plane is to avoid flying into the ground? The speed of the plane is 1300 km/h.

Problem 3-2 (*HRW37, 19P*): How far does the runner whose velocity versus time graph is shown in Figure P3-2 travel in 16 s?

Problem 3-3 (*HRW37, 22E*): An object moves in a straight line as described by the velocity versus time graph of Figure P3-3. Sketch a graph that represents the acceleration of the object as a function of time.

Problem 3-4 (*HRW37, 23E, 24E*): The graphs of position versus time in Figures P3-4a and b are for particles in straight-line motion.

4.3°   35 m

FIGURE P3-1

[5]These designations reference Halliday, Resnick, and Walker. For example, "HRW36, 14P" denotes "Halliday, Resnick, and Walker; page 36, problem 14P."

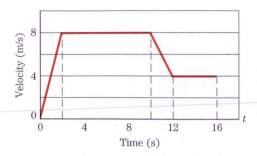

FIGURE P3-2

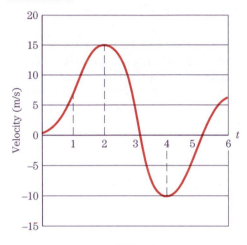

Time (s)

FIGURE P3-3

(a) State, for each of the intervals $AB, BC, CD,$ and DE, whether the velocity v is positive, negative or zero and whether the acceleration a is positive, negative or zero. (Ignore end points of the intervals.)

(b) From the curves, is there any interval over which the acceleration is obviously not constant?

(c) If the axes are shifted upward together such that the time axes end up running along the dashed lines, does any of your answers change?

Problem 3-5 (*HRW38, 25E*): A particle moves along the x axis with $x(t)$ as shown in Figure P3-5. Make rough sketches of velocity versus time and acceleration versus time for this motion.

Category II: Intermediate

Problem 3-6: A particle moves in a straight line with a constant acceleration a and initial velocity v_0. Determine the following:

(a) The velocity of the particle as a function of time;

(b) The distance the particle travels as a function of time; and

(c) If the acceleration is negative, the distance the particle travels before it stops momentarily.

Problem 3-7: With its brakes fully engaged, a car comes to a complete stop from an initial speed of 55 mi/hr in a distance of 120 ft (called the *stopping distance*). Given the same constant acceleration, what would be the *stopping distance* if the initial speed were 75 mi/hr?

Problem 3-8: A police car starts from rest at point A 2 seconds after a speeding car passes that point, as sketched in Figure P3-8. The speeding car is moving at 75 mi/hr. If the police car accelerates at the rate

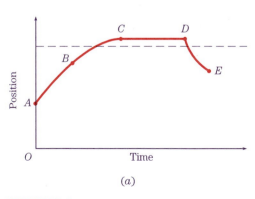

(a)

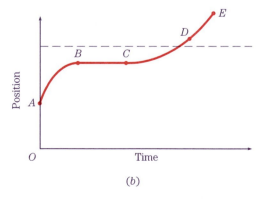

(b)

FIGURE P3-4

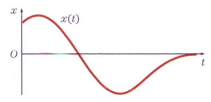

FIGURE P3-5

of 20 ft/sec^2 until it reaches its maximum achievable speed of 90 mi/hr, which is maintained thereafter, determine the distance between point A and the point at which the police car overtakes the speeding car.

Problem 3-9: Two cars travel along a straight section of highway at the same constant speed V_0. Suddenly, the driver in the front car begins an emergency stop by applying a constant acceleration of -20 ft/sec^2. After a 0.1 second of response delay time, the driver in the rear car applies a steadily increasing force to brake the car so that the acceleration of the rear car changes linearly from 0 to -20 ft/sec^2 within 1 second. The acceleration versus time graphs for both cars are sketched in Figures P3-9a and b, respectively. Determine, in order to avoid a collision,

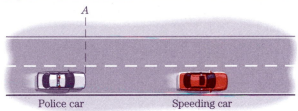

FIGURE P3-8

the minimum distance the drivers should maintain between their cars when they are driving at the same constant speed. Obtain numerical values for an initial speed V_0 of 50 mi/hr, 60 mi/hr, and 70 mi/hr.

Geometrical constraints: Problems 3-10 through 3-15

Problem 3-10: Explain how every point on the smaller disk in Figure 3-11 can have the same velocity despite the fact that some points are farther from the axis of rotation than others.

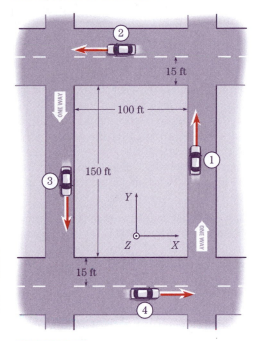

FIGURE P3-11

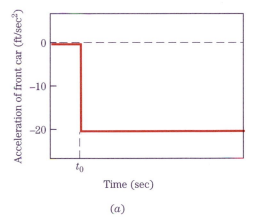
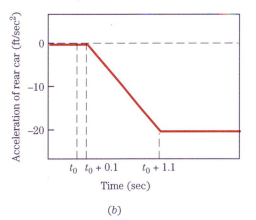

FIGURE P3-9

Problem 3-11: An automobile traverses a city's streets, as sketched in Figure P3-11. Describe the rotation of the automobile about the Z axis as it goes to each successive location. Assuming the trip from location ① back to location ① takes 100 seconds, find the average velocity and the average angular velocity of the automobile.

Problem 3-12: Explain intuitively why the answer is 4 in Example 3-2.

Problem 3-13: Assume that the nonslipping disk in Figure 3-12 rolls *inside* the stationary disk as sketched in Figure P3-13. How many complete rotations does the smaller disk make in one complete transit around the inside of the larger disk?

Problem 3-14: Gears are important mechanisms for transferring rotational motion from one axis to another. Show that the ratio of the angular velocity of a pair of gears is the reciprocal of the ratio of their radii (called the *gear ratio*). Use the result to determine the speed at which the weight W in the system sketched in Figure P3-14 is being lifted when the motor turns at 50 rev/s.

Problem 3-15: Astronomers sometimes use a time-keeping system called *sidereal time* (sidereal means "by the stars"), which is not exactly the same as the time system we use in our daily lives, known as *solar time*. Kinematically, the difference in times is caused by the use of two different reference frames with respect to which the Earth's rotation is measured. A *sidereal day* is the length of time that a given "fixed star" in the sky takes to return to the same direction in the sky. A *solar day* is the length of time that the

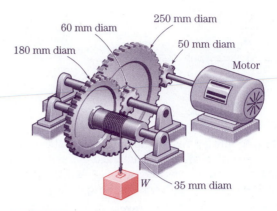

FIGURE P3-14

Sun takes to return to the same direction in the sky. Such a difference is illustrated in Figure P3-15.

Assume we are located at point A at noon of Day 1 where both the Sun and the "fixed star" are directly above us. In one sidereal day, we must move to point A', at which time the "fixed star" is again directly above us, however, the Sun is not. In one solar day, we must move to A'' where the Sun is directly above us, but now the "fixed star" is not. According to our daily timekeeping, the time at which A arrives at A'' would be noon of Day 2.

Given that a solar year is approximately $365\frac{1}{4}$ solar days and that a solar day is defined to be 24 hours, determine the length of a sidereal day. Also, how many sidereal days are there in a solar year?

Use of differential operator $\dfrac{d}{dt} = \left(\dfrac{\partial}{\partial t}\right)_{\text{rel}} + \omega\times$:
Problems 3-16 through 3-18

Problem 3-16: A disk is spinning on shaft AB and rotates at an angular velocity $\omega_2 = 15$ rev/s with respect to shaft AB, as sketched in Figure P3-16. Shaft AB is mounted on a cone-shaped base of conical angle of 25°, which is rotating about its axis CD at an angular velocity $\omega_1 = 200$ rev/s, with respect to ground. Find the angular acceleration of the disk (with respect to ground).[6]

Problem 3-17: As sketched in Figure P3-17, a ship is rolling at a constant angular velocity ω_1 about the X direction, with respect to ground, and is also pitching

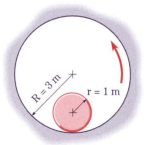

FIGURE P3-13

---

[6]The phrase "with respect to ground" is generally superfluous because essentially all kinematical analyses seek these quantities with respect to ground or a "fixed" reference. Nevertheless, in order to emphasize this point, we shall use this phrase for the remainder of the problems in Category II.

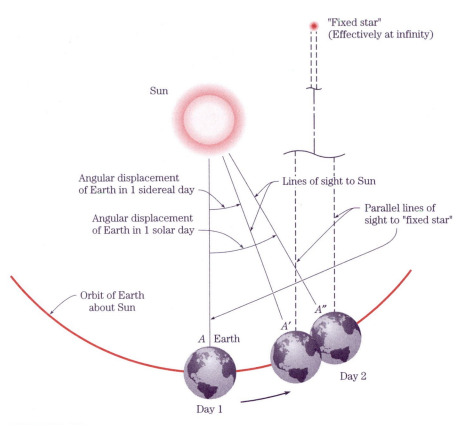

"Fixed star"
(Effectively at infinity)

Sun

Angular displacement
of Earth in 1 sidereal day

Lines of sight to Sun

Angular displacement
of Earth in 1 solar day

Parallel lines of
sight to "fixed star"

A''

Orbit of Earth
about Sun

A'

A Earth

Day 2

Day 1

FIGURE P3-15

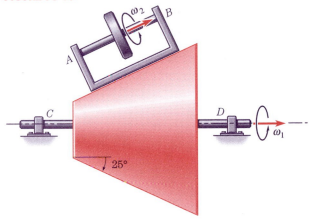

ω_2

B

A

C

D

ω_1

$25°$

FIGURE P3-16

at a constant angular velocity ω_2 about the Y direction, with respect to ground. The ship is level at the instant shown. The radar antenna located on the ship is rotating at a constant angular velocity ω_0 about the Z direction, with respect to the pitching and rolling

ship. Find the angular velocity and angular acceleration of the radar antenna (with respect to ground) at the instant shown.

Problem 3-18: A rod of length ℓ with a wheel of radius r attached to one of its ends is rotating about

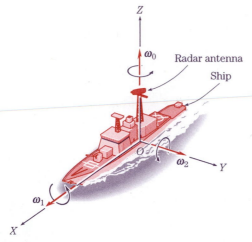

FIGURE P3-17

a vertical axis AA at constant angular velocity Ω, as sketched in Figure P3-18. The rod is in a horizontal plane and the disk is in a vertical plane. The wheel is rolling without slip on the horizontal supporting surface, as sketched in Figure P3-18. Find the angular velocity and angular acceleration of the wheel (with respect to ground).

Use of Eqs. (3-23) and (3-26) utilizing single intermediate frame: Problems 3-19 through 3-29

Although the first few problems below are sufficiently simple that readers are likely capable of writing the answer by inspection, we strongly encourage the use of Eqs. (3-23) and (3-26). To encourage this approach, a set of potentially useful reference frames

has been provided in each problem. Of course, readers may choose any other reference frames they prefer, but such frames must be *clearly* defined.

Problem 3-19: A boy is spinning a bucket of water such that the bucket moves in a circle of radius R in the vertical plane at a constant speed v, as sketched in Figure P3-19. Note that gravity acts. Determine the acceleration of the bucket (with respect to ground) at locations A and B; that is, the highest and lowest points of the path.

The reference frames defined in Figure P3-19 are as follows: $OXYZ$ is fixed to ground; $Oxyz$ is fixed to the rope.

Problem 3-20: An amusement park ride consists of swinging arms (only two of which are shown) pinned to a rotating column, as sketched in Figure P3-20. As the column rotates at a constant rate, the arms swing out. Each swinging arm has an effective length of 50 ft. At the instant shown, the column is rotating at 0.225 rev/sec and the arms form an angle of 50° with respect to the vertical. Note that gravity acts. Find the velocity and acceleration of the crouched rider inside the cage labeled A, at the instant shown (with respect to ground).

The reference frames defined in Figure P3-20 are as follows: $OXYZ$ is fixed to ground; $Oxyz$ is fixed to the swinging arm.

Problem 3-21: Three passengers, each having a mass of 90 kg, are located 90° apart on a large Ferris wheel of radius 20 m as sketched in Figure P3-21. Passenger ① is located at the highest point of the wheel

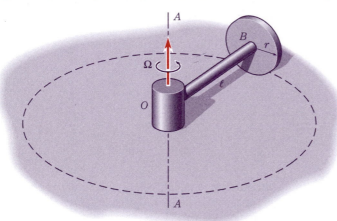

FIGURE P3-18

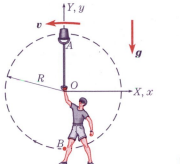

FIGURE P3-19

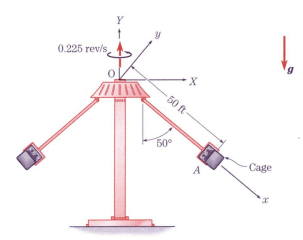

FIGURE P3-20

at the instant shown. The wheel's angular velocity is 1 rad/s counterclockwise and its angular acceleration is -0.1 rad/s^2. Note that gravity acts. Find the velocity and acceleration of each passenger, at the instant shown (with respect to ground).

The reference frames defined in Figure P3-21 are as follows: $OXYZ$ is fixed to ground; $Oxyz$ is fixed to the Ferris wheel.

Problem 3-22: A centrifuge is often used for astronaut training. A simplified version is sketched in Figure P3-22, in which a cockpit is rigidly attached to an arm that is driven by a motor. If in a particular test, starting from rest, the driving motor supplies a constant angular acceleration of 0.5 rad/s^2, and the distance from the astronaut's navel to the rotation axis is 10 m, find the acceleration of the astronaut's navel as a function of time (with respect to ground). Note that gravity acts.

The reference frames defined in Figure P3-22 are as follows: $OXYZ$ is fixed to ground; $Oxyz$ is fixed to the arm.

Problem 3-23: A ladder of length ℓ moves in contact with the wall and the floor as sketched in Figure P3-23. The angle θ describes the location of the ladder completely. In terms of θ and its derivatives, find the velocity and acceleration of points A, B, and C on the ladder (with respect to ground) as it falls while remaining in contact with the wall and the floor. Note that gravity acts.

The reference frames defined in Figure P3-23 are as follows: $OXYZ$ is fixed to ground; $Cxyz$ is fixed to the ladder.

Problem 3-24: A commuter train is moving toward the right along a straight rail at the speed v_0 and acceleration a_0. A boy in the train runs toward the forward end of the car, along the center aisle at the speed v_1 and acceleration a_1, both with respect to the train, as sketched in Figure P3-24. Find the velocity and acceleration of the boy (with respect to ground).

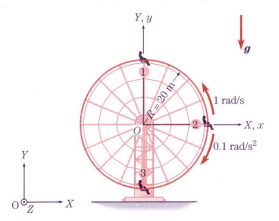

FIGURE P3-21

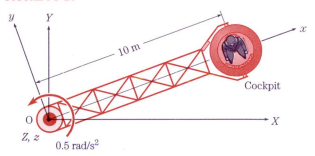

FIGURE P3-22

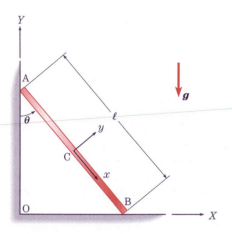

FIGURE P3-23

The reference frames defined in Figure P3-24 are as follows: $OXYZ$ is fixed to ground; $oxyz$ is fixed to the train.

Problem 3-25: A ferry is trying to cross a 500 m wide river by starting from point A on one bank of the river with the following four possible destinations on the other bank of the river as sketched in Figure P3-25.

(a) Point B, which is directly across the river from point A;

(b) Point C, which is 200 m upstream from point B;

(c) Point D, which is 200 m downstream from point B; and

(d) Anywhere on the other bank but should take the minimum time to cross the river.

The boat is powered to travel 25 km/h in still water and the water in the river runs at 13 km/h. For each of the four destinations, determine the time required to complete the crossing.

The reference frames defined in Figure P3-25 are as follows: $OXYZ$ is fixed to ground with the Y axis along the river; $Oxyz$ is fixed to the water.

Problem 3-26: Rotor blades of a helicopter are of radius 5.2 m and rotating at a constant angular velocity of 5 rev/s. The helicopter is flying horizontally in a straight line at a speed of 5 m/s and an acceleration of 0.5 m/s^2, as sketched in Figure P3-26. Determine the velocity and acceleration of point A, which is the outermost tip of one of the blades, (with respect to ground) when θ is 0°, 90°, and 180°.

The reference frames defined in Figure P3-26 are as follows: $OXYZ$ is fixed to ground; $Oxyz$ is fixed to the translating helicopter.

Problem 3-27: A test chamber sketched in Figure P3-27 is used to study vehicular ride quality parameters. A passenger's head of mass m is located at point P. At the instant shown, the chamber is moving outward at speed v_0 and acceleration a_0, with respect to the rotating base, which is rotating at angular velocity ω and angular acceleration $\dot{\omega}$, both with respect to ground. Note that gravity acts. Find the velocity and acceleration of the passenger's head (with respect to ground).

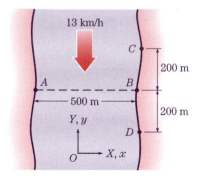

FIGURE P3-25

FIGURE P3-24

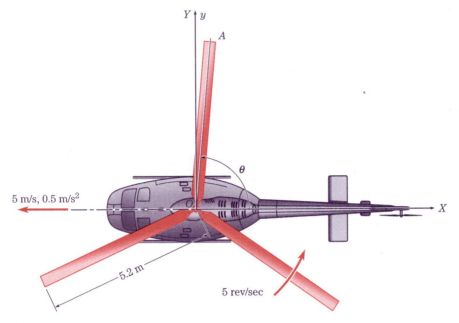

FIGURE P3-26

The reference frames defined in Figure P3-27 are as follows: $OXYZ$ is fixed to ground; $Oxyz$ is fixed to the rotating base.

Problem 3-28: A skater is spinning on the ice about the vertical axis of her body with angular velocity ω_0, with respect to ground. She has training weights (mass m_0 each) in her hands, as sketched in Figure P3-28. At the instant shown, her arms are nearly fully extended, and she is moving her hands radically inward toward her body. Due to the angular momentum principle, which will be discussed in Chapters 4 and 6, the motion of the weights causes her body to rotate faster, resulting in an angular acceleration denoted by $\dot{\omega}_0$, with respect to ground. At the instant shown, each of her arms extends to a radius of ℓ about the axis of her body. Note that gravity acts. Find the velocity and acceleration of each weight (with respect to ground).

The reference frames defined in Figure P3-28 are as follows: $OXYZ$ is fixed to ground; $Oxyz$ is fixed to the body of the skater.

Problem 3-29: A sportsman is skeet shooting. As sketched in Figure P3-29, as the shot of mass m is exiting the barrel of the rifle, the sportsman is rotating the rifle about a vertical axis that passes through point O, at an angular velocity $\dot{\theta}$ and angular acceleration $\ddot{\theta}$, both with respect to ground. The exit velocity and acceleration of the shot relative to the barrel are $\dot{r}$ and $\ddot{r}$, respectively. Note that gravity acts. Find the velocity and acceleration of the shot in the barrel (with respect to ground) when it is a distance r from point O.

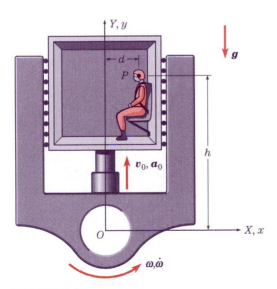

FIGURE P3-27

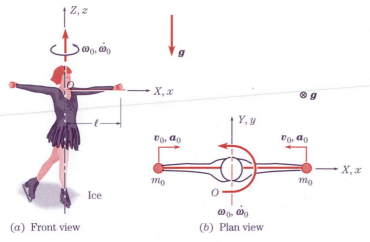

(a) Front view
(b) Plan view

FIGURE P3-28

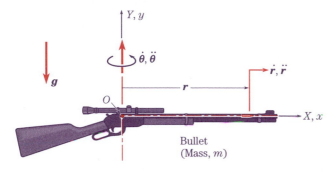

Bullet
(Mass, m)

FIGURE P3-29

The reference frames defined in Figure P3-29 are as follows: $OXYZ$ is fixed to ground; $Oxyz$ is fixed to the gun barrel.

Category III: More Difficult

- In general, the problems in this category involve three-dimensional motion and require the use of multiple intermediate frames. In Problems 3-30 through 3-53, a set of potentially useful reference frames has been provided. It may not be necessary to use all the reference frames provided. Alternatively, you may choose any other reference frames you prefer, but they must be *clearly* defined.

Problem 3-30: A ride at an amusement park consists of four symmetrically located seats driven to rotate about the vertical A axis at a constant rate of 10 rev/min, with respect to the supporting arm AB which is 6 m long. The vertical B axis is driven by another motor at a constant rate of 6 rev/min, with respect

to ground. All four seats are 2 m from the A axis, and the A and B axes are parallel, as sketched in Figure P3-30. The mass of each passenger is 75 kg. Note that gravity acts. Find the velocities and accelerations of all four passengers at the instant shown.

The reference frames defined in Figure P3-30 are as follows: $BXYZ$ is fixed to ground; $Bx_1y_1z_1$ is fixed to the arm AB; and $Ax_2y_2z_2$ is fixed to the frame that carries the seats.

Problem 3-31: A robot built by Chris·Dana·Gina, Inc. for handling rare gems is moving in the plane of the sketch with velocity v_0 and acceleration a_0, both with respect to the showroom floor, as sketched in Figure P3-31. Link AB is rotating at ω_1, with respect to the floor, and link BC is rotating at ω_2 and $\dot{\omega}_2$, both with respect to link AB. A delicate gem of mass m is being transported; the gem's geometric dimensions are negligible compared with l and l_0. Note that gravity acts. Find the velocity and acceleration of the gem.

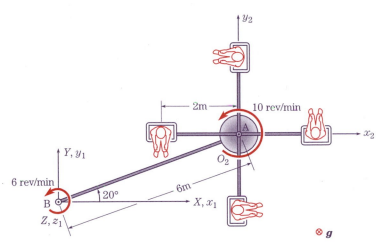

FIGURE P3-30

The reference frames defined in Figure P3-31 are as follows: $AXYZ$ is fixed to ground; $Ax_1y_1z_1$ is fixed in link AB; and $Bx_2y_2z_2$ is fixed in link BC.

Problem 3-32: A carnival ride consists of two platforms as sketched in Figure P3-32. The smaller platform of radius r rotates at $\boldsymbol{\omega}_2$ with respect to a larger platform of radius R which rotates at $\boldsymbol{\omega}_1$ with respect to ground. A rider P of mass m is located along the X axis, at the instant shown. Note that gravity acts. Find the velocity and acceleration of rider P.

The reference frames defined in Figure P3-32 are as follows: $OXYZ$ is fixed to ground; $Ox_1y_1z_1$ is fixed to the larger platform; and $ox_2y_2z_2$ is fixed to the smaller platform.

Problem 3-33: A carnival ride consists of two platforms of radii r and R rotating at angular velocities $\boldsymbol{\omega}_1$ and $\boldsymbol{\omega}_2$, respectively, as sketched in Figure P3-32.

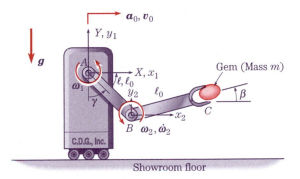

FIGURE P3-31

The platforms are driven by independent sets of gears fixed to ground and thus the angular velocities, $\boldsymbol{\omega}_1$ and $\boldsymbol{\omega}_2$, are both defined with respect to ground. A rider P of mass m is located along the X axis, at the instant shown. Note that gravity acts. Find the velocity and acceleration of rider P.

The reference frames defined in Figure P3-32 are as follows: $OXYZ$ is fixed to ground; $Ox_1y_1z_1$ is fixed to the larger platform; and $ox_2y_2z_2$ is fixed to the smaller platform.

Problem 3-34: A large disk rotates at a constant angular velocity Ω, with respect to ground, about a central shaft through O. A small disk rotates about Q at a constant angular velocity ω with respect to the large disk. A point P is located on the perimeter of the small disk such that line PQ is normal to line OQ, as sketched in Figure P3-34. Find the relation between ω and Ω for which the acceleration of point P would be parallel to line OQ.

The reference frames defined in Figure P3-34 are as follows: $OXYZ$ is fixed to ground; $Ox_1y_1z_1$ is fixed to the large disk; and $Qx_2y_2z_2$ is fixed to the small disk.

Problem 3-35: A ride at an amusement park is sketched in Figure P3-35. The cockpit containing two passengers rotates at an angular velocity $\omega_1 = 3$ rad/s, relative to the main arm, and the main arm rotates at an angular velocity $\omega_2 = 1$ rad/s relative to ground. Point A, which corresponds to an eye of a passenger, is located 0.8 m off the centerline of the

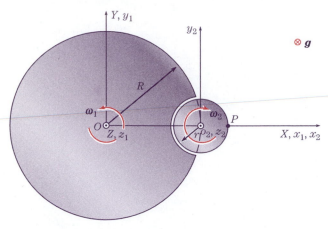

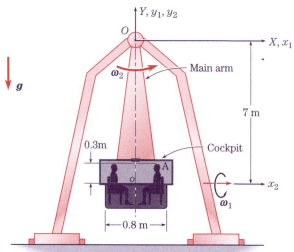

FIGURE P3-32

main arm and 0.3 m above the cockpit's rotating axis, which is also 7 m from the main arm's rotating axis. Note that gravity acts. For the instant shown, find the velocity and acceleration of point A.

The reference frames defined in Figure P3-35 are as follows: $OXYZ$ is fixed to ground; $Ox_1y_1z_1$ is fixed to the main arm; and $ox_2y_2z_2$ is fixed to the cockpit.

Problem 3-36: Curtis Lee pitches a curve ball toward home plate with velocity 120 ft/sec. The pitch curves at a radius of curvature of 2500 ft if viewed from directly above the field, as sketched in Figure P3-36a. In addition, he puts a spin of 20 rev/sec on the ball about the vertical direction, with respect to ground. Note that gravity acts. Find the acceleration of a pine tar glob, which weighs 0.2 oz and is attached to the ball as sketched in Figure P3-36b.

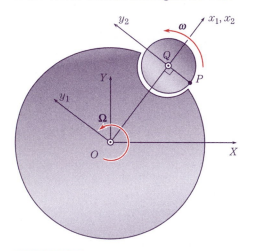

FIGURE P3-35

The reference frames defined in Figure P3-36 are as follows: $OXYZ$ is fixed to ground; $Ox_1y_1z_1$ is defined such that the z_1 axis coincides with the Z axis while the ball remains on the x_1 axis at all times; and the origin of $ox_2y_2z_2$ is at the center of the ball with the x_2 axis along the x_1 axis while the y_2 axis and z_2 axis are parallel to the y_1 axis and z_1 axis, respectively.

Problem 3-37: A pendulum of length $\ell = 1$ m is attached to a moving cart as sketched in Figure P3-37. When the pendulum is vertical, it has an angular velocity $\omega_1 = 2$ rad/s and an angular acceleration $\dot{\omega}_1 = 5$ rad/s^2, and the cart moves at a velocity $v_0 = 0.2$ m/s and an acceleration $a_0 = 0.1$ m/s^2, all

FIGURE P3-34

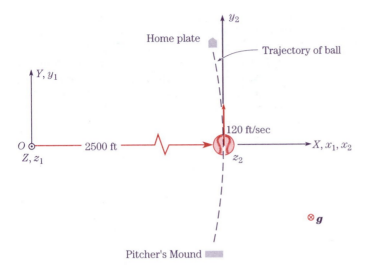

(a) View from above field.

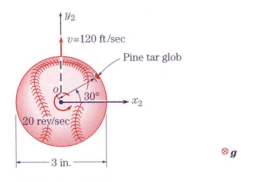

(b) Close-up of ball.

FIGURE P3-36

with respect to ground. A point P, which is 5 cm from the center of the bob C, is moving radically toward the center of the bob at a velocity $v_1 = -0.1$ m/s and an acceleration $a_1 = -0.2$ m/s^2, both with respect to the bob. Note that gravity acts. Find the velocity and acceleration of point P.

The reference frames defined in Figure P3-37 are as follows: $OXYZ$ is fixed to ground; $Ox_1y_1z_1$ is fixed to the translating cart and does not rotate; and $Ox_2y_2z_2$ is fixed to the pendulum.

Problem 3-38: A wheel rotates at an angular velocity $\omega_2 = 5$ rad/s with respect to a platform that rotates at an angular velocity $\omega_1 = 10$ rad/s with respect to ground. A bead of mass 20 g moves along a spoke of the wheel at a velocity of 6 m/s and an acceleration of 3 m/s^2, both downward along and with respect to

the spoke. At the instant sketched in Figure P3-38, the spoke is in the Z direction and the bead is located at 0.25 m from the shaft axis of the wheel. Note that gravity acts. Find the velocity and acceleration of the bead at the instant shown.

The reference frames defined in Figure P3-38 are as follows: $OXYZ$ is fixed to ground; $Ox_1y_1z_1$ is fixed to the platform; and $ox_2y_2z_2$ is fixed to the wheel.

Problem 3-39: A duck is flying at velocity $\boldsymbol{v}_b$ and acceleration $\boldsymbol{a}_b$ (both in negative Y direction), and diving at angular velocity $\boldsymbol{\omega}_1$ (about X direction) and angular acceleration $\dot{\boldsymbol{\omega}}_1$ (about negative X direction), all with respect to ground. In order to maintain control, the duck is rotating its wings, as if they were straight rigid links, at angular velocity $\boldsymbol{\omega}_2$ and angular acceleration $\dot{\boldsymbol{\omega}}_2$ (about Y direction), both with

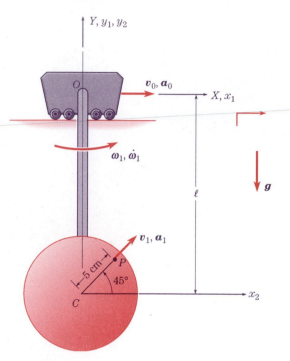

FIGURE P3-37

respect to its body, as sketched in Figure P3-39. Note that gravity acts. Find the velocity and acceleration of the duck's right wing tip, which is located by vector **s**, at the instant shown.

The reference frames defined in Figure P3-39 are as follows: $OXYZ$ is fixed to ground; $Ox_1y_1z_1$ is fixed to the duck's body; and $Ox_2y_2z_2$ is fixed to the duck's right wing.

Problem 3-40: A space shuttle is maneuvering a payload M of mass 100 kg into position, as sketched in Figure P3-40. At the instant shown, the base A is

rotating about the Y axis at an angular velocity $\omega_2 = 0.1$ rad/s and an angular acceleration $\dot{\omega}_2 = 0.2$ rad/s^2, both with respect to the shuttle. The shuttle is moving at a velocity of 2000 m/s and an acceleration of 1 m/s^2 toward the viewer (that is, in the Z direction), both with respect to a fixed reference frame $AXYZ$. At the instant shown, $\alpha = 45°$ and $\beta = 30°$. Arm BC is rotating at an angular velocity $\omega_1 = 0.2$ rad/s and an angular acceleration $\dot{\omega}_1 = 0.8$ rad/s^2, both with respect to arm AB. The lengths of arms AB and BC are 10 m and 3 m, respectively. Find the velocity and acceleration of the payload M.

The reference frames defined in Figure P3-40 are as follows: $AXYZ$ is fixed to "ground" or "fixed space"; $Ax_1y_1z_1$ is fixed to arm AB; and $Bx_2y_2z_2$ is fixed to arm BC.

Problem 3-41: An astronaut A weighs W and is on a platform that is rotating at an angular velocity ω_1 with respect to ground. Also, he is walking on a table which, at the instant shown in Figure P3-41, is in the plane of the platform. His velocity and acceleration are v_0 and a_0, respectively, both with respect to the rotating table and in a direction that forms an angle θ with respect to the rotating axis of the platform. The rotating table has an angular velocity ω_2 with respect to the platform. At the instant shown, the astronaut is a radial distance b from the axis of the rotation of the table. Find the velocity and acceleration which the astronaut is experiencing at the instant shown.

The reference frames defined in Figure P3-41 are as follows: $OXYZ$ is fixed to ground; $Ox_1y_1z_1$ is fixed to the platform; and $Ox_2y_2z_2$ is fixed to the table.

Problem 3-42: A conveyer cart moves down an inclined plane, which is inclined at an angle of 30° with

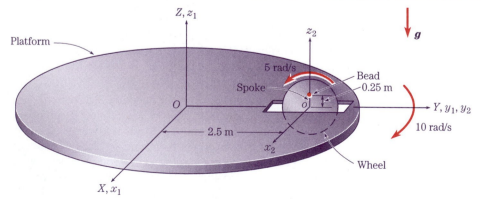

FIGURE P3-38

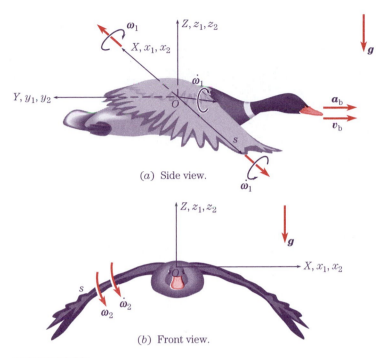

(a) Side view.

(b) Front view.

FIGURE P3-39

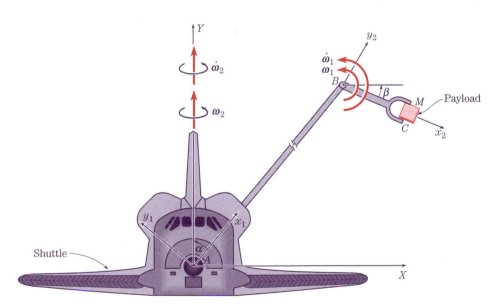

FIGURE P3-40

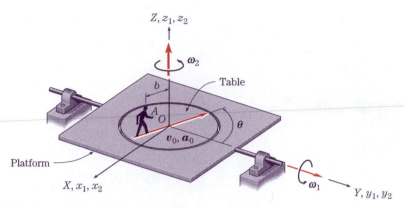

FIGURE P3-41

respect to the horizontal, at a speed $v_0 = 5$ ft/sec and an acceleration $a_0 = 0.5$ ft/sec^2. The cart moves in the YZ plane, which is the plane of the sketch. A 10 ft long shaft AB swings in the YZ plane at an angular velocity $\omega_1 = 1$ rad/sec. At the instant shown in Figure P3-42, shaft AB is vertical. A lower platform of radius 1 ft, which is perpendicular to shaft AB and contains a particle of weight 10 lb fixed at point D, has an angular velocity $\omega_2 = 0.5$ rad/sec with respect to the shaft AB. Find the velocity and acceleration of point D at the instant shown. Note that gravity acts, and that point D is located on the perimeter of the platform and, at the instant shown, in the YZ plane.

The reference frames defined in Figure P3-42 are as follows: $AXYZ$ is fixed to ground; $Ax_1y_1z_1$ is fixed to shaft AB; and $Bx_2y_2z_2$ is fixed to the platform that contains point D.

Problem 3-43: A robot arm is manipulating a delicate workpiece P of mass m along an assembly line, as sketched in Figure P3-43. At the instant shown, all members are in the plane of the sketch, that is, the XY plane. The gripper, which is pin-jointed at point B, is rotating about the z_2 axis at an angular velocity ω_1 and an angular acceleration $\dot{\omega}_1$, both with respect to arm AB. The centerline of the gripper forms an angle θ with respect to the X axis. The arm AB is rotat-

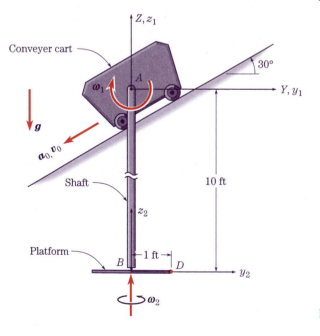

FIGURE P3-42

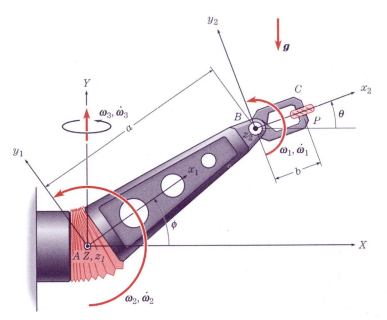

FIGURE P3-43

ing about the Z axis at an angular velocity ω_2 and an angular acceleration $\dot{\omega}_2$, both with respect to ground. The arm AB is also rotating about the Y axis at an angular velocity ω_3 and an angular acceleration $\dot{\omega}_3$, both with respect to ground. Arm AB forms an angle ϕ with respect to the X axis. The lengths of arm AB and gripper BC are a and b, respectively. Note that gravity acts. Find the velocity and acceleration of the workpiece P at the instant shown.

The reference frames defined in Figure P3-43 are as follows: $AXYZ$ is fixed to ground; $Ax_1y_1z_1$ is fixed to arm AB; and $Bx_2y_2z_2$ is fixed to gripper BC.

Problem 3-44: A carnival ride is modeled as two huge disks, a larger one of radius d and a smaller one of radius a. At the instant sketched in Figure P3-44, a rider whose weight is W and whose size is negligible is located at point B as shown. The angular velocity components of the smaller disk are ω_1 and ω_2, both with respect to the larger disk; and the angular velocity and angular acceleration of the larger disk are ω_0 and $\dot{\omega}_0$, respectively, both with respect to ground. Note that ω_0, $\dot{\omega}_0$, and ω_1 are defined in the Z direction, and ω_2 is defined in the Y direction. Note that gravity acts. Find the velocity and acceleration of the rider at B at the instant shown.

The reference frames defined in Figure P3-44 are as follows: $CXYZ$ is fixed to ground; $Cx_1y_1z_1$ is fixed to the larger disk; and $o_2x_2y_2z_2$ is fixed to the smaller disk.

Problem 3-45: An airliner is making an approach to an airport, as sketched in Figure P3-45. At the instant shown, the aircraft is circling in a horizontal plane and at a radius of 2000 m at an angular velocity $\omega_1 = 1$ rev/3 min while the aircraft is descending at $\omega_3 = 0.5$ rev/min, both with respect to ground. To accomplish the turning, the aircraft is also banking about its longitudinal axis at $\omega_2 = 2$ rev/min with respect to ground. A passenger P is walking down a side aisle at a distance $d = 2$ m from the centerline toward the back of the aircraft at a constant speed of 0.75 m/s with respect to the plane. Note that gravity acts. Find the velocity and acceleration of this passenger.

The reference frames defined in Figure P3-45 are as follows: $OXYZ$ is fixed to ground; $Ox_1y_1z_1$ rotates at ω_1 with respect to $OXYZ$ so its x_1 axis always passes through the point C about which the rotations of the aircraft occur; and $Cx_2y_2z_2$ is fixed to the aircraft.

Problem 3-46: An astronaut training system consists of a telescoping tube and a rotating platform, as sketched in Figure P3-46. One end of the tube is connected to ground where it is forced to rotate at

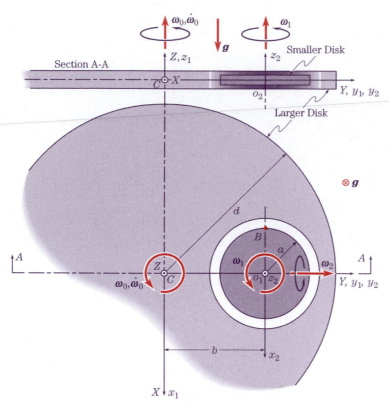

FIGURE P3-44

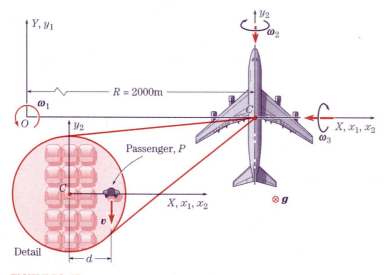

FIGURE P3-45

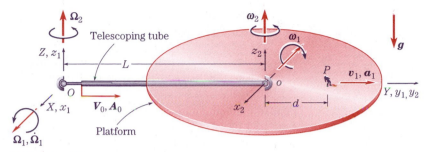

FIGURE P3-46

angular velocities Ω_1 and Ω_2 and angular acceleration $\dot{\Omega}_1$, all with respect to ground. The other end of the tube is connected to the platform, which rotates at ω_1 and ω_2, both with respect to ground, as sketched. At the instant shown, an astronaut P, weight W and located along OY a distance d from point o as sketched, is moving at velocity v_1 and acceleration a_1, both with respect to the platform. Further, the tube is being telescoped at velocity V_0 and acceleration A_0, both with respect to ground. Note that Ω_1 and $\dot{\Omega}_1$ are in the X direction; Ω_2 is in the Z direction; Ω_1 is in the negative x_2 direction; and ω_2 is in the z_2 direction. Note that gravity acts. Find the velocity and acceleration of the astronaut at the instant shown.

The reference frames defined in Figure P3-46 are as follows: $OXYZ$ is fixed to ground; $Ox_1y_1z_1$ is fixed to the telescoping tube; and $ox_2y_2z_2$ is fixed to the platform.

Problem 3-47: A robot welder is sketched in Figure P3-47. Welding is accomplished by moving the welding gun BC so that the point C is located properly in relation to the parts to be welded. However, the quality of the weld also depends on the velocity and acceleration of point C with respect to the parts to be welded. At the instant shown, the robot arm AB is rotating at ω_1, ω_2, and $\dot{\omega}_2$, all with respect to ground. The arm AB contains an actuator that is being telescoped at a constant speed of v_0 with respect to ground. The instantaneous length of arm AB is L_1. Also, arm AB makes an angle θ with respect to the horizontal as shown. The welding gun BC is rotating at ω_3 with respect to the robot arm AB. The length of BC is L_2. The welding gun is horizontal as the instant shown. Both the arm AB and the welding gun BC are in the plane of the sketch. Note that gravity acts. Find the velocity and acceleration of point C relative

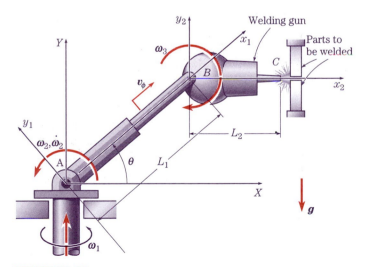

FIGURE P3-47

to the parts to be welded (ground) at the instant shown.

The reference frames defined in Figure P3-47 are as follows: $AXYZ$ is fixed to ground; $Ax_1y_1z_1$ is fixed to arm AB; and $Bx_2y_2z_2$ is fixed to gun BC.

Problem 3-48: A truck, which is designed to make concrete on its way to the construction site by mixing water, sand, cement, and pebbles, is being driven along a straight horizontal road at linear velocity v_1 and linear acceleration a_1. Because the road surface is not well maintained, the truck is experiencing rolling and pitching at angular velocities ω_1 and ω_2, respectively, both with respect to ground, as sketched in Figure P3-48. The concrete container of conical shape is rotating at angular velocity ω_3 with respect to the truck. Find the velocity and acceleration of point P on the container at the instant shown. Note that gravity acts, and that, at the instant shown, P is located at the lowest, rearmost point on the container.

The reference frames defined in Figure P3-48 are as follows: $OXYZ$ is fixed to ground; $Ox_1y_1z_1$ is fixed to the truck; and $o_2x_2y_2z_2$ is fixed to the container.

Problem 3-49: A robot, sketched in Figure P3-49, is moving a workpiece of mass m from the site of process 1 to the site of process 2. The column is moving with respect to the fixed base at linear velocity v_0 and angular velocity ω_1. The arm is rotating at ω_2 with respect to the column. The gripper, which holds the workpiece, is moving at zero velocity and an acceleration a_0 and at angular velocity ω_3 and angular acceleration α_3, all with respect to the arm. Note that

gravity acts. Find the velocity and acceleration of the workpiece at the instant shown.

The reference frames defined in Figure P3-49 are as follows: $OXYZ$ is fixed to ground; $Ox_1y_1z_1$ is fixed to the column; $Ox_2y_2z_2$ is fixed to the arm; and $Ox_3y_3z_3$ is fixed to the gripper.

Problem 3-50: An inverted pendulum is attached to a cart that is accelerating on the horizontal surface at a_0 as sketched in Figure P3-50. When the pendulum is vertical, it has an angular velocity ω_2 and angular acceleration $\dot{\omega}_2$, both about the Z axis and both with respect to ground. At the instant shown, a body P of mass m slides along rod AB with an outward radial velocity v_0 with respect to rod AB. The rod is one of many radial rods (others are not sketched) rigidly connected to a ring of 3 in. radius, and body P is 2 inches radically from the center A of the ring. At the instant shown, rod AB is in the plane of the sketch and forms an angle of $\pi/3$ with respect to the horizontal, the ring is rotating at ω_0 and ω_1, both with respect to the pendulum. Note that gravity acts. Find the velocity and acceleration of body P.

The reference frames defined in Figure P3-50 are as follows: $OXYZ$ is fixed to ground; $Ox_1y_1z_1$ is fixed to the translating cart and does not rotate; $Ax_2y_2z_2$ is fixed to the pendulum; and $Ax_3y_3z_3$ is fixed to the ring.

Problem 3-51: In the hi-tech filming of star ship flight sequences in science fiction movies, the camera is often moved by computer control around a stationary scaled model of the star ship. (The star ship model is filmed against a blue color screen. Later, by

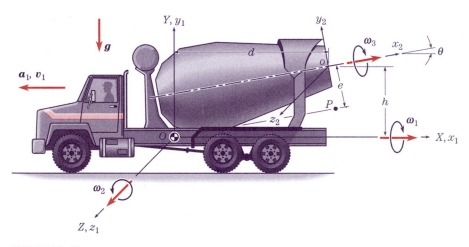

FIGURE P3-48

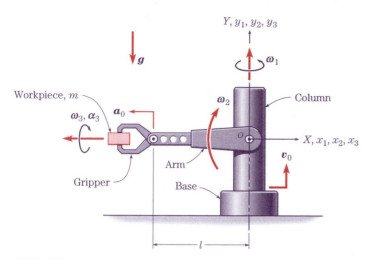

FIGURE P3-49

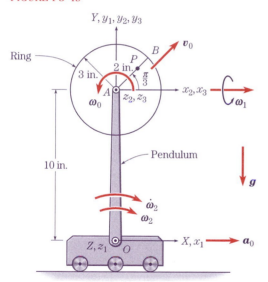

FIGURE P3-50

photographic and/or computer graphic techniques, other film sequences are superimposed according to whether they belong to the foreground or the background. Such an effort can produce fascinating images such as a star ship navigating in a meteor field.)

The mechanism sketched in Figure P3-51 shows one such filming sequence. The star ship model E is fixed in space. The camera D of mass m is rigidly mounted on link CD. Link CD rotates at angular velocity ω_3 with respect to link BC. Link BC rotates at angular velocity ω_2 with respect to link AB. Link AB

rotates at angular velocity ω_1 with respect to ground. The magnitudes of the angular velocities ω_1, ω_2, and ω_3 are all 0.2 rev/s. Furthermore, links AB, BC, and CD are extensible and each is being telescoped at the same relative speed of 0.1 m/s; that is, link AB relative to ground, link BC relative to link AB, and link CD relative to link BC. At the instant shown, links AB, BC, and CD are all of length 1 m. Note that gravity acts. In order to design the "motion" of star ship recorded on film, it is necessary to compute the motion of the star ship with respect to the camera. Determine the velocity and acceleration of the star ship E *with respect to the camera D* at the instant shown.

The reference frames defined in Figure P3-51 are as follows: $AXYZ$ is fixed to ground; $Ax_1y_1z_1$ is fixed to the link AB; $Bx_2y_2z_2$ is fixed to the link BC; and $Cx_3y_3z_3$ is fixed to the link CD. $Dx_4y_4z_4$ is fixed to the camera.

Problem 3-52: Find the motion of point P in Figure 3-28 with respect to the "fixed" reference frame $OXYZ$, by using the strategy of "moving" from frame $o_2x_2y_2z_2$ "inward" toward frame $OXYZ$, for the case in which both ω_1 and ω_2 are defined with respect to ground.

Problem 3-53: Find the motion of point P in Figure 3-28 with respect to the "fixed" reference frame $OXYZ$, by using the strategy of "moving" from frame $OXYZ$ "outward" toward frame $o_2x_2y_2z_2$, for the case in which ω_1 is defined with respect to ground and ω_2 is defined with respect to frame $o_1x_1y_1z_1$.

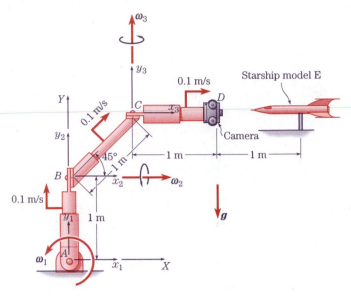

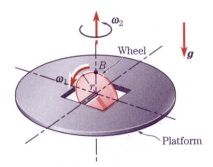

FIGURE P3-51

● In Problems 3-54 through 3-60, reference frames are *not* provided. Choose any reference frames deemed convenient and define them *clearly*.

Problem 3-54: Indicate in Figure P3-54 the location of the point on the surface of the Earth that a particle has the *maximum* magnitudes of velocity and acceleration caused by the rotation of the Earth about its own axis and about the axis of the Sun. Find the velocity and acceleration of a particle at that location relative to a fixed reference frame at the center of the Sun, $OXYZ$.

It is assumed that the equators of the Sun and the Earth are in the plane of the paper. The radius of the Earth's orbit may be assumed to be 1.50×10^{11} m and the radius of the Earth may be assumed to be

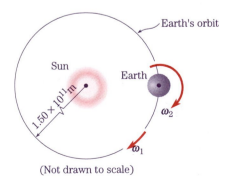

FIGURE P3-54

FIGURE P3-55

6.365×10^6 m. The constant angular velocity ω_1, relative to the $OXYZ$ reference frame, can be calculated because the Earth rotates about the Sun once a year; that is, once every 365.25 days. The constant angular velocity ω_2, relative to the Earth's axis, can also be calculated because the Earth rotates about its own axis once a day; that is, once every 24 hours.

Problem 3-55: A wheel of radius r is rotating at ω_1 with respect to a platform. The platform is rotating at ω_2 with respect to ground, as sketched in Figure P3-55. Find the velocity and acceleration of point B at the instant shown. Note that gravity acts. Let B be a tiny bug of mass m which lies along a vertical radius of the wheel, at the instant shown.

Problem 3-56: During takeoff, an airplane is pitching at an angle θ above the horizontal ground at an angular velocity $\dot{\theta}$ and an angular acceleration $\ddot{\theta}$, both

FIGURE P3-56

with respect to ground and about an axis perpendicular to the body of the aircraft through its ground contact point A. The velocity and acceleration of the plane are v_0 and a_0, respectively, both with respect to ground and in the horizontal direction, as sketched in Figure P3-56. A flight attendant, who weighs W and is located at point P in the airplane, is walking down the center aisle of the airplane at a speed v_1 and an acceleration a_1, both with respect to the airplane and along the aisle; that is, along the direction that forms an angle θ with respect to the horizontal. Point P is located a distance h above the ground and a distance ℓ (that is, PB) along the center aisle such that the angle PBA is a right angle. Note that gravity acts. Find the velocity and acceleration of this flight attendant.

Problem 3-57: A stationary truck is carrying a worker in its cockpit to replace street lights. At the instant sketched in Figure P3-57, all members are in the plane of the sketch, and the base A is rotating about a vertical axis at an angular velocity $\omega_1 = 0.25$ rad/s and an angular acceleration $\dot{\omega}_1 = 0.15$ rad/s^2, both with respect to ground. The main arm AB forms

an angle of 40° with respect to the vertical. The forearm BC is rotating at an angular velocity $\omega_2 = 0.20$ rad/s and an angular acceleration $\dot{\omega}_2 = 0.35$ rad/s^2, both with respect to the main arm AB, and about an axis that is orthogonal to the plane of the sketch, and forms an angle of 50° with respect to the vertical. The lengths of the main arm and the forearm are 4.5 m and 2.5 m, respectively. A worker, weighing 200 lb, is standing in the cockpit, and can be modeled as being at point C. Note that gravity acts. Find the velocity and acceleration of the worker at the instant shown.

Problem 3-58: A car travels at a constant speed v_0 along a horizontal circular track of radius R. Figure P3-58a shows a frontal view of the car windshield.

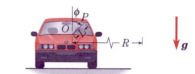

(a) View of windshield from in front of car.

(b) Sideview.

FIGURE P3-58

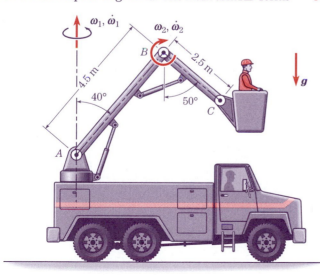

FIGURE P3-57

The wiper blade has length L and oscillates according to $\phi = \phi_0 \sin \Omega t$. As sketched in Figure P3-58b, the windshield is inclined at an angle α with respect to the vertical. Note that gravity acts. Find an expression for the acceleration of the tip of the wiper point P at an arbitrary instant of time t. Assume that R is much greater than all other lengths in this problem; thus, as indicated in Figure P3-58a, you may assume that the base of the blade O travels in a circle of radius R.

Problem 3-59: A clown pushes a cart E in a circus act as sketched in Figure P3-59. The cart is moving at a velocity v_1 and an acceleration a_1, both with respect to ground, as sketched. A rotating shaft DE is mounted on the cart and rotates at angular velocity ω_1 with respect to the cart. The bar CA of length ℓ is hinged at point C of the rotating shaft, which is a distance d from the rotating axis, and is swinging at angular velocity ω_2 and angular acceleration $\dot{\omega}_2$, both with respect to the shaft DE. Bar CA forms an angle θ with respect to the vertical. The acrobat ① is strapped down tightly at point A on bar CA. Acrobat ① pulls acrobat ② inward with a velocity v_2 radically along bar CA, with respect to acrobat ①. Note that gravity acts. Find the velocity and acceleration of acrobat ②. Length ℓ is much greater than heights of acrobats.

Problem 3-60: A centrifuge is often used to simulate flight conditions for training pilots and astronauts. In the simulation sketched in Figure P3-60, the arm of the centrifuge was rotating at an angular velocity ω_2 and an angular acceleration $\dot{\omega}_2$, both with respect to ground. The cockpit had angular velocity ω_1 and angular acceleration of $\dot{\omega}_1$, both with respect to the arm. In this particular test, the astronaut was not strapped-in well and at the instant shown was sliding at ω_3 and $\dot{\omega}_3$ with respect to the seat. In order to assess the situation, find the velocity and acceleration of the astronaut's head with respect to ground. Note that the profile of the seat is a portion of a spherical surface that is centered at point C, through which the arm passes. The bottom of the seat A is a distance h below the arm and a distance L from the axis of rotation. At the instant shown, the astronaut's head is a distance b above the bottom of the seat and a horizontal distance d from point C. Also note that gravity acts.

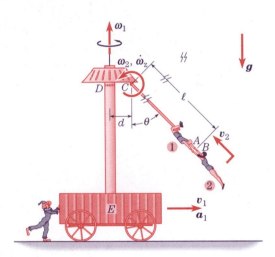

FIGURE P3-59

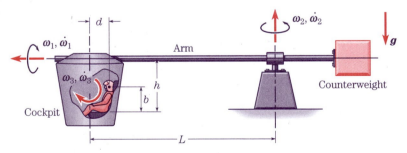

FIGURE P3-60

Sonnets from the Portuguese (Especially XII & XLIII)
  —Elizabeth Barrett Browning

CHAPTER FOUR

MOMENTUM FORMULATION FOR SYSTEMS OF PARTICLES

4-1   INTRODUCTION

In this chapter we present the *momentum approach* for the formulation of the equations of motion for dynamic systems consisting of bodies that may be modeled as particles. This approach, which is also called *vectorial mechanics* as well as the *direct approach,* takes as its fundamental quantities *forces* and *momenta.* We shall emphasize the *foundations* of the underlying physics and the *structure* of the requirements for the formulation of the equations of motion.

By virtue of his *Principia,* Isaac Newton must be considered to be the premier author of the direct approach to dynamics. The *Principia* is noteworthy for its deductive mathematical character, its technique of working from the general to the particular, and its unification of terrestrial and celestial mechanics, notwithstanding the earlier ruminations of Kepler and Descartes regarding such unification.

By the standards of rigor of his day (but not by modern standards of rigor), Newton carefully defined concepts such as force and mass. Building upon the work primarily of Galilei and Kepler, but also that of many others, Newton stated his famous *axioms,* or *principles* or *laws of motion,* and then proceeded to solve for the first time in history major problems in celestial and terrestrial mechanics. Even though Newton almost certainly used the calculus in obtaining some of his major results, no such formulations appear in the *Principia.* Indeed, the *Principia* is so ponderous in its *geometrical* constructions that there are few people who have carefully read it in its entirety.

Here we note that a *fundamental law* or *principle* of science as used above and throughout denotes a statement, quantitative or qualitative, about a governing relationship between two or more definable quantities. Such statements are given the status of law or principle after a long history of failure to disprove them. A fundamental law is a generally sweeping statement based on empirical observations. A fundamental law or principle can be neither derived nor deduced from purely theoretical considerations or from other laws. A law can never be proved theoretically or experimentally; it can only be disproved, in which case, it is no longer a law.

In our opinion, the current techniques for formulating and solving mechanics problems are due more to Leonhard Euler than any other single individual in history. Furthermore, considering his achievements in organizing, unifying, applying, and creating mathematical concepts, calculations, and theories—most of such achievements accomplished in the context of mechanics—the contributions of Euler place him among the preeminent mathematicians in history, irrespective of how elite one chooses to forge the guild. Although the origin of the concept of the *free body diagram* is obscure, it was Euler who brought it into common explicit use in mechanics. For this reason, it is sometimes called *Euler's cut principle.* In his monumental work, *Mechanica sive Motus Scientia Analytice Exposita* (Mechanics, or the Science of Motion Set Forth Analytically), generally simply called *Mechanica,* which he published in 1736, Euler abandoned the geometrical methods of Newton and adopted the more powerful mathematics of the calculus and the differential notation initiated by Leibniz. (It is interesting to imagine what mechanics would look like today if Euler had been English, and thus likely politically impelled to adopt Newton's notation.) In particular, Euler illustrated the *direct approach* by deriving the differential equations of motion for a particle, and then integrated those equations to obtain the particle's motion. In this way, Euler laid the path for the subsequent contributions of Lagrange, as well as his own, in the development of the *indirect approach.*

Furthermore, although the relations $F_x = ma_x$, $F_y = ma_y$, and $F_z = ma_z$ (where F is force, m is mass, and a is acceleration) are generally called Newton's equations, it was Euler who in 1752 *first* articulated these equations as general expressions, applicable to every part of every solid or fluid system, discrete or continuous. To suggest today that these equations represent a trivial consequence of Newton's second law is to ignore the decades of research by the world's scholars who did not recognize them as a general principle. Also, as we shall emphasize in Chapter 6, it was Euler who played the major role in identifying the *angular momentum principle* for extended bodies as an independent law of mechanics.

Finally, as two of history's greatest mathematicians and mechanicians, both Newton and Euler may be cited as prime examples of the interplay between mathematics and mechanics, displayed by so many scientists before and after them, thus obviating any further imperative to demonstrate the historical union between mathematics and mechanics.

4-2   THE FUNDAMENTAL PHYSICS

4-2.1   Newton's Laws of Motion

The necessary physical principles for a quantitative study of the dynamics of mechanical systems were first stated by Isaac Newton in his *Principia* of 1687. Newton's laws of motion are statements of *empirical fact,* and are the cumulative products of millennia of speculation, theory, and experimental observation of both terrestrial and celestial motion. Newton's laws form the basis for the direct approach to the formulation of the equations of motion for particles. A typical modern statement of these laws is as follows:

First Law:

A particle remains at rest or continues to move in a straight line with constant velocity if there is no resultant force acting on it.

Second Law:

A particle acted upon by a resultant force moves in such a manner that the time rate of change of its linear momentum is equal to the force.

Third Law:

Forces result from the interaction of particles and such forces between two particles are equal in magnitude, opposite in direction, and collinear.

The above statements of Newton's laws are not verbatim but are given in terms of a modern interpretation of their meanings. Even so, there are important elucidations, limitations, and ambiguities regarding these laws that have been explored by many of history's greatest scientific thinkers. Newton left much to be done by succeeding generations. During the past three centuries, much has been written about these laws. Indeed, it is likely that no set of physical laws has ever received more comment or analysis. Thus, it is equally likely that much that has been written is repetitive, and that some of it is wrong.

It is important to keep in mind that these laws were written as the governing principles for particles, the concept of which we shall discuss in Subsection 4-2.2. In writing the *Principia,* Newton's primary interest in mechanics lay in the motions of celestial bodies. He treated such bodies as particles, their movements due only to universal gravitation and their own momenta. In order to justify the modeling of such immense bodies as the Moon, Earth, and Sun as particles, he proved that the attractive force between two homogeneous spheres (which was, of course, a modeling decision since these bodies are neither homogeneous nor spheres) is directed along the line connecting their centers and is independent of their diameters, a result that is rederived in many elementary college physics textbooks.

In explaining his first law, Newton wrote: "A top, whose parts by their cohesion are continually drawn aside from rectilinear motions, does not stop spinning, except insofar as it is slowed by the air." Thus, it could be reasoned that Newton was in some ways concerned with extended or rigid bodies and angular momentum, neither of which is true. The lack of such analyses in the *Principia* supports this assertion.

It has been said by various authors that the first law lacks content or is trivial or is meaningless. Well, it is true that in a reference frame where the second law holds, the first law is a consequence of it for zero force. But, first, to dismiss the first law in this way is to ignore its historical significance: The idea that velocity does not have to be sustained by any agent or force, but that a *change in velocity* requires a cause, eluded mankind for thousands of years. Second, it is a lack of understanding of the first law that is at the core of the concoction called *centrifugal force.* This fictitious and fallacious entity is a fabrication to explain the tendency of a rotating body to fly away from the axis of rotation. A straightforward recourse to the first law in such cases will reveal neither the existence of nor the need for inventing centrifugal forces.[1] Third, and of most practical consequence, the question remains as to how such a reference frame in which the second law holds is to be established. The reference frame with respect to which the first law's "constant velocity" is to be measured establishes the concept of an *inertial reference frame.* The first law asserts that such reference frames exist. Thus, the first law identifies the reference frames in which the second law can be expected to hold.

4-2.2   A Particle

A *particle*—sometimes called a *mass-point*—is a point endowed with fixed mass m. Because no physical body can be a point, we immediately see that a particle is an idealized concept. It is a model. Also, a particle preserves its identity, as its mass does not change with time or space. The velocity of a particle is uniquely defined by the velocity of the point that locates it. Further, all of the applied forces on a particle act at a single point—the location of the particle—and can be summed vectorially to give a single resultant applied force.

As a practical matter, a body may be modeled as a particle whenever its dimensions are irrelevant to the analysis of its motion or the forces acting on it. In dynamics the question of whether a body may be reasonably modeled as a particle often reduces to the significance of rotational causes or effects (loading, motion, or energy) in comparison with translational

[1]See Problems 4-18 and 4-19.

causes or effects (loading, motion, or energy). Specifically, when the rotational causes and effects are negligible in comparison with the translational causes and effects, the body may be reasonably modeled as a particle.

In the context of classical mechanics, a particle, by virtue of its infinitesimal size, cannot be said to deform or rotate or sustain torques about an axis through it. Thus, the application of Newton's laws to a body modeled as a particle can tell us nothing about the rotation or deformation of the body. To obtain information of this type, it is necessary to model bodies or systems as interconnected swarms of particles or continuous media, rigid or deformable, held together by internal forces. Particles are *not* meant to be thought of as atoms, but as elements of real bodies or as real bodies themselves, small enough to allow the application of Newton's laws with satisfactory accuracy.

4-2.3   Linear Momentum and Force

If relativistic effects are not modeled, the linear momentum $\boldsymbol{p}$ of a particle is defined as the product of the mass and the velocity of the particle, namely,

$$\boldsymbol{p} = m\boldsymbol{v} \tag{4-1}$$

Equation (4-1) is the *constitutive relationship* governing the particle of mass m, where m is a positive constant. Common units of mass are the slug (lb-sec^2/ft) and kilogram (kg). Bodies whose velocity–momentum behavior is governed by Eq. (4-1) are called *newtonian particles*. The linear momentum is a vector quantity. *Mass* is the quantitative measure of the resistance of a body to be accelerated. Mass is also a measure of that property that produces gravitational attraction of a body. It is Albert Einstein's *principle of equivalence* that demonstrates the equality of the so-called *inertial* mass and the so-called *gravitational* mass. As the reader may recall from elementary physics, a synopsis of the principle of equivalence may be expressed as follows: In a closed system from which an outside reference frame cannot be consulted, it is not possible to distinguish between an acceleration due to gravity and an acceleration due to a resultant force that is equal to the gravitational force.

The resultant force $\boldsymbol{f}$ is then related to the particle's linear momentum by Newton's second law as

$$\boldsymbol{f} = \frac{d\boldsymbol{p}}{dt} \tag{4-2}$$

Equation (4-2) is the *force–dynamic relationship* governing the particle. Force is the vector interaction between bodies and is commonly classified as a *contact* force (such as due to spring or dashpot elements) or a *field* force (such as due to gravity or electromagnetism). If in any system of units, the entities length, time, and mass are taken as the base units, then the corresponding unit of force is determined in accordance with Newton's second law. Common units of force are the pound (lb) and newton (N). Whereas force can be defined in accordance with Eq. (4-2), more appropriately for statics, force can be defined in terms of its effect to produce deformation in a standard spring. Extensive discussions on the nature and concept of force are important in philosophical discourses of mechanics but cannot be gainfully explored here.

The analysis of many problems in dynamics can be achieved in the form of conservation theorems. In general, conservation theorems specify the conditions under which one or more quantities remain constant in time. Equation (4-2) provides the first of several such conservation theorems: *conservation of linear momentum of a particle*.

- If the resultant force on a particle is zero, then $\dfrac{d\boldsymbol{p}}{dt} = 0$, and thus the particle's linear momentum $\boldsymbol{p}$ is conserved; that is, $\boldsymbol{p}$ remains constant.

Being a vector, linear momentum can be conserved in one direction if the resultant force in that direction is zero, independently of all other directions.

Conservation of linear momentum is not a law but a theorem that is a direct consequence of Newton's second law. Like Newton's laws, this consequence has broad, but not universal applicability. (For example, newtonian mechanics is not valid for the interaction of charged particles moving at large fractions of the speed of light, or for quantum mechanics.) Therefore, as a consequence of Newton's laws themselves, the conservation of linear momentum must be subject to the verification of experiment, with the results providing the measure of their confirmation. Nevertheless, if Newton's laws hold, so does the conservation of linear momentum for free (i.e., zero resultant force) particles.

4-2.4   Inertial Reference Frames

Without a reference frame, neither motion nor dynamics makes sense. If you were the only object in the universe and you were floating freely through space, you could not tell whether you were moving or stationary, whether you were upside-down or downside-up (though these might be the least of your problems). Such concepts as "moving" or "upside-down" would be nonsensical without a reference by which to decide.

Experience has shown that, in the absence of relativistic effects, the mass of a particle and the force applied to a particle do not depend on the reference frame with respect to which they are measured. Yet, the value of $\dfrac{d\boldsymbol{p}}{dt}$ *does* depend on the reference frame from which the motion is observed. This dilemma is operationally resolved by declaring that there are only *some* reference frames in which $\boldsymbol{f} = \dfrac{d\boldsymbol{p}}{dt}$, and by defining these favored frames as *inertial reference frames*.

By virtue of this operational definition, the rate of change of the linear momentum $\dfrac{d\boldsymbol{p}}{dt}$ must be the same when a particle is observed from any inertial reference frame; thus, the linear momentum of a particle as seen from different inertial reference frames can differ at most by a constant. So, the velocity of a particle as seen from two inertial reference frames must also differ at most by a constant. Hence, any two inertial reference frames are at most *moving relative to each other with a constant velocity*. Thus, an inertial reference is a nonaccelerating, whether due to translation or rotation, reference frame.

An absolute inertial reference frame is an idealization. Such an idealization would require an absolutely "fixed" reference frame located at some gravitationally averaged point of all the matter in the universe—otherwise the unbalanced gravitational force would produce an acceleration. No such location is known. In this regard, our continued use of a "fixed" reference throughout Chapter 3 may not be fully justifiable. Nevertheless, if such questionable terminology assisted the reader in grasping the kinematic concepts, we feel vindicated in having made that choice.

In the analysis of a specific dynamics problem, the choice of an inertial reference frame depends on the required accuracy. In celestial mechanics, a reference frame located at the center of the solar system (near the center of the Sun[2]) and that does not rotate with respect

---

[2]The Sun contains more than 99 percent of the mass in the entire solar system.

to our galaxy has been found to provide very accurate predictions. Such a reference frame is also generally required for those meteorological calculations that model the motions of weather systems over days, or for analyses of satellites. These calculations could not be made with sufficient accuracy by using an Earth-based reference frame because of the Earth's acceleration with respect to the center of the solar system as well as because of the Earth's daily rotation about its own axis.

In many terrestrial dynamics problems, a reference frame that is fixed in the Earth is often found to be a sufficiently accurate inertial reference frame for short periods of observations, of the order of minutes. However, even a perfect billiard ball that was struck along a perfect line and thus appearing to travel along a straight path would be found to curve (with respect to the table and the Earth) if measured with sufficient accuracy.

Ultimately, confirmation of the accuracy of an inertial reference frame must be experimentally verifiable. In all subsequent analyses in this text, we shall assume the specified inertial reference frames to be fixed, within satisfactory accuracy.

4-2.5 The Universal Law of Gravitation

Gravity is a naturally occurring force of mutual attraction between any two masses in the universe. One of Newton's major contributions to the development of mechanics is his law of gravitation for particles, which, if written in terms of the force *on* mass m, may be expressed as

$$\boldsymbol{f} = -\frac{Gm_0m}{r^2}\boldsymbol{u}_r \qquad (4\text{-}3)$$

where, referring to Figure 4-1, $\boldsymbol{u}_r$ is the unit vector from m_0 toward m, m_0 is the mass of the attracting particle and m is the mass of attracted particle, r is the separation of the particles, and G is a universal constant called the *constant of gravitation,* which is independent of the locations of the particles in space. In 1798, in the most significant contribution to the study of gravitation since Newton, Henry Cavendish* (1731–1810) performed a historic experiment to measure G. Currently, the accepted value of the gravitation constant is

$G \approx 66.73 \times 10^{-12} \dfrac{\text{m}^3}{\text{kg} \cdot \text{s}^2}.$

---

*(Cavendish's measurement of $67.54 \times 10^{-12} \dfrac{\text{m}^3}{\text{kg} \cdot \text{s}^2}$ was within $1\frac{1}{2}$% of the currently accepted value. The chronicle of the measurement of G provides an extended and fascinating history.)

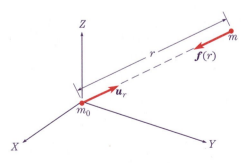

FIGURE 4-1 Central force field in which mass m experiences attractive force $\boldsymbol{f}$ toward mass m_0 at origin.

As expressed by Eq. (4-3), $\boldsymbol{f}$ is the attractive force that m experiences due to the presence of m_0. The mass m_0 also experiences a force that is equal in magnitude and oppositely directed along the line joining the two particles.

The form of Newton's law of gravitation as expressed by Eq. (4-3) applies strictly to particles. If either or both of the particles are replaced by bodies of spatial extent, it is necessary to make an additional hypothesis in order for Eq. (4-3) to remain valid. The required additional hypothesis is that the gravitational force field is a linear field. That is, the net gravitational force on a particle due to many other particles is the vector sum of all the individual forces, in which case, r is the distance between the mass centers of the interacting bodies.

The *gravitational field vector*[*] $\boldsymbol{g}$ is defined as the vector that represents the force-per-unit mass exerted on the particle m by (the field of) the particle m_0. So,

$$\boldsymbol{g} = \frac{\boldsymbol{f}}{m} = -G\frac{m_0}{r^2}\boldsymbol{u}_r \tag{4-4}$$

---

[*]The gravitational symbol g was introduced into mechanics by John Bernoulli, though obviously not as a vector, since vectors had not yet been invented.

The gravitational field vector has the dimensions of force per unit mass, or acceleration. Because the Earth is a rotating oblate spheroid that is flattened at the poles,[3] $\boldsymbol{g}$ varies slightly over the Earth's surface—from about 9.78 m/s^2 (32.09 ft/sec^2) at the equator to about 9.83 m/s^2 (32.26 ft/sec^2) at the poles. In most engineering analyses of problems on the Earth's surface, this variation is neglected, and the widely accepted values of 9.81 m/s^2 in SI units and 32.17 ft/sec^2 in British or U.S. customary units are used.

The Earth's gravitational force $\boldsymbol{f}_g$ of attraction on a body is sometimes called the true weight $\boldsymbol{W}$, since a body's weight with respect to some other field—say, such as that due to the Moon—will be different. Nevertheless, we shall simply use the word *weight* when referring to $\boldsymbol{W}$. Being a force, weight is a vector, and from Eq. (4-4) may be written as

$$\boldsymbol{f}_g = \boldsymbol{W} = m\boldsymbol{g} \tag{4-5}$$

Because of the effects that cause $\boldsymbol{g}$ to vary over the Earth's surface, $\boldsymbol{W}$ will also vary accordingly. However, as indicated above, we shall neglect such effects for all analyses at or near the Earth's surface. Whereas mass is an absolute quantity since it can be measured at any location, weight is not an absolute quantity since it must be measured with respect to a gravitational field.

4-3   TORQUE AND ANGULAR MOMENTUM FOR A PARTICLE

Newton's second law has been stated in Eq. (4-2). In this section, we simply present an *alternative* form of Eq. (4-2). One might ask why, if there is no new dynamic information, the discussion of it deserves an entire section. It is precisely because of that fact and the misunderstandings associated with it that we emphasize these topics in this section. Let's proceed.

---

[3]The polar radius is about 0.3% less than the equatorial radius.

A useful consequence of and alternative to the force–linear momentum relation is obtained if we take moments of both sides of Eq. (4-2) with respect to a point B. We shall see that the derived result simplifies for two specific classes of points B. Initially, we consider point B in Figure 4-2 to be an arbitrary point located by the position vector $\mathbf{R}_B$ and having velocity $\mathbf{v}_B = \dfrac{d\mathbf{R}_B}{dt}$, both with respect to the inertial reference frame $OXYZ$. The particle m in Figure 4-2 has linear momentum $\mathbf{p}$ with respect to the inertial frame $OXYZ$, is subjected to the resultant force $\mathbf{f}$, and is located with respect to B by $\mathbf{r}$.

If, with respect to B, the vector cross-product of $\mathbf{r}$ is taken on both sides of Eq. (4-2), Eq. (4-2) becomes

$$\mathbf{r} \times \mathbf{f} = \mathbf{r} \times \frac{d\mathbf{p}}{dt} \tag{4-6}$$

The rule for differentiation of a product such as $\mathbf{r} \times \mathbf{p}$ is

$$\frac{d}{dt}(\mathbf{r} \times \mathbf{p}) = \frac{d\mathbf{r}}{dt} \times \mathbf{p} + \mathbf{r} \times \frac{d\mathbf{p}}{dt}$$

so

$$\mathbf{r} \times \frac{d\mathbf{p}}{dt} = \frac{d}{dt}(\mathbf{r} \times \mathbf{p}) - \frac{d\mathbf{r}}{dt} \times \mathbf{p} \tag{4-7}$$

Replacement of the right-hand side of Eq. (4-6) by the right-hand side of Eq. (4-7) gives

$$\begin{aligned}
\mathbf{r} \times \mathbf{f} &= \frac{d}{dt}(\mathbf{r} \times \mathbf{p}) - \frac{d\mathbf{r}}{dt} \times \mathbf{p} \\
&= \frac{d}{dt}(\mathbf{r} \times \mathbf{p}) - \left(\frac{d\mathbf{R}}{dt} - \frac{d\mathbf{R}_B}{dt}\right) \times \mathbf{p} \\
&= \frac{d}{dt}(\mathbf{r} \times \mathbf{p}) - (\mathbf{v} - \mathbf{v}_B) \times \mathbf{p}
\end{aligned} \tag{4-8}$$

where reference to Figure 4-2 should prove helpful, particularly noting that $\mathbf{R} = \mathbf{R}_B + \mathbf{r}$ has been used.

The term $\mathbf{v} \times \mathbf{p}$ on the right-hand side of Eq. (4-8) is zero because the particle's velocity and momentum are always collinear. The quantity $\mathbf{r} \times \mathbf{f}$ is called the *torque,* or *moment,* of the force $\mathbf{f}$ about B and is denoted by $\boldsymbol{\tau}_B$. The quantity $\mathbf{r} \times \mathbf{p}$ is called the *angular momentum,* or *moment of momentum,* of the particle about B and is denoted by $\mathbf{h}_B$. In terms of these definitions, Eq. (4-8) becomes

$$\boldsymbol{\tau}_B = \frac{d\mathbf{h}_B}{dt} + \mathbf{v}_B \times \mathbf{p} \tag{4-9}$$

The term $\mathbf{v}_B \times \mathbf{p}$ is not always zero, but will be zero if either

1. B is *fixed* in the inertial reference frame; that is, $\mathbf{v}_B = 0$; or
2. The velocity of B is always *parallel* to the velocity of the particle P.

If either of these conditions holds, Eq. (4-9) reduces to the more commonly quoted form of the angular momentum principle,

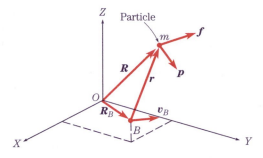

FIGURE 4-2 Particle with momentum $\boldsymbol{p}$ and subjected to force $\boldsymbol{f}$.

$$\boldsymbol{\tau}_B = \frac{d\boldsymbol{h}_B}{dt} \tag{4-10}$$

Thus, whenever B is either a fixed point or a point that moves parallel to the particle, *the resultant torque applied to a particle equals the time rate of change of the particle's angular momentum*, which is a statement that is often called the *angular momentum principle* or the *moment of momentum principle*. We shall consider either of Eqs. (4-9) or (4-10) as the angular momentum principle, noting the appropriate restrictions represented by conditions 1 and 2 when Eq. (4-10) is used.

Recall from Subsection 4-2.2 that a particle cannot deform or rotate or sustain torques about an axis through it. Thus, in the context of Eqs. (4-9) and (4-10), torques relate to the torque of a force as indicated in Figure 4-2.

In analogy with Eq. (4-2), Eq. (4-10) provides a second conservation theorem: *conservation of angular momentum of a particle*.

- If the resultant torque about a point B of all the forces on a particle is zero, then $\dfrac{d\boldsymbol{h}_B}{dt} = 0$, and thus the particle's angular momentum $\boldsymbol{h}_B$ is conserved; that is, $\boldsymbol{h}_B$ remains constant.

In stating the conservation of angular momentum of a particle as expressed above, it is understood that one of the two conditions immediately preceding Eq. (4-10) pertains to the point B about which the torque and angular momentum are calculated. In the case of a particle, point B almost always refers to a point, say O, that is fixed in inertial space; but clearly, this does not have to be the case since the second condition may apply.

Now we return to the issue raised in the first paragraph of this section. Often students are confronted with the question of whether to use torque and angular momentum, or force and linear momentum, or both. Well, the crux is that "both" is never an appropriate answer for a particle. This is because no new dynamic information is introduced by the torque and angular momentum equation for a particle and thus the use of "both" would result in the duplication of information. It follows that *either* the force and linear momentum (vector) equation *or* the torque and angular momentum (vector) equation can be chosen; but not both.

It is interesting, however, to note that there are other appropriate choices. Since the force and linear momentum equation (Eq. (4-2)) and the torque and angular momentum equation (Eq. (4-10)) are both vector equations, each of Eqs. (4-2) and (4-10) may be expanded as three scalar equations, for example, in the X, Y, and Z directions shown in Figure 4-2. Therefore, using either Eq. (4-2) or Eq. (4-10) means using the corresponding three scalar equations. This decomposition of vector equations enables the visualization of other possible choices than the rather obvious choice of either exclusively Eq. (4-2) or exclusively Eq. (4-10). The idea is that it may be possible to choose *some* three of these six scalar equations. Indeed, it can be shown[4] that *certain* choices of three out of the six scalar equations constitute a necessary and sufficient set for formulating the dynamic equations of a particle, including the set of three scalar equations exclusively from Eq. (4-2) and the set of three scalar equations exclusively from Eq. (4-10).[5] The most convenient choice among these alternatives can be made, depending upon the particular problem to be studied. It is the emphasis of this point that has led us to present this alternative formulation in a separate section.

4-4   FORMULATION OF EQUATIONS OF MOTION: EXAMPLES

Having reviewed newtonian particle dynamics, we are now prepared to formulate the governing equations of motion for lumped-parameter systems whose mass may be modeled as one or more particles. For mechanical systems (as opposed to electrical or electromechanical systems, which we shall consider later), the requirements for the formulation of the equations of motion are similar to the requirements for systems in mechanical equilibrium, but are more extensive. As stated in Eq. (2-1), they are

1. Geometric requirements on the motions (kinematic requirements);

2. Dynamic requirements on the forces (force–dynamic requirements); and

3. Constitutive requirements for all the system elements and fields.   (4-11)

In order to emphasize the underlying structure of these formulations, we dignify these three requirements with an additional equation number. Requirement 1 is essentially the satisfaction of the kinematics. Requirement 2 is essentially Newton's second law. Requirement 3 is the collection of the constitutive relations that express the velocity-momentum relations for the masses, relate forces and motion, and render the constitutive character of the fields. Although these requirements must be satisfied for both the direct approach and the indirect approach, the equations of motion will be formulated via the *direct approach*.

It is interesting, though not necessary, to categorize these problems into broad classes. Although there is no generally accepted classification of all the types of problems in particle dynamics, two types of problems arise most often. The *first kind* is where the motion is known and the forces are desired (see Figure 4-3). The *second kind* is where the forces are known and the motion is desired (see Figure 4-4).

$x_i(t)$ Known system $F_j(t)$

(Known inputs)   (Computed outputs)   FIGURE 4-3   Dynamics problem of the *first kind*.

[4]See Problem 4-20.

[5]As we shall see in Chapter 6, this is not the case for extended bodies, which may be modeled as continua or as systems of particles; that is, all six scalar equations must be used because of their independence. See Section 6-2 or Appendix C for an extensive discussion of momentum principles.

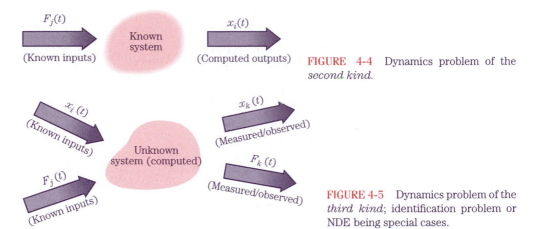

FIGURE 4-4 Dynamics problem of the *second kind.*

FIGURE 4-5 Dynamics problem of the *third kind*; identification problem or NDE being special cases.

An obvious type of problem that does not fall into either of the above two classes is the *identification problem,* in which some combination of forces and motions is known and it is desired to characterize the system (see Figure 4-5). We shall call this type of problem the *third kind.* One of the active research areas in identification is *nondestructive evaluation* (NDE), where known inputs and measured outputs are combined to infer some characteristic(s) of the system; this is to identify the system or some characteristics of the system, such as structural flaws or elastic modulus.

Another obvious problem of particle dynamics is the *hybrid problem,* which is a mixture of two or more kinds, for instance, the first and second kinds. As an example, consider a robot that is designed to place delicate objects onto a flat surface. Over a portion of the object's trajectory, the robot's driving motors, producing specified forces or torques, may be fully engaged. As the object approaches the surface, it is likely that a desirable soft landing will result in a specified "safe" velocity. Thus, over the first portion of the trajectory, the problem may be characterized as a problem of the second kind (forces specified), and over the second part of the trajectory, as a problem of the first kind (motion specified).

By way of examples, we shall briefly explore a few problems of the first and second kinds. The analyses will likely be viewed as increasingly difficult, but the structure of the approach will remain identical throughout. It should be understood, however, that our presentation of newtonian dynamics is far from complete. Since the material in this chapter is primarily in the context of a review, except for a brief discussion in Chapter 6, further pursuit of newtonian dynamics will go unexplored.

4-4.1 Problems of Particle Dynamics of the First Kind

In problems of particle dynamics of the first kind, the motion of the system is specified for all time and it is desired to find the corresponding forces.

■ **Example 4-1:** Antenna Deployed by Maneuvering Airplane

We reconsider Example 3-8, where a maneuvering airplane is deploying an antenna as sketched in Figure 4-6a. The acceleration $\boldsymbol{a}_A$ of the antenna, which is modeled as a particle of weight 2 lb, was calculated for the parameters given in the problem statement.

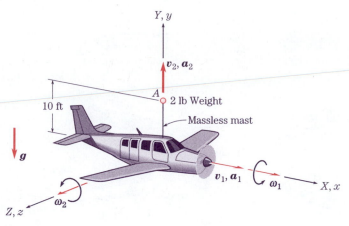

(a) Maneuvering airplane and antenna A.

(b) Free-body diagram of antenna A.   (c) Kinetic diagram of antenna A.

FIGURE 4-6   Airplane deploying antenna at tip of massless mast.

For the instant shown, we want to find the force applied by the massless mast on the antenna A.

□   **Solution:**   Here we emphasize the structure of the formulation requirements defined in Eq. (4-11).

1. Kinematic requirements: The geometric requirements on the motion of the antenna A have been addressed; in particular, refer to Eq. (e) in Example 3-8. The important result from that analysis is

$$\boldsymbol{a}_A = \frac{d\boldsymbol{v}_A}{dt} = \frac{d^2\boldsymbol{R}_0}{dt^2} + \boldsymbol{a}_{\text{rel}} + 2\boldsymbol{\omega} \times \boldsymbol{v}_{\text{rel}} + \dot{\boldsymbol{\omega}} \times \boldsymbol{r} + \boldsymbol{\omega} \times (\boldsymbol{\omega} \times \boldsymbol{r})$$

$$= [99.6\boldsymbol{I} + 0.064\boldsymbol{J} + 0.6\boldsymbol{K}] \frac{\text{ft}}{\text{sec}^2} \tag{a}$$

where $\boldsymbol{I}, \boldsymbol{J}$, and $\boldsymbol{K}$ are the X, Y, and Z unit vectors, respectively.

2. Force–dynamic requirements: The dynamic requirements on the forces are satisfied by Newton's second law

$$\sum_{i=1}^{q} f_i = \frac{d\boldsymbol{p}}{dt} \qquad \text{(b)}$$

where q is the total number of individual forces. For this analysis, Eq. (b) becomes

$$\boldsymbol{f}_m + \boldsymbol{f}_g = \frac{d\boldsymbol{p}_A}{dt} \qquad \text{(c)}$$

where $\boldsymbol{f}_m$ is the force exerted by the mast and $\boldsymbol{f}_g$ is the force exerted by gravity, both on antenna A; and where $\boldsymbol{p}_A$ is the linear momentum of the antenna. Although they are not necessary, the *free-body diagram* and the so-called *kinetic diagram* of the antenna are sketched in Figures 4-6b and 4-6c, respectively. The vector directions of $\boldsymbol{f}_m$ and $\frac{d\boldsymbol{p}_A}{dt}$ in Figures 4-6b and 4-6c are arbitrary. The free-body diagram should be quite familiar to the reader from work in statics. The kinetic diagram is simply a sketch that represents the resulting dynamic effects of Eq. (c), or more generally Newton's second law. These diagrams are very useful when working directly with the individual vector components, in which case, the diagrams should also contain the coordinate axes with clearly indicated positive directions. When working in vector notation and when all the forces pass through a single point as is the case here, these diagrams are less useful, though they may nonetheless be helpful in some analyses.

3. *Constitutive requirements:*   From Eqs. (4-1) and (4-5), the constitutive relations for this problem are

$$\boldsymbol{p}_A = m\boldsymbol{v}_A \qquad \text{(d)}$$

$$\boldsymbol{f}_g = m\boldsymbol{g} = -mg\boldsymbol{J} \qquad \text{(e)}$$

where the momentum–velocity relation in Eq. (d) has been written for a newtonian particle of mass m.

Combining the requirements in Eqs. (c) and (d) gives

$$\boldsymbol{f}_m + \boldsymbol{f}_g = m\boldsymbol{a}_A \qquad \text{(f)}$$

Substitution of Eqs. (a) and (e) into Eq. (f) leads to

$$\boldsymbol{f}_m = m\boldsymbol{a}_A - m\boldsymbol{g}$$

$$= \left[\frac{2}{32.17}(99.6\boldsymbol{I} + 0.064\boldsymbol{J} + 0.6\boldsymbol{K}) - \frac{2}{32.17}(-32.17\boldsymbol{J}) \right] \text{lb} \qquad \text{(g)}$$

or

$$\boldsymbol{f}_m = [6.192\boldsymbol{I} + 2.004\boldsymbol{J} + 0.037\boldsymbol{K}] \text{ lb}. \qquad \text{(h)}$$

Probably the most striking observation that can be made about this solution is the observation that can be made about dynamics in general: In most dynamics problems, kinematics is the most difficult step.

Note that when the airplane is not maneuvering, the force in Eq. (g) is $\boldsymbol{f}_m = 2\boldsymbol{J}$ (lb), which indicates that the mast applies an upward (positive Y direction) force of 2 lb to the antenna. This force, of course, is simply due to the weight of the antenna. Also, from Eq.

(h), we note that during the maneuvering of the airplane, the force on the antenna in the X direction is greater than three times the antenna's weight.

Finally, it is important not to lose sight of why such dynamic analyses are conducted. In engineering the ultimate goal is design, in the broadest sense. From this analysis, numerous design issues can be explored. We shall examine only a few of such considerations.

As an instrument or a set of instruments, the antenna may be delicate or sensitive to applied forces. From Eq. (h), we see that the antenna must be capable of withstanding a "force of $3g$'s" in the X direction. If it cannot withstand such a force, it may be subject to damage or the maneuvering of the airplane may have to be curtailed.

Also, the mast (as well as its connections with the antenna and the airplane) must be designed to perform its deployment function. By Newton's third law, a force that is equal in magnitude and opposite in direction to the force given in Eq. (h) is applied *to* the mast. Thus, the mast is a structural element that must be designed to be *strong* enough to withstand the maneuvering loads, and *stiff* enough not to undergo intolerable deflections, including flexure and buckling.

So, although we shall not explore these design issues in this text, it is important to appreciate the purposes of the dynamic analyses and the subsequent benefit and use of their results.

■ **Example 4-2:**   Bird on Mobile

We reconsider Example 3-9, where a bird is running along a member of a moving mobile as sketched in Figures 4-7a and 4-7b. The acceleration $\boldsymbol{a}_B$ of the bird, which is modeled as a particle, was calculated for the parameters given in the problem statement.

For the instant shown, we want to find the force exerted on the bird by the leg of the mobile along which it is running.

❑ **Solution:**   Again, we emphasize the structure of the formulation requirements defined in Eq. (4-11).

1. Kinematic requirements: The geometric requirements on the motions of the bird have been addressed; in particular, refer to Eq. (f) in Example 3-9. The important result from that analysis is

$$\boldsymbol{a}_B = \frac{d\boldsymbol{v}_B}{dt} = -(2\omega_1 v_0 + R\omega_1^2)\boldsymbol{I} + (a_0 - d\omega_1^2 - d\omega_2^2)\boldsymbol{J} + (2v_0\omega_2)\boldsymbol{K} \qquad (a)$$

where $\boldsymbol{I}$, $\boldsymbol{J}$, and $\boldsymbol{K}$ are the X, Y, and Z unit vectors, respectively.

2. Force–dynamic requirements: The dynamic requirements on the forces are satisfied by Newton's second law

$$\sum_{i=1}^{q} \boldsymbol{f}_i = \frac{d\boldsymbol{p}}{dt} \qquad (b)$$

where q is the total number of individual forces. For this analysis, Eq. (b) becomes

$$\boldsymbol{f}_m + \boldsymbol{f}_g + \boldsymbol{f}_w = \frac{d\boldsymbol{p}_B}{dt} \qquad (c)$$

where $\boldsymbol{f}_m, \boldsymbol{f}_g$, and $\boldsymbol{f}_w$ are the forces exerted on the bird by the mobile, gravity, and the wind, respectively, and $\boldsymbol{p}_B$ is the linear momentum of the bird. Also, the free-body diagram and

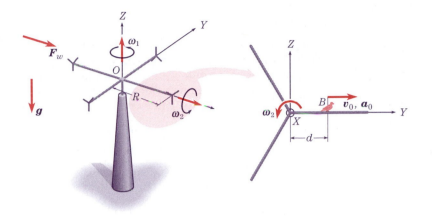

(*a*) Mobile structure.   (*b*) Detail of *Y* and bird.

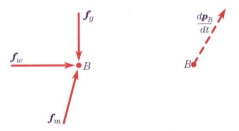

(*c*) Free-body diagram of bird *B*.   (*d*) Kinetic diagram of bird *B*.

FIGURE 4-7 Bird running along moving mobile.

the kinetic diagram of the bird are sketched in Figures 4-7*c* and 4-7*d*, respectively, where the vector directions of $\boldsymbol{f}_m$ and $\dfrac{d\boldsymbol{p}_B}{dt}$ are arbitrary.

3. Constitutive requirements: Again, from Eqs. (4-1) and (4-5), the constitutive relations for this problem are

$$\boldsymbol{p}_B = m\boldsymbol{v}_B \tag{d}$$

$$\boldsymbol{f}_g = m\boldsymbol{g} = -mg\boldsymbol{K} \tag{e}$$

where the momentum–velocity relation in Eq. (d) has been written for a newtonian particle of mass m.

Combining the requirements in Eqs. (c) and (d) gives

$$\boldsymbol{f}_m + \boldsymbol{f}_g + \boldsymbol{f}_w = m\boldsymbol{a}_B \tag{f}$$

Substitution of Eqs. (a) and (e) into Eq. (f) leads to

$$\begin{aligned} \boldsymbol{f}_m &= m\boldsymbol{a}_B - m\boldsymbol{g} - f_w\boldsymbol{I} \\ &= -(f_w + 2m\omega_1 v_0 + mR\omega_1^2)\boldsymbol{I} + m(a_0 - d\omega_1^2 - d\omega_2^2)\boldsymbol{J} \\ &\quad + m(2v_0\omega_2 + g)\boldsymbol{K} \end{aligned} \tag{g}$$

where, from the problem statement, we have used the relation $\boldsymbol{f}_w = \boldsymbol{F}_w = f_w\boldsymbol{I}$.

■ **Example 4-3:**   **Pilot in Airplane on Aircraft Carrier**

We reconsider Example 3-11, where an airplane is taking off from a maneuvering aircraft carrier as sketched in Figure 4-8*a*. The acceleration $\boldsymbol{a}_P$ of the pilot, who is modeled as a particle of weight W, was calculated for the parameters given in the problem statement.

For the instant shown, we want to find the force exerted on the pilot by the seat in the airplane, assuming that the seat is the only contact between the pilot and the airplane.

☐ **Solution:**   Again, we emphasize the structure of the formulation requirements defined in Eq. (4-11).

1. Kinematic requirements: The geometric requirements on the motion of the pilot have been addressed; in particular, refer to Eq. (k) in Example 3-11. The important result from that analysis is

$$\boldsymbol{a}_P = \frac{d\boldsymbol{v}_P}{dt} = (-R\omega_1^2 - 2\omega_1 v_0 + 2b\omega_2\omega_3)\boldsymbol{I}$$
$$+ (a_0 - b\omega_3^2 - b\omega_1^2 + h\omega_1\omega_2)\boldsymbol{J}$$
$$+ (ba_3 + b\omega_1\omega_2 - h\omega_2^2)\boldsymbol{K} \tag{a}$$

where $\boldsymbol{I}$, $\boldsymbol{J}$, and $\boldsymbol{K}$ are the X, Y, and Z unit vectors, respectively.

2. Force–dynamic requirements: The dynamic requirements on the forces are satisfied by Newton's second law

$$\sum_{i=1}^{q} \boldsymbol{f}_i = \frac{d\boldsymbol{p}}{dt} \tag{b}$$

where q is the total number of individual forces. For this analysis, Eq. (b) becomes

$$\boldsymbol{f}_s + \boldsymbol{f}_g = \frac{d\boldsymbol{p}_P}{dt} \tag{c}$$

where $\boldsymbol{f}_s$ and $\boldsymbol{f}_g$ are the forces exerted on the pilot by the airplane's seat and gravity, respectively; and $\boldsymbol{p}_P$ is the linear momentum of the pilot. The free-body diagram and the kinetic diagram of the pilot are sketched in Figures 4-8*b* and 4-8*c*, respectively, where the vector directions of $\boldsymbol{f}_s$ and $\frac{d\boldsymbol{p}_P}{dt}$ are arbitrary.

3. Constitutive requirements: Again, from Eqs. (4-1) and (4-5), the constitutive relations for this problem are

$$\boldsymbol{p}_P = m\boldsymbol{v}_P \tag{d}$$

and

$$\boldsymbol{f}_g = m\boldsymbol{g} = -mg\boldsymbol{K} \tag{e}$$

where the momentum–velocity relation in Eq. (d) has been written for a newtonian particle of mass m.

Combining the requirements in Eqs. (c) and (d) gives

$$\boldsymbol{f}_s + \boldsymbol{f}_g = m\boldsymbol{a}_P \tag{f}$$

Substitution of Eqs. (a) and (e) into Eq. (f) leads to

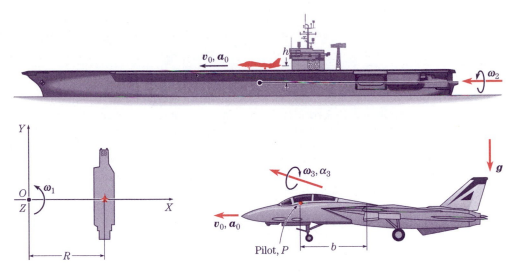

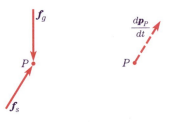

(a) Maneuvering aircraft carrier with airplane taking off.

(b) Free-body diagram of pilot P.   (c) Kinetic diagram of pilot P.

FIGURE 4-8   Aircraft carrier and airplane.

$$\boldsymbol{f}_s = m\boldsymbol{a}_P - m\boldsymbol{g}$$

$$= \frac{W}{g}(2b\omega_2\omega_3 - 2\omega_1 v_0 - R\omega_1^2)\boldsymbol{I}$$

$$+ \frac{W}{g}(a_0 - b\omega_3^2 - b\omega_1^2 + h\omega_1\omega_2)\boldsymbol{J}$$

$$+ \frac{W}{g}(b\alpha_3 + b\omega_1\omega_2 - h\omega_2^2 + g)\boldsymbol{K} \tag{g}$$

where we have used the scalar relationship from Eq. (4-5) that $m = W/g$. Several design questions may be explored at this point. We leave these for the reader's consideration.

4-4.2   Problems of Particle Dynamics of the Second Kind

In problems of particle dynamics of the second kind, the forces on the system are specified for all time and it is desired to find the corresponding motion that will satisfy the constraints and the initial conditions. In general, for dynamics problems involving the same system, problem solutions of the second kind are often more difficult than those of the first kind.

■ **Example 4-4:**   **Package-Transfer System**

We reconsider Example 2-1 in Chapter 2, in particular, the solution obtained via the direct approach. Referring to the solution there, it may be observed that the solution was obtained for each of three regions, designated as the regions between points A and B, points B and C, and points C and D. In each of these regions, the force–dynamic requirements and the constitutive requirements were combined and then used to integrate the kinematic equations to calculate the displacements. In a strict sense, this example may not be considered as a problem of the second kind since some of the forces are a function of the kinematic variable. However, since such strict delineation of these classifications is not central to our interests, we shall not pursue this issue further. Thus, this example constitutes a problem of particle dynamics of the second kind.

■ **Example 4-5:**   **Multistory Buildings**

Engineers who are concerned with the response of multistory buildings to natural phenomena such as winds and earthquakes are required to develop mathematical models for the dynamic analysis of these structures. Many models have been developed for this purpose. One type of model that is frequently used is illustrated in Figure 4-9, where a four-story building is modeled.

In the model in Figure 4-9, the mass of the entire building is assumed to be concentrated in the floors, which are assumed to undergo only horizontal lateral displacements in the plane of the sketch. The vertical walls in the building model provide the elastic stiffness such that the resisting spring force is proportional to the relative displacement of adjacent

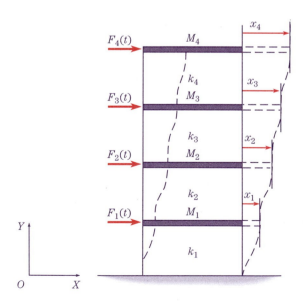

FIGURE 4-9   Model of four-story building.

floors, the proportionality constants being the spring constants k_1 through k_4. The forces $F_1(t)$ through $F_4(t)$ denote the known (by physical measurement, theory, or assumption) time-varying forces. We want to find the equations of motion for this model.

☐   Solution: The formulation regimen given in Eq. (4-11) still applies.

1. Kinematic requirements: The kinematics of this model are relatively simple—indeed one dimensional—because all the displacements are along a single reference direction. Because of this one-dimensional character of the problem, vectors are not needed in the analysis. The positions of the four masses relative to their undisplaced configurations are indicated in Figure 4-9 and are written as

$$x_1(t) \qquad x_2(t) \qquad x_3(t) \qquad x_4(t) \tag{a}$$

Because there is a single reference frame, the indicated inertial reference frame, without ambiguity we may write the corresponding velocities as

$$\dot{x}_1(t) \qquad \dot{x}_2(t) \qquad \dot{x}_3(t) \qquad \dot{x}_4(t) \tag{b}$$

which may be written more succinctly as

$$\dot{x}_j(t) \qquad j = 1, 2, 3, 4$$

The corresponding accelerations are

$$\ddot{x}_1(t) \qquad \ddot{x}_2(t) \qquad \ddot{x}_3(t) \qquad \ddot{x}_4(t) \tag{c}$$

which may be written more succinctly as

$$\ddot{x}_j(t) \qquad j = 1, 2, 3, 4$$

2. Force–dynamic requirements: The dynamic requirements on the forces are satisfied by Newton's second law

$$\sum_{i=1}^{q} f_i = \frac{dp_j}{dt} \qquad j = 1, 2, 3, 4 \tag{d}$$

where q is the total number of individual forces for each j. For this analysis, with the aid of the free-body diagrams in Figure 4-10, Eq. (d) becomes

$$F_1(t) - F_{s_1} + F_{s_2} = \dot{p}_1 \tag{e}$$
$$F_2(t) - F_{s_2} + F_{s_3} = \dot{p}_2 \tag{f}$$
$$F_3(t) - F_{s_3} + F_{s_4} = \dot{p}_3 \tag{g}$$
$$F_4(t) - F_{s_4} = \dot{p}_4 \tag{h}$$

where Eqs. (e) through (h) correspond to masses M_1 through M_4, respectively. It should be understood that the floors sketched in Figure 4-10 are modeled as particles; they are simply shown as extended bodies to assist the relational correspondence with the model in Figure 4-9.

3. Constitutive requirements: The constitutive relations for this system are[6]

---

[6]Sometimes, the manner in which the subscripts in Eqs. (i) and (j) appear might suggest a summation. No summation is implied here.

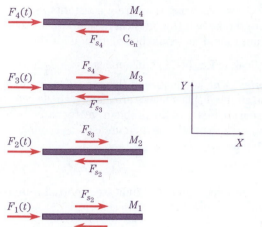

$$p_j = M_j \dot{x}_j \qquad j = 1, 2, 3, 4 \tag{i}$$

and

$$F_{s_j} = k_j e_j \qquad j = 1, 2, 3, 4 \tag{j}$$

where e is the relative (interstory) displacement of the two ends of the spring elements, k_j are the indicated interstory spring constants, and F_{s_j} are the corresponding spring forces.

Substitution of Eq. (a) into Eq. (j) gives

$$F_{s_1} = k_1 x_1 \tag{k}$$
$$F_{s_2} = k_2(x_2 - x_1) \tag{l}$$
$$F_{s_3} = k_3(x_3 - x_2) \tag{m}$$
$$F_{s_4} = k_4(x_4 - x_3) \tag{n}$$

Then, substitution of Eqs. (k) through (n) into the left-hand sides of Eqs. (e) through (h) gives

$$F_1(t) - k_1 x_1 + k_2(x_2 - x_1) = M_1 \ddot{x}_1 \tag{o}$$
$$F_2(t) - k_2(x_2 - x_1) + k_3(x_3 - x_2) = M_2 \ddot{x}_2 \tag{p}$$
$$F_3(t) - k_3(x_3 - x_2) + k_4(x_4 - x_3) = M_3 \ddot{x}_3 \tag{q}$$
$$F_4(t) - k_4(x_4 - x_3) = M_4 \ddot{x}_4 \tag{r}$$

where Eqs. (c) and (i) have been used. Thus,

$$M_1 \ddot{x}_1 + (k_1 + k_2)x_1 - k_2 x_2 = F_1(t) \tag{s}$$
$$M_2 \ddot{x}_2 - k_2 x_1 + (k_2 + k_3)x_2 - k_3 x_3 = F_2(t) \tag{t}$$
$$M_3 \ddot{x}_3 - k_3 x_1 + (k_3 + k_4)x_3 - k_4 x_4 = F_3(t) \tag{u}$$
$$M_2 \ddot{x}_2 - k_4 x_3 + k_4 x_4 = F_4(t) \tag{v}$$

FIGURE 4-10   Free-body diagrams of floors in four-story building model.

Equations (s) through (v) are the differential equations of motion for the model of the four-story building. In order to determine the system's response (that is, the time history of the displacement of each floor), the external loads $F_1(t)$ through $F_4(t)$ and the initial conditions of each mass must be specified. Then, the equations can be integrated. We shall return to such an analysis in Chapter 8.

For readers who are familiar with matrix notation (as well as readers who are not), Eqs. (s) through (v) may be rewritten as

$$[M]\{\ddot{x}\} + [K]\{x\} = \{F(t)\} \tag{w}$$

where

$$[M] = \begin{bmatrix} M_1 & 0 & 0 & 0 \\ 0 & M_2 & 0 & 0 \\ 0 & 0 & M_3 & 0 \\ 0 & 0 & 0 & M_2 \end{bmatrix} \quad [K] = \begin{bmatrix} k_1 + k_2 & -k_2 & 0 & 0 \\ -k_2 & k_2 + k_3 & -k_3 & 0 \\ 0 & -k_3 & k_3 + k_4 & -k_4 \\ 0 & 0 & -k_4 & k_4 \end{bmatrix} \tag{x}$$

and

$$\{x\} = \begin{Bmatrix} x_1 \\ x_2 \\ x_3 \\ x_4 \end{Bmatrix} \quad \{\ddot{x}\} = \begin{Bmatrix} \ddot{x}_1 \\ \ddot{x}_2 \\ \ddot{x}_3 \\ \ddot{x}_4 \end{Bmatrix} \quad \{F(t)\} = \begin{Bmatrix} F_1(t) \\ F_2(t) \\ F_3(t) \\ F_4(t) \end{Bmatrix} \tag{y}$$

Expressed in matrix notation, the equations of motion display obvious symmetries, which we shall also explore in Chapter 8. The names given to these matrices are as follows:

$$[M] \equiv \text{Mass matrix} \qquad [K] \equiv \text{Stiffness matrix}$$

$$\{x\} \equiv \text{Displacement vector} \quad \{\ddot{x}\} \equiv \text{Acceleration vector} \quad \{F(t)\} \equiv \text{Force vector}$$

■ **Example 4-6:**   **Geosynchronously Orbiting Satellite**

In the modern era, satellites play increasingly important roles in our everyday lives, particularly in information transfer. One type of satellite is the so-called *geosynchronously orbiting satellite,* which orbits the Earth in a circular manner such that it remains above the same location on the equator of the Earth at all times. We should like to determine the radius of an equatorial geosynchronous orbit.

We may model both the satellite and the Earth as particles (refer to Section 4-2) that remain in the plane of the sketch, as illustrated in Figure 4-11. The gravitational field is the only source of forces acting on the satellite. Furthermore, we assume the center of the rotating Earth is fixed; while such an assumption would not be valid in determining the long-term motion of the satellite, it is satisfactory for the purpose at hand.

❒ **Solution:**   The formulation regimen given in Eq. (4-11) still applies. We shall find the equations of motion for the satellite, which will be used to find the geosynchronous orbit radius.

1. Kinematic requirements: We fix the origin of an inertial reference frame $OXYZ$ at the center of the rotating Earth. Then the position vector of the rotating satellite is $\boldsymbol{r}$ as sketched in Figure 4-11. By definition, motions of the satellite are

$$\boldsymbol{v} = \frac{d\boldsymbol{r}}{dt} \qquad \boldsymbol{a} = \frac{d^2\boldsymbol{r}}{dt^2} \tag{a}$$

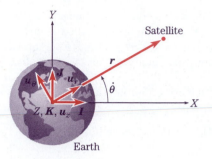

FIGURE 4-11 Satellite orbiting the Earth, as viewed from the North Pole. Both satellite and Earth are modeled as particles.

Since r is rotating with respect to $OXYZ$, use of the differential operator in Eq. (3-19) once gives

$$v = \left(\frac{\partial r}{\partial t}\right)_{rel} + \boldsymbol{\omega} \times r \tag{b}$$

and use of Eq. (3-19) a second time gives

$$a = \left(\frac{\partial^2 r}{\partial t^2}\right)_{rel} + 2\boldsymbol{\omega} \times \left(\frac{\partial r}{\partial t}\right)_{rel} + \dot{\boldsymbol{\omega}} \times r + \boldsymbol{\omega} \times [\boldsymbol{\omega} \times r] \tag{c}$$

where $\boldsymbol{\omega}$ is the angular velocity of the intermediate frame having unit vectors u_r, u_θ, and u_z for the radial, circumferential, and Z directions, respectively. The mathematical operations leading to Eqs. (b) and (c) are identical to those performed in Example 3-5.

In terms of unit vectors of the intermediate frame, the individual terms in Eqs. (b) and (c) are given by

$$r = ru_r \qquad \left(\frac{\partial r}{\partial t}\right)_{rel} = \dot{r}u_r \qquad \left(\frac{\partial^2 r}{\partial t^2}\right)_{rel} = \ddot{r}u_r \tag{d}$$

$$\boldsymbol{\omega} = \dot{\theta}u_z \qquad \dot{\boldsymbol{\omega}} = \ddot{\theta}u_z \tag{e}$$

Substitution of Eqs. (d) and (e) into Eq. (c) yields the acceleration of the satellite, defined in the intermediate reference frame, with respect to $OXYZ$, as

$$a = (\ddot{r} - \dot{\theta}^2 r)u_r + (2\dot{\theta}\dot{r} + \ddot{\theta}r)u_\theta \tag{f}$$

2. Force–dynamic requirements: The dynamic requirements on the forces are satisfied by Newton's second law

$$\sum_{i=1}^{q} f_i = \frac{dp}{dt} \tag{g}$$

where q is the total number of individual forces.

3. Constitutive requirements: The only force acting on the satellite is the gravitational force and is given by Eq. (4-3) as

$$f_g = -G\frac{m_0 m}{r^2}u_r \tag{h}$$

where m_0 and m are the masses of the Earth and the satellite, respectively. Also, the linear momentum p of the satellite is defined by

$$\boldsymbol{p} = m\boldsymbol{v} \tag{i}$$

Combining Eqs. (g) and (i) gives

$$\boldsymbol{f}_g = m\boldsymbol{a} \tag{j}$$

Substitution of Eqs. (f) and (h) into Eq. (j) gives

$$-G\frac{m_0}{r^2}\boldsymbol{u}_r = (\ddot{r} - \dot{\theta}^2)\boldsymbol{u}_r + (2\dot{\theta}\dot{r} + \ddot{\theta}r)\boldsymbol{u}_\theta \tag{k}$$

Or Eq. (k) can be written in terms of its scalar components. In the radial direction,

$$-G\frac{m_0}{r^2} = \ddot{r} - \dot{\theta}^2 r \tag{l}$$

and in circumferential direction,

$$0 = 2\dot{\theta}\dot{r} + \ddot{\theta}r \tag{m}$$

Equations (l) and (m) are the component equations of motion for the satellite. If the satellite is orbiting in a circular orbit, then r is a constant and $\dot{r}$ and $\ddot{r}$ are zero. Since r is a constant and $\ddot{r}$ is zero, Eq. (l) requires that $\dot{\theta}$ is a constant. Equation (m) also yields the fact that the angular velocity of the satellite is a constant; since $r \neq 0$ and $\dot{r} = 0$, $\ddot{\theta}r = 0$ requires that $\dot{\theta} = $ constant. Equation (l) gives the relation between the angular velocity and the orbiting radius as

$$\dot{\theta}^2 = \frac{Gm_0}{r^3} \tag{n}$$

Based on Eq. (n), we can find the radius of the geosynchronous orbit. It is known that the mass of the Earth is 5.976×10^{24} kg, and the period of the geosynchronously orbiting satellite is 24 hours. Then,

$$\dot{\theta} = \frac{2\pi\left(\dfrac{\text{rad}}{\text{day}}\right)}{24\left(\dfrac{\text{hr}}{\text{day}}\right) \cdot 3600\left(\dfrac{\text{s}}{\text{hr}}\right)} = 7.272 \times 10^{-5}\,\frac{\text{rad}}{\text{s}} \tag{o}$$

Thus,

$$r = \sqrt[3]{\frac{Gm_0}{\dot{\theta}^2}} = \sqrt[3]{\frac{6.673 \times 10^{-11}\,(\text{m}^3/\text{kg}\cdot\text{s}^2) \times 5.976 \times 10^{24}\,(\text{kg})}{7.272^2 \times 10^{-10}\,(\text{rad}^2/\text{s}^2)}} = 4.225 \times 10^7\,\text{m} \tag{p}$$

which is the required radius of a geosynchronous orbit. By the way, the radius of the Earth is about 6.365×10^6 m.

■ **Example 4-7:** Space Litterbug

Suppose a passenger in a geosynchronously orbiting spacecraft discards a banana peel out of a porthole or airlock, with velocity v vertically "downward" toward the Earth, expecting the banana peel to fall until it incinerates in the atmosphere. The spacecraft is orbiting in the geosynchronous circular orbit of radius r_0 given by Eq. (p) in Example 4-6. Is the expectation of this space litterbug fulfilled? Find the motion of the banana peel.

☐ **Solution:**   The formulation regimen given in Eq. (4-11) still applies.

In essence, the equations of motion for the banana peel are the same as those for the satellite since, in both cases, the only force is the gravitation of the Earth. As we can see from the previous example, the equations of motion are nonlinear and difficult to integrate. Here we shall simply find the motion of the banana peel with respect to the orbiting spacecraft. By doing this, we shall be able to exploit the results obtained in the previous example.

1. Kinematic requirements: We define the position vector of the banana peel as

$$\boldsymbol{r} = r\boldsymbol{u}_r = (r_0 + s)\boldsymbol{u}_r \tag{a}$$

where s is the radial distance between the banana peel and the spacecraft orbit and, as specified in the problem statement, r_0 is the radius of the spacecraft's geosynchronous circular orbit (given by Eq. (p) in Example 4-6). The unit vector $\boldsymbol{u}_r$ is along the radial direction from the center of the Earth to the banana peel. Also, we define the angular velocity of the banana peel's position vector as[7]

$$\boldsymbol{\omega} = (\omega_0 + \omega_1)\boldsymbol{u}_z \tag{b}$$

where ω_0 is the angular velocity of the spacecraft's position vector with respect to the Earth (given by Eq. (o) in Example 4-6) and ω_1 is the angular velocity of the banana peel's position vector with respect to the spacecraft.

By these definitions, the following initial conditions for the problem may be observed: at $t = 0$, the time at which the banana peel is expelled,

$$s = 0 \qquad \dot{s} = -v \qquad \omega_1 = 0 \tag{c}$$

where the initial condition on $\dot{s}$ is in accordance with the problem statement.

Then, referring to Eq. (f) in Example 4-6, the acceleration of the banana peel is

$$\boldsymbol{a} = \left[\ddot{s} - (\omega_0 + \omega_1)^2(r_0 + s)\right]\boldsymbol{u}_r + [2(\omega_0 + \omega_1)\dot{s} + \dot{\omega}_1(r_0 + s)]\boldsymbol{u}_\theta \tag{d}$$

where Eqs. (a) and (b) and the conditions $\dot{r}_0 = \ddot{r}_0 = \dot{\omega}_0 = 0$ have been used, since the spacecraft is moving in a circular orbit at a constant angular velocity.

2. Force–dynamic requirements: The dynamic requirements on the forces are satisfied by Newton's second law

$$\sum_{i=1}^{q} \boldsymbol{f}_i = \frac{d\boldsymbol{p}}{dt} \tag{e}$$

where q is the total number of individual forces.

3. Constitutive requirements: The only force acting on the banana peel is the gravitational force and is given by Eq. (4-3) as

$$\boldsymbol{f}_g = -G\frac{m_0 m}{r^2}\boldsymbol{u}_r \tag{f}$$

where m_0 and m are the masses of the Earth and the banana peel, respectively. In making this assumption, we are neglecting gravitational interactions between the spacecraft and the banana peel. Also, the linear momentum $\boldsymbol{p}$ of the banana peel is defined by

---

[7]The angular velocity $\boldsymbol{\omega}$ might be commonly called the "angular velocity of the banana peel." However, as a particle, the banana peel cannot possess an angular velocity, despite the fact that in the context of Eq. (4-8) we may define its angular momentum.

$$\boldsymbol{p} = m\boldsymbol{v}. \tag{g}$$

Similar to the developments in Example 4-6, by substituting Eqs. (d), (f), and (g) into Eq. (e), and writing the results in scalar component form, the scalar equations of motion for the banana peel are obtained in the radial direction as

$$-G\frac{m_0}{(r_0 + s)^2} = \ddot{s} - (\omega_0 + \omega_1)^2(r_0 + s) \tag{h}$$

and in the circumferential direction as

$$2(\omega_0 + \omega_1)\dot{s} + \dot{\omega}_1(r_0 + s) = 0 \tag{i}$$

Equations (h) and (i) for s and ω_1 are nonlinear and thus somewhat difficult to integrate. We want to linearize them.

Recalling the initial conditions as expressed in Eqs. (c), we note that shortly after the banana peel is expelled, s and ω_1 are very small compared with r_0 and ω_0, respectively. They may gradually increase during the course of the banana peel's motion; but for now, we investigate the motion of the banana peel at this early stage. Thus,

$$s \ll r_0 \qquad\qquad \omega_1 \ll \omega_0 \tag{j}$$

In fact, as long as these conditions are satisfied, the solutions that we shall obtain will be valid.

With the assumptions in Eq. (j), a Taylor series expansion is obtained for the left-hand side of Eq. (h) as

$$-G\frac{m_0}{(r_0 + s)^2} = -\frac{Gm_0}{r_0^2} + \frac{2Gm_0 s}{r_0^3} - \frac{3Gm_0 s^2}{r_0^4} + \cdots \tag{k}$$

By retaining at most first-order terms in s and ω_1, Eq. (h) becomes

$$\ddot{s} - \omega_0^2 r_0 - 2\omega_1\omega_0 r_0 - \omega_0^2 s = -\frac{Gm_0}{r_0^2} + \frac{2Gm_0 s}{r_0^3} \tag{l}$$

where Eq. (k) has been used to replace the left-hand side of Eq. (h). Since the spacecraft is moving in a circular orbit, recall Eq. (n) in Example 4-6, which gives the relation between the orbiting radius r_0 and the angular velocity ω_0 as

$$\frac{Gm_0}{r_0^2} = \omega_0^2 r_0 \tag{m}$$

Equation (m) can be used to show that in Eq. (l) the second term on the left-hand side and the first term on the right-hand side cancel. Equation (m) can also be used to show that in Eq. (l) the fourth term on the left-hand side and the second term on the right-hand side may be combined to produce $-3\omega_0^2 s$ on the left-hand side of Eq. (l). In that case, Eq. (l) simplifies to

$$\ddot{s} - 3\omega_0^2 s - 2\omega_1\omega_0 r_0 = 0 \tag{n}$$

Similarly, by retaining s and ω_1, including products containing their derivatives, up to the first order, Eq. (i) simplifies to

$$2\omega_0\dot{s} + \dot{\omega}_1 r_0 = 0 \tag{o}$$

Equation (o) can be integrated directly to give

$$2\omega_0 s + \omega_1 r_0 = C \tag{p}$$

where C is a constant of integration that can be evaluated in accordance with the initial conditions in Eq. (c). At $t = 0$, $s = 0$ and $\omega_1 = 0$, which lead to $C = 0$. Thus,

$$\omega_1 = -\frac{2\omega_0 s}{r_0} \tag{q}$$

Substitution of Eq. (q) into Eq. (n) to eliminate ω_1 yields a second-order linear ordinary differential equation in s as

$$\ddot{s} + \omega_0^2 s = 0 \tag{r}$$

The general solution to Eq. (r) is well known as[8]

$$s = A \cos \omega_0 t + B \sin \omega_0 t \tag{s}$$

where A and B are constants of integration that may be evaluated in accordance with the initial conditions in Eq. (c). At $t = 0$, $s = 0$ and $\dot{s} = -v$. Thus

$$A = 0 \qquad B = -\frac{v}{\omega_0} \tag{t}$$

Therefore, substitution of Eqs. (t) into Eq. (s) gives

$$s = -\frac{v}{\omega_0} \sin \omega_0 t \tag{u}$$

Also, the angular velocity of the banana peel can be calculated from Eq. (q) as

$$\omega_1 = -\frac{2\omega_0 s}{r_0} = \frac{2v}{r_0} \sin \omega_0 t \tag{v}$$

where Eq. (u) has been used. Equation (v) can be integrated to obtain the circumferential angle of the banana peel with respect to the spacecraft, denoted as ϕ. If we assume that $\phi = 0$ is the initial circumferential angle, then integration of Eq. (v) from $t = 0$ to $t = t$ gives

$$\phi = \frac{2v}{\omega_0 r_0}(1 - \cos \omega_0 t) \tag{w}$$

Together, Eqs. (u) and (w) define the motion of the banana peel with respect to the spacecraft. Trajectories of the banana peel, as viewed relative to the spacecraft, are sketched in Figure 4-12, using the numerical values of ω_0 and r_0 given in Eqs. (o) and (p) in Example 4-6, respectively, for various values of the expulsion velocity v as indicated.

Viewed relative to the litterbug in the spacecraft, the banana peel moves in an elliptical path with exactly the same period at which the spacecraft orbits the Earth. In other words, the banana peel will arrive back at the porthole of the spacecraft one orbit of the spacecraft later.

Also, Figure 4-13 shows the orbits of both the banana peel (solid curve) and of the spacecraft (dot-dash curve) for an expulsion velocity of 200 m/s, as viewed from a nonrotating reference frame located at the center of the Earth. The locations of both the banana peel and the spacecraft at different times during one geosynchronous orbit are also sketched. Note that the banana peel "leads" the spacecraft throughout their orbits, until it returns to the expulsion porthole.

Finally, to complete the solution, we reexamine the validity of our linearization assumptions. The solutions in Eqs. (u) and (v) state that s and ω_1 vary periodically with maximum magnitudes

[8]Such equations as Eq. (r) and their solutions are discussed in Chapter 8.

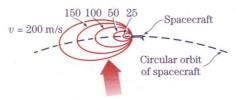

150 100 50 25

$v = 200$ m/s

Spacecraft

Circular orbit of spacecraft

Trajectories of banana peel relative to spacecraft

Earth

FIGURE 4-12 Trajectories of banana peel relative to litterbug in orbiting geosynchronous spacecraft, for various expulsion velocities of banana peel. (All orbits are drawn to scale with respect to Earth.)

$$|s|_{\max} = \frac{v}{\omega_0} \quad \text{and} \quad |\omega_1|_{\max} = \frac{2v}{r_0} \tag{x}$$

Dividing both sides of the first expression in Eqs. (x) by r_0 gives

$$\frac{|s|_{\max}}{r_0} = \frac{v}{\omega_0 r_0} = \frac{v}{V_{\text{spacecraft}}} \tag{y}$$

where $V_{\text{spacecraft}}$ is the orbiting velocity of the spacecraft, the value of which can be easily calculated. Similarly, dividing both sides of the second expression in Eqs. (x) by ω_0 gives

$$\frac{|\omega_1|_{\max}}{\omega_0} = \frac{1}{\omega_0}\frac{2v}{r_0} = 2\frac{v}{V_{\text{spacecraft}}} \tag{z}$$

Thus, since we can expect that $v \ll V_{\text{spacecraft}}$, both of our linearization assumptions in Eqs. (j) are satisfied throughout the time of consideration; hence the solutions in Eqs. (u) and (w) are valid approximations to the dynamic behavior of the banana peel.

In conclusion, we note that although this example was slightly extended, much was accomplished. First, the formulation of the problem was moderately straightforward, following the regimen of Eq. (4-11), so the length of the analysis did not stem from the formulation. Second, the solution of this problem was moderately sophisticated, requiring the specification of initial conditions, order-of-magnitude approximations, a Taylor series expansion, the linearization of nonlinear differential equations, the solution of the resulting linearized differential equations, the interpretation of the obtained solutions, and, of course, a final check to authenticate our assumptions and approximations. Finally, as depicted in Figures 4-12 and 4-13, we also obtained a fascinating result. Wow, we accomplished a great deal!

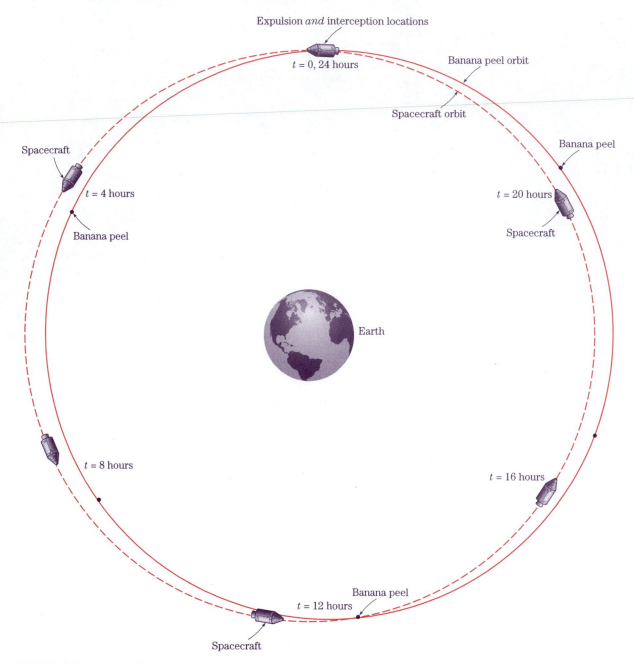

FIGURE 4-13 Orbits of banana peel and geosynchronous spacecraft, with respect to nonrotating reference frame located at center of Earth, when banana peel expulsion velocity is 200 m/s. (All orbits are drawn to scale with respect to Earth.)

Problems for Chapter 4

Category I: Back to Physics

Generally problems of first kind: Problems 4-1 through 4-6

Problem 4-1 *(HRW127, 64P)*[9]: Figure P4-1 shows a section of an alpine cable-car system. The maximum permissible mass of each car with occupants is 2,800 kg. The cars, riding on a support cable, are pulled by a second cable attached to each pylon. What is the difference in tension between adjacent sections of pull cable if the cars are at maximum mass and are being accelerated up the 35° incline at 0.81 m/s^2?

Problem 4-2 *(HRW129, 73P)*: Figure P4-2 shows a man sitting in a bosun's chair that dangles from a massless rope, which runs over a massless frictionless pulley and back down to his hands. The combined mass of the man and the chair is 95 kg.

(a) With what force must the man pull the rope for him to rise at constant speed?

(b) What force is needed for an upward acceleration of 1.3 m/s^2?

(c) Suppose, instead, that the rope on the right is held by a person on the ground. Repeat parts (a) and (b) for this new situation.

(d) In each of the four cases, what is the force exerted on the ceiling by the pulley system?

Problem 4-3 *(HRW284, 82)*: A 0.30 kg softball has a velocity of 12 m/s at an angle of 35° below the horizontal just before making contact with the bat. The ball leaves the bat 2.0 ms later with a vertical velocity of 10 m/s as sketched in Figure P4-3. What is the magnitude of the average force acting on the ball while it is in contact with the bat?

Problem 4-4 *(HRW253, 56P)*: Two long barges are moving in the same direction in still water, one with a speed of 10 km/h and the other with a speed of 20 km/h. While they are passing each other, coal is shoveled from the slower to the faster one at a rate of 1000 kg/min; see Figure P4-4. How much additional force must be provided by the driving engines of each barge if neither is to change speed? Assume that the shoveling is always perfectly sideways and that the frictional forces between the barges and the water do not depend on the weight of the barge.

Problem 4-5 *(HRW215, 31P)*: Tarzan, who weighs 688 N, swings from a cliff at the end of a convenient

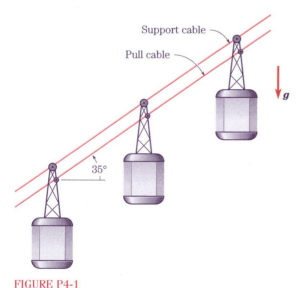

FIGURE P4-1

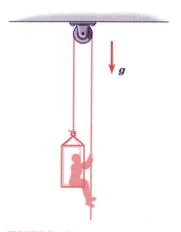

FIGURE P4-2

[9]These designations reference Halliday, Resnick, and Walker. For example, "HRW127, 64P" denotes "Halliday, Resnick, and Walker; page 127, problem 64P."

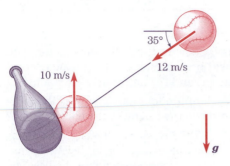

35°

10 m/s

12 m/s

g

FIGURE P4-3

vine that is 18 m long, see Figure P4-5. From the top of the cliff to the bottom of the swing, he descends by 3.2 m. The vine will break if the force on it exceeds 950 N. Does the vine break?

Problem 4-6 *(HRW216, 37P):* A boy is seated on the top of a hemispherical mound of ice, as sketched in Figure P4-6. He is given a very small push and starts sliding down the ice. Show that he leaves the ice (loses contact) at a point whose height is $2R/3$ if the ice is frictionless.

Generally problems of second kind: Problems 4-7 through 4-12

Problem 4-7 *(HRW156, 65P):* An airplane is flying in a horizontal circle at a speed of 480 km/h. If the wings of the plane are tilted 40° to the horizontal, as sketched in Figure P4-7, what is the radius of the circle in which the plane is flying? Assume that the required force is provided entirely by an "aerodynamic lift" that is perpendicular to the wing surface.

Problem 4-8 *(HRW135, Sample Problem 6-2):* If a car's wheels are "locked" (kept from rolling) during emergency braking, the car slides along the road. Ripped-off bits of tire and small melted sections of road form the "skid marks" that reveal the cold-welding during the slide. The record for the longest skid marks on a public road was reportedly set in 1960 by a Jaguar on the M1 highway in England—the marks were 290 m long! Assuming that the kinetic coefficient of friction was 0.6, how fast was the car going when the wheels locked?

Problem 4-9 *(HRW89, 17E):* A dart is thrown horizontally toward the bull's-eye point P on the dart board as sketched in Figure P4-9 with an initial speed

of 10 m/s. It hits point Q on the rim, vertically below P, 0.19 s later.

(a) What is the distance PQ?

(b) How far away from the dart board did the dart thrower stand?

Problem 4-10 *(HRW90, 36P):* In Galileo's *Two New Sciences*, the author states that "for elevations (angles of projection) which exceed or fall short of 45° by equal amounts, the ranges (distance between projection and landing points) are equal...." Prove this statement. (See Figure P4-10.)

Problem 4-11 *(HRW140, Sample Problem 6-6):* During the motion of projectiles, air resistance effects may be neglected if the velocity is low and the *time of flight* (the duration of the motion) is short. However, in many instances, they play an important role. The force of air resistance is often called the *drag force* and is frequently expressed as

$$D = \frac{1}{2}C\rho Av^2$$

where D the drag force, C the nondimensional *drag coefficient*, ρ the air density, A the effective cross-sectional area (the projected area perpendicular to the velocity), and v the velocity of the projectile define the parameters.

A raindrop with radius $R = 1.5$ mm falls from a cloud that is at height $h = 1200$ m above the ground. The drag coefficient C for the raindrop is 0.60. Assume the raindrop is spherical throughout its fall, the density of water ρ_w is 1000 kg/m^3, and the density of air ρ_a is 1.2 kg/m^3.

(a) What is the terminal speed of the raindrop?

(b) What would have been the speed just before impact if there had been no drag force?

Problem 4-12 *(HRW250, 15E):* A man of mass m clings to a rope ladder suspended below a balloon of mass M; see Figure P4-12. The balloon is stationary with respect to the ground.

(a) If the man begins to climb the ladder at speed v (with respect to the ladder), in what direction and with what speed (with respect to the Earth) will the balloon move?

(b) What is the state of the motion after the man stops climbing?

FIGURE P4-4

FIGURE P4-5

FIGURE P4-6

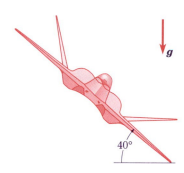

FIGURE P4-7

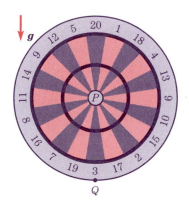

FIGURE P4-9

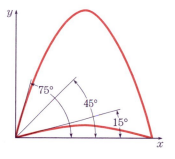

FIGURE P4-10

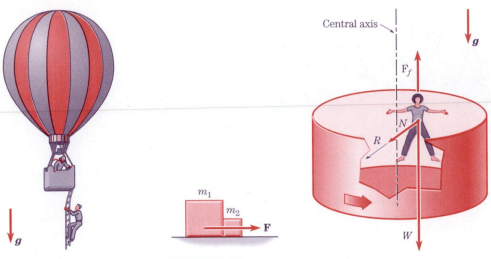

FIGURE P4-12

FIGURE P4-13

FIGURE P4-15

Generally hybrid problems: Problems 4-13 through 4-17

Problem 4-13 *(HRW125, 40P):* Two blocks are in contact on a frictionless table. A horizontal force is applied to one block, as sketched in Figure P4-13.

(a) if $m_1 = 2.3$ kg, $m_2 = 1.2$ kg, and $F = 3.2$ N, find the force of contact between the two blocks.

(b) Show that if the same force F is applied to m_2 rather than m_1, the force of contact between the blocks is 2.1 N, which is not the same value derived in (a). Explain the difference.

Problem 4-14 *(HRW126, 49P):* Three blocks are connected via a cable on a horizontal frictionless table and pulled to the right by a force $T_3 = 65.0$ N, as sketched in Figure P4-14. If $m_1 = 12.0$ kg, $m_2 = 24.0$ kg, and $m_3 = 31.0$ kg, calculate

(a) the acceleration of the system; and

(b) the tension T_1 and T_2.

Problem 4-15 *(HRW146, Sample Problem 6-12):* Even some seasoned rollercoaster riders blanch at the thought of riding the Rotor, which is essentially a large hollow cylinder that rotates rapidly around its central axis, as sketched in Figure P4-15. Before the ride begins, a rider enters the cylinder through a door on the side and stands on a floor against a canvas-covered wall. The door is closed, and as the cylinder begins to turn, the rider, wall, and floor move in unison. When the rider's speed reaches some predetermined value, the floor abruptly and alarmingly falls away. The rider does not fall with it but instead is pinned to the wall while the cylinder rotates, as if an unseen (and somewhat unfriendly) agent is pressing the body to the wall. Later, the floor is eased back to the rider's feet, the cylinder slows and the rider sinks a few centimeters to regain footing on the floor. (Some riders consider all this to be fun.)

Suppose the static coefficient of friction between the rider's clothing and the canvas is 0.4 and that the cylinder's radius is 2.1 m.

FIGURE P4-14

(a) What minimum speed must the cylinder and rider have if the rider is not to fall when the floor drops?

(b) If the rider's mass is 49 kg, what is the magnitude of the so-called "centripetal force" on her?

Problem 4-16 *(HRW126, 54P)*: A new 26-ton Navy jet requires an airspeed of 280 ft/sec for takeoff. Its engine develops a maximum thrust (force) of 24,000 lb.

(a) What is the minimum length of the runway required for the plane to take off from a ground airport?

(b) On an aircraft carrier, the runway is only 300 ft long. What is the minimum thrust (force) (assumed constant throughout the takeoff) needed from the catapult that is used to help launch the jet?[10]

Problem 4-17 *(HRW152, 25P)*: A 3.5 kg block is pushed along a horizontal floor by a force $F = 15$ N that makes an angle $\theta = 40°$ with the horizontal, as sketched in Figure P4-17. The kinetic coefficient of friction between the block and the floor is 0.25.

(a) Calculate the magnitude of the frictional force exerted on the block.

(b) Calculate the acceleration of the block.

(c) If the force is pulling the block, instead of pushing, what are the answers for parts (a) and (b) in this new situation?

Category II: Intermediate

Conceptual problems: Problems 4-18 through 4-21

Problem 4-18: You and your friend are riding in a car along a straight section of highway and, as you enter a curve toward left, you notice that the box of dynamics books on the back seat slides from the center toward the right. See Figure P4-18. Your friend suggests that the two of you have just witnessed "centrifugal force" in action. What qualitative response might you offer, since he is your friend?

Problem 4-19: Late one night, during your return to college with a friend who is driving, in the back seat you awaken from a deep sleep and observe that the speedometer is reading a steady 135 mi/hr. You also notice that the medal hanging from the rearview mirror on a 6 in. long chain is at a steady angle of 15° with respect to the vertical (See Fig. P4-19).

(a) You have heard that such medals are pulled to the side by "centrifugal forces." Is this so? Why do you suppose the medal is hanging at an angle?

(b) Assuming you are on a level section of the highway (indeed, on the highway at all), find the radius of curvature of the highway, as

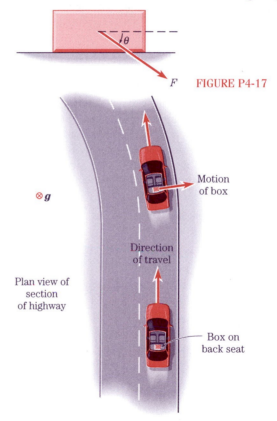

F FIGURE P4-17

Motion of box

⊗ g

Direction of travel

Plan view of section of highway

Box on back seat

FIGURE P4-18

---

[10]Takeoffs and landings of aircraft on carriers are high-risk activities. The landing is often characterized as a "controlled crash" since, for example, a 25-ton F-14 hooks one of four cables stretched across the carrier's deck and is brought from about 125 mi/hr to a "dead" stop in less than 200 ft.

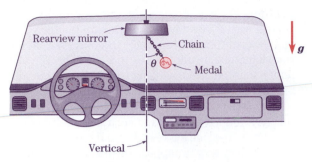

Rearview mirror — Chain — Medal — θ — g — Vertical

FIGURE P4-19

seen in a plane view. Does the highway turn left or right at that location?

Problem 4-20: Find all the necessary and sufficient sets of the scalar equations, obtained by decomposing the vector expressions in Eqs. (4-2) and (4-10), for the momentum principle of a particle.

Problem 4-21: One of several bolts that hold two rotating plates together is sketched in Figure P4-21. The head of the bolt of interest shears off instantaneously in brittle fracture, at the instant shown. The angular velocity ω_0 is constant. The mass of the bolt head is m_0, the mass of the threaded portion of the bolt is m_1, and the diameter of the threaded portion is d. The friction between the bolt head and the upper plate is negligible.

 (a) What is the fracture stress of the bolt material?

 (b) Does the bolt head fly off along trajectory A, B, or C?

 (c) Is the trajectory selected in part (b) due to centrifugal force?

 (d) If your answer to part (c) is yes, calculate the centrifugal force. If your answer to part (c) is no, then what force or principle can you cite to explain your answer?

Generally problems of first kind: Problems 4-22 through 4-43

• Note: In Problems 4-22 through 4-38, the system description can be found in the referenced problems in Chapter 3, where the velocity and acceleration of the body of concern were calculated.

Problem 4-22: In Problem 3-19, determine the tension in the rope at positions A and B.

Problem 4-23: In Problem 3-21, determine the force exerted by the seat on each passenger.

Problem 4-24: In Problem 3-27, determine the force exerted on the head of the passenger by his neck.

Problem 4-25: In Problem 3-28, determine the force applied to each weight by the skater's hands.

Problem 4-26: In Problem 3-31, determine the force exerted on the gem by the robot.

Problem 4-27: In Problem 3-32, determine the force exerted on the rider by the seat.

Problem 4-28: In Problem 3-33, determine the force exerted on the rider by the seat.

Problem 4-29: In Problem 3-36, determine the force exerted on the ball by the pine tar glob.

Problem 4-30: In Problem 3-38, determine the force exerted on the bead by the spoke.

Problem 4-31: In Problem 3-40, determine the force exerted on the payload by the arm.

Problem 4-32: In Problem 3-41, determine the force exerted on the platform by the astronaut.

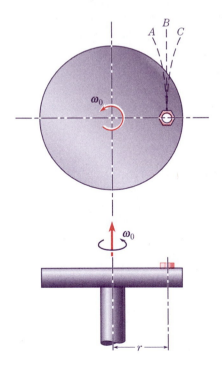

FIGURE P4-21

Problem 4-33: In Problem 3-43, determine the force exerted on the workpiece by the robot arm.

Problem 4-34: In Problem 3-44, determine the force exerted on the rider by the seat.

Problem 4-35: In Problem 3-45, determine the force exerted on the floor by the passenger.

Problem 4-36: In Problem 3-46, determine the force exerted on the astronaut by the platform.

Problem 4-37: In Problem 3-49, determine the force exerted on the workpiece by the robot.

Problem 4-38: In Problem 3-56, determine the force exerted on the floor by the passenger.

Problem 4-39: A baseball traveling horizontally strikes a rigid vertical wall at a speed of 35 m/s and returns at a speed of 30 m/s, as sketched in Figure P4-39a. It is of interest to find the maximum force exerted on the wall by the impact. Only the time interval during which the ball and the wall are in contact

is known as 0.025 s. So, as a rough estimation, the force versus time plot is approximated as a parabola as sketched in Figure P4-39b. *Estimate* the maximum force F_{max}.

Problem 4-40: In the design of couplers—the hook-like linking mechanisms—between the cars of a train, extreme situations are often considered to assess limiting requirements. Suppose a locomotive of mass 30,000 kg pulls 80 cars each of mass 30,000 kg. At an acceleration of 0.5 m/s^2, how much load is transferred by the coupler between the locomotive and the first car; between the 79th and 80th cars?

Problem 4-41: In testing the bonding strength of a new adhesive, a small test specimen of mass m is attached to a horizontal rotating platform by use of this adhesive, as sketched in Figure P4-41. The specimen is a distance r from the center, and has a bonding area A. If the adhesive ruptures when the platform is rotating at angular velocity Ω, determine its bonding strength.

Problem 4-42: In testing the bonding strength of a new adhesive, a small test specimen of mass m is attached to a small horizontal platform that is *translating* along a circular path, as sketched in Figure P4-42. (Refer to Figure 3-6b for the meaning of *curvilinear translation.*) The arm holding the platform rotates at a constant angular velocity. The specimen is a distance r from the center, and has a bonding area A. If the adhesive ruptures when the platform is rotating at angular velocity Ω, determine its bonding strength.

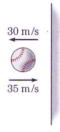

30 m/s

35 m/s

(a) The impact.

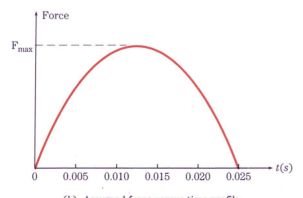

Force

F_{max}

0   0.005   0.010   0.015   0.020   0.025 $t(s)$

(b) Assumed force versus time profile.

FIGURE P4-39

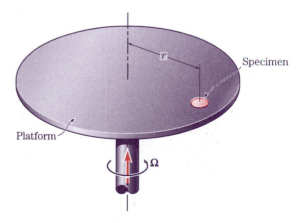

r Specimen

Platform

Ω

FIGURE P4-41

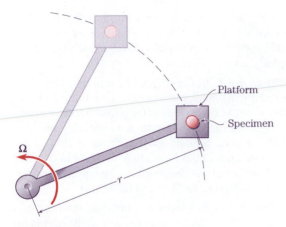

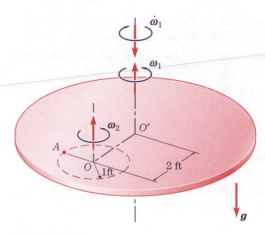

FIGURE P4-42

FIGURE P4-43

Problem 4-43: A mass A weighing 4 oz is made to rotate at a constant angular velocity $\omega_2 = 15$ rad/sec with respect to a platform. This motion is in the plane of the platform which, at the instant sketched in Figure P4-43, is rotating at an angular velocity $\omega_1 = 10$ rad/sec and an angular acceleration $\dot{\omega}_1 = -4$ rad/sec^2, both with respect to ground. Note that mass A does not contact the platform, and that gravity acts. If the mass of the rod supporting A is neglected, determine the axial and shear forces in the rod at point o at the instant shown.

Generally problems of second kind: Problems 4-44 through 4-59

Problem 4-44: As illustrated in Figure P4-44, *rectilinear motion* is a special type of motion such that when a rectilinear reference frame (cartesian coordinate system) is appropriately chosen for the dynamics of a particle, the motions along the three rectilinear directions are independent of each other and yet the combined motions along the three rectilinear directions define the motion of the particle. Assume a particle of mass m and initial velocity $\boldsymbol{v}_0$ is in rectilinear motion with respect to an $OXYZ$ reference frame under the applied arbitrary force $\boldsymbol{F}$. Further, the initial position of the particle is the point whose cartesian coordinates are (x_0, y_0, z_0), as sketched in Figure P4-44a.

(a) Find the position of a particle subjected to this force. Express the results as the cartesian coordinates of the particle as functions of time.

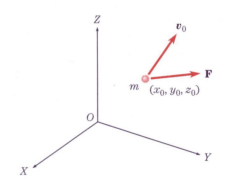

(a) Particle m in rectilinear motion subjected to force $\boldsymbol{F}$.

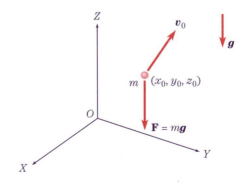

(b) Particle m in rectilinear motion in gravitational field; motion of projectile.

FIGURE P4-44

(b) When the applied force is due to gravity, as sketched in Figure P4-44b, the motion is usually referred to as the *motion of a projectile*. Specialize the results in (a) for this case.

Problem 4-45: For the *motion of a projectile*, it is always possible to find a vertical plane within which the motion evolves, as sketched in Figure 4-45a. Thus the motion of a projectile is usually treated as two-dimensional—one dimension in the horizontal direction and one dimension in the vertical—where the vertical direction is defined as the same as the direction of $\boldsymbol{g}$, as sketched in Figure 4-45b. If the projectile at position (x_0, y_0) in the xy coordinate system is given the initial velocity v_0 in a direction that forms an angle θ_0 with respect to the horizontal as sketched in Figure 4-45b, find the equation for the planar curve that describes the trajectory of the projectile. The reader may recall such a planar curve is a *parabola*.

Problem 4-46: When a bullet is fired from a rifle, the plot of the pressure in the barrel versus the time between the firing and the bullet's exit from the barrel is sketched in Figure P4-46. Assuming the friction between the barrel and the bullet is negligible and noting that the bullet weighs 0.025 lb with a diameter of 0.35 in., estimate the speed at which the bullet exits the barrel.

Problem 4-47: The four-seater racing scull is poised for a start, as sketched in Figure P4-47. At the sound of the gun, each oarsman exerts a 30 lb constant push on the water with his oar in the forward direction of the boat. Find the speed of the scull after 1 sec. The scull weighs 400 lb, and each oarsman weighs 150 lb. Each oarsman is moving forward in the direction of the scull's motion and relative to the boat at 1 ft/sec. The weight of entrained water and friction may be neglected.

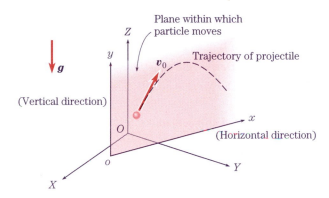

(a) Perspective of motion of projectile in cartesian coordinate system $OXYZ$.

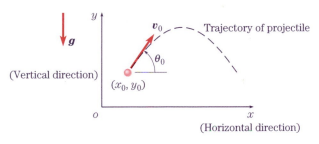

(b) Motion of projectile in vertical plane oxy.

FIGURE P4-45

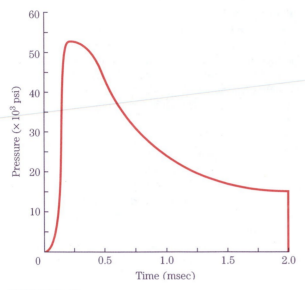

FIGURE P4-46

Problem 4-48: A barge of mass M carrying a car of mass m is still and afloat a distance d from the pier. The car is parked at the far end of the barge, which is free to move in the water, as sketched in Figure P4-48. Can the barge be moved closer to the pier only by driving the car on the barge? If yes, how? If not, why? Ignore the frictional effects between the barge and the water.

Problem 4-49: Two barges of mass M and $2M$ are connected by a cable. The cable is shortened to one-half of its original length ℓ by turning a windlass on one of the barges. Neglecting any frictional effects between the barges and the water, find the distance moved by the barge of mass $2M$.

Problem 4-50: A young man is standing in a canoe awaiting a young lady. The young man weighs 200 lb and is at the far end of a 20 ft canoe; the other end of the canoe is at the end of the pier. See Figure P4-50. The canoe also weighs 200 lb and its center of mass is located at its mid-length. When the young lady appears, he scrambles forward to greet her, but when he has moved the 20 ft to the other end of the canoe, to his surprise he finds that he cannot reach her. How far from the pier is the near tip of the canoe after he has made his 20 ft dash? The canoe is not tied to the pier and there are no currents in the water. Neglect all frictional effects between the canoe and the water.

Problem 4-51: The equation of motion of a simple plane pendulum is well known as

$$\ddot{\theta} + \frac{g}{\ell} \sin \theta = 0$$

where ℓ is the length of the pendulum and θ is the angle between the pendulum and the vertical, as indicated in Figure P4-51. Note that gravity acts.

    (*a*) Derive this equation via the direct approach.

    (*b*) Derive this equation using the *principle of angular momentum.*

Problem 4-52: Johannes Kepler devised his three laws of planetary motion after years of observation and arduous calculations. Today many undergraduates could tell him that his second law is a direct consequence of the angular momentum principle.

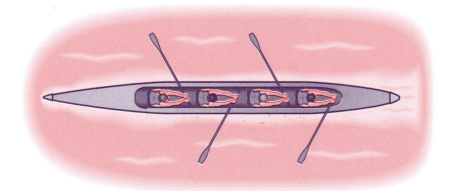

FIGURE P4-47

FIGURE P4-48

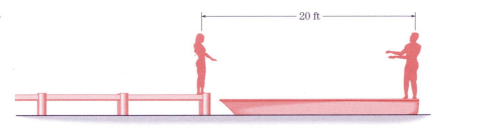

FIGURE P4-50

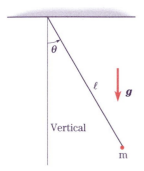

FIGURE P4-51

(a) Design the optimal banking angle so that no lateral frictional force is required to help the car to make the turn when it travels at the speed limit.

(b) Given the bank angle in part (a), if the static coefficient of friction between the road surface and tires is 0.5 for a dry day, 0.25 for a rainy day and 0.05 during icy conditions, what is the maximum speed such that a car will not slip for each of these driving conditions?

Kepler's second law states that an orbiting planet sweeps out equal areas in equal times. (Refer to Figure 1-13.) Prove this statement.

Problem 4-53: A portion of a highway turns at a radius of $R = 2000$ ft, as sketched in Figure P4-53. The standard speed limit for this portion of the highway is 55 mi/hr. Usually, the road surface is designed to bank at some angle to improve safety and ease of driving. Note that gravity acts.

Problem 4-54: Two masses m_1 and m_2 are connected by a massless inextensible strap that runs through a massless and frictionless pulley, as sketched in Figure P4-54. The strap does not slip on the rotating pulley. Each mass is then connected to the floor by a spring. The spring constants are k_1 and k_2. Note that gravity acts, and assume the strap remains taut at all times. Derive the equation(s) of motion.

Problem 4-55: Two masses m_1 and m_2 are connected by an inextensible massless strap that runs through a massless frictionless pulley as sketched in Figure P4-55. The mass on the horizontal surface, m_1, is connected to a rigid wall by a spring of spring constant k. The mass m_2 can move frictionlessly along the ramp, which is inclined at angle θ. Note that gravity acts, and assume the strap remains taut at all times. Derive the equation(s) of motion.

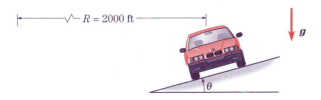

FIGURE P4-53

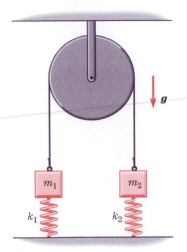

FIGURE P4-54

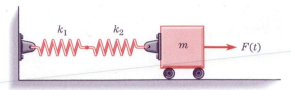

FIGURE P4-56

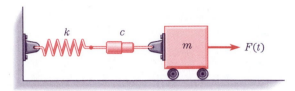

FIGURE P4-57

Problem 4-56: In the mass spring system sketched in Figure P4-56, mass m moves without friction on a horizontal floor. It is restrained by two springs connected in series to a rigid wall. The spring constants are k_1 and k_2. A force $F(t)$ acts on the mass. Derive the equation(s) of motion.

Problem 4-57: In the mass spring dashpot system sketched in Figure P4-57, mass m moves without friction on a horizontal floor. It is restrained by a spring of spring constant k and a dashpot of dashpot constant c connected in series to a rigid wall. A force $F(t)$ acts on the mass. Derive the equation(s) of motion.

Problem 4-58: In the mass spring dashpot system sketched in Figure P4-58, two masses m_1 and m_2 move without friction on a horizontal floor. Mass m_1

is restrained to a rigid wall by a spring of spring constant k_1. Mass m_2 is restrained to the same rigid wall by a dashpot of dashpot constant c_1. The two masses are connected by a spring of spring constant k_2 and a dashpot of dashpot constant c_2. An external force $F(t)$ acts on mass m_2. Derive the equation(s) of motion.

Problem 4-59: In the mass spring dashpot system sketched in Figure P4-59, two masses m_1 and m_2 move without friction on a horizontal floor. The two masses are connected by a spring of spring constant k_2 and a dashpot of dashpot constant c_2. Each of the two masses is connected to a fixed wall by another spring, of spring constants k_1 and k_3, and another dashpot, of dashpot constants c_1 and c_3, all as sketched in Figure P4-59. An external force $F(t)$ acts on mass m_1. Derive the equation(s) of motion.

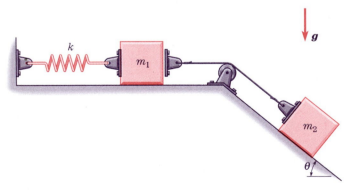

FIGURE P4-55

Category III: More Difficult

Generally problems of first kind: Problems 4-60 through 4-62

Problem 4-60: A mass m slides frictionlessly on a semicircular track of radius R, as sketched in Figure P4-60. Determine the contact force as a function of θ if the velocity of the mass is zero when $\theta = 0$. Note that gravity acts.

Problem 4-61: A particle of mass m initially rests at the inside bottom of a cylindrical pipe of interior radius R, as sketched in Figure P4-61. The mass can slide on the interior surface of the pipe without friction. The mass is suddenly given a velocity v_0 in the tangential direction such that it starts to slide up the pipe. At what angle θ and for what possible initial velocities v_0 does the mass m lose contact with the pipe? Note that gravity acts.

Problem 4-62: Reconsider the mast-deploying antenna in Examples 3-8 and 4-1. Suppose the mast has a circular cross-section of radius 0.2 in., and is made of an aluminum alloy having a modulus of elasticity of 10,000 psi. The mast is deployed through a guide tube such that during flexure of the mast there is no slope at its base, as sketched in Figure P4-62.

    (a) Find the stress in the mast at the instant shown in Figure 4-6a.

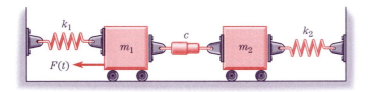

FIGURE P4-58

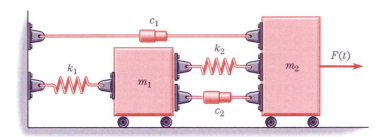

FIGURE P4-59

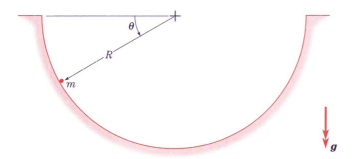

FIGURE P4-60

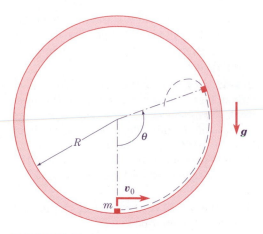

FIGURE P4-61

(b) Will the mast buckle at the instant shown in Figure 4-6a?

(c) At what length L will buckling occur?

Ignore the effects of lateral loads on the enhancement of buckling.[11]

Generally problems of second kind: Problems 4-63 through 4-66

Problem 4-63: A double-pulley, consisting of two massless frictionless pulleys of different radii r_1 and

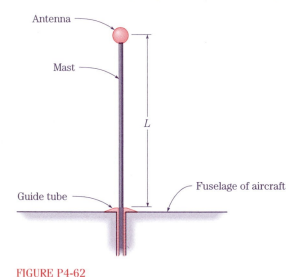

FIGURE P4-62

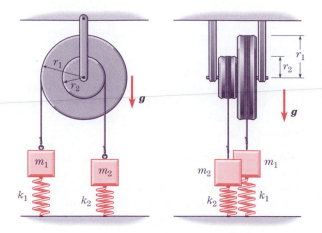

(a) Front view.   (b) Side view.

FIGURE P4-63

r_2 and welded together concentrically, supports two masses m_1 and m_2, as sketched in Figure P4-63. A spring connects each mass to the floor. The spring constants are k_1 and k_2. The connecting strap is inextensible and massless and does not slip on the rotating pulley. Note that gravity acts, and assume the strap remains taut at all times. Derive the equation(s) of motion.

Problem 4-64: A simple plane pendulum of mass m is attached to a cart of mass M, which moves without friction on a horizontal rail (Refer to Fig. P4-64). The pendulum has length ℓ. Note that gravity acts. Derive the equation(s) of motion.

Problem 4-65: A shock absorber is shown schematically in Figure P4-65. It consists of two identical

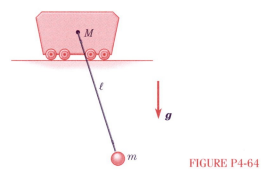

FIGURE P4-64

[11]We emphasize that this formulation is for a constant load. The analysis for a harmonically varying load is more difficult, yet the results of such an analysis can be applied in a straightforward manner. For example, see *Theory of Elastic Stability,* 2nd ed., by Stephen P. Timoshenko and James M. Gere, McGraw-Hill, 1961, pp. 158–162.

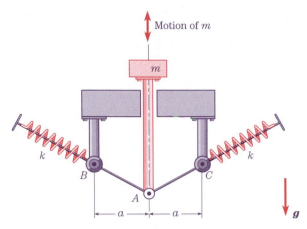

FIGURE P4-65

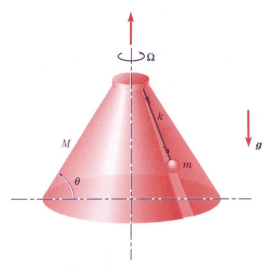

FIGURE P4-66

massless rigid rods of length ℓ that are pin-joined at point A and that can frictionlessly slide through two guides B and C, which themselves are free to rotate so as to adjust their guiding direction. B and C are a distance $2a$ apart. Two springs are precompressed and mounted between the guides and the free ends of the rods. A mass m is supported at point A by a valve that can slide along a vertical slot. A dashpot of dashpot constant c is incorporated to model the losses due to the sliding of the valve along the slot. When in equilibrium, each arm forms an angle θ_0 with respect to the ceiling. Note that gravity acts. Derive the equation(s) of motion for *small motions* about the equilibrium position.

Problem 4-66: A mass m slides frictionlessly in a track cut into the surface of a cone of mass M, which has a conical angle θ, as sketched in Figure P4-66. The cone rotates at a constant angular velocity Ω about its axis of symmetry. The mass is restrained by a spring of spring constant k and unstretched length ℓ_0, the other end of the spring being at the axis of rotation. Note that gravity acts. Derive the equation(s) of motion.

Generally hybrid problems: Problems 4-67 through 4-70

Problem 4-67: A space station rotates relative to an inertial reference frame $BXYZ$ at a constant angular velocity ω_1 in order to simulate Earth's gravity in the living quarters located on the rim, such as at point A, as sketched in Figure P4-67.

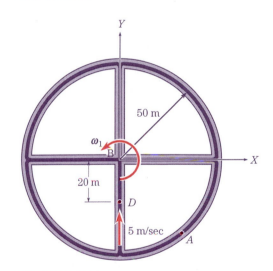

FIGURE P4-67

(a) Find the value of ω_1.

(b) A 10 kg package D is being transported to the center B at a constant velocity of 5 m/s along and relative to a spoke of the space station. Find the force vector acting on the package D, at the instant shown.

Problem 4-68: During landing, an aircraft weighing 50,000 lb has a touchdown speed of 150 ft/sec. A parachute is deployed and the engine is shut off at the instant of touchdown. The drag force generated by the parachute is $f = 0.6v^2$ where f is in lb and v is the instantaneous speed of the aircraft in ft/sec. Determine the length of the runway required in order to reduce the speed of the aircraft to 50 ft/sec, after which the brakes are to be applied to bring the aircraft to a complete stop.

Problem 4-69: A segment of chain of mass per unit length ρ and total length ℓ is placed on a table. The kinetic coefficient of friction between the chain and the table is μ_k. The entire chain is fully extended and one of its ends initially dangles over the table edge a length ℓ_0, as sketched in Figure P4-69. The dangled length, when released, is sufficient to drag the entire chain over the edge due to gravity. Determine the dangled length of the chain as a function of time after its release. Note that gravity acts.

Problem 4-70: A small ring, which may be modeled as a particle, travels along a circular hoop at an initial speed of v_0, as sketched in Figure P4-70. The hoop has a radius R and lies in a horizontal plane with respect to gravity. The kinetic coefficient of friction between the hoop and the ring is μ_k. Note that gravity acts. Determine the distance traveled by the ring before it stops.

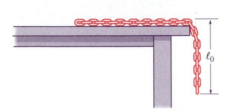

FIGURE P4-69

FIGURE P4-70

c	Campanella			Aaron	
1st	McCovey			Carlton	
2nd	Morgan			Musial	
3rd	Schmidt	755 HR		Palmer	Most difficult omissions
ss	Banks	5714 SO		Robinson, F.	
of	Mantle			Rose	
of	Mays			Ryan	
of	Williams	J. Robinson		Yastrzemski	

rhp	Seaver	Career	rhp	Gibson	One game	
lhp	Spahn		lhp	Koufax		

CHAPTER FIVE

VARIATIONAL FORMULATION FOR SYSTEMS OF PARTICLES

5-1 · INTRODUCTION

In this chapter we begin the formal presentation of the *indirect approach,* the approach based on the scalar concepts of energy and work. For reasons that will become quite apparent, we shall generally refer to this approach variously as *lagrangian dynamics* or as *variational dynamics.* (*Lagrangian dynamics* will be commonly used when we limit our interest to lumped-parameter systems. Furthermore, *analytical dynamics* is another frequently used synonym for this approach; however, in general, we shall not use this term.) We shall be particularly deliberate in our presentations. In this chapter we shall be concerned exclusively with mechanical systems where all of the mass is contained in particles. Also, in this chapter, as well as throughout this book, we shall restrict our analyses to *holonomic* systems, a term that we shall define in Subsection 5-4.2.

The development of variational dynamics for mechanical systems is intimately associated with the contributions of Leibniz, John Bernoulli, Euler, d'Alembert, and Lagrange in the eighteenth century, and Hamilton in the nineteenth century. We shall take Hamilton's principle as the fundamental law for deriving the equations of motion of dynamic systems via the indirect approach. While this perspective is not unique, it is unconventional, since many writers designate Newton's laws as the fundamental principles of mechanical dynamics, irrespective of the approach. Indeed, many writers begin with Newton's laws and then proceed to "derive" Hamilton's principle, a process we feel is philosophically and technically inappropriate. For mechanical particulate systems, Hamilton's principle is an alternative to Newton's laws. We espouse the perspective that Hamilton's principle is a fundamental law of science, and that as such it cannot be derived. Furthermore, we adopt the view that Hamilton's principle is arguably more fundamental than Newton's laws, as it is also applicable to mechanical systems containing extended bodies (Chapter 6) as well as to nonmechanical systems (Chapter 7).

As expressed in his *Principia* (1687), Newton presented particle dynamics in terms of the quantities of force and momentum. In a series of letters and papers written between 1686 and 1695, Leibniz chose to advocate the alternative quantities of what we now call *energy* and *work.* The concept of imagining allowable neighboring configurations was exploited in John Bernoulli's 1717 letter announcing for statics the *principle of virtual work,* which was extended to dynamics by d'Alembert in a 1742 letter. From his *Mechanics* (1736) and numerous mathematical and mechanical contributions throughout much of the eighteenth century, Euler furnished analyses that would blossom into Lagrange's *calculus of variations* (ca. 1755–60) and *Méchanique Analytique* (1788), one of the most renowned books in mechanics. It was Lagrange who introduced into dynamics the important concept of potential energy, though not the name, which appears to be due to George Green (1793–1841) of Cambridge in 1828; recall that Leibniz's *vis viva* had plowed the way for kinetic energy. Furthermore, it was Thomas Young, of hieroglyphics decipherment and "Young's modulus" notoriety, who appears to have been the first person to use the word "energy" to mean *kinetic energy* in lieu of *vis viva* or "living force." In two papers in 1834 and 1835, Hamilton presented the principle under which all variational dynamics (including electricity, magnetism, fluids, optics, and quantum mechanics, although obviously not so generally envisioned by him) could be embraced.

It will be seen that Lagrange's equations provide a powerful tool for formulating the equations of motion for lumped-parameter systems, leading to ordinary differential equations (ODEs). Hamilton's principle may be used for this purpose also, but not as efficiently. Hamilton's principle will provide the comparably powerful tool for formulating the equations of motion for continuous-parameter systems, leading to partial differential equations (PDEs). Thus, in this chapter we shall go far in constructing an immense scheme capable of addressing a broad range of problems, and as the reader shall see, substantially more efficiently than via the direct approach.

5-2 FORMULATION OF EQUATIONS OF MOTION

We begin our formal presentation of variational mechanics by reemphasizing the basic requirements for formulating the equations of motion for mechanical systems:

1. Geometric requirements on the motions;

2. Dynamic requirements on the forces; and   (5-1)

3. Constitutive requirements for all the system elements and fields.

The importance of this collection of statements is emphasized (again) by giving it not only an equation number but an entire section in this chapter.

In formulating the equations of motion for a specific model, we begin by defining the geometric constraints on the system. As we shall define in more specific terms later in this chapter, a geometric constraint is a requirement that restricts or imposes the system's spatial motion. Adherence to these geometric constraints will limit consideration to geometric configurations that do not violate these restrictions. Configurations of the system that satisfy the geometric constraints are called *geometrically admissible states*. By defining geometric coordinates that are compatible with the geometrically admissible (that is, allowable) states, we shall satisfy Requirement 1. However, the dynamic requirements on the forces that act throughout the system will not necessarily be satisfied by all the spatial motions that are compatible with the geometrically admissible states. Because a particle *may* go to a particular point in space does not mean that it *will* go to that point when subjected to the specified forces.

In Chapter 4, we stated that in the direct approach (momentum approach), Requirement 2 for a particle is satisfied by Newton's linear momentum principle. In the indirect approach (variational approach), Requirement 2 will be satisfied by Hamilton's principle. In essence, Hamilton's principle will be expressed as an equation that will vanish (that is, equal zero) if the proposed geometrically admissible state also satisfies Requirement 2. We shall find that for lumped-parameter systems, Hamilton's principle reduces to Lagrange's equations, which generally provide the most straightforward route to obtaining the equations of motion.

In order to apply Hamilton's principle or Lagrange's equations to a specific system, it is necessary to introduce the constitutive information characterizing the system's elements and fields. This is accomplished in terms of worklike and energylike expressions, enunciated in accordance with the constitutive relations, which are rendered in terms of the geometrically admissible states. Thus, Requirement 3 will be satisfied by the proper definition of these worklike and energylike expressions.

Consider a particle that is confined to a room by its walls, ceiling, and floor. In this case, the walls, ceiling, and floor impose the constraints, and the resulting geometrically admissible states consist of all locations within the room. Hamilton's principle will be satisfied when the particle undergoes only motions (within the room) that also are consistent with the applied forces. The connective information between the forces and the motions is provided via the constitutive worklike and energylike expressions in which Hamilton's principle is expressed.

Hamilton's principle may not be immediately comprehensible to the reader at first encounter. The reason for this is likely to be because, *in form*, it is different from other principles the reader has studied. Thus, it is suggested not to think of it in terms of *other* principles that have been encountered. Think of it as it truly is—as a new principle about the way nature behaves. As a principle, like all laws of nature, it cannot be derived and it cannot be proved. It can only be repeatedly tested against the physical reality to which it pertains.

5-3 WORK AND STATE FUNCTIONS

The instantaneous constitutive characterization of a system depends on the system parameters (in mechanical systems, for example, mass or stiffness element models) and system variables (in mechanical systems, for example, velocity or position at specified locations throughout the system). Functions that provide an instantaneous constitutive characterization of some property of the system in terms of these parameters and variables are called *state functions*. For mechanical systems, we shall be concerned primarily with two types of state functions—kinetic and potential state functions—both of which will be defined in terms of work.

We begin with formal definitions of *work, kinetic state functions,* which are functions of either a system's momenta or velocities, and *potential state functions,* which are functions of either a system's deformations of conservative elements or forces in conservative elements as well as positions of bodies (often particles) within conservative force fields. A dynamic system element or force field is called *conservative* if the net work done on it around any closed path is zero. Kinetic and potential *state functions* are alternative and complete representations of the constitutive characterizations of conservative elements and conservative force fields. The specific value of a state function depends only on the instantaneous value of the independent variable(s). The importance of these alternative constitutive characterizations—that is, state functions—is that they play a central role in formulations via the indirect approach. The appropriate evaluations of the work and the kinetic and potential state functions will satisfy Requirement 3 in Eq. (5-1).

5-3.1 Work

By definition, the *work* done by a force $\boldsymbol{F}$ acting on a probe as it moves along a path from $\boldsymbol{R}_0$ to $\boldsymbol{R}$ is

$$W = \int_{\boldsymbol{R}_0}^{\boldsymbol{R}} \boldsymbol{F} \cdot d\boldsymbol{R} \tag{5-2}$$

where the vector dot product indicates that work is done only by the component of the force along the trajectory. In general, the force may be written as $\boldsymbol{F}(\boldsymbol{R}, \boldsymbol{v}, t)$, indicating that it may depend on kinematic variables such as the probe's position $\boldsymbol{R}$ and velocity $\boldsymbol{v}$, as well as the time t. The force may be also a function of the probe's mass and electrical charge, due to fields—gravitational, electric, or magnetic—in which the probe is immersed.

By definition as well as by analogy with Eq. (5-2), the work done by a moment or torque $\boldsymbol{\tau}$ acting on a nonparticulate body (particles cannot sustain torques) as it undergoes a rotation from θ_0 to θ is

$$W = \int_{\theta_0}^{\theta} \boldsymbol{\tau} \cdot d\theta \tag{5-3}$$

where the torque may be written as $\boldsymbol{\tau}(\theta, \boldsymbol{\omega}, t)$, indicating that it may be a function of angular position θ, angular velocity $\boldsymbol{\omega}$ and time t.

■ **Example 5-1:**   Work Done by Forces

Find the work done by the constant forces acting on the body undergoing one-dimensional translation as sketched in Figure 5-1.

❏ **Solution:**   By definition, for cartesian coordinates xyz, Eq. (5-2) may be rewritten as

$$W = \int_{1}^{2} (F_x\, dx + F_y\, dy + F_z\, dz) \tag{a}$$

where the integration limits 1 and 2 indicate the end points of the range over which the work is computed, and F_x, F_y, and F_z are the components of $\boldsymbol{F}$ in the x, y, and z directions, respectively. For the problem defined in Figure 5-1, Eq. (a) becomes

$$W = \int_{x=0}^{x_0} (F_1 \cos\theta - F_3)\, dx + \int_{y=0}^{0} (-F_1 \sin\theta - F_2)\, dy + \int_{z=0}^{0} (F_4)\, dz \tag{b}$$

or

$$W = (F_1 \cos\theta)x_0 - F_3 x_0 \tag{c}$$

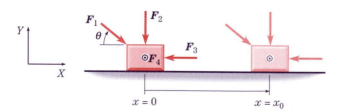

FIGURE 5-1   Body subjected to constant forces and undergoing one-dimensional translation.

Clearly,

1. Only force components—such as $F_1 \cos \theta$ or F_3—along the direction of the motion contribute to the work (or, equivalently, forces, such as F_2, which are orthogonal to the motion do no work).
2. Force components, such as F_3, which are oppositely directed to the motion, do negative work.

Next, the concept of work will be used in assisting in the definitions of several important concepts for mechanical systems. In Chapter 7, we shall consider comparable concepts for electrical and electromechanical systems.

5-3.2   Kinetic State Functions

By definition, the work done on a particle in increasing its momentum from 0 to $\boldsymbol{p}$ is the *kinetic energy function* $T(\boldsymbol{p})$, or

$$T(\boldsymbol{p}) = \int_0^{\boldsymbol{p}} \boldsymbol{v} \cdot d\boldsymbol{p} \tag{5-4}$$

Equation (5-4) is independent of the velocity–momentum relation for the particle. Figure 5-2 is a plot of a nonlinear velocity–momentum constitutive relation, such as that for a relativistic particle. Equation (5-4) could have been derived by noting that the work increment due to the resultant force $\boldsymbol{f}$ on a particle is

$$dW = \boldsymbol{f} \cdot d\boldsymbol{R} = \frac{d\boldsymbol{p}}{dt} \cdot d\boldsymbol{R} = \frac{d\boldsymbol{p}}{dt} \cdot \boldsymbol{v} \, dt = \boldsymbol{v} \cdot d\boldsymbol{p} \tag{5-5}$$

where Newton's linear momentum principle has been used. The integration of Eq. (5-5) gives Eq. (5-4).

A complementary kinetic state function can be defined as the *kinetic coenergy function* and can be obtained directly from Figure 5-2 as

$$T^*(\boldsymbol{v}) = \int_0^{\boldsymbol{v}} \boldsymbol{p} \cdot d\boldsymbol{v} \tag{5-6}$$

which is independent of the velocity–momentum relation for the particle.

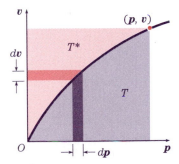

FIGURE 5-2   Kinetic state functions for particles, illustrating their complementary character.

The complementary character of $T(\boldsymbol{p})$ and $T^*(\boldsymbol{v})$ is illustrated in Figure 5-2, where it is also noted that the areas below and above the curve are not equal. Equations (5-4) and (5-6) describe the same constitutive information, but in terms of different independent variables. If one of the kinetic state functions is known, it is straightforward to determine its complementary kinetic state function. For example, given the operating point $(\boldsymbol{p}, \boldsymbol{v})$ and the kinetic energy $T(\boldsymbol{p})$, then $T^*(\boldsymbol{v})$ may be evaluated from

$$T^*(\boldsymbol{v}) = \boldsymbol{p} \cdot \boldsymbol{v} - T(\boldsymbol{p}) \tag{5-7}$$

where it is understood that Eq. (5-7) is valid only at points along the constitutive relation in the $\boldsymbol{p}$–$\boldsymbol{v}$ plane. Although we shall not pursue this further, the definition of $T^*(\boldsymbol{v})$ in Eq. (5-7) is a simple example of a *Legendre transformation*. In essence, this Legendre transformation is a method for changing from the independent variable $\boldsymbol{p}$ in $T(\boldsymbol{p})$ to the independent variable $\boldsymbol{v}$ in $T^*(\boldsymbol{v})$, without loss of constitutive information. (See Problem 5-11.)

If the particle is newtonian, $\boldsymbol{p} = m\boldsymbol{v}$, and Eq. (5-4) gives the kinetic energy as

$$T(\boldsymbol{p}) = \int_0^{\boldsymbol{p}} \frac{1}{m} \boldsymbol{p} \cdot d\boldsymbol{p} = \frac{1}{2m} \boldsymbol{p} \cdot \boldsymbol{p} \tag{5-8}$$

Also, for a newtonian particle, Eq. (5-6) gives the kinetic coenergy as

$$T^*(\boldsymbol{v}) = \int_0^{\boldsymbol{v}} m\boldsymbol{v} \cdot d\boldsymbol{v} = \frac{1}{2}m\boldsymbol{v} \cdot \boldsymbol{v} \tag{5-9}$$

Note that we leave the right-hand sides of Eqs. (5-8) and (5-9) in their vector forms. Although $T(\boldsymbol{p})$ and $T^*(\boldsymbol{v})$ are scalar functions, as we shall emphasize (see Example 5-23), it is useful to keep in mind the forms of the right-hand sides of Eqs. (5-8) and (5-9). Except for cases involving one-dimensional motion, the retention of the vector dependence of $T(\boldsymbol{p})$ and $T^*(\boldsymbol{v})$ will be shown to be important in properly evaluating these kinetic state functions. Although the magnitudes of $T(\boldsymbol{p})$ and $T^*(\boldsymbol{v})$ are equal for a newtonian particle, their independent arguments are clearly different. Furthermore, for a relativistic particle, the magnitudes of $T(\boldsymbol{p})$ and $T^*(\boldsymbol{v})$ are different, as shown in Problem 5-12.

In accordance with our earlier considerations, we shall continue to restrict our interest to nonrelativistic effects. In this case, the constitutive relation in Figure 5-2 is a straight line and, for any operating point $(\boldsymbol{p}, \boldsymbol{v})$, the area below the line is equal to the area above the line. So, although the magnitude of T and T^* are equal in this case, for reasons that we shall discuss in Subsection 5-3.4, we shall maintain the distinction in their independent variables as expressed in Eqs. (5-8) and (5-9); T is a function of $\boldsymbol{p}$ and T^* is a function of $\boldsymbol{v}$.

System Kinetic State Functions   Equations (5-4) and (5-6) were written for a single particle. For a system of N particles, the system kinetic state functions are simply the sums of the respective state functions for the individual particles. Thus, the system kinetic energy function T and the system kinetic coenergy function T^* may be written, respectively, as

$$T = \sum_{i=1}^{N} T_i$$
$$T^* = \sum_{i=1}^{N} T_i^* \tag{5-10}$$

where for an individual particle T_i and T_i^* are the kinetic energy function and kinetic coenergy function, respectively.

5-3.3 Potential State Functions

Ideal Two-Force Conservative Elements A two-force element is a model of a mechanical component that has contact force interactions with its environment at only two points, typically sketched at the end nodes of the schematic of the element. Simple extensional springs, as represented in the schematic in Figure 5-3a, are one type of mechanical component that is often modeled as an ideal two-force conservative element. In our terminology, *ideal* denotes that the element is *massless,* and *conservative* denotes that the net work on the element vanishes during any closed cycle of loading that returns the element to its initial configuration. Because the element is massless, it would undergo infinite accelerations if it experienced an unbalanced force or torque. Thus, the two forces on any ideal two-force element must be equal, opposite, and collinear, at all times.

The unloaded length of the ideal spring is often denoted by ℓ_0. If its elongation is e, the force f acting on the spring depends only on the elongation. For sufficiently small elongations, the constitutive relation can often be treated as *linear,* thus

$$f = ke \tag{5-11}$$

where k is called the *spring constant.* The phrase "sufficiently small elongations" can be established only in the context of a specific spring whose constitutive behavior can be modeled by Eq. (5-11).

By definition, for an ideal two-force conservative element, the work done in deforming it from 0 to e is the *potential energy function $V(e)$,* or

$$V(e) = \int_0^e f \, de \tag{5-12}$$

Strictly, integrals such as Eq. (5-12) are a function of the upper limit e and should be written as

$$V(e) = \int_0^e f(\phi) \, d\phi$$

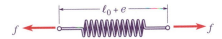

(a) Schematic of an ideal spring.

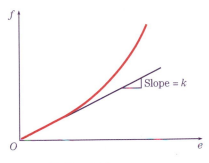

(b) Constitutive relation.

FIGURE 5-3 Schematic and constitutive relation for an ideal two-force conservative spring.

where ϕ is called a *dummy variable* because the value of the integral is independent of the choice of this variable. In general, we shall continue to write such integrals as written in Eq. (5-12) unless doing so leads to ambiguity.

Figure 5-4 is a plot of a nonlinear force–elongation constitutive relation, illustrating the potential energy function $V(e)$ and the *potential coenergy function* $V^*(f)$, which, from Figure 5-4, can be written as

$$V^*(f) = \int_0^f e\,df \tag{5-13}$$

Both Eqs. (5-12) and (5-13) are valid, independent of the force–elongation relation for the element. Also, both Eqs. (5-12) and (5-13) describe the same constitutive information, but in terms of different independent variables. In addition to Eq. (5-13), $V^*(f)$ can be evaluated via a Legendre transformation as

$$V^*(f) = ef - V(e) \tag{5-14}$$

for the operating point (e, f), where it is understood that Eq. (5-14) is valid only at points along the constitutive relation in the e–f plane.

If the constitutive relation is linear, Eqs. (5-11) and (5-12) give the potential energy function as

$$V(e) = \int_0^e ke\,de = \frac{1}{2}ke^2 \tag{5-15}$$

Force Fields   Here, we consider force fields for which the force on a body in the field depends only on the body's position; that is, we restrict our consideration to fields having forces of the form $f(R)$. In accordance with Eq. (5-2), the work done on a body by a conservative or a nonconservative force field as the body is moved from the datum R_0 to the position R is

$$W = \int_{R_0}^{R} f(R) \cdot dR \tag{5-16}$$

If the value of this work integral is the same for all paths between the datum R_0 and R, the value of the integral in Eq. (5-16) will depend only on the instantaneous position of the body. In this case, the net work around any closed path must be zero. If the net work around any closed path is zero, the force field $f(R)$ is characterized as *conservative*. So, if the force field is conservative, the integral in Eq. (5-16) is a *state function*.

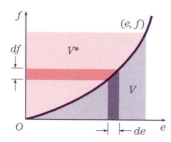

FIGURE 5-4   Potential state functions for an ideal two-force element, illustrating their complementary character.

By definition, the work done on a body in a conservative force field as the body moves from $\boldsymbol{R}_0$ to $\boldsymbol{R}$ is the *potential energy function* of the body and is defined as

$$V(\boldsymbol{R}) = -\int_{\boldsymbol{R}_0}^{\boldsymbol{R}} \boldsymbol{f}(\boldsymbol{R}) \cdot d\boldsymbol{R} \tag{5-17}$$

Thus, the potential energy $V(\boldsymbol{R})$ of the body is clearly a state function. The negative sign in Eq. (5-17) denotes that the potential energy of the body decreases if the field does work on the body.

■ **Example 5-2:**   Gravitational Potential Energy Function for Particle

In the vicinity of the Earth's surface, the gravitational force field constitutive relation for a particle of mass m is given by $\boldsymbol{f} = m\boldsymbol{g}$, where $\boldsymbol{g}$ is the vector pointing toward the Earth's center and having a magnitude equal to the local acceleration due to gravity. Find the gravitational potential energy function for the particle sketched in Figure 5-5.

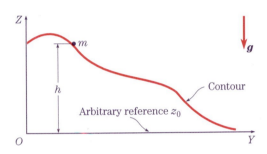

FIGURE 5-5   Particle on contour above arbitrary fixed reference z_0.

☐ **Solution:**   Substitution of the gravitational force field constitutive relation into Eq. (5-17) gives

$$V = -\int_{\boldsymbol{R}_0}^{\boldsymbol{R}} \boldsymbol{f}(\boldsymbol{R}) \cdot d\boldsymbol{R} = -\int_{\boldsymbol{R}_0}^{\boldsymbol{R}} (-mg\boldsymbol{k}) \cdot (dx\boldsymbol{i} + dy\boldsymbol{j} + dz\boldsymbol{k}) = \int_{z_0}^{z} mg\,dz = mg(z - z_0) \tag{a}$$

The result in Eq. (a) is a general expression for particles in the vicinity of the Earth's surface. In particular, when the height of the particle is h and the reference z_0 is arbitrarily defined as 0 at the Earth's surface, Eq. (a) reduces to

$$V = mgh \tag{b}$$

The Legendre transformation, which can be used to determine the complementary functions for the kinetic state functions for particles and the potential state functions for springs, may not in general be used for force fields. In general, transformations such as Eqs. (5-7) or (5-14) do not exist if the constitutive relation $f = f(x)$ cannot be inverted to obtain a unique relation of the form $x = x(f)$. Such a unique inversion does not exist when the constitutive relation is $f = c$, where c is a constant. As illustrated in Figure 5-6, if x is given, f can be found uniquely; but if f is given as $f = c$, x cannot be determined uniquely. In this case, $V = cx$ can be found via Eq. (5-12), but a unique nontrivial V^* cannot be found. The constitutive relation for a particle in a uniform gravitational field is of the form illustrated in Figure 5-6; this can be seen if c and x are replaced by mg and z, respectively.

If the force field $f(R)$ is conservative, it immediately follows that

$$\oint_C f(R) \cdot dR = 0 \tag{5-18}$$

where C represents any closed contour. Also, recall from vector analysis that Stokes' theorem,* which is valid for *any* vector field, relates the path integral around a closed contour C to a surface integral over an arbitrary surface S bounded by C as

$$\oint_C f \cdot dR = \oiint_S (\nabla \times f) \cdot dS \tag{5-19}$$

---

*This is a theorem that was initially proved in an 1847 letter by the Irishman William Thomson (1824–1907), later Lord Kelvin. The Irish recipient of the letter, George Gabriel Stokes (1819–1903), first published the result as a mathematical tripos question in 1854, which was taken by, among others, the Scotsman James Clerk Maxwell (1831–1879). It was Maxwell who in 1870 associated the terms *gradient, curl,* and *divergence* with their modern mathematical interpretations. Thomson, Stokes, and Maxwell were all Cantabrigian wranglers.

---

where ∇ is the differential vector operator *del* or *nabla,* and where $\nabla \times f$ is the curl of f, often written as "curl f." Since for a conservative force field, the left-hand side of Eq. (5-19) vanishes in accordance with Eq. (5-18), this requires that

$$\nabla \times f = 0 \tag{5-20}$$

Equations (5-18) and (5-20) provide two techniques for establishing the conservative character of a force field $f(R)$. Thus, if either of Eqs. (5-18) or (5-20) holds for a force field $f(R)$, the field is conservative. Further, it can be shown[1] that Eq. (5-17) is equivalent to

$$f = -\nabla V \tag{5-21}$$

If Eq. (5-21) is written in cartesian coordinates as

$$f_x i + f_y j + f_z k = -\left(\frac{\partial V}{\partial x} i + \frac{\partial V}{\partial y} j + \frac{\partial V}{\partial z} k\right) \tag{5-22}$$

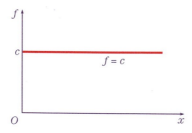

FIGURE 5-6 Constitutive relation for which $V = cx$, but V^* does not exist.

[1] See Problem 5-9.

we note that, given the potential energy function, the conservative force components may be found as

$$f_x = -\frac{\partial V}{\partial x} \qquad f_y = -\frac{\partial V}{\partial y} \qquad f_z = -\frac{\partial V}{\partial z} \tag{5-23}$$

since the corresponding components on both sides of a vector expression must be equal. Note that the gravitational force field constitutive relation for a particle can be recovered by substituting Eq. (a) in Example 5-2 into Eq. (5-23).

System Potential Energy Function   For a system consisting of any number of particles, interconnected by conservative springs and acted on by one or more conservative force fields, the system potential energy is

$$V = \sum_{i=1}^{N} V_i \tag{5-24}$$

where the $V_i(i = 1, 2, \cdots, N)$ represent the individual potential energies of each particle in the force fields and of each conservative spring, N being the total number of such individual contributions.

Finally, note that comparable—not identical—statements could be made about V^* as those made about V in regard to Eq. (5-24). However, there is no general one-to-one correspondence between V and V^* for force fields, as illustrated in Figure 5-6. Thus, although V^* is a valid concept in general, in practical terms it is much less useful than V.

5-3.4   Energy and Coenergy

Here, a few words about the choice to use energy and coenergy complete this section. In maintaining the distinction between energy and coenergy functions, we continue a tradition that has been adopted by a significant group of engineers, but rejected by many.

Most often it is argued by those who do *not* make the distinction between energy and coenergy that because in newtonian dynamics the magnitudes of T and T^* are equal, it is irrelevant to make such a delineation. Such an argument is true but insufficient. We shall list several reasons for retaining the distinction between energy and coenergy state functions.

First, the underlying theoretical foundations of mechanics can be presented more clearly by retaining the energy–coenergy distinction. Second, many of the results obtained here can be extended directly to relativistic dynamics, but only if the energy–coenergy distinction is maintained. Third, in other areas of mechanics such as nonlinear elasticity, this energy–coenergy distinction is broadly used. Fourth, any attempt to incorporate nonlinear springs into the problems considered here can be easily and unambiguously accomplished. Fifth, and perhaps pragmatically most important, is the fact that many modern treatments of electromechanical systems and control use this complementarity of state functions, and we shall study such mixed mechanical and electrical systems. Sixth, there is an unambiguous basis for illustrating that the kinetic energy T depends *naturally* on momentum and that the kinetic coenergy depends *naturally* on velocity, and that a failure to make this distinction may lead to a loss of information. See, for example, Problems 5-10 and 5-11.

Thus, besides being the analytically correct approach, the retention of the energy–coenergy distinction has practical significance, especially, though not necessarily only, when nonlinear constitutive relations pertain.

5-4   GENERALIZED VARIABLES AND VARIATIONAL CONCEPTS

In this section we introduce a number of fundamental concepts in variational mechanics. The subjects discussed in this section can be quite mathematical in their presentation. Here, however, we focus on the physical interpretation of the concepts.

5-4.1   Generalized Coordinates

In formulating the equations of motion for a mechanical system, we shall have to satisfy the kinematic requirements. We shall use the phrase *geometrically admissible motions* (often simply stated as *admissible motions*) to denote a spatial configuration that is always kinematically compatible with the problem statement. Thus, if the motion of a system is modeled as planar, a geometrically admissible motion will not admit motion out of that plane. These *geometrically admissible motions* will be described by *generalized coordinates*.* And, in formulating the equations of motion for mechanical systems, the geometric requirements on the motions—that is, Requirement 1 in Eq. (5-1)—will be satisfied by the selection of a *complete* and *independent* set of *generalized coordinates*. We now proceed to explore the meanings of "generalized coordinates," "completeness," and "independence."

> *The term *generalized coordinates,* as well as the forthcoming terms *generalized velocities* and *generalized momenta,* first appeared in 1867 in the *Treatise on Natural Philosophy* by the Irishman William Thomson, later Lord Kelvin, (1824–1907) and the Scotsman Peter G. Tait (1831–1901).

• *Generalized coordinates* are variables that locate dynamic systems with respect to a reference frame.

Although the individual generalized coordinates may be inertial or noninertial, ultimately, as a set, they must be capable of representing the inertial velocity of each particle in the system. (For instance, see the alternative solutions in Example 5-21.)

■ **Example 5-3:**   Bead on Rigid Rod

The bead B in Figure 5-7, which is modeled as a particle, slides along the rigid rod. Find a set of generalized coordinate(s) for this system.

❑ **Solution:**   We simply ask the question "How many geometric quantities or variables do we need to locate the system, and what might they be?" One possible answer is "x". We express this formally as follows:

$$\xi_j : x; \text{ where } x \text{ is the position of the bead along the } X$$
$$\text{axis, with respect to the left-hand wall.}$$

The Greek letter ξ_j will be used to denote generalized coordinates where j is the number of generalized coordinates, j obviously being equal to one in this example. Note that the directed line for x has a magnitude (indicated by the length of the line), a direction (indicated by the arrowhead), and a reference with respect to the left-hand wall. (Note that x is not intended to be a vector, however.) Also, note that we shall emphasize the importance of ex-

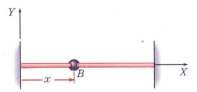

FIGURE 5-7 Bead on rigid rod, illustrating generalized coordinate x.

pressing in words precisely what is intended as the meaning of each generalized coordinate. Failure to do so may lead to ambiguity in the resulting equations of motion.

■ **Example 5-4:** Bead in Plane

The bead B in Figure 5-8, which is modeled as a particle, is constrained to remain in the XY plane. Find a set of generalized coordinate(s) for this system.

▢ **Solution:** We simply ask the question "How many geometric quantities or variables do we need to locate the system, and what might they be?" One possible answer is "x and y." We express this formally as follows:

> $\xi_j : x, y$; where x and y are the cartesian coordinates of the bead along the X and Y axes, respectively, with respect to the $OXYZ$ reference frame.

We note that another set of generalized coordinates could have been selected. This alternative set is "r and θ." We express this formally as follows:

> $\xi_j : r, \theta$; where r and θ are the polar coordinates of the bead in the XY plane, with respect to the $OXYZ$ reference frame.

There are other conceivable choices of generalized coordinates; indeed, there is an infinite number.

Again, we emphasize the importance of stating precisely what is intended by the choice of generalized coordinates. Also, note the directed lines, indicating the positive sense and the reference frame with respect to which each of the generalized coordinates is defined.

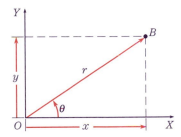

FIGURE 5-8 Bead in plane, illustrating two alternative sets of generalized coordinates.

■ **Example 5-5:**   **Rod in Plane**

The slender rod of length l in Figure 5-9 is constrained to remain in the XY plane. Geometrically, a slender rod is the one-dimensional extension of a particle; it has only length. Note that the two ends of the rod are different. One end of the rod is flat and the other end is rounded.[2] The attribution of these end features to the rod will assist our discussion. Find a set of generalized coordinate(s) for this system.

□ **Solution:**   We simply ask the question "How many geometric quantities or variables do we need to locate the system, and what might they be?" Figure 5-9 illustrates two of many choices that often come to mind during the initial encounter with this problem. They may be formally expressed as follows:

In Figure 5-9*a*:

$$\xi_j : x_1, x_2, y_1, y_2; \text{ where } x_1 \text{ and } y_1 \text{ are the cartesian}$$
coordinates of the flat end of the rod and x_2 and y_2 are the cartesian coordinates of the rounded end of the rod, both with respect to the $OXYZ$ reference frame.

In Figure 5-9*a*, if x_1, y_1, and x_2 are selected, we need y_2 also. For example, if x_1, y_1, and x_2 are defined, the system is not unambiguously fixed or located since it could be in either the solid-sketched configuration or the dashed-sketched configuration.

In Figure 5-9*b*:

$$\xi_j : x, y, \theta; \text{ where } x \text{ and } y \text{ are the cartesian coordi-}$$
nates of the flat end of the rod with respect to the $OXYZ$ reference frame, and θ is the angle measured counterclockwise from the X axis to a directed line segment from the flat end of the rod to its rounded end.

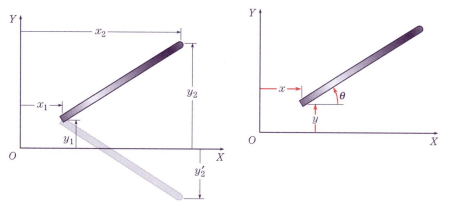

FIGURE 5-9   Rod in plane, illustrating two proposed sets of generalized coordinates.

[2]Although the rod has no cross-sectional dimensions, the assignment of a flat end and a round end might promote a challenge from the strident reader. Our response: no problem, this is a *model*, not a real thing.

Clearly, other conceivable choices of sets of generalized coordinates could have been selected.

With Examples 5-3, 5-4, and 5-5 as background, we can proceed with our definition of *completeness*.

- A set of generalized coordinates is *complete* if it is capable of locating all parts of the system at all times.

■ **Example 5-6:**   Completeness of Generalized Coordinates

Determine which of the sets of generalized coordinates in Examples 5-3, 5-4, and 5-5 are *complete*.

❑   **Solution:**   All the generalized coordinates in Examples 5-3, 5-4, and 5-5 (Figures 5-7, 5-8, and 5-9) are *complete*. Each set is capable of fixing (that is, locating) all parts of the system at all times.

We may also proceed with our definition of *independence*.

- A set of generalized coordinates is *independent* if, when all but one of the generalized coordinates are fixed, there remains a continuous range of values for that one coordinate.

The test for independence requires us to conduct a thought experiment. Further, we note that it may not always be possible to go from one admissible configuration to another simply by varying only the one remaining coordinate. For independence, we simply require that a continuous range of admissible motions involving that coordinate is accessible. (In isolation, this statement is not easily understood. It is particularly relevant in cases such as that discussed in part 2 of Example 5-10, and should be read again during the study of that example.)

■ **Example 5-7:**   Independence of Generalized Coordinates

Determine which of the sets of the generalized coordinates in Examples 5-3, 5-4, and 5-5 are *independent*.

❑   **Solution:**   **1.**   Example 5-3 is essentially a trivial case. Because there is only one coordinate, we cannot fix all but one coordinate, while simultaneously imagining another coordinate to assume a continuous range of values. Thus, the set $\xi_j : x$ is a *complete* (as shown in Example 5-6) and *independent* set of *generalized coordinates*.

2.   In Example 5-4, first consider the set $\xi_j : x, y$. Now, if we fix x, we can conduct a thought experiment by imagining y to assume a continuous range of admissible values. This thought experiment envisions the bead moving along a "vertical" line in the plane, parallel to the Y axis, at constant x. Then, if we fix y and conduct a second thought experiment, we imagine x to assume a continuous range of admissible values. This second thought experiment envisions the bead moving along a "horizontal" line in the plane, parallel to the X axis, at constant y. Thus, the set $\xi_j : x, y$ satisfies the criteria for both completeness and independence, so we conclude that it is a *complete* and *independent* set of *generalized coordinates*.

Next, consider the set $\xi_j : r, \theta$. If we fix r and conduct a thought experiment, we imagine θ to assume a continuous range of admissible values. This thought experiment envisions the bead moving along a circular arc of radius r in the plane, centered at the origin. Then, if we fix θ and conduct a second thought experiment, we imagine r to assume a continuous range of admissible values. This thought experiment envisions the bead moving along a radial line through the origin and in the plane, at constant θ. Thus, the set $\xi_j : r, \theta$ satisfies the criteria for both completeness and independence, and so we conclude that it is a *complete* and *independent* set of *generalized coordinates*.

3. In Example 5-5, first consider the set $\xi_j : x_1, y_1, x_2, y_2$. Note that if we fix any three coordinates, there does *not* remain a continuous range of values for that lone remaining coordinate. For instance, as sketched in Figure 5-9a, if x_1, y_1, and x_2 are fixed, there is one other possible value of y_2, namely y_2'; but there is not a continuous range of values of y_2. A similar result is obtained if any different three of the coordinates are fixed, with one remaining not fixed. Thus, the set $\xi_j : x_1, y_1, x_2, y_2$ is complete, as determined in Example 5-6, but is not independent.

   Next, consider the set $\xi_j : x, y, \theta$. If we fix x and y we can imagine θ to assume a continuous range of admissible values; then fix x and θ, and imagine y to assume a continuous range of admissible values; and finally fix y and θ, and imagine x to assume a continuous range of admissible values. Except for the fact that there are three generalized coordinates here, the envisioned changes during the thought experiments are quite similar to case 2 (the particle in the plane) immediately above. Thus, we conclude that the set $\xi_j : x, y, \theta$ satisfies the criteria for both completeness and independence, and is therefore a *complete* and *independent* set of *generalized coordinates*.

Case 2 in Example 5-7 illustrates an important characteristic of generalized coordinates. Because both (x, y) and (r, θ) represent complete and independent sets, we conclude that sets of complete and independent generalized coordinates are *not unique*. However, the *number* of generalized coordinates in a complete and independent set of generalized coordinates is unique. Thus, in Examples 5-3, 5-4, and 5-5(b) above, although there may exist an infinite number of possible generalized coordinates, any complete and independent set of generalized coordinates must consist of 1, 2, and 3 generalized coordinates, respectively.

Finally, the thought experiments that envisioned small allowable changes in the generalized coordinates resulted in what we shall call *admissible variations*. The small changes were allowable in the sense that they were envisioned in accordance with the kinematic constraints given in problem statements. The bead in Figure 5-7, for example, was not allowed to move in the Y or Z directions. The special symbol δ will be used in conjunction with the generalized coordinates to denote an admissible variation; for example, δx denotes the envisioned admissible variation of x in Figure 5-7. The symbol δ was introduced by Lagrange to emphasize the hypothetical character of the displacements resulting from the thought experiments, as contrasted with the symbol d, denoting actual differentials of displacements that occur in the time interval dt, resulting from real motion or forces. In contrast, δx is an imagined displacement that occurs *at an instant in time*; in this sense, δx is sometimes said to be a *contemporaneous admissible variation*. For example, consider the bead in Figure 5-8. If it were sitting at the origin O with no velocity and if then it were acted upon by a net force in the x direction only, it would undergo some dx as a function of time. On the other hand, we understand that as possibilities, δx and δy exist as conceivable and allowable changes that the bead *might* undergo without violating the requirement that the bead remains in the XY plane.

5-4.2 Admissible Variations, Degrees of Freedom, Geometric Constraints, and Holonomicity

The concepts of *completeness* and *independence*, which apply to *generalized coordinates*, also apply to *admissible variations*.

- A set of admissible variations is *complete* if it is capable of representing all the geometric variations of the system at all times.
- A set of admissible variations is *independent* if, when all but one of the admissible variations are fixed, there remains a continuous range of values for that one admissible variation.

These admissible variations occur at a fixed instant in time. Although the system is dynamic, it is during this *fixed instant in time* that we *imagine* the various incremental compatible displacements that the system *may* undergo, consistent with the definition of the problem.

■ **Example 5-8:**   Bead in Plane

We reconsider the bead in a plane as sketched in Figure 5-8, where the complete and independent set of generalized coordinates is, say, x and y. We want to investigate the completeness and the independence of the admissible variations for this system.

❑ **Solution:**   We conduct our thought experiment by considering all admissible variations of the generalized coordinates. As indicated in Subsection 5-4.1, we shall introduce the notation $\delta\xi_j$ to indicate admissible variations, that is, the hypothetical or imagined small allowable changes of the generalized coordinates ξ_j. (We shall explore the mathematical behavior of δ in Section 5-4.3.) Practically, for this example we imagine the changes sequentially by imagining δx, and then δy. Note that these changes are taking place in the respective positive x and y directions, and that they are kinematically allowable in that they are made consistent with the statement that the bead remains in the plane; that is, $\delta z = 0$. These comments are illustrated in Figure 5-10. Thus, we conclude that for the bead in the plane:

$$\xi_j : x, y \qquad \text{and} \qquad \delta\xi_j : \delta x, \delta y$$

where δx and δy comprise a complete and independent set of admissible variations.

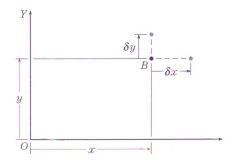

FIGURE 5-10   Bead in plane, illustrating admissible variations δx and δy.

Note that an equally valid consideration of this problem could have resulted in the following:

$$\xi_j : r, \theta \qquad \text{and} \qquad \delta\xi_j : \delta r, \delta\theta$$

where δr and $\delta\theta$ comprise a complete and independent set of admissible variations.

■ **Example 5-9:**   Rod in Plane

We reconsider the slender rod in a plane as sketched in Figure 5-9*b*, with the complete and independent set of generalized coordinates x, y, and θ. We want to investigate the completeness and the independence of the admissible variations for this system.

❑ **Solution:**   We conduct our thought experiment by considering all admissible variations of the generalized coordinates. Practically, we imagine the changes sequentially by imagining δx (with no δy or $\delta\theta$), then δy (with no δx or $\delta\theta$), and then $\delta\theta$ (with no δx or δy), all taken in the positive sense of the respective generalized coordinates x, y, and θ. These admissible variations are made kinematically consistent with the statement that the rod remains in the plane. These admissible variations are illustrated in Figure 5-11. Thus, we conclude that for the rod in the plane:

$$\xi_j : x, y, \theta \qquad \text{and} \qquad \delta\xi_j : \delta x, \delta y, \delta\theta$$

where δx, δy, and $\delta\theta$ comprise a complete and independent set of admissible variations.

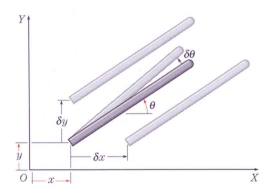

FIGURE 5-11   Rigid rod in plane, illustrating admissible variations δx, δy, and $\delta\theta$.

■ **Example 5-10:**   Disk on Plane

Figure 5-12 shows a thin disk of radius r that is constrained to remain in contact with the XY plane. ("Thin" indicates that the disk is modeled to have negligible thickness.) The plane of the disk is always vertical; yet the disk may rotate or spin about its vertical (Z direction) diameter, irrespective of whether it is allowed to slip on the XY plane. We want to investigate the completeness and independence of the generalized coordinates and the admissible variations for this system.

❑ **Solution:**   Some modest considerations reveal that we are further confronted with the question of whether the disk is allowed to slip as it moves on the XY plane. Indeed, it is useful to consider both cases: slipping allowed and no slipping allowed.

1. **Slipping Allowed:** If slipping is allowed, we need four generalized coordinates to fix the location of the disk. As sketched in Figure 5-12, these are chosen as x and y, which locate the contact point on the XY plane or, equivalently, the center of the disk above the XY plane; ϕ, which is the heading of the disk defined by the angle between the plane of the disk and the XZ plane; and θ, which is the angle of rotation of the disk about an axis through its center and perpendicular to its plane. Further, if all but any one of these coordinates are fixed, there remains a continuous range of admissible values of that one remaining coordinate. Thus, a complete and independent set of generalized coordinates for the slipping disk is $\xi_j : x, y, \phi, \theta$. The variations δx, δy, $\delta \phi$, and $\delta \theta$ comprise a complete set of admissible variations. Further, if all but any one of these admissible variations are fixed, there remains a continuous range of values for that one remaining admissible variation. So, δx, δy, $\delta \phi$, and $\delta \theta$ also comprise an independent set of admissible variations. Thus, for this case of the vertical disk on the plane we conclude the following:

$$\xi_j : x, y, \phi, \theta \qquad \text{and} \qquad \delta \xi_j : \delta x, \delta y, \delta \phi, \delta \theta$$

where the ξ_j and the $\delta \xi_j$ comprise complete and independent sets of generalized coordinates and admissible variations, respectively.

2. **No Slipping Allowed:** If slipping is not allowed, the need for four generalized coordinates to fix the location of the disk remains. For convenience, we choose as the complete set the same four as above: x, y, ϕ, and θ. Further, these four generalized coordinates continue to comprise an independent set. This independence is illustrated in Figure 5-13, where, if three of the coordinates are fixed, say $x = x_0$, $\phi = \phi_0$, and $\theta = \theta_0$, there remains a continuous range of admissible values for the fourth coordinate, in this case y. This continuous range of admissible values of y exists and is accessible, even though it is not possible to pass from one configuration to another by simply varying y alone. The same discussion can be composed for x, if y, ϕ, and θ are fixed; for ϕ, if x, y, and θ are fixed; and for θ, if x, y, and ϕ are fixed. Thus, a complete and independent set of generalized coordinates for the nonslipping disk is $\xi_j : x, y, \phi, \theta$, the same as for the slipping disk.

   In considering the admissible variations, we encounter a new situation. The variations δx, δy, $\delta \phi$, and $\delta \theta$ constitute a *complete* set of admissible variations; however, they are not independent. The requirement that the disk must roll without slipping implies one or more constraining relations between the four admissible variations. For

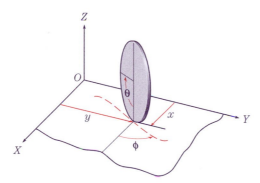

FIGURE 5-12   Disk constrained to remain vertical as it moves on XY plane.

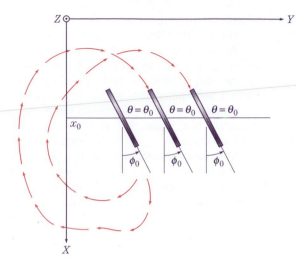

FIGURE 5-13 Configurations of disk with fixed x, ϕ, and θ but different y, illustrating possible nonslipping admissible paths between admissible configurations.

example, in this problem, if we let $\delta\phi$ and $\delta\theta$ be independent, from Figure 5-12 we see that the nonslipping requirement implies that

$$\delta x = r\cos\phi\,\delta\theta$$
$$\delta y = r\sin\phi\,\delta\theta \tag{a}$$

Note that we could have chosen δx and $\delta\phi$ to be independent and written constraining relations for δy and $\delta\theta$, and similarly we could have chosen δy and $\delta\phi$ to be independent and written constraining relations for δx and $\delta\theta$. Then, there are only two independent admissible variations for the nonslipping disk. Thus, for the nonslipping disk we conclude the following:

$$\xi_j : x, y, \phi, \theta \qquad \text{and} \qquad \delta\xi_k : \delta\phi, \delta\theta$$

where ξ_j and $\delta\xi_k$ comprise *independent* sets of generalized coordinates and admissible variations, respectively; but $\xi_j : x, y, \phi, \theta$ and $\delta\xi_j : \delta x, \delta y, \delta\phi, \delta\theta$ comprise *complete* sets of generalized coordinates and admissible variations, respectively.

Based on the discussion in Examples 5-8, 5-9, and 5-10, we are prepared to define several additional concepts for use in our variational formulation.

- The number of independent admissible variations in a complete set of admissible variations is the number of *degrees of freedom*.

Thus, we see that the bead in the plane has *two* degrees of freedom; the rod in the plane has *three* degrees of freedom; the disk on the plane for which slipping is allowed has *four* degrees of freedom; and the disk on the plane for which slipping is not allowed has *two* degrees of freedom.

- A *geometric constraint* is any requirement that reduces the number of degrees of freedom.

We have already encountered several forms of geometric constraints. For example, an unconstrained bead, which we continue to model as a particle, has three generalized coordinates, say x, y, and z, and three degrees of freedom characterized by the three associated admissible variations δx, δy, and thus δz. By restricting such a particle to the XY plane, we require that $z = 0$ and $\delta z = 0$. Thus, the geometric constraint of limiting the motion to the XY plane reduces the number of degrees of freedom from three to two. We may also note that in restricting the bead to the plane, there is a one-to-one correspondence between a restriction on the generalized coordinates (x, y, z go to x, y) and a restriction on the admissible variations (δx, δy, δz go to δx, δy); each is reduced by one. Constraints that act in this way are called *holonomic constraints.*

Now, let's consider the disk on a plane. With slipping allowed, the disk has four complete and independent generalized coordinates and four complete and independent admissible variations. By restricting the disk to roll without slipping, the number of independent admissible variations is reduced from four to two; yet the number of complete and independent generalized coordinates remains at four. Constraints that do not impose a one-to-one correspondence between a restriction on the generalized coordinates and a restriction on the admissible variations are called *nonholonomic constraints.*

- If the number of generalized coordinates is equal to the number of degrees of freedom, the system is said to be *holonomic.*

Note again that for holonomic constraints, there is a one-to-one correspondence between a restriction on the generalized coordinates and a restriction on the admissible variations. The bead in the plane, the rod in the plane, and the disk with slipping allowed are all holonomic systems.

- If the number of generalized coordinates is not equal to the number of degrees of freedom, the system is said to be *nonholonomic.*

Note that in nonholonomic systems, the number of degrees of freedom will always be *less* than the number of generalized coordinates in a complete and independent set. The disk that is constrained to roll on the plane without slipping is a nonholonomic system.*

---

*The terms *holonomic* and *nonholonomic* were introduced into dynamics around 1894 by the German physicist Heinrich R. Hertz (1857–1894), although he was not the first to formulate either of these types of problems. "Holonomic" has a Greek root variously translated as "altogether lawful" or "whole law."

In a strict sense, the bead on the rod, say of length l, considered in Example 5-3 is a nonholonomic system if we consider the possibility of the walls at the ends of the rod constraining the bead's motion. Systems that are holonomic over a range of their coordinates, but nonholonomic beyond that range, are formally called *piecewise holonomic.* Such cases are considered in the more mathematical presentations of dynamics where constraints involving inequalities (for example, for the bead on the rod, $0 < x < l$) are treated. We do not propose to study such cases. Our purpose in mentioning this issue here is that in problems such as Example 5-3 (and there are many such cases), generally a wall is intended to provide a reference, not a constraint. Thus, unless stated otherwise, we shall assume that such walls merely act as references, and we shall assume that all geometric trajectories of interest never come into contact with the walls. Thus, we shall think of such systems simply as *holonomic.*

Finally, in completing our discussion of geometric constraints, we consider a time-varying constraint in which *motion* is imposed upon the system. This is variously called *forced motion* or *prescribed motion* or, as we shall call it, *specified motion*.

■ **Example 5-11:**   Specified Motion

Consider Figure 5-14 where the system model, consisting of m_1, m_2, and k, is connected to an external drive that produces a time-varying displacement $x_0(t)$. The *specified motion $x_0(t)$*, including all its time derivatives, is fully prescribed by the external drive, independently of the motion of m_2 or any other system characteristics. As indicated in Figure 5-14, *specified motion* is also called a *time-varying constraint*. We want to define the generalized coordinates, the degrees of freedom, and the characterization of the system's constraints as holonomic or nonholonomic.

❏ **Solution:**   From the previous examples, we might be inclined to define x_1 and x_2 as a complete set of coordinates where x_1 and x_2 are the displacements of m_1 and m_2, respectively, from their respective equilibrium positions. However, the introduction of the specified motion requires that

$$x_1 = x_0(t) \tag{a}$$

Since $x_0(t)$ is completely specified, so too is $x_1(t)$ via Eq. (a). Thus, there exists only one generalized coordinate in a complete set, namely x_2. In this case, Eq. (a) provides an *auxiliary condition* that enables us to locate the system completely. As we shall see in Section 5-5, Eq. (a) will be important in deriving the equations of motion. Also, in our thought experiment, we cannot imagine x_1 to remain fixed as x_2 undergoes a continuous range of values. To imagine x_1 as fixed would be a violation of the constraint in the specification of the problem as represented by Eq. (a). So, x_2 represents the only independent coordinate and, indeed, constitutes the independent, as well as complete, set of generalized coordinates for the system sketched in Figure 5-14.

Further, as already suggested, an externally imposed *time-varying constraint* such as $x_0(t)$ cannot be arbitrarily varied; it is geometrically specified for all time. Thus, by definition,

$$\delta x_0 = 0 \tag{b}$$

and the combination of Eqs. (a) and (b) implies that

$$\delta x_1 = 0 \tag{c}$$

Therefore, we conclude that $\xi_j : x_2$ is a complete *and* independent set of generalized coordinates, with Eq. (a) as an *auxiliary condition;* and that $\delta\xi_j : \delta x_2$ is a complete *and*

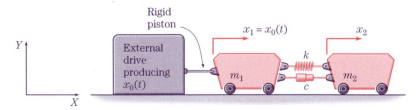

FIGURE 5-14   Time-varying constraint of specified motion $x_1 = x_0(t)$.

independent set of admissible variations. Because there is one admissible variation, there is *one* degree of freedom. Hence, because the number of generalized coordinates is equal to the number of degrees of freedom, we conclude that the system is holonomic.

Equation (a) represents a different type of constraint from those discussed previously. It is a time-varying constraint, as opposed to a purely *spatial* restriction, such as a nonslipping constraint or a requirement to remain in a plane. Nevertheless, provided that time-varying constraints are such that there remains a one-to-one correspondence between a restriction on the prospective generalized coordinates and a restriction on the prospective admissible variations, the system will be holonomic.

Without further discussion, we note that *holonomic constraints* involving time, such as Eq. (a), are sometimes called *rheonomic;* while holonomic constraints not involving time (such as those in Examples 5-8 and 5-9) are sometimes called *scleronomic.* *

> *The terms *rheonomic* and *scleronomic* were introduced by the Austrian physicist Ludwig Boltzmann (1844–1906) around 1904. Both terms have Greek roots, "rheonomic" meaning "flow" and "scleronomic" meaning "rigid."

5-4.3   Variational Principles in Mechanics[3]

In Subsections 5-3.2 and 5-3.3 we studied kinetic and potential state functions; these will satisfy a substantial portion of Requirement 3 in Eq. (5-1). In Subsections 5-4.1 and 5-4.2 we studied generalized coordinates, degrees of freedom, and constraints; these will satisfy a substantial portion of Requirement 1 in Eq. (5-1). In this and the next subsection we preview the unifying concepts, and in the next subsection we conduct the task primarily of characterizing nonconservative dynamic forces in preparation for satisfying Requirement 2 in Eq. (5-1). The fundamental principles and the regimen for equation formulation via variational dynamics will be enunciated and repeatedly demonstrated in Section 5-5.

The fundamental idea of a mechanical variational principle is that by examining neighboring *geometrically admissible states,* a *natural state* can be identified. The process of examining these neighboring geometrically admissible states is called a variation. These concepts were introduced in Subsection 5-4.2 in conjunction with degrees of freedom. Variations are hypothetical small changes from one geometrically admissible state to a neighboring geometrically admissible state. In most textbooks, these hypothetical or imagined geometrical variations are called *virtual displacements.* As defined in Subsection 5-4.2, we shall call them *admissible variations.*

In the mechanics of *particles, natural states* are geometrically admissible states that also correspond to one of the following:

- In statics: $\sum \boldsymbol{f} = 0$ or $\sum \boldsymbol{\tau} = 0$

- In dynamics: $\sum \boldsymbol{f} = \dfrac{d\boldsymbol{p}}{dt}$ or $\sum \boldsymbol{\tau} = \dfrac{d\boldsymbol{h}}{dt}$

---

[3]This subsection is concerned primarily with perspective and may be read cursorily. In Chapter 7, we shall explore variational principles in electrical networks and electromechanical systems.

In statics, a *natural state* for a particle is characterized by the equilibrium requirement on the forces (f), or the *equivalent* equilibrium requirement on the torques (τ). In dynamics, a *natural state* for a particle is characterized by the linear momentum principle, or the *equivalent* angular momentum principle. (These momentum principles were discussed in Chapter 4, and their interrelations are presented in detail in Appendix C.)

In the mechanics of *extended bodies,* rigid or flexible, *natural states* are geometrically admissible states that also correspond to one of the following:

- In statics: $\sum f = 0$ and $\sum \tau = 0$

- In dynamics: $\sum f = \dfrac{d\boldsymbol{P}}{dt}$ and $\sum \tau = \dfrac{d\boldsymbol{H}}{dt}$

where $\boldsymbol{P}$ and $\boldsymbol{H}$ are the yet-to-be-defined linear momentum and angular momentum of the extended body. In statics, a *natural state* for an extended body is characterized by the equilibrium requirements on the forces (f) *and* the torques (τ). In dynamics, a *natural state* for an extended body is characterized by the linear momentum principle *and* the angular momentum principle. (These momentum principles will be discussed in Chapter 6, and their interrelations are presented in detail in Appendix C.)

Thus, given a set of laws or principles that govern the system under consideration, we would expect a statics variational principle to deliver the equilibrium equations for the system, and a dynamics variational principle to deliver the dynamics equations for the system.

As introduced in Subsection 5-4.2, the symbol δ is used to denote the variation of a quantity. The variational operator δ and the differential operator d behave the same mathematically, but they are not the same conceptually. For example, assume that x represents a dependent variable in a dynamics problem where time is an independent variable. In such a case, the expression dx would designate a real change occurring during an increment of time dt. On the other hand, the expression δx would designate a hypothetical change *occurring at an instant in time* (that is, contemporaneously), not during an increment in time. The rules governing the mathematical operations involving δ are discussed in Appendix D; as indicated already, these rules are the same for δ and d.

For illustrative purposes, we digress into statics via two examples. These examples may be omitted without loss of continuity in our pursuit of a variational formulation for dynamic systems. Without detailed discussion, we state the following principle for a mechanical system of rigid bodies in equilibrium in which the work of all the constraint forces is zero:

A geometrically admissible configuration of the system is an equilibrium configuration if, and only if, the variational work increment vanishes for arbitrary geometrically admissible variations; thus

$$\delta W = \sum_{i=1} f_i \cdot \delta \boldsymbol{R}_i = 0 \tag{5-25}$$

where δW is the variational work increment generated by the *real* (or actual) externally applied forces f_i acting at the ith location and moving through $\delta \boldsymbol{R}_i$, which is the admissible variation at that same i^{th} location.

In summary, the variational principle in Eq. (5-25) requires that (1) the system be in equilibrium, (2) the admissible variations be compatible with the geometric constraints, and (3) the work of the constraint forces must be zero; for example, pinned connections must be frictionless. Equation (5-25) is often called the *principle of virtual work.*

■ **Example 5-12:**   Rigid Hinged Bar

Consider a rigid bar, hinged by a frictionless pin at O as sketched in Figure 5-15a. The bar is in equilibrium and is constrained to rotate about the longitudinal axis of the pin. The bar is loaded by a known force F and an unknown reaction R as shown. Find R.

□   **Solution:**   In accordance with the hinged constraint, we *imagine* a small rotation of the bar about the pivot O. This small rotation is imagined because the bar that is in equilibrium, by definition, does not move in any manner. This admissible variation is shown as $\delta\theta$ in Figure 5-15b, where the underlying angle θ represents a complete and independent set of generalized coordinates. Also, at the load sites, there are the corresponding displacement variations

$$\delta y_1 = a\,\delta\theta \qquad \text{and} \qquad \delta y_2 = \ell\,\delta\theta \tag{a}$$

where the linearized relationships in Eq. (a) are in accordance with the requirement that admissible variations are *small* hypothetical changes. Use of Eq. (5-25) gives

$$\begin{aligned}
\delta\mathcal{W} &= \sum_i \boldsymbol{f}_i \cdot \delta\boldsymbol{R}_i = 0 \\
&= (-R)(\delta y_1) + (F)(\delta y_2) \\
&= (-R)(a\,\delta\theta) + F(\ell\,\delta\theta) = 0
\end{aligned} \tag{b}$$

where the negative sign on R indicates its direction in the negative Y direction. Since $\delta\theta$ is nonzero but arbitrary, we may divide both sides of Eq. (b) by $\delta\theta$. Thus,

$$R = \left(\frac{\ell}{a}\right)F \tag{c}$$

Clearly, this result could have been easily obtained by the "direct approach" of taking the sum of the moments about O. Nevertheless, we have illustrated that there is this alternative approach—the "indirect approach" or the "variational approach"—which under some circumstances might offer significant advantages.

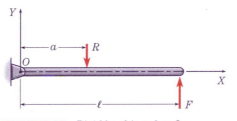
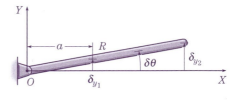

FIGURE 5-15   Rigid bar hinged at O.

Let's consider another example.

■ **Example 5-13:**   Two-bar Linkage Compactor

A two-bar linkage compactor is sketched in Figure 5-16a. The known load F acting at B is in equilibrium with the unknown reaction load R, which produces compaction at C. The bar-pin connections at A, B, and C are frictionless, as are the interfaces between the piston and ground. Find R.

❏ **Solution:**   This problem is more complicated than the previous example, and thus is capable of re-vealing more details of the analytical technique. As depicted in Figure 5-16b, the angle θ represents a set of complete and independent generalized coordinates. Also, as shown in Figure 5-16b, the admissible variation is $\delta\theta$. Note that the variation $\delta\theta$ is always taken in the positive sense of its underlying coordinate; that is, both $\delta\theta$ and θ are positive in the counterclockwise direction. Now, rather than attempt to write δx and δy by inspection, we derive them in a more general manner, which we shall have occasion to use in future prob-lems. Guided by our need to obtain the admissible variations at the load-application sites, we write

$$\text{at } B, \quad y = \ell \sin \theta$$

and   (a)

$$\text{at } C, \quad x = 2\ell \cos \theta$$

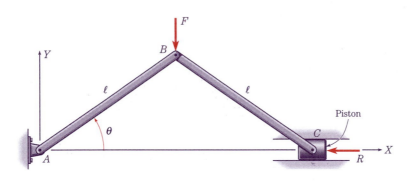

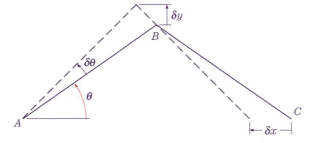

FIGURE 5-16   Two-bar linkage compactor.

where we have written y at B and x at C in accordance with the directions of F at B and R at C, respectively. As we shall see next, these are just the quantities we need. Using the rule for obtaining the variations of these functions,[4] Eqs. (a) give

$$\text{at } B, \qquad \delta y = \ell \cos \theta \, \delta \theta$$

and (b)

$$\text{at } C, \qquad \delta x = -2\ell \sin \theta \, \delta \theta$$

Use of Eq. (5-25) gives

$$\delta \mathcal{W} = \sum_i \boldsymbol{f}_i \cdot \delta \boldsymbol{R}_i$$

$$= (-F)(\ell \cos \theta \, \delta \theta) + (-R)(-2\ell \sin \theta \, \delta \theta) = 0$$ (c)

where the negative signs on F and R indicate their directions in the negative Y and X directions, respectively. Since $\delta \theta$ is nonzero but arbitrary, we may divide both sides of Eq. (c) by $\delta \theta$. Thus Eq. (c) gives

$$R = \frac{F}{2 \tan \theta}$$ (d)

Even though this was not a difficult problem, the *variational approach* appears to offer advantages over the *direct approach*. The direct approach would have required a free-body diagram analysis of each member in the compactor, including the interactions at each pin. Then a set of simultaneous equations would have been derived and solved, ultimately resulting in the answer for R. Furthermore, suppose a similar configuration consisting of *ten* bars instead of two bars had existed. In that case, the variational approach would have required essentially the same analysis! The direct approach, however, would have required a tedious analysis involving free-body diagrams for each of the ten bars and the piston, followed by the solution of the resulting simultaneous equations; not a pretty thought.

5-4.4   Generalized Velocities and Generalized Forces for Holonomic Systems

It is important to emphasize that the variational analyses that we now proceed to develop are limited to holonomic systems. Fortunately, most systems of engineering interest are holonomic. If in examining a system we determine that it is nonholonomic, we must simply halt in our application of the procedures we are about to delineate. In such a case we could return to the *direct approach* (momentum approach) or we could consider slightly more advanced variational techniques that are capable of addressing nonholonomic systems; however, we shall not pursue those more advanced variational techniques. Despite the fact that we shall restrict our attention to holonomic systems, we judge the practice of establishing the character of a system's constraints as important, and we shall emphasize and illustrate this practice throughout the remainder of the book. That is, for each system analyzed, we shall address the question "Is it holonomic or nonholonomic?"

---

[4]Note that δ behaves the same as d. Appendix D gives a summary of such rules.

We begin by considering a holonomic system consisting of N particles m_i ($i = 1, 2, 3, \ldots, N$) acted upon by N resultant forces $\boldsymbol{f}_i$ ($i = 1, 2, 3, \ldots, N$), as represented in Figure 5-17. Let all the admissible configurations of the systems be represented by the *complete* and *independent* set of n *generalized coordinates* ξ_j ($j = 1, 2, 3, \ldots, n$), and let all the *admissible variations* of the system be represented by the associated *complete* and *independent* set of n *admissible variations* $\delta \xi_j$ ($j = 1, 2, 3, \ldots, n$). Henceforth, unless we make a statement to the contrary, "generalized coordinates" will generally denote a "complete and independent set of generalized coordinates" and "admissible variations" will generally denote a "complete and independent set of admissible variations." The generalized coordinates may be inertial or noninertial. As indicated in Figure 5-17, the position of the ith particle m_i can be located by the position vector

$$\boldsymbol{R}_i = \boldsymbol{R}_i(\xi_1, \xi_2, \ldots, \xi_n, t) \tag{5-26}$$

where $\boldsymbol{R}_i$ is a function of the generalized coordinates and, if there are time-varying specified motions (or time-varying constraints), $\boldsymbol{R}_i$ is also a function of the time t.

The inertial velocity of the ith particle, $\boldsymbol{v}_i$, is

$$
\begin{aligned}
\boldsymbol{v}_i = \frac{d\boldsymbol{R}_i}{dt} &= \sum_{j=1}^{n} \frac{\partial \boldsymbol{R}_i}{\partial \xi_j} \frac{d\xi_j}{dt} + \frac{\partial \boldsymbol{R}_i}{\partial t} \\
&= \sum_{j=1}^{n} \frac{\partial \boldsymbol{R}_i}{\partial \xi_j} \dot{\xi}_j + \frac{\partial \boldsymbol{R}_i}{\partial t}
\end{aligned}
\tag{5-27}
$$

where we have used the kinematic relation of Eq. (3-1) and the chain rule of differentiation. As noted in Eq. (5-27), the time derivatives of the generalized coordinates are $\dfrac{d\xi_j}{dt}$, which by convention we shall write henceforth as $\dot{\xi}_j$, and which are called the *generalized velocities*. The last term on the right-hand side of Eq. (5-27) is the velocity due to time-varying specified motions that may be imposed on the system. (Refer to Example 5-11, where $\dot{x}_1 = \dot{x}_0$ corresponds to $\dfrac{\partial \boldsymbol{R}_i}{\partial t}$ in Eq. (5-27).) Although Eq. (5-27) suggests that the actual particle velocities are complicated combinations of the generalized coordinates and time, in most elementary problems these velocities are quite easily calculated. In fact, we note that such velocities may be readily calculated using Eqs. (3-1) or (3-23). Thus, Chapter 3 is an important resource in the variational approach as well as in the momentum approach.

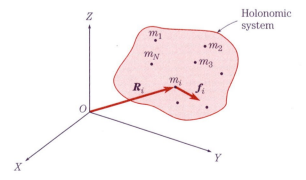

FIGURE 5-17 Holonomic system consisting of N particles.

The substitution of the term $\dot{\xi}_j$ for $\dfrac{d\xi_j}{dt}$ introduces no ambiguity into Eq. (5-27), although it is sometimes thought to do so. The generalized coordinates for lumped-parameter systems are scalar quantities, having time as their only dependent variable. Thus, irrespective of whether the generalized coordinates are inertial or noninertial, as scalar quantities their time rate of change is identical when viewed from all frames. So, $\dfrac{d\xi_j}{dt}$, $\dfrac{\partial \xi_j}{\partial t}$ and $\dot{\xi}_j$ are all identical, always.

Next, we separate the total force $\boldsymbol{f}_i$ acting on a given particle m_i into three categories according to

$$\boldsymbol{f}_i = \boldsymbol{f}_i^{\mathrm{c}} + \boldsymbol{f}_i^{\mathrm{nc}} + \boldsymbol{f}_i^{\mathrm{cst}} \tag{5-28}$$

where $\boldsymbol{f}_i^{\mathrm{c}}$ is a conservative force, $\boldsymbol{f}_i^{\mathrm{nc}}$ is a nonconservative force, and $\boldsymbol{f}_i^{\mathrm{cst}}$ is a workless constraint force. For example, conservative forces are forces such as those in conservative elements, conservative force fields, or any forces that can be represented by a potential function such as defined in Eq. (5-17); spring forces and gravitational forces are in this category. Nonconservative forces are forces that dissipate energy or introduce energy into the system; such forces are not representable by a potential function. Dashpot forces and most externally applied or so-called driving forces are in this category. If there is doubt regarding whether a force is conservative or nonconservative or if the potential energy function of a known conservative force cannot be constructed, incorporating such forces into the formulation as nonconservative forces is an option that nevertheless will lead to the correct equations of motion. Workless constraint forces are forces that confine the motion of the system in accordance with imposed constraints and do no work during an admissible variation. Forces in the Z direction that keep the particle in the XY plane in Example 5-4 or keep the rod in the XY plane in Example 5-5 are in this category. Such forces, no matter how large they may be, do no work because there is no accompanying displacement in the Z direction.

Now consider the total variational work increment done by all the individual forces $\boldsymbol{f}_i$ that act on all the individual particles m_i. This total variational work increment is evaluated by considering the total work done by all the forces as the system undergoes an admissible variation. (This is analogous to Eq. (5-25), except now we are interested in dynamics where the unbalanced forces give rise to accelerations.) Thus,

$$\delta\mathcal{W} = \sum_{i=1}^{N} \boldsymbol{f}_i \cdot \delta\boldsymbol{R}_i \tag{5-29}$$

Substitution of Eq. (5-28) into Eq. (5-29) gives

$$\delta\mathcal{W} = \sum_{i=1}^{N} \boldsymbol{f}_i^{\mathrm{c}} \cdot \delta\boldsymbol{R}_i + \sum_{i=1}^{N} \boldsymbol{f}_i^{\mathrm{nc}} \cdot \delta\boldsymbol{R}_i + \sum_{i=1}^{N} \boldsymbol{f}_i^{\mathrm{cst}} \cdot \delta\boldsymbol{R}_i \tag{5-30}$$

By definition of the admissible variations, the variational work increment due to the workless constraint forces during an admissible variation is zero. So,

$$\sum_{i=1}^{N} \boldsymbol{f}_i^{\mathrm{cst}} \cdot \delta\boldsymbol{R}_i = 0 \tag{5-31}$$

Also, in accordance with Eqs. (5-21) and (5-22), the variational work increment due to the conservative forces during an admissible variation, temporarily expressed in cartesian coordinates for illustration, is

$$\sum_{i=1}^{N} \boldsymbol{f}_i^c \cdot \delta \boldsymbol{R}_i = -\nabla V \cdot \delta \boldsymbol{R}_i$$

$$= -\left(\frac{\partial V}{\partial x}\boldsymbol{i} + \frac{\partial V}{\partial y}\boldsymbol{j} + \frac{\partial V}{\partial z}\boldsymbol{k}\right) \cdot (\delta x \boldsymbol{i} + \delta y \boldsymbol{j} + \delta z \boldsymbol{k}) \tag{5-32}$$

$$= -\left(\frac{\partial V}{\partial x}\delta x + \frac{\partial V}{\partial y}\delta y + \frac{\partial V}{\partial z}\delta z\right)$$

$$= -\delta V$$

where the rules of vector and differential calculus (or Appendix D) have been used.

Now, in considering the variational work increment due to the nonconservative forces $\boldsymbol{f}_i^{\text{nc}}$, we note that the variation of the position vector $\boldsymbol{R}_i \ (\xi_1, \xi_2, \ldots, \xi_n, t)$ can be obtained by the rules of differential calculus (or Appendix D) as

$$\delta \boldsymbol{R}_i = \sum_{j=1}^{n} \frac{\partial \boldsymbol{R}_i}{\partial \xi_j} \delta \xi_j \tag{5-33}$$

where time does not appear in Eq. (5-33) because the admissible variations occur at an instant in time; that is, the variations are said to be *contemporaneous*. Use of Eqs. (5-30) and (5-33) gives the variational work increment due exclusively to the nonconservative forces as

$$\delta \mathcal{W}^{\text{nc}} = \sum_{i=1}^{N} \boldsymbol{f}_i^{\text{nc}} \cdot \delta \boldsymbol{R}_i = \sum_{i=1}^{N} \boldsymbol{f}_i^{\text{nc}} \cdot \sum_{j=1}^{n} \frac{\partial \boldsymbol{R}_i}{\partial \xi_j} \delta \xi_j$$

$$= \sum_{j=1}^{n} \left(\sum_{i=1}^{N} \boldsymbol{f}_i^{\text{nc}} \cdot \frac{\partial \boldsymbol{R}_i}{\partial \xi_j}\right) \delta \xi_j \tag{5-34}$$

where the summations over i and j have been interchanged.

The terms in parentheses in Eq. (5-34) are somewhat complicated combinations of vector dot products of the nonconservative forces and the position vectors projected along the direction of a generalized coordinate. Fortunately, as we shall see in the examples that follow, these terms are not difficult to evaluate. The important idea in evaluating these terms is to recognize that the product of these sets of parentheses with their associated admissible variations gives the variational work increment $\delta \mathcal{W}^{\text{nc}}$. Because of this important idea, these combinations in parentheses are called *generalized forces,* and they are denoted by $\Xi_j \ (j = 1, 2, \ldots, n)$; thus,

$$\Xi_j = \sum_{i=1}^{N} \boldsymbol{f}_i^{\text{nc}} \cdot \frac{\partial \boldsymbol{R}_i}{\partial \xi_j} \tag{5-35}$$

Substitution of Eq. (5-35) into Eq. (5-34) gives

$$\delta \mathcal{W}^{\text{nc}} = \sum_{i=1}^{N} \boldsymbol{f}_i^{\text{nc}} \cdot \delta \boldsymbol{R}_i = \sum_{j=1}^{n} \Xi_j \delta \xi_j \tag{5-36}$$

Equation (5-36) is a statement that the variational work increment done by the real nonconservative forces during an admissible variation of the system's configuration is equal to the variational work increment done by the generalized forces during the admissible variations of each generalized coordinate. Equation (5-36) is *very important* in the variational formulation because it is the expression that enables us to account for the contributions by all the nonconservative forces to the system dynamics; that is, it is the means by which we evaluate the insertion of energy into the system or the extraction of energy from the system. Note that Eq. (5-36) is only that portion of Eq. (5-30) that accounts for the nonconservative forces. Thus, in calculating the generalized forces Ξ_j, the summation over N in Eq. (5-36) is effectively a summation over all the nonconservative forces in the system.

■ **Example 5-14:**   **Bead in Plane**

We consider the bead, modeled as a particle, which is constrained to remain in the XY plane, and which is subjected to the force $f(t)$, as sketched in Figure 5-18. The angle defining the direction of $f(t)$ with respect to the X axis is ψ, which is a constant. (More generally, the angle ψ may be a function of time, but must be prescribed by an external agent to the system; in this case, we simply state that the prescribed value of ψ is a constant.) Find the generalized force(s) for this system.

❑ **Solution:**   Referring to Example 5-8, a complete and independent set of generalized coordinates and a complete and independent set of the associated admissible variations are

$$\xi_j : x, y \qquad \delta\xi_j : \delta x, \delta y \tag{a}$$

From Eq. (a), we note that because there are *two* admissible variations, δx and δy, there are *two* degrees of freedom. Furthermore, there are *two* generalized coordinates, x and y. So, the number of degrees of freedom and the number of generalized coordinates are equal. Thus, the system is *holonomic;* so we are allowed to proceed with the analyses developed for holonomic systems.

To find the generalized forces, we use Eq. (5-36), into which we substitute the appropriate variables in this example:

$$\delta \mathcal{W}^{\mathrm{nc}} = f(t)\cos\psi\,\delta x + f(t)\sin\psi\,\delta y = \Xi_x\,\delta x + \Xi_y\,\delta y \tag{b}$$

The middle terms in Eq. (b) correspond to the middle expression in Eq. (5-36), and the right-hand terms in Eq. (b) correspond to the right-hand expression in Eq. (5-36). The thought experiment for generating the middle terms in Eq. (b) is as follows: Imagine the variational work done by the real forces during the sequential tweaking (imagined perturbation or admissible variation) of each generalized coordinate. So, if we imagine the work done during an admissible variation of x, we find that the component of $f(t)$ in the x direction is the

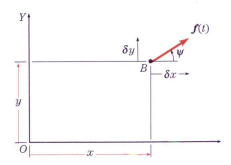

FIGURE 5-18   Bead in plane subjected to force $f(t)$ at prescribed angle ψ.

work-producing component; that is, $f(t) \cos \psi$ times the admissible variation δx produces the associated variational work increment. Similarly, for y, we find that $f(t) \sin \psi$ times the admissible variation δy produces the associated variational work increment.

Focusing only on the middle and right-hand portions of Eq. (b), we observe that all terms multiplied by δx in the middle portion must be equal to Ξ_x in the right-hand portion, and all terms multiplied by δy in the middle portion must be equal to Ξ_y in the right-hand portion. Therefore,

$$\Xi_x = f(t) \cos \psi \qquad \Xi_y = f(t) \sin \psi \qquad (c)$$

where Ξ_x and Ξ_y are the generalized forces for the system sketched in Figure 5-18.

■ **Example 5-15:** "Floating Dashpot"

The viscous extensional dashpot is an ideal two-force dissipative mechanical element model, having a *force–elongation rate* constitutive relation given by $f = cv$, where c is the positive *dashpot constant* relating force and elongation rate v (velocity) across the ends or the terminals of the dashpot. The linear dashpot schematic is depicted in Figure 5-19a.

Two carts of unknown mass are sketched in Figure 5-19b and are connected by an element we call a *floating dashpot*. A floating dashpot is simply a dashpot with neither end fixed in an inertial reference frame. Find the generalized force(s) for this system.

❑ **Solution:**   Referring to Figure 5-19b, we assume the system is constrained to remain in the plane of the sketch and the massive carts are constrained to remain on the flat bed. The adjective "massive" means "mass-bearing," not "large" or "heavy." We select a complete and independent set of generalized coordinates and a complete and independent set of the associated admissible variations as

$$\xi_j : x_1, x_2 \qquad \delta \xi_j : \delta x_1, \delta x_2 \qquad (a)$$

where x_1 and x_2 are the displacements of carts 1 and 2, respectively, with respect to their undisplaced positions, defined in an inertial reference frame. From Eq. (a), we note that because there are *two* admissible variations, δx_1 and δx_2, there are *two* degrees of freedom. Furthermore, there are *two* generalized coordinates. So, the number of degrees of freedom and the number of generalized coordinates are equal. Thus, the system is *holonomic;* so we are allowed to proceed with the analyses developed for holonomic systems.

To find the generalized forces, we use Eq. (5-36), where we emphasize that the middle portion of Eq. (5-36) represents the admissible work increment done by the real forces during an admissible variation of the system's configuration. The admissible variation of the system's configuration is accomplished through a sequential variation of each of the admissible variations. So, considering *only* the middle portion of Eq. (5-36) for the system sketched in Figure 5-19b gives

$$\sum_{i=1}^{N} \boldsymbol{f}_i^{\mathrm{nc}} \cdot \delta \boldsymbol{R}_i = c(\dot{x}_2 - \dot{x}_1) \, \delta x_1 - c(\dot{x}_2 - \dot{x}_1) \, \delta x_2 \qquad (b)$$

where δx_1 and δx_2 are positive toward the right, the same as their underlying generalized coordinates x_1 and x_2, respectively. Let us consider the analysis leading to Eq. (b). Although it is not necessary, assume for the moment that in Eq. (b), $\dot{x}_2 > \dot{x}_1 > 0$. Then, for c, $\dot{x}_2$ and $\dot{x}_1 > 0$, the real force in the dashpot will give the indicated variational work

FIGURE 5-19   Linear dashpot schematic and "floating dashpot" between two carts.

increments shown in Eq. (b). In other words, when $\dot{x}_2 > \dot{x}_1 > 0$, the force *in the dashpot* will act in the same direction as δx_1, resulting in a positive sign on the first term; and the force *in the dashpot* will act in the opposite direction from δx_2, resulting in a negative sign on the second term. Alternatively, assume that in Eq. (b), $\dot{x}_1 > \dot{x}_2 > 0$. Then, for c, $\dot{x}_2$ and $\dot{x}_1 > 0$, the real force in the dashpot again will give the indicated variational work increments shown in Eq. (b). The gist is that for any thought experiment involving the dashpot forces and the admissible variations, Eq. (b) will result.

Substitution of the results in Eq. (b) into Eq. (5-36) gives

$$\delta \mathcal{W}^{\mathrm{nc}} = c(\dot{x}_2 - \dot{x}_1)\delta x_1 - c(\dot{x}_2 - \dot{x}_1)\delta x_2 = \Xi_{x_1}\delta x_1 + \Xi_{x_2}\delta x_2 \qquad (c)$$

The middle terms in Eq. (c) correspond to the middle expression in Eq. (5-36), and the right-hand terms in Eq. (c) correspond to the right-hand expression in Eq. (5-36). From Eq. (c) we conclude that

$$\Xi_{x_1} = c(\dot{x}_2 - \dot{x}_1) = -c(\dot{x}_1 - \dot{x}_2) \qquad (d)$$

$$\Xi_{x_2} = -c(\dot{x}_2 - \dot{x}_1) \qquad (e)$$

where in Eq. (d) we have changed the order of the terms by factoring the negative sign forward to suggest consistency with our understanding that dashpots are dissipative elements.

As an alternative to the above method of deriving the generalized forces, we have devised a mnemonic for treating *floating* dashpots. *It's simple and it's worth remembering.* Using this mnemonic as explained below, we may rewrite Eq. (c) as follows:

$$\delta \mathcal{W}^{\mathrm{nc}} = -c(\dot{x}_2 - \dot{x}_1)(\delta x_2 - \delta x_1) = \Xi_{x_1}\delta x_1 + \Xi_{x_2}\delta x_2 \qquad (f)$$

In Eq. (f) we have written a minus sign on the dashpot term to reflect its dissipative character and we have written the velocity term to express the relative velocity of the two ends of the dashpot. Then the variational terms for the dashpot are written such that (i) the order of the coordinates in the velocity and the varied terms is the same and (ii) the sign between the velocities and the sign between the varied terms are the same. So, in Eq. (f) the subscripted order 2 then 1 is the same in the velocity and the varied terms, and the sign (minus) is the same between the velocity terms and between the varied terms. Equation (f) gives the identical generalized forces as concluded in Eqs. (d) and (e).

In order to apply the above mnemonic, remember that the minus sign, reflecting the dissipative character of the dashpot, and the velocity across the terminals of the dashpot must be used. Otherwise, we emphasize that we have designed this rule to produce the correct answer, irrespective of whether the generalized coordinates are in the same or opposite directions, inertial or noninertial. These points are illustrated by the two solutions given in Example 5-21, and again, in Example 5-22.

■ **Example 5-16:**   Force on Rigid Link

A rigid link, pivoted from above and loaded by a time-varying force $F(t)$ that remains vertical, is constrained to remain in the plane as sketched in Figure 5-20a. The length of the rigid link is ℓ and its undisplaced position is vertically downward (negative Y direction). Find the generalized force(s) for this system.

❑ **Solution:**   The generalized coordinate and the associated admissible variation are

$$\xi_j : \theta \qquad\qquad \delta\xi_j : \delta\theta$$

where θ is the angle of the link from its undisplaced position in an inertial reference frame, measured positive in a counterclockwise sense. From Eq. (a) we note that because there is *one* admissible variation, there is *one* degree of freedom. Furthermore, there is *one* generalized coordinate. So, the number of degrees of freedom and the number of generalized coordinates are equal. Thus, the system is *holonomic;* so we may proceed with the analyses developed for holonomic systems.

In accordance with Eq. (5-36), we want to evaluate the middle expression first; that is, we want to determine the variational work increment done by the real force(s) during an admissible variation in the system's configuration. By tweaking each (in this case, one) of the coordinates sequentially and observing the work done by the real forces (in this case, one), we can evaluate the middle term. In Figure 5-20b, an admissible variation of the system's configuration due to the admissible variation $\delta\theta$ is shown. Note that, in accordance with our convention, $\delta\theta$ is positive in the positive θ direction. Note also that because $F(t)$ is always vertical, the work-producing displacement component due to the admissible variation of the configuration is the vertical line of the hypothetical triangle at the tip of the link. This is

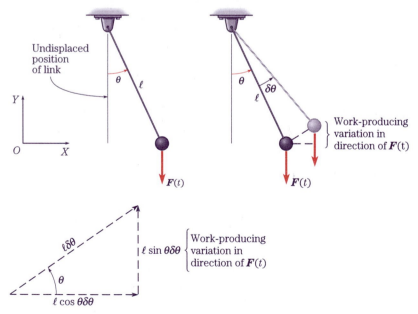

FIGURE 5-20   Rigid link loaded by vertical force, and associated geometry for admissible variation.

illustrated in Figure 5-20*b*, and the expanded view in Figure 5-20*c* shows that the work-producing variation is $\ell \sin \theta \, \delta\theta$. Thus, Eq. (5-36) may be evaluated as

$$\delta\mathcal{W}^{\mathrm{nc}} = -F(t)\ell \sin \theta \, \delta\theta = \Xi_\theta \delta\theta$$

where the middle term in Eq. (b) corresponds to the middle expression in Eq. (5-36), and the right-hand term in Eq. (b) corresponds to the right-hand expression in Eq. (5-36). The negative sign in Eq. (b) accounts for the fact that for positive $\delta\theta$, the work-producing variation $\ell \sin \theta \, \delta\theta$ and the force $F(t)$ are in opposite directions; or simply that $\ell \sin \theta \, \delta\theta$ and $F(t)$ are in the positive and negative Y directions, respectively.

Focusing only on the middle and right-hand portions of Eq. (b), we observe that all terms multiplied by $\delta\theta$ in the middle portion must be equal to Ξ_θ. Therefore,

$$\Xi_\theta = -F(t)\ell \sin \theta$$

which is the generalized force associated with the generalized coordinate θ.

From Examples 5-14, 5-15, and 5-16, several observations can be made. These observations, though made in the context of these examples, are general. All of the remarks are made for holonomic systems, however, and therefore may not hold for nonholonomic systems.

The number of generalized forces is equal to the number of generalized coordinates (which, of course, is equal to the number of degrees of freedom). One or more of the generalized forces may be zero, but even in such cases, they exist and should be written as being equal to zero. The product of each generalized force and its associated admissible variation must be a variational work increment. Thus, if the generalized coordinate has the dimension of length, the generalized force must have the dimension of force (Examples 5-14 and 5-15); or if the generalized coordinate is an angle, the generalized force must be a torque or moment (Example 5-16).

In the broadest sense, the types of real forces and torques that can be incorporated into generalized forces can be arbitrary. That is, generalized forces can be used to account for any set of adequately characterized forces: internal forces, external forces, conservative forces, or nonconservative forces. However, as indicated in our derivation of Eq. (5-36), in our formulations we shall reserve generalized forces to account for the *nonconservative* forces and torques that may be either internal (such as due to dashpots) or external (such as due to applied force and torque excitation).

5-5   EQUATIONS OF MOTION FOR HOLONOMIC MECHANICAL SYSTEMS VIA VARIATIONAL PRINCIPLES

We are now prepared to state a specific *regimen* for the formulation of the equations of motion for holonomic mechanical systems of arbitrary complexity. The foundation of our formulation will be based on the unifying law called *Hamilton's principle*. We shall find that for lumped-parameter systems, the formulation can be more directly obtained via Lagrange's equations, which are derivable from Hamilton's principle, though obviously Lagrange's equations were not originally derived via Hamilton's principle.

In Section E-1 of Appendix E, a more extensive discussion of Hamilton's principle and Lagrange's equations is presented. As usual, it is not necessary to read Appendix E in order to proceed. In this section, in order to facilitate a more direct and expeditious presentation of these formulations, only a brief summary is provided.

The formulation of the equations of motion for holonomic mechanical systems via variational principles requires the satisfaction of the three general requirements for equation derivation as emphasized in Section 5-2, and summarized in Eq. (5-1).

1. Geometric requirements: Select a complete and independent set of generalized coordinates; ξ_j $(j = 1, 2, \ldots, n)$.

2. Force–dynamic requirements:

$$\text{Hamilton's Principle—} \quad \text{V.I.} = \int_{t_1}^{t_2} [\delta\mathcal{L} + \sum_{j=1}^{n} \Xi_j \delta\xi_j]\, dt \qquad (5\text{-}37)$$

or

$$\text{Lagrange's Equations—} \quad \frac{d}{dt}\left(\frac{\partial\mathcal{L}}{\partial\dot{\xi}_j}\right) - \frac{\partial\mathcal{L}}{\partial\xi_j} = \Xi_j \qquad (5\text{-}38)$$

where V.I. in Eq. (5-37) designates a *variational indicator* and $\mathcal{L}$ is called the *lagrangian function,* or more simply the *lagrangian,* defined as

$$\mathcal{L} = T^* - V \qquad (5\text{-}39)$$

3. Constitutive requirements: $T^*(\xi_j, \dot{\xi}_j, t)$ for inertia (mass-bearing) elements,[5]
   $V(\xi_j)$ for conservative forces, and
   $\Xi_j(\xi_j, \dot{\xi}_j)$ for nonconservative forces.

Hamilton's principle for a holonomic system of n degrees of freedom may be stated as follows:

• An admissible motion of the system between specified configurations at t_1 and t_2 is a natural motion if, and only if, the variational indicator vanishes for arbitrary admissible variations.

In essence, Hamilton's principle in Eq. (5-37) states that if the geometric requirements (here called "admissible motions") are satisfied by ξ_j and the constitutive requirements are satisfied by T^*, V, Ξ_j, the force–dynamic requirements (here called "natural motions") will be satisfied by the vanishing (that is, equals zero) of the variational indicator (V.I.). The name "variational indicator" is used to denote that the integral on the right-hand side of Eq. (5-37) will vanish for the indicated arbitrary variations if, and only if, the force–dynamic requirements are satisfied.

At this point, the reader may be wondering whence Hamilton's principle. Well, we adopt the philosophical position that Hamilton's principle is a fundamental law of nature, and like all fundamental laws it cannot be derived or proved; it is an empirical statement of fact. It is equivalent to Newton's laws for mechanical problems that are governed by Newton's laws, but it also is capable of addressing other types of systems (see Chapter 7). In this sense, Hamilton's principle is more comprehensive and thus arguably more fundamental than Newton's laws.

---

[5] T^* is an explicit function of t only when there are time-varying constraints; for example, such a system is depicted in Figure 5-14.

The general relations in Eq. (5-38) are Lagrange's equations, and for lumped-parameter systems they are equivalent to Hamilton's principle. The combination $T^* - V$, which appears often in variational dynamics, is given the symbol $\mathcal{L}$ and is called the *lagrangian function,* or simply the *lagrangian.* For lumped-parameter systems, Lagrange's equations generally provide the most straightforward route to the equations of motion. After satisfying the geometric requirements in terms of the generalized coordinates and then the constitutive requirements in terms of T^*, V, and Ξ_j, Lagrange's equations are used to satisfy the force–dynamic requirements. In deriving the equations of motion via Lagrange's equations, the following steps are undertaken:

1. Select a complete and independent set of generalized coordinates ξ_j.
2. Account for all nonconservative forces by deriving the generalized forces Ξ_j.
3. Derive T^* and V to construct the lagrangian $\mathcal{L}$.
4. Substitute the results from steps 1, 2, and 3 into Lagrange's equations.

This procedure is summarized in Table 5-1, where it is emphasized that this formulation is for holonomic systems. The remainder of this chapter will focus on the implementation of the regimen in Table 5-1.

Several questions might arise as a consequence of the formulation presented as Eqs. (5-37) through (5-39) and summarized in Table 5-1. How can we be certain the various steps fit together? What guarantees that these steps assure the satisfaction of Eq. (5-1)? Why is the lagrangian expressed here as $\mathcal{L} = T^* - V$ rather than as it is expressed in most books as $\mathcal{L} = T - V$? The answer to the first and second of the questions is that the formulation flows from Hamilton's principle as a fundamental law of science, which accordingly assures the satisfaction of Eq. (5-1). Some of the relationships between Hamilton's principle and various direct approaches are explored in Appendix E. In particular, it is shown in Appendix E that for a class of problems governed by Newton's laws, Hamilton's principle provides an alternative route to the equations of motion. But Hamilton's principle is shown to embrace much more. The answer to the third question also flows from Hamilton's principle, where the appropriate kinetic state function is a function of generalized coordinates and generalized velocities, not momentum variables. Thus, from Subsection 5-3.2, a kinetic state function that is a function of geometric variables is the *kinetic coenergy function,* not the kinetic energy function.

TABLE 5-1 Structure of Equation of Motion Formulation via Lagrange's Equation for Holonomic Mechanical Systems

- Generalized Coordinates: ξ_j $(j = 1, 2, \ldots, n)$
  Admissible Variations: $\delta\xi_j$ $(j = 1, 2, \ldots, n')$
  If holonomic, then note $n' = n$ and proceed. (If nonholonomic, seek alternative formulation.)

- Generalized Forces: $\delta\mathcal{W}^{\mathrm{nc}} = \displaystyle\sum_{i=1}^{N} \boldsymbol{f}_i^{\mathrm{nc}} \cdot \delta\boldsymbol{R}_i = \sum_{j=1}^{n} \Xi_j \delta\xi_j$

- Lagrangian: $\mathcal{L} = T^*(\xi_j, \dot{\xi}_j) - V(\xi_j)$

- Lagrange's Equations: $\dfrac{d}{dt}\left(\dfrac{\partial\mathcal{L}}{\partial\dot{\xi}_j}\right) - \dfrac{\partial\mathcal{L}}{\partial\xi_j} = \Xi_j \qquad (j = 1, 2, \ldots, n)$

■ **Example 5-17:**   Bead on Rigid Rod

The bead of mass m in Figure 5-21a, which is modeled as a particle, slides along the rigid rod without friction. A known force $\boldsymbol{f}(t)$ acts on the bead at a constant angle θ, with respect to the rod. Find the equation(s) of motion for the bead.

❏ **Solution:**   It is assumed that we are concerned only with the behavior of the system as the bead moves between the two walls, or we may think of the support walls as being so far away as to have no influence on the problem. Otherwise, we are confronted with piecewise holonomic dynamic behavior. While we make such an observation here in order to indicate that such considerations may be of interest, in general, we shall omit such discursive discussions.

We simply want to emphasize the four steps listed in Table 5-1 for deriving the equations of motion via Lagrange's equations. We refer to Example 5-3.

1. Generalized coordinate(s):   The generalized coordinate and admissible variation are

$$\xi_j : x \qquad \delta\xi_j : \delta x \tag{a}$$

where x is the inertial position of the bead, relative to the left-hand wall. (We use the unconventional phrase "inertial position" to emphasize that the position x is relative to an inertial reference frame. We shall continue to emphasize the importance of expressing in words precisely what is intended as the meaning of each generalized coordinate. As we shall see, failure to do so may lead to ambiguity in the resulting equations of motion.) Because there is one admissible variation, there is one degree of freedom. Since there is also one generalized coordinate, the system is holonomic.

$$\text{Holonomic!} \tag{b}$$

Thus we may proceed with our lagrangian formulation.

2. Generalized force: In the absence of specific information to the contrary, we take the force $\boldsymbol{f}(t)$ to be nonconservative. So, the variational work expression in Eq. (5-36) becomes

$$\delta\mathcal{W}^{\text{nc}} = \sum_{i=1}^{N} \boldsymbol{f}_i^{\text{nc}} \cdot \delta\boldsymbol{R}_i = \sum_{j=1}^{n} \Xi_j \delta\xi_j$$

or

$$= f(t)\cos\theta\,\delta x = \Xi_x\,\delta x$$

$$\Xi_x = f(t)\cos\theta \tag{c}$$

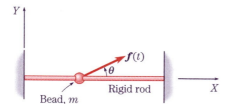

(a) Bead on rigid rod.

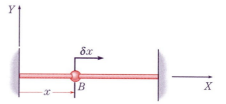

(b) Generalized coordinate and admissible variation.

FIGURE 5-21   Bead, modeled as particle, on rigid rod and subjected to force $\boldsymbol{f}(t)$ at constant angle θ.

3. Lagrangian: Use of Eq. (5-9) gives

$$T^* = \frac{1}{2}m\boldsymbol{v}\cdot\boldsymbol{v} = \frac{1}{2}m\dot{x}^2 \tag{d}$$

and consideration of Eqs. (5-12) and (5-17) gives

$$V = 0 \tag{e}$$

Thus, in accordance with Eq. (5-39),

$$\mathcal{L} = T^* - V = \frac{1}{2}m\dot{x}^2 \tag{f}$$

Note that in writing T^* we maintain the general form of Eq. (5-9), where $\boldsymbol{v}\cdot\boldsymbol{v}$ was written in Eq. (d). We shall return to this point in more complex problems such as Example 5-23. Equation (e) suggests that no conservative forces act on m. Also, note that even after writing Eqs. (d) and (e), we chose to write Eq. (f). Experience has shown that for subsequent mathematical manipulations, forming Eq. (f) is a worthwhile substitution in spite of the apparent simplicity of the operation.

4. Lagrange's equation: Equation (5-38) gives

$$\frac{d}{dt}\left(\frac{\partial\mathcal{L}}{\partial\dot{x}}\right) - \left(\frac{\partial\mathcal{L}}{\partial x}\right) = \Xi_x \tag{g}$$

where Eq. (a) has been used. Substitution of Eqs. (c) and (f) into Eq. (g) gives

$$\frac{d}{dt}(m\dot{x}) - (0) = f(t)\cos\theta \tag{h}$$

Thus,

$$m\ddot{x} = f(t)\cos\theta \tag{i}$$

which is the equation of motion.

Equation (i) is a second-order linear differential equation governing the bead's dynamics along the rod. The force $f(t)\cos\theta$ is the force applied to the bead along its direction of motion. There is also a force $f(t)\sin\theta$ applied to the bead, but this force does not appear in the formulation; it does no work as it is always precisely counteracted by a constraint force imposed by the rigid rod.

Note that in stating the problem, we sought the "equation(s)" of motion. We now see that there is only *one* equation of motion, corresponding to the single degree of freedom. It will always be the case that for holonomic systems, the number of equations of motion will be equal to the number of generalized coordinates, which, of course, is equal to the number of degrees of freedom. Also, in the definition of this problem, θ was constant. Actually, the formulation above would be identical for a time-varying θ, assuming the time-varying behavior was prescribed (imposed) by an external agent.

■ **Example 5-18:** Single Mass Spring Dashpot System

A single mass spring dashpot system subjected to an external force $f(t)$ is sketched in Figure 5-22a. The mass m, linear spring constant k, and linear dashpot constant c are shown on the

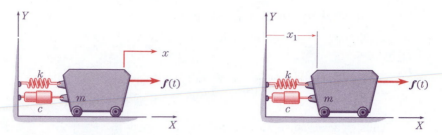

(a) System and generalized coordinate x.   (b) System and alternative generalized coordinate x_1.

FIGURE 5-22   Single mass spring dashpot system.

respective elements. The undeformed length of the spring is ℓ_0. Derive the equation(s) of motion.

☐ **Solution:**

1. Generalized coordinate(s):   The generalized coordinate and admissible variation are

$$\xi_j : x \qquad \delta\xi_j : \delta x \tag{a}$$

where, as sketched in Figure 5-22a, x is the inertial displacement in the X direction of the mass m from its equilibrium position. Note that the equilibrium position of the mass corresponds to its undisturbed location when $f(t) = 0$ and the spring is undeformed. Because there is one admissible variation, there is one degree of freedom. Since there is also one generalized coordinate, the system is holonomic.

$$\text{Holonomic!} \tag{b}$$

Thus we may proceed with our lagrangian formulation.

2. Generalized force: In the absence of specific information to the contrary, we take the force $f(t)$ to be nonconservative. We know that the dissipative dashpot force is nonconservative. (There is also a spring force but because it is conservative, it will be accommodated by the potential energy function in the lagrangian.) So, use of Eq. (5-36) gives

$$\delta\mathcal{W}^{\text{nc}} = \sum_{i=1}^{N} \boldsymbol{f}_i^{\text{nc}} \cdot \delta\boldsymbol{R}_i = \sum_{j=1}^{n} \Xi_j \delta\xi_j$$

$$= f(t)\delta x - c\dot{x}\delta x = \Xi_x \, \delta x \tag{c}$$

or

$$\Xi_x \, \delta x = [f(t) - c\dot{x}]\delta x$$

Thus, the generalized force is

$$\Xi_x = f(t) - c\dot{x} \tag{d}$$

Consider two observations. First, in determining the signs on the forces in Eq. (c), we consider the variational work increment done during an admissible variation δx, always remembering that δx is taken in the same direction as the positive direction of the underlying coordinate x. Because $f(t)$ and δx are in the same direction, the associated sign is positive; and because the dashpot force (for positive $\dot{x}$) and δx are in opposite directions, the associated sign is negative. Second, although there are two nonconservative forces acting on the mass, there is only one generalized force, corresponding to the single degree of freedom.

3. Lagrangian: Equation (5-9) gives

$$T^* = \frac{1}{2}m\boldsymbol{v} \cdot \boldsymbol{v} = \frac{1}{2}m\dot{x}^2 \tag{e}$$

and Eq. (5-15) gives

$$V = \frac{1}{2}kx^2 \tag{f}$$

Thus, in accordance with Eq. (5-39),

$$\mathcal{L} = T^* - V = \frac{1}{2}m\dot{x}^2 - \frac{1}{2}kx^2 \tag{g}$$

Note that in writing T^*, we maintain the general form of Eq. (5-9), and that the conservative force in the spring has now been taken into account via the potential energy function.

4. Lagrange's equation: Equation (5-38) gives

$$\frac{d}{dt}\left(\frac{\partial \mathcal{L}}{\partial \dot{x}}\right) - \frac{\partial \mathcal{L}}{\partial x} = \Xi_x \tag{h}$$

where Eq. (a) has been used. Substitution of Eqs. (d) and (g) into Eq. (h) gives

$$\frac{d}{dt}(m\dot{x}) - (-kx) = f(t) - c\dot{x}$$

or

$$m\ddot{x} + c\dot{x} + kx = f(t) \tag{i}$$

which is the equation of motion.

For discussion only, let's assume that $c = f(t) = 0$. Then, Eq. (i) reduces to

$$m\ddot{x} + kx = 0 \tag{j}$$

Equation (j) indicates that if the mass is against the wall $(x = -\ell_0)$,[6] the spring is fully compressed, resulting in a force $k\ell_0$ which accelerates the mass in the positive X direction. Thus, at $x = -\ell_0$

$$m\ddot{x} = k\ell_0 \tag{k}$$

Also, Eq. (j) indicates that at $x = 0$, the spring is undeformed and no forces act on the mass. Thus, at $x = 0$

$$m\ddot{x} = 0 \tag{l}$$

So, $x = 0$ is the equilibrium position of the mass; and assuming it had no velocity $(\dot{x} = 0)$ when $x = 0$, the mass would remain at its equilibrium position.

---

[6]In a real system, $x = -\ell_0$ may not be physically realizable, and almost certainly the proposed model would have long since broken down. However, in the context of the assumed *model*, $x = -\ell_0$ is a valid value of x for examination.

❑ **Alternative Solution:**

Consider an alternative solution where the generalized coordinate is sketched in Figure 5-22b.

1. Generalized coordinate(s):   The generalized coordinate and admissible variation are

$$\xi_j : x_1 \qquad \delta\xi_j : \delta x_1 \tag{m}$$

where x_1 is the inertial position of the mass relative to the left-hand wall.

$$\text{Holonomic!} \tag{n}$$

2. Generalized force: Equation (5-36) gives

$$\delta\mathcal{W}^{\text{nc}} = \sum_{i=1}^{N} \boldsymbol{f}_i^{\text{nc}} \cdot \delta\boldsymbol{R}_i = \sum_{j=1}^{n} \Xi_j \delta\xi_j$$
$$= f(t)\delta x_1 - c\dot{x}_1 \delta x_1 = \Xi_x \delta x_1$$

Thus, the generalized force is

$$\Xi_{x_1} = f(t) - c\dot{x}_1 \tag{o}$$

3. Lagrangian: Equation (5-9) gives

$$T^* = \frac{1}{2}m\boldsymbol{v} \cdot \boldsymbol{v} = \frac{1}{2}m\dot{x}_1^2 \tag{p}$$

and Eq. (5-15) gives

$$V = \frac{1}{2}k(x_1 - \ell_0)^2 \tag{q}$$

Thus, in accordance with Eq. (5-39),

$$\mathcal{L} = T^* - V = \frac{1}{2}m\dot{x}_1^2 - \frac{1}{2}k(x_1 - \ell_0)^2 \tag{r}$$

In the potential energy function we note the first significant difference between the previous solution and this alternative solution. This difference is due to the fact that the potential energy stored in the spring is still due to its *change* in length, but now that deformation is (indeed, must be) expressed in terms of the present generalized coordinate.

4. Lagrange's equation: Equation (5-38) gives

$$\frac{d}{dt}\left(\frac{\partial\mathcal{L}}{\partial\dot{x}_1}\right) - \frac{\partial\mathcal{L}}{\partial x_1} = \Xi_{x_1} \tag{s}$$

where Eq. (m) has been used. Substitution of Eqs. (o) and (r) into Eq. (s) gives

$$m\ddot{x}_1 + c\dot{x}_1 + k(x_1 - \ell_0) = f(t) \tag{t}$$

which is the equation of motion.

Again for discussion only, let's assume that $c = f(t) = 0$. Then, Eq. (t) reduces to

$$m\ddot{x}_1 + kx_1 = k\ell_0 \tag{u}$$

At $x_1 = 0$,[7] Eq. (u) reduces to

$$m\ddot{x}_1 = k\ell_0 \tag{v}$$

Equation (v) indicates that if the mass is against the wall, the spring is fully compressed, resulting in a force $k\ell_0$ which accelerates the mass in the positive X direction. At $x_1 = \ell_0$, Eq. (u) reduces to

$$m\ddot{x}_1 = 0 \tag{w}$$

Equation (w) indicates that if the mass is at $x_1 = \ell_0$, the spring is undeformed and no forces act on the mass. So, $x_1 = \ell_0$ is the equilibrium position of the mass, and assuming it had no velocity when $x_1 = \ell_0$, the mass would remain at its equilibrium position.

We conclude this example with two comments about our alternative analyses of the system sketched in Figure 5-22. And, although these comments are made in the context of this particular example, they are broadly applicable.

First, it is useful to note that in selecting x and alternatively x_1 as generalized coordinates for the system, different equations of motion were obtained. Compare Eqs. (i) and (t). Different generalized coordinates, in general (but not always), will result in different equations of motion. Thus, as stated earlier, it is very important to be clear and explicit in defining the meanings of the selected generalized coordinates.

Second, although the equations of motion for the two choices of generalized coordinates are different, both equations of motion contain the same physical information. In particular, although Eqs. (j), (k), and (l) are different from Eqs. (u), (v), and (w), respectively, as discussed above, each corresponding pair (that is, (j) and (u), (k) and (v), and (l) and (w)) captures the same physical phenomena.

■ **Example 5-19:**   Simple Plane Pendulum

A simple plane pendulum—defined as "simple" because the mass m is concentrated in the bob, the massless link is rigid, and the pendulum remains in the plane of the figure—is sketched in Figure 5-23a. The simple plane pendulum is a model, of course. The upper end of the rigid massless link is supported by a frictionless joint. Derive the equation(s) of motion.

☐ **Solution:**

1. Generalized coordinate(s):   The generalized coordinate and admissible variation are

$$\xi_j : \theta \qquad \delta\xi_j : \delta\theta \tag{a}$$

where θ is the inertial counterclockwise angular displacement of the pendulum from its downward-hanging equilibrium position, as sketched in Figure 5-23a. Because there is one admissible variation, there is one degree of freedom. Since there is also one generalized coordinate, the system is holonomic.

$$\text{Holonomic!} \tag{b}$$

Thus we may proceed with our lagrangian formulation.

---

[7]Again, in a real system, $x_1 = 0$ may not be physically realizable; however, in the context of the assumed *model*, $x_1 = 0$ is a valid value of x_1 for examination.

2. Generalized force: There are only two forces acting on the mass m: the gravitational force and the constraint force applied by the rigid link. The gravitational force is a conservative force, thus we shall account for it in the potential energy function. The force due to the rigid link is a *workless constraint force* since it is always orthogonal to the displacement of the mass. (From statics, the reader may recall that because the massless rigid link is a two-force member, its forces must be directed along its length. Indeed, the forces must be equal, opposite, and collinear.) In accordance with Eq. (5-2) and Example 5-1, forces such as the force along the rigid link, which are orthogonal to the displacement at the point of their application, do no work. Alternatively, a nonconservative torque could exist at the joint, but in the absence of friction between the link and the joint pin, it must also be zero. So, use of Eq. (5-36) gives

$$\delta \mathcal{W}^{\mathrm{nc}} = \sum_{i=1}^{N} \boldsymbol{f}_i^{\mathrm{nc}} \cdot \delta \boldsymbol{R}_i = \sum_{j=1}^{n} \Xi_j \delta \xi_j$$

$$= 0 = \Xi_\theta \delta \theta$$

Thus, the generalized force is

$$\Xi_\theta = 0 \tag{c}$$

Note that we were not finished here until we expressed Eq. (c). That is, Eq. (c) is what we sought, not the statement that $\delta \mathcal{W}^{\mathrm{nc}} = 0$. The fact that $\Xi_\theta = 0$ does not mean that Ξ_θ does not exist; Ξ_θ does exist and it is equal to zero.

3. Lagrangian: From Figure 5-23b, the inertial velocity of the mass is $\ell\dot{\theta}$. So, Eq. (5-9) gives

$$T^* = \frac{1}{2} m \boldsymbol{v} \cdot \boldsymbol{v} = \frac{1}{2} m (\ell\dot{\theta})^2 \tag{d}$$

The potential energy contribution is due to the change in elevation of the mass of the pendulum in the gravitational field, relative to a fixed reference. Any *fixed* reference may be used, with an emphasis on the word "fixed." In this problem, and in general, there are an infinite number of possibilities, but two stand out: the downward-hanging equilibrium

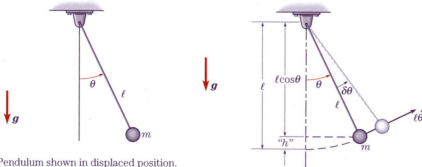

(a) Pendulum shown in displaced position.

(b) Geometry for energy calculations.

FIGURE 5-23   Simple plane pendulum.

position of the mass m, and the support joint at the upper end of the link. From Figure 5-23b, note that

$$\text{“}h\text{”} = \ell - \ell \cos \theta = \ell(1 - \cos \theta) \tag{e}$$

So, relative to the downward-hanging equilibrium position of m, use of Eq. (5-17) or, more directly, Eq. (b) in Example 5-2, gives

$$V_1 = mg \text{“}h\text{”} = mg\ell(1 - \cos \theta) \tag{f}$$

where V_1 is the corresponding potential energy function. Alternatively, relative to the support joint

$$\begin{aligned} V_2 &= -mg\ell + mg \text{“}h\text{”} \\ &= -mg\ell + mg\ell - mg\ell \cos \theta \\ &= -mg\ell \cos \theta \end{aligned} \tag{g}$$

where V_2 is the corresponding potential energy function. Also, Eq. (g) could have been written directly from Figure 5-23b by noting that the mass m is a distance of $\ell \cos \theta$ below the support joint. Thus, in accordance with Eq. (5-39), the lagrangian is

$$\mathcal{L}_1 = T^* - V_1 = \frac{1}{2}m(\ell\dot{\theta})^2 + mg\ell \cos \theta - mg\ell \tag{h}$$

or

$$\mathcal{L}_2 = T^* - V_2 = \frac{1}{2}m(\ell\dot{\theta})^2 + mg\ell \cos \theta \tag{i}$$

where $\mathcal{L}_1$ and $\mathcal{L}_2$ are the lagrangians corresponding to V_1 and V_2, respectively.

4. Lagrange's equation: Equation (5-38) gives

$$\frac{d}{dt}\left(\frac{\partial \mathcal{L}}{\partial \dot{\theta}}\right) - \frac{\partial \mathcal{L}}{\partial \theta} = \Xi_\theta \tag{j}$$

where Eq. (a) has been used. In preparing to substitute either Eq. (h) or Eq. (i) into Eq. (j), we observe that the resulting equation of motion will be the same for either of Eqs. (h) or (i). Equations (h) and (i) differ only by a constant, and because of the differentiation operations in Eq. (j), these constants do not influence the resulting equations of motion. Thus, we make the interesting observation that *any* constant whatever may be added to the lagrangian without affecting the equations of motion. Therefore, substitution of Eq. (c) and either of Eqs. (h) or (i) into Eq. (j) gives

$$m\ell^2 \ddot{\theta} + mg\ell \sin \theta = 0 \tag{k}$$

which is the equation of motion.

The equations of motion in the two preceding examples (Examples 5-17 and 5-18) were linear, whereas Eq. (k) here is nonlinear. As we shall see in Chapter 8, the closed form analytical solution techniques for solving linear differential equations are well developed; the same cannot be said for nonlinear equations, as most nonlinear differential equations cannot be solved in closed form.

In many instances, although the underlying phenomena and the resulting equations of motion are nonlinear, we shall settle for understanding the "small" motion or linearized behavior of systems in the vicinity of one or more equilibrium positions. Here we briefly explore the linearization of the nonlinear differential equation of motion in Eq. (k) in the vicinity of its downward-hanging equilibrium position.

It is the presence of the $\sin \theta$ term in Eq. (k) that makes the equation of motion non-linear. By appropriately truncating the Taylor series expansion of $\sin \theta$, we can linearize Eq. (k). Thus,* in the vicinity of $\theta = 0$, the series

$$\sin \theta = \theta - \frac{\theta^3}{3!} + \frac{\theta^5}{5!} - \cdots \tag{l}$$

---

*Strictly, the Taylor series given in Eq. (l) is called the *Maclaurin series* because it is expanded about the special point $\theta = 0$. Nevertheless, we shall generally refer to all such series expansions as Taylor series, named in honor of the Englishman Brook Taylor (1685–1731) who published the more general form of the series in 1715. The term *Taylor series* was not associated with this result until the mid-1780s.

---

is truncated at the first term for linearized motion and substituted into Eq. (k), giving

$$m\ell^2 \ddot{\theta} + mg\ell\theta = 0 \tag{m}$$

Alternatively, we note that Lagrange's equations require a single derivative in either ξ_j or $\dot{\xi}_j$. So, if we want the resulting equations of motion to be linear, this suggests that the la-grangian should be *quadratic*. This is indeed the case, as we conclude that when a quadratic lagrangian is substituted into Lagrange's equations, the linear equations of motion will be obtained. By returning to Eq. (f) or Eq. (g), expanding the $\cos \theta$ term into its Taylor series, and then truncating it at the quadratic term, Eq. (m) could be obtained via this alternative procedure. We shall illustrate this technique in Example 5-20.

Finally, we have suggested the need for amplification by the quotation marks on the word "small" in conjunction with our discussion leading to the linearized equation (m). As it so happens, a closed-form analytical solution does exist for Eq. (k). This solution is given in terms of an elliptic integral of the first kind, a subject that is beyond the scope of our interests. Although elliptic integrals will not be considered here, we state the result, correct to fourth order, as

$$\mathcal{T} = 2\pi \sqrt{\frac{\ell}{g}} \left[1 + \frac{1}{16} \theta_0^2 + \frac{11}{3072} \theta_0^4 + \cdots \right] \tag{n}$$

where $\mathcal{T}$ is the period or time for a complete oscillation of the pendulum and θ_0 is the maximum angular displacement of the pendulum. So, we see that while the period $\mathcal{T}_0 = 2\pi \sqrt{\frac{\ell}{g}}$, which is the well-known solution for Eq. (m), is not a function of the angular displacement, $\mathcal{T}$ in Eq. (n), which is the solution for Eq. (k), is a function of the angular displacement.

More directly to the point of "small" are the tabulated results in Table 5-2. From Table 5-2, we can see that "small" means that even for $\theta_0 = 30°$, Eq. (m) approximates Eq. (k) to within about 2% and that even at $\theta_0 = 60°$, the error introduced by using Eq. (m) instead of Eq. (k) is only about 8%. Compared with most problems, the plane pendulum allows for a very generous interpretation of the word "small."

The equation for the period during small angular displacements was derived in 1673 by Christiaan Huygens. Although Huygens was well aware of large-angle nonlinearities in pen-dula, the first significant treatment of the pendulum undergoing large angular displacements

TABLE 5-2   Numerical Results from Elliptic
Integral for Plane Pendulum

Angular Displacement, θ_0	Normalized Period, $\mathcal{T}/\mathcal{T}_0$
0°	1.00
30°	1.02
60°	1.08
90°	1.18
120°	1.37
150°	1.76
170°	2.44
180°	∞

$\mathcal{T}_0 = 2\pi\sqrt{\ell/g}$ = Period for $\theta_0 \approx 0°$, which is solution to Eq. (m).

was due to Leonhard Euler (1736), and substantial work on elliptic integrals was published
by Adrien M. Legendre in 1825 and 1826.

■ **Example 5-20:**   Restrained Plane Pendulum

A plane pendulum, restrained by a linear spring of spring constant k and a linear dashpot
of dashpot constant c, is sketched in Figure 5-24. The pendulum's length ℓ and mass m are
also indicated in Figure 5-24. The upper end of the rigid massless link is supported by a
frictionless joint. Derive the *linear* equation(s) of motion.

☐ **Solution:**   Note that the problem statement requests the *linear* equation(s) of motion. This is equivalent to assuming that the system should be analyzed for small excursions from equilibrium
for which the variational work increment may be written as a function of the admissible
variations in the vicinity of the equilibrium, and the lagrangian may be written as a quadratic
function of the generalized coordinates and generalized velocities.

1. Generalized coordinate(s):   The generalized coordinate and admissible variation are

$$\xi_j : \theta \qquad \delta\xi_j : \delta\theta \tag{a}$$

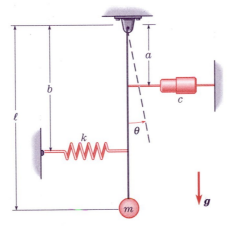

FIGURE 5-24   Plane pendulum restrained by linear
spring and dashpot.

where θ is the inertial counterclockwise angular displacement of the pendulum from its downward-hanging equilibrium position. Because there is one admissible variation, there is one degree of freedom. Since there is also one generalized coordinate, the system is holonomic.

$$\text{Holonomic!} \tag{b}$$

Thus we may proceed with our lagrangian formulation.

2. Generalized force: The only nonconservative force that acts in the system is the dashpot force; all other forces (spring and gravity) are conservative or workless constraint (rigid link) forces. So, use of Eq. (5-36) gives

$$\delta W^{\text{nc}} = \sum_{i=1}^{N} f_i^{\text{nc}} \cdot \delta R_i = \sum_{j=1}^{n} \Xi_j \delta \xi_j$$
$$= -c(a\dot\theta)(a\,\delta\theta) = \Xi_\theta\,\delta\theta \tag{c}$$

Note that Eq. (c) has been written for small angular displacements, with the goal of a linear equation of motion. In the middle portion of Eq. (c), the terms in parentheses are $(a\dot\theta)$, which is the velocity across the dashpot, and $(a\,\delta\theta)$, which is the variational displacement across the dashpot during an admissible variation $\delta\theta$. From Eq. (c), it follows that

$$\Xi_\theta = -ca^2\dot\theta \tag{d}$$

where Ξ_θ is equal to every term in the multiplicand of $\delta\theta$ in the middle portion of Eq. (c). Note also that since the generalized coordinate is an angle, the generalized force is a torque or moment.

3. Lagrangian: As in Example 5-19, the inertial velocity of the mass is $\ell\dot\theta$. So, Eq. (5-9) gives

$$T^* = \frac{1}{2}mv \cdot v = \frac{1}{2}m(\ell\dot\theta)^2 \tag{e}$$

Further, the potential energy function accounts for contributions due to the gravitational force and the spring force, both of which are conservative. Considering gravity only, as in Example 5-19,

$$V = mg\ell(1 - \cos\theta) \tag{f}$$

However, since we want to derive the linear equation of motion, as we indicated in Example 5-19, we want to retain quadratic terms in θ and $\dot\theta$ in the lagrangian. Note that Eq. (e) is already quadratic in $\dot\theta$. Now we note that Eq. (f) can be rewritten as

$$V = mg\ell\left[1 - \left(1 - \frac{\theta^2}{2!} + \frac{\theta^4}{4!} - \cdots\right)\right]$$
$$V \approx mg\ell\frac{\theta^2}{2} \tag{g}$$

where we have used the Taylor series expansion for $\cos\theta$ in the vicinity of $\theta = 0$ and truncated the series at the quadratic term. Thus, the quadratic potential energy function for the system is

$$V = mg\ell\frac{\theta^2}{2} + \frac{1}{2}k(b\theta)^2 \tag{h}$$

where, in accordance with Eq. (5-15), the second term on the right-hand side of Eq. (h) accounts for the spring, which undergoes extension $b\theta$, where $b\theta$ is the approximation for small motions. So, substitution of Eqs. (e) and (h) into Eq. (5-39) gives the lagrangian as

$$\mathcal{L} = T^* - V = \frac{1}{2}m(\ell\dot{\theta})^2 - mg\ell\frac{\theta^2}{2} - \frac{1}{2}k(b\theta)^2 \tag{i}$$

4. Lagrange's equation: Equation (5-38) gives

$$\frac{d}{dt}\left(\frac{\partial\mathcal{L}}{\partial\dot{\theta}}\right) - \frac{\partial\mathcal{L}}{\partial\theta} = \Xi_\theta \tag{j}$$

where Eq. (a) has been used. Substituting Eqs. (d) and (i) into Eq. (j) gives

$$m\ell^2\ddot{\theta} + ca^2\dot{\theta} + (kb^2 + mg\ell)\theta = 0 \tag{k}$$

which is the *linear* equation of motion for the system.

■ **Example 5-21:**   **Mass Spring Dashpot Subsystem in Falling Container**

A mass spring dashpot subsystem in a falling container of mass m_1 is sketched in Figure 5-25. The system is subject to constraints (not shown) that confine its motion to the vertical direction only. The mass m_2, linear spring of undeformed length ℓ_0 and spring constant k, and linear dashpot of dashpot constant c of the internal subsystem are also indicated in Figure 5-25. Derive the equation(s) of motion for the system.

❏ **Solution:**   In Example 5-18 we considered an alternative set of generalized coordinates: one with respect to the equilibrium position of the mass, and one with respect to ground; both nevertheless being inertial. In this example, there are two masses, suggesting two generalized coordinates in a complete and independent set; and as for all problems, there is a variety of possible choices of generalized coordinates. However, in this example, where there is no meaningful equilibrium position, at least one of the generalized coordinates *must* be with respect to ground. The relevance of this requirement should be noted when considering both T^* and V in the solutions. For example, recall that the evaluation of T^* requires the *inertial velocity.* We shall consider two possible choices of sets of generalized coordinates. As usual, without an additional specific question or issue under consideration, neither set is intrinsically preferable. (For example, an additional specific question might be "What is the relative displacement of m_2 with respect to m_1?" In that case, y_1 and y_2 *might* be preferable to x_1 and x_2 since y_2 addresses the additional specific question directly, where reference to Figure 5-25 suggests the meanings of $x_1, x_2, y_1,$ and y_2.)

1. Generalized coordinate(s):   First we select x_1 and x_2 as a complete and independent set of generalized coordinates, where x_1 and x_2 are the inertial positions of m_1 and m_2, respectively, both with respect to the horizontal ground reference sketched in Figure 5-25. Thus,

$$\xi_j : x_1, x_2 \qquad \delta\xi_j : \delta x_1, \delta x_2 \tag{a}$$

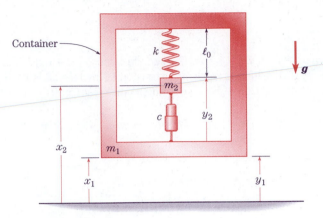

FIGURE 5-25   Container and mass spring dashpot subsystem, constrained to move in vertical direction only; constraints not shown.

Because there are two admissible variations, there are two degrees of freedom. Since there are also two generalized coordinates, the system is holonomic.

$$\text{Holonomic!} \tag{b}$$

Thus we may proceed with our lagrangian formulation.

2. Generalized forces: The dashpot is the only element that contributes to the noncon-servative variational work increment. In terms of the generalized coordinates x_1 and x_2, the dashpot is identified as a "floating dashpot." So, use of Eq. (5-36) gives

$$\delta\mathcal{W}^{\text{nc}} = \sum_{i=1}^{N} \boldsymbol{f}_i^{\text{nc}} \cdot \delta\boldsymbol{R}_i = \sum_{j=1}^{n} \Xi_j \delta\xi_j$$

$$= -c(\dot{x}_2 - \dot{x}_1)(\delta x_2 - \delta x_1) = \Xi_{x_1}\delta x_1 + \Xi_{x_2}\delta x_2 \tag{c}$$

where both $\dot{x}_1$ and $\dot{x}_2$ are inertial velocities. Thus, the generalized forces are

$$\Xi_{x_1} = -c(\dot{x}_1 - \dot{x}_2) \tag{d}$$

$$\Xi_{x_2} = -c(\dot{x}_2 - \dot{x}_1) \tag{e}$$

where reference to Example 5-15 may be beneficial.

3. Lagrangian: In accordance with Eq. (5-10), the system kinetic coenergy function is

$$T^* = T^*_{m_1} + T^*_{m_2} \tag{f}$$

where $T^*_{m_1}$ and $T^*_{m_2}$ are the kinetic coenergies of m_1 and m_2, respectively. By use of Eq. (5-9), Eq. (f) becomes

$$T^* = \sum_{i=1}^{2} \frac{1}{2} m_i \boldsymbol{v}_i \cdot \boldsymbol{v}_i = \frac{1}{2} m_1 \dot{x}_1^2 + \frac{1}{2} m_2 \dot{x}_2^2 \tag{g}$$

where again it is worth noting that $\dot{x}_1$ and $\dot{x}_2$ are inertial velocities. Using Eqs. (5-24) and (5-15) and Eq. (b) in Example 5-2 for the potential energy function gives

$$V = \frac{1}{2} k(x_1 - x_2)^2 + m_1 g x_1 + m_2 g x_2 \tag{h}$$

Since both x_1 and x_2 are positive in the same direction, the potential energy function for the spring is a function of their difference, which gives the spring deformation. Furthermore, the fact that the deformation is squared results in a positive value of the potential energy, regardless of the sign of the deformation. In this sense, the state functions for masses and springs are said to be *positive definite* since they are always nonnegative, and moreover they are zero only when every velocity (for masses in T^*) or every deformation (for springs in V) is zero. However, it should be noted that when V is a linear function of the generalized coordinate(s), such as due to gravity in Eq. (h), it may be positive or negative, depending on the choice of reference. Such potential energy functions result in constants in the equations of motion, which simply change the equilibrium about which the dynamics occur. (See Problem 5-19 or Example 6-12.)

Thus, the lagrangian may be written by substituting Eqs. (g) and (h) into Eq. (5-39) as

$$\mathcal{L} = T^* - V$$

$$= \frac{1}{2}m_1\dot{x}_1^2 + \frac{1}{2}m_2\dot{x}_2^2 - \frac{1}{2}k(x_1 - x_2)^2 - m_1gx_1 - m_2gx_2 \tag{i}$$

4. Lagrange's equations: Equation (5-38) gives

$$\frac{d}{dt}\left(\frac{\partial \mathcal{L}}{\partial \dot{x}_j}\right) - \frac{\partial \mathcal{L}}{\partial x_j} = \Xi_{x_j} \qquad j = 1, 2 \tag{j}$$

where Eq. (a) has been used. Substitution of Eqs. (d), (e) and (i) into Eqs. (j) gives

$$x_1\text{-equation:} \qquad m_1\ddot{x}_1 + c\dot{x}_1 - c\dot{x}_2 + kx_1 - kx_2 + m_1g = 0 \tag{k}$$

and

$$x_2\text{-equation:} \qquad m_2\ddot{x}_2 - c\dot{x}_1 + c\dot{x}_2 - kx_1 + kx_2 + m_2g = 0 \tag{l}$$

Equations (k) and (l) are a pair of coupled linear ordinary differential equations of motion.

☐ Alternative Solution:

We now consider an alternative solution where the generalized coordinates are the second set of coordinates sketched in Figure 5-25. Although the same dynamic information will be contained in this alternative formulation, the form of the equations of motion will change significantly due to this second choice of generalized coordinates.

1. Generalized coordinate(s):   We select y_1 and y_2 as a complete and independent set of generalized coordinates, where y_1 is the inertial position of m_1 with respect to the horizontal ground reference; and y_2 is the position of m_2 relative to m_1 and from the unstretched length of the spring. (See Problem 6-12.) So, we observe that y_2 is noninertial. Thus,

$$\xi_j : y_1, y_2 \qquad \delta\xi_j : \delta y_1, \delta y_2 \tag{m}$$

Because there are two admissible variations, there are two degrees of freedom. Since there are also two generalized coordinates, the system is holonomic.

$$\text{Holonomic!} \tag{n}$$

Thus we may proceed with our lagrangian formulation. Note that at the outset of this alternative solution we knew that we should select only two generalized coordinates. The reason is that while generalized coordinates are not unique, the number of generalized coordinates in a complete and independent set is unique; in this case that number is *two*.

2. Generalized forces: The dashpot is the only element that contributes to the nonconservative variational work increment; however, in this alternative solution the character of its contribution is significantly different. Use of Eq. (5-36) gives

$$\delta W^{\mathrm{nc}} = \sum_{i=1}^{N} \boldsymbol{f}_i^{\mathrm{nc}} \cdot \delta \boldsymbol{R}_i = \sum_{j=1}^{n} \Xi_j \delta \xi_j$$

$$= -c(\dot{y}_2)\delta y_2 = \Xi_{y_1} \delta y_1 + \Xi_{y_2} \delta y_2 \tag{o}$$

which may be obtained by sequentially tweaking each of the generalized coordinates. Thus, the generalized forces are

$$\Xi_{y_1} = 0 \tag{p}$$

$$\Xi_{y_2} = -c\dot{y}_2 \tag{q}$$

Notice that Eq. (o) is a degenerate form of the "floating dashpot" mnemonic. Nevertheless, the rules we established in Example 5-15 do not require alteration. Still, because there are two degrees of freedom, there are (indeed, must be) two generalized forces. Equations (p) and (q) should be contrasted with Eqs. (d) and (e).

3. Lagrangian: Again, in accordance with Eqs. (5-9) and (5-10), the system kinetic coenergy function is

$$T^* = \sum_{i=1}^{2} \frac{1}{2} m_i \boldsymbol{v}_i \cdot \boldsymbol{v}_i = \frac{1}{2} m_1 \dot{y}_1^2 + \frac{1}{2} m_2 (\dot{y}_1 + \dot{y}_2)^2 \tag{r}$$

Equation (r) should be contrasted with Eq. (g), where it is noted that in Eq. (r) the inertial velocity of m_2 is (indeed, must be) used. Use of Eqs. (5-24) and (5-15) and Eq. (b) in Example 5-2 for the potential energy function gives

$$V = \frac{1}{2} k y_2^2 + m_1 g y_1 + m_2 g (y_1 + y_2) \tag{s}$$

Equation (s) should be contrasted with Eq. (h).

Thus, the lagrangian may be written by substituting Eqs. (r) and (s) into Eq. (5-39) as

$$\mathcal{L} = T^* - V$$

$$= \frac{1}{2} m_1 \dot{y}_1^2 + \frac{1}{2} m_2 (\dot{y}_1 + \dot{y}_2)^2 - \frac{1}{2} k y_2^2 - m_1 g y_1 - m_2 g (y_1 + y_2) \tag{t}$$

4. Lagrange's equations: Equation (5-38) gives

$$\frac{d}{dt} \left(\frac{\partial \mathcal{L}}{\partial \dot{y}_j} \right) - \frac{\partial \mathcal{L}}{\partial y_j} = \Xi_{y_j} \qquad j = 1, 2 \tag{u}$$

where Eq. (m) has been used. Substitution of Eqs. (p), (q), and (t) into Eqs. (u) gives

$$y_1\text{-equation:} \qquad (m_1 + m_2)\ddot{y}_1 + m_2 \ddot{y}_2 + m_1 g + m_2 g = 0 \tag{v}$$

and

$$y_2\text{-equation:} \qquad m_2 \ddot{y}_1 + m_2 \ddot{y}_2 + c\dot{y}_2 + k y_2 + m_2 g = 0 \tag{w}$$

Equations (v) and (w) are the equations of motion. Equations (v) and (w) are a pair of coupled linear ordinary differential equations and should be compared with Eqs. (k) and (l).

For convenience we rewrite both sets of equations in matrix form as

$$\begin{bmatrix} m_1 & 0 \\ 0 & m_2 \end{bmatrix} \begin{Bmatrix} \ddot{x}_1 \\ \ddot{x}_2 \end{Bmatrix} + \begin{bmatrix} c & -c \\ -c & c \end{bmatrix} \begin{Bmatrix} \dot{x}_1 \\ \dot{x}_2 \end{Bmatrix} + \begin{bmatrix} k & -k \\ -k & k \end{bmatrix} \begin{Bmatrix} x_1 \\ x_2 \end{Bmatrix} = -g \begin{Bmatrix} m_1 \\ m_2 \end{Bmatrix} \tag{x}$$

and

$$\begin{bmatrix} m_1 + m_2 & m_2 \\ m_2 & m_2 \end{bmatrix} \begin{Bmatrix} \ddot{y}_1 \\ \ddot{y}_2 \end{Bmatrix} + \begin{bmatrix} 0 & 0 \\ 0 & c \end{bmatrix} \begin{Bmatrix} \dot{y}_1 \\ \dot{y}_2 \end{Bmatrix} + \begin{bmatrix} 0 & 0 \\ 0 & k \end{bmatrix} \begin{Bmatrix} y_1 \\ y_2 \end{Bmatrix} = -g \begin{Bmatrix} m_1 + m_2 \\ m_2 \end{Bmatrix} \tag{y}$$

where Eq. (x) corresponds to Eqs. (k) and (l), and Eq. (y) corresponds to Eqs. (v) and (w).

The three constant coefficient matrices in the order in which they appear in both Eqs. (x) and (y) are often designated by $[m]$, $[c]$, and $[k]$ and are called the *mass matrix, damping matrix,* and *stiffness matrix,* respectively. The three time-varying 2×1 matrices in the order in which they appear in Eq. (x) (or Eq. (y)) are designated by $\{\ddot{x}(t)\}$, $\{\dot{x}(t)\}$, and $\{x(t)\}$ (or $\{\ddot{y}(t)\}$, $\{\dot{y}(t)\}$ and $\{y(t)\}$) and are called the *acceleration vector, velocity vector,* and *displacement vector,* respectively. The differences in the $[m]$, $[c]$, and $[k]$ matrices in Eqs. (x) and (y) are due to the choices of generalized coordinates used in deriving each set of equations. Indeed, the use of *noninertial* generalized coordinates (such as y_2) may lead to unsymmetric matrices. The pair of equations in Eq. (x) and the pair of equations in Eq. (y) are coupled through the off-diagonal terms in the matrices. If the off-diagonal terms in all of the three square matrices vanish, the individual equations of motion in each formulation will be uncoupled and independent.

Finally, note that throughout this example, the potential energy functions have been written for a spring element k whose undeformed length does not enter the formulation. In reconsidering this system in Problem 5-46, we select a set of generalized coordinates that result in a potential energy function in which the spring's undeformed length appears.

■ **Example 5-22:** System Driven by Specified Motion

The system sketched in Figure 5-26, consisting of masses m_1, m_2, and m_3, two linear springs of spring constants k_1 and k_2, and two linear dashpots of dashpot constants c_1 and c_2, is connected to an external drive that produces a time-varying displacement $x_0(t)$. The *specified motion $x_0(t)$* is fully prescribed by the external drive, independently of any motion or characteristics of the system. The system is constrained to remain in the plane of the sketch. Derive the equation(s) of motion for the system.

❏ **Solution:**

1. Generalized coordinate(s): From Example 5-11, we note that the system can be completely located by the coordinates x_1, x_2, and x_3, which are the inertial displacements of m_1,

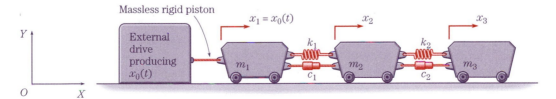

FIGURE 5-26 Dynamic system subjected to time-varying constraint of specified motion $x_0(t)$.

m_2, and m_3, respectively, from their equilibrium positions. However, because of the imposed *time-varying* specified motion, that is,

$$x_1 = x_0(t) \tag{a}$$

there remain only two coordinates in a complete and independent set, namely, x_2 and x_3. Furthermore, the time-varying specified motion, by definition, requires that

$$\delta x_0(t) = 0 \tag{b}$$

which in accordance with Eq. (a) requires that

$$\delta x_1 = 0 \tag{c}$$

Thus, we conclude that a complete *and* independent set of generalized coordinates and a complete *and* independent set of admissible variations are expressed by

$$\xi_j : x_2, x_3 \qquad \delta\xi_j : \delta x_2, \delta x_3 \tag{d}$$

Because there are two admissible variations, there are two degrees of freedom. Since there are also two generalized coordinates, the system is holonomic.

$$\text{Holonomic!} \tag{e}$$

Thus we may proceed with our lagrangian formulation.

2. Generalized forces: Because there are two dissipative linear dashpots and no other sources of nonconservative forces, we conclude that there are two elements that contribute to the variational work increment expression. Both dashpots are "floating dashpots," so in accordance with Eq. (5-36) and Example 5-15,

$$\delta\mathcal{W}^{\text{nc}} = \sum_{i=1}^{N} \boldsymbol{f}_i^{\text{nc}} \cdot \delta\boldsymbol{R}_i = \sum_{j=1}^{n} \Xi_j \delta\xi_j$$

$$= -c_1(\dot{x}_2 - \dot{x}_1)(\delta x_2 - \delta x_1) - c_2(\dot{x}_3 - \dot{x}_2)(\delta x_3 - \delta x_2)$$

$$= \Xi_{x_1}\delta x_1 + \Xi_{x_2}\delta x_2 + \Xi_{x_3}\delta x_3 \tag{f}$$

However, in accordance with Eqs. (a) and (c), Eq. (f) should be immediately rewritten as

$$-c_1(\dot{x}_2 - \dot{x}_0)\delta x_2 - c_2(\dot{x}_3 - \dot{x}_2)(\delta x_3 - \delta x_2) = \Xi_{x_2}\delta x_2 + \Xi_{x_3}\delta x_3 \tag{g}$$

Now, Eq. (g) leads directly to the generalized forces as

$$\Xi_{x_2} = -c_1(\dot{x}_2 - \dot{x}_0) - c_2(\dot{x}_2 - \dot{x}_3) \tag{h}$$

$$\Xi_{x_3} = -c_2(\dot{x}_3 - \dot{x}_2) \tag{i}$$

where every term on the left-hand side of Eq. (g) that has δx_2 as its multiplicand contributes to Ξ_{x_2} and every term on the left-hand side of Eq. (g) that has δx_3 as its multiplicand contributes to Ξ_{x_3}. Because there are two degrees of freedom, as expected, there are two generalized forces. Further, we note that although $x_1(t)$, or equivalently $x_0(t)$, is not a generalized coordinate, $x_0(t)$ nevertheless enters the formulation in Eq. (h) via the imposed time-varying specified motion given in Eq. (a).

3. Lagrangian: In accordance with Eq. (5-10), the system kinetic coenergy function may be written as

$$T^* = T^*_{m_1} + T^*_{m_2} + T^*_{m_3} \tag{j}$$

where the three terms on the right-hand side of Eq. (j) account for the kinetic coenergies of m_1, m_2, and m_3, as indicated. Use of Eq. (5-9) gives

$$T^* = \sum_{i=1}^{3} \frac{1}{2} m_i \boldsymbol{v}_i \cdot \boldsymbol{v}_i$$

$$= \frac{1}{2} m_1 \dot{x}_0^2 + \frac{1}{2} m_2 \dot{x}_2^2 + \frac{1}{2} m_3 \dot{x}_3^2 \tag{k}$$

where Eq. (a) has been used. The first term on the right-hand side of both Eqs. (j) and (k) is frequently not written, for a reason that may already be apparent, yet which we shall note in the equation derivation.

The system potential energy function may be written in accordance with Eq. (5-24) as

$$V = V_{k_1} + V_{k_2} \tag{l}$$

where the two terms on the right-hand side of Eq. (l) account for the potential energies of k_1 and k_2. The use of Eq. (5-15) gives

$$V = \frac{1}{2} k_1 (x_2 - x_0)^2 + \frac{1}{2} k_2 (x_3 - x_2)^2 \tag{m}$$

where Eq. (a) has been used in expressing the potential energy of the spring k_1. Thus, the lagrangian may be written by substituting Eqs. (k) and (m) into Eq. (5-39) as

$$\mathcal{L} = T^* - V$$

$$= \frac{1}{2} m_1 \dot{x}_0^2 + \frac{1}{2} m_2 \dot{x}_2^2 + \frac{1}{2} m_3 \dot{x}_3^2 - \frac{1}{2} k_1 (x_2 - x_0)^2 - \frac{1}{2} k_2 (x_3 - x_2)^2 \tag{n}$$

4. Lagrange's equations: Equation (5-38) gives

$$\frac{d}{dt} \left(\frac{\partial \mathcal{L}}{\partial \dot{x}_j} \right) - \frac{\partial \mathcal{L}}{\partial x_j} = \Xi_{x_j} \qquad j = 2, 3 \tag{o}$$

where Eq. (d) has been used. Substitution of Eqs. (h), (i), and (n) into Eqs. (o) gives

x_2-equation: $m_2 \ddot{x}_2 + (c_1 + c_2) \dot{x}_2 - c_2 \dot{x}_3 + (k_1 + k_2) x_2 - k_2 x_3 = c_1 \dot{x}_0 + k_1 x_0$ \qquad (p)

and

x_3-equation: $m_3 \ddot{x}_3 - c_2 \dot{x}_2 + c_2 \dot{x}_3 - k_2 x_2 + k_3 x_3 = 0$ \qquad (q)

Equations (p) and (q) are the equations of motion. Note that in deriving the equations of motion, the first term on the right-hand side of Eq. (n) simply vanishes as j assumes either the value of 2 or 3 in Eq. (o). Physically, this indicates that regardless of the value of m_1, the time-varying specified motion in Eq. (a) can be thought to drive the system via imposed motion on the spring k_1 and the dashpot c_1. The mass m_1 plays no role in the equations of motion. Indeed, as written, Eq. (p) suggests this idea, since the x_2 coordinate is "driven" by the force $c_1 \dot{x}_0 + k_1 x_0$. The x_2 and x_3 coordinates communicate via the *coupling* terms associated with the spring k_2 and the dashpot c_2. These *coupling* elements produce terms containing x_3 and $\dot{x}_3$ in the x_2 equation, and x_2 and $\dot{x}_2$ in the x_3 equation.

■ **Example 5-23:**   Mass and Plane Pendulum Dynamic System

A simple plane pendulum of mass m_0 and length ℓ is suspended from a cart of mass m as sketched in Figure 5-27. The motion of the cart is restrained by a spring of spring constant k and a dashpot of dashpot constant c; and the angle of the pendulum is restrained by a torsional spring of spring constant k_t and a torsional dashpot of dashpot constant c_t. Note that the constitutive relations for the torsional spring and the torsional dashpot are linear expressions $\tau = k_t\theta$ and $\tau = c_t\dot{\theta}$, respectively, where τ is the torque and θ is the change in angle across the element from its undeformed configuration. The torsional spring is undeformed when the pendulum is in its downward-hanging equilibrium position. Also, m_0 is acted upon by a known harmonic force $F_o \sin \omega t$, which remains horizontal, regardless of the angle θ and the motion of the cart. The system is constrained to remain in the plane of the sketch and the cart remains on the bed throughout its motion. Derive the equation(s) of motion for the system.

❑ **Solution:**

1. Generalized coordinate(s):   The coordinates x and θ sketched in Figure 5-27 represent a complete and independent set of generalized coordinates, where x is the inertial displacement of the cart from its equilibrium position and θ is the inertial counterclockwise angular displacement of the pendulum from its downward-hanging equilibrium position. Further, consistent with the constraints on the system, we can define δx and $\delta\theta$ as a complete and independent set of admissible variations. Thus,

$$\xi_j : x, \theta \qquad \delta\xi_j : \delta x, \delta\theta \tag{a}$$

Because there are two admissible variations, there are two degrees of freedom. Since there are also two generalized coordinates, the system is holonomic.

$$\text{Holonomic!} \tag{b}$$

Thus we may proceed with our lagrangian formulation.

2. Generalized forces: Because there are two dashpots and an externally applied force, we conclude that there are three contributions to the variational work increment. So, use of Eq. (5-36) gives

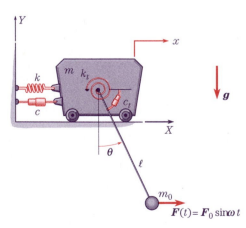

FIGURE 5-27   Dynamic system of mass with suspended simple plane pendulum.

$$\delta \mathcal{W}^{nc} = \sum_{i=1}^{N} f_i^{nc} \cdot \delta R_i = \sum_{j=1}^{n} \Xi_j \delta \xi_j$$

$$= -c\dot{x}\, \delta x - c_t \dot{\theta}\, \delta \theta + F_0 \sin \omega t (\delta x + \ell \cos \theta\, \delta \theta)$$

$$= \Xi_x\, \delta x + \Xi_\theta\, \delta \theta \tag{c}$$

where the notation f_i^{nc} is interpreted to incorporate both nonconservative forces and non-conservative torques. The expressions for the dashpots are straightforward; the expression for the externally applied force may not appear to be so. First, we note that although the force $F(t)$ is applied at m_0, even when $\delta \theta = 0$, $F(t)$ does work during an admissible variation δx. Further, $F(t)$ also does work during an admissible variation $\delta \theta$, where the admissible displacement at m_0 can be expressed by referring to Example 5-16 and, particularly, to Figure 5-20c, where the work-producing variation in the direction of $F(t)$ here is $\ell \cos \theta\, \delta \theta$. Thus, from Eq. (c), the generalized forces are

$$\Xi_x = F_0 \sin \omega t - c\dot{x} \tag{d}$$

$$\Xi_\theta = F_0 \ell \cos \theta \sin \omega t - c_t \dot{\theta} \tag{e}$$

3. Lagrangian: In accordance with Eq. (5-10), the system kinetic coenergy function may be written as

$$T^* = T_m^* + T_{m_0}^* \tag{f}$$

where the two terms on the right-hand side of Eq. (f) account for the kinetic coenergies of m and m_0, respectively. The kinetic coenergy function for the mass m is

$$T_m^* = \frac{1}{2} m v \cdot v = \frac{1}{2} m \dot{x}^2 \tag{g}$$

and the kinetic coenergy function for m_0 is

$$T_{m_0}^* = \frac{1}{2} m_0 v \cdot v = \frac{1}{2} m_0 \left[(\dot{x})^2 + (\ell\dot{\theta})^2 + 2\dot{x}\ell\dot{\theta} \cos \theta \right] \tag{h}$$

where Eq. (5-9) has been used to write both Eqs. (g) and (h), and where the inertial velocity of m and m_0 has been used in their respective equations.

Our persistence in writing T^* as $\frac{1}{2} m v \cdot v$ instead of the more common form as $\frac{1}{2} m v^2$ has been finally rewarded and can now be explored in the context of Eq. (h). The velocity of m_0 is due to both generalized velocities $\dot{x}$ and $\dot{\theta}$, and their respective contributions are neither collinear nor orthogonal. So, we shall examine several methods of evaluating $v \cdot v$ for m_0.

First, the noncollinearity and the nonorthogonality of $\dot{x}$ and $\ell\dot{\theta}$ are depicted in Figure 5-28. In Figure 5-28a, we note that the *law of cosines* may be used to evaluate $v \cdot v$. Note that the law of cosines is expressed in terms of the acute angle θ rather than the more commonly used obtuse angle; nevertheless, we shall continue to refer to Figure 5-28a as the law of cosines.

Second, we recall that in cartesian coordinates, $v \cdot v$ may be expressed as

$$v \cdot v = v_x^2 + v_y^2 + v_z^2 \tag{i}$$

where here we immediately note that $v_z = 0$ in accordance with the planar constraint of the problem and the inertial axes given in Figure 5-27. So, according to Figure 5-28b, we

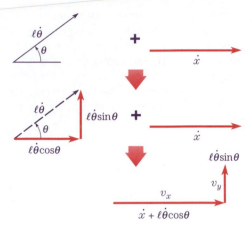

$$v^2 = (\dot{x})^2 + \left(\ell\dot{\theta}\right)^2 + 2\dot{x}\ell\dot{\theta}\cos\theta$$

(a) Law of cosines.

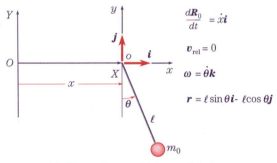

(b) Orthogonal vector components.

$$\frac{d\mathbf{R}_0}{dt} = \dot{x}\mathbf{i}$$

$$\mathbf{v}_{\text{rel}} = 0$$

$$\boldsymbol{\omega} = \dot{\theta}\mathbf{k}$$

$$\mathbf{r} = \ell\sin\theta\mathbf{i} - \ell\cos\theta\mathbf{j}$$

(c) Kinematic use of intermediate frame.

FIGURE 5-28 Three techniques for determining kinematics for m_0 of suspended plane pendulum.

determine the vector components along the X and Y orthogonal directions, and find from Eq. (i)

$$
\begin{aligned}
\mathbf{v} \cdot \mathbf{v} &= (\dot{x} + \ell\dot{\theta}\cos\theta)^2 + (\ell\dot{\theta}\sin\theta)^2 + 0 \\
&= \dot{x}^2 + 2\dot{x}\ell\dot{\theta}\cos\theta + (\ell\dot{\theta})^2(\cos^2\theta + \sin^2\theta)
\end{aligned}
\tag{j}
$$

which, because $\sin^2\theta + \cos^2\theta = 1$, reduces to the result in the square brackets in Eq. (h) or Figure 5-28a. Here, we emphasize that although we established components along the inertial X and Y directions, *any* orthogonal directions in the plane of the sketch would have been equally valid. This approach, where the velocities are resolved along any convenient set of orthogonal axes and then the equivalent of Eq. (i) is applied, is probably the easiest to visualize.

A third alternative is to recognize that the pendulum's location can be expressed in a noninertial intermediate reference frame attached to the cart, and to use the kinematics provided by Eq. (3-23). Figure 5-28c illustrates this third alternative where the substitution of the terms indicated in Figure 5-28c into Eq. (3-23) leads directly to the anticipated result. The details of this approach are similar, though not identical, to part (1) in Example 3-10.

The system potential energy function may be written as

$$V = \frac{1}{2}kx^2 + \frac{1}{2}k_t\theta^2 + m_0g\ell(1 - \cos\theta) \tag{k}$$

where Eqs. (5-15) and (5-24), and Eq. (f) in Example 5-19 have been used. Note also that in analogy with Eq. (5-12) and Figure 5-4, for the torsional spring

$$V = \int_0^\theta \tau\, d\theta = \int_0^\theta k_t\theta\, d\theta = \frac{1}{2}k_t\theta^2 \tag{l}$$

Thus, the lagrangian may be written by substituting Eqs. (f), (g), (h), and (k) into

$$\mathcal{L} = T^* - V$$
$$= \frac{1}{2}m\dot{x}^2 + \frac{1}{2}m_0\left[\dot{x}^2 + 2\dot{x}\ell\dot{\theta}\cos\theta + (\ell\dot{\theta})^2\right]$$
$$-\frac{1}{2}kx^2 - \frac{1}{2}k_t\theta^2 + m_0g\ell\cos\theta - m_0g\ell \tag{m}$$

where Eq. (5-39) has been used.

4. Lagrange's equations: Substitution of Eqs. (a), (d), (e), and (m) into Lagrange's equations

$$\frac{d}{dt}\left(\frac{\partial\mathcal{L}}{\partial\dot{\xi}_j}\right) - \frac{\partial\mathcal{L}}{\partial\xi_j} = \Xi_j \tag{n}$$

gives the two equations of motion.

In general, as up to now, we shall continue to leave to the reader the differentiation in obtaining the equations of motion. However, in completing the examples in this section, we want to emphasize that, although straightforward, the direct substitutions indicated in Eq. (n) require care in their execution. With this advice in mind, we proceed to derive the equations of motion in some detail. (It might be instructive for the reader to obtain the equations of motion prior to reading the next two paragraphs. Go on, try it. The effort and the achievement may be beneficial to innermost recesses of yourself you may not know exist.)

In accordance with Eq. (n), the equation of motion for the x generalized coordinate is

$$\frac{d}{dt}\left(\frac{\partial\mathcal{L}}{\partial\dot{x}}\right) - \frac{\partial\mathcal{L}}{\partial x} = \Xi_x \tag{o}$$

Substitution of Eqs. (d) and (m) into Eq. (o) gives

$$\frac{d}{dt}\left[m\dot{x} + m_0\dot{x} + m_0\ell\dot{\theta}\cos\theta\right] - [-kx] = F_0\sin\omega t - c\dot{x} \tag{p}$$

The third term in the first set of brackets in Eq. (p) contains a product of $\dot{\theta}$ and $\cos\theta$, both of which are variable. Applying the ordinary rules for differentiating a product to this term as well as conducting the indicated differentiation on the other terms lead to

$$m\ddot{x} + m_0\ddot{x} + m_0\ell\ddot{\theta}\cos\theta - m_0\ell\dot{\theta}\sin\theta\,\dot{\theta} + kx = F_0\sin\omega t - c\dot{x} \tag{q}$$

or

$$(m + m_0)\ddot{x} + c\dot{x} + kx + m_0\ell(\ddot{\theta}\cos\theta - \dot{\theta}^2\sin\theta) = F_0\sin\omega t \tag{r}$$

In accordance with Eq. (n), the equation of motion for the θ generalized coordinate is

$$\frac{d}{dt}\left(\frac{\partial\mathcal{L}}{\partial\dot{\theta}}\right) - \frac{\partial\mathcal{L}}{\partial\theta} = \Xi_\theta \tag{s}$$

Substitution of Eqs. (e) and (m) into Eq. (s) gives

$$\frac{d}{dt}\left[m_0\ell\dot{x}\cos\theta + m_0\ell^2\dot{\theta}\right] - \left[m_0\dot{x}\ell\dot{\theta}(-\sin\theta) - k_t\theta - m_0g\ell\sin\theta\right]$$

$$= F_0\ell\cos\theta\sin\omega t - c_t\dot{\theta} \tag{t}$$

where we observe that in the first set of brackets, there is a product term, $\dot{x}\cos\theta$, similar to the term noted in Eq. (p); and in the second set of brackets, the $(-\sin\theta)$ term is due to the differentiation of the $\cos\theta$ term in Eq. (h). Performing the remaining operations leads to

$$m_0\ell^2\ddot{\theta} + c_t\dot{\theta} + k_t\theta + m_0g\ell\sin\theta + m_0\ddot{x}\ell\cos\theta = F_0\ell\cos\theta\sin\omega t \tag{u}$$

Clearly the pair of equations of motion represented by Eqs. (r) and (u) is nonlinear. (How did you do in your efforts to derive them?)

5-6   WORK–ENERGY RELATION

When a system is not subjected to any time-varying constraints, there exists an important and useful relation between the *total mechanical energy* of the system and the work done by all the nonconservative forces. The *work–energy relation* can be derived directly from Lagrange's equations, as shown in Section E-4 of Appendix E. Here, only the result for a mechanical system is given and is discussed briefly.

When a system is free of time-varying constraints, the time rate of the system's *total energy* is equal to the time rate of the work done on the system by all the nonconservative forces. Here, the *total energy* is defined as the sum of the kinetic energy T (not kinetic coenergy T^*) and the potential energy V. That is, the following expression holds:

$$\frac{dE}{dt} = \frac{d\mathcal{W}^{\text{nc}}}{dt} \tag{5-40}$$

where

$$E(t) = T + V \tag{5-41}$$

and where the nonconservative work is given, using Eq. (5-2), as

$$\mathcal{W}^{\text{nc}} = \sum_{i=1}^{N}\int_{\mathbf{R}_{i0}}^{\mathbf{R}_i} \boldsymbol{f}_i^{\text{nc}} \cdot d\mathbf{R}_i = \sum_{i=1}^{N}\int_0^t \boldsymbol{f}_i^{\text{nc}} \cdot \boldsymbol{v}_i\,dt \tag{5-42}$$

where $\boldsymbol{R}_{i0}$ is the initial position vector of the ith particle and where the definition $\boldsymbol{v}_i \equiv \dfrac{d\boldsymbol{R}_i}{dt}$ has been used. Substituting Eq. (5-42) into Eq. (5-40) gives

$$\frac{dE}{dt} = \sum_{i=1}^{N} \boldsymbol{f}_i^{\text{nc}} \cdot \boldsymbol{v}_i \tag{5-43}$$

where the right-hand side of Eq. (5-43) is the power input into the system by the nonconservative forces.

Integration of Eq. (5-43) with respect to time yields

$$E(t) = E_0 + \mathcal{W}^{\text{nc}}(t) \tag{5-44}$$

where Eq. (5-42) has been used and where E_0 is a constant of integration, which we recognize as the system's total mechanical energy at time $t = 0$. Thus, Eq. (5-44) states that the system's total mechanical energy at time t is equal to the initial total energy E_0 plus the nonconservative work done on the system from the initial time to time t.

One case of particular interest is when there are *no nonconservative forces* and thus no nonconservative work. In this case the second term on the right-hand side of Eq. (5-44) vanishes, yielding

$$E(t) = E_0 \tag{5-45}$$

Equation (5-45) states that when a system has neither time-varying constraints nor nonconservative forces, the system's total mechanical energy is conserved; in this instance, Eq. (5-45) is often cited as a statement of the principle of the *conservation of mechanical energy*. Motions of a dynamic system that are governed by Eq. (5-45) are called *conservative motions*. And, if all the motions of a system are conservative, the system is called a *conservative system*.

Furthermore, for conservative mechanical systems, Eqs. (5-41) and (5-45) can be combined to give

$$T + V = E_0 \tag{5-46}$$

which is valid for linear or nonlinear constitutive relations. Recall from Figures 5-2 and 5-4 that for linear constitutive relations, coenergies are equal in magnitude to their corresponding energies. In this case, for newtonian particles where $T^* = T$, a corollary of Eq. (5-46) is

$$T^* + V = E_0 \tag{5-47}$$

Expressions such as Eqs. (5-44) through (5-47) are particularly useful when it is of interest to study a system at specific locations or times, such as at extremes where the displacements or deformations are maximum or minimum. In this regard, for example, Eq. (5-47) is sufficient by itself to establish the generalized velocity $\dot{\xi}$ as a function of the generalized coordinate ξ. However, such a one-to-one correspondence exists only for single-degree-of-freedom systems. If the system has n degrees of freedom where $n > 1$, although the conservation of energy principle holds, it is necessary to find $n - 1$ additional independent expressions in order to completely fix the generalized velocities as functions of the generalized coordinates.

■ **Example 5-24:**   Package Transfer System

Reconsider Example 2-1 via the work–energy relation. Figure 5-29 shows a model of a package transfer system where a package or container, released at A, is transferred to a second

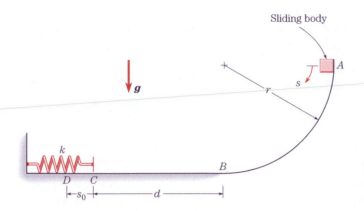

FIGURE 5-29 Body sliding along surface of package-transfer system.

location at C. Such a concept is used in warehouses where packages, released from storage or inventory at A, are distributed to a dock or customer pickup at C.

In the situation of interest, the package is modeled as a sliding body of negligibly small geometric dimensions. The sliding body of mass m is released from rest (height of r above horizontal surface BD) along a frictionless circular surface AB of radius r. It then slides along the horizontal surface, having a coefficient of kinetic friction with respect to the body of μ, and encounters a spring at point C. The spring constant is k and the distance BC is d. Note that gravity acts. How much will the spring be compressed?

☐ **Solution:**   The generalized coordinate and admissible variation are

$$\xi_j : s \qquad \delta\xi_j : \delta s \tag{a}$$

where, as sketched in Figure 5-29, s is the distance traveled by the package. Thus, the system has a single degree of freedom.

Since the package is a newtonian body, its kinetic energy is equal to its kinetic coenergy. Therefore,

$$T = T^* = \frac{1}{2}m\boldsymbol{v}\cdot\boldsymbol{v} = \frac{1}{2}m\dot{s}^2 \tag{b}$$

The potential energy of the system consists of two contributions: one due to gravity and the other due to the elastic spring. That is,

$$V = V_g + V_{spring} \tag{c}$$

where V_g is the potential energy due to gravity and V_{spring} is the potential energy due to the elastic spring. Note that during the motion between location A and location C, there is no potential energy in the spring (that is, $V_{spring} = 0$). Note also that during the motion between location B and location D, there is no change in the potential energy due to gravity (that is, $V_g = $ constant). Further, if the reference for V_g is taken as the horizontal surface BD, $V_g = 0$ between locations B and D.

By using Eq. (5-46) and Eqs. (b) and (c), the initial energy of the system can be expressed as

$$E_0 = E(0) = T(0) + V(0) = T(0) + V_g(0) + V_{spring}(0) = 0 + mg \text{ "}h\text{"} + 0 = mgr \tag{d}$$

where $T(0) = 0$ because the package is at rest at $t = 0$ (that is, $\dot{s} = 0$ at $t = 0$) and where $V_{spring} = 0$ at $t = 0$ as discussed above. Denoting as t_1 the time at which the package compresses the spring to its maximum deformation and denoting as s_0 the corresponding maximum deformation of the spring, the total energy at $t = t_1$ can be expressed as

$$E(t_1) = T(t_1) + V(t_1) = T(t_1) + V_g(t_1) + V_{spring}(t_1) = 0 + 0 + \frac{1}{2}ks_0^2 = \frac{1}{2}ks_0^2 \quad \text{(e)}$$

where $T(t_1) = 0$ because the instantaneous velocity of the package must be zero at $t = t_1$ and where $V_g = 0$ as discussed above.

The nonconservative work done during the motion between locations A and D is due solely to the frictional force acting along the horizontal surface (that is, between locations B and D, or equivalently, in the range $\frac{\pi r}{2} \le s \le \frac{\pi r}{2} + d + s_0$). Therefore, by using Eq. (5-42), the nonconservative work can be expressed as

$$\mathcal{W}^{nc}(t_1) = \int_{\boldsymbol{R}_0}^{\boldsymbol{R}_1} \boldsymbol{f}^{nc} \cdot d\boldsymbol{R} = \int_{\frac{\pi r}{2}}^{\frac{\pi r}{2}+d+s_0} (-\mu mg)\, ds = -\mu mg\, (d + s_0) \quad \text{(f)}$$

where the fact that the frictional force has a magnitude of μmg and acts in the direction opposite to s has been used.

Thus, substitution of Eqs. (d) through (f) into Eq. (5-44), for $t = t_1$, yields

$$E(t_1) = E_0 + \mathcal{W}^{nc}(t_1) \quad \text{(g)}$$

or

$$\frac{1}{2}ks_0^2 = mgr - \mu mg(d + s_0) \quad \text{(h)}$$

which can be rearranged to give

$$ks_0^2 + 2\mu mgs_0 - 2mg(r - \mu d) = 0 \quad \text{(i)}$$

Note that Eq. (i) is identical to Eq. (w) or Eq. (ad) in Example 2-1; thus, the solution of Eq. (h) gives the same solution as found in Eq. (x) of Example 2-1, if the numerical values of the parameters in Example 2-1 are used here.

5-7 NATURE OF LAGRANGIAN DYNAMICS

Some of the language introduced in this chapter is slightly formal; but this is intended to be comprehensive in scope, and is not designed to be unintuitive or pompous. For example, in this chapter we have introduced the phrase "admissible motion" to mean a "geometrically compatible motion that is consistent with the constraints," and the phrase "natural motion" to mean a "geometrically compatible motion that satisfies the *force–dynamic requirements.*" So, for example, why not use "geometrically compatible motion" instead of "admissible motion," if geometrically compatible motion is intended? The major practical reason is that by using more familiar mechanical terms, we limit our scope to mechanical concepts, whereas one of our goals is to emphasize the breadth of the principles explored. For example, we shall reinterpret the phrases "admissible motions" and "natural motions" in the context of electrical and electromechanical concepts in Chapter 7. Furthermore, these concepts can be extended to other disciplines, including fluid mechanics, quantum mechanics, or statistical mechanics.

Then, there is the matter of the energy-coenergy duality. This duality may be treated with indifference in the analysis of exclusively linear newtonian mechanical systems, with little practical cost (despite the intellectual misdemeanor for doing so); but the neglect of this duality in run-of-the-mill electrical and electromechanics problems may lead the analyst to a very simple reward: the wrong answer.

Variational dynamics is the formal manifestation of the indirect approach, where in mechanical terms, the fundamental quantities of energy and work replace the fundamental quantities of force and momentum of the direct approach. Philosophically, Hamilton's principle is rooted in the concept that *Nature* will ensure efficiency as a system moves freely along a path from one configuration to some other configuration; of all the paths that a system might traverse, the path chosen by Nature will always be the path that minimizes *something*. Historically, that *something* has been called the *action*. Various definitions have been used for the *action*. In some definitions of Hamilton's principle where all the generalized forces vanish, the action is the integral of the lagrangian from time t_1 to time t_2, where these two times are arbitrary. By use of the calculus of variations, the necessary condition that this action be *minimum* results in Lagrange's equations. A tremendous amount has been written about the philosophical roots and implications of a physical theory that is capable of selecting the one path among all possibilities that minimizes a specific quantity.

We have adopted the unconventional, though not unique, perspective of introducing Hamilton's principle as the unifying tenet of our variational formulations. This clearly inverts the historical order, since we derive Lagrange's equations as merely *one* consequence of Hamilton's principle. In Section E-1 of Appendix E, we present *an* analytical overview of a connection between Newton's second law, Lagrange's equations, and Hamilton's principle. There is likely to be more in Appendix E than many readers of this book care to know; yet there is equally likely to be something in Appendix E that most readers will find of interest.

It is somewhat fascinating that the vectorial character of the system's motion is retained in the scalar lagrangian. From a practical perspective, scalars are generally easier to manipulate than vectors. Further, the requirement to compute accelerations in the direct approach is a major drawback as compared with the requirement to compute velocities in the indirect approach.

In light of the present chapter, we encourage the reader to revisit Table 2-1 where we outlined some of the most significant differences between the direct approach and the indirect approach. In the variational approach, the concepts of *generalized coordinates* and *generalized forces* are central. By use of generalized coordinates, the introduction of kinematics into the equation formulation becomes more precise than in typical expositions of the direct approach. (Think about this point; in the direct approach it was not always crystal clear whether all the kinematic variables that were needed had in fact been defined.) In the variational approach the kinematic variables are not unique because generalized coordinates are not unique; but *precision* is ensured since a complete and independent set of generalized coordinates is just large enough to describe every allowable configuration of the system. The generalized coordinates are "generalized" in the sense that no adherence to a prior set of coordinates (such as cartesian or cylindrical coordinates) is imposed. Generalized coordinates may be any convenient combination of other coordinates. The generalized forces are "generalized" in the sense that each generalized force represents the sum of all the work-producing components of the nonconservative internal and external forces and torques along the direction of its corresponding generalized coordinate. In this work-producing context, if a generalized coordinate is a length, its corresponding generalized force will be a force; or if a generalized coordinate is an angle, its corresponding generalized force will be a torque or moment. The constraint forces, which often need to be considered in the direct approach, do no work along the directions of the generalized coordinates and therefore need not be considered in the variational approach. The vanishing of effects due to the constraint forces is a major aspect of the relative ease of implementation of variational techniques over the direct approach techniques in mechanics.

Most postnewtonian theories of the dynamics of mechanical systems exploit the concept that effects due to the workless constraint forces can be ignored. This is a centuries-old idea that has been periodically revisited, for the past two hundred years, mostly in the context of "Lagrange's form of d'Alembert's principle." To this day there continue to be some differences of opinion regarding both the relations between the variously named theories–methods–equations–approaches and their relative efficacy. We believe that, in general, the formulation of the equations of motion of lumped-parameter holonomic mechanical systems via Lagrange's equations is more efficient than via all the other various methods flowing from Lagrange's form of d'Alembert's principle. Lagrange's form of d'Alembert's principle is discussed in Appendix F. The choice of method to be used in any technical undertaking resides ultimately with the analyst.

Finally, readers who elect to pursue in depth several of the references on these topics will find themselves swept up in an intellectual elation for which, like a magnificent symphony or an insightful poem or a great basketball game, they may find little "practical" value.

Problems for Chapter 5

Category I: Back to Physics

Work and energy: Problems 5-1 through 5-6

Problem 5-1 (*HRW185, 38P*)[8]: A 250-g block is dropped onto a vertical spring of spring constant $k = 2.5$ N/cm (Figure P5-1). The block becomes attached to the spring, and the spring compresses 12 cm before momentarily stopping. While the spring is being compressed, what work is done

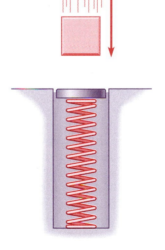

FIGURE P5-1

  (*a*) By the block's weight and

  (*b*) By the spring?

  (*c*) What was the speed of the block just before it hit the spring?

  (*d*) If the speed at impact is doubled, what is the maximum compression of the spring? Assume the friction is negligible.

Problem 5-2 (*HRW212, 7E*): A frictionless roller-coaster car tops the first hill in Figure P5-2 with speed v_0. What is its speed at

---

[8]These designations reference Halliday, Resnick, and Walker. For example, "HRW185, 38P" denotes "Halliday, Resnick and Walker; page 185, problem 38P".

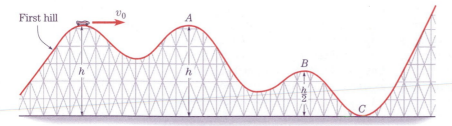

FIGURE P5-2

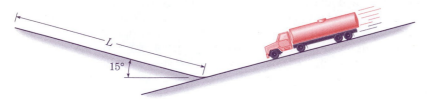

FIGURE P5-3

(a) Point A,

(b) Point B, and

(c) Point C?

(d) How high will it go on the last hill, which is too high for it to cross?

Problem 5-3 (*HRW212, 8E*): A runaway truck with failed brakes is moving downgrade at 80 mi/hr. Fortunately, there is an emergency escape ramp near the bottom of the hill. The inclination of the ramp is 15° (see Figure P5-3). What minimum length must it have to make certain of bringing the truck to rest, at least momentarily? Real escape ramps are often covered with a thick layer of sand or gravel. Why?

Problem 5-4 (*HRW422, Sample Problem 15-8*): If a projectile is fired upward, usually it will slow, stop momentarily, and return to Earth. There is, however, a certain initial speed that will cause it to move upward forever, theoretically coming to rest only at infinity. This initial speed is called the *escape speed*.

Assuming the Earth is a uniform sphere of radius 6.37×10^6 m and mass 5.98×10^{24} kg, find the escape speed from the Earth. Ignore the effects caused by air drag and the Earth's rotation.

Problem 5-5 (*HRW220, 81P*): A particle slides along a track with elevated ends and a flat central part, as sketched in Figure P5-5. The flat part has length L. The curved portions of the track are fric-

tionless, but for the flat part the kinetic coefficient of friction is $\mu_k = 0.20$. The particle is released at point A, which is a height $h = \dfrac{L}{2}$ above the flat part of the track. Where does the particle finally come to rest?

Problem 5-6 (*HRW221, 89P*): A *governor* consists of two 200 g masses attached by massless, rigid 10 cm rods to a vertical rotating axle (Figure P5-6). The rods are hinged so that the masses swing out from the axle as they rotate with it. However, when the angle θ is 45°, the masses encounter the wall of the cylinder in which the governor is rotating.

(a) What is the minimum rate of rotation, in revolutions per minute, required for the masses to touch the wall?

(b) If the kinetic coefficient of friction between the masses and the wall is 0.35, at what rate does the frictional force dissipate mechanical energy as a result of the masses rubbing against the wall when the mechanism is rotating at 300 rev/min?

FIGURE P5-5

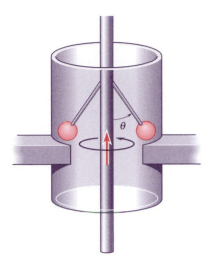

FIGURE P5-6

Category II: Intermediate

Work and energy: Problems 5-7 and 5-8

Problem 5-7: An automobile (weight 1500 lb) is being transported in a railway car as sketched in Figure P5-7. The railway car is moving at a constant velocity of 10 mi/hr and strikes an abutment, which brings the railway car to an "immediate" stop. The automobile's transmission is in neutral and its parking brake is disengaged.

(a) Assuming the stiffness of the automobile bumper may be modeled as a linear spring of spring constant of 10,000 lb/in., find the maximum force applied to the automobile by the bumper.

(b) Assuming that the maximum force that may be applied to the bumper-automobile connections without sustaining any damage is 15,000 lb and noting that there are two such

connections, determine whether the velocity v will cause damage to the connections.

Problem 5-8: A double spring bumper is used to stop packages at the end of an automated package transport line. There are two prospective designs of the bumper as sketched in Figures P5-8a and b. In each design, the two springs are identical of spring constant $k = 500$ N/cm and the face plates of the bumper have negligible masses. The kinetic coefficient of friction between the floor and the package is $\mu_k = 0.4$. For each of the two designs, find the maximum displacement of a package of mass $m = 30$ kg after it comes into contact with the bumper at the speed $v = 3$ m/s. Assuming that without causing damage the maximum force that may be applied to the variety of packages transported by the system is 2500 N, determine whether the proposed designs are safe. If either or both are not safe, suggest changes that will make them safe.

Conceptual questions: Problems 5-9 through 5-12

Problem 5-9: For a conservative force field, the *potential energy function* is the work done by the force field on a particle from a datum position $\boldsymbol{R}_0$ to a new position $\boldsymbol{R}$, which is *defined* as

$$V(\boldsymbol{R}) = -\int_{\boldsymbol{R}_0}^{\boldsymbol{R}} \boldsymbol{f}(\boldsymbol{R}) \cdot d\boldsymbol{R}$$

Show that this is equivalent to

$$\boldsymbol{f}(\boldsymbol{R}) = -\nabla V(\boldsymbol{R})$$

Problem 5-10: When the *kinetic energy function* is expressed as a function of momentum, the complete velocity–momentum relation can be recovered from the kinetic energy function; so it is thus said that kinetic energy depends *naturally* on the momentum. Show this by considering a one-dimensional

FIGURE P5-7

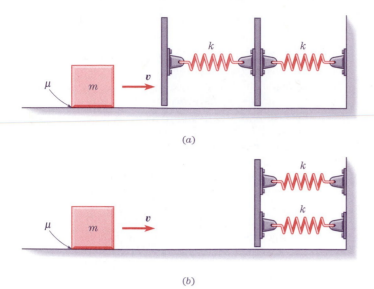

(a)

(b)

FIGURE P5-8

case. Also, show that if the kinetic energy function is expressed as a function of velocity, the velocity–momentum relation cannot be fully recovered.

Problem 5-11: When the *kinetic coenergy function* is expressed as a function of velocity, the complete velocity–momentum relation can be recovered from the kinetic coenergy function; so it is thus said that kinetic coenergy depends *naturally* on the velocity. Show this by considering a one-dimensional case. Also, show that if the kinetic coenergy function is expressed as a function of momentum, the velocity–momentum relation cannot be fully recovered.

Problem 5-12: The momentum for a *relativistic particle* is defined as

$$p = \frac{mv}{\sqrt{1 - v^2/c^2}}$$

where v is the speed of the particle and c is the speed of light. Derive the expressions for the kinetic energy function and kinetic coenergy function for such a relativistic particle.

Holonomicity: Problems 5-13 through 5-18

For the systems described in the problem statements 5-13 through 5-18, do the following:

(a) Determine the number of degrees of freedom;

(b) Determine whether the system is holonomic or nonholonomic; and

(c) If the system is holonomic, define a set of complete and independent generalized coordinates.

Problem 5-13: A rigid circular ring must remain vertical and remain in contact with and roll without slip on a horizontal plane, as sketched in Figure P5-13.

Problem 5-14: The rigid circular ring in Problem 5-13 is allowed to slip.

Problem 5-15: A rigid thin disk rolls without slip on a horizontal surface, as sketched in Figure P5-15. In addition to rolling, the disk is allowed to tip with respect to the vertical; thus, the line joining the center of the disk and the contact point does not necessarily remain vertical.

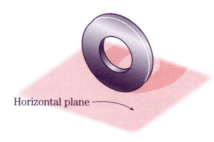

Horizontal plane

FIGURE P5-13

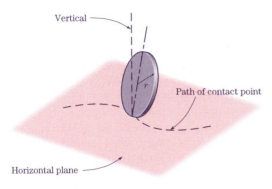

Vertical

Path of contact point

Horizontal plane

FIGURE P5-15

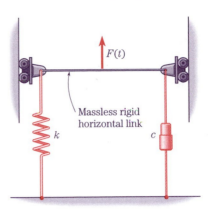

$F(t)$

Massless rigid
horizontal link

k

c

FIGURE P5-20

Problem 5-16: The rigid thin disk in Problem 5-15 is allowed to slip.

Problem 5-17: A rigid sphere of radius R rolls on a horizontal plane without slip.

Problem 5-18: The rigid sphere in Problem 5-17 is allowed to slip.

Equation(s) of motion—systems of 1 degree of freedom: Problems 5-19 through 5-37

Problem 5-19: Consider the mass spring system sketched in Figure P5-19. Note that gravity acts. Construct the lagrangian and obtain the equation(s) of motion for the system, as follows

> (a) In terms of the displacement η of the mass from the unstretched length ℓ_0 of the spring; and
>
> (b) In terms of the displacement ξ of the mass from its (sagged) equilibrium position.

What role does gravity play in the motion of the mass? (This is an important concept.)

Problem 5-20: The system sketched in Figure P5-20 consists of a spring of spring constant k and a dashpot of dashpot constant c, arranged in parallel. Assume that the upper ends of the dashpot and spring are connected by a massless rigid link that is constrained to remain horizontal at all times. A force $F(t)$ is applied vertically to the horizontal massless link. Derive the equation(s) of motion for the system.

Problem 5-21: The system sketched in Figure P5-21 consists of a mass m and dashpot c, connected in series. An external force $f(t)$ acts on the mass m. Derive the equation(s) of motion for the system.

Problem 5-22: Derive the equation(s) of motion for the system described in Problem 4-54.

Problem 5-23: Derive the equation(s) of motion for the system described in Problem 4-55.

Problem 5-24: Derive the equation(s) of motion for the system described in Problem 4-63.

Problem 5-25: An inverted simple pendulum of length ℓ and mass m is restrained by two springs of spring constants k_1 and k_2, each of which is attached

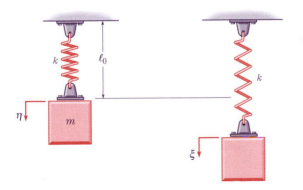

k ℓ_0

k

η m

ξ

FIGURE P5-19

g

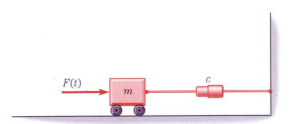

$F(t)$ m c

FIGURE P5-21

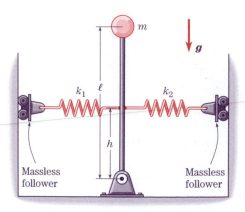

FIGURE P5-25

to the massless rigid link at height h at one end and a massless follower at the other end. The massless followers maintain the springs in their horizontal orientation and are constrained to move without friction along the vertical walls, as sketched in Figure P5-25. The pendulum's equilibrium position is vertical and the unstretched length of the springs is ℓ_0. Note that gravity acts. Derive the equation(s) of motion for the system.

Problem 5-26: An inverted simple pendulum is restrained by a spring of spring constant k and a dashpot of dashpot constant c to a massless follower, which is constrained to move without friction along a horizontal slot, as sketched in Figure P5-26. The pendulum's equilibrium position is vertical and the un-

stretched length of the spring is ℓ_0. Note that gravity acts. Derive the equation(s) of motion for the system.

Problem 5-27: An inverted simple pendulum consists of a mass m and a massless T-shaped link, and is restrained to ground by two springs of spring constant k, as sketched in Figure P5-27. The vertical bar of the T-shaped link is of length ℓ, the horizontal bar is of length b, and the unstretched length of the springs is ℓ_0. The massless followers, which are constrained to move without friction along a horizontal slot, maintain the springs in their vertical orientation. The pendulum's equilibrium position is such that the bar of length ℓ is vertical. Note that gravity acts. Derive the equation(s) of motion for the system.

Problem 5-28: A massless rigid link of length L is pivoted about point O, as sketched in Figure P5-28. Two masses m_1 and m_2 are attached to the ends of the link. Note that gravity acts. Derive the equation(s) of motion for the system.

Problem 5-29: Three identical particles of mass m are mounted on a massless rigid link at a distance ℓ apart, as sketched in Figure P5-29. The link is pivoted about point O. The middle particle is restrained by a spring of spring constant k and a dashpot of dashpot constant c, which connect this particle to a massless follower that slides without friction along a vertical slot. The spring is unstretched when the link is in its vertical equilibrium orientation. Note that gravity acts. Derive the equation(s) of motion for the system.

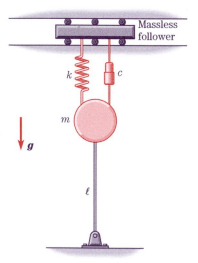

FIGURE P5-26

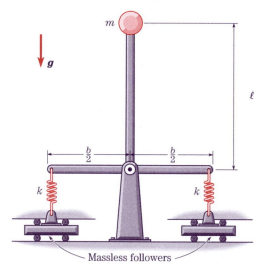

FIGURE P5-27

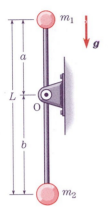

FIGURE P5-28

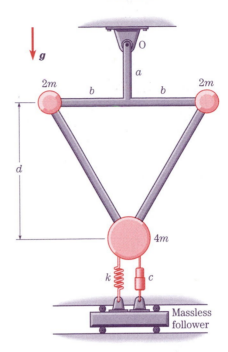

FIGURE P5-30

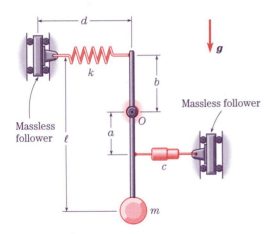

FIGURE P5-29

pendulum is restrained by a spring of spring constant k and a dashpot of dashpot constant c. Each of two massless followers moves without friction along its vertical slot, thus maintaining a horizontal orientation of the spring and the dashpot. The spring is un-

Problem 5-30: A plane pendulum consists of three particles attached to a massless rigid frame, which is pivoted about point O, as sketched in Figure P5-30. The bottom mass is connected to a massless follower by a spring of spring constant k and a dashpot of dashpot constant c. The follower can move without friction along a horizontal slot, thus maintaining a vertical orientation of the spring and the dashpot. The spring is unstretched when the pendulum is in its vertical equilibrium orientation shown in the sketch. Note that gravity acts. Derive the equation(s) of motion for the system.

Problem 5-31: An inverted plane pendulum consists of three particles, each of mass m, and a T-shaped massless rigid link, as sketched in Figure P5-31. The

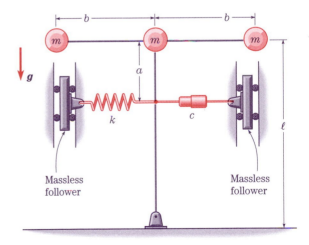

FIGURE P5-31

stretched when the link is in its vertical equilibrium orientation. Note that gravity acts. Derive the equation(s) of motion for the system.

Problem 5-32: A system has been modeled as sketched in Figure P5-32. The rigid link AB is massless. Each of two massless followers moves without friction along its vertical slot, thus maintaining a horizontal orientation of the spring and the dashpot. The spring is unstretched when the link is in its vertical equilibrium orientation. Note that gravity acts. Derive the equation(s) of motion for the system.

Problem 5-33: A massless rigid link ABC is pivoted about point O and is restrained by two springs of spring constant k_1 and k_2 and a dashpot of dashpot constant c. A mass m is suspended from one end of the link ABC at point C and is constrained to move without friction along a vertical slot, as sketched in Figure P5-33. The spring is unstretched when the link ABC is in its horizontal orientation. Assume *small motion* so that changes in the orientations of the springs and dashpot are negligible. Note that gravity acts. Derive the equation(s) of motion for the system.

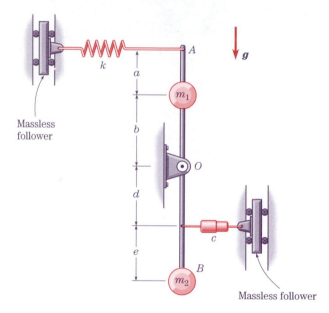

Massless follower

FIGURE P5-32

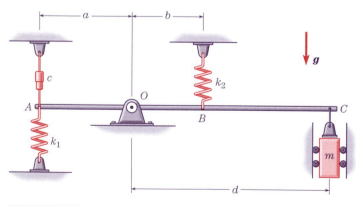

FIGURE P5-33

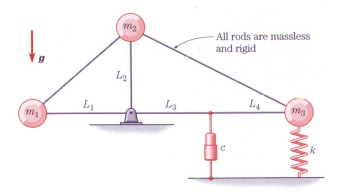

FIGURE P5-34

Problem 5-34: Three particles are mounted on a massless rigid frame, which is pivoted about point O, as sketched in Figure P5-34. The frame is restrained by a spring of spring constant k and a dashpot of dashpot constant c. Masses m_1 and m_3 lie along a horizontal line when the system is in equilibrium. Assume *small motion* about the equilibrium position so that the rotation of the frame about point O does not change the orientations of the spring and the dashpot. Note that gravity acts. Derive the equation(s) of motion for the system.

Problem 5-35: The mechanism sketched in Figure P5-35 consists of four identical massless rigid links of length ℓ that form a rhombus. Across its horizontal diagonal, a spring of spring constant k and a dashpot of dashpot constant c are attached. One end of

the vertical diagonal is pivoted about point O and the other end supports a mass m. The undeformed length of the spring is also ℓ. Note that gravity acts. Derive the equation(s) of motion for the system.

Problem 5-36: The mechanism sketched in Figure P5-36 consists of four identical massless rigid links of length ℓ that form a rhombus. On each end of its horizontal diagonal, a torsional spring of spring constant k_t is attached to the two adjacent links. One end of the vertical diagonal is pivoted about point O and the other end supports a mass m. When in equilibrium, the mechanism is a square. Note that gravity acts. Derive the equation(s) of motion for the system.

Problem 5-37: Derive the equation(s) of motion for the shock-absorber system described in Problem 4-65.

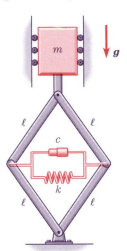

FIGURE P5-35

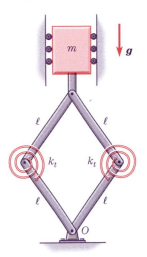

FIGURE P5-36

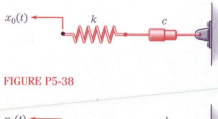

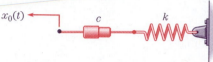

FIGURE P5-39

Equation(s) of motion—systems of 1 degree of freedom with specified motion: Problems 5-38 through 5-45

Problem 5-38: The system sketched in Figure P5-38 consists of a dashpot of dashpot constant c and a spring of spring constant k connected in series. The displacement of the left-hand node is specified as $x_0(t)$. Derive the equation(s) of motion for the system.

Problem 5-39: The system sketched in Figure P5-39 is identical to the one in Figure P5-38 except that the locations of the spring and the dashpot have been reversed. The displacement of the left-hand node is specified as $x_0(t)$. Derive the equation(s) of motion for the system.

Problem 5-40: A bead of mass m slides along a semicircular hoop of radius R, which is rotating at a constant angular velocity Ω about its axis of geometrical symmetry, as sketched in Figure P5-40. The friction between the bead and the hoop has been modeled as viscous damping having an equivalent dashpot constant c. Note that gravity acts. Derive the equation(s) of motion for the bead.

Problem 5-41: A vibration testing setup in which a delicate instrument of mass m is packed in a crate of mass M is sketched in Figure P5-41. The packing material has been modeled as a spring of spring constant k and a dashpot of dashpot constant c. The crate is placed directly on the vibration table, which moves vertically according to $y(t) = a \sin \Omega t$, where y is the displacement of the top surface of the table

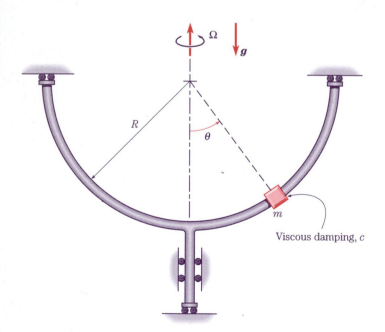

FIGURE P5-40

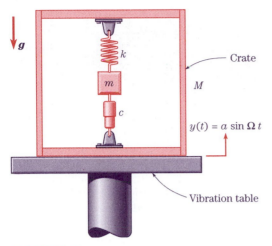

FIGURE P5-41

measured from its initial position and where a and Ω are specified constants. Note that gravity acts. Derive the equation(s) of motion for the system.

Problem 5-42: A picture frame of mass M is being rotated at a specified angular velocity $\Omega(t)$. A simple plane pendulum of length ℓ and mass m is suspended from the picture frame and is constrained to remain within the plane of the picture frame, as sketched in Figure P5-42. A dashpot of dashpot constant c connects the mass m to a massless follower, which moves without friction along a horizontal slot and maintains

the dashpot in a vertical orientation. Note that gravity acts. Derive the equation of motion for the system.

Problem 5-43: Two simple plane pendulums, both of length ℓ but of masses m_1 and m_2, are joined by a spring of spring constant k and a dashpot of dashpot constant c via their massless rigid links. The left-hand pendulum is driven according to a specified angular displacement $\theta_1(t)$ with respect to the vertical. The pendulum on the right-hand side is excited by an external force $F(t)$, which remains horizontal at all times, as sketched in Figure P5-43. The spring is unstretched when the links of the pendulums are in their downward hanging equilibrium positions.

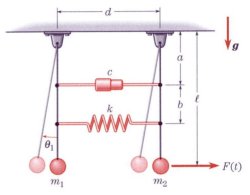

FIGURE P5-43

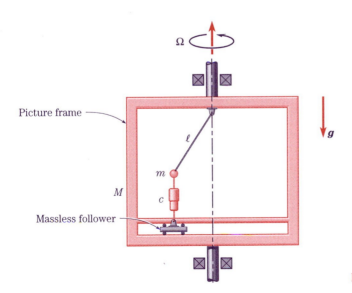

FIGURE P5-42

Assume *small motions* and note that gravity acts. Derive the equation(s) of motion for the system.

Problem 5-44: A turntable rotates in a horizontal plane (with respect to gravity) about a vertical axis passing through its center O, at a specified angular velocity $\Omega(t)$. A particle of mass m is attached to the center of the turntable by a spring of spring constant k and slides without friction along a radial track on the turntable, as sketched in Figure P5-44. The unstretched length of the spring is ℓ_0. Derive the equation(s) of motion for the system.

Problem 5-45: Derive the equation(s) of motion for the system described in Problem 4-66.

Equation(s) of motion—systems of 2 degrees of freedom: Problems 5-46 through 5-67

Problem 5-46: For the system in Example 5-21, assume the unstretched length of the spring is ℓ_0. Derive the equation(s) of motion for the system using x_1 and x_2 as the generalized coordinates, where x_1 and x_2 are inertial positions as sketched in Figure P5-46.

Problem 5-47: The system sketched in Figure P5-47 consists of a dashpot of dashpot constant c and a mass m. A horizontal force $F(t)$ is applied at the free end of the dashpot. Derive the equation(s) of motion for the system.

Problem 5-48: Derive the equation(s) of motion for the system described in Problem 4-56.

Problem 5-49: Derive the equation(s) of motion for the system described in Problem 4-57.

Problem 5-50: Two springs, of identical unstretched length ℓ_1 but of different spring constants k_1 and k_2, are connected in parallel. They are further connected in series to a third spring of spring constant

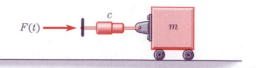

FIGURE P5-47

k_3 and unstretched length ℓ_3. The massless followers, which are constrained to move without friction along horizontal slots, maintain a massless rigid link in its vertical orientation. The three springs connect mass M to a rigid wall, as sketched in Figure P5-50. The mass moves without friction on a horizontal surface and is excited by an external force $F(t) = F_0 \sin\Omega t$. Derive the equation(s) of motion for the system.

Problem 5-51: A system of two blocks of masses m_1 and m_2 connected to each other by a spring of spring constant k_1 strikes a buffer of spring constant k_2 attached to a rigid wall, as sketched in Figure P5-51. The motion of the blocks during the period that m_2 is in contact with the buffer is considered. Derive the equation(s) of motion for the system.

Problem 5-52: Derive the equation(s) of motion for the system described in Problem 4-58.

Problem 5-53: Derive the equation(s) of motion for the system described in Problem 4-59.

Problem 5-54: A forging machine is modeled as sketched in Figure P5-54, where the hammer is

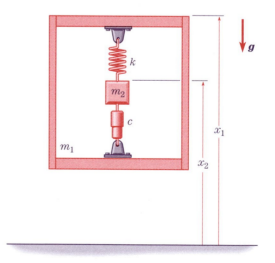

FIGURE P5-46

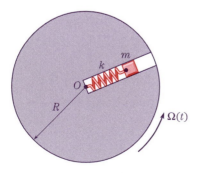

FIGURE P5-44

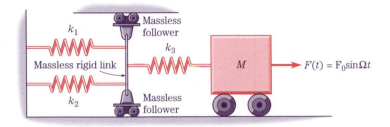

FIGURE P5-50

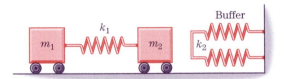

FIGURE P5-51

dropped onto an anvil supported by a shock absorber, which in turn is mounted on a foundation block on the ground. The shock absorber between the anvil and the foundation block exhibits stiffness and damping. Also, due to the large impact force, the ground immediately below the foundation block deforms; thus, the damping and stiffness of the soil beneath the foundation block are also modeled. Let the masses of the hammer, anvil, and foundation block be m_0, m_1, and

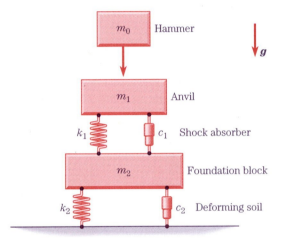

FIGURE P5-54

m_2, respectively; and, the spring constants and dashpot constants of the shock absorber and the soil beneath the foundation block be k_1 and c_1, and k_2 and c_2, respectively, as sketched in Figure P5-54. Assuming the hammer remains in contact with the anvil after impact, the motion of the system following the impact of the hammer is considered. Note that gravity acts. Derive the equation(s) of motion for the system.

Problem 5-55: Three masses are connected by massless pulleys and inextensible cables, as sketched in Figure P5-55. Mass m_1 is excited by an external force $F(t)$. All the masses move without friction with respect to their constraining surfaces. Assume *small motions* about the equilibrium configuration of the system and assume $F(t)$ is much smaller than the weight of m_1 such that all the cables remain taut at all times, and note that gravity acts. Derive the equa-

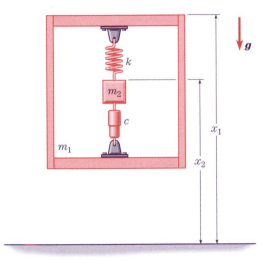

FIGURE P5-53

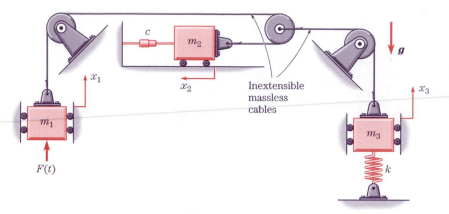

FIGURE P5-55

tion(s) of motion for the system. Begin by selecting a complete and independent set of generalized coordinates, starting with $x_1(t)$, then $x_2(t)$ and $x_3(t)$, as necessary. Note that the meanings of these coordinates remain to be clearly defined.

Problem 5-56: A particular vibration absorber can be modeled as sketched in Figure P5-56. A massive platform of mass M is mounted on a horizontally flexible foundation that exerts a horizontal restoring force Kd when the platform moves a distance d horizontally. Also, a mass spring system sketched in Figure

P5-56 moves without friction along the horizontal surface of the platform. Derive the equation(s) of motion for the system. Begin by selecting a complete and independent set of generalized coordinates, starting with $x_1(t)$ and then $x_2(t)$, as necessary. Note that the meanings of these coordinates remain to be clearly defined.

Problem 5-57: A model of a two-story building is sketched in Figure P5-57. The mass of the building is connected in the two floors of masses m and $\dfrac{m}{2}$, and the stiffness of the building is represented by two

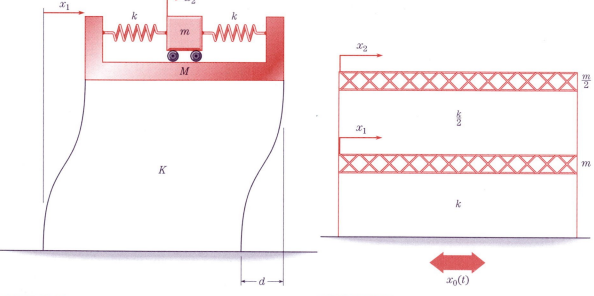

FIGURE P5-56   FIGURE P5-57

interstory springs of spring constants k and $\dfrac{k}{2}$, as shown. Such a model is sometimes used for seismic structural analyses. It is desired to analyze the motion of the building with respect to the ground, when subjected to a specified ground motion $x_0(t)$. Derive the equation(s) of motion for the system. Begin by selecting a complete and independent set of generalized coordinates, starting with $x_1(t)$ and then $x_2(t)$, as necessary. Note that the meanings of these coordinates remain to be clearly defined.

Problem 5-58: The system consists of two masses m and $2m$ attached to a massless rigid rod that is supported by two identical springs of spring constant k, as sketched in Figure P5-58. The right-hand mass is loaded by a vertical force $F(t)$. Assume that no gravity acts on the system and that the springs are undeformed when the rod is horizontal (with respect to the ground) and unloaded.

Show that for *small vertical motions* in the plane of the sketch the equations of motion for the system are given by

$$
\begin{bmatrix} 3m & 0 \\ 0 & \dfrac{2}{3}ml^2 \end{bmatrix} \begin{Bmatrix} \ddot{y}_c \\ \ddot{\theta} \end{Bmatrix} + \begin{bmatrix} 2k & -\dfrac{kl}{3} \\ -\dfrac{kl}{3} & \dfrac{13kl^2}{72} \end{bmatrix} \begin{Bmatrix} y_c \\ \theta \end{Bmatrix}
$$

$$
= \begin{Bmatrix} 1 \\ \dfrac{l}{3} \end{Bmatrix} F(t)
$$

where $y_c(t)$ is the vertical displacement of the system's center of mass and $\theta(t)$ is the angle of the rod, both defined with respect to the unloaded horizontal equilibrium position of the system.

Problem 5-59: A simple plane pendulum of mass m_1 and length L is pivoted about point O. A mass m_2 slides frictionlessly along the massless rigid link of the pendulum and is connected to point O by a spring of spring constant k, as sketched in Figure P5-59. The unstretched length of the spring is ℓ_0 ($\ell_0 < L$). Note that gravity acts. Derive the equation(s) of motion for the system.

Problem 5-60: An extensible plane pendulum is modeled as a mass spring dashpot system, as sketched in Figure P5-60. The assembly of the spring of spring constant k and the dashpot of dashpot constant c forms the straight and massless (extensible) link of the pendulum. The mass of the bob is m. The unstretched length of the pendulum, and thus the spring, is ℓ_0. Note that gravity acts. Derive the equation(s) of motion for the system.

Problem 5-61: Figure P5-61 shows a pendulum consisting of a mass m and a massless rigid link of length ℓ. The pendulum is pivoted, via a frictionless pin joint, to a massless vertical shaft that can rotate about its axis but is otherwise fixed. An external torque $T(t)$ is applied to the massless shaft. Due to the pin joint coupling, the pendulum is constrained to remain in the rotating plane, which passes through the axis of the vertical shaft and the massless link. Note that gravity acts. Derive the equation(s) of motion for the system.

Problem 5-62: A composite pendulum consists of a box of mass M and a massless rigid link of length ℓ. Within the box, a small mass m is attached to the wall of the box via a spring of spring constant k. Two configurations are sketched in Figures P5-62a and b. The masses m and M have negligible dimensions. Note that gravity acts. Derive the equation(s) of motion for both configurations.

FIGURE P5-58

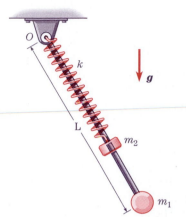

FIGURE P5-59

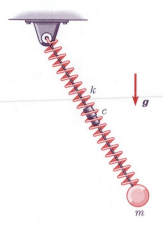

FIGURE P5-60

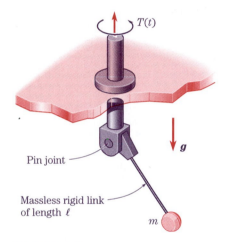

FIGURE P5-61

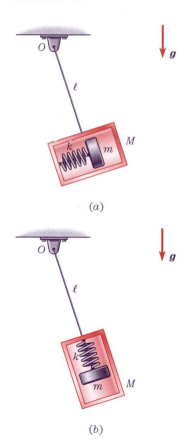

(a)

(b)

FIGURE P5-62

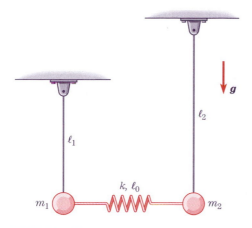

FIGURE P5-63

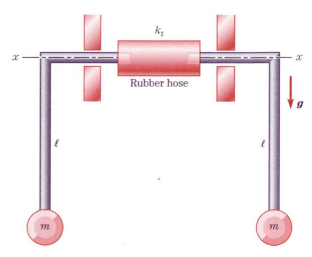

FIGURE P5-64

Problem 5-63: The system sketched in Figure P5-63 consists of two simple plane pendulums, masses m_1 and m_2 and respective lengths ℓ_1 and ℓ_2. The two masses are coupled via a spring of spring constant k and unstretched length ℓ_0. Assume *small motions* and note that gravity acts. Derive the equation(s) of motion for the system.

Problem 5-64: Figure P5-64 shows two identical simple pendulums of mass m that rotate about the x-x axis and are coupled by a rubber hose of torsional spring constant k_t. Assume that the rods of length ℓ of the pendulums, the two shafts along the x-x axis, and the rubber hose are all massless. Note that gravity acts. Derive the equation(s) of motion for the system.

Problem 5-65: As sketched in Figure P5-65, at the ends of a massless rigid link of length ℓ are attached two identical masses, each of mass m. The link is pivoted about its midlength at point O. The two masses at its ends are connected to ground via springs and a mass $2m$. Assume small motions and ignore gravity. Derive the equation(s) of motion for the system.

Problem 5-66: The system shown in Figure P5-66 consists of two masses m_1 and m_2 carried by a weightless string subjected to a large initial tension T. Consider small transverse motions of the masses; and ignore gravity. Derive the equation(s) of motion for the system. (*Hint:* For the derivation of the potential

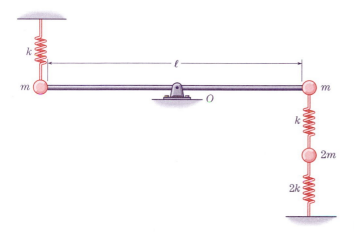

FIGURE P5-65

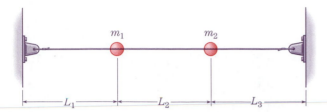

FIGURE P5-66

energy function for such a string model, see Example K-4 in Appendix K.)

Problem 5-67: A particle of mass m moves without friction on a table top. The particle is connected to a string that runs through a small hole in the center of the table and hangs vertically. A force $F(t)$ is applied to the free end of the string, which always remains taut, as sketched in Figure P5-67. Derive the equation(s) of motion.

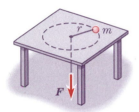

FIGURE P5-67

Category III: More Difficult

Equation(s) of motion—systems with nonorthogonal velocity components: Problems 5-68 through 5-74

Problem 5-68: In a proposed space station, the crew is to live in a torus-shaped station that rotates at a constant angular velocity Ω. (Refer to Figure P5-68.) It is argued that the "resulting field" will simulate the gravity field to which the crew is accustomed. In the space station, a pendulum is suspended from point O at radius R. For the following two cases, obtain the equation of motion of the mass m:

    (a) the pendulum is swinging in the XY plane; and

    (b) the pendulum is swinging in the XZ plane.

Problem 5-69: A massless container has a simple plane pendulum of mass m and length ℓ suspended from its ceiling, as sketched in Figure P5-69. A torsional spring of spring constant k_t connects the

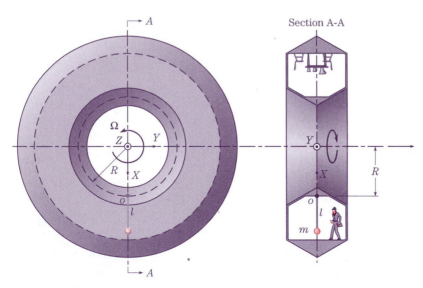

FIGURE P5-68

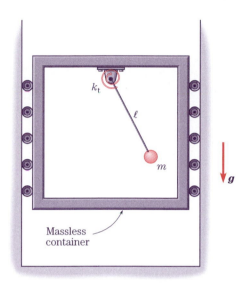

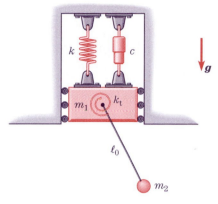

FIGURE P5-70

FIGURE P5-69

massless rigid link of the pendulum to the ceiling. The spring is undeformed when the pendulum hangs vertically downward. The container is constrained to move vertically along a channel. Note that gravity acts. Derive the equation(s) of motion for the system.

Problem 5-70: A mass m_1 is constrained to move in a vertical channel and is restrained by a spring of spring constant k and a dashpot of dashpot constant c. A simple plane pendulum of length ℓ_0 and mass m_2 is suspended from m_1, as sketched in Figure P5-70. A torsional spring of spring constant k_t restrains the rotation of the pendulum and is undeformed when the pendulum hangs vertically downward. Note that gravity acts. Derive the equation(s) of motion for the system.

Problem 5-71: A block of mass m moves without friction along a horizontal surface and is connected to rigid walls via two identical springs of spring constant k, as sketched in Figure P5-71. Also, a simple plane pendulum of mass m (which is identical to the mass of the block) and length L is pivoted about point O in the block. Note that gravity acts.

Show that, when the generalized coordinates are chosen to be x and θ as indicated in Figure P5-71, the linearized equation(s) of motion for the system are given by

$$2m\ddot{x} + mL\ddot{\theta} + 2kx = 0$$
$$mL\ddot{x} + mL^2\ddot{\theta} + mgL\theta = 0$$

where x is the displacement of the block from its equilibrium positions and θ is the counterclockwise angular displacement of the pendulum from its downward-hanging equilibrium position.

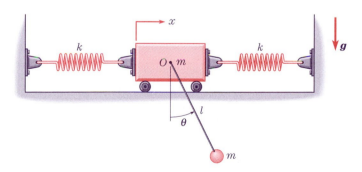

FIGURE P5-71

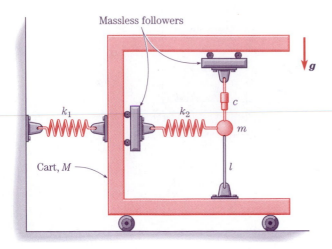

FIGURE P5-72

Problem 5-72: A cart of mass M is moving along a horizontal surface, restrained by a spring of spring constant k_1 to a rigid wall. On the cart is an inverted pendulum of length ℓ and mass m that is connected to a dashpot of dashpot constant c and a spring of spring constant k_2, as sketched in Figure P5-72. The massless followers, which are constrained to move without friction, are designed so that, at all times, the dashpot c remains vertical and the spring k_2 remains horizontal. Note that gravity acts. Derive the equation(s) of motion for the system.

Problem 5-73: A system consists of four masses (m_0, m_1, m_2, and m_3), three extensional springs (k_0, k_1, and k_2), one torsional spring k_t, three dashpots (c_0, c_1, and c_2), and massless rigid links and massless followers (A, B, C, and D), as sketched in Figure P5-73. The massless followers, which are constrained to move without friction, maintain the attached springs and dashpots in either a vertical or horizontal orientation, as sketched in Figure P5-73. The link connecting m_2 and m_3 is of total length ($a + b$) and may undergo large rotations about its pivot, which translates with m_1. The link ($a + b$) is vertical *and* all springs are undeformed when the system is in its equilibrium configuration. Note that gravity acts. Derive the equation(s) of motion for the system.

Problem 5-74: A mass m is suspended from the ceiling by two identical springs each of spring constant k. The mounting points are a distance $2a$ apart, as sketched in Figure P5-74. The mass is constrained to remain within the plane of the sketch. When in equilibrium, each spring forms an angle θ_0 with respect to the ceiling. Note that gravity acts. Derive the equation(s) of motion for *small motions* about the equilibrium position.

Equation(s) of motion—systems of 3 or more degrees of freedom: Problems 5-75 through 5-86

Problem 5-75: A block of mass m is connected to a spring of spring constant k, which in turn is connected to point O at the center of a massless rigid link of length ℓ, as sketched in Figure P5-75. The link may rotate about its center at point O, which is supported by a massless follower that can translate horizontally without friction. The ends of the link are attached to two springs of spring constants $2k$ and $3k$. The link is vertical when the $2k$ and $3k$ springs are undeformed and undergoes only small rotations. Derive the equation(s) of motion for the system.

Problem 5-76: A massless triangular frame contains 3 masses, m_0, m_1, and m_2, supported by spring dashpot suspensions, as sketched in Figure P5-76. The unstretched length of both springs k_1 and k_2 is ℓ_0. Also, there are eight spring elements designed to absorb ground impact of the frame. The masses undergo only small vertical motions relative to the frame in the plane of the sketch; thus, the massless link of length ℓ may undergo small vertical motion relative to the frame and small rotations in the plane. Note that only the motions prior to ground impact are of interest,

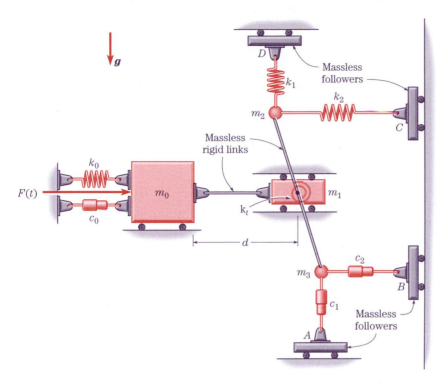

FIGURE P5-73

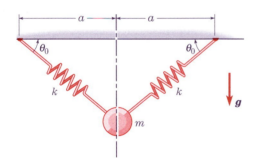

FIGURE P5-74

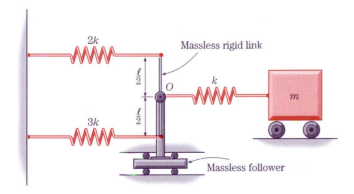

FIGURE P5-75

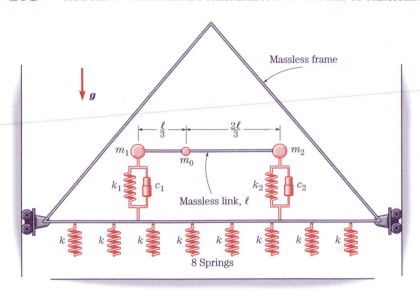

FIGURE P5-76

and that gravity acts. Derive the equation(s) of motion for the system.

Problem 5-77: Two masses m_1 and m_2 are connected by a cable that runs over a massless pulley without slip, as sketched in Figure P5-77. The cable is extensible and thus has been modeled as a section of inextensible cable combined with a spring of spring constant k_0 connected in series at each of its ends. Each of the masses is connected to ground by a spring of spring constants k_1 or k_2 and a dashpot of dashpot constant c_1 or c_2, as shown. Note that gravity acts. Derive the equation(s) of motion for the system.

Problem 5-78: The model for a vehicle is sketched in Figure P5-78. The chassis of mass M_0 traverses a flat horizontal terrain. The cabin of mass M and the combined passenger and seat of mass m are constrained to move only in the vertical direction relative to the chassis. All bodies remain in the plane of the sketch. Note that gravity acts. Derive the equation(s) of motion for the system.

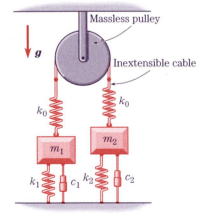

FIGURE P5-77

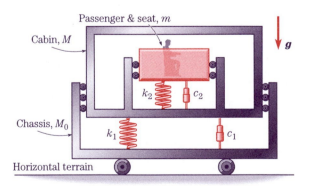

FIGURE P5-78

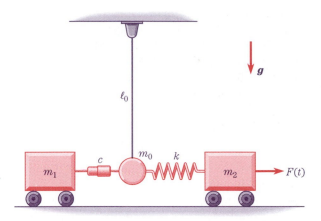

FIGURE P5-79

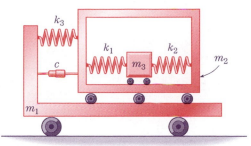

FIGURE P5-80

Problem 5-79: The bob of a simple plane pendulum of mass m_0 and length ℓ_0 is connected to two blocks, one of mass m_1 by a dashpot of dashpot constant c, and the other of mass m_2 by a spring of spring constant k and unstretched length ℓ_1, as sketched in Figure P5-79. An external force $F(t)$ acts on m_2. Assume *small motions* so that changes in the orientations of the spring and the dashpot caused by the angular displacement of the pendulum are negligible. Note that gravity acts. Derive the equation(s) of motion for the system.

Problem 5-80: Sketched in Figure P5-80 is a model for a vehicle of mass m_1 transporting a crate of mass m_2 that contains a delicate instrument of mass m_3. The instrument is restrained in the crate by two springs of spring constants k_1 and k_2. The crate is restrained on the vehicle by a spring of spring constant k_3 and a dashpot of dashpot constant c. Derive the equation(s) of motion for the system.

Problem 5-81: A model of a vehicle of mass m_1 transporting two crates of masses m_2 and m_3 is sketched in Figure P5-81. The crates are connected to each other by a spring of spring constant k_1. The crate m_2 is attached to the vehicle by a spring of spring constant k_2. The dissipation due to the relative motion between the crates and the vehicle can be modeled as viscous damping having an equivalent dashpot constant c. The vehicle's power plant generates a force $F(t)$ that moves the system. Derive the equation(s) of motion for the system.

Problem 5-82: Sketched in Figure P5-82 is a model for a high-precision testbed. The testbed of mass M is supported by two identical but independent suspension systems, each of mass m and each connected to the ground by a spring of spring constant k_1 and a dashpot of dashpot constant c_1. Each suspension system is connected to the testbed by a spring of spring constant k_2 and a dashpot of dashpot constant c_2. The testbed is subjected to an external force $F(t)$. Note that gravity acts. Derive the equation(s) of motion for the system.

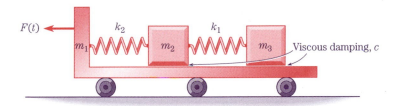

FIGURE P5-81

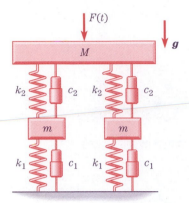

FIGURE P5-82

Problem 5-83: In a produce processing plant, a box of canned vegetables is dropped from above onto an individual mechanism that is modeled as sketched in Figure P5-83. The mechanism, whose model consists of masses (M, m_1, and m_2) and several springs (k_1, k_1, k_2, and k_2), moves without friction along the horizontal floor. The box does not bounce from the platform, and the mass of the box (m_0) is negligible when compared with M, m_1 or m_2. It is desired to analyze the motion of the mechanism after the box is dropped onto the platform. Note that gravity acts. Derive the equation(s) of motion for the system.

Problem 5-84: A mass spring dashpot system consists of three masses (m_1, m_2, and m_3), four springs (k_1, k_2, k_3, and k_4) and one dashpot (c). The masses are connected to each other and then to the ceiling by the springs and dashpot, as sketched in Figure P5-84. An external force $F(t)$ excites the system by acting on mass m_3. Note that gravity acts. Derive the equation(s) of motion for the system.

Problem 5-85: Mass m_1 is restrained on a cart of mass M by a spring of spring constant k_1. Mass m_1 also drags mass m_2 via a second spring of spring constant k_2. Dissipation between the indicated contacting surfaces can be modeled as viscous damping having equivalent dashpot constants c_1, c_2, and c_3, as sketched in Figure P5-85. Cart M is pulled by an external force $F(t)$. Derive the equation(s) of motion for the system.

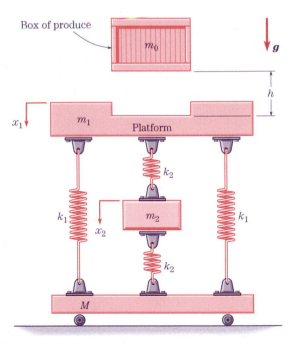

FIGURE P5-83

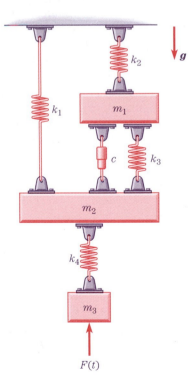

$F(t)$ FIGURE P5-84

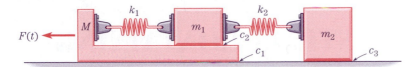

FIGURE P5-85

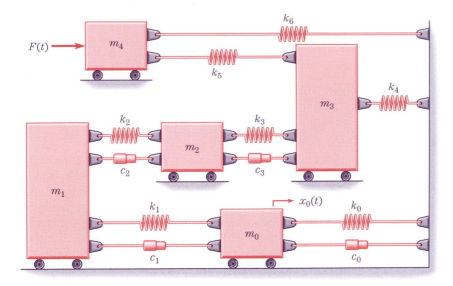

FIGURE P5-86

Problem 5-86: A mass spring dashpot system consists of five masses (m_0 through m_4), seven springs (k_0 through k_6) and four dashpots (c_0 through c_3), as sketched in Figure P5-86. All masses move without friction on their supporting horizontal surfaces. Mass m_0 moves according to a *specified motion* $x_0(t)$ with respect to ground. An external force $F(t)$ acts on mass m_4. Derive the equation(s) of motion for the system.

Baylor; Bird; Johnson; Jordan; Russell
Abdul-Jabbar; Chamberlain; Erving; Robertson; West

Chamberlain
55 rebounds vs. Boston (1960)
100 points vs. New York (1962)
50.4 pts/gm, 1961–62 season

Russell
11 rings in 13 years

Maravich
44.2 pts/gm, Collegiate Career

CHAPTER SIX

DYNAMICS OF SYSTEMS CONTAINING RIGID BODIES

6-1 INTRODUCTION

In this chapter we enlarge the application of Lagrange's equations from systems of particles as presented in Chapter 5 to systems containing rigid bodies, that is, extended bodies in which each constituent element does not move with respect to a coordinate system fixed, say, in the body. A major step in this development is the analysis of the mass distribution properties of rigid bodies, which, via the concept of angular momentum, leads to the inertia tensor. We shall restrict our lagrangian formulation to rigid bodies in plane motion. Although our focus is clearly the lagrangian formulation, for perspective we shall begin this chapter with a review of equation formulation via momentum principles; that is, via the direct approach.

It is taught in college freshman physics courses that the dynamics of a rigid body are governed by the linear momentum principle and the angular momentum principle,[1] which may be stated, respectively, as

$$\boldsymbol{F} = \frac{d\boldsymbol{P}}{dt} \quad \text{and} \quad \boldsymbol{\tau} = \frac{d\boldsymbol{H}}{dt}$$

where $\boldsymbol{F}$ is the resultant external force, $\boldsymbol{P}$ is the body's linear momentum, $\boldsymbol{\tau}$ is the resultant external torque, $\boldsymbol{H}$ is the body's angular momentum, and t is time. Readers who have not explored the origins of these equations are likely to believe, by virtue of the manner in which they are typically presented, that they are due to Newton. Well, only the linear momentum principle may be so attributed. The angular momentum principle was not clearly formulated until the latter half of the eighteenth century.

In *The Principia* (1687), Newton gave the principle of linear momentum (for particles) but did not state the principle of angular momentum. Various contributions to the angular momentum principle were made throughout Newton's lifetime and beyond by Christiaan Huygens (1629–1695), whose investigations of clock pendulums in 1673 led him to the first mathematical discussions of a dynamics problem beyond the dynamics of particles; Jacques I (Anglicized, James or Germanic, Jakob) Bernoulli (1654–1705), who, in applying "the principle of the lever" (that is, the equilibrium of moments) in 1686, corrected in 1703, gave an unclear statement of the angular momentum principle; Daniel I Bernoulli (1700–1782), who

---

[1]This is sometimes called the *moment of momentum principle*.

in 1744 derived the principle of angular momentum from the principle of linear momentum, and who had earlier (1733) given the first of many statements by himself and by others of the *principle of superposition;* and Leonhard Euler, who used both the linear and angular momentum principles as independent laws in 1744 and concluded that the principle of angular momentum was independent of the principle of linear momentum in 1775.

6-2   MOMENTUM PRINCIPLES FOR RIGID BODIES[2]

In this section we review the roles of forces and torques in analyzing the dynamics of extended solids via the *momentum* or *direct approach.* By extended bodies, we mean solid objects whose dynamics cannot be adequately modeled by considering them to be particles. These extended solids may be further modeled as systems of particles or continua, rigid or deformable. In particular, in this chapter, we consider these extended solids to be rigid bodies, and both systems of particles and continuum models are investigated; in Chapter 9, we consider extended solids that are modeled as elastic continua. After reviewing the requirements on forces and torques for solids in equilibrium, we briefly present two concepts of rigid bodies—systems of particles and continua—and then summarize the dynamic requirements on forces and torques for each of these models.

It is important to emphasize that, in this chapter as throughout the book, our focus will be the *variational* or *indirect approach.* In this section, we review the direct approach. Although it is likely that the reader has studied rigid bodies via the direct approach, the statement of these principles here is likely to be philosophically different from that which has been encountered.

In most dynamics textbooks, Newton's second law is stated for a particle where it is properly understood that particles cannot sustain torques. Then, by using Newton's third law (whether explicitly stated or simply in essence), the force–linear momentum and the torque–angular momentum equations are derived for rigid bodies from Newton's second law. Occasionally, there are some reservations expressed about this approach by the authors; but they then proceed to apply these principles to extended bodies, whether such bodies are considered as systems of particles or continua. Indeed, this approach should justifiably give rise to several major issues and questions. For example, it ought to be clear to those who do not explicitly cite Newton's third law that some new information beyond mathematical manipulations must be added to the force–linear momentum principle for a single particle in order to derive the force–linear momentum and torque–angular momentum principles for a system of particles. Further, the use of Newton's third law is, in effect, an assumption regarding the nature of the force interactions between the constituents of the rigid body.

These issues and questions may not be sufficient to drive the immediate interests of some readers of this book. On the other hand, at some point, in more relaxed times, these issues and questions may generate a bud of curiosity. It is in this context that Appendix C is provided, and for no other reason.

The course taken here is not typical; the force–linear momentum principle and the torque–angular momentum principle are asserted as independent principles that govern the dynamics of extended bodies—rigid or deformable. The former of these is called the *linear*

---

[2]This section is an *overview* of momentum principles for use in the direct approach, and therefore, except for Subsection 6-2.2, this section may be skimmed (or even omitted) without significant loss of continuity if the primary goal is a lagrangian presentation.

momentum principle; the latter of these is called the *angular momentum principle* (also sometimes called the *moment of momentum principle*). These are two independent fundamental laws of mechanics; neither is a corollary of the other. Further, from these two laws as launching sites, the nature of the internal force and internal torque interactions between the object's constituents can be derived for various models of extended bodies. For rigid bodies modeled as systems of particles, the internal force interactions are indeed found to satisfy Newton's third law. However, for rigid bodies modeled as continua, the internal force and internal torque interactions are found not to be governed by Newton's third law. Here, we shall simply outline the underlying connections between these principles, omitting most of the detailed mathematics. These omitted details are neither recommended (during a first encounter) nor required for proceeding in this chapter. Nevertheless, the mathematical details supporting the significant results for systems of particles may be found in Appendix C.

Finally, the question of whether an object should be modeled as an extended body can be answered only in the context of a specific problem. The *qualitative* means by which such a specific assessment can be made are the subject of Chapter 2. The *quantitative* means by which such a specific assessment can be made are contained within this chapter. So, if the effects associated with modeling the object as an extended body are not negligible in comparison with the effects associated with modeling the object as a particle, then the extended-body character of the object is significant. This statement may have the appearance of circularity, but it is not idly circuitous. This is the nature of the modeling process. After engaging this chapter, the reader should be armed with the quantitative means by which such assessments can be accomplished.

6-2.1 Review of Solids in Equilibrium and Particle Dynamics

In the study of the mechanics of rigid and deformable solids in equilibrium, the force–equilibrium requirement represents one of the three major requirements in the formulation of the system equations. For a particle, this requirement may be expressed as

$$\boldsymbol{f} = 0 \tag{6-1}$$

where $\boldsymbol{f}$ is the resultant force acting on the particle. For an extended body, rigid or deformable, this requirement may be expressed as

$$\boldsymbol{F} = 0 \tag{6-2}$$

$$\boldsymbol{\tau} = 0 \tag{6-3}$$

where $\boldsymbol{F}$ and $\boldsymbol{\tau}$ are the resultant force and the resultant torque about an arbitrary point, respectively, acting on the extended body.

Further, recall from Chapter 4 that the momentum principles governing particles in motion are

$$\boldsymbol{f} = \frac{d\boldsymbol{p}}{dt} \tag{6-4}$$

and

$$\boldsymbol{\tau}_B = \frac{d\boldsymbol{h}_B}{dt} + \boldsymbol{v}_B \times \boldsymbol{p} \tag{6-5}$$

where f, the resultant force; p, the linear momentum; τ_B, the resultant torque about an arbitrary point B; h_B, the angular momentum about the same point B; and v_B, the velocity of point B, define the variables in Eqs. (6-4) and (6-5). In discussing these principles, we made the significant conclusion that Eqs. (6-4) and (6-5) are *not* independent principles for a particle. Thus, as indicated in Chapter 4, certain combinations of three scalar equations among the six implied by the two vector equations of Eqs. (6-4) and (6-5)—including the obvious choices of three exclusively from Eq. (6-4) or three exclusively from Eq. (6-5)—may be used in the formulation of the equations of motion for a particle, but not more than three such scalar equations are necessary.

If the particle is modeled as newtonian, its linear momentum is defined as

$$p = mv \tag{6-6}$$

and its angular momentum about point B is defined as

$$h_B = r \times mv \tag{6-7}$$

where m is the particle's mass, r is the position of the particle with respect to the point B, and v is the velocity of the particle, all with respect to an inertial reference frame.

6-2.2 Models of Rigid Bodies

Just as the concept of a particle provides a useful mathematical model for a certain class of problems (see Subsection 4–2.2), the idealization of a perfectly rigid body also provides a useful model for the gross dynamic behavior of extended bodies that are modeled to undergo negligible deformation. The two most widely adopted models of rigid bodies are (1) a distribution of a very large number of discrete particles that are rigidly bound together and (2) a distribution of continuous rigid mass. The former of these is called a *system of particles;* the latter is called a *continuous body.* Conceptually, an integral over a *continuous body* may be considered to be the limiting extension of a summation over a corresponding *system of particles.* Such a limiting extension poses no theoretical problems when the goal is to compute mass distribution properties or overall dynamic behavior of the body. However, if the nature of the internal forces and internal torques of the body's constituents is desired, the simple conversion from a system of particles to a continuous body may require additional considerations, as indicated in Appendix C.

Indeed, as we shall see, both models are governed by the identical principles. However, as we shall also see, the intrinsic character of the internal forces and internal torques that must exist in each of these models is quite different. While this point may prove to be of primarily theoretical value, it is just this point that is central to an appreciation of the theoretical relationship between the linear momentum and the angular momentum principles.

Sketches of rigid bodies modeled as a *system of particles* and a *continuum* are shown in Figures 6-1a and 6-1b, respectively. The $oxyz$ frame in each model is assumed to be rigidly attached to the model. So, the kinematics of the rigid body are equivalent to the kinematics of the $oxyz$ frame. Such a rigidly attached frame is commonly called a *body-coordinate frame* (or *body-coordinate axes* or *body-coordinate system*). In particular, given the position of a single point in the $oxyz$ body-coordinate frame and the angular orientation of $oxyz$, with respect to the reference frame $OXYZ$, the location of the attached rigid body is completely determined.

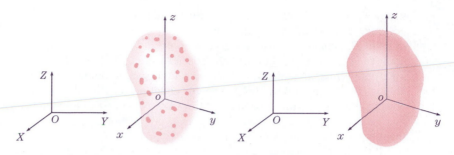

(a) System of particles model.   (b) Continuum model.

FIGURE 6-1   Rigid bodies modeled as (a) system of large number of particles or (b) continuum, both rigidly attached to body-coordinate frame $oxyz$.

6-2.3   Momentum Principles for Extended Bodies: The Newton-Euler Equations

The governing momentum principles for extended bodies are the linear momentum principle, due substantially to Isaac Newton, and the angular momentum principle, due to a lesser extent to Christiaan Huygens and James Bernoulli and to a greater extent to Leonhard Euler. We assert that for extended bodies, modeled as a system of particles or a continuum, deformable or rigid, these two principles are *independent* laws of mechanics.

The linear momentum principle may be stated as follows:

- The resultant force acting on a body or any subset thereof is equal to the time rate of change of its linear momentum with respect to an inertial reference frame. Or,

$$\boldsymbol{F} = \frac{d\boldsymbol{P}}{dt} \tag{6-8}$$

where $\boldsymbol{F}$ is the resultant force, and $\boldsymbol{P}$ is the linear momentum with respect to the inertial reference frame. We shall designate Eq. (6-8) as *Newton's linear momentum principle.*

Equation (6-8) provides a conservation theorem: *conservation of linear momentum of an extended body.*

- If the resultant force on an extended body is zero, then $\frac{d\boldsymbol{P}}{dt} = 0$, and thus the body's linear momentum $\boldsymbol{P}$ is conserved; that is, $\boldsymbol{P}$ remains constant.

Being a vector, linear momentum can be conserved in one direction if the resultant force in that direction is zero, independently of all other directions.

The angular momentum principle may be stated as follows:

- The resultant torque acting on a body or any subset thereof about a fixed point O in an inertial reference frame is equal to the time rate of change of its angular momentum with respect to the inertial reference frame. Or,

$$\boldsymbol{\tau}_O = \frac{d\boldsymbol{H}_O}{dt} \tag{6-9}$$

where $\boldsymbol{\tau}_O$ is the resultant torque about the fixed point O, and $\boldsymbol{H}_O$ is the angular momentum of the body about the same point O. We shall designate Eq. (6-9) and all its various forms for extended bodies (that is, Eqs. (6-12) and (6-15)) as *Euler's angular momentum principle*.

In analogy with Eq. (6-8), Eq. (6-9) provides another conservation theorem: *conservation of angular momentum of an extended body*.

- If the resultant torque on an extended body about a fixed point O is zero, then $\dfrac{d\boldsymbol{H}_O}{dt} = 0$, and thus the body's angular momentum $\boldsymbol{H}_O$ about the point O is conserved; that is, $\boldsymbol{H}_O$ remains constant.

We shall designate the combination of relations such as Eqs. (6-8) and (6-9) as the *Newton-Euler equations*. These equations are fundamental tenets of mechanics; they can neither be derived nor proved.

6-2.4 Momentum Principles for Rigid Bodies Modeled as Systems of Particles

We consider a rigid body modeled as a system of N particles as depicted in Figure 6-2. Equations (6-8) and (6-9) are the applicable linear momentum and angular momentum principles for such a model. If the N particles are newtonian, the linear momentum and the angular momentum are given, by definition, as

$$\boldsymbol{P} = \sum_{i=1}^{N} m_i \boldsymbol{v}_i \tag{6-10}$$

and

$$\boldsymbol{H}_O = \sum_{i=1}^{N} \boldsymbol{R}_i \times m_i \boldsymbol{v}_i \tag{6-11}$$

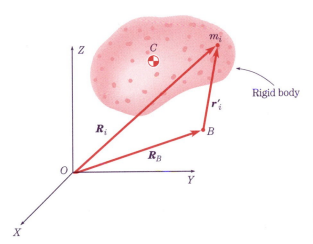

FIGURE 6-2 Rigid body modeled as system of particles, showing center of mass C and arbitrary point B.

where m_i and $\boldsymbol{v}_i$ are the mass and inertial velocity of the ith particle, respectively, and $\boldsymbol{R}_i$ is the position of the ith particle with respect to the $OXYZ$ inertial reference frame.

Given, as we have just declared, that the linear momentum principle and the angular momentum principle for extended bodies are two independent laws, we may use this annunciation to explore the nature of the mutual internal force interactions between the constituents of such bodies. The strategy by which we conduct such an investigation is to derive the linear momentum principle for an extended body from the linear momentum principle for a particle, and then explore the character of the mutual force interactions between the constituents of the body, which is required to insure that both the *asserted* and the *derived* principles hold. Also, we derive the angular momentum principle for an extended body from the angular momentum principle for a particle, and then explore the character of the mutual force interactions between the constituents of the body, which is required to insure that both the *asserted* and the *derived* principles hold.

It can be shown (Appendix C) that the linear momentum principle for a system of particles is derivable from the linear momentum principle for a single particle if the pairwise mutual internal forces between the system's particles are equal and opposite. Further, it can be shown that the angular momentum principle for a system of particles is derivable from the angular momentum principle for a single particle if additionally the pairwise mutual internal forces between the system's particles act along the line connecting any pair of particles; that is, the pairwise mutual internal forces are *collinear*. Thus, the combination of these two requirements on the mutual internal forces between any two particles is that they are of the same magnitude, in opposite directions, and collinear. Therefore, the Newton-Euler equations for a system of particles require that the mutual internal forces between any two particles must satisfy Newton's third law.

Finally, it can be shown (Appendix C) that the angular momentum principle for a system of particles may be expressed in the alternative form

$$\boldsymbol{\tau}_B = \frac{d\boldsymbol{H}_B}{dt} + \boldsymbol{v}_B \times \boldsymbol{P} \tag{6-12}$$

where B is an arbitrary point whose velocity with respect to an inertial reference frame is $\boldsymbol{v}_B$, and $\boldsymbol{P}$ is the linear momentum of the system of particles in accordance with Eq. (6-10). From Eq. (6-11) and Figure 6-2,

$$\boldsymbol{H}_B = \sum_{i=1}^{N} \boldsymbol{r}_i' \times m_i \boldsymbol{v}_i \tag{6-13}$$

where $\boldsymbol{r}_i'$ is the position of m_i with respect to point B. Note that Eq. (6-12) reduces to Eq. (6-9) if

1. Point B has zero velocity with respect to the inertial reference frame (that is, $\boldsymbol{v}_B = 0$); or

2. Point B has a velocity $\boldsymbol{v}_B$ that is parallel to the linear momentum $\boldsymbol{P}$; thus, $\boldsymbol{v}_B \times \boldsymbol{P} = 0$. The most important case in this category is when point B coincides with the center of mass[3] of the system of particles.

---

[3]We assume the reader is familiar with the concept of the *center of mass*. Nevertheless, referring to Eqs. (6-41) may be helpful. In general, we shall use the phrases "center of mass" and "central axes" when referring to the mass center and its associated axes, respectively. We shall reserve the terms "centroid" and "centroidal" to refer to the geometric midpoint of an area or a volume of a body.

Since the total linear momentum of the system of particles can be expressed as

$$\boldsymbol{P} = M\boldsymbol{v}_C \tag{6-14}$$

where M is the total mass of the system of particles and $\boldsymbol{v}_C$ is the velocity of the center of mass of the system, $\boldsymbol{v}_B \times \boldsymbol{P}$ vanishes if $\boldsymbol{v}_B$ is either parallel to $\boldsymbol{v}_C$ or is itself $\boldsymbol{v}_C$. Thus, Eq. (6-9) is valid not only for a fixed point O but also for the center of mass of a system of particles. Therefore, we can write, additionally,

$$\boldsymbol{\tau}_C = \frac{d\boldsymbol{H}_C}{dt} \tag{6-15}$$

where C denotes the center of mass of the system of particles.

6-2.5   Momentum Principles for Rigid Bodies Modeled as Continua

Next we consider the rigid body modeled as a continuum as depicted in Figure 6-3. Equations (6-8) and (6-9) are the applicable linear momentum and angular momentum principles for such a model. If the continuum is modeled as a newtonian medium, the linear momentum and the angular momentum are given, by definition, as

$$\boldsymbol{P} = \int_M \boldsymbol{v}\, dm \tag{6-16}$$

and

$$\boldsymbol{H}_O = \int_M \boldsymbol{R} \times \boldsymbol{v}\, dm \tag{6-17}$$

where $\boldsymbol{v}$ is the inertial velocity of an infinitesimal mass element dm, $\boldsymbol{R}$ is the position of dm with respect to the $OXYZ$ inertial reference frame, and M is the total mass of the body. In some branches of mechanics, in particular quantum mechanics, the infinitesimal mass element dm is endowed with the *intrinsic angular momentum* (also called the *spin angular momentum*). In that case, the intrinsic angular momentum, commonly denoted as $\boldsymbol{l}$, must be added to the integrand in Eq. (6-17). We shall, however, disregard the intrinsic angular momentum due to the rare acknowledgment of its existence in classical mechanics.

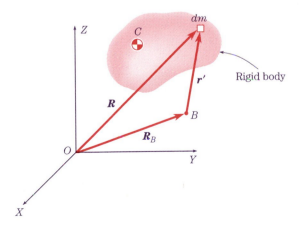

FIGURE 6-3   Rigid body modeled as continuum, showing center of mass C and arbitrary point B.

Again, an exploration of the character of the internal interactions is instructive. This exploration is based on the necessary conditions to bring the momentum principles for a particle and the corresponding momentum principles for an extended rigid body into a consistent framework. As in the preceding subsection, the strategy is to derive the linear momentum principle for an extended body from the linear momentum principle for a mass element of the body, and then explore the character of the mutual force and torque interactions between the constituents of the body, which is required to insure that both the *asserted* and the *derived* principles hold. Also, we derive the angular momentum principle for an extended body from the angular momentum principle for a mass element of the body, and then explore the character of the mutual force and torque interactions between the constituents of the body, which is required to insure that both the *asserted* and the *derived* principles hold.

In the preceding subsection we found that for an extended body modeled as a system of particles, the necessary conditions on the internal interactions are satisfied by Newton's third law. On the other hand, when an extended body is modeled as a continuum, as shown in Appendix C, we find that the satisfaction of Newton's third law by the internal forces between the mass elements requires a Poisson's ratio of $\frac{1}{4}$ for an isotropic elastic solid, a result that is generally not observed in real materials.

It can be shown that the linear momentum principle for an extended body modeled as a continuum is derivable from the linear momentum principle for each mass element if the total internal force acting mutually between all mass elements is zero. Further, it can be shown that the angular momentum principle for an extended body modeled as a continuum is derivable from the angular momentum principle for each mass element if the total internal torque acting mutually between all mass elements is zero. The conditions that are imposed on the internal interactions of rigid bodies modeled as continua are somewhat more complex than the conditions that are imposed on the internal interactions for rigid bodies modeled as systems of particles.

Finally it can be shown that the angular momentum principle for a continuum may be expressed in the alternative form

$$\boldsymbol{\tau}_B = \frac{d\boldsymbol{H}_B}{dt} + \boldsymbol{v}_B \times \boldsymbol{P} \tag{6-18}$$

where B is an arbitrary point whose velocity with respect to an inertial reference frame is $\boldsymbol{v}_B$, and $\boldsymbol{P}$ is the linear momentum of the continuous rigid body in accordance with Eq. (6-16). From Eq. (6-17) and Figure 6-3,

$$\boldsymbol{H}_B = \int_M \boldsymbol{r}' \times \boldsymbol{v} \, dm \tag{6-19}$$

where $\boldsymbol{r}'$ is the position of dm with respect to point B.

Again, it is noted that the form of Eq. (6-18) reduces to the form of Eq. (6-9) if

1. Point B has zero velocity with respect to the inertial reference frame (that is, $\boldsymbol{v}_B = 0$); or

2. Point B has a velocity $\boldsymbol{v}_B$ that is parallel to the linear momentum $\boldsymbol{P}$; thus, $\boldsymbol{v}_B \times \boldsymbol{P} = 0$. The most important case in this category is when point B coincides with the center of mass of the continuous rigid body.

TABLE 6-1 Summary of Momentum Principles for Particles and Extended—Rigid or Deformable—Bodies

	Linear Momentum Principle		Angular Momentum Principle	
	Dynamic Requirements on Forces		Dynamic Requirements on Torques	Character of Internal Force Interactions
			Case 1 (O): $v_B = 0$ Case 2 (C): $v_B \times p = 0$ or $v_B \times P = 0$	
Body Modeled as Particle	$f = \dfrac{dp}{dt}$	or $\Updownarrow$	$\tau_B = \dfrac{dh_B}{dt} + v_B \times p$ $\qquad$ $\tau_O = \dfrac{dh_O}{dt}$ Not Meaningful	Not Meaningful
Body Modeled as System of Particles	$F = \dfrac{dP}{dt}$	and $\Updownarrow$	$\tau_B = \dfrac{dH_B}{dt} + v_B \times P$ $\qquad$ Case 1: $\tau_O = \dfrac{dH_O}{dt}$ Case 2: $\tau_C = \dfrac{dH_C}{dt}$	Satisfied by Newton's Third Law (forces are equal, opposite, and collinear)
Body Modeled as Continuum	$F = \dfrac{dP}{dt}$	and $\Updownarrow$	$\tau_B = \dfrac{dH_B}{dt} + v_B \times P$ $\qquad$ Case 1: $\tau_O = \dfrac{dH_O}{dt}$ Case 2: $\tau_C = \dfrac{dH_C}{dt}$	Internal Body Forces are Not Collinear

Since the total linear momentum of the continuous body can be expressed as

$$\boldsymbol{P} = M\boldsymbol{v}_C \qquad (6\text{-}20)$$

where M is the total mass of the body and $\boldsymbol{v}_C$ is the velocity of the center of mass of the body, $\boldsymbol{v}_B \times \boldsymbol{P}$ vanishes if $\boldsymbol{v}_B$ is either parallel to $\boldsymbol{v}_C$ or is itself $\boldsymbol{v}_C$. Thus, Eq. (6-9) is valid not only for a fixed point O, but also for the center of mass of the continuous body. Therefore, we can write, additionally,

$$\boldsymbol{\tau}_C = \frac{d\boldsymbol{H}_C}{dt} \qquad (6\text{-}21)$$

where C denotes the center of mass of the continuous body.

A summary of the momentum principles is given in Table 6-1. A few points summarized in Table 6-1 are worth repeating.

1. The linear momentum principle and the angular momentum principle for a particle are *not* independent. Certain combinations of three scalar equations among the six implied by the two vector momentum principles must be satisfied in equation formulation; the satisfaction of these three ensures the satisfaction of the other three.

2. The linear momentum principle and the angular momentum principle for extended bodies are independent principles, not derivable from one another. Both must be satisfied independently in equation formulation.

3. The momentum principles for extended bodies apply equally to deformable and rigid bodies. Rigid bodies are simply idealizations of real (deformable) bodies in which the internal deformations are assumed to be negligibly small.

Although we shall deemphasize the application of momentum principles here, Table 6-1 is nevertheless worthy of both philosophical and practical note. The relatively innocuous simplicity of the entries in Table 6-1 belies both their enormously broad applicability as well as the subtleties and complexities in their implementation. We have barely penetrated the subject of the direct approach for rigid bodies here.

Finally, based on the foregoing presentations in Chapter 4 and in this section, the mechanical requirements on the forces and torques on particles and extended bodies for a formulation via the direct approach may be summarized as given in Table 6-2. Both the static

TABLE 6-2   Mechanical Requirements (Natural Motions) on Forces and Torques for Particles and Extended (Rigid or Deformable) Bodies

	Statics	Dynamics
Particle	$\sum \boldsymbol{f} = 0$ $\left(\text{or, if convenient, } \sum \boldsymbol{\tau} = 0\right)$	$\sum \boldsymbol{f} = \dfrac{d\boldsymbol{p}}{dt}$ $\left(\text{or, if convenient, } \sum \boldsymbol{\tau}_0 = \dfrac{d\boldsymbol{h}_0}{dt}\right)$
Extended Body (Rigid or Flexible)	$\sum \boldsymbol{F} = 0$ *and* $\sum \boldsymbol{\tau} = 0$	$\sum \boldsymbol{F} = \dfrac{d\boldsymbol{P}}{dt}$ *and* $\sum \boldsymbol{\tau}_{0,C} = \dfrac{d\boldsymbol{H}_{0,C}}{dt}$

and dynamic requirements are given in order to display the similarities and differences between statics and dynamics as well as between particles and extended bodies. The dynamic requirements on the forces and torques are in accordance with Requirement 2 in Eqs. (2-1), (4-11), or (5-1). The reader is expected to have encountered the statics requirements on forces and torques, as expressed by the statics column in Table 6-2.

6-3 DYNAMIC PROPERTIES OF RIGID BODIES

In analyzing the dynamics of rigid bodies, it is clearly important to be able to calculate the angular momentum. In this section we explore such calculations. The outcome will be a convenient form that separates the mass distribution properties of the body from the angular motion of the body. The calculation of these mass distribution properties is required for formulations via both the direct and the indirect approaches.

6-3.1 The Inertia Tensor

In calculating the angular momentum of rigid bodies, we are required to consider relations such as Eqs. (6-11) or (6-17). In the explicit forms given, Eqs. (6-11) and (6-17) are not convenient for evaluation. For example, the evaluation of Eq. (6-11) requires the position and velocity vectors of each of a large number of particles, taking their vector cross products, and summing over all the particles, for each instant in time, a substantial undertaking in all but the most elementary of applications. While such an enormous series of operations is clearly possible, any option for simplifying the calculation of the angular momentum vector is potentially attractive.

Further, it is worth emphasizing that in discussing Eqs. (6-8) through (6-21) as general principles for rigid bodies, all the kinematic variables were defined in terms of inertial unit vectors. However, as we saw in our discussion of kinematics in Chapter 3, in the course of making specific calculations it is often most convenient to express kinematic variables in terms of the unit vectors of an intermediate reference frame. As emphasized in Chapter 3, the velocities and accelerations of most interest are the inertial values; however, they are often conveniently expressed in terms of the unit vectors of the intermediate reference frame. Similarly, in the analyses that follow, the kinematic variables will be generally expressed in terms of the unit vectors of an intermediate reference frame. In particular, these will be the unit vectors of a reference frame attached to the rigid body, variously called the *body-coordinate axes* or the *body-coordinate system* or, as we shall generally call it, the *body-coordinate frame.*

We consider a rigid body consisting of a very large number of particles, and which is undergoing angular velocity $\boldsymbol{\omega}$ as sketched in Figure 6-4. The body is attached to the $oxyz$ *body-coordinate frame,* where the origin o of this frame is fixed in inertial space. In accordance with Eq. (3-23), the velocity of the ith particle is

$$\boldsymbol{v}_i = \frac{d\boldsymbol{R}_o}{dt} + \boldsymbol{v}_{\text{rel}} + \boldsymbol{\omega} \times \boldsymbol{r}_i$$

$$= \boldsymbol{\omega} \times \boldsymbol{r}_i \tag{6-22}$$

where the term $\dfrac{d\boldsymbol{R}_o}{dt}$ vanishes because point o is fixed, the term $\boldsymbol{v}_{\text{rel}}$ vanishes because the body is rigid, and $\boldsymbol{r}_i$ is the position vector of m_i with respect to o. Substitution of Eq. (6-22) into Eq. (6-11) gives

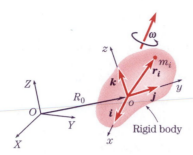

FIGURE 6-4 Rigid body attached to $oxyz$ body-coordinate frame, undergoing angular velocity $\boldsymbol{\omega}$ about fixed point o.

$$\boldsymbol{H}_o = \sum_{i=1}^{N} \boldsymbol{r}_i \times m_i \boldsymbol{v}_i$$

$$= \sum_{i=1}^{N} \boldsymbol{r}_i \times m_i (\boldsymbol{\omega} \times \boldsymbol{r}_i)$$

$$= \sum_{i=1}^{N} m_i \boldsymbol{r}_i \times (\boldsymbol{\omega} \times \boldsymbol{r}_i) \tag{6-23}$$

where N is the very large number of particles which comprise the rigid body.[4] Note that in accordance with Eq. (6-22) and Figure 6-4, $\boldsymbol{H}_O$ in Eq. (6-11) has become $\boldsymbol{H}_o$ in Eq. (6-23); that is, Eq. (6-23) is the angular momentum of the body with respect to the body-coordinate frame, not necessarily originating from any special point in the body, such as the center of mass of the body.

The vector $\boldsymbol{r}_i$ is conveniently expressed in terms of the $oxyz$ unit vectors. Also, as is common, the angular velocity vector $\boldsymbol{\omega}$ is expressed in terms of the $oxyz$ unit vectors. Thus, $\boldsymbol{H}_o$ is also expressed in terms of the same unit vectors. So, in general, we may write these vectors in terms of their respective components as

$$\boldsymbol{H}_o = H_x \boldsymbol{i} + H_y \boldsymbol{j} + H_z \boldsymbol{k}$$
$$\boldsymbol{r}_i = x_i \boldsymbol{i} + y_i \boldsymbol{j} + z_i \boldsymbol{k} \tag{6-24}$$
$$\boldsymbol{\omega} = \omega_x \boldsymbol{i} + \omega_y \boldsymbol{j} + \omega_z \boldsymbol{k}$$

If $\boldsymbol{A}$, $\boldsymbol{B}$, and $\boldsymbol{C}$ are arbitrary vectors, recall that a vector triple product may be written equivalently as

$$\boldsymbol{A} \times (\boldsymbol{B} \times \boldsymbol{C}) = (\boldsymbol{A} \cdot \boldsymbol{C})\boldsymbol{B} - (\boldsymbol{A} \cdot \boldsymbol{B})\boldsymbol{C} \tag{6-25}$$

Use of the vector identity in Eq. (6-25) to rewrite Eq. (6-23) gives

$$\boldsymbol{H}_o = \sum_{i=1}^{N} m_i [(\boldsymbol{r}_i \cdot \boldsymbol{r}_i)\boldsymbol{\omega} - (\boldsymbol{r}_i \cdot \boldsymbol{\omega})\boldsymbol{r}_i] \tag{6-26}$$

where from Eq. (6-25) we have identified $\boldsymbol{A}$ with the first $\boldsymbol{r}_i$, $\boldsymbol{B}$ with $\boldsymbol{\omega}$ and $\boldsymbol{C}$ with the second $\boldsymbol{r}_i$ in Eq. (6-23). Expansion of Eq. (6-26), with the use of Eqs. (6-24) and (6-25), gives

---

[4]By the phrase "the very large number" of particles, we imply that N is sufficiently large that the particle distribution is capable of representing the physical body in a selected calculation to any desired accuracy.

$$\boldsymbol{H}_o = \sum_{i=1}^{N} m_i \left[(x_i^2 + y_i^2 + z_i^2)(\omega_x \boldsymbol{i} + \omega_y \boldsymbol{j} + \omega_z \boldsymbol{k}) \right.$$
$$\left. - (x_i \omega_x + y_i \omega_y + z_i \omega_z)(x_i \boldsymbol{i} + y_i \boldsymbol{j} + z_i \boldsymbol{k}) \right] \quad (6\text{-}27)$$

Examination of Eq. (6-27) clearly shows $\boldsymbol{i}, \boldsymbol{j}$, and $\boldsymbol{k}$ components on the right-hand side. By definition, and in accordance with the first of Eqs. (6-24), the sum of all the x components on the right-hand side of Eq. (6-27) is equal to H_x. That is,

$$H_x = \sum_{i=1}^{N} m_i \left[x_i^2 \omega_x + y_i^2 \omega_x + z_i^2 \omega_x - x_i^2 \omega_x - x_i y_i \omega_y - x_i z_i \omega_z \right] \quad (6\text{-}28)$$

By factoring out the angular velocity components in Eq. (6-28), H_x may be written as

$$H_x = \left[\sum_{i=1}^{N} m_i (y_i^2 + z_i^2) \right] \omega_x + \left[-\sum_{i=1}^{N} m_i (x_i y_i) \right] \omega_y + \left[-\sum_{i=1}^{N} m_i (x_i z_i) \right] \omega_z \quad (6\text{-}29)$$

Equation (6-27) contains analogous terms for H_y and H_z, which we may identify as

$$H_y = \left[-\sum_{i=1}^{N} m_i (y_i x_i) \right] \omega_x + \left[\sum_{i=1}^{N} m_i (x_i^2 + z_i^2) \right] \omega_y + \left[-\sum_{i=1}^{N} m_i (y_i z_i) \right] \omega_z \quad (6\text{-}30)$$

and

$$H_z = \left[-\sum_{i=1}^{N} m_i (z_i x_i) \right] \omega_x + \left[-\sum_{i=1}^{N} m_i (z_i y_i) \right] \omega_y + \left[\sum_{i=1}^{N} m_i (x_i^2 + y_i^2) \right] \omega_z \quad (6\text{-}31)$$

Equations (6-29), (6-30) and (6-31) may be collected in the form

$$H_x = I_{xx} \omega_x + I_{xy} \omega_y + I_{xz} \omega_z$$
$$H_y = I_{yx} \omega_x + I_{yy} \omega_y + I_{yz} \omega_z \quad (6\text{-}32)$$
$$H_z = I_{zx} \omega_x + I_{zy} \omega_y + I_{zz} \omega_z$$

where

$$I_{xx} = \sum_{i=1}^{N} m_i \left(y_i^2 + z_i^2 \right) \qquad I_{yy} = \sum_{i=1}^{N} m_i \left(x_i^2 + z_i^2 \right) \qquad I_{zz} = \sum_{i=1}^{N} m_i \left(x_i^2 + y_i^2 \right) \quad (6\text{-}33)$$

and

$$I_{xy} = -\sum_{i=1}^{N} m_i x_i y_i = I_{yx} \qquad I_{xz} = -\sum_{i=1}^{N} m_i x_i z_i = I_{zx} \qquad I_{yz} = -\sum m_i y_i z_i = I_{zy}$$
$$(6\text{-}34)$$

The terms in Eqs. (6-33) are called the *moments of inertia* and the terms in Eqs. (6-34) are called the *products of inertia.*[*]

---

[*]The terms *moment of inertia* and *product of inertia* were composed by Euler (1760), although several authors, dating from Huygens in 1673, had encountered equivalent quantities. During the 1760s, Euler proceeded to calculate these quantities for a number of homogeneous bodies.

Equations (6-32), (6-33), and (6-34) are the equivalent of Eq. (6-23); and although it may not appear at all obvious why this collection of equations is preferable to the single expression in Eq. (6-23), this collection is generally more useful for calculation. The reason for this preference lies in the form of Eqs. (6-32), where the mass distribution properties of the body as represented by the inertia terms are expressed separately from the motion of the body as represented by the angular velocity components. Also, the dimensions (length squared times mass) of the inertia components reveal the reason why they are sometimes said to be an "integral of the second moment of the body's mass" about an axis.

Equations (6-32) may be expressed in matrix notation as

$$\{H\}_o = [I]_o\{\omega\} \tag{6-35}$$

where

$$\{H\}_o = \begin{Bmatrix} H_x \\ H_y \\ H_z \end{Bmatrix}_o \quad \text{and} \quad \{\omega\} = \begin{Bmatrix} \omega_x \\ \omega_y \\ \omega_z \end{Bmatrix} \tag{6-36}$$

are *column matrices* (also called *column vectors* because they are vectors) and

$$[I]_o = \begin{bmatrix} I_{xx} & I_{xy} & I_{xz} \\ I_{yz} & I_{yy} & I_{yz} \\ I_{zx} & I_{zy} & I_{zz} \end{bmatrix}_o \tag{6-37}$$

is a *square matrix*. As already defined, $\{H\}$ and $\{\omega\}$ are the angular momentum vector and the angular velocity vector, respectively. The new quantity $[I]$ is called the *inertia matrix* or the *inertia tensor* because, as we shall see, like the familiar quantities of stress and strain, $[I]$ has the rotational transformation properties of a second-order tensor.[*] The individual components or terms in these matrices are interchangeably called *elements* or *entries*. The moments of inertia and products of inertia for the central axes of some *uniform* (that is, constant density) solids are given in Table 6-3.

---

[*]The word *tensor* as it is presently used to denote quantities such as stress, strain and the inertia matrix was introduced into the mechanics vocabulary by the German Woldemar Voigt(1850–1919) in 1898.

The inertia tensor describes the manner in which the mass is distributed with respect to the selected *oxyz* body-coordinate frame. So, although the rigid body may rotate and translate, the inertia tensor, with respect to the body-coordinate frame, is invariant with time. If, on the other hand, the *oxyz* axes were to rotate with respect to the body, in general, the inertia tensor would be a function of time, thus introducing significant complexity into the angular momentum calculations. An exception to this observation is the case when the rigid body is spinning about an axis for which all the associated products of inertia vanish.

The analyses leading to the results for the inertia tensor have been obtained for an arbitrary location of the fixed origin *o*, relative to the body. Thus, we conclude that an inertia tensor may be calculated for each and every point in a rigid body. Furthermore, this is also true for every point outside the body since the fixed origin *o* need not be inside the body. Also, the results given in Eqs. (6-32) or (6-35) represent the angular momentum of the body not only with respect to a fixed point *o*, but more generally for any point that is either

TABLE 6-3 Central Moments and Products of Inertia of Some Uniform Solids. (The mass of the body is m and its center of mass is located at c.)

Sphere

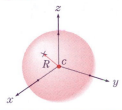

$$I_{xx} = I_{yy} = I_{zz} = \frac{2}{5}mR^2$$

$$I_{xy} = I_{xz} = I_{yz} = 0$$

$$\mathcal{V} = \frac{4}{3}\pi R^3$$

Hemisphere

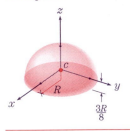

$$I_{xx} = I_{yy} = \frac{83}{320}mR^2$$

$$I_{zz} = \frac{2}{5}mR^2$$

$$I_{xy} = I_{xz} = I_{yz} = 0$$

Cylinder

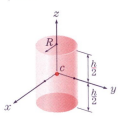

$$I_{xx} = I_{yy} = \frac{1}{12}m(3R^2 + h^2)$$

$$I_{zz} = \frac{1}{2}mR^2$$

$$I_{xy} = I_{xz} = I_{yz} = 0$$

$$\mathcal{V} = \pi R^2 h$$

Semicylinder

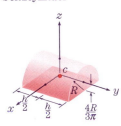

$$I_{xx} = 0.0699mR^2 + \frac{1}{12}mh^2$$

$$I_{yy} = 0.320mR^2$$

$$I_{zz} = \frac{1}{12}m(3R^2 + h^2)$$

$$I_{xy} = I_{xz} = I_{yz} = 0$$

Slender rod

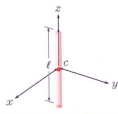

$$I_{xx} = I_{yy} = \frac{1}{12}m\ell^2$$

$$I_{zz} = 0$$

$$I_{xy} = I_{xz} = I_{yz} = 0$$

TABLE 6-3 (continued)

Circular disk or plate

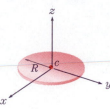

$$I_{xx} = I_{yy} = \frac{1}{4}mR^2$$

$$I_{zz} = \frac{1}{2}mR^2$$

$$I_{xy} = I_{xz} = I_{yz} = 0$$

Thin ring or hoop

$$I_{xx} = I_{yy} = \frac{1}{2}mR^2$$

$$I_{zz} = mR^2$$

$$I_{xy} = I_{xz} = I_{yz} = 0 \ (R = \text{Ring radius})$$

Cone

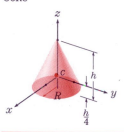

$$I_{xx} = I_{yy} = \frac{3}{80}m(4R^2 + h^2)$$

$$I_{zz} = \frac{3}{10}mR^2$$

$$I_{xy} = I_{xz} = I_{yz} = 0$$

$$\mathcal{V} = \frac{1}{3}\pi R^2 h$$

Rectangular prism

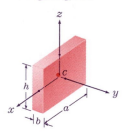

$$I_{xx} = \frac{1}{12}m(b^2 + h^2)$$

$$I_{yy} = \frac{1}{12}m(a^2 + h^2)$$

$$I_{zz} = \frac{1}{12}m(a^2 + b^2)$$

$$I_{xy} = I_{xz} = I_{yz} = 0$$

$$\mathcal{V} = abh$$

Rectangular plate

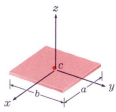

$$I_{xx} = \frac{1}{12}mb^2$$

$$I_{yy} = \frac{1}{12}ma^2$$

$$I_{zz} = \frac{1}{12}m(a^2 + b^2)$$

$$I_{xy} = I_{xz} = I_{yz} = 0$$

1. Fixed in inertial space, or

2. The center of mass of the rigid body.

Points corresponding to case 1 will be denoted by a subscript o. This case is the direct consequence of Eq. (6-11), which was used in the derivations above. Points corresponding to case 2 will be denoted by a subscript c. The subscripts o and c on the angular momentum vector and the inertia tensor denote the origins from which the body-coordinate frame emanates. By rewriting Eq. (6-11) about the body's center of mass, case 2 can be verified, as demonstrated in Problem 6-26.

In three-dimensional rigid body motion, the angular momentum and the angular velocity are generally not parallel, a fact that leads to much of the complexity of rigid body dynamics. As we shall see in Section 6-6, this situation simplifies considerably for rigid bodies undergoing plane motion. Although we shall restrict our attention primarily to rigid bodies undergoing plane motion, it is beneficial to investigate some of the properties of the three-dimensional inertia tensor. By doing so, we shall be better able to appreciate the character of the inertia tensor itself as well as the conditions under which a three-dimensional dynamics problem reduces to a planar or two-dimensional dynamics problem.

■ **Example 6-1:** In General, $\boldsymbol{\omega}$ and $\boldsymbol{H}$ are not in the Same Direction

Suppose a rigid body with its attached $oxyz$ body-coordinate frame is being driven at a constant angular velocity $\boldsymbol{\omega} = \omega_z \boldsymbol{k}$. For specificity, assume point o is fixed in inertial space, although such an assumption is not necessary. Such a body and the angular velocity vector are sketched in Figure 6-5. For an arbitrary body, according to Eqs. (6-32), the angular momentum vector is

$$\boldsymbol{H}_o = \begin{Bmatrix} H_x \\ H_y \\ H_z \end{Bmatrix}_o = I_{xz}\omega_z \mathbf{i} + I_{yz}\omega_z \mathbf{j} + I_{zz}\omega_z \mathbf{k} \tag{a}$$

which is also sketched in Figure 6-5. (For convenience only, we have assumed that both I_{xz} and I_{yz} are greater than zero.) The important conclusion resulting from Eq. (a) is that, in general, the angular velocity and the angular momentum are not in the same direction.

Furthermore, since the body is rotating at $\boldsymbol{\omega}$ and since the angular momentum vector is expressed in terms of the body-coordinate frame, $\boldsymbol{H}_o$ is also rotating with respect to an inertial space. According to Eq. (6-9), if $\boldsymbol{H}_o$ is changing with respect to inertial space, there must be a net external torque applied to the body. In many rotating machines, such a torque is undesirable as it must be delivered to the body by dynamic forces on the bearings and

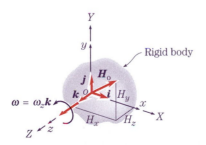

FIGURE 6-5   Angular momentum $\boldsymbol{H}_o$ and angular velocity $\boldsymbol{\omega}$ for arbitrary rigid body, with attached body-coordinate frame, are not in same direction.

the supporting structure. The forces and(or) moments corresponding to these time-varying torques may lead to undesirable stresses, noise, and structural fatigue.

■ **Example 6-2:**   **Dynamic Balancing**

Figure 6-6 shows a rotor that, for the discussion to follow, is modeled as a rigid body driven at ω about the z axis. We assume the $oxyz$ axes in Figure 6-6 to be a body-coordinate frame. As indicated in Example 6-1, a time-varying angular momentum vector requires a corresponding torque, which for the rotor sketched in Figure 6-6 must be delivered by forces on the bearings. The change in angular momentum is due to a change in its direction, not its magnitude, as it sweeps out a cone as suggested in Figure 6-6.

If the center of mass of the rotor does not lie on the axis of rotation, the linear momentum of the rotor will change constantly. This is due to the centripetal acceleration of the center of mass. If such a rotor is not spinning, in a gravitational field it will hang "heavy side down." This condition of the rotor, in which the center of mass does not lie along the axis of rotation, is called *static unbalance*. A statically unbalanced rotor can be statically balanced either by adding (hopefully) small counterbalancing weights to its periphery or by removing small amounts of material from the rotor itself in such a manner that its center of mass is brought onto its axis of rotation.

Even if a rotor is brought into static balance, in general, the angular momentum will not also be brought into alignment with the angular velocity. Thus, both time-varying angular momentum and the associated external torques are likely to persist, even for a statically balanced rotor. If the angular momentum is brought into alignment with the angular velocity, then as the rotor spins about its rotation axis (the z axis in Figure 6-6), the angular momentum vector will remain constant—in both magnitude and direction—in the z direction, implying zero net external torque. Such a rotor is said to be *dynamically balanced*.

Assuming static balance has been achieved, the additional requirement for achieving dynamic balance can be met via the following argument. If Eq. (a) in Example 6-1 is considered, it becomes clear that if the products of inertia associated with the z axis (I_{xz} and I_{yz}) are eliminated from the rotor, the angular momentum vector will be in the z direction as is the angular velocity vector.

By examining Eqs. (6-32), it is clear that analogous statements can be made regarding the x and y axes in an arbitrary body. That is, for example, if a body rotates about its y axis, the angular momentum will also be about the y axis if the associated products of inertia (I_{zy} and I_{xy}) vanish or are eliminated. Thus, we conclude that if the products of inertia associated with the axis of rotation are eliminated, a rotor will be dynamically balanced for rotation about that axis. (These concepts are explored further in Subsections 6-3.3 and 6-3.4.)

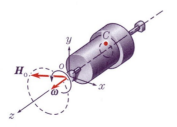

FIGURE 6-6   Unbalanced rotor.

■ **Example 6-3:** $[I]_c$ for Uniform Sphere

Given the uniform (that is, constant-density) rigid sphere sketched in Figure 6-7a, find the inertia tensor for a set of axes having its origin at the center of mass of the sphere (mass M and radius R). Such an inertia tensor is required in order to conduct a scientific investigation of the game of pool; sorry, we meant "pocket billiards."

❏ **Solution:**

The components of the inertia tensor are expressed in Eq. (6-37) and may be calculated using Eqs. (6-33) and (6-34). Since the sphere is a uniform continuous distribution of mass, the summations over a large number of discrete particles in Eqs. (6-33) and (6-34) may be replaced by integrals.

A differential volume element is identified in Figure 6-7a and is shown expanded in Figure 6-7b. From Figure 6-7a, the location of this element in cartesian coordinates is

$$x = r \cos \theta \sin \phi \qquad y = r \sin \theta \sin \phi \qquad z = r \cos \phi \tag{a}$$

and from Figure 6-7b, the differential volume dV is

$$dV = (r\,d\phi)(r \sin \phi\,d\theta)(dr) = r^2 \sin \phi\,d\phi\,d\theta\,dr \tag{b}$$

Further, the constant density ρ is

$$\rho = \frac{M}{\text{Volume}} = \frac{M}{\dfrac{4}{3}\pi R^3} \tag{c}$$

and the corresponding differential mass dm of the volume element is

$$dm = \rho\,dV \tag{d}$$

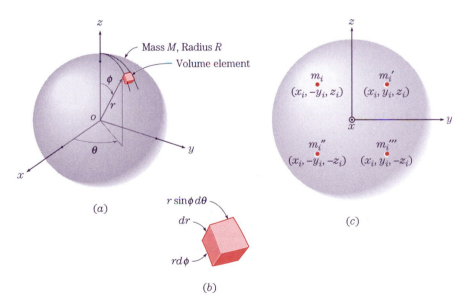

(a)

(b)

(c)

FIGURE 6-7 Uniform sphere showing central axes, volume element, and cross-section at arbitrary x.

Arbitrarily selecting I_{yy} as one of the moments of inertia for calculation gives

$$I_{yy} = \sum_{i=1}^{N} m_i(x_i^2 + z_i^2) = \int_M (x^2 + z^2)\, dm = \int_V (x^2 + z^2)\rho\, dV \tag{e}$$

where the second expression in Eqs. (6-33) has been used and rewritten in its integral form. Substitution of Eqs. (a) and (b) into Eq. (e) gives

$$I_{yy} = \int_{r=0}^{R} \int_{\theta=0}^{2\pi} \int_{\phi=0}^{\pi} \left[(r\cos\theta\sin\phi)^2 + (r\cos\phi)^2 \right] \rho r^2 \sin\phi\, d\phi\, d\theta\, dr \tag{f}$$

where the limits on r, θ, and ϕ have been indicated.

Integration of Eq. (f) over ϕ gives

$$I_{yy} = \int_0^R \int_0^{2\pi} \left[r^2 \cos^2\theta \left(\frac{4}{3}\right) + r^2 \left(\frac{2}{3}\right) \right] \rho r^2\, d\theta\, dr \tag{g}$$

where the elementary calculus evaluations

$$\int_0^\pi \sin^3\phi\, d\phi = \frac{4}{3} \qquad \int_0^\pi \cos^2\phi \sin\phi\, d\phi = \frac{2}{3} \tag{h}$$

have been used. Integration of Eq. (g) over θ gives

$$I_{yy} = \int_0^R \left[r^4 \left(\frac{4}{3}\right)(\pi) + r^4 \left(\frac{2}{3}\right)(2\pi) \right] \rho\, dr \tag{i}$$

where the elementary calculus evaluations

$$\int_0^{2\pi} \cos^2\theta\, d\theta = \pi \qquad \int_0^{2\pi} d\theta = 2\pi \tag{j}$$

have been used. Integration of Eq. (i) over r gives

$$I_{yy} = \frac{2}{5} MR^2 \tag{k}$$

where Eq. (c) has been used to express the result in terms of the given parameters, M and R. Symmetry arguments, which may be confirmed by formal integration, lead to

$$I_{xx} = I_{yy} \qquad \text{and} \qquad I_{zz} = I_{yy} \tag{l}$$

Integration of any of the products of inertia reveals the presence of integrals of the type

$$\int_0^{n\pi} \sin\alpha \cos\alpha\, d\alpha = 0 \qquad n = 1, 2, \ldots, p \tag{m}$$

where α is either ϕ or θ, n is either 1 or 2 here, and p is an arbitrary integer. Thus, in accordance with Eq. (m), all the products of inertia are zero:

$$I_{xy} = I_{yz} = I_{zx} = 0 \tag{n}$$

Note that the same results for the products of inertia can be reached via symmetry arguments in conjunction with Figure 6-7c, which is a cross-section of the sphere taken at an arbitrary value of x. At the arbitrary value of x, four differential elements are sketched in

Figure 6-7c. For example, in accordance with the constant density requirement, all differential elements of the same volume have the same mass; for the four elements shown in Figure 6-7c, these are

$$m_i = m_i' = m_i'' = m_i'''$$ (o)

Thus, referring to Eqs. (6-34), when computing products of inertia such as I_{xy}, the contributions of m_i and m_i' (as well as those of m_i'' and m_i''') cancel. For example, the contribution to I_{xy} due exclusively to m_i and m_i' is $I_{xy} = -[m_i(x_i)(-y_i) + m_i'(x_i)(y_i)] = 0$. Similarly, when computing products of inertia such as I_{xz}, the contributions of m_i and m_i'' (as well as those of m_i' and m_i''') cancel.

Therefore, the final result for the uniform sphere may be written as

$$[I]_c = \begin{bmatrix} \frac{2}{5}MR^2 & 0 & 0 \\ 0 & \frac{2}{5}MR^2 & 0 \\ 0 & 0 & \frac{2}{5}MR^2 \end{bmatrix} = \frac{2}{5}MR^2 \begin{bmatrix} 1 & 0 & 0 \\ 0 & 1 & 0 \\ 0 & 0 & 1 \end{bmatrix}$$ (p)

where Eqs. (k), (l), and (n) have been expressed in the form of Eq. (6-37), while also noting the symmetry of $[I]_c$ in accordance with Eqs. (6-34).

■ **Example 6-4:** $[I]_c$ for Uniform Cube

Find the inertia tensor for a uniform (that is, constant-density) rigid cube of mass M and edge length L about its center of mass for the central axes sketched in Figure 6-8.

❏ **Solution:**   The components of the inertia tensor are shown in Eq. (6-37) and may be calculated using Eqs. (6-33) and (6-34). Since the cube is a uniform continuous distribution of mass, the summations in Eqs. (6-33) and (6-34) may be replaced by integrals.

The differential mass of a volume element is the product of the constant density $\frac{M}{L^3}$ and the differential volume $dx\,dy\,dz$, and may be written as

$$dm = \frac{M}{L^3}dx\,dy\,dz$$ (a)

Arbitrarily selecting I_{zz} as one of the moments of inertia for calculation, we find

$$I_{zz} = \int_M (x^2 + y^2)dm = \int_{-\frac{L}{2}}^{\frac{L}{2}} \int_{-\frac{L}{2}}^{\frac{L}{2}} \int_{-\frac{L}{2}}^{\frac{L}{2}} (x^2 + y^2)\frac{M}{L^3}\,dx\,dy\,dz = \frac{1}{6}ML^2$$ (b)

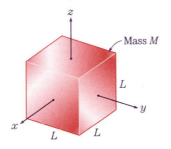

FIGURE 6-8   Uniform cube of mass M and edge length L, with central axes parallel to edges.

where Eqs. (6-33) and Eq. (a) have been used. I_{xx} and I_{yy}, which are both equal to I_{zz}, may be computed in a similar manner. Further, arbitrarily selecting I_{xz} as one of the products of inertia for calculation leads to

$$I_{xz} = -\int_M xz \, dm = -\int_{-\frac{L}{2}}^{\frac{L}{2}} \int_{-\frac{L}{2}}^{\frac{L}{2}} \int_{-\frac{L}{2}}^{\frac{L}{2}} xz \frac{M}{L^3} \, dx \, dy \, dz = 0 \tag{c}$$

where Eqs. (6-34) and Eq. (a) have been used. The remaining products of inertia, which are all equal to I_{xz}, may be computed in a similar manner. Therefore,

$$[I]_c = \frac{ML^2}{6} \begin{bmatrix} 1 & 0 & 0 \\ 0 & 1 & 0 \\ 0 & 0 & 1 \end{bmatrix} \tag{d}$$

The similarity between Eq. (p) in Example 6-3 and Eq. (d) is interesting. We shall return to this point in Examples 6-5 and 6-8.

6-3.2 Parallel-Axes Theorem

Given the inertia matrix with respect to one reference frame, we may want to determine the inertia matrix for the two cases, as follows:

1. A second reference frame, having a different origin from the first, but where the respective axes of the two reference frames are parallel; and

2. A second reference frame, having the same origin as the first, but rotated with respect to the first reference frame.

In this subsection, we examine the first of these two cases; in the next subsection, we consider the second. The results of this subsection will be useful for calculating the inertia matrix of the same body with respect to different reference frames having respective parallel axes, the inertia matrix of composite bodies, and the inertia matrix of bodies having cutouts.

We consider the case illustrated in Figure 6-9, where the rigid body has two attached reference frames, one with its origin at A and the other with its origin at B. It is worth noting that the origins of these reference frames do not have to be located within the body. The axes of the two reference frames are parallel such that x and x', y and y', and z and z' are parallel, respectively. As indicated in Figure 6-9, points A and B are displaced with respect to each other such that the transformations between their respective axes are

$$x' = x + a \qquad y' = y + b \qquad z' = z + c \tag{6-38}$$

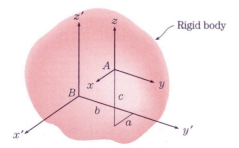

FIGURE 6-9   Two sets of reference frames with parallel axes and origins at A and B, attached to same rigid body.

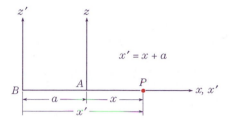

FIGURE 6-10 Location of arbitrary point P in Axyz is simply x, while location of P in B$x'y'z'$ is x', which is equal to $x + a$.

The validity of these expressions is easily seen by considering the one-dimensional case when point A is placed along any one of the primed axes, and an arbitrary point is located with respect to each reference frame. This is illustrated in Figure 6-10 for the x and x' axes.

So, here's the problem. Given the inertia tensor $[I]_A$ for the Axyz reference frame, we seek the inertia tensor $[I]_B$ for the B$x'y'z'$ reference frame where the subscripts A and B denote the origins of the respective frames. The resulting expression for accomplishing this task is called the *parallel-axes theorem*. As before, for convenience we continue to assume that the rigid body consists of N particles, where N is a very large number, although the results will be equally valid for continuous bodies by allowing the summations to become integrals. We provide the summation presentation rather than the integral presentation simply because the summation presentation appears to be slightly easier to visualize during an initial encounter.

In a step-by-step fashion, we shall calculate each of the matrix entries in Eq. (6-37) where the point o there is selected conveniently to correspond to the point B here. The entries of $[I]_B$ can be computed by substituting Eqs. (6-38) into each of the appropriate expressions in Eqs. (6-33) and (6-34). For example, we arbitrarily compute $I_{z'z'}$, the entry in the lower right-hand corner of $[I]_B$, using the third expression in Eqs. (6-33), and Eqs. (6-38).

$$
\begin{aligned}
I_{z'z'} &= \sum_{i=1}^{N} m_i ({x'_i}^2 + {y'_i}^2) \\
&= \sum_{i=1}^{N} m_i \left[(x_i + a)^2 + (y_i + b)^2 \right] \\
&= \sum_{i=1}^{N} m_i \left(x_i^2 + 2ax_i + a^2 + y_i^2 + 2by_i + b^2 \right) \\
&= \sum_{i=1}^{N} m_i \left(x_i^2 + y_i^2 \right) + \sum_{i=1}^{N} m_i \left(a^2 + b^2 \right) + \sum_{i=1}^{N} 2m_i (ax_i + by_i)
\end{aligned}
\tag{6-39}
$$

Recall from elementary mechanics that

$$
M = \sum_{i=1}^{N} m_i
\tag{6-40}
$$

and

$$
x_c = \frac{\sum_{i=1}^{N} m_i x_i}{M} \qquad y_c = \frac{\sum_{i=1}^{N} m_i y_i}{M} \qquad z_c = \frac{\sum_{i=1}^{N} m_i z_i}{M}
\tag{6-41}
$$

where Eq. (6-40) denotes that the total mass of the body M is the sum of the masses of all the individual particles, and the three expressions in Eqs. (6-41) are simply the x, y, and z coordinates of the center of mass of the body. It is important to note that x_c, y_c, and z_c are relative to the origin A of the xyz reference frame; this is because of the x_i, y_i, and z_i on the right-hand sides of the expressions in Eqs. (6-41). For centers of mass relative to the origin B, x_i', y_i', and z_i' must be used in Eqs. (6-41) instead of x_i, y_i, and z_i, respectively.

Note that the first summation in Eq. (6-39) is simply I_{zz}, which is given, namely, the entry in the lower right-hand corner of $[I]_A$. The second summation in Eq. (6-39) combined with Eq. (6-40) gives

$$\sum_{i=1}^{N} m_i(a^2 + b^2) = M(a^2 + b^2) \tag{6-42}$$

The third summation in Eq. (6-39) combined with Eqs. (6-41) gives

$$\sum_{i=1}^{N} 2m_i[ax_i + by_i] = 2M(ax_c + by_c) \tag{6-43}$$

Thus, substitution of these results (I_{zz} from $[I]_A$ and Eqs. (6-42) and (6-43)) into Eq. (6-39) gives

$$I_{z'z'} = I_{zz} + M(a^2 + b^2) + 2M(ax_c + by_c) \tag{6-44}$$

Similarly, we arbitrarily select the product of inertia $I_{x'y'}$ for computation, where use of Eqs. (6-34) and (6-38), and then Eqs. (6-40) and (6-41), gives

$$
\begin{aligned}
I_{x'y'} &= -\sum_{i=1}^{N} m_i x_i' y_i' \\
&= -\sum_{i=1}^{N} m_i(x_i y_i + ab + bx_i + ay_i) \\
&= I_{xy} - Mab - M(bx_c + ay_c) \tag{6-45}
\end{aligned}
$$

We may proceed in a similar manner to compute the remaining seven entries of the matrix $[I]_B$; actually, there are only four more independent entries because of the symmetry of the inertia tensor as depicted in Eq. (6-34). Upon completion of these computations, a general expression relating $[I]_B$ to $[I]_A$, M, and the given and calculated dimensions is obtained. Thus,

$$
\begin{bmatrix} I_{xx} & I_{xy} & I_{xz} \\ I_{yx} & I_{yy} & I_{yz} \\ I_{zx} & I_{zy} & I_{zz} \end{bmatrix}_B = \begin{bmatrix} I_{xx} & I_{xy} & I_{xz} \\ I_{yx} & I_{yy} & I_{yz} \\ I_{zx} & I_{zy} & I_{zz} \end{bmatrix}_A + M \begin{bmatrix} b^2 + c^2 & -ab & -ac \\ -ab & a^2 + c^2 & -bc \\ -ac & -bc & a^2 + b^2 \end{bmatrix}
$$
$$
+ M \begin{bmatrix} 2(by_c + cz_c) & -(bx_c + ay_c) & -(cx_c + az_c) \\ -(bx_c + ay_c) & 2(cz_c + ax_c) & -(cy_c + bz_c) \\ -(cx_c + az_c) & -(cy_c + bz_c) & 2(ax_c + by_c) \end{bmatrix} \tag{6-46}
$$

Equation (6-46) is the *general form of the parallel-axes theorem*. It applies to any pair of origins A and B, provided the respective axes are parallel as indicated in Figure 6-9.

If point A is at the center of mass of the rigid body, $x_c = y_c = z_c = 0$. (Refer to Eqs. (6-41).) In this case, the third matrix on the right-hand side of Eq. (6-46) vanishes, and Eq. (6-46) reduces to the simpler form

$$[I]_{\mathrm{B}} = [I]_{\mathrm{c}} + M \begin{bmatrix} b^2 + c^2 & -ab & -ac \\ -ab & c^2 + a^2 & -bc \\ -ac & -bc & a^2 + b^2 \end{bmatrix} \tag{6-47}$$

Equation (6-47) is the more commonly expressed form of the *parallel-axes theorem*. So, given the central inertia matrix $[I]_{\mathrm{c}}$ and the dimensions a, b, and c, $[I]_{\mathrm{B}}$ can be easily computed from Eq. (6-47).

■ **Example 6-5:** $[I]_{\mathrm{c}}$ and $[I]_{\mathrm{B}}$ for Cube with Hole

A uniform (that is, constant-density) cube, except for a centroidally located cubic hole, is sketched in Figure 6-11. The mass of a corresponding uniform cube of identical outer dimensions but without a hole is 27 m, where both cubes have the same edge lengths 3ℓ. The length of each edge of the cubic hole is ℓ. Find the inertia tensors for the two sets of parallel axes located at c, the center of the cube, and at B, the midlength of an edge, as sketched in Figure 6-11.

❑ **Solution:**   Since both the cube without the hole and the hole itself have the same centrally located axes $cxyz$, we may write

$$[I_1]_{\mathrm{c}} = [I]_{\mathrm{c}} + [I_2]_{\mathrm{c}} \tag{a}$$

where $[I_1]_{\mathrm{c}}$ is the inertia tensor of the cube without the hole, $[I]_{\mathrm{c}}$ is the inertia tensor of the cube with the hole, and $[I_2]_{\mathrm{c}}$ is the inertia tensor of a uniform cube of mass m and edge lengths ℓ. Note that Eq. (a) is, in reality, an identity that must be true for inertia tensors referenced to the same axes. For a given set of reference axes, the whole must equal the sum of the parts, since there is no means by which the whole could be otherwise. Equation (a) leads to

$$[I]_{\mathrm{c}} = [I_1]_{\mathrm{c}} - [I_2]_{\mathrm{c}} \tag{b}$$

which is less obvious but must be true via Eq. (a), and which is a *very useful* result for cutouts.

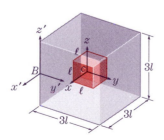

FIGURE 6-11   Cube of uniform density except for centrally located cubic hole.

From Example 6-4 we may write

$$[I_1]_c = \frac{(27m)(3\ell)^2}{6}\begin{bmatrix}1 & 0 & 0 \\ 0 & 1 & 0 \\ 0 & 0 & 1\end{bmatrix} = \frac{243m\ell^2}{6}\begin{bmatrix}1 & 0 & 0 \\ 0 & 1 & 0 \\ 0 & 0 & 1\end{bmatrix} \tag{c}$$

and

$$[I_2]_c = \frac{m\ell^2}{6}\begin{bmatrix}1 & 0 & 0 \\ 0 & 1 & 0 \\ 0 & 0 & 1\end{bmatrix} \tag{d}$$

Substitution of Eqs. (c) and (d) into Eq. (b) gives

$$[I]_c = \frac{242m\ell^2}{6}\begin{bmatrix}1 & 0 & 0 \\ 0 & 1 & 0 \\ 0 & 0 & 1\end{bmatrix} \tag{e}$$

which is within 0.5 percent of $[I_1]_c$. The reader should reflect upon the reasons for this small difference.

To calculate $[I]_B$ for the cube with the hole, we note that Eq. (6-47) is appropriate where for the axes sketched in Figure 6-11

$$a = -\frac{3\ell}{2} \qquad b = \frac{3\ell}{2} \qquad c = 0 \tag{f}$$

where reference to Figure 6-9 may be helpful. Thus, substitution of Eqs. (e) and (f) into Eq. (6-47) gives

$$[I]_B = \frac{242m\ell^2}{6}\begin{bmatrix}1 & 0 & 0 \\ 0 & 1 & 0 \\ 0 & 0 & 1\end{bmatrix} + 26m\begin{bmatrix}\left(\frac{3\ell}{2}\right)^2 & \left(\frac{3\ell}{2}\right)^2 & 0 \\ \left(\frac{3\ell}{2}\right)^2 & \left(\frac{3\ell}{2}\right)^2 & 0 \\ 0 & 0 & \left(\frac{3\ell}{2}\right)^2 + \left(\frac{3\ell}{2}\right)^2\end{bmatrix} \tag{g}$$

or

$$[I]_B = \frac{m\ell^2}{6}\begin{bmatrix}593 & 351 & 0 \\ 354 & 593 & 0 \\ 0 & 0 & 944\end{bmatrix} \tag{h}$$

Could the zeroes in Eq. (h) have been predicted? The answer to this question lies within Subsection 6-3.4.

❏ **Interesting Observation:**

The answer in Eq. (e) suggests a fascinating result. (Actually, the following discussion is more than a suggestion; it is true.) Note that the results in Eq. (p) in Example 6-3, Eq. (d) in Example 6-4, and Eq. (e) here can be written as

$$[I]_c = \text{Constant}\begin{bmatrix}1 & 0 & 0 \\ 0 & 1 & 0 \\ 0 & 0 & 1\end{bmatrix} \tag{i}$$

that is, a constant times the *identity matrix*. Equation (i) indicates that the central dynamics of a uniform cube, with or without a symmetric centroidal hole, and the central dynamics of a uniform sphere are essentially the same. By "central dynamics," in this context, we mean all calculations involving either of Eqs. (6-15) or (6-21). Thus, if the central dynamics of a cube were being monitored without visual access to the cube, those dynamics could not be distinguished from the corresponding dynamics of a sphere. Equations (6-15) and (6-35) support this observation. We shall revisit this issue in Example 6-8.

■ **Example 6-6:** Snowman Toy

The toy sketched in Figure 6-12 consists of three uniform disks (masses m_1, m_2, and m_3; radii r_1, r_2, and r_3; and constant thickness d; two identical slender rods of mass m_4 and length ℓ, each; and a massive particle m_5 as shown. (Note that a slender rod is the one-dimensional analogy of a particle.)

a. Find the center of mass of the toy in terms of the $OXYZ$ reference frame indicated in the sketch, where it may be assumed that the XY plane is located at the midthickness of the toy.

b. Find the I_{zz} component of the inertia tensor for the toy, about its center of mass. For this calculation, assume that the value of the parameters in part (a) are such that the center of mass is located at $3r_3/4$ above the contact point O.

❑ **Solution:**

Part (a)

The center of mass of the toy is located at (X_C, Y_C, Z_C), where X_C, Y_C, and Z_C are defined in Eqs. (6-41). Note that in terms of the $OXYZ$ reference frame sketched in Figure 6-12, due to symmetry,

$$X_C = 0 \quad \text{and} \quad Z_C = 0 \tag{a}$$

where the $Z_C = 0$ is due to the location of the XY plane at the midthickness of the toy. Also, from the second expression of Eqs. (6-41):

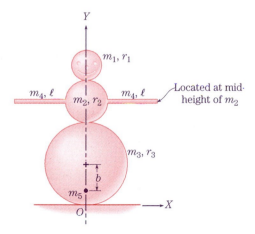

FIGURE 6-12 Snowman toy.

$$Y_C = \frac{\sum_{i=1}^{5} m_i y_i}{M}$$

$$= \frac{m_1(r_1 + 2r_2 + 2r_3) + (m_2 + 2m_4)(r_2 + 2r_3) + m_3(r_3) + m_5(r_3 - b)}{m_1 + m_2 + m_3 + 2m_4 + m_5} \tag{b}$$

Thus the center of mass of the toy, with respect to O, is given by the coordinates in Eqs. (a) and (b).

Part (b)

Equation (6-46) and Table 6-3 may be used to find the I_{zz} component of the inertia tensor of the toy about its center of mass. For this calculation, note that in Figure 6-9 as well as in Eq. (6-46) the origin B there corresponds to the toy's center of mass here; and the origin A there corresponds to the centers of mass of the five individual elements here. So, the third matrix on the right-hand side of Eq. (6-46) vanishes, reducing the required expression to Eq. (6-47). Furthermore, we are concerned only with the z component of the inertia tensor in Eq. (6-47). Thus, indicating the toy's center of mass by the subscript C_O and using Eq. (6-47), we may write

$$(I_{zz})_{C_O} = \sum_{i=1}^{5} \left[I_{ci} + m_i(a_i^2 + b_i^2) \right]$$

$$= \left[\frac{m_1 r_1^2}{2} + m_1 \left(r_1 + 2r_2 + \frac{5}{4} r_3 \right)^2 \right] + \left[\frac{m_2 r_2^2}{2} + m_2 \left(r_2 + \frac{5}{4} r_3 \right)^2 \right]$$

$$+ \left[\frac{m_3 r_3^2}{2} + m_3 \left(\frac{r_3}{4} \right)^2 \right] + 2 \left\{ \frac{m_4 \ell^2}{12} + m_4 \left[\left(r_2 + \frac{5}{4} r_3 \right)^2 + \left(r_2 + \frac{\ell}{2} \right)^2 \right] \right\}$$

$$+ m_5 \left(b - \frac{r_3}{4} \right)^2 \tag{c}$$

where Table 6-3 (circular plate for the disks and slender rods) has been used to provide the central moment of inertia components for the disks and rods. Note that except for the slender rods, all the a_i in Eq. (c) are zero.

6-3.3   Principal Directions and Principal Moments of Inertia

As stated in Subsections 6-3.1 and 6-3.2, for each reference frame originating at *every* point inside (as well as outside) of a rigid body, a corresponding inertia matrix exists. If at a given point, there is a continuous change in the orientation of the body-coordinate frame centered at that point, there will be a corresponding continuous change in the elements in the matrix. According to Eqs. (6-34) and (6-46), the inertia matrix will be symmetric for every body-coordinate frame centered at every point and in every orientation. Further, in most orientations, the matrix will contain nine (six independent) nonzero elements. While the moments of inertia cannot be negative, the products of inertia may be positive, zero, or negative.

In particular, a change in the orientation of the body-coordinate axes relative to the body will generally result in changes in the magnitudes as well as in the signs of the products of inertia. For example, assuming nonzero products of inertia, a rotation of the axes of π radians about the x axis will reverse the signs of I_{xy} and I_{xz}, leaving I_{yz} unchanged. This happens because such a rotation, relative to the body, reverses the directions of the positive y and z body-coordinate axes. Conversely, a rotation of $\frac{\pi}{2}$ radians about the x axis will reverse the sign of I_{yz}. (The visualization of these various sign reversals may require some hand-sketching by the reader; see Problem 6-21.) Because these sign reversals are the result of continuously varying values, the values of these products of inertia must pass through zero. Indeed, it is always possible to find a set of body-coordinate axes for which all the products of inertia are zero simultaneously, thus resulting in an inertia matrix which is *diagonal*, such as*

$$[I] = \begin{bmatrix} I_1 & 0 & 0 \\ 0 & I_2 & 0 \\ 0 & 0 & I_3 \end{bmatrix} \tag{6-48}$$

*Proof of this theorem is substantially due to the German Johann A. Segner (1704–1777) in 1755. Segner's paper proved to be very important to Euler's subsequent work in rigid body dynamics.

The directions of the body-coordinate axes in this special orientation are defined as the *principal directions*, called 1, 2, and 3, and the associated I_1, I_2, and I_3 diagonal elements in Eq. (6-48) are called the *principal moments of inertia*. Such a diagonalized matrix exists for body-coordinate axes at every point inside (and outside) the body.

An important consequence of Eq. (6-48) is that when the body rotates about a principal direction, the angular momentum vector will also be in that same direction; that is, the angular momentum vector and the angular velocity vector will be parallel. For example, suppose the principal directions, with their associated unit vectors $\boldsymbol{u}_1, \boldsymbol{u}_2$ and $\boldsymbol{u}_3$, have been found, resulting in an inertia matrix corresponding to Eq. (6-48). If the angular velocity vector were $\boldsymbol{\omega} = \omega_1 \boldsymbol{u}_1$, the corresponding angular momentum vector according to Eq. (6-35) would be

$$\{H\} = \begin{bmatrix} I_1 & 0 & 0 \\ 0 & I_2 & 0 \\ 0 & 0 & I_3 \end{bmatrix} \begin{Bmatrix} \omega_1 \\ 0 \\ 0 \end{Bmatrix} = \begin{Bmatrix} I_1\omega_1 \\ 0 \\ 0 \end{Bmatrix} \tag{6-49}$$

or

$$\boldsymbol{H} = I_1\omega_1\boldsymbol{u}_1 \tag{6-50}$$

where Eq. (6-32), written for directions 1, 2, and 3, could also have been used. Similarly, if $\boldsymbol{\omega} = \omega_2\boldsymbol{u}_2$, then $\boldsymbol{H} = I_2\omega_2\boldsymbol{u}_2$; and if $\boldsymbol{\omega} = \omega_3\boldsymbol{u}_3$, then $\boldsymbol{H} = I_3\omega_3\boldsymbol{u}_3$. Thus, unlike the general case where the angular velocity and the angular momentum are not parallel (as shown in Example 6-1), when the body rotates about a principal direction, the angular momentum and the angular velocity are in the same direction. Referring to Example 6-2, we see that in order for a rotor to be dynamically balanced, its angular velocity must be along one of its principal directions.

The demonstration that, given an arbitrary inertia tensor, a set of principal directions and the associated inertia tensor can always be found reduces to a three-dimensional *eigenvalue problem.*[*] The terms "eigenvalues" and "eigenvectors" are generic designations, which, in this context, are the "principal moments of inertia" and the "principal directions," respectively. In engineering and many other fields, eigenvalue problems of this type are commonplace and the procedures for obtaining their solutions are straightforward. We choose to forgo a formal eigenvalue presentation because, as we shall see shortly, such an analysis of many bodies requires only a two-dimensional eigenvalue solution. From studies of the technical mechanics of solids, the reader is likely to be familiar with a two-dimensional eigenvalue solution. We shall review briefly such a solution next. Furthermore, we shall return to a more detailed discussion of the eigenvalue problem in the context of obtaining the solutions of linear(ized) equations of motion, a subject commonly called *vibration*.

---

[*]The development of eigenvalue analysis attracted the interests of numerous great mathematicians of the nineteenth and twentieth centuries. Perhaps most significant among these were the Russian Vladimir A. Steklov (1864–1926), the Swede Ivar Fredholm (1866–1927), and the German David Hilbert (1862–1943), to whom appears to go credit for the modern terms *eigenvalues* and *eigenfunctions* in 1904.

6-3.4 Uses of Mass Symmetry

Although the procedures for calculating the principal directions and the principal moments of inertia are straightforward, for most three-dimensional problems the numerical evaluation of the result by hand is burdensome. It is fortunate, however, that in most practical problems, the use of symmetry provides a means for reducing the three-dimensional problem to a two-dimensional problem, by inspection.

We consider the rigid body having a single plane of *mass symmetry* (not necessarily geometric symmetry) as sketched in Figure 6-13a. Note that the requirement of mass symmetry with respect to the symmetry plane admits potentially complicated geometric variations as well as mass variations along the coordinate axes. (It may be convenient, although certainly not necessary, to think of the body in Figure 6-13a as having constant density.) For convenience, we shall consider the rigid body to be a system of N uniformly distributed identical particles. In essence, for such a system of particles, the existence of the plane of mass symmetry, here the xy plane, suggests that for every particle having coordinates (x_i, y_i, z_i) there is a particle of equal mass having coordinates $(x_i, y_i, -z_i)$.

For the body sketched in Figures 6-13a and 6-13b, where the xy plane is the plane of mass symmetry,

$$I_{xz} = -\sum_{i=1}^{N} m_i x_i z_i = I_{zx} = 0 \tag{6-51}$$

and

$$I_{yz} = -\sum_{i=1}^{N} m_i y_i z_i = I_{zy} = 0 \tag{6-52}$$

In performing the summations in Eqs. (6-51) and (6-52), particles such as those labeled ① and ② must have the same mass, and thus represent canceling pairs in both I_{xz} and

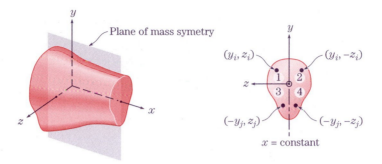

(a) Perspective view of body.

(b) Section through body at arbitrary value of x.

FIGURE 6-13   Rigid body having one plane of mass symmetry.

I_{yz}. (For example, note that ① and ② produce $I_{xz} = -[m_i(x_i)(z_i) + m_i(x_i)(-z_i)] = 0$.) Similarly, particles ③ and ④ represent canceling pairs. Extending this idea throughout the entire body over all the N particles leads to the results in Eqs. (6-51) and (6-52). (Recall that this idea was used in Example 6-3 to deduce the vanishing of the products of inertia of the uniform sphere.) Thus, the inertia matrix corresponding to Figure 6-13, in its most general form, becomes

$$[I] = \begin{bmatrix} I_{xx} & I_{xy} & 0 \\ I_{yx} & I_{yy} & 0 \\ 0 & 0 & I_{zz} \end{bmatrix} \tag{6-53}$$

Since, for the inertia tensor in Eq. (6-53), an angular velocity in the z direction produces an angular momentum that is also in the z direction, the z direction is a principal direction and I_{zz} is a principal moment of inertia. Furthermore, for any frame originating anywhere within the symmetry plane, the direction perpendicular to that plane continues to be a principal direction, and the moment of inertia associated with that direction is a principal moment of inertia. Given this new insight, could the zeroes in Eq. (h) of Example 6-5 have been predicted?

A scan of Figure 6-14 reveals that mass symmetry is quite common among dynamic systems. Thus, the analysis relating to Figure 6-13 is quite useful. For systems sketched in Figure 6-14, suppose the inertia matrix is known for a set of axes whose origin lies in the symmetry plane, with one direction perpendicular to that symmetry plane. Then, the analysis above shows that the eigenvalue analysis to determine the principal directions and principal moments of inertia would be immediately reduced from three dimensions to two dimensions. (Again, we caution that the requirement here is for mass symmetry, not the obviously observed geometric symmetry displayed in Figure 6-14.)

In solving the two-dimensional eigenvalue problem, both transformation equations and graphical techniques are widely used. Both of these techniques are summarized next; the transformation relations are explored more generally in Problem 6-89.

Suppose we wish to diagonalize the inertia matrix in Eq. (6-53), thus finding all the corresponding principal moments of inertia and principal directions. We may use Mohr's

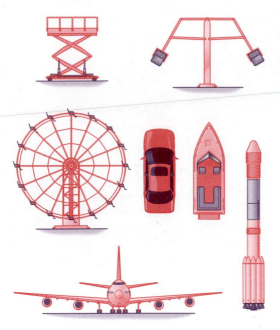

FIGURE 6-14 A small percentage of common examples of systems having at least one plane of (at least, approximate) mass symmetry.

circle* for representing the inertia components associated with the xy plane, namely I_{xx}, I_{xy}, and I_{yy}. Although the reader is likely to have encountered Mohr's circle during the study of such other tensors as stress and strain, we briefly review the procedures for determining the principal moments of inertia and the principal directions using this graphical technique.

---

*The concept of representing two-dimensional stress states by a circle was apparently invented by the German Karl Culmann (1821–1881) in 1866. The development and a more complete study of this simplifying graphical device were performed by the German Otto Mohr (1835–1918) in 1882, for whom this technique has been named.

Figure 6-15 is a graphical plot that is constructed for the xy plane, with the moments of inertia plotted along the abscissa labeled I_{ii} and the products of inertia plotted along the ordinate labeled I_{ij}. All plots must be in the right half-plane because I_{ii} must be nonnegative. After locating I_{xx} and I_{yy} along the abscissa, positive products of inertia are plotted downward along the x direction and upward along the y direction. Negative products of inertia are plotted upward along the x direction and downward along the y direction. Points labeled x and y in Figure 6-15 are thus located. The intersection of a straight line joining x and y with the abscissa locates the center C of the Mohr's circle. With C as the center and line Cx or Cy as a radius, Mohr's circle can be drawn. These sign conventions for plotting Mohr's circle preserve the sense of rotation between the physical plane and the Mohr's circle plane; so clockwise directions in the physical plane correspond to clockwise directions in the Mohr's circle plane, and vice versa.

For specificity, we have assumed that I_{xx} is greater than I_{yy}, which is greater than zero; and I_{xy} is greater than zero. Figure 6-15 gives the resulting Mohr's circle where the principal moments of inertia correspond to points 1 and 2, and are called I_{max} and I_{min}, respectively.

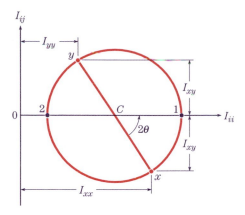

FIGURE 6-15   Mohr's circle for the xy plane, assuming $I_{xx} > I_{yy} > 0$ and $I_{xy} > 0$.

The angle between the x direction and the 1 direction in the physical plane is θ, in the same sense as the 2θ in the Mohr's circle plane. Permutation of the axes in this discussion provides the sign conventions for the yz plane and the zx plane.

Mohr's circle can provide physical insight into the manner in which the inertia components change as θ changes. Mohr's circle can be drawn to scale and physically measured to obtain numerical results graphically, or it can be sketched to serve as a mnemonic for obtaining analytical results. In this latter instance, from Figure 6-15, the specific—*not* general—relations for the principal moments of inertia can be written as

$$I_1, I_2 = \frac{1}{2}(I_{xx} + I_{yy}) \pm \sqrt{\frac{1}{4}(I_{xx} - I_{yy})^2 + I_{xy}^2} \tag{6-54}$$

where the plus sign of the $\pm$ symbol corresponds to I_1 and the minus sign corresponds to I_2, and

$$\tan 2\theta = \frac{I_{xy}}{\frac{1}{2}(I_{xx} - I_{yy})} \tag{6-55}$$

Note that on the right-hand side of Eq. (6-54), the first term is the location of the center C of the Mohr's circle and the second term is the radius of the Mohr's circle. Thus, the value of I_1 is the location of the center plus the radius, and the value of I_2 is the location of the center minus the radius.

■ **Example 6-7:**   **Airplane Inertia Properties**

An airplane sketched in Figure 6-16 is modeled as a rigid body that is mass-symmetric with respect to the yz plane. The central inertia tensor, with respect to the $cxyz$ body-coordinate frame shown, is

$$[I]_c = \begin{bmatrix} 6{,}000 & 0 & 0 \\ 0 & 10{,}000 & 1{,}000 \\ 0 & 1{,}000 & 4{,}000 \end{bmatrix} \text{ kg-m}^2 \tag{a}$$

Find the principal directions and the associated principal moments of inertia.

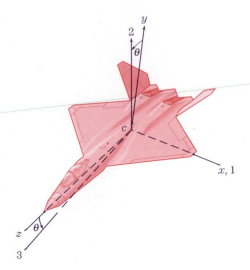

FIGURE 6-16   Airplane with body-coordinate frame and principal directions.

☐ **Solution:**   By symmetry, the x direction is a principal direction that can be concluded by either (i) the zero values of the I_{xy} and I_{xz} products of inertia in Eq. (a) or (ii) the fact that the x axis is perpendicular to the yz (mass) symmetry plane.

Using the analogous forms of Eqs. (6-54) and (6-55) for transformation in the yz plane, obtained by permutation of the axes, gives

$$I_2, I_3 = \frac{1}{2}(I_{yy} + I_{zz}) \pm \sqrt{\frac{1}{4}(I_{yy} - I_{zz})^2 + I_{yz}^2} \qquad \text{(b)}$$

and

$$\tan 2\theta = \frac{I_{yz}}{\frac{1}{2}(I_{yy} - I_{zz})} \qquad \text{(c)}$$

Substitution of the appropriate values of the moments of inertia and the products of inertia from Eq. (a) in the problem statement into Eqs. (b) and (c) gives

$$I_2 = 10{,}162 \text{ kg-m}^2 \qquad I_3 = 3{,}838 \text{ kg-m}^2 \qquad \theta = 9.22° = 9°13' \qquad \text{(d)}$$

Furthermore, $I_1 = I_{xx} = 6{,}000$ kg-m^2. The principal directions are also sketched in Figure 6-16.

■ **Example 6-8:**   Mohr's Circle for $[I]_c$ for Uniform Sphere and Cube

Sketch the Mohr's circle for the central inertia tensors of a uniform sphere and a uniform cube.

☐ **Solution:**   From Examples 6-3 and 6-4, the Mohr's circles for the central inertia tensors of a uniform sphere and a uniform cube are sketched in Figures 6-17a and 6-17b, respectively.

From Figures 6-17a and 6-17b, we observe that the Mohr's circles for the central axes of a uniform sphere and a uniform cube are both points, at some constant distance along

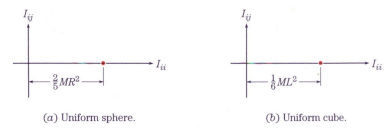

(*a*) Uniform sphere.   (*b*) Uniform cube.

FIGURE 6-17   Mohr's circles for central axes of uniform sphere and uniform cube. (I_{ii} denotes any moment of inertia and I_{ij} denotes any product of inertia.)

the I_{ii} axis. (Recall that comparable statements may be made about the uniform cube with a symmetric centroidally located hole.) Thus, irrespective of the orientation of the central axes, the Mohr's circles remain as points, implying that for all orientations, all the moments of inertia are equal and all the products of inertia are zero. While this is not surprising for a uniform sphere, it may be unexpected for a uniform cube. Thus, the remark in Example 6-5 that the central dynamics of a uniform cube or a uniform cube with a symmetric centroidally located hole are indistinguishable from the central dynamics of a uniform sphere is reinforced.

Indeed, this idea may be extended to include any object for which the central inertia tensor may be written as

$$[I]_c = C' \begin{bmatrix} 1 & 0 & 0 \\ 0 & 1 & 0 \\ 0 & 0 & 1 \end{bmatrix} \qquad \text{(a)}$$

where C' is an arbitrary constant. Such an arbitrary rigid body is indicated in Figure 6-18.

6-4   DYNAMICS OF RIGID BODIES VIA DIRECT APPROACH

Before proceeding to our formulations of the equations of motions via the indirect approach, we consider two examples of rigid body dynamics using the direct approach. These examples are only two of a potentially large set of examples and they are presented primarily for illustration; our major goal in this book is to explore analyses via the indirect approach.

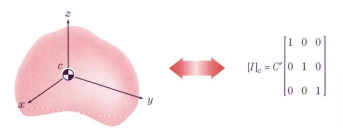

$$[I]_c = C' \begin{bmatrix} 1 & 0 & 0 \\ 0 & 1 & 0 \\ 0 & 0 & 1 \end{bmatrix}$$

FIGURE   6-18   Arbitrary rigid body having central dynamic properties that are indistinguishable from a uniform sphere.

Beyond these two examples, we shall return to the indirect approach and illustrate in some detail the methods of equation formulation associated with variational rigid body dynamics.

■ **Example 6-9:**   **Robot Forces and Torques**

We reconsider the robot sketched in Figure 6-19 and discussed in Example 3-10. We seek the forces and moments applied to the 1 m long work piece, which we assume can be modeled as a uniform slender rod, as the robot and its links move as specified in Example 3-10. (The reader may find it instructive to refer to the structure of the examples in Chapter 4.)

◻ **Solution:**

1. Kinematic requirements: The kinematics of the work piece for this analysis were computed in Example 3-10 and they are repeated as

$$\boldsymbol{a}_C = (-14.66\boldsymbol{I} + 2.951\boldsymbol{J})\left(\frac{\text{m}}{\text{s}}\right) \tag{a}$$

$$\dot{\boldsymbol{\omega}} = \dot{\boldsymbol{\omega}}_2 = 0.5\boldsymbol{K}\left(\frac{\text{rad}}{\text{s}^2}\right) \tag{b}$$

where $\boldsymbol{a}_C$ is the acceleration of the center of mass of the work piece and $\dot{\boldsymbol{\omega}}$ is the angular acceleration of the work piece.

2. Dynamic requirements on forces and torques: From Eq. (6-8) and Eqs. (6-15) or (6-21), the dynamic requirements on the forces and torques are represented by the Newton-Euler equations as

$$\boldsymbol{F} = \frac{d\boldsymbol{P}}{dt} \tag{c}$$

$$\boldsymbol{\tau}_C = \frac{d\boldsymbol{H}_C}{dt} \tag{d}$$

Equations (c) and (d) are also summarized in Table 6-2.

3. Constitutive requirements: From Eqs. (6-14) or (6-20) and the central form of Eq. (6-35), we write

$$\boldsymbol{P} = M\boldsymbol{v}_C \tag{e}$$

$$\{H\}_C = [I]_C\{\omega\} \tag{f}$$

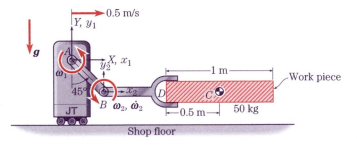

FIGURE 6-19   Robot manipulating work piece.

and, in accordance with Eq. (4-5), the gravitational force $\boldsymbol{F}_g$ is

$$\boldsymbol{F}_g = M\boldsymbol{g} \tag{g}$$

The gravitational resultant force $\boldsymbol{F}_g$, which acts at the center of the work piece, is a result from elementary statics, typically discussed under the heading of "resultants of distributed loads." (Also see Problem 6-28.)

Now we shall manipulate the equations contained within the three basic requirements above. We note that in accordance with the planar character of the problem,

$$F_Z = 0 \qquad \tau_X = 0 \qquad \tau_Y = 0 \tag{h}$$

Via formal calculations, the first expression in Eqs. (h) follows from Eq. (a), where $a_Z = 0$; and the second and third expressions in Eqs. (h) follow from Eq. (b), where $\dot{\omega}_X = \dot{\omega}_Y = 0$. Then, combining Eqs. (a) through (g) and noting the free-body diagrams in Figure 6-20 give

$$\sum F_X = ma_X : \qquad F_1 = M(-14.66) \tag{i}$$

$$\sum F_Y = ma_Y : \qquad F_2 - M_g = M(2.951) \tag{j}$$

$$\sum \tau_Z = I_{ZZ}\dot{\omega}_Z : \qquad \tau_Z - F_2\left(\frac{\ell}{2}\right) = I_{ZZ}(0.5) \tag{k}$$

where

$$M = 50 \text{ (kg)} \qquad g = 9.81 (\text{m/s}^2) \qquad \ell = 1 \text{ (m)} \qquad I_{ZZ} = \frac{M\ell^2}{12} = 4.167 \text{ (kg-m}^2\text{)} \tag{l}$$

The solution of Eqs. (i), (j), and (k), consistent with the numerical values in Eqs. (l), gives

$$F_1 = -733 \text{ (N)} \qquad F_2 = 638 \text{ (N)} \qquad \tau_Z = 321 \text{ (N-m)} \tag{m}$$

The solutions given by the relations in Eq. (m) provide the forces and moment that *must* be applied to the work piece in order to move it in the specified manner. If the work piece were very delicate, these forces and moment could be of interest because they might represent crushing or buckling loads. The reverse of these forces and moment represent reactions that are applied *to* the robot. Such forces might be useful in designing torque motors located at joints A and B, in assessing whether the robot could be tipped by these loads, or for calculating the stresses in the links AB and BD. Remember, the purpose of our analyses is some aspect of design; we are now in possession of the forces and moments to assess a range of design considerations.

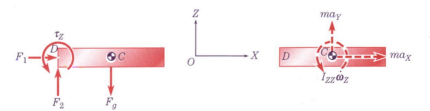

(a) Force and moment free body diagram.   (b) Kinetic diagram.

FIGURE 6-20   Free-body diagrams of work piece modeled as uniform slender rod.

■ **Example 6-10:** Gyroscopes—A Simplified Perspective

A very interesting and practical example of rigid body dynamics is the gyroscope. In a general sense, any rotating rigid body, symmetric or unsymmetric, that is suspended such that the orientation of its axis of rotation can be altered may be considered a gyroscope. In order to exhibit the most interesting and practical characteristics of gyroscopic motion, however, the gyroscope should possess a relatively large spin about an axis of symmetry. A thorough description of gyroscopic behavior requires considerable complexity; thus only a simplified physical presentation is offered here. Nevertheless, this simplified discussion should provide essential insight into the major dynamic gyroscopic phenomenon, which is described analytically by the relationship between torque and the *rate of change* of angular momentum. Rather than emphasizing the three basic requirements of equation formulation, we focus on a general discussion of the dynamic gyroscopic phenomenon.

A sketch of a simple gyroscope appears in Figure 6-21*a*, where the *xyz* rectangular frame is a *body-coordinate frame,* which, in this example, does not undergo the so-called *spin* **ω** of the gyroscope, but is otherwise constrained to move with the rotor. The basic element of the gyroscope is a symmetrical rotor, spinning with a large angular velocity **ω** about the *z* axis, resulting in a large angular momentum along the *z* axis; as may be seen below, the larger the angular velocity and the associated angular momentum, the larger the desired gyroscopic effects.

If the gyroscopic rotor is mounted such that it is free of external torques, then, in accordance with Eqs. (6-15) or (6-21), its angular momentum will remain constant. In this case, the *z* axis will retain a fixed direction in inertial space, regardless of the motion of the structure or vehicle in which the gyroscope is located. In this manner, the gyroscope may be used for *inertial guidance.* Because it is free of external torques, its axis of rotation will continue to point in any inertial direction in which it is initially set. Such a (near) torque-free environment can be accomplished by mounting the rotor in settings called *gimbal rings.* It is in the application of inertial guidance of aircraft and spacecraft that gyroscopes—though not in this torque-free configuration—may be ranked among the most important devices developed during the twentieth century. The next time you fly in an airplane through clouds and/or at night, perhaps the significance of the gyroscope will be sharpened in your mind; that is, assuming you want to arrive at the city whose name is printed on your ticket.

In most practical applications, the gyroscope is not torque-free, but held in bearings that are fixed in a vehicle. Then, when the orientation of the vehicle changes, the angular momentum vector of the attached gyroscope must also change, not in magnitude but in direction. According to the angular momentum principle, any change in angular momentum must be accompanied by a torque. If the vehicle in which the bearings are fixed turns such that the angular momentum rotates in the *xz* plane about the *y* axis, the change of this angular momentum will be in the *x* direction, shown incrementally as $\Delta \boldsymbol{H}_c$ in Figure 6-21*a*. According to the angular momentum principle, the torque produced by the rate of this change must also be in the *x* direction as sketched in Figure 6-21*a*. Such a torque gives rise to bearing forces that can be detected, measured, and thus related back to the *rate of angular displacement* (or the angular velocity) of the vehicle's motion. Configured in this application, the gyroscope is commonly called a *rate gyro:* The faster the imposed turning about the *y* axis, the larger the $\dfrac{d\boldsymbol{H}_c}{dt}$, and thus, the larger the resulting torque output.

Precision rate gyros are instruments that are capable of enormous sensitivity. The unit of precision of state-of-the-art gyroscopic devices is the meru (*milli earth rate unit*), or

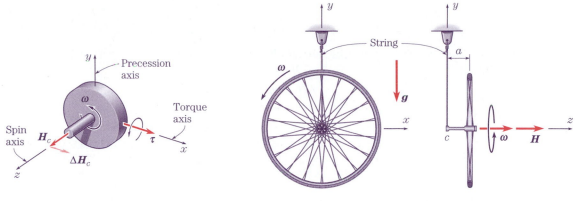

(a) Gyroscope with axes.

(b) Bicycle wheel supported by string.

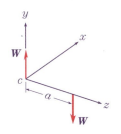

(c) Perspective of forces on bicycle wheel.

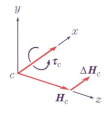

(d) Torque and angular momentum, showing τ and ΔH in same direction.

FIGURE 6-21 Gyroscope and bicycle wheel with related force, torque, and angular momentum diagrams.

10^{-3} of the Earth's rotation rate. For example, note that the rotation rate corresponding to 1 meru is $(360°/\text{day} \div 24 \text{ hr/day}) \cdot 10^{-3} = 0.015°/\text{hr}$. Commercial aircraft are piloted on instruments with a precision of the order of 10^{-2} meru; that is, a capability of detecting 1 revolution approximately every 300 years. Rate gyros used for lunar landings had a precision of the order of 10^{-4} meru; or equivalently, the capability of detecting rotation rates of 1 revolution approximately every 30,000 years! (Many navigational requirements for vehicles on or near the surface of the Earth are being met increasingly by the Global Positioning System (GPS), which is a system of orbiting satellites. Capable of being utilized in two distinct modes, the GPS's civilian mode precision is comparable to 10^{-2} meru and the GPS's military mode precision is comparable to 10^{-4} meru.)

A vivid demonstration of the nonintuitive phenomenon of the gyroscopic effect can be provided by a bicycle wheel such as sketched in Figure 6-21b. When the bicycle wheel is held with its axle horizontal, spun to produce the highest convenient $\boldsymbol{\omega}$, and then held only by the string as indicated in Figure 6-21b, the z-axis of the wheel does not fall due to gravity, but moves slowly in a horizontal plane. As illustrated in Figure 6-21c, the weight W acts downward; hence the force in the string must be W upward. So, although the net force on the wheel is zero, the net torque is not zero. That is,

$$\sum \boldsymbol{F} = W\boldsymbol{j} - W\boldsymbol{j} = 0 \tag{a}$$

and

$$\sum \boldsymbol{\tau}_c = \boldsymbol{a} \times \boldsymbol{W} = (a\boldsymbol{k}) \times (-W\boldsymbol{j}) = aW\boldsymbol{i} \tag{b}$$

In accordance with the angular momentum principle, during an infinitesimal increment of time Δt

$$\Delta \boldsymbol{H} = \boldsymbol{\tau}_c \, \Delta t = aW \, \Delta t\boldsymbol{i} \tag{c}$$

where Eq. (b) has been used in Eq. (c). Thus, $\Delta \boldsymbol{H}$ is in the same direction as $\boldsymbol{\tau}_c$ (see Figure 6-21d), indicating that the axle of the wheel rotates toward the x axis as it rotates about the string in a horizontal plane. (Note that reversing the direction of the spin $\boldsymbol{\omega}$ would result in a $\Delta \boldsymbol{H}$ in the negative x direction.) Thus, all the forces occur in the $\pm y$ directions while the resulting motions occur in the $\pm x$ directions: The applied forces are *orthogonal* to the resulting motion!

This rotation (that is, rate of angular displacement) about the string will be relatively slow when compared with $\boldsymbol{\omega}$. This much slower rotation about the y axis is called *precession*, a term that had its origin in astronomy to describe the 26,000 year precession of the Earth's equinoxes; a phenomenon noted by Hipparchus as early as ca. 150 B.C. and accurately calculated by Newton. (Also, see Figure 1-10.) Now, referring back to Figure 6-21a, we see the bases for the names associated with the x, y, and z axes; namely, the torque, precession and spin axes, respectively.

Finally, in addition to inertial guidance systems, toys, and demonstration models, gyroscopes are useful as stabilizing devices, especially on ships that may have one or more gyroscopes of *several tons* as part of the vessel's antiroll system. And, in addition to the potentially dangerous nonintuitive behavior of hand-held high-speed rotary tools, the gyroscopic effect is a very important consideration in the design of bearings for the shafts of many rotating elements and systems such as propellers, turbines, and engines in watercraft, aircraft, and landcraft.

6-5   LAGRANGIAN FOR RIGID BODIES

In order to be able to use the indirect formulation for systems containing rigid bodies, we need only calculate the lagrangian for a rigid body; otherwise, all the analyses in Chapter 5 remain in effect. For convenience, we continue to model the rigid bodies as systems of N particles. Again, the analogous relations for rigid bodies modeled as continua could be derived simply by substituting integrals for summations and then proceeding identically.

6-5.1   Kinetic Coenergy Function for Rigid Body

We consider the rigid body sketched in Figure 6-22, where the ith particle m_i is located by $\boldsymbol{r}_i$ with respect to a point o that is fixed in inertial space. Also, the center of mass c as well as a point b that has an arbitrary velocity are shown. $OXYZ$ is an inertial reference frame.

The kinetic coenergy of the rigid body is the sum of the kinetic coenergies of the individual particles, namely,

$$T^* = \sum_{i=1}^{N} \frac{1}{2} m_i \boldsymbol{v}_i \cdot \boldsymbol{v}_i \tag{6-56}$$

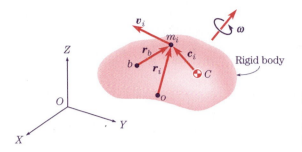

FIGURE 6-22 Rigid body undergoing angular velocity $\boldsymbol{\omega}$, where point c is center of mass, point o has zero velocity, and point b has arbitrary velocity.

First, we consider T^* in terms of the velocity of the particles with respect to the point o, which is fixed with respect to the inertial reference frame. In this case, the velocity of the ith particle is given by Eq. (6-22). Substitution of Eq. (6-22) into Eq. (6-56) gives

$$T^* = \frac{1}{2}\sum_{i=1}^{N} m_i(\boldsymbol{\omega} \times \boldsymbol{r}_i) \cdot (\boldsymbol{\omega} \times \boldsymbol{r}_i) \tag{6-57}$$

If $\boldsymbol{A}$, $\boldsymbol{B}$, and $\boldsymbol{C}$ are arbitrary vectors, recall that

$$\boldsymbol{A} \cdot (\boldsymbol{B} \times \boldsymbol{C}) = \boldsymbol{B} \cdot (\boldsymbol{C} \times \boldsymbol{A}) \tag{6-58}$$

Comparing Eqs. (6-57) and (6-58) and identifying $\boldsymbol{A}$ with the first $(\boldsymbol{\omega} \times \boldsymbol{r}_i)$, $\boldsymbol{B}$ with the second $\boldsymbol{\omega}$, and $\boldsymbol{C}$ with the second $\boldsymbol{r}_i$ lead to

$$T^* = \frac{1}{2}\boldsymbol{\omega} \cdot \sum_{i=1}^{N} m_i \boldsymbol{r}_i \times (\boldsymbol{\omega} \times \boldsymbol{r}_i) \tag{6-59}$$

The summations in Eqs. (6-59) and (6-23) are identical. Thus, Eq. (6-59) reduces to

$$T^* = \frac{1}{2}\boldsymbol{\omega} \cdot \boldsymbol{H}_o \tag{6-60}$$

By introducing the matrix notation in Eq. (6-35) into Eq. (6-60), the kinetic coenergy for a rigid body with respect to an inertial reference frame becomes

$$T^* = \frac{1}{2}\{\boldsymbol{\omega}\}^T[I]_o\{\boldsymbol{\omega}\} \tag{6-61}$$

The matrices in Eq. (6-61) are $\{\boldsymbol{\omega}\}^T$, a row matrix that is the transpose of $\{\boldsymbol{\omega}\}$; $[I]_o$, the inertia matrix of the rigid body with respect to a body-coordinate frame whose origin is the fixed point o; and $\{\boldsymbol{\omega}\}$, a column matrix of the components of the angular velocity of the body defined with respect to the inertial reference frame. The reader may recall that the transpose of a column matrix is a row matrix of the same elements. The reader should select the more familiar or convenient of the two mathematical forms of Eqs. (6-60) or (6-61); the two equations are physically identical.

An alternative form of Eqs. (6-60) and (6-61) can be obtained by expanding the velocity of the ith particle in terms of $\boldsymbol{v}_c$, the velocity of the center of mass. For this case

$$\boldsymbol{v}_i = \boldsymbol{v}_c + \boldsymbol{\omega} \times \mathbf{c}_i \tag{6-62}$$

where $\mathbf{c}_i$ is the position vector of the ith particle with respect to the center of mass. (The reader may think of Eq. (6-62) as being equivalent to Eq. (3-23) where $\dfrac{d\boldsymbol{R}_0}{dt}$, $\boldsymbol{v}_{\text{rel}}$, $\boldsymbol{\omega}$, and $\boldsymbol{r}$ in

Eq. (3–23) correspond to $\boldsymbol{v}_c$, zero, $\boldsymbol{\omega}$, and $\boldsymbol{c}_i$ in Eq. (6-62), respectively.) It can be shown[5] that substitution of Eq. (6-62) into Eq. (6-56) leads to

$$T^* = \frac{1}{2}M\boldsymbol{v}_c \cdot \boldsymbol{v}_c + \frac{1}{2}\boldsymbol{\omega} \cdot \boldsymbol{H}_c \tag{6-63}$$

Similar to Eq. (6-61), Eq. (6-63) can be expressed in matrix representation as

$$T^* = \frac{1}{2}M\{v_c\}^T\{v_c\} + \frac{1}{2}\{\omega\}^T[I]_c\{\omega\} \tag{6-64}$$

where $[I]_c$ is the inertia matrix of the rigid body with respect to a body-coordinate frame whose origin is the center of mass of the body; and $\{\omega\}$ and $\{\omega\}^T$ are the column and row matrices containing the components of the inertial angular velocity of the body.

Equations (6-63) and (6-64) are equivalent forms of a theorem that we state as follows:

- The kinetic coenergy of a system is equal to the sum of (i) the kinetic coenergy due to a (fictitious) particle having a mass equal to the total mass of the system and moving at the velocity of the system's center of mass and (ii) the kinetic coenergy due to the motion of the system relative to its center of mass.*

---

*This is a theorem substantially due to the German Johann S. Koenig (1712–1757), published in 1751.

Alternatively stated, Eqs. (6-63) and (6-64) express the concept that the kinetic coenergy is the sum of a translational coenergy (which is equivalent to the kinetic coenergy of a particle of mass M moving with velocity $\boldsymbol{v}_c$) plus a rotational kinetic coenergy (which is equivalent to the kinetic coenergy due to $\boldsymbol{\omega}$ only; that is, if $\boldsymbol{v}_c$ were zero). Again, since Eqs. (6-63) and (6-64) are physically identical, the reader should select the more familiar or convenient mathematical form.

Finally, it is also possible to derive still another form of T^* by expanding the velocity of the ith particle in terms of $\boldsymbol{v}_b$, the velocity of an arbitrary point fixed in the body, as required by Problem 6-42.

6-5.2 Potential Energy Function for Rigid Body

The expression for the calculation of the potential energy of a probe in a conservative force field given as Eq. (5–17) and used for particles in Chapter 5 continues to be valid for rigid bodies. Recall from Eq. (5–17) that

$$V = -\int_{\boldsymbol{R}_O}^{\boldsymbol{R}} \boldsymbol{f}(\boldsymbol{R}) \cdot d\boldsymbol{R} \tag{6-65}$$

where $\boldsymbol{f}(\boldsymbol{R})$ represents the spatially dependent force of the field. In particular, from Eq. (4-5) for a particle m_i in a gravitational field $\boldsymbol{g}$ near the Earth's surface,

$$\boldsymbol{f}_i = m_i\boldsymbol{g} \tag{6-66}$$

---

[5]See Problem 6-41.

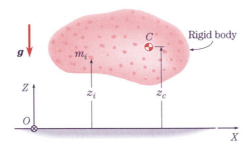

FIGURE 6-23   Rigid body modeled as system of particles in gravitational field.

If we consider the rigid body of total mass M to be modeled as a system of particles as sketched in Figure 6-23, then by use of Eqs. (6-65) and (6-66) where arbitrarily $\boldsymbol{R}_0 = \boldsymbol{z}_0 = 0$ (also refer to Example 5-2), the potential energy due to gravity may be written as

$$V = -\sum_{i=1}^{N} \boldsymbol{f}_i \cdot \boldsymbol{z}_i = -\sum_{i=1}^{N}(-m_i g \boldsymbol{k}) \cdot z_i \boldsymbol{k} = \sum_{i=1}^{N} m_i g z_i = M g z_c \qquad (6\text{-}67)$$

where, from Eq. (6-41), z_c is the location of the center of mass of the rigid body with respect to the indicated reference.

6-6   EQUATIONS OF MOTION FOR SYSTEMS CONTAINING RIGID BODIES IN PLANE MOTION

If the velocities of all the points of a rigid body remain parallel to a plane that is fixed in inertial space, then the body may be modeled as undergoing *plane motion*. Bodies modeled as thin slabs having their motion confined to the plane of the slab may be treated as undergoing plane motion. Further, bodies of appreciable dimensions normal to a plane of motion may also be treated as undergoing plane motion. In all cases, it is to be understood that the plane of motion contains the center of mass and that the velocities of all points of the body remain parallel to the plane of motion throughout the duration of the analysis; only the angular velocity and its derivatives may be normal to the plane of motion. Indeed, all angular velocities and their derivatives *must* be normal to the plane of motion. Many problems of practical interest may be idealized to the category of plane motion.

The fundamental concepts and analytical techniques for deriving the equations of motion via the indirect approach have been presented in Chapter 5. Restricting our attention to rigid bodies in plane motion, we now apply the same concepts and techniques, which are summarized in Table 5-1. Examples 6-12 through 6-17 are conducted using Lagrange's equations. The principles and procedures for each example are the same, yet the subtleties in their application are often enlightening. These examples have been carefully selected in an attempt to reveal those subtleties, specific details that appear repeatedly in the analysis and design of *many* dynamic systems. First, however, in Example 6-11 we derive a very useful simplification for the kinetic coenergy function for rigid bodies undergoing plane motion.

■ **Example 6-11:**   Simplification of T^* for Plane Motion

Assume the motion of a rigid body is confined to the xy plane. Find the corresponding most general forms for T^* resulting from Eqs. (6-60) or (6-61) and (6-63) or (6-64).

☐ **Solution:**   The equations for the kinetic coenergy appear to be somewhat difficult to evaluate, and indeed this is sometimes the case for fully three-dimensional motion. For plane motion, however, these equations simplify significantly. For example, when the motion of a rigid body is confined to the xy plane, then the only possibly nonzero component of the angular velocity is ω_z. So, for motion confined to the xy plane, at most,

$$\boldsymbol{\omega} = \omega_z \boldsymbol{k} \tag{a}$$

From Eq. (6-60)

$$T^* = \frac{1}{2} \boldsymbol{\omega} \cdot \boldsymbol{H}_o \tag{b}$$

where from Eq. (6-32)

$$\begin{aligned}
\boldsymbol{H}_o = {} & (I_{xx}\omega_x + I_{xy}\omega_y + I_{xz}\omega_z)\boldsymbol{i} \\
& + (I_{yx}\omega_x + I_{yy}\omega_y + I_{yz}\omega_z)\boldsymbol{j} \\
& + (I_{zx}\omega_x + I_{zy}\omega_y + I_{zz}\omega_z)\boldsymbol{k}
\end{aligned} \tag{c}$$

In accordance with Eq. (a), Eq. (c) reduces to

$$\boldsymbol{H}_o = I_{xz}\omega_z \boldsymbol{i} + I_{yz}\omega_z \boldsymbol{j} + I_{zz}\omega_z \boldsymbol{k} \tag{d}$$

Then, substitution of Eqs. (a) and (d) into Eq. (b) gives

$$T^* = \frac{1}{2}\omega_z \boldsymbol{k} \cdot (I_{xz}\omega_z \boldsymbol{i} + I_{yz}\omega_z \boldsymbol{j} + I_{zz}\omega_z \boldsymbol{k}) = \frac{1}{2}(I_{zz})_o \omega_z^2 \tag{e}$$

where the subscript "o" on I_{zz} is present to remind us that the relevant moment of inertia is about an inertially fixed point. Equation (e) is a *general* result for rigid bodies confined to the xy plane. Thus, Eq. (e) represents a substantial simplification of Eqs. (6-60) and (6-61) for planar rigid body motion.

Analogously, Eqs. (6-63) and (6-64) reduce to

$$T^* = \frac{1}{2}M\boldsymbol{v}_c \cdot \boldsymbol{v}_c + \frac{1}{2}(I_{zz})_c \omega_z^2 \tag{f}$$

where the subscript "c" on I_{zz} is present to remind us that the relevant moment of inertia is about the center of mass of the rigid body. Equations (e) and (f) are *very useful* reductions of their more general counterparts.

Equations (e) and (f) in Example 6-11 are so frequently cited that they are summarized in Table 6-4. They have been written on the assumption that the angular velocity occurs in the z direction and the motion is confined to the xy plane. A simple change of subscripts on the moment of inertia and angular velocity provides the corresponding expressions for plane motion within any plane. Further, we have assigned these expressions equation numbers for citation purposes. Nevertheless, it is important to appreciate the more general equations from which they were extracted; namely, Eq. (6-68) is the reduced form of Eqs. (6-60) or (6-61), and Eq. (6-69) is the reduced form of Eqs. (6-63) or (6-64).

Therefore, we reemphasize that the regimen for the formulation of the equations of motion presented in Chapter 5 for systems of particles may be used now for systems containing

TABLE 6-4 T^* for Rigid Bodies Undergoing Plane Motion[†]

	Kinetic Coenergy Function	Equation Number
Fixed-Axis Rotation About o	$T^* = \dfrac{1}{2}(I_{zz})_o \omega_z^2$	(6-68)
General Plane Motion	$T^* = \dfrac{1}{2}M\boldsymbol{v}_c \cdot \boldsymbol{v}_c + \dfrac{1}{2}(I_{zz})_c \omega_z^2$	(6-69)

[†] Equations are written on the assumption that motion occurs in the xy plane.

rigid bodies. The only modification required for rigid bodies undergoing plane motion, which is the focus of our interest, is that T^* for rigid bodies should be calculated using either Eq. (6-68) or (6-69), and V for the gravitational potential energy should be calculated using Eq. (6-67).

■ **Example 6-12:** Pulley and Strap System

Figure 6-24 shows a pulley of mass m_o that is modeled as a disk, supported by an axle (which itself has bearings but is not shown) through its center. An ideal (that is, massless) inextensible strap over the disk supports a mass m at one of its ends and at the other end is restrained by an elastic element that is characterized by a linear spring of spring constant k. The disk does not slip relative to the strap. Derive the equation(s) of motion for the system for two different generalized coordinates, namely, x and x_1 where

a. x is the downward displacement of m from the unstretched length of the spring; and

b. x_1 is the downward displacement of m from its equilibrium position (that is, from its assembled equilibrium position where the system rests under the action of gravity).

❑ **Solution:**

Part (a)

1. Generalized coordinate(s): The coordinate x sketched in Figure 6-24 constitutes a complete and independent set of generalized coordinates. In accordance with the problem

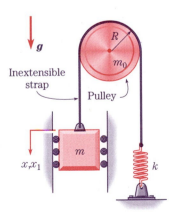

FIGURE 6-24 Pulley mass spring system.

statement in part (a), x is the (inertial) displacement of the mass m from the unstretched length of the spring, measured positive downward. So,

$$\xi_j : x \qquad \delta\xi_j : \delta x \tag{a}$$

The system is *holonomic;* thus we may proceed to use our lagrangian formulation.

2. Generalized force: There are no nonconservative forces acting in the system. So,

$$\delta\mathcal{W}^{nc} = \sum_{i=1}^{N} f_i^{nc} \cdot \delta R_i = \sum_{j=1}^{n} \Xi_j\, \delta\xi_j$$

$$= 0 = \Xi_x\, \delta x$$

Thus, the generalized force is

$$\Xi_x = 0 \tag{b}$$

As emphasized previously, even though $\delta\mathcal{W}^{nc}$ is zero, this portion of our solution is incomplete until we write Eq. (b) explicitly.

3. Lagrangian: There is a single rigid body, the pulley, which we model as a disk. In accordance with Eq. (5-10), the system kinetic coenergy function may be written as

$$T^* = T^*_{\text{mass}} + T^*_{\text{disk}} \tag{c}$$

where we may consider using either of Eqs. (6-68) or (6-69) for T^*_{disk}. Note that since $v_c = 0$ for the disk, the distinction between points c (center of mass) and o (inertially fixed point) vanishes; thus Eqs. (6-68) and (6-69) become the same expression. Then, in preparation of using Eq. (c), we may write

$$T^*_{\text{mass}} = \frac{1}{2}m\boldsymbol{v}\cdot\boldsymbol{v} = \frac{1}{2}m\dot{x}^2$$

$$T^*_{\text{disk}} = \frac{1}{2}(I_{zz})_c\omega^2 = \frac{1}{2}\left(\frac{m_0 R^2}{2}\right)\omega^2$$

where, in addition to the use of Eq. (5-9), the use of $(I_{zz})_c$ implies the motion is occurring in the (otherwise undefined) XY plane. Or

$$T^* = \frac{1}{2}m\dot{x}^2 + \frac{1}{2}\left(\frac{m_0 R^2}{2}\right)\omega^2 \tag{d}$$

where in this problem I_c and I_o for this disk are the same and Table 6-3 has been appropriately used. Note that Eq. (d) is not an appropriate final result for T^* because it is expressed in terms of kinematic variables that are not exclusively generalized coordinates or time derivatives of generalized coordinates. Thus, we should immediately use the kinematic requirement that

$$\omega = \dot{\theta} = \frac{\dot{x}}{R} \tag{e}$$

to convert Eq. (d) into

$$T^* = \frac{1}{2}m\dot{x}^2 + \frac{1}{2}\left(\frac{m_0 R^2}{2}\right)\left(\frac{\dot{x}}{R}\right)^2 \tag{f}$$

where, unlike Eq. (d), Eq. (f) is expressed kinematically exclusively in terms of the generalized velocity $\dot{x}$.

In accordance with Eq. (5-24), the system potential energy function may be written as

$$V = V_{\text{spring}} + V_{\text{mass}}$$

$$= \frac{1}{2}kx^2 - mgx \tag{g}$$

where Eq. (5-15) and Example 5-2 have been used. In writing Eq. (g), we have observed that the potential energy of the spring is zero when $x = 0$, in accordance with the definition of the generalized coordinate. Also, the gravitational potential energy of the mass m has been set to zero when $x = 0$. This is an arbitrary choice; and, of course, we could add a constant to Eq. (g), representing the gravitational potential energy at $x = 0$, with no effect on the resulting equation of motion.

Thus, the lagrangian may be written by substituting Eqs. (f) and (g) into

$$\mathcal{L} = T^* - V \tag{h}$$

4. Lagrange's equation: Substitution of Eqs. (b), (f), (g) and (h) into

$$\frac{d}{dt}\left(\frac{\partial \mathcal{L}}{\partial \dot{x}}\right) - \frac{\partial \mathcal{L}}{\partial x} = \Xi_x \tag{i}$$

gives

$$\left(m + \frac{m_0}{2}\right)\ddot{x} + kx = mg \tag{j}$$

Part (b)

1. Generalized coordinate(s):   The coordinate x_1 sketched in Figure 6-24 constitutes a complete and independent set of generalized coordinates. In accordance with the problem statement in part (b), x_1 is the (inertial) displacement of the mass m from its equilibrium position, measured positive downward. So,

$$\xi_j : x_1 \qquad \delta\xi_j : \delta x_1 \tag{k}$$

As in part (a), the system is observed to be *holonomic*.

2. Generalized force: There are no nonconservative forces acting in the system. So, as in part (a),

$$\delta W^{\text{nc}} = \sum_{i=1}^{N} f_i^{\text{nc}} \cdot \delta R_i = \sum_{j=1}^{n} \Xi_j \, \delta\xi_j$$

$$= 0 = \Xi_{x_1} \delta x_1$$

Thus, the generalized force is

$$\Xi_{x_1} = 0 \tag{l}$$

3. Lagrangian: Following the analogous discussion as in part (a) leads to

$$T^* = \frac{1}{2}m\dot{x}_1^2 + \frac{1}{2}\left(\frac{m_0 R^2}{2}\right)\left(\frac{\dot{x}_1}{R}\right)^2 \tag{m}$$

The calculation of the potential energy function leads to an apparently different result from that in part (a). Here

$$V = V_{\text{spring}} + V_{\text{mass}} = \frac{1}{2}k(x_1 + e)^2 - mgx_1 \qquad (n)$$

where e, the initial elongation in the spring when m is in its equilibrium position, can be determined from the equilibrium condition

$$mg = ke \qquad (o)$$

which gives

$$e = \frac{mg}{k} \qquad (p)$$

Thus, the lagrangian may be written by substituting Eqs. (m) and (n) into

$$\mathcal{L} = T^* - V \qquad (q)$$

4. Lagrange's equation: Substitution of Eqs. (l), (m), (n), and (q) into

$$\frac{d}{dt}\left(\frac{\partial \mathcal{L}}{\partial \dot{x}_1}\right) - \frac{\partial \mathcal{L}}{\partial x_1} = \Xi_{x_1} \qquad (r)$$

gives

$$\left(m + \frac{m_0}{2}\right)\ddot{x}_1 + kx_1 + ke - mg = 0 \qquad (s)$$

However, in accordance with Eq. (p), the last two terms on the left-hand side of Eq. (s) cancel, giving

$$\left(m + \frac{m_0}{2}\right)\ddot{x}_1 + kx_1 = 0 \qquad (t)$$

Several comments should be made in response to the likely question of "Why present two solutions to this relatively simple problem?" As indicated in Chapter 5 and as illustrated in Example 5-18, first of all, the choice of different generalized coordinates in general will lead to different equations of motion. Thus, we reemphasize the importance of being explicit in defining the meanings of generalized coordinates. Second, both formulations contain the same physical information, and for identical physical inputs to the system, the dynamic response of the system obtained from each formulation will be identical. This is suggested, though not proved, by a reexamination of the two equations of motion. Although in Chapter 8 we shall explore the explicit solution to equations such as Eqs. (j) and (t), we note that for $\ddot{x} = 0$ and $\ddot{x}_1 = 0$, the two equations give the same physical position of the system. That is, $x_1 = 0$ in Eq. (t) is physically identical to $x = \dfrac{mg}{k}$ in Eq. (j), as suggested by Eq. (p). Third, note that gravity does not appear in the equation of motion for part (b), that is, Eq. (t). Thus, we are led to conclude that if gravity enters the lagrangian in at most a linear fashion, it may be ignored if the system is restrained[6] *and* if the equations of motion are formulated for motion about the system's equilibrium configuration. Therefore, these three

[6]Refer to Example 5-21 for a system that is "unrestrained" in this context.

observations, which are *general*, although illustrated in the context of this simple example, provide the justification for presenting two solutions.

■ **Example 6-13: Disk Supported by Strap**

A pulley of mass m that is modeled as a disk is supported by an ideal (that is, massless) inextensible strap, which is connected to an elastic element having a spring constant k, as sketched in Figure 6-25a. The disk is assumed to move only vertically in the plane of the sketch. As it moves up and down, the disk is not allowed to slip relative to the strap. Derive the equation(s) of motion.

☐ **Solution:**

1. Generalized coordinate(s):   As we have repeatedly emphasized, the clear definition of the generalized coordinates is important in obtaining an unambiguous set of equations of motion. In this solution, we select the coordinate x, as sketched in Figure 6-25a, as a complete and independent set of generalized coordinates, where x is the vertical displacement of the center of mass of the disk from the unstretched length of the spring, measured positive downward. So,

$$\xi_j : x \qquad \delta\xi_j : \delta x \tag{a}$$

The system is *holonomic*; thus we may proceed to use our lagrangian formulation.

2. Generalized force: There are no nonconservative forces acting in the system. So,

$$\delta\mathcal{W}^{nc} = \sum_{i=1}^{N} \boldsymbol{f}_i^{nc} \cdot \delta\boldsymbol{R}_i = \sum_{j=1}^{n} \Xi_j \, \delta\xi_j$$
$$= 0 = \Xi_x \, \delta x$$

Thus, the generalized force is

$$\Xi_x = 0 \tag{b}$$

3. Lagrangian: One of the interesting aspects of this example is that the kinetic coenergy may be calculated using either Eq. (6-68) or (6-69). For comparison and illustration, we shall use both.

Initially, use of Eq. (6-69) may be easier to visualize. That is, the kinetic coenergy for the disk is

$$T^* = \frac{1}{2}M\boldsymbol{v}_c \cdot \boldsymbol{v}_c + \frac{1}{2}(I_{zz})_c \omega_z^2 \tag{c}$$

which reduces to

$$T^* = \frac{1}{2}m\dot{x}^2 + \frac{1}{2}(I_{zz})_c \omega_z^2$$
$$= \frac{1}{2}m\dot{x}^2 + \frac{1}{2}\left(\frac{mR^2}{2}\right)\left(\frac{\dot{x}}{R}\right)^2$$
$$= \frac{1}{2}\left(\frac{3m}{2}\right)\dot{x}^2 \tag{d}$$

where $(I_{zz})_c$ from Table 6-3 and the kinematic requirement

$$\omega = \dot{\theta} = \frac{\dot{x}}{R} \tag{e}$$

have been used.

Alternatively, we may use Eq. (6-68). First, in accordance with Figure 3-3, we note that the instantaneous point identified as A in Figure 6-25a—the tangent contact point of the disk at the strap—has zero velocity! This is an interesting, useful, and subtle concept. First, note that all points along the strap from the ceiling to point A' are fixed in inertial space, where A' is a point on the strap coincident with point A on the disk. Because the disk does not slip relative to the strap, the velocity of point A must be the same as the velocity of point A'; that is, zero. This is a major result of Example 3-1, where the point P on the perimeter was shown to have zero velocity whenever it was in contact with the stationary roadbed. Then, Eq. (6-68) gives

$$T^* = \frac{1}{2}(I_{zz})_o\omega_z^2 \tag{f}$$

where the fixed point o specifically denotes point A here.

Use of the parallel-axes theorem in Eq. (6-47) gives

$$(I_{zz})_o = (I_{zz})_c + mR^2 = \frac{mR^2}{2} + mR^2 = \frac{3mR^2}{2} \tag{g}$$

Based upon Example 6-11 and the discussions leading to Eqs. (6-68) and (6-69), Eq. (g) is clearly all that is required of the parallel-axes theorem in Eq. (6-47). Equation (g) indicates that only the lower right-hand element in each of the three matrices in Eq. (6-47) plays a role in this analysis. The term $(I_{zz})_c$ in the first matrix on the right-hand side of Eq. (6-47) is provided in Table 6-3, or could be calculated using Eq. (6-33). The term in the second matrix on the right-hand side of Eq. (6-47) requires a and b to be evaluated. In this analysis, for a second reference frame centered at point A', $a = 0$ and $b = R$. Such analyses requiring the use of the parallel-axes theorem as in Eq. (g) should become familiar and be quickly performed by the reader.

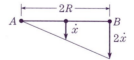

(b) Kinematics of horizontal diameter AB.

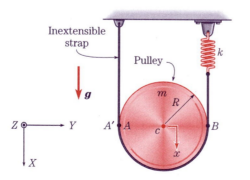

(a) System configuration.

FIGURE 6-25   Disk supported by ideal inextensible strap connected to spring element.

Substitution of Eqs. (e) and (g) into Eq. (f) gives

$$T^* = \frac{1}{2}\left(\frac{3m}{2}\right)\dot{x}^2 \tag{h}$$

which is the same as Eq. (d).

In accordance with Eq. (5–24), the system potential energy may be written as

$$V = V_{\text{spring}} + V_{\text{disk}}$$
$$= \frac{1}{2}k(2x)^2 - mgx \tag{i}$$

where the kinematic condition sketched in Figure 6-25b has been used. Note that because point A is instantaneously fixed, the kinematics of the diameter AB are as indicated in Figure 6-25b, where the integrated value of the velocity at point B is used in Eq. (i).

Thus, the lagrangian may be written by substituting Eqs. (d) or (h) and (i) into

$$\mathcal{L} = T^* - V \tag{j}$$

4. *Lagrange's equation:*   Substitution of Eqs. (b), (d) or (h), (i), and (j) into

$$\frac{d}{dt}\left(\frac{\partial \mathcal{L}}{\partial \dot{x}}\right) - \frac{\partial \mathcal{L}}{\partial x} = \Xi_x \tag{k}$$

gives

$$\frac{3}{2}m\ddot{x} + 4kx = mg \tag{l}$$

■ **Example 6-14:**   **Coupled Rotors**

Two thin cylindrical rotors (masses m_1 and m_2, and radii R_1 and R_2, respectively) are mounted on two identical shafts (length $2L$, shear modulus G and polar moment of the cross-sectional area about the shaft axis J). Refer to Figure 6-26. The radii and masses of the shafts are sufficiently small that their moments of inertia about their longitudinal axes (sometimes called their mass rotary inertias) may be neglected. Also, the two rotors are coupled by a linear spring, having a spring constant k, which is attached as shown. The right-hand rotor is driven by a harmonic torque $\tau(t)$ of known amplitude (τ_0) and frequency (Ω).

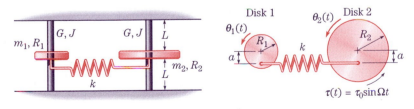

FIGURE 6-26   Coupled rotors excited by harmonic torque, shown in their equilibrium configurations.

Prove that for small angular displacements of the rotors, the equations of motion of the system are

$$
\begin{bmatrix} \dfrac{1}{2}m_1R_1^2 & 0 \\[2mm] 0 & \dfrac{1}{2}m_2R_2^2 \end{bmatrix} \begin{Bmatrix} \ddot{\theta}_1 \\[1mm] \ddot{\theta}_2 \end{Bmatrix} + \begin{bmatrix} \dfrac{2GJ}{L} + ka^2 & -ka^2 \\[2mm] -ka^2 & \dfrac{2GJ}{L} + ka^2 \end{bmatrix} \begin{Bmatrix} \theta_1 \\[1mm] \theta_2 \end{Bmatrix} = \begin{Bmatrix} 0 \\ 1 \end{Bmatrix} \tau_0 \sin \Omega t
$$

where, as shown in Figure 6-26, $\theta_1(t)$ and $\theta_2(t)$ are the counterclockwise angular displacements of the left-hand and right-hand rotors, respectively, both measured from the equilibrium configuration of the system.

☐ **Solution:**

1. Generalized coordinate(s): The coordinates θ_1 and θ_2 defined in the problem and sketched in Figure 6-26 constitute a complete and independent set of generalized coordinates. So,

$$
\xi_j : \theta_1, \theta_2 \qquad\qquad \delta\xi_j : \delta\theta_1, \delta\theta_2 \tag{a}
$$

The system is *holonomic*; thus we may proceed to use our lagrangian formulation.

2. Generalized forces: The variational work done by all the nonconservative forces during an admissible variation of the system is

$$
\delta\mathcal{W}^{\mathrm{nc}} = \sum_{i=1}^{N} \boldsymbol{f}_i^{\mathrm{nc}} \cdot \delta\boldsymbol{R}_i = \sum_{j=1}^{n} \Xi_j\, \delta\xi_j
$$

$$
= \tau(t)\,\delta\theta_2 = \Xi_{\theta_1}\,\delta\theta_1 + \Xi_{\theta_2}\,\delta\theta_2 \tag{b}
$$

where, as always, the notation $\boldsymbol{f}_i^{\mathrm{nc}}$ is interpreted to incorporate both nonconservative forces and torques. Thus, from Eq. (b), the generalized forces are

$$
\Xi_{\theta_1} = 0 \qquad\qquad \Xi_{\theta_2} = \tau(t) \tag{c}
$$

3. Lagrangian: Since the longitudinal axis of each of the rotors is fixed, we may use either of Eqs. (6-68) or (6-69) since they are equivalent for this analysis. We model the rotors as uniform disks and define $(I_{zz_1})_c$ and $(I_{zz_2})_c$ as the moments of inertia of the disks about the longitudinal axes of shafts 1 and 2, respectively. Then, according to Eq. (5–10), the kinetic coenergy function is

$$
T^* = T^*_{\mathrm{disk1}} + T^*_{\mathrm{disk2}}
$$

$$
= \frac{1}{2}(I_{zz_1})_c \omega_1^2 + \frac{1}{2}(I_{zz_2})_c \omega_2^2
$$

$$
= \frac{1}{2}\left(\frac{m_1R_1^2}{2}\right)\dot{\theta}_1^2 + \frac{1}{2}\left(\frac{m_2R_2^2}{2}\right)\dot{\theta}_2^2 \tag{d}
$$

where Table 6-3, $\omega_1 = \dot{\theta}_1$, and $\omega_2 = \dot{\theta}_2$ have been used.

Recall from the technical mechanics of solids that for a circular cylindrical linearly elastic shaft undergoing small torsional deformation as sketched in Figure 6-27, the constitutive relation between the *twist angle* ϕ and the *twisting moment* τ_t is

$$
\phi = \frac{\tau_t L}{GJ} \tag{e}
$$

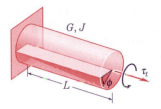

FIGURE 6-27   Circular cylindrical elastic shaft.

where L, G, and J are the length, shear modulus, and polar moment of the cross-sectional area about the shaft axis, respectively. Equation (e) may be used to define a corresponding torsional spring constant as

$$k_t = \frac{\tau_t}{\phi} = \frac{GJ}{L} \tag{f}$$

where the analogy with $k = \dfrac{F}{\delta}$ for a linear extensional spring has been observed.

Thus, from Eqs. (5-15) and (5-24), the potential energy function of the shafting and the coupling spring is

$$V = 2\left[\frac{1}{2}k_t\theta_1^2\right] + 2\left[\frac{1}{2}k_t\theta_2^2\right] + \frac{1}{2}ke^2 \tag{g}$$

where the first term accounts for the two (upper and lower) segments of shafting that act as torsional springs attached to disk 1, the second term accounts for the two segments of shafting that act as torsional springs attached to disk 2, and e is the extension in the linear coupling spring. For small angular displacements about the equilibrium configuration, as specified by the problem statement, the spring extension may be written as

$$e = a\theta_2 - a\theta_1 \tag{h}$$

Substitution of Eqs. (f) and (h) into Eq. (g) gives

$$V = 2\left[\frac{1}{2}\left(\frac{GJ}{L}\right)\theta_1^2\right] + 2\left[\frac{1}{2}\left(\frac{GJ}{L}\right)\theta_2^2\right] + \frac{1}{2}k(a\theta_2 - a\theta_1)^2 \tag{i}$$

The lagrangian may be written by substituting Eqs. (d) and (i) into

$$\mathcal{L} = T^* - V \tag{j}$$

4. Lagrange's equations: Substitution of Eqs. (a), (c), (d), (i) and (j) into

$$\frac{d}{dt}\left(\frac{\partial \mathcal{L}}{\partial \dot{\theta}_j}\right) - \frac{\partial \mathcal{L}}{\partial \theta_j} = \Xi_{\theta_j} \qquad j = 1, 2 \tag{k}$$

leads directly to

θ_1-equation: $\dfrac{m_1 R_1^2}{2}\ddot{\theta}_1 + 2\dfrac{GJ}{L}\theta_1 + ka^2(\theta_1 - \theta_2) = 0 \tag{l}$

and

θ_2-equation: $\dfrac{m_2 R_2^2}{2}\ddot{\theta}_2 + 2\dfrac{GJ}{L}\theta_2 + ka^2(\theta_2 - \theta_1) = \tau_0 \sin \Omega t \tag{m}$

Equations (l) and (m) may be rewritten in matrix form as given in the problem statement. Furthermore, as noted above, "small angular displacements" in the problem statement provided the basis for adopting both the linear constitutive model in Eq. (e) as well as the linearized spring extension model in Eq. (h). Note that the two rotors are truly coupled by the "coupling" spring k since it is only through this spring term that θ_2 appears in Eq. (l) and that θ_1 appears in Eq. (m).

■ **Example 6-15:** Mechanical Mechanism

A mechanism consists of a bar of mass m_0 that is pivoted about its upper end and restrained by a torsional spring of spring constant k_t, a spur pinion gear of mass m, and a viscous torsional dashpot of dashpot constant c_t, as sketched in Figure 6-28. The spring is connected between the bar and ground (or the stationary gear), and the dashpot is connected between the bar and the spur pinion gear. Assume that the spring is undeformed when the bar is vertical, and its constitutive relation is $\tau = k_t \theta_0$ where τ is torque and θ_0 is the angular displacement across its terminals. Derive the equation(s) of motion for the system. Do *not* assume small motions.

❑ **Solution:**

1. Generalized coordinate(s): The coordinates ϕ and θ shown in the expanded view in Figure 6-29 constitute a complete set of generalized coordinates where θ is the *inertial* angular displacement of the bar and ϕ is the *inertial* angular displacement of the pinion gear, both with respect to a *vertical line* locating their undisturbed configurations. It is very important to note that both ϕ and θ are measured from a *vertical* line that is *inertial*.

Referring to Figure 6-29, in position 1 or the undisturbed configuration, point D on the stationary gear and point A on the rolling gear are coincident. So the angle ϕ represents the inertial rotation of line CA. The presence of the gear teeth imposes the constraint that any change in θ requires a corresponding change in ϕ, and vice versa. This constraint may be derived by noting that

$$DB = AB \tag{a}$$

or

$$R\theta = r(\phi - \theta) \tag{b}$$

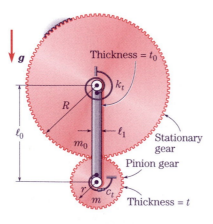

FIGURE 6-28   Two-gear mechanical mechanism.

Thus,

$$\phi = \left(\frac{R+r}{r}\right)\theta \tag{c}$$

which gives

$$\dot{\phi} = \left(\frac{R+r}{r}\right)\dot{\theta} \tag{d}$$

Equation (c) implies that

$$\delta\phi = \left(\frac{R+r}{r}\right)\delta\theta \tag{e}$$

The combination of Eqs. (c) and (e) serves to reduce the complete and independent set of generalized coordinates to either ϕ or θ, and the corresponding complete and independent set of admissible variations to either $\delta\phi$ or $\delta\theta$, respectively. Selecting θ as the choice of generalized coordinate leads to

$$\xi_j : \theta \qquad\qquad \delta\xi_j : \delta\theta \tag{f}$$

The system is *holonomic*; thus we may proceed to use our lagrangian formulation.

2. Generalized force: The calculation of the generalized force is one of several interesting aspects of this problem. The only contribution to the nonconservative variational work increment is due to the viscous torsional dashpot c_t. Note that the dashpot c_t is a "floating dashpot," so in accordance with our discussion in Example 5-15,

$$\delta\mathcal{W}^{\text{nc}} = \sum_{i=1}^{N} \boldsymbol{f}_i^{\text{nc}} \cdot \delta\boldsymbol{R}_i = \sum_{j=1}^{n} \Xi_j\,\delta\xi_j$$

$$= -c_t(\dot{\phi} - \dot{\theta})(\delta\phi - \delta\theta) = \sum_{j=1}^{n} \Xi_j\,\delta\xi_j \tag{g}$$

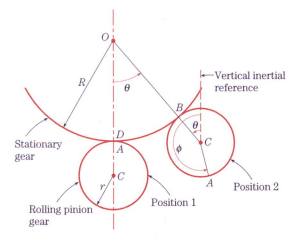

FIGURE 6-29 Geometric variables and parameters for two-gear mechanical mechanism.

where, as always, the notation $\boldsymbol{f}_i^{\mathrm{nc}}$ is interpreted to incorporate both nonconservative forces and torques. Further, Eq. (g) is consistent with the fact that the ends of the dashpot c_t are fixed to the bar and the pinion gear. Noting that $\delta\phi$ and $\delta\theta$ as well as $\dot{\phi}$ and $\dot{\theta}$ in Eq. (g) are not independent, substitution of Eqs. (d) and (e) into Eq. (g) gives

$$
\begin{aligned}
\delta W^{\mathrm{nc}} &= -c_t\left(\frac{R+r}{r}\dot{\theta} - \dot{\theta}\right)\left(\frac{R+r}{r}\delta\theta - \delta\theta\right) = \Xi_\theta\,\delta\theta \\
&= -c_t\left(\frac{R+r}{r} - 1\right)^2 \dot{\theta}\,\delta\theta = \Xi_\theta\,\delta\theta \\
&= -c_t\left(\frac{R}{r}\right)^2 \dot{\theta}\delta\theta = \Xi_\theta\delta\theta
\end{aligned}
\tag{h}
$$

Thus, from Eq. (h)

$$
\Xi_\theta = -c_t\left(\frac{R}{r}\right)^2 \dot{\theta}
\tag{i}
$$

which is a result made considerably more accessible via our concept of the "floating dashpot."

3. Lagrangian: In accordance with Eq. (5-10), the system kinetic coenergy function may be written as

$$
T^* = T^*_{\mathrm{gear}} + T^*_{\mathrm{bar}}
\tag{j}
$$

For the pinion gear, of the two forms for T^* in Table 6-4, Eq. (6-69) is easier to visualize although both forms are appropriate.

$$
\begin{aligned}
T^*_{\mathrm{gear}} &= \frac{1}{2}m\boldsymbol{v}_c \cdot \boldsymbol{v}_c + \frac{1}{2}(I_{zz})_c\omega^2_{\mathrm{gear}} \\
&= \frac{1}{2}m(\ell_0\dot{\theta})^2 + \frac{1}{2}\left(\frac{mr^2}{2}\right)\left(\frac{R+r}{r}\right)^2 \dot{\theta}^2
\end{aligned}
\tag{k}
$$

where the use of $(I_{zz})_c$ implies all motion is occurring in the (otherwise undefined) XY plane, and where Table 6-3 (circular disk or plate), and Eq. (d) have been used in conjunction with $\omega_{\mathrm{gear}} = \dot{\phi}$. Recognizing that $R + r = \ell_0$ enables Eq. (k) to be reduced to

$$
T^*_{\mathrm{gear}} = \frac{3}{4}m\ell_0^2\dot{\theta}^2
\tag{l}
$$

For the bar, either of Eqs. (6-68) or (6-69) may be conveniently used. Use of Eq. (6-68) gives

$$
T^*_{\mathrm{bar}} = \frac{1}{2}(I_{zz})_o\omega^2_{\mathrm{bar}} = \frac{1}{2}\left[\frac{m_0}{12}(\ell_0^2 + \ell_1^2) + m_0\left(\frac{\ell_0}{2}\right)^2\right]\dot{\theta}^2
\tag{m}
$$

where Table 6-3 (rectangular prism) and the parallel-axes theorem in Eq. (6-47) have been used for $(I_{zz})_o$, and $\omega_{\mathrm{bar}} = \dot{\theta}$ has been applied. Equation (m) reduces to

$$
T^*_{\mathrm{bar}} = \frac{1}{2}m_0\left(\frac{\ell_0^2}{3} + \frac{\ell_1^2}{12}\right)\dot{\theta}^2
\tag{n}
$$

Thus, substitution of Eqs. (l) and (n) into Eq. (j) gives

$$T^* = \frac{1}{2}\left(\frac{3}{2}m\ell_0^2 + \frac{1}{3}m_0\ell_0^2 + \frac{1}{12}m_0\ell_1^2\right)\dot{\theta}^2 \tag{o}$$

In accordance with Eq. (5-24), the system potential energy function may be written as

$$V = V_{\text{spring}} + V_{\text{gear}} + V_{\text{bar}}$$

$$= \frac{1}{2}k_t\theta^2 + mg\ell_0(1 - \cos\theta) + m_0 g\frac{\ell_0}{2}(1 - \cos\theta)$$

$$= \frac{1}{2}k_t\theta^2 + (m\ell_0 + \frac{1}{2}m_0\ell_0)g(1 - \cos\theta) \tag{p}$$

where an equation analogous to Eq. (5-15), but for a torsional spring, has been used, and the combination of Eq. (6-67) and Eq. (f) in Example 5-19 has been used for both the gear and the bar.

Thus, the lagrangian may be written by substituting Eqs. (o) and (p) into

$$\mathcal{L} = T^* - V \tag{q}$$

4. Lagrange's equation: Substitution of Eqs. (i), (o), (p) and (q) into

$$\frac{d}{dt}\left(\frac{\partial\mathcal{L}}{\partial\dot{\theta}}\right) - \frac{\partial\mathcal{L}}{\partial\theta} = \Xi_\theta \tag{r}$$

gives

$$\left(\frac{3}{2}m\ell_0^2 + \frac{1}{3}m_0\ell_0^2 + \frac{1}{12}m_0\ell_1^2\right)\ddot{\theta} + c_t\left(\frac{R}{r}\right)^2\dot{\theta} + k_t\theta + \left(m\ell_0 + \frac{1}{2}m_0\ell_0\right)g\sin\theta = 0 \tag{s}$$

■ **Example 6-16:** Disk with Internal Piston

The planar system sketched in Figure 6-30 consists of a circular disk of thickness d, radius R, and density ρ. The disk is free to rotate about a pivot at its inertially fixed center O. An $a \times 4a \times d$ rectangular slot is cut into the disk, and a rectangular solid piston, which is modeled as a block constrained by a spring of spring constant k, slides without friction in the slot. The piston has mass m and dimensions $a \times a \times t$ where t is its thickness in the Z direction. The spring is unstretched when $x = 0$; that is, the unstretched spring length is $\frac{3a}{2}$. Reference frame $OXYZ$ is inertial and reference frame $ox_1y_1z_1$ is attached to the disk. Derive the equation(s) of motion.

□ **Solution:**

1. Generalized coordinate(s): The coordinates θ and x sketched in Figure 6-30 constitute a complete and independent set of generalized coordinates, where θ is the angular displacement of the disk with respect to its indicated inertial reference and x is the displacement of the mass m along the slot with respect to the unstretched length of the spring. It is important to note that x is not inertial. So,

$$\xi_j : \theta, x \qquad \delta\xi_j : \delta\theta, \delta x \tag{a}$$

The system is *holonomic*; thus we may proceed to use our lagrangian formulation.

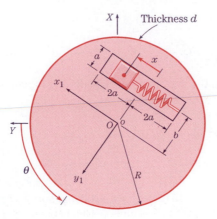

FIGURE 6-30   Plan view of rotating disk with internal piston modeled as block.

2. Generalized forces: There are no nonconservative forces acting in the system. So,

$$\delta W^{\text{nc}} = \sum_{i=1}^{N} \boldsymbol{f}_i^{\text{nc}} \cdot \delta \boldsymbol{R}_i = \sum_{j=1}^{N} \Xi_j \, \delta \xi_j$$

$$= 0 = \Xi_\theta \, \delta\theta + \Xi_x \, \delta x$$

Thus, the generalized forces are

$$\Xi_\theta = 0 \qquad\qquad \Xi_x = 0 \tag{b}$$

3. Lagrangian: In accordance with Eq. (5-10), the kinetic coenergy function for the system is

$$T^* = T_{\text{disk}}^* + T_{\text{block}}^* \tag{c}$$

Because point O of the disk is fixed in the inertial reference frame, Eq. (6-68) may be used to give

$$T_{\text{disk}}^* = \frac{1}{2}(I_{\text{disk}})_o \omega_{\text{disk}}^2 \tag{d}$$

where

$$\omega_{\text{disk}} = \dot{\theta} \tag{e}$$

Further, the moment of inertia in Eq. (d) is given by

$$(I_{\text{disk}})_o = (I_{zz} \text{ of circular disk about its center})$$
$$- (I_{zz} \text{ of rectangular slot of disk material about center of disk}) \tag{f}$$

where reference to Eq. (b) in Example 6-5 may be beneficial. Use of Table 6-3 to evaluate both terms on the right-hand side of Eq. (f), and the additional use of the parallel-axes theorem in Eq. (6-47) to evaluate the second term on the right-hand side of Eq. (f) give

$$(I_{\text{disk}})_o = \frac{MR^2}{2} - \left\{ \frac{m'}{12}[(4a)^2 + a^2] + m'b^2 \right\} \tag{g}$$

where

$$M = \rho \pi R^2 d \quad \text{and} \quad m' = \rho(4a)(a)(d) \tag{h}$$

That is, the second term (including its minus sign) on the right-hand side of each of Eqs. (f) and (g) is the moment of inertia of the slot with respect to the fixed point O, where the term $m'b^2$ is the contribution due to the parallel-axes theorem.

For the block (piston), Eq. (6-69) will be used, namely,

$$T^*_{block} = \frac{1}{2}m\boldsymbol{v}_c \cdot \boldsymbol{v}_c + \frac{1}{2}(I_{block})_c \omega^2_{block} \tag{i}$$

where

$$\omega_{block} = \dot{\theta} \tag{j}$$

since the block is constrained to rotate with the disk (refer to Figure 3-10) and where use of Table 6-3 gives

$$(I_{block})_c = \frac{m}{12}(a^2 + a^2) \tag{k}$$

Further, recall that $\boldsymbol{v}_c$ in Eq. (i) must be the inertial velocity of the center of mass of the block. Then, since x is defined with respect to the disk and is therefore noninertial, Eq. (3–23) may be used to write

$$\boldsymbol{v}_c = \frac{d\boldsymbol{R}_o}{dt} + \boldsymbol{v}_{rel} + \boldsymbol{\omega} \times \boldsymbol{r}$$

$$= 0 + \dot{x}\boldsymbol{i} + \dot{\theta}\boldsymbol{k} \times (x\boldsymbol{i} - b\boldsymbol{j})$$

$$= (\dot{x} + b\dot{\theta})\boldsymbol{i} + x\dot{\theta}\boldsymbol{j} \tag{l}$$

where Eq. (j) has been used, noting that $\dfrac{d\boldsymbol{R}_o}{dt} = 0$. Thus, Eq. (l) represents the inertial velocity of the block's center of mass, defined in $ox_1y_1z_1$ but with respect to $OXYZ$.

The potential energy function for the system is simply

$$V = \frac{1}{2}kx^2 \tag{m}$$

Then, the lagrangian may be written by substituting Eqs. (c) through (m) into

$$\mathcal{L} = T^* - V \tag{n}$$

4. *Lagrange's equations:*   Substitution of Eqs. (a) through (n) into

$$\frac{d}{dt}\left(\frac{\partial \mathcal{L}}{\partial \dot{\xi}_j}\right) - \frac{\partial \mathcal{L}}{\partial \xi_j} = \Xi_j \tag{o}$$

gives

θ-equation: $\dfrac{d}{dt}\left\{[(I_{disk})_o + (I_{block})_c]\dot{\theta} + m[(b^2 + x^2)\dot{\theta} + b\dot{x}]\right\} = 0 \tag{p}$

and

x-equation: $m(\ddot{x} + b\ddot{\theta} - x\dot{\theta}^2) + kx = 0 \tag{q}$

Equation (p) has been left in a slightly unusual form, providing us the opportunity to make an interesting, though for our goals not particularly important, observation. In general, the lagrangian is a function of all the generalized coordinates ξ_j ($j = 1, 2, \ldots, n$) and the associated generalized velocities $\dot{\xi}_j$ ($j = 1, 2, \ldots, n$). In some analyses of dynamics problems (1) a particular generalized coordinate ξ_k does not appear explicitly in the lagrangian *and* (2) the corresponding generalized force Ξ_k is also zero. When both of these conditions are satisfied, Eq. (o) reveals that the kth equation of motion integrates directly to

$$\frac{\partial \mathcal{L}}{\partial \dot{\xi}_k} = \Pi_k = \text{Constant} \tag{r}$$

where Π_k is called the kth *generalized momentum*. (Note that if $\dfrac{d}{dt}\{\Pi_k\} = 0$, then $\Pi_k =$ constant.) The significance of Eq. (r) is that it may be noted that Π_k is *conserved* in the analysis throughout the duration of the dynamics. The associated generalized coordinate ξ_k is called a *cyclic* or *ignorable* coordinate.[*]

---

[*]The terms *cyclic* or *ignorable* were introduced into dynamics by the German Hermann L.F. von Helmholtz (1821–1894) in 1884. Also, as indicated in Chapter 5, the term *generalized momentum* was invented by William Thomson and Peter G. Tait in 1867.

The θ coordinate in this example is a cyclic or ignorable coordinate and the associated generalized momentum Π_θ is conserved. Thus, from Eq. (p), we may write

$$\Pi_\theta = \left\{ [(I_{\text{disk}})_o + (I_{\text{block}})_c]\dot{\theta} + m[(b^2 + x^2)\dot{\theta} + b\dot{x}] \right\} = \text{Constant} \tag{s}$$

Equation (s) indicates that the angular momentum of the system in Figure 6-30 is conserved.

■ **Example 6-17:** Rocking Snowman Toy

The snowman toy sketched in Figure 6-31 has a total mass M and a central moment of inertia I_C in the Z direction. (See Example 6-6.) The circular radius of the bottom section is r as shown. Assume that the toy is constrained to rock in the plane of the figure and that there is no slip between the toy and the horizontal inertial surface. If the center of mass C is located at $3r/4$ above point O prior to rocking, derive the equation(s) of motion for the rocking snowman toy. Note that $OXYZ$ is an inertial reference frame.

❑ **Solution:**

1. Generalized coordinate(s): The toy in its undisplaced configuration is sketched in Figure 6-32a, where d is defined as the distance from A to C, and where A is the geometrical center of the bottom section as sketched in Figure 6-31. As a compound rigid body subject to a nonslipping constraint, the toy may be located by the single generalized coordinate θ, where θ is the angular displacement of the toy with respect to an inertial vertical line, as sketched in Figure 6-32b. So,

$$\xi_j : \theta \qquad \delta\xi_j : \delta\theta \tag{a}$$

The system is *holonomic*; thus we may proceed to use our lagrangian formulation.

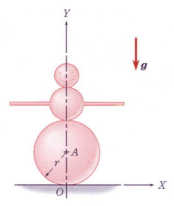

FIGURE 6-31   Rocking snowman toy.

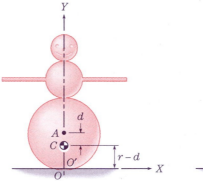

(*a*) Undisplaced configuration.

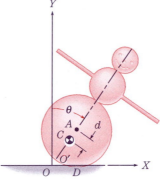

(*b*) Displaced configuration.

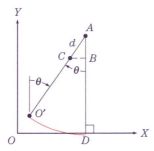

(*c*) Expanded view of displaced centerline $AO' = r$.

FIGURE 6-32   Geometric variables and parameters for rocking snowman toy.

2. Generalized force: There are no nonconservative forces acting in the system. So,

$$\delta W^{nc} = \sum_{i=1}^{N} \boldsymbol{f}_i^{nc} \cdot \delta \boldsymbol{R}_i = \sum_{j=1}^{n} \Xi_j \, \delta \xi_j$$

$$= 0 = \Xi_\theta \, \delta \theta$$

Thus, the generalized force is

$$\Xi_\theta = 0 \tag{b}$$

3. Lagrangian: The kinetic coenergy function can be calculated using either Eqs. (6-68) or (6-69). Equation (6-69) is likely to be easier to visualize initially.[7] So,

$$T^* = \frac{1}{2}M\boldsymbol{v}_c \cdot \boldsymbol{v}_c + \frac{1}{2}(I_{zz})_c \omega_z^2 \tag{c}$$

Unlike many of the previous problems, which were sufficiently simple that the velocity components could be written directly or with little effort from the problem statement, the inertial velocity $\boldsymbol{v}_c$ in this analysis is not so easily determined. In this example, a more elementary initiation point in the analysis is beneficial. In particular, in order to find the velocity of the center of mass of the toy, we shall begin by writing the *position* of the center of mass and then differentiate it with respect to time to obtain velocity. This approach is more general and indeed the reader should appreciate the fact that this approach has been used throughout the book in essentially *every* kinematic analysis, albeit implicitly in most simpler problems. This is a subtle yet important concept. Furthermore, this is just the approach that was emphasized and presented in detail in Example 3-1. Thus, in the analysis of mechanical systems, we have come full circle in our kinematic analyses!

In Figure 6-32b, point D is the instantaneous contact point between the toy and the inertial horizontal surface. In the initial position of the toy, point O' on the toy is coincident with the origin O of the inertial reference frame. An expanded view of the displaced centerline AO' is sketched in Figure 6-32c. From Figures 6-32b and 6-32c we note that

$$OD = O'D = r\theta \qquad AD = r \qquad AC = d \tag{d}$$

$$AB = d\cos\theta \qquad CB = d\sin\theta \tag{e}$$

So, the components of the position vector to point C, with respect to $OXYZ$, are

$$x_C = OD - CB = r\theta - d\sin\theta \tag{f}$$

$$y_C = AD - AB = r - d\cos\theta \tag{g}$$

where Eqs. (d) and (e) have been used. Then, by time differentiation of Eqs. (f) and (g), the components of the inertial velocity of point C are

$$\dot{x}_C = r\dot{\theta} - d\cos\theta\,\dot{\theta} \tag{h}$$

$$\dot{y}_C = d\sin\theta\,\dot{\theta} \tag{i}$$

where $(\dot{\square})$ has been used to denote $\dfrac{d\square}{dt}$. Further,

$$\boldsymbol{v}_c \cdot \boldsymbol{v}_c = \dot{x}_C^2 + \dot{y}_C^2 = (r^2 - 2dr\cos\theta + d^2)\dot{\theta}^2 \tag{j}$$

---

[7]See Problem 6-43 for an application of Eq. (6-68) in this regard.

where $\sin^2 \theta + \cos^2 \theta = 1$ has been used. Thus, substitution of Eq. (j) into Eq. (c) and note of the fact that $\omega_z = \dot{\theta}$ give the kinetic coenergy function as

$$T^* = \frac{1}{2} \left[M(r^2 - 2dr \cos \theta + d^2) + I_C \right] \dot{\theta}^2 \tag{k}$$

where $(I_{zz})_C$ in Eq. (c) has been set equal to I_C, which is given in the problem statement. The potential energy function is

$$V = Mgy_C = Mg(r - d \cos \theta) \tag{l}$$

where Eqs. (6-67) and (g) have been used.

Thus, the lagrangian may be written by substituting Eqs. (k) and (l) into

$$\mathcal{L} = T^* - V \tag{m}$$

4. Lagrange's equation: Substitution of Eqs. (b), (k), (l), and (m) into

$$\frac{d}{dt} \left(\frac{\partial \mathcal{L}}{\partial \dot{\theta}} \right) - \frac{\partial \mathcal{L}}{\partial \theta} = \Xi_\theta \tag{n}$$

gives

$$\left\{ I_C + Mr^2 \left(\frac{17}{16} - \frac{1}{2} \cos \theta \right) \right\} \ddot{\theta} + \frac{Mr^2}{4} \dot{\theta}^2 \sin \theta + \frac{Mgr}{4} \sin \theta = 0 \tag{o}$$

where d has been replaced by its assigned value of $\dfrac{r}{4}$.

As we shall discuss in Chapter 8 in considering *stability*, for small displacements from equilibrium, the toy represented by Eq. (o) will rock without tipping over. On the other hand, if the center of mass of the toy had been located above point A, the equation of motion would change in a significant manner and the dynamic behavior of the toy would also change greatly. In this second case, any disturbance of the toy would cause it to undergo a large tipping motion. (Indeed, the purpose of the mass m_5 sketched in Figure 6-12 is to deter such large "unstable" tipping by ensuring that the center of mass of the toy is below point A.)

■ **Example 6-18:** Rollercoaster—Use of Conservation of Energy

A twenty-car rollercoaster has a uniform velocity of 80 km/h on a horizontal section of its track, as shown in Figure 6-33. Neglect all losses due to wind resistance and friction, and note that gravity acts. The vertical dimension of the cars above the track may be neglected.

Find the velocity of the rollercoaster as its center car passes over the top of the circular track of radius 20 m. All twenty cars have identical mass and the total length of the rollercoaster is 30 m. Further, it may be assumed that the mass is uniformly distributed along the length of the rollercoaster.

❑ **Solution:** The generalized coordinate and admissible variation are

$$\xi_j : s \qquad \delta\xi_j : \delta s \tag{a}$$

where, as sketched in Figure 6-33, s is the distance traveled by the center car, with respect to an arbitrary inertial reference. Thus, the system has a single degree of freedom.

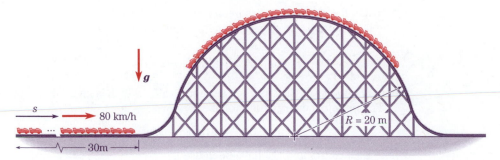

FIGURE 6-33   Rollercoaster running along horizontal and circular tracks.

Since the rollercoaster is a newtonian body, its kinetic energy is equal to its kinetic coenergy. Therefore,

$$T = T^* = \frac{1}{2}M\boldsymbol{v} \cdot \boldsymbol{v} = \frac{1}{2}M\dot{s}^2 \tag{b}$$

where M is the total mass of the rollercoaster. The potential energy of the system is due solely to gravity and may be expressed as

$$V = V_g = Mg\text{"}h\text{"} \tag{c}$$

where V_g is the potential energy due to gravity and where "h" is the height of the center of mass of the rollercoaster above an appropriate reference. Because there are no nonconservative forces in the system, the system is *conservative*.

The initial energy E_0 can be found, via Eq. (5-46) or Eq. (5-47), as

$$E_0 = E(0) = T^*(0) + V(0) = \frac{1}{2}M\dot{s}_0^2 + 0 = \frac{1}{2}M\dot{s}_0^2 \tag{d}$$

where Eq. (b) has been used, the reference for the potential energy has been chosen as the horizontal section of the track, and $\dot{s}_0$ is the velocity of the rollercoaster along the horizontal section of the track. Denoting the time at which the center car passes over the top of the circular section of the track as t_1, the kinetic coenergy at $t = t_1$ can be written as

$$T^*(t_1) = \frac{1}{2}M\dot{s}_1^2 \tag{e}$$

where $\dot{s}_1$ is the velocity of the rollercoaster at $t = t_1$, $\dot{s}_1$ being the quantity we seek. The potential energy at $t = t_1$ can be expressed as

$$V(t_1) = Mg\text{"}h_1\text{"} \tag{f}$$

where "h_1" is the height of the center of mass of the rollercoaster above the horizontal track at $t = t_1$.

Due to the assumption of uniform mass distribution, the mass per unit length of the rollercoaster can be defined as

$$\rho_0 = \frac{M}{l_0} \tag{g}$$

where l_0 is the total length of the rollercoaster. By using the definition in Eqs. (6-41) for the center of mass and Figure 6-34, "h_1" in Eq. (f) can be expressed as[8]

$$
"h_1" = \frac{1}{M} \int_{-\frac{l_0}{2}}^{\frac{l_0}{2}} y \rho_0 \, du = \int_{-\frac{l_0}{2}}^{\frac{l_0}{2}} \frac{\rho_0 r}{M} \cos \frac{u}{r} \, du = \frac{\rho_0 r^2}{M} \sin \frac{u}{r} \Big|_{-\frac{l_0}{2}}^{\frac{l_0}{2}}
$$

$$
= \frac{2\rho_0 r^2}{M} \sin \frac{l_0}{2r} = \frac{2r^2}{l_0} \sin \frac{l_0}{2r} \tag{h}
$$

where u is a dummy variable along the circular path as sketched in Figure 6-34; y is the vertical distance of the differential element du; and, Eq. (g) has been used in the last equality of Eq. (h). Thus, substituting Eq. (h) into Eq. (f) gives

$$
V(t_1) = \frac{2Mgr^2}{l_0} \sin \frac{l_0}{2r} \tag{i}
$$

Therefore, substituting Eqs. (d),(e), and (i) into Eq. (5-47) yields

$$
T^* + V = E_0 \tag{j}
$$

or

$$
\frac{1}{2}M\dot{s}_1^2 + \frac{2Mgr^2}{l_0} \sin \frac{l_0}{2r} = \frac{1}{2}M\dot{s}_0^2 \tag{k}
$$

Rearranging Eq. (k) and substituting the numerical values for the parameters yield

$$
\dot{s}_1^2 = \dot{s}_0^2 - \frac{4gr^2}{l_0} \sin \frac{l_0}{2r}
$$

$$
= \left(\frac{80 \ (\text{km/h}) \cdot 1000 \ (\text{m/km})}{3600 \ (\text{s/h})} \right)^2 - \frac{4 \cdot 9.81 \ (\text{m/s}^2) \cdot (20 \ (\text{m}))^2}{30 \ (\text{m})} \sin \frac{30 \ (\text{m})}{2 \cdot 20 \ (\text{m})}
$$

$$
= 137.2 \ (\text{m/s})^2 \tag{l}
$$

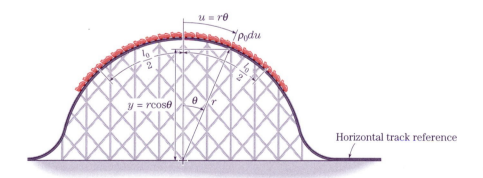

FIGURE 6-34   Rollercoaster modeled as uniform mass distribution and its potential energy.

[8]See Problem 6-65 for an alternative analysis that may be easier to visualize.

or

$$\dot{s}_1 = 11.7 \, \text{m/s} = 42.2 \, \text{km/h} \qquad \text{(m)}$$

which is the velocity of the rollercoaster as its center car passes through the top of the circular path, and which is independent of the mass of the rollercoaster.

As an additional elementary calculation or a reconsideration of Eq. (h) will show (see Problem 6-65), it is interesting to note that the center of mass of the rollercoaster never reaches the top of the 20 m circular section!

Problems for Chapter 6

Category I: Back to Physics

Problem 6-1 *(HRW250, 11P)*[9]: The Great Pyramid of Khufu at Giza, Egypt, had height $H = 147$ m before its topmost stone fell. Its base is a square with edge length $L = 230$ m (see Figure P6-1). Assuming that it had uniform density, find the original height of its center of mass above the base.

Problem 6-2 *(HRW250, 10P)*: Two pieces of sheet metal, each in the shape of a right triangle having height $H = 2.0$ cm and length $L = 3.5$ cm, are shown in Figure P6-2.

(a) Find the coordinates of the center of mass of the composite system.

(b) If each piece is reversed left-for-right so that the 2.0 cm sides are against each other, find the coordinates of the center of mass of the composite system.

Problem 6-3 *(HRW250, 12P):* A cylindrical can of uniform thickness, mass M, and height H is initially

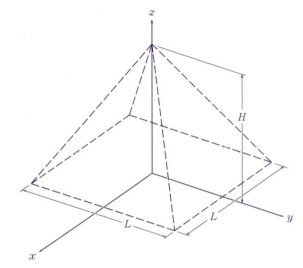

FIGURE P6-1

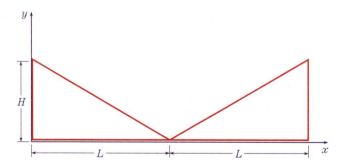

FIGURE P6-2

[9]These designations reference Halliday, Resnick, and Walker. For example, "HRW250, 11P" denotes "Halliday, Resnick, and Walker; page 250, problem 11P."

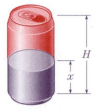

FIGURE P6-3

filled with pop of mass m, as shown in Figure P6-3. We punch small holes in the top and bottom to drain the pop. Consider the varying height h of the center of mass of the can and any pop within it.

    (a) What is h initially?

    (b) What is h when all the pop has drained?

    (c) How does h change during the draining of the pop?

    (d) If x is the height of the remaining pop at any given instant, find x in terms of M, H, and m when the center of mass reaches its lowest point.

Problem 6-4 *(6:HRW296, Fig11-12):* Two uniform plastic tubes are twisted back and forth rapidly about their midpoints. Inside the first tube are two fillers, each of mass m and length l, fixed at the ends of the tube (see Figure P6-4a). The second tube contains a larger filler, made by welding two fillers, which are the same as those in the first tube of Figure P6-4a, at the

midpoint of the tube (see Figure P6-4b). Which tube is easier to wiggle rotationally? Explain your answer.

Problem 6-5 *(HRW350, 60P):* A toy train track is mounted on a large wheel that is free to turn with negligible friction about a vertical axis, as shown in Figure P6-5. A toy train of mass m is placed on the track and, with the system initially at rest, the electrical power is turned on. The train reaches a steady state speed v_0 with respect to the track. What is the angular velocity ω_0 of the wheel, if its mass is M and its radius is R? Assume that the wheel may be treated as a hoop and that its spokes and hub are massless.

Problem 6-6 *(HRW351, 64P):* The particle of mass m in Figure P6-6 slides down the frictionless surface and collides with the uniform vertical rod, sticking to it. The rod pivots about the point O through an angle θ_0 before momentarily coming to rest. Find the angle θ_0 in terms of other parameters given in the figure.

Problem 6-7 *(HRW408, 66E, part1):* A pendulum consists of a uniform disk of radius 10.0 cm and mass 500 g attached to a uniform rod of length 50.0 cm and mass 270 g; see Figure P6-7.

    (a) Find the distance between the pivot and the center of mass of the pendulum.

    (b) Find the moment of inertia of the pendulum about the pivot.

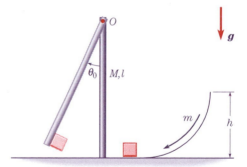

FIGURE P6-6

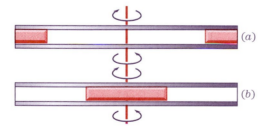

FIGURE P6-4

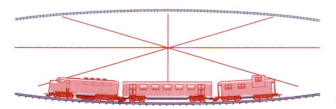

FIGURE P6-5

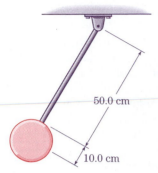

50.0 cm

10.0 cm

FIGURE P6-7

Problem 6-8 *(HRW317, 86P):* A uniform spherical shell of mass M and radius R rotates about a vertical axis on frictionless bearings, as shown in Figure P6-8. A massless cord passes around the equator of the shell, over a pulley of rotational inertia I and radius r, and is attached to a small object of mass m that is otherwise free to fall under the influence of gravity. There is no friction on the pulley's axle and the cord does not slip on the pulley. Find the speed of the object after it has fallen a distance h' from rest. *(Hint:* Use the work-energy approach.)

Category II: Intermediate

Inertia Tensor: Problems 6-9 through 6-25

Problem 6-9: Derive the moment of inertia about the longitudinal axis of a uniform cylinder of mass M, height h and radius R.

Problem 6-10: Derive the moment of inertia about a diameter of a uniform thin disk of mass M and radius R, for centroidally located axes.

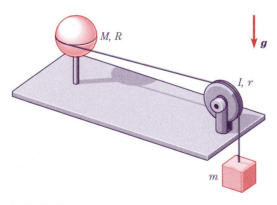

FIGURE P6-8

Problem 6-11: Find the inertia tensor of the uniform rectangular block of mass M sketched in Figure P6-11, with respect to the $oxyz$ axes shown.

Problem 6-12: The uniform octant of a sphere of radius R is sketched in Figure P6-12 with two sets of parallel axes $oxyz$ and $Cx'y'z'$. Derive the three moments of inertia of the octant of mass M with respect to the axes $Cx'y'z'$ through its center of mass, clearly indicating the location of the center of mass.

Problem 6-13: The revolving door sketched in Figure P6-13 consists of four rectangular sections, each of which can be modeled as a uniform rectangular plate of weight 70 lb, height 8 ft, width 4 ft. Using Table 6-3, find the moment of inertia of the revolving door with respect to its axis of rotation.

Problem 6-14: A cylinder of radius R and height h has a square through hole along its longitudinal axis, as sketched in Figure P6-14. The sides of the square hole are $\dfrac{R}{4}$ and the mass of the cylinder with the hole is M. Find the inertia tensor $[I]_C$ of the cylinder.

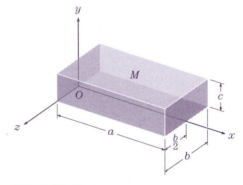

FIGURE P6-11

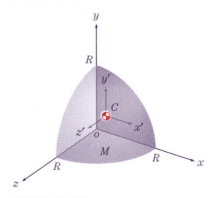

FIGURE P6-12

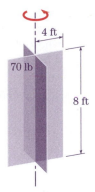

FIGURE P6-13

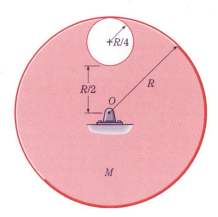

FIGURE P6-15

Problem 6-15: A uniform circular plate of radius R and mass M having a circular hole of radius $\dfrac{R}{4}$, as sketched in Figure P6-15, is pivoted at its center O.

(a) Find the location of the center of mass of the plate.

(b) Find the moment of inertia of the plate for the axis through O and perpendicular to the plane of sketch.

Problem 6-16: A uniform composite body consists of a cone (mass m, base radius R and height h) and a hemisphere (mass m_0 and radius R) as sketched in Figure P6-16. Find the inertia tensor $[I]_c$ of the composite body about its center of mass, clearly indicating the location of the composite body's center of mass.

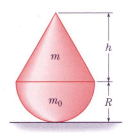

FIGURE P6-16

Problem 6-17: The circular cylindrical vessel of mass m_0 contains grain of mass $2m_0$ as sketched in Figure P6-17. The inertia tensor of the vessel with the grain about the point o located at its bottom edge is $[I]_o$ with respect to $oxyz$. Find the inertia tensor of the

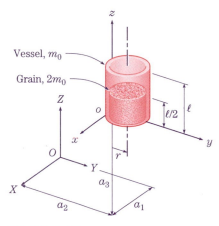

FIGURE P6-17

vessel with the grain with respect to the $OXYZ$ axes, which are respectively parallel to the $oxyz$ axes.

Problem 6-18: The composite pendulum sketched in Figure P6-18 is made of a massless rectangular frame of height l_2 and width $2l_1$. Two cubes of mass m with edges of length l_0 are attached to the upper corners of the frame. The cubes are positioned at an

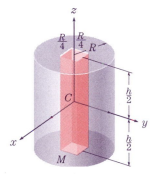

FIGURE P6-14

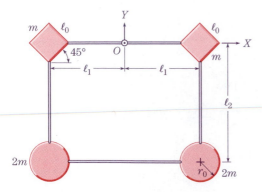

FIGURE P6-18

angle of 45° with respect to the frame, as shown. Two spheres of mass $2m$ and radius r_0 are attached to the lower corners of the frame. Find the moment of inertia I_{ZZ} of this pendulum, about its pivot point O.

Problem 6-19: The flywheel sketched in Figure P6-19 consists of a rim made of cast-iron (density $\rho_i = 13.7$ slug/ft^3), a hub made of aluminum (density $\rho_a = 5.28$ slug/ft^3), and six spokes made of copper (density $\rho_c = 17.4$ slug/ft^3). Using Table 6-3, find the moment of inertia of the composite flywheel with respect to its axis of rotation. Also, find the total mass of the composite flywheel.

Problem 6-20: The moment of inertia of a body about a specified axis is often expressed in handbooks in terms of a quantity called the *radius of gyration*. The radius of gyration has the unit of length and is defined as

$$\kappa = \sqrt{\frac{I}{m}}$$

where I is the moment of inertia of the body about the specified axis and m is the mass of the body. Note

that since the above definition yields $I = m\kappa^2$, the radius of gyration κ gives the radial distance from the specified axis of an equivalent particle whose mass and moment of inertia are identical to those of the body, respectively.

(a) Determine the radius of gyration of a sphere of radius R about its diameter.

(b) Determine the radius of gyration of the rectangular block in Problem 6-11 about the z axis.

(c) Determine the radius of gyration of the flywheel in Problem 6-19 about its axis of rotation.

Problem 6-21: The inertia tensor of the rigid body sketched in Figure P6-21 with respect to a body-coordinate frame $Axyz$ is given as

$$[I]_A = \begin{bmatrix} I_{xx} & I_{xy} & I_{xz} \\ I_{xy} & I_{yy} & I_{yz} \\ I_{xz} & I_{yz} & I_{zz} \end{bmatrix}$$

Find the inertia tensor of the body with respect to $Ax'y'z'$ in the following cases.

(a) When the axes $x'y'z'$ are rotated by π radians about the x axis.

(b) When the axes $x'y'z'$ are rotated by $\frac{\pi}{2}$ radians about the x axis.

(c) When the axes $x'y'z'$ are rotated by π radians about the y axis.

(d) When the axes $x'y'z'$ are rotated by $\frac{\pi}{2}$ radians about the y axis.

(e) When the axes $x'y'z'$ are rotated by π radians about the z axis.

(f) When the axes $x'y'z'$ are rotated by $\frac{\pi}{2}$ radians about the z axis.

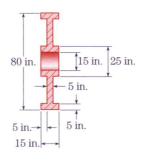

FIGURE P6-19

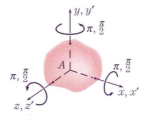

FIGURE P6-21

Problem 6-22: For the $oxyz$ axes shown, the inertia tensor of the uniform cone sketched in Figure P6-22 is given by

$$[I] = m \begin{bmatrix} \dfrac{12a^2 + 3h^2}{80} & 0 & 0 \\ 0 & \dfrac{12a^2 + 3h^2}{80} & 0 \\ 0 & 0 & \dfrac{3a^2}{10} \end{bmatrix}$$

where m is the mass of the cone, a is its base radius and h is its height.

(a) If $a = \dfrac{h}{2}$, find the inertia tensor in terms of a and m only.

(b) Sketch the Mohr's circle for the inertia tensor in part (a).

(c) Sketch the Mohr's circle for the $ox'y'z'$ axes where $ox'y'z'$ may be reached simply by rotating $oxyz$ about the x axis by $60°$, in a positive sense, as indicated in Figure P6-22.

(d) Write the inertia tensor for the $ox'y'z'$ axes.

Problem 6-23: Find the principal directions and the principal moments of inertia of the rectangular block in Problem 6-11, for the $oxyz$ axes shown. Assume $M = 10$ kg, $a = 30$ cm, $b = 10$ cm, and $c = 15$ cm.

Problem 6-24: The centroidal inertia tensor of the cube sketched in Figure P6-24 with respect to the $cxyz$ frame is

$$[I]_c = \frac{mL^2}{6} \begin{bmatrix} 1 & 0 & 0 \\ 0 & 1 & 0 \\ 0 & 0 & 1 \end{bmatrix}$$

A new reference frame $cx'y'z'$ is obtained by rotating $cxyz$ through a fixed angle $\Delta\theta$ about the vector cA where A is at a corner of the cube as shown. This ro-

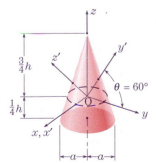

FIGURE P6-22

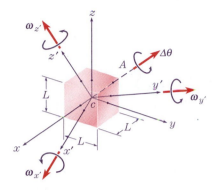

FIGURE P6-24

tation results in the angles α between x and x', β between y and y', and γ between z and z'. The motion of the cube is prescribed by the angular velocity vector $\{\omega\}^T = \{\omega_{x'}, \omega_{y'}, \omega_{z'}\}$, as sketched. Find the angular momentum vector $\boldsymbol{H}_c$. (It is probably most convenient to give your answer in terms of the $cx'y'z'$ reference frame.)

Problem 6-25: Briefly answer the following question: Can the central dynamics of each of the following bodies be qualitatively indistinguishable from that of a uniform sphere? If so, under what condition? If not, why?

(a) A uniform cylinder having mass m, radius R, and height h.

(b) A uniform slender rod having mass m and length l.

(c) A uniform cone having mass m, base radius R, and height h.

Direct approach: Problems 6-26 through 6-40

Problem 6-26: By rewriting Eq. (6-11) about the center of mass of a uniform rigid body and following the procedure in Subsection 6-3.1, show that

$$\{H\}_c = [I]_c\{\omega\}$$

where $\{H\}_c$ is the angular momentum vector about the body's center of mass, $[I]_c$ is the inertia tensor about the body's center of mass, and $\{\omega\}$ is the body's angular velocity vector.

Problem 6-27: As shown in Figure P6-27, a rectangular parallelepiped container of mass M is moving in a circular path in a circular torus in outer space. There

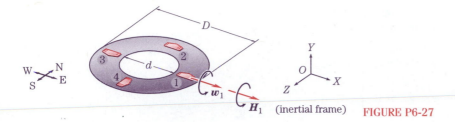

FIGURE P6-27

are three monkeys, each having mass m, in the container. The environment is torque free. The geometry of the container and the torus are given in Figure P6-27.

At point 1, it is known that the monkeys are asleep and that the angular momentum of the container (including the monkeys) is $\boldsymbol{H}_o = H_X\boldsymbol{I}$ (that is, due East).

(a) What do you suppose the monkeys were doing between points 1 and 2?

(b) What do you suppose the monkeys were doing between points 2 and 3?

(c) Sketch the angular momentum vector of the container-monkey system at points 2, 3, and 4.

Problem 6-28: Show that the gravitational resultant force $\boldsymbol{F}_g$ on an extended body is

$$\boldsymbol{F}_g = M\boldsymbol{g}$$

where M is the mass of the body and $\boldsymbol{g}$ is the gravitational field vector and that the force $\boldsymbol{F}_g$ acts at the center of mass of the body.

Problem 6-29: As sketched in Figure P6-29, a uniform slender rod of mass M and length L is suspended via two inextensible cables at locations A and B. Find the tension in the cable at B at the instant the other cable at A ruptures suddenly. Assume the rod is initially at rest and note that gravity acts on the system.

Problem 6-30: A body of mass M and centroidal moment of inertia I_c pivots about a fixed point O, as sketched in Figure P6-30. The body is initially at rest, with its center of mass located a distance a directly below O. The body is suddenly struck by a constant horizontal force $\boldsymbol{F}_o$ at a point A located vertically downward a distance d from point O, as indicated in Figure P6-30. Find the distance d such that there is no horizontal reaction force of the pin at O. Note that

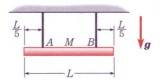

FIGURE P6-29

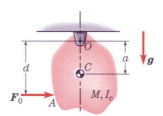

FIGURE P6-30

gravity acts on the system. (Such a point as A is called the *center of percussion*.)

Problem 6-31: The overhead monorail sketched in Figure P6-31 is driven by either of its two wheels, A or B. The center of mass of the monorail is located at C as indicated. The static coefficient of friction between the wheels and the overhead rail is μ_s. Determine which of the wheels, A or B, should be driven in order to accelerate the car faster toward the right; and find that maximum value of acceleration, assuming the wheels do not slip. Assume the mass of the wheels is negligible and note that gravity acts on the system.

Problem 6-32: The automobile in Figure P6-32 weighs 3000 lb and its center of mass is located at C as indicated. The static coefficient of friction between the wheels and the pavement is 0.75. Note that gravity acts and assume the mass of the wheels is negligible.

Determine the minimum time required for the automobile to start from rest and accelerate to 55 mi/hr for parts (a) through (e), assuming in all cases that the wheels do not slip.

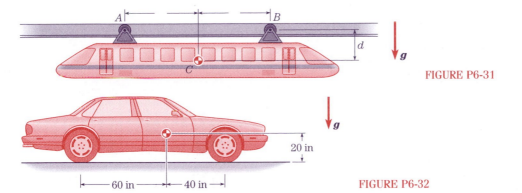

FIGURE P6-31

FIGURE P6-32

(a) When the road is level and the automobile is rear-wheel drive.

(b) When the road is level and the automobile is front-wheel drive.

(c) When the road is level and the automobile is four-wheel drive.

(d) Repeat parts (a) through (c) when the road is inclined 10° upward.

(e) Repeat parts (a) through (c) when the road is inclined 10° downward.

Determine the minimum time required for the automobile to decelerate from 55 mi/hr to a full stop for parts (f) through (h), assuming the braking system operates on all four wheels and the wheels do not slip.

(f) When the road is level.

(g) When the road is inclined 10° upward.

(h) When the road is inclined 10° downward.

Problem 6-33: The four-wheel drive truck sketched in Figure P6-33 has a mass of 3000 kg and its center of mass is located 2.5 m behind the front axle and 0.8 m above ground. The truck is carrying a crate of mass 800 kg on its flat bed. The static coefficient of friction between the truck's bed and the crate is 0.3. Find the maximum acceleration of the truck without causing the crate to slip or tip. Also, find the minimum static

coefficient of friction between the wheels and ground for the truck to attain such an acceleration. Assume that the mass of the wheels is negligible and that the wheels do not slip. Note that gravity acts.

Problem 6-34: As sketched in Figure P6-34, a uniform plate of mass 20 kg is being carried by a pickup truck with its tailgate locked upright. The front end of the plate A is attached to the truck's bed via a horizontal pin joint. Note that gravity acts on the system and ignore the friction between the plate and tailgate at B.

(a) If the truck accelerates at 4 m/s², find the force exerted on the pin joint at A and the normal force exerted on the plate by the tailgate at B.

(b) If the pin joint fails when the force acting on it exceeds 100 N, find the maximum allowable acceleration of the truck without breaking the pin joint.

Problem 6-35: A conveyor belt transports a uniform cube having sides of length 1 m. (See Figure P6-35.) Suppose the belt starts from rest and accelerates such that its speed in m/s is given by $v = 2t^2$, t being the time in seconds. Find the distance traveled by the cube before it starts to tip over. Assume that the friction between the cube and the belt is sufficiently

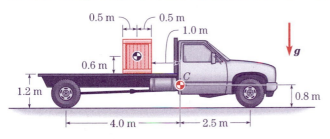

FIGURE P6-33

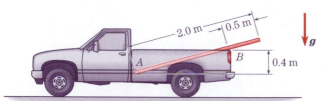

FIGURE P6-34

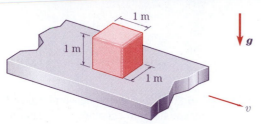

FIGURE P6-35

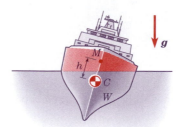

FIGURE P6-37

large to prevent slipping. Note that gravity acts on the system.

Problem 6-36: A uniform billiard ball of mass m and radius r is struck by a cue stick, as sketched in Figure P6-36. Due to the cue stick's downward glancing blow, the ball starts to move toward the right with an initial counterclockwise angular velocity ω_1 and an initial centroidal velocity v_1. The kinetic coefficient of friction between the ball and the surface is μ_k. Note that gravity acts. Find the time at which the billiard ball starts to return toward its original position. Also, find the number of revolutions that the ball has undergone until it starts to return.

Problem 6-37: The rocking motion of a ship at sea is considered. The weight of the ship is W and the center of mass is located at C as indicated in Figure P6-37. The centroidal moment of inertia about a longitudinal axis of the ship is I_c. The intersection between the line of action of the buoyant forces and the ship's vertical center line is called the *metacenter* of the ship, as denoted by M in Figure P6-37. Throughout small rocking motions of the ship, the location of the meta-

center can be assumed to depend only on the shape of the hull. The distance between the metacenter and the center of mass of the ship (often called the *metacentric height*) is denoted by h in Figure P6-37. Derive the equation(s) of motion for the system. Note that gravity acts on the system and assume small motions.

Problem 6-38: A cylinder of radius r and mass m rests between two massless rigid rods of length l that are hinged below at a fixed point O, as sketched in Figure P6-38. The cylinder-rod interface is frictionless. A force P is applied to the upper end of each rod in the horizontal direction. Note that gravity acts. Derive the equation(s) of motion.

Problem 6-39: The thin hoop of thickness t as depicted in Figure P6-39 is made of a material of density ρ and its width and mean radius are w and r_0, respectively. Suppose the hoop is rotating about its vertical axis at a constant angular velocity ω_o.

    (*a*) Find the tension T in the hoop and the corresponding mean hoop stress.

    (*b*) Suppose the hoop is welded through a vertical section and the fracture circumferential (hoop) stress at the weld is σ_{frac}. Find the maximum constant angular velocity that will not cause the hoop to fracture.

Problem 6-40: An overhead crane is lifting a crate of weight of 500 lb, as sketched in Figure P6-40a. The total weight of the crane cabin and the rigid horizontal

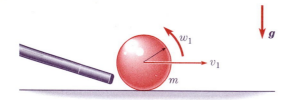

FIGURE P6-36

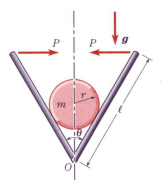

FIGURE P6-38

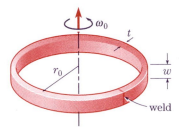

FIGURE P6-39

beam is 2000 lb. Note that gravity acts on the system and that there are four identical columns which support equal load. The cross-section of each column is square with dimensions shown in Figure P6-40b. Assume in all cases that the horizontal beam and all four columns remain in static equilibrium.

(a) Find the force and the corresponding stress in each of the columns when the crate is being lifted with an acceleration of 10 ft/sec^2.

(b) Find the force and the corresponding stress in each of the columns when the crate is being lifted with a constant velocity of 1 ft/sec.

(c) Suppose the columns are made of steel whose yield strength is 60,000 psi. Find the maximum acceleration of the crate without causing yielding of the columns.

(d) Suppose the columns are made of steel whose modulus of elasticity is $E = 29 \times 10^5$ psi. Assuming both ends of the columns are clamped, find the maximum acceleration of the crate without causing the columns to buckle.

Indirect approach: Problems 6-41 through 6-64

Problem 6-41: Show that substitution of Eq. (6-62) into Eq. (6-56) leads to Eq. (6-63).

Problem 6-42: Derive an expression for the kinetic coenergy function for a rigid body by expanding the velocity of the ith particle in terms of $\boldsymbol{v}_b$, the velocity of an arbitrary point, fixed in the body.

Problem 6-43: Derive an expression for the kinetic coenergy function for the rocking snowman toy in Example 6-17 via Eq. (6-68), noting and using the fact that the varying contact point between the toy and the horizontal surface is instantaneously fixed.

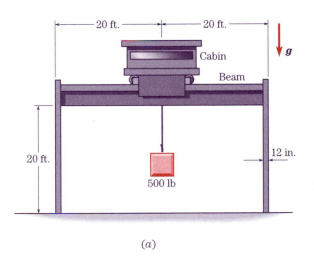

(a)

(b)

FIGURE P6-40

Problem 6-44: Calculate the kinetic coenergy of the smaller disk A in both Figures 3-10 and 3-11. Assume that the smaller disk has mass m and radius r and the larger disk has radius R. (This problem should assist in further understanding the issues addressed in Example 3-2.)

Problem 6-45: A uniform disk of mass m and radius R is pivoted at O about its axis, as sketched in Figure P6-45. A spring of spring constant k is attached to the disk at a radial distance r from the pivot O. Assume the spring is undeformed when its attachment point is directly above O. Assuming small motions, derive the equation(s) of motion for the system.

Problem 6-46: A general plane pendulum of arbitrary shape and mass M pivots about a fixed point O, as sketched in Figure P6-46. The distance between the point O and the pendulum's center of mass is h and the moment of inertia of the pendulum about the pivot point O is I_o. Derive the equation(s) of motion for the system. Note that gravity acts on the system.

Problem 6-47: By simply using the results from Problems 6-7 and 6-46, write the equation of motion for the pendulum in Problem 6-7.

Problem 6-48: By simply using the results from Problems 6-15 and 6-46, write the equation of motion for the pendulum in Problem 6-15.

Problem 6-49: A torsional pendulum as sketched in Figure P6-49 undergoes small angular oscillations

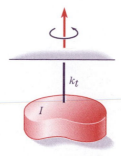

FIGURE P6-49

about its suspension axis, constrained by a torsional spring of torsional spring constant k_t. The moment of inertia of the pendulum about its axis of rotation is I. Derive the equation(s) of motion of the system.

Problem 6-50: The system sketched in Figure P6-50 consists of two blocks of masses m_1 and m_2, interconnected by a massless flexible inextensible cord and a uniform cylindrical pulley of mass m_0 and radius r_0. The friction in the bearings of the pulley is negligible whereas the friction between mass m_1 and the horizontal surface is significant and is modeled as viscous damping having an equivalent dashpot constant c. Initially, both masses are at rest and the cord is taut. It is desired to study the motion of the masses when the support of mass m_2 is suddenly removed. Derive the equation(s) of motion for the system. Note that gravity acts on the system.

Problem 6-51: A uniform rigid rod of mass m and length L slides without friction in a slot along a rigid body AB, as sketched in Figure P6-51. The rigid body AB is pivoted without friction about a fixed point O so that it can rotate in the plane of the sketch. The moment of inertia of the rigid body AB about the point O is given as I_o. Derive the equation(s) of motion for the system. Note that gravity acts on the system.

Problem 6-52: A uniform cylinder of mass m and radius R rolls without slip along a flat bed. A spring ele-

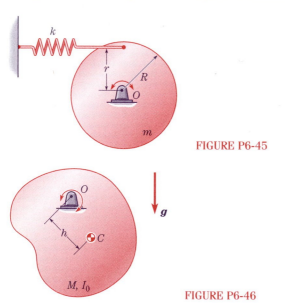

FIGURE P6-45

FIGURE P6-46

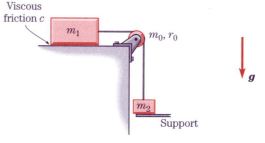

FIGURE P6-50

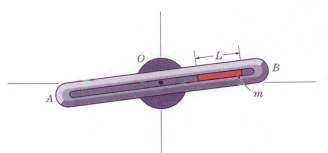

ment of spring constant k and a linear dashpot of dashpot constant c are attached as sketched in Figure P6-52. The system is in equilibrium when the attachment point of the spring and the dashpot is directly above the center of the cylinder. Derive the equation(s) of motion for the system. Assume small motions about the equilibrium position.

Problem 6-53: The cart of mass m_0, which has an inclined surface, carries a disk of mass m and radius R, as sketched in Figure P6-53. The disk rolls without slip on the inclined surface and it is restrained via the spring of spring constant k, which is attached at the point A. The spring is unstretched when the point A is diametrically opposite the contact point of the disk on the surface. Derive the equation(s) of motion for the system. Assume small motions of the disk about the unstretched length of the spring and note that gravity acts on the system.

Problem 6-54: Consider the system sketched in Figure P6-54, where one end of a rigid bar of mass m and length l is pivoted and the other end is supported by a spring of spring constant k. The system is initially in equilibrium with the bar in the horizontal position. Suppose a particle of small mass m_0 ($m_0 \ll m$) strikes the bar at a point B, as shown in Figure P6-54, and remains stuck to the bar thereafter. Derive the equation(s) of small motion of the bar about its horizontal equilibrium position after the collision. Note that gravity acts on the system.

Problem 6-55: In the lumped-parameter torsional system sketched in Figure P6-55, it is assumed that all the mass is in the uniform disk and all the compliance is in the shafts. The geometric and physical properties of the elements are indicated in Figure P6-55. Derive the equation(s) of motion for the system.

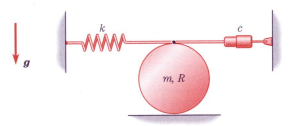

FIGURE P6-52

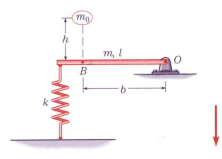

FIGURE P6-54

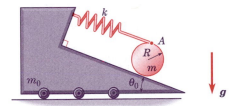

FIGURE P6-53

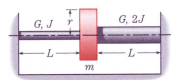

FIGURE P6-55

Problem 6-56: The rotary system shown in Figure P6-56 consists of two rotors (axial moments of inertia I_1 and I_2) connected by a single shaft (length L, shear modulus G and polar moment of the cross-sectional area about the shaft axis J). Let $\theta_1(t)$ and $\theta_2(t)$ be the inertial angular displacements with respect to a common reference of the rotors as shown, which are also the generalized coordinates. Prove that the equations of motion for the system are

$$I_1\ddot{\theta}_1 + \frac{GJ}{L}\theta_1 - \frac{GJ}{L}\theta_2 = 0$$

$$I_2\ddot{\theta}_2 - \frac{GJ}{L}\theta_1 + \frac{GJ}{L}\theta_2 = 0$$

Problem 6-57: Consider the torsional system model sketched in Figure P6-57, which consists of two identical circular disks, each having axial moment of inertia I_0, mounted on a circular shaft having negligible axial moment of inertia. The shaft consists of two identical uniform segments, each having length l, polar moment of the cross-sectional area about the shaft axis J, and shear modulus G. Also, the two disks are subjected to applied torques $T_1(t)$ and $T_2(t)$, as sketched in Figure P6-57. Suppose the generalized coordinates of the system are chosen as the angular displacements of the disks with respect to their equilibrium positions, where $\theta_1(t)$ and $\theta_2(t)$ are as sketched in Figure P6-57. Then show that the equations of motion for the system are given as

$$I_0\ddot{\theta}_1 + \frac{2GJ}{l}\theta_1 - \frac{GJ}{l}\theta_2 = T_1(t)$$

$$I_0\ddot{\theta}_2 - \frac{GJ}{l}\theta_1 + \frac{GJ}{l}\theta_2 = T_2(t)$$

Problem 6-58: Two uniform cylindrical rotors (masses m_1 and m_2, and radii r_1 and r_2) are connected to each other and rigid walls via three shafts,

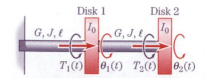

FIGURE P6-57

as sketched in Figure P6-58. Around the perimeter of the m_2 rotor is wrapped a flexible inextensible massless cord, which is attached to a mass m_0. The mass m_0 is constrained to move vertically only, without friction. Derive the equation(s) of motion for the system. Note that gravity acts on the system. Assume that the shafts are supported by frictionless rigid bearings (not shown) such that their bending is negligible.

Problem 6-59: As sketched in Figure P6-59, a long constant-width inextensible fabric is wrapped onto two identical uniform rolls, each of mass m and radius r. The upper roll rotates without friction about its fixed axis, and the lower roll is located immediately below and in contact with the upper roll. After the lower roll is released from rest from this configuration, derive the equation(s) of motion for the system. Assume the mass of the fabric is negligible and

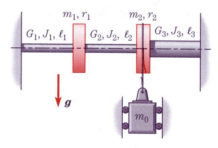

FIGURE P6-58

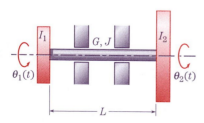

FIGURE P6-56

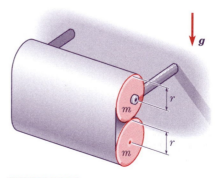

FIGURE P6-59

the lower roll's axis is constrained to move vertically only. Note that gravity acts on the system.

Problem 6-60: A horizontal force of constant direction and constant magnitude F_0 acts on a uniform rod of mass m and length l, as sketched in Figure P6-60. The rod is connected to a fixed support by a pin joint, where a torsional spring of spring constant k_t provides a restoring torque due to the rotation of the rod. Derive the equation(s) of motion for the system. Ignore gravity and assume that the torsional spring is undeformed when the rod is horizontal.

Problem 6-61: The system sketched in Figure P6-61 consists of a thin uniform rod of mass m and length L, and is pivoted about O, located at $\dfrac{L}{3}$ from the right-hand end of the rod. Also, a viscous dashpot having dashpot constant c is located at the right-hand end. At the other end, there is a spring having spring constant k and an applied force $f(t)$. The spring is undeformed when the rod is horizontal. Derive the equation(s) of motion for the system. Assume that the rod undergoes only small motions and ignore gravity.

Problem 6-62: In the system sketched in Figure P6-62, the cart has mass m and slides without friction on the horizontal surface. A uniform cylinder of mass m and radius r is on the cart. The cylinder rolls without slip on the cart and is elastically restrained by the spring having spring constant k. The cart is acted upon by a horizontal force $f_1(t)$. The position of the system can be fully described by the two inertial displacements, $x_1(t)$ and $x_2(t)$, as shown in Figure P6-62, and which are measured from a reference of equilibrium in which the spring is unstretched. Prove that the equations of motion for the system are given by

$$\begin{bmatrix} \dfrac{3}{2}m & -\dfrac{1}{2}m \\ -\dfrac{1}{2}m & \dfrac{3}{2}m \end{bmatrix} \begin{Bmatrix} \ddot{x}_1 \\ \ddot{x}_2 \end{Bmatrix} + \begin{bmatrix} k & -k \\ -k & k \end{bmatrix} \begin{Bmatrix} x_1 \\ x_2 \end{Bmatrix} = \begin{Bmatrix} f_1(t) \\ 0 \end{Bmatrix}$$

Problem 6-63: As sketched in Figure P6-63, three disks with rotary inertias I_1, I_2, and I_3 are connected by shafts of torsional spring constants k_1, k_2, k_3, and k_4 and torsional damping constants c_1, c_2, c_3, and c_4, due to the material behavior in the shafts. The disks are excited by the external torques $T_1(t), T_2(t)$, and $T_3(t)$, as shown. Derive the equation(s) of motion for the system.

Problem 6-64: The pendulum, which consists of a thin uniform rod and a thin uniform disk, pivots about the point O, as sketched in Figure P6-64. For this problem, it remains in a plane that is orthogonal to the plane of the sketch. That is, it swings back and forth in the yz plane. Derive the equation(s) of motion for the system. Note that gravity acts on the system.

Work and energy: Problems 6-65 through 6-77

Problem 6-65: Recall that the centroid C of a line segment L may be found from

$$x_c = \frac{\int x \, dL}{L} \qquad y_c = \frac{\int y \, dL}{L} \qquad z_c = \frac{\int z \, dL}{L}$$

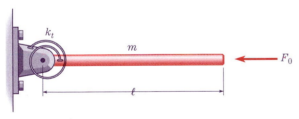

FIGURE P6-60

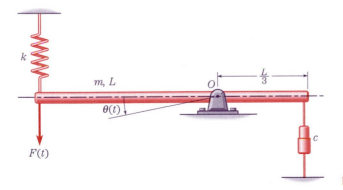

FIGURE P6-61

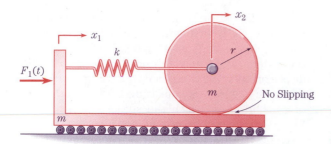

FIGURE P6-62

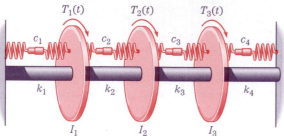

FIGURE P6-63

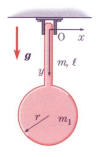

FIGURE P6-64

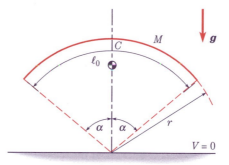

FIGURE P6-65

(a) Using these results, locate the centroid of the circular arc shown in Figure P6-65. Express your answer both in terms of r and α and in terms of r and l_0.

(b) Does the centroid of the circular arc in part (a) reach its top?

(c) Let the line segment in part (a) have uniform density and total mass M. Find the potential energy V for the reference indicated in the figure, both in terms of r and α and in terms of r and l_0.

Problem 6-66: In the late 1960s, Richard ("Dick") Fosbury (1947–) revolutionized the *high jump* event in track and field by deciding not to jump as did the top-ranked jumpers in the world (Figure P6-66a) but to jump as illustrated in Figure P6-66b. In a pre-

Fosbury jump, the athlete jumped (approximately) sideways face down by elevating his or her entire body simultaneously across the bar. In the "Fosbury flop," the athlete leaped head-first face up.

In 1968, Fosbury, who was not expected by many to earn a place even on the American National Team, in fact, went on to win the high jump gold medal at the Olympics held in Mexico City.

Suppose an athlete of mass M attempts to jump over the bar located a height h above the ground. For simplicity, assume that a high jumper's body may be represented by a uniform density line segment of mass M and length l_0—straight in the case of a pre-Fosbury jumper, curved with radius of curvature R in the case of a post-Fosbury jumper. For the parame-

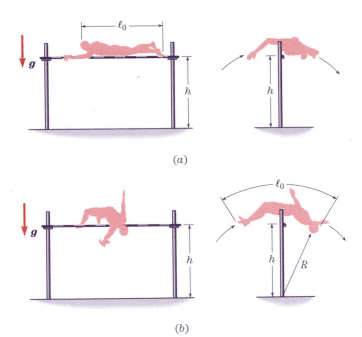

(a)

(b)

FIGURE P6-66

ters given, compare the required vertical velocity with which the jumpers of the two techniques must leap.

Note that the center of mass of a successful Fosbury flopper may never cross the bar!

Problem 6-67: Due to the torsional spring attached at each hinge, the rectangular door sketched in Figure P6-67 closes automatically when released from an open position. The spring constant of each torsional spring is $k_t = 100$ N·m/rad and the mass of the door is 50 kg. The door can be modeled as a uniform rect-

angular plate having the dimensions indicated in Figure P6-67. Suppose the door is released from rest at a 45° open position. Find the angular speed of the door immediately before it fully closes.

Problem 6-68: Figure P6-68 shows a toy car that is driven by a torsional spring attached at the axle of the rear wheels. The spring constant of the torsional spring is $k_t = 0.02$ N·m/rad. The mass of the body of the car is 0.3 kg. Each of the rear wheels can be modeled as a circular uniform cylinder of mass 0.5 kg and radius 5 cm; and the front wheels can each be modeled as a circular uniform cylinder of mass 0.1 kg and radius 2 cm. Suppose the torsional spring is wound up 5 revolutions and the car is released from rest. Find the maximum speed of the car, assuming that the wheels roll without slip and that the spring disengages when it is fully unwound.

Problem 6-69: The yo-yo sketched in Figure P6-69 is climbing with an angular velocity of $\omega_1 = 50$ rad/s

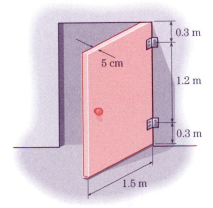

FIGURE P6-67

FIGURE P6-68

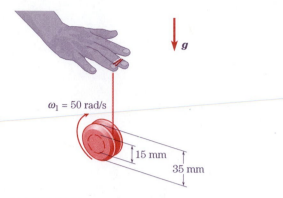

FIGURE P6-69

at the instant shown. The yo-yo has a mass of 0.1 kg and a radius of gyration of $\kappa = 17$ mm about its centroidal axis. (See Problem 6-20 for the definition of the radius of gyration.) How high will the yo-yo climb before it comes to rest? Note that gravity acts.

Problem 6-70: The braking system on a rotor sketched in Figure P6-70 is being tested. The rotor and shaft assembly has a mass of 30 kg and a centroidal radius of gyration of 20 mm. (See Problem 6-20 for the definition of the radius of gyration.) The rotor is initially rotating at a constant angular speed of 1000 rpm. The braking system, when engaged, exerts a constant braking torque T. Assume that the shaft is rigid and that its bearing friction is negligible.

(a) Find the number of revolutions of the rotor before it comes to a complete stop if $T = 1.0$ N·m.

(b) Find the constant braking torque T required to stop the rotor in two revolutions.

Problem 6-71: In Figure P6-71, the uniform disk of weight 20 lb and radius 5 in. can rotate about its centroidal axis without friction. The disk is initially at rest before a flexible inextensible cable wrapped around it is pulled vertically by a constant force of 15 lb. If the force on the cable is released after the disk has rotated 5 revolutions, find the final angular velocity of the disk.

Problem 6-72: As sketched in Figure P6-72, a uniform thin rod of mass m and length l is free to rotate about its pivot O at its lower end. The rod is initially vertical and is released from rest such that it falls over onto the horizontal surface. Find the angular velocity of the rod as a function of the angle between the rod and the vertical. Note that gravity acts on the system.

Problem 6-73: The uniform sphere of mass m and radius r is at rest at the top of a fixed hemispherical surface of radius R, as sketched in Figure P6-73. The sphere is released from rest and starts to roll down the hemisphere without slip. Find the angular and linear velocities of the sphere, as functions of the angle θ shown in Figure P6-73, before the sphere loses contact with the surface. Note that gravity acts.

Problem 6-74: The uniform cylinder of mass 10 kg and radius 5 cm, as sketched in Figure P6-74, rolls without slip on a 30° inclined surface. A spring of spring constant 100 N/m is attached to a massless flexible inextensible strap that is wrapped around the cylinder. The spring is initially stretched 0.5 m. Suppose the system is released from rest with this initial stretch of the spring. Note that gravity acts on the system.

(a) Find the speed of the centroid of the cylinder and the angular speed of the cylinder when the spring is completely unstretched.

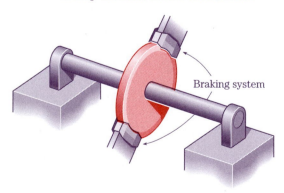

FIGURE P6-70

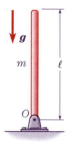

FIGURE P6-71

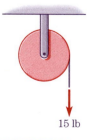

FIGURE P6-72

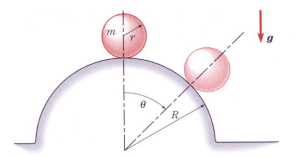

FIGURE P6-73

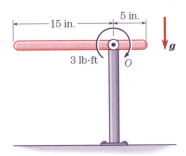

FIGURE P6-75

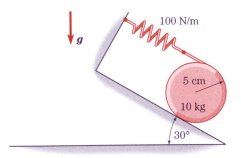

FIGURE P6-74

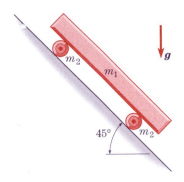

FIGURE P6-76

(*b*) Find the stretch of the spring when the speed of the cylinder becomes zero.

Problem 6-75: A uniform rod of weight 2 lb and length 20 in. pivots about the point O, as sketched in Figure P6-75. Suppose the rod is initially horizontal and at rest. If the rod is released and at the same time is acted upon by a constant counter torque of 3 lb·ft, find its angular velocity when it becomes vertical. Note that gravity acts on the system.

Problem 6-76: As sketched in Figure P6-76, a rectangular block of mass m_1 is placed on two identical uniform cylinders, each of mass m_2 and radius r, which roll on a 45° inclined surface. No slipping occurs between any contacting surfaces. Note that gravity acts on the system. Suppose the system is released from rest. Find the speed of the cylinders and the block when the block has moved down a distance x along the inclined surface.

Problem 6-77: A ship of mass M and centroidal moment of inertia I_c is being turned by two tugboats, as sketched in Figure P6-77. Each tugboat exerts a constant force F on the ship in a direction normal to the ship's centerline at all times. Suppose the ship is ini-

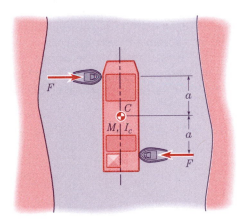

FIGURE P6-77

tially at rest and the resistance of the water against the ship's rotation is negligible. Find the angular velocity of the ship after it has been turned 90°.

Review question: Problem 6-78

Problem 6-78: Briefly answer the following questions.

(a) A rotor is statically unbalanced. What must be done to overcome this condition?

(b) A rotor is dynamically unbalanced. What must be done to overcome this condition?

(c) An angular velocity is imposed about an axis through a rigid body. The resulting angular momentum is not about this same axis. How can the axis of rotation be altered in order to ensure that the angular momentum vector and the angular velocity vector are in the same direction?

(d) Suppose a system, whose total energy may be changing, is operating in a torque-free environment. What can be said about the system's momenta?

Category III: More Difficult

Inertia tensor: Problems 6-79 through 6-80

Problem 6-79: Consider the two rectangular coordinate systems $oxyz$ and $ox'y'z'$ shown in Figure P6-79, one rotated with respect to the other. The unit vectors associated with all axes are also shown in Figure P6-79.

(a) The position vector of the ith particle of a system of particles can be written either as $\mathbf{r}_i = \{x_i\ y_i\ z_i\}^T$ with respect to the unprimed coordinate system or as $\mathbf{r}'_i = \{x'_i\ y'_i\ z'_i\}^T$ with respect to the primed coordinate system, the superscript T denoting the transpose of the row matrices, yielding

column vectors. Show that the components of $\mathbf{r}_i$ and $\mathbf{r}'_i$ are related by the expression

$$\begin{Bmatrix} x'_i \\ y'_i \\ z'_i \end{Bmatrix} = [C] \begin{Bmatrix} x_i \\ y_i \\ z_i \end{Bmatrix}$$

where $[C]$ is a square matrix called the rotational-transformation matrix, or more commonly, the rotation matrix, whose elements are the direction cosines of the angles between the primed and unprimed axes; that is,

$$[C] = \begin{bmatrix} c_{x'x} & c_{x'y} & c_{x'z} \\ c_{y'x} & c_{y'y} & c_{y'z} \\ c_{z'x} & c_{z'y} & c_{z'z} \end{bmatrix}$$

where $c_{pq} = \cos\theta_{pq} = \mathbf{u}_p \cdot \mathbf{u}_q = c_{qp}$, θ_{pq} being the angle between the p-axis and the q-axis.

(b) Show that the rotation matrix $[C]$ has the property

$$[C][C]^T = [1]$$

where $[C]^T$ denotes the transpose of $[C]$ and $[1]$ is the identity matrix. All matrices having this property are called orthogonal matrices.

(c) Show that the inertia tensor with respect to the unprimed axes $oxyz$, whose elements are defined by Eqs. (6-33) and (6-34), can be written as

$$[I]_o = \sum_{i=1}^{N} m_i (r_i^2[1] - \mathbf{r}_i \mathbf{r}_i^T)$$

where $\mathbf{r}_i$ is the position vector of the ith particle of the system, and r_i is the magnitude of the position vector $\mathbf{r}_i$.

(d) Using parts (a), (b) and (c), show that the inertia tensor with respect to the primed axes $[I']_o$ can be written in terms of the inertia tensor with respect to the unprimed axes $[I]$ and the rotation matrix $[C]$ as

$$[I']_o = [C][I]_o[C]^T$$

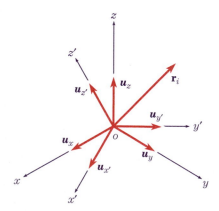

Note that all tensors obey this transformation rule; indeed, the inertia tensor is named so only because it obeys such a rule.
(*Hint:* Write $[I']_o$ by using the expression in part (*c*); substitute the expression in part (*a*) and its transpose into the expression for $[I']_o$; then reduce the resulting expression for $[I']_o$ into the sought form by using the equation in part (*b*).)

(*e*) Show that the relationship between $[I']_o$ and $[I]_o$ given in part (*d*) also holds for a continuous body, where the position vector of a point may be written as either $\boldsymbol{r} = \{x \ y \ z\}^T$ with respect to the unprimed coordinate system or as $\boldsymbol{r}' = \{x' \ y' \ z'\}^T$ with respect to the primed coordinate system.

Problem 6-80: The inertia tensor for a continuous body having mass symmetry about the xy plane is given in the form of Eq. (6-53), from which it was noted that the z direction is a principal direction. The four elements of the inertia tensor associated with x and y axes are considered in an attempt to find the principal directions in the xy plane and their associated principal moments of inertia. To this end, rotation of the axes about z direction is considered, as sketched in Figure P6-80. As a result of this rotation of axes, the z' axis is the same as the z axis, and the angle between the x' and x axes is equal to the angle between the y' and y axes, denoted by θ as indicated.

(*a*) By using the general transformation relation given in part (*d*) of Problem 6-79, show that the elements of the inertia tensor associated with the x' and y' axes can be written as

$$I_{x'x'} = I_{xx} \cos^2 \theta + I_{yy} \sin^2 \theta$$
$$+ 2I_{xy} \sin \theta \cos \theta$$
$$I_{y'y'} = I_{yy} \cos^2 \theta + I_{xx} \sin^2 \theta$$
$$- 2I_{xy} \sin \theta \cos \theta$$
$$I_{x'y'} = I_{xy} \cos^2 \theta - I_{xy} \sin^2 \theta$$
$$- (I_{xx} - I_{yy}) \sin \theta \cos \theta$$

(*b*) By using trigonometric identities, show that the equations in part (*a*) can be rewritten as

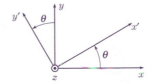

FIGURE P6-80

$$I_{x'x'} = \frac{1}{2}(I_{xx} + I_{yy}) + \frac{1}{2}(I_{xx} - I_{yy}) \cos 2\theta$$
$$+ I_{xy} \sin 2\theta$$

$$I_{y'y'} = \frac{1}{2}(I_{xx} + I_{yy}) - \frac{1}{2}(I_{xx} - I_{yy}) \cos 2\theta$$
$$- I_{xy} \sin 2\theta$$

$$I_{x'y'} = I_{xy} \cos 2\theta - \frac{1}{2}(I_{xx} - I_{yy}) \sin 2\theta$$

(*c*) Using the equations in part (*a*) or those in part (*b*), plot $I_{x'x'}$, $I_{y'y'}$ and $I_{x'y'}$ versus θ for a case where $I_{xx} > I_{yy} > 0$ and $I_{xy} > 0$, by assuming arbitrary values for I_{xx}, I_{yy} and I_{xy}. Compare these plots with the Mohr's circle shown in Figure P6-15.

(*d*) In particular, perform the task in part (*c*) for the rectangular block of Problem 6-23. Also, find the principal directions and principal moments of inertia from these plots and compare them with those found in Problem 6-23.

(*e*) Using the equations in part (*b*), derive Eqs. (6-54) and (6-55) for the principal moments of inertia and the principal directions, respectively.

Indirect approach: Problems 6-81 through 6-119

Problem 6-81: A thin rectangular plate (thickness t) is made of a material having density ρ and it has two identical circular holes (radius R), as sketched in Figure P6-81. The plate pivots about the fixed point O. Assume that all the dimensions shown are much greater than the uniform thickness t. Derive the equation(s) of motion for the system. Do not assume small motions and note that gravity acts on the system.

Problem 6-82: The rectangular frame sketched in Figure P6-82 is pivoted about point O. The frame is constructed from four slender rods having lengths a

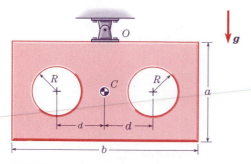

FIGURE P6-81

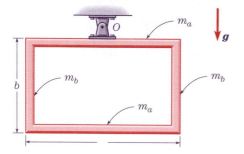

FIGURE P6-82

and b, and masses m_a and m_b as shown. Derive the equation(s) of motion for the system. Do not assume small motions and note that gravity acts on the system.

Problem 6-83: The system sketched in Figure P6-83 consists of slender rods of given masses and lengths,

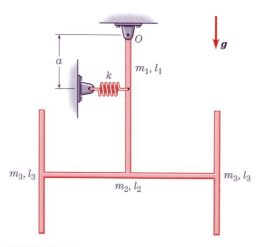

FIGURE P6-83

along with a spring of spring constant k. For small oscillations from the vertical equilibrium configuration, derive the equation(s) of motion for the system. Note that gravity acts on the system.

Problem 6-84: A special-purpose scale is designed as sketched in Figure P6-84, where the weighing dish of mass m_0 is constrained to move vertically only and it is linked via a flexible inextensible massless cord and a system of massless rigid levers to the pointer. The lever system is restrained by a spring of spring constant k and a dashpot of dashpot constant c, as shown. Assume that all the levers are horizontal when the system is at equilibrium with no object placed on the weighing dish. For small motions about this equilibrium, derive the equation(s) of motion for the system. Note that gravity acts on the system.

Problem 6-85: In Figure P6-85, a mass m_0 is supported in a gravitational field via a flexible inextensible massless cable that wraps around two uniform cylindrical pulleys via frictionless fixed guides. The mass m_0 is constrained to move vertically only and the friction between the mass m_0 and the walls of the vertical guide may be modeled as viscous damping having a dashpot constant c. The two pulleys are connected to ground via two springs of spring constants k_1 and k_2. Assume that the m_1 pulley is constrained to move horizontally only whereas the m_2 pulley is constrained to move vertically only. Derive the equation(s) of motion for the system. Note that gravity acts on the system.

Problem 6-86: An artistic mobile is suspended via a spring of spring constant k attached at point O, as sketched in Figure P6-86. The mass of the mobile is m and its center of mass is located a distance d horizontally from the point O, at equilibrium. The moment of inertia of the mobile about the point O is given as I_o and its rotational motion is restrained by a torsional spring k_t. Assume the motion of the mobile is confined to the plane of sketch and the linear spring k remains vertical at all times. Derive the equation(s) of motion for the system. Note that gravity acts on the system and do not assume small motions.

Problem 6-87: A simple model for a truck suspension is sketched in Figure P6-87. The location of the center of mass of the truck is shown. The truck's mass and moment of inertia about its center of mass are

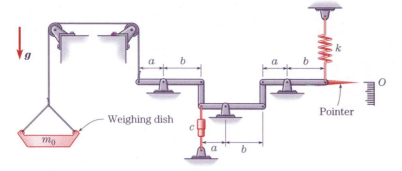

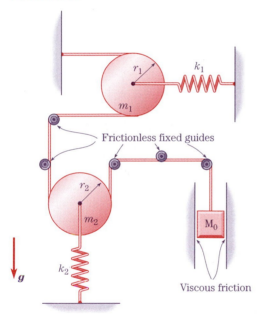

FIGURE P6-84

FIGURE P6-85

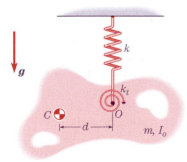

FIGURE P6-86

M and I, respectively. The suspension systems on the rear and front of the truck are modeled by springs k_1 and k_2, respectively. For small motions in the plane of the figure, $y(t)$ and $\theta(t)$ constitute a convenient set of generalized coordinates as shown in Figure P6-87. Note that gravity acts on the system and assume that $y = 0$ and $\theta = 0$ at equilibrium. Show that the equations of motion for the truck are

$$
\begin{bmatrix} M & 0 \\ 0 & I \end{bmatrix} \begin{Bmatrix} \ddot{y} \\ \ddot{\theta} \end{Bmatrix} + \begin{bmatrix} (k_1 + k_2) & (k_2 l_2 - k_1 l_1) \\ (k_2 l_2 - k_1 l_1) & (k_2 l_2^2 + k_1 l_1^2) \end{bmatrix} \begin{Bmatrix} y \\ \theta \end{Bmatrix} = \begin{Bmatrix} 0 \\ 0 \end{Bmatrix}
$$

Problem 6-88: To simulate the effect of an earthquake on very low-rise buildings, a rigid building model has been devised as shown in Figure P6-88, where the base is connected to the ground through a linear spring of spring constant k, and the rotational motion of the building is restrained by the rotational spring of spring constant k_t connected to the base. The rigid building is uniform and has mass m, height L, and centroidal moment of inertia I_c. It is desired to determine the response of the building subjected to horizontal earthquakes represented by the specified ground motion $x_0(t)$. Derive the equation(s) of motion for the system. Note that gravity acts on the system and assume small rotations of the building.

Problem 6-89: A torsional system, consisting of two identical thin disks and three identical massless stiffness elements, has been modeled as shown in Figure P6-89. It is desired to study the longitudinal (in the x

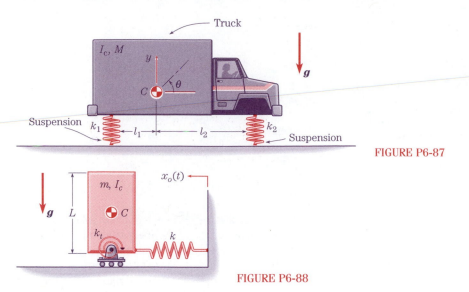

FIGURE P6-87

FIGURE P6-88

direction) motion as well as the torsional (about the x axis) motion of this system. Assume that all the stiffness elements are undeformed in the configuration shown.

(a) For the longitudinal motion only, derive the equation(s) of motion for the system.

(b) For the torsional motion only, derive the equation(s) of motion for the system.

Problem 6-90: As sketched in Figure P6-90, a wheel of mass m and radius a rolls without slip on a horizontal surface. The spokes of the wheel are assumed to be massless. At the center of the wheel is pivoted a body of mass $2m$ whose center of mass is located a distance a from the pivot and whose moment of inertia about its center of mass is $I_c = 4ma^2$. Also, the center of the wheel is connected to a fixed wall via a spring of spring constant k. Note that gravity acts on the system.

(a) Without assuming small motions, derive the equation(s) of motion for the system.

(b) By linearizing the equation(s) of motion found in part (a), derive the equation(s) of small motion for the system.

Problem 6-91: As sketched in Figure P6-91, two uniform cylindrical pulleys of masses m_1 and m_2 and radii R_1 and R_2 are welded together. This composite pulley rotates as a unit and without friction about a fixed point O. The larger pulley is also connected to the ground via a spring of spring constant k and a dashpot of dashpot constant c, both wrapped without slipping around its periphery. A block of mass m_0 is attached to the smaller pulley via an inextensible massless strap wrapped without slipping around the smaller pulley. Assume that the block is constrained to move only vertically. Derive the equation(s) of motion for the system. Note that gravity acts on the system.

Problem 6-92: As sketched in Figure P6-92, two blocks of masses m_1 and m_2 move on horizontal surfaces without frictional losses. The blocks are attached to springs of spring constants k_1 and k_2, respectively, and are also connected to the ends of a pivoted rigid slender rod of mass m_0 and length $(a + b)$. At the pivot of the rod is a rotational spring of spring constant k_t. Assuming small motions and $a > b$, derive the equation(s) of motion for the system. Also, assume that all springs are undeformed when the rod is vertical and note that gravity acts on the system.

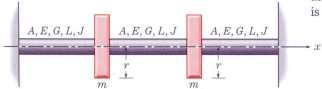

FIGURE P6-89

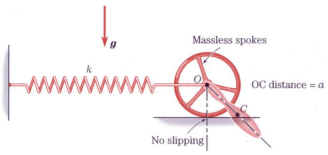

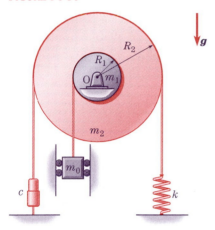

FIGURE P6-91

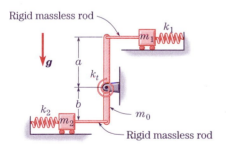

FIGURE P6-92

Problem 6-93: The dial of a torque meter is modeled as a uniform slender rigid rod of mass m_1 and length l_1, which is welded to a uniform rigid cylindrical shaft of mass m_2 and radius r_2. As shown in Figure P6-93, the shaft is supported by two bearings whose friction is modeled as viscous damping having equivalent rotational dashpot constants c_{t1} and c_{t2}. Also, the shaft is restrained by a torsional spring of spring constant k_t, and the dial is restrained by two linear springs of

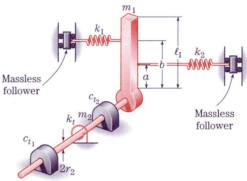

FIGURE P6-93

spring constants k_1 and k_2. The linear springs k_1 and k_2 are connected to frictionless massless followers so that they remain horizontal at all times. Derive the equation(s) of motion for the system. Ignore gravity and assume that all springs are undeformed when the dial is vertical.

Problem 6-94: The ends of the slender rod (mass m and length l) sketched in Figure P6-94 are free to slide on the cylindrical surface of radius R without friction. The motion is constrained to remain in the plane of the sketch. Derive the equation(s) of motion. Note that gravity acts on the system.

Problem 6-95: As sketched in Figure P6-95, a rectangular container of length l and width w rocks without slip on a fixed semicylinder of radius R. The container is filled with sand of mass density ρ, to a constant

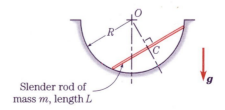

FIGURE P6-94

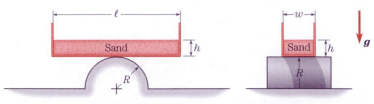

Front view   Side view   FIGURE P6-95

depth h. Assume that the mass of the container itself is negligible compared with the mass of the sand and that when the container is horizontal, its midpoint is directly above the top of the semicylinder. Also, assume that the depth of the sand remains constant at all times during the system's rocking motion. Note that gravity acts on the system.

(a) Without assuming small motions, derive the equation(s) of motion for the system.

(b) By linearizing the equation(s) of motion obtained in part (a), find the equation(s) of small motion for the system about its equilibrium position.

Problem 6-96: A homogeneous solid semicylinder rocks without slip on a horizontal surface, as sketched in Figure P6-96. Note that gravity acts on the system.

(a) Derive the equation(s) of motion for the system for rocking motion of the semicylinder on the horizontal surface.

(b) By linearizing the equation(s) of motion obtained in part (a), find the equation(s) of small motion for the system about its equilibrium position.

Problem 6-97: A disk of mass m and radius r rolls without slip on a circular surface of radius R, as sketched in Figure P6-97. Note that gravity acts.

(a) Derive the equation(s) of motion for the system.

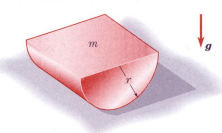

FIGURE P6-96

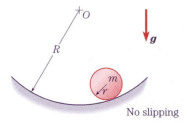

No slipping

FIGURE P6-97

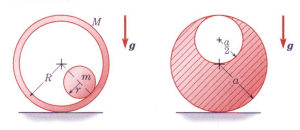

FIGURE P6-98   FIGURE P6-99

(b) By linearizing the equation(s) of motion obtained in part (a), find the equation(s) of small motion for the system about its equilibrium position.

Problem 6-98: As sketched in Figure P6-98, a uniform disk of mass m and radius r rolls without slip inside a large hoop of mass M and radius R, which rolls without slip along a horizontal track. Derive the equation(s) of motion for the system. Note that gravity acts on the system.

Problem 6-99: The solid circular cylinder of radius a has a hole of radius $\frac{a}{2}$ drilled through it parallel to and $\frac{a}{2}$ from its axis, as sketched in Figure P6-99. The cylinder is placed on a horizontal plane on which it rolls in the plane of the sketch without slipping. Derive the equation(s) of motion for the system. Note that gravity acts on the system.

Problem 6-100: As sketched in Figure P6-100, a uniform cylinder of mass m and radius r rolls without slip on a horizontal surface. A rigid thin hoop of mass M and radius R ($R > r$) rolls without slip on the cylinder. Derive the equation(s) of motion for the system. Note that gravity acts on the system.

Problem 6-101: The rocking chair sketched in Figure P6-101 is rocking in the plane of the sketch, sliding along the horizontal frictionless floor, and subjected to the force $\boldsymbol{F}$ as shown. The force is always parallel to the floor. The mass of the chair is m and the moment of inertia about its center of mass is I_c. OCB is a line of mass symmetry of the chair and is vertical when the chair is unforced and at equilibrium. Note that gravity acts on the system.

(a) Derive the equation(s) of motion for the rocking chair.

(b) Suppose a person of mass M is seated on the chair. It is assumed that the person can be modeled as a rigid body and does not move with respect to the chair. The person's center of mass is along the line OCB at a distance H above the floor ($h < H < L$), and the moment of inertia of the person about his

or her own center of mass is I_p. Derive the equation(s) of motion for the chair-person system.

Problem 6-102: The uniform rectangular plate sketched in Figure P6-102 has sides of length L and mass m. It is constrained to remain in the plane of the sketch while one of its corners slides without friction along a horizontal surface. Derive the equation(s) of motion for the system. Note that gravity acts on the system.

Problem 6-103: As sketched in Figure P6-103, two carts of mass M and M_0, connected via a spring and a dashpot, move on a horizontal surface without frictional losses. The cart of mass M has a circular cylindrical surface of radius R on which a small disk of mass m and radius r rolls without slip. Also, the cart of mass M is acted on by a horizontal force $F(t)$. Derive the equation(s) of motion for the system. Note that gravity acts on the system.

Problem 6-104: Figure P6-104 shows a simple model of a tank truck with suspension for dynamic analysis. The cab and platform of the truck are modeled as uniform rectangular blocks, and the tank is modeled as a uniform solid cylinder, with the dimensions given in Figure P6-104. The entire truck may be treated as a single rigid body.

(a) Find the location of the centroid of the truck.

(b) Find the moment of inertia of the truck I_{zz} with respect to its centroid.

(c) Derive the equation(s) of motion of the truck for small motions, ignoring gravity.

Problem 6-105: As sketched in Figure P6-105, two uniform thin rods (masses m_0 and m_1, lengths l_0 and l_1) and a uniform sphere (mass m and radius r) are attached to form a rigid body that is constrained to remain in the plane of the sketch. A linear spring and dashpot are attached to each end of l_0. Consider small motions of the system relative to its horizontal equilibrium position, while assuming that the system is

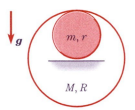

FIGURE P6-100

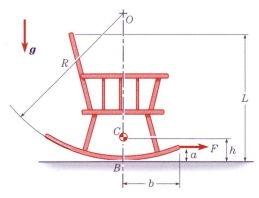

FIGURE P6-101

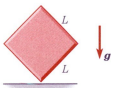

FIGURE P6-102

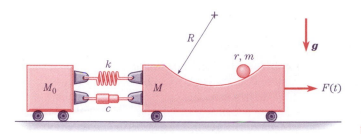

FIGURE P6-103

(Depth into page = b_1)

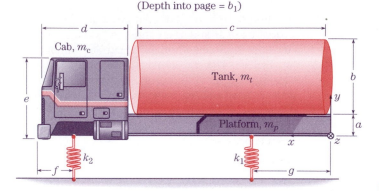

FIGURE P6-104

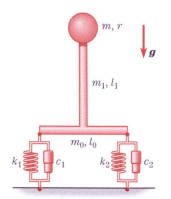

FIGURE P6-105

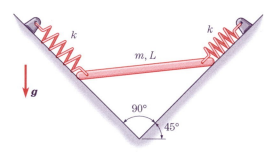

FIGURE P6-106

constrained against horizontal motion. (That is, it can move vertically and rotate.) Derive the equation(s) of motion for the system. Note that gravity acts on the system.

Problem 6-106: As sketched in Figure P6-106, a uniform rod of mass m and length L slides without friction, in the plane of the sketch, in a trough. The springs are unstressed when the rod is horizontal. Derive the equation(s) of motion for the system. Note that gravity acts on the system.

Problem 6-107: In the Kendall Square subway station in Cambridge, Massachusetts, there are bells to entertain passengers waiting on the platform. The bells are struck by hammers such as the one sketched in Figure P6-107.

The hammer can be modeled as a rectangular block of mass m_2, width b, and height a, rigidly attached to a rod of mass m_1 and length l. The hammer pivots freely about point O, in the plane of the figure. A passenger sets the hammer in motion by pulling on a lever (not shown), which imparts a specified motion $x_0(t)$ to a mass m_0 containing the pivot point O.

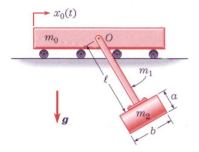

FIGURE P6-107

Derive the equation(s) of motion for the hammer system. Note that gravity acts on the system.

Problem 6-108: The cart of mass M sketched in Figure P6-108 rolls without frictional losses along a track on the horizontal surface. Two identical thin rods of mass m and length l are attached to the cart by pinned joints. A third rod of mass m_0 and length l_0 is attached to the pendulous rods by pinned joints. Derive the equation(s) of motion for the system. Note that gravity acts on the system.

Problem 6-109: A rack and pinion system is sketched in Figure P6-109. The axis of the pinion is fixed in frictionless bearings, and the flexible inextensible massless cord wrapped around the circular

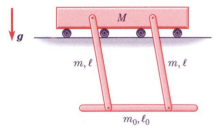

FIGURE P6-108

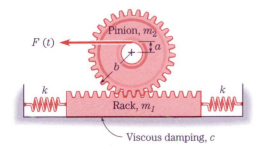

FIGURE P6-109

pulley of radius b is pulled by an external force $f(t)$, which remains horizontal at all times. Assume that the pinion can be modeled as a uniform cylinder of mass m_2 and radius b and that the friction between the rack and the horizontal surface can be modeled as viscous damping having a dashpot constant c. Derive the equation(s) of motion for the system.

Problem 6-110: In the dynamic system sketched in Figure P6-110, the uniform cylinder of mass M and radius a rolls without slipping on the lower surface. The bar of mass m is geared to the cylinder and held horizontally by frictionless rollers as indicated. The bar is acted upon by a force $F(t)$, which is always horizontal. Derive the equation(s) of motion for the system.

Problem 6-111: Figure P6-111 shows a schematic of a scale. Objects to be weighed are placed on the weighing table (m_1). The numbers on the indicator (m_4) appear at a window (not shown) according to the weight of the object being weighed. Derive the equation(s) of motion for the system. Note that gravity acts on the system and assume that the torsional spring is unstressed when the linear spring is unstressed.

Problem 6-112: The system sketched in Figure P6-112 consists of two pendulums. Each of the pendulums consists of a weight (mass m_1 or m_2), a massless rigid rod (length l_1 or l_2), and a gear (radius r_1 or r_2 and centroidal moment of inertia I_1 or I_2), as sketched in Figure P6-112. Derive the equation(s) of motion for the system. Note that gravity acts on the system.

Problem 6-113: As sketched in Figure P6-113, two uniform sector gears m_1 and m_2 are rigidly attached to the ends of two uniform links m_3 and m_4, respectively. The sector gears rotate about fixed points O and O'. The centroidal moment of inertia of m_1 is I_1 and the centroidal moment of inertia of m_2 is I_2. The system contains two torsional springs, two extensional springs, and a dashpot, and is acted upon by the force $\mathbf{F}(t)$, which is always horizontal. The longitudinal centerline of m_4 may be assumed to pass through point O'. The longitudinal centerline of m_3 passes through points O and C_1. At equilibrium, the three springs of spring constants k_1, k_2, k_{t1} are undeformed, and the lines OC_1 and $O'C_2$ are vertical. Assuming small angular displacements, derive the equation(s) of motion for the system. Note that gravity acts on the system.

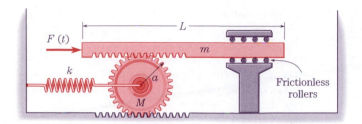

FIGURE P6-110

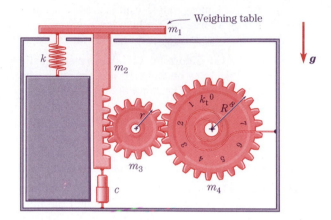

FIGURE P6-111

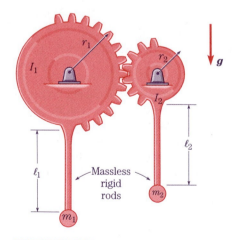

FIGURE P6-112

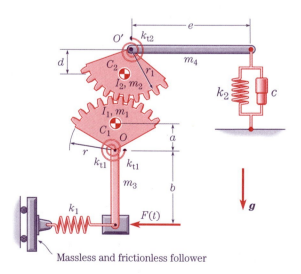

Massless and frictionless follower

FIGURE P6-113

Problem 6-114: The system sketched in Figure P6-114 consists of a mass m_1 carried inside a cart of mass m_0 that is rigidly connected via a massless rigid link to another mass m_3, which can translate only horizontally. A slender rod of mass m_2 and length l pivots on m_3 and is rotationally restrained via a torsional spring k_t, and its ends are attached to springs and dashpots as sketched in Figure P6-114. The massless followers maintain the attached springs and dashpots in either a vertical or horizontal orientation as shown. The slender rod connecting masses m_4 and m_5 is of total length $(a + b)$ and may undergo large rotations about its pivot, which translates with m_3, where a is smaller than b. The rod $(a + b)$ is vertical *and* all springs are undeformed when the system is in its equilibrium configuration. Define a complete and independent set of generalized coordinates, find the generalized forces, and evaluate the kinetic coenergy function and the potential energy function of the system. Do not assume small motions and note that gravity acts on the system.

Problem 6-115: As sketched in Figure P6-115, a uniform bar of length $2r_0$ and mass m is pivoted at the end of a spring of unstretched length l_0 and spring constant k. As it changes its length, the spring is constrained to remain straight, while the bar may pivot

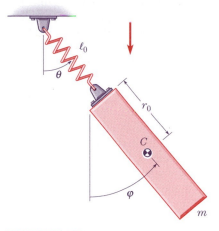

FIGURE P6-115

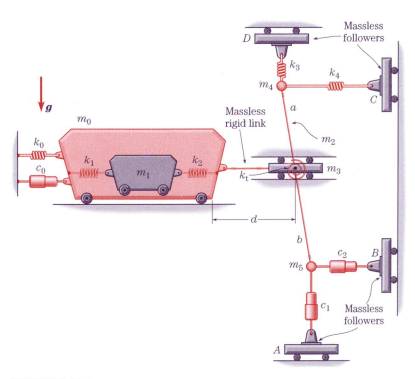

FIGURE P6-114

about the end of the spring. Assuming that the motion remains in the plane of the sketch, derive the equation(s) of motion for the system. Note that gravity acts on the system.

Problem 6-116: A system of two welded rigid rods (masses m_2 and m_3, length L and h) is connected to the ground via ball joints, as sketched in Figure P6-116. The mass m_1 attached to the end of the horizontal rod is acted upon by a harmonic force $F(t) = f_0 \sin \omega t$, which is always horizontal and orthogonal to L. Derive the equation(s) of motion for the system. Note that gravity acts on the system.

Problem 6-117: A disk of radius R has a rectangular cutout ($d \times b$) as sketched in Figure P6-117. The mass of the disk with the cutout is m_0. The center of the disk is supported by two identical shafts of negligible density, shear modulus G, polar moment of cross-sectional area about their axes J and length L. Also, the disk is being driven by an external torque

$\tau(t)$. Further, a small disk of mass m and radius r rolls freely and without slip on the surface of the cutout. The smaller disk is constrained to remain in contact with the cutout surface and it is attached to the larger disk via a spring of spring constant k and a dashpot of dashpot constant c. The rectangular block removed from the larger disk to produce the cutout is of mass m_1 and has centroidal moment of inertia I_1 about an axis perpendicular to the plane of the disk. Assume the spring k is unstretched when the center of the smaller disk is midway along the surface of the cutout and that the two shafts are both unstressed at equilibrium. Derive the equation(s) of motion for the system. Assume that no gravity acts on the system.

Problem 6-118: A uniform cylinder (mass M and radius R) rolls without slip on the bottom section of a pivoted frame (mass m), as sketched in Figure P6-118. The frame is pivoted at its bottom face. The moment of inertia of the frame about its pivot O is given as I_o. The centroid of the frame is located at a distance d from the frame's pivot. The cylinder is connected to a spring (spring constant k) and a dashpot (dashpot constant c), such that the spring is unstretched when the center of the cylinder is directly above the pivot point O of the frame. There is a vertical spring (spring constant k_1) that connects a follower (mass m_1) to the frame and is unstretched when the frame's bottom section is horizontal. The spring constant k_1 is for longitudinal extension or compression as this spring is sufficiently stiff laterally to remain vertical at all times. Derive the equation(s)

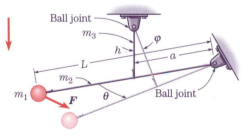

$$F = f_0 \sin \omega t \text{ (always horizontal)}$$

FIGURE P6-116

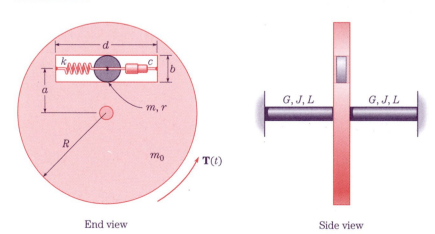

End view   Side view

FIGURE P6-117

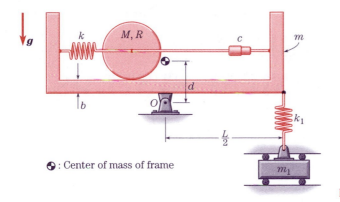

: Center of mass of frame

FIGURE P6-118

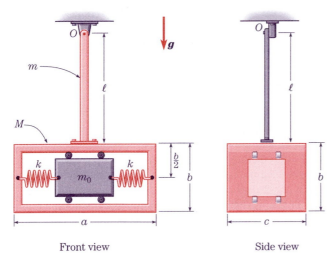

Front view   Side view   FIGURE P6-119

of motion for the system. Do not assume small motions and note that gravity acts on the system.

Problem 6-119: The plane pendulum sketched in Figure P6-119, which consists of a slender uniform rod (mass m and length l) and a crate (mass M and dimensions $a \times b \times c$), can swing about the fixed point O. The crate carries a cart (mass m_0) that is restrained by springs as shown. The centroidal moments of inertia, about the axis normal to the sketch, of the crate and the cart are I and I_0, respectively. The springs are unstretched when the cart is halfway along the inside surface of the crate. Derive the equation(s) of motion for the system. Do not assume small motions and note that gravity acts on the system.

Casablanca
Dr. Strangelove
Raging Bull
Star Wars
Swept Away   (The Three Stooges)

Robert DeNiro's performance in
Raging Bull

Gene Kelly's signature dance in
Singing in the Rain

CHAPTER SEVEN

DYNAMICS OF ELECTRICAL AND ELECTROMECHANICAL SYSTEMS

7-1 INTRODUCTION

In this chapter we consider the application of Lagrange's equations to the formulation of the equations of motion for lumped-parameter electrical networks and electromechanical systems. The use of electrical and electromechanical devices and systems is omnipresent throughout modern society. The generation, transmission, and use of energy; communication systems including telephone, radio, and television; scientific, entertainment, and household devices and instruments; and transportation and industrial equipment are only a few of many broad categories involving electrical and electromechanical systems, and by which humankind has expanded its innately limited ranges of functioning. Indeed, we are so dependent on—yet ignorant of the detailed functioning of—most of these systems that the collapse of a few of them would lead to major dislocations within modern society.

A detailed understanding of many of the systems cited above requires substantial knowledge of electromagnetic theory. But fortunately for equation formulation, the variational approach can be used to circumvent many of the intricacies of electromagnetic theory. Furthermore, many of these systems operate at frequencies at which the electrical and magnetic effects substantially decouple, resulting in problems that can be satisfactorily modeled as lumped-parameter networks. With Maxwell's equations as the starting point, the theoretical bases for *quasistatic models,* as electromagnetic lumped-parameter dynamic models are called, are explored in Appendix G. Models of this type can be characterized by a finite number of generalized electrical coordinates (combined with a finite number of mechanical coordinates in the case of electromechanics) by which energy functions, coenergy functions, and work expressions can be derived for use in a lagrangian formulation. Compared with the direct approaches of circuit analysis and field theory, the variational approach is enormously accessible. For those electrical circuit elements commanding our interest, Table 7-2 summarizes all the equations required to formulate the equations of motion for electrical networks via the variational approach; the reader should get to it as soon as practicable. Beyond the examples following Table 7-2, we consider electromechanical systems.

We note, and show in Appendix G, that in the analysis of lumped-parameter electrical networks, electric and magnetic effects substantially decouple; indeed, this is a requirement for the validity of the lumped-parameter approximation. From such approximations, the concepts of circuit theory, including "Kirchhoff's laws," can be deduced from Maxwell's equations. Thus, Kirchhoff's laws are not independent laws of physics; they are simply convenient *approximate* consequences of Maxwell's equations for physical systems in which the electric and magnetic field effects substantially decouple. Thus we shall refer to them as *Kirchhoff's rules*.

The initial observations of electricity and magnetism occurred in antiquity, and were associated with such natural phenomena as (1) atmospheric lightning, (2) St. Elmo's fire—a pale glow on the tips of pointed objects during stormy weather, (3) torpedo fish or electric eels stunning their prey, (4) the amber effect—the characteristic of amber, when rubbed, to attract objects, and (5) lodestone's ability to attract iron objects. Records from antiquity from the Near East and the Far East chronicle various of these natural phenomena.

It has become somewhat commonplace to attribute the discovery of the amber effect to Thales of Miletus, but such an attribution is problematic since there is no extant document verifying that Thales, like Pythagoras, ever wrote any technical manuscript. Nevertheless, like so many western technological terms, the words *electricity* and *magnetism* are of Greek origin. Our word "electron" is derived from the Greek word for amber and the word "magnetism" was probably derived from Magnesia, a district in Thessaly in Greece, where lodestone was found in the West; lodestone had already been discovered and noted in the Far East.

Unlike astronomy, mathematics, and mechanics, which were already being developed in antiquity, electricity and magnetism as sciences were substantially created in the seventeenth and eighteenth centuries. While there are numerous scientists who can be cited for efforts during this period, in forgoing such a list it seems appropriate to bracket these two centuries by two renowned events: (1) the publication in London of *On the Magnet* in 1600 by the Englishman William Gilbert (1544–1603) and (2) the monumental invention of a method to produce a continuous electric current via a *voltaic pile* by the Italian Alessandro Volta (1745–1827), as read before the Royal Society on June 26, 1800 and subsequently published in the *Transactions of the Royal Society* later that year. Throughout the seventeenth and eighteenth centuries, the two sciences of electricity and magnetism developed as electrostatics and magnetostatics, entirely independent of one another.

The development of electromagnetic theory includes the names of many of the great scientific experimentalists and theoreticians of the nineteenth century. In this abridged summary, however, we shall cite three critical accomplishments that represent paramount contributions to this development.

First, until the early nineteenth century, it was generally believed that electricity and magnetism were two independent phenomena. In 1820, after more than a decade of experiments, the Danish scientist Hans Christian Oersted (1771–1851) demonstrated that electric currents can exert forces on a compass needle; that is, *electric currents can be used to create magnetic fields!* The discipline of electromagnetism had been born. Within a year, the Frenchmen André Marie Ampere (1775–1836), Jean Baptiste Biot (1774–1862), and Felix Savant (1791–1841) had deduced the laws for forces between two conductors and between a conductor and a magnet.

Second, in juxtaposition with the experiments of Oersted, the English experimentalist Michael Faraday (1791–1867) in 1831 demonstrated that a time-varying magnetic field can produce an electric current; that is, *magnetic fields can be used to create electric currents!* Along the way, Faraday conceived the concepts of electric and magnetic fields. Faraday encountered substantial difficulty in expressing his ideas quantitatively, however, because he had little formal education and knew little mathematics. Although Faraday's concept of *fields* was by no means readily accepted by the scientific community, it was one of the important contributions that would lead Maxwell to write the monumental reciprocal fact: electricity in motion produces the same effect as magnets at rest, while magnets in motion produce the same effects as electricity at rest.

Third, the experiments of Oersted (electric currents produce magnetic fields) and Faraday (time-varying magnetic fields produce electric currents) led the Scottish physicist James Clerk Maxwell (1831–1879)—like Newton, of Trinity College, Cambridge—to predict the existence of *electromagnetic waves*. Earlier experiments by the Frenchman Charles Augustin Coulomb (1736–1806) had shown that the attractive electric force[1] $\boldsymbol{F}_e$ on charge q_2 exerted by an opposite charge q_1 is

$$\boldsymbol{F}_e = -K_e \frac{q_1 q_2}{r^2} \boldsymbol{u}_r$$

where $\boldsymbol{u}_r$ is the unit vector from q_1 toward q_2, r is their separation and, by measurement,

$$K_e = 9 \times 10^9 \frac{(\text{newton})(\text{meter})^2}{(\text{coulomb})^2}$$

And, experiments by the Englishman John Michell (1724–1793) had shown that the attractive magnetic force[1] $\boldsymbol{F}_m$ on pole p_2 exerted by an opposite pole p_1 is

$$\boldsymbol{F}_m = -K_m \frac{p_1 p_2}{r^2} \boldsymbol{u}_r$$

where $\boldsymbol{u}_r$ is the unit vector from p_1 toward p_2, r is their separation and, by measurement,

$$K_m = 1 \times 10^{-7} \frac{(\text{newton})(\text{second})^2}{(\text{coulomb})^2}$$

Note that the ratio of $\dfrac{K_e}{K_m}$ has units of $\dfrac{m^2}{s^2}$; that is, the square of a speed. But, as Maxwell was to ask, is this a speed of any special significance?

Clearly,

$$\sqrt{\frac{K_e}{K_m}} = 3 \times 10^8 \ \frac{\text{m}}{\text{s}}$$

is the speed of light in a vacuum, which had been measured sufficiently accurately prior to Maxwell's work that he concluded that light is an electromagnetic wave. Continuing, Maxwell made a correction to Ampere's law consistent with this finding and the revealed coupling of electric and magnetic fields, gathered the work of the above authors and others, and compiled a summary of electromagnetic theory, thus laying the foundations for radio, television, and an enormous number of modern devices and systems. Within a quarter century of Maxwell's 1864 publication of his theoretical results, the German Heinrich Hertz (1857–1894) experimentally demonstrated radio waves (1887). And ultimately, Oliver Heaviside, who had cast the work of Grassman and Gibbs into modern vector notation, would do the same for the ponderous equations that Maxwell left.

---

[1] Refer to Eq. (4-3) and Figures 1-16 and 4-1 for interesting comparisons in form and explanations.

7-2 FORMULATION OF EQUATIONS OF MOTION FOR ELECTRICAL NETWORKS

Electrical networks are representations of the interconnections of passive and active electromagnetic elements that constitute a system or subsystem. The electrical variables that describe the behavior of the system are constrained by the physical interconnections and the constitutive relations of the individual elements. The interconnection constraints on electrical networks are represented by *Kirchhoff's rules.**

> *During 1846, while he was still a student, the German Robert G. Kirchhoff (1824–1887) derived the rules commonly called "Kirchhoff's laws," which govern the current and voltage at the nodes in an electrical network.

Kirchhoff's current rule (KCR) applies to every node in an electrical network. It is a statement of the conservation of charge, indicating that an unneutralized charge cannot accumulate at a node in the network. *Kirchhoff's current rule states that the algebraic sum of the currents entering any node must be zero.* A consequence of KCR is that elements connected in series always conduct the same current.

Kirchhoff's voltage rule (KVR) applies to every closed loop in an electrical network. It is a statement of the fact that the voltage between any two points in a network is independent of the path through the network from one of such points to the other. *Kirchhoff's voltage rule states that the algebraic sum of the voltage drops around any closed loop in a network must be zero.* A consequence of KVR is that elements connected in parallel always have the same voltage across them.

With Kirchhoff's interconnection rules as the framework, the requirements that must be satisfied in the formulation of the equations of motion for electrical networks may be stated as follows:

1. Requirements on the currents, which encompass KCR and the relationships between charge variables q_k and current variables i_k given by

$$\frac{dq_k}{dt} = i_k \tag{7-1}$$

2. Requirements on the voltages, which encompass KVR and the relationships between the flux linkage variables λ_k and voltage variables e_k given by

$$\frac{d\lambda_k}{dt} = e_k \tag{7-2}$$

3. Constitutive relations for all the passive and active elements.

The fundamental quantities and the corresponding units for the electrical variables used in these requirements are summarized in Table 7-1. The subscript k in Eqs. (7-1) and (7-2) is simply a counter that indicates a particular variable. The importance of this collection of statements is emphasized by giving it essentially an entire section in this chapter. These statements are analogous to Eq. (2-1) or (4-11) or (5-1) for mechanical systems.

The concepts of completeness and independence that were defined in conjunction with generalized geometric coordinates in Chapter 5 have their counterparts for electrical variables. The definitions of completeness and independence given in Chapter 5 also apply to charge coordinates and flux linkage coordinates. By definition, the independent loop currents, which satisfy KCR and which are capable of defining the current in every network

TABLE 7-1 Quantities and MKS Units of Basic Electrical Variables

Quantity	Variable	Unit
Charge	q	coulomb
Current	i	ampere
Flux Linkage	λ	weber
Voltage	e	volt

element, are the time derivatives of a *complete* and *independent* set of *generalized charge coordinates* q_k. Also, by definition, the independent node voltages, which satisfy KVR and which are capable of defining the voltage across every network element, are the time derivatives of a *complete* and *independent* set of *generalized flux linkage coordinates* λ_k.

The concept of "degrees of freedom" is not generally used in conjunction with electrical networks. There is no electrical network that can be meaningfully regarded as nonholonomic. To the extent that it might be desirable to so designate them, *all electrical networks are holonomic.* Thus, in a charge formulation, the number of "degrees of freedom" is equal to the number of generalized charge coordinates q_k in a complete and independent set, the number of degrees of freedom being the number of δq_k. And in a flux linkage formulation, the number of "degrees of freedom" is equal to the number of generalized flux linkage coordinates λ_k in a complete and independent set, the number of degrees of freedom being the number of $\delta \lambda_k$.

In the *direct approach* for formulating the equations of motion for electrical networks, the analysis consists of selecting a set of generalized variables, writing equations that satisfy each of the three general requirements, and then combining these equations, eliminating variables wherever possible, ideally with the use of an established procedure or protocol. For example, in a procedure that is sometimes called "loop analysis," a set of complete and independent charge variables q_k is defined, and Requirement 1 (KCR) is used to express all currents and the remaining (dependent) charges in terms of the q_k. Then Requirement 3 is used to express the voltages or flux linkages across all elements in terms of the q_k. Finally, Requirement 2 (KVR) is imposed on the voltages and flux linkages, resulting in a set of equations in terms of the q_k.

Alternatively, in a procedure that is sometimes called "node analysis," a set of complete and independent flux linkage variables λ_k is defined, and Requirement 2 (KVR) is used to express all voltages and the remaining (dependent) flux linkages in terms of the λ_k. Then Requirement 3 is used to express the currents or charges through all elements in terms of the λ_k. Finally, Requirement 1 (KCR) is imposed on the currents and charges, resulting in a set of equations in terms of the λ_k.

Except for the summaries given immediately above, we shall not pursue the direct approach for electrical networks. It is likely that most readers have already encountered the direct approach for electrical networks. The remainder of our electrical network analyses will focus on the *indirect* or *variational approach*.

In developing the indirect approach, we shall be concerned with the flow of energy. The rate of electrical energy transfer is called the *power* $\mathcal{P}$, and is defined as the product of voltage and current as

$$\mathcal{P} = ei \qquad (7\text{-}3)$$

The unit of power is the *watt*, which is one volt-ampere.

Our purpose in presenting the indirect approach for electrical networks is twofold. First, although the direct approach for analysis of electrical networks is well established and is the preferred approach of many engineers, the indirect approach may often prove to be more efficient and easier to apply in circuits containing many elements. Second, in presenting the indirect approach for electrical networks, we lay the foundations for the analysis of electromechanical systems, where energetic concepts and variational techniques are broadly used. Furthermore, it is worth noting that, except where otherwise indicated, the concepts established here are equally valid for systems containing nonlinear as well as linear elements.

7-3 CONSTITUTIVE RELATIONS FOR CIRCUIT ELEMENTS

The physical devices in electrical systems, as compared with those in mechanical systems, are *relatively* easy—not *easy*—to model because the ideal network elements, called resistors, capacitors, inductors, amplifiers, and voltage and current sources, correspond rather well to the physical devices having those same names. In this section we consider the constitutive descriptions of several of the most common ideal electrical network elements. These elements are *ideal* in the sense that each of them models only a single electromagnetic characteristic; resistance, capacitance, or inductance for passive elements, and other single characteristics to be specified for active elements. Although the accurate modeling of many physical devices requires the use of more than a single ideal element, the resulting network model nevertheless contains only combinations of these common ideal elements. For example, the modeling of physical devices called resistors may require, in addition to an ideal element having the same resistance as the physical device, other resistors in series with it to model the increase in resistance due to bonded connector resistance, or other resistors in parallel with it to model the decrease in resistance due to leakage effects. However, in general, we shall not explore such modeling strategies, as we simply present the constitutive descriptions of ideal electrical elements, which we categorize broadly as passive or active network elements.

Furthermore, for the first time[2] we note that the variational techniques presented here can be extended to such modern electronic elements as operational amplifiers and transistors. We present operational amplifiers, voltage sources, and current sources as ideal active elements, and forgo transistors in this chapter.

7-3.1 Passive Elements

Passive electrical elements are those that do not possess an energy source and therefore are incapable of delivering more electrical energy to the network than was previously supplied to them. We now consider the three most common lumped-parameter passive electrical elements: capacitors, inductors, and resistors. The capacitor and the inductor are capable of storing energy and delivering stored energy back to the network; the resistor dissipates energy primarily as radiation, mostly in the form of heat, but also in other forms such as electric lighting.

---

[2]We write "for the first time" because we can find no examples in the literature where this has been accomplished.

The Capacitor: A capacitor is an electrical network element whose characteristics can be measured statically by applying a known charge and measuring the corresponding voltage. Capacitors are made in many sizes and shapes, but they always consist of two conductors, generally separated by a dielectric material. Dielectrics, commonly called insulators, are materials in which an electric field can be established without permitting a significant flow of charge through them. The capacitor is said to be uncharged when both plates, or conductor terminations, contain the same quantity of electric charge. The capacitor is charged to q when q units of charge have been added to one plate *and* q units of charge have been removed from the other plate. The network schematic for a capacitor is sketched in Figure 7-1*a*.

Via static measurement, the constitutive relation for an *ideal capacitor* can be determined in the form of an equation, $e = e(q)$; or inversely, $q = q(e)$. A sketch of such a relation appears as the curve in Figure 7-1*b*. If e is constant, q is constant. In order to conduct current, the capacitor must experience $\dfrac{dq}{dt}$, which implies $\dfrac{de}{dt}$. Because current flows only when e is changing, the capacitor is an *open circuit* for direct current for which e is a constant. Thus, it may be shown[3] that the capacitor stores energy via its *electric field*. Therefore, an alternative constitutive description of a capacitor can be given in terms of a pair of electrical complementary state functions as follows: the electrical energy function W_e and the electrical coenergy function W_e^*.

The *electrical energy function*[4] $W_e(q)$ for a capacitor is the work done on it in changing its charge from a datum of no charge to charge q. This work can be computed as the time integral of the electrical power delivered to the capacitor. Combining Eqs.(7-1) and (7-3) gives

$$W_e(q) = \int_0^t \mathcal{P}\,dt = \int_0^t ei\,dt = \int_0^q e\,dq \qquad (7\text{-}4)$$

The unit of electrical energy is the *joule,* which is one watt-second. The electrical energy associated with a specific operating point (q, e) is represented by the area below the curve in Figure 7-1*b*. When the electrical energy function is known, the voltage across the capacitor can be recovered by differentiating Eq. (7-4) as

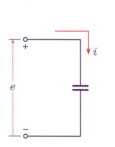

(*a*) Network schematic.

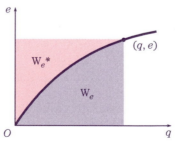

(*b*) Constitutive relation.

FIGURE 7-1 Network schematic and constitutive relation for capacitor.

---

[3]See Section G-4 in Appendix G.

[4]Although $W_e(q)$ is more properly called the *electrical energy function,* for brevity we shall often refer to it simply as the *electrical energy.*

$$\frac{\partial W_e}{\partial q} = e \tag{7-5}$$

Using the concept of the Legendre transformation, which was introduced in Chapter 5, the *electrical coenergy function* $W_e^*(e)$ for the capacitor may be written as

$$W_e^*(e) = qe - W_e(q) \tag{7-6}$$

The electrical coenergy function associated with a specific operating point is represented by the area above the curve in Figure 7-1*b*, and thus can be written alternatively as

$$W_e^*(e) = \int_0^e q\, de \tag{7-7}$$

When the electrical coenergy is known, the charge on the capacitor can be obtained by differentiating Eq. (7-7) as

$$\frac{\partial W_e^*}{\partial e} = q \tag{7-8}$$

For most capacitors, the constitutive relation is nearly linear over the operating range of its variables. If the capacitor can be modeled as linear, the relation between q and e can be written as

$$q = Ce \tag{7-9}$$

where the proportionality constant C is called the *capacitance*. The unit of capacitance is the *farad,* which is one coulomb per volt. For a linear capacitor, substitution of Eq. (7-9) into Eq. (7-4) and integration of the resulting expression give the electrical energy as

$$W_e(q) = \int_0^q \frac{q}{C}\, dq = \frac{q^2}{2C} \tag{7-10}$$

It is interesting to note that differentiation of Eq. (7-10) in accordance with Eq. (7-5) gives Eq. (7-9) since

$$\frac{\partial W_e}{\partial q} = \frac{q}{C} = e \tag{7-11}$$

The recovery of the constitutive relation in this manner is generally not used for electrical networks but does prove to be beneficial in some analyses of electromechanical systems. For a capacitor that can be modeled by the linear constitutive relation in Eq. (7-9), the electrical coenergy is found via Eq. (7-7) as

$$W_e^*(e) = \int_0^e Ce\, de = \frac{1}{2}Ce^2 \tag{7-12}$$

and so Eq. (7-8) becomes

$$\frac{\partial W_e^*}{\partial e} = Ce = q \tag{7-13}$$

The Inductor: According to Olmsted's classic experiment, when current flows through a conductor, a magnetic field is established (induced) in the space or material around the conductor, the strength of the field being proportional to the current. According to *Faraday's* *law,* when a conductor encloses a region containing a magnetic field, a voltage is induced in the conductor whenever the magnetic field linking the conductor changes.[5] To some extent, all conductors exhibit these induction effects; if the conductor is a coil of many closely packed turns, induction effects are significantly enhanced. Complementary to the $q-e$ characteristics of a capacitor, an *ideal inductor* is an electrical network element in which the flux linkage λ depends only on the instantaneous current i. A network schematic of an inductor is sketched in Figure 7-2a.

The inductance of a coil in air may be generally modeled as linear. In many real inductors, ferromagnetic cores such as iron, nickel, or cobalt are added to increase the coil inductance by several orders of magnitude. The addition of such cores, however, generally leads to nonlinear inductance and sometimes to undesirable hysteretic effects where the flux linkage may depend on the history of the current.

By its definition, if the current in an ideal inductor is constant, the flux linkage is also constant. Then, for constant current, the voltage across an ideal inductor must be zero since $e = \dfrac{d\lambda}{dt} = 0$. So, for direct current, the ideal inductor behaves like a perfect conductor, a *short circuit.* Thus, it may be shown[6] that the inductor stores energy via its *magnetic field.* Therefore, a constitutive description of an ideal inductor can be given in terms of a pair of magnetic complementary state functions as follows: the magnetic energy function W_m and the magnetic coenergy function W_m^*.

The *magnetic energy function*[7] $W_m(\lambda)$ for an ideal inductor is the work done on it in changing its flux linkage from a datum of no flux linkage to a flux linkage λ. This work can be computed as the time integral of the magnetic power delivered to the inductor. Combining Eqs. (7-2) and (7-3) gives

$$W_m(\lambda) = \int_0^t \mathcal{P}\, dt = \int_0^t ei\, dt = \int_0^\lambda i\, d\lambda \tag{7-14}$$

The unit of magnetic energy is the *joule* since via Eqs. (7-4) and (7-14), W_m and W_e must have the same units. The magnetic energy associated with a specific operating point (λ, i) is represented by the area below the curve in Figure 7-2b. When the magnetic energy function is known, the current in the inductor can be recovered by differentiating Eq. (7-14) as

$$\frac{\partial W_m}{\partial \lambda} = i \tag{7-15}$$

Using the concept of the Legendre transformation, the *magnetic coenergy function* $W_m^*(i)$ for the inductor may be written as

$$W_m^*(i) = \lambda i - W_m(\lambda) \tag{7-16}$$

The magnetic coenergy function associated with a specific operating point is represented by the area above the curve in Figure 7-2b, and thus can be written alternatively as

$$W_m^*(i) = \int_0^i \lambda\, di \tag{7-17}$$

[5]See Problem 7-1 for an elementary interpretation of flux linkage.

[6]See Section G-4 in Appendix G.

[7]Although $W_m(\lambda)$ is more properly called the *magnetic energy function,* for brevity we shall often refer to it simply as the *magnetic energy.*

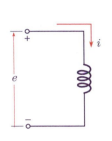

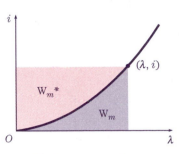

(a) Network schematic.   (b) Constitutive relation.

FIGURE 7-2 Network schematic and constitutive relation for inductor.

When the magnetic coenergy is known, the flux linkage associated with the inductor can be obtained by differentiating Eq. (7-17) as

$$\frac{\partial W_m^*}{\partial i} = \lambda \tag{7-18}$$

If the inductor can be modeled as linear, the relation between λ and i can be written as

$$\lambda = Li \tag{7-19}$$

where the proportionality constant L is called the *inductance*. The unit of inductance is the *henry,* which is one weber per ampere or one volt-second per ampere. For a linear inductor, substitution of Eq. (7-19) into Eq. (7-14) and integration of the resulting expression give the magnetic energy as

$$W_m(\lambda) = \int_0^\lambda \frac{\lambda}{L}\, d\lambda = \frac{\lambda^2}{2L} \tag{7-20}$$

It is interesting to note that differentiation of Eq. (7-20) in accordance with Eq. (7-15) gives Eq. (7-19) since

$$\frac{\partial W_m}{\partial \lambda} = \frac{\lambda}{L} = i \tag{7-21}$$

The recovery of the constitutive relation in this manner is generally not used for electrical networks but does prove to be beneficial in some analyses of electromechanical systems. For an inductor that can be modeled by the linear constitutive relation in Eq. (7-19), the magnetic coenergy is found via Eq. (7-17) as

$$W_m^*(i) = \int_0^i Li\, di = \frac{1}{2}Li^2 \tag{7-22}$$

and so Eq. (7-18) becomes

$$\frac{\partial W_m^*}{\partial i} = Li = \lambda \tag{7-23}$$

The Resistor: Many dissipative devices may be modeled as resistive electrical network elements for which their constitutive relations can be plotted via a voltage-current graph. The linear constitutive relation for a resistive element is called *Ohm's law,*

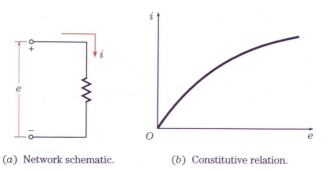

(a) Network schematic.   (b) Constitutive relation.

FIGURE 7-3   Network schematic and constitutive relation for resistor.

$$e = Ri \qquad (7\text{-}24)$$

where the proportionality constant R is the *resistance*. The unit of resistance is the *ohm*, which is one volt per ampere. Ohm's law may be expressed alternatively as

$$i = Ge \qquad (7\text{-}25)$$

where the proportionality constant $G = R^{-1}$ is the *conductance*. The unit of conductance is the *siemens*, which is one ampere per volt (ohm^{-1}). The *mho* is an antiquated name for the siemens.

Resistive elements dissipate electrical energy, which is usually transformed into thermal energy by a process called *joule heating*. This heat might be useful (such as in electric stoves, toasters, or safety fuses) or harmful (such as the overheating of devices). Many electronic devices such as diodes and transistors behave as nonlinear resistive elements.

Resistors are electrical resistive elements that are designed and built into devices to provide the useful functioning of the device or the system. Resistors are manufactured to have specific values of resistance. The network schematic for a resistor is sketched in Figure 7-3a and its constitutive relation is sketched in Figure 7-3b.

Furthermore, many resistors obey Ohm's linear constitutive relation over a broad portion of their operating ranges. In such resistors, the major cause of deviation from Ohm's law is excessive heating during electrical energy dissipation, which may cause the resistor to burn or melt.

7-3.2   Active Electrical Elements

An *active* electrical element is a device that is capable of controlling the flow of electrical energy from a source to an electrical load. Voltage sources, current sources, and operational amplifiers are three such active devices that we shall consider. Our goal is not to discuss the detailed design or internal functioning of these devices, but to summarize their constitutive or operating characteristics in order to analyze their dynamic behaviors in electrical networks.

Ideal Voltage Source: An *ideal voltage source* is an electrical network element that produces a prescribed time history of voltage difference $E(t)$ across it, independent of the magnitude or direction of the current flowing through it. Thus, an ideal voltage source is called an *independent source*. The network schematic for an ideal voltage source is sketched in Figure 7-4. Within their operating limits, fully charged batteries can be modeled as ideal voltage sources of nearly constant voltage.

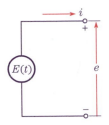

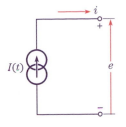

FIGURE 7-4   Network schematic for ideal voltage source $E(t)$.

FIGURE 7-5   Network schematic for ideal current source $I(t)$.

Ideal Current Source: An *ideal current source* is an electrical network element that produces a prescribed time history of current $I(t)$ through it, independent of the voltage difference across its terminals. Thus, an ideal current source is also called an *independent source.* The network schematic for an ideal current source is sketched in Figure 7-5. There is no single simple physical device, analogous to the battery as an ideal voltage source, which approximates an ideal current source. Nevertheless, the current source is a useful device for modeling some electrical networks, particularly those containing transistors and power supplies that have been designed specifically to approximate a constant current source.

Operational Amplifier*:   The *operational amplifier* is an active device and is one of the most versatile components in modern electronics, frequently used in conjunction with feedback and control. The *op-amp,* as it is usually called, may be a complex integrated circuit consisting of dozens of transistors, resistors, and capacitors, but as an electrical network element it has a relatively simple input–output relationship. Under normal operating conditions, an op-amp behaves generally as a high-gain linear voltage amplifier; that is, its output voltage is a large multiple of its input voltage, the input–output correspondence being linear

---

*Numerous developments of twentieth century electronics were achieved in the industry of the United States. Much of the earliest development of operational amplifiers was conducted from the 1940s onward at the Bell Telephone Laboratories (Bell Labs) by the Americans Harold S. Black, Hendrik W. Bode, Clarence A. Lovell, and David B. Parkinson. Also, George A. Philbrick, while at the National Defense Research Council, was an early significant contributor to the development and later to the commercialization of operational amplifiers. The longer history of the design of modern op-amps can be constituted to include the 1904 *vacuum diode* of the Englishman John A. Fleming (also the inventor of the "right-hand rule" which we have used); the 1907 *vacuum triode* of the American Lee DeForest; the late-1940s *transistors* of the Americans John Bardeen, Walter H. Brattain, and William B. Shockley (all at Bell Labs); and the late-1950s *integrated circuit* of Americans Jack S. Kilby (Texas Instruments) and Robert N. Noyce (Fairchild Semiconductor Corporation). The name "operational amplifier" appears to have been devised by John R. Ragazzini and his colleagues R.H. Randall and F.A. Russell in 1947.

---

over the operating range. For example, a typical use of an operational amplifier is to amplify relatively low levels of electrical signals to sufficiently high levels that are capable of driving electromechanical devices such as audio speakers or instruments or actuators, perhaps, for the purpose of feedback control. Other uses include assisting in active signal filtering and performing analog mathematical operations, the latter application being the one in which the term "operational amplifier" originated—for instance, see Example 7-6.

The basic op-amp schematic is sketched in Figure 7-6a. The five terminals shown comprise the minimum set required to describe its functioning. Many op-amps have one or two additional external terminals. The terminals connected to the supply voltages $+E_s$ and $-E_s$

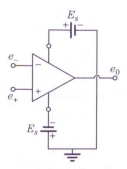

(*a*) Schematic of op-amp, showing connections
to direct current supplies $+E_s$ and $-E_s$.

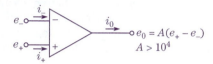

(*b*) Common schematic of op-amp, with supply terminals not shown.

FIGURE 7-6   Schematics of op-amps, indicating node voltage and
current notations.

of the device emphasize that the op-amp is an active device. The unconventional use of the subscript "s" denotes "supply" on the supply voltages. The primary purpose of these voltage supplies is to provide all the power that may be needed for the op-amp to function as described below. The input voltages e_- and e_+ and the output voltage e_0 are network voltages defined in accordance with the op-amp's location in an electrical circuit. Their subscripts designate the particular op-amp terminal with which they are associated, and are not suggestions of algebraic signs.

The more common schematic for an op-amp is sketched in Figure 7-6*b*, where the supply terminals have been omitted since, as we shall see, they do not participate in the network connections of the circuit in which the device is one of a number of elements. This more common schematic causes no confusion when deriving the equations of motion, since these equations of motion will be derived based upon idealized behavior, which will be outlined below. Note that in Figure 7-6*b*, we have added input currents, i_+ and i_-, and an input–output relationship, each of which we shall now address.

Op-amps are designed to function almost exclusively within their linear operating range. Within this linear range, their behavior is simple and can be characterized by two features: (1) they have an infinite input resistance and (2) they have a zero output resistance. The infinite input resistance effectively signifies that the input currents, i_+ and i_-, are both essentially zero. The zero output resistance effectively signifies that the op-amp can be modeled as a dependent voltage source of voltage e_0, where the functional dependence of e_0 on the difference in input voltages, $(e_+ - e_-)$, is a linear one. The constant of linear proportionality is denoted by A, which is called the *open-loop gain,* and which typically ranges from 10^4 to 10^9. For example, for $A = 10^4$ and for a supply voltage of 10 volts, the width of the linear region of allowable inputs is ± 1 millivolt, as sketched in Figure 7-7. For higher open-loop gains and the same supply voltage, the linear region would be correspondingly narrower.

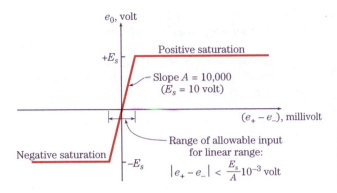

FIGURE   7-7 Input–output relationship for op-amp, indicating range of allowable input for linear range when open-loop gain is 10^4.

Therefore, in their linear range of operation, op-amps may be characterized by the following set of modeling approximations:

1. Input currents, i_+ and i_-, are zero.
2. Input voltages must be approximately equal.
3. Op-amp is modeled as a dependent voltage source.
4. Open loop gain is a constant, for all input frequencies.

(7-26)

Characteristic 1 means that

$$i_- = 0 \qquad i_+ = 0 \tag{7-27}$$

and characteristics 2, 3, and 4 plus Figure 7-7 lead to a device that may be modeled as a high-gain dependent voltage source having an input–output relationship expressible as

$$e_0 = A(e_+ - e_-) \tag{7-28}$$

where e_+ and e_- must be approximately equal and where Eq. (7-28) is valid for all input frequencies. The phrase "dependent voltage source" means that we can treat the op-amp just as we treat any other voltage source; it's just that the specific value of the source voltage now *depends* on the difference between the two input voltages e_+ and e_-.

In the analysis of idealized op-amps, Eq. (7-28) implies something about A, the open-loop gain, which is not generally stated, but which we have expressed explicitly. The voltages e_0, e_+, and e_- may be functions of frequency. As typically written, however, A is not considered to be a function of frequency, although in any real device this assumption must

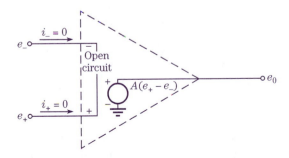

FIGURE   7-8 Idealized equivalent circuit model for op-amp.

ultimately break down. Thus, as we stated in characteristic 4 in Eq. (7-26), in the analysis of ideal op-amps, the open-loop gain is assumed to be a constant for all input frequencies.

The characteristic behavior described above leads to the idealized circuit model sketched in Figure 7-8 for an op-amp operating within its linear range. The dashed triangular outline of the circuit model in Figure 7-8 simply indicates that the op-amp is modeled as a single element and, as we shall note in the examples later in this chapter, without loss of information the dashed line may be omitted from all equivalent circuits under analysis.

From Figure 7-8, we see that the idealized equivalent circuit model for an op-amp is a combination of an open circuit plus a dependent voltage source having the input–output relationship given in Eq. (7-28). As such, as a circuit element, the dependent voltage source portion of the op-amp should be treated precisely the same as an ideal voltage source as sketched in Figure 7-4.

An interesting, though unconventional, interpretation of the idealized equivalent circuit model in Figure 7-8 is that the voltage supplies (E_s and $-E_s$) have been incorporated into the dependent voltage source portion of the model. In this manner, during linear operation, the level of current that is drawn from the supplies is whatever current is required in order to satisfy the voltage transfer relation given in Eq. (7-28); but that current does not affect the choice of generalized charge coordinates q_k. Nevertheless, the significant feature of the dependent voltage source portion of the idealized equivalent circuit model is that it should be treated the same as an ideal voltage source. For the two input terminals of the idealized equivalent circuit model in Figure 7-8, characteristics 1 and 2 in Eq. (7-26) apply.

Finally, whereas all the above discussion on op-amps has been based on linear operation, which is typical in the vast majority of op-amp applications, our *assumption* of linearity does not ensure linearity. Thus, with op-amp models, as with all linear models, it is important to verify that linearity is a valid assumption. Perhaps the most common nonlinear occurrence in op-amps is saturation, which fortunately is easy to detect. In particular, from Figure 7-7 it is clear that whenever the op-amp output e_0 approaches the supply voltage E_s, nonlinearity is also approached. Typically, in physical devices saturation can be observed within a volt or two of the supply voltage.

7-4 HAMILTON'S PRINCIPLE AND LAGRANGE'S EQUATIONS FOR ELECTRICAL NETWORKS

As in the case of mechanical systems, the equations of motion for electrical networks may be formulated via the direct approach or the indirect approach. Irrespective of the approach used for equation formulation, the three requirements given in Section 7-2 must be satisfied. In both the direct approach and the indirect approach, either a charge variable formulation or a flux linkage variable formulation may be used. It is likely that most readers have had some experience with the direct approach, as reviewed briefly in Section 7-2. No further discussion of the direct approach for electrical networks will be considered, however, as we turn exclusively to formulations via the indirect approach.

7-4.1 Generalized Charge Variables

In a charge variable formulation, the admissibility requirements (that is, admissible motions) are satisfied by the selection of a complete and independent set of loop currents

i_k $(k = 1, 2, \ldots, n_e)$, which satisfies Kirchhoff's current rule, consistent with the relations $\dot{q}_k = i_k$. Then, Hamilton's principle for electrical networks may be expressed as

$$\text{V.I.} = \int_{t_1}^{t_2} \left[\delta(W_m^* - W_e) + \sum_{k=1}^{n_e} \mathcal{E}_k \delta q_k \right] dt \tag{7-29}$$

where n_e is the number of complete and independent generalized charge coordinates. Equation (7-29), which is an extended form of the classical Hamilton's principle, may be expressed as follows:

- An admissible motion of the system between specified configurations at t_1 and t_2 is a natural motion if, and only if, the variational indicator vanishes for arbitrary admissible variations.

It should be noticed immediately that this statement of Hamilton's principle is the same as that given in Chapter 5 for mechanical systems. Indeed, the principle has not changed; we need only interpret its terminology in the context of the charge formulation that we are considering. Also, as in Chapter 5, we adopt Eq. (7-29) as a fundamental law of nature, not derivable from any other law or principle.

Here, an "admissible motion" is simply an electrically dynamic requirement on the charges that satisfies KCR in accordance with the choice of a complete and independent set of loop currents. This satisfies Requirement 1 for equation formulation, Eq. (7-1). In a charge formulation, the "natural motion" is equivalent to KVR. This satisfies Requirement 2 for equation formulation, Eq. (7-2). The constitutive relations (Requirement 3) are incorporated into the formulation via

1. W_m^*, which represents the sum of the magnetic coenergies of all the inductance in the network;

2. W_e, which represents the sum of the electrical energies of all the capacitance in the network; and

3. The generalized voltages $\mathcal{E}_k$, which represent the contributions of all the nonconservative elements in the network; nonconservative elements being those that supply energy to the network or dissipate energy from the network.

The summation over k in Eq. (7-29) accounts for the nonconservative elements that do work under an admissible variation of the charges. The work expression containing these generalized voltages will be discussed in Subsection 7-4.3. Thus, in this manner, the three requirements for the formulation of the equations of motion are satisfied.

In taking Hamilton's principle in Eq. (7-29) as a fundamental law of nature, we observe that it cannot be proved. Nevertheless, just as we demonstrated the equivalence of Hamilton's principle and Newton's second law along with kinematic and constitutive requirements for mechanical systems (Section E-1 of Appendix E), Hamilton's principle may be shown to incorporate the three requirements for the formulation of the equations of motion for electrical networks. For a charge variable formulation, this equivalence is shown in Section E-2 of Appendix E.

For electrical networks containing only lumped-parameter elements, Lagrange's equations generally provide the most direct path for deriving the equations of motion. As shown in Section E-2 of Appendix E, the necessary conditions for the vanishing of the variational indicator of Hamilton's principle in Eq. (7-29) are Lagrange's equations in the form

$$\frac{d}{dt}\left(\frac{\partial \mathcal{L}}{\partial \dot{q}_k}\right) - \frac{\partial \mathcal{L}}{\partial q_k} = \mathcal{E}_k \qquad k = 1, \dots, n_e \tag{7-30}$$

where the lagrangian is

$$\mathcal{L}(q_k, \dot{q}_k) = W_m^* - W_e \tag{7-31}$$

and the generalized voltages $\mathcal{E}_k$ corresponding to the generalized charge coordinates are derived from the variational work expression

$$\delta \mathcal{W}^{nc} = \sum_{j=1}^{M} e_j^{nc} \delta q_j = \sum_{k=1}^{n_e} \mathcal{E}_k \delta q_k \tag{7-32}$$

In Eq. (7-32) the term $\sum_{j=1}^{M} e_j^{nc} \delta q_j$ is the variational work increment done by the M non-conservative elements during an admissible variation in the charge coordinates. Also, as indicated in conjunction with Eq. (7-29), n_e is the number of complete and independent generalized charge coordinates. As in Chapter 5, the "nc" superscript denotes "nonconservative." (Note the correspondence of Eqs. (7-29) through (7-32) with those in Table 5-1.)

7-4.2   Generalized Flux Linkage Variables

In a flux linkage formulation, the admissibility requirements (that is, admissible motions) are satisfied by the selection of a complete and independent set of node potentials $e_k (k = 1, 2, \dots, n_e')$, which satisfies Kirchhoff's voltage rule, consistent with the relations $\dot{\lambda}_k = e_k$. Then, Hamilton's principle for electrical networks may be expressed as

$$\text{V.I.} = \int_{t_1}^{t_2} \left[\delta(W_e^* - W_m) + \sum_{k=1}^{n_e'} \mathcal{I}_k \, \delta \lambda_k \right] dt \tag{7-33}$$

where n_e' is the number of complete and independent generalized flux linkage coordinates. Equation (7-33), which is another extended form of the classical Hamilton's principle, may be expressed as follows:

- An admissible motion of the system between specified configurations at t_1 and t_2 is a natural motion if, and only if, the variational indicator vanishes for arbitrary admissible variations.

Again, we note that the statement of Hamilton's principle remains unaltered; we need only interpret its terminology in the context of the flux linkage formulation under consideration. Also, as in Chapter 5 and as in Subsection 7-4.1, we adopt Eq. (7-33) as a fundamental law of nature, not derivable from any other law or principle.

Here, an "admissible motion" is an electrically dynamic requirement on the flux linkages that satisfies KVR in accordance with the choice of a complete and independent set of node voltages. This satisfies Requirement 2 for equation formulation, Eq. (7-2). In a flux linkage formulation, the "natural motion" is equivalent to KCR. This satisfies Requirement 1 for equation formulation, Eq. (7-1). The constitutive relations (Requirement 3) are incorporated into the formulation via

1. W_e^*, which represents the sum of the electrical coenergies of all the capacitance in the network;

2. W_m, which represents the sum of the magnetic energies of all the inductance in the network; and

3. The generalized currents $\mathcal{I}_k$, which represent the contributions of all the nonconservative elements in the network; nonconservative elements being those that supply energy to the network or dissipate energy from the network.

The summation over k in Eq. (7-33) accounts for those nonconservative elements that do work under an admissible variation of the flux linkages. The work expression containing these generalized currents will be discussed in Subsection 7-4.3. Thus, in this manner the three requirements for the formulation of the equations of motion are satisfied.

In taking Hamilton's principle in Eq. (7-33) as a fundamental law of nature, we observe that it cannot be proved. Nevertheless, just as we demonstrated the equivalence of Hamilton's principle and Newton's second law along with kinematic and constitutive requirements for mechanical systems (Section E-1 of Appendix E), Hamilton's principle may be shown to incorporate the three requirements for the formulation of the equations of motion for electrical networks. For a flux linkage variable formulation, this equivalence is shown in Section E-3 of Appendix E.

For electrical networks containing only lumped-parameter elements, Lagrange's equations generally provide the most direct path for deriving the equations of motion. As shown in Section E-3 of Appendix E, the necessary conditions for the vanishing of the variational indicator of Hamilton's principle in Eq. (7-33) are Lagrange's equations in the form

$$\frac{d}{dt}\left(\frac{\partial \mathcal{L}}{\partial \dot{\lambda}_k}\right) - \frac{\partial \mathcal{L}}{\partial \lambda_k} = \mathcal{I}_k \qquad k = 1, \ldots, n'_e \tag{7-34}$$

where the lagrangian is

$$\mathcal{L}\left(\lambda_k, \dot{\lambda}_k\right) = W^*_e - W_m \tag{7-35}$$

and the generalized currents $\mathcal{I}_k$ corresponding to the generalized flux linkage coordinates are derived from the variational work expression

$$\delta \mathcal{W}^{nc} = \sum_{j=1}^{M'} i_j^{nc} \delta \lambda_j = \sum_{k=1}^{n'_e} \mathcal{I}_k \delta \lambda_k \tag{7-36}$$

In Eq. (7-36) the term $\sum_{j=1}^{M'} i_j^{nc} \delta \lambda_j$ is the variational work increment done by the M' nonconservative elements during an admissible variation of the flux linkage coordinates. Also, as indicated in conjunction with Eq. (7-33), n'_e is the number of complete and independent generalized flux linkage coordinates. As in Chapter 5, the "nc" superscript denotes "nonconservative." (Note the correspondence of Eqs. (7-33) through (7-36) with Eqs. (7-29) through (7-32), respectively. Also, refer to Table 5-1.)

7-4.3   Work Expressions

In both the charge formulation and the flux linkage formulation it is necessary to account for the contributions of each nonconservative element in the network. In a charge formulation the contributions of these elements are computed via Eq. (7-32), and in a flux linkage formulation the contributions of these elements are computed via Eq. (7-36). To clarify these work expressions, we consider a single arbitrary nonconservative network element as illustrated in Figure 7-9. The nonconservative element might be a resistor, an op-amp, an ideal voltage source, or an ideal current source. The directions of current flow i *through* the element and the voltage increase e *across* the element are shown for power delivered *to* the network.

In accordance with Eq. (7-3), the power delivered to the network by the element is ei, and the work increment delivered during time dt is

$$ei\,dt = e\,dq = i\,d\lambda \tag{7-37}$$

where Eqs. (7-1) and (7-2) have been used. Thus, the work increment done by the element during a differential change in q is $e\,dq$, or the work increment done by the element during a differential change in λ is $i\,d\lambda$. Analogously, the variational work increment done by the element during an admissible variation of δq or $\delta\lambda$ is $e\,\delta q$ or $i\,\delta\lambda$, respectively. These variational work increments are positive if the element supplies energy *to* the network during a positive variation of the generalized variable and are negative if the element absorbs energy *from* the network during a positive variation of the generalized variable.

Resistor: If the resistor sketched in Figure 7-3 is linear, in terms of the sign convention indicated in Figure 7-9, the constitutive relation may be rewritten as

$$e = -Ri \tag{7-38}$$

where Eq. (7-24) has been used. The change in sign between Eqs. (7-24) and (7-38) is due to the reversal of the direction of current flow in Figures 7-3a and 7-9. The physical meaning of Eq. (7-38) in conjunction with Figure 7-9 is that if the directions of current flow and voltage increase are the same, the nonconservative element delivers energy to the network and thus acts as a source. If the directions of current flow and voltage increase are opposite, which is the case for the resistor (Figure 7-3a), the nonconservative element extracts energy from the network and thus acts as a dissipator. So, for the resistor, Eq. (7-24) corresponds to Figure 7-3a and Eq. (7-38) corresponds to Figure 7-9, where Figure 7-9 is the basis for calculating the work expressions associated with the nonconservative elements in the networks.

For a charge formulation, substitution of Eq. (7-38) into Eq. (7-32) gives

$$\delta\mathcal{W}^{\text{nc}} = e^{\text{nc}}\,\delta q = -Ri\,\delta q = -R\dot{q}\,\delta q \tag{7-39}$$

where Eq. (7-1) has been used. Thus, in accordance with Eq. (7-32), the generalized voltage for a resistor in a charge formulation for a single admissible variation δq is

$$\mathcal{E} = -R\dot{q} \tag{7-40}$$

For a flux linkage formulation, substitution of Eq. (7-38) into Eq. (7-36) gives

$$\delta\mathcal{W}^{\text{nc}} = i^{\text{nc}}\,\delta\lambda = -R^{-1}e\,\delta\lambda = -R^{-1}\dot{\lambda}\,\delta\lambda \tag{7-41}$$

where Eq. (7-2) has been used. Thus, in accordance with Eq. (7-36), the generalized current for a resistor in a flux linkage formulation for a single admissible variation $\delta\lambda$ is

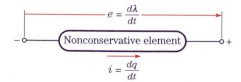

FIGURE 7-9 Arbitrary nonconservative network element, showing sign convention for power delivered *to* network in work calculations.

$$\mathcal{I} = -R^{-1}\dot{\lambda} \qquad (7\text{-}42)$$

Furthermore, in both the charge and flux linkage formulations, it is worthwhile to recall the analysis of the "floating dashpot" as presented in Example 5-15. This point will be illustrated for resistors in Examples 7-1, 7-5, 7-6, 7-7, and 7-10.

Ideal Voltage Source: Next consider the ideal voltage source $E(t)$ represented in Figure 7-4. For a charge formulation, the use of Eq. (7-32) gives

$$\delta\mathcal{W}^{\text{nc}} = e^{\text{nc}}\,\delta q = E(t)\,\delta q \qquad (7\text{-}43)$$

Note that the directions of current flow and voltage increase in Figures 7-4 and 7-9 are identical, resulting in a positive sign on the right-hand side term of Eq. (7-43). Thus, in accordance with Eq. (7-32), the generalized voltage for an ideal voltage source in a charge formulation is

$$\mathcal{E} = E(t) \qquad (7\text{-}44)$$

For a flux linkage formulation, the use of Eq. (7-36) for an ideal voltage source contributes nothing to the work expression. The voltage across an ideal voltage source at each instant in time is *specified* or *prescribed*, and therefore the corresponding flux linkage coordinate cannot be arbitrarily varied during an admissible variation of the flux linkages. Thus, for a flux linkage formulation, the ideal voltage source does not appear in the work expression. The ideal voltage source does enter the equation formulation, however, as it becomes a part of the admissibility requirements on the flux linkage coordinates. Such an admissibility requirement will be illustrated in Example 7-3.

Ideal Current Source: Now consider the ideal current source $I(t)$ represented in Figure 7-5. For a charge formulation, the use of Eq. (7-32) for an ideal current source contributes nothing to the work expression. The current through an ideal current source at each instant in time is *specified* or *prescribed*, and therefore the corresponding charge coordinate cannot be arbitrarily varied during an admissible variation of the charges. Thus, for a charge formulation, the ideal current source does not appear in the work expression. The ideal current source does enter the equation formulation, however, as it becomes a part of the admissibility requirements on the charge coordinates. Such an admissibility requirement will be illustrated in Example 7-4.

For a flux linkage formulation, the use of Eq. (7-36) gives

$$\delta\mathcal{W}^{\text{nc}} = i^{\text{nc}}\,\delta\lambda = I(t)\,\delta\lambda \qquad (7\text{-}45)$$

Note that the directions of current flow and voltage increase in Figures 7-5 and 7-9 are identical, resulting in a positive sign on the right-hand side term of Eq. (7-45). Thus, in accordance with Eq. (7-36), the generalized current for an ideal current source in a flux linkage formulation is

$$\mathcal{I} = I(t) \qquad (7\text{-}46)$$

Operational Amplifier: Since the modeled characteristic behavior of an operational amplifier is as a dependent voltage source, the work expression for an operational amplifier is essentially the same as that of an ideal voltage source. In the course of incorporating operational amplifiers into the analysis of electrical networks, it is also necessary to include each of the four modeling characteristics given in Eq. (7-26) as well as the op-amp input–output relationship given in Eq. (7-28). The analysis of electrical networks containing operational amplifiers will be illustrated in Examples 7-5 through 7-10.

TABLE 7-2   *Lumped-Parameter Offering of Variational Electricity*

	Charge Formulation	Flux Linkage Formulation
• *Generalized Coordinates*	$q_k \quad k = 1, 2, \ldots, n_e$	$\lambda_k \quad k = 1, 2, \ldots, n'_e$
• *Lagrangian*	$\mathcal{L}(q_k, \dot{q}_k) = W^*_m - W_e$	$\mathcal{L}(\lambda_k, \dot{\lambda}_k) = W^*_e - W_m$
• *State Functions*	Linear: $\quad W^*_m = \dfrac{1}{2}L\dot{q}^2$	Linear: $\quad W^*_e = \dfrac{1}{2}C\dot{\lambda}^2$
	$W_e = \dfrac{q^2}{2C}$	$W_m = \dfrac{\lambda^2}{2L}$
	Nonlinear: $\quad W^*_m = \displaystyle\int_0^{\dot{q}} \lambda\, d\dot{q}$	Nonlinear: $\quad W^*_e = \displaystyle\int_0^{\dot{\lambda}} q\, d\dot{\lambda}$
	$W_e = \displaystyle\int_0^{q} \dot{\lambda}\, dq$	$W_m = \displaystyle\int_0^{\lambda} \dot{q}\, d\lambda$
• *Work Expressions*	$\delta\mathcal{W}^{\text{nc}} = \displaystyle\sum_{j=1}^{M} e^{\text{nc}}_j \delta q_j = \sum_{k=1}^{n_e} \mathcal{E}_k \delta q_k$	$\delta\mathcal{W}^{\text{nc}} = \displaystyle\sum_{j=1}^{M'} i^{\text{nc}}_j \delta\lambda_j = \sum_{k=1}^{n'_e} \mathcal{I}_k \delta\lambda_k$
Resistor (Linear)	$-R\dot{q}\,\delta q$	$-R^{-1}\dot{\lambda}\,\delta\lambda$
Voltage Source	$E(t)\,\delta q$	No Work Expression: KVR Admissibility
Current Source	No Work Expression: KCR Admissibility	$I(t)\,\delta\lambda$
Operational Amplifier	Akin to Voltage Source Plus Idealized Characteristics	Akin to Voltage Source Plus Idealized Characteristics
• *Equations of Motion*	$\dfrac{d}{dt}\left(\dfrac{\partial\mathcal{L}}{\partial\dot{q}_k}\right) - \dfrac{\partial\mathcal{L}}{\partial q_k} = \mathcal{E}_k$ $\quad k = 1, 2, \ldots, n_e$	$\dfrac{d}{dt}\left(\dfrac{\partial\mathcal{L}}{\partial\dot{\lambda}_k}\right) - \dfrac{\partial\mathcal{L}}{\partial\lambda_k} = \mathcal{I}_k$ $\quad k = 1, 2, \ldots, n'_e$

• *Linear Constitutive Relations*
  Capacitor: $q = C\lambda$ Inductor: $\lambda = L\dot{q}$ Resistor: $\dot{\lambda} = R\dot{q}$
• *Linear Op-Amp*
  Voltage-Transfer Characteristic: $e_0 = A(e_+ - e_-)$

7-4.4   Summary of Lumped-Parameter Offering of Variational Electricity

The entire collection of results obtained in this chapter for deriving the equations of motion for lumped-parameter electrical networks is summarized in Table 7-2. The use of Table 7-2 enables one to derive the equations of motion for lumped-parameter electrical networks consisting of linear and nonlinear elements, given the appropriate constitutive relations. In the examples below, only linear elements are considered in illustrating the various types of network problems that might be encountered.

7-4.5   Examples

■ **Example 7-1:**   Illustration of Charge Formulation

Consider the electrical network sketched in Figure 7-10. This network is designed to illustrate the use of a charge formulation. As such, note that the circuit contains one of each

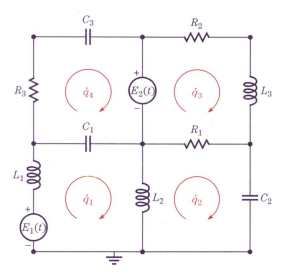

FIGURE 7-10 Electrical network for illustrating charge formulation.

type of element (capacitor, inductor, resistor, and voltage source) on the perimeter and one of each type of element on the interior. Using a charge formulation, derive the equation(s) of motion.

☐ **Solution:** For each loop in the network a current is sketched as indicated. By "loop" we mean an elemental mesh which recognizes the smallest closed circuit.[8] For consistency and convenience these are all sketched as positive in the clockwise direction. However, these directions are arbitrary as they may be all counterclockwise or mixed; except for convenience, it doesn't matter. This set of loop currents ($\dot{q}_1$, $\dot{q}_2$, $\dot{q}_3$ and $\dot{q}_4$) is complete and independent: Every element has some defined current and every defined current (or an equivalent) is needed to ensure completeness and independence. So, as a starting point in a charge formulation, simply place a loop current in every (elemental) loop of the network!

The net current flowing through any element is the sum of all the loop currents passing through that element. In general, for elements subject to two currents, we shall take *currents which flow upward or toward the right as positive*, although this too is arbitrary. For example, the upward current through L_2 is ($\dot{q}_2 - \dot{q}_1$) or the current toward the right through C_1 is ($\dot{q}_1 - \dot{q}_4$).

1. Generalized coordinate(s): The generalized charge coordinates are defined by q_1, q_2, q_3, and q_4. Then, referring to the charge formulation column in Table 7-2, we can proceed in a surprisingly straightforward manner.[9]

$$q_k : q_1 \ q_2 \ q_3 \ q_4 \tag{a}$$

2. Lagrangian:

$$\mathcal{L}(q_k, \dot{q}_k) = W_m^* - W_e \tag{b}$$

where

---

[8]See Problem 7-15, where an unusual set of loop currents is selected for this circuit.

[9]The similarity of this regimen with the procedure outlined in Table 5-1 should be observed by the reader.

$$W_m^* = \frac{1}{2}L_1\dot{q}_1^2 + \frac{1}{2}L_2(\dot{q}_2 - \dot{q}_1)^2 + \frac{1}{2}L_3\dot{q}_3^2 \tag{c}$$

$$W_e = \frac{(q_1 - q_4)^2}{2C_1} + \frac{q_2^2}{2C_2} + \frac{q_4^2}{2C_3} \tag{d}$$

and where we observe that W_m^* accounts for all the inductance in the circuit and W_e accounts for all the capacitance.

3. Generalized voltages: Further, the nonconservative variational work expression becomes

$$\delta W^{nc} = E_1(t)\delta q_1 + E_2(t)(\delta q_3 - \delta q_4)$$
$$- R_1(\dot{q}_2 - \dot{q}_3)(\delta q_2 - \delta q_3) - R_2\dot{q}_3\delta q_3 - R_3\dot{q}_4\delta q_4$$
$$= \sum_{k=1}^{4} \mathcal{E}_k \, \delta q_k \tag{e}$$

where for $E_2(t)$ and R_1 we are reminded of the "floating dashpot" in Example 5-15. So, from Eq. (e), the generalized voltages are

$$\mathcal{E}_1 = E_1(t) \qquad \mathcal{E}_2 = -R_1(\dot{q}_2 - \dot{q}_3)$$
$$\mathcal{E}_3 = E_2(t) - R_1(\dot{q}_3 - \dot{q}_2) - R_2\dot{q}_3 \qquad \mathcal{E}_4 = -E_2(t) - R_3\dot{q}_4 \tag{f}$$

4. Lagrange's equations: Substitution of Eqs. (a) through (d) and Eqs. (f) into Lagrange's equations

$$\frac{d}{dt}\left(\frac{\partial \mathcal{L}}{\partial \dot{q}_k}\right) - \frac{\partial \mathcal{L}}{\partial q_k} = \mathcal{E}_k \qquad k = 1, 2, 3, 4 \tag{g}$$

gives

q_1-equation: $\qquad (L_1 + L_2)\ddot{q}_1 - L_2\ddot{q}_2 + \dfrac{q_1}{C_1} - \dfrac{q_4}{C_1} = E_1(t)$ $\qquad$ (h)

q_2-equation: $\qquad -L_2\ddot{q}_1 + L_2\ddot{q}_2 + R_1(\dot{q}_2 - \dot{q}_3) + \dfrac{q_2}{C_2} = 0$ $\qquad$ (i)

q_3-equation: $\qquad L_3\ddot{q}_3 + (R_1 + R_2)\dot{q}_3 - R_1\dot{q}_2 = E_2(t)$ $\qquad$ (j)

q_4-equation: $\qquad R_3\dot{q}_4 + \left(\dfrac{C_1 + C_3}{C_1 C_3}\right)q_4 - \dfrac{q_1}{C_1} = -E_2(t)$ $\qquad$ (k)

The reader should note that the identical equations of motion would have been obtained if the function W_m^* for L_2 had been written as $\frac{1}{2}L_2(\dot{q}_1 - \dot{q}_2)^2$ or the function W_e for C_1 had been written as $\dfrac{(q_4 - q_1)^2}{2C_1}$. That is, Eqs. (h) through (k) would have been identical if we had used a different sign convention for the currents. Remember, these energy functions are positive definite; that is, they are nonnegative and moreover they are zero only when the net independent variable (net current or net charge) is zero. Similarly, we could have written $-R_1(\dot{q}_3 - \dot{q}_2)(\delta q_3 - \delta q_2)$ instead of the corresponding term in the work expression in Eq. (e). Our "floating resistor" rule, which we define in analogy with our "floating dashpot" rule, is versatile and forgiving.

■ **Example 7-2:** Illustration of Flux Linkage Formulation

Consider the electrical network sketched in Figure 7-11. In this case, the purpose is to illustrate the use of a flux linkage formulation. Using a flux linkage formulation, derive the equation(s) of motion.

❑ **Solution:** For each node (except for the ground, which represents the reference) in the network, a flux linkage coordinate is sketched as indicated. (If no node is grounded, then any node may be selected as the datum for which we could arbitrarily set $\lambda = 0$.)[10] By definition, all flux linkage coordinates are defined with respect to ground (or an arbitrary datum). So, as a starting point in a flux linkage formulation, simply place a flux linkage variable at every ungrounded node!

1. Generalized coordinate(s): A complete and independent set of generalized flux linkage coordinates is defined by λ_1, λ_2 and λ_3. Then, referring to the flux linkage formulation column in Table 7-2, again we can proceed in a surprisingly straightforward manner.

$$\lambda_k : \quad \lambda_1 \ \lambda_2 \ \lambda_3 \tag{a}$$

2. Lagrangian:

$$\mathcal{L}(\lambda_k, \dot{\lambda}_k) = W_e^* - W_m \tag{b}$$

where

$$W_e^* = \frac{1}{2}C_1(\dot{\lambda}_2 - \dot{\lambda}_1)^2 + \frac{1}{2}C_2(\dot{\lambda}_3 - \dot{\lambda}_2)^2 \tag{c}$$

$$W_m = \frac{\lambda_1^2}{2L} \tag{d}$$

and where we observe that W_e^* accounts for all the capacitance in the circuit and W_m accounts for all the inductance.

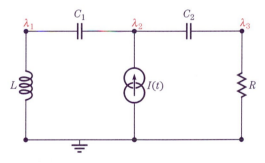

FIGURE 7-11 Electrical network for illustrating flux linkage formulation.

[10] See Problem 7-18.

3. *Generalized currents:*   Further, the nonconservative variational work expression becomes

$$\delta \mathcal{W}^{nc} = -\frac{\dot{\lambda}_3}{R} \delta \lambda_3 + I(t)\, \delta \lambda_2 = \sum_{k=1}^{3} \mathcal{I}_k\, \delta \lambda_k \tag{e}$$

So, from Eq. (e), the generalized currents are

$$\mathcal{I}_1 = 0 \qquad \mathcal{I}_2 = I(t) \qquad \mathcal{I}_3 = -\frac{\dot{\lambda}_3}{R} \tag{f}$$

4. *Lagrange's equations:*   Substitution of Eqs. (a) through (d) and Eqs. (f) into Lagrange's equations

$$\frac{d}{dt}\left(\frac{\partial \mathcal{L}}{\partial \dot{\lambda}_k}\right) - \frac{\partial \mathcal{L}}{\partial \lambda_k} = \mathcal{I}_k \qquad k = 1, 2, 3 \tag{g}$$

gives

λ_1-equation:
$$C_1\left(\ddot{\lambda}_1 - \ddot{\lambda}_2\right) + \frac{\lambda_1}{L} = 0 \tag{h}$$

λ_2-equation:
$$-C_1\ddot{\lambda}_1 + (C_1 + C_2)\,\ddot{\lambda}_2 - C_2\ddot{\lambda}_3 = I(t) \tag{i}$$

λ_3-equation:
$$-C_2\ddot{\lambda}_2 + C_2\ddot{\lambda}_3 + \frac{\dot{\lambda}_3}{R} = 0 \tag{j}$$

■ **Example 7-3:**   **Illustration of KVR Admissibility**

Consider the electrical network sketched in Figure 7-12, which we use to illustrate the flux linkage admissibility for an independent voltage source. Derive the equation(s) of motion via a flux linkage formulation.

❑ **Solution:**   As in Example 7-2, for each ungrounded node in the network, a flux linkage coordinate is sketched as indicated. Also, as before, all flux linkage coordinates are defined with respect to ground. From Table 7-2, because of the presence of a voltage source in a flux linkage formulation, we note that in lieu of a work expression for the voltage source we must consider admissibility. Specifically, to address the issue of admissibility, we note that at the node where the source feeds into the network, designated as node ① in Figure 7-12, the voltage is *specified* or *prescribed* for all time by the voltage source. So, λ_0 cannot be arbitrarily varied. Thus, at node ①

$$\dot{\lambda}_0 = E(t) \qquad \delta \lambda_0 = 0 \tag{a}$$

1. *Generalized coordinate(s):*   Given Eq. (a), a complete and independent set of generalized flux linkage coordinates is defined by λ_1 and λ_2. Then referring to the flux linkage formulation column in Table 7-2, we may proceed.

$$\lambda_k: \ \lambda_1 \ \lambda_2 \tag{b}$$

2. *Lagrangian:*

$$\mathcal{L}(\lambda_k, \dot{\lambda}_k) = W_e^* - W_m \tag{c}$$

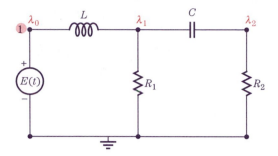

FIGURE 7-12 Electrical network requiring KVR admissibility for flux linkage formulation.

where

$$W_e^* = \frac{1}{2}C(\dot{\lambda}_2 - \dot{\lambda}_1)^2 \tag{d}$$

$$W_m = \frac{(\lambda_1 - \lambda_0)^2}{2L} \tag{e}$$

3. Generalized currents: Further, the nonconservative variational work expression becomes

$$\delta \mathcal{W}^{\mathrm{nc}} = -\frac{\dot{\lambda}_1}{R_1}\delta\lambda_1 - \frac{\dot{\lambda}_2}{R_2}\delta\lambda_2 = \sum_{k=1}^{2}\mathcal{I}_k\,\delta\lambda_k \tag{f}$$

So, from Eq. (f), the generalized currents are

$$\mathcal{I}_1 = -\frac{\dot{\lambda}_1}{R_1} \qquad \mathcal{I}_2 = -\frac{\dot{\lambda}_2}{R_2} \tag{g}$$

4. Lagrange's equations: Substitution of Eqs. (b) through (e) and Eqs. (g) into Lagrange's equations

$$\frac{d}{dt}\left(\frac{\partial \mathcal{L}}{\partial \dot{\lambda}_k}\right) - \frac{\partial \mathcal{L}}{\partial \lambda_k} = \mathcal{I}_k \qquad\qquad k = 1, 2 \tag{h}$$

gives

$$\lambda_1\text{-equation:} \quad C\ddot{\lambda}_1 - C\ddot{\lambda}_2 + \frac{\dot{\lambda}_1}{R_1} + \frac{\lambda_1}{L} = \frac{\lambda_0}{L} \quad \text{where } \dot{\lambda}_0 = E(t) \tag{i}$$

$$\lambda_2\text{-equation:} \quad -C\ddot{\lambda}_1 + C\ddot{\lambda}_2 + \frac{\dot{\lambda}_2}{R_2} = 0 \tag{j}$$

where the admissibility constraint on $\dot{\lambda}_0$ is indicated in the adjoining expression in Eq. (i). So, although λ_0 is not a generalized flux linkage coordinate, λ_0 does enter the formulation via the admissibility constraint expressed in Eq. (a) or adjoining Eq. (i).

■ **Example 7-4:** Illustration of KCR Admissibility

Consider the electrical network sketched in Figure 7-13a which we use to illustrate the charge admissibility for an independent current source. Derive the equation(s) of motion via a charge formulation.

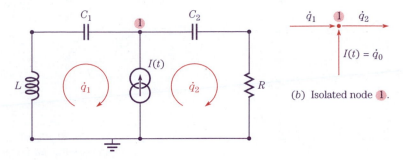

(a) Network showing loop currents.

FIGURE 7-13 Electrical network requiring KCR admissibility for charge formulation, and isolated node where $I(t)$ feeds into network.

❐ **Solution:** As in Example 7-1, for each loop in the network, a current is sketched. From Table 7-2, because of the presence of a current source in a charge formulation, we note that in lieu of a work expression for the current source we must consider admissibility. Specifically, to address the issue of admissibility, we consider the circuit node where the current source feeds into the network, which has been designated as node ①. For convenience, node ① has been isolated in Figure 7-13b where the associated currents are also sketched. Imposing KCR at node ① requires that

$$\dot{q}_2 = \dot{q}_1 + \dot{q}_0 \tag{a}$$

where $I(t) = \dot{q}_0$ has been defined. Integrating Eq. (a) gives

$$q_2 = q_1 + q_0 \tag{b}$$

where we have omitted any constant of integration since there can be no initial charge in the network at node ①. Taking the variation of Eq. (b) (which would eliminate any constants of integration even if they had been written in Eq. (b)) gives

$$\delta q_2 = \delta q_1 \tag{c}$$

because $\delta q_0 = 0$ since the source cannot be arbitrarily varied. Thus, there is only one generalized charge in a complete and independent set, which we may take as either q_1 or q_2; we shall take q_1.

1. Generalized coordinate(s): We now proceed to implement the remainder of Table 7-2 for the network in Figure 7-13a.

$$q_k : q_1 \tag{d}$$

2. Lagrangian:

$$\mathcal{L}(q_k, \dot{q}_k) = W_m^* - W_e \tag{e}$$

where

$$W_m^* = \frac{1}{2}L(\dot{q}_1^2) \tag{f}$$

$$W_e = \frac{q_1^2}{2C_1} + \frac{(q_1 + q_0)^2}{2C_2} \tag{g}$$

and where Eq. (b) has been used instead of q_2 for the electrical energy in C_2.

3. Generalized voltage: Further, the nonconservative variational work expression becomes

$$\delta \mathcal{W}^{\text{nc}} = -R\dot{q}_2 \, \delta q_2 = -R(\dot{q}_1 + \dot{q}_0)\delta q_1 = \mathcal{E}_1 \, \delta q_1 \tag{h}$$

where Eqs. (a) and (c) have been used to rewrite $\dot{q}_2$ and δq_2, respectively, in terms of the generalized charge $\dot{q}_1$ and its variation. So, the generalized voltage is

$$\mathcal{E}_1 = -R(\dot{q}_1 + \dot{q}_0) \tag{i}$$

4. Lagrange's equation: Substitution of Eqs. (d) through (g) and Eq. (i) into Lagrange's equation

$$\frac{d}{dt}\left(\frac{\partial \mathcal{L}}{\partial \dot{q}_k}\right) - \frac{\partial \mathcal{L}}{\partial q_k} = \mathcal{E}_k \qquad k = 1 \tag{j}$$

gives

$$L\ddot{q}_1 + R\dot{q}_1 + \left(\frac{C_1 + C_2}{C_1 C_2}\right)q_1 = -R\dot{q}_0 - \frac{q_0}{C_2} \qquad \text{where } \dot{q}_0 = I(t) \tag{k}$$

and where $I(t)$ has been indicated as an admissibility constraint on $\dot{q}_0$.

Observations: With the exception of the op-amp, all the circuit elements that appear in Table 7-2 have been considered in the four examples above. The comprehensive character of these examples should allow the reader to be fully confident in the formulation of electrical network problems of any complexity involving these elements. As the Beatles might have been overheard to say: In order to derive equations of motion for electrical networks of any complexity, all you need is $\mathcal{LOVE}$; $\mathcal{LOVE}$ is all you need.

It should be clear that on any circuit, the analyst has the choice of formulating the equations via either charges or flux linkages. If the goal is to formulate the equations of motion in the fewer generalized coordinates of the two choices, then the choice is clear: If the network has fewer independent loops than independent nodes, a charge formulation should be the choice; and vice versa.

The reader should feel truly empowered at this point. Problems 7-6 through 7-18 should fall with a very modest push. Several of these problems should be attempted before the reader does anything else; *anything*. Then the reader will be empowered.

From our discussion of op-amps, recall that the complexity of these devices does not result in a simple constitutive relation but in four major characteristics of their operation in their linear range, given as Eq. (7-26). The equation embodiment of these four major characteristics of ideal op-amps is given substantially in Eqs. (7-27) and (7-28). But these are not constitutive relations in the sense used elsewhere, relating charge variables (or their derivatives) to flux linkage variables (or their derivatives). Equation (7-27) states that the currents through an op-amp are essentially zero; and Eq. (7-28) is called a *voltage transfer characteristic,* relating output voltage to the *difference* in input voltages. So, despite the fact that Eqs. (7-26) through (7-28) represent an enormous simplification of a class of complex devices, we are nevertheless dealing with an uncharacteristic circuit element.

Thus, satisfying the admissibility requirements for op-amps, as indicated in Table 7-2, requires more thoughtful analysis than any of the other entries in the table. As the reader will observe, the examples involving op-amp circuits will be extended slightly because, beyond equations formulation, we shall take the additional steps of exploring the physical response characteristics of these networks.

Now that several of Problems 7-6 through 7-18 have been successfully completed, we proceed to incorporate circuits containing op-amps into various lagrangian formulation examples. The competency that the reader feels is very real; and this empowerment is just what is needed to consider the next set of examples on operational amplifiers.

■ **Example 7-5:** Inverting Amplifier

The op-amp circuit sketched in Figure 7-14a represents an *inverting amplifier.* The reason for this name will become apparent by virtue of the result obtained in this example. Such a circuit is one of the major components used in analog computers, some of the others being the integrator (Example 7-6), and various noninverting, follower, addition, and subtraction amplifiers. The element $e_s(t)$ represents an input *source voltage* to the circuit, irrespective of whether this voltage comes from a device or another network. Through its *voltage transfer characteristics,* the function of the op-amp is to convert the *source voltage* into a different (generally amplified, but conceptually diminished) voltage for subsequent use.

In accordance with Figure 7-8, the equivalent circuit for the inverting amplifier is sketched in Figure 7-14b, but without the unnecessary dashed triangular outline. Derive the equation(s) of motion for this circuit and then examine its input–output voltage transfer characteristic for operation within its linear range.

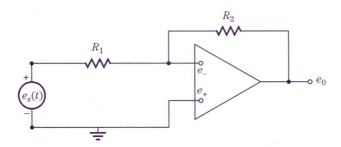

(a) Schematic of op-amp network model.

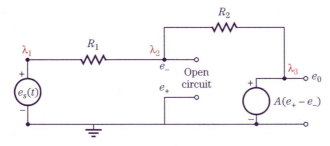

(b) Idealized equivalent circuit, showing flux linkage variables.

FIGURE 7-14 Schematics of inverting amplifier and its idealized equivalent circuit for flux linkage formulation.

☐ **Solution:** From Figure 7-14b, we make the following observations.

1. $e_+ = 0$ because the "+" terminal of the op-amp is connected to ground.

2. For linear operation, characteristic 2 in Eq. (7-26) requires that the maximum value of e_- is ϵ, where ϵ is a voltage that is much less than the other voltages in the circuit.

3. For a flux linkage formulation, as effected in Examples 7-2 and 7-3 above, λ_1, λ_2 and λ_3 are variously identified with each of the (ungrounded) nodes. Further,

$$\dot{\lambda}_1 = e_s(t) \qquad \text{and} \qquad \delta\lambda_1 = 0 \qquad \text{(a)}$$

$$\dot{\lambda}_2 = e_- = \epsilon(t) \qquad \text{and} \qquad \delta\lambda_2 \neq 0 \qquad \text{(b)}$$

$$\dot{\lambda}_3 = e_0(t) \qquad \text{and} \qquad \delta\lambda_3 = 0 \qquad \text{(c)}$$

where the variations $\delta\lambda_1$ and $\delta\lambda_3$ are zero because they are associated with the voltage sources. (Refer to Example 7-3.) The variation $\delta\lambda_2$ is arbitrary.

This set of three statements comprises the KVR admissibility conditions plus the idealized op-amp characteristics as identified in Table 7-2 for operational amplifier networks. While not difficult, they do require mental engagement. Here they have been summarized in a manner that can be repeatedly used in numerous op-amp equation formulations.

1. Generalized coordinate(s): Use of the flux linkage formulation column in Table 7-2 leads to

$$\lambda_k : \lambda_2 \qquad \text{(d)}$$

where, in accordance with Eqs. (a), (b) and (c), λ_2 represents a complete and independent set of generalized flux linkage coordinates.

2. Lagrangian: The lagrangian is

$$\mathcal{L}(\lambda_k, \dot{\lambda}_k) = W_e^* - W_m \qquad \text{(e)}$$

where

$$W_e^* = 0 \qquad \text{and} \qquad W_m = 0 \qquad \text{(f)}$$

3. Generalized current: The nonconservative variational work expression is

$$\delta\mathcal{W}^{nc} = -\frac{1}{R_1}(\dot{\lambda}_2 - \dot{\lambda}_1)(\delta\lambda_2 - \delta\lambda_1)$$

$$-\frac{1}{R_2}(\dot{\lambda}_3 - \dot{\lambda}_2)(\delta\lambda_3 - \delta\lambda_2) = \mathcal{I}_2\,\delta\lambda_2 \qquad \text{(g)}$$

However, Eqs. (a) and (c) are used to discard $\delta\lambda_1$ and $\delta\lambda_3$; so,

$$\mathcal{I}_2 = -\frac{1}{R_1}(\dot{\lambda}_2 - \dot{\lambda}_1) + \frac{1}{R_2}(\dot{\lambda}_3 - \dot{\lambda}_2) \qquad \text{(h)}$$

4. Lagrange's equation: Substitution of Eqs. (d) through (f) and Eq. (h) into Lagrange's equation

$$\frac{d}{dt}\left(\frac{\partial\mathcal{L}}{\partial\dot{\lambda}_k}\right) - \frac{\partial\mathcal{L}}{\partial\lambda_k} = \mathcal{I}_k \qquad k = 2 \qquad \text{(i)}$$

gives

$$0 = -\frac{1}{R_1}(\dot{\lambda}_2 - \dot{\lambda}_1) + \frac{1}{R_2}(\dot{\lambda}_3 - \dot{\lambda}_2) \tag{j}$$

Equation (j) is the equation of motion for the network model sketched in Figure 7-14a.

Now we proceed to examine the input–output voltage transfer characteristic of the network. By virtue of observation 2 or Eq. (b), λ_2, which is equal to ϵ, may be discarded in Eq. (j) since it is negligible in comparison with λ_1 and λ_3. Also, if Eqs. (a) and (c) are substituted into Eq. (j), the input–output voltage transfer characteristic becomes

$$e_0 = -\left(\frac{R_2}{R_1}\right)e_s \tag{k}$$

The sign of the output e_0 has been inverted from that of the input e_s, thus the name "inverting amplifier." Furthermore, it is also for this reason that the op-amp terminal into which e_s was fed is generally called the "inverting input" terminal. As illustrated in Example 7-10, for analogous reasons, the other terminal is called the "noninverting input" terminal. If $\dfrac{R_2}{R_1} > 1$, the device acts as an amplifier. If $\dfrac{R_2}{R_1} = 1$, the device simply inverts the sign of the input, a characteristic that provides a useful function in analog computers and other hardware.

■ **Example 7-6:** Analog Integrator

The op-amp circuit sketched in Figure 7-15a represents an *integrator.* It is another of the important devices for analog computation. As before, $e_s(t)$ is an input source voltage to

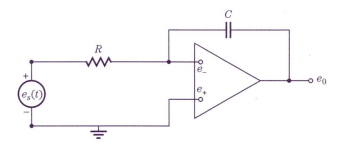

(a) Schematic of op-amp network model.

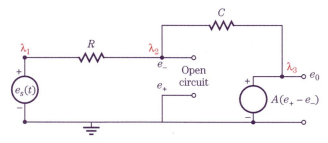

(b) Idealized equivalent circuit, showing flux linkage variables.

FIGURE 7-15 Schematics of integrator and its idealized equivalent circuit for flux linkage formulation.

the circuit, which may be due to a device or another network. Derive the equation(s) of motion for this circuit and then examine its input–output voltage transfer characteristic for operation within its linear range.

☐ **Solution:** From Figure 7-15b, which is the idealized equivalent circuit, we make the following observations.

1. $e_+ = 0$ because the "+" terminal of the op-amp is connected to ground.

2. For linear operation, characteristic 2 in Eq. (7-26) requires that the maximum value of e_- is ϵ, where ϵ is a voltage that is much less than the other voltages in the circuit.

3. For a flux linkage formulation, as effected in Examples 7-2, 7-3, and 7-5, λ_1, λ_2, and λ_3 are variously identified with each of the (ungrounded) nodes. Further,

$$\dot{\lambda}_1 = e_s(t) \quad \text{and} \quad \delta\lambda_1 = 0 \tag{a}$$

$$\dot{\lambda}_2 = e_- = \epsilon(t) \quad \text{and} \quad \delta\lambda_2 \neq 0 \tag{b}$$

$$\dot{\lambda}_3 = e_0(t) \quad \text{and} \quad \delta\lambda_3 = 0 \tag{c}$$

where the variations $\delta\lambda_1$ and $\delta\lambda_3$ are zero because they are associated with the voltage sources. (Refer to Examples 7-3 and 7-5.) The variation $\delta\lambda_2$ is arbitrary.

This set of three statements comprises the KVR admissibility conditions plus the idealized op-amp characteristics as identified in Table 7-2 for operational amplifier networks.

1. Generalized coordinate(s): Use of the flux linkage formulation column in Table 7-2 leads to

$$\lambda_k : \lambda_2 \tag{d}$$

where, in accordance with Eqs. (a), (b), and (c), λ_2 represents a complete and independent set of generalized flux linkage coordinates.

2. Lagrangian: The lagrangian is

$$\mathcal{L}(\lambda_k, \dot{\lambda}_k) = W_e^* - W_m \tag{e}$$

where

$$W_e^* = \frac{1}{2}C(\dot{\lambda}_3 - \dot{\lambda}_2)^2 \tag{f}$$

and

$$W_m = 0 \tag{g}$$

3. Generalized current: The nonconservative variational work expression is

$$\delta\mathcal{W}^{nc} = -\frac{1}{R}(\dot{\lambda}_2 - \dot{\lambda}_1)(\delta\lambda_2 - \delta\lambda_1) = \mathcal{I}_2\,\delta\lambda_2 \tag{h}$$

However, Eq. (a) is used to discard $\delta\lambda_1$; so,

$$\mathcal{I}_2 = -\frac{1}{R_1}(\dot{\lambda}_2 - \dot{\lambda}_1) \tag{i}$$

4. Lagrange's equation: Substitution of Eqs. (d) through (g) and Eq. (i) into Lagrange's equation

$$\frac{d}{dt}\left(\frac{\partial \mathcal{L}}{\partial \dot{\lambda}_k}\right) - \frac{\partial \mathcal{L}}{\partial \lambda_k} = \mathcal{I}_k \qquad k = 2 \tag{j}$$

gives

$$C(\ddot{\lambda}_3 - \ddot{\lambda}_2) = \frac{1}{R}(\dot{\lambda}_2 - \dot{\lambda}_1) \tag{k}$$

Equation (k) is the equation of motion for the network model sketched in Figure 7-15a.

Now we proceed to examine the input–output voltage transfer characteristic of the network. In performing the limiting process of allowing λ_2 (that is, e_-) to go to zero, we *must* be careful because we also observe the presence of $\ddot{\lambda}_2$ in Eq. (k). As we know, the fact that any function, say λ_2, goes to zero does not mean that the derivative of that function, $\dot{\lambda}_2$, also go to zero. So, the limiting-process arguments regarding terms in Eq. (k) go (carefully) as below. It is important to understand that neither λ_2 nor $\dot{\lambda}_2$ ever becomes zero; the limiting analysis that follows is for the purpose of neglecting $\dot{\lambda}_2$ in comparison with $\dot{\lambda}_1$ and $\ddot{\lambda}_2$ in comparison with $\ddot{\lambda}_3$. The limiting process of letting λ_2 go to zero for any $\dot{\lambda}_2$ is due to the fact that A effectively is (goes to) infinity. (Refer to the meaning of A in Figures 7-7 and 7-15.)

Let's reexamine the behavior of λ_2. In particular, from Figure 7-15, Eq. (7-28) effectively reduces to $e_0 = A(e_+ - e_-) = A(0 - \epsilon) = A(-\dot{\lambda}_2)$. Since e_0 must remain finite, $\dot{\lambda}_2$ must go to zero as A effectively goes to infinity; or more accurately, $\dot{\lambda}_2$ must be very small if A is very large. The practical consequence of $\dot{\lambda}_2$ being very small is that $\dot{\lambda}_2$ may be neglected in comparison with $\dot{\lambda}_1$. Now let's examine the behavior of $\ddot{\lambda}_2$. Recall that A is independent of frequency (characteristic 4 in Eq. (7-26)). From Figure 7-15, for this problem Eq. (7-28) effectively reduces to $\dot{e}_0 = A(\dot{e}_+ - \dot{e}_-) = A(0 - \dot{\epsilon}_-) = A(-\ddot{\lambda}_2)$. Since $\dot{e}_0$ must remain finite (which is true because e_0 must remain finite), $\ddot{\lambda}_2$ must go to zero as A effectively goes to infinity; or more accurately, $\ddot{\lambda}_2$ must be very small if A is very large. The practical consequence of $\ddot{\lambda}_2$ being very small is that $\ddot{\lambda}_2$ may be neglected in comparison with $\ddot{\lambda}_3$. Thus, Eq. (k) becomes

$$C\ddot{\lambda}_3 = -\frac{\dot{\lambda}_1}{R} \tag{l}$$

If Eqs. (a) and (c) are substituted into Eq. (l), we get

$$\dot{e}_0 = -\frac{1}{RC}e_s \tag{m}$$

or

$$e_0 = -\frac{1}{RC}\int_{-\infty}^{t} e_s \, dt \tag{n}$$

which indicates the reason the circuit in Figure 7-15a is called an *integrator.* By incorporating a switch into the element that initializes the integration at zero time, no initial stored energy will exist in the circuit, allowing the lower limit on the integral to be set to zero. Also, it is noted that by interchanging the capacitor and the resistor in the circuit in Figure 7-15, the op-amp will become a differentiator. However, differentiators are intrinsically less stable than integrators, and decisions to use them should take into account this inherent characteristic.

■ **Example 7-7:** Active Band-Pass Filter

The op-amp circuit sketched in Figure 7-16a represents an *active band-pass filter*. In this context, a filter is an electrical network that selectively passes signals having a specific minimum, maximum, or range of frequencies. Low-pass filters pass signals below a specified value of frequency, high-pass filters pass signals above a specified value of frequency, and band-pass filters pass signals within a specified range. (Mechanical filters—or strainers—are mechanical analogs of electrical filters. Most mechanical filters allow particles smaller than a specified size to pass through them. In this sense, most mechanical filters are "low-pass filters.") The circuit in Figure 7-16a is said to be *active* because it uses an op-amp, which is an active element. Derive the equation(s) of motion for this circuit, and obtain an expression for the input–output voltage transfer characteristic of the device.

☐ **Solution:** From Figure 7-16b, which is the idealized equivalent circuit of the active band-pass filter, we make the following observations.

1. $e_+ = 0$ because the "+" terminal of the op-amp is connected to ground.

2. For linear operation, characteristic 2 in Eq. (7-26) requires that the maximum value of e_- is ϵ, where ϵ is a voltage that is much less than the other voltages in the circuit.

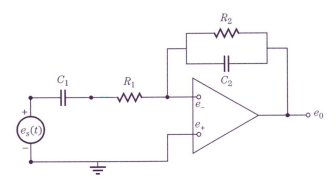

(a) Schematic of op-amp network model.

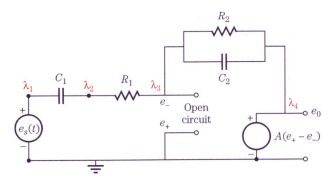

(b) Idealized equivalent circuit, showing flux linkage variables.

FIGURE 7-16 Schematics of active band-pass filter and its idealized equivalent circuit for flux linkage formulation.

3. For a flux linkage formulation, as in Examples 7-2, 7-3, 7-5, and 7-6, λ_1, λ_2, λ_3, and λ_4 are variously identified with each of the (ungrounded) nodes. Further,

$$\dot{\lambda}_1 = e_s(t) \quad \text{and} \quad \delta\lambda_1 = 0 \tag{a}$$

$$\dot{\lambda}_3 = e_- = \epsilon(t) \quad \text{and} \quad \delta\lambda_3 \neq 0 \tag{b}$$

$$\dot{\lambda}_4 = e_0(t) \quad \text{and} \quad \delta\lambda_4 = 0 \tag{c}$$

where the variations $\delta\lambda_1$ and $\delta\lambda_4$ are zero because they are associated with the voltage sources. As in Examples 7-5 and 7-6, Eq. (b) denotes that the variation $\delta\lambda_2$ is arbitrary.

This set of three statements comprises the KVR admissibility conditions plus the idealized op-amp characteristics as identified in Table 7-2 for operational amplifier networks.

1. Generalized coordinate(s): Use of the flux linkage formulation column in Table 7-2 leads to

$$\lambda_k : \lambda_2 \ \lambda_3 \tag{d}$$

where, in accordance with Eqs. (a), (b), and (c), λ_2 and λ_3 represent a complete and independent set of generalized flux linkage coordinates.

2. Lagrangian: The lagrangian is

$$\mathcal{L}(\lambda_k, \dot{\lambda}_k) = W_e^* - W_m \tag{e}$$

where

$$W_e^* = \frac{1}{2}C_1(\dot{\lambda}_2 - \dot{\lambda}_1)^2 + \frac{1}{2}C_2(\dot{\lambda}_4 - \dot{\lambda}_3)^2 \tag{f}$$

and

$$W_m = 0 \tag{g}$$

3. Generalized currents: The nonconservative variational work expression is

$$\delta\mathcal{W}^{nc} = -\frac{1}{R_1}(\dot{\lambda}_3 - \dot{\lambda}_2)(\delta\lambda_3 - \delta\lambda_2) - \frac{1}{R_2}(\dot{\lambda}_4 - \dot{\lambda}_3)(\delta\lambda_4 - \delta\lambda_3) = \sum_{k=2}^{3} \mathcal{I}_k \, \delta\lambda_k \tag{h}$$

However, Eq. (c) is used to discard $\delta\lambda_4$. So, from Eq. (h)

$$\mathcal{I}_2 = -\frac{1}{R_1}(\dot{\lambda}_2 - \dot{\lambda}_3) \tag{i}$$

and

$$\mathcal{I}_3 = -\frac{1}{R_1}(\dot{\lambda}_3 - \dot{\lambda}_2) - \frac{1}{R_2}(\dot{\lambda}_3 - \dot{\lambda}_4) \tag{j}$$

4. Lagrange's equations: Substitution of Eqs. (d) through (g), and Eqs. (i) and (j) into Lagrange's equations

$$\frac{d}{dt}\left(\frac{\partial \mathcal{L}}{\partial \dot{\lambda}_k}\right) - \frac{\partial \mathcal{L}}{\partial \lambda_k} = \mathcal{I}_k \qquad k = 2, 3 \tag{k}$$

gives

λ_2-equation: $$-C_1\ddot{\lambda}_1 + C_1\ddot{\lambda}_2 + \frac{1}{R_1}\dot{\lambda}_2 - \frac{1}{R_1}\dot{\lambda}_3 = 0 \tag{l}$$

and

$$\lambda_3\text{-equation:}\quad C_2\ddot{\lambda}_3 - C_2\ddot{\lambda}_4 - \frac{1}{R_1}\dot{\lambda}_2 + \left(\frac{1}{R_1} + \frac{1}{R_2}\right)\dot{\lambda}_3 - \frac{1}{R_2}\dot{\lambda}_4 = 0 \tag{m}$$

By virtue of observation 2 and Eq. (b), $\dot{\lambda}_3$ may be neglected in comparison with $\dot{\lambda}_2$ and $\dot{\lambda}_4$; and by virtue of the arguments presented in Example 7-6, $\ddot{\lambda}_3$ may be neglected in comparison with $\ddot{\lambda}_4$. So, Eqs. (l) and (m) become

$$C_1\ddot{\lambda}_2 + \frac{\dot{\lambda}_2}{R_1} = C_1\dot{e}_s \tag{n}$$

$$R_1C_2\dot{e}_0 + \frac{R_1}{R_2}e_0 + \dot{\lambda}_2 = 0 \tag{o}$$

where Eqs. (a) and (c) have been used to replace $\dot{\lambda}_1$ and $\dot{\lambda}_4$ with $e_s(t)$ and $e_0(t)$, respectively. Equations (n) and (o) are the equations of motion for the network model in Figure 7-16a.

In general, the primary interest in the op-amp network is its input–output voltage transfer characteristic. In this regard, Eqs. (n) and (o) can be used to solve for e_0 in terms of e_s. If Eq. (o) and its time-differentiated form are used to obtain $\dot{\lambda}_2$ and $\ddot{\lambda}_2$, then $\dot{\lambda}_2$ and $\ddot{\lambda}_2$ can be eliminated from Eq. (n) by use of the two so-obtained expressions. Thus,

$$C_1\left(-R_1C_2\ddot{e}_0 - \frac{R_1}{R_2}\dot{e}_0\right) + \frac{1}{R_1}\left(-R_1C_2\dot{e}_0 - \frac{R_1}{R_2}e_0\right) = C_1\dot{e}_s$$

or

$$(R_1R_2C_1C_2)\ddot{e}_0 + (R_1C_1 + R_2C_2)\dot{e}_0 + e_0 = -R_2C_1\dot{e}_s \tag{p}$$

the solution of which we shall consider in Chapter 8.

Observations: Two significant observations can be made about Examples 7-5 through 7-7. First, all the variational formulations have been conducted using generalized flux linkages. One might be interested in the variational techniques for formulating the corresponding equations using generalized charges. Well, being *voltage transfer devices,* op-amps are most conveniently formulated using flux linkages, more directly yielding the desired voltage transfer relations. In general, charge formulations for any but the simplest op-amps do not lead directly to these voltage transfer relations. Nevertheless, we shall illustrate charge formulations in Examples 7-8 and 7-9 below. Second, in each of Examples 7-5 through 7-7, one of the op-amp terminals was connected to ground. This was done intentionally since this is, by far, the most frequent topological configuration for op-amps. Nevertheless, in Example 7-10, we shall illustrate the equation formulation for an alternative configuration in which neither op-amp terminal is grounded.

■ **Example 7-8:** Inverting Amplifier via Charge Formulation

For illustration, we reconsider the inverting amplifier of Example 7-5. Derive the equation(s) of motion and ultimately the voltage transfer relation via a generalized charge coordinate

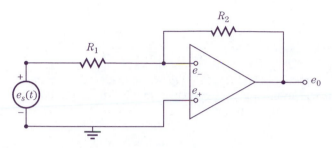

(a) Schematic of op-amp network model.

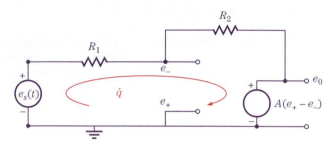

(b) Idealized equivalent circuit, showing loop current $\dot{q}$.

FIGURE 7-17 Schematics of inverting amplifier and its idealized equivalent circuit for charge formulation.

formulation.[11] The schematic of the device is sketched in Figure 7-17a and its idealized equivalent circuit is sketched in Figure 7-17b, where a loop current is sketched.

☐ **Solution:** From Figure 7-17b, we make the following observations.

1. $e_+ = 0$ because the "+" terminal of the op-amp is connected to ground.

2. For linear operation, characteristic 2 in Eq. (7-26) requires that the maximum value of e_- is ϵ, where ϵ is a voltage that is much less than the other voltages in the circuit.

3. In accordance with Eq. (7-28),

$$e_0 = A(e_+ - e_-) \tag{a}$$

where A, the *open-loop gain*, is typically 10^4 to 10^9.

This set of three statements comprises the idealized op-amp characteristics as identified in Table 7-2 for operational amplifier networks.

1. Generalized coordinate(s): Use of the charge formulation column in Table 7-2 leads to

$$q_k : q \tag{b}$$

where, in accordance with Figure 7-17b, q represents a complete and independent set of generalized charge coordinates, thus satisfying the KCR admissibility conditions.

---

[11]Because of the simplicity of this device, the direct method—as found in most textbooks on op-amps— may be easier to apply here. However, this is not generally the case for devices with modestly increased complexity.

2. Lagrangian: The lagrangian is

$$\mathcal{L}(q_k, \dot{q}_k) = W_m^* - W_e \tag{c}$$

where

$$W_m^* = 0 \quad \text{and} \quad W_e = 0 \tag{d}$$

3. Generalized voltage: The nonconservative variational work expression is

$$\delta \mathcal{W}^{nc} = e_s(t)\delta q - R_1 \dot{q} \delta q - R_2 \dot{q} \delta q - A(e_+ - e_-)\delta q = \mathcal{E}\delta q \tag{e}$$

where the sign on each of the terms in the middle portion of Eq. (e) is consistent with the sign convention defined in Figure 7-9. In particular, the current through the source $e_s(t)$ flows from negative to positive, resulting in a plus sign on the corresponding term in Eq. (e); and the current through the source $A(e_+ - e_-)$ flows from positive to negative, resulting in a minus sign on the corresponding term in Eq. (e). Thus, Eq. (e) gives

$$\mathcal{E} = e_s(t) - R_1 \dot{q} - R_2 \dot{q} - A(e_+ - e_-) \tag{f}$$

4. Lagrange's equation: Substitution of Eqs. (b) through (d) and Eq. (f) into Lagrange's equation

$$\frac{d}{dt}\left(\frac{\partial \mathcal{L}}{\partial \dot{q}}\right) - \frac{\partial \mathcal{L}}{\partial q} = \mathcal{E} \tag{g}$$

gives

$$e_s(t) - R_1 \dot{q} - R_2 \dot{q} = A(e_+ - e_-) \tag{h}$$

Further, substitution of Eq. (a) for the right-hand side of Eq. (h) gives

$$e_s(t) - (R_1 + R_2)\dot{q} = e_0 \tag{i}$$

which is the equation of motion for the network model sketched in Figure 7-17*a*, for a charge formulation.

Although Eq. (i) is the equation of motion for the charge formulation of the inverting amplifier, it is not in a useful form where its voltage transfer characteristic is apparent. Equation (i) can be converted into the form given in Eq. (k) of Example 7-5 as follows. Note from Ohm's law and Figure 7-17*b*,

$$\dot{q} = R_1^{-1}(e_s(t) - e_-) \tag{j}$$

From observation 2 at the beginning of this solution, we observe that

$$e_- \ll e_s(t) \tag{k}$$

and substitution of Eq. (k) into Eq. (j) effectively gives

$$\dot{q} = R_1^{-1}e_s(t) \tag{l}$$

Thus, substitution of Eq. (l) into Eq. (i) yields

$$e_s(t) - R_1[R_1^{-1}e_s(t)] - R_2[R_1^{-1}e_s(t)] = e_0 \tag{m}$$

or

$$e_0 = -\left(\frac{R_2}{R_1}\right)e_s \tag{n}$$

which is the anticipated voltage transfer characteristic of the circuit.

■ **Example 7-9:** Active Band-Pass Filter via Charge Formulation

We reconsider the active band-pass filter of Example 7-7. Derive the equation(s) of motion via a generalized charge coordinate formulation. The schematic of the device is sketched in Figure 7-18a and its idealized equivalent circuit appears in Figure 7-18b, where two loop currents are sketched.

□ **Solution:**

1. Generalized coordinate(s): From Figure 7-18b, we define a loop current in each loop of the network, where the open-circuit characteristic of the op-amp has been noted. This set of loop currents is complete and independent. Thus, q_1 and q_2 represent a complete and independent set of generalized charge coordinates that satisfy the KCR admissibility conditions.

$$q_k : q_1 \; q_2 \tag{a}$$

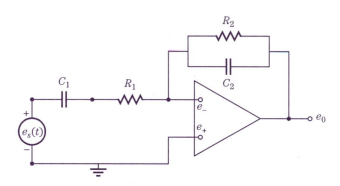

(a) Schematic of op-amp network model.

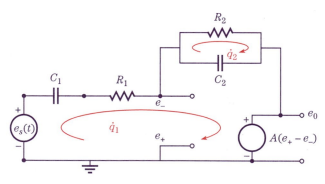

(b) Idealized equivalent circuit, showing loop currents $\dot{q}_1$ and $\dot{q}_2$.

FIGURE 7-18 Schematics of active band-pass filter and its idealized equivalent circuit for charge formulation.

2. Lagrangian: In accordance with the charge formulation column in Table 7-2, the lagrangian is

$$\mathcal{L}(q_k, \dot{q}_k) = W_m^* - W_e \tag{b}$$

where from Figure 7-18*b*

$$W_m^* = 0 \quad \text{and} \quad W_e = \frac{q_1^2}{2C_1} + \frac{(q_1 - q_2)^2}{2C_2} \tag{c}$$

3. Generalized voltages: The nonconservative variational work expression is

$$\delta\mathcal{W}^{\text{nc}} = e_s(t)\delta q_1 - R_1\dot{q}_1\delta q_1 - R_2\dot{q}_2\delta q_2$$

$$- A(e_+ - e_-)\delta q_1 = \sum_{k=1}^{2} \mathcal{E}_k\delta q_k \tag{d}$$

where the sign on each of the terms in the middle portion of Eq. (d) is consistent with the sign convention defined in Figure 7-9. The generalized voltages are found from Eq. (d) as

$$\mathcal{E}_1 = e_s(t) - R_1\dot{q}_1 - A(e_+ - e_-)$$
$$= e_s(t) - R_1\dot{q}_1 - e_0 \tag{e}$$

where Eq. (7-28) has been used, and

$$\mathcal{E}_2 = -R_2\dot{q}_2 \tag{f}$$

4. Lagrange's equations: Substitution of Eqs. (a), (b), (c), (e), and (f) into Lagrange's equations

$$\frac{d}{dt}\left(\frac{\partial\mathcal{L}}{\partial\dot{q}_k}\right) - \frac{\partial\mathcal{L}}{\partial q_k} = \mathcal{E}_k \quad k = 1, 2 \tag{g}$$

gives

q_1-equation: $$R_1\dot{q}_1 + \left(\frac{C_1 + C_2}{C_1 C_2}\right)q_1 - \frac{q_2}{C_2} = e_s - e_0 \tag{h}$$

and

q_2-equation: $$R_2\dot{q}_2 - \frac{q_1}{C_2} + \frac{q_2}{C_2} = 0 \tag{i}$$

Equations (h) and (i) are the equations of motion for the charge formulation of the active band-pass filter; however, they do not illustrate the useful voltage transfer characteristics of the filter. It can be shown (see Problem 7-23) that Eqs. (h) and (i) can be reduced to Eq. (p) in Example 7-7.

■ **Example 7-10:** Noninverting Amplifier with Ungrounded Op-Amp

The op-amp device sketched in Figure 7-19 can be used for amplification, without the sign inversion that occurs for the circuit in Examples 7-5 or 7-8. Thus, it is called a *noninverting amplifier.* Derive the equation(s) of motion for this circuit and then find its input–output voltage transfer characteristic for operation within its linear range.

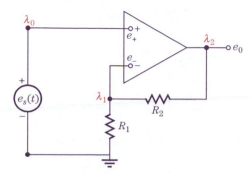

FIGURE 7-19 Schematic of noninverting amplifier.

□ **Solution:** For a flux linkage formulation, flux linkage variables are located at each of the (ungrounded) nodes, as sketched in Figure 7-19. Now, we make the following observations.

1.

$$e_+(t) = e_s(t) = \dot{\lambda}_0 \quad \text{and} \quad \delta\lambda_0 = 0 \tag{a}$$

$$e_-(t) = \dot{\lambda}_1 \quad \text{and} \quad \delta\lambda_1 \neq 0 \tag{b}$$

$$e_0 = \dot{\lambda}_2 \quad \text{and} \quad \delta\lambda_2 = 0 \tag{c}$$

where the variations $\delta\lambda_0$ and $\delta\lambda_2$ are zero because they are associated with the voltage sources. (The reader may want to sketch an idealized equivalent circuit for Figure 7-19, corresponding to those in Figures 7-14 through 7-18.)

2. In accordance with Eq. (7-28),

$$e_0 = A(e_+ - e_-) \tag{d}$$

where A is typically of the order of 10^4 to 10^9.

This set of two statements comprises the KVR admissibility conditions plus the idealized op-amp characteristics as identified in Table 7-2 for operational amplifier networks.

1. Generalized coordinate(s): Use of the flux linkage formulation column in Table 7-2 leads to

$$\lambda_k : \lambda_1 \tag{e}$$

where, in accordance with Figure 7-19 and Eqs. (a), (b), and (c), λ_1 represents a complete and independent set of generalized flux linkage coordinates.

2. Lagrangian: The lagrangian is

$$\mathcal{L}(\lambda_k, \dot{\lambda}_k) = W_e^* - W_m \tag{f}$$

where

$$W_e^* = 0 \quad \text{and} \quad W_m = 0 \tag{g}$$

3. Generalized current: The nonconservative variational work expression is

$$\delta\mathcal{W}^{\text{nc}} = -\frac{1}{R_1}(\dot{\lambda}_1)\delta\lambda_1 - \frac{1}{R_2}(\dot{\lambda}_2 - \dot{\lambda}_1)(\delta\lambda_2 - \delta\lambda_1) = \mathcal{I}_1\delta\lambda_1 \tag{h}$$

and noting that, in accordance with Eq. (c), $\delta\lambda_2 = 0$. So, Eq. (h) gives

$$\mathcal{I}_1 = -\frac{\dot{\lambda}_1}{R_1} - \frac{1}{R_2}(\dot{\lambda}_1 - \dot{\lambda}_2) \tag{i}$$

4. Lagrange's equation: Substitution of Eqs. (e), (f), (g), and (i) into Lagrange's equation

$$\frac{d}{dt}\left(\frac{\partial \mathcal{L}}{\partial \dot{\lambda}_1}\right) - \frac{\partial \mathcal{L}}{\partial \lambda_1} = \mathcal{I}_1 \tag{j}$$

gives

$$\left(\frac{1}{R_1} + \frac{1}{R_2}\right)\dot{\lambda}_1 - \frac{\dot{\lambda}_2}{R_2} = 0 \tag{k}$$

where Eq. (k) is the equation of motion for the circuit sketched in Figure 7-19.

The second goal in this example is to obtain the voltage transfer relation for the circuit. So, substitution of Eqs. (a) and (b) into Eq. (d) yields

$$e_0 = A(e_s - \dot{\lambda}_1) \tag{l}$$

which gives

$$\dot{\lambda}_1 = e_s - \frac{e_0}{A} \tag{m}$$

Substitution of Eqs. (c) and (m) into Eq. (k) in order to eliminate $\dot{\lambda}_1$ and $\dot{\lambda}_2$, after a bit of algebraic manipulation, gives

$$\left(\frac{1}{AR_1} + \frac{1}{AR_2} + \frac{1}{R_2}\right)e_0 = \left(\frac{1}{R_1} + \frac{1}{R_2}\right)e_s \tag{n}$$

For *very* large values of A, Eq. (n) reduces approximately to

$$e_0 = \left(1 + \frac{R_2}{R_1}\right)e_s \tag{o}$$

which is the voltage transfer relation for the circuit sketched in Figure 7-19. Notice that if $R_1 = R_2$, this op-amp device simply has a doubling transfer characteristic, without sign inversion.

7-5 CONSTITUTIVE RELATIONS FOR TRANSDUCERS

We now turn our attention to composite systems that function on the basis of the interaction of both mechanical and electrical variables. Devices that convert mechanical energy to electrical energy or electrical energy to mechanical energy are known as *electromechanical transducers.* Examples of such devices include motors, generators, computer printers, telephones, electric typewriters, video and audio tape decks, loudspeakers, microphones, accelerometers, and many metering, switching, and monitoring devices. Here we shall limit our consideration to two broad categories of electromechanical transducers: those based upon the movable-plate capacitor and the movable-core inductor.

Electromechanical transducers that conserve energy are often called *lossless.* We shall instead continue to use the word *conservative* to describe energy-conserving transducers. Conservative transducers may be subdivided into two classes: those that *store* energy and then subsequently release it, and those that only *transfer* energy.

An electromechanical *energy storage* transducer may store energy when either mechanical work or electrical work is done on it. The stored energy can be recovered later as either mechanical or electrical work, irrespective of the form in which the energy was originally inserted. In this manner, energy conversion can be performed by inserting energy in one form, mechanical or electrical, and extracting it in the other form.

A conservative *energy transfer* transducer is a device that does not store energy; the instantaneous power output always equals the instantaneous power input. Energy conversion in a conservative electromechanical energy-transfer transducer is accomplished instantaneously.

In considering lumped-parameter systems, we are continuing to operate under the restrictions of electromagnetic quasistatics. In essence, electromagnetic quasistatics will be valid if $\dfrac{L}{\lambda} \ll 1$, where L is the length of the device or circuit under consideration and λ is the electromagnetic wavelength. Under these quasistatic assumptions, the fields that produce forces in a given element are either electrical or magnetic, but not both. Therefore, our analyses will consider separately the forces due to electric fields and the forces due to magnetic fields. The theoretical bases for these operating assumptions are given in Appendix G.

The capacitive and inductive transducers for which we shall derive electromechanical constitutive relations are *ideal*. That is, we shall be concerned exclusively with the coupling character between either the electrical and mechanical variables or the magnetic and mechanical variables. Beyond this section, in order to model real electromechanical transducers, it will be necessary to account for electromagnetic and mechanical losses, complementary electrical inductance or capacitance, and mechanical effects due to mass and stiffness. However, these complicating analytical requirements are suspended temporarily as we proceed to develop the constitutive relations for the ideal movable-plate capacitor and the ideal movable-core inductor.

7-5.1   Ideal Movable-Plate Capacitor

An ideal movable-plate capacitor sketched in Figure 7-20 is a conservative energy storage transducer. The charge on the capacitor is q, which is one-half the difference of the charge on the two plates. The voltage across the plates is e. The displacement of the movable plate from its datum is x. And the external force required to hold the movable plate in equilibrium at any value of x, against the electrical attraction of the fixed plate, is f.

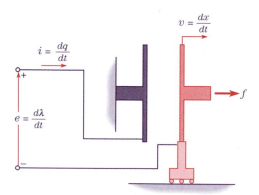

FIGURE 7-20   Ideal movable-plate capacitor.

By definition, an *ideal* movable-plate capacitor has negligible electrical inductance and resistance, as well as negligible mechanical properties such as mass, stiffness, and damping. To the extent that a real movable-plate capacitor may exhibit properties due to inductance, resistance, mass, stiffness, or damping, these characteristics can be introduced external to the capacitor in the electromechanical model of the entire physical system.

The constitutive relations for an ideal movable-plate capacitor may be written as equations for the voltage and force in terms of the displacement and charge as

$$e = e(x, q) \qquad f = f(x, q) \tag{7-47}$$

The form of the first of Eqs. (7-47) follows directly from electromagnetic theory. That is, the voltage across a movable-plate capacitor depends on the charge and the separation between the plates. The form of the second of Eqs. (7-47) is an assumption, but one that follows directly from our definition of *ideal;* that is, since the voltage is a function of x and q, the force can be at most a function of x and q. For most devices, it is necessary to determine both of Eqs. (7-47) by conducting physical experiments, assuming the device exists, or by conducting numerical analyses in the design phase of the device. For simple devices such as the parallel movable-plate capacitor sketched in Figure 7-20, these constitutive relations can be derived theoretically from electromagnetic field theory. The theoretical analysis that follows is often the simplest route to determining Eqs. (7-47).

When there is no charge on the plates of the movable-plate capacitor, there is no electric field, and thus no attractive force between the plates. So, an important auxiliary expression to the constitutive relations in Eqs. (7-47) is

$$f(x, 0) = 0 \tag{7-48}$$

That is, when the charge q is zero, the force f must also be zero for all values of x.

The electrical power delivered to the movable-plate capacitor is ei, and the mechanical power delivered to it is fv. Thus, the work increment done on the capacitor in time dt is

$$d\text{Work} = ei\,dt + fv\,dt = e\,dq + f\,dx = dW_e(x, q) \tag{7-49}$$

As indicated by Eq. (7-49), for a conservative movable-plate capacitor, this work is equal to the increase in the stored electrical energy dW_e in the transducer. Although both electrical and mechanical work increments are represented in the work increment in Eq. (7-49), the phrase *electrical energy* is used for the stored energy W_e because the energy that is stored in an *ideal* movable-plate capacitor is due to the electric field. That is, there is no mechanical energy because in the ideal movable–plate capacitor the mass and stiffness are zero and, as shown in Subsection G-4.1 of Appendix G, the electromagnetic energy is stored in the capacitor's electric field.

The total *electrical energy function* $W_e(x, q)$ with respect to a datum configuration is obtained by integrating Eq. (7-49) along *any* path from the datum to the operating point (x, q). Once the energy function $W_e(x, q)$ is known, the constitutive relations in Eqs. (7-47) can be recovered by differentiating Eq. (7-49) according to

$$\frac{\partial W_e}{\partial x} = f \qquad \frac{\partial W_e}{\partial q} = e \tag{7-50}$$

An alternative state function for the movable-plate capacitor is the *electrical coenergy function* $W_e^*(x, e)$, defined by a Legendre transformation as

$$W_e^*(x, e) = eq - W_e(x, q) \tag{7-51}$$

where the electrical coenergy is a function of the complementary electrical variable, the voltage e, and the same mechanical variable, the displacement x. When the electrical coenergy function is known, the constitutive relations for the transducer can be obtained by differentiating Eq. (7-51) according to

$$\frac{\partial W_e^*}{\partial x} = -\frac{\partial W_e}{\partial x} = -f \qquad \frac{\partial W_e^*}{\partial e} = q \tag{7-52}$$

where Eq. (7-50) has been used in obtaining the first of Eqs. (7-52).

7-5.2   Electrically Linear Movable-Plate Capacitor

For many movable-plate capacitors operating within their design range, the transducer may be modeled as *electrically linear.* That is, to within a multiplicative constant, there is a one-to-one linear correspondence between the electrical variables e and q. Thus, if electrical linearity can be assumed, the first of the constitutive relations in Eqs. (7-47) can be written as

$$e = \frac{q}{C(x)} \tag{7-53}$$

where the form of Eq. (7-9) has been adopted and where $C(x)$ is the capacitance when the displacement of the movable plate is x. In general, $C(x)$ is a nonlinear function of x. The assumption of electrical linearity is generally valid if *fringing* effects can be neglected. (Around the edges of a parallel-plate capacitor in the vicinity of the gap between the plates, the field lines may "bulge" outwardly in rounded contours. This effect may lead to nonlinearities and is known as *fringing.*) In real transducers where the plate perimeter dimensions are much larger than the gap dimension, the effects of fringing are usually negligible or, if necessary, may be incorporated empirically or by theoretical analysis. Thus, as we shall proceed to demonstrate, having made only the frequently valid assumption of *electrical linearity* enables us to derive explicit expressions for both of the electrical state functions W_e and W_e^* for the transducer.

One very useful feature of a *conservative* element, transducer, or system is that the change in stored energy between any two points in its variables space is independent of the path in going from the initial state to the final state. Indeed, this is a major characteristic of *state functions.* Thus, for integrating Eq. (7-49), we can select *any* integration path that is computationally convenient. In Figure 7-21, if we integrate Eq. (7-49) from a datum configuration of $x = 0$ and $q = 0$ along any path to reach a final configuration (x, q), the stored electrical energy must be the same. Thus, integration along any of paths Ⓐ, Ⓑ or Ⓒ in Figure 7-21 will result in the identical final stored electrical energy. A particularly convenient path is the successive two-step contour indicated as ① and ② in Figure 7-21. From Eq. (7-49) we note that

$$W_e(x, q) = \int_0^q e \, dq + \int_0^x f \, dx \tag{7-54}$$

Along path ①, $q = 0$, making the first integral in Eq. (7-54) zero; and because of Eq. (7-48), $f = 0$ also, making the second integral in Eq. (7-54) zero. Thus, along path ①, $W_e(x, q) = 0$. Along path ②, x is a constant; so $dx = 0$, making the second integral in Eq. (7-54) zero. Thus, Eq. (7-54) reduces to an integration along path ②, which is simply

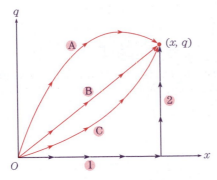

FIGURE 7-21 Integration paths for electrical energy of movable-plate capacitor.

$$W_e(x, q) = \int_0^q e \, dq = \int_0^q \frac{q \, dq}{C(x)} = \frac{q^2}{2C(x)} \tag{7-55}$$

where Eq. (7-53) has been used to obtain the result in Eq. (7-55).

Physically, the process of integration along path ① means that the capacitor is mechanically positioned into its datum configuration at $x = 0$ when no charge is on the plates. Then, since the force is also zero and remains zero as long as q remains zero, locating the movable plate into its final position x can be accomplished with no work; and thus no stored energy. Then, integration along path ② can be interpreted to mean that the movable plate is "locked" into its position x and then the capacitor is charged to its final value of q. Even though the force on the movable plate changes during the electrical charging, there is no displacement of the plate, and thus no work is done by this force.

Since the electrical energy function $W_e(x, q)$ is known, the constitutive relations corresponding to Eqs. (7-47) can be derived. Substituting the result from Eq. (7-55) into Eqs. (7-50) gives

$$f = \frac{\partial W_e}{\partial x} = \frac{q^2}{2} \frac{\partial}{\partial x}[C^{-1}(x)] = -\frac{q^2}{2C^2(x)} \frac{\partial C(x)}{\partial x} \tag{7-56}$$

and

$$e = \frac{\partial W_e}{\partial q} = \frac{\partial}{\partial q}\left[\frac{q^2}{2C(x)}\right] = \frac{q}{C(x)} \tag{7-57}$$

Alternatively, the electrical coenergy function for the electrically linear movable-plate capacitor may be obtained by substituting Eqs. (7-53) and (7-55) into Eq. (7-51) as

$$W_e^*(x, e) = eq - W_e(x, q)$$

$$= e[eC(x)] - \frac{e^2C^2(x)}{2C(x)} = \frac{1}{2}C(x)e^2 \tag{7-58}$$

Now, since the electrical coenergy function $W_e^*(x, e)$ is known, the constitutive relations can also be derived by substituting Eqs. (7-53) and (7-58) into Eqs. (7-52), giving

$$f = -\frac{\partial W_e^*}{\partial x} = -\frac{e^2}{2} \frac{\partial}{\partial x}[C(x)] = -\frac{1}{2}\left[\frac{q^2}{C^2(x)}\right] \frac{\partial C(x)}{\partial x} \tag{7-59}$$

and

$$q = \frac{\partial W_e^*}{\partial e} = \frac{\partial}{\partial e}\left[\frac{1}{2}C(x)e^2\right] = C(x)e \tag{7-60}$$

Finally, it is interesting to observe that the frequently valid assumption of electrical linearity for the conservative movable-plate capacitor is sufficient to provide all results contained in Eqs. (7-55) through (7-60).

7-5.3   Ideal Movable-Core Inductor

Another conservative energy storage transducer is the movable-core inductor. The ideal movable-core inductor is the magnetic counterpart of the ideal movable-plate capacitor, and this magnetic-electrical complementarity will be emphasized in the analysis below. The most common configuration of the ideal movable-core inductor is sketched in Figure 7-22. The flux linkage of the coil is λ, and the current through it is i. The displacement of the movable core from its datum is x, and the external force required to hold the core in equilibrium at any value of x, against the magnetic attraction of the solenoid, is f.

By definition, an *ideal* movable-core inductor has negligible electrical capacitance and resistance, as well as negligible mechanical properties such as mass, stiffness, and damping. To the extent that a real movable-core inductor may exhibit properties due to capacitance, resistance, mass, stiffness, or damping, these characteristics can be introduced external to the movable-core inductor in the electromechanical model of the entire physical system.

The constitutive relations for an ideal movable-core inductor may be written as equations for the current and force in terms of the displacement and flux linkage as

$$i = i(x, \lambda) \qquad f = f(x, \lambda) \tag{7-61}$$

The form of the first of Eqs. (7-61) follows directly from electromagnetic theory. That is, the current through a movable-core inductor depends on the flux linkage and the displacement of the core. We neglect magnetic hysteresis, so Eqs. (7-61) are single-valued functions. The form of the second of Eqs. (7-61) is an assumption but follows directly from our definition of *ideal*; that is, since the current is a function of x and λ, the force can be at most a function of x and λ. For most devices, it is necessary to determine both of Eqs. (7-61) by conducting physical experiments, assuming the device exists, or by conducting numerical analyses in the design phase of the device. For simple devices such as the movable-core inductor sketched in Figure 7-22, these constitutive relations can be derived theoretically from electromagnetic field theory. The theoretical analysis that follows is often generally the simplest route to determining Eqs. (7-61).

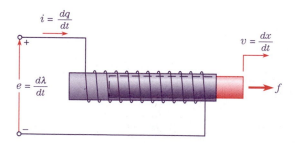

FIGURE 7-22   Ideal movable-core inductor.

An important characteristic of the movable-core inductor is that when $\lambda = 0$, there is no magnetic field, and thus no magnetic attraction between the solenoid and the movable core. So, the force f is zero for any x when $\lambda = 0$. Thus, an important auxiliary expression to the constitutive relations in Eqs. (7-61) is

$$f(x, 0) = 0 \tag{7-62}$$

That is, when the flux linkage is zero, the force f must also be zero for all values of x.

The electrical power delivered to the movable-core inductor is ei, and the mechanical power delivered to it is fv. Thus, the work increment done *on* the movable-core inductor in time dt is

$$d\text{Work} = ei \, dt + fv \, dt = i \, d\lambda + f \, dx = dW_m(x, \lambda) \tag{7-63}$$

As indicated by Eq. (7-63), for a conservative movable-core inductor this work is equal to the increase in the stored magnetic energy dW_m in the transducer. Although both magnetic and mechanical work increments are represented in the work increment in Eq. (7-63), the phrase *magnetic energy* is used for the stored energy W_m because the energy that is stored in an *ideal* movable-core inductor is due to the magnetic field. That is, there is no mechanical energy because in the ideal movable-core inductor the mass and stiffness are zero and, as shown in Subsection G-4.2 of Appendix G, the electromagnetic energy is stored in the inductors magnetic field.

The total *magnetic energy function* $W_m(x, \lambda)$ with respect to a datum configuration is obtained by integrating Eq. (7-63) along *any* path from the datum to the operating point (x, λ). Once the energy function $W_m(x, \lambda)$ is known, the constitutive relations in Eqs. (7-61) can be recovered by differentiating Eq. (7-63) according to

$$\frac{\partial W_m}{\partial x} = f \qquad \frac{\partial W_m}{\partial \lambda} = i \tag{7-64}$$

An alternative state function for the movable-core inductor is the *magnetic coenergy function* $W_m^*(x, \lambda)$, defined by a Legendre transformation as

$$W_m^*(x, i) = i\lambda - W_m(x, \lambda) \tag{7-65}$$

where the magnetic coenergy is a function of the complementary magnetic variable, the current i, and the same mechanical variable, the displacement x. When the magnetic coenergy function is known, the constitutive relations for the transducer can be obtained by differentiating Eq. (7-65) according to

$$\frac{\partial W_m^*}{\partial x} = -\frac{\partial W_m}{\partial x} = -f \qquad \frac{\partial W_m^*}{\partial i} = \lambda \tag{7-66}$$

where Eq. (7-64) has been used in obtaining the first of Eqs. (7-66).

7-5.4   Magnetically Linear Movable-Core Inductor

For many movable-core inductors operating within their design range, the constitutive relation between current and flux linkage may be modeled as *magnetically linear.* That is, to within a multiplicative constant, there is a one-to-one linear correspondence between the magnetic variables i and λ. Thus, if magnetic linearity can be assumed, the first of the constitutive relations in Eqs. (7-61) can be written as

$$i = \frac{\lambda}{L(x)} \tag{7-67}$$

where the form of Eq. (7-19) has been adopted and where $L(x)$ is the inductance when the displacement of the core is x. In general, $L(x)$ is a nonlinear function of x. Although the requirements for magnetic linearity are satisfied less often than those for electrical linearity, these requirements are frequently met by devices functioning in their design operating ranges. Thus, as we shall proceed to demonstrate, having made only the frequently valid assumption of *magnetic linearity* enables us to derive explicit expressions for the magnetic state functions W_m and W_m^* for the transducer.

One very useful feature of a *conservative* element, transducer, or system is that the change in stored energy between any two points in its variables space is independent of the path in going from the initial state to the final state. Thus, for integrating Eq. (7-63), we can select *any* integration path that is computationally convenient. In Figure 7-23, if we integrate Eq. (7-63) from a datum configuration of $x = 0$ and $\lambda = 0$ along any path to reach a final configuration (x, λ), the stored magnetic energy must be the same. Thus, integration along any of paths (A), (B), or (C) in Figure 7-23 all result in the identical final stored magnetic energy. A particularly convenient path is the successive two-step contour indicated as (1) and (2) in Figure 7-23. From Eq. (7-63) we note that

$$W_m(x, \lambda) = \int_0^\lambda i \, d\lambda + \int_0^x f \, dx \tag{7-68}$$

Along path (1), $\lambda = 0$, making the first integral in Eq. (7-68) zero; and because of Eq. (7-62), $f = 0$ also, making the second integral in Eq. (7-68) zero. Thus, along path (1), $W_m(x, \lambda) = 0$. Along path (2), x is a constant; so $dx = 0$, making the second integral in Eq. (7-68) zero. Thus, Eq. (7-68) reduces to an integration along path (2), which is simply

$$W_m(x, \lambda) = \int_0^\lambda i \, d\lambda = \int_0^\lambda \frac{\lambda d\lambda}{L(x)} = \frac{\lambda^2}{2L(x)} \tag{7-69}$$

where Eq. (7-67) has been used to obtain the result in Eq. (7-69).

Physically, the process of integration along path (1) means that the core is mechanically positioned into its datum configuration at $x = 0$ when the flux linkage λ is zero. Then, since the force is also zero and remains zero as long as λ remains zero, locating the movable core into its final position x can be accomplished with no work; and thus no stored energy. Then,

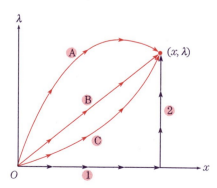

FIGURE 7-23   Integration paths for magnetic energy of movable-core inductor.

integration along path ② can be interpreted to mean that the movable core is "locked" into its position x, and then the flux linkage is increased to its final value of λ. Even though the force on the movable core changes as the flux linkage is increased to its final value, there is no displacement of the core, and thus no work is done by this force.

Since the magnetic energy function $W_m(x, \lambda)$ is known, the constitutive relations corresponding to Eqs. (7-61) can be derived. Substituting the result from Eq. (7-69) into Eqs. (7-64) gives

$$f = \frac{\partial W_m}{\partial x} = \frac{\lambda^2}{2} \frac{\partial}{\partial x}[L^{-1}(x)] = -\frac{\lambda^2}{2L^2(x)} \frac{\partial L(x)}{\partial x} \tag{7-70}$$

and

$$i = \frac{\partial W_m}{\partial \lambda} = \frac{\partial}{\partial \lambda}\left[\frac{\lambda^2}{2L(x)}\right] = \frac{\lambda}{L(x)} \tag{7-71}$$

Alternatively, the magnetic coenergy function for the magnetically linear movable-core inductor may be obtained by substituting Eqs. (7-67) and (7-69) into Eq. (7-65) as

$$W_m^*(x, i) = i\lambda - W_m(x, \lambda)$$

$$= i[iL(x)] - \frac{i^2 L^2(x)}{2L(x)} = \frac{1}{2}L(x)i^2 \tag{7-72}$$

Now, since the magnetic coenergy function $W_m^*(x, i)$ is known, the constitutive relations can also be derived by substituting Eqs. (7-67) and (7-72) into Eqs. (7-66), giving

$$f = \frac{\partial W_m^*}{\partial x} = -\frac{i^2}{2} \frac{\partial}{\partial x}[L(x)] = -\frac{1}{2}\left[\frac{\lambda^2}{L^2(x)}\right]\frac{\partial L(x)}{\partial x} \tag{7-73}$$

and

$$\lambda = \frac{\partial W_m^*}{\partial i} = \frac{\partial}{\partial i}\left[\frac{1}{2}L(x)i^2\right] = L(x)i \tag{7-74}$$

Finally, it is interesting to observe that the frequently valid assumption of magnetic linearity for the conservative movable-core inductor is sufficient to provide all the results contained in Eqs. (7-69) through (7-74). The parallels between the analytical developments for the ideal movable-plate capacitor and the ideal movable-core inductor should be striking.

7-6 HAMILTON'S PRINCIPLE AND LAGRANGE'S EQUATIONS FOR ELECTROMECHANICAL SYSTEMS

As in the cases of mechanical systems and electrical systems, the equations of motion for electromechanical systems may be formulated via the direct approach or the indirect approach. Electromechanical systems are by definition both electrical and mechanical in character. Thus, irrespective of the approach used for equation formulation, it is necessary to satisfy all the dynamic requirements for electrical networks (Section 7-2) plus all the dynamic requirements for mechanical systems (Section 5-2). The coupling between the electrical variables and the mechanical variables is provided by the constitutive relation(s) of the electromechanical transducer(s) in the system.

The formulation of the equations of motion for electromechanical systems via the *direct approach* involves expressing each of the individual requirements for all the electrical and mechanical variables in analytical form, combining the resulting equations, and then ideally systematically eliminating excessive variables to obtain a set of simultaneous differential equations of motion. For example, suppose there are n_e complete and independent electrical variables and n_m complete and independent mechanical variables. Then, imposing all remaining electrical requirements will lead to n_e differential equations, and imposing all remaining mechanical requirements will lead to n_m differential equations. Due to the electromechanical transducer(s) in the system, some mechanical variables will be present in the n_e electrical equations and some electrical variables will be present in the n_m mechanical equations, thus resulting in a set of $n_e + n_m$ coupled differential equations of motion.

No further discussion of the direct approach for electromechanical systems will be considered here. We now turn exclusively to equation formulation via the indirect approach, which is of substantial utility and is broadly used, though not generally in the form presented here.

7-6.1   Displacement–Charge Variables Formulation

In a displacement–charge variables formulation, the admissibility requirements are satisfied by the selection of a complete and independent set of generalized geometric coordinates ξ_j ($j = 1, 2, \ldots, n_m$) and the selection of a complete and independent set of loop currents i_k ($k = 1, 2, \ldots, n_e$), consistent with the relations $\dot{q}_k = i_k$. Then, Hamilton's principle for electromechanical systems using displacement and charge variables may be defined as

$$\text{V.I.} = \int_{t_1}^{t_2} \left[\delta(T^* - V + W_m^* - W_e) + \sum_{j=1}^{n_m} \Xi_j \delta \xi_j + \sum_{k=1}^{n_e} \mathcal{E}_k \delta q_k \right] dt \qquad (7\text{-}75)$$

Equation (7-75), which is an extended form of the classical Hamilton's principle, may be defined as follows:

- An admissible motion of the system between specified configurations at t_1 and t_2 is a natural motion if, and only if, the variational indicator vanishes for arbitrary admissible variations.

Just as for electrical networks, this statement of Hamilton's principle for electromechanical systems is the same as that defined for mechanical systems. We need only interpret its terminology in the context of the displacement–charge formulation for electromechanical systems.

Here, an "admissible motion" is represented by both the generalized geometric coordinates ξ_j that satisfy the kinematic requirements and the generalized charge coordinates q_k that satisfy Kirchhoff's current rule in accordance with a complete and independent set of loop currents. The "natural motion" is an admissible motion that also satisfies both the dynamic force requirements in accordance with Newton's second law and the dynamic voltage requirements in accordance with Kirchhoff's voltage rule. The constitutive relations for all the mechanical elements and electrical elements, including their coupling, are incorporated into the formulation via the proper representation of the four state functions (T^*, V, W_m^* and W_e), the generalized forces (Ξ_j), and the generalized voltages ($\mathcal{E}_k$). Thus, in this way, all three mechanical and all three electrical requirements are satisfied.

In adopting Hamilton's principle in Eq. (7-75) as a fundamental law of nature, we conclude that it cannot be *derived;* it is an unprovable empirical fact. Still, we can show the equivalence of Hamilton's principle in Eq. (7-75) with the underlying mechanical and electrical requirements and the alternative laws and rules associated with the direct approach for electromechanical systems. For a displacement–charge variables formulation, this equivalence is demonstrated in Section E-2 of Appendix E.

For electromechanical systems containing only lumped-parameter elements, Lagrange's equations generally provide the most direct procedure for deriving the equations of motion. As shown in Section E-2 of Appendix E, the necessary conditions for the vanishing of the displacement–charge variational indicator in Eq. (7-75) are Lagrange's equations in the form of two sets:

$$\frac{d}{dt}\left(\frac{\partial \mathcal{L}}{\partial \dot{\xi}_j}\right) - \frac{\partial \mathcal{L}}{\partial \xi_j} = \Xi_j \qquad j = 1, 2, \ldots, n_m \tag{7-76}$$

$$\frac{d}{dt}\left(\frac{\partial \mathcal{L}}{\partial \dot{q}_k}\right) - \frac{\partial \mathcal{L}}{\partial q_k} = \mathcal{E}_k \qquad k = 1, 2, \ldots, n_e \tag{7-77}$$

where the lagrangian is

$$\mathcal{L}(\xi_j, \dot{\xi}_j, q_k, \dot{q}_k) = T^* - V + W_m^* - W_e \tag{7-78}$$

and the generalized forces Ξ_j and the generalized voltages $\mathcal{E}_k$ are derived from the nonconservative variational work expression

$$\delta \mathcal{W}^{nc} = \sum_{h=1}^{N} \boldsymbol{f}_h^{nc} \cdot \delta \boldsymbol{R}_h + \sum_{i=1}^{M} e_i^{nc} \delta q_i = \sum_{j=1}^{n_m} \Xi_j \delta \xi_j + \sum_{k=1}^{n_e} \mathcal{E}_k \delta q_k \tag{7-79}$$

In Eq. (7-79) the term $\sum_{h=1}^{N} \boldsymbol{f}_h^{nc} \cdot \delta \boldsymbol{R}_h$ is the variational work increment done by the N nonconservative mechanical forces and the term $\sum_{i=1}^{M} e_i^{nc} \delta q_i$ is the variational work increment done by the M nonconservative electrical elements, both effected during an admissible variation of the system. Also, as indicated in conjunction with Eq. (7-75), n_m is the number of complete and independent generalized geometric coordinates and n_e is the number of complete and independent generalized charge coordinates.

7-6.2   Displacement–Flux Linkage Variables Formulation

In a displacement–flux linkage variables formulation, the admissibility requirements are satisfied by the selection of a complete and independent set of generalized geometric coordinates ξ_j ($j = 1, 2, \ldots, n_m$) and the selection of a complete and independent set of node voltages e_k ($k = 1, 2, \ldots, n_e'$), consistent with the relations $\dot{\lambda}_k = e_k$. Then, Hamilton's principle for electromechanical systems using displacement and flux linkage variables may be expressed as

$$\text{V.I.} = \int_{t_1}^{t_2} \left[\delta(T^* - V + W_e^* - W_m) + \sum_{j=1}^{n_m} \Xi_j \delta \xi_j + \sum_{k=1}^{n_e'} \mathcal{I}_k \delta \lambda_k \right] dt \tag{7-80}$$

Equation (7-80), which is an extended form of the classical Hamilton's principle, may be expressed as follows:

- An admissible motion of the system between specified configurations at t_1 and t_2 is a natural motion if, and only if, the variational indicator vanishes for arbitrary admissible variations.

Again, just as for electrical networks and the preceding displacement–charge variables formulation, this statement of Hamilton's principle for electromechanical systems is the same as that defined for mechanical systems. We need only interpret its terminology in the context of the displacement–flux linkage formulation for electromechanical systems.

Here, an "admissible motion" is represented by both the generalized geometric coordinates ξ_j that satisfy the kinematic requirements and the generalized flux linkage coordinates λ_k that satisfy Kirchhoff's voltage rule in accordance with a complete and independent set of node voltages. The "natural motion" is an admissible motion that also satisfies both the dynamic force requirements in accordance with Newton's second law and the dynamic current requirements in accordance with Kirchhoff's current rule. The constitutive relations for all the mechanical elements and electrical elements, including their coupling, are incorporated into the formulation via the proper representation of the four state functions (T^*, V, W_e^* and W_m), the generalized forces (Ξ_j), and the generalized currents ($\mathcal{I}_k$). Thus, in this way, all three mechanical and all three electrical requirements are satisfied.

In adopting Hamilton's principle in Eq. (7-80) as a fundamental law of nature, again we conclude that it cannot be *derived;* it is an unprovable empirical fact. Nevertheless, we can show the equivalence of Hamilton's principle in Eq. (7-80) with the underlying mechanical and electrical requirements and the alternative laws and rules associated with the direct approach for electromechanical systems. For a displacement–flux linkage variables formulation, this equivalence is demonstrated in Section E-3 of Appendix E.

For electromechanical systems containing only lumped-parameter elements, Lagrange's equations generally provide the most direct procedure for deriving the equations of motion. As shown in Section E-3 of Appendix E, the necessary conditions for the vanishing of the displacement–flux linkage variational indicator in Eq. (7-80) are Lagrange's equations in the form of two sets:

$$\frac{d}{dt}\left(\frac{\partial \mathcal{L}}{\partial \dot{\xi}_j}\right) - \frac{\partial \mathcal{L}}{\partial \xi_j} = \Xi_j \qquad j = 1, 2, \ldots, n_m \tag{7-81}$$

$$\frac{d}{dt}\left(\frac{\partial \mathcal{L}}{\partial \dot{\lambda}_k}\right) - \frac{\partial \mathcal{L}}{\partial \lambda_k} = \mathcal{I}_k \qquad k = 1, 2, \ldots, n_e' \tag{7-82}$$

where the lagrangian is

$$\mathcal{L}(\xi_j, \dot{\xi}_j, \lambda_k, \dot{\lambda}_k) = T^* - V + W_e^* - W_m \tag{7-83}$$

and the generalized forces Ξ_j and the generalized currents $\mathcal{I}_k$ are derived from the nonconservative variational work expression

$$\delta \mathcal{W}^{nc} = \sum_{h=1}^{N} \boldsymbol{f}_h^{nc} \cdot \delta \boldsymbol{R}_h + \sum_{i=1}^{M'} i_i^{nc} \delta \lambda_i = \sum_{j=1}^{n_m} \Xi_j \delta \xi_j + \sum_{k=1}^{n_e'} \mathcal{I}_k \delta \lambda_k \tag{7-84}$$

In Eq. (7-84) the term $\sum_{h=1}^{N} \boldsymbol{f}_h^{nc} \cdot \delta \boldsymbol{R}_h$ is the variational work increment done by the N nonconservative mechanical forces and the term $\sum_{i=1}^{M'} i_i^{nc} \delta \lambda_i$ is the variational work increment done by the M' nonconservative electrical elements, both effected during an admissible variation of the system. Also, as indicated in conjunction with Eq. (7-80), n_m is the number of complete and independent generalized geometric coordinates and n'_e is the number of complete and independent generalized flux linkage coordinates.

It is instructive to note the parallels between Eqs. (7-80) through (7-84) with Eqs. (7-75) through (7-79), as well as with the corresponding equations in Table 5-1.

7-6.3   Examples

■ **Example 7-11:**   Butterfly Condenser

A butterfly condenser consists of a rotor of axial moment of inertia I_0 and a stator that is grounded (mechanically and electrically) as sketched in Figure 7-24. The rotor rotates in the fixed bearings without friction. The capacitance of the condenser varies with rotor angle according to the constitutive relation $C(\theta) = C_0 + C_1 \cos 2\theta$, where C_0 and C_1 are positive constants. Assuming electrical linearity, find the equation(s) of motion for the rotor when an ideal constant voltage source E_0 is connected across the condenser.

❐ **Solution:**   To model the physical butterfly capacitor, we consider the capacitor to be a combination of an ideal electrical element, mechanical element, and an electromechanical transducer element. This is a concept that is broadly used, and may be (but is not) used in all the examples that appear here on electromechanics. This concept is sketched in Figure 7-25a, where the electrical element consists of the ideal constant-voltage source E_0 (on the left), the mechanical element consists of the rotor, having moment of inertia I_0 (on the right), and the electromechanical transducer is in the middle. The symbolic representation of the electromechanical transducer is sketched with the electrically linear constitutive relation indicated.

Displacement–flux linkage formulation

1. Generalized coordinate(s):   From Figure 7-25b, we note that the generalized geometric coordinate and flux linkage admissibility requirement may be defined as

$$\xi_j : \theta \qquad \delta\xi_j : \delta\theta \tag{a}$$

and

$$\dot{\lambda}_0 = E_0 \qquad \delta\lambda_0 = 0 \tag{b}$$

where θ is the angular displacement of the rotor with respect to the stator, and we note that the system is holonomic. Equation (a) is a straightforward statement of the single degree of freedom of the rotor and the holonomic character of the "mechanical side" of the transducer. Equation (b) simply indicates that at the output terminal labeled ① of the voltage source,

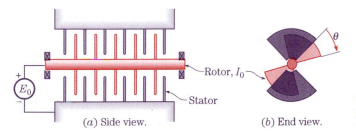

(a) Side view.   (b) End view.

Rotor, I_0
Stator
E_0
θ

FIGURE 7-24   Butterfly condenser, showing angular displacement θ of rotor in end view.

the flux linkage is uniquely prescribed. (Refer to Table 7-2 and Example 7-3.) Thus, θ is the only generalized coordinate.

2. Generalized force: Consideration of Eq. (7-84), namely,

$$\delta \mathcal{W}^{nc} = \sum_{h=1}^{N} \boldsymbol{f}_h^{nc} \cdot \delta \boldsymbol{R}_h + \sum_{i=1}^{M'} i_i^{nc} \delta \lambda_i = \sum_{j=1}^{n_m} \Xi_j \delta \xi_j + \sum_{k=1}^{n_e'} \mathcal{I}_k \delta \lambda_k$$

leads immediately to

$$\Xi_\theta = 0 \qquad \mathcal{I}_k \text{ does not apply} \tag{c}$$

indicating that there are no nonconservative mechanical torques and the generalized current does not apply since there is no generalized flux linkage coordinate.

3. Lagrangian: Use of Eq. (7-83) gives the lagrangian as

$$\mathcal{L}(\theta, \dot{\theta}) = T^* - V + W_e^* - W_m$$

$$= \frac{1}{2} I_0 \dot{\theta}^2 - 0 + \frac{1}{2} C(\theta) E_0^2 - 0 \tag{d}$$

where Eq. (7-58), incorporating the condition of electrical linearity of the transducer, has been used. If the capacitance constitutive relation is substituted into Eq. (d), the lagrangian becomes

$$\mathcal{L} = \frac{1}{2} I_0 \dot{\theta}^2 + \frac{1}{2} [C_0 + C_1 \cos 2\theta] E_0^2 \tag{e}$$

(*a*) Model indicating electrically linear constitutive relation.

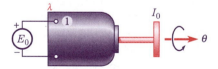

(*b*) Model for displacement–flux linkage formulation.

(*c*) Model for displacement–charge formulation.

FIGURE 7-25   Electromechanical models of butterfly condenser.

4. Lagrange's equation: Substitution of Eq. (a) into Eq. (7-81) gives

$$\frac{d}{dt}\left(\frac{\partial \mathcal{L}}{\partial \dot{\theta}}\right) - \frac{\partial \mathcal{L}}{\partial \theta} = \Xi_\theta \tag{f}$$

and substitution of Eqs. (c) and (e) into Eq. (f) gives

$$I_0 \ddot{\theta} + E_0^2 C_1 \sin 2\theta = 0 \tag{g}$$

which is a single nonlinear differential equation describing the behavior of the butterfly condenser.

Displacement–charge formulation

1. Generalized coordinate(s):   Alternatively, reference to Figure 7-25c indicates that for a displacement–charge formulation, the generalized coordinates are

$$\xi_j : \theta \qquad \delta \xi_j : \delta \theta \tag{h}$$

$$q_k : q \qquad \delta q_k : \delta q \tag{i}$$

Thus, both θ and q are generalized coordinates, and the system is holonomic.

2. Generalized force and generalized voltage: Consideration of Eq. (7-79) gives

$$\delta \mathcal{W}^{nc} = \sum_{h=1}^{N} \boldsymbol{f}_h^{nc} \cdot \delta \boldsymbol{R}_h + \sum_{i=1}^{M} e_i^{nc} \delta q_i = \sum_{j=1}^{n_m} \Xi_j \delta \xi_j + \sum_{k=1}^{n_e} \mathcal{E}_k \delta q_k$$

$$= (0)\delta \theta + E_0 \delta q = \Xi_\theta \delta \theta + \mathcal{E}_q \delta q \tag{j}$$

Or,

$$\Xi_\theta = 0 \qquad \mathcal{E}_q = E_0 \tag{k}$$

3. Lagrangian: Use of Eq. (7-78) gives the lagrangian as

$$\mathcal{L}(\theta, \dot{\theta}, q, \dot{q}) = T^* - V + W_m^* - W_e$$

$$= \frac{1}{2} I_0 \dot{\theta}^2 - 0 + 0 - \frac{q^2}{2C(\theta)} \tag{l}$$

where Eq. (7-55), incorporating the condition of electrical linearity of the transducer, has been used.

4. Lagrange's equations: Substitution of Eqs. (h) and (i) into Eqs. (7-76) and (7-77), respectively, gives

$$\frac{d}{dt}\left(\frac{\partial \mathcal{L}}{\partial \dot{\theta}}\right) - \frac{\partial \mathcal{L}}{\partial \theta} = \Xi_\theta \tag{m}$$

and

$$\frac{d}{dt}\left(\frac{\partial \mathcal{L}}{\partial \dot{q}}\right) - \frac{\partial \mathcal{L}}{\partial q} = \mathcal{E}_q \tag{n}$$

Then substitution of Eqs. (k) and (l) into Eqs. (m) and (n) leads to

θ-equation: $$I_0 \ddot{\theta} - \frac{q^2}{2}[C(\theta)]^{-2}\frac{dC(\theta)}{d\theta} = 0 \tag{o}$$

and

q-equation: $$\frac{q}{C(\theta)} = E_0 \tag{p}$$

We immediately observe that Eq. (p) is the electromechanical transduction equation and if it is substituted into Eq. (o) along with the constitutive relation for $C(\theta)$, we get

$$I_0\ddot{\theta} + E_0^2 C_1 \sin 2\theta = 0 \tag{q}$$

Equation (q) is the same as Eq. (g). The fact that Eqs. (g) and (q) are identical is a consequence of the simplicity of the system since, in general, a displacement–flux linkage formulation will not lead to the same equations of motion as a displacement–charge formulation. In Chapter 8 (Example 8-20), we shall return to this example to explore some of the features of the dynamic behavior of this system.

■ **Example 7-12:**   Condenser Microphone

The condenser microphone sketched in Figure 7-26 is an electric field transducer that senses pressure. (If the same transducer were used as a pressure *source,* it would then be configured as a speaker.) The microphone consists of a flexible diaphragm that is mounted parallel to a fixed back plate, both within a case. In this simplified model, the flexible diaphragm consists of a movable plate of mass m and its circumferential support, having an effective stiffness modeled by a linear spring with spring constant k. The effective mechanical damping of the microphone is due to air motion around the fixed plate and through the holes in it and is modeled by a linear dashpot with dashpot constant c, associated with the velocity of the diaphragm.

The time-varying acoustic pressure is due to the sound field and produces motion of the flexible diaphragm. The motion $x(t)$ of the diaphragm changes the capacitance between the movable and fixed plates, and produces currents that the amplifier magnifies in order to drive a recorder or a loudspeaker (not shown). The capacitance function $C(x)$ is given by

$$C(x) = C_0 \frac{d_0}{d_0 + x}$$

where d_0 is the separation between the diaphragm and the fixed plate when $x = 0$, and C_0 is the capacitance of the microphone when $x = 0$. The acoustic pressure acting over the

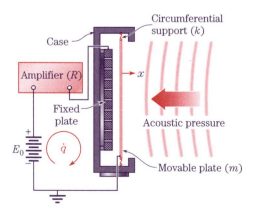

FIGURE 7-26   Condenser microphone, showing coordinates x and q.

area of the diaphragm exerts a time-varying force $F(t)$ on the movable plate. The amplifier may be modeled by a resistance and the battery that powers the amplifier is modeled as an ideal constant voltage source.

Obtain the equation(s) of motion for the microphone. Electrical linearity may be assumed.

◻ **Solution:**

1. Generalized coordinate(s):   We consider the variables x and q as generalized coordinates as indicated in Figure 7-26. Thus

$$\xi_j : x \qquad \delta\xi_j : \delta x \tag{a}$$

$$q_k : q \qquad \delta q_k : \delta q \tag{b}$$

Therefore, x and q are the generalized coordinates, and the system is holonomic.

2. Generalized force and generalized voltage: Consideration of Eq. (7-79) gives

$$\delta \mathcal{W}^{\mathrm{nc}} = \sum_{h=1}^{N} \boldsymbol{f}_h^{\mathrm{nc}} \cdot \delta \boldsymbol{R}_h + \sum_{i=1}^{M} e_i^{\mathrm{nc}} \delta q_i = \sum_{j=1}^{n_m} \Xi_j \delta\xi_j + \sum_{k=1}^{n_e} \mathcal{E}_k \delta q_k$$

$$= F(t)\delta x - c\dot{x}\delta x + E_0\delta q - R\dot{q}\delta q = \Xi_x\delta x + \mathcal{E}_q\delta q \tag{c}$$

Or,

$$\Xi_x = F(t) - c\dot{x} \qquad \mathcal{E}_q = E_0 - R\dot{q} \tag{d}$$

3. Lagrangian: Use of Eq. (7-78) gives the lagrangian as

$$\mathcal{L}(x, \dot{x}, q, \dot{q}) = T^* - V + W_m^* - W_e$$

$$= \frac{1}{2}m\dot{x}^2 - \frac{1}{2}kx^2 + 0 - \frac{q^2}{2C(x)} \tag{e}$$

where Eq. (7-55), incorporating the condition of electrical linearity of the transducer, has been used.

4. Lagrange's equation: Substitution of Eqs. (a) and (b) into Eqs. (7-76) and (7-77), respectively, gives

$$\frac{d}{dt}\left(\frac{\partial\mathcal{L}}{\partial\dot{x}}\right) - \frac{\partial\mathcal{L}}{\partial x} = \Xi_x \tag{f}$$

and

$$\frac{d}{dt}\left(\frac{\partial\mathcal{L}}{\partial\dot{q}}\right) - \frac{\partial\mathcal{L}}{\partial q} = \mathcal{E}_q \tag{g}$$

Then substitution of Eqs. (d) and (e) into Eqs. (f) and (g) leads to

$$m\ddot{x} + kx - \frac{q^2 C'(x)}{2C^2(x)} = F(t) - c\dot{x} \tag{h}$$

and

$$\frac{q}{C(x)} = E_0 - R\dot{q} \tag{i}$$

where

$$C'(x) \equiv \frac{\partial C(x)}{\partial x} = (C_0 d_0)\frac{\partial}{\partial x}(d_0 + x)^{-1} = (C_0 d_0)(-1)(d_0 + x)^{-2}$$

Note that

$$-\frac{C'(x)}{C^2(x)} = -\frac{-\dfrac{C_0 d_0}{(d_0 + x)^2}}{\dfrac{(C_0 d_0)^2}{(d_0 + x)^2}} = \frac{1}{C_0 d_0} \tag{j}$$

Substitution of Eq. (j) into Eq. (h) and substitution of the capacitance function for $C(x)$ into Eq. (i) give

x-equation:
$$m\ddot{x} + c\dot{x} + kx + \frac{q^2}{2C_0 d_0} = F(t) \tag{k}$$

and

q-equation:
$$R\dot{q} + \frac{q(d_0 + x)}{C_0 d_0} = E_0 \tag{l}$$

as the equations of motion for the condenser microphone, given the stated capacitance function for $C(x)$.

■ **Example 7-13:** **Magnetic Suspension System**

The magnetic suspension system depicted in Figure 7-27a is proposed for levitating model aircraft during wind tunnel testing. The model of mass m is magnetizable and is supported against gravity by the magnetic field induced by the coil current I, which is a constant. (Designs of low friction magnetic bearings are also based on similar concepts.)

The magnetic inductance $L(x)$ of the system can be measured as a function of the position x of the model, producing a curve as sketched in Figure 7-27b. The inductance assumes its maximum value when the model is closest to the coil and decreases toward a constant as the model moves off toward very large values of x. The inductance function can be approximated by

$$L(x) = L_1 + \frac{L_0}{1 + \dfrac{x}{a}}$$

where L_1, L_0, and a are positive constants.

Obtain the equation(s) of motion for the system shown. Magnetic linearity may be assumed.

□ **Solution:**

1. Generalized coordinate(s): As prospective generalized coordinates, x and q as indicated in Figure 7-27a are considered. Thus

$$\xi_j : x \qquad \delta\xi_j : \delta x \tag{a}$$
$$\dot{q} : I \qquad \delta q : 0 \tag{b}$$

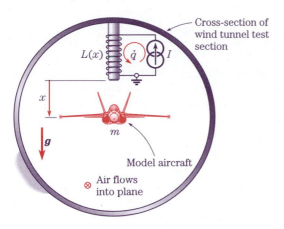

(a) Suspension system with model aircraft.

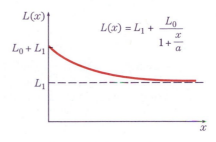

(b) Magnetic inductance function.

FIGURE 7-27   Magnetic suspension system, showing (a) position x and current $\dot{q}$ and (b) inductance function.

where in Eq. (b), we note that $\delta q = 0$ since $\dot{q} = I$ is a constant and thus δq cannot be arbitrarily varied. So, x is the only generalized coordinate, and the system is holonomic.

2. *Generalized force:*   Consideration of Eq. (7-79) gives

$$\delta \mathcal{W}^{\mathrm{nc}} = \sum_{h=1}^{N} \boldsymbol{f}_h^{\mathrm{nc}} \cdot \delta \boldsymbol{R}_h + \sum_{i=1}^{M} e_i^{\mathrm{nc}} \delta q_i = \sum_{j=1}^{n_m} \Xi_j \delta \xi_j + \sum_{k=1}^{n_e} \mathcal{E}_k \delta q_k$$

which leads immediately to

$$\Xi_x = 0 \qquad \mathcal{E}_k \text{ does not apply} \tag{c}$$

Equation (c) indicates that there are no nonconservative mechanical forces and the generalized voltage does not apply since there is no generalized charge coordinate.

3. *Lagrangian:*   Use of Eq. (7-78) gives the lagrangian as

$$\mathcal{L}(x, \dot{x}) = T^* - V + W_m^* - W_e$$

$$= \frac{1}{2}m\dot{x}^2 + mgx + \frac{1}{2}L(x)\dot{q}^2 - 0 \tag{d}$$

where Eq. (7-72), incorporating the condition of electrical linearity of the transducer, has been used.

4. *Lagrange's equation:*   Substitution of Eq. (a) into Eq. (7-76) gives

$$\frac{d}{dt}\left(\frac{\partial \mathcal{L}}{\partial \dot{x}}\right) - \frac{\partial \mathcal{L}}{\partial x} = \Xi_x \tag{e}$$

Then substitution of Eqs. (c) and (d) into Eq. (e) leads to

$$m\ddot{x} - mg - \frac{1}{2}L'(x)\dot{q}^2 = 0 \tag{f}$$

where $L'(x) \equiv \partial L(x)/\partial x$. Substitution of the inductance function into Eq. (f) gives

$$m\ddot{x} + \frac{L_0}{2a}\frac{1}{\left(1 + \dfrac{x}{a}\right)^2}I^2 = mg \tag{g}$$

where $\dot{q} = I$ has been used. Equation (g) is the equation of motion for the magnetic suspension system, given the stated inductance function for $L(x)$. In Problem 8-102 we shall return to this example to explore some of the features of the dynamic behavior of this system, as well as a corresponding system when feedback control is introduced (Problem 8-103).

■ **Example 7-14:**   Electromagnetic Plunger

A movable plunger of mass m is driven by an electromagnet as sketched in Figure 7-28. This is a basic configuration used for tripping circuit breakers, operating valves, and other applications in which a relatively large force is applied to a member that moves a relatively small distance.

The gap below the plunger is maintained at x_0 by a linear spring with spring constant k when the current in the coil of the electromagnet is zero. When the switch is closed, the current in the coil produces a magnetic force that pulls the plunger down to the stop at $x = -x_0$. The effective damping is modeled by a linear dashpot with dashpot constant c and represents losses in the mechanical system. The electromagnet has an inductance function

$$L(x) = \frac{L_0}{1 + \dfrac{x_0 + x}{h}}$$

where L_0, x_0, and h are positive constants. The displacement of the mass from its equilibrium position is x. The resistor R accounts for electromagnetic losses, including the resistance of the winding and any additional resistance of the electrical circuit.

Obtain the equation(s) of motion for the electromagnetic plunger. Magnetic linearity may be assumed; that is, fringing of the magnetic fields may be neglected and the magnetic material may be assumed to be infinitely permeable.

❑ **Solution:**

1. Generalized coordinate(s):   We consider the variables x and q as generalized coordinates as indicated in Figure 7-28. Thus

$$\xi_j : x \qquad \delta\xi_j : \delta x \tag{a}$$

$$q_k : q \qquad \delta q_k : \delta q \tag{b}$$

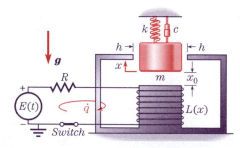

FIGURE   7-28   Electromagnetic plunger, showing displacement x and current $\dot{q}$.

where x is the displacement of m from its equilibrium position, and $\dot{q}$ is the current in the electrical circuit. So, gravity may be ignored in constructing the potential function. (See Example 6-12.) Therefore, x and q are the generalized coordinates, and the system is holonomic.

2. Generalized force and generalized voltage: Consideration of Eq. (7-79) gives

$$\delta W^{\mathrm{nc}} = \sum_{h=1}^{N} \boldsymbol{f}_h^{\mathrm{nc}} \cdot \delta \boldsymbol{R}_h + \sum_{i=1}^{M} e_i^{\mathrm{nc}} \delta q_i = \sum_{j=1}^{n_m} \Xi_j \delta \xi_j + \sum_{k=1}^{n_e} \mathcal{E}_k \delta q_k$$

$$= -c\dot{x}\delta x - R\dot{q}\delta q + E(t)\delta q = \Xi_x \delta x + \mathcal{E}_q \delta q \tag{c}$$

Or,

$$\Xi_x = -c\dot{x} \qquad \mathcal{E}_q = E(t) - R\dot{q} \tag{d}$$

3. Lagrangian: Use of Eq. (7-78) gives the lagrangian as

$$\mathcal{L}(x, \dot{x}, q, \dot{q}) = T^* - V + W_m^* - W_e$$

$$= \frac{1}{2}m\dot{x}^2 - \frac{1}{2}kx^2 + \frac{1}{2}L(x)\dot{q}^2 - 0 \tag{e}$$

where Eq. (7-72), incorporating the condition of electrical linearity of the transducer, has been used.

4. Lagrange's equations: Substitution of Eqs. (a) and (b) into Eqs. (7-76) and (7-77), respectively, gives

$$\frac{d}{dt}\left(\frac{\partial \mathcal{L}}{\partial \dot{x}}\right) - \frac{\partial \mathcal{L}}{\partial x} = \Xi_x \tag{f}$$

and

$$\frac{d}{dt}\left(\frac{\partial \mathcal{L}}{\partial \dot{q}}\right) - \frac{\partial \mathcal{L}}{\partial q} = \mathcal{E}_q \tag{g}$$

Then substitution of Eqs. (d) and (e) into Eqs. (f) and (g) leads to

$$m\ddot{x} + kx - \frac{1}{2}L'(x)\dot{q}^2 = -c\dot{x} \tag{h}$$

and

$$\frac{d}{dt}(L(x)\dot{q}) - 0 = E(t) - R\dot{q} \tag{i}$$

where $L'(x) \equiv \dfrac{\partial L(x)}{\partial x}$. Noting that

$$L'(x) = -\frac{L_0}{h\left(1 + \dfrac{x_0 + x}{h}\right)^2} \tag{j}$$

and that

$$\frac{d}{dt}(L(x)\dot{q}) = L(x)\ddot{q} + \frac{\partial L(x)}{\partial x}\frac{dx}{dt}\dot{q} = L(x)\ddot{q} + L'(x)\dot{x}\dot{q} \tag{k}$$

then substitution of Eqs. (j) and (k) into Eqs. (h) and (i) gives

x-equation:
$$m\ddot{x} + c\dot{x} + kx + \frac{L_0\dot{q}^2}{2h\left(1 + \dfrac{x_0 + x}{h}\right)^2} = 0 \qquad (l)$$

and

q-equation:
$$\frac{L_0}{1 + \dfrac{x_0 + x}{h}}\ddot{q} + R\dot{q} - \frac{L_0}{h\left(1 + \dfrac{x_0 + x}{h}\right)^2}\dot{x}\dot{q} = E(t) \qquad (m)$$

Equations (l) and (m) are the equations of motion for the electromagnetic plunger, given the stated inductance function for $L(x)$. In Chapter 8 (Example 8-21), we shall return to this example to explore some of the features of the dynamic behavior of this system.

7-7   ANOTHER LOOK AT LAGRANGIAN DYNAMICS

In conjunction with Section 5-7, we now take another look at the nature and structure of lagrangian dynamics; this time with a broader perspective that encompasses electrical and electromechanical analyses. The equations of motion for a variety of electrical networks and electromechanical systems have been derived. In the next chapter we shall return to several of these equations of motion in order to examine the dynamic behavior represented by them. The solutions to some of the linear equations of motion will be obtained. Also in Chapter 8, some of the nonlinear equations of motion will be examined by exploring their linearized behavior in the vicinity of states of special significance called *equilibrium configurations*.

It should be exceedingly clear to the reader that a failure either to define or to maintain the energy–coenergy distinction would have rendered the analyses in this chapter less coherent and awkward. Thus, once established, this energy–coenergy complementarity cannot justifiably be treated cavalierly or ignored.

We want to conclude this chapter with the discussion of two points that may have occurred to the reader during the development of the electrical and electromechanical formulations. These are (1) the issue of holonomic or nonholonomic constraints for electrical networks and electromechanical systems and (2) the existence of variational formulations beyond the displacement–charge variables and the displacement–flux linkage variables presentations considered here.

First, as indicated at the beginning of Section 7-2, the concept of holonomicity, which is of importance in analyzing mechanical systems, does not arise in conjunction with electrical systems. There does not appear to be an electromagnetic equivalent of a nonholonomic constraint. Thus, the rules governing the equation formulation for holonomic mechanical systems apply also to the formulation of the equations of motion for *all* electrical networks. Regarding electromechanical systems, the concepts and requirements of holonomic versus nonholonomic constraints must continue to be a consideration for the "mechanical side" of the system; however, as already indicated, these concepts and requirements do not arise for the "electrical side" of the system.

The presence of an ideal current source in an electrical network that is analyzed using charge coordinates reduces the number of independent charge variables in an otherwise complete set. However, as we have shown, the use of admissibility in the form of KCR provides a corresponding decrease in the number of charge variables in the complete set; thus resulting in an equality in the number of charge coordinates in an independent set *and* in a complete set. (Recall that this equality is a major feature of a holonomic system.) The analogous situation is found to hold for an electrical network containing an ideal voltage source and which is analyzed via flux linkage coordinates that are subject to KVR.

Second, we have seen that complementary forms of Hamilton's principle for electrical networks exist and may be beneficially alternatively used. That is, the lagrangian has been beneficially written as either $\mathcal{L}(q_k, \dot{q}_k) = W_m^* - W_e$ or $\mathcal{L}(\lambda_k, \dot{\lambda}_k) = W_e^* - W_m$. Well, comparably useful complementary variational principles for static mechanical systems are also commonly applied. For dynamic mechanical systems, however, a complementary form of Hamilton's principle exists, but its usefulness is somewhat limited.[12]

In the complementary form of Hamilton's principle for dynamic mechanical systems, the lagrangian is $\mathcal{L}(\mathbf{p}_k, \dot{\mathbf{p}}_k) = V^* - T$ and the admissibility conditions are expressed via restrictions on forces ($\dot{\mathbf{p}}_k$, that is, time rates of change of momenta) and momenta ($\mathbf{p}_k$). Because this complementary form of Hamilton's principle requires that the force and momentum admissibility conditions must be independent of displacements, many problems involving large motions cannot be addressed directly via this complementary principle. In mechanical systems, geometric compatibility can always be expressed in geometric variables (and time); but, except in one-dimensional problems, force and momentum requirements generally cannot be expressed exclusively in terms of forces and momenta without reference to geometry.

We simply note in closing that, in analyzing lumped-parameter electrical networks, Kirchhoff's current rule and Kirchhoff's voltage rule are always uncoupled. This fact establishes the underlying reason for the comparable utility and ease of application of both the charge and the flux linkage forms of Hamilton's principle and Lagrange's equations for lumped-parameter electrical networks.

Problems for Chapter 7

Problem 7-1: By reviewing elementary physics, recall your acquaintance with magnetic *flux linkage*. In particular, consider Figure P7-1, where the elements of electromagnetic induction are illustrated. A conductor, connected to a galvanometer (a device for measuring weak currents or voltages), is moved through the field of a permanent magnet. Recall and write the following:

(a) Define magnetic flux.

(b) In accordance with Faraday's law of electromagnetic induction, define the output voltage for a single loop of coil passing through the magnetic field.

(c) Repeat part (b) for a coil of N loops.

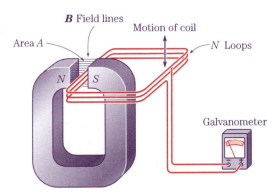

FIGURE P7-1

---

[12]See Appendix D of S. H. Crandall et al. [1968].

(d) Define flux linkage in this elementary demonstration.

Category I: Back to Physics

Problem 7-2 (*HRW759, 3E; HRW760, 16E; HRW 760, 17E*)[13]: The electrical networks sketched in Figure P7-2 correspond to circuits analyzed in HRW (Chapter 27; Problems 3E, 16E, and 17E). Using a charge formulation, derive the equation(s) of motion for each network.

Problem 7-3 (*HRW810, 8E; HRW811, 29E; HRW 812, 37E*): The electrical networks sketched in Figure P7-3 correspond to circuits analyzed in HRW (Chapter 29; Problems 8E, 29E, and 37E). Using a charge formulation, derive the equation(s) of motion for each network.

Problem 7-4 (*HRW915, 25P; HRW915, 26P; HRW 916, 28P*): The electrical networks sketched in Figure P7-4 correspond to circuits analyzed in HRW (Chapter 33; Problems 25P, 26P, and 28P). Using a

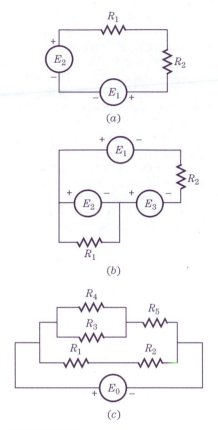

FIGURE P7-3

charge formulation, derive the equation(s) of motion for each network.

Problem 7-5 (*HRW760, 17E; HRW812, 37E; HRW 916, 28P*): Consider the electrical networks sketched in Figures 7-2c, 7-3c, and 7-4c. Using a flux linkage formulation, derive the equation(s) of motion for each network.

Categories II & III: Intermediate & More Difficult

RLC networks: Problems 7-6 through 7-18

Problem 7-6: Consider the electrical network sketched in Figure P7-6, and perform the tasks below.

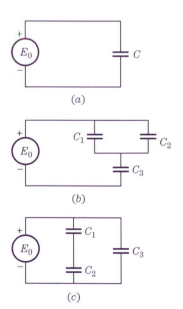

FIGURE P7-2

---

[13]These designations reference Halliday, Resnick, and Walker. For example, "HRW759, 3E" denotes "Halliday, Resnick, and Walker; page 759, problem 3E."

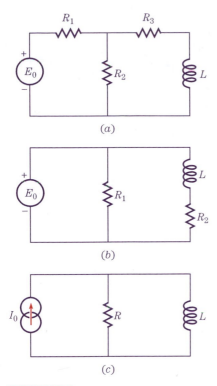

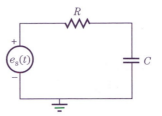

FIGURE P7-6

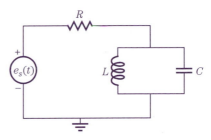

FIGURE P7-7

(a)

(b)

(c)

FIGURE P7-4

(a) Select a complete and independent set of generalized charge coordinates. State each admissibility condition.

(b) Also, select a complete and independent set of generalized flux linkage coordinates. State each admissibility condition.

For the flux linkage coordinates in part (b), do the following:

(c) Derive the generalized currents.

(d) Derive the electrical coenergy function.

(e) Derive the magnetic energy function.

(f) Derive the equation(s) of motion.

Problem 7-7: Perform the same tasks listed in Problem 7-6 for the electrical network sketched in Figure P7-7.

Problem 7-8: Perform the same tasks listed in Problem 7-6 for the electrical network sketched in Figure P7-8.

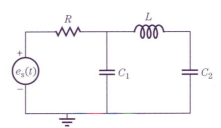

FIGURE P7-8

Problem 7-9: Perform the same tasks listed in Problem 7-6 for the electrical network sketched in Figure P7-9.

Problem 7-10: Consider the electrical network sketched in Figure P7-10, and perform the tasks below.

(a) Select a complete and independent set of generalized flux linkage coordinates. State each admissibility condition.

(b) Also, select a complete and independent set of generalized charge coordinates. State each admissibility condition. (Use

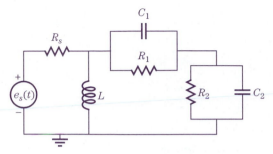

FIGURE P7-9

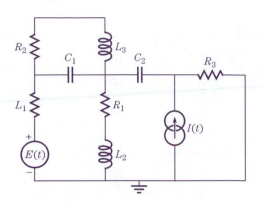

FIGURE P7-11

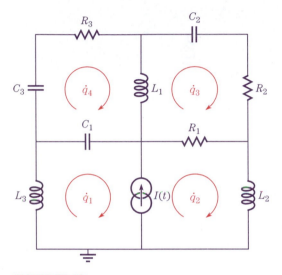

FIGURE P7-10

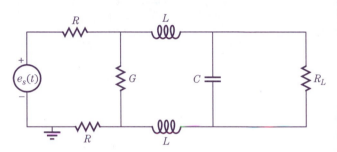

FIGURE P7-12

the loop currents shown in the schematic, and select q_1 as one of your generalized coordinates.)

For the charge coordinates in part (b), do the following:

    (a) Derive the generalized voltages.

    (b) Derive the magnetic coenergy function.

    (c) Derive the electrical energy function.

    (d) Derive the equation(s) of motion.

Problem 7-11: Perform the same tasks listed in Problem 7-10 for the electrical network sketched in Figure P7-11.

Problem 7-12: Perform the same tasks listed in Problem 7-10 for the electrical network sketched in Figure P7-12.

Problem 7-13: Perform the same tasks listed in Problem 7-10 for the electrical network sketched in Figure P7-13.

Problem 7-14: Perform the same tasks listed in Problem 7-10 for the electrical network sketched in Figure P7-14.

Problem 7-15: Reconsider the electrical network sketched in Figure 7-10. An unorthodox—but valid—set of loop currents is sketched in Figure P7-15. Such a set of alternative loop currents illustrates the nonuniqueness of generalized charge coordinates. Derive the equation(s) of motion for the network in Figure P7-15.

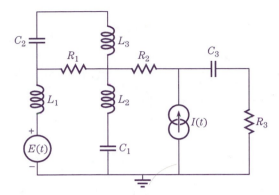

FIGURE P7-13

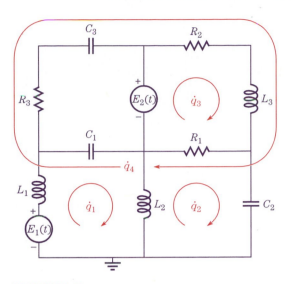

FIGURE P7-15

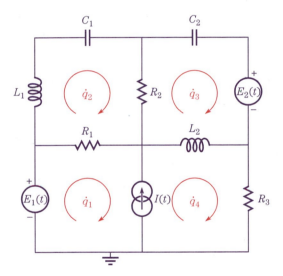

FIGURE P7-14

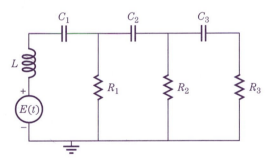

FIGURE P7-16

Problem 7-18: For the electrical network sketched in Figure P7-18, derive the equation(s) of motion as follows:

    (a) For a charge variables formulation; and

    (b) For a flux linkage variables formulation.

Problem 7-16: For the electrical network sketched in Figure P7-16, derive the equation(s) of motion as follows:

    (a) For a charge variables formulation; and

    (b) For a flux linkage variables formulation.

Problem 7-17: For the electrical network sketched in Figure P7-17, derive the equation(s) of motion as follows:

    (a) For a charge variables formulation; and

    (b) For a flux linkage variables formulation.

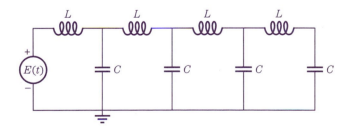

FIGURE P7-17

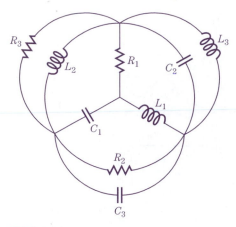

FIGURE P7-18

Operational amplifier networks: Problems 7-19 through 7-23

Problem 7-19: For the op-amp network sketched in Figure P7-19,

  (*a*) Derive the equation(s) of motion using a flux linkage formulation; and

  (*b*) Derive the equation(s) of motion using a charge formulation.

Problem 7-20: The op-amp network sketched in Figure P7-20 is called a *summer.* The two source voltages $e_1(t)$ and $e_2(t)$ are specified inputs. Using a flux linkage formulation, derive the equation(s) of motion. Then obtain an expression for the input–output voltage transfer characteristic of the network for operation within its linear range.

Problem 7-21: The op-amp network sketched in Figure P7-21 is called a *compensator amplifier.* The source voltage $e_s(t)$ is a specified input. Using a flux

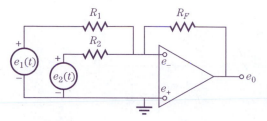

FIGURE P7-20

linkage formulation, derive the equation(s) of motion. Then obtain an expression for the input–output voltage transfer characteristic of the network for operation within its linear range.

Problem 7-22: For the op-amp network sketched in Figure P7-22, derive the equation(s) of motion using a flux linkage formulation. Then obtain an expression for the input–output voltage transfer characteristic of the network for operation within its linear range.

Problem 7-23: Show that Eqs. (h) and (i) in Example 7-9 can be reduced to Eq. (p) in Example 7-7.

Electromechanical systems: Problems 7-24 through 7-31

Problem 7-24: A proposed design for an electromagnetic hammer is sketched in Figure P7-24. The hammer driving element has mass m and is restrained

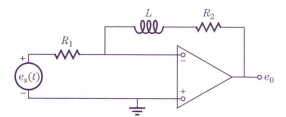

FIGURE P7-21

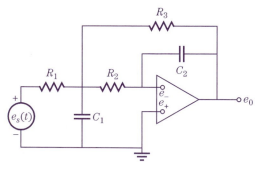

FIGURE P7-19

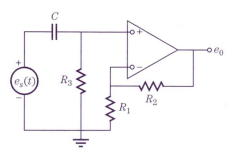

FIGURE P7-22

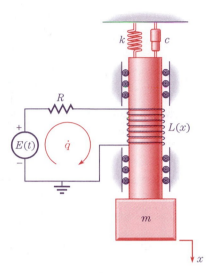

FIGURE P7-24

by a linear spring of spring constant k. The linear dashpot of dashpot constant c represents losses in the mechanical system due to damping. The electromagnet has an inductance function

$$L(x) = \frac{L_0}{\left(1 + \dfrac{x_0 + x}{h}\right)}$$

where L_0, x_0, and h are known constants. The linear resistor R represents the resistance of the circuit. Ignore gravity and assume magnetic linearity. Select $x(t)$ and $q(t)$ as a complete and independent set of generalized coordinates, as shown in Figure P7-24. Using a displacement–charge variables formulation, show that the equations of motion are

$$m\ddot{x} + c\dot{x} + kx + \frac{L_0}{2h\left(1 + \dfrac{x_0 + x}{h}\right)^2}\left(\frac{dq}{dt}\right)^2 = 0$$

and

$$\frac{L_0}{\left(1 + \dfrac{x_0 + x}{h}\right)}\frac{d^2q}{dt^2} - \frac{L_0}{h\left(1 + \dfrac{x_0 + x}{h}\right)^2}\frac{dx}{dt}\frac{dq}{dt}$$

$$= -R\frac{dq}{dt} + E(t)$$

Problem 7-25: The device sketched in Figure P7-25 is suggested for measuring liquid levels in storage

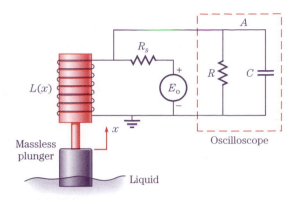

FIGURE P7-25

tanks. A massless plunger floats on the surface of the liquid. The level of the liquid is measured by the displacement x of the plunger. The plunger is made of magnetizable material and forms the core of a movable-core inductor. The inductance of the inductor changes with the position of the plunger and is given by $L(x)$. The inductor is driven by a constant voltage source E_0 through a series resistor R_s. The response of the inductor is displayed on an oscilloscope that has been modeled by the resistor R and the capacitor C, as shown. (Note that the oscilloscope actually forms a portion of the electrical network.) The display on the oscilloscope screen corresponds to the voltage (or the time derivative of the flux linkage) at point A. In order to study the behavior of this device, derive the equation(s) of motion for the system using a displacement–flux linkage variables formulation. Magnetic linearity may be assumed. The plunger inductance function is

$$L(x) = L_0\left(1 + \frac{x}{x_0}\right)$$

where L_0 and x_0 are known constants.

Problem 7-26: A mechanical vibration monitoring system is sketched in Figure P7-26. A movable-core

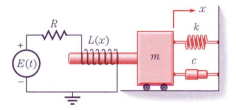

FIGURE P7-26

inductor uses a portion of the vibrating mass as its core to monitor the system's motion. The inductance function of the movable-core inductor is

$$L(x) = \frac{L_0}{\left(1 - \dfrac{x}{a_0}\right)}$$

where L_0 and a_0 are known constants. Derive the equation(s) of motion using a displacement–charge variables formulation. Magnetic linearity may be assumed.

Problem 7-27: A bead of mass m slides without friction on top of a horizontal table. See Figure P7-27. A taut inextensible but flexible string connects the mass m through a hole in the center of the table to the upper plate of a capacitor that is free to move up and down under the combined influence of gravity, the tension in the string, and the electrostatic attraction between the capacitor plates. The upper plate of capacitor has mass M. The bead's polar coordinates

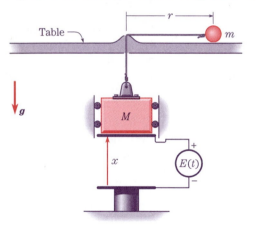

(a) Cross-sectional view

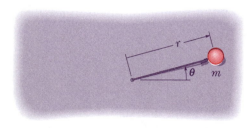

(b) Top view

FIGURE P7-27

with respect to the hole are denoted by r and θ as indicated. The capacitance between the two plates is given by

$$C(x) = \frac{B}{x + a_0}$$

where B and a_0 are known constants and x is the separation of the two plates. The string length is adjusted such that $x = r$. A time-varying voltage $E(t)$ is applied across the capacitor.

(a) Derive the equation(s) of motion using a displacement–flux linkage variables formulation. Electrical linearity may be assumed.

(b) Determine which of the coordinates of θ and r is cyclic or ignorable, and find the corresponding constant of the motion. Give a physical interpretation of that constant of the motion.

Problem 7-28: The model of an electromechanical system is sketched in Figure P7-28. The respective functions for the movable-plate capacitor and the movable-core inductor are

$$C(x_2) = \frac{C_0 d_0}{d_0 - x_2} \quad \text{and} \quad L(x_1) = \frac{L_0}{1 + \left(\dfrac{x_1}{x_0}\right)^2}$$

where C_0, d_0, L_0, and x_0 are known constants. Derive the equation(s) of motion using a displacement–charge variables formulation. Electrical linearity may be assumed.

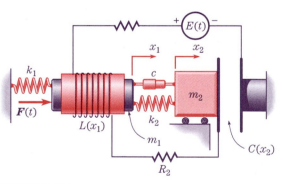

FIGURE P7-28

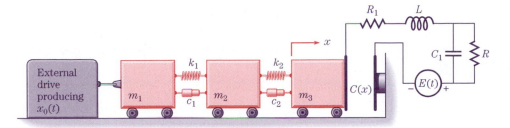

FIGURE P7-29

Problem 7-29: A vibration monitoring system is sketched in Figure P7-29. A movable-plate capacitor uses one of the masses (m_3) as its movable plate to monitor the mechanical system's motion at that location. The capacitance function of the movable-plate capacitor is

$$C(x) = \frac{C_0 d_0}{d_0 - x}$$

where C_0 and d_0 are known constants. Derive the equation(s) of motion using a displacement–charge variables formulation. Electrical linearity may be assumed.

Problem 7-30: The model of the electromechanical system as sketched in Figure P7-30 consists of a movable-plate capacitor, a galvanometer, several electrical elements, and several mechanical (torsional and translational) elements. The movable-plate capacitance function is

$$C(x) = \frac{A}{s - x}$$

where A and s are known constants, and where the displacement x is measured from the unstretched length of the spring k. The magnetic coenergy function for the galvanometer is

$$W_m^*(\theta, i) = Ti(\theta - \theta_0)$$

where T and θ_0 are known constants. I is the moment of inertia of the galvanometer's moving coil about its rotational longitudinal axis. The system is powered by two batteries E_1 and E_2 as shown. Derive the equation(s) of motion using a displacement–charge variables formulation. Electrical linearity may be assumed.

Problem 7-31: An electromechanical transducer, as sketched in Figure P7-31, has unknown transduction characteristics; so the following two measurements were performed.

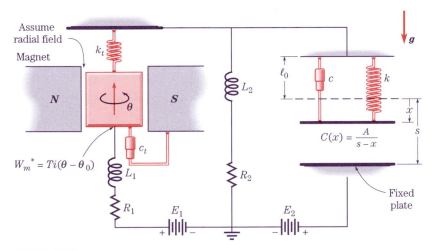

FIGURE P7-30

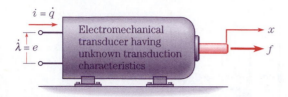

FIGURE P7-31

(1) For arbitrary displacement x and current i, the first measurement revealed that

$$\lambda = Ai^3(x + l)$$

where A and l are constants, now determined.

(2) In the second measurement, the current was turned off, revealing that the force f may be expressed as

$$f = f(x, i) = kx$$

when $i = 0$.

   (a) Find the general expression for $f(x, i)$.

   (b) Also, find the complementary function $f(x, \lambda)$.

CHAPTER EIGHT

VIBRATION OF LINEAR LUMPED-PARAMETER SYSTEMS

8-1   INTRODUCTION

In this chapter we examine the dynamic response of some of the system models for which we have derived equations of motion. Primarily, we shall consider the closed form solutions to equations for one- and two-degree-of-freedom linear system models, which resulted in first- and second-order linear differential equations. Furthermore, for system models represented by nonlinear differential equations, we shall examine the behavior of these systems for small excursions about states known as *equilibrium configurations*. Such examinations lead to the concept of *stability*. We shall find that for small excursions of systems residing in a *stable equilibrium configuration* the motion of the system can be analyzed via the linear analyses presented early in the chapter. On the other hand, small excursions of systems residing in an *unstable equilibrium configuration* lead to larger excursions that cannot be analyzed via linear analyses.

In this chapter we shall make extensive use of the well known solutions of linear ordinary differential equations having constant coefficients. It has been stated by some historians that the solutions of ordinary differential equations can be said to have begun at the very instant that Newton's teacher Isaac Barrow* recognized the inverse relationship between differentiation and integration. It is more widely accepted that if Barrow ever had such a revelation, he never appreciated its significance. Both Newton and Leibniz certainly did, however, and each exploited it. In examining the behavior of several nonlinear system models for small excursions about equilibrium configurations, we shall use the powerful Taylor series expansion published by Brook Taylor (1685–1731) in 1715, the special case being the so-called Maclaurin series published by Colin Maclaurin (1698–1746) in 1742. Furthermore, the "Taylor series" for several functions were known to James Gregory (1638–1675) by 1670, and by mathematicians in India prior to 1550.

---

*Isaac Barrow (1630–1677) was the first holder of the Lucasian Professorship at Cambridge University—established by Henry Lucas (ca. 1610–1663) in 1663 via his bequest of "lands to the annual value of $100"—which he held from 1664 to 1669 when he was succeeded by Isaac Newton.

The name of nearly every great mathematician of the eighteenth century can be associated in one context or another with solutions of differential equations. Nevertheless, in our opinion, Euler was perhaps the most significant and influential individual in the history of the solutions of differential equations. While credit for many of his ideas must be shared with his contemporaries, the distinction between homogeneous and nonhomogeneous linear differential equations, the distinction between the homogeneous (complementary) and the particular solutions, the use of integrating factors, and the systemic techniques for solving linear ordinary differential equations having constant coefficients are all substantially due to Leonhard Euler. Furthermore, by first publishing the notation or symbols e to represent the base of the system of natural logarithms (1736), $f(x)$ for "a function of x" (1740), Δx for a finite difference of x (1755), $\sum$ to denote summation (1755), and i to represent $\sqrt{-1}$ (1794), as well as popularizing the symbol π (which, however, had been used by William Jones in 1706), Euler placed his imprimatur on the solutions of differential equations for all future generations.

8-2   SINGLE-DEGREE-OF-FREEDOM FIRST-ORDER SYSTEMS

Many single-degree-of-freedom systems are modeled by differential equations that are first order. The equation of motion for some of these systems may be written as

$$\tau\frac{d\nu}{dt} + \nu = f(t) \tag{8-1}$$

where $\nu(t)$ is the output or response variable, t is time, $f(t)$ is the input or forcing excitation, and τ is a positive coefficient, which for dimensional consistency must have units of time. In the most general case, τ may be constant or variable, positive or negative; however, we shall consider only the case where τ is a positive constant. For example, for the system sketched in Figure 8-1a, $\nu = v$, v being the velocity or the time derivative of the generalized coordinate x, and $\tau = \dfrac{m}{c}$. Examples of single-degree-of-freedom first-order systems are numerous, occurring throughout the social sciences, the biological sciences, the physical sciences, as well as engineering. Note that Eq. (8-1) is a generic equation of motion that governs many single-degree-of-freedom first-order systems. For example, Eq. (8-1) describes the dynamics of all the systems sketched in Figure 8-1, when ν and τ are replaced by the corresponding quantities indicated in Figure 8-1.

In order to determine the solution to Eq. (8-1), according to the theory of ordinary differential equations, only one initial condition is needed. Therefore, the initial condition is specified as

$$\nu(0) = \nu_0 \tag{8-2}$$

where ν_0 is the value of $\nu(t)$ at $t = 0$.

In accordance with the theory of ordinary differential equations, the complete solution of Eq. (8-1) can be found as the sum of the *particular solution* and the *homogeneous solution*. The particular solution depends on the excitation $f(t)$ on the right-hand side of Eq. (8-1). As a special case, when $f(t) = 0$ in Eq. (8-1), the particular solution is zero and therefore the complete solution consists of the homogeneous solution only; this solution represents the *free response* of the system. When $f(t) \neq 0$ in Eq. (8-1), the complete solution consists of the sum of the particular and homogeneous solutions.

The general procedure to obtain the complete solution in the case of a nonzero excitation is shown in Table 8-1, consisting of four main steps. The procedure in Table 8-1 applies not only to single-degree-of-freedom first-order systems but also to all linear dynamic systems of higher order and higher degrees of freedom. For example, we shall use the procedure in Table 8-1 to obtain harmonic responses of single-degree-of-freedom second-order systems (Section 8-3) and two-degree-of-freedom second-order systems (Section 8-4).

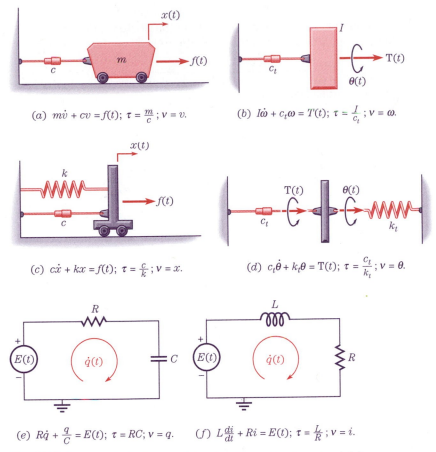

(a) $m\dot{v} + cv = f(t)$; $\tau = \frac{m}{c}$; $v = v$.

(b) $I\dot{\omega} + c_t\omega = T(t)$; $\tau = \frac{I}{c_t}$; $v = \omega$.

(c) $c\dot{x} + kx = f(t)$; $\tau = \frac{c}{k}$; $v = x$.

(d) $c_t\dot{\theta} + k_t\theta = T(t)$; $\tau = \frac{c_t}{k_t}$; $v = \theta$.

(e) $R\dot{q} + \frac{q}{C} = E(t)$; $\tau = RC$; $v = q$.

(f) $L\frac{di}{dt} + Ri = E(t)$; $\tau = \frac{L}{R}$; $v = i$.

FIGURE 8-1   Some single-degree-of-freedom first-order dynamic models.

Next, we consider the response of the generic single-degree-of-freedom first-order system when $f(t)$ is a step input, ramp input, or harmonic input, as well as when $f(t) = 0$. In all cases that follow, it is assumed that the observation of the system response commences at $t = 0$.

8-2.1   Free Response

For the *free response,* by definition, we are concerned with the response of the system in the absence of external excitation. Thus, for the free response, we consider the first-order homogeneous equation of the generic form

$$\tau\frac{dv}{dt} + v = 0 \tag{8-3}$$

which can be obtained from Eq. (8-1) by setting $f(t) = 0$. The system is subjected to the initial condition (I.C.)

$$v(0) = v_0 \tag{8-4}$$

TABLE 8-1   General Procedure to Obtain Complete Solution for Linear Dynamic System Subjected to Nonzero Excitation and Initial Condition(s)

Step 1
Establish equation(s) of motion with specified excitation function(s) and initial condition(s).

$\downarrow$

Step 2
Assume particular solution according to theory of ODEs; determine coefficient(s) of particular solution such that it satisfies equation of motion.

$\downarrow$

Step 3
Establish complete solution by summing particular solution found in Step 2 and homogeneous solution (found according to theory of ODEs) with unknown coefficient(s).

$\downarrow$

Step 4
Impose initial condition(s) onto complete solution found in Step 3; evaluate coefficient(s) of homogeneous solution. Complete solution found.

The solution to Eq. (8-3) is often called the homogeneous or complementary solution since Eq. (8-3) is a homogeneous ordinary differential equation (ODE).

In the theory of ordinary differential equations, it is *assumed* that the solution to Eq. (8-3) is of the form

$$\nu(t) = Ce^{\lambda t} \tag{8-5}$$

where C and λ are scalar constants and e is the base of the system of natural logarithms, equal to $2.71828\ldots$. Substitution of Eq. (8-5) into Eq. (8-3) gives

$$(\tau\lambda + 1)Ce^{\lambda t} = 0 \tag{8-6}$$

which, because $Ce^{\lambda t}$ cannot be zero for a nontrivial solution, requires

$$\tau\lambda + 1 = 0 \tag{8-7}$$

Equation (8-7) is called the *characteristic equation,* having the solution

$$\lambda = -\frac{1}{\tau} \tag{8-8}$$

Thus, by substituting Eq. (8-8) into Eq. (8-5), the solution becomes

$$\nu(t) = Ce^{-\frac{t}{\tau}} \tag{8-9}$$

where the *constant of integration* C can be determined in accordance with the initial condition in Eq. (8-4).

Combining Eqs. (8-4) and (8-9) at $t = 0$ leads to

$$\nu(0) = \nu_0 = Ce^{-\frac{0}{\tau}} = C(1) \tag{8-10}$$

or

$$C = \nu_0 \qquad (8\text{-}11)$$

Thus, substitution of Eq. (8-11) into Eq. (8-9) gives

$$\nu(t) = \nu_0 e^{-\frac{t}{\tau}} \qquad t \geq 0 \qquad (8\text{-}12)$$

where we have written Eq. (8-12) with $t \geq 0$ to remind us that the solution does not hold for $t < 0$.

In the second column of Table 8-2 the values of the free response normalized by the initial value of the response variable are listed; that is, $\dfrac{\nu(t)}{\nu_0} = e^{\frac{-t}{\tau}}$ from Eq. (8-12), for times in multiples of τ, τ being called the system *time constant;* that is, $t = n\tau$ where $n = 0, 1, 2, \ldots$. It can be seen from the tabulated values that the response decreases by approximately 63% (more precisely, $1 - e^{-1}$) of the initial value, when the time one τ is elapsed. Moreover, this 63% decrease of the previous value occurs for any and every time increment of τ. So the value of ν at $t = 2\tau$ will be decreased by approximately 63% when $t = 3\tau$, and the value of ν at, say, $t = 2.6\tau$ will be decreased by approximately 63% when $t = 3.6\tau$, or $t = 2.6\tau + \tau$. Also, note that the response $\nu(t)$ is within 2% of its final value (which is zero in this case) by $t = 4\tau$. For this reason, such a system's response is often said to be consummated in about four time constants, or $t = 4\tau$. So, this means that systems having smaller τ's reach their final state faster than those having larger τ's.

The generic character of the free response given by Eq. (8-12) is sketched in Figure 8-2. Note that the response is nonoscillatory since $\nu(t)$ decreases monotonically and approaches zero asymptotically for all positive values of τ. We make the following observations concerning the free response curve sketched in Figure 8-2:

1. The slope of the response is always negative, as its magnitude decreases monotonically with time and ultimately approaches zero.
2. Quantitatively, the slope of the response at any time is such that if a tangent to the response curve were drawn at a particular time, such an extended straight line would intersect the final value (which is zero in this case) in one τ beyond that time. This is indicated in Figure 8-2 for various multiples of τ.[1]
3. The entire response curve can be generated from two pieces of information: ν_0 and τ.

TABLE 8-2 Normalized Free and Step Response for Single-Degree-of-Freedom First-Order System at Various Times

Time t	Normalized Free Response $\dfrac{\nu}{\nu_0}$ in Equation (8-12): $e^{\frac{-t}{\tau}}$	Normalized Step Response $\dfrac{\nu}{\nu_f}$ in Equation (8-22): $(1 - e^{\frac{-t}{\tau}})$
0	1	0
τ	0.368	0.632
2τ	0.135	0.865
3τ	0.050	0.950
4τ	0.018	0.982
5τ	0.007	0.993
6τ	0.002	0.998

[1]This may be shown mathematically; see Problem 8-14.

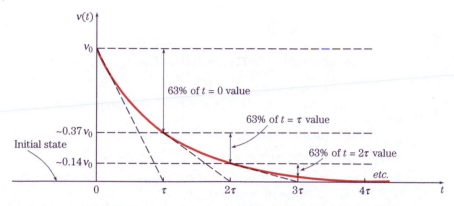

FIGURE 8-2   Free response $\nu_0 e^{\frac{-t}{\tau}}$ versus τ, illustrating generic decaying behavior.

As indicated in the observations above, the time constant τ plays a dominant role in the free response of first-order systems modeled by Eq. (8-1). In order to illustrate further the effect of the time constant on the system response, the solution in Eq. (8-12) is plotted in Figure 8-3 for several values of τ. Note that the smaller the (positive) time constant, the faster $\nu(t)$ approaches zero, as may be concluded from Figure 8-3 or, as already indicated, in Table 8-2 and Figure 8-2.

8-2.2   Step Response

The exponential solution $e^{\frac{-t}{\tau}}$ also arises when the final value of the solution is nonzero. For example, assume the input $f(t)$ is a step function. Then, the step response analysis may be formulated as

$$\tau \frac{d\nu}{dt} + \nu = f(t) \tag{8-13}$$

where

$$f(t) = \begin{cases} 0 & t < 0 \\ \nu_f & t \geq 0 \end{cases} \tag{8-14}$$

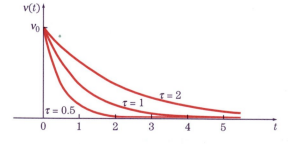

FIGURE 8-3   Response of equation $\tau \dot{\nu} + \nu = 0$, subjected to initial condition $\nu(0) = \nu_0$ for various time constants τ.

where ν_f is a constant. Figure 8-4a shows the forcing function given by Eq. (8-14), which is called the *step input*. Further, suppose the initial condition for $\nu(t)$ is given by

$$\nu(0) = \nu_0 \tag{8-15}$$

This completes Step 1 of Table 8-1. Continuing with the remaining steps in Table 8-1 will give the complete solution.

As indicated in Step 2 of Table 8-1, the particular solution $\nu_p(t)$ is obtained next. From the theory of ordinary differential equations, it is *assumed* that the form for $\nu_p(t)$, when $f(t)$ is given by Eq. (8-14), is

$$\nu_p(t) = C_0 \tag{8-16}$$

where C_0 is the undetermined coefficient to be determined by requiring that Eq. (8-16) satisfies Eqs. (8-13) and (8-14). Therefore, substitution of Eq. (8-16) into Eq. (8-13) gives

$$C_0 = \nu_f \tag{8-17}$$

where Eq. (8-14) has been used. Thus, the particular solution may be obtained by substituting Eq. (8-17) into Eq. (8-16). This completes Step 2 of Table 8-1.

As indicated in Step 3 of Table 8-1, the complete solution is established next. Recall that the general homogeneous solution $\nu_h(t)$ to Eq. (8-13) is given by Eq. (8-9). Thus, combining Eqs. (8-9), (8-16), and (8-17) gives the complete solution as

$$\nu(t) = \nu_h(t) + \nu_p(t) = C'e^{-\frac{t}{\tau}} + \nu_f \qquad t \geq 0 \tag{8-18}$$

This completes Step 3 of Table 8-1.

The only unknown constant in Eq. (8-18) is C', which, as indicated in Step 4 of Table 8-1, can be evaluated by imposing the initial condition. Substitution of Eq. (8-15) into Eq. (8-18) at $t = 0$ requires that

$$C' = \nu_0 - \nu_f \tag{8-19}$$

Therefore, substitution of Eq. (8-19) into Eq. (8-18) gives the complete solution as

$$\nu(t) = (\nu_0 - \nu_f)e^{-\frac{t}{\tau}} + \nu_f \qquad t \geq 0 \tag{8-20}$$

This completes Step 4 of Table 8-1, as Eq. (8-20) is the complete solution.

Consider the special case of zero initial condition such that Eq. (8-15) becomes

$$\nu(0) = \nu_0 = 0 \tag{8-21}$$

Substitution of Eq. (8-21) into Eq. (8-20) gives

$$\nu(t) = \nu_f \left(1 - e^{-\frac{t}{\tau}}\right) \qquad t \geq 0 \tag{8-22}$$

As indicated by the common exponential function $e^{\frac{-t}{\tau}}$, the free response in Eq. (8-12) and the step response in Eq. (8-22) share some of the characteristics discussed in the previous subsection. In the third column of Table 8-2 the values of the response given by Eq. (8-22) are shown, normalized by ν_f for several integer values of τ. As in the free response case, the normalized response is within 2% of its final value (which is unity in this case) by four time constants, or $t = 4\tau$. Also, as sketched in Figure 8-4b, the instantaneous slope of the step response curve is such that if extended, the tangent line at any particular time $t = t_0$ would intersect the final value in one additional time constant, that is, at $t = t_0 + \tau$.

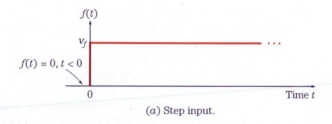

(a) Step input.

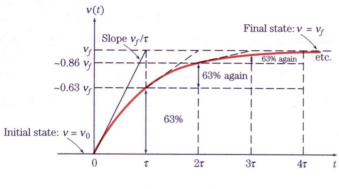

(b) Exponential response.

FIGURE 8-4   Step input and corresponding exponential response of single-degree-of-freedom first-order system.

8-2.3   Ramp Response

Now, consider the response of the generic single-degree-of-freedom first-order system to a forcing function that is a linear function of time for nonnegative time ($t \geq 0$). That is, we seek the solution to the equation

$$\tau \frac{d\nu}{dt} + \nu = f(t) \tag{8-23}$$

where

$$f(t) = \begin{cases} 0 & t < 0 \\ a_0 t & t \geq 0 \end{cases} \tag{8-24}$$

in which a_0 is a positive constant slope of the forcing function. Figure 8-5 shows the forcing function given by Eq. (8-24), which is called a *ramp input*. Also, suppose the initial condition for $\nu(t)$ is given by

$$\nu(0) = \nu_0 \tag{8-25}$$

This completes Step 1 of Table 8-1.

As indicated in Step 2 of Table 8-1, the particular solution $\nu_p(t)$ is obtained next. From the theory of ordinary differential equations, it is *assumed* that the form for $\nu_p(t)$, when $f(t)$ is given by Eq. (8-24), is

$$\nu_p(t) = C_1 t + C_0 \tag{8-26}$$

where C_1 and C_0 are the undetermined coefficients to be determined by requiring that Eq. (8-26) satisfies Eqs. (8-23) and (8-24). Therefore, substitution of Eq. (8-26) into Eq. (8-23) gives

$$\tau C_1 + C_1 t + C_0 = a_0 t \qquad t \geq 0 \tag{8-27}$$

where Eq. (8-24) has been used. In order for Eq. (8-27) to hold for all positive values of time t, like coefficients of time on both sides of Eq. (8-27) must be equal, giving

$$\tau C_1 + C_0 = 0 \tag{8-28}$$
$$C_1 = a_0 \tag{8-29}$$

or, upon solution for C_1 and C_0,

$$C_1 = a_0 \tag{8-30}$$
$$C_0 = -\tau a_0 \tag{8-31}$$

So, by substitution of Eqs. (8-30) and (8-31) into Eq. (8-26), the particular solution becomes

$$\nu_p(t) = a_0 t - a_0 \tau \qquad t \geq 0 \tag{8-32}$$

This completes Step 2 of Table 8-1.

As indicated in Step 3 of Table 8-1, the complete solution is established next. Recall that the general homogeneous solution $\nu_h(t)$ to Eq. (8-23) is given by Eq. (8-9) as

$$\nu_h(t) = C' e^{-\frac{t}{\tau}} \tag{8-33}$$

where C' is an unknown coefficient. Thus, combining Eqs. (8-32) and (8-33) gives the complete solution as

$$\nu(t) = \nu_h(t) + \nu_p(t) = C' e^{-\frac{t}{\tau}} + a_0 t - a_0 \tau \qquad t \geq 0 \tag{8-34}$$

This completes Step 3 of Table 8-1.

The only unknown constant in Eq. (8-34) is C', which, as indicated in Step 4 of Table 8-1, can be evaluated by imposing the initial condition. Substitution of Eq. (8-25) into Eq. (8-34) at $t = 0$ requires that

$$\nu_0 = C' - a_0 \tau \tag{8-35}$$

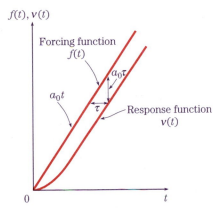

FIGURE 8-5   Ramp input and corresponding response of single-degree-of-freedom first-order system for zero initial condition.

or,

$$C' = \nu_0 + a_0\tau \tag{8-36}$$

Therefore, substitution of Eq. (8-36) into Eq. (8-34) gives the complete solution as

$$\nu(t) = (\nu_0 + a_0\tau)e^{-\frac{t}{\tau}} + a_0t - a_0\tau \qquad t \geq 0 \tag{8-37}$$

This completes Step 4 of Table 8-1, as Eq. (8-37) is the complete solution.

There is a subtle but important observation that should be made here. It is sometimes written in textbooks (1) that the homogeneous solution $\nu_h(t)$ is called the *transient solution,* which is due to the system's initial conditions; and (2) that the particular solution $\nu_p(t)$ is called the *steady-state solution,* which is due to the forcing function and will persist after the transient solution has disappeared. It is sometimes further stated that the complete solution is the superposition of the homogeneous solution and the particular solution. Such statements are potentially misleading, with potentially significant consequences.

The subtlety lies in the manner in which the unknown coefficient in the homogeneous solution is determined. The comments in the preceding paragraph suggest that C' in the homogeneous solution should be determined via the initial condition by combining Eqs. (8-25) and (8-33) directly. If this were so, C' would be the same as C in Eq. (8-11) found via the initial condition in Eq. (8-4). But the C' found in Eq. (8-36) is not equal to the C found in Eq. (8-11). The important conclusion here is that the unknown constant C' associated with the homogeneous solution may be properly determined only within the context of the complete solution (Eq. (8-34), in this case), that is, only *after* the complete solution has been constructed in an expression corresponding to Eq. (8-34), containing the evaluated constants of integration from the particular solution. In this way, the complete solution will not only satisfy the particular response but will also satisfy the initial condition(s) imposed onto the system. Therefore, while the complete solution may be generically thought to be the superposition of a homogeneous solution and a particular solution, the constants of integration of the homogeneous solution must be evaluated (1) only *after* the particular solution has been fully evaluated and (2) only within the context of the complete solution.

Furthermore, regarding some terminology, the *transient response* should be considered as that portion of the response that dies out, period; and the *steady-state* or *long-term solution* should be considered as that portion of the response that persists for unlimited time, period. For example, if a system having no damping were not subjected to any external disturbance except its nonzero initial condition(s), its homogeneous solution would never decay, and thus this motion cannot be considered to be transient. Therefore, the phrases *transient solution* and *steady-state solution* should not be viewed as synonyms for homogeneous solution and particular solution, respectively.

Now, consider the special case of zero initial condition; that is, $\nu_0 = 0$. The solution in this case is easily found from Eq. (8-37) as

$$\nu(t) = a_0\tau\left(e^{-\frac{t}{\tau}} - 1\right) + a_0t \tag{8-38}$$

(Note that although the initial condition vanishes—that is, is equal to zero—the homogeneous solution's constant C' does not vanish.) Figure 8-5 contains a sketch of the output ramp response $\nu(t)$ given by Eq. (8-38) for zero initial condition, along with the input forcing function $f(t)$ given by Eq. (8-24). For large time t, the exponential term on the right-hand side of Eq. (8-38) vanishes; and, therefore, the remaining term ($a_0t - a_0\tau$) describes the long-term response, as sketched in Figure 8-5. In the long term, the forcing function minus the steady-state response function is a constant, (Forcing) $-$ (Response) $= a_0t - (a_0t - a_0\tau) = a_0\tau$, as indicated in Figure 8-5.

8-2.4   Harmonic Response

We shall now consider the response of the generic single-degree-of-freedom first-order system to a forcing function that is harmonic for nonnegative time $(t \geq 0)$. That is, we seek the solution to the equation

$$\tau \frac{d\nu}{dt} + \nu = f(t) \tag{8-39}$$

where

$$f(t) = \begin{cases} 0 & t < 0 \\ \sum_{i=1}^{m} a_i \cos \Omega_i t + b_i \sin \Omega_i t & t \geq 0 \end{cases} \tag{8-40}$$

where the a_i's and b_i's are the constant amplitudes of the cosinusoidal and sinusoidal forcing functions, respectively, having m distinct (radian) frequencies Ω_i's. The function $f(t)$ in Eq. (8-40) accommodates the most general harmonic forcing function because it contains both cosinusoidal and sinusoidal functions having all possible forcing frequencies and, further because any cosinusoidal or sinusoidal function having nonzero phase angle may be expanded into a linear combination of both cosinusoidal and sinusoidal functions (that is, for example, $c_i \cos(\Omega_i t + \varphi_i) = c_i \cos \varphi_i \cos \Omega_i t - c_i \sin \varphi_i \sin \Omega_i t \equiv a_i \cos \Omega_i t + b_i \sin \Omega_i t$, where the c_i's are the constant amplitudes and the φ_i's are the phase angles). Also, suppose the initial condition for $\nu(t)$ is given by

$$\nu(0) = \nu_0 \tag{8-41}$$

To facilitate the analysis, we shall simplify the problem. That is, the excitation given by Eq. (8-40) is reduced to

$$f(t) = \begin{cases} 0 & t < 0 \\ a \cos \Omega t + b \sin \Omega t & t \geq 0 \end{cases} \tag{8-42}$$

which retains only a pair of single frequency cosinusoidal and sinusoidal functions, and therefore the subscript i has been dropped from Eq. (8-40). Because the system model governed by Eq. (8-39) is linear, the principle of superposition is valid in its analysis; thus, the particular solution for a sum of various forcing functions is the same as the sum of the particular solutions for each of the forcing functions. Therefore, no generality is lost due to the simplification of the forcing function from Eq. (8-40) to Eq. (8-42).

Thus, we shall consider the problem whose equation of motion is given by Eq. (8-39) where $f(t)$ is given by Eq. (8-42) and the initial condition is given by Eq. (8-41). This completes Step 1 of Table 8-1.

Two alternative techniques for obtaining the particular solution are presented in Subdivision 8-2.4.1 via real variables exclusively and Subdivision 8-2.4.2 via complex variables (Step 2 of Table 8-1). In Subdivision 8-2.4.3, the complete solution is established by adding the particular solution obtained in either of Subdivisions 8-2.4.1 or 8-2.4.2 and the homogeneous solution obtained in Subsection 8-2.1 (Step 3 of Table 8-1); and, the coefficient of the homogeneous solution is evaluated in accordance with the initial condition, yielding the complete solution (Step 4 of Table 8-1). As a summary, the procedures for finding the complete solution are tabulated in Subdivision 8-2.4.4. Further, various aspects of the particular solution are examined more closely in Subdivision 8-2.4.5.

8-2.4.1 *Particular Solution*   Now, consider the particular solution as indicated in Step 2 of Table 8-1. According to the theory of ordinary differential equations, the particular solution for the forcing function given by Eq. (8-42) is *assumed* to be of the form

$$\nu_p(t) = A \cos \Omega t + B \sin \Omega t \qquad t \geq 0 \tag{8-43}$$

where A and B are the undetermined coefficients and where it should be noted that the response frequency is the same as the excitation frequency Ω. By requiring that the assumed particular solution satisfies the equation of motion, the unknown coefficients in the particular solution may be evaluated. Substituting Eqs. (8-42) and (8-43) into Eq. (8-39) yields, for positive time,

$$\tau(-\Omega A \sin \Omega t + \Omega B \cos \Omega t) + A \cos \Omega t + B \sin \Omega t = a \cos \Omega t + b \sin \Omega t \tag{8-44}$$

Grouping separately terms containing $\cos \Omega t$ and those containing $\sin \Omega t$ in Eq. (8-44) leads to

$$(A + \tau\Omega B - a) \cos \Omega t + (-\tau\Omega A + B - b) \sin \Omega t = 0 \qquad t \geq 0 \tag{8-45}$$

In order for Eq. (8-45) to hold for all nonnegative time, each of the coefficients of $\cos \Omega t$ and $\sin \Omega t$ in Eq. (8-45) must be zero. Thus,

$$A + \tau\Omega B - a = 0 \tag{8-46}$$

$$-\tau\Omega A + B - b = 0 \tag{8-47}$$

Equations (8-46) and (8-47) may be rewritten in matrix form as

$$\begin{bmatrix} 1 & \tau\Omega \\ -\tau\Omega & 1 \end{bmatrix} \begin{Bmatrix} A \\ B \end{Bmatrix} = \begin{Bmatrix} a \\ b \end{Bmatrix} \tag{8-48}$$

Equation (8-48) can be solved via Cramer's rule for A and B as

$$A = \frac{\begin{vmatrix} a & \tau\Omega \\ b & 1 \end{vmatrix}}{\begin{vmatrix} 1 & \tau\Omega \\ -\tau\Omega & 1 \end{vmatrix}} = \frac{a - \tau\Omega b}{1 + \tau^2\Omega^2} \tag{8-49}$$

$$B = \frac{\begin{vmatrix} 1 & a \\ -\tau\Omega & b \end{vmatrix}}{\begin{vmatrix} 1 & \tau\Omega \\ -\tau\Omega & 1 \end{vmatrix}} = \frac{b + \tau\Omega a}{1 + \tau^2\Omega^2} \tag{8-50}$$

Clearly, Eqs. (8-46) and (8-47) could have been solved via either Cramer's rule* or any alternative procedure.

---

*This procedure for solving simultaneous linear algebraic equations is named for the Swiss physicist Gabriel Cramer (1704–1752), who clearly articulated it in 1750, although Colin Maclaurin (1698–1746) had published the same rule in 1748 and Leibniz had described essentially the same technique in a 1693 letter to l'Hospital. It appears that Cramer's work has been given precedence primarily because of his more accessible notation and because Euler gave credit for this rule to him. Furthermore, although it is not clear when this method came into use by Chinese scholars, it is evident that they knew and used such a rule hundreds of years before any of the above authors.

Substituting Eqs. (8-49) and (8-50) into Eq. (8-43) gives

$$\nu_p(t) = \frac{1}{1 + \tau^2\Omega^2}\{(a - \tau\Omega b)\cos\Omega t + (b + \tau\Omega a)\sin\Omega t\}$$

$$= \frac{1}{1 + \tau^2\Omega^2}\{a(\cos\Omega t + \tau\Omega\sin\Omega t) + b(\sin\Omega t - \tau\Omega\cos\Omega t)\} \quad (8\text{-}51)$$

Alternatively, Eq. (8-51) may be rewritten as[2]

$$\nu_p(t) = \frac{1}{1 + \tau^2\Omega^2}\left\{a\sqrt{1 + \tau^2\Omega^2}\cos(\Omega t - \phi) + b\sqrt{1 + \tau^2\Omega^2}\sin(\Omega t - \phi)\right\}$$

$$= \frac{1}{\sqrt{1 + \tau^2\Omega^2}}\{a\cos(\Omega t - \phi) + b\sin(\Omega t - \phi)\} \quad (8\text{-}52)$$

where the angle ϕ is defined by

$$\phi = \tan^{-1}\tau\Omega \qquad 0 \le \phi < \frac{\pi}{2} \quad (8\text{-}53)$$

Note that the range of the angle ϕ indicated in Eq. (8-53) follows from the physical restrictions that $\tau > 0$ and $\Omega \ge 0$. That is, under these restrictions, when $\Omega = 0$, $\phi = \tan^{-1}0 = 0$; and, as Ω increases toward ∞, ϕ monotonically increases approaches $\pi/2$.[3] This completes Step 2 of Table 8-1.

Here, we want to be somewhat deliberate in our definitions because the reader will find a variety of meanings associated with the terms phase angle, phase lag, and phase lead. Referring to Eq. (8-52) we define *phase angle* as the quantity in parentheses beyond Ωt, in this case $-\phi$. So, the phase angle would have been $+\phi$ if the term in parentheses had been $\Omega t + \phi$. Furthermore, if the phase angle when evaluated is a positive quantity, it is called a *phase lead,* whereas if the phase angle when evaluated is a negative quantity, its absolute value is called a *phase lag.* That is, when evaluated, the phase angle will produce the specific result $(\Omega t \pm C_0)$, where C_0 is a known positive constant. If the sign in this result is the plus sign, C_0 is a phase lead. If the sign in this result is the negative sign, C_0 is called a phase lag. Therefore, since the ϕ determined via Eq. (8-53) must be positive, $-\phi$ is the *phase angle* in Eq. (8-52) and ϕ is the *phase lag* in Eq. (8-52).

It is to be emphasized that phase denotes a *relative* behavior of two or more quantities. In particular, the terminology we have adopted provides that the response variable $\nu_p(t)$ will trail the harmonic forcing function by C_0 when C_0 is a phase lag, and will precede the harmonic forcing function by C_0 when C_0 is a phase lead. Physically realizable systems do *not* exhibit response phase leads because the system response cannot occur before the input.[4]

Note that either Eq. (8-51) or Eqs. (8-52) *and* (8-53) also provide the particular solution for the special cases where the excitation is either purely cosinusoidal or purely sinusoidal. When the excitation is purely cosinusoidal (that is, $b = 0$ in Eq. (8-42)), Eq. (8-52) reduces immediately to

---

[2]See Section H-6 in Appendix H; also, see Problem 8-18.

[3]Theoretically, the condition $\tau < 0$ is possible, leading to "unstable" systems. Some economic, sociological, and active control models may possess negative τ and thus exhibit inherent instability. However, we shall not consider such cases here although Appendix H allows for such possibilities.

[4]This is often called the *causality condition.*

$$\nu_p(t) = \frac{a}{\sqrt{1 + \tau^2 \Omega^2}} \cos(\Omega t - \phi) \tag{8-54}$$

And, when the excitation is purely sinusoidal (that is, $a = 0$ in Eq. (8-42)), Eq. (8-52) reduces immediately to

$$\nu_p(t) = \frac{b}{\sqrt{1 + \tau^2 \Omega^2}} \sin(\Omega t - \phi) \tag{8-55}$$

It is straightforward to see that the sum of Eqs. (8-54) and (8-55) is identical to Eq. (8-52). This is a consequence of the linearity of the system. That is, the particular solution of a linear system model for an excitation consisting of more than one forcing function is identical to the sum of its particular solutions for each of those forcing functions.

Similarly, the principle of superposition can be applied to find the particular solution for the general excitation in Eq. (8-40) from the particular solution for the simplified excitation in Eq. (8-42). That is, by adding the particular solution given by Eq. (8-52) for all forcing frequencies in the excitation, the particular solution for the general excitation of Eq. (8-40) may be found as

$$\nu_p(t) = \sum_{i=1}^{m} \frac{1}{\sqrt{1 + \tau^2 \Omega_i^2}} \{a_i \cos(\Omega_i t - \phi_i) + b_i \sin(\Omega_i t - \phi_i)\} \tag{8-56}$$

where

$$\phi_i = \tan^{-1} \tau \Omega_i \qquad 0 \le \phi_i < \frac{\pi}{2} \tag{8-57}$$

Therefore, as indicated earlier, no generality was sacrificed due to the simplification of the general excitation in Eq. (8-40) to the apparently simpler one in Eq. (8-42).

8-2.4.2 Alternative Technique for Obtaining Particular Solution Using Complex Variables

The particular solution found in Subdivision 8-2.4.1 may be derived in a slightly more concise fashion by utilizing complex variables.[5] In this technique, the harmonic excitation $f(t)$ on the right-hand side of Eq. (8-39) is written in the form of a complex function as

$$f(t) = \begin{cases} 0 & t < 0 \\ \nu_f e^{i\Omega t} & t \ge 0 \end{cases} \tag{8-58}$$

where ν_f is a real constant representing the amplitude of the harmonic excitation at the (radian) frequency Ω, where $i \equiv \sqrt{-1}$, and where ν_f will be identified subsequently with a or b in Eq. (8-42). The complex forcing function $\nu_f e^{i\Omega t}$ in Eq. (8-58) has no physical meaning! It is simply a useful mathematical tool for deriving the particular responses to sinusoidal and cosinusoidal excitations simultaneously and somewhat efficiently.[6]

In accordance with the theory of ordinary differential equations, it is *assumed* that the particular solution $\nu_p(t)$ of Eq. (8-39) due to the function $f(t)$ in Eq. (8-58) is of the form

$$\nu_p(t) = C_1 e^{i\Omega t} \tag{8-59}$$

[5]See Appendix H.
[6]See Section H-5 in Appendix H.

where C_1 is an undetermined coefficient. Substituting Eqs. (8-58) and (8-59) into Eq. (8-39), canceling the common $e^{i\Omega t}$ from both sides of the resulting expression, and solving for C_1 give

$$C_1 = \frac{\nu_f}{1 + i\tau\Omega} \tag{8-60}$$

Multiplying the numerator and the denominator of the right-hand side of Eq. (8-60) by the complex conjugate of the denominator, that is, $(1 - i\tau\Omega)$, and then separating the real and the imaginary parts of the resulting expression give[7]

$$C_1 = \frac{\nu_f}{1 + \tau^2\Omega^2} - i\frac{\tau\Omega\nu_f}{1 + \tau^2\Omega^2} \tag{8-61}$$

Equation (8-61) may be rewritten as[8]

$$C_1 = \frac{\nu_f}{\sqrt{1 + \tau^2\Omega^2}}e^{-i\phi} \tag{8-62}$$

where the angle ϕ is given by

$$\phi = \tan^{-1}\tau\Omega \qquad 0 \le \phi < \frac{\pi}{2} \tag{8-63}$$

The physical meaning of the angle ϕ defined in Eq. (8-63) will be discussed shortly. Substitution of Eqs. (8-62) and (8-63) into Eq. (8-59) yields the particular solution as

$$\nu_p(t) = \frac{\nu_f}{\sqrt{1 + \tau^2\Omega^2}}e^{i(\Omega t - \phi)}$$

$$= \frac{\nu_f}{\sqrt{1 + \tau^2\Omega^2}}\{\cos(\Omega t - \phi) + i\sin(\Omega t - \phi)\} \tag{8-64}$$

where Euler's formula has been used.

As discussed in Section H-5 in Appendix H, the real part and the imaginary part of the complex particular solution given in Eq. (8-64) represent the particular solutions due to the forcing functions given by the real part and the imaginary part, respectively, of Eq. (8-58) for nonnegative time. Therefore, when the forcing function $f(t)$ is specified by[9] $f(t) = \text{Re}\{\nu_f e^{i\Omega t}\} = \nu_f\cos\Omega t$ (purely cosinusoidal), the particular solution is given by the real part of Eq. (8-64); that is,

$$\nu_p(t) = \frac{\nu_f}{\sqrt{1 + \tau^2\Omega^2}}\cos(\Omega t - \phi) \tag{8-65}$$

Similarly, when the forcing function $f(t)$ is specified by[10] $f(t) = \text{Im}\{\nu_f e^{i\Omega t}\} = \nu_f\sin\Omega t$ (purely sinusoidal), the particular solution is given by the imaginary part of Eq. (8-64); that is,

---

[7]See, for example, Eq. (H-18) in Appendix H.

[8]See Section H-4 in Appendix H, in particular Example H-2.

[9]The notation "Re{□}" denotes "the real part of □."

[10]The notation "Im{□}" denotes "the imaginary part of □."

$$\nu_p(t) = \frac{\nu_f}{\sqrt{1 + \tau^2\Omega^2}} \sin(\Omega t - \phi) \tag{8-66}$$

Note here that the angle $-\phi$ in Eqs. (8-65) and (8-66) is called the phase angle because it represents the shift of the response in time relative to the excitation. Furthermore, since ϕ is nonnegative as defined in Eq. (8-63), the angle ϕ is a phase lag.

Via the principle of superposition, as indicated in Subdivision 8-2.4.1, the particular solution due to the forcing function $f(t) = a\cos\Omega t + b\sin\Omega t$ can be obtained by summing the particular solution for $f(t) = a\cos\Omega t$ and the particular solution for $f(t) = b\sin\Omega t$. Note that the particular solution for $f(t) = a\cos\Omega t$ can be obtained from Eq. (8-65) by setting $\nu_f = a$ and the particular solution for $f(t) = b\sin\Omega t$ can be obtained from Eq. (8-66) by setting $\nu_f = b$. Therefore, the particular solution due to $f(t) = a\cos\Omega t + b\sin\Omega t$ can be obtained as

$$\nu_p(t) = \frac{1}{\sqrt{1 + \tau^2\Omega^2}}\{a\cos(\Omega t - \phi) + b\sin(\Omega t - \phi)\} \tag{8-67}$$

where we recall that ϕ is defined by Eq. (8-63).

Note that as expected, due to the identity of Eqs. (8-67) and (8-63) with Eqs. (8-52) and (8-53), respectively, the particular solution obtained via the complex variable technique is identical to the particular solution obtained without using complex variables. The only advantage of the complex variable technique lies in its relatively shorter calculation when compared with the technique used in Subdivision 8-2.4.1.

8-2.4.3 Complete Solution
Now, as indicated in Step 3 of Table 8-1, the complete solution is established next. Since the homogeneous solution $\nu_h(t)$ is the solution to Eq. (8-3), it is given by Eq. (8-9) as

$$\nu_h(t) = Ce^{-\frac{t}{\tau}} \tag{8-68}$$

where C is a coefficient to be evaluated in accordance with the initial condition. Therefore, summing Eqs. (8-52) and (8-68) gives the complete solution as

$$\begin{aligned}\nu(t) &= \nu_p(t) + \nu_h(t) \\ &= \frac{1}{\sqrt{1 + \tau^2\Omega^2}}\{a\cos(\Omega t - \phi) + b\sin(\Omega t - \phi)\} + Ce^{-\frac{t}{\tau}}\end{aligned} \tag{8-69}$$

This completes Step 3 of Table 8-1.

As indicated in Step 4 of Table 8-1, imposing the initial condition in Eq. (8-41) onto Eq. (8-69) gives

$$\nu_0 = \frac{1}{\sqrt{1 + \tau^2\Omega^2}}(a\cos\phi - b\sin\phi) + C \tag{8-70}$$

or

$$C = \nu_0 - \frac{a\cos\phi - b\sin\phi}{\sqrt{1 + \tau^2\Omega^2}} \tag{8-71}$$

Substituting Eq. (8-71) into Eq. (8-69) gives the complete solution as

$$\nu(t) = \frac{1}{\sqrt{1 + \tau^2 \Omega^2}} \{a \cos(\Omega t - \phi) + b \sin(\Omega t - \phi)\}$$

$$+ \left\{ \nu_0 - \frac{a \cos \phi - b \sin \phi}{\sqrt{1 + \tau^2 \Omega^2}} \right\} e^{-\frac{t}{\tau}} \quad (8\text{-}72)$$

where

$$\phi = \tan^{-1} \tau \Omega \qquad 0 \leq \phi < \frac{\pi}{2} \quad (8\text{-}73)$$

This completes Step 4 of Table 8-1. Thus, Eqs. (8-72) and (8-73) give the complete solution for the excitation in Eq. (8-42) and the initial condition in Eq. (8-41).

Furthermore, when the excitation is given by the general forcing function in Eq. (8-40), the complete solution can be obtained as the sum of Eqs. (8-56) and (8-68). It can be shown[11] that imposing the initial condition in Eq. (8-41) onto the complete solution and evaluating C in the homogeneous solution lead to the complete solution for the general excitation of Eq. (8-40) as

$$\nu(t) = \sum_{i=1}^{m} \frac{1}{\sqrt{1 + \tau^2 \Omega_i^2}} \{a_i \cos(\Omega_i t - \phi_i) + b_i \sin(\Omega_i t - \phi_i)\}$$

$$+ \left\{ \nu_0 - \sum_{i=1}^{m} \frac{a_i \cos \phi_i - b_i \sin \phi_i}{\sqrt{1 + \tau^2 \Omega_i^2}} \right\} e^{-\frac{t}{\tau}} \quad (8\text{-}74)$$

where

$$\phi_i = \tan^{-1} \tau \Omega_i \qquad 0 \leq \phi_i < \frac{\pi}{2} \quad (8\text{-}75)$$

8-2.4.4 Summary of Approaches for Complete Solution   In summary, the procedure for finding the complete solution to Eq. (8-39), where $f(t)$ is given by Eq. (8-42) and $\nu(t)$ is subjected to the initial condition in Eq. (8-41), is listed in Table 8-3. That is, Table 8-3 presents the procedure for finding the complete solution used in Subdivisions 8-2.4.1 and 8-2.4.3, without the use of complex variables. Table 8-4 shows the alternative procedure using complex variables for finding the particular solution only, as presented in Subdivision 8-2.4.2. In both Tables 8-3 and 8-4, the steps corresponding to Steps 1 through 4 of Table 8-1 are indicated. The purpose in providing Tables 8-3 and 8-4 is fourfold:

1. To present the reader with summaries that may be helpful for inculcating the solution techniques;

2. To reinforce the idea that complex functions provide a mathematical tool that should be used with flexibility and at the analyst's convenience;

3. To emphasize that superposition of dynamic responses to multiple inputs (here, $a \cos \Omega t$ and $b \cos \Omega t$) is valid for linear systems; and

---

[11]See Problem 8-19.

TABLE 8-3 Procedure to Obtain Complete Solution for Single-Degree-of-Freedom First-Order System Subjected to Harmonic Excitation, $f(t) = a\cos\Omega t + b\sin\Omega t$, and Initial Condition

---

Step 1

Establish equation of motion as

$$\tau\frac{d\nu}{dt} + \nu = a\cos\Omega t + b\sin\Omega t$$

and specify initial condition as

$$\nu(0) = \nu_0$$

Step 2

Assume particular solution as

$$\nu_p(t) = A\cos\Omega t + B\sin\Omega t$$

and evaluate A and B such that $\nu_p(t)$ satisfies equation of motion; particular solution found.

Step 3

Recall general homogeneous solution as

$$\nu_h(t) = Ce^{-t/\tau}$$

and establish complete solution as

$$\nu(t) = \nu_p(t) + \nu_h(t).$$

Step 4

Impose initial condition onto complete solution and evaluate C of homogeneous solution. Complete solution found.

---

4. To note that the concepts illustrated in Tables 8-3 and 8-4, which are presented for relatively simple linear systems here, can be extended to more complicated linear dynamic systems of higher order and higher degrees of freedom.

8-2.4.5 Closer Look at Particular Solution Note that the second term in Eq. (8-72) (or Eq. (8-74)) decays exponentially with time. Hence, for sufficiently large values of t, the complete response is described by the first term in Eq. (8-72), that is, the particular solution that represents the sum of both the cosinusoidal and sinusoidal oscillations. Therefore, in the harmonic response of single-degree-of-freedom first-order systems, the steady-state response *happens* to coincide with the particular solution. This is an exceptional case to the cautionary comment that precedes Eq. (8-38). One example of a complete harmonic response is sketched in Figure 8-6 for arbitrary values of system parameters and initial condition. In Figure 8-6, the particular or steady-state response is shown to prevail after a time lapse where the exponential term has become negligible.

TABLE 8-4 Alternative Procedure Using Complex Variables to Obtain Particular Solution for Single-Degree-of-Freedom First-Order System Subjected to Harmonic Excitation, $f(t) = a\cos\Omega t + b\sin\Omega t$

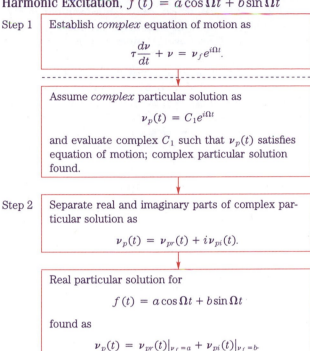

Step 1 | Establish *complex* equation of motion as

$$\tau\frac{d\nu}{dt} + \nu = \nu_f e^{i\Omega t}.$$

Assume *complex* particular solution as

$$\nu_p(t) = C_1 e^{i\Omega t}$$

and evaluate complex C_1 such that $\nu_p(t)$ satisfies equation of motion; complex particular solution found.

Step 2 | Separate real and imaginary parts of complex particular solution as

$$\nu_p(t) = \nu_{pr}(t) + i\nu_{pi}(t).$$

Real particular solution for

$$f(t) = a\cos\Omega t + b\sin\Omega t$$

found as

$$\nu_p(t) = \nu_{pr}(t)|_{\nu_f = a} + \nu_{pi}(t)|_{\nu_f = b}.$$

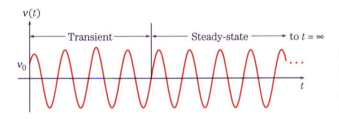

$v(t)$

← Transient → ← Steady-state → to $t = \infty$

v_0

t

FIGURE 8-6 Complete response of single-degree-of-freedom first-order system subjected to harmonic excitation and initial condition ν_0.

Here we examine the particular solution in some detail. Consider the case where the harmonic excitation is given by either a cosinusoidal or sinusoidal function. Then, when $f(t) = \nu_f \cos\Omega t$ (ν_f being the amplitude of the harmonic excitation function), the particular solution is given by Eq. (8-54) as

$$\nu_p(t) = \frac{\nu_f}{\sqrt{1 + \tau^2\Omega^2}}\cos(\Omega t - \phi) \tag{8-76}$$

where we have set $a = \nu_f$, the amplitude of the excitation function. Also, when $f(t) = \nu_f \sin\Omega t$ (ν_f being the amplitude of the harmonic excitation function), the particular solution is given by Eq. (8-55) as

$$\nu_p(t) = \frac{\nu_f}{\sqrt{1 + \tau^2\Omega^2}} \sin(\Omega t - \phi) \qquad (8\text{-}77)$$

where we have set $b = \nu_f$, the amplitude of the excitation function. The phase lag ϕ in both Eqs. (8-76) and (8-77) is given by Eq. (8-53). For example, in Figure 8-7, a sinusoidal input and the corresponding steady-state responses for different values of $\tau\Omega$ are sketched. Similar plots can be shown for a cosinusoidal input and the corresponding steady-state responses.

Note that the amplitudes as well as the phase angles of the particular solutions given by Eqs. (8-76) and (8-77) are identical to each other. Therefore, the same observations can be made regarding the amplitude and phase angle of the particular solutions, whether the input harmonic function is cosinusoidal or sinusoidal. Below, we make a series of observations about the amplitude and phase angle of the steady-state harmonic response given by either Eq. (8-76) or Eq. (8-77); that is, when the forcing function is either $f(t) = \nu_f \cos\Omega t$ or $\nu_f \sin\Omega t$:

1. While the input amplitude is ν_f, the output amplitude is diminished according to
    $$\frac{\nu_f}{\sqrt{1 + \tau^2\Omega^2}}.$$

2. The response frequency is the same as the excitation frequency, a characteristic of all harmonically excited linear systems.

3. The phase angle of the response is $-\phi = -\tan^{-1}(\tau\Omega)$, which means that the output *lags* the input by ϕ. Thus, the output peaks *after* the input peaks by a *phase lag* ϕ, or equivalently by a time $t' = \phi/\Omega$.

From the observations 1 through 3 immediately above, it should be noted that the steady-state harmonic response depends upon two parameters: the amplitude of the input ν_f and the quantity $\tau\Omega$. The linear dependence of the output (response) amplitude upon the input (excitation) amplitude ν_f is an inherent characteristic of linear systems; for example, when the input amplitude is doubled, the output amplitude is also doubled. As indicated in observations 1 and 3 above and also sketched in Figure 8-7, the system parameter $\tau\Omega$ affects both the amplitude and the phase angle of the output. So, as $\tau\Omega$ increases, the output amplitude decreases while the phase lag (or, equivalently, the time delay of the output) increases. For example, when $\tau\Omega$ becomes very large (that is, the time constant or the forcing frequency or both become very large), the output amplitude approaches zero and the phase lag approaches 90°; whereas when $\tau\Omega$ becomes very small, the output

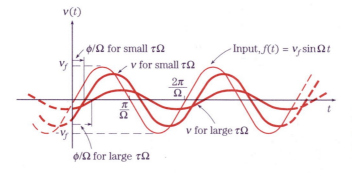

FIGURE 8-7 Sinusoidal input and steady-state response of single-degree-of-freedom first-order system, for two values of $\tau\Omega$.

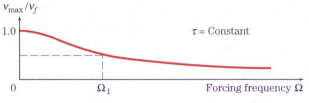

(a) Output/input amplitude ratio, $v_{max}/v_f = 1/\sqrt{1+\tau^2\Omega^2}$.

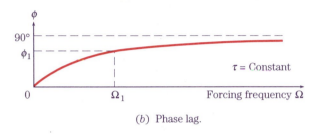

(b) Phase lag.

FIGURE 8-8   Amplitude and phase lag characteristics of steady-state harmonic response of single-degree-of-freedom first-order system subjected to $f(t) = v_f \cos \Omega t$ or $v_f \sin \Omega t$.

amplitude approaches the input amplitude and the phase lag approaches 0°. So, when $\tau\Omega$ is very small, the output follows the input nearly faithfully; whereas when $\tau\Omega$ is very large, the output amplitude becomes small relative to the input amplitude and the output follows the input with a time delay of $t' \approx \dfrac{\pi}{2\Omega}$.

A graphical method is commonly used to present the steady-state harmonic response in the frequency domain (that is, as functions of forcing frequency Ω rather than time t). We shall illustrate this graphical method for the steady-state harmonic response, given by either Eq. (8-76) or Eq. (8-77), or when the harmonic input is specified as either $f(t) = v_f \cos \Omega t$ or $f(t) = v_f \sin \Omega t$. In Figure 8-8a and b are plotted the output/input amplitude ratio and the phase lag, respectively, as functions of the forcing frequency for a given τ. As shown in Figure 8-8, for a given excitation frequency Ω_1, $\dfrac{v_{max}}{v_f}\left(= \dfrac{1}{\sqrt{1 + \tau^2\Omega_1^2}}\right)$ and ϕ may be determined from the plots.

8-2.5   Summary of Responses for Single-Degree-of-Freedom First-Order Systems

In Subsections 8-2.1 through 8-2.4, the responses of the generic single-degree-of-freedom first-order system have been obtained for various forms of excitations. In summary, the complete responses for all these cases are listed in Table 8-5, for the initial condition $v(0) = v_0$.

In the first column of Table 8-5, the forcing function $f(t)$ for positive time is specified for each case; and in the second column, the particular solution is given for each case; in the third column, the homogeneous solution is given for each case. Therefore, the complete solution may be easily obtained by summing the second and third columns for each case.

Further, since all the unknown coefficients of both the particular solution and the homogeneous solution have been evaluated in Table 8-5, any specific particular, homogeneous, and complete solutions can be obtained, simply by substituting the system parameter (in this case, τ), the excitation parameter(s) (for example, v_f in the step input), and the initial condition (in this case, v_0) into the specific formula(s) of interest.

TABLE 8-5 Summary of Responses for Single-Degree-of-Freedom First-Order System $\tau\dot{v} + v = f(t)$ Subjected to Various Excitations and Initial Condition $v(0) = v_0$.

Forcing function, $f(t), t \geq 0$	Particular solution, $v_p(t)$	Homogeneous solution, $v_h(t)$
0 (No input)	—	$v_0 e^{\frac{-t}{\tau}}$
v_f (constant) (Step input)	v_f	$(v_0 - v_f)e^{\frac{-t}{\tau}}$
$a_0 t (a_0$: constant) (Ramp input)	$a_0 t - \tau a_0$	$(v_0 + \tau a_0)e^{\frac{-t}{\tau}}$
$a \cos \Omega t + b \sin \Omega t$ (a, b, Ω: constant) (Harmonic input)	$\dfrac{a\cos(\Omega t - \phi) + b\sin(\Omega t - \phi)}{\sqrt{1 + \tau^2\Omega^2}}$ where $\phi = \tan^{-1}\tau\Omega,$ $0 \leq \phi < \dfrac{\pi}{2}$	$\left\{ v_0 - \dfrac{a\cos\phi - b\sin\phi}{\sqrt{1 + \tau^2\Omega^2}} \right\} e^{\frac{-t}{\tau}}$ where $\phi = \tan^{-1}\tau\Omega,$ $0 \leq \phi < \dfrac{\pi}{2}$

8-3 SINGLE-DEGREE-OF-FREEDOM SECOND-ORDER SYSTEMS

The equation of motion for the mass spring dashpot system sketched in Figure 8-9 is

$$m\ddot{x} + c\dot{x} + kx = f(t) \tag{8-78}$$

where $x(t)$ is the generalized coordinate representing the displacement of the mass from its equilibrium position. Although the form of Eq. (8-78) is general in the sense that it describes the behavior of models in many disciplines, as indicated by the specific parameters used, we shall discuss this equation in the context of a mechanical system. Equation (8-78), or its equivalent various forms, is the most ubiquitous equation in vibration applications, perhaps commanding the attention of vibration engineers more than all other equations of motion combined. Thus, any presentation of this equation in a single section of a single chapter is subject to copious truncation. It is in this limited context that we briefly explore the free response and the harmonic response of systems modeled by Eq. (8-78).

According to the theory of ordinary differential equations, in order to determine the solution to Eq. (8-78) for a given $f(t)$, two initial conditions are needed. These initial conditions in their most general form may be expressed as

$$x(0) = x_0 \tag{8-79}$$

$$\dot{x}(0) = v_0 \tag{8-80}$$

where x_0 and v_0 are the initial displacement and initial velocity, respectively.

In Subsection 8-3.1, the free response is found; in Subsections 8-3.2 through 8-3.4, several features of the free response are studied. In Subsection 8-3.5, the harmonic response is found and several of its characteristics are investigated. In finding the (complete) harmonic response, as emphasized earlier, the procedure in Table 8-1 is used.

8-3.1 Free Response

For the free (unforced) response, the equation of motion given by Eq. (8-78) for the system in Figure 8-9 reduces to a homogeneous ordinary differential equation

$$m\ddot{x} + c\dot{x} + kx = 0 \tag{8-81}$$

where $f(t)$ in Eq. (8-78) has been set to zero. By dividing Eq. (8-81) by m, it may be rewritten in a somewhat generic form as

$$\ddot{x} + 2\zeta\omega_n\dot{x} + \omega_n^2 x = 0 \tag{8-82}$$

where the nondimensional parameter ζ is called the *damping factor* or damping ratio and ω_n is called the *natural frequency*. From Eqs. (8-81) and (8-82), it is noted that for the system in Figure 8-9

$$\zeta = \frac{c}{2m\omega_n} = \frac{c}{c_c} \tag{8-83}$$

and

$$\omega_n = \sqrt{\frac{k}{m}} \tag{8-84}$$

where by definition c_c is equal to $2m\omega_n$ and is called the *critical damping coefficient*. Both ζ and ω_n will be prominent in the solution of Eq. (8-81) or Eq. (8-82), and we shall interpret the physical significance of each, including that of c_c. The system is assumed to be subjected to the following initial conditions (I.C.) at $t = 0$:

$$x(0) = x_0 \tag{8-85}$$
$$\dot{x}(0) = v_0 \tag{8-86}$$

From the theory of ordinary differential equations, it is *assumed* that the solution to Eq. (8-82) is of the exponential form

$$x(t) = Ce^{\lambda t} \tag{8-87}$$

where C and λ are unknown constants. Substitution of Eq. (8-87) into Eq. (8-82) yields

$$(\lambda^2 + 2\zeta\omega_n\lambda + \omega_n^2)Ce^{\lambda t} = 0 \tag{8-88}$$

Equation (8-88) is a product of two terms, one of which must be zero. If $Ce^{\lambda t}$ were zero, then via Eq. (8-87), $x(t)$ would also be zero, resulting in a trivial solution. So, $Ce^{\lambda t}$ cannot be zero for a nontrivial solution. Thus, Eq. (8-88) requires that

$$\lambda^2 + 2\zeta\omega_n\lambda + \omega_n^2 = 0 \tag{8-89}$$

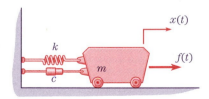

FIGURE 8-9 Simple model for single-degree-of-freedom second-order systems.

which is called the *characteristic equation* of the system. So, Eq. (8-87) is a solution to Eq. (8-82) if λ is a root to Eq. (8-89) such that

$$\lambda_1 = \left[-\zeta + \sqrt{\zeta^2 - 1} \right] \omega_n \qquad \lambda_2 = \left[-\zeta - \sqrt{\zeta^2 - 1} \right] \omega_n \qquad (8\text{-}90)$$

where the roots λ_1 and λ_2 are called the *characteristic values* or more commonly the *eigenvalues*. Here, we emphasize that we are limiting our consideration to systems in which $m > 0, c \geq 0$, and $k > 0$, thus requiring that $0 \leq \zeta < \infty$; so the radicand $(\zeta^2 - 1)$ may be positive, negative, or zero. Thus, from the theory of ordinary differential equations, the general solution to Eq. (8-82) is

$$x(t) = C_1 e^{\lambda_1 t} + C_2 e^{\lambda_2 t} \qquad (8\text{-}91)$$

where C_1 and C_2 are arbitrary constants and λ_1 and λ_2 are given in Eqs. (8-90).

The values of the constants C_1 and C_2 can be found by introducing the initial conditions, Eqs. (8-85) and (8-86), into Eq. (8-91), giving

$$x(0) = C_1 + C_2 = x_0 \qquad (8\text{-}92)$$

$$\dot{x}(0) = C_1 \lambda_1 + C_2 \lambda_2 = v_0 \qquad (8\text{-}93)$$

Solution of Eqs. (8-92) and (8-93) yields the values for C_1 and C_2 as

$$C_1 = \frac{x_0 \lambda_2 - v_0}{\lambda_2 - \lambda_1} \qquad (8\text{-}94)$$

$$C_2 = \frac{v_0 - x_0 \lambda_1}{\lambda_2 - \lambda_1} \qquad (8\text{-}95)$$

The solution to Eq. (8-82), Eq. (8-91) with Eqs. (8-90), (8-94), and (8-95), may be summarized in four cases. The consideration of four cases is suggested by the change in character of ζ as the system parameters m, c, and k vary over their permissible ranges.

CASE 1: Undamped ($\zeta = 0$; or, via Eq. (8-83), $c = 0$)

This case is labeled *undamped* because $c = 0$, and its corresponding equation of motion may be expressed as

$$\ddot{x} + \omega_n^2 x = 0 \qquad (8\text{-}96)$$

which is subjected to the initial conditions in Eqs. (8-85) and (8-86). From Eqs. (8-90)

$$\lambda_1 = i\omega_n \qquad \lambda_2 = -i\omega_n \qquad (8\text{-}97)$$

where $i = \sqrt{-1}$. Then, Eq. (8-91) becomes

$$x(t) = C_1 e^{i\omega_n t} + C_2 e^{-i\omega_n t} \qquad (8\text{-}98)$$

Using Euler's formula,[12] $e^{\pm i\omega t} = \cos \omega t \pm i \sin \omega t$, Eq. (8-98) can be written as

$$x(t) = (C_1 + C_2) \cos \omega_n t + (C_1 - C_2) i \sin \omega_n t \quad (8\text{-}99)$$

Then, substitution of Eq. (8-97) into Eqs. (8-94) and (8-95) gives

$$C_1 = \frac{-ix_0 \omega_n - v_0}{-2i\omega_n} = \frac{x_0 \omega_n - iv_0}{2\omega_n} \qquad (8\text{-}100)$$

$$C_2 = \frac{v_0 - ix_0 \omega_n}{-2i\omega_n} = \frac{x_0 \omega_n + iv_0}{2\omega_n} \qquad (8\text{-}101)$$

Substitution of Eqs. (8-100) and (8-101) into Eq. (8-99) yields

$$x(t) = x_0 \cos \omega_n t + \frac{v_0}{\omega_n} \sin \omega_n t \qquad (8\text{-}102)$$

Briefly, we make two observations. First, the physical significance of ω_n should be clear. It is the inherent fre-

---

[12]See Eq. (H-12) in Appendix H.

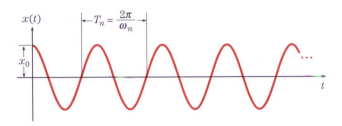

(a) $x(0) = x_0$ and $\dot{x}(0) = 0$

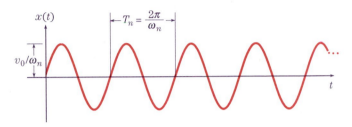

(b) $x(0) = 0$ and $\dot{x}(0) = v_0$.

FIGURE 8-10 Free response of undamped single-degree-of-freedom second-order system for (a) zero initial velocity and (b) zero initial displacement.

quency at which a free undamped single-degree-of-freedom second-order system will oscillate, and for this reason it is called the *natural frequency*, expressed in radians per second. The number of complete cycles per unit time is often called the *natural cyclic frequency* and is given by $f_n = \dfrac{\omega_n}{2\pi}$, expressed in hertz (1 hertz (Hz) is 1 cycle per second, or 1 revolution per second). The time required for one complete cycle is called the *natural period* and is given by $T_n = \dfrac{1}{f_n} = \dfrac{2\pi}{\omega_n}$, expressed in seconds. Second, the eerie little $i = \sqrt{-1}$ quietly and conveniently appeared and just as quietly and conveniently disappeared. The use of complex numbers in this analysis simply provided a shorthand round-trip from the real variables (Eq. (8-96)) through complex variables (Eqs. (8-97) through (8-101)) back to real variables (Eq. (8-102)). The trip was simply a mathematical ruse, taken exclusively for convenience. There was no physics associated with the trip, so no physical interpretations need be attributed to it.

The response represented by Eq. (8-102) is sketched in Figure 8-10 for the two simplest sets of initial conditions: an initial displacement and zero velocity *or* an initial velocity and zero displacement. In accordance with the principle of superposition, combinations of nonzero initial displacement and nonzero initial velocity will result in combinations of the responses of Figures 8-10a and b.

Finally, two alternative forms of Eq. (8-102) that are often used by vibration analysts are

$$x(t) = C_0 \sin(\omega_n t + \psi_1) \qquad (8\text{-}103)$$

or

$$x(t) = C_0 \cos(\omega_n t + \psi_2) \qquad (8\text{-}104)$$

where C_0, ψ_1, and ψ_2 are constants that may be determined from the initial conditions[13] or directly from Eq. (8-102) as shown in Section H-6 in Appendix H.[14]

CASE 2: Overdamped ($\zeta > 1$; or, via Eq. (8-83), $c > c_c = 2m\omega_n$)

This case is labeled *overdamped* because the damping coefficient is greater than its critical value; that is, $c > c_c$ in Eq. (8-83) for $\zeta > 1$. For this case, the equation of motion is Eq. (8-82), subjected to the initial con-

ditions in Eqs. (8-85) and (8-86), and its solution is Eq. (8-91),

$$x(t) = C_1 e^{\lambda_1 t} + C_2 e^{\lambda_2 t} \qquad (8\text{-}105)$$

[13] See Problem 8-29.

[14] See Problem 8-30.

where from Eqs. (8-90),

$$\lambda_{1,2} = \left[-\zeta \pm \sqrt{\zeta^2 - 1}\right]\omega_n \qquad (8\text{-}106)$$

Further, from Eqs. (8-94), (8-95), and (8-106),

$$C_1 = x_0\left[\frac{\zeta}{2\sqrt{\zeta^2 - 1}} + \frac{1}{2}\right] + \frac{v_0}{2\sqrt{\zeta^2 - 1}\,\omega_n} \qquad (8\text{-}107)$$

and

$$C_2 = -x_0\left[\frac{\zeta}{2\sqrt{\zeta^2 - 1}} - \frac{1}{2}\right] - \frac{v_0}{2\sqrt{\zeta^2 - 1}\,\omega_n} \qquad (8\text{-}108)$$

From Eq. (8-106) we see that the eigenvalues λ_1 and λ_2 are distinct, real, and negative.

The solution is given by substituting Eqs. (8-106) through (8-108) into Eq. (8-105). That is,

$$x(t) =$$

$$\left\{\frac{x_0}{2}\left(\frac{\zeta}{\sqrt{\zeta^2 - 1}} + 1\right) + \frac{v_0}{2\omega_n\sqrt{\zeta^2 - 1}}\right\}e^{(-\zeta\omega_n + \sqrt{\zeta^2-1}\,\omega_n)t}$$

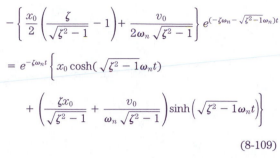

$$-\left\{\frac{x_0}{2}\left(\frac{\zeta}{\sqrt{\zeta^2 - 1}} - 1\right) + \frac{v_0}{2\omega_n\sqrt{\zeta^2 - 1}}\right\}e^{(-\zeta\omega_n - \sqrt{\zeta^2-1}\,\omega_n)t}$$

$$= e^{-\zeta\omega_n t}\left\{x_0\cosh(\sqrt{\zeta^2 - 1}\,\omega_n t)\right.$$

$$\left. + \left(\frac{\zeta x_0}{\sqrt{\zeta^2 - 1}} + \frac{v_0}{\omega_n\sqrt{\zeta^2 - 1}}\right)\sinh\left(\sqrt{\zeta^2 - 1}\,\omega_n t\right)\right\}$$

$$(8\text{-}109)$$

The response represented by Eq. (8-109) is sketched in Figure 8-11 for several combinations of generic initial conditions. The motion expressed by Eq. (8-109) is generally nonoscillatory—although sufficiently large initial velocities toward the equilibrium position can produce a single zero-crossing—and decays to zero for sufficiently long times t. The two curves labeled C in Figure 8-11a depict the two possibilities where for $\dot{x} < 0$ the response may or may not result in a single zero-crossing.

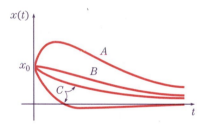

(a) Positive initial displacement.

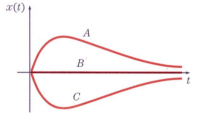

(b) Zero initial displacement.

FIGURE 8-11   Response of overdamped single-degree-of-freedom second-order system for various values of initial velocity and for (a) positive initial displacement x_0 and (b) zero initial displacement. For both (a) and (b), the individual curves are denoted by A: $\dot{x}(0) > 0; B : \dot{x}(0) = 0; C : \dot{x}(0) < 0$.

CASE 3: Critically Damped ($\zeta = 1$; or, via Eq. (8-83), $c = c_c = 2m\omega_n$)

This case is labeled *critically damped* because the damping coefficient is equal to its so-called critical value; that is, $c = c_c = 2m\omega_n$ in Eq. (8-83), giving $\zeta = 1$. As in Case 2, the equation of motion is Eq. (8-82), subjected to the initial conditions in Eqs. (8-85) and (8-86), and its solution is Eq. (8-91). However, from Eqs. (8-90), the eigenvalues λ_1 and λ_2 are equal,

$$\lambda_1 = \lambda_2 = -\omega_n \qquad (8\text{-}110)$$

Thus, the trial solution in Eq. (8-87) gives only a partial solution.

From the theory of ordinary differential equations, for repeated roots, it can be verified by substitution that a second solution is represented by the trial solution

$$x(t) = Cte^{-\omega_n t} \qquad (8\text{-}111)$$

Thus, the general solution for this case may be expressed as

$$x(t) = C_3 e^{-\omega_n t} + C_4 t e^{-\omega_n t} \qquad (8\text{-}112)$$

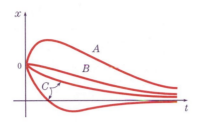

(a) Positive initial displacement.

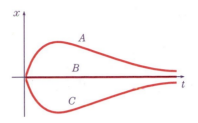

(b) Zero initial displacement.

FIGURE 8-12 Response of critically damped single-degree-of-freedom second-order system for various values of initial velocity and for (a) positive initial displacement x_0 and (b) zero initial displacement. For both (a) and (b), the individual curves are denoted by $A : \dot{x}(0) > 0; B : \dot{x}(0) = 0; C : \dot{x}(0) < 0$.

Substitution of the initial conditions in Eqs. (8-85) and (8-86) into Eq. (8-112) gives

$$C_3 = x_0 \qquad (8\text{-}113)$$

$$-\omega_n C_3 + C_4 = v_0 \qquad (8\text{-}114)$$

whereupon solving for C_3 and C_4 yields

$$C_3 = x_0 \qquad (8\text{-}115)$$

$$C_4 = \omega_n x_0 + v_0 \qquad (8\text{-}116)$$

Therefore, the critically damped motion of the system is expressed by Eqs. (8-112), (8-115) and (8-116). That is,

$$x(t) = e^{-\omega_n t}\{x_0 + (\omega_n x_0 + v_0)t\} \qquad (8\text{-}117)$$

Once again, the motion is generally nonoscillatory and decays to zero for sufficiently long time t. The response represented by the solution in Eq. (8-117) is sketched in Figure

8-12 for several combinations of generic initial conditions. Again, the two curves labeled as C in Figure 8-12a indicate that for various values of $\dot{x} < 0$, the response may or may not exhibit a single zero-crossing. The strong similarity between Figures 8-11 and 8-12 should be noted. This strong similarity is due to the fact that Case 3 is the limiting condition of Case 2. When excited by the identical initial conditions of displacement and velocity, a critically damped system will approach equilibrium faster than will an otherwise identical overdamped system. Furthermore, now we may interpret the meaning of c_c, the critical damping coefficient. If the damping constant c is greater than c_c, the system will be overdamped. If the damping constant c is equal to c_c, the system will be critically damped. And if the damping constant c is less than c_c, the system will be underdamped; this is the case that we consider next.

CASE 4: Underdamped ($0 < \zeta < 1$; or, via Eq. (8-83), $0 < c < c_c = 2m\omega_n$)

This case is labeled *underdamped* because the damping coefficient c is positive but less than its critical value; that is, $0 < c < c_c$. Once again, the solution to Eq. (8-82), subjected to the initial conditions in Eqs. (8-85) and (8-86), is represented by Eq. (8-91). Note, however, that $(\zeta^2 - 1)$ is negative, so Eqs. (8-90) become

$$\lambda_1 = \left[-\zeta + i\sqrt{1 - \zeta^2}\right]\omega_n \qquad \lambda_2 = \left[-\zeta - i\sqrt{1 - \zeta^2}\right]\omega_n \qquad (8\text{-}118)$$

where $i = \sqrt{-1}$. It is now convenient to define a new variable ω_d as

$$\omega_d = \sqrt{1 - \zeta^2}\,\omega_n \qquad (8\text{-}119)$$

where ω_d is called the *damped natural frequency*, expressed in radians per second.

In analogy with the (undamped) natural frequency, the damped natural frequency will be shown to be the frequency at which a free underdamped single-degree-of-freedom second-order system will oscillate. The corresponding number of complete cycles per unit time is often called the *damped natural cyclic frequency* and is given by $f_d = \dfrac{\omega_d}{2\pi}$, expressed in hertz (1 hertz (Hz) is 1 cycle per second). The corresponding time required for one complete cycle is called the *damped natural period* and is given by $T_d = \dfrac{1}{f_d} = \dfrac{2\pi}{\omega_d}$, expressed in seconds. The nondimensional damped natural frequency, $\dfrac{\omega_d}{\omega_n} = \sqrt{1 - \zeta^2}$ from Eq. (8-119), is plotted in Figure 8-13 as a function of ζ.

Recalling that $e^{(a+b)} = e^a e^b$ and using Eqs. (8-118) and (8-119), we may rewrite Eq. (8-91) as

$$x(t) = e^{-\zeta\omega_n t}\left[C_1 e^{i\omega_d t} + C_2 e^{-i\omega_d t}\right] \qquad (8\text{-}120)$$

Use of Euler's formula, $e^{\pm i\omega_d t} = \cos\omega_d t \pm i\sin\omega_d t$, allows Eq. (8-120) to be written as

$$\begin{aligned} x(t) &= e^{-\zeta\omega_n t}\left[C_1(\cos\omega_d t + i\sin\omega_d t)\right.\\ &\quad \left. + C_2(\cos\omega_d t - i\sin\omega_d t)\right]\\ &= e^{-\zeta\omega_n t}\left[(C_1 + C_2)\cos\omega_d t + i(C_1 - C_2)\sin\omega_d t\right] \end{aligned}$$
$$(8\text{-}121)$$

The coefficients C_1 and C_2 may be found by substituting Eqs. (8-118) into Eqs. (8-94) and (8-95), which yields

$$C_1 = x_0\left[\frac{\zeta\omega_n}{2i\omega_d} + \frac{1}{2}\right] + \frac{v_0}{2i\omega_d} \qquad (8\text{-}122)$$

$$C_2 = -x_0\left[\frac{\zeta\omega_n}{2i\omega_d} - \frac{1}{2}\right] - \frac{v_0}{2i\omega_d} \qquad (8\text{-}123)$$

which can be combined to show that

$$C_1 + C_2 = x_0 \qquad (8\text{-}124)$$

$$\begin{aligned} i(C_1 - C_2) &= i\left(\frac{\zeta\omega_n x_0}{i\omega_d} + \frac{v_0}{i\omega_d}\right)\\ &= \frac{\zeta\omega_n x_0 + v_0}{\omega_d} \end{aligned} \qquad (8\text{-}125)$$

Substituting Eqs. (8-124) and (8-125) into Eq. (8-121) gives

$$x(t) = e^{-\zeta\omega_n t}\left(x_0\cos\omega_d t + \frac{v_0 + \zeta\omega_n x_0}{\omega_d}\sin\omega_d t\right) \qquad (8\text{-}126)$$

Equation (8-126) is the solution to Eq. (8-82) for free underdamped vibration. Once again, note that the use of complex variables in Eqs. (8-118) through (8-125) was simply for mathematical convenience and the final form of the solution in Eq. (8-126) is, as it must be, in terms of real functions.

Alternative forms of the solution in Eq. (8-126) may be found by rewriting the expression in parentheses in the forms of Eqs. (8-103) and (8-104). That is,

$$x(t) = Ce^{-\zeta\omega_n t}\sin(\omega_d t + \psi_1) \qquad (8\text{-}127)$$

or

$$x(t) = Ce^{-\zeta\omega_n t}\cos(\omega_d t - \psi_2) \qquad (8\text{-}128)$$

where C, ψ_1, and ψ_2 are constants that may be determined from the initial conditions[15] or directly from Eq. (8-126), as shown in Section H-6 in Appendix H.[16]

The motion in this underdamped case, Case 4, is oscillatory and periodic as in Case 1. However, unlike Case 1, the amplitude of the displacement decreases with increasing time, and the frequency of the displacement (as well as the other kinematic variables) is equal to the damped natural frequency, rather than the natural frequency. The motion in Case 4 given by Eq. (8-126) is sketched in Figure 8-14 for an arbitrary set of initial conditions. We make two observations from Figure 8-14. First, as expressed in Eq. (8-127), the motion is represented by the product of a decaying exponential and a harmonic function. As such, the motion is bounded by two envelopes $\pm Ce^{-\zeta\omega_n t}$, where C may be found via the solution scheme summarized above. Second, the period of the oscillation (that is, the time interval between any two consecutive peaks, such as $(t_2 - t_1)$) is equal to the damped natural period, $T_d = \dfrac{1}{f_d} = \dfrac{2\pi}{\omega_d}$.

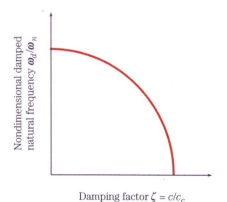

FIGURE 8-13   Nondimensional damped natural frequency versus the damping factor.

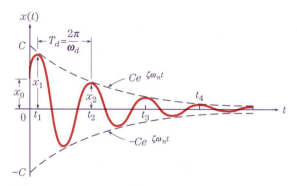

FIGURE 8-14   Free vibration response of underdamped ($\zeta < 1$) single-degree-of-freedom second-order system subjected to arbitrary initial conditions.

---

[15] See Problem 8-29.
[16] See Problem 8-30.

8-3.2 Natural Frequency via Static Deflection

Suppose we were asked to find the natural frequency of a massive machine mounted on flexible bearings. We were given the undeformed (that is, unloaded or freestanding) height of the bearings, and further we were allowed to have access only to a length-measuring device (such as a ruler, meter stick, or micrometer). Such a system is sketched in Figure 8-15. Although this is a slightly peculiar limitation for such an important request, a response to the request leads to a simple and fascinating result.

Sketches of a single-degree-of-freedom model of the system are shown in Figure 8-16 where the flexible bearings represented by the linear spring and the dashpot are shown before and after the machine is mounted. Note that we do *not* know the specific values of m, k, or c, and that we are concerned exclusively with statics, not dynamics. The elongation Δ in the bearings is due to the machine's weight W, which we also do not know and which can be modeled as a static load, as sketched in Figure 8-16.

From the free-body diagram in Figure 8-16c,

$$k\Delta = W = mg \tag{8-129}$$

Dividing both sides of Eq. (8-129) by the product $m\Delta$ gives

$$\frac{k}{m} = \frac{g}{\Delta} \tag{8-130}$$

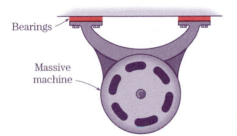

Bearings

Massive
machine

FIGURE 8-15 Model of massive machine supported by flexible bearings.

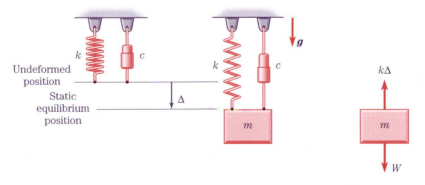

Undeformed
position

Static
equilibrium
position

k c Δ g m $k\Delta$ W

(*a*) Undeformed bearing.   (*b*) Mounted machine.   (*c*) Free-body diagram.

FIGURE 8-16 Single-degree-of-freedom model and free-body diagram of massive machine mounted on flexible bearings.

Recognizing the left-hand side of Eq. (8-130) as ω_n^2, we conclude that

$$\omega_n = \sqrt{\frac{g}{\Delta}} \tag{8-131}$$

Thus, by knowing only the *static* deflection, we may compute the (dynamic) natural frequency!

Finally, we make three observations. First, the systems in both Figures 8-15 and 8-16 could have been rotated by 180° within the plane of sketch, resulting in compression instead of extension of the bearings. Such a rotation, however, would not have changed the result in Eq. (8-131). Second, because the analysis is concerned exclusively with statics, the dashpot plays no role in the evaluation; so c may just as well have been set to zero for this specific exposition. Third, in the course of designing mountings for machinery, engineers encounter several performance tables in bearing catalogs that are based exclusively on Eq. (8-131).

■ **Example 8-1:**   Natural Frequency of Instrument Isolator System

To decrease the structural vibratory motion transmitted to an instrument package (total mass = 20 kg) in an aircraft, the package is mounted on isolators. If the isolator, having negligible damping, deflects 3 mm under the weight of the instrument package, find the natural frequency of the system that comprises the instrument package and the isolator.

□ **Solution:**   Direct substitution of $\Delta = 3$ mm $= 0.003$ m and $g = 9.81 \frac{m}{s^2}$ into Eq. (8-131) gives the natural frequency of the system as

$$\omega_n = \sqrt{\frac{g}{\Delta}} = \sqrt{\frac{9.81 \left(\frac{m}{s^2}\right)}{0.003 \; (m)}} = 57.18 \; (rad/s)$$

8-3.3   Logarithmic Decrement

In the preceding subsection, we considered the natural frequency of a single-degree-of-freedom second-order system. In this section, we consider the second of the intrinsic properties of such systems as represented in Eq. (8-82), the damping factor ζ.

Considering the important underdamped case, we note that there is no generally reliable purely theoretical technique for determining ζ. After all, in most physical systems there is no identifiable component that can be associated primarily with energy dissipation. (Of course, sometimes a shock absorber or comparable device is specifically designed for energy dissipation. In such cases, there is likely to be a set of estimating calculations to assess the performance of the device. We are not considering such specific devices here. The damping with which we are concerned is due to a broad range of unmodeled mechanisms, including material damping or damping due to relative motion of components or damping due to surrounding fluid.) Here, we present the most common technique for determining ζ, which requires a combination of experiment and theory. This technique requires exciting the system by initial conditions, recording the resulting displacement x versus time t (such x versus t representations are sometimes called the *system ringdown*), and using the analyses developed thus far to derive an expression for ζ in terms of the experimental ringdown measurements.

Suppose a set of experiments provides the motion of an underdamped single-degree-of-freedom second-order system as sketched in Figure 8-14. From Figure 8-14 we note that the times when the motion peaks are designated arbitrarily as $t_1, t_2, t_3, \ldots$ and the displacements at those times are designated as $x_1, x_2, x_3, \ldots$, respectively. Consider the ratio of any two successive displacements, say x_1 and x_2, which are measured experimentally. This ratio of x_1 to x_2 is given theoretically by Eq. (8-127) as

$$\frac{x_1}{x_2} = \frac{Ce^{-\zeta\omega_n t_1}\sin(\omega_d t_1 + \psi_1)}{Ce^{-\zeta\omega_n t_2}\sin(\omega_d t_2 + \psi_1)} \tag{8-132}$$

Note that $t_2 = t_1 + T_d$, T_d being the damped natural period given by

$$T_d = \frac{2\pi}{\omega_d} = \frac{2\pi}{\sqrt{1 - \zeta^2}\,\omega_n} \tag{8-133}$$

Note that $\sin(\omega_d t_2 + \psi_1) = \sin(\omega_d t_1 + \omega_d T_d + \psi_1) = \sin(\omega_d t_1 + 2\pi + \psi_1) = \sin(\omega_d t_1 + \psi_1)$, which when used in Eq. (8-132) gives

$$\frac{x_1}{x_2} = \frac{e^{-\zeta\omega_n t_1}}{e^{-\zeta\omega_n(t_1 + T_d)}} = e^{\zeta\omega_n T_d} \tag{8-134}$$

where in the denominator we have used the fact that $e^{-\zeta\omega_n(t_1 + T_d)} = e^{-\zeta\omega_n t_1}e^{-\zeta\omega_n T_d}$.

The *logarithmic decrement* δ is defined as the natural logarithm of the ratio of any two successive displacement amplitudes, so that by taking the natural logarithm of both sides of Eq. (8-134), we find that

$$\delta \equiv \ln\left(\frac{x_j}{x_{j+1}}\right) = \zeta\omega_n T_d = \zeta\omega_n\frac{2\pi}{\sqrt{1 - \zeta^2}\,\omega_n} = \frac{2\pi\zeta}{\sqrt{1 - \zeta^2}} \tag{8-135}$$

where the subscript j is an arbitrary integer and where Eq. (8-133) has been used. Equation (8-135) may be rewritten for ζ as

$$\zeta = \frac{\delta}{\sqrt{(2\pi)^2 + \delta^2}} \tag{8-136}$$

Note from Eq. (8-135) that if $\zeta \ll 1$ (that is, very low damping, which quantitatively means $c \ll c_c$), $\sqrt{1 - \zeta^2} \approx 1$ and thus

$$\delta \approx 2\pi\zeta \tag{8-137}$$

Figure 8-17 gives a comparison of Eqs. (8-135) and (8-137) versus the damping factor ζ.

Because few systems behave perfectly in accordance with Eq. (8-135), it is useful to conduct the experiment for δ several times, using different pairs of successive amplitudes. In addition and particularly for very low damping where x_i and x_{i+1} are nearly the same, it is also beneficial to consider the ratio of nonsuccessive amplitudes, which has the effect of "averaging" the value of δ over a longer portion of the system ringdown. Then, for the amplitudes x_1 and x_{n+1}, where n is an integer, we observe that

$$\frac{x_1}{x_{n+1}} = \frac{x_1}{x_2}\frac{x_2}{x_3}\frac{x_3}{x_4}\frac{x_4}{x_5}\cdots\frac{x_n}{x_{n+1}} \tag{8-138}$$

Taking the natural logarithm of both sides of Eq. (8-138) gives

$$\ln\left(\frac{x_1}{x_{n+1}}\right) = \ln\left(\frac{x_1}{x_2}\right) + \ln\left(\frac{x_2}{x_3}\right) + \cdots + \ln\left(\frac{x_n}{x_{n+1}}\right) \tag{8-139}$$

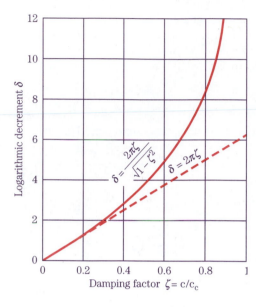

FIGURE 8-17 Logarithmic decrement versus damping factor for exact and approximate expressions.

And, because $\delta = \ln\left(\dfrac{x_i}{x_{i+1}}\right)$, Eq. (8-139) can be rewritten as

$$\ln\left(\frac{x_1}{x_{n+1}}\right) = \delta + \delta + \cdots + \delta = n\delta \qquad (8\text{-}140)$$

So,

$$\delta = \frac{1}{n}\ln\left(\frac{x_1}{x_{n+1}}\right) \qquad (8\text{-}141)$$

or alternatively,

$$\delta = \frac{1}{n-1}\ln\left(\frac{x_1}{x_n}\right) \qquad (8\text{-}142)$$

where the amplitude of x_n is used instead of x_{n+1}. Therefore, in cases where the damping is very low or in cases where the frequency during ringdown is high, the use of Eq. (8-141) or Eq. (8-142) can facilitate the calculation of the logarithmic decrement δ.

In Table 8-6, typical values of logarithmic decrement for several systems are summarized.

TABLE 8-6 Logarithmic Decrement for Various Types of Structures

Type of Structure	Approximate Range of Logarithmic Decrement, δ
Multistory Steel Buildings	0.02–0.10
Steel Bridges	0.05–0.15
Multistory Concrete Buildings	0.10–0.20
Concrete Bridges	0.10–0.30
Machinery Foundations	0.40–0.60

■ **Example 8-2:**   **Logarithmic Decrement and Damping Coefficient**

The following data are given for the mass spring dashpot model sketched in Figure 8-9: mass $= 5$ kg, spring constant $= 2$ kN$/$m. It is found experimentally that the motion of the system is periodically oscillatory and the amplitude of the oscillation on the fifth cycle is one-third of its initial value. Find the dashpot constant c of the system.

☐ **Solution:**   Because the amplitude x_5 is equal to $\dfrac{x_1}{3}$, substituting $\dfrac{x_1}{x_5} = 3$ and $n = 5$ into Eq. (8-142) gives

$$\delta = \frac{1}{4}\ln 3 = 0.275 \tag{a}$$

The damping factor ζ can be found by substituting the result in Eq. (a) into Eq. (8-136) as

$$\zeta = \frac{\delta}{\sqrt{(2\pi)^2 + \delta^2}} = \frac{0.275}{\sqrt{4\pi^2 + 0.275^2}} = 0.0437 \tag{b}$$

The dashpot constant c can be found from Eqs. (8-83) and (8-84) as

$$c = c_c\zeta = 2m\omega_n\zeta = 2m\sqrt{\frac{k}{m}}\,\zeta = 2\zeta\sqrt{\frac{km^2}{m}} = 2\zeta\sqrt{km} \tag{c}$$

Substituting the given values for m and k and the computed value for ζ into Eq. (c) gives

$$c = 2(0.0437)\sqrt{(2000)(5)} = 8.74 \ (\text{N} \cdot \text{s}/\text{m}) \tag{d}$$

which is the sought-after dashpot constant.

8-3.4   Energy Loss of Free Vibration

For cases in which the damping is low, the energy loss or dissipation in underdamped cases can be expressed in terms of the logarithmic decrement. At the maximum displacement during a given cycle, the velocity of the mass is zero, so all the energy is stored in the spring element. If V_j is defined as the mechanical energy in the system at the maximum displacement of the jth cycle,

$$V_j = \frac{1}{2}kx_j^2 \tag{8-143}$$

Further, the system mechanical energy at the maximum amplitude of the $j + 1^{\text{st}}$ cycle is

$$V_{j+1} = \frac{1}{2}kx_{j+1}^2 \tag{8-144}$$

The energy loss between the jth and $j + 1^{\text{st}}$ cycles is

$$\Delta V = V_j - V_{j+1} = \frac{1}{2}k(x_j^2 - x_{j+1}^2) \tag{8-145}$$

which can be written as

$$\Delta V = \frac{1}{2}k(x_j - x_{j+1})(x_j + x_{j+1}) \tag{8-146}$$

The *specific energy loss* is defined from Eqs. (8-143) and (8-146) as

$$
\begin{aligned}
\frac{\Delta V}{V_j} &\equiv \frac{\frac{1}{2}k(x_j - x_{j+1})(x_j + x_{j+1})}{\frac{1}{2}kx_j^2} \\
&= \left(1 - \frac{x_{j+1}}{x_j}\right)\left(1 + \frac{x_{j+1}}{x_j}\right) \\
&= (1 - e^{-\delta})(1 + e^{-\delta}) = 1 - e^{-2\delta}
\end{aligned} \tag{8-147}
$$

where Eq. (8-135) has been used. (Or, note from Eq. (8-134) that $\left(\dfrac{x_j}{x_{j+1}}\right) = (e^{\delta})$ which leads directly to $\left(\dfrac{x_j}{x_{j+1}}\right)^{-1} = (e^{\delta})^{-1} = e^{-\delta}$.) Equation (8-147) expresses the facts that (1) the specific energy loss increases as the logarithmic decrement increases and (2) if the damping is zero, δ is zero and the specific energy loss is also zero. Equation (8-147) is used frequently in industrial vibration applications where it is common to express system or component specifications in terms of specific energy loss.

8-3.5 Harmonic Response

Consider the system sketched in Figure 8-9 for the case when the applied force $f(t)$ is harmonic for nonnegative time ($t \geq 0$). Since $f(t)$ is nonzero, each step of the general procedure of Table 8-1 for the complete solution must be addressed explicitly. The equation of motion for the system is given by

$$m\ddot{x} + c\dot{x} + kx = f(t) \tag{8-148}$$

where

$$f(t) = \begin{cases} 0 & t < 0 \\ \sum_{j=1}^{m} a_j \cos \Omega_j t + b_j \sin \Omega_j t & t \geq 0 \end{cases} \tag{8-149}$$

where the a_j's and b_j's are the constant amplitudes of the cosinusoidal and sinusoidal forcing functions, respectively, having m distinct (radian) frequencies Ω_j's. As indicated in the discussion immediately following Eq. (8-40) in Subsection 8-2.4, the excitation $f(t)$ in Eq. (8-149) accommodates all possible harmonic excitations. Also, we assume the initial conditions are given by Eqs. (8-79) and (8-80) as

$$x(0) = x_0 \tag{8-150}$$

$$\dot{x}(0) = v_0 \tag{8-151}$$

As in Subsection 8-2.4, we shall focus our attention on a simpler problem where the excitation $f(t)$ is abridged to

$$f(t) = \begin{cases} 0 & t < 0 \\ a \cos \Omega t + b \sin \Omega t & t \geq 0 \end{cases} \tag{8-152}$$

Thus, the problem is to find the complete solution to Eq. (8-148) where $f(t)$ is given by Eq. (8-152) and the initial conditions are given by Eqs. (8-150) and (8-151). This satisfies Step 1 of Table 8-1.

In the first two subdivisions below (Subdivisions 8-3.5.1 and 8-3.5.2), two alternative approaches for the particular solution are presented (Step 2 of Table 8-1); in the third subdivision (Subdivision 8-3.5.3), the complete solution is established (Step 3 of Table 8-1); and then the coefficients of the homogeneous solution are evaluated in accordance with the initial conditions, yielding the complete solution (Step 4 of Table 8-1). As a summary, the procedures for finding the complete solution are tabulated in Subdivision 8-3.5.4. Further, the particular solution is examined more closely in Subdivision 8-3.5.5.

8-3.5.1 Particular Solution

In accordance with the theory of ordinary differential equations, the particular solution $x_p(t)$ for the $f(t)$ given by Eq. (8-152) is *assumed* to be of the form

$$x_p(t) = A\cos\Omega t + B\sin\Omega t \qquad t \geq 0 \tag{8-153}$$

where A and B are the undetermined coefficients and where it should be noted that the response frequency is the same as the excitation frequency Ω. By requiring that the assumed particular solution satisfies the equation of motion, the undetermined coefficients in the particular solution may be evaluated. Substituting Eqs. (8-152) and (8-153) into Eq. (8-148) yields, for nonnegative time,

$$-m\Omega^2 A\cos\Omega t - m\Omega^2 B\sin\Omega t - c\Omega A\sin\Omega t + c\Omega B\cos\Omega t$$
$$+ kA\cos\Omega t + kB\sin\Omega t = a\cos\Omega t + b\sin\Omega t \tag{8-154}$$

Grouping terms containing $\cos\Omega t$ and those containing $\sin\Omega t$ separately in Eq. (8-154) leads to

$$\{(k - m\Omega^2)A + c\Omega B - a\}\cos\Omega t + \{-c\Omega A + (k - m\Omega^2)B - b\}\sin\Omega t = 0 \tag{8-155}$$

In order that Eq. (8-155) holds for all nonnegative time, each set of the coefficients in the flower brackets must be zero. This condition of zero coefficients in Eq. (8-155) can be written in matrix form as

$$\begin{bmatrix} (k - m\Omega^2) & c\Omega \\ -c\Omega & (k - m\Omega^2) \end{bmatrix} \begin{Bmatrix} A \\ B \end{Bmatrix} = \begin{Bmatrix} a \\ b \end{Bmatrix} \tag{8-156}$$

Equation (8-156) can be solved via Cramer's rule for A and B as

$$A = \frac{\begin{vmatrix} a & c\Omega \\ b & (k - m\Omega^2) \end{vmatrix}}{\begin{vmatrix} (k - m\Omega^2) & c\Omega \\ -c\Omega & (k - m\Omega)^2 \end{vmatrix}} = \frac{(k - m\Omega^2)a - c\Omega b}{(k - m\Omega^2)^2 + (c\Omega)^2} \tag{8-157}$$

$$B = \frac{\begin{vmatrix} (k - m\Omega^2) & a \\ -c\Omega & b \end{vmatrix}}{\begin{vmatrix} (k - m\Omega^2) & c\Omega \\ -c\Omega & (k - m\Omega^2) \end{vmatrix}} = \frac{(k - m\Omega^2)b + c\Omega a}{(k - m\Omega^2)^2 + (c\Omega)^2} \tag{8-158}$$

Substituting Eqs. (8-157) and (8-158) into Eq. (8-153) gives

$$x_p(t) = \frac{\{(k - m\Omega^2)a - c\Omega b\}\cos\Omega t + \{(k - m\Omega^2)b + c\Omega a\}\sin\Omega t}{(k - m\Omega^2)^2 + (c\Omega)^2} \tag{8-159}$$

The function in the numerator of Eq. (8-159) may be rewritten as[17]

$$\{(k - m\Omega^2)a - c\Omega b\}\cos\Omega t + \{(k - m\Omega^2)b + c\Omega a\}\sin\Omega t$$
$$= a\{(k - m\Omega^2)\cos\Omega t + c\Omega\sin\Omega t\} + b\{(k - m\Omega^2)\sin\Omega t - c\Omega\cos\Omega t\}$$
$$= \sqrt{(k - m\Omega^2)^2 + (c\Omega)^2}\{a\cos(\Omega t - \phi) + b\sin(\Omega t - \phi)\} \qquad (8\text{-}160)$$

where the angle ϕ is defined by

$$\phi = \tan^{-1}\frac{c\Omega}{k - m\Omega^2} \qquad 0 \le \phi < \pi \qquad (8\text{-}161)$$

Note that the range of the angle ϕ indicated in Eq. (8-161) follows from the physical restrictions that $m, k > 0$ and $c, \Omega \ge 0$. That is, under these conditions, when $\Omega = 0$, $\phi = \tan^{-1}0 = 0$; as Ω increases up to $\Omega = \omega_n = \sqrt{\dfrac{k}{m}}$, ϕ monotonically increases up to $\dfrac{\pi}{2}$; and, finally, as Ω increases further and beyond $\sqrt{\dfrac{k}{m}}$, ϕ increases further up toward π.

Substitution of Eq. (8-160) into Eq. (8-159) yields

$$x_p(t) = \frac{1}{\sqrt{(k - m\Omega^2)^2 + (c\Omega)^2}}\{a\cos(\Omega t - \phi) + b\sin(\Omega t - \phi)\} \qquad (8\text{-}162)$$

where ϕ is given in Eq. (8-161). We note that the angle $-\phi$ in Eq. (8-162) represents the shift of the response in time with respect to the excitation and thus the angle $-\phi$ is the phase angle in accordance with the definition following Eq. (8-53). Furthermore, since ϕ as defined in Eq. (8-161) will always be nonnegative, the ϕ is a phase lag, which means that the system response trails, does not lead, the excitation.

When the excitation is purely cosinusoidal (that is, $f(t) = a\cos\Omega t$ as obtained from Eq. (8-152) by setting $b = 0$), the particular solution can be obtained from Eq. (8-162) by setting $b = 0$ as

$$x_p(t) = \frac{a}{\sqrt{(k - m\Omega^2)^2 + (c\Omega)^2}}\cos(\Omega t - \phi) \qquad (8\text{-}163)$$

Similarly, when the excitation is purely sinusoidal (that is, $f(t) = b\sin\Omega t$ as obtained from Eq. (8-152) by setting $a = 0$), the particular solution can be obtained from Eq. (8-162) by setting $a = 0$ as

$$x_p(t) = \frac{b}{\sqrt{(k - m\Omega^2)^2 + (c\Omega)^2}}\sin(\Omega t - \phi) \qquad (8\text{-}164)$$

Furthermore, the sum of Eqs. (8-163) and (8-164) for those two harmonic excitation cases restores Eq. (8-162)—a demonstration of the principle of superposition.

It should be noted that the denominator of Eq. (8-162) becomes zero when the system is undamped ($c = 0$) *and* the forcing frequency coincides with the natural frequency of the system, that is, when $\Omega = \omega_n = \sqrt{\dfrac{k}{m}}$. In this special case, the expression in Eq. (8-162) becomes infinite. In fact, the solution form assumed in Eq. (8-153) for the particular solution

---

[17]See Section H-6 in Appendix H; see, also, Problem 8-54.

does not provide a nontrivial solution to the equation of motion because in this special case ($c = 0$ and $k - m\Omega^2 = 0$) an attempt to use this assumed particular solution would reduce Eq. (8-155) to $\{-a\}\cos\Omega t + \{-b\}\sin\Omega t = 0$ for all positive time, which means that both a and b must be zero; that is, no excitation. In this special case, the equation of motion, Eq. (8-148), becomes

$$m\ddot{x} + kx = a\cos\omega_n t + b\sin\omega_n t \tag{8-165}$$

where $\omega_n^2 = \dfrac{k}{m}$.

In accordance with the theory of ordinary differential equations, the particular solution to Eq. (8-165) must be *assumed* in the form of

$$x_p(t) = At\cos\omega_n t + Bt\sin\omega_n t \tag{8-166}$$

Substitution of Eq. (8-166) into Eq. (8-165) yields

$$m\{(-\omega_n^2 At\cos\omega_n t - \omega_n^2 Bt\sin\omega_n t) - 2\omega_n A\sin\omega_n t + 2\omega_n B\cos\omega_n t\}$$
$$+ k(At\cos\omega_n t + Bt\sin\omega_n t) = a\cos\omega_n t + b\sin\omega_n t \tag{8-167}$$

Note that the quantities in the two sets of parentheses cancel each other because $m\omega_n^2 = m\left(\dfrac{k}{m}\right) = k$. Therefore, after some rearrangement, Eq. (8-167) becomes

$$(2m\omega_n B - a)\cos\omega_n t - (2m\omega_n A + b)\sin\omega_n t = 0 \tag{8-168}$$

Setting the coefficients of $\cos\omega_n t$ and $\sin\omega_n t$ in Eq. (8-168) separately equal to zero yields

$$A = -\frac{b}{2m\omega_n} \tag{8-169}$$

$$B = \frac{a}{2m\omega_n} \tag{8-170}$$

Substituting Eqs. (8-169) and (8-170) into Eq. (8-166) gives

$$x_p(t) = -\frac{bt}{2m\omega_n}\cos\omega_n t + \frac{at}{2m\omega_n}\sin\omega_n t \qquad \Omega = \omega_n, c = 0 \tag{8-171}$$

Note that the particular solution given by Eq. (8-171) increases linearly and indefinitely with time t, a condition nontechnically called *resonance,* which occurs when the system is (approximately) undamped and the forcing frequency coincides (approximately) with the system's natural frequency. Equation (8-171) indicates the manner in which $x_p(t)$ goes to infinity, information which is not contained in Eq. (8-162).

Therefore, in summary, the particular solution for Eq. (8-148) where $f(t)$ is defined by Eq. (8-152) is given by Eq. (8-162), except for the special case of $c = 0$ and $\Omega = \omega_n$ for which the particular solution is given by Eq. (8-171). This completes Step 2 of Table 8-1.

Now, the particular solution for the general excitation given in Eq. (8-149) is found, via the principle of superposition, from Eqs. (8-162) and (8-171). That is, if the system is damped or if the system is undamped but none of the forcing frequencies $\Omega_j, j = 1, 2, \ldots, m$ coincides with the system's natural frequency ω_n, the particular solution for the excitation of Eq. (8-149) is given by

$$x_p(t) = \sum_{j=1}^{m} \frac{1}{\sqrt{(k - m\Omega_j^2)^2 + (c\Omega_j)^2}}\{a_j\cos(\Omega_j t - \phi_j) + b_j\sin(\Omega_j t - \phi_j)\} \tag{8-172}$$

where ϕ_j is given by

$$\phi_j = \tan^{-1} \frac{c\Omega_j}{k - m\Omega_j^2} \qquad 0 \leq \phi_j < \pi \tag{8-173}$$

If the system is undamped *and* one of the forcing frequencies, say Ω_p, coincides with the system's natural frequency ω_n, the particular solution is given by the combination of Eqs. (8-162) and (8-171) as

$$x_p(t) = \sum_{j=1, j\neq p}^{m} \frac{1}{\sqrt{\left(k - m\Omega_j^2\right)^2 + (c\Omega_j)^2}} \{a_j \cos(\Omega_j t - \phi_j) + b_j \sin(\Omega_j t - \phi_j)\}$$

$$- \frac{b_p t}{2m\omega_n} \cos \omega_n t + \frac{a_p t}{2m\omega_n} \sin \omega_n t \qquad \Omega_p = \omega_n = \sqrt{\frac{k}{m}} \tag{8-174}$$

Equation (8-174) elucidates the interesting fact that if an undamped system is forced by several forces, each having a different frequency, and if one of those frequencies corresponds to ω_n, the system will selectively respond to each of the excitation frequencies. So, even if the amplitude (a_p or/and b_p) of the force at ω_n is small, the long-term persistence of such an excitation can cause the system to respond (resonate) with increasing, and probably damaging, deformation.

8-3.5.2 Alternative Technique for Obtaining Particular Solution Using Complex Variables

The particular solution found in Subdivision 8-3.5.1 may be obtained in a more concise fashion by utilizing complex variables.[18] In this approach, the harmonic excitation $f(t)$ on the right-hand side of Eq. (8-148) is written in the form of a complex function as

$$f(t) = \begin{cases} 0 & t < 0 \\ f_0 e^{i\Omega t} & t \geq 0 \end{cases} \tag{8-175}$$

where f_0 is a real constant representing the amplitude of the harmonic force excitation at the (radian) frequency Ω and f_0 can be identified with a or b in Eq. (8-152). The complex forcing function $f_0 e^{i\Omega t}$ in Eq. (8-175) has no physical meaning. It is simply a useful mathematical tool for deriving the particular responses to sinusoidal and cosinusoidal excitations simultaneously and somewhat efficiently.[19]

In accordance with the theory of ordinary differential equations, the particular solution $x_p(t)$ of Eq. (8-148) due to the $f(t)$ in Eq. (8-175) is *assumed* to be of the form

$$x_p(t) = X_1 e^{i\Omega t} \tag{8-176}$$

where X_1 is unknown but may be determined by requiring that Eq. (8-176) satisfies Eq. (8-148) for nonnegative time. Therefore, substitution of Eqs. (8-175) and (8-176) into Eq. (8-148) for $t \geq 0$, cancellation of the common $e^{i\Omega t}$ term from both sides of the resulting expression, and the subsequent solution for X_1 give

$$X_1 = \frac{f_0}{(k - m\Omega^2) + ic\Omega} \tag{8-177}$$

---

[18]See Appendix H.

[19]See Section H-5 in Appendix H.

Multiplying the numerator and the denominator of the right-hand side of Eq. (8-177) by $[(k - m\Omega^2) - ic\Omega]$, which is the complex conjugate of the denominator, and separating the real and the imaginary parts of the resulting expression give

$$X_1 = f_0 \left[\frac{(k - m\Omega^2)}{(k - m\Omega^2)^2 + (c\Omega)^2} - \frac{ic\Omega}{(k - m\Omega^2)^2 + (c\Omega)^2} \right] \tag{8-178}$$

Note that if Eq. (8-178) is substituted into Eq. (8-176), an equation comparable to Eq. (8-159) will be obtained.

Equation (8-178) can be rewritten as[20]

$$X_1 = \frac{f_0}{\sqrt{(k - m\Omega^2)^2 + (c\Omega)^2}} e^{-i\phi} \tag{8-179}$$

where the angle ϕ is given by

$$\phi = \tan^{-1} \frac{c\Omega}{k - m\Omega^2} \qquad 0 \le \phi < \pi \tag{8-180}$$

The physical meaning of ϕ will be discussed shortly. Substitution of Eq. (8-179) into Eq. (8-176) gives

$$x_p(t) = \frac{f_0}{\sqrt{(k - m\Omega^2)^2 + (c\Omega)^2}} e^{i(\Omega t - \phi)}$$

$$= \frac{f_0}{\sqrt{(k - m\Omega^2)^2 + (c\Omega)^2}} \{\cos(\Omega t - \phi) + i\sin(\Omega t - \phi)\} \tag{8-181}$$

where Euler's formula has been used.

As discussed in Section H-5 in Appendix H, the real part and the imaginary part of the complex particular solution given in Eq. (8-181) represent the particular solutions due to the forcing functions given by the real part and the imaginary part, respectively, of Eq. (8-175) for nonnegative time. Therefore, when the forcing function $f(t)$ is specified by[21] $f(t) = \text{Re}\{f_0 e^{i\Omega t}\} = f_0 \cos \Omega t$ (purely cosinusoidal), the particular solution is given by the real part of Eq. (8-181); that is,

$$x_p(t) = \frac{f_0}{\sqrt{(k - m\Omega^2)^2 + (c\Omega)^2}} \cos(\Omega t - \phi) \tag{8-182}$$

Similarly, when the forcing function $f(t)$ is specified by[22] $f(t) = \text{Im}\{f_0 e^{i\Omega t}\} = f_0 \sin \Omega t$ (purely sinusoidal), the particular solution is given by the imaginary part of Eq. (8-181); that is,

$$x_p(t) = \frac{f_0}{\sqrt{(k - m\Omega^2)^2 + (c\Omega)^2}} \sin(\Omega t - \phi) \tag{8-183}$$

Note here that the angle $-\phi$ in Eqs. (8-182) and (8-183) is called the phase angle because it represents the shift of the response in time relative to the excitation. Furthermore, since ϕ as defined in Eq. (8-180) will always be nonnegative, the angle ϕ is a phase lag.

---

[20]See Section H-4 in Appendix H, in particular Example H-2.

[21]The notation "Re{□}" denotes "the real part of □."

[22]The notation "Im{□}" denotes "the imaginary part of □."

Via the principle of superposition, the particular solution due to the forcing function $f(t) = a\cos\Omega t + b\sin\Omega t$ can be obtained by summing the particular solution for $f(t) = a\cos\Omega t$ and the particular solution for $f(t) = b\sin\Omega t$. Note that the particular solution for $f(t) = a\cos\Omega t$ can be obtained from Eq. (8-182) by setting $f_0 = a$ and the particular solution for $f(t) = b\sin\Omega t$ can be obtained from Eq. (8-183) by setting $f_0 = b$. Therefore, the particular solution due to $f(t) = a\cos\Omega t + b\sin\Omega t$ can be obtained as

$$x_p(t) = \frac{1}{\sqrt{(k - m\Omega^2)^2 + (c\Omega)^2}}\{a\cos(\Omega t - \phi) + b\sin(\Omega t - \phi)\} \qquad (8\text{-}184)$$

where we recall that ϕ is defined by Eq. (8-180).

Note that as expected, due to the identity of Eqs. (8-184) and (8-180) with Eqs. (8-161) and (8-162), respectively, the particular solution obtained via the complex variable technique is identical to the particular solution obtained without using complex variables.

Furthermore, for the special case of undamped resonance ($c = 0$ and $\Omega = \omega_n$), the particular solution found in Subdivision 8-3.5.1 as Eq. (8-171) can be obtained[23] via the complex variable technique by assuming a particular solution in the form of

$$x_p(t) = X_1 t e^{i\omega_n t} \qquad (8\text{-}185)$$

and finding X_1 such that Eq. (8-185) satisfies Eq. (8-148), when $\Omega = \omega_n$ in $f(t)$ in Eq. (8-175). The only advantage of the complex variable technique lies in its relatively shorter calculation when compared with the approach used in Subdivision 8-3.5.1.

8-3.5.3 Complete Solution
Now, as indicated in Step 3 of Table 8-1, the complete solution is established as the sum of the particular solution and the homogeneous solution. Recall that the form of the homogeneous solution depends on the damping factor ζ. The general forms of the homogeneous solution for each of four cases of damping were found in Subsection 8-3.1 and they are relisted as

$$x_h(t) = C_1\cos\omega_n t + C_2\sin\omega_n t \qquad \zeta = 0 \qquad (8\text{-}186)$$
$$x_h(t) = e^{-\zeta\omega_n t}(C_1\cos\omega_d t + C_2\sin\omega_d t) \qquad 0 < \zeta < 1 \qquad (8\text{-}187)$$
$$x_h(t) = C_1 e^{-\omega_n t} + C_2 t e^{-\omega_n t} \qquad \zeta = 1 \qquad (8\text{-}188)$$
$$x_h(t) = C_1 e^{(-\zeta + \sqrt{\zeta^2 - 1})\omega_n t} + C_2 e^{(-\zeta - \sqrt{\zeta^2 - 1})\omega_n t} \qquad \zeta > 1 \qquad (8\text{-}189)$$

where C_1 and C_2 are the coefficients to be evaluated in accordance with the initial conditions. The complete solution is established by summing the particular solution $x_p(t)$ and the homogeneous solution given by *one* of Eqs. (8-186) through (8-189).

A Word of Caution: The unknown coefficients C_1 and C_2 in Eqs. (8-186) through (8-189) for the complete solution are *not* given by the various C_1 and C_2 coefficients found in Subsection 8-3.1; but C_1 and C_2 for the complete solution must remain unevaluated until the initial conditions are imposed onto the *complete solution*. The reason for this requirement is that the C_1 and C_2 found in Subsection 8-3.1 are valid only when the homogeneous solution alone satisfies the initial conditions. So, if the homogeneous solution and the complete solution were identical (that is, if $f(t) = 0$), the various C_1 and C_2 coefficients in Subsection 8-3.1 would satisfy the complete solution. On the other hand, a complete solution consisting of a homogeneous solution and a nonzero particular solution would not satisfy the initial conditions if the C_1 and C_2 coefficients from Subsection 8-3.1 were used. Therefore,

---

[23]See Problem 8-86.

the unknown coefficients in Eqs. (8-186) through (8-189) must be evaluated only *after* the complete solution has been established, yielding different values for C_1 and C_2 from those found in Subsection 8-3.1. This is an important point, often misunderstood.

The following discussion will be focused on the undamped case, subjected to harmonic excitation. Nevertheless, the identical procedure would yield the complete solution for all the other cases. For undamped systems ($c = 0$), the general (sinusoidal and cosinusoidal) particular solution in Eq. (8-162) for $\Omega \neq \omega_n$ reduces to

$$x_p(t) = \frac{1}{k - m\Omega^2}(a\cos\Omega t + b\sin\Omega t) \qquad \Omega \neq \omega_n \qquad (8\text{-}190)$$

where we have used $c = 0$ and we have used $\phi = 0$ from Eq. (8-161) because $c = 0$. Also, recall that when the forcing frequency coincides with the system's natural frequency ω_n, the particular solution for the undamped system is given by Eq. (8-171) as

$$x_p(t) = -\frac{bt}{2m\omega_n}\cos\omega_n t + \frac{at}{2m\omega_n}\sin\omega_n t \qquad \Omega = \omega_n, c = 0 \qquad (8\text{-}191)$$

Therefore, the complete undamped solution for $\Omega \neq \omega_n$ is given by the sum of Eqs. (8-186) and (8-190) as

$$x(t) = \frac{a\cos\Omega t + b\sin\Omega t}{k - m\Omega^2} + C_1\cos\omega_n t + C_2\sin\omega_n t \qquad \Omega \neq \omega_n, c = 0 \quad (8\text{-}192)$$

and the complete undamped solution for $\Omega = \omega_n$ is given by the sum of Eqs. (8-186) and (8-191) as

$$x(t) = \left(C_1 - \frac{bt}{2m\omega_n}\right)\cos\omega_n t + \left(C_2 + \frac{at}{2m\omega_n}\right)\sin\omega_n t \qquad \Omega = \omega_n, c = 0 \quad (8\text{-}193)$$

This completes Step 3 of Table 8-1.

The coefficients C_1 and C_2 can be determined by imposing the initial conditions in Eqs. (8-150) and (8-151) onto each of the *complete solutions* in Eqs. (8-192) and (8-193). First, consider the case of $\Omega \neq \omega_n$. Imposing the initial condition in Eq. (8-150) onto Eq. (8-192) at $t = 0$ and imposing the initial condition in Eq. (8-151) onto the time derivative of Eq. (8-192) at $t = 0$ give, respectively,

$$x_0 = \frac{a}{k - m\Omega^2} + C_1 \qquad (8\text{-}194)$$

$$v_0 = \frac{b\Omega}{k - m\Omega^2} + C_2\omega_n \qquad (8\text{-}195)$$

Solving Eqs. (8-194) and (8-195) for C_1 and C_2 and substituting the results into Eq. (8-192) give the complete solution as

$$x(t) = \frac{a\cos\Omega t + b\sin\Omega t}{k - m\Omega^2} + \left(x_0 - \frac{a}{k - m\Omega^2}\right)\cos\omega_n t$$

$$+ \left(\frac{v_0}{\omega_n} - \frac{b\Omega}{\omega_n(k - m\Omega^2)}\right)\sin\omega_n t \qquad \Omega \neq \omega_n, c = 0 \quad (8\text{-}196)$$

Similarly, when $\Omega = \omega_n$, imposing the initial condition in Eq. (8-150) onto Eq. (8-193) at $t = 0$ and imposing the initial condition in Eq. (8-151) onto the time derivative of Eq. (8-193) at $t = 0$ give

$$x_0 = C_1 \tag{8-197}$$

$$v_0 = -\frac{b}{2m\omega_n} + C_2\omega_n \tag{8-198}$$

Solving Eqs. (8-197) and (8-198) for C_1 and C_2 and substituting the results into Eq. (8-193) give the complete solution as

$$x(t) = \left(x_0 - \frac{bt}{2m\omega_n}\right)\cos\omega_n t + \left(\frac{v_0}{\omega_n} + \frac{b}{2m\omega_n^2} + \frac{at}{2m\omega_n}\right)\sin\omega_n t \qquad \Omega = \omega_n, c = 0$$

$$\tag{8-199}$$

This completes Step 4 of Table 8-1. Equations (8-196) and (8-199) give the complete solutions for the undamped system subjected to the harmonic excitation in Eq. (8-152), for the cases of $\Omega \neq \omega_n$ and $\Omega = \omega_n$, respectively.

As indicated earlier, the complete solutions for all cases of nonzero damping can also be found via the identical procedure.[24]

An important point emphasized following Eq. (8-37) in Subsection 8-2.3 should be reemphasized here. The point is regarding the distinction of various terminology: "transient solution," "homogeneous solution," "steady-state solution," and "particular solution." It is clear that the homogeneous solutions in all of Eqs. (8-187) through (8-189) for damped systems decay exponentially with time and ultimately vanish. After these damped homogeneous solutions dampen out, the system response is represented solely by the particular solution. It is for this reason that the homogeneous solution is often called the *transient response,* and the particular solution is often called the *steady-state* or *long-term response.* These names are not always appropriate, however, because in undamped systems the homogeneous solutions persist without decaying amplitudes. So, in undamped models, the steady-state solution is given by the sum of both the homogeneous and the particular solutions, as illustrated by the complete solutions in Eqs. (8-196) and (8-199). Therefore, in general, the phrase "transient solution" should not be used interchangeably with the phrase "homogeneous solution," and the phrase "steady-state solution" should not be used interchangeably with the phrase "particular solution." In problems where the external excitation is of finite duration, this point is of even stronger significance, irrespective of whether the system is damped or undamped.

8-3.5.4 *Summary of Approaches for Complete Solution*

In summary, Table 8-7 shows the alternative procedure using complex variables for finding the particular solution only, as presented in Subdivision 8-3.5.2. Also, the procedure for finding the complete solution to Eq. (8-148), where $f(t)$ is given by Eq. (8-152) and $x(t)$ is subjected to the initial conditions in Eqs. (8-150) and (8-151), is listed in Table 8-8. That is, Table 8-8 shows the procedure used in Subdivisions 8-3.5.1 and 8-3.5.3 for finding the complete solution, without the use of complex variables. In both Tables 8-7 and 8-8, the steps corresponding to Steps 1 through 4 of Table 8-1 are indicated. As with Tables 8-3 and 8-4, the purpose in providing Tables 8-7 and 8-8 is fourfold. The reader is referred to the discussions of Tables 8-3 and 8-4 in Subdivision 8-2.4.4.

8-3.5.5 *Closer Look at Particular Solution*

The harmonic excitation of the single-degree-of-freedom second-order system is the most frequently applied model in vibration analysis. Furthermore, since all physical systems exhibit some damping, the response of

---

[24]See Problem 8-87; also, see Table 8-9 for the results.

TABLE 8-7 Alternative Procedure Using Complex Variables to Obtain Particular Solution for Single-Degree-of-Freedom Second-Order System Subjected to Harmonic Excitation, $f(t) = a \cos \Omega t + b \sin \Omega t$

Step 1	Establish *complex* equation of motion as
	$$m\ddot{x} + c\dot{x} + kx = f_0 e^{i\Omega t}.$$

	Assume *complex* particular solution as
	$$x_p(t) = X_1 e^{i\Omega t}$$
	and evaluate complex X_1 such that $x_p(t)$ satisfies equation of motion; complex particular solution found.

Step 2	Separate real and imaginary parts of complex particular solution as
	$$x_p(t) = x_{pr}(t) + i x_{pi}(t).$$

	Real particular solution for
	$$f(t) = a \cos \Omega t + b \sin \Omega t$$
	found as
	$$x_p(t) = x_{pr}(t)\big|_{f_0=a} + x_{pi}(t)\big|_{f_0=b}.$$

such systems will be dominated by the particular solution after some time beyond initialization. Here, we examine the particular solution of underdamped systems subjected to harmonic excitation, reformatting it in nondimensional form. Next, we present several of the frequently encountered applications related to these results.

Consider the case where the harmonic excitation is given by either a cosinusoidal or sinusoidal function. Then, when $f(t) = f_0 \cos \Omega t$ (f_0 being the amplitude of the harmonic input force), the particular solution is given by Eq. (8-163) as

$$x_p(t) = \frac{f_0}{\sqrt{(k - m\Omega^2)^2 + (c\Omega)^2}} \cos(\Omega t - \phi) \qquad (8\text{-}200)$$

where we have set $a = f_0$, and where

$$\phi = \tan^{-1} \frac{c\Omega}{k - m\Omega^2} \qquad 0 \le \phi < \pi \qquad (8\text{-}201)$$

Also, when $f(t) = f_0 \sin \Omega t$ (f_0 being the amplitude of the harmonic input force), the particular solution is given by Eq. (8-164) as

$$x_p(t) = \frac{f_0}{\sqrt{(k - m\Omega^2)^2 + (c\Omega)^2}} \sin(\Omega t - \phi) \qquad (8\text{-}202)$$

where we have set $b = f_0$, and where ϕ in Eq. (8-202) is given by Eq. (8-201).

TABLE 8-8 Procedure to Obtain Complete Solution for Single-Degree-of-Freedom Second-Order System Subjected to Harmonic Excitation, $f(t) = a \cos \Omega t + b \sin \Omega t$, and Initial Conditions

Step 1
Establish equation of motion as
$$m\ddot{x} + c\dot{x} + kx = a \cos \Omega t + b \sin \Omega t$$
and specify initial conditions as
$$x(0) = x_0 \qquad \dot{x}(0) = v_0.$$

$\downarrow$

Step 2
Assume particular solution as
$$x_p(t) = A \cos \Omega t + B \sin \Omega t$$
and evaluate A and B such that $x_p(t)$ satisfies equation of motion; particular solution found.

$\downarrow$

Step 3
Recall general homogeneous solution as
$x_h(t) = C_1 \cos \omega_n t + C_2 \sin \omega_n t, \qquad\qquad \zeta = 0$
$x_h(t) = e^{-\zeta \omega_n t}(C_1 \cos \omega_d t + C_2 \sin \omega_d t), \qquad 0 < \zeta < 1$
$x_h(t) = C_1 e^{-\zeta \omega_n t} + C_2 t e^{-\zeta \omega_n t}, \qquad\qquad \zeta = 1$
$x_h(t) = C_1 e^{(-\zeta + \sqrt{\zeta^2 - 1})\omega_n t} + C_2 e^{(-\zeta - \sqrt{\zeta^2 - 1})\omega_n t}, \qquad \zeta > 1;$
and establish complete solution as
$$x(t) = x_p(t) + x_h(t).$$

$\downarrow$

Step 4
Impose initial conditions onto complete solution and evaluate C_1 and C_2 of homogeneous solution. Complete solution found.

Note that the amplitudes as well as the phase angles of the particular solutions given by Eqs. (8-200) and (8-202) are identical. Therefore, the same observations can be made regarding the amplitude and phase angle of the particular solutions, whether the input harmonic function is cosinusoidal or sinusoidal. Before we make several of these observations, it is convenient to rewrite the amplitude and phase of the particular solutions in nondimensional forms.

Denoting the amplitude of the particular solutions by X, Eqs. (8-200) and (8-202) can be rewritten, respectively, as

$$x_p(t) = X \cos(\Omega t - \phi) \tag{8-203}$$

$$x_p(t) = X \sin(\Omega t - \phi) \tag{8-204}$$

where

$$X = \frac{f_0}{\sqrt{(k - m\Omega^2)^2 + (c\Omega)^2}} \tag{8-205}$$

$$\phi = \tan^{-1} \frac{c\Omega}{k - m\Omega^2} \qquad 0 \le \phi < \pi \tag{8-206}$$

Dividing the numerators and the denominators of both Eqs. (8-205) and (8-206) by k gives

$$X = \frac{\dfrac{f_0}{k}}{\sqrt{\left(1 - \dfrac{m\Omega^2}{k}\right)^2 + \left(\dfrac{c\Omega}{k}\right)^2}} \tag{8-207}$$

and

$$\phi = \tan^{-1} \frac{\dfrac{c\Omega}{k}}{1 - \dfrac{m\Omega^2}{k}} \tag{8-208}$$

We may define

$$\frac{m\Omega^2}{k} = \frac{\Omega^2}{\omega_n^2} \equiv r^2 \tag{8-209}$$

which further leads to

$$\frac{c\Omega}{k} = \frac{c\Omega}{k} \cdot \frac{2m\omega_n}{2m\omega_n} = 2 \cdot \frac{c}{2m\omega_n} \cdot \frac{\Omega}{1} \cdot \frac{\omega_n}{1} \cdot \frac{m}{k}$$

$$= 2 \cdot \frac{c}{c_c} \cdot \frac{\Omega}{\omega_n} = 2\zeta\left(\frac{\Omega}{\omega_n}\right) = 2\zeta r \tag{8-210}$$

where Eqs. (8-83) and (8-84) have been used, and r is the excitation frequency divided by the natural frequency and is called the *frequency ratio*. Using Eqs. (8-209) and (8-210) in Eqs. (8-207) and (8-208) gives

$$\frac{X}{X_0} = \frac{1}{\sqrt{\left[1 - \left(\dfrac{\Omega}{\omega_n}\right)^2\right]^2 + \left[2\zeta\dfrac{\Omega}{\omega_n}\right]^2}} = \frac{1}{\sqrt{(1 - r^2)^2 + (2\zeta r)^2}} \tag{8-211}$$

and

$$\phi = \tan^{-1} \frac{2\zeta\left(\dfrac{\Omega}{\omega_n}\right)}{1 - \left(\dfrac{\Omega}{\omega_n}\right)^2} = \tan^{-1} \frac{2\zeta r}{1 - r^2} \qquad 0 \le \phi < \pi \tag{8-212}$$

where $X_0 \equiv \dfrac{f_0}{k}$ is called the *zero-frequency deflection* (or the *static deflection* due to f_0).

The ratio $\dfrac{X}{X_0}$ is called the *magnification factor* (or the *amplification factor* or the *system receptance*).

Therefore, when the harmonic excitation is given by the forcing function $f(t) = f_0 \cos \Omega t$ or $f(t) = f_0 \sin \Omega t$, the constant amplitude X of the particular solution is given by Eq. (8-211) and the constant phase lag ϕ of the particular solution is given by Eq. (8-212). We make the following observations about the amplitude and the phase lag ϕ of the particular solution when $f(t) = f_0 \cos \Omega t$ or $f(t) = f_0 \sin \Omega t$.

1.   When the input (force) amplitude is f_0, the output (displacement) amplitude is given by $\dfrac{f_0}{k\sqrt{(1-r^2)^2 + (2\zeta r)^2}}$. We note here that when $\zeta \approx 0$ and $r \approx 1$ (or, $\Omega \approx \omega_n$), the output becomes very large. Clearly, the modeling assumptions that led to the linear model in Figure 8-9 and Eq. (8-78) will require recurrent reassessment as the output displacement amplitude becomes increasingly large. As indicated following Eq. (8-171), *resonance* is a nontechnical word associated with the condition of large amplitudes when $\Omega \approx \omega_n$.

2.   The response frequency is the same as the excitation frequency.

3.   The phase angle of the response is $-\phi = -\tan^{-1}\left\{\dfrac{2\zeta r}{1-r^2}\right\}$, which means that the output *lags* the input by ϕ radians. Thus, the output peaks *after* the input peaks by a *phase lag* ϕ or equivalently by a time $t' = \dfrac{\phi}{\Omega}$.

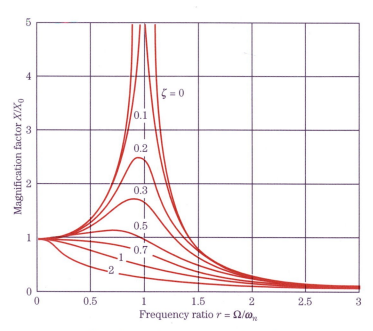

FIGURE 8-18   Magnification factor spectrum (magnification factor versus frequency ratio) for generic system sketched in Figure 8-9 with $f(t) = f_0 \sin \Omega t$ or $f(t) = f_0 \cos \Omega t$ and $X_0 = \dfrac{f_0}{k}$.

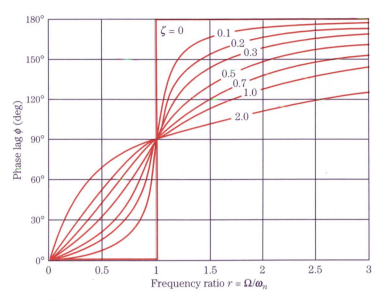

FIGURE 8-19 Phase lag versus frequency ratio for generic system sketched in Figure 8-9 with $f(t) = f_0 \sin \Omega t$ or $f(t) = f_0 \cos \Omega t$.

From the three observations above, it can be stated that the particular solution for the harmonic response depends upon three parameters: the zero-frequency deflection, the frequency ratio r, and the damping factor ζ. The linear dependence of the output upon the input amplitude f_0 is an inherent characteristic of linear systems: for example, when the input amplitude is doubled, the output amplitude is also doubled, or when the input amplitude is halved, the output amplitude is also halved. As indicated in observations 1 and 3 above, the frequency ratio r affects both the amplitude and the phase lag of the response. The effects of r on the amplitude and phase lag of the response, as given by Eqs. (8-211) and (8-212), are depicted graphically in Figures 8-18 and Figure 8-19, respectively, for several values of ζ. In particular, the graph of $\dfrac{X}{X_0}$ as a function of the frequency ratio r (Figure 8-18) is called the *magnification factor spectrum* (or the *amplification factor spectrum*). Note in Figure 8-19 that the phase lag ϕ ranges from 0 to π as specified in Eq. (8-212).

■ **Example 8-3:** Numerical Example of Particular Solution of Single-Degree-of-Freedom Second-Order System Subjected to Harmonic Force

Consider the single-degree-of-freedom second-order system sketched in Figure 8-9 whose parameters are given by $m = 10$ kg, $k = 1000$ N/m, $c = 50$ N $\cdot$ s/m. Find the particular solution for this system subjected to a harmonic force $f(t) = 100 \sin 12t$ N.

❑ **Solution:** Since the forcing function is a sinusoidal function, the particular solution is given by Eqs. (8-204), (8-211), and (8-212). In order to find X and ϕ from Eqs. (8-211) and (8-212), the frequency ratio, zero-frequency deflection and damping factor must be evaluated. The natural frequency of the system can be found from Eq. (8-84) as

$$\omega_n = \sqrt{\frac{k}{m}} = \sqrt{\frac{1000}{10}} = \sqrt{100} = 10\left(\frac{\text{rad}}{\text{s}}\right) \qquad \text{(a)}$$

The damping factor of the system can be calculated from Eq. (8-83) as

$$\zeta = \frac{c}{2m\omega_n} = \frac{50}{(2)(10)(10)} = 0.25 \qquad \text{(b)}$$

Therefore, the frequency ratio is

$$r = \frac{\Omega}{\omega_n} = \frac{12}{10} = 1.2 \qquad \text{(c)}$$

Substituting Eqs. (b) and (c) into Eq. (8-211) gives

$$\frac{X}{X_0} = \frac{1}{\sqrt{(1 - r^2)^2 + (2\zeta r)^2}} = \frac{1}{\sqrt{(1 - 1.2^2)^2 + ((2)(0.25)(1.2))^2}} = 1.34 \qquad \text{(d)}$$

Note that the zero-frequency deflection is given by

$$X_0 = \frac{f_0}{k} = \frac{100}{1000} = 0.1 \text{ (m)} \qquad \text{(e)}$$

Substitution of Eq. (e) into Eq. (d) gives

$$X = (1.34)(0.1) = 0.134 \text{ (m)} \qquad \text{(f)}$$

Also, substituting Eqs. (b) and (c) into Eq. (8-212) gives

$$\phi = \tan^{-1}\frac{2\zeta r}{1 - r^2}$$

$$= \tan^{-1}\frac{(2)(0.25)(1.2)}{1 - (1.2)^2} = \tan^{-1}\frac{0.6}{-0.44} = \tan^{-1}(-1.364) = 126.3° \qquad \text{(g)}$$

where the angle ϕ has been evaluated within the range specified in Eq. (8-212). Therefore, the particular solution can be expressed by substituting Eqs. (f) and (g) into Eq. (8-204) as

$$x_p(t) = X\sin(\Omega t - \phi) = 0.134\sin(12t - 126.3°) \text{ (m)} \qquad \text{(h)}$$

■ **Example 8-4:** **Is the System Static or Dynamic?**

Consider the system model in Figure 8-9. Under what harmonic loading conditions can a static analysis of the system be assessed to provide an adequate representation of the system's dynamic response?

❏ **Solution:** Equations (8-163) and (8-164) or Eqs. (8-182) and (8-183) or Eqs. (8-200) and (8-202) or Figures 8-18 and 8-19 are equivalent pairs of results that enable a quantitative assessment of the subtle question "Is the system static or dynamic?" In order to exploit the information in these equations and figures, it may have been necessary to have engaged in a substantial amount of modeling. That is, getting from the physical system to the model in Figure 8-9 may be difficult. Having arrived at such a model, quantitative evaluations of ω_n, ζ, and Ω must then be determined.

For models such as shown in Figure 8-9, it is frequently said that dynamics occurs when "inertia forces" (by which is meant $m\ddot{x}_{max}$) become comparable with "elastic forces" (by which is meant kx_{max}), where $\ddot{x}_{max}$ and x_{max} represent the maximum acceleration and displacement, respectively. For example, if for a model represented by $m\ddot{x} + kx = f(t)$ the accelerations are vanishingly small such that $|m\ddot{x}| \ll |kx|$, the characterization of the model clearly reduces to quasistatics; that is, the model is time-varying but effectively static. Indeed, this is substantially true; however, the presence of damping complicates such a simplistic characterization. So, let's consider the simpler undamped case in an attempt to address the question at hand.

The crux of the analysis is contained in Eqs. (8-211) and (8-212), which are reformulations of the results cited in the first paragraph of this solution. For $\zeta = 0$ we may rewrite Eq. (8-211) as

$$\frac{X}{X_0} = \frac{1}{1 - \left(\dfrac{\Omega}{\omega_n}\right)^2} = \frac{1}{1 - r^2} \tag{a}$$

where X is the amplitude of the dynamic response and X_0 is the amplitude of the static response. Then, the magnification factor $\dfrac{X}{X_0}$ in Figure 8-18 is a measure of the relative magnitude of the dynamic system displacement to the static displacement, the difference between X and X_0 being that portion of the response due to dynamics or "inertia effects." Also, the phase lag ϕ (Figure 8-19 and Eq. (8-212)) corresponds to a timewise system displacement response relative to the applied loading.

All time-varying loads, no matter how slowly they are applied, will produce some inertia effects. The practical issue is associated with the level at which we choose to consider those inertia effects to be significant. If, for example, an acceptable assessment based exclusively on statics is required to be within 10% of the corresponding dynamic analysis, Eq. (a) and Figure 8-18 deliver the criterion when $\zeta = 0$. Equation (a) or Figure 8-18 informs us that $\dfrac{X}{X_0} \leqslant 1.1$ when $r \leqslant 0.3$. That is, if the forcing frequency of $f(t)$ never exceeds $0.3\omega_n$, X of the dynamic analysis will not exceed X_0, the static displacement, by more than 10%; and thus the dynamic f_{max} in the spring will never exceed the static f_0 by more than 10%. If, on the other hand, the analytical and modeling considerations allow for the overlooking of dynamic effects provided the *undamped* system faithfully "follows" the forcing function, then Figure 8-19 informs us that $0 < r < 1$ (that is, Ω up to ω_n) is an allowable forcing frequency range.

Damping, of course, will modify all the discussions above. No longer can inertia and elastic forces be considered exclusively, as damping forces must also be reckoned. Nevertheless, Figures 8-18 and 8-19 (or Eqs. (8-211) and (8-212)) provide the manner in which damping alters the discussion. For example, as displayed in Figure 8-18, for $\zeta = 0.5$, the 10% acceptability in the previous paragraph would extend the frequency range of $r \approx 0.47$. That is, when $\zeta = 0.5$, a static analysis of the system will give displacements and spring forces that are within 10% of the corresponding dynamic displacements and forces, provided the excitation frequency is less than approximately $0.47\omega_n$.

Further, beyond $r = 1$ and for all values of damping factor, the question may appear to become substantially more complicated, but in reality it becomes trivial. The question may appear to be more complicated because despite the fact that there are substantial inertia forces acting within the system, $\dfrac{X}{X_0} < 1$ for sufficiently large r. Indeed, as shown in Figure 8-18, regardless of the level of damping, when $r \gg 1$, $\dfrac{X}{X_0}$ becomes much less than unity. However, even though $\dfrac{X}{X_0}$ may be small, the inertia forces, given by $m\ddot{x}_{max} = m\Omega^2 X$, are large because of the large values of Ω. Thus, the presence of substantial inertia forces, implying that a significant fraction of the system's energy is at some times kinetic, is sufficient to conclude that the system is dynamic. After all, such energy has the capacity of being converted into useful or damaging work. In any event, the question becomes trivial because despite the frequent desirability of operating at $r > 1$ where $\dfrac{X}{X_0} < 1$, it is always necessary to get to that high value of r by passing through the region around $r = 1$, the region where the dynamic response of the system can do enormous harm. Thus, irrespective of the low values of $\dfrac{X}{X_0}$ for $r \gg 1$, such cases must be considered as dynamic with all the associated design concerns.

In summary, the question "Is the system static or dynamic?" may be answered as follows. The system may be considered to be statically excited in the frequency range $0 < r < r_0$, where $r_0 < 1$ *and* r_0 is determined in accordance with an accuracy criterion due to the neglect of dynamic effects. More importantly, this example makes the point that all moving systems do not have to be treated as dynamical; such time-varying analyses where inertia effects can be neglected are called *quasistatic*.

8-3.5.6 Transmissibility and Isolation

In many engineering systems, it is desirable to diminish the transmission of force or motion from one region of the system to other regions. For example, the transmission of the "vibration" or "vibratory motion" of a piece of machinery or an engine into its supporting structure (building, ships, or automobiles) may be highly undesirable. Or, the transmission of the "vibration" of a vehicle during its traverse of a cobble roadway or irregular track to its passengers may be uncomfortable. The former example is associated with *transmissibility;* the latter example is associated with *isolation.* Figure 8-20 illustrates the concepts of transmissibility and isolation, for harmonic excitation.

In Figure 8-20a a mass spring dashpot model is excited by a harmonic force $f(t) = f_0 \sin \Omega t$. The details of the force $F_T(t)$ that is transmitted into the foundation may not be of interest; often only a single measure is necessary for an adequate vibration assessment. Such a measure is the transmissibility (TR), defined as the ratio of the maximum transmitted force $(F_T)_{max}$ to the maximum excitation force f_0. In Figure 8-20b a mass spring dashpot model is excited by a base harmonic excitation $x_0(t) = A_0 \sin \Omega t$. Again, the details of the motion $x(t)$ that is transmitted to the structure (or passenger in a vehicle) may not be of interest; often only a single measure is necessary for an adequate vibration assessment. The appropriate measure in this case is the isolation (IS), defined as the steady-state ratio of the maximum transmitted displacement $(x)_{max}$ to the maximum excitation displacement A_0.

As revealed in the analyses to follow, the analytical expressions for transmissibility and isolation are identical. Perhaps this is the reason why these two important yet different concepts are sometimes interchangeably and confusingly cited.

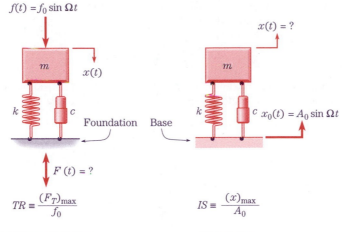

(a) Transmissibility.

(b) Isolation.

FIGURE 8-20   Concepts of transmissibility and isolation illustrated for harmonic excitation.

■ **Example 8-5:**   Transmissibility (of Force)

Consider the mass spring dashpot model sketched in Figure 8-20a, where the mass is excited by a harmonic force $f(t) = f_0 \sin \Omega t$. Find the transmissibility (TR) for this system; that is, the ratio of the maximum force transmitted into the foundation in steady state to the amplitude of the excitation force f_0.

❏ **Solution:**   The transmitted force is due to the combined forces in the spring k and the dashpot c. That is,

$$F_T(t) = kx + c\dot{x} \tag{a}$$

Because the system is damped, the steady-state response of the system to the harmonic forcing $f(t) = f_0 \sin \Omega t$ is given by Eqs. (8-161) and (8-162) with $a = 0$ and $b = f_0$. That is,

$$x(t) = X \sin(\Omega t - \phi) \tag{b}$$

where

$$X = \frac{f_0}{\sqrt{(k - m\Omega^2)^2 + (c\Omega)^2}} \tag{c}$$

and

$$\phi = \tan^{-1} \frac{c\Omega}{k - m\Omega^2} \qquad 0 \le \phi < \pi \tag{d}$$

Substituting Eq. (b) into Eq. (a) gives

$$F_T = kX \sin(\Omega t - \phi) + c\Omega X \cos(\Omega t - \phi)$$
$$= \sqrt{(kX)^2 + (c\Omega X)^2} \sin(\Omega t - \phi + \alpha) \tag{e}$$

where the two trigonometric functions in Eq. (e) have been combined to yield a single trigonometric function[25] and where

$$\alpha = \tan^{-1} \frac{c\Omega}{k} \qquad 0 \le \alpha < \frac{\pi}{2} \tag{f}$$

Thus, the amplitude or maximum magnitude of the transmitted force is given by the coefficient of the sinusoidal function in Eq. (e), and substituting Eq. (c) into this coefficient gives

$$(F_T)_{\max} = \sqrt{k^2 + (c\Omega)^2}\, X = \frac{\sqrt{k^2 + (c\Omega)^2}\, f_0}{\sqrt{(k - m\Omega^2)^2 + (c\Omega)^2}} \tag{g}$$

Dividing both the numerator and the denominator in Eq. (g) by the spring constant k gives

$$(F_T)_{\max} = \frac{\sqrt{1 + \left(\dfrac{c\Omega}{k}\right)^2}\, f_0}{\sqrt{\left(1 - \dfrac{m\Omega^2}{k}\right)^2 + \left(\dfrac{c\Omega}{k}\right)^2}} = \frac{\sqrt{1 + (2\zeta r)^2}}{\sqrt{(1 - r^2)^2 + (2\zeta r)^2}}\, f_0 \tag{h}$$

where Eqs. (8-209) and (8-210) have been used.

The steady-state ratio of the maximum transmitted force to the maximum excitation force f_0 can be found from Eq. (h) as

$$\mathrm{TR} \equiv \frac{(F_T)_{\max}}{f_0} = \frac{\sqrt{1 + (2\zeta r)^2}}{\sqrt{(1 - r^2)^2 + (2\zeta r)^2}} \tag{i}$$

which is defined as the *transmissibility,* and denoted by TR. The transmissibility given by Eq. (i) is a function of r and ζ only. Figure 8-21 shows plots of TR for different values of ζ as a function of r. It immediately follows from Figure 8-21 that the condition $r > \sqrt{2}$ must be met in order that TR < 1, which means the transmitted force amplitude is less than the excitation force amplitude. Therefore, in order to achieve transmitted force reduction—often called "suppression"—it is important to design the spring constant k such that ω_n satisfies the condition that $r = \dfrac{\Omega}{\omega_n} > \sqrt{2} \left(\text{or, } \omega_n = \sqrt{\dfrac{k}{m}} < \dfrac{\Omega}{\sqrt{2}}\right)$ for a given mass m and a specified dominant forcing frequency Ω.

■ **Example 8-6:** A Numerical Example of Transmissibility

Suppose a machine is mounted on an elastic bearing, which in turn sits on a rigid foundation. The bearing damping is negligible. In operation the machine generates a harmonic force having a frequency of 1000 rpm. If the mass of the machine is 50 kg, find the condition on the equivalent spring constant k of the elastic bearing for suppression of the transmitted force. Also, find the percentage of the dynamic force, generated by the machine, that is transmitted into the foundation if the stiffness of the bearing is $k = 200$ kN/m.

---

[25]See Section H-6 in Appendix H.

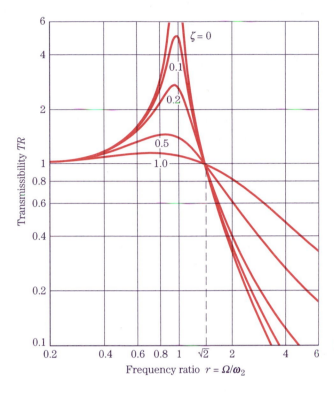

FIGURE 8-21 Transmissibility as function of frequency ratio, for various values of damping factor.

☐ **Solution:** The condition for suppression of the transmitted force is given by $r = \dfrac{\Omega}{\omega_n} > \sqrt{2}$ or

$$\sqrt{\frac{k}{m}} < \frac{\Omega}{\sqrt{2}} \tag{a}$$

The forcing frequency is $\Omega = 1000 \text{ rpm} = (1000)(2\pi)/(60) = 104.7 \text{ rad/s}$. Substituting the values of Ω and m into Eq. (a) gives

$$\sqrt{\frac{k}{m}} < \frac{104.7}{\sqrt{2}} = 74.05 \text{ (rad/s)} \tag{b}$$

or

$$k < m(74.05)^2 = (50)(74.05)^2 = 274 \text{ (kN/m)} \tag{c}$$

which is the desired condition on the stiffness of the bearing.

When the stiffness of the foundation is $k = 200 \text{ kN/m}$, the natural frequency of the machine-bearing system is given by

$$\omega_n = \sqrt{\frac{k}{m}} = \sqrt{\frac{200000}{50}} = 63.2 \text{ (rad/s)} \tag{d}$$

Therefore, the frequency ratio is $r = \dfrac{\Omega}{\omega_n} = \dfrac{(104.7)}{(63.2)} = 1.657$. Substituting this value of r and $\zeta = 0$ into Eq. (i) in Example 8-5 yields

$$\mathrm{TR} = \frac{1}{\sqrt{(1 - r^2)^2}} = \frac{1}{\sqrt{(1 - 1.657^2)^2}} = 0.573 \tag{e}$$

Therefore, 57.3% of the machine-generated dynamic force is transmitted into the foundation. An assessment of whether this is an adequate reduction must be based upon information not provided in the problem statement.

■ **Example 8-7:** Isolation (of Motion)

Consider the mass spring dashpot system sketched in Figure 8-20b, where the mass of the system is connected to a base via a spring and dashpot and the displacement of the base has a constant magnitude and is time harmonic. Derive an expression for the isolation (IS) for this system.

❑ **Solution:** Denoting the generalized coordinate for this system as the inertial displacement of the mass as given by $x(t)$ and denoting the inertial specified displacement of the base by $x_0(t)$, as sketched in Figure 8-20b, the equation of motion for the model is

$$m\ddot{x} + c(\dot{x} - \dot{x}_0) + k(x - x_0) = 0 \tag{a}$$

Upon rearrangement, Eq. (a) becomes

$$m\ddot{x} + c\dot{x} + kx = c\dot{x}_0 + kx_0 \tag{b}$$

We note that the displacement $x_0(t)$ is defined by

$$x_0(t) = A_0 \sin \Omega t \tag{c}$$

Substituting Eq. (c) into Eq. (b) gives

$$m\ddot{x} + c\dot{x} + kx = c\Omega A_0 \cos \Omega t + kA_0 \sin \Omega t \tag{d}$$

Note that the equation of motion in Eq. (d) represents a single-degree-of-freedom second-order system subjected to a harmonic forcing in the form of Eq. (8-152). Therefore, the steady-state motion of this damped system is given by Eqs. (8-161) and (8-162) with $a = c\Omega A_0$ and $b = kA_0$. That is,

$$x(t) = \frac{1}{\sqrt{(k - m\Omega^2)^2 + (c\Omega)^2}}\{c\Omega A_0 \cos(\Omega t - \phi) + kA_0 \sin(\Omega t - \phi)\} \tag{e}$$

where

$$\phi = \tan^{-1}\frac{c\Omega}{k - m\Omega^2} \qquad 0 \le \phi < \pi \tag{f}$$

Equation (e) can be rewritten[26] as

$$x(t) = X \sin(\Omega t - \phi + \alpha) \tag{g}$$

---

[26]See Section H-6 in Appendix H.

where

$$X = \frac{A_0 \sqrt{k^2 + (c\Omega)^2}}{\sqrt{(k - m\Omega^2)^2 + (c\Omega)^2}} \tag{h}$$

and

$$\alpha = \tan^{-1} \frac{c\Omega}{k} \qquad 0 \le \alpha < \frac{\pi}{2} \tag{i}$$

Thus, the magnitude or the maximum displacement of the steady-state response is X given by Eq. (h). Dividing both the numerator and the denominator in Eq. (h) by the spring constant k and using Eqs. (8-209) and (8-210) give

$$X = \frac{A_0 \sqrt{1 + \left(\frac{c\Omega}{k}\right)^2}}{\sqrt{\left(1 - \frac{m\Omega^2}{k}\right)^2 + \left(\frac{c\Omega}{k}\right)^2}} = \frac{\sqrt{1 + (2\zeta r)^2}}{\sqrt{(1 - r^2)^2 + (2\zeta r)^2}} A_0 \tag{j}$$

Therefore, the steady-state ratio of the magnitude of the transmitted displacement to the magnitude of the excitation displacement is given by

$$\text{IS} \equiv \frac{(x)_{\max}}{A_0} = \frac{\sqrt{1 + (2\zeta r)^2}}{\sqrt{(1 - r^2)^2 + (2\zeta r)^2}} \tag{k}$$

which is defined as the *isolation,* denoted by IS.

The isolation given by Eq. (k) is a function of r and ζ only, and its expression is indeed identical to the expression for TR given by Eq. (i) in Example 8-5. Due to the identity of the expressions for IS and TR, the condition for suppression of motion transmission is the same as that for suppression of force transmission. That is, the condition $r > \sqrt{2}$ must be met in order that IS < 1, which means that the transmitted displacement amplitude is less than the excitation displacement amplitude. Therefore, in order to achieve transmitted displacement reduction, the spring constant k needs to be designed such that ω_n satisfies the condition that $r = \dfrac{\Omega}{\omega_n} > \sqrt{2}\left(\text{or, } \omega_n = \sqrt{\dfrac{k}{m}} < \dfrac{\Omega}{\sqrt{2}}\right)$ for a given mass m and a specified dominant excitation frequency Ω.

■ **Example 8-8:**   Isolation of Traversing Vehicle

Consider the vehicle suspension system modeled in Figure 8-22. The vertical motion of the vehicle of mass m relative to the wheel is restrained by the supporting spring of spring constant k and a viscous shock absorber of dashpot constant c. The wheel follows an idealized surface, assumed to be spatially harmonic with amplitude A_0 and wavelength l (crest-to-crest distance). We want to find the maximum vertical displacement of the vehicle m relative to the maximum undulation of the surface A_0, when the vehicle is moving over this idealized surface at a constant forward speed v.

❏ **Solution:**   If the motion of the wheel is denoted by $x_0(t)$, it follows that

$$x_0(t) = A_0 \sin \Omega t \tag{a}$$

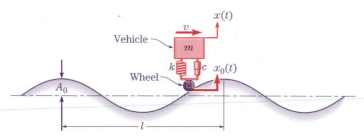

FIGURE 8-22 Idealized model of vehicle moving over undulating surface.

where we note that for the idealized surface

$$\Omega t = (2\pi f)t = \left(2\pi \frac{v}{l}\right)t \tag{b}$$

because the wheel passes successive crests (hills) at the rate or cyclic frequency $f = \frac{v}{l}$. Therefore, the forcing frequency is $\Omega = 2\pi f = \frac{2\pi v}{l}$. The ratio of the maximum vertical displacement of the vehicle relative to the maximum undulation of the surface A_0 is found by Eq. (k) in Example 8-7 when the following equivalent parameters are used:

$$r = \frac{\Omega}{\omega_n} = \frac{\frac{2\pi v}{l}}{\sqrt{\frac{k}{m}}} \tag{c}$$

$$\zeta = \frac{c}{2\sqrt{mk}} \tag{d}$$

where Eq. (8-83) has been used in writing Eq. (d).

This particular example is characteristic of the most common type of elementary isolation analysis. Clearly, such issues as passenger "ride quality" are associated with the concept of isolation.

■ **Example 8-9:** A Numerical Example of Isolation

Reconsider the mounted instrument package of Example 8-1. If the aircraft engines are operating at 3000 rpm, find the percentage of the displacement generated by the engines in the structure of the aircraft at the site of the instruments that is transmitted to the instrument package.

❏ **Solution:** From Example 8-1 the natural frequency of the mounted instrument package is $57.18\,\mathrm{rad/s}$. The forcing frequency is $\Omega = 3000$ rpm $= \dfrac{(3000)(2\pi)}{60} = 314.2\,\mathrm{rad/s}$, and the frequency ratio is $r = \dfrac{\Omega}{\omega_n} = \dfrac{314.2}{57.18} = 5.495$. Therefore, noting that $\zeta = 0$, Eq. (k) in Example 8-7 gives

$$\mathrm{IS} = \frac{1}{\sqrt{(1-r^2)^2}} = 0.034 \tag{a}$$

That is, only approximately 3.4% of the displacement caused by the engines in the aircraft at the mounting site will be transmitted to the instrument package.

8-3.5.7 Harmonic Response of Electrical System
Now, consider an electrical system. It will be demonstrated that the forced response of an electrical system to a harmonic input can be obtained via the same procedures used for mechanical systems.

■ **Example 8-10:** Harmonic Response of Active Band-Pass Filter

Consider the active band-pass filter model presented in Example 7-7. From the formulation in Example 7-7, the output voltage $e_0(t)$ is governed by the equation of motion

$$(R_1 R_2 C_1 C_2)\ddot{e}_0 + (R_1 C_1 + R_2 C_2)\dot{e}_0 + e_0 = -R_2 C_1 \dot{e}_s \tag{a}$$

where the R_i's and C_i's are the resistances and capacitances sketched in Figure 7-16 and $e_s(t)$ is the input voltage.

Find the forced response of the output voltage $e_0(t)$ when the input voltage is harmonic. For such a device, the primary interest is in its input–output characteristics; thus, the input–output characteristics should be sketched.

☐ **Solution:**

First, note that Eq. (a) is of the same form as Eq. (8-148), which is emphasized by rewriting Eq. (a) as

$$m_{eq}\ddot{e}_0 + c_{eq}\dot{e}_0 + k_{eq}e_0 = f_{eq}(t) \tag{b}$$

where m_{eq}, c_{eq}, k_{eq}, and f_{eq} are given by

$$m_{eq} = R_1 R_2 C_1 C_2 \tag{c}$$

$$c_{eq} = R_1 C_1 + R_2 C_2 \tag{d}$$

$$k_{eq} = 1 \tag{e}$$

$$f_{eq}(t) = -R_2 C_1 \dot{e}_s \tag{f}$$

where the subscript "eq" stands for "equivalent." The particular solution of Eq. (b) can be found by following the procedure of Subdivision 8-3.5.2.[27] When the input voltage is harmonic, in analogy with Eq. (8-175) for nonnegative time, let

$$e_s(t) = E_s e^{i\Omega t} \tag{g}$$

where E_s is the constant amplitude of the harmonic input voltage having frequency Ω.

So, substituting Eq. (g) into Eq. (f) and then substituting that result into Eq. (b) yield

$$m_{eq}\ddot{e}_0 + c_{eq}\dot{e}_0 + K_{eq}e_0 = -i\Omega R_2 C_1 E_s e^{i\Omega t} \tag{h}$$

The particular solution of Eq. (h) can be found by *assuming* the form of the output voltage as

$$e_{0p}(t) = E_0 e^{i\Omega t} \tag{i}$$

---

[27]The procedure utilizing complex functions is adopted here because they are more commonly used in the analysis of electrical circuits. However, the procedure of Subdivision 8-3.5.1 can also be used.

Substituting Eq. (i) into Eq. (h) and canceling the common $e^{i\Omega t}$ term yield

$$(k_{eq} - m_{eq}\Omega^2)E_0 + ic_{eq}\Omega E_0 = -i\Omega R_2 C_1 E_s \qquad (j)$$

or

$$E_0 = \frac{-i\Omega R_2 C_1}{(k_{eq} - m_{eq}\Omega^2) + ic_{eq}\Omega} E_s \qquad (k)$$

Multiplying the numerator and the denominator of Eq. (k) by the complex conjugate of the denominator, $[(k_{eq} - m_{eq}\Omega^2) - ic_{eq}\Omega]$, and separating the resulting expression into its real and imaginary parts give

$$E_0 = -\Omega R_2 C_1 E_s \left[\frac{c_{eq}\Omega}{(k_{eq} - m_{eq}\Omega^2)^2 + c_{eq}^2\Omega^2} + \frac{i(k_{eq} - m_{eq}\Omega^2)}{(k_{eq} - m_{eq}\Omega^2)^2 + c_{eq}^2\Omega^2} \right] \qquad (l)$$

Equation (l) can be rewritten as[28]

$$E_0 = \frac{-\Omega R_2 C_1 E_s}{\sqrt{(k_{eq} - m_{eq}\Omega^2)^2 + (c_{eq}\Omega)^2}} e^{i\phi} \qquad (m)$$

where

$$\phi = \tan^{-1}\frac{k_{eq} - m_{eq}\Omega^2}{c_{eq}\Omega} \qquad \frac{\pi}{2} \geq \phi > -\frac{\pi}{2} \qquad (n)$$

where the range of the angle ϕ follows from the conditions $m_{eq}, k_{eq}, c_{eq} > 0$ and $\Omega \geq 0$. That is, when $\Omega = 0$, $\phi = \tan^{-1}\infty = \frac{\pi}{2}$; as Ω increases to $\Omega = \sqrt{\frac{k_{eq}}{m_{eq}}}$, ϕ decreases monotonically to 0 (when $\Omega = \sqrt{\frac{k_{eq}}{m_{eq}}}$, $\phi = \tan^{-1} 0 = 0$); and, as Ω increases further toward ∞, ϕ decreases monotonically further toward $-\frac{\pi}{2}$ $\left(\text{as } \Omega \to \infty, \phi \to \tan^{-1}(-\infty) = -\frac{\pi}{2} \right)$. Therefore, substituting Eq. (m) into Eq. (i) gives

$$e_{0p}(t) = \frac{-\Omega R_2 C_1 E_s}{\sqrt{(k_{eq} - m_{eq}\Omega^2)^2 + (c_{eq}\Omega)^2}} e^{i(\Omega t + \phi)} \qquad (o)$$

where ϕ is given by Eq. (n).

In order to examine the input–output characteristics of the active band-pass filter, the amplitude ratio of the output voltage to the input voltage is considered in detail. The ratio of the output to the input may be obtained by dividing Eq. (o) by Eq. (g); that is,

$$\frac{e_{0p}(t)}{e_s(t)} \equiv H(\Omega)$$

$$= \frac{-\Omega R_2 C_1 E_s e^{i(\Omega t + \phi)}}{\sqrt{(k_{eq} - m_{eq}\Omega^2)^2 + (c_{eq}\Omega)^2}} \bigg/ \left(E_s e^{i\Omega t} \right)$$

$$= \frac{-\Omega R_2 C_1 e^{i\phi}}{\sqrt{(k_{eq} - m_{eq}\Omega^2)^2 + (c_{eq}\Omega)^2}} \qquad (p)$$

[28]See Section H-4 in Appendix H. Also, see Problem 8-54.

where the conventional notation $H(\Omega)$ for harmonic output–input ratio has been used. The amplitude ratio of the output to the input may be obtained by taking the magnitude of the complex quantity $H(\Omega)$ in Eq. (p). That is,

$$
|H(\Omega)| = \left| \frac{-\Omega R_2 C_1}{\sqrt{(k_{eq} - m_{eq}\Omega^2)^2 + (c_{eq}\Omega)^2}} \right| \cdot \left| e^{i\phi} \right|
$$

$$
= \frac{\Omega R_2 C_1}{\sqrt{(k_{eq} - m_{eq}\Omega^2)^2 + (c_{eq}\Omega)^2}} \tag{q}
$$

where the identity $|a \cdot b| = |a| \cdot |b|$ for two complex quantities a and b and, in accordance with Eqs. (H-12) and (H-21) in Appendix H, $\left| e^{i\phi} \right| = 1$ have been used.

The radicand of the denominator in Eq. (q) can be simplified via Eqs. (c) through (e). That is,

$$
\begin{aligned}
(k_{eq} - m_{eq}\Omega^2)^2 &+ (c_{eq}\Omega)^2 \\
&= k_{eq}^2 + \left(c_{eq}^2 - 2m_{eq}k_{eq} \right)\Omega^2 + m_{eq}^2\Omega^4 \\
&= 1 + \left\{ (R_1C_1 + R_2C_2)^2 - 2(R_1C_1R_2C_2) \right\}\Omega^2 + R_1^2R_2^2C_1^2C_2^2\Omega^4 \\
&= 1 + \left\{ R_1^2C_1^2 + R_2^2C_2^2 \right\}\Omega^2 + R_1^2R_2^2C_1^2C_2^2\Omega^4 \\
&= \left(1 + \Omega^2R_1^2C_1^2 \right)\left(1 + \Omega^2R_2^2C_2^2 \right)
\end{aligned} \tag{r}
$$

Substitution of Eq. (r) into Eq. (q) gives

$$
\begin{aligned}
|H(\Omega)| &= \frac{\Omega R_2 C_1}{\sqrt{\left(1 + \Omega^2R_1^2C_1^2 \right)\left(1 + \Omega^2R_2^2C_2^2 \right)}} \\
&= \frac{R_2}{R_1} \frac{\Omega R_1 C_1}{\sqrt{1 + \Omega^2R_1^2C_1^2}} \frac{1}{\sqrt{1 + \Omega^2R_2^2C_2^2}} \\
&= \frac{R_2}{R_1} \left[\frac{\Omega\tau_1}{\sqrt{1 + (\Omega\tau_1)^2}} \right]\left[\frac{1}{\sqrt{1 + (\Omega\tau_2)^2}} \right]
\end{aligned} \tag{s}
$$

where new symbols, $\tau_1 \equiv R_1C_1$ and $\tau_2 \equiv R_2C_2$, have been introduced for simplification. Note that τ_1 and τ_2 have the unit of time.

Now, we examine the spectral (that is, frequency) characteristics of the output–input ratio by examining separately the behavior of the two terms in the square brackets in Eq. (s). The term in the first set of square brackets in Eq. (s) exhibits the spectral characteristic shown in the log-log plot of Figure 8-23, which can be approximated by considering the low-frequency ($\Omega\tau_1 \ll 1$) and high-frequency ($\Omega\tau_1 \gg 1$) asymptotes. That is, when $\Omega\tau_1 \ll 1$, the term is approximately equal to $\Omega\tau_1$ whereas when $\Omega\tau_1 \gg 1$, it is approximately equal to unity. Such characteristics as sketched in Figure 8-23 are typical of a *high-pass filter* because it passes (with very modest or negligible attenuation) only high-frequency signals and suppresses low-frequency signals. Similarly, the term in the second set of square brackets in Eq. (s) exhibits the spectral characteristic shown in the log-log plot of Figure 8-24, which can be approximated by considering the low-frequency ($\Omega\tau_2 \ll 1$) and high-frequency ($\Omega\tau_2 \gg 1$) asymptotes. That is, when $\Omega\tau_2 \ll 1$, the term is approximately equal to unity whereas when $\Omega\tau_2 \gg 1$, it is approximately equal to $\frac{1}{(\Omega\tau_2)}$. Such characteristics as sketched in Figure 8-24 are typical of a *low-pass filter* because it passes (with very

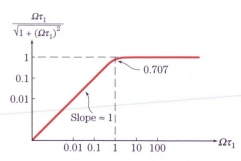

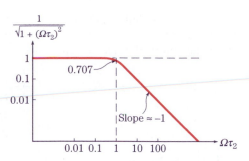

FIGURE 8-23   High-pass filter characteristic of $\dfrac{\Omega\tau_1}{\sqrt{1+(\Omega\tau_1)^2}}$.

FIGURE 8-24   Low-pass filter characteristic of $\dfrac{1}{\sqrt{1+(\Omega\tau_2)^2}}$.

modest or negligible attenuation) only low-frequency signals and suppresses high-frequency signals.

If the circuit were designed such that $\left(\dfrac{1}{\tau_2}\right) \gg \left(\dfrac{1}{\tau_1}\right)$, the output–input amplitude ratio given by Eq. (s) would exhibit the characteristics of a *band-pass filter*. These spectral characteristics can be found by combining Figures 8-23 and 8-24, as sketched in Figure 8-25. The flat region between $\dfrac{1}{\tau_1}$ and $\dfrac{1}{\tau_2}$ is called the *passband* because the filter (approximately) uniformly passes only those signals in this frequency range and suppresses those outside this frequency range. Note from Eq. (s) that if $\left(\dfrac{R_2}{R_1}\right) > 1$, this filter has a gain that is greater than unity within the passband. Therefore, this active filter circuit can be used as a frequency-selective amplifier.

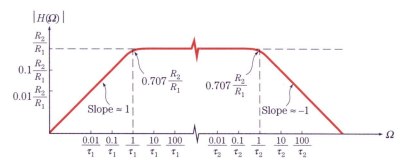

FIGURE 8-25   Band-pass filter characteristic of $|H(\Omega)|$.

8-3.6   Summary of Responses for Single-Degree-of-Freedom Second-Order Systems

In Subsections 8-3.1 and 8-3.5, the free and harmonic responses, respectively, of the generic single-degree-of-freedom second-order system were obtained. In summary, the free and harmonic responses are listed in Table 8-9.

TABLE 8-9 Summary of Free and Harmonic Responses for Single-Degree-of-Freedom Second-Order System, $m\ddot{x} + c\dot{x} + kx = f(t)$, Subject to Initial Conditions, $x(0) = x_0$ and $\dot{x}(0) = v_0$

Forcing function, $f(t)$, $t \geq 0$	Particular solution, $x_p(t)$	Homogeneous solution, $x_h(t)$
0 (No input)	—	*Undamped* ($\zeta = 0$): $$x_0 \cos \omega_n t + \left(\frac{v_0}{\omega_n}\right) \sin \omega_n t$$ *Underdamped* ($0 < \zeta < 1$): $$e^{-\zeta \omega_n t}\left(x_0 \cos \omega_d t + \frac{v_0 + \zeta \omega_n x_0}{\omega_d} \sin \omega_d t\right)$$ *Critically damped* ($\zeta = 1$): $$e^{-\omega_n t}\{x_0 \cos \omega_d t + (w_n x_0 + v_0)t\}$$ *Overdamped* ($\zeta > 1$): $$e^{-\zeta \omega_n t}\left\{x_0 \cosh\left(\sqrt{\zeta^2 - 1}\,\omega_n t\right) + \left(\frac{\zeta \omega_n x_0 + v_0}{\omega_n \sqrt{\zeta^2 - 1}}\right) \sinh\left(\sqrt{\zeta^2 - 1}\,\omega_n t\right)\right\}$$
$a \cos \Omega t + b \sin \Omega t$ (a, b, Ω: constant) (Harmonic input)	*Undamped* ($\zeta = 0$) *and* $\Omega \neq \omega_n$: $$\frac{a \cos \Omega t + b \sin \Omega t}{k - m\Omega^2}$$ *Undamped* ($\zeta = 0$) *and* $\Omega = \omega_n$: $$-\frac{bt}{2m\omega_n} \cos \omega_n t + \frac{at}{2m\omega_n} \sin \omega_n t$$ *Damped* ($\zeta > 0$): $$\frac{a \cos(\Omega t - \phi) + b \sin(\Omega t - \phi)}{\sqrt{(k - m\Omega^2)^2 + (c\Omega)^2}}$$	*Undamped* ($\zeta = 0$) *and* $\Omega \neq \omega_n$: $$\left\{x_0 - \frac{a}{k - m\Omega^2}\right\} \cos \omega_n t + \left\{\frac{v_0}{\omega_n} - \frac{b\Omega}{\omega_n(k - m\Omega^2)}\right\} \sin \omega_n t$$ *Undamped* ($\zeta = 0$) *and* $\Omega = \omega_n$: $$x_0 \cos \omega_n t + \left(\frac{v_0}{\omega_n} - \frac{b}{2m\omega_n^2}\right) \sin \omega_n t$$ *Underdamped* ($0 < \zeta < 1$): $$e^{-\zeta \omega_n t}\left[\left(x_0 - \frac{a\cos\phi - b\sin\phi}{\sqrt{(k-m\Omega^2)^2 + (c\Omega)^2}}\right)\cos\omega_d t\right.$$ $$\left.+ \frac{1}{\omega_d}\left\{v_0 + \zeta\omega_n x_0 - \frac{a(\Omega\sin\phi + \zeta\omega_n\cos\phi) - b(\zeta\omega_n\sin\phi - \Omega\cos\phi)}{\sqrt{(k-m\Omega^2)^2 + (c\Omega)^2}}\right\}\sin\omega_d t\right]$$ *Critically damped* ($\zeta = 1$): $$e^{-\omega_n t}\left[\left(x_0 - \frac{a\cos\phi - b\sin\phi}{\sqrt{(k-m\Omega^2)^2 + (c\Omega)^2}}\right)\right.$$ $$\left.+ \left\{v_0 + \omega_n x_0 - \frac{a\cos\phi - b\sin\phi}{\sqrt{(k-m\Omega^2)^2 + (c\Omega)^2}} - \frac{a(\Omega\sin\phi + \omega_n\cos\phi) - b(\zeta\omega_n\sin\phi - \Omega\cos\phi)}{\sqrt{(k-m\Omega^2)^2 + (c\Omega)^2}}\right\}t\right]$$ *Overdamped* ($\zeta > 1$): $$e^{-\zeta\omega_n t}\left[\left(x_0 - \frac{a\cos\phi - b\sin\phi}{\sqrt{(k-m\Omega^2)^2 + (c\Omega)^2}}\right)\cosh\left(\sqrt{\zeta^2-1}\,\omega_n t\right)\right.$$ $$+ \frac{1}{\omega_n\sqrt{\zeta^2-1}}\left\{v_0 + \zeta\omega_n x_0 \right.$$ $$\left.\left.- \frac{a(\Omega\sin\phi + \zeta\omega_n\cos\phi) - b(\zeta\omega_n\sin\phi - \Omega\cos\phi)}{\sqrt{(k-m\Omega^2)^2 + (c\Omega)^2}}\right\}\sinh\left(\sqrt{\zeta^2-1}\,\omega_n t\right)\right]$$

Note that $\omega_n = \sqrt{\dfrac{k}{m}}$; $\zeta = \dfrac{c}{2m\omega_n}$; $\omega_d = \omega_n\sqrt{1 - \zeta^2}$; $\phi = \tan^{-1}\left[\dfrac{c\Omega}{(k - m\Omega^2)}\right]$, $0 \leq \phi < \pi$.

In the first column of Table 8-9, the forcing function $f(t)$ for nonnegative time is specified for each case; in the second column, the particular solution is given for the harmonic response; and, in the third column, the homogeneous solution is given for each of the free and harmonic responses for the various categories of system damping. Therefore, the complete solution for the harmonic response may be easily obtained by summing the corresponding formulas in the second and third columns.

Recall that the complete solution for a harmonic excitation was derived in Subsection 8-3.5 only for the undamped case whereas the free (homogeneous) response was derived in Subsection 8-3.1 for all categories of system damping. The harmonic response solutions for all categories of *damped* systems are listed in Table 8-9 and the reader is urged to derive them by the procedure in Subsection 8-3.5.[29]

Further, since all the coefficients of both the particular solutions and the homogeneous solutions have been evaluated for general initial conditions and substituted into the formulas of Table 8-9, the particular, homogeneous, and complete solutions can be obtained for a specific system model by substituting the system parameters, excitation parameters, and initial conditions into the specific formula(s) of interest.

8-4 TWO-DEGREE-OF-FREEDOM SECOND-ORDER SYSTEMS

The model that we adopt for our discussion is sketched in Figure 8-26. If the generalized coordinates are selected as $x_1(t)$ and $x_2(t)$ which denote the inertial displacements from equilibrium of m_1 and m_2, respectively, the equations of motion are easily derived[30] as

$$m_1\ddot{x}_1(t) + (c_1 + c_2)\dot{x}_1(t) - c_2\dot{x}_2(t) + (k_1 + k_2)x_1(t) - k_2x_2(t) = F_1(t) \quad (8\text{-}213)$$

$$m_2\ddot{x}_2(t) - c_2\dot{x}_1(t) + (c_2 + c_3)\dot{x}_2(t) - k_2x_1(t) + (k_2 + k_3)x_2(t) = F_2(t) \quad (8\text{-}214)$$

Equations (8-213) and (8-214) are simultaneous *coupled* second-order linear differential equations. These equations are coupled because $x_2(t)$ and $\dot{x}_2(t)$ appear in Eq. (8-213), and $x_1(t)$ and $\dot{x}_1(t)$ appear in Eq. (8-214). Terms such as $-c_2\dot{x}_2(t)$ and $-k_2x_2(t)$ in Eq. (8-213) and $-c_2\dot{x}_1(t)$ and $-k_2x_1(t)$ in Eq. (8-214) are called *coupling terms*. Here, coupling suggests that the motion of m_1 affects the motion of m_2 and the motion of m_2 affects the motion of m_1, in both cases through the k_2 and c_2 *coupling elements*.

The equations of motion in Eqs. (8-213) and (8-214) can be rewritten in the slightly abbreviated and more general form for two-degree-of-freedom systems as

$$m_1\ddot{x}_1 + c_{11}\dot{x}_1 + c_{12}\dot{x}_2 + k_{11}x_1 + k_{12}x_2 = F_1(t) \quad (8\text{-}215)$$

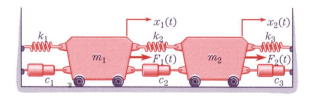

FIGURE 8-26 Model of two-degree-of-freedom system.

[29]Also see Problem 8-87.

[30]Or, see Problem 5-53.

$$m_2\ddot{x}_2 + c_{21}\dot{x}_1 + c_{22}\dot{x}_2 + k_{21}x_1 + k_{22}x_2 = F_2(t) \tag{8-216}$$

where, for the specific system in Figure 8-26, we can identify the new terms in Eqs. (8-215) and (8-216) as

$$c_{11} = c_1 + c_2 \qquad c_{22} = c_2 + c_3 \qquad c_{12} = c_{21} = -c_2 \tag{8-217}$$

$$k_{11} = k_1 + k_2 \qquad k_{22} = k_2 + k_3 \qquad k_{12} = k_{21} = -k_2 \tag{8-218}$$

In accordance with the theory of ordinary differential equations, in order to determine the solution to Eqs. (8-215) and (8-216), four initial conditions are needed. Those initial conditions may be most generally chosen as the initial displacements and initial velocities of the two generalized coordinates and they are given as

$$x_1(0) = x_{10} \tag{8-219}$$

$$x_2(0) = x_{20} \tag{8-220}$$

$$\dot{x}_1(0) = v_{10} \tag{8-221}$$

$$\dot{x}_2(0) = v_{20} \tag{8-222}$$

where the constants x_{10}, v_{10}, x_{20}, and v_{20} are known values of the corresponding system variables at time t equals zero.

In the following subsections, the free response and harmonic response of the system governed by Eqs. (8-215) and (8-216) are considered with emphasis on undamped systems where all the c_{ij}'s are zero. For the free response, we develop in Subsection 8-4.2 the concept of natural modes, which provide the general form for the free undamped response solution with unknown coefficients; coefficients which in turn are determined in Subsection 8-4.3 in accordance with the initial conditions. In Subsection 8-4.4, we develop solution techniques for the harmonic response and explore the dynamic characteristics of two-degree-of-freedom systems subjected to harmonic excitations. Again, the emphasis will be on undamped systems, but not exclusively so.

8-4.1   Natural Modes of Vibration

We examine the *free undamped* equations of motion. As indicated previously, in this context, *free* means $F_1(t) = F_2(t) = 0$, and *undamped* means $c_{11} = c_{12} = c_{21} = c_{22} = 0$ in Eqs. (8-215) and (8-216). The outcome of this examination will be the calculation of intrinsic properties of the system, somewhat analogous to the natural frequency for single-degree-of-freedom systems but distinctly more elaborate. We shall find that a two-degree-of-freedom system not only has *two natural frequencies,* but such a system also has *two natural mode shapes* that describe the spatial configuration of motion of the system at each of the two natural frequencies. The concept of a mode shape does not arise in single-degree-of-freedom systems and it is likely to be a new idea for most readers; however, it is an important concept, comparable in significance to a natural frequency. A *natural mode of vibration* will be expressed quantitatively as the combination of a natural frequency and its corresponding natural mode shape.

Substantially for reasons cited in the preceding paragraph, the analysis of two-degree-of-freedom systems is significantly more difficult than the analysis of single-degree-of-freedom systems. Indeed, because of issues involving natural mode shapes, multiple natural frequencies, and coupling (as described below), two-degree-of-freedom systems are conceptually more like twenty- or thirty- or even higher degree-of-freedom systems than one-degree-of-freedom systems. So, in exploring two-degree-of-freedom systems, we introduce the major concepts that are applicable in the analyses of systems having an arbitrary number of degrees of freedom. Thus, in the limited presentation of this single section, we restrict our interests to only a few—but very important—aspects of the vibration of two-degree-of-freedom systems.

In the absence of both external forces and damping, Eqs. (8-215) and (8-216) reduce to

$$m_1 \ddot{x}_1(t) + k_{11} x_1(t) + k_{12} x_2(t) = 0 \qquad (8\text{-}223)$$
$$m_2 \ddot{x}_2(t) + k_{21} x_1(t) + k_{22} x_2(t) = 0 \qquad (8\text{-}224)$$

Equations (8-223) and (8-224) are the free undamped equations of motion, which will be the form used in our analyses of the natural modes of vibration.

We say that a system is vibrating in a *natural mode* when during free undamped vibration its coordinates $x_1(t)$ and $x_2(t)$ execute the identical timewise periodic motion—motion that is sometimes called *synchronous motion*. This is a characterization of a natural mode of vibration for which we shall be greatly rewarded. Accepting this characterization, we make a series of interrelated observations.

1. Equations (8-223) and (8-224) are homogeneous ordinary differential equations.
2. If $x_1(t)$ and $x_2(t)$ are solutions to Eqs. (8-223) and (8-224), $C_0 x_1(t)$ and $C_0 x_2(t)$ are also solutions, where C_0 is an arbitrary constant.
3. Thus, solutions to Eqs. (8-223) and (8-224) can be found only to within a multiplicative constant.
4. Furthermore, if a natural mode of vibration (in which $x_1(t)$ and $x_2(t)$ execute synchronous motion) exists, the ratio $\dfrac{x_2(t)}{x_1(t)}$ must be time-independent. That is, whenever either $x_1(t)$ or $x_2(t)$ experiences its maximum (or minimum) value, so does the other.
5. Thus, when the system is vibrating in a natural mode, the geometric or spatial configuration of the system remains constant, but its amplitude changes. So, only the ratio $\dfrac{x_2(t)}{x_1(t)}$ can be determined explicitly.

One or more of these five observations may appear to be moderately obscure although each follows from purely mathematical considerations of Eqs. (8-223) and (8-224). Nevertheless, there are also distinct physical interpretations associated with these observations and we shall indicate them in the analyses and examples in this section.

Based on the above observations, solutions to Eqs. (8-223) and (8-224) may be expressed as

$$x_1(t) = u_1 g(t) \qquad (8\text{-}225)$$
$$x_2(t) = u_2 g(t) \qquad (8\text{-}226)$$

where u_1 and u_2 are constant amplitudes and $g(t)$ represents the common timewise function. It can be shown that during a natural mode of vibration (or simply, a *natural mode*), the function $g(t)$ must be harmonic,[31] and that Eqs. (8-225) and (8-226) can be rewritten as

---

[31]See Eq. (I-18) in Appendix I. There is an exceptional case where $g(t)$ is not harmonic; this case will be discussed later in this subsection, for instance, in Example 8-13 and Subdivision 8-4.2.2.

$$x_1(t) = u_1 \cos(\omega t - \psi) \tag{8-227}$$

$$x_2(t) = u_2 \cos(\omega t - \psi) \tag{8-228}$$

where ω is the harmonic frequency of the motion and ψ is a constant angle. Both ω and ψ are the same for both coordinates $x_1(t)$ and $x_2(t)$, and later in this section both will be interpreted physically.

Substitution of Eqs. (8-227) and (8-228) into Eqs. (8-223) and (8-224) and cancellation of the common cosine term give

$$(k_{11} - m_1\omega^2)u_1 + k_{12}u_2 = 0 \tag{8-229}$$

$$k_{12}u_1 + (k_{22} - m_2\omega^2)u_2 = 0 \tag{8-230}$$

Equations (8-229) and (8-230) are simultaneous homogeneous linear algebraic equations in the unknown amplitudes u_1 and u_2. If Eqs. (8-229) and (8-230) have solutions, then Eqs. (8-227) and (8-228) are solutions to Eqs. (8-223) and (8-224).

An attempt to solve Eqs. (8-229) and (8-230) by Cramer's rule for either u_1 or u_2 would lead to a zero in the numerator divided by the determinant of the coefficients of Eqs. (8-229) and (8-230). Readers should convince themselves of this point. Thus, in order for Eqs. (8-229) and (8-230) to represent a solution other than the trivial solution $u_1 = u_2 = 0$, the determinant of the coefficients of Eqs. (8-229) and (8-230) (that is, the denominator in Cramer's rule) must be zero; so

$$\Delta(\omega^2) = \det \begin{bmatrix} k_{11} - m_1\omega^2 & k_{12} \\ k_{12} & k_{22} - m_2\omega^2 \end{bmatrix} = 0 \tag{8-231}$$

where $\Delta(\omega^2)$ is called the *characteristic determinant*. Expanding the characteristic determinant leads to

$$\Delta(\omega^2) = m_1 m_2 \omega^4 - (m_1 k_{22} + m_2 k_{11})\omega^2 + k_{11}k_{22} - k_{12}^2 = 0 \tag{8-232}$$

which is called the *characteristic equation* and which is a quadratic equation in ω^2. The roots of the characteristic equation are given by

$$\begin{matrix} \omega_1^2 \\ \omega_2^2 \end{matrix} = \frac{m_1 k_{22} + m_2 k_{11}}{2 m_1 m_2} \mp \frac{1}{2}\sqrt{\left(\frac{m_1 k_{22} + m_2 k_{11}}{m_1 m_2}\right)^2 - 4\frac{k_{11}k_{22} - k_{12}^2}{m_1 m_2}} \tag{8-233}$$

where we emphasize that ω_1^2 is associated with the negative sign on the square root and ω_2^2 is associated with the positive sign. This choice is by convention, making ω_1^2 less than (or equal to) ω_2^2.

Note that it is possible for the last term in the square root of Eq. (8-233) to vanish; that is, $k_{11}k_{22} - k_{12}^2 = 0$, leading to $\omega_1 = 0$. The natural modes of vibration for such a case are explored in Example 8-13 and in the discussion following Example 8-13. Note also that it is possible for the square root to vanish, leading to *repeated roots;* however, our limited presentation does not allow us to pursue this primarily analytical topic. Furthermore, repeated roots are modeling artifacts that can be designed out of models of (real) physical systems by the introduction of small changes in the idealized elemental properties that would otherwise give rise to the repeated roots.

From Eq. (8-233) we conclude that when the system vibrates in a natural mode, only two frequencies are possible: ω_1 and ω_2, which are called the *natural frequencies.* The natural frequencies ω_1 and ω_2 are unique intrinsic properties of the system. (The quantities ω_1^2 and ω_2^2 obtained from Eq. (8-233) are called *eigenvalues*.)

Returning to consideration of the amplitudes u_1 and u_2, we note from Eqs. (8-229) and (8-230) that they depend on the natural frequencies ω_1 and ω_2. We designate $u_1^{(1)}$ and $u_2^{(1)}$ as the values of u_1 and u_2 that correspond to ω_1, and $u_1^{(2)}$ and $u_2^{(2)}$ as the values of u_1 and u_2 that correspond to ω_2. The superscripts in parentheses on u_1 and u_2 do not represent powers of exponentiation, but simply designate the natural frequency to which they correspond. Also, recall from the fifth observation following Eq. (8-224) that only the ratios $\dfrac{u_2^{(1)}}{u_1^{(1)}}$ and $\dfrac{u_2^{(2)}}{u_1^{(2)}}$ can be determined explicitly. Again, this follows directly from the fact that Eqs. (8-223) and (8-224) are homogeneous.

Substituting ω_1^2 and ω_2^2 into Eqs. (8-229) and (8-230) and defining the corresponding *amplitude ratios* as r_1 and r_2, respectively, give

$$r_1 \equiv \frac{u_2^{(1)}}{u_1^{(1)}} = -\frac{k_{11} - m_1\omega_1^2}{k_{12}} = -\frac{k_{12}}{k_{22} - m_2\omega_1^2} \tag{8-234}$$

$$r_2 \equiv \frac{u_2^{(2)}}{u_1^{(2)}} = -\frac{k_{11} - m_1\omega_2^2}{k_{12}} = -\frac{k_{12}}{k_{22} - m_2\omega_2^2} \tag{8-235}$$

In writing Eqs. (8-234) and (8-235), we have used both of Eqs. (8-229) and (8-230), for each ω separately. That is, to calculate r_1, either of Eqs. (8-229) or (8-230) can be used, and indeed both are shown in Eq. (8-234). The same value of r_1 will be obtained from either of the expressions given in Eq. (8-234). Similarly, either expression in Eq. (8-235) may be used to obtain r_2. Thus, r_1 gives the relative amplitude of u_2 and u_1 when the system is vibrating at ω_1; and r_2 gives the relative amplitude of u_2 and u_1 when the system is vibrating at ω_2. The geometric configuration of the system at ω_1 is often written in matrix form as

$$\{u^{(1)}\} = \left\{ \begin{array}{c} u_1^{(1)} \\ u_2^{(1)} \end{array} \right\} = u_1^{(1)} \left\{ \begin{array}{c} 1 \\ r_1 \end{array} \right\} \tag{8-236}$$

and the geometric configuration of the system at ω_2 is often written in matrix form as

$$\{u^{(2)}\} = \left\{ \begin{array}{c} u_1^{(2)} \\ u_2^{(2)} \end{array} \right\} = u_1^{(2)} \left\{ \begin{array}{c} 1 \\ r_2 \end{array} \right\} \tag{8-237}$$

where $\{u^{(1)}\}$ and $\{u^{(2)}\}$ are called the *mode shapes* (or *modal vectors* or *eigenvectors*). In all terms in Eqs. (8-236) and (8-237), the superscripts indicate the *natural mode* number. Unlike the natural frequencies that are unique for a system, the modal vectors are not unique. While the spatial configuration of the system vibrating at either of the natural frequencies is unique, different generalized coordinates in which such a configuration is expressed will generally give rise to different modal vectors.

We summarize by emphasizing that for a two-degree-of-freedom system, the system is characterized by two natural modes of vibration, each consisting of its natural frequency and its mode shape. We abridge our results as follows:

$$\begin{array}{lll} \text{Natural Mode 1:} & \omega_1, & \{u^{(1)}\} \\ \text{Natural Mode 2:} & \omega_2, & \{u^{(2)}\} \end{array} \tag{8-238}$$

We emphasize that the results contained in Eq. (8-238) are *intrinsic properties* of the system model. Nothing in Eq. (8-238) implies anything about the forcing of the system, the system's initial conditions, or any other possible excitation of the system; Eq. (8-238) contains only information describing intrinsic properties of the system model. The natural frequencies are unique for *any* set of generalized coordinates; the mode shapes are unique for *any specific* set of generalized coordinates, but, in general, depend on the choice of generalized coordinates. We shall see that the calculation of these properties will be enormously beneficial in understanding the vibration response of the system, in general, and especially the free response due to initial conditions. Furthermore, as we shall explore in Subsection 8-4.2, any free undamped response of the system can be obtained as a combination of these two modes. This statement is supported mathematically because the two modes in Eq. (8-238) both satisfy the equations of motion, Eqs. (8-223) and (8-224).

■ **Example 8-11:**   Natural Modes of Mass Spring System

Find the natural modes of vibration for the two-degree-of-freedom mass spring system sketched in Figure 8-27.

□ **Solution:**   The equations of motion for the system are given by

$$2m\ddot{x}_1 + 3kx_1 - kx_2 = 0 \tag{a}$$

$$m\ddot{x}_2 - kx_1 + 2kx_2 = 0 \tag{b}$$

where the generalized coordinates, $x_1(t)$ and $x_2(t)$, are the inertial displacements from the equilibrium positions of masses $2m$ and m, respectively. Equations (a) and (b) may be obtained either by the techniques of Chapter 5 or by particularizing Eqs. (8-213) and (8-214) to the system model in Figure 8-27. Comparing Eqs. (a) and (b) with Eqs. (8-223) and (8-224), respectively, gives

$$m_1 = 2m \qquad m_2 = m \tag{c}$$

and

$$k_{11} = 3k \qquad k_{12} = k_{21} = -k \qquad k_{22} = 2k \tag{d}$$

Substitution of Eqs. (c) and (d) into Eq. (8-233) gives

$$\begin{aligned}
\omega_1^2 \atop \omega_2^2 &= \frac{(2m)(2k) + (m)(3k)}{2(2m)(m)} \mp \frac{1}{2}\sqrt{\left[\frac{(2m)(2k) + (m)(3k)}{(2m)(m)}\right]^2 - 4\frac{(3k)(2k) - (-k)^2}{(2m)(m)}} \\
&= \left\{ \frac{7}{4} \mp \frac{1}{2}\sqrt{\left(\frac{7}{2}\right)^2 - 10} \right\} \frac{k}{m}
\end{aligned} \tag{e}$$

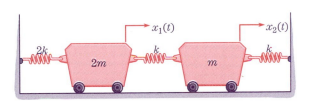

FIGURE 8-27   Model of free undamped two-degree-of-freedom system.

Then

$$\omega_1^2 = \frac{k}{m} \qquad \omega_2^2 = \frac{5k}{2m} \qquad \text{(f)}$$

or

$$\omega_1 = \sqrt{\frac{k}{m}} \qquad \omega_2 = 1.58\sqrt{\frac{k}{m}} \qquad \text{(g)}$$

where ω_1 and ω_2 are the natural frequencies of the system.

Substitution of Eqs. (c), (d), and (f) into Eqs. (8-234) and (8-235) yields

$$r_1 \equiv \frac{u_2^{(1)}}{u_1^{(1)}} = -\frac{3k - (2m)\left(\dfrac{k}{m}\right)}{-k} = -\frac{-k}{2k - (m)\left(\dfrac{k}{m}\right)} = 1 \qquad \text{(h)}$$

$$r_2 \equiv \frac{u_2^{(2)}}{u_1^{(2)}} = -\frac{3k - (2m)\left(\dfrac{5k}{2m}\right)}{-k} = -\frac{-k}{2k - (m)\left(\dfrac{5k}{2m}\right)} = -2 \qquad \text{(i)}$$

Therefore, the modal vectors, according to Eqs. (8-236) and (8-237), are

$$\{u^{(1)}\} = \left\{ \begin{array}{c} u_1^{(1)} \\ u_2^{(1)} \end{array} \right\} = u_1^{(1)} \left\{ \begin{array}{c} 1 \\ r_1 \end{array} \right\} = \left\{ \begin{array}{c} 1 \\ 1 \end{array} \right\} \qquad \text{(j)}$$

$$\{u^{(2)}\} = \left\{ \begin{array}{c} u_1^{(2)} \\ u_2^{(2)} \end{array} \right\} = u_1^{(2)} \left\{ \begin{array}{c} 1 \\ r_2 \end{array} \right\} = \left\{ \begin{array}{c} 1 \\ -2 \end{array} \right\} \qquad \text{(k)}$$

where $u_1^{(1)}$ and $u_1^{(2)}$ have been arbitrarily set equal to unity. Recall that the modal vectors can be determined only to within a multiplicative constant; so by arbitrarily setting $u_1^{(1)}$ and $u_1^{(2)}$ to unity, we use that fact to circumvent having to carry these arbitrary constants through the subsequent discussions. The natural modes are summarized in Figure 8-28, in which the mode shapes are sketched in plots having the linear extent of the system drawn along the horizontal axis and the horizontal displacements of the masses drawn along the vertical axis. Note that the "linear extent" of models such as that in Figure 8-27 is an indeterminate quantity as such models have no defined dimensions; nevertheless, horizontal "linear extent" is a useful concept for interpreting mode shape diagrams.

Discussion

In preparation for the discussion of the results obtained in this example, we recall that the force-deformation relation for a linearly elastic rod subjected to longitudinal loading is

$$F = \frac{AE}{L}\delta \qquad \text{(l)}$$

where F is longitudinal force (positive for tension), δ is elongation (for tensile force), A is cross-sectional area, E is modulus of elasticity, and L is undeformed length. Equation (l) can be written in the form of a constitutive relation for a linear spring as

$$F = k\delta \qquad \text{(m)}$$

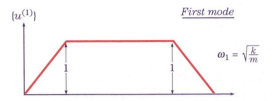

(a) Mode shape and natural frequency for first mode.

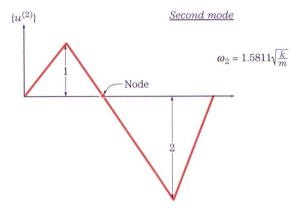

(b) Mode shape and natural frequency for second mode.

FIGURE 8-28 Natural modes of vibration for system in Figure 8-27.

where the spring constant is given by

$$k = \frac{AE}{L} \tag{n}$$

As sketched in Figure 8-28a, the first natural mode, or simply the *first mode,* consists of a mode shape *and* a natural frequency. The mode shape is a schematic of the spatial configuration of the system when it is vibrating freely at ω_1. As indicated in Eq. (j) where $u_1^{(1)}$ was arbitrarily set to unity, the amplitudes of the two masses are normalized with respect to the mass located by $x_1(t)$. Again, such an arbitrary normalization is consistent with the third observation following Eq. (8-224). Furthermore, as indicated by the fourth and fifth observations following Eq. (8-224), the ratio of $\dfrac{x_2(t)}{x_1(t)}$ as given by $\{u^{(1)}\}$ remains constant at unity as the two masses oscillate back and forth at ω_1 in the first mode.

Pause and reflect upon the physical meaning of the first mode. What is the coupling (middle) spring doing during the first mode of vibration? Other than going along for the ride, it is doing nothing. As such, it could be replaced by a massless rigid rod (Figure 8-29a) or simply removed (Figure 8-29b). Thus, we note that the two parts of the system in Figure 8-29b may be thought to be single-degree-of-freedom subsystems that do not communicate, and the subsystem natural frequency for each part of the system must be consistent with ω_1 in Eq. (f). Clearly, the natural frequency of each subsystem in Figure 8-29b is ω_1. Furthermore, from Fig 8-29a, the two masses oscillate in unison, apparently as a single mass of $3m$ and working against the two outer springs having a combined stiffness of $3k$. Again, as a hypothetical single-degree-of-freedom system, this leads to $\omega_n^2 = \omega_1^2 = \dfrac{3k}{3m} = \dfrac{k}{m}$.

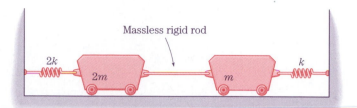

(a) Coupling spring replaced by massless rigid rod.

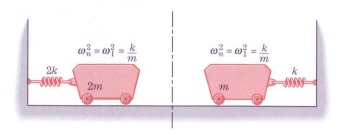

(b) Coupling spring removed.

FIGURE 8-29 Schematics of hypothetical models for system in Figure 8-27 vibrating in first mode.

When the system is vibrating freely in the *second mode,* its frequency will be ω_2 and its spatial configuration will be in accordance with the corresponding mode shape in Figure 8-28b. As indicated in Eq. (k) where $u_1^{(2)}$ was arbitrarily set to unity, the amplitudes of the two masses are normalized with respect to the mass located by $x_1(t)$. As expressed in Eq. (k) and Figure 8-28b, the ratio of $\dfrac{x_2(t)}{x_1(t)}$ as given by $\{u^{(2)}\}$ remains constant at minus two as the two masses oscillate back and forth at ω_2 in the second mode.

What is the coupling spring doing during the second mode of vibration? As indicated by the mode shape (Figure 8-28b) and the modal vector (Eq. (k)), the two masses move in opposition such that there is a node in the coupling spring. The node—a point which does not move—is located one-third of the way between the mass $2m$ and the mass m. (This distance may be determined from similar triangles in the mode shape of Figure 8-28b.) Thus, a hypothetical model for the system may be sketched as in Figure 8-30. As indicated in Eq. (n), the stiffness of a specific spring whose length is altered is inversely proportional to its length. So the stiffness of a segment of a spring of length $\dfrac{L}{3}$ where the original stiffness and length were k and L, respectively, is $3k$. Similarly, the stiffness of a segment of length $\dfrac{2L}{3}$ of an original spring k of length L is $\dfrac{3k}{2}$. The corresponding hypothetical single-degree-of-freedom subsystems in Figure 8-30 both give $\omega_n^2 = \omega_2^2 = \dfrac{5k}{2m}$, which is consistent with ω_2^2 in Eq. (f).

While the physical discussions that led to the hypothetical single-degree-of-freedom subsystem models are not necessary in analyzing such systems, they are beneficial as an aid to understanding the numerical results. Furthermore, such physical insight can be useful in the rapid evaluation of systems possessing some symmetries, and for the approximate checking of numerical results.

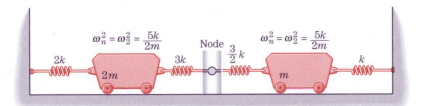

FIGURE 8-30 Schematic of hypothetical model for system in Figure 8-27 vibrating in second mode.

■ **Example 8-12:** Natural Modes of Lumped-Parameter Model of Multistory Building

In many engineering applications, multistory buildings may be modeled as lumped-parameter mass spring systems. For example, in one of the most frequently used models, a four-story building may be modeled as sketched in Figure 8-31a. The elements m_i's represent the masses of the floors plus walls, and the springs k_i's represent the interstory lateral stiffness due primarily to the walls. Such models are often used in the earthquake and wind analyses of multistory buildings. (See Example 4-5.)

For simplicity, consider such a model for a two-story building as sketched in Figure 8-31b. Find the natural modes of vibration for the system when $m_2 = \dfrac{m_1}{2} = \dfrac{m}{2}$ and $k_2 = \dfrac{k_1}{2} = \dfrac{k}{2}$.

□ **Solution:** The equations of motion for the system are given by

$$m\ddot{x}_1 + \frac{3}{2}kx_1 - \frac{k}{2}x_2 = 0 \tag{a}$$

$$\frac{m}{2}\ddot{x}_2 - \frac{k}{2}x_1 + \frac{k}{2}x_2 = 0 \tag{b}$$

where $x_1(t)$ and $x_2(t)$ are the generalized coordinates representing the horizontal inertial displacements from the equilibrium positions of the masses m_1 and m_2, respectively.

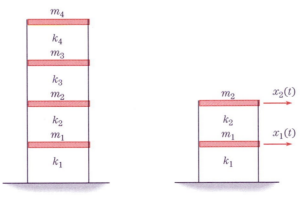

(a) Model of four-story building.   (b) Model of two-story building.

FIGURE 8-31 Lumped-parameter models of multistory buildings.

Equations (a) and (b) may be obtained by the techniques of Chapter 5 or by appropriately reducing the equations of motion in Example 4-5. Furthermore, comparing Eqs. (a) and (b) with Eqs. (8-223) and (8-224), respectively, gives

$$m_1 = m \qquad m_2 = \frac{m}{2} \tag{c}$$

and

$$k_{11} = \frac{3k}{2} \qquad k_{12} = k_{21} = -\frac{k}{2} \qquad k_{22} = \frac{k}{2} \tag{d}$$

Substitution of Eqs. (c) and (d) into Eq. (8-233) gives

$$
\begin{aligned}
\omega_1^2 \atop \omega_2^2 &= \frac{(m)\left(\frac{k}{2}\right) + \left(\frac{m}{2}\right)\left(\frac{3k}{2}\right)}{2(m)\left(\frac{m}{2}\right)} \mp \frac{1}{2}\sqrt{\left[\frac{(m)\left(\frac{k}{2}\right) + \left(\frac{m}{2}\right)\left(\frac{3k}{2}\right)}{(m)\left(\frac{m}{2}\right)}\right]^2 - 4\frac{\left(\frac{3k}{2}\right)\left(\frac{k}{2}\right) - \left(-\frac{k}{2}\right)^2}{(m)\left(\frac{m}{2}\right)}} \\
&= \left\{\frac{5}{4} \mp \frac{1}{2}\sqrt{\left(\frac{5}{4}\right)^2 - 4}\right\}\frac{k}{m}
\end{aligned}
\tag{e}
$$

Then

$$\omega_1^2 = \frac{k}{2m} \qquad \omega_2^2 = \frac{2k}{m} \tag{f}$$

or

$$\omega_1 = 0.707\sqrt{\frac{k}{m}} \qquad \omega_2 = 1.414\sqrt{\frac{k}{m}} \tag{g}$$

where ω_1 and ω_2 are the natural frequencies of the system.

Substitution of Eqs. (c), (d), and (f) into Eqs. (8-234) and (8-235) yields

$$r_1 \equiv \frac{u_2^{(1)}}{u_1^{(1)}} = -\frac{\frac{3k}{2} - (m)\left(\frac{k}{2m}\right)}{-\left(\frac{k}{2}\right)} = -\frac{-\left(\frac{k}{2}\right)}{\frac{k}{2} - \left(\frac{m}{2}\right)\left(\frac{k}{2m}\right)} = 2 \tag{h}$$

$$r_2 \equiv \frac{u_2^{(2)}}{u_1^{(2)}} = -\frac{\frac{3k}{2} - (m)\left(\frac{2k}{m}\right)}{-\left(\frac{k}{2}\right)} = -\frac{-\left(\frac{k}{2}\right)}{\frac{k}{2} - \left(\frac{m}{2}\right)\left(\frac{2k}{m}\right)} = -1 \tag{i}$$

Therefore, the modal vectors, according to Eqs. (8-236) and (8-237), are

$$\{u^{(1)}\} = \left\{ \begin{matrix} u_1^{(1)} \\ u_2^{(1)} \end{matrix} \right\} = u_1^{(1)}\left\{ \begin{matrix} 1 \\ r_1 \end{matrix} \right\} = \left\{ \begin{matrix} 1 \\ 2 \end{matrix} \right\} \tag{j}$$

$$\{u^{(2)}\} = \left\{ \begin{matrix} u_1^{(2)} \\ u_2^{(2)} \end{matrix} \right\} = u_1^{(2)}\left\{ \begin{matrix} 1 \\ r_2 \end{matrix} \right\} = \left\{ \begin{matrix} 1 \\ -1 \end{matrix} \right\} \tag{k}$$

where, as discussed in Example 8-11, $u_1^{(1)}$ and $u_1^{(2)}$ have been arbitrarily set equal to unity. The natural modes are summarized in Figure 8-32. The mode shapes in Figure 8-32 are sketched such that the vertical extent of the system is plotted vertically and the horizontal displacements of the system are plotted horizontally. The vertical line in both Figures 8-32a and 8-32b may be thought to represent the undeformed vertical centerline of the model, or actually any vertical undeformed line of the model.

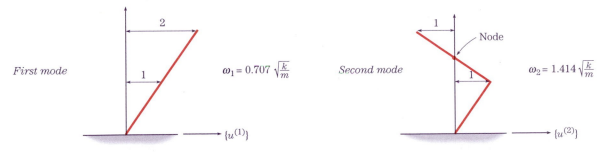

First mode $\omega_1 = 0.707\sqrt{\frac{k}{m}}$ *Second mode* $\omega_2 = 1.414\sqrt{\frac{k}{m}}$

(a) Mode shape and natural frequency for first mode.   (b) Mode shape and natural frequency for second mode.

FIGURE 8-32   Natural modes of vibration for system in Figure 8-31.

■ **Example 8-13:**   ***Semidefinite*** Mass Spring System

Consider the mass spring system sketched in Figure 8-33. Find the natural modes of vibration for the system.

❑ **Solution:**   The equations of motion for the system are given by

$$2m\ddot{x}_1 + kx_1 - kx_2 = 0 \qquad\text{(a)}$$
$$m\ddot{x}_2 - kx_1 + kx_2 = 0 \qquad\text{(b)}$$

where the generalized coordinates, $x_1(t)$ and $x_2(t)$, are the inertial displacements from the equilibrium positions of masses $2m$ and m, respectively. Comparing Eqs. (a) and (b) with Eqs. (8-223) and (8-224), respectively, gives

$$m_1 = 2m \qquad m_2 = m \qquad\text{(c)}$$

and

$$k_{11} = k \qquad k_{12} = k_{21} = -k \qquad k_{22} = k \qquad\text{(d)}$$

Substitution of Eqs. (c) and (d) into Eq. (8-233) gives

$$\begin{aligned}\omega_1^2 \\ \omega_2^2\end{aligned} = \frac{(2m)(k) + (m)(k)}{2(2m)(m)} \mp \frac{1}{2}\sqrt{\left[\frac{(2m)(k) + (m)(k)}{(2m)(m)}\right]^2 - 4\frac{(k)(k) - (-k)^2}{(2m)(m)}}$$

$$= \left\{\frac{3}{4} \mp \frac{1}{2}\sqrt{\left(\frac{3}{2}\right)^2 - 0}\right\}\frac{k}{m} \qquad\text{(e)}$$

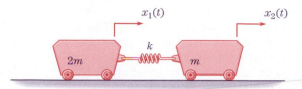

FIGURE 8-33   Semidefinite system.

Then

$$\omega_1^2 = 0 \qquad \omega_2^2 = \frac{3k}{2m} \tag{f}$$

or

$$\omega_1 = 0 \qquad \omega_2 = 1.225\sqrt{\frac{k}{m}} \tag{g}$$

where ω_1 and ω_2 are the natural frequencies of the system. Note that one of the two natural frequencies is zero ($\omega_1 = 0$), the physical meaning of which will be discussed shortly. Nevertheless, the procedure for finding the modal vectors also remains unchanged.

Substitution of Eqs. (c), (d), and (f) into Eqs. (8-234) and (8-235) yields

$$r_1 \equiv \frac{u_2^{(1)}}{u_1^{(1)}} = -\frac{k - (2m)(0)}{-k} = -\frac{-k}{k - (m)(0)} = 1 \tag{h}$$

$$r_2 \equiv \frac{u_2^{(2)}}{u_1^{(2)}} = -\frac{k - (2m)\left(\dfrac{3k}{2m}\right)}{-k} = -\frac{-k}{k - (m)\left(\dfrac{3k}{2m}\right)} = -2 \tag{i}$$

Therefore, the modal vectors, according to Eqs. (8-236) and (8-237), are

$$\{u^{(1)}\} = \begin{Bmatrix} u_1^{(1)} \\ u_2^{(1)} \end{Bmatrix} = u_1^{(1)} \begin{Bmatrix} 1 \\ r_1 \end{Bmatrix} = \begin{Bmatrix} 1 \\ 1 \end{Bmatrix} \tag{j}$$

$$\{u^{(2)}\} = \begin{Bmatrix} u_1^{(2)} \\ u_2^{(2)} \end{Bmatrix} = u_1^{(2)} \begin{Bmatrix} 1 \\ r_2 \end{Bmatrix} = \begin{Bmatrix} 1 \\ -2 \end{Bmatrix} \tag{k}$$

where, as discussed in Example 8-11, $u_1^{(1)}$ and $u_1^{(2)}$ have been arbitrarily set equal to unity. The modes are summarized in Figure 8-34, where the mode shapes are sketched in plots having the linear extent of the system drawn along the horizontal axis and the horizontal displacements of the masses drawn along the vertical axis. The dashed lines in the mode shapes in Figure 8-34 are intended to emphasize that no element exists outwardly of the masses, and thus, the dashed lines are sketched for reference only.

In the first mode, the coupling spring undergoes no deformation, as the two masses simply translate together as if they were a single rigid body. This point is similar to the first mode of Example 8-11, where the coupling spring remained undeformed. However, unlike Example 8-11, there is no restoring force in the first mode here because the coupling spring that does not deform is the only stiffness element that could exert a restoring force. Therefore, we cannot expect harmonic motion of this system in the first mode because the

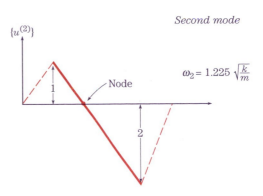

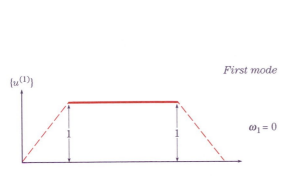

(a) Mode shape and natural frequency for first mode.   (b) Mode shape and natural frequency for second mode.

FIGURE 8-34   Natural modes of vibration for system in Figure 8-33.

masses are not subjected to restoring forces during their first mode motion. Then, what kind of motion can we expect in the first mode of the system? The answer—though perhaps obvious—will be given in the discussion following this example.

In the second mode, the two masses vibrate in opposition, with a node at one-third of the spring's length from the mass $2m$. The reader may wish to sketch the hypothetical model analogous to Figure 8-30 for this system.

Discussion

As indicated, the first mode in Example 8-13 has zero natural frequency and unity for all components of the modal vector and, as suggested above, it is not harmonic with time. In fact, the first mode motion in this case is simply a translation of the system as a whole. Mathematically, when a natural frequency is zero, the corresponding time response function is not given as indicated in Eqs. (8-227) and (8-228), but it is given by[32]

$$g(t) = C_1 + C_2 t \tag{l}$$

Therefore, when $\omega_1 = 0$, the motion of the system in that natural mode is obtained by substituting Eq. (l) into Eqs. (8-225) and (8-226), giving

$$x_1(t) = u_1^{(1)}(C_1 + C_2 t) = C_1 + C_2 t \tag{m}$$

$$x_2(t) = u_2^{(1)}(C_1 + C_2 t) = r_1 u_1^{(1)}(C_1 + C_2 t) = C_1 + C_2 t \tag{n}$$

where $u_2^{(1)} = r_1 u_1^{(1)} = u_1^{(1)}$ (since $r_1 = 1$) has been used; and $u_1^{(1)}$ has been arbitrarily set equal to unity. As discussed in Example 8-13, this modal motion is nonoscillatory, unlike those having nonzero natural frequencies; and, the actual modal motion represents an indefinite translation of the system as a whole. The following generic observations are made regarding the modal motion corresponding to a zero natural frequency, such as that described by Eqs. (m) and (n).

---

[32]See Eq. (I-22) in Appendix I.

1. A modal motion having a zero natural frequency represents a uniform motion of the entire system as a whole, exhibiting no oscillations. That is, there is no relative motion within the system in such a mode; the system behaves as if it were a single rigid body. Therefore, a mode corresponding to zero natural frequency is often called a *rigid-body mode*. Recalling that the natural modes are arranged in ascending order of natural frequencies, a mode having zero natural frequency is sometimes called a *zeroth mode*.

2. At least one rigid-body mode exists in all unrestrained systems because such systems can move freely as a rigid body, that is, without relative motion of the system elements or components.

3. In order to visualize a rigid-body mode, note that during rigid-body modal motion, no elastic energy is stored in the stiffness elements. So, during rigid-body modal motion, there is no restoring force due to the potential energy of the elastic elements, which will return the system to some previous state. Thus, the system can move freely and indefinitely without relative displacements within it.

4. Systems having a rigid-body mode are called *semidefinite systems*.

8-4.2   Response to Initial Conditions

In the previous subsection we found that two-degree-of-freedom undamped systems have two natural modes of vibration and that any free undamped vibration of the system *must* consist of these two modes. In terms of the displacement amplitudes u_1 and u_2 in Eqs. (8-227) and (8-228), we may express these results as

$$u_1 = u_1^{(1)}@\omega_1 + u_1^{(2)}@\omega_2 \tag{8-239}$$

$$u_2 = u_2^{(1)}@\omega_1 + u_2^{(2)}@\omega_2 \tag{8-240}$$

where we interpret Eq. (8-239) as stating that the total amplitude of u_1 consists of the portion of u_1 at ω_1 to be $u_1^{(1)}$ and the portion of u_1 at ω_2 to be $u_1^{(2)}$. A similar statement may be made for u_2 in Eq. (8-240). It is important to appreciate the fact that Eqs. (8-239) and (8-240) are the conceptual results of Subsection 8-4.1. We shall now find the free undamped response of various two-degree-of-freedom systems subjected to initial conditions.

8-4.2.1 Systems Having Nonzero Natural Frequencies   First, we consider the case when both natural frequencies are nonzero. Thus, combining Eqs. (8-227) and (8-228) with Eqs. (8-239) and (8-240) gives

$$x_1(t) = u_1^{(1)} \cos(\omega_1 t - \psi_1) + u_1^{(2)} \cos(\omega_2 t - \psi_2) \tag{8-241}$$

$$x_2(t) = u_2^{(1)} \cos(\omega_1 t - \psi_1) + u_2^{(2)} \cos(\omega_2 t - \psi_2) \tag{8-242}$$

where the first terms on the right-hand sides of Eqs. (8-241) and (8-242) are the first mode components, and the second terms are the second mode components. If we define $C_1 = u_1^{(1)}$ and $C_2 = u_1^{(2)}$, and use Eqs. (8-234) and (8-235), we may also write

$$u_2^{(1)} = u_1^{(1)} r_1 = C_1 r_1 \tag{8-243}$$

$$u_2^{(2)} = u_1^{(2)} r_2 = C_2 r_2 \tag{8-244}$$

Using Eqs. (8-243) and (8-244), Eqs. (8-241) and (8-242) may be rewritten as

$$\{x(t)\} = \left\{ \begin{array}{c} x_1(t) \\ x_2(t) \end{array} \right\} = \left\{ \begin{array}{c} x_1^{(1)}(t) \\ x_2^{(1)}(t) \end{array} \right\} + \left\{ \begin{array}{c} x_1^{(2)}(t) \\ x_2^{(2)}(t) \end{array} \right\}$$

$$= C_1 \left\{ \begin{array}{c} 1 \\ r_1 \end{array} \right\} \cos(\omega_1 t - \psi_1) + C_2 \left\{ \begin{array}{c} 1 \\ r_2 \end{array} \right\} \cos(\omega_2 t - \psi_2) \quad (8\text{-}245)$$

where C_1, C_2, ψ_1, and ψ_2 are constants, the four of which may be determined by the initial conditions on the system. Note that we have introduced the intermediate results that $x_1(t) = x_1^{(1)}(t) + x_1^{(2)}(t)$ and $x_2(t) = x_2^{(1)}(t) + x_2^{(2)}(t)$, which emphasize that $x_1(t)$ and $x_2(t)$ are each simply a combination of the system's response at each of the two natural frequencies. That is, for example, the free (unforced) response of $x_1(t)$ is the sum of the free response of x_1 due to ω_1 (written as $x_1^{(1)}(t)$) plus the free response of x_1 due to ω_2 (written as $x_1^{(2)}(t)$). This statement is analogous to Eq. (8-239) and identical to Eq. (8-241). Comparable statements about $x_2(t)$ can be made.

We consider the general[33] set of initial conditions (I.C.) given by Eqs. (8-219) through (8-222) as

$$x_1(0) = x_{10} \quad (8\text{-}246)$$
$$x_2(0) = x_{20} \quad (8\text{-}247)$$
$$\dot{x}_1(0) = v_{10} \quad (8\text{-}248)$$
$$\dot{x}_2(0) = v_{20} \quad (8\text{-}249)$$

Substitution of Eqs. (8-246) and (8-247) into Eq. (8-245) at $t = 0$ and substitution of Eqs. (8-248) and (8-249) into the time derivative of Eq. (8-245) at $t = 0$ give

$$x_1(0) = x_{10} = C_1 \cos \psi_1 + C_2 \cos \psi_2 \quad (8\text{-}250)$$
$$x_2(0) = x_{20} = C_1 r_1 \cos \psi_1 + C_2 r_2 \cos \psi_2 \quad (8\text{-}251)$$
$$\dot{x}_1(0) = v_{10} = C_1 \omega_1 \sin \psi_1 + C_2 \omega_2 \sin \psi_2 \quad (8\text{-}252)$$
$$\dot{x}_2(0) = v_{20} = C_1 \omega_1 r_1 \sin \psi_1 + C_2 \omega_2 r_2 \sin \psi_2 \quad (8\text{-}253)$$

Equations (8-250) through (8-253) constitute a set of four algebraic equations in four unknowns, apparently requiring a straightforward solution technique. However, as shown in the examples below, care must be exercised in solving these equations for the constants C_1, C_2, ψ_1, and ψ_2. A perfunctory attempt to solve Eqs. (8-250) through (8-253) may lead to singularities for some combinations of initial conditions; such singularities can be avoided by use of an appropriate solution scheme.

The meanings of C_1 and C_2 in Eq. (8-245) are fairly clear; that is, C_1 and C_2 represent the $x_1(t)$ amplitudes of modes 1 and 2, respectively, in the subsequent motion due to the imposed initial conditions. The meanings of ψ_1 and ψ_2 in Eq. (8-245) are less obvious; however, ψ_1 and ψ_2 simply ensure shifts in time of modes 1 and 2 in the subsequent motion due to the imposed initial conditions. Note that we do *not* refer to ψ_1 and ψ_2 as *phase angles,* a term that we reserve exclusively for forced responses.

---

[33]Our use of the word "general" as an adjective for this set of initial conditions is to convey only the fact that none of $x_1(0)$, $x_2(0)$, $\dot{x}_1(0)$ or $\dot{x}_2(0)$ is zero. Nevertheless, it is always necessary and sufficient to define all four quantities, even if one or more of them are zero.

It is important to appreciate that a procedure to evaluate C_1, C_2, ψ_1, and ψ_2 is simply a means to properly distribute the energy associated with the initial conditions between the two modes of vibrations. Once this energy partitioning is achieved, the energy associated with each mode remains in that mode forever; that is, as long as the system is undergoing a free undamped response due to those initial conditions only. Any subsequent motion of the system, no matter how complicated it may appear to visual inspection, will be simply a superposition of the motions of the two modes. Furthermore, since the initial energy in each mode remains in that mode, the modes do not "communicate" with each other. In this context, they are said to be *orthogonal* to each other; thus they are called *orthogonal modes*.[34]

■ **Example 8-14:** Scheme for Finding Constants from Initial Conditions—Case of Nonzero Natural Frequencies

Find C_1, C_2, ψ_1, and ψ_2 from Eqs. (8-250) through (8-253) for all choices of initial conditions when both natural frequencies are nonzero.

☐ **Solution:** In order to eliminate C_2 and ψ_2 from Eqs. (8-250) and (8-251), multiply Eq. (8-250) by r_2 and subtract Eq. (8-251) from the resulting expression, yielding

$$r_2 x_{10} - x_{20} = (r_2 - r_1) C_1 \cos \psi_1 \tag{a}$$

or

$$C_1 \cos \psi_1 = \frac{r_2 x_{10} - x_{20}}{r_2 - r_1} \tag{b}$$

where it is assumed that $r_2 \neq r_1$, which follows from the earlier exclusion of cases having repeated natural frequencies. Also, multiplying Eq. (8-252) by r_2 and subtracting Eq. (8-253) from the resulting expression yield

$$r_2 v_{10} - v_{20} = (r_2 - r_1) C_1 \omega_1 \sin \psi_1 \tag{c}$$

or

$$C_1 \sin \psi_1 = \frac{r_2 v_{10} - v_{20}}{\omega_1 (r_2 - r_1)} \tag{d}$$

First, suppose the numerator of Eq. (b) is nonzero (that is, $r_2 x_{10} - x_{20} \neq 0$). Then, dividing Eq. (d) by Eq. (b) gives

$$\tan \psi_1 = \frac{r_2 v_{10} - v_{20}}{\omega_1 (r_2 x_{10} - x_{20})} \tag{e}$$

or

$$\psi_1 = \tan^{-1} \frac{r_2 v_{10} - v_{20}}{\omega_1 (r_2 x_{10} - x_{20})} \tag{f}$$

---

[34]See Problem 8-72.

If the numerator and the denominator on the right-hand side of Eq. (f) are both nonzero (that is, $r_2 x_{10} - x_{20} \neq 0$ and $r_2 v_{10} - v_{20} \neq 0$), neither $\cos \psi_1$ nor $\sin \psi_1$ is zero; therefore, from Eqs. (b) and (d),

$$C_1 = \frac{r_2 x_{10} - x_{20}}{(r_2 - r_1) \cos \psi_1} = \frac{r_2 v_{10} - v_{20}}{\omega_1 (r_2 - r_1) \sin \psi_1} \tag{g}$$

where ψ_1 is given by Eq. (f).

If on the right-hand side of Eq. (f) the denominator is nonzero and the numerator is zero (that is, $r_2 x_{10} - x_{20} \neq 0$ and $r_2 v_{10} - v_{20} = 0$), Eq. (f) yields $\psi_1 = 0$; then, Eq. (d) becomes a trivial identity and Eq. (b) yields

$$C_1 = \frac{r_2 x_{10} - x_{20}}{(r_2 - r_1) \cos(0)} = \frac{r_2 x_{10} - x_{20}}{r_2 - r_1} \tag{h}$$

Now, suppose the numerator of Eq. (b) is zero (that is, $r_2 x_{10} - x_{20} = 0$). Then, Eq. (b) requires that either $C_1 = 0$ or $\psi_1 = \dfrac{\pi}{2}$. The case when $C_1 = 0$ is possible only when the numerator on the right-hand side of Eq. (d) is also zero (that is, $r_2 x_{10} - x_{20} = 0$ and $r_2 v_{10} - v_{20} = 0$), in which case the angle ψ_1 is undetermined. If the numerator on the right-hand side of Eq. (d) is not zero (that is, $r_2 x_{10} - x_{20} = 0$ and $r_2 v_{10} - v_{20} \neq 0$), C_1 cannot be zero; therefore, $\psi_1 = \dfrac{\pi}{2}$ must hold for Eq. (b) to vanish, in which case C_1 can be found from Eq. (d) as

$$C_1 = \frac{r_2 v_{10} - v_{20}}{\omega_1 (r_2 - r_1) \sin \dfrac{\pi}{2}} = \frac{r_2 v_{10} - v_{20}}{\omega_1 (r_2 - r_1)} \tag{i}$$

In summary, C_1 and ψ_1 can be determined for all combinations of initial conditions by adhering to the solution scheme synopsized below.

1.   If $r_2 x_{10} - x_{20} \neq 0$ and $r_2 v_{10} - v_{20} \neq 0$,

$$\psi_1 = \tan^{-1} \frac{r_2 v_{10} - v_{20}}{\omega_1 (r_2 x_{10} - x_{20})}; \qquad C_1 = \frac{r_2 x_{10} - x_{20}}{(r_2 - r_1) \cos \psi_1} = \frac{r_2 v_{10} - v_{20}}{\omega_1 (r_2 - r_1) \sin \psi_1}$$

2.   If $r_2 x_{10} - x_{20} \neq 0$ and $r_2 v_{10} - v_{20} = 0$,

$$\psi_1 = 0; \qquad C_1 = \frac{r_2 x_{10} - x_{20}}{r_2 - r_1}$$

3.   If $r_2 x_{10} - x_{20} = 0$ and $r_2 v_{10} - v_{20} \neq 0$,

$$\psi_1 = \frac{\pi}{2}; \qquad C_1 = \frac{r_2 v_{10} - v_{20}}{\omega_1 (r_2 - r_1)}$$

4.   If $r_2 x_{10} - x_{20} = 0$ and $r_2 v_{10} - v_{20} = 0$,

$$\psi_1: \text{arbitrary}; \qquad C_1 = 0$$

Performing a similar procedure, the solution scheme for C_2 and ψ_2 can be developed and is synopsized below.

1.   If $r_1 x_{10} - x_{20} \neq 0$ and $r_1 v_{10} - v_{20} \neq 0$,

$$\psi_2 = \tan^{-1} \frac{r_1 v_{10} - v_{20}}{\omega_2(r_1 x_{10} - x_{20})}; \qquad C_2 = \frac{r_1 x_{10} - x_{20}}{(r_1 - r_2)\cos\psi_2} = \frac{r_1 v_{10} - v_{20}}{\omega_2(r_1 - r_2)\sin\psi_2}$$

2. If $r_1 x_{10} - x_{20} \neq 0$ and $r_1 v_{10} - v_{20} = 0$,

$$\psi_2 = 0; \qquad C_2 = \frac{r_1 x_{10} - x_{20}}{r_1 - r_2}$$

3. If $r_1 x_{10} - x_{20} = 0$ and $r_1 v_{10} - v_{20} \neq 0$,

$$\psi_2 = \frac{\pi}{2}; \qquad C_2 = \frac{r_1 v_{10} - v_{20}}{\omega_2(r_1 - r_2)}$$

4. If $r_1 x_{10} - x_{20} = 0$ and $r_1 v_{10} - v_{20} = 0$,

$$\psi_2: \text{arbitrary}; \qquad C_2 = 0$$

An important observation can be made regarding the two solution schemes summarized above. In the fourth case of each scheme, the coefficient C_1 or C_2 vanishes, indicating that the corresponding mode does not appear in the response to the initial conditions. In these cases, the corresponding ψ_j (ψ_1 for C_1 and ψ_2 for C_2) may be anything; that is, as required by Eq. (8-245), they are arbitrary. Thus, for example, the first mode is simply not excited ($C_1 = 0$) by the initial conditions satisfying

$$r_2 x_{10} - x_{20} = 0 \tag{j}$$

and

$$r_2 v_{10} - v_{20} = 0 \tag{k}$$

The condition in Eq. (j) requires either that both x_{10} and x_{20} are zero or that $\dfrac{x_{20}}{x_{10}} = r_2$; and, similarly, the condition in Eq. (k) requires either that both v_{10} and v_{20} are zero or that $\dfrac{v_{20}}{v_{10}} = r_2$. Therefore, the first mode motion is not excited when the initial displacements are both zero or they are selfsame with the second mode shape *and* the initial velocities are both zero or they are selfsame with the second mode. In this context, "selfsame" for displacements means the displacements are identical with the second mode shape, and "selfsame" for the velocities means the velocities are such that they produce and/or preserve displacements that are identical with the second mode shape. Similarly, from the fourth case of the solution scheme for C_2 and ψ_2, the second mode is not excited when the initial displacements are both zero or they are selfsame with the first mode *and* the initial velocities are both zero or they are selfsame with the first mode.

Finally, as readers may verify for themselves, attempts to solve Eqs. (8-250) through (8-253) in a standard fashion in analyzing many problems will lead to abortive singularities. Although the schemes for calculating C_1, C_2, ψ_1, and ψ_2 may appear to be slightly complicated or entwined, they are straightforward to implement. Furthermore, these schemes are *essentially* foolproof; we emphasize "essentially" as no scheme is fully foolproof.

■ **Example 8-15:**   **Response of Two-Story Building to Initial Conditions**

Consider the two-story building model of Example 8-12. We analyze the result of an accident in which a helicopter of mass m_0, in attempting to land on the roof, struck the second story of a building with a horizontal velocity of v_0, and fell vertically to ground, as sketched in Figure 8-35. (Fortunately, no one was injured, not even scratched.) Find the motion of the building due to this collision.

☐ **Solution:** Since the natural frequencies found in Example 8-12 are both nonzero, the solution is given by Eq. (8-245), whose unknown constants can be found by the solution scheme given in Example 8-14. Recall from Example 8-12 that

$$\omega_1 = 0.707\sqrt{\frac{k}{m}} \qquad \omega_2 = 1.414\sqrt{\frac{k}{m}} \tag{a}$$

and

$$r_1 = 2 \qquad r_2 = -1 \tag{b}$$

where the specific parameter values used in calculating these modes were $m_2 = \dfrac{m_1}{2} = \dfrac{m}{2}$ and $k_2 = \dfrac{k_1}{2} = \dfrac{k}{2}$. (If the natural modes of vibration had not been previously calculated, it would have been necessary to do so prior to proceeding.)

The initial conditions for this system can be found via the principle of linear momentum. That is, the linear momentum in the horizontal direction of the entire system, consisting of the building and the helicopter, must be the same before and immediately after the collision. That is,

$$(m_2)(0) + (m_0)(-v_0) = (m_2)(v_{20}) + (m_0)(0) \tag{c}$$

The left-hand side of Eq. (c) is the pre-collision linear momentum of the entire system with m_2 at rest (also, m_1 is at rest both before and immediately after the collision), and the helicopter of mass m_0 having a negative velocity $-v_0$. The right-hand side of Eq. (c) is the immediate post-collision linear momentum of the entire system with m_2 having unknown horizontal velocity v_{20} and the helicopter m_0 having zero horizontal velocity. From Eq. (c), the velocity v_{20} of m_2 immediately after the collision can be found as

$$v_{20} = -\frac{m_0}{m_2}v_0 = -\frac{2m_0}{m}v_0 \tag{d}$$

where the specific value $m_2 = \dfrac{m}{2}$ has been used. All other initial conditions are zero; that is,

$$x_{10} = x_{20} = v_{10} = 0 \tag{e}$$

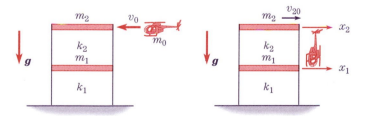

(a) Immediately pre-collision.   (b) Immediately post-collision.

FIGURE 8-35 Model of two-story building subjected to collision by helicopter.

Combining the second of Eqs. (b) with Eqs. (d) and (e) yields $r_2 x_{10} - x_{20} = 0$ and $r_2 v_{10} - v_{20} \neq 0$. Therefore, C_1 and ψ_1 can be found via the third case in the first solution scheme developed in Example 8-14. That is,

$$\psi_1 = \frac{\pi}{2} \tag{f}$$

and

$$C_1 = \frac{r_2 v_{10} - v_{20}}{\omega_1 (r_2 - r_1)} = \frac{(-1)(0) + \dfrac{2m_0}{m} v_0}{0.707 \sqrt{\dfrac{k}{m}}(-1-2)} = \frac{2}{(0.707)(-3)} \frac{m_0 v_0}{m} \sqrt{\frac{m}{k}} = -0.943 \frac{m_0 v_0}{\sqrt{mk}} \tag{g}$$

Similarly, combining the first of Eqs. (b) with Eqs. (d) and (e) yields $r_1 x_{10} - x_{20} = 0$ and $r_1 v_{10} - v_{20} \neq 0$. Therefore, C_2 and ψ_2 can be found via the third case in the second solution scheme developed in Example 8-14. That is,

$$\psi_2 = \frac{\pi}{2} \tag{h}$$

and

$$C_2 = \frac{r_1 v_{10} - v_{20}}{\omega_2 (r_1 - r_2)} = \frac{(2)(0) + \dfrac{2m_0}{m} v_0}{1.414 \sqrt{\dfrac{k}{m}}(2+1)} = \frac{2}{(1.414)(3)} \frac{m_0 v_0}{m} \sqrt{\frac{m}{k}} = 0.471 \frac{m_0 v_0}{\sqrt{mk}} \tag{i}$$

Therefore, substituting Eqs. (f) through (i) into Eq. (8-245) gives

$$
\begin{aligned}
\{x(t)\} = \left\{ \begin{array}{c} x_1(t) \\ x_2(t) \end{array} \right\} &= -0.943 \frac{m_0 v_0}{\sqrt{mk}} \left\{ \begin{array}{c} 1 \\ 2 \end{array} \right\} \cos\left(0.707 \sqrt{\frac{k}{m}} t - \frac{\pi}{2} \right) \\
&\quad + 0.471 \frac{m_0 v_0}{\sqrt{mk}} \left\{ \begin{array}{c} 1 \\ -1 \end{array} \right\} \cos\left(1.414 \sqrt{\frac{k}{m}} t - \frac{\pi}{2} \right) \\
&= -0.943 \frac{m_0 v_0}{\sqrt{mk}} \left\{ \begin{array}{c} 1 \\ 2 \end{array} \right\} \sin\left(0.707 \sqrt{\frac{k}{m}} t \right) \\
&\quad + 0.471 \frac{m_0 v_0}{\sqrt{mk}} \left\{ \begin{array}{c} 1 \\ -1 \end{array} \right\} \sin\left(1.414 \sqrt{\frac{k}{m}} t \right) \tag{j}
\end{aligned}
$$

which is the response of the building due to the helicopter collision.

Figure 8-36 shows the system response given by Eq. (j) where each response plot is normalized by the convenient coefficient $\dfrac{m_0 v_0}{\sqrt{mk}}$. Figure 8-36a shows $x_1(t)$ and $x_2(t)$ as computed from Eq. (j). Figure 8-36b shows those portions of $x_1(t)$ and $x_2(t)$ due exclusively to the first mode, namely, $x_1^{(1)}(t)$ and $x_2^{(1)}(t)$, respectively. Figure 8-36c shows those portions of $x_1(t)$ and $x_2(t)$ due exclusively to the second mode, namely, $x_1^{(2)}(t)$ and $x_2^{(2)}(t)$, respectively. Figure 8-36d shows the total response of $x_1(t)$, along with its modal components, and the total response of $x_2(t)$, along with its modal components. So, the heavy lines in Figure 8-36d are identical to the corresponding lines in Figure 8-36a. Figure 8-36d illustrates that the system response is established from the superposition of the individual modal responses.

In this example, since the second natural frequency ω_2 happens to be twice the first natural frequency ω_1, as shown in Eq. (a), $\left(\text{or, the period of the first mode response } T_1 = \dfrac{2\pi}{\omega_1} \text{ is}\right.$

twice the period of second mode response $T_2 = \left.\dfrac{2\pi}{\omega_2}\right)$, the system response is periodic with period T_1 which is the least common multiple of T_1 and T_2. Note that the energy introduced into the system via the initial conditions in Eqs. (d) and (e) has been distributed between the two modes. This energy partitioning remains unchanged with time.

The fact that the helicopter did not become a part of the post-collision vibratory system meant that the natural modes of vibration obtained in Example 8-12 could be used in this analysis. If the helicopter had remained on the roof, for example, the natural modes of vibration would have to be found for the alternative system where m_2 would be replaced by $(m_2 + m_0)$.

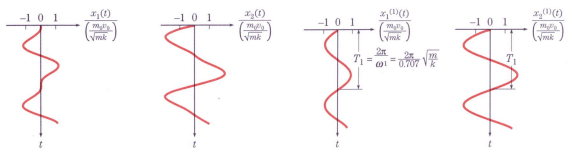

(a) System response.

(b) First mode response only.

(c) Second mode response only.

FIGURE 8-36 a,b,c Response of two-story building model in Figure 8-35.

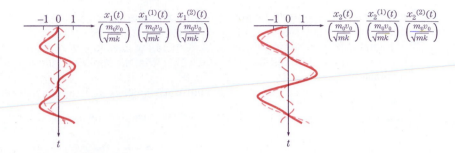

(*d*) Both first and second mode responses(dashed lines), *plus* system response (solid lines).

FIGURE 8-36*d*

8-4.2.2 Systems Having a Rigid-Body Mode—Semidefinite Systems

Thus far, in considering two-degree-of-freedom system response to initial conditions, we have assumed that the natural frequencies were both nonzero. Now we consider semidefinite systems having a rigid-body mode associated with a zero natural frequency. In this case, as indicated in Eqs. (m) and (n) in Example 8-13, the rigid-body modal motion is given by

$$\{x^{(1)}(t)\} = \begin{Bmatrix} x_1^{(1)}(t) \\ x_2^{(1)}(t) \end{Bmatrix} = C_1 \begin{Bmatrix} 1 \\ 1 \end{Bmatrix} + C_2 \begin{Bmatrix} 1 \\ 1 \end{Bmatrix} t \tag{8-254}$$

where the superscript "(1)" on the generalized coordinates indicates that they are associated with the first mode, which is the rigid-body mode. Recall that the modal motion corresponding to the other (or nonzero) natural frequency ω_2 is of the form in Eq. (8-245). Therefore, combining the two modal motions yields the general form for the free vibration of semidefinite systems as

$$\{x(t)\} = \begin{Bmatrix} x_1(t) \\ x_2(t) \end{Bmatrix} = \begin{Bmatrix} x_1^{(1)}(t) \\ x_2^{(1)}(t) \end{Bmatrix} + \begin{Bmatrix} x_1^{(2)}(t) \\ x_2^{(2)}(t) \end{Bmatrix}$$

$$= C_1 \begin{Bmatrix} 1 \\ 1 \end{Bmatrix} + C_2 \begin{Bmatrix} 1 \\ 1 \end{Bmatrix} t + C_3 \begin{Bmatrix} 1 \\ r_2 \end{Bmatrix} \cos(\omega_2 t - \psi) \tag{8-255}$$

where ω_2 is the nonzero natural frequency of the second mode and r_2 is the corresponding amplitude ratio, and where C_1, C_2, C_3, and ψ are unknown constants to be determined in accordance with the initial conditions. Substitution of the displacement initial conditions, Eqs. (8-246) and (8-247), into Eq. (8-255) at $t = 0$ and substitution of the velocity initial conditions, Eqs. (8-248) and (8-249), into the time derivative of Eq. (8-255) at $t = 0$ give

$$x_1(0) = x_{10} = C_1 + C_3 \cos \psi \tag{8-256}$$

$$x_2(0) = x_{20} = C_1 + C_3 r_2 \cos \psi \tag{8-257}$$

$$\dot{x}_1(0) = v_{10} = C_2 + C_3 \omega_2 \sin \psi \tag{8-258}$$

$$\dot{x}_2(0) = v_{20} = C_2 + C_3 \omega_2 r_2 \sin \psi \tag{8-259}$$

which can be solved for the unknown constants C_1, C_2, C_3, and ψ in terms of the given initial conditions. All the precautions regarding the solution of Eqs. (8-250) through (8-253) are advised here. Thus, we shall devise a scheme for reliably calculating these constants for two-degree-of-freedom semidefinite systems.

■ **Example 8-16:** Scheme for Finding Constants from Initial Conditions—Case of Semidefinite Systems

Find C_1, C_2, C_3, and ψ from Eqs. (8-256) through (8-259) for all choices of initial conditions.

❐ **Solution:** In order to eliminate C_3 and ψ from Eqs. (8-256) and (8-257), multiply Eq. (8-256) by r_2 and subtract Eq. (8-257) from the resulting expression, yielding

$$r_2 x_{10} - x_{20} = (r_2 - 1)C_1 \tag{a}$$

or

$$C_1 = \frac{r_2 x_{10} - x_{20}}{r_2 - 1} \tag{b}$$

Also, multiplying Eq. (8-258) by r_2 and subtracting Eq. (8-259) from the resulting expression yield

$$r_2 v_{10} - v_{20} = (r_2 - 1)C_2 \tag{c}$$

or

$$C_2 = \frac{r_2 v_{10} - v_{20}}{r_2 - 1} \tag{d}$$

Therefore, the unknown constants, C_1 and C_2 in Eq. (8-255), can be determined from Eqs. (b) and (d), respectively, for all initial conditions without encountering any singularity, since $r_2 \neq 1$. (Recall that $r_1 = 1$; so $r_2 \neq 1$ simply signifies that the system has a single rigid-body mode.)

Now, in order to find C_3 and ψ, subtracting Eq. (8-257) from Eq. (8-256) gives

$$C_3 \cos \psi = \frac{x_{10} - x_{20}}{1 - r_2} \tag{e}$$

and, subtracting Eq. (8-259) from Eq. (8-258) gives

$$C_3 \sin \psi = \frac{v_{10} - v_{20}}{\omega_2(1 - r_2)} \tag{f}$$

where we note that $\omega_2 \neq 0$. (Recall that $\omega_1 = 0$.) Following the procedure used in Example 8-14, the solution scheme for C_3 and ψ can be developed from Eqs. (e) and (f) and is given below.

1. If $x_{10} \neq x_{20}$ and $v_{10} \neq v_{20}$,

$$\psi = \tan^{-1} \frac{v_{10} - v_{20}}{\omega_2(x_{10} - x_{20})}; \qquad C_3 = \frac{x_{10} - x_{20}}{(1 - r_2)\cos \psi} = \frac{v_{10} - v_{20}}{\omega_2(1 - r_2)\sin \psi}$$

2. If $x_{10} \neq x_{20}$ and $v_{10} = v_{20}$,

$$\psi = 0; \qquad C_3 = \frac{x_{10} - x_{20}}{1 - r_2}$$

3. If $x_{10} = x_{20}$ and $v_{10} \neq v_{20}$,

$$\psi = \frac{\pi}{2}; \qquad C_3 = \frac{v_{10} - v_{20}}{\omega_2(1 - r_2)}$$

4. If $x_{10} = x_{20}$ and $v_{10} = v_{20}$,

$$\psi: \text{ arbitrary}; \qquad C_3 = 0$$

Therefore, the unknown constants C_1 and C_2 in Eq. (8-255) are determined by Eqs. (b) and (d), respectively, regardless of the character of initial conditions; and, the unknown constants C_3 and ψ in Eq. (8-255) are determined via the solution scheme summarized above. Observations comparable to those made in Example 8-14 can be made regarding semidefinite systems. In particular, the modal motion of one natural mode does not appear in the system response when the initial conditions do not contain energy in that mode. Also, as required by Eq. (8-255), if $C_3 = 0$ as in the fourth case above, ψ does not appear in the system response, making it arbitrary.

■ **Example 8-17:** Response of Semidefinite System to Initial Conditions

Consider the semidefinite system of Example 8-13 (Figure 8-33) with $m = 1$ kg and $k = 10$ kN/m. Find the system responses to each of the following sets of initial conditions.

(1) $x_{10} = 1$ (m); $x_{20} = -2$ (m); $v_{10} = -50$ (m/s); $v_{20} = 100$ (m/s).

(2) $x_{10} = x_{20} = -1$ (m); $v_{10} = v_{20} = 200$ (m/s).

❑ **Solution:** From Example 8-13, recall that $\omega_1 = 0$, $r_1 = 1$ and

$$\omega_2 = 1.225\sqrt{\frac{k}{m}} = 122.5 \text{ (rad/s)}; \qquad r_2 = -2 \tag{a}$$

where ω_2 has been evaluated for the parameter values in this example.

Part (1)

In this part, substitution of the given initial conditions and $r_2 = -2$ into Eqs. (b) and (d) of Example 8-16 gives

$$C_1 = \frac{r_2 x_{10} - x_{20}}{r_2 - 1} = \frac{(-2)(1) - (-2)}{-2 - 1} = 0 \tag{b}$$

$$C_2 = \frac{r_2 v_{10} - v_{20}}{r_2 - 1} = \frac{(-2)(-50) - 100}{-2 - 1} = 0 \tag{c}$$

which indicate that the rigid-body mode motion does not appear in the system response. Since $x_{10} \neq x_{20}$ and $v_{10} \neq v_{20}$, via the first case of the solution scheme developed in Example 8-16,

$$\psi = \tan^{-1}\frac{v_{10} - v_{20}}{\omega_2(x_{10} - x_{20})} = \tan^{-1}\frac{-50 - 100}{(122.5)(1 - (-2))} = \tan^{-1}\frac{-150}{367.5} = -22.2° \tag{d}$$

and

$$C_3 = \frac{x_{10} - x_{20}}{(1 - r_2)\cos\psi} = \frac{1 - (-2)}{(1 - (-2))\cos(-22.2°)}$$

$$= \frac{v_{10} - v_{20}}{\omega_2(1 - r_2)\sin\psi} = \frac{-50 - 100}{(122.5)(1 - (-2))\sin(-22.2°)}$$

$$= 1.08 \tag{e}$$

Therefore, according to Eq. (8-255), the system response to the initial conditions is

$$\{x(t)\} = \left\{ \begin{matrix} x_1(t) \\ x_2(t) \end{matrix} \right\} = 1.08 \left\{ \begin{matrix} 1 \\ -2 \end{matrix} \right\} \cos(122.5t + 22.2°) \text{ (m)} \tag{f}$$

Note that the solution of Eq. (d) may have been obtained as $-22.2° + 180° = 157.8°$ (or equivalently $-22.2° - 180° = -202.2°$) due to the π periodicity of the tangent function. However, this alternative angle does *not* result in a different physical system response. Indeed, it may be shown that if ψ in Eq. (d) had been found as $\psi = 157.8°$, Eq. (e) would result in $C_3 = -1.08$, yielding the same numerical result as in Eq. (f).[35] These consequences are true for all cases in Examples 8-14 and 8-16 where inverse tangent equations need to be evaluated. Therefore, it is not necessary to define a range of the angle for the inverse tangent equations when evaluating the unknown constants for initial condition responses. This is in contrast with the evaluation of the phase angles of harmonic responses in Subsections 8-2.4 and 8-3.5, where the physically acceptable ranges of the phase angles must be specified.

As indicated earlier, the response in Eq. (f) has no rigid-body motion because the given initial conditions are selfsame with the second mode (that is, $\frac{x_{20}}{x_{10}} = -2 = r_2$ and $\frac{v_{20}}{v_{10}} = -2 = r_2$). The response given by Eq. (f) is shown in Figure 8-37a. It may be noted from Figure 8-37a that the solution indeed represents exclusively the second-mode motion with a period of $T_2 = \frac{2\pi}{\omega_2}$.

Note from Figure 8-37a that the two masses oscillate at the same frequency ω_2 and that they oscillate in spatial opposition. That is, the mass $2m$ reaches its maximum positive amplitude at the same instant that the mass m reaches its maximum negative amplitude, and vice versa. Thus, the angle ψ does *not* represent a geometric or spatial phase angle between the two masses. Indeed, both $x_1(t)$ and $x_2(t)$ are associated with the same angle ψ as indicated in Eq. (f). Therefore, we see that the angle ψ is simply that value which must be incorporated into the harmonic function to ensure that the displacements and velocities of the masses at $t = 0$ and beyond are precisely consistent with the imposed initial conditions. That is, in this example, the maximum (or minimum) displacements and velocities do not occur at $t = 0$, and it is the angle ψ which mathematically captures this fact.

Part (2)
Similarly, substitution of the given initial conditions and $r_2 = -2$ into Eqs. (b) and (d) of Example 8-16 gives

$$C_1 = \frac{r_2 x_{10} - x_{20}}{r_2 - 1} = \frac{(-2)(-1) - (-1)}{-2 - 1} = -1 \tag{g}$$

---

[35]See Problem 8-73.

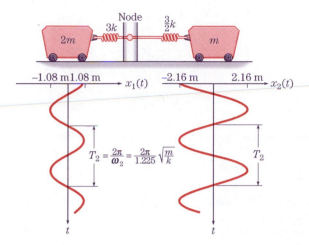

(a) Response to I.C. in part 1, with hypothetical spring elements in lieu of original spring.

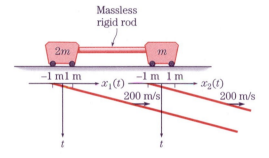

(b) Response to I.C. in part 2, with hypothetical massless rigid rod in lieu of original spring.

FIGURE 8-37   Responses of system in Figure 8-33 to two sets of initial conditions.

$$C_2 = \frac{r_2 v_{10} - v_{20}}{r_2 - 1} = \frac{(-2)(200) - 200}{-2 - 1} = 200 \qquad \text{(h)}$$

Since $x_{10} = x_{20}$ and $v_{10} = v_{20}$, via the fourth case of the solution scheme developed in Example 8-16,

$$\psi : \text{arbitrary}; \qquad C_3 = 0 \qquad \text{(i)}$$

which indicate that the second-mode motion does not appear in the system response. Therefore, according to Eq. (8-255), the response to the initial conditions is

$$\{x(t)\} = \left\{ \begin{array}{c} x_1(t) \\ x_2(t) \end{array} \right\} = -\left\{ \begin{array}{c} 1 \\ 1 \end{array} \right\} + 200 \left\{ \begin{array}{c} 1 \\ 1 \end{array} \right\} t = \left\{ \begin{array}{c} -1 + 200t \\ -1 + 200t \end{array} \right\} \text{ (m)} \qquad \text{(j)}$$

which represents a pure rigid-body motion. The system response in Eq. (j) reflects the fact that the initial conditions are selfsame with the rigid-body mode (that is, $\dfrac{x_{20}}{x_{10}} = 1 = r_1$ and

$\frac{v_{20}}{v_{10}} = 1 = r_1$). The response given by Eq. (j) is shown in Figure 8-37b. The motion shown in Figure 8-37b is indeed the first mode rigid-body motion. That is, the two masses move at the same velocity as if they were portions of the same rigid body, and the velocity is equal to the specified initial velocity.

8-4.3 Harmonic Response

The response of a linear two-degree-of-freedom system subjected to harmonic excitation and initial conditions can be found conceptually by the same procedures as those described for single-degree-of-freedom systems in Subsections 8-2.4 and 8-3.5 and summarized in Tables 8-3, 8-4, 8-7, and 8-8. All these procedures are within the context of the general technique for obtaining the complete solution as given in Table 8-1.

We shall consider the response of the generic two-degree-of-freedom system to an excitation that is harmonic for nonnegative time ($t \geq 0$). That is, we seek the solution to the equations of motion given by Eqs. (8-215) and (8-216) as

$$m_1 \ddot{x}_1 + c_{11} \dot{x}_1 + c_{12} \dot{x}_2 + k_{11} x_1 + k_{12} x_2 = F_1(t) \tag{8-260}$$

$$m_2 \ddot{x}_2 + c_{21} \dot{x}_1 + c_{22} \dot{x}_2 + k_{21} x_1 + k_{22} x_2 = F_2(t) \tag{8-261}$$

where

$$F_1(t) = \begin{cases} 0 & t < 0 \\ \displaystyle\sum_{j=1}^{m} f_{1j} \cos \Omega_j t + g_{1j} \sin \Omega_j t & t \geq 0 \end{cases} \tag{8-262}$$

and

$$F_2(t) = \begin{cases} 0 & t < 0 \\ \displaystyle\sum_{j=1}^{m} f_{2j} \cos \Omega_j t + g_{2j} \sin \Omega_j t & t \geq 0 \end{cases} \tag{8-263}$$

where the f_{kj}'s and g_{kj}'s are the constant amplitudes of the corresponding harmonic forcing functions having distinct (radian) frequencies Ω_j's. As indicated in the discussion immediately following Eq. (8-40) in Subsection 8-2.4, the forcing functions in Eqs. (8-262) and (8-263) accommodate all possible combinations of harmonic excitations. Also, we observe that the initial conditions are given by Eqs. (8-219) through (8-222) as

$$x_1(0) = x_{10} \tag{8-264}$$

$$x_2(0) = x_{20} \tag{8-265}$$

$$\dot{x}_1(0) = v_{10} \tag{8-266}$$

$$\dot{x}_2(0) = v_{20} \tag{8-267}$$

As in Subsection 8-2.4 or Subsection 8-3.5, we shall focus our interest on a simpler problem where the excitations $F_j(t)$'s are simplified to

$$F_1(t) = \begin{cases} 0 & t < 0 \\ f_1 \cos \Omega t + g_1 \sin \Omega t & t \geq 0 \end{cases} \tag{8-268}$$

and

$$F_2(t) = \begin{cases} 0 & t < 0 \\ f_2 \cos \Omega t + g_2 \sin \Omega t & t \geq 0 \end{cases} \tag{8-269}$$

which retain only pairs of single-frequency cosinusoidal and sinusoidal functions and, therefore, the subscript j has been dropped from Eqs. (8-262) and (8-263).

Due to the linearity of the system governed by Eqs. (8-260) and (8-261), the principle of superposition is valid and thus the particular solution for the forcing functions in Eqs. (8-262) and (8-263) can be obtained as the superposition of the particular solutions for the forcing functions in Eqs. (8-268) and (8-269) where, in going from the specific to the general, $\Omega = \Omega_j$, $f_1 = f_{1j}$, $g_1 = g_{1j}$, $f_2 = f_{2j}$ and $g_2 = g_{2j}$, $j = 1, 2, \ldots, m$. Therefore, no generality is lost by the simplification of the excitation from Eqs. (8-262) and (8-263) to Eqs. (8-268) and (8-269).

Thus, the problem is to find the solution to Eqs. (8-260) and (8-261) where the $F_j(t)$'s are given by Eqs. (8-268) and (8-269) and the initial conditions are given by Eqs. (8-264) through (8-267). This completes Step 1 of Table 8-1.

Recall from the general procedure in Table 8-1 that the complete solution of a system subjected to both nonzero excitations and initial conditions is given as the sum of the particular solution and the homogeneous solution. Since we have not derived the general homogeneous solution for the damped case (which is beyond the scope of our interests), we shall concentrate on undamped systems. Nevertheless, in Subdivision 8-4.3.1, the procedure to obtain the particular solution for general damped systems is described via the procedure of Subdivision 8-2.4.1 or Subdivision 8-3.5.1; and, in Subdivision 8-4.3.2, an alternative procedure to obtain the particular solution for general damped systems is described via the procedure of Subdivision 8-2.4.2 or Subdivision 8-3.5.2 (Step 2 of Table 8-1). In Subdivision 8-4.3.3, the procedures presented in Subdivisions 8-4.3.1 and 8-4.3.2 are applied to the special case of undamped systems to find the particular solution for undamped systems in detail (Step 2 of Table 8-1 for undamped systems); and the solution scheme is demonstrated in an example of an undamped two-degree-of-freedom torsional system. In Subdivision 8-4.3.4, the complete solution for an undamped system is established by summing the general homogeneous solution found in Subsection 8-4.1 and the particular solution found in Subdivision 8-4.3.3 (Step 3 of Table 8-1); and the procedure for determining the coefficients of the homogeneous solution in accordance with the initial conditions is described (Step 4 of Table 8-1).

8-4.3.1 Particular Solution for General Damping Case

Consider the particular solution as indicated in Step 2 of Table 8-1. According to the theory of ordinary differential equations, the particular solutions for the harmonic excitations in Eqs. (8-268) and (8-269) are *assumed* to be of the form

$$x_{1p}(t) = A_1 \cos \Omega t + B_1 \sin \Omega t \tag{8-270}$$

$$x_{2p}(t) = A_2 \cos \Omega t + B_2 \sin \Omega t \tag{8-271}$$

where A_1, A_2, B_1, and B_2 are undetermined constants.

Substitution of Eqs. (8-270) and (8-271) into the equations of motion, Eqs. (8-260) and (8-261), along with the harmonic forcing functions in Eqs. (8-268) and (8-269), yields

$$-m_1 \Omega^2 A_1 \cos \Omega t - m_1 \Omega^2 B_1 \sin \Omega t$$
$$- c_{11} \Omega A_1 \sin \Omega t + c_{11} \Omega B_1 \cos \Omega t - c_{12} \Omega A_2 \sin \Omega t + c_{12} \Omega B_2 \cos \Omega t$$
$$+ k_{11} A_1 \cos \Omega t + k_{11} B_1 \sin \Omega t + k_{12} A_2 \cos \Omega t + k_{12} B_2 \sin \Omega t$$
$$= f_1 \cos \Omega t + g_1 \sin \Omega t \tag{8-272}$$

$$-m_2\Omega^2 A_2 \cos \Omega t - m_2 \Omega^2 B_2 \sin \Omega t$$
$$- c_{21}\Omega A_1 \sin \Omega t + c_{21}\Omega B_1 \cos \Omega t - c_{22}\Omega A_2 \sin \Omega t + c_{22}\Omega B_2 \cos \Omega t$$
$$+ k_{21}A_1 \cos \Omega t + k_{21}B_1 \sin \Omega t + k_{22}A_2 \cos \Omega t + k_{22}B_2 \sin \Omega t$$
$$= f_2 \cos \Omega t + g_2 \sin \Omega t \quad (8\text{-}273)$$

Since Eqs. (8-272) and (8-273) must hold for all positive time, grouping the terms containing $\cos \Omega t$ and those containing $\sin \Omega t$ separately, and setting the coefficients of $\cos \Omega t$ and of $\sin \Omega t$ separately equal to zero in both Eqs. (8-272) and (8-273) yield four equations in A_1, B_1, A_2, and B_2:

$$(k_{11} - m_1\Omega^2)A_1 + c_{11}\Omega B_1 + k_{12}A_2 + c_{12}\Omega b_2 = f_1 \quad (8\text{-}274)$$
$$-c_{11}\Omega A_1 + (k_{11} - m_1\Omega^2)B_1 - c_{12}\Omega A_2 + k_{12}B_2 = g_1 \quad (8\text{-}275)$$
$$k_{21}A_1 + c_{21}\Omega B_1 + (k_{22} - m_2\Omega^2)A_2 + c_{22}\Omega B_2 = f_2 \quad (8\text{-}276)$$
$$-c_{21}\Omega A_1 + k_{21}B_1 - c_{22}\Omega A_2 + (k_{22} - m_2\Omega^2)B_2 = g_2 \quad (8\text{-}277)$$

Equations (8-274) through (8-277) may be rewritten, in matrix form, as

$$\begin{bmatrix} (k_{11} - m_1\Omega^2) & c_{11}\Omega & k_{12} & c_{12}\Omega \\ -c_{11}\Omega & (k_{11} - m_1\Omega^2) & -c_{12}\Omega & k_{12} \\ k_{21} & c_{21}\Omega & (k_{22} - m_2\Omega^2) & c_{22}\Omega \\ -c_{21}\Omega & k_{21} & -c_{22}\Omega & (k_{22} - m_2\Omega^2) \end{bmatrix} \begin{Bmatrix} A_1 \\ B_1 \\ A_2 \\ B_2 \end{Bmatrix} = \begin{Bmatrix} f_1 \\ g_1 \\ f_2 \\ g_2 \end{Bmatrix} \quad (8\text{-}278)$$

The matrix equation in Eq. (8-278) can be solved via Cramer's rule; however, due to the protracted algebra, the solutions to Eq. (8-278) will not be given here. If the solutions for A_1, B_1, A_2, and B_2 were found, substituting them into Eqs. (8-270) and (8-271) would yield the particular solution. This completes Step 2 of Table 8-1 for general damped systems.

8-4.3.2 Alternative Technique for Obtaining Particular Solution for General Damping Case Using Complex Variables

There is an alternative technique for obtaining the particular solution to that used in Subdivision 8-4.3.1. Similar to Subdivisions 8-2.4.2 and 8-3.5.2, this alternative technique is via complex functions,[36] and the computation involved in using it is relatively short compared with that in Subdivision 8-4.3.1. In this technique, the harmonic excitations $F_j(t)$'s on the right-hand sides of Eqs. (8-260) and (8-261) are written in the form of complex functions as

$$F_1(t) = \begin{cases} 0 & t < 0 \\ F_1 e^{i\Omega t} & t \geq 0 \end{cases} \quad (8\text{-}279)$$

and

$$F_2(t) = \begin{cases} 0 & t < 0 \\ F_2 e^{i\Omega t} & t \geq 0 \end{cases} \quad (8\text{-}280)$$

where F_1 and F_2 are real constants representing the amplitudes of the harmonic excitations at the (radian) frequency Ω and these constants will be identified shortly with f_1 and f_2, respectively, or with g_1 and g_2, respectively, in the real forcing functions in Eqs. (8-268)

[36]See Appendix H.

and (8-269). The complex forcing functions in Eqs. (8-279) and (8-280) have no physical meaning; they are simply useful mathematical tools for deriving the particular responses to sinusoidal and cosinusoidal excitations simultaneously and somewhat efficiently.[37]

In accordance with the theory of ordinary differential equations, we *assume* the particular solutions for $x_1(t)$ and $x_2(t)$ denoted by $x_{1p}(t)$ and $x_{2p}(t)$, respectively, to be of the form

$$x_{1p}(t) = X_1 e^{i\Omega t} \tag{8-281}$$

$$x_{2p}(t) = X_2 e^{i\Omega t} \tag{8-282}$$

where X_1 and X_2 are undetermined complex amplitudes. Substituting Eqs. (8-281) and (8-282) and Eqs. (8-279) and (8-280) into Eqs. (8-260) and (8-261) for nonnegative time and canceling the common $e^{i\Omega t}$ term give

$$(-m_1\Omega^2 + ic_{11}\Omega + k_{11})X_1 + (ic_{12}\Omega + k_{12})X_2 = F_1 \tag{8-283}$$

$$(ic_{21}\Omega + k_{21})X_1 + (-m_2\Omega^2 + ic_{22}\Omega + k_{22})X_2 = F_2 \tag{8-284}$$

Cramer's rule may be used to solve Eqs. (8-283) and (8-284) for the complex amplitudes X_1 and X_2. Thus,

$$X_1 = \frac{\begin{vmatrix} F_1 & (ic_{12}\Omega + k_{12}) \\ F_2 & (-m_2\Omega^2 + ic_{22}\Omega + k_{22}) \end{vmatrix}}{\begin{vmatrix} (-m_1\Omega^2 + ic_{11}\Omega + k_{11}) & (ic_{12}\Omega + k_{12}) \\ (ic_{21}\Omega + k_{21}) & (-m_2\Omega^2 + ic_{22}\Omega + k_{22}) \end{vmatrix}} \tag{8-285}$$

and

$$X_2 = \frac{\begin{vmatrix} (-m_1\Omega^2 + ic_{11}\Omega + k_{11}) & F_1 \\ (ic_{21}\Omega + k_{21}) & F_2 \end{vmatrix}}{\begin{vmatrix} (-m_1\Omega^2 + ic_{11}\Omega + k_{11}) & (ic_{12}\Omega + k_{12}) \\ (ic_{21}\Omega + k_{21}) & (-m_2\Omega^2 + ic_{22}\Omega + k_{22}) \end{vmatrix}} \tag{8-286}$$

Because of their extensive character, the numerators and denominators in Eqs. (8-285) and (8-286) will not be expanded by multiplication. However, if they were expanded, the complex amplitudes X_1 and X_2 could be expressed in terms of their real and imaginary parts. Designating the real and imaginary parts by the second subscripts r and i, respectively, X_1 and X_2 could be written as

$$X_1 = X_{1r} + iX_{1i} \tag{8-287}$$

$$X_2 = X_{2r} + iX_{2i} \tag{8-288}$$

Equations (8-287) and (8-288) can be rewritten as[38]

$$X_1 = Z_1 e^{i\phi_1} \tag{8-289}$$

$$X_2 = Z_2 e^{i\phi_2} \tag{8-290}$$

---

[37]See Section H-5 in Appendix H.
[38]See Section H-4 in Appendix H.

where

$$Z_1 = \sqrt{X_{1r}^2 + X_{1i}^2} \qquad (8\text{-}291)$$

$$\phi_1 = \tan^{-1}\frac{X_{1i}}{X_{1r}} \qquad (8\text{-}292)$$

$$Z_2 = \sqrt{X_{2r}^2 + X_{2i}^2} \qquad (8\text{-}293)$$

$$\phi_2 = \tan^{-1}\frac{X_{2i}}{X_{2r}} \qquad (8\text{-}294)$$

and where Z_1, ϕ_1, Z_2, and ϕ_2 are all real quantities. The physical significance of ϕ_1 and ϕ_2 will be made clear shortly. Substituting Eqs. (8-289) and (8-290) into Eqs. (8-281) and (8-282) gives the complex particular solution as

$$x_{1p}(t) = Z_1 e^{i(\Omega t + \phi_1)} \qquad (8\text{-}295)$$

$$x_{2p}(t) = Z_2 e^{i(\Omega t + \phi_2)} \qquad (8\text{-}296)$$

As indicated in Subdivisions 8-2.4.2 and 8-3.5.2,[39] the real parts of Eqs. (8-295) and (8-296) give the particular solution for the forcing functions of the real parts of Eqs. (8-279) and (8-280) (that is, purely cosinusoidal forcing functions); and, the imaginary parts of Eqs. (8-295) and (8-296) give the particular solution for the forcing functions of the imaginary parts of Eqs. (8-279) and (8-280) (that is, purely sinusoidal forcing functions). Therefore, as emphasized in Subdivisions 8-2.4.2 and 8-3.5.2, the introduction of complex forcing functions simplifies the derivation of the particular solutions for both cosinusoidal and sinusoidal excitations simultaneously.

Thus, the particular solution for the excitation in Eqs. (8-268) and (8-269) can be found as the superposition of the real and imaginary parts of Eqs. (8-295) and (8-296). So, the particular solution due to the cosinusoidal excitation (namely, $F_1(t) = f_1 \cos\Omega t$ and $F_2(t) = f_2 \cos\Omega t$ in Eqs. (8-268) and (8-269)) is given by the real parts of Eqs. (8-295) and (8-296); that is,

$$x_{1p}(t) = \mathrm{Re}\left\{ Z_1 e^{i(\Omega t + \phi_1)} \right\} = Z_1 \cos(\Omega t + \phi_1) \qquad (8\text{-}297)$$

$$x_{2p}(t) = \mathrm{Re}\left\{ Z_2 e^{i(\Omega t + \phi_2)} \right\} = Z_2 \cos(\Omega t + \phi_2) \qquad (8\text{-}298)$$

where Z_1, ϕ_1, Z_2, and ϕ_2 are evaluated via Eqs. (8-285) through (8-294) with $F_1 = f_1$ and $F_2 = f_2$. Similarly, the particular solution due to the sinusoidal excitation (namely, $F_1(t) = g_1 \sin\Omega t$ and $F_2(t) = g_2 \sin\Omega t$ in Eqs. (8-268) and (8-269)) is given by the imaginary parts of Eqs. (8-295) and (8-296); that is,

$$x_{1p}(t) = \mathrm{Im}\left\{ Z_1 e^{i(\Omega t + \phi_1)} \right\} = Z_1 \sin(\Omega t + \phi_1) \qquad (8\text{-}299)$$

$$x_{2p}(t) = \mathrm{Im}\left\{ Z_2 e^{i(\Omega t + \phi_2)} \right\} = Z_2 \sin(\Omega t + \phi_2) \qquad (8\text{-}300)$$

[39] Also, see Section H-5 in Appendix H.

where Z_1, ϕ_1, Z_2, and ϕ_2 are evaluated via Eqs. (8-285) through (8-294) with $F_1 = g_1$ and $F_2 = g_2$. Note that the angles ϕ_1 and ϕ_2 in Eqs. (8-297) through (8-300) represent the shifts of the system response in time with respect to the excitation. Therefore, ϕ_1 and ϕ_2 are phase angles of the generalized coordinates $x_1(t)$ and $x_2(t)$, respectively. Once Eqs. (8-297) through (8-300) are evaluated with these specific settings of F_1 and F_2, the particular solution for the harmonic excitation in Eqs. (8-268) and (8-269) is given by the sum of Eqs. (8-297) and (8-299) for $x_{1p}(t)$ and by the sum of Eqs. (8-298) and (8-300) for $x_{2p}(t)$. Although we shall not evaluate Eqs. (8-297) through (8-300), we want to underscore the straightforward character of the procedure for obtaining them as well as for evaluating them, if desired.

8-4.3.3 Particular Solution for Undamped System

Now, we consider the particular solution for an undamped system in some detail, using the two approaches presented in Subdivisions 8-4.3.1 and 8-4.3.2.

First, we derive the particular solution for undamped systems by following the procedure in Subdivision 8-4.3.1 where the particular solution for damped systems was not explicitly evaluated, only because of the protracted algebra associated with it; there exists no conceptual hurdle in finding the damped response. In the special case of undamped systems, however, the required algebraic calculations are reduced significantly, so the details will be presented.

Omitting all the damping terms in Eqs. (8-274) through (8-277) gives

$$(k_{11} - m_1\Omega^2)A_1 + k_{12}A_2 = f_1 \tag{8-301}$$

$$(k_{11} - m_1\Omega^2)B_1 + k_{12}B_2 = g_1 \tag{8-302}$$

$$k_{21}A_1 + (k_{22} - m_2\Omega^2)A_2 = f_2 \tag{8-303}$$

$$k_{21}B_1 + (k_{22} - m_2\Omega^2)B_2 = g_2 \tag{8-304}$$

Note that Eqs. (8-301) and (8-303) contain only A_1 and A_2 whereas Eqs. (8-302) and (8-304) contain only B_1 and B_2; that is, the two pairs of equations are uncoupled. Therefore, Eqs. (8-301) and (8-303) can be solved for A_1 and A_2 via Cramer's rule as

$$A_1 = \frac{\begin{vmatrix} f_1 & k_{12} \\ f_2 & (k_{22} - m_2\Omega^2) \end{vmatrix}}{\begin{vmatrix} (k_{11} - m_1\Omega^2) & k_{12} \\ k_{21} & (k_{22} - m_2\Omega^2) \end{vmatrix}} = \frac{(k_{22} - m_2\Omega^2)f_1 - k_{12}f_2}{(k_{11} - m_1\Omega^2)(k_{22} - m_2\Omega^2) - k_{12}^2} \tag{8-305}$$

$$A_2 = \frac{\begin{vmatrix} (k_{11} - m_1\Omega^2) & f_1 \\ k_{21} & f_2 \end{vmatrix}}{\begin{vmatrix} (k_{11} - m_1\Omega^2) & k_{12} \\ k_{21} & (k_{22} - m_2\Omega^2) \end{vmatrix}} = \frac{(k_{11} - m_1\Omega^2)f_2 - k_{12}f_1}{(k_{11} - m_1\Omega^2)(k_{22} - m_2\Omega^2) - k_{12}^2} \tag{8-306}$$

where in the second equalities of Eqs. (8-305) and (8-306), we have used $k_{12} = k_{21}$ from the third of Eqs. (8-218). We note that the denominators of Eqs. (8-305) and (8-306) are essentially the same as the characteristic equation in Eq. (8-232). (In Eqs. (8-305) and (8-306) the value of Ω is the known forcing frequency; whereas in Eq. (8-232) the roots of the characteristic equation were the eigenvalues ω_1^2 and ω_2^2.) Thus, the response amplitudes A_1 and A_2 become infinite if the excitation frequency Ω of f_1 or f_2 coincides with either of the natural frequencies ω_1 or ω_2. The natural frequencies are sometimes called the *resonant frequencies* since the system is said to resonate if it is driven at ω_1 or ω_2.

Similarly, applying Cramer's rule to solve Eqs. (8-302) and (8-304) for B_1 and B_2 yields

$$B_1 = \frac{\begin{vmatrix} g_1 & k_{12} \\ g_2 & (k_{22} - m_2\Omega^2) \end{vmatrix}}{\begin{vmatrix} (k_{11} - m_1\Omega^2) & k_{12} \\ k_{21} & (k_{22} - m_2\Omega^2) \end{vmatrix}} = \frac{(k_{22} - m_2\Omega^2)g_1 - k_{12}g_2}{(k_{11} - m_1\Omega^2)(k_{22} - m_2\Omega^2) - k_{12}^2} \qquad (8\text{-}307)$$

$$B_2 = \frac{\begin{vmatrix} (k_{11} - m_1\Omega^2) & g_1 \\ k_{21} & g_2 \end{vmatrix}}{\begin{vmatrix} (k_{11} - m_1\Omega^2) & k_{12} \\ k_{21} & (k_{22} - m_2\Omega^2) \end{vmatrix}} = \frac{(k_{11} - m_1\Omega^2)g_2 - k_{12}g_1}{(k_{11} - m_1\Omega^2)(k_{22} - m_2\Omega^2) - k_{12}^2} \qquad (8\text{-}308)$$

where, again, we have used $k_{12} = k_{21}$. Note that the denominator of Eqs. (8-307) and (8-308) is identical to that of Eqs. (8-305) and (8-306). Therefore, when the excitation frequency Ω of g_1 or g_2 coincides with either of the natural frequencies of the system, the amplitudes B_1 and B_2 will also become infinite.

Thus, substitution of Eqs. (8-305) through (8-308) into Eqs. (8-270) and (8-271) gives the particular solution for the harmonic excitation in Eqs. (8-268) and (8-269) as

$$x_{1p}(t) = A_1 \cos\Omega t + B_1 \sin\Omega t$$
$$= \frac{(k_{22} - m_2\Omega^2)f_1 - k_{12}f_2}{(k_{11} - m_1\Omega^2)(k_{22} - m_2\Omega^2) - k_{12}^2} \cos\Omega t$$
$$+ \frac{(k_{22} - m_2\Omega^2)g_1 - k_{12}g_2}{(k_{11} - m_1\Omega^2)(k_{22} - m_2\Omega^2) - k_{12}^2} \sin\Omega t \qquad (8\text{-}309)$$

$$x_{2p}(t) = A_2 \cos\Omega t + B_2 \sin\Omega t$$
$$= \frac{(k_{11} - m_1\Omega^2)f_2 - k_{12}f_1}{(k_{11} - m_1\Omega^2)(k_{22} - m_2\Omega^2) - k_{12}^2} \cos\Omega t$$
$$+ \frac{(k_{11} - m_1\Omega^2)g_2 - k_{12}g_1}{(k_{11} - m_1\Omega^2)(k_{22} - m_2\Omega^2) - k_{12}^2} \sin\Omega t \qquad (8\text{-}310)$$

Note that when the excitation is purely cosinusoidal (that is, when $g_1 = g_2 = 0$ in Eqs. (8-268) and (8-269)), B_1 and B_2 both become zero as shown in Eqs. (8-307) and (8-308). Thus, in this case, the particular solution in Eqs. (8-309) and (8-310) becomes purely cosinusoidal. Similarly, when the excitation is purely sinusoidal (that is, when $f_1 = f_2 = 0$ in Eqs. (8-268) and (8-269)), A_1 and A_2 both become zero as shown in Eqs. (8-305) and (8-306). Thus, in this case, the particular solution in Eqs. (8-309) and (8-310) becomes purely sinusoidal. Therefore, an important observation here is that the particular solution for undamped systems subjected to cosinusoidal excitation is purely cosinusoidal with no sinusoidal terms; and, the particular solution for undamped systems subjected to sinusoidal excitation is purely sinusoidal with no cosinusoidal terms.

In summary, the entire procedure for obtaining the particular solution, without using complex functions, for an undamped system subjected to harmonic excitations can be summarized as follows:

1. Establish the equations of motion including the harmonic forcing functions.

2. If the excitation is specified to be cosinusoidal (that is, exclusively containing $\cos\Omega t$, Ω being the forcing frequency), assume a particular solution as purely cosinusoidal at the forcing frequency; that is, $x_{1p}(t) = A_1 \cos\Omega t$ and $x_{2p}(t) = A_2 \cos\Omega t$, where $x_1(t)$

and $x_2(t)$ are the generalized coordinates. If the excitation is specified to be sinusoidal (that is, exclusively containing $\sin \Omega t$, Ω being the forcing frequency), assume a particular solution as purely sinusoidal at the forcing frequency; that is, $x_{1p}(t) = B_1 \sin \Omega t$ and $x_{2p}(t) = B_2 \sin \Omega t$. If the excitation is specified as combinations of cosinusoidal and sinusoidal functions (that is, containing both $\cos \Omega t$ and $\sin \Omega t$, Ω being the forcing frequency), assume a particular solution as a combination of both cosinusoidal and sinusoidal functions at the forcing frequency; that is, $x_{1p}(t) = A_1 \cos \Omega t + B_1 \sin \Omega t$ and $x_{2p}(t) = A_2 \cos \Omega t + B_2 \sin \Omega t$.

3. Substitute the assumed particular solution into the equations of motion and factor out the terms $\cos \Omega t$ and/or $\sin \Omega t$, depending on the forcing function.

4. Set the coefficients of the terms $\cos \Omega t$ and/or $\sin \Omega t$ equal to zero and, via Cramer's rule, evaluate the undetermined amplitudes (A_1, A_2, and/or B_1, B_2) from the resulting equations.

5. Obtain the particular solutions by substituting the thus-found coefficients into the assumed solutions.

Alternatively, the particular solutions for the undamped case can be derived[40] by the procedure in Subdivision 8-4.3.2. That is, first, the complex particular solution for the complex excitation in Eqs. (8-279) and (8-280) can be found as Eqs. (8-281) and (8-282) where the coefficients X_1 and X_2 are obtained from Eqs. (8-285) and (8-286), which simplify for the undamped case due to the vanishing of all the damping constants c_{ij}. Then, when the (real) forcing functions are given as purely cosinusoidal, the particular solution is given by the real part of the complex particular solution as given in Eqs. (8-297) and (8-298); when the (real) forcing functions are given as purely sinusoidal, the particular solution is given by the imaginary part of the complex particular solution as given in Eqs. (8-299) and (8-300); and when the (real) forcing functions are given as a combination of both cosinusoidal and sinusoidal functions, the particular solution is given by the combination of the real and imaginary parts of the complex particular solution as discussed in the last paragraph of Subdivision 8-4.3.2.

The particular solution is often called the *forced response* of the system because the particular solution is generally associated with the forcing functions. In order to demonstrate the procedure summarized above, we shall consider the forced response of an undamped system subjected to harmonic excitation.

■ **Example 8-18:** Forced Response of Torsional System to Harmonic Excitation

Consider the torsional system model sketched in Figure 8-38, which consists of two identical circular disks, each having axial rotary inertia I_0, mounted on a circular shaft having negligible rotary inertia. The shaft consists of two identical uniform segments, each having

FIGURE 8-38 Model of two-degree-of-freedom torsional system subjected to external torques.

40See Problem 8-96.

equilibrium length l, polar moment of the cross-sectional area about the shaft axis J, and shear modulus G. Also, the two disks are subjected to applied torques $T_1(t)$ and $T_2(t)$ as sketched in Figure 8-38.

When the generalized coordinates of the system are chosen as the angular displacements of the disks with respect to their equilibrium positions, $\theta_1(t)$ and $\theta_2(t)$ as sketched in Figure 8-38, the equations of motion for the system are given as[41]

$$I_0 \ddot{\theta}_1 + \frac{2GJ}{l}\theta_1 - \frac{GJ}{l}\theta_2 = T_1(t) \tag{a}$$

$$I_0 \ddot{\theta}_2 - \frac{GJ}{l}\theta_1 + \frac{GJ}{l}\theta_2 = T_2(t) \tag{b}$$

Also, the natural frequencies of the system are given as[42]

$$\begin{matrix} \omega_1^2 \\ \omega_2^2 \end{matrix} = \left\{\frac{3}{2} \mp \frac{\sqrt{5}}{2}\right\}\frac{GJ}{I_0 l} \tag{c}$$

Thus,

$$\omega_1 = 0.61803\sqrt{\frac{GJ}{I_0 l}} \qquad \omega_2 = 1.61803\sqrt{\frac{GJ}{I_0 l}} \tag{d}$$

Further, the modal vectors are given by[42]

$$\{u^{(1)}\} = u_1^{(1)}\left\{\begin{matrix} 1 \\ \dfrac{1+\sqrt{5}}{2} \end{matrix}\right\} = \left\{\begin{matrix} 1 \\ 1.61803 \end{matrix}\right\} \tag{e}$$

$$\{u^{(2)}\} = u_1^{(2)}\left\{\begin{matrix} 1 \\ \dfrac{1-\sqrt{5}}{2} \end{matrix}\right\} = \left\{\begin{matrix} 1 \\ -0.61803 \end{matrix}\right\} \tag{f}$$

where $u_1^{(1)}$ and $u_1^{(2)}$ are the amplitudes of $\theta_1(t)$ in the first and second modes, respectively, and have been arbitrarily set equal to unity.

Find the forced response of the system when $T_1(t) = 0$ and $T_2(t) = T_2 \cos \Omega t$.

Solution: The procedure summarized above can be used to obtain the forced response (or particular solution) of the system. The equations of motion for the given excitations are

$$I_0 \ddot{\theta}_1 + \frac{2GJ}{l}\theta_1 - \frac{GJ}{l}\theta_2 = 0 \tag{g}$$

$$I_0 \ddot{\theta}_2 - \frac{GJ}{l}\theta_1 + \frac{GJ}{l}\theta_2 = T_2 \cos \Omega t \tag{h}$$

Since the excitation has only a cosinusoidal function, *assume* the forced response in the form

$$\theta_{1p}(t) = \Theta_1 \cos \Omega t \tag{i}$$

$$\theta_{2p}(t) = \Theta_2 \cos \Omega t \tag{j}$$

[41]See Problem 6-57.
[42]See Problem 8-63.

where Θ_1 and Θ_2 are the undetermined amplitudes of $\theta_{1p}(t)$ and $\theta_{2p}(t)$, respectively.

Substituting Eqs. (i) and (j) into Eqs. (g) and (h) and cancelling the common $\cos \Omega t$ term yield

$$\left(\frac{2GJ}{l} - I_0\Omega^2\right)\Theta_1 - \frac{GJ}{l}\Theta_2 = 0 \tag{k}$$

$$-\frac{GJ}{l}\Theta_1 + \left(\frac{GJ}{l} - I_0\Omega^2\right)\Theta_2 = T_2 \tag{l}$$

Solving Eqs. (k) and (l) for Θ_1 and Θ_2 via Cramer's rule gives

$$\Theta_1 = \frac{\begin{vmatrix} 0 & -\dfrac{GJ}{l} \\ T_2 & \left(\dfrac{GJ}{l} - I_0\Omega^2\right) \end{vmatrix}}{\begin{vmatrix} \left(\dfrac{2GJ}{l} - I_0\Omega^2\right) & -\dfrac{GJ}{l} \\ -\dfrac{GJ}{l} & \left(\dfrac{GJ}{l} - I_0\Omega^2\right) \end{vmatrix}}$$

$$= \frac{T_2\dfrac{GJ}{l}}{\left(2\dfrac{GJ}{l} - I_0\Omega^2\right)\left(\dfrac{GJ}{l} - I_0\Omega^2\right) - \left(\dfrac{GJ}{l}\right)^2} \tag{m}$$

$$\Theta_2 = \frac{\begin{vmatrix} \left(\dfrac{2GJ}{l} - I_0\Omega^2\right) & 0 \\ -\dfrac{GJ}{l} & T_2 \end{vmatrix}}{\begin{vmatrix} \left(\dfrac{2GJ}{l} - I_0\Omega^2\right) & -\dfrac{GJ}{l} \\ -\dfrac{GJ}{l} & \left(\dfrac{GJ}{l} - I_0\Omega^2\right) \end{vmatrix}}$$

$$= \frac{T_2\left(2\dfrac{GJ}{l} - I_0\Omega^2\right)}{\left(2\dfrac{GJ}{l} - I_0\Omega^2\right)\left(\dfrac{GJ}{l} - I_0\Omega^2\right) - \left(\dfrac{GJ}{l}\right)^2} \tag{n}$$

Note that Eqs. (m) and (n) could have been obtained directly by substituting the equivalent m's, k's and f_{ij}'s into Eqs. (8-305) and (8-306): $m_1 \equiv I_0$, $m_2 \equiv I_0$, $k_{11} \equiv \dfrac{2GJ}{l}$, $k_{12} = k_{21} \equiv -\dfrac{GJ}{l}$, $k_{22} \equiv \dfrac{GJ}{l}$, $f_1 = 0$, $f_2 \equiv T_2$ and $g_1 = g_2 = 0$.

Note that the common denominator of Eqs. (m) and (n) can be rearranged as

$$\left(2\frac{GJ}{l} - I_0\Omega^2\right)\left(\frac{GJ}{l} - I_0\Omega^2\right) - \left(\frac{GJ}{l}\right)^2$$

$$= 2\left(\frac{GJ}{l}\right)^2 - 2\frac{GJI_0}{l}\Omega^2 - \frac{GJI_0}{l}\Omega^2 + I_0^2\Omega^4 - \left(\frac{GJ}{l}\right)^2$$

$$= I_0^2\left[\Omega^4 - 3\frac{GJ}{I_0l}\Omega^2 + \left(\frac{GJ}{I_0l}\right)^2\right]$$

$$= I_0^2\left[\Omega^2 - \frac{3 - \sqrt{5}}{2}\frac{GJ}{I_0l}\right]\left[\Omega^2 - \frac{3 + \sqrt{5}}{2}\frac{GJ}{I_0l}\right]$$

$$= I_0^2(\Omega^2 - \omega_1^2)(\Omega^2 - \omega_2^2) \tag{o}$$

where ω_1 and ω_2 are the natural frequencies of the system whose squares are given in Eq. (c). As indicated following Eqs. (8-306) and (8-308), the rearrangement in Eq. (o) is possible due to the fact that the determinant in the common denominator of Eqs. (m) and (n) has the same form as the characteristic equation.[43]

Replacing the denominators of Eqs. (m) and (n) by Eq. (o) leads to

$$\Theta_1 = \frac{T_2}{I_0}\frac{\dfrac{GJ}{I_0l}}{(\Omega^2 - \omega_1^2)(\Omega^2 - \omega_2^2)} \tag{p}$$

$$\Theta_2 = \frac{T_2}{I_0}\frac{2\dfrac{GJ}{I_0l} - \Omega^2}{(\Omega^2 - \omega_1^2)(\Omega^2 - \omega_2^2)} \tag{q}$$

where in both Eqs. (p) and (q) the numerator and denominator have been divided by I_0. Substituting Eqs. (p) and (q) into Eqs. (i) and (j) gives the forced response of the system.

Discussion

In Figure 8-39, the amplitudes of the forced response are plotted as functions of the forcing frequency Ω. In both plots, Θ_1 and Θ_2 have been normalized by the value of Θ_1 when $\Omega = 0$; that is, from Eq. (p)

$$\Theta_{10} \equiv \Theta_1\Big|_{\Omega=0} = \frac{\dfrac{T_2}{I_0}\cdot\dfrac{GJ}{I_0l}}{\omega_1^2\cdot\omega_2^2} = \frac{\dfrac{T_2}{I_0}\cdot\dfrac{GJ}{I_0l}}{\dfrac{3 - \sqrt{5}}{2}\dfrac{GJ}{I_0l}\cdot\dfrac{3 + \sqrt{5}}{2}\dfrac{GJ}{I_0l}}$$

$$= \frac{\dfrac{T_2}{I_0}\cdot\dfrac{GJ}{I_0l}}{\dfrac{1}{4}(9 - 5)\left(\dfrac{GJ}{I_0l}\right)^2} = \frac{T_2}{I_0}\frac{I_0l}{GJ}$$

$$= \frac{T_2 l}{GJ} \tag{r}$$

---

[43] Also see Problem 8-74.

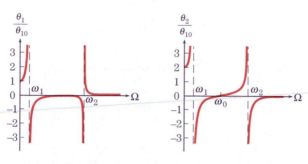

(a) Normalized angular displacement amplitude of disk 1.

(b) Normalized angular displacement amplitude of disk 2.

FIGURE 8-39 Normalized amplitudes of forced response of system in Figure 8-38.

where Eq. (c) has been used for ω_1^2 and ω_2^2. The quantity given in Eq. (r) is the *static deflection* of the left-hand disk because the harmonic excitation having zero forcing frequency (or, equivalently, infinite forcing period) denotes a static or constant load. Accordingly, Eq. (r) could have been written directly from statics or via Eq. (e) in Example 6-14.

From the plots of Figure 8-39, several observations are made. First, when the forcing frequency approaches either of the natural frequencies, ω_1 or ω_2, the forced response amplitudes of both disks approach infinity. This is the phenomenon called resonance. Second, normalized amplitudes of Θ_1 and Θ_2 at $\Omega = 0$ are consistent with the values that would be determined from statics. For a static torque on disk 2, the angular displacement of disk 2 should be twice that of disk 1. Third, to the extent that "phase" exists in this system, it is automatically captured in Figure 8-39 where negative values indicate a phase of π radians with respect to the applied torque. By π radians, we mean that, at the frequency of interest, the output (angular displacement) achieves its negative-most value when the input (torque) is at its positive-most value. Note that the "phase" here is simply a change of sign in the amplitude, its source being seen in Eqs. (p) and (q) as Ω increases or decreases.

The fourth observation is rather intriguing. In Figure 8-39b, there exists a forcing frequency at which the amplitude of disk 2 vanishes while that of disk 1 is finite. This means that when the system is forced at this particular frequency, disk 1 oscillates at the forcing frequency, but disk 2 remains stationary even though it is disk 2 that is being forced! Also, disk 1 is π radians out of phase with $T_2(t)$. This particular frequency, say ω_0, can be found by setting the numerator of Eq. (q) equal to zero. Thus,

$$\omega_0^2 = 2\frac{GJ}{I_0 l} \qquad \text{or} \qquad \omega_0 = \sqrt{2\frac{GJ}{I_0 l}} \tag{s}$$

Since at the forcing frequency $\Omega = \omega_0$ disk 2 is stationary, a hypothetical model of the system may be considered to consist of only one disk (disk 1), mounted on two identical shafts that are fixed at their far ends as sketched in Figure 8-40. The equation of motion for the hypothetical degenerate single-degree-of-freedom system of Figure 8-40 is

$$I_0 \ddot{\theta}_1 + 2\frac{GJ}{l}\theta_1 = 0 \tag{t}$$

whose natural frequency is

$$\omega_n = \sqrt{2\frac{GJ}{I_0 l}} \tag{u}$$

which is identical to ω_0 given by Eq. (s). Recall that, by definition, the natural frequency ω_n for the hypothetical degenerate system is the frequency at which the degenerate system (which is equivalent to the original system in Figure 8-38 with disk 2 fixed) will oscillate freely. Therefore, when the excitation frequency Ω of the original system coincides with the natural frequency of the degenerate system, the system will oscillate in a particular fashion that can be modeled as sketched in Figure 8-40.

The fifth observation is related to the asymptotic behavior of the amplitude spectra. As the forcing frequency becomes much larger than ω_2, the forced response amplitudes of both disks approach zero. This means that if excitation of very high frequency is applied, the system is not able to respond to it with large amplitudes. Indeed, the preceding sentence can be turned inside-out by observing that a quantitative measure of a very high frequency is the frequency beyond ω_2 at which the system response is negligible.

Finally, as indicated earlier, the signs of Θ_1 and Θ_2 change as Ω changes. Regarding the signs of Θ_1 and Θ_2, the forcing frequency can be divided into four ranges. That is, when $0 < \Omega < \omega_1$, Θ_1 and Θ_2 are both positive, and so both disks rotate in the same direction as the excitation; when $\omega_1 < \Omega < \omega_0$, Θ_1 and Θ_2 are both negative, and so both disks rotate in the opposite direction of the excitation; when $\omega_0 < \Omega < \omega_2$, Θ_1 is negative and Θ_2 is positive, and so disk 1 rotates in opposition to the excitation and disk 2 rotates in the same direction as the excitation; and when $\omega_2 < \Omega$, Θ_1 is positive and Θ_2 is negative, and so disk 1 rotates in the same direction as the excitation and disk 2 rotates in opposition to the excitation. Note that only when $\Omega > \omega_0$, Θ_1 and Θ_2 have different signs from each other. Therefore, when $\Omega > \omega_0$, there must be a location, between the two disks, where for each such value of Ω a cross-section of the shaft remains stationary at all times. As indicated previously, this stationary location is called a *node*. Unlike the node discussed in Subsection 8-4.1 where the node is an inherent characteristic of a mode shape, here both the existence of and the location of the node are dependent upon the excitation frequency.

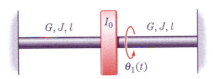

FIGURE 8-40   Hypothetical degenerate system having natural frequency equal to ω_0 as depicted in Figure 8-39.

8-4.3.4 Complete Solution for Undamped Two-Degree-of-Freedom Systems   Since the homogeneous solution for undamped two-degree-of-freedom systems has been derived in Subsection 8-4.1, the complete solution for undamped systems can be established. Therefore, we shall briefly outline the procedure to obtain the complete solution for undamped two-degree-of-freedom systems subjected to harmonic excitation *and* initial conditions.

As indicated in Step 3 of Table 8-1, the complete solution can be established as the sum of the homogeneous and particular solutions. First, consider the case where the natural frequencies of the system are both nonzero. In this case, the general homogeneous solution for the undamped case is given by Eq. (8-245) and the particular solution is given by Eqs. (8-309) and (8-310). Therefore, the complete solution can be written as the sum of Eq. (8-245) and Eqs. (8-270) and (8-271) as

$$\{x(t)\} = \begin{Bmatrix} x_1(t) \\ x_2(t) \end{Bmatrix}$$

$$= C_1 \begin{Bmatrix} 1 \\ r_1 \end{Bmatrix} \cos(\omega_1 t - \psi_1) + C_2 \begin{Bmatrix} 1 \\ r_2 \end{Bmatrix} \cos(\omega_2 t - \psi_2)$$

$$+ \begin{Bmatrix} A_1 \\ A_2 \end{Bmatrix} \cos \Omega t + \begin{Bmatrix} B_1 \\ B_2 \end{Bmatrix} \sin \Omega t \quad (8\text{-}311)$$

where A_1, A_2, B_1, and B_2 are known and given in Eqs. (8-305) through (8-308). This completes Step 3 of Table 8-1 when the system is *not* semidefinite.

Imposing the initial conditions, Eqs. (8-264) through (8-267), onto Eq. (8-311) yields

$$x_1(0) = x_{10} = C_1 \cos \psi_1 + C_2 \cos \psi_2 + A_1 \quad (8\text{-}312)$$

$$x_2(0) = x_{20} = C_1 r_1 \cos \psi_1 + C_2 r_2 \cos \psi_2 + A_2 \quad (8\text{-}313)$$

$$\dot{x}_1(0) = v_{10} = C_1 \omega_1 \sin \psi_1 + C_2 \omega_2 \sin \psi_2 + B_1 \Omega \quad (8\text{-}314)$$

$$\dot{x}_2(0) = v_{20} = C_1 \omega_1 r_1 \sin \psi_1 + C_2 \omega_2 r_2 \sin \psi_2 + B_2 \Omega \quad (8\text{-}315)$$

By defining new constants $x'_{10} \equiv x_{10} - A_1$, $x'_{20} \equiv x_{20} - A_2$, $v'_{10} \equiv v_{10} - B_1 \Omega$ and $v'_{20} \equiv v_{20} - B_2 \Omega$, Eqs. (8-312) through (8-315) can be rewritten as

$$x'_{10} = C_1 \cos \psi_1 + C_2 \cos \psi_2 \quad (8\text{-}316)$$

$$x'_{20} = C_1 r_1 \cos \psi_1 + C_2 r_2 \cos \psi_2 \quad (8\text{-}317)$$

$$v'_{10} = C_1 \omega_1 \sin \psi_1 + C_2 \omega_2 \sin \psi_2 \quad (8\text{-}318)$$

$$v'_{20} = C_1 \omega_1 r_1 \sin \psi_1 + C_2 \omega_2 r_2 \sin \psi_2 \quad (8\text{-}319)$$

Then, Eqs. (8-316) through (8-319) constitute a set of simultaneous equations having the same form as the set of Eqs. (8-250) through (8-253). Therefore, the unknown constants, C_1, C_2, ψ_1, and ψ_2, can be found by the procedures summarized in Example 8-14. Therefore, substituting the thus-found unknown constants into Eq. (8-311) yields the complete solution. This completes Step 4 of Table 8-1 when the system is *not* semidefinite.

Now, consider the case of semidefinite systems. In this case, the general homogeneous solution for the undamped case is given by Eq. (8-255). Therefore, the complete solution may be written as the sum of Eq. (8-255) and Eqs. (8-270) and (8-271) as

$$\{x(t)\} = \begin{Bmatrix} x_1(t) \\ x_2(t) \end{Bmatrix}$$

$$= C_1 \begin{Bmatrix} 1 \\ 1 \end{Bmatrix} + C_2 \begin{Bmatrix} 1 \\ 1 \end{Bmatrix} t + C_3 \begin{Bmatrix} 1 \\ r_2 \end{Bmatrix} \cos(\omega_2 t - \psi)$$

$$+ \begin{Bmatrix} A_1 \\ A_2 \end{Bmatrix} \cos \Omega t + \begin{Bmatrix} B_1 \\ B_2 \end{Bmatrix} \sin \Omega t \quad (8\text{-}320)$$

where A_1, A_2, B_1, and B_2 are known and given in Eqs. (8-305) through (8-308). This completes Step 3 of Table 8-1 for semidefinite systems.

Imposing the initial conditions, Eqs. (8-264) through (8-267), onto Eq. (8-320) yields

$$x_1(0) = x_{10} = C_1 + C_3 \cos \psi + A_1 \tag{8-321}$$

$$x_2(0) = x_{20} = C_1 + C_3 r_2 \cos \psi + A_2 \tag{8-322}$$

$$\dot{x}_1(0) = v_{10} = C_2 + C_3 \omega_2 \sin \psi + B_1 \Omega \tag{8-323}$$

$$\dot{x}_2(0) = v_{20} = C_2 + C_3 \omega_2 r_2 \sin \psi + B_2 \Omega \tag{8-324}$$

By defining new constants $x'_{10} \equiv x_{10} - A_1$, $x'_{20} \equiv x_{20} - A_2$, $v'_{10} \equiv v_{10} - B_1 \Omega$ and $v'_{20} \equiv v_{20} - B_2 \Omega$, Eqs. (8-321) through (8-324) can be rewritten as

$$x'_{10} = C_1 + C_3 \cos \psi \tag{8-325}$$

$$x'_{20} = C_1 + C_3 r_2 \cos \psi \tag{8-326}$$

$$v'_{10} = C_2 + C_3 \omega_2 \sin \psi \tag{8-327}$$

$$v'_{20} = C_2 + C_3 \omega_2 r_2 \sin \psi \tag{8-328}$$

Then, Eqs. (8-325) through (8-328) constitute a set of simultaneous equations having the same form as the set of Eqs. (8-256) through (8-259). Therefore, the unknown constants, C_1, C_2, C_3, and ψ, can be found by the procedures summarized in Example 8-16. Therefore, substituting the thus-found unknown constants into Eq. (8-320) yields the complete solution. This completes Step 4 of Table 8-1 for semidefinite systems.

8-5 STABILITY OF NONLINEAR SYSTEMS

So far in this chapter, we have developed vibration theories for *linear* lumped-parameter systems having one or two degrees of freedom. Thus, the elements of such systems were assumed to behave linearly on the condition that the excursion of the generalized coordinates, measured from their equilibrium values or some other appropriate reference, was of sufficiently small amplitude that linear equations of motion pertained.

All physical systems are inherently nonlinear, with various degrees of nonlinearity. Therefore, the most accurate general equations of motion for all real systems must be nonlinear. If it is desired to analyze the system behavior over a broad or full range of possible system motion—sometimes called the *global motion*—the nonlinear equations of motion must be considered. There are several theoretical and graphical methods for such global analyses, but these will not be explored in this section. Nevertheless, this section will seek to address several important questions: Is the assumption of small amplitude motion sufficient to apply the linear vibration theories developed in the previous sections? If not, what condition(s) should be met for the linear vibration theories to be applicable? What is the notion of stability, and what are the various types of stability?

In order to address these questions, it is useful to clarify the concept of an *equilibrium position* or *equilibrium configuration*. For convenience, let's limit our preliminary discussions to mechanical systems. From the discipline of statics we note that it is commonly stated that for a particle the necessary and sufficient condition for any configuration—that is, any combination of generalized coordinates—to be in equilibrium is that the sum of all forces must vanish. We note that what's implied in statics, but rarely stated, is that in addition to the vanishing of the sum of all the forces, the velocity of such a particle is assumed to vanish. In our discussions of equilibrium configurations, we state that an equilibrium configuration is any combination of generalized coordinates for which the accelerations *and* the velocities vanish. (Note that via the linear momentum principle the vanishing of the acceleration is equivalent to the vanishing of the sum of the forces, which is more commonly stated.)

We shall find that nonlinear equations of motion frequently possess more than one equilibrium configuration. Then, we shall investigate the linearized behavior of the model in the vicinity of each equilibrium configuration. We shall note that some equilibrium configurations will be *stable* and some *unstable,* and that the linear vibration analyses developed previously in this chapter can be applied to small amplitude oscillations in the vicinity of stable equilibrium configurations. Even though a more precise discussion of stability is given in Appendix J, the concepts of stable and unstable can be beneficially roughly explored here.

An equilibrium configuration is said to be *stable* if when the system is perturbed slightly from that equilibrium configuration, the system oscillates about or converges to the equilibrium configuration. Equilibrium configurations that are not stable are said to be *unstable*. When the system is slightly (no matter how slightly) perturbed from an unstable equilibrium configuration, the system will depart (perhaps indefinitely far) away from the equilibrium configuration. Therefore, the linear vibration theories developed in the previous sections cannot be applied to motion about unstable equilibrium configurations. In summary, the linear vibration theories can be applied to small-amplitude oscillations only in the vicinity of stable equilibrium configurations.

What remains to be achieved is a method for finding the equilibrium configurations of nonlinear systems and for examining the character of their stabilities. In this section, the equilibrium positions and their stabilities are found in three specific examples according primarily to physical arguments; however, a distinct procedure will emerge. Here, we recognize the unorthodox approach of presenting examples without the benefit of a foregoing theoretical framework, but obviously perceive value in doing so. As a prelude to *control theory,* and for specificity, a comparable *state-space* procedure is presented in Appendix J in the context of an example of an inverted pendulum. Finally, the *potential energy method* that can be applied to nonlinear conservative systems is also presented in Appendix J.

■ **Example 8-19:** Inverted Pendulum

Consider the inverted pendulum sketched in Figure 8-41*a* consisting of a massless rigid rod of length L and a mass m at the tip of the rod. The mass is attached to a linear spring of spring constant k, the other end of which is connected to a massless follower, which in turn can move vertically without friction. Note that the massless follower maintains the spring in a horizontal orientation. Note also that gravity acts.

When the generalized coordinate is chosen as the angle θ of the rod measured from the vertical inertial reference as shown, the equation of motion for the system is given by[44]

$$mL^2\ddot{\theta} + kL^2 \sin\theta \cos\theta - mgL \sin\theta = 0 \tag{a}$$

which is a nonlinear equation in θ, due to the sine and cosine terms. The first term in Eq. (a) represents the time rate of change of the angular momentum of the system about the pivot point O; the second term is the negative of the torque exerted by the spring; and, the third term is the negative of the torque exerted by gravity. Note that the nonlinear equation of motion holds for all values of θ between $\pm\dfrac{\pi}{2}$ radians, which is the allowable range of θ given by the geometric constraint of the floor at the pivot.

---

[44]See Problem 5-25, which simplifies to this model.

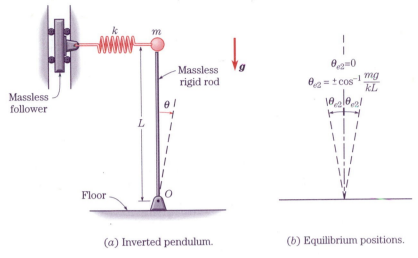

k m

Massless
rigid rod

g

Massless
follower

θ

L

Floor

O

$\theta_{e2}=0$

$\theta_{e2} = \pm\cos^{-1}\dfrac{mg}{kL}$

$|\theta_{e2}|\theta_{e2}|$

(a) Inverted pendulum. (b) Equilibrium positions.

FIGURE 8-41 Inverted pendulum and equilibrium positions.

For arbitrary values of m, k and L, except that mg may not equal kL, perform the following tasks:

1. Find the equilibrium positions (or equilibrium configurations).
2. Examine the stability of all equilibrium positions found in part (1).
3. Find the frequencies for small free oscillations about the stable equilibrium positions found in part (2).

❑ **Solution:**

Part (1)

In general, the (static) equilibrium positions may be found by setting the (angular) velocity and acceleration equal to zero in the equation of motion. Therefore, setting $\ddot{\theta} = 0$ (and $\dot{\theta} = 0$, if such a term had existed) gives

$$kL^2 \sin\theta_e \cos\theta_e - mgL\sin\theta_e = 0 \qquad (b)$$

or upon factoring $\sin\theta_e$ out of each term gives

$$\sin\theta_e(kL^2\cos\theta_e - mgL) = 0 \qquad (c)$$

where the subscript "e" designates an "equilibrium" value of θ.

Equation (c) is satisfied either when $\sin\theta_e = 0$ or when the quantity in parentheses equals zero. The first case of $\sin\theta_e = 0$ leads to

$$\theta_{e1} = 0 \qquad (d)$$

where the mathematical possibility of $\theta_e = \pi$ (meaning that the mass is directly below the pivot point) is excluded because of the geometric constraint of the floor. The second case in Eq. (c) requires that

$$kL^2\cos\theta_{e2} - mgL = 0 \qquad (e)$$

which leads to

$$\cos\theta_{e2} = \frac{mg}{kL} \tag{f}$$

or

$$\theta_{e2} = \pm\cos^{-1}\frac{mg}{kL} \tag{g}$$

where we note that the two equilibrium positions in Eq. (g) exist only when $mg < kL$ because the cosine function cannot exceed unity. (The condition that $mg \leq kL$ may be most easily seen from Eq. (f), and the equality of mg and kL is excluded by the problem statement.) Furthermore, since the cosine function is an even function, θ_{e2} represents two values of θ that lie within the allowable range of $\pm\frac{\pi}{2}$ radians. These three equilibrium positions (one given by Eq. (d) and two given by Eq. (g)) are indicated in Figure 8-41b, and each of them should appeal to the reader's intuition. When θ_e corresponds to Eq. (d), the spring is unstretched and the mass m is statically balanced by the vertical rod. When θ_e corresponds to Eq. (g), the torque about the pivot O caused by gravity acting on the mass is precisely balanced by the torque about the pivot produced by the spring force.

Part (2)

In order to examine the stability of the three equilibrium positions shown in Figure 8-41b, the nonlinear equation of motion in Eq. (a) is linearized about the equilibrium positions. To this end, we first consider a perturbation (a small change) of θ about a generic equilibrium position θ_e; that is, without specifying the first, second, or third equilibrium position, consider small excursions about θ_e. Thus, let

$$\theta = \theta_e + \epsilon(t) \tag{h}$$

where the time-varying perturbation $\epsilon(t)$ is small; that is, $\epsilon^2(t) \ll \epsilon(t)$. Note the requirement that $\epsilon^2(t) \ll \epsilon(t)$ is a quantitative characterization of the "small" change or perturbation. Substitution of Eq. (h) into Eq. (a) gives

$$mL^2\ddot{\epsilon} + kL^2\sin(\theta_e + \epsilon)\cos(\theta_e + \epsilon) - mgL\sin(\theta_e + \epsilon) = 0 \tag{i}$$

where we note that $\ddot{\theta} = \ddot{\epsilon}$ for Eq. (h) since θ_e is a constant.

Using common trigonometric identities for the sine and cosine of the sum of two angles yields

$$\sin(\theta_e + \epsilon) = \sin\theta_e\cos\epsilon + \cos\theta_e\sin\epsilon$$
$$\approx \sin\theta_e + \epsilon\cos\theta_e \tag{j}$$
$$\cos(\theta_e + \epsilon) = \cos\theta_e\cos\epsilon - \sin\theta_e\sin\epsilon$$
$$\approx \cos\theta_e - \epsilon\sin\theta_e \tag{k}$$

where, in noting that ϵ is small, we have used $\cos\epsilon \approx 1$ and $\sin\epsilon \approx \epsilon$. Note that Eqs. (j) and (k) could have been equally easily found by Taylor series expansions of $\sin\theta$ and $\cos\theta$ in the neighborhood of θ_e; that is,

$$\sin(\theta_e + \epsilon) = \sin\theta|_e + \cos\theta|_e\epsilon + \text{(higher-order terms)}$$
$$\approx \sin\theta_e + \epsilon\cos\theta_e \tag{l}$$

$$\cos(\theta_e + \epsilon) = \cos \theta|_e - \sin \theta|_e \epsilon + \text{(higher-order terms)}$$
$$\approx \cos \theta_e - \epsilon \sin \theta_e \qquad \text{(m)}$$

where the notation "$|_e$" designates evaluation at the equilibrium position θ_e.

Substitution of Eqs. (j) and (k) (or, of course, Eqs. (l) and (m)) into Eq. (i) yields

$$mL^2 \ddot{\epsilon} + kL^2(\sin \theta_e + \epsilon \cos \theta_e)(\cos \theta_e - \epsilon \sin \theta_e) - mgL(\sin \theta_e + \epsilon \cos \theta_e) = 0 \quad \text{(n)}$$

or

$$mL^2 \ddot{\epsilon} + \underline{kL^2 \sin \theta_e \cos \theta_e} - \epsilon kL^2 \sin^2 \theta_e + \epsilon kL^2 \cos^2 \theta_e - \epsilon^2 kL^2 \sin \theta_e \cos \theta_e$$
$$\underline{- mgL \sin \theta_e} - \epsilon mgL \cos \theta_e = 0 \qquad \text{(o)}$$

Note that the combination of the two underlined terms in Eq. (o) vanishes because of Eq. (b). Such a cancellation based on the equilibrium condition should always be anticipated; furthermore, if such a cancellation does not occur, the calculations should be checked and corrected. Also in Eq. (o), neglecting the term in ϵ^2 since such a term is much less than terms in ϵ, leads to

$$mL^2 \ddot{\epsilon} - \epsilon kL^2 \sin^2 \theta_e + \epsilon kL^2 \cos^2 \theta_e - \epsilon mgL \cos \theta_e = 0 \qquad \text{(p)}$$

Upon division of Eq. (p) by mL^2 throughout and grouping terms, it is found that

$$\ddot{\epsilon} + \left[\frac{k}{m}(\cos^2 \theta_e - \sin^2 \theta_e) - \frac{g}{L} \cos \theta_e \right] \epsilon = 0 \qquad \text{(q)}$$

Use of the identity $\cos^2 \theta_e + \sin^2 \theta_e = 1$ to eliminate $\sin^2 \theta_e$ in Eq. (q) leads to

$$\ddot{\epsilon} + \left[\frac{k}{m}(2 \cos^2 \theta_e - 1) - \frac{g}{L} \cos \theta_e \right] \epsilon = 0 \qquad \text{(r)}$$

which is the linear equation of motion governing the small motion ϵ about the generic equilibrium position θ_e.

Note that Eq. (r) in ϵ is of the same form as the equation of motion, Eq. (8-96), for linear undamped single-degree-of-freedom second-order systems. Therefore, the frequency of the free oscillation of ϵ as governed by Eq. (r) is given by the square root of

$$\tilde{\omega}_n^2 = \left[\frac{k}{m}(2 \cos^2 \theta_e - 1) - \frac{g}{L} \cos \theta_e \right] \qquad \text{(s)}$$

It is important to note that $\tilde{\omega}_n^2$ in Eq. (s) can be positive or negative, depending upon the system parameters and the specific equilibrium position being considered. The crux of the sign dependence of $\tilde{\omega}_n^2$ is this: If the sign of $\tilde{\omega}_n^2$ is *positive,* the system will be *stable* and will oscillate at $\tilde{\omega}_n$ for small excursions about θ_e; on the other hand, if the sign of $\tilde{\omega}_n^2$ is *negative,* the system will be *unstable* and will not oscillate at $\tilde{\omega}_n$ for small excursions about θ_e, but will move away from θ_e.

The significance of different signs of $\tilde{\omega}_n^2$ can be examined mathematically by rewriting Eq. (r) as

$$\ddot{\epsilon} + \tilde{\omega}_n^2 \epsilon = 0 \qquad \text{(t)}$$

and recalling from the theory of differential equations that the solution of Eq. (t) is of the form

$$\epsilon(t) = e^{\lambda t} \qquad \text{(u)}$$

which, when substituted into Eq. (t), gives

$$\lambda^2 + \tilde{\omega}_n^2 = 0 \tag{v}$$

or

$$\lambda^2 = -\tilde{\omega}_n^2 \tag{w}$$

When $\tilde{\omega}_n^2$ is *positive*, Eq. (w) yields $\lambda = \pm i\tilde{\omega}_n$ which, when substituted into Eq. (u), leads to an oscillatory motion of $\epsilon(t)$ about $\epsilon = 0$ (or, equivalently, an oscillatory motion of $\theta(t)$ about $\theta = \theta_e$). This case was considered in the linear vibration theory of Subsection 8-3.1. Since the system oscillates freely about the equilibrium position and does not move away from the equilibrium position, the system is stable about this equilibrium configuration. The equilibrium position about which the system is stable is called a *stable equilibrium*. While nonlinear systems in general do not possess a single entity that may be called a natural frequency, when the equilibrium configuration is stable, $\tilde{\omega}_n$ behaves just like a natural frequency for the *linearized model* represented by Eqs. (r) or (t).

When $\tilde{\omega}_n^2$ is *negative*, however, Eq. (w) yields $\lambda = \pm \sqrt{-\tilde{\omega}_n^2}$ (here, the radicand $-\tilde{\omega}_n^2$ is positive), which are positive and negative real values. Since the exponential function in Eq. (u) having a positive exponent grows indefinitely with time, $\epsilon(t)$ (or equivalently $\theta(t)$) exhibits exponential growth. In this case, due to the indefinite departure of the system from the equilibrium position, the system is unstable about the equilibrium position. The equilibrium about which the system is unstable is called an *unstable equilibrium*.

Part (3)

Now, having developed the criterion for stability of equilibrium positions, each of the three equilibrium positions, θ_{e1} and $\pm\theta_{e2}$, will be considered. For small oscillations about the first equilibrium position, $\theta_{e1} = 0$, Eq. (s) reduces to

$$\tilde{\omega}_{n1}^2 = \left[\frac{k}{m}(2-1) - \frac{g}{L}(1)\right]$$
$$= \frac{k}{m} - \frac{g}{L} \tag{x}$$

Considering the sign of Eq. (x), the system is stable about $\theta_{e1} = 0$ only when $\left(\frac{k}{m}\right) > \left(\frac{g}{L}\right)$; and it is unstable when $\left(\frac{k}{m}\right) < \left(\frac{g}{L}\right)$. Thus, when the system is stationary at $\theta_{e1} = 0$ *and* $\left(\frac{k}{m}\right) > \left(\frac{g}{L}\right)$, if the system is subjected to small disturbances, it will oscillate freely at a frequency $\tilde{\omega}_{n1}$ given by Eq. (x). On the other hand, when the system is stationary at $\theta_{e1} = 0$ *and* $\left(\frac{k}{m}\right) < \left(\frac{g}{L}\right)$, if someone in Nebraska sneezes, the pendulum will fall without oscillation, or at least move significantly from the position $\theta_{e1} = 0$. Without analyzing the original nonlinear equation, Eq. (a), we cannot determine precisely the manner in which it moves away from $\theta_{e1} = 0$.

Now, consider the two equilibrium positions given by Eq. (g), keeping in mind that these equilibrium positions exist only when $\frac{(mg)}{(kL)} < 1$. Substituting Eq. (f) into Eq. (s) gives

$$\tilde{\omega}_{n2}^2 = \frac{k}{m}\left\{2\left(\frac{mg}{kL}\right)^2 - 1\right\} - \frac{g}{L}\left(\frac{mg}{kL}\right)$$

$$= \frac{k}{m}\left\{2\left(\frac{mg}{kL}\right)^2 - 1 - \left(\frac{mg}{kL}\right)^2\right\}$$

$$= \frac{k}{m}\left\{\left(\frac{mg}{kL}\right)^2 - 1\right\} \tag{y}$$

which is negative because of the necessary condition $\frac{(mg)}{(kL)} < 1$ for the existence of these equilibrium positions. Therefore, the system is always unstable about the two equilibrium positions given by $\pm\theta_{e2} = \cos^{-1}\left\{\frac{mg}{kL}\right\}$.

In summary, it has been found that (1) when $mg < kL$, there are three equilibrium positions ($\theta_{e1} = 0$ and $\pm\theta_{e2} = \cos^{-1}\left\{\frac{mg}{kL}\right\}$); and θ_{e1} is a stable equilibrium position while $\pm\theta_{e2}$ are unstable equilibrium positions; and furthermore (2) when $mg > kL$, there is only one equilibrium position ($\theta_e = 0$), which is unstable. Therefore, for this system the linear vibration theories can be applied only to small-amplitude oscillations about the equilibrium position $\theta_{e1} = 0$ when $mg < kL$.

■ **Example 8-20:** Butterfly Condenser

Consider the butterfly condenser sketched in Figure 8-42 whose equation of motion was derived in Example 7-11. When the generalized coordinate is θ, which is the angular displacement of the rotor from the vertical, and with respect to the stator as indicated in Figure 8-42, the equation of motion is

$$I_0\ddot{\theta} + E_0^2 C_1 \sin 2\theta = 0 \tag{a}$$

which is a nonlinear differential equation in θ, due to the presence of the sine term. The first term in Eq. (a) represents the time rate of change of the angular momentum of the rotor about its axis and the second term in Eq. (a) represents the negative of the torque due to the electrical interactions between the stator and rotor that are charged by the constant voltage source E_0. Note that the nonlinear equation of motion given by Eq. (a) holds for all values of θ.

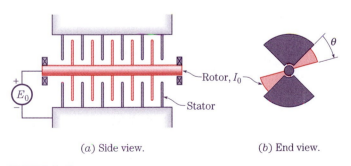

(a) Side view.   (b) End view.

FIGURE 8-42   Butterfly condenser.

For this system, perform the following tasks:

1. Find the equilibrium positions.
2. Examine the stability of all equilibrium positions.
3. Find the frequencies for small free oscillations about the stable equilibrium positions.

☐ **Solution:**

Part (1)
The (static) equilibrium positions may be found by setting $\ddot{\theta} = 0$ in Eq. (a). (Note that if an angular velocity term had been present in Eq. (a), it too would have been set to zero.) Thus, Eq. (a) reduces to

$$E_0^2 C_1 \sin 2\theta_e = 0 \tag{b}$$

which leads to

$$\theta_{e1} = 0, \qquad \theta_{e2} = \frac{\pi}{2}, \qquad \theta_{e3} = \pi, \qquad \theta_{e4} = \frac{3\pi}{2} \tag{c}$$

which are the equilibrium positions of the rotor. If the rotor is at any of these four positions, all the electrostatic torques on the rotor are completely balanced.

Parts (2) and (3)
The nonlinear equation of motion, Eq. (a), can be linearized about a generic equilibrium position θ_e by assuming a perturbation (small change) of θ in the form of

$$\theta = \theta_e + \epsilon(t) \tag{d}$$

where the time-varying perturbation $\epsilon(t)$ is small; that is, $\epsilon^2(t) \ll \epsilon(t)$. Again, note that the requirement that $\epsilon^2(t) \ll \epsilon(t)$ is a quantitative characterization of the word "small" associated with the perturbation. Substitution of Eq. (d) into Eq. (a) gives

$$I_0 \ddot{\epsilon} + E_0^2 C_1 \sin(2\theta_e + 2\epsilon) = 0 \tag{e}$$

where we note that $\ddot{\theta} = \ddot{\epsilon}$ from Eq. (d) since θ_e is a constant.

Using a trigonometric identity for the sine of the sum of two angles yields

$$\sin(2\theta_e + 2\epsilon) = \sin 2\theta_e \cos 2\epsilon + \cos 2\theta_e \sin 2\epsilon$$
$$\approx \sin 2\theta_e + 2\epsilon \cos 2\theta_e \tag{f}$$

where, in noting that ϵ is small, we have used $\cos 2\epsilon \approx 1$ and $\sin 2\epsilon \approx 2\epsilon$. (As in Example 8-19, a Taylor series expansion of $\sin 2\theta$ in the neighborhood of θ_e could also be used here.) Substitution of Eq. (f) into Eq. (e) gives

$$I_0 \ddot{\epsilon} + E_0^2 C_1 (\sin 2\theta_e + 2\epsilon \cos 2\theta_e) = 0 \tag{g}$$

which, upon recalling from Eq. (b) that $\sin 2\theta_e = 0$, reduces to

$$I_0 \ddot{\epsilon} + 2 E_0^2 C_1 \epsilon \cos 2\theta_e = 0 \tag{h}$$

Again, note that the equilibrium condition in Eq. (b) ($\sin 2\theta_e = 0$) has dropped out of the equation of motion, Eq. (g), as a consequence of the linearization. Dividing Eq. (h) by I_0 gives

$$\ddot{\epsilon} + \left[\frac{2 E_0^2 C_1}{I_0} \cos 2\theta_e \right] \epsilon = 0 \tag{i}$$

which is a linearized equation of motion in the vicinity of a generic equilibrium θ_e.

For the same reasons discussed in Example 8-19, Eq. (i) represents a system that may undergo stable oscillation if the quantity in square brackets is positive; further, in that case, the quantity in square brackets is equal to the square of the frequency of the stable oscillation. So, we shall examine the sign of the quantity in square brackets of Eq. (i) for each of the four equilibrium positions given in Eq. (c).

Because $\cos 2\theta_e = 1$ for the first and third equilibrium positions (that is, $\theta_{e1} = 0$ and $\theta_{e3} = \pi$), the system is stable about those equilibrium positions. That is, the system will oscillate freely about the equilibrium positions $\theta_{e1} = 0$ and $\theta_{e3} = \pi$ with a frequency

$$\tilde{\omega}_n = \sqrt{\frac{2E_0^2 C_1}{I_0}} \tag{j}$$

For the other two equilibrium positions $\left(\text{that is, } \theta_{e2} = \dfrac{\pi}{2} \text{ and } \theta_{e4} = \dfrac{3\pi}{2}\right)$, $\cos 2\theta_e = -1$; so, the system is unstable about these two equilibrium positions. Therefore, these two equilibrium positions are unstable. Small disturbances to the system at either of these equilibrium positions will result in large amplitudes of the rotor away from these locations. Again, considering the fact that the electrostatic forces between the stator and the rotor are attractive, the stability/instability of the four equilibrium positions should appeal to the intuition of the reader.

The results above can be interpreted easily based on the physics of the system. Note that the stator and rotor have opposite charge due to the constant voltage source E_0. Also, note that the stable equilibrium positions (that is, $\theta_{e1} = 0$ and $\theta_{e3} = \pi$) represent the case when the rotor is vertical and the stator and rotor overlap. When the rotor position is disturbed a small amount from this overlapped configuration, all forces due to the electrical attraction between the oppositely charged stator and rotor would cause the rotor to move back toward the equilibrium position. Therefore, these equilibrium positions are stable. Alternatively, consider the two unstable equilibrium positions $\left(\text{that is, } \theta_{e2} = \dfrac{\pi}{2} \text{ and } \theta_{e4} = \dfrac{3\pi}{2}\right)$. Note that these two equilibrium positions represent the case when the alignment of the rotor plates is horizontal and midway between the stator plates. In this configuration the electrical attractive forces exerted on each rotor plate by the two adjacent stator plates *are* precisely balanced. However, any perturbation of the rotor from either of these equilibrium positions would cause unbalanced electrical attractive forces, which would accelerate the rotor in the same direction of the perturbation, away from the equilibrium position. Therefore, these equilibrium positions are unstable.

In summary, there are four equilibrium positions: $\theta_{e1} = 0$, $\theta_{e2} = \dfrac{\pi}{2}$, $\theta_{e3} = \pi$, and $\theta_{e4} = \dfrac{3\pi}{2}$. Only two of these four equilibrium positions are stable; they are $\theta_{e1} = 0$, $\theta_{e3} = \pi$, with an associated free oscillation frequency given by Eq. (j). Therefore, the linear vibration theories can be applied to small-amplitude oscillations only about these two stable equilibrium positions.

■ **Example 8-21:** Plunger Switch

Consider the plunger switch shown in Figure 8-43, whose equations of motion were derived in Example 7-14 via a displacement-charge formulation, in which the generalized

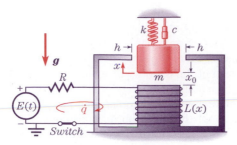

FIGURE 8-43   Electromagnetic plunger.

coordinates were x, the displacement of plunger with respect to its reference position, and q, the charge in the electrical circuit. The equations of motion are

$$m\ddot{x} + c\dot{x} + kx + \frac{L_0 \dot{q}^2}{2h\left(1 + \dfrac{x_0 + x}{h}\right)^2} = 0 \tag{a}$$

$$\frac{L_0}{\left(1 + \dfrac{x_0 + x}{h}\right)}\ddot{q} - \frac{L_0}{h\left(1 + \dfrac{x_0 + x}{h}\right)^2}\dot{x}\dot{q} + R\dot{q} = E(t) \tag{b}$$

which are nonlinear, due to the presence of the $\dot{q}^2$ term in Eq. (a), the $\dot{x}\dot{q}$ term in Eq. (b), and the x-dependent coefficients in both equations.

The first three terms in Eq. (a) represent the equation of motion for a simple mass spring dashpot system and the fourth term in Eq. (a) represents the negative of the electrical force exerted by the electromagnet. The first two terms in Eq. (b) represent the voltage drop in the coil of the electromagnet; in particular, the x-dependence of the coefficients of the first and second terms is due to the x-dependent inductance function,

$$L(x) = \frac{L_0}{1 + \dfrac{x_0 + x}{h}} \tag{c}$$

The third term in Eq. (b) is the voltage drop in the resistor R and the right-hand side of Eq. (b) is the voltage provided by the voltage supply. Note that the nonlinear equations of motion given by Eqs. (a) and (b) hold for all values of x and q where, of course, x is constrained by the upper and lower stops.

When $E(t) = E_0$ (a constant) and when $h \ll x_0$, perform the followings tasks:

1. Establish the equation(s) that must be analyzed in order to find the equilibrium positions of the mass m.

2. Analyze and discuss the stability for the equilibrium positions associated with the equation(s) obtained in part (1).

☐ **Solution:**

Part (1)

The (static) equilibrium positions may be found by setting $\ddot{x} = \dot{x} = 0$ in Eq. (a) and setting $\ddot{q} = 0$ (that is, $\dot{q}$ = constant, meaning a constant current in the circuit)[45] in Eq. (b). Therefore, Eqs. (a) and (b) become, respectively,

$$-kx_e = \frac{L_0 \dot{q}_e^2}{2h\left(1 + \dfrac{x_0 + x_e}{h}\right)^2} \tag{d}$$

$$\dot{q}_e = \frac{E_0}{R} \tag{e}$$

where x_e and $\dot{q}_e$ are the equilibrium values of the displacement and the current, respectively. Equation (e) can be used to eliminate q_e from Eq. (d). Then,

$$-kx_e = \frac{L_0 E_0^2}{2R^2 h\left(1 + \dfrac{x_0 + x_e}{h}\right)^2} \tag{f}$$

Writing out Eq. (f) and rearranging it in powers of x_e yield a cubic equation in x_e:

$$x_e^3 + 2(h + x_0)x_e^2 + (h + x_0)^2 x_e + \frac{hL_0 E_0^2}{2kR^2} = 0 \tag{g}$$

which is the equation whose solution gives the equilibrium positions x_e of the mass m.

Part (2)

The cubic equation of Eq. (g) can be solved analytically for x_e. This analytical method will be described only briefly here, without finding the closed-form analytical solution because of its complexity. Equation (g) can be rewritten as

$$x_e^3 + px_e^2 + qx_2 + r = 0 \tag{h}$$

where the coefficients of Eq. (g) are

$$p = 2(h + x_0) \tag{i}$$
$$q = (h + x_0)^2 \tag{j}$$
$$r = \frac{hL_0 E_0^2}{2kR^2} \tag{k}$$

Then, the roots of the general cubic equation, Eq. (h), are given in standard references as

$$x_e = \begin{cases} A + B - \dfrac{p}{3} \\[2mm] -\dfrac{A+B}{2} + \dfrac{A-B}{2}\sqrt{-3} - \dfrac{p}{3} \\[2mm] -\dfrac{A+B}{2} - \dfrac{A-B}{2}\sqrt{-3} - \dfrac{p}{3} \end{cases} \tag{l}$$

[45]The equilibrium conditions for electrical components may be given as follows: (1) For an inductive system whose constitutive relation is $\lambda = L\dot{q}$, $\dot{q}$ = constant or λ = constant, implying a constant current, depending on the specific formulation used; (2) for a capacitive system whose constitutive relation is $q = C\lambda$, q = constant or $\dot{\lambda}$ = constant, implying a constant voltage, depending on the specific formulation used.

where

$$A = \sqrt[3]{-\frac{b}{2} + \sqrt{\frac{b^2}{4} + \frac{a^3}{27}}} \tag{m}$$

$$B = \sqrt[3]{-\frac{b}{2} - \sqrt{\frac{b^2}{4} + \frac{a^3}{27}}} \tag{n}$$

and where

$$a = \frac{1}{3}(3q - p^2) \tag{o}$$

$$b = \frac{1}{27}(2p^3 - 9pq + 27r) \tag{p}$$

The roots given by Eq. (l) are real or imaginary depending on the sign of the radicand of the square roots in Eqs. (m) and (n). That is,

Case 1: If $\dfrac{b^2}{4} + \dfrac{a^3}{27} > 0$, there will be a real root and two conjugate imaginary roots.

Case 2: If $\dfrac{b^2}{4} + \dfrac{a^3}{27} = 0$, there will be three real roots of which at least two are equal.

Case 3: If $\dfrac{b^2}{4} + \dfrac{a^3}{27} < 0$, there will be three real and unequal roots.

In this problem, the roots of Eq. (g) can be found by finding a and b from Eqs. (o) and (p), using Eqs. (i) through (k); finding A and B using Eqs. (m) and (n); and finally, substituting the thus found A, B, and p into Eq. (l). Because of the protracted character of these nevertheless straightforward calculations, we shall not find the closed-form solution.

The conditions for the three different root compositions can be found, however, in accordance with the conditions following Eq. (p). In this regard, if all system parameters except E_0 were fixed, it could be found[46] that

Case 1: Equation (g) has one real and two conjugate imaginary roots when $E_0 > E_m$;

Case 2: Equation (g) has three real roots, two of which are equal when $E_0 = E_m$; and

Case 3: Equation (g) has three real unequal roots when $E_0 < E_m$;

where in all cases

$$E_m^2 = \frac{8}{27}\frac{kR^2}{hL_0}(h + x_0)^3 \tag{q}$$

These different cases are illustrated in Figure 8-44.

In Figure 8-44 are plotted the negative of the spring force, that is, $f_s = -kx$ as given by the left-hand side of Eq. (f), and the electrical force f_e as given by the right-hand side of Eq. (f), both versus the displacement x of the mass m. Note that the electrical force f_e shifts vertically in accordance with the value of E_0. Therefore, three typical plots are

---

[46]See Problem 8-101.

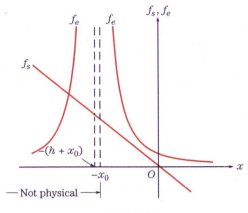

(a) $E_0 > E_m$.

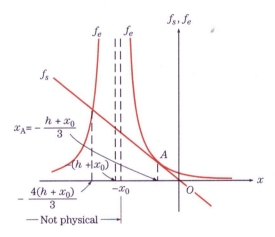

(b) $E_0 = E_m$.

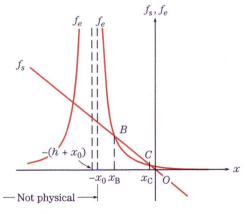

(c) $E_m < E_0$.

FIGURE 8-44 Equilibrium positions as intersections of spring force and electrical force for various values of E_0.

shown in Figure 8-44a, b, and c representing Cases 1, 2, and 3, respectively. Because the equilibrium condition given by Eq. (f) requires $f_s = f_e$, each of the intersections of the two curves for f_s and f_e represents an equilibrium position and the corresponding abscissa of each intersection gives an equilibrium position x_e for the mass m.

When E_0 is larger than E_m (Case 1 above), there is only one mathematical intersection between the f_s-curve and the f_e-curve, as shown in Figure 8-44a. Note that due to the geometrical constraint, the displacement x cannot be more negative than $-x_0$ and thus the region $x < -x_0$ is denoted as being "not physical" in Figure 8-44. Therefore, the mathematical intersection of the two curves in Figure 8-44a is not physical and thus there is no physically possible equilibrium in this case. When E_0 is equal to E_m (Case 2 above), there are two intersections between the f_s-curve and the f_e-curve as shown in Figure 8-44b[47]:

$$x_e = -\frac{4(h + x_0)}{3} \text{ and a double root } x_e = -\frac{(h + x_0)}{3}. \text{ Of these two mathematical equilib-}$$

rium positions, only one is physically possible; that is, $x_e = -\dfrac{(h + x_0)}{3}$. When E_0 is smaller than E_m (Case 3 above), there are three intersections between the f_s-curve and the f_e-curve as shown in Figure 8-44c. Because one of these three mathematical intersections is in the "not-physical" region and therefore should be discarded, there are only two physically possible equilibrium positions: one between $-(h + x_0)$ and $-\dfrac{(h + x_0)}{3}$, the other between $-\dfrac{(h + x_0)}{3}$ and 0. Here we recall that the region $x < -x_0$ is not physical; so, the equilibrium position between $-(h + x_0)$ and $-\dfrac{(h + x_0)}{3}$ will not be physical if it falls into the range $-(h + x_0) < x_e < -x_0$. However, we shall not consider this case further because of the assumption that $h \ll x_0$ in the problem statement. Thus, in the discussions that follow, we shall assume that both of the physically possible equilibrium positions for Case 3 do occur.

In summary, (1) when $E_0 > E_m$, there is no equilibrium position; (2) when $E_0 = E_m$, there is one equilibrium position; (3) when $E_0 < E_m$, there are two equilibrium positions.

The stability of these equilibrium positions could be explored analytically if the closed-form roots to Eq. (g) had been found for the equilibrium positions x_e. That is, the equations of motion given by Eqs. (a) and (b) could be linearized about the equilibrium values x_e and $\dot{q}_e$, by replacing x by $x_e + \epsilon_x(t)$ and by replacing $\dot{q}$ by $\dot{q}_e + \epsilon_{\dot{q}}(t)$, as in Examples (8-19) and (8-20). The equilibrium conditions would produce canceling terms, which would drop out of the linearized equations of motion. These linearized equations of motion would be coupled in $\epsilon_x(t)$ and $\epsilon_{\dot{q}}(t)$. The stability of *each* equilibrium position could be found by assuming exponential time functions for $\epsilon_x(t)$ and $\epsilon_{\dot{q}}(t)$, substituting these assumed solutions into the linearized equations, and examining the sign of the time function exponents that satisfy the linearized equations of motion. Since this entire process is protracted, we shall not pursue this analytical method for the equilibrium positions and their stability.

As an alternative to closed-form equilibrium and stability analyses, the equilibrium positions and their stability can be analyzed by a graphical method and physical arguments, as already initiated above. For instance, in this example, the analysis starts with several observations regarding Eq. (f). First, the left-hand side of Eq. (f) is the negative of the spring force and the right-hand side is the electrical force, both exerted on the plunger

---

[47]See Problem 8-101.

mass. Therefore, equilibrium positions are the positions where the spring force and the electrical force are balanced. Second, since the right-hand side of Eq. (f) is always numerically positive, the left-hand side must be positive, which means that the equilibrium position x_e must be negative. Third, when x is negative, the spring force pulls the plunger upward (that is, positive in accordance with the sign convention of x defined in Figure 8-43) while the electrical force pulls the plunger downward (that is, negative in accordance with the sign convention of x).

Recall from Figure 8-44 that equilibrium positions exist only when $E_0 \leq E_m$. Thus, the case of $E_0 = E_m$ in Figure 8-44b and the case of $E_0 < E_m$ in Figure 8-44c are considered. When $E_0 = E_m$, there exists only one equilibrium labeled A in Figure 8-44b, whose x-coordinate (say, x_A) is the equilibrium position of the mass and is given by $x_A = -\dfrac{(h + x_0)}{3}$. In order to determine the stability of this equilibrium position, consider a perturbation or small departure of x from the equilibrium position x_A. When the plunger is moved upward a little (that is, $x > x_A$), it may be noted from the curves in Figure 8-44b that the (downward) electrical force exceeds the (upward) spring force, resulting in a net downward force and consequently a downward motion of the plunger toward the equilibrium position. Therefore, the equilibrium position x_A seems to be stable for an upward perturbation of the plunger. However, when a downward perturbation of the plunger is considered, the curves in Figure 8-44b reveal that the (downward) electrical force exceeds the (upward) spring force, resulting in a net downward force and a further downward motion of the plunger. Thus, the equilibrium position x_A is unstable for a downward perturbation of the plunger. Furthermore, even when the plunger is given an upward perturbation, the resulting downward motion may take the plunger below the equilibrium position, "overshooting" the equilibrium position, especially if the damping of the system is less than critical. Then, the net force would act downward and, therefore, the plunger would continue to move downward, away from the equilibrium position. That is, the stability of the equilibrium position x_A concluded above for an upward perturbation is not valid in general. Therefore, the equilibrium position x_A is unstable.

When $E_0 < E_m$, there exist two equilibrium positions labeled B and C in Figure 8-44c, whose x-coordinates (say, x_B and x_C) are the equilibrium positions of the mass. To examine the stability of the equilibrium point B first, consider perturbations of x from the equilibrium position x_B. When the plunger is perturbed upward slightly (that is, $x > x_B$), the upward spring force exceeds the downward electrical force, resulting in a net upward force and consequently further upward motion of the plunger. Similarly, when the plunger is perturbed downward (that is, $x < x_B$), the downward electrical force exceeds the upward spring force, resulting in a net downward force and consequently a further downward motion of the plunger. Therefore, the equilibrium position x_B is an unstable equilibrium. Now, consider the equilibrium position x_C. When the plunger is perturbed upward (that is, $x > x_C$), the downward electrical force exceeds the upward spring force, resulting in a net downward force and consequently a downward motion of the plunger toward the equilibrium position x_C. Similarly, when the plunger is perturbed downward (that is, $x < x_C$), the upward spring force exceeds the downward electrical force, resulting in a net upward force and consequently an upward motion of the plunger toward the equilibrium position x_C. Therefore, the equilibrium position x_C is a stable equilibrium.

Thus, the vibration theories developed in the previous sections can be applied to small amplitude oscillations about the equilibrium position x_C, which exists only if $E_0 < E_m$.

In Examples 8-19 through 8-21, the equilibrium positions were found and the stability of each equilibrium position was examined. In Examples 8-19 and 8-20, the solution of the linearized equations of motion was assumed to be of the form $e^{\lambda t}$ and, by examining the λ's that satisfied the linearized equations, the stability of each equilibrium position was explored. Note that since the analyses were based on the linearized equations, the resulting stability described the linearized systems, not the original nonlinear systems. In Example 8-21, however, due to the complexity of the problem, a graphical method was used to explore the stability, based on the relative magnitudes of opposing forces acting during perturbations about the equilibrium positions. Note that the graphical technique was based on the nonlinear equations and therefore the resulting stability described the original nonlinear system, not the linearized system. This is a simple example of *nonlinear stability analysis,* a topic that is explored briefly in Appendix J.

The analyses in Examples 8-19 through 8-21 were based substantially on physical arguments; yet, it is likely that the reader has detected some suggestion of an underlying structure or procedure. Well, there was a procedure. The procedure used in Examples 8-19 and 8-20 and described in Example 8-21 for finding equilibrium configurations and their corresponding linearized stability is summarized in Table 8-10. Here we use the phrase *equilibrium configuration* as opposed to *equilibrium position* to emphasize the generality of the concept and the possibility that the variables of interest may not be geometric. First, establish the nonlinear equation(s) of motion; let's say there is one such equation of motion. Second, from the equation of motion, find the equilibrium configurations by setting, for example in mechanical systems, the velocity and acceleration to zero. Third, linearize the nonlinear equation of motion about the equilibrium configurations. Fourth, assume an exponential form solution for the linearized equation of motion and find the exponents of the solution that satisfies the linearized equation. Fifth, according to the exponents found for each equilibrium configuration, determine the stability for that equilibrium. Appendix J contains a brief and more formal presentation of the procedure summarized in Table 8-10. For illustrative purposes, the inverted pendulum in Example 8-19 is reexamined in parallel with that presentation.

TABLE 8-10 Summary of Procedure for Finding Equilibrium Configurations and Their Corresponding Linearized Stability

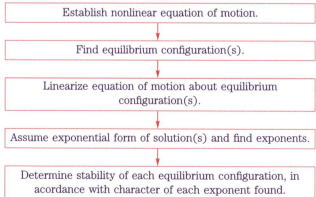

Problems for Chapter 8

Category I: Back to Physics

Problem 8-1 (*HRW407, 59E*):[48] In Figure P8-1 are shown two oscillating systems: a mass spring system and a simple pendulum, both under gravity. There is an interesting relation between these two systems. Suppose you hang a mass on the end of a spring, and when the mass is at rest, the spring is stretched a distance h. Show that the natural frequency of this mass spring system is the same as that of a simple pendulum having the same mass and length h.

Problem 8-2 (*HRW410, 85P*): Assume that you are examining the mechanical characteristics of the suspension system of a 2000 kg automobile. The suspension "sags" 10 cm when the weight of the entire automobile is placed on it. In addition, the amplitude of oscillation decreases by 50% during one complete cycle of free oscillation. Estimate the values of the spring constant and dashpot constant for the spring and shock absorber system of one wheel, assuming each wheel supports 500 kg.

Problem 8-3 (*HRW407, 56E*): A 2500 kg demolition ball swings from the end of a crane, under gravity (Figure P8-3). The length of the swinging segment of cable is 17 m.

    (*a*) Find the period of swing, assuming that the system can be treated as a simple pendulum.

    (*b*) Does the period depend on the ball's mass?

Problem 8-4 (*HRW405, 36P*): A block of weight 14.0 N, which slides without friction on a 40.0° incline, is connected to the top of the incline by a massless spring of unstretched length 0.45 m and spring con-

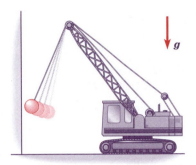

FIGURE P8-3

stant 120 N/m, as shown in Figure P8-4*a*. Note that gravity acts on the system.

    (*a*) Find the equilibrium position of the block.

    (*b*) If the block is pulled slightly down the incline from the equilibrium position, and released, what is the period of the ensuing oscillations?

    (*c*) If the mass spring system is now hung vertically as shown in Figure P8-4*b*, what is the period of the free oscillations of the system? Compare the result with part (*b*), and explain.

Problem 8-5 (*HRW405, 33P, 34P*): Two springs k_1 and k_2 are attached to a block of mass m and to fixed walls as shown in Figure P8-5. The block can move along the horizontal surface without friction.

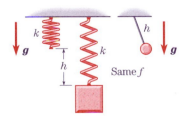

FIGURE P8-1

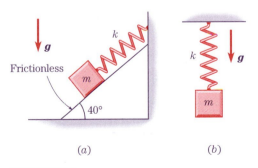

    (*a*)   (*b*)

FIGURE P8-4

[48]These designations reference Halliday, Resnick, and Walker. For example, "HRW407, 59E" denotes "Halliday, Resnick and Walker; page 407, problem 59E."

FIGURE P8-5

(a) Show that the natural cyclic frequency (in hertz) of the block's oscillations is

$$f_n = \frac{1}{2\pi}\sqrt{\frac{k_1 + k_2}{m}}$$

(b) Show that the natural cyclic frequency in part (a) may be written as

$$f_n = \sqrt{f_{n1}^2 + f_{n2}^2}$$

where f_{n1} and f_{n2} are the natural cyclic frequencies (in hertz) at which the block would oscillate if connected only to spring k_1 or only to spring k_2, respectively.

Problem 8-6 (*HRW409, 76P*): Consider the rotating disk in Problem 6-45.

(a) Find the natural frequency for small rotational oscillations of the disk.

(b) For what value of r ($0 \le r \le R$) will the natural frequency become maximum? What is that maximum natural frequency?

(c) What is the natural frequency when $r = 0$? Will the disk still oscillate?

Problem 8-7 (*HRW392, Eqn.(14-32)*): Using the equation of motion derived in Problem 6-46, show that the natural frequency for small motion about the equilibrium position of a general plane pendulum is

$$\omega_n = \sqrt{\frac{Mgh}{I_o}}$$

where M is the mass of the pendulum; g is the gravitational acceleration; h is the distance between the pivot point and the center of mass of the pendulum; and I_o is the moment of inertia of the pendulum about the pivot point.

Problem 8-8 (*HRW408, 66E*): Find the natural frequency of the composite pendulum sketched in Problem 6-7 by substituting the solutions of Problem 6-7 into the equation given in Problem 8-7.

Problem 8-9 (*HRW409, 78P*): A pendulum has two possible pivot points A and B; point A has a fixed

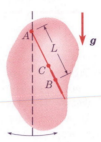

FIGURE P8-9

position and B is adjustable along the length of the pendulum, as shown in Figure P8-9. The period of the pendulum when suspended from A is found to be T. The pendulum is then reversed and suspended from B, which is moved until the pendulum again has period T. Using the equation given in Problem 8-7, show that the gravitational acceleration g is given by

$$g = \frac{4\pi^2 L}{T^2}$$

in which L is the distance between A and B for equal periods T. (Note that g can be measured in this way without knowing the moment of inertia of the pendulum or any of its dimensions except L.)

Problem 8-10 (*HRW389, Eqn.(14-25)*): Using the equation of motion derived in Problem 6-49, show that the natural frequency for small angular oscillations of a general torsional pendulum is given by

$$\omega_n = \sqrt{\frac{k_t}{I}}$$

where k_t is the spring constant of the torsional spring; I is the moment of inertia of the pendulum about its axis of rotation.

Problem 8-11 (*HRW405, 38P*): Three 10,000 kg ore cars are held at rest on a 30° incline on a mine rail-

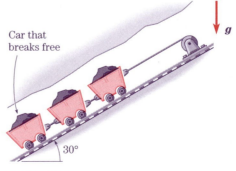

Car that breaks free

30°

FIGURE P8-11

way using a cable that is parallel to the incline (Figure P8-11). The cable is observed to stretch 15 cm just before the coupling between the middle and lowest cars breaks, detaching the lowest car. Assume that the cable's stiffness can be modeled as a linear spring. Note that gravity acts on the system.

(a) Find the frequency of the resulting oscillations of the remaining two cars.

(b) Find the amplitude of the resulting oscillations of the remaining two cars.

Problem 8-12 (*HRW406, 45E*): A block of mass M, at rest on a horizontal frictionless table, is attached to a rigid support by a spring of spring constant k. A bullet of mass m and velocity v_0 strikes the block as shown in Figure P8-12. The bullet remains embedded in the block.

(a) Determine the velocity of the block immediately after the collision.

(b) Determine the amplitude of the resulting simple harmonic motion of the block.

Problem 8-13 (*HRW404, 25P*): Two blocks ($m = 1$ kg and $M = 10$ kg) and a spring ($k = 200$ N/m) are arranged on a horizontal frictionless surface as shown in Figure P8-13. The coefficient of static friction between the two blocks is 0.40. What is the maximum possible amplitude of the simple harmonic motion if no slippage is to occur between the two blocks?

Category II: Intermediate

Single-degree-of-freedom first-order systems: Problems 8-14 through 8-26

Problem 8-14: Show that in the free response of single-degree-of-freedom first-order systems, the time-axis intercept of a tangent straight line to the response curve at any time t is simply given by $t + \tau$ where τ is the time constant of the system.

Problem 8-15: The step response of a single-degree-of-freedom first-order system (Subsection 8-2.2) can be obtained in an alternative manner by utilizing the free response solution. Toward this end, define a new variable $\tilde{v}(t) \equiv v(t) - v_f$.

(a) Show that the equation of motion can be written in terms of the new variable $\tilde{v}(t)$ as

$$\tau \frac{d\tilde{v}}{dt} + \tilde{v} = 0$$

which is in the same form as the equation of motion for the free response in Eq. (8-3).

(b) Using the result of part (a) above with the corresponding initial condition $\tilde{v}(0) = v_0 - v_f$, find the solution $\tilde{v}(t)$ from the free response solution in Eq. (8-12).

(c) From the result of part (b) above, find the step response $v(t)$. Compare the result with Eq. (8-20).

Problem 8-16: Consider the rotational system sketched in Figure P8-16, where the rotational spring constant and the rotational dashpot constant are given as $k_t = 100$ N·m/rad and $c_t = 100$ N·m·s/rad, respectively. When the left-hand end of the rotational spring is suddenly rotated by 1 radian and held there, find the angular motion $\theta(t)$ of the right-hand end of the spring as a function of time.

Problem 8-17: It can be shown that the response of a single-degree-of-freedom first-order system subjected to a *general* force $f(t)$, governed by

$$\tau \frac{dv}{dt} + v = f(t)$$

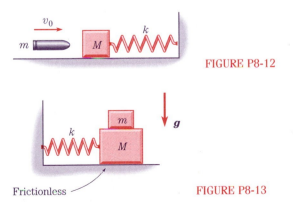

FIGURE P8-12

FIGURE P8-13

Frictionless

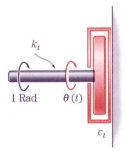

1 Rad $\theta(t)$

k_t

c_t

FIGURE P8-16

and subjected to the initial condition $v(0) = v_0$, can be found as

$$v(t) = v_0 e^{\frac{-t}{\tau}} + e^{\frac{-t}{\tau}} \int_0^t \frac{f(t')}{\tau} e^{\frac{t'}{\tau}} dt'$$

where t' is a dummy variable. (See, for example, G. Strang, *Introduction to Applied Mathematics*, Wellesley-Cambridge Press, Wellesley, MA, 1986, pp. 472–473.)

Using this general solution, find the ramp response for $f(t) = a_0 t, t \geq 0$ and the initial condition $v(0) = v_0$. Compare the result with Eq. (8-37).

Problem 8-18: Show that Eq. (8-51) can be rewritten as Eqs. (8-52) and (8-53). (*Hint:* See Section H-6 in Appendix H.)

Problem 8-19: Show that the complete response for a single-degree-of-freedom first-order system to multiple harmonic excitations can be found as Eqs. (8-74) and (8-75), by (1) superposing the particular solution in Eq. (8-56) and the homogeneous solution in Eq. (8-68) and (2) imposing the initial condition in Eq. (8-41) to determine the coefficient C of the homogeneous solution.

Problem 8-20: Consider the system in Problem 5-20. If the vertical displacement of the horizontal link is given as sketched in Figure P8-20, find the required force $F(t)$ as a function of time. Assume $c = kt_0$.

Problem 8-21: In the spring dashpot system of Problem 5-38, find the forced response (that is, only the particular solution without considering the effects of initial conditions) of the displacement of the node between the dashpot and spring, as functions of time in the following cases.

(a) When the specified displacement of the left node $x_0(t)$ is as sketched in Figure P8-21a.

(b) When the specified displacement of the left node $x_0(t)$ is as sketched in Figure P8-21b.

(c) When the specified displacement of the left node $x_0(t)$ is as sketched in Figure P8-21c.

(d) Can you obtain the solution of part (c) exclusively from the solutions of parts (a) and (b)? If so, explain how and why.

Problem 8-22: Consider the system in Problem 5-39. Find the forced response (that is, only the particular solution without considering the effects of initial conditions) of the displacement of the node between the dashpot and spring, as functions of time in the following cases.

(a) When the specified displacement of the left node $x_0(t)$ is as sketched in Figure P8-22a.

(b) When the specified displacement of the left node $x_0(t)$ is as sketched in Figure P8-22b.

(c) When the specified displacement of the left node $x_0(t)$ is as sketched in Figure P8-22c.

(d) Can you obtain the solution of part (c) exclusively from the solutions of parts (a) and (b)? If so, explain how and why.

Problem 8-23: Find the velocity of the masses sketched in Problem 6-50, as a function of time, after the support of mass m_2 is suddenly removed.

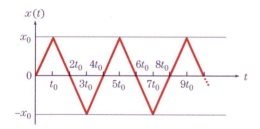

FIGURE P8-20

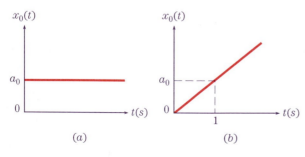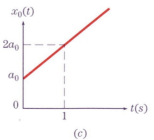

FIGURE P8-21

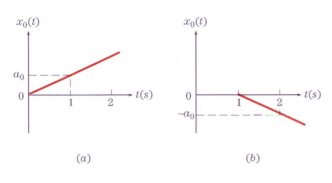

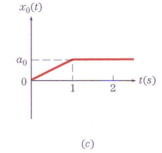

(a)   (b)   (c)   **FIGURE P8-22**

Problem 8-24: Consider the mechanical system sketched in Problem 5-21. Let $m = 10$ kg and $c = 10$ N·s/m. When the force $f(t) = 100 \cos 20t$ N is suddenly applied at $t = 0$ and the initial velocity of the mass is 0.1 m/s toward the left, find the velocity and displacement of the mass, as functions of time. Find the steady-state velocity and steady-state displacement of the mass.

Problem 8-25: In the electrical circuit sketched in Figure 8-1e, find the charge across the capacitor C as a function of time after the voltage source $E(t)$ is suddenly increased from zero to E_0 and remains constant. Assume that the initial charge across the capacitor is q_0.

Problem 8-26: Consider the mechanical system sketched in Figure 8-1c. The harmonic response of the system was recorded for the applied forces $f(t) = 10 \sin \Omega t$ newtons at two different forcing frequencies $\Omega_1 = 3$ rad/s and $\Omega_2 = 9$ rad/s. The amplitudes of the steady-state displacement of the system were measured as 0.02 m and 0.01 m for the two different forcing frequencies Ω_1 and Ω_2, respectively. Based on these measurements, evaluate the spring constant and the dashpot constant of the system.

Single-degree-of-freedom second-order systems: Problems 8-27 through 8-54

Problem 8-27: Determine the constants C_0, ψ_1, and ψ_2 in Eqs. (8-103) and (8-104) by imposing the initial conditions in Eqs. (8-85) and (8-86).

Problem 8-28: Determine the constants C_0, ψ_1, and ψ_2 in Eqs. (8-103) and (8-104) by rewriting Eq. (8-102). (*Hint:* See Section H-6 in Appendix H.)

Problem 8-29: Determine the constants C, ψ_1, and ψ_2 in Eqs. (8-127) and (8-128) by imposing the initial conditions in Eqs. (8-85) and (8-86).

Problem 8-30: Determine the constants C, ψ_1 and ψ_2 in Eqs. (8-127) and (8-128) by rewriting Eq. (8-126). (*Hint:* See Section H-6 in Appendix H.)

As a result, show that the free vibration of a single-degree-of-freedom *underdamped* second-order system subjected to the initial displacement $x(0) = x_0$ and initial velocity $\dot{x}(0) = v_0$ may be expressed alternatively as

$$x(t) = Ae^{-\zeta\omega_n t} \sin(\omega_d t + \psi_1)$$

or

$$x(t) = Ae^{-\zeta\omega_n t} \cos(\omega_d t - \psi_2)$$

where

$$A = \frac{\sqrt{(x_0\omega_d)^2 + (v_0 + \zeta\omega_n x_0)^2}}{\omega_d}$$

$$\psi_1 = \tan^{-1} \frac{x_0\omega_d}{v_0 + \zeta\omega_n x_0}$$

$$\psi_2 = \tan^{-1} \frac{v_0 + \zeta\omega_n x_0}{x_0\omega_d}$$

These are very convenient forms for this response, worth committing to memory.

Problem 8-31: A very lightly damped flexible floor system deflects downward by 6 cm at the location

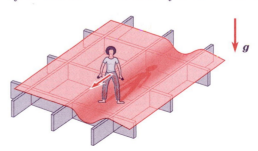

FIGURE P8-31

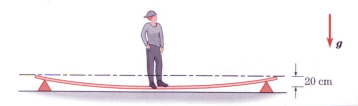

20 cm

FIGURE P8-32

where a man stands still, due to the man's weight (Figure P8-31). If he wants to avoid excessive vibration of the floor system caused by his walking (or running) on it, how many steps should he not take in a minute?

Problem 8-32: A woman stands at the midspan of a very lightly damped flexible board as sketched in Figure P8-32. The center of the board has been measured to sag by 20 cm due to her weight. If the woman flexes her knees harmonically, what frequency of her knee-flexing motion will cause the greatest deflection of the board?

Problem 8-33: A heavy crate of unknown mass is suspended from a rigid crane by an extensible cable as sketched in Figure P8-33 and is observed to undergo vertical oscillations of period 0.5 seconds. Find the static elongation of the cable when the crate is suspended by the cable and stays still.

Problem 8-34: The front bumper of a car is being tested by driving the car forward against a rigid wall at a speed of 10 km/h. (See Figure P8-34.) The car weighs 1 ton and the equivalent spring constant of the bumper is 5×10^6 N/m. Assume that the mass and effective damping of the bumper are negligible for such tests.

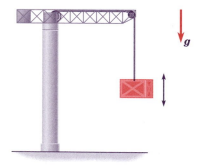

FIGURE P8-33

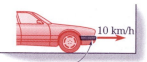

10 km/h

Bumper, $k = 5 \times 10^6$ N/m   FIGURE P8-34

(a) Determine the velocity and displacement of the car as functions of time during the period the bumper is in contact with the wall.

(b) If the bumper deforms more than 5 cm, plastic deformation remains even after unloading. Find the maximum initial velocity of the car without causing plastic deformation of the bumper.

Problem 8-35: The shafting and propeller of a propulsion system in a ship are connected to the turbine, as sketched in Figure P8-35. The mass and the axial rotary inertia of the propeller are given as $m = 200$ kg and $I_0 = 1000$ kg $\cdot$ m^2, respectively. The propeller is connected to a turbine, rotating at a constant angular velocity, by a shaft consisting of two sections made of the same material having a shear modulus of $G = 1.0 \times 10^{11}$ N/m^2. The two sections of the shaft have polar moments of the cross-sectional area about the shaft axis of $J_1 = 8 \times 10^{-7}$ m^4 and $J_2 = 4 \times 10^{-6}$ m^4 and, lengths 1 m and 1.5 m, respectively, as shown in Figure P8-35. For this analysis, the rotating inertia of all the shafts may be assumed to be negligible compared with the rotary inertia of the

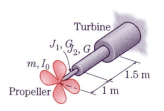

Turbine

J_1, G, J_2, G

m, I_0

Propeller

1.5 m

1 m

FIGURE P8-35

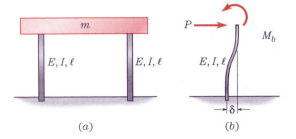

(a)

(b)

FIGURE P8-36

turbine. Find the natural frequency for torsional vibration of the propeller-shaft system.

Problem 8-36: A simple model for a one-story building is sketched in Figure P8-36a. The roof of the building is supported by four columns, each of which is clamped into the roof and the ground at the upper and lower ends, respectively, so that rotation is not allowed at either end. Such column or beam models are sometimes called "guided cantilevers." When a column is subjected to such boundary conditions where a transverse force P and the corresponding moment M_b at one end are as shown in Figure P8-36b, the transverse deflection δ of the column at the tip of the applied force is

$$\delta = \frac{Pl^3}{12EI}$$

where l is the length of the column, E is the modulus of elasticity of the column, and I is the second moment of the cross-sectional area of the column about its neutral axis. Find the natural frequency for the small horizontal vibration of the roof, assuming the four columns all deform only in the configuration shown in Figure P8-36b and the mass of the columns is negligible. Assume gravity is negligible.

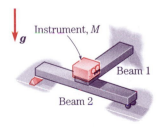

FIGURE P8-37

Problem 8-37: As sketched in Figure P8-37, an instrument of mass M is mounted on a beam system that consists of a cantilever beam (Beam 1) and a simply supported beam (Beam 2). The length, modulus of elasticity, and the second moment of the cross-sectional area about the neutral axis are given as l_1, E_1, and I_1 for Beam 1 and l_2, E_2, and I_2 for Beam 2. Note that the instrument is mounted at the center of Beam 2 and that gravity acts on the system. In this analysis, the beams have negligible mass.

It is known that the tip deflection of a cantilever beam subjected to a transverse force P applied at its tip is given by

$$\delta = \frac{Pl^3}{3EI}$$

and that the deflection at the midspan of a simply supported beam subjected to a transverse force P applied at its midspan is given by

$$\delta = \frac{Pl^3}{48EI}$$

(a) Find the static deflection of the instrument.

(b) Find the natural frequency of the system during small vertical vibration of the instrument.

Problem 8-38: Figure P8-38 shows two systems where a mass m_0 falls due to gravity from rest from a height h above a platform and moves with the platform after the collision.

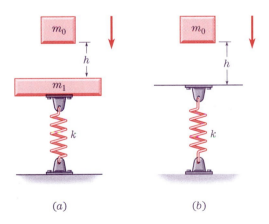

(a)

(b)

FIGURE P8-38

(a) When the platform has mass m_1 (Figure P8-38a), obtain the motion of the platform after the collision.

(b) When the platform is massless (Figure P8-38b), obtain the motion of the platform after the collision.

Problem 8-39: Consider the system sketched in Problem 6-54, where a particle of small mass m_0 is dropped from a height h above the bar and it strikes and sticks to the bar at the point B located a distance b from the pivot point O.

(a) Find the motion of the system after the collision of the particle with the bar.

(b) Suppose the maximum load that the spring can carry is $f_{\max}$. Find the condition on the distance b for which the spring does not rupture.

Problem 8-40: In a junkyard, a trashed car is being moved by a magnetic crane. (See Figure P8-40.) Once it has been moved to the desired location, the magnet of the crane is deactivated so that the car is dropped. The spring constant of each of the four independent suspension systems of the car is 20 kN/m including the stiffness of tires, and the mass of the car is 1200 kg. The vertical oscillation of the car after being dropped onto the ground is considered, assuming the tires do not bounce from the ground and the body undergoes only vertical translational motion due to the even distribution of its weight among the four wheels. Note that gravity acts on the system.

(a) Find the natural frequency for the vertical vibration of the car.

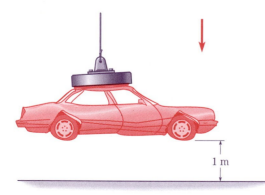

FIGURE P8-40

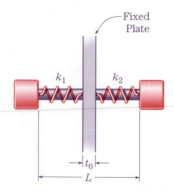

FIGURE P8-41

(b) If the bottoms of the tires are 1 m above ground prior to the drop, find the motion of the car as a function of time. Assume that the suspensions are neither stretched nor compressed before the drop.

Problem 8-41: As sketched in Figure P8-41, two springs of different spring constants k_1 and k_2, respectively, support a rigid plunger of mass m. No part of the two springs is welded to the plunger or the fixed plate structure; yet, the ends of each spring remain in contact either with a plunger head or the plate because they were designed to be in compression at all times. The distance between the plunger heads is L and the thickness of the plate is t_0, as sketched in Figure P8-41. Also, the undeformed length of each spring is l_0.

(a) Find the condition on the dimensions L, t_0, and l_0 for the springs to be in compression at all times.

(b) Find the equilibrium position of the plunger.

(c) Find the natural frequency for the free vibration of the plunger-spring system about the equilibrium position.

Problem 8-42: Find the natural frequency for small vibration of the system sketched in Problem 6-92.

Problem 8-43: Consider the motion of the pulley system sketched in Problem 6-91.

(a) Find the natural frequency for the rotational oscillation of the composite pulley.

(b) If the strap connecting the mass m_0 suddenly breaks when the system is at rest in equilibrium, find the ensuing motion. Assume that the system is underdamped.

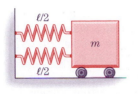 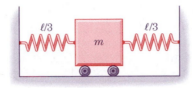

(a)   (b)   **FIGURE P8-44**

Problem 8-44: A spring of stiffness k and length l is cut into halves and a mass m is connected to the two parts as sketched in Figure P8-44a. The natural period T_n of this system is found to be 0.5 second.

If a spring that is identical to the original spring of stiffness k and length l is now cut such that one part is $\dfrac{l}{3}$ and the other part is $\dfrac{2l}{3}$, and the same mass m is connected as shown in Figure P8-44b, find the natural period of this new system.

Problem 8-45: Find the natural frequency for small motions of the system sketched in Problem 6-83.

Problem 8-46: Find the natural frequency of the lumped-parameter torsional system sketched in Problem 6-55.

Problem 8-47: Find the natural frequency for small rotational motions of the torque meter dial sketched in Problem 6-93.

Problem 8-48: Find the natural frequency for small motions of the system sketched in Problem 6-112.

Problem 8-49: The system sketched in Figure P8-49 is dropped due to gravity from rest from a height h to a rigid surface. Prior to being dropped,

the system was at equilibrium, while being supported by its massless base. The variable x is the displacement of the mass from its equilibrium position prior to being dropped. Assume that the system is critically damped.

   (a) Find the motion of m resulting from the impact of the system with the surface.

   (b) If the maximum load in the spring without its rupture is f_{max}, find the maximum height h from which the system can be dropped without the spring's rupture.

Problem 8-50: The springs of a single-axle automobile trailer are compressed four inches under its weight. Find the critical speed—that is, the speed at which the trailer will undergo maximum vertical displacement—when the trailer is traveling over an undulating road with a profile approximated by a sine wave of wavelength 48 feet. Assume the damping of the trailer's suspension system is negligible.

Problem 8-51: Consider small rocking motions of the semicylinder sketched in Problem 6-96. Assume the radius of the semicylinder is $r = 10$ cm.

   (a) Find its natural frequency for the small motion.

   (b) If the semicylinder is held at rest, tilted by $5°$ from its equilibrium position and released, find the ensuing motion.

Problem 8-52: Consider the weighing scale sketched in Problem 6-84, where $m_0 = 1$ kg, $k = 400$ N/m, $a = 5$ cm, $b = 10$ cm, and the viscous damping factor is known to be $\zeta = 0.8$.

   (a) Find the natural frequency of the system.

   (b) An object weighing 3 kg is removed after the scale reading indicated and remained at "3 kg". Find the time required for the scale reading to return to "0" for the first time.

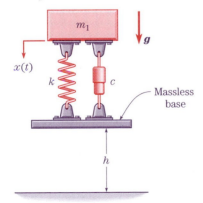

FIGURE P8-49

Problem 8-53: Show that the numerator in Eq. (8-159) can be rewritten as Eqs. (8-160) and (8-161). (*Hint:* See Section H-6 in Appendix H.)

Problem 8-54: Show that Eq.(1) in Example 8-10 can be rewritten as Eqs. (m) and (n). (*Hint:* See Section H-4 in Appendix H.)

Two-degree-of-freedom systems: Problems 8-55 through 8-74

Problem 8-55: Consider the vibration absorber model sketched in Problem 5-56. Assuming $M = 5m$ and $K = 9k$, find the natural frequencies and mode shapes of the system.

Problem 8-56: Consider the pulley system sketched in Problem 6-85. Assuming $m_1 = m_2 = m_0 = m$ and $k_1 = k_2 = k$, find the natural frequencies and mode shapes of the system.

Problem 8-57: Find the natural frequencies and mode shapes of the double pendulum system sketched in Problem 5-63. Assume $m_1 = m_2 = m$, $l_1 = l, l_2 = 2l$ and $mg = kl$.

Problem 8-58: Find the natural frequencies and mode shapes of the system sketched in Problem 5-65.

Problem 8-59: Assuming small motions, find the natural frequencies and mode shapes of the system sketched in Problem 6-86. Assume $m_0 = m, mg = kl$ and $k_t = kl^2$.

Problem 8-60: Assuming small motions and letting $m_0 = m, mg = kl$ and $k_t = kl^2$, find the natural frequencies and mode shapes of the system of Example 5-23.

Problem 8-61: For the system sketched in Problem 5-71, find the natural frequencies and mode shapes, assuming $mg = \dfrac{kl}{2}$.

Problem 8-62: For the torsional system sketched in Problem 6-58, find the natural frequencies and mode shapes, assuming $m_0 = m$, $m_1 = m_2 = 2m$, $G_1 = G_2 = G_3 = G, l_1 = l_2 = l_3 = l, J_1 = J, J_2 = 2J$, $J_3 = 3J$ and $r_1 = r_2 = r$.

Problem 8-63: Find the natural frequencies and mode shapes of the torsional system of Example 8-18.

Problem 8-64: Assuming $m_1 = m, m_2 = 2m, R_1 = R, R_2 = 2R, \dfrac{GJ}{L} = ka^2$ and $a = \dfrac{R}{2}$, find the natural frequencies and mode shapes of the system of Example 6-14.

Problem 8-65: A linear dynamic system is described by the equations of motion

$$\begin{bmatrix} 4m & m \\ m & 4m \end{bmatrix} \begin{Bmatrix} \ddot{x}_1 \\ \ddot{x}_2 \end{Bmatrix} + \begin{bmatrix} 2k & -k \\ -k & 2k \end{bmatrix} \begin{Bmatrix} x_1 \\ x_2 \end{Bmatrix} = 0$$

where m denotes mass, k denotes stiffness and x denotes displacement. Find the natural frequencies and modal vectors of the system. (*Hint:* Direct use of Eqs. (8-229) through (8-235) is not possible because the equations of motion given above are not in the form of Eqs. (8-223) and (8-224). The modes for this type of system can be found by assuming the solutions in the form of Eqs. (8-227) and (8-228) and following the same procedure as in Subsection 8-4.1.)

Problem 8-66: Figure Ṗ8-66 shows a simple model for an overhead crane. The masses of the crane cabin and the load are $5m$ and m, respectively. The supporting beam of the crane is modeled as simply supported. The span, modulus of elasticity, and the second moment of the cross-sectional area about the neutral axis of the beam are L, E, and I, respectively. In this model, the mass of the beam is negligible. From elementary technical mechanics of solids, the midspan

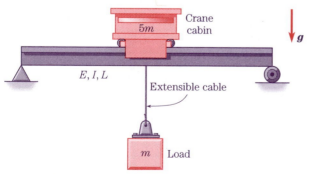

FIGURE P8-66

deflection δ for such a simply supported beam subjected to a vertical static load P at its midspan is known as

$$\delta = \frac{PL^3}{48EI}$$

A simple static test shows that the lifting cable of undeformed length l_0 elongates δ_0 when the load m is suspended by it. Note that gravity acts on the system.

Determine the natural frequency (or, frequencies and modal vectors) of the system in each of the following cases. Assume $\frac{EI}{L^3} = k_0$ and $\delta_0 = \frac{mg}{k_0}$, where k_0 is a constant having the same dimension as a spring constant.

(a) The crane cabin is midway along the span of the beam and the undeformed length of the suspended lifting cable is l_0.

(b) The crane cabin is midway along the span of the beam and the load has been hoisted fully up to the cabin.

(c) The crane cabin is at either end of the beam and the undeformed length of the suspended lifting cable is $\frac{l_0}{2}$.

(d) The crane cabin is at either end of the beam and the load has been hoisted fully up to the cabin.

Problem 8-67: An automobile of mass m_1 is towing a cart of mass m_2 as sketched in Figure P8-67. The stiffness of the link between the automobile and the cart is modeled as a linear spring of spring constant k.

(a) When $m_2 = 0.5m_1$, find the natural frequencies and mode shapes of the system.

(b) Suppose both the automobile and cart have been at rest until another car strikes the cart from the rear, imparting an initial forward velocity v_0 to the cart. Find the ensuing motion of the system. Assume that the brakes

of the automobile are not engaged and its transmission is in neutral.

(c) Find the force in the link as a function of time during the motion in part (b).

(d) Suppose the link ruptures when subjected to an axial force larger than f_{max}. Find the maximum initial speed v_0 in part (b) that will not rupture the link.

Problem 8-68: Find the response of the rotary system sketched in Problem 6-56 for the following set of initial conditions:

$$\theta_1(0) = \alpha I_2 \qquad \dot{\theta}_1(0) = 0$$
$$\theta_2(0) = -\alpha I_1 \qquad \dot{\theta}_2(0) = 0$$

where α is a nonzero constant.

Problem 8-69: Consider the system of two blocks hitting a buffer as sketched in Problem 5-51. Assume that prior to the collision with the buffer, both blocks have the same velocity v_0 and all springs are undeformed and that the blocks remain in contact with the buffer springs after the collision. Find the ensuing motion of the system. Let $m_1 = 2m_0$, $m_2 = m_0$ and $k_1 = k_2 = k_0$.

Problem 8-70: The two-degree-of-freedom system in Example 8-11 is at rest. Then, both blocks are struck sharp blows simultaneously, imparting to them the same initial velocity v_0 (toward the right).

(a) Find the subsequent motion of the system.

(b) Suppose the maximum allowed load in all three springs is the same, which is f_{max}. Find the maximum value of v_0 that will not cause any spring to rupture. Which spring will break first when v_0 exceeds this critical value?

Problem 8-71: Consider the motion of the forging machine model sketched in Problem 5-54, after the impact of the hammer onto the anvil. Assume that $m_1 = 19m_0$, $m_2 = 30m_0$, $k_2 = 49k_1$ and $c_1 = c_2 = 0$.

v_0 ←

m_2

m_1

k

(a) Find the natural frequencies and mode shapes of the system.

(b) If the velocity of the hammer immediately prior to the impact is v_0, find the ensuing motion of the anvil and the foundation block. Assume that the equilibrium positions of the anvil and the foundation block after the impact remain the same as before the impact because of the relatively small mass of the hammer.

(c) Suppose the shock absorber fails when it is subjected to a force larger than F_Y. Find the maximum allowable initial velocity of hammer v_0 that does not cause failure of the shock absorber.

Problem 8-72: The equations of motion of undamped and unforced linear lumped-parameter systems may be expressed in matrix form as (see Problem 8-65)

$$[M]\{\ddot{x}\} + [K]\{x\} = 0$$

where $\{x\}$ is the vector whose elements are the generalized coordinates; $[M]$ and $[K]$ are the symmetric mass matrix and stiffness matrix, respectively, which are 2×2 for two-degree-of-freedom systems. Substituting the modal response functions in Eqs. (8-227) and (8-228) for two-degree-of-freedom systems into the above matrix equation yields

$$[K]\{u\} = \omega^2 [M]\{u\}$$

Since all natural frequencies and modal vectors satisfy the immediately preceding equation, it can be written for the two distinct modes that

$$[K]\{u^{(1)}\} = \omega_1^2[M]\{u^{(1)}\} \quad [K]\{u^{(2)}\} = \omega_2^2[M]\{u^{(2)}\}$$

Do the following:

(a) Premultiply the first and second of the immediately preceding equations by the transposes of the second and first modal vectors (that is, $\{u^{(2)}\}^T$ and $\{u^{(1)}\}^T$), respectively.

(b) Take the transpose of the first of the resulting equations in part (a); and subtract it from the second of the resulting equations in part (a).

(c) Using the result of part (b) and the symmetry of the $[M]$ and $[K]$ matrices, show that

$$\{u^{(1)}\}^T[M]\{u^{(2)}\} = 0$$

for $\omega_1 \neq \omega_2$. This equation is said to state the *orthogonality* of modal vectors.

Problem 8-73: Show that evaluating the angle ψ in Eq. (d) in Example 8-17 alternatively as $\psi = 157.8° = -22.2° + 180°$ or $\psi = -202.2° = -22.2° - 180°$ does not affect the system response in Eq. (f).

Problem 8-74: Show that, in general, a quadratic polynomial equation $x^2 + ax + b = 0$ having two roots α and β can be factored into $(x - \alpha)(x - \beta) = 0$. (Therefore, the characteristic equation for two-degree-of-freedom systems, $\omega^4 + a\omega^2 + b = 0$ having roots ω_1^2 and ω_2^2, can be factored into $(\omega^2 - \omega_1^2)(\omega^2 - \omega_2^2) = 0$.)

State-space representation: Problem 8-75

Problem 8-75: The equation of motion for the system sketched in Figure P8-75 is given as

$$m\ddot{x} + c\dot{x} + kx = f(t)$$

This equation of motion can be rewritten in a matrix form by defining a set of independent variables called *state variables*. (See Appendix K for the definition of state variables.) Also, the vector whose elements are state variables is called the *state vector*. Though the choice of state variables is not unique, for lumped-parameter dynamic systems, the generalized coordinates and the generalized velocities constitute a convenient set of state variables. Therefore, for the system sketched in Figure P8-75, the state variables can be defined as $x_1 \equiv x$ and $x_2 \equiv \dot{x}$, yielding the state vector

$$x = \begin{Bmatrix} x_1 \\ x_2 \end{Bmatrix} = \begin{Bmatrix} x \\ \dot{x} \end{Bmatrix}$$

Finding the time derivatives of the state variables and arranging them in a standard matrix form give an alternative representation of the equation(s) of motion, called the *state-space representation*. For the choice

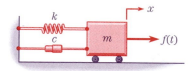

FIGURE P8-75

of the state vector above, $\dot{x}_1 = x_2$; and from the equation of motion, $\dot{x}_2 = -\dfrac{k}{m}x_1 - \dfrac{c}{m}x_2 + \dfrac{1}{m}f(t)$. Arranging these in matrix form gives

$$\dot{\boldsymbol{x}} = \begin{Bmatrix} \dot{x}_1 \\ \dot{x}_2 \end{Bmatrix} = \begin{bmatrix} 0 & 1 \\ -\dfrac{k}{m} & -\dfrac{c}{m} \end{bmatrix} \begin{Bmatrix} x_1 \\ x_2 \end{Bmatrix} + \begin{Bmatrix} 0 \\ \dfrac{1}{m} \end{Bmatrix} f(t)$$

which is the state-space representation of the equation of motion of the system sketched in Figure P8-75. For general linear lumped-parameter dynamic systems, the state-space representation of the equations of motion is in the form of

$$\dot{\boldsymbol{x}} = A\boldsymbol{x} + B\boldsymbol{u}$$

where $\boldsymbol{x}$ is the state vector, A and B are matrices, and $\boldsymbol{u}$ is the input vector to the system.

Express the equations of motion for the following systems in state-space representation.

(a) The system sketched in Figure 8-1b.

(b) The system in Problem 5-63.

Stability: Problems 8-76 through 8-80

Problem 8-76: Consider the small rocking motion of the sand container on a semicylinder, about its horizontal position, sketched in Problem 6-95. Find the condition on the depth h of sand in the container for the small rocking motion of the container to be stable.

Problem 8-77: Consider the small rocking motion of the ship sketched in Problem 6-37.

(a) Find the condition on the location of the metacenter M with respect to the center of mass C, for stable rocking motion of the ship.

(b) Find the natural frequency for such stable rocking motion of the ship.

Problem 8-78: Consider the rod system sketched in Problem 6-60. Find the critical value of the applied force F_0 for which the small motion of the rod about its horizontal position becomes unstable.

Problem 8-79: Consider the pendulum sketched in Problem 5-28.

(a) Find the equilibrium positions.

(b) Linearize the equations of motion about the equilibrium positions.

(c) Identify the stable and unstable equilibrium positions and find the frequency of small oscillation in the vicinity of each stable equilibrium position, if any.

Problem 8-80: The friction coefficient μ between two surfaces in mutual contact is known to depend generally on the relative velocity of the two surfaces as sketched in Figure P8-80a, where μ_0 is the static coefficient of friction and V is the relative velocity of the contacting surfaces.

Consider a block of mass m being pushed along a flat surface by a spring of spring constant k whose other end is attached to a cart moving at a constant speed V_0, as sketched in Figure P8-80b. Suppose the coefficient of friction between the block and the flat surface depends on the velocity of the block, as depicted in Figure P8-80a. It is desired to study the relative motion of the block with respect to the cart.

(a) Find the equilibrium position of the block relative to the cart.

(b) Linearize the equation of motion for the relative displacement of the block to the cart about the equilibrium position, for both cases of $0 < V_0 < V_1$ and $V_0 > V_1$, V_1 being indicated in Figure P8-80a.

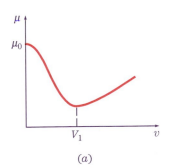

(a)

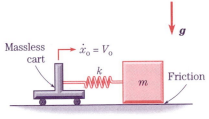

(b)

FIGURE P8-80

(a)

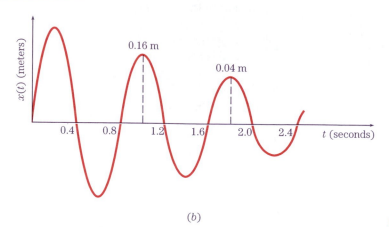

(b)

FIGURE P8-82

(c) Determine the stability of the small relative motion of the block in the vicinity of the equilibrium position, for both cases of $0 < V_0 < V_1$ and $V_0 > V_1$.

Category III: More Difficult

Single-degree-of-freedom second-order systems: Problems 8-81 through 8-87

Problem 8-81: Consider the small rotational motion of the rod sketched in Problem 6-61.

(a) Find the natural frequency of the system.

(b) If the dashpot constant is given by $c = 2\sqrt{mk}$, show that the system is underdamped.

(c) When the initial conditions on the system are given by

$$\theta(0) = 0 \quad \dot{\theta}(0) = 1$$

derive an expression for the free (unforced) response of the system. For this response, let $c = 2\sqrt{mk}$ as in part (b) above. Sketch the response, clearly indicating the parameters on your plot, including the response en-

velope, its θ-intercept at time $t = 0$ (that is, the value of the envelope at $t = 0$), and the period of oscillation.

Problem 8-82: The system sketched in Figure P8-82a is a setup to determine the weight of a tracked vehicle. The vehicle collides with a massless bumper and couples with it. The bumper displacement $x(t)$ from its equilibrium position versus time is recorded as sketched in Figure P8-82b. The spring constant of the bumper is known to be $k = 200{,}000$ N/m. Determine the value of the damping factor ζ and the mass of the vehicle in kilograms.

*Problem 8-83: When a motor is mounted on the midspan of a beam of negligible mass, the center of the beam is observed to deflect statically 0.1 meters. (See Figure P8-83 and assume the gravitational acceleration is $g \approx 10.0$ m/s².) Further, to determine the damping characteristics of the motor-beam system, the motor-beam system is deflected, released, and allowed to vibrate freely. It is observed that during free vibration, the motor's amplitude of vibration is reduced by a factor of 3 during each cycle. Furthermore, because of an unbalance in the motor, a harmonic force in the vertical direction is generated.

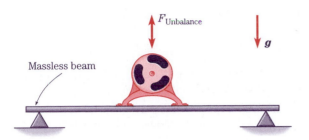

Massless beam

FIGURE P8-83

When the motor is operating at 300 rpm, the amplitude of the harmonic unbalance force is 100 newtons.

(a) What is the amplitude of the force transmitted to each support at the ends of the beam due to the harmonic unbalance force?

(b) Suppose the maximum force that each support can withstand is $f_{max} = 10$ N. Will the support break in part (a) due to the harmonic unbalance force? Assume that the static force in the springs due to the weight of the motor is negligible compared with the dynamic harmonic force due to the unbalance of the motor.

(c) If there were no damping in the system, what operational speed of the motor would be most damaging to the motor-beam structure?

Problem 8-84: Find the natural frequency of the scale sketched in Problem 6-111. How does the natural frequency change when an object of mass m_0 is placed on the weighing table?

Problem 8-85: The system consists of three thin disks (each of mass m and radius r) rigidly connected as sketched in Figure P8-85, and two shaft-beam-rod elastic elements having circular cross-sections of area A, shear modulus G, modulus of elasticity E, polar moment of cross-sectional area about the longitudinal axis J, and second moment of cross-sectional area about the neutral axis I. (Specifically, one disk each lies in the x-y, y-z, and z-x planes, and the center of each disk coincides with the origin of the xyz axes.) Ignore gravity and note that the shaft-beam-rod elastic elements are clamped in the rigid wall at the ends 1 and 2 indicated in Figure P8-85. If the mass of each shaft-beam-rod elastic element is negligible compared with that of the disks and if $r \ll L$, estimate the following:

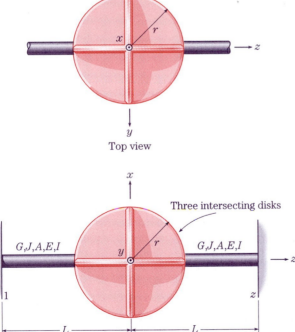

FIGURE P8-85

(a) Translational natural frequency in the x direction.

(b) Translational natural frequency in the y direction.

(c) Translational natural frequency in the z direction.

(d) Torsional natural frequency about the z axis.

Problem 8-86: Using the complex variables technique, obtain the particular solution for an undamped single-degree-of-freedom second-order system subjected to a combination of cosinusoidal and sinusoidal excitations of forcing frequency equal to the system's natural frequency (that is, $f(t) = a \cos \omega_n t + b \sin \omega_n t$). Begin by assuming a complex particular solution (Eq. (8-185))

$$x_p(t) = X_1 t e^{i \omega_n t}$$

and then find X_1 such that the complex solution satisfies the equation of motion. Compare the result with Eq. (8-171).

Problem 8-87: By following the procedure in Subsection 8-3.5, obtain the complete response of a damped single-degree-of-freedom second-order system subjected to a harmonic excitation $f(t) = a \cos \Omega t + b \sin \Omega t$ and initial conditions $x(0) = x_0$ and $\dot{x}(0) = v_0$, for various system damping factors:

(a) $0 < \zeta < 1$ (underdamped).

(b) $\zeta = 1$ (critically damped).

(c) $\zeta > 1$ (overdamped).

Compare the results with those listed in Table 8-9.

Two-degree-of-freedom systems: Problems 8-88 through 8-98

Problem 8-88: Consider the string-mass system sketched in Problem 5-66. Let $m_1 = m_2 = m$ and $L_1 = L_2 = L_3 = L$.

(a) Find the natural frequencies and mode shapes. Plot the mode shapes.

(b) Obtain the system response when the system is released from rest with the left-hand mass displaced upward and the right-hand mass downward by the same amount, equal to a unit length. In one or two sentences, explain your results.

Problem 8-89: Consider the model of a truck suspension sketched in Problem 6-87. For simplicity, assume that $k_1 = k_2 = k$, $l_1 = \dfrac{l_2}{2} = l$, and $I = Ml^2$.

(a) Calculate the modes of vibration of the truck, and sketch the mode shapes.

(b) Give a set of initial conditions that will excite only the first mode of vibration of the truck. Briefly explain your answer.

Problem 8-90: Consider the seismic structural analysis model for a two-story building sketched in Problem 5-57.

(a) Find the natural frequencies and mode shapes of the building.

(b) Suppose the building is subjected to an earthquake, modeled as an impulsive acceleration of the ground $\ddot{x}_0(t) = V_0 \delta(t)$, where $\delta(t)$ is the unit *delta function* defined such that

$$\delta(t) = \begin{cases} \infty & t = 0 \\ 0 & t \neq 0 \end{cases}$$

and

$$\int_{-\alpha}^{\alpha} \delta(t) = 1 \qquad \text{for all } \alpha > 0$$

Find the subsequent motion of the building with respect to the ground after such a shock.

Problem 8-91: Consider the coupled pendulums sketched in Problem 5-64. Assume small motions.

(a) Find the natural frequencies and mode shapes of the system.

(b) If the pendulums are released from rest with the left-hand pendulum (Pendulum 1) vertical and the right-hand one (Pendulum 2) rotated by a small angle θ_0, it is observed that Pendulum 2 oscillates for a while, then slows down as Pendulum 1 begins to oscillate, the process continuing with transferral of energy back and forth between the two pendulums. This phenomenon is called "beating" and it occurs when the coupling spring (the rubber hose in this problem) is relatively soft compared with the other springs (gravity in this problem). If $l = 19.3$ in, $mg = 3.86$ lb, and $k_t = 2.0$ lb·in/rad, determine the beating period (from Pendulum 1 to 2 and back to Pendulum 1) for such a motion.

Problem 8-92: Consider the vertical motion of the produce processing system model sketched in Problem 5-83. Assume that $m_1 = 2m$, $m_2 = m$, $m_0 \ll m$ and $k_1 = k_2 = k$.

(a) Find the dynamic response of the system after the box is dropped due to gravity from a height h above the platform m_1.

(b) Suppose the maximum load that the spring connecting m_1 and m_2 can withstand is f_{max}; find the minimum height h that causes this particular spring to fail.

Problem 8-93: Consider the system sketched in Problem 6-90. Assume that $mg = ka$.

(a) Find the natural frequencies and mode shapes of the system.

(b) Obtain the subsequent motion of the system when the system is released from rest with the spring undeformed and the pivoted body rotated by a small initial angle θ_0. (See Example 5-19 for an appropriate definition of "small" in this context.)

Problem 8-94: Consider the system sketched in Problem 6-62.

(a) Find the natural frequencies and mode shapes of the system.

(b) When the system is released from rest with the spring elongated by x_0, find the subsequent motion of the system. Based on the result, find the motion of the center of mass. Explain your answer.

(c) Suppose the system is acted upon by an external force $f(t) = f_0 \cos \Omega t$ applied horizontally to the cart and positive toward the right. Find the harmonic forced motion of the system. Based on the result, find the motion of the center of mass. Again, explain your answer.

Problem 8-95: Consider the rigid building model for earthquake analyses sketched in Problem 6-88. If the ground is given a harmonic motion $x_0(t) = X_0 \sin \Omega t$, for what value(s) of Ω would one expect large am-

plitude responses to occur? Give an equation for such Ω.

Problem 8-96: Using the complex variables technique described in Subdivision 8-4.3.2, obtain the particular solution for an undamped two-degree-of-freedom system subjected to a combination of cosinusoidal and sinusoidal excitations (that is, $F_1(t) = f_1 \cos \Omega t + g_1 \sin \Omega t$ and $F_2(t) = f_2 \cos \Omega t + g_2 \sin \Omega t$). Begin by assuming complex particular solutions (Eqs. (8-281) and (8-282)) as

$$x_{1_p}(t) = X_1 e^{i\Omega t} \qquad x_{2_p}(t) = X_2 e^{i\Omega t}$$

and find X_1 and X_2 from Eqs. (8-285) and (8-286), with all c_{ij} set to zero. Show that the result is identical to Eqs. (8-309) and (8-310).

Problem 8-97: Consider the system sketched in Problem 5-58, when the applied force is given as $F(t) = F_0 \sin \Omega t$.

(a) Find the forced response of the system.

(b) Is there a frequency at which the system can be driven, resulting in only vertical motion of the center of mass? If so, find it.

(c) Is there a frequency at which the system can be driven, resulting in only rotational motion about the center of mass? If so, find it.

(d) At what frequency or frequencies will the driven system resonate?

Problem 8-98: Consider the forced rotary system sketched in Problem 6-63. If $T_1(t) = T_3(t) = 0$, $T_2(t) = T_0 \sin \Omega t$, and $c_1 = c_2 = c_3 = c_4 = 0$, find the nonzero values of Ω such that the middle disk I_2 will not move.

Stability: Problems 8-99 through 8-103

Problem 8-99: Consider the rotating planar pendulum sketched in Problem 5-61. Suppose the angular velocity of the vertical shaft is regulated to be a constant Ω_0.

(a) Find the equilibrium position(s) of the pendulum.

(b) Determine the stability of small motions about each equilibrium position.

(c) Find the frequency of the system for small free oscillations about each stable equilibrium position.

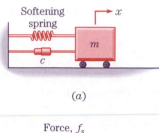

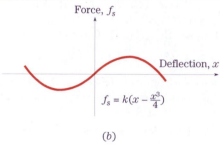

(a)

(b)

FIGURE P8-100

Problem 8-100: A single-degree-of-freedom system sketched in Figure P8-100a consists of a mass, a linear dashpot, and a spring where the spring is not linear but a so-called *softening spring*. The spring force versus deflection is given by the expression $f_s = k\left(x - \dfrac{x^3}{4}\right)$ (as opposed to a linear spring force $f_s = kx$), as sketched in Figure P8-100b.

(a) Write the equation of motion for the unforced system.

(b) Obtain the equilibrium position(s) of the system.

(c) Linearize the equation of motion about each equilibrium position.

(d) Determine the stability of each equilibrium position.

(e) Neglecting damping of the system, find the frequency for small free oscillations about each stable equilibrium position.

Problem 8-101: It is known that a general cubic polynomial equation $x^3 + px^2 + qx + r = 0$ can be solved by defining a set of new parameters as

$$A = \sqrt[3]{-\frac{b}{2} + \sqrt{\frac{b^2}{4} + \frac{a^3}{27}}}$$

$$B = \sqrt[3]{-\frac{b}{2} - \sqrt{\frac{b^2}{4} + \frac{a^3}{27}}}$$

where

$$a = \frac{1}{3}(3q - p^2) \qquad b = \frac{1}{27}(2p^3 - 9pq + 27r)$$

Then, the roots of the cubic equation are given as

$$x = \begin{cases} A + B - \dfrac{p}{3} \\[2mm] -\dfrac{A+B}{2} + \dfrac{A-B}{2}\sqrt{-3} - \dfrac{p}{3} \\[2mm] -\dfrac{A+B}{2} - \dfrac{A-B}{2}\sqrt{-3} - \dfrac{p}{3} \end{cases}$$

These roots are real or imaginary depending on the sign of the radicand of the square roots in the equations above defining the parameters A and B. That is,

Case 1: If $\dfrac{b^2}{4} + \dfrac{a^3}{27} > 0$, there will be a real root and two conjugate imaginary roots.

Case 2: If $\dfrac{b^2}{4} + \dfrac{a^3}{27} = 0$, there will be three real roots of which at least two are equal.

Case 3: If $\dfrac{b^2}{4} + \dfrac{a^3}{27} < 0$, there will be three real and unequal roots.

Using the solution scheme above, the roots of Eq. (h) in Example 8-21 are investigated.

(a) Assuming all parameters but E_0 in Eqs. (i), (j), and (k) are fixed, find the conditions on E_0 for the roots of Eq. (h) to be real and complex in the three cases above.

(b) In part (a), particularly when Eq. (h) has two repeated real roots and a distinct root (Case 2), find all the three roots.

Problem 8-102: Consider the magnetic suspension system in Example 7-13.

(a) Find the equilibrium position(s) of the model aircraft.
    (*Hint:* After eliminating the acceleration in the equation of motion, separately plot the opposing gravitational and electrical forces.)

(b) Determine the stability of each equilibrium position.

Problem 8-103: Because the system in Example 7-13 has been shown to be unstable (Problem 8-102), a feedback control system has been designed to attempt to stabilize it.

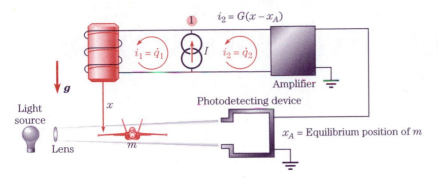

$i_2 = G(x - x_A)$

$i_1 = \dot{q}_1$ I $i_2 = \dot{q}_2$

Amplifier

Photodetecting device

Light source

Lens

x

m

g

x_A = Equilibrium position of m

FIGURE P8-103

As sketched in Figure P8-103, an optical system detects the position of the model. The signal from the photodetecting device is fed into an amplifier that produces a current i_2 that is proportional to the model's deviation from its equilibrium position x_A. Note that $i_2 = G(x - x_A)$ where G is a constant, frequently called the *feedback loop gain*.

(a) By considering x, q_1, and q_2 as variables, show that these reduce to x as the only generalized coordinate for the system sketched in Figure P8-103.

(b) Derive the equation of motion for the system sketched in Figure P8-103.

(c) Linearize the equation of motion about x_A.

(d) Show that in order to ensure stability, the feedback loop gain must satisfy the following criterion:

$$G > \frac{I}{a + x_A}$$

(e) What is the frequency of small oscillation about x_A if the feedback loop gain satisfies the criterion in part (d)?

CHAPTER NINE

DYNAMICS OF CONTINUOUS SYSTEMS

9-1   INTRODUCTION

In this chapter we analyze systems that can be modeled as one-dimensional continuous mechanical or electrical media. In particular, we consider continuum models of strings, rods, shafts, beams, and electrical transmission lines, systems that by their continuous nature are said to possess an infinite number of degrees of freedom. Hamilton's principle is used to derive the equations of motion, with the significant benefit of providing the *natural* (force-dynamic) *boundary conditions*. We forgo consideration of the direct approach in this chapter; it can be found in most other textbooks that consider continuum models.

It is very beneficial for the reader to appreciate the parallelism between Section 8-4, where we studied two-degree-of-freedom systems, and this chapter, where we study systems having an infinite number of degrees of freedom. Following the derivation of the equations of motion in each instance, we consider the *natural modes of vibration* (Subsection 8-4.1 and Section 9-3), *response to initial conditions* (Subsection 8-4.2 and Section 9-4), and *response to harmonic excitation* (Subsection 8-4.3 and Section 9-5). From these correspondences, it should be clear that two-degree-of-freedom systems are in many ways analogous to systems having an infinite number of degrees of freedom. Such a recognition has led to such vernacular idioms as "one, two, infinity" or "one, two, too many," both depicting the fact that the intrinsic character of two-degree-of-freedom systems is the same as that of three-degree-of-freedom systems, four, ..., infinity. Thus, having studied Chapter 8, the reader may wish to refer to Section 8-4 throughout the study of Chapter 9, which by necessity is somewhat more mathematically protracted.

While one-dimensional lumped-parameter dynamics models usually lead to ordinary differential equations (ODEs) of motion, one-dimensional continuous dynamics models lead to partial differential equations (PDEs) of motion. In particular, partial differential equations arise when the generalized coordinate is a function of two (or more) variables, in this case, one spatial variable plus time. Although there are several well-developed techniques for the solution of partial differential equations, we shall explore only one such method: *separation of variables*. This is a powerful technique in applications where the domain of interest is finite. The attractive feature of this solution technique for our application to one-dimensional continua is that it transforms a more difficult PDE problem into a pair of simpler ODE problems.

In deriving the equations of motion for systems containing continuous elements, we need no new theory; we shall simply apply Hamilton's principle. In applying Hamilton's principle, we must define the terms in the appropriate variational indicator and then use the calculus of variations to obtain the equations of motion. In this use, the calculus of variations reduces essentially to integration by parts.*

---

*In some respects, it could be said that a significant underlying feature of integration by parts was known at the instant humankind understood the geometrical postulate that the sum of the areas of the parts of a plane figure is equal to the area of the entire figure. This idea was known to the Ancient Egyptians and it inspired the more theoretical postulations of the Greeks. Other theoretical notions associated with integration by parts were anticipated by the Frenchman Blaise Pascal (1623–1662)— though obviously before the invention of calculus—and contributions to the technique may be thinly attributed to John Bernoulli. Also, integration by parts was one of the most opportunistically used techniques of Lagrange. Nevertheless, priority for the invention of integration by parts perhaps should be given to the Englishman Brook Taylor for his 1715 publication on the subject.

Recall that the variational indicator for mechanical systems is

$$\text{V.I.} = \int_{t_1}^{t_2} \left[\delta(T^* - V) + \sum_{j=1}^{n} \Xi_j \, \delta\xi_j \right] dt \tag{9-1}$$

and the variational indicator for electrical systems using a charge formulation is

$$\text{V.I.} = \int_{t_1}^{t_2} \left[\delta(W_m^* - W_e) + \sum_{k=1}^{n_e} \mathcal{E}_k \, \delta q_k \right] dt \tag{9-2}$$

and the variational indicator for electrical systems using a flux linkage formulation is

$$\text{V.I.} = \int_{t_1}^{t_2} \left[\delta(W_e^* - W_m) + \sum_{k=1}^{n_e'} \mathcal{I}_k \, \delta\lambda_k \right] dt \tag{9-3}$$

Eqs. (9-1), (9-2), and (9-3) are representations of Hamilton's principle which has been stated as follows:

- An admissible motion of the system between specified configurations at t_1 and t_2 is a natural motion if, and only if, the variational indicator vanishes for arbitrary admissible variations.

Whereas Eqs. (9-1), (9-2), and (9-3) could have been used to formulate equations of motion for lumped-parameter systems (see Problem 9-1), we found Lagrange's equations—which are derivable from Hamilton's principle—to provide the simpler technique for such formulations. For systems containing continuous elements, of the variational approaches, Hamilton's principle provides the most straightforward route to the equations of motion.

Considering, as we do, several of the most frequently used one-dimensional structural models (strings, rods, shafts, and beams), a historical perspective of the development of both the models themselves and the solutions to the resulting equations of motion would be far beyond a reasonable undertaking. Not only does the simple immensity of such a perspective support this view, but the history itself is a convolution of hits and misses, followed by incremental improvements, the recounting of which is configured to thwart the best-intentioned equitable attribution. Without question, the Ancient Egyptians developed a collection of empirical codes for *strength of materials*. Otherwise, it would have been impossible for them to achieve their great structural accomplishments. A chronological review of their pyramids, for example, reveals technical developments and modifications that are reflective of an evolving set of structural codes. And later, during the reign of Hammurabi (ca. 1792–1650 B.C.), a strict Mesopotamian code of laws pledged death to the builders of inadequate structures whose collapse caused the death of their occupants, a dependable means of discouraging lackadaisical adherence to the reliable rules of construction.

Modern notions of linear elasticity can be associated with the Englishman Robert Hooke (1635–1703) in 1678 and the Frenchman Edme Mariotte (1620–1684) in 1680 for their publications citing a proportionality between force and elongation in many materials. That deforming solids are subjected to *internal forces* was noted by Leibniz in 1684 and James Bernoulli in 1691. In 1705 Bernoulli specified that a proper means of representing such internal forces was in terms of force per unit area, namely stress, as a function of elongation per unit length, namely strain. The relationship between stress (σ) and strain (ϵ) was explored by Thomas Young in 1807, for whom the proportionality constant E in the relation $\sigma = E\epsilon$ is often called Young's modulus.

In the technical mechanics of solids, nearly every historical figure mentioned thus far in this textbook could claim a contribution. In addition to those individuals, (most notably Euler and James and Daniel Bernoulli for numerous contributions, especially for those relating to the bending of beams), probably the most significant contributors to these developments were Charles A. Coulomb (1736–1806)—primarily on torsion but also on bending—and Claude-Louis-Marie-Henri Navier (1785–1836)—primarily on bending but also as a transitional figure between corpuscular strength of materials and continuum theory of elasticity. Also, Marie-Sophie Germain (1776–1831) made influential contributions to the analysis of models of vibration of thin flat elastic plates, for which in 1816 she won a 3,000 franc prize, so established for that goal by the First Class of the Institute—the French Science Academy's elite group in the mathematical and physical sciences. In the subsequent evolution of the mechanics of deformable solids, the notable event was the presentation by Augustin-Louis Cauchy (1789–1857) on September 30, 1822, of the formulation for the mathematical theory of continuum mechanics. Even so, significant research on the elastic and inelastic mechanics of many of these structural elements proceeded beyond this period, and indeed, many notable developments in plates and shells occurred in the twentieth century.

9-2   EQUATIONS OF MOTION

In this section, we consider several one-dimensional continuous systems, and their equations of motion are found by applying Hamilton's principle. In the first four subsections, we consider examples in which we derive the equations of motion for several one-dimensional continuous systems. In Subsection 9-2.5, where we summarize this section, we list the boundary conditions for various physical terminations at each end of the various continua, along with the governing equations of motion.

In analyzing structural members, we limit our analyses to the "technical mechanics of solids." Thus, we restrict our analyses to slender structural members typically called rods (which undergo uniaxial deformation due to longitudinal forces), shafts (which undergo

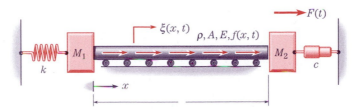

FIGURE 9-1   Model for longitudinal vibration of propulsion system.

twisting due to torsional moments), and beams (which undergo transverse deflection due to bending moments).

9-2.1   Longitudinal Motion of System Containing Rod

We consider the system containing a rod undergoing longitudinal motion as sketched in Figure 9-1. The rod is modeled as a one-dimensional continuum of (mass) density ρ, cross-sectional area A, modulus of elasticity E, and equilibrium length l. The product EA, which often appears in the mechanical analyses of rods undergoing longitudinal deformation, is called the *extensional rigidity*. The entire extent of the rod ($0 < x < l$) is subjected to a distributed longitudinal force per unit length of the rod $f(x, t)$. The left-hand end of the rod ($x = 0$) is attached to a mass M_1 and an elastic element having a spring constant k. The right-hand end of the rod ($x = l$) is subjected to a prescribed force $F(t)$ and is also attached to a mass M_2 and a dissipative element characterized by a linear dashpot constant c. Without explanation, we note that such a model has proved to be useful in analyzing longitudinal vibration and wave propagation in ship propulsion systems, such as those in ocean liners and aircraft carriers.

We want to derive the equation of motion and the boundary conditions for the system model in Figure 9-1. In addition to formulating the variational indicator for Hamilton's principle, we must ensure that the geometric boundary conditions are appropriately incorporated; Hamilton's principle will deliver not only the equation of motion but also the *natural* (force-dynamic) *boundary conditions*.

The generalized coordinate for the rod is denoted by $\xi(x, t)$, where $\xi(x, t)$ is the longitudinal displacement of the section whose equilibrium position is x. The choice of $\xi(x, t)$ is guided by our understanding of lumped-parameter systems where a generalized coordinate would be a time-dependent function of the longitudinal displacement of a specific undisturbed or equilibrium position in the system. Here, because of the selection of a continuous model, we observe that the generalized coordinate is not only a function of time but is also a function of space. Thus we resolve that

$$\xi_j : \xi(x, t) \qquad \text{and} \qquad \delta\xi_j : \delta\xi(x, t) \tag{9-4}$$

The distributed force $f(x, t)$, the localized force $F(t)$, and the dashpot force provide the contributions to the nonconservative work increment. In accordance with Eq. (5-36), these nonconservative force contributions give

$$\delta \mathcal{W}^{nc} = \sum_{i=1}^{N} \boldsymbol{f}_i \cdot \delta \boldsymbol{R}_i = \sum_{j=1}^{n} \Xi_j \, \delta \xi_j$$

$$= f(x,t)\,\delta\xi(x,t) + F(t)\,\delta\xi(l,t) - c\frac{\partial\xi(l,t)}{\partial t}\,\delta\xi(l,t) = \sum_{j=1}^{n} \Xi_j \, \delta\xi_j \qquad (9\text{-}5)$$

where we emphasize that the force $F(t)$ and the dashpot force both act at $x = l$ as noted in Eq. (9-5).

The *kinetic coenergy function* consists of contributions due to the masses M_1 and M_2 and the mass ρA (per unit length) of the rod. No new theory is required, as we simply note that

$$T^* = \frac{1}{2}M_1\left[\frac{\partial\xi(0,t)}{\partial t}\right]^2 + \int_0^l \frac{1}{2}\rho A\left[\frac{\partial\xi(x,t)}{\partial t}\right]^2 dx + \frac{1}{2}M_2\left[\frac{\partial\xi(l,t)}{\partial t}\right]^2 \qquad (9\text{-}6)$$

where in the first and third terms in T^* we emphasize that the velocities of M_1 and M_2 are indicated at $x = 0$ and $x = l$, respectively. Note that the second term in T^* has the correct form of $\frac{1}{2}$ (mass)(velocity squared), but it must be written as an integral to reconcile the fact that the velocity in the rod varies continuously throughout the rod from $x = 0$ to $x = l$. Note also that both ρ and A may be functions of x.

The *potential energy function* consists of contributions due to the spring k and the strain energy of the elastic rod. The potential energy function may be written as

$$V = \frac{1}{2}k[\xi(0,t)]^2 + \int_0^l \frac{1}{2}EA\left[\frac{\partial\xi(x,t)}{\partial x}\right]^2 dx \qquad (9\text{-}7)$$

where in the first term we emphasize that the deformation of the spring element is due to the rod displacement at $x = 0$. The second term in Eq. (9-7) is the strain energy function for the continuous elastic medium of the rod and may be derived as illustrated in Example K–1 of Appendix K.

Substitution of Eqs. (9-4) through (9-7) into Eq. (9-1) gives the variational indicator of Hamilton's principle as

$$\text{V.I.} = \int_{t_1}^{t_2}\left[\!\left[\delta\left\{\overbrace{\frac{1}{2}M_1\left[\frac{\partial\xi(0,t)}{\partial t}\right]^2}^{①} + \overbrace{\int_0^l \frac{1}{2}\rho A\left[\frac{\partial\xi(x,t)}{\partial t}\right]^2 dx}^{②} + \overbrace{\frac{1}{2}M_2\left[\frac{\partial\xi(l,t)}{\partial t}\right]^2}^{③}\right.\right.$$

$$\overbrace{-\frac{1}{2}k\,[\xi(0,t)]^2}^{④} \overbrace{- \int_0^l \frac{1}{2}EA\left[\frac{\partial\xi(x,t)}{\partial x}\right]^2 dx}^{⑤}\Bigg\}$$

$$\overbrace{+ f(x,t)\delta\xi(x,t) + \left[F(t) - c\frac{\partial\xi(l,t)}{\partial t}\right]\delta\xi(l,t)}^{⑥}\Bigg]\!\Bigg] dt \qquad (9\text{-}8)$$

At first sight, even though we should be comfortable with the source of each of the terms in the variational indicator, as a whole Eq. (9-8) may appear to be somewhat daunting. Yet, as we shall see, the careful consideration of each term, followed by the application of Hamilton's principle, will reveal both the enormous amount of information contained in Eq. (9-8) as well as the straightforward manner of extracting that information.

First, we note from Figure 9-1 that there are no geometric restrictions on the motions of the rod at either end. As we shall see in Subsection 9-2.4, the existence of geometric restrictions leads directly to *geometric boundary conditions*. So, for the system sketched in Figure 9-1, there are no geometric boundary conditions. (There are force-dynamic requirements on the ends of the rod that will be delivered by Hamilton's principle, but no geometric constraints.)

With the exception of the last two terms in Eq. (9-8), the variation in Eq. (9-8) is on the lagrangian contained within the flower brackets. In order to apply Hamilton's principle to the variational indicator, we must express the V.I. in terms of variations on the generalized coordinate. Although the operations for this conversion are provided in Appendix E, we shall proceed to illustrate these operations in detail for this problem.

In conducting the conversion of the variational indicator from the form in Eq. (9-8) (where the variation is on the lagrangian) to the final form below (where the variation is on the generalized coordinate), we shall need to apply two sets of operations. First, we shall use the fact that the variation operation is commutative with both space and time differentiation. Thus, as discussed in Appendix D,

$$\delta\left(\frac{\partial \xi}{\partial t}\right) = \frac{\partial(\delta \xi)}{\partial t} \tag{9-9}$$

and

$$\delta\left(\frac{\partial \xi}{\partial x}\right) = \frac{\partial(\delta \xi)}{\partial x} \tag{9-10}$$

Second, in instances where we encounter terms such as Eq. (9-9) we shall use timewise integration by parts, and in instances where we encounter terms such as Eq. (9-10) we shall use spacewise integration by parts.

In order to convert the variational indicator in a somewhat methodical manner, the terms in Eq. (9-8) are designated as ① through ⑥, with the understanding that each term includes the variation δ (except for term ⑥) as well as the timewise integration $\int dt$. We shall consider each of these terms in sequence.

Term ①:

Term ① contains a derivative that is analogous to Eq. (9-9); thus we shall integrate it in a timewise sense. So, it becomes

$$\int_{t_1}^{t_2} \delta\left\{\frac{1}{2}M_1\left[\frac{\partial \xi(0,t)}{\partial t}\right]^2\right\} dt = \int_{t_1}^{t_2} \frac{1}{2}\cdot M_1\cdot 2\left[\frac{\partial \xi(0,t)}{\partial t}\right]\delta\left[\frac{\partial \xi(0,t)}{\partial t}\right] dt \tag{9-11}$$

where we have used the fact that δ behaves mathematically the same as d. So, $\delta(\dot{\xi})^2 = 2(\dot{\xi})\,\delta(\dot{\xi})$. Use of Eq. (9-9) in Eq. (9-11) gives

$$\int_{t_1}^{t_2} \delta\left\{\frac{1}{2}M_1\left[\frac{\partial \xi(0,t)}{\partial t}\right]^2\right\} dt = \int_{t_1}^{t_2} M_1\left[\frac{\partial \xi(0,t)}{\partial t}\right]\left[\frac{\partial \delta \xi(0,t)}{\partial t}\right] dt \tag{9-12}$$

Recall that integration by parts is often expressed as

$$\int u\,dv = uv - \int v\,du \tag{9-13}$$

where we have omitted any constants of integration.[1] Identifying corresponding terms in Eqs. (9-12) and (9-13) yields[2]

$$\int_{t_1}^{t_2} \delta\left\{\frac{1}{2}M_1\left[\frac{\partial\xi(0,t)}{\partial t}\right]^2\right\}dt$$

$$= M_1\left[\frac{\partial\xi(0,t)}{\partial t}\right]\delta\xi(0,t)\Big|_{t_1}^{t_2} - \int_{t_1}^{t_2}\frac{\partial}{\partial t}\left\{M_1\left[\frac{\partial\xi(0,t)}{\partial t}\right]\right\}\delta\xi(0,t)\,dt \tag{9-14}$$

In accordance with Hamilton's principle that the variations vanish at t_1 and t_2 (refer to Figure 9-2), the first term on the right-hand side of Eq. (9-14) vanishes. Thus, conversion of term ① gives

$$\int_{t_1}^{t_2} \delta\left\{\frac{1}{2}M_1\left[\frac{\partial\xi(0,t)}{\partial t}\right]^2\right\}dt = -\int_{t_1}^{t_2} M_1\left[\frac{\partial^2\xi(0,t)}{\partial t^2}\right]\delta\xi(0,t)\,dt \tag{9-15}$$

where, in going from Eq. (9-14) to Eq. (9-15), we have assumed that M_1 does not change with time. Otherwise, in going from Eq. (9-14) to Eq. (9-15), M_1 would not have passed through $\frac{\partial}{\partial t}$ unaffected. Thus, we have converted the variation on the left-hand side of Eq. (9-15) from one on (a portion of) the lagrangian to one on the generalized coordinate on the right-hand side of Eq. (9-15).

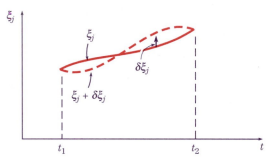

FIGURE 9-2   A trajectory and its neighboring admissible trajectory, emphasizing that $\delta\xi_j = 0$ at t_1 and t_2.

[1]Note that Eq. (9-13) is analogous to the Legendre transformations, which we have used most notably in Chapters 5 and 7. Further, when written as $uv = \int v\,du + \int u\,dv$, Eq. (9-13) is simply a statement that in the $u-v$ plane, the whole is equal to the sum of the parts. Actually, a more general form of integration by parts, which is more generally appropriate for these analyses, may be expressed as $\int f(\eta)g'(\eta)d\eta = f(\eta)g(\eta) - \int g(\eta)f'(\eta)d\eta$ where $(\ )' \equiv \frac{\partial(\ )}{\partial\eta}$; however, we shall restrict our notation to the form in Eq. (9-13).

[2]This identification is made easier by *imagining* the cancellation of the second ∂t and dt on the right-hand side of Eq. (9-12). Mathematically, no such cancellation ever occurs, however. Nevertheless, contriving such an imagined operation and considering Eqs. (9-12) and (9-13), we may identify the following analogous terms:

$$u = M_1\frac{\partial\xi(0,t)}{\partial t} \qquad dv = \partial\delta\xi(0,t) \qquad du = \partial\left[M_1\frac{\partial\xi(0,t)}{\partial t}\right] \qquad v = \delta\xi(0,t).$$

Term ②:

Term ② also contains a derivative that is analogous to Eq. (9-9); thus we shall integrate it in a timewise sense. So, it becomes

$$\int_{t_1}^{t_2} \delta \left\{ \int_0^l \frac{1}{2}\rho A \left[\frac{\partial \xi(x,t)}{\partial t} \right]^2 dx \right\} dt = \int_0^l dx \int_{t_1}^{t_2} \frac{1}{2} \cdot \rho A \cdot 2 \left[\frac{\partial \xi(x,t)}{\partial t} \right] \delta \left[\frac{\partial \xi(x,t)}{\partial t} \right] dt \quad (9\text{-}16)$$

where we have used the fact that δ behaves mathematically the same as d (for example, $\delta(\dot\xi)^2 = 2(\dot\xi)\,\delta(\dot\xi)$) and where we have extracted the space integral forward for aesthetics only. (It should be noted that the location of the integral $\int_0^l dx$ on the right-hand side of Eq. (9-16) does not alter the meaning of the expression; it can be pulled forward, as shown, or placed at the rear, or located anyplace—it doesn't matter.) Use of Eq. (9-9) in Eq. (9-16) gives

$$\int_{t_1}^{t_2} \delta \left\{ \int_0^l \frac{1}{2}\rho A \left[\frac{\partial \xi(x,t)}{\partial t} \right]^2 dx \right\} dt = \int_0^l dx \int_{t_1}^{t_2} \rho A \left[\frac{\partial \xi(x,t)}{\partial t} \right] \left[\frac{\partial \delta \xi(x,t)}{\partial t} \right] dt \quad (9\text{-}17)$$

Analogous to term ①, identifying corresponding terms in Eqs. (9-13) and (9-17) yields

$$\int_{t_1}^{t_2} \delta \left\{ \int_0^l \frac{1}{2}\rho A \left[\frac{\partial \xi(x,t)}{\partial t} \right]^2 dx \right\} dt$$

$$= \int_0^l dx \left\{ \rho A \frac{\partial \xi(x,t)}{\partial t} \delta \xi(x,t) \Big|_{t_1}^{t_2} - \int_{t_1}^{t_2} \frac{\partial}{\partial t} \left[\rho A \frac{\partial \xi(x,t)}{\partial t} \right] \delta \xi(x,t)\, dt \right\} \quad (9\text{-}18)$$

In accordance with Hamilton's principle that the variations vanish at t_1 and t_2 (refer to Figure 9-2), the first term on the right-hand side of Eq. (9-18) vanishes. Thus, conversion of term ② gives

$$\int_{t_1}^{t_2} \delta \left\{ \int_0^l \frac{1}{2}\rho A \left[\frac{\partial \xi(x,t)}{\partial t} \right]^2 dx \right\} dt = -\int_{t_1}^{t_2} dt \int_0^l \rho A \left[\frac{\partial^2 \xi(x,t)}{\partial t^2} \right] \delta \xi(x,t)\, dx \quad (9\text{-}19)$$

where, in going from Eq. (9-18) to Eq. (9-19), we have assumed that ρA does not change with time. Thus, we have converted the variation on the left-hand side of Eq. (9-19) from one on (a portion of) the lagrangian to one on the generalized coordinate on the right-hand side of Eq. (9-19).

Term ③:

Term ③ is very similar to term ①, the only difference being that term ① captures information at $x = 0$ whereas term ③ captures information at $x = l$. Thus, via an identical procedure used for term ①, the conversion of term ③ becomes

$$\int_{t_1}^{t_2} \delta \left\{ \frac{1}{2}M_2 \left[\frac{\partial \xi(l,t)}{\partial t} \right]^2 \right\} dt = -\int_{t_1}^{t_2} M_2 \left[\frac{\partial^2 \xi(l,t)}{\partial t^2} \right] \delta \xi(l,t)\, dt \quad (9\text{-}20)$$

where as before we have assumed that M_2 does not change with time, and where Eq. (9-20) represents a conversion from a variation on (a portion of) the lagrangian to a variation on the generalized coordinate.

Term ④:

The conversion of term ④ is straightforward, involving the elementary operation quoted immediately after both Eqs. (9-11) and (9-16). Thus, term ④ becomes

$$\int_{t_1}^{t_2} \delta\left\{-\frac{1}{2}k\left[\xi(0,t)\right]^2\right\} dt = \int_{t_1}^{t_2} -\frac{1}{2}\cdot k\cdot 2\cdot \xi(0,t)\delta\xi(0,t)\, dt$$

$$= -\int_{t_1}^{t_2} k\xi(0,t)\delta\xi(0,t)\, dt \qquad (9\text{-}21)$$

where the operation in Eq. (9-21) has converted the variation from one on (a portion of) the lagrangian to one on the generalized coordinate.

Term ⑤:

Term ⑤ contains a derivative that is analogous to Eq. (9-10); thus we shall integrate it in a spacewise sense. So, it becomes

$$\int_{t_1}^{t_2} \delta\left\{-\int_0^l \frac{1}{2}EA\left[\frac{\partial\xi(x,t)}{\partial x}\right]^2 dx\right\} dt = \int_{t_1}^{t_2} dt \int_0^l -\frac{1}{2}\cdot EA\cdot 2\left[\frac{\partial\xi(x,t)}{\partial x}\right]\delta\left[\frac{\partial\xi(x,t)}{\partial x}\right] dx \qquad (9\text{-}22)$$

where we have used the fact that δ behaves mathematically the same as d $\Big($for example,

$\delta(\xi')^2 = 2(\xi')\,\delta(\xi')$ where $\xi' \equiv \dfrac{\partial\xi}{\partial x}\Big)$ and where we have extracted the entire time integral

forward for aesthetics only. Use of Eq. (9-10) in Eq. (9-22) gives

$$\int_{t_1}^{t_2} \delta\left\{-\int_0^l \frac{1}{2}EA\left[\frac{\partial\xi(x,t)}{\partial x}\right]^2 dx\right\} dt = \int_{t_1}^{t_2} dt \int_0^l -EA\left[\frac{\partial\xi(x,t)}{\partial x}\right]\left[\frac{\partial\delta\xi(x,t)}{\partial x}\right] dx \qquad (9\text{-}23)$$

Identifying corresponding terms in Eqs. (9-13) and (9-23) yields[3]

$$\int_{t_1}^{t_2} \delta\left\{-\int_0^l \frac{1}{2}EA\left[\frac{\partial\xi(x,t)}{\partial x}\right]^2 dx\right\} dt$$

$$= \int_{t_1}^{t_2} dt\left\{-EA\frac{\partial\xi(x,t)}{\partial x}\,\delta\xi(x,t)\Big|_0^l + \int_0^l \frac{\partial}{\partial x}\left[EA\frac{\partial\xi(x,t)}{\partial x}\right]\delta\xi(x,t)\, dx\right\}$$

$$= \int_{t_1}^{t_2} dt\left\{-EA\frac{\partial\xi(l,t)}{\partial x}\,\delta\xi(l,t) + EA\frac{\partial\xi(0,t)}{\partial x}\,\delta\xi(0,t)\right.$$

$$\left. + \int_0^l \frac{\partial}{\partial x}\left[EA\frac{\partial\xi(x,t)}{\partial x}\right]\delta\xi(x,t)\, dx\right\} \qquad (9\text{-}24)$$

---

[3]This identification is made easier by *imagining* the cancellation of the second ∂x and the dx on the right-hand side of Eq. (9-23). Mathematically, no such cancellation ever occurs, however. Nevertheless, contriving such an imagined operation and considering Eqs. (9-13) and (9-23), we may identify the following terms:

$$u = -EA\frac{\partial\xi(x,t)}{\partial x} \qquad dv = \partial\delta\xi(x,t) \qquad du = -\partial\left[EA\frac{\partial\xi(x,t)}{\partial x}\right] \qquad v = \delta\xi(x,t).$$

Thus, we have converted the variation on the left-hand side of Eq. (9-24) from one on (a portion of) the lagrangian to one on the generalized coordinate on the right-hand side of Eq. (9-24).

Term ⑥:

Term ⑥ already contains a variation on the generalized coordinate, and thus requires no conversion.

Now, we are prepared to collect all of the converted terms back into the variational indicator in Eq. (9-8). In doing this, it is convenient to aggregate terms containing alike variations on the generalized coordinate. Accordingly, all terms containing $\delta\xi(x,t)$ should be grouped together, all terms containing $\delta\xi(0,t)$ should be grouped together, and all terms containing $\delta\xi(l,t)$ should be grouped together. Thus, the variational indicator becomes

$$
\begin{aligned}
\text{V.I.} = \int_{t_1}^{t_2} \Bigg[& \int_0^l \left\{ -\rho A\left[\frac{\partial^2 \xi(x,t)}{\partial t^2} \right] + \frac{\partial}{\partial x}\left[EA\frac{\partial \xi(x,t)}{\partial x} \right] + f(x,t) \right\} \delta\xi(x,t)\, dx \\
& + \left\{ -M_1\left[\frac{\partial^2 \xi(0,t)}{\partial t^2} \right] - k\xi(0,t) + EA\frac{\partial \xi(0,t)}{\partial x} \right\} \delta\xi(0,t) \\
& + \left\{ -M_2\left[\frac{\partial^2 \xi(l,t)}{\partial t^2} \right] + F(t) - c\frac{\partial \xi(l,t)}{\partial t} - EA\frac{\partial \xi(l,t)}{\partial x} \right\} \delta\xi(l,t) \Bigg] dt \qquad (9\text{-}25)
\end{aligned}
$$

Hamilton's principle requires that the variational indicator in Eq. (9-25) vanishes for arbitrary admissible variations of the generalized coordinate. Thus, the necessary conditions that Eq. (9-25) vanishes for arbitrary values of $\delta\xi(x,t)$ throughout $0 < x < l$, for arbitrary values of $\delta\xi(0,t)$ at $x = 0$, and for arbitrary values of $\delta\xi(l,t)$ at $x = l$ are that each of the terms in flower brackets in Eq. (9-25) vanishes.[4] Thus,

$$
\rho A\frac{\partial^2 \xi}{\partial t^2} = \frac{\partial}{\partial x}\left(EA\frac{\partial \xi}{\partial x} \right) + f(x,t) \qquad 0 < x < l \qquad (9\text{-}26)
$$

which is the governing partial differential equation (PDE) for the longitudinal motion of the rod. Furthermore,

$$
M_1\frac{\partial^2 \xi}{\partial t^2} + k\xi = EA\frac{\partial \xi}{\partial x} \qquad x = 0 \qquad (9\text{-}27)
$$

and

$$
M_2\frac{\partial^2 \xi}{\partial t^2} + c\frac{\partial \xi}{\partial t} + EA\frac{\partial \xi}{\partial x} = F(t) \qquad x = l \qquad (9\text{-}28)
$$

where Eqs. (9-27) and (9-28) are called the *natural boundary conditions*. Note that the natural boundary conditions are statements of the dynamic requirements on the forces at each end of the rod. Because $\dfrac{\partial \xi}{\partial x}$ is simply the longitudinal strain *in* the rod, we see that the right-hand side of Eq. (9-27) is simply the force *in* the rod at $x = 0$. Thus, Eq. (9-27) simply indicates that the force *in* the rod at $x = 0$ dynamically interacts with M_1 through its

---

[4]This issue is discussed in Section D-2 in Appendix D although, as usual, it is not necessary to refer to this additional narrative.

acceleration and k through its extension. Similarly, Eq. (9-28) indicates that the externally applied force $F(t)$ dynamically interacts with M_2, c, and the force *in* the rod at $x = l$.

Not only has Hamilton's principle produced the partial differential equation governing the dynamic behavior of the rod continuum; it has also furnished the *natural* (force-dynamic) *boundary conditions*. Herein lies one of the valuable features of variational methods in mechanical dynamics as well as in other disciplines. In many systems without simple boundary conditions, it may be nontrivial to ensure the proper signs on the various terms that are present in boundary conditions such as represented in Eqs. (9-27) and (9-28); Hamilton's principle automatically gives these complete dynamic relationships. Finally, in order to solve Eq. (9-26) in conjunction with Eqs. (9-27) and (9-28) in a particular problem, it is necessary to prescribe the system's *initial conditions* and the external forces $f(x, t)$ and $F(t)$. For the rod continuum in this example, these initial conditions would be represented by the complete specification of

$$\xi(x, 0) \qquad \text{and} \qquad \frac{\partial \xi(x, 0)}{\partial t} \qquad 0 < x < l \qquad (9\text{-}29)$$

which are the longitudinal displacement and longitudinal velocity throughout the rod at time $t = 0$.

9-2.2   Twisting Motion of System Containing Shaft

We consider the system containing a shaft undergoing twisting motion as sketched in Figure 9-3. The circular cylindrical shaft is modeled as a one-dimensional continuum of (mass) density ρ, polar moment of the cross-sectional area about the axis of the shaft J, moment of inertia per unit length of shaft about its axis I_p,[5] shear modulus G, and equilibrium length l. The product GJ, which often appears in the mechanical analyses of shafts undergoing torsional deformation, is called the *torsional rigidity*. The entire extent of the shaft ($0 < x < l$) is subjected to a distributed torque per unit length of the shaft $\tau(x, t)$. The left-hand end of the shaft ($x = 0$) is attached to a disk of rotary inertia I_1 (that is, the moment of inertia of the disk about the shaft's axis) and an elastic element having a torsional spring constant k_t. The right-hand end of the shaft ($x = l$) is subjected to a prescribed torque, or moment, $T(t)$ and is attached to a disk of rotary inertia I_2 (that is, the moment of inertia of the disk about the shaft's axis) and a dissipative element that is characterized by a torsional dashpot constant c_t. Without explanation, we note that such a model has proved to be useful in analyzing the torsional vibration and wave propagation in ship propulsion systems, such as those in ocean liners and aircraft carriers. As we shall see, this torsional system behaves similarly to the system in Figure 9-1, which models the longitudinal wave propagation in propulsion systems.

We want to derive the equation of motion and the boundary conditions for the system model in Figure 9-3. In addition to formulating the variational indicator for Hamilton's principle, we must ensure that the geometric boundary conditions are appropriately incorporated; Hamilton's principle will deliver not only the equation of motion but also the *natural* (force-dynamic) *boundary conditions*.

---

[5]It can be shown that $I_p = \rho J$; see Problem 9-14.

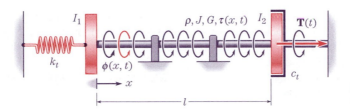

FIGURE 9-3 Model for twisting vibration of propulsion system.

The generalized coordinate for the shaft is denoted by $\varphi(x, t)$, where $\varphi(x, t)$ is the angular displacement of the section whose position is x. The choice of $\varphi(x, t)$ is guided by our understanding of lumped-parameter systems where generalized coordinates are time-dependent functions of the translational or angular displacement of a specific undisturbed or equilibrium position in the system. Here, because of the selection of a continuous model, we observe that the generalized coordinate is not only a function of time but also a function of space. Thus we resolve that

$$\xi_j : \varphi(x, t) \qquad \text{and} \qquad \delta\xi_j : \delta\varphi(x, t) \tag{9-30}$$

The distributed torque $\tau(x, t)$, the localized torque $\mathrm{T}(t)$ applied on I_2, and the dashpot moment provide the contributions to the nonconservative work increment. In accordance with Eq. (5-36), these nonconservative torque contributions give

$$\delta\mathcal{W}^{nc} = \sum_{i=1}^{N} \boldsymbol{f}_i \cdot \delta\boldsymbol{R}_i = \sum_{j=1}^{n} \Xi_j \, \delta\xi_j$$

$$= \tau(x, t)\,\delta\varphi(x, t) + \mathrm{T}(t)\,\delta\varphi(l, t) - c_t \frac{\partial\varphi(l, t)}{\partial t}\,\delta\varphi(l, t) = \sum_{j=1}^{n} \Xi_j \, \delta\xi_j \tag{9-31}$$

where we emphasize that the torque $\mathrm{T}(t)$ and the dashpot torque both act at $x = l$ as noted in Eq. (9-31).

The *kinetic coenergy function* consists of contributions due to the rotary inertias I_1 and I_2 and the rotary inertia ρJ (per unit length) of the shaft. No new theory is required, as we simply note that

$$T^* = \frac{1}{2}I_1 \left[\frac{\partial\varphi(0, t)}{\partial t}\right]^2 + \int_0^l \frac{1}{2}\rho J \left[\frac{\partial\varphi(x, t)}{\partial t}\right]^2 dx + \frac{1}{2}I_2 \left[\frac{\partial\varphi(l, t)}{\partial t}\right]^2 \tag{9-32}$$

where in the first and third terms in T^* we emphasize that the angular velocities of I_1 and I_2 are indicated at $x = 0$ and $x = l$, respectively, and where in the second term ρJ has been used in place of the more obvious I_p. (See Problem 9-14.) Further, note that the second term in T^* has the correct form of $\frac{1}{2}$(rotary inertia)(angular velocity squared), but it must be written as an integral to reconcile the fact that the angular velocity in the rod varies continuously throughout the rod from $x = 0$ to $x = l$. Note also that both ρ and J may be functions of x.

The *potential energy function* consists of contributions due to the torsional spring k_t and the strain energy of the elastic shaft. The potential energy function may be written as

$$V = \frac{1}{2}k_t[\varphi(0, t)]^2 + \int_0^l \frac{1}{2}GJ \left[\frac{\partial\varphi(x, t)}{\partial x}\right]^2 dx \tag{9-33}$$

where in the first term we emphasize that the deformation of the torsional spring is due to the shaft angular displacement at $x = 0$. The second term in Eq. (9-33) is the strain energy function for the continuous elastic medium of the shaft and may be derived as illustrated in Example K-2 of Appendix K.

Substitution of Eqs. (9-30) through (9-33) into Eq. (9-1) gives the variational indicator of Hamilton's principle as

$$
\begin{aligned}
\text{V.I.} = \int_{t_1}^{t_2} \Bigg[\Bigg[\delta \bigg\{ \frac{1}{2} I_1 \bigg[\frac{\partial \varphi(0,t)}{\partial t} \bigg]^2 + \int_0^l \frac{1}{2} \rho J \bigg[\frac{\partial \varphi(x,t)}{\partial t} \bigg]^2 dx + \frac{1}{2} I_2 \bigg[\frac{\partial \varphi(l,t)}{\partial t} \bigg]^2 \\
- \frac{1}{2} k_t [\varphi(0,t)]^2 - \int_0^l \frac{1}{2} G J \bigg[\frac{\partial \varphi(x,t)}{\partial x} \bigg]^2 dx \bigg\} \\
+ \tau(x,t)\, \delta\varphi(x,t) + \bigg[\mathrm{T}(t) - c_t \frac{\partial \varphi(l,t)}{\partial t} \bigg] \delta\varphi(l,t) \Bigg] \Bigg] dt \quad (9\text{-}34)
\end{aligned}
$$

As we shall see, the careful consideration of each term followed by the application of Hamilton's principle will reveal the enormous amount of information contained in Eq. (9-34).

First, we note from Figure 9-3 that there are no geometric restrictions on the motions of the shaft at either end. As we shall see in Subsection 9-2.4, the existence of geometric restrictions leads directly to *geometric boundary conditions*. So, for the system sketched in Figure 9-3, there are no geometric boundary conditions.

With the exception of the last two terms in Eq. (9-34), the variation in Eq. (9-34) is on the lagrangian contained within the flower brackets. In order to apply Hamilton's principle to the variational indicator, we must express the V.I. in terms of variations on the generalized coordinate.

Note that Eq. (9-34) is an exact analogy with Eq. (9-8). Equation (9-34) indeed becomes identical to Eq. (9-8) if we replace φ in Eq. (9-34) by ξ, I_1 by M_1, I_2 by M_2, ρJ by ρA, k_t by k, GJ by EA, $\tau(x,t)$ by $f(x,t)$, $\mathrm{T}(t)$ by $F(t)$, and c_t by c. Due to this analogy between Eqs. (9-8) and (9-34), the conversion of the terms in Eq. (9-34), in order to obtain variations on the generalized coordinate, may be conducted via the identical procedures used for the rod undergoing longitudinal motion in Subsection 9-2.1.

Thus, by rewriting Eq. (9-25) with the parameters and generalized coordinate replaced as noted above, the variational indicator for the torsional system sketched in Figure 9-3 becomes

$$
\begin{aligned}
\text{V.I.} = \int_{t_1}^{t_2} \Bigg[\Bigg[\int_0^l \bigg\{ -\rho J \bigg[\frac{\partial^2 \varphi(x,t)}{\partial t^2} \bigg] + \frac{\partial}{\partial x} \bigg[GJ \frac{\partial \varphi(x,t)}{\partial x} \bigg] + \tau(x,t) \bigg\} \delta\varphi(x,t)\, dx \\
+ \bigg\{ -I_1 \bigg[\frac{\partial^2 \varphi(0,t)}{\partial t^2} \bigg] - k_t \varphi(0,t) + GJ \frac{\partial \varphi(0,t)}{\partial x} \bigg\} \delta\varphi(0,t) \\
+ \bigg\{ -I_2 \bigg[\frac{\partial^2 \varphi(l,t)}{\partial t^2} \bigg] + \mathrm{T}(t) - c_t \frac{\partial \varphi(l,t)}{\partial t} - GJ \frac{\partial \varphi(l,t)}{\partial x} \bigg\} \delta\varphi(l,t) \Bigg] \Bigg] dt \quad (9\text{-}35)
\end{aligned}
$$

Hamilton's principle requires that the variational indicator in Eq. (9-35) vanishes for arbitrary admissible variations of the generalized coordinate. Thus, the necessary conditions that Eq. (9-35) vanishes for arbitrary values of $\delta\varphi(x,t)$ throughout $0 < x < l$, for arbitrary values of $\delta\varphi(0,t)$ at $x = 0$, and for arbitrary values of $\delta\varphi(l,t)$ at $x = l$ are that each of the terms in flower brackets in Eq. (9-35) vanishes. Thus,

$$\rho J \frac{\partial^2 \varphi}{\partial t^2} = \frac{\partial}{\partial x}\left(GJ\frac{\partial \varphi}{\partial x}\right) + \tau(x,t) \qquad 0 < x < l \qquad (9\text{-}36)$$

which is the governing partial differential equation (PDE) for the twisting motion of the shaft. Furthermore,

$$I_1 \frac{\partial^2 \varphi}{\partial t^2} + k_t \varphi = GJ\frac{\partial \varphi}{\partial x} \qquad x = 0 \qquad (9\text{-}37)$$

and

$$I_2 \frac{\partial^2 \varphi}{\partial t^2} + c_t \frac{\partial \varphi}{\partial t} + GJ\frac{\partial \varphi}{\partial x} = \mathrm{T}(t) \qquad x = l \qquad (9\text{-}38)$$

where Eqs. (9-37) and (9-38) are called the *natural boundary conditions*. Note that the natural boundary conditions are statements of the dynamic requirements on the torques (or moments) at each end of the shaft. Because the right-hand side of Eq. (9-37) is simply the torque *in* the shaft at $x = 0$ according to the technical mechanics of solids, Eq. (9-37) simply indicates that the torque in the shaft at $x = 0$ dynamically interacts with I_1 through its angular acceleration and k_t through its angular displacement. Similarly, Eq. (9-38) indicates that the externally applied torque $\mathrm{T}(t)$ dynamically interacts with I_2, c_t, and the torque *in* the shaft at $x = l$.

Not only has Hamilton's principle produced the partial differential equation governing the dynamic behavior of the shaft continuum; it has also furnished the *natural* (force-dynamic or torque-dynamic) *boundary conditions*. Again, herein lies one of the valuable characteristics of variational methods in mechanical dynamics as well as in other disciplines. In many systems without simple boundary conditions, it may be rather difficult to ensure the proper signs on the various terms that are present in boundary conditions such as represented in Eqs. (9-37) and (9-38); Hamilton's principle automatically gives these complete dynamic relationships. Finally, in order to solve Eq. (9-36) in conjunction with Eqs. (9-37) and (9-38) in a particular problem, it is necessary to prescribe the system's *initial conditions* and the external torques $\tau(x,t)$ and $\mathrm{T}(t)$. For the shaft continuum in this example, these initial conditions would be represented by the complete specification of

$$\varphi(x,0) \qquad \text{and} \qquad \frac{\partial \varphi(x,0)}{\partial t} \qquad 0 < x < l \qquad (9\text{-}39)$$

which are the angular displacement and angular velocity throughout the shaft at time $t = 0$.

9-2.3   Electric Transmission Line

We consider a long electric transmission line as sketched in Figure 9-4. The transmission line of length l is modeled as a one-dimensional continuum having distributed series inductance $L(x)$ per unit length and distributed shunt capacitance $C(x)$ per unit length (Figure 9-5). The left-hand end of the transmission line ($x = 0$) is attached to a loop containing an inductor of inductance L_1 and a capacitor of capacitance C_1. The right-hand end of the transmission line ($x = l$) is attached to a loop containing an inductor of inductance L_2, a resistor of resistance R, and an ideal voltage source providing a prescribed voltage $E(t)$. We shall note below that as far as its dynamic behavior is concerned, this electric transmission line is approximately analogous to the mechanical systems sketched in Figures 9-1 and 9-3.

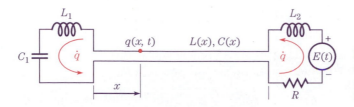

FIGURE 9-4 Electric transmission line having same dynamic characteristics as system sketched in Figure 9-1 or Figure 9-3.

We want to derive the equation of motion and the boundary conditions for the system model of Figure 9-4 by applying Hamilton's principle. The equation of motion for this system can be most conveniently derived via the charge formulation scheme presented in Subsection 7-4.1. Thus, a convenient generalized coordinate for the transmission line is chosen to be $q(x, t)$, where $q(x, t)$ is the charge on the positive line at the section whose position is x, as indicated in Figure 9-4. Here, because of the selection of a continuous model, we observe that the generalized coordinate is not only a function of time but also a function of space. Thus we resolve that

$$q_k : q(x, t) \quad \text{and} \quad \delta q_k : \delta q(x, t) \tag{9-40}$$

The voltage source $E(t)$ and the resistor R are the only nonconservative elements in the system. In accordance with Eq. (7-32), the work expression for these nonconservative elements is written as

$$\delta \mathcal{W}^{nc} = \sum_{j=1}^{M} e_j^{nc} \, \delta q_j = \sum_{k=1}^{n_e} \mathcal{E}_k \, \delta q_k$$

$$= E(t) \, \delta q(l, t) - R \frac{\partial q(l, t)}{\partial t} \, \delta q(l, t) = \sum_{k=1}^{n_e} \mathcal{E}_k \, \delta q_k \tag{9-41}$$

where we have used Eqs. (7-43) and (7-39) and where we emphasize that the work associated with both nonconservative elements is due to the charge and current at $x = l$.

Before we write the magnetic coenergy function and the electrical energy function, we need to consider the differential element of the transmission line dx sketched in Figure 9-5.

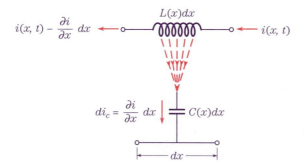

FIGURE 9-5 Current flow through differential element of electric transmission line.

The differential element dx consists of the series inductance $L(x)\,dx$ and the shunt capacitance $C(x)\,dx$. Assume that the current $i(x,t)$ in the top conductor flows from right to left as indicated in Figure 9-5. A portion of the current that flows through the top conductor "leaks" through the shunt capacitance $C(x)\,dx$ and flows into the bottom conductor. If the current entering the top conductor element is denoted by $i(x,t)$, the current exiting the top conductor may be expressed by $i(x,t) - \left(\dfrac{\partial i}{\partial x}\right)dx$,[6] as sketched in Figure 9-5. Then, since the element $L(x)\,dx$ is passive and conservative, the current outflow from it must be the same as the current inflow to it, in accordance with KCR (Kirchhoff's current rule). Thus, the differential "leakage" current through the shunt capacitance $C\,dx$ must be given by

$$di_c = \frac{\partial i(x,t)}{\partial x}\,dx = \frac{\partial}{\partial x}\left[\frac{\partial q(x,t)}{\partial t}\right]dx \tag{9-42}$$

where, after applying KCR to the differential element, we have used the relationship between charge and current, $i = \dfrac{\partial q}{\partial t}$. Combining Eq. (9-42) and the relationship $i = \dfrac{\partial q}{\partial t}$ indicates that the differential leakage charge dq_c on the capacitance $C(x)\,dx$ must satisfy

$$di_c = \frac{\partial}{\partial x}\left[\frac{\partial q(x,t)}{\partial t}\right]dx = \frac{\partial}{\partial t}\left[\frac{\partial q(x,t)}{\partial x}\right]dx = \frac{\partial}{\partial t}(dq_c) \tag{9-43}$$

where we have used the fact that the timewise and spacewise differentiations are commutative and where we have used the observation that the time derivative of the differential leakage charge must be the differential leakage current. Integrating the last two expressions in Eq. (9-43) with respect to time gives

$$dq_c = \frac{\partial q(x,t)}{\partial x}\,dx \tag{9-44}$$

where we have omitted any integration constant on the assumption that there is no initial charge. Note that Eq. (9-44) resulted from the consideration of charge conservation or, equivalently, KCR and that this is the only compatibility requirement for the generalized charge coordinate $q(x,t)$.

The *magnetic coenergy function* of the system consists of contributions due to the inductors at both ends L_1 and L_2 and the distributed series inductance $L(x)\,dx$ contained in the transmission line. The form of the magnetic coenergy function for inductors is $\dfrac{1}{2}$(inductance)(current squared), as given by Eq. (7–22). Thus, the magnetic coenergy function W_m^* can be written as

$$W_m^* = \frac{1}{2}L_1\left[\frac{\partial q(0,t)}{\partial t}\right]^2 + \int_0^l \frac{1}{2}L(x)\left[\frac{\partial q(x,t)}{\partial t}\right]^2 dx + \frac{1}{2}L_2\left[\frac{\partial q(L,t)}{\partial t}\right]^2 \tag{9-45}$$

---

[6]This result follows simply from the definition of the spacewise partial derivative of $i(x,t)$, which gives

$$\frac{\partial i(x,t)}{\partial x} = \lim_{\Delta x \to 0}\frac{i(x,t) - i(x - \Delta x, t)}{\Delta x} = \frac{i(x,t) - i(x - dx, t)}{dx}$$

where Δx is an incremental length. The above equation may be rearranged to give

$$i(x - dx, t) = i(x,t) - \frac{\partial i(x,t)}{\partial x}\,dx$$

where in the first and third terms on the right-hand side of Eq. (9-45) we emphasize that the currents through L_1 and L_2 are indicated at $x = 0$ and $x = l$, respectively. Note that we obtained the second term in W_m^* using the form $\frac{1}{2}$(inductance)(current squared) for each differential element dx (that is, $\frac{1}{2}(L(x)\,dx)i^2 = \frac{1}{2}L(x)\left(\frac{\partial q}{\partial t}\right)^2 dx$) and then integrating the resulting expression along the line from $x = 0$ to $x = l$ to reconcile the fact that the inductance and current in the transmission line vary continuously throughout the line.

The *electrical energy function* consists of contributions due to the capacitor C_1 and the distributed shunt capacitance in the transmission line. The form of the electrical energy function for capacitors is $\frac{1}{2}$(capacitance)$^{-1}$(charge squared), as given by Eq. (7-10). Thus, the electrical energy function W_e can be written as

$$W_e = \frac{1}{2C_1}[q(0,t)]^2 + \int_0^l \frac{1}{2C(x)}\left[\frac{\partial q(x,t)}{\partial x}\right]^2 dx \qquad (9\text{-}46)$$

where in the first term we emphasize that the charge on the capacitor C_1 is due to the charge $q(0,t)$ at $x = 0$. Note that we obtained the second term in W_e using the form $\frac{1}{2}$(capacitance)$^{-1}$(charge squared) for the shunt capacitance $C(x)\,dx$ associated with each differential element dx and then integrating the resulting expression along the line from $x = 0$ to $x = l$ to reconcile the fact that the capacitance and charge in the transmission line vary continuously throughout the line. That is, the electrical energy for $C(x)\,dx$ is given by

$$\frac{1}{2}\frac{(\text{charge})^2}{(\text{capacitance})} = \frac{(dq_c)^2}{2C(x)\,dx} = \frac{\left(\frac{\partial q}{\partial x}\,dx\right)^2}{2C(x)\,dx} = \frac{1}{2C(x)}\left[\frac{\partial q(x,t)}{\partial x}\right]^2 dx \qquad (9\text{-}47)$$

where Eq. (9-44) has been used to express the differential charge dq_c on $C(x)\,dx$ in terms of the generalized coordinate $q(x,t)$. Upon integration with respect to x, Eq. (9-47) yields the second term of Eq. (9-46).

Substitution of Eqs. (9-40), (9-41), (9-45), and (9-46) into Eq. (9-2) gives the variational indicator of Hamilton's principle as

$$\text{V.I.} = \int_{t_1}^{t_2}\left[\left[\delta\left\{\frac{1}{2}L_1\left[\frac{\partial q(0,t)}{\partial t}\right]^2 + \int_0^l \frac{1}{2}L(x)\left[\frac{\partial q(x,t)}{\partial t}\right]^2 dx + \frac{1}{2}L_2\left[\frac{\partial q(l,t)}{\partial t}\right]^2\right.\right.\right.$$

$$\left.\left. - \frac{1}{2C_1}[q(0,t)]^2 - \int_0^l \frac{1}{2C(x)}\left[\frac{\partial q(x,t)}{\partial x}\right]^2 dx\right\}\right.$$

$$\left.\left. + \left[E(t) - R\frac{\partial q(l,t)}{\partial t}\right]\delta q(l,t)\right]dt \qquad (9\text{-}48)\right.$$

First, we note from Figure 9-4 that there are no restrictions on the admissible variations on the coordinate $q(x,t)$ at either end. (Refer to Table 7-2.) With the exception of the term $\delta q(l,t)$ in Eq. (9-48), the variation is on the lagrangian contained within the flower brackets. In order to apply Hamilton's principle to the variational indicator, we must express the V.I. in terms of variations on the generalized coordinate.

Note that Eq. (9-48) is a near-exact analogy with Eq. (9-8) or Eq. (9-34) (only the absence of a term analogous to the $f(x,t)$ term in Eq. (9-8) or the $\tau(x,t)$ term in Eq. (9-34) is different). Except for absence of a distributed loading term, Eq. (9-48) becomes identical to Eq. (9-8) if we replace q in Eq. (9-48) by ξ, L_1 by M_1, L_2 by M_2, $L(x)$ by ρA, $\dfrac{1}{C_1}$ by k, $\dfrac{1}{C(x)}$ by EA, $E(t)$ by $F(t)$, and R by c. Due to this analogy between Eqs. (9-8) and (9-48), the conversion of the terms in Eq. (9-48), in order to obtain variations on the generalized coordinate, may be conducted via the identical procedures used for the rod undergoing longitudinal motion in Subsection 9-2.1.

Thus, by rewriting Eq. (9-25) with the parameters and generalized coordinate replaced as noted above, the variational indicator for the electric transmission line sketched in Figure 9-4 becomes

$$
\begin{aligned}
\text{V.I.} = \int_{t_1}^{t_2} \Bigg[\!\!\int_0^l & \left\{ -L(x)\left[\frac{\partial^2 q(x,t)}{\partial t^2}\right] + \frac{\partial}{\partial x}\left[\frac{1}{C(x)}\frac{\partial q(x,t)}{\partial x}\right] \right\} \delta q(x,t)\,dx \\
& + \left\{ -L_1\left[\frac{\partial^2 q(0,t)}{\partial t^2}\right] - \frac{1}{C_1}q(0,t) + \frac{1}{C(x)}\frac{\partial q(0,t)}{\partial x} \right\} \delta q(0,t) \\
& + \left\{ -L_2\left[\frac{\partial^2 q(l,t)}{\partial t^2}\right] + E(t) - R\frac{\partial q(l,t)}{\partial t} - \frac{1}{C(x)}\frac{\partial q(l,t)}{\partial x} \right\} \delta q(l,t) \Bigg]\,dt
\end{aligned}
\tag{9-49}
$$

Hamilton's principle requires that the variational indicator in Eq. (9-49) vanishes for arbitrary admissible variations of the generalized coordinate. Thus, the necessary conditions that Eq. (9-49) vanishes for arbitrary values of $\delta q(x,t)$ throughout $0 < x < l$, for arbitrary values of $\delta q(0,t)$ at $x = 0$, and for arbitrary values of $\delta q(l,t)$ at $x = l$ are that each of the terms in flower brackets in Eq. (9-49) vanishes. Thus,

$$
L(x)\frac{\partial^2 q}{\partial t^2} = \frac{\partial}{\partial x}\left(\frac{1}{C(x)}\frac{\partial q}{\partial x}\right) \qquad 0 < x < l
\tag{9-50}
$$

which is the governing partial differential equation (PDE) for the electric transmission line. Furthermore,

$$
L_1\frac{\partial^2 q}{\partial t^2} + \frac{1}{C_1}q = \frac{1}{C(x)}\frac{\partial q}{\partial x} \qquad x = 0
\tag{9-51}
$$

and

$$
L_2\frac{\partial^2 q}{\partial t^2} + R\frac{\partial q}{\partial t} + \frac{1}{C(x)}\frac{\partial q}{\partial x} = E(t) \qquad x = l
\tag{9-52}
$$

where Eqs. (9-51) and (9-52) are equivalent to the natural boundary conditions in the mechanical systems studied in the two previous subsections. We note that for this charge formulation the "natural boundary conditions" for the electric transmission line are simply statements of KVR (Kirchhoff's voltage rule) at the ends of the transmission line. To illustrate this point, we find the voltage drop across the top and bottom conductors (that is, across the shunt capacitance) in the transmission line. Recall that the differential charge dq_c on $C(x)\,dx$ is given by Eq. (9-44) and that the constitutive relation for linear capacitors is $q = Ce$. Thus, the voltage difference $e_c(x,t)$ across $C(x)\,dx$ is given by

$$
e_c(x,t) = \frac{dq_c}{C(x)\,dx} = \frac{\left(\dfrac{\partial q}{\partial x}\right)dx}{C(x)\,dx} = \frac{1}{C(x)}\frac{\partial q}{\partial x}
\tag{9-53}
$$

where Eq. (9-44) has been used in writing the differential charge dq_c on the shunt capacitance $C(x)\,dx$ in terms of the generalized coordinate $q(x, t)$. Therefore, the right-hand side of Eq. (9-51) is simply the voltage difference across the gap between the top and bottom conductors in the transmission line at $x = 0$. Thus, Eq. (9-51) simply indicates that the sum of the voltage drops across the inductor L_1 and across the capacitor C_1 balances the voltage increase in the top conductor relative to the bottom conductor in the transmission line at $x = 0$. This zero voltage difference around a closed loop is described by KVR, as noted in Chapter 7. Similarly, Eq. (9-52) indicates that the externally supplied voltage of the voltage source $E(t)$ is balanced by the sum of the voltage drops across the elements L_2 and R and the voltage drop in the transmission line at $x = l$.

As mentioned earlier, because the governing equation of motion and the boundary conditions for the electric transmission line have essentially identical forms as those for the systems sketched in Figures 9-1 and 9-3, all three of these systems behave in the same fashion as far as their dynamic characteristics are concerned.

Not only has Hamilton's principle produced the partial differential equation governing the dynamic behavior of the electric transmission continuum, it has also furnished the "natural (KVR) boundary conditions." Herein lies one of the significant benefits of variational methods in electrical dynamics as well as in mechanical dynamics, as shown in previous subsections. In many systems without simple boundary conditions, it may be rather difficult to ensure the proper signs on the various terms that are present in boundary conditions such as represented in Eqs. (9-51) and (9-52); Hamilton's principle automatically gives these complete dynamic relationships. Finally, in order to solve Eq. (9-50) in conjunction with Eqs. (9-51) and (9-52) in a particular problem, it is necessary to prescribe the system's *initial conditions* and the voltage supply $E(t)$. For the electric transmission continuum in this example, these initial conditions would be represented by the complete specification of

$$q(x, 0) \quad \text{and} \quad \frac{\partial q(x, 0)}{\partial t} \quad\quad 0 < x < l \quad\quad\quad (9\text{-}54)$$

which are the charge and current throughout the transmission line at time $t = 0$.

9-2.4   Flexural Motion of System Containing Beam

We consider the system containing a Bernoulli-Euler beam undergoing flexural motion as sketched in Figure 9-6. The beam is modeled as a one-dimensional continuum of (mass) density ρ, cross-sectional area A, modulus of elasticity E, second moment of the cross-sectional area about the neutral axis I, and equilibrium length l. Because in many mechanics formulations the second moment of the cross-sectional area appears with the modulus of elasticity

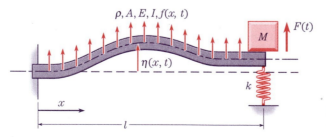

FIGURE 9-6   Model for flexural vibration of Bernoulli-Euler beam.

as the product EI, EI is often called the *flexural rigidity* (or *bending modulus*) of the beam. The entire extent of the beam $(0 < x < l)$ is subjected to a distributed transverse (lateral) force $f(x, t)$ per unit length of the beam. The left-hand end of the beam $(x = 0)$ is clamped or built into an immovable rigid wall. The right-hand end of the beam $(x = l)$ is subjected to a prescribed vertical force $F(t)$, and is also attached to the mass M and is restrained vertically by a linear spring having a spring constant k.

In preparation for the analysis of beams, we review several models of energy storage and the commonly defined boundary conditions for such structures. Table 9-1 contains a summary of four energy storage mechanisms in beam dynamics; Table 9-2 contains a summary of the four most commonly assigned boundary conditions in beam structures known as Bernoulli-Euler beams. Entries such as those in both Tables 9-1 and 9-2 as always are *models*, idealizations of real bodies or systems. No physical boundary conditions behave precisely as any of the entries in Table 9-2.

As indicated in Table 9-1, we consider the motion and deformation of an arbitrary longitudinal element of a dynamically deforming beam. Also, the sign convention for a set of positive bending moments (M_b) and shear forces (S) is shown on an element in equilibrium. As illustrated, the beam element may store kinetic coenergy by the transverse translation of its mass ($^\#1$) and (or) by the rotation of its rotary inertia about an axis perpendicular to the plane of the sketch ($^\#2$), and it may store strain energy by its bending deformation

TABLE 9-1 Beam Models for Determining State Functions

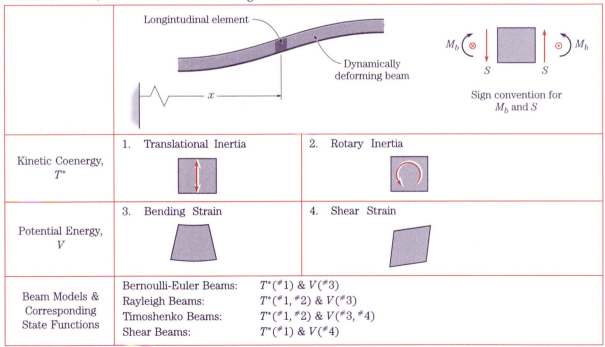

Kinetic Coenergy, T^*	1. Translational Inertia	2. Rotary Inertia
Potential Energy, V	3. Bending Strain	4. Shear Strain

Beam Models & Corresponding State Functions	Bernoulli-Euler Beams: $T^*(^\#1)$ & $V(^\#3)$ Rayleigh Beams: $T^*(^\#1, ^\#2)$ & $V(^\#3)$ Timoshenko Beams: $T^*(^\#1, ^\#2)$ & $V(^\#3, ^\#4)$ Shear Beams: $T^*(^\#1)$ & $V(^\#4)$

TABLE 9-2 Boundary Conditions for Bernoulli-Euler Beams

Schematic of End Condition	Analytical Boundary Condition		
Clamped	Displacement:	$\eta(0,t) = 0$	$\triangledown$
	Slope:	$\dfrac{\partial \eta(0,t)}{\partial x} = 0$	$\triangledown$
Free	Bending moment:	$M_b(0,t) = 0$	$\triangle$
	Shear force:	$S(0,t) = 0$	$\triangle$
Simply Supported	Displacement:	$\eta(0,t) = 0$	$\triangledown$
	Bending moment:	$M_b(0,t) = 0$	$\triangle$
Sliding–Rotationally Restrained	Slope:	$\dfrac{\partial \eta(0,t)}{\partial x} = 0$	$\triangledown$
	Shear force:	$S(0,t) = 0$	$\triangle$

$\triangledown$: Geometric Boundary Condition—Responsibility of Analyst
$\triangle$: Natural Boundary Condition—Provided by Hamilton's Principle

($^{\#}3$) and (or) its shearing deformation ($^{\#}4$). Several beam models that account for various combinations of these energy storage mechanisms have been used by analysts to account for specific aspects of a dynamic phenomenon. Four of the most widely used models are given in Table 9-1; clearly, other models could be devised from different combinations of these energy storage mechanisms. The choice of one model over another is generally guided by the modeling of the structure under consideration, including the analyst's desire for simplicity. Of these models, the most commonly used is the Bernoulli-Euler beam, for which it is assumed that all the kinetic coenergy is due to translation ($^{\#}1$) and all the potential (strain) energy is due to bending deformation ($^{\#}3$). Accordingly, we shall restrict our presentation to Bernoulli-Euler beams.

In Table 9-2, we consider the most commonly defined boundary conditions for Bernoulli-Euler beams. In addition to the schematics of the various end conditions and the corresponding analytical expressions for the boundary conditions (B.C.), we distinguish between those B.C. that are characterized as *geometric* and those B.C. that are characterized as *natural*. In the course of applying Hamilton's principle, it is necessary for the analyst to define explicitly all geometric boundary conditions; Hamilton's principle will intrinsically deliver all the natural boundary conditions.

We want to derive the equation of motion and all the boundary conditions for the system model in Figure 9-6. In addition to formulating the variational indicator for Hamilton's principle, we must ensure that the geometric boundary conditions are appropriately incorporated; Hamilton's principle will deliver not only the equation of motion but also the natural (force-dynamic) boundary conditions.

The generalized coordinate for the beam is denoted by $\eta(x, t)$, where $\eta(x, t)$ is the transverse (or lateral) displacement of the section whose position is x. The choice of $\eta(x, t)$ is guided by our understanding of lumped-parameter systems where generalized coordinates are time-dependent functions of the translational or angular displacement of a specific undisturbed or equilibrium position in the system. Here, because of the selection of a continuous model, we observe that the generalized coordinate is not only a function of time but also a function of space. Thus we resolve that for the Bernoulli-Euler beam model

$$\xi_j : \eta(x, t) \qquad \text{and} \qquad \delta\xi_j : \delta\eta(x, t) \tag{9-55}$$

The distributed force $f(x, t)$ and the localized force $F(t)$ provide the only contributions to the nonconservative work increment. In accordance with Eq. (5-36), these nonconservative force contributions give

$$\delta\mathcal{W}^{nc} = \sum_{i=1}^{N} \boldsymbol{f}_i \cdot \delta\boldsymbol{R}_i = \sum_{j=1}^{n} \Xi_j \, \delta\xi_j$$

$$= f(x, t) \, \delta\eta(x, t) + F(t) \, \delta\eta(l, t) = \sum_{j=1}^{n} \Xi_j \, \delta\xi_j \tag{9-56}$$

where we emphasize that the force $F(t)$ acts at $x = l$ as noted in Eq. (9-56).

The *kinetic coenergy function* consists of contributions due to the mass M and the mass of the beam. No new theory is required, as we simply note that

$$T^* = \frac{1}{2}M\left[\frac{\partial\eta(l, t)}{\partial t}\right]^2 + \int_0^l \frac{1}{2}\rho A\left[\frac{\partial\eta(x, t)}{\partial t}\right]^2 dx \tag{9-57}$$

where in the first term in T^* we emphasize that the velocity of M is indicated at $x = l$. Note that the second term in T^* has the correct form of $\frac{1}{2}$(mass)(velocity squared), but it must be written as an integral to reconcile the fact that the transverse velocity in the beam varies continuously throughout the beam from $x = 0$ to $x = l$. Here, as indicated above, we are restricting our consideration to Bernoulli-Euler beams, so the integral term in Eq. (9-57) corresponds to item $^\#1$ in Table 9-1. Note also that both ρ and A may be functions of x.

The *potential energy function* consists of contributions due to the spring k and the strain energy of the elastic beam. This may be written as

$$V = \frac{1}{2}k[\eta(l, t)]^2 + \int_0^l \frac{1}{2}EI\left[\frac{\partial^2\eta(x, t)}{\partial x^2}\right]^2 dx \tag{9-58}$$

where in the first term we emphasize that the deformation of the spring element is due to the transverse beam displacement at $x = l$. The second term in Eq. (9-58) is the strain energy function for the continuous elastic medium of the Bernoulli-Euler beam (item $^\#3$ in Table 9-1) and may be derived as illustrated in Example K-3 of Appendix K.

Substitution of Eqs. (9-55) through (9-58) into Eq. (9-1) gives the variational indicator of Hamilton's principle as

$$
\text{V.I.} = \int_{t_1}^{t_2} \left[\delta \left\{ \overbrace{\frac{1}{2}M\left[\frac{\partial \eta(l,t)}{\partial t}\right]^2}^{\textcircled{7}} + \overbrace{\int_0^l \frac{1}{2}\rho A\left[\frac{\partial \eta(x,t)}{\partial t}\right]^2 dx}^{\textcircled{8}} \right. \right.
$$

$$
\left. \overbrace{-\frac{1}{2}k[\eta(l,t)]^2}^{\textcircled{9}} \quad \overbrace{- \int_0^l \frac{1}{2}EI\left[\frac{\partial^2 \eta(x,t)}{\partial x^2}\right]^2 dx}^{\textcircled{10}} \right\}
$$

$$
\left. + \overbrace{f(x,t)\delta\eta(x,t) + F(t)\delta\eta(l,t)}^{\textcircled{11}} \right] dt \quad (9\text{-}59)
$$

Once again, as we shall see, the careful consideration of each term followed by the application of Hamilton's principle will reveal the enormous amount of information contained in Eq. (9-59).

First, we note from Figure 9-6 that there are two geometric restrictions on the motions of the beam, both of them at the left-hand end ($x = 0$). These geometric restrictions are due to the boundary conditions of the clamped end. (Refer to Table 9-2.) The appropriate boundary conditions for such a clamped end are given by the vanishing of both the transverse displacement and the beam's (longitudinal) slope at $x = 0$. That is,

$$
\eta(0,t) = 0 \quad \text{and} \quad \frac{\partial \eta(0,t)}{\partial x} = 0 \quad (9\text{-}60)
$$

These are *geometric boundary conditions* as opposed to natural boundary conditions because these boundary conditions are the result of geometric restrictions imposed on the motion of the beam by the clamped-end condition at $x = 0$. Therefore, the geometrically admissible motion must satisfy

$$
\delta\eta(0,t) = 0 \quad \text{and} \quad \delta\left[\frac{\partial \eta(0,t)}{\partial x}\right] = 0 \quad (9\text{-}61)
$$

That is, we cannot "imagine" a nonzero deflection or nonzero slope at $x = 0$ without violating the geometric constraints on the system.

In order to apply Hamilton's principle to the variational indicator, we must express the V.I. in terms of variations on the generalized coordinates. In order to indicate clearly this conversion procedure for each term in the V.I., the terms in Eq. (9-59) are designated as ⑦ through ⑪, with the understanding that terms ⑦ through ⑩ include the variational operator δ and all terms include the timewise integration $\int \, dt$. It is useful to note that terms ⑦, ⑧, and ⑨ in Eq. (9-59) are of the same forms as ①, ②, and ④, respectively, in Eq. (9-8). Also, we note that term ⑪ already has a variation on the generalized coordinate and, therefore, no conversion is needed, just as in the case of term ⑥ in Eq. (9-8). We shall write the converted expressions for these terms without repeating the mathematical operations, and we shall consider in detail only the conversion of term ⑩, a type of term that we have not previously encountered.

Terms ⑦ through ⑨:

Terms ⑦ through ⑨ are written below by appropriately modifying Eqs. (9-15), (9-19), and (9-21), respectively. That is, term ⑦ becomes

$$\int_{t_1}^{t_2} \delta \left\{ \frac{1}{2} M \left[\frac{\partial \eta(l,t)}{\partial t} \right]^2 \right\} dt = -\int_{t_1}^{t_2} M \left[\frac{\partial^2 \eta(l,t)}{\partial t^2} \right] \delta \eta(l,t)\, dt \tag{9-62}$$

Term ⑧ becomes

$$\int_{t_1}^{t_2} \delta \left\{ \int_0^l \frac{1}{2} \rho A \left[\frac{\partial \eta(x,t)}{\partial t} \right]^2 dx \right\} dt = -\int_{t_1}^{t_2} dt \int_0^l \rho A \left[\frac{\partial^2 \eta(x,t)}{\partial t^2} \right] \delta \eta(x,t)\, dx \tag{9-63}$$

And term ⑨ becomes

$$\int_{t_1}^{t_2} \delta \left\{ \frac{1}{2} k \left[\eta(l,t) \right]^2 \right\} dt = -\int_{t_1}^{t_2} k \eta(l,t) \delta \eta(l,t)\, dt \tag{9-64}$$

Note that we have converted the variation of terms ⑦ through ⑨, from a variation on (a portion of) the lagrangian to a variation on the generalized coordinate.

Terms ⑩:

Term ⑩ is similar to term ⑤ in Eq. (9-8), the significant difference being that it contains a second derivative in a spacewise sense. This difference results in term ⑩ requiring two successive executions of integration by parts in a spacewise sense, as will be shown below.

So, term ⑩ becomes

$$\int_{t_1}^{t_2} \delta \left\{ -\int_0^l \frac{1}{2} EI \left[\frac{\partial^2 \eta(x,t)}{\partial x^2} \right]^2 dx \right\} dt = \int_{t_1}^{t_2} dt \int_0^l -\frac{1}{2} \cdot EI \cdot 2 \left[\frac{\partial^2 \eta(x,t)}{\partial x^2} \right] \delta \left[\frac{\partial^2 \eta(x,t)}{\partial x^2} \right] dx \tag{9-65}$$

where we have used the operation quoted immediately after both Eqs. (9-11) and (9-16) and where we have extracted the entire time integral forward for aesthetics only. Using Eq. (9-10) yields

$$\delta \left[\frac{\partial^2 \eta(x,t)}{\partial x^2} \right] = \delta \left[\frac{\partial}{\partial x} \left(\frac{\partial \eta(x,t)}{\partial x} \right) \right] = \frac{\partial}{\partial x} \delta \left[\frac{\partial \eta(x,t)}{\partial x} \right] \tag{9-66}$$

Thus, substituting Eq. (9-66) into Eq. (9-65) gives

$$\int_{t_1}^{t_2} \delta \left\{ -\int_0^l \frac{1}{2} EI \left[\frac{\partial^2 \eta(x,t)}{\partial x^2} \right]^2 dx \right\} dt = \int_{t_1}^{t_2} dt \int_0^l -EI \left[\frac{\partial^2 \eta(x,t)}{\partial x^2} \right] \frac{\partial}{\partial x} \delta \left[\frac{\partial \eta(x,t)}{\partial x} \right] dx \tag{9-67}$$

Identifying corresponding terms in Eqs. (9-13) and (9-67) yields

$$\int_{t_1}^{t_2} \delta \left\{ -\int_0^l \frac{1}{2} EI \left[\frac{\partial^2 \eta(x,t)}{\partial x^2} \right]^2 dx \right\} dt$$

$$= \int_{t_1}^{t_2} dt \left\{ -EI \frac{\partial^2 \eta(x,t)}{\partial x^2} \delta \left[\frac{\partial \eta(x,t)}{\partial x} \right] \Big|_0^l + \int_0^l \frac{\partial}{\partial x} \left[EI \frac{\partial^2 \eta(x,t)}{\partial x^2} \right] \delta \left[\frac{\partial \eta(x,t)}{\partial x} \right] dx \right\}$$

$$= \int_{t_1}^{t_2} dt \left\{ -EI \frac{\partial^2 \eta(l,t)}{\partial x^2} \delta \left[\frac{\partial \eta(l,t)}{\partial x} \right] + EI \frac{\partial^2 \eta(0,t)}{\partial x^2} \delta \left[\frac{\partial \eta(0,t)}{\partial x} \right] \right.$$

$$\left. + \int_0^l \frac{\partial}{\partial x} \left[EI \frac{\partial^2 \eta(x,t)}{\partial x^2} \right] \delta \left[\frac{\partial \eta(x,t)}{\partial x} \right] dx \right\} \tag{9-68}$$

where, in writing the first two terms in the last expression of Eq. (9-68), the commutativity of variation and differentiation as stated in Eq. (9-10) has been used. By virtue of the second of the geometric admissibility requirements in Eq. (9-61), the second term in the last expression of Eq. (9-68) vanishes. This is an important consequence of having stated the *geometric boundary conditions,* one of which has now been imposed onto the formulation. Further, we note that the last term on the right-hand side of Eq. (9-68) continues to possess a spacewise integration having a variation on the spacewise derivative of the generalized coordinate. This term must be converted further in order to obtain a variation on the generalized coordinate. Therefore, Eq. (9-68) becomes

$$\int_{t_1}^{t_2} \delta \left\{ -\int_0^l \frac{1}{2} EI \left[\frac{\partial^2 \eta(x,t)}{\partial x^2} \right]^2 dx \right\} dt$$

$$= \int_{t_1}^{t_2} dt \left\{ -EI \frac{\partial^2 \eta(l,t)}{\partial x^2} \delta \left[\frac{\partial \eta(l,t)}{\partial x} \right] + \int_0^l \frac{\partial}{\partial x} \left[EI \frac{\partial^2 \eta(x,t)}{\partial x^2} \right] \frac{\partial \delta \eta(x,t)}{\partial x} dx \right\} \quad (9\text{-}69)$$

where in writing the last term in Eq. (9-69), we have again used the commutativity of variation and differentiation. Thus, a second integration by parts in a spacewise sense can be conveniently performed. Identifying corresponding terms in Eq. (9-13) and the last term in Eq. (9-69) yields

$$\int_{t_1}^{t_2} \delta \left\{ -\int_0^l \frac{1}{2} EI \left[\frac{\partial^2 \eta(x,t)}{\partial x^2} \right]^2 dx \right\} dt$$

$$= \int_{t_1}^{t_2} dt \left\{ -EI \frac{\partial^2 \eta(l,t)}{\partial x^2} \delta \left[\frac{\partial \eta(l,t)}{\partial x} \right] + \frac{\partial}{\partial x} \left[EI \frac{\partial^2 \eta(x,t)}{\partial x^2} \right] \delta \eta(x,t) \Big|_0^l \right.$$

$$\left. -\int_0^l \frac{\partial^2}{\partial x^2} \left[EI \frac{\partial^2 \eta(x,t)}{\partial x^2} \right] \delta \eta(x,t) \, dx \right\}$$

$$= \int_{t_1}^{t_2} dt \left\{ -EI \frac{\partial^2 \eta(l,t)}{\partial x^2} \delta \left[\frac{\partial \eta(l,t)}{\partial x} \right] + \frac{\partial}{\partial x} \left[EI \frac{\partial^2 \eta(l,t)}{\partial x^2} \right] \delta \eta(l,t) \right.$$

$$\left. - \frac{\partial}{\partial x} \left[EI \frac{\partial^2 \eta(0,t)}{\partial x^2} \right] \delta \eta(0,t) - \int_0^l \frac{\partial^2}{\partial x^2} \left[EI \frac{\partial^2 \eta(x,t)}{\partial x^2} \right] \delta \eta(x,t) \, dx \right\} \quad (9\text{-}70)$$

By virtue of the first of the geometric admissibility requirements in Eq. (9-61), the third term in the last expression of Eq. (9-70) vanishes. Again, as indicated following Eq. (9-68), this is an important consequence of having stated the *geometric boundary conditions,* the other of which has now been imposed onto the formulation. Therefore, term ⑩ becomes

$$\int_{t_1}^{t_2} \delta \left\{ -\int_0^l \frac{1}{2} EI \left[\frac{\partial^2 \eta(x,t)}{\partial x^2} \right]^2 dx \right\} dt$$

$$= \int_{t_1}^{t_2} dt \left\{ -EI \frac{\partial^2 \eta(l,t)}{\partial x^2} \delta \left[\frac{\partial \eta(l,t)}{\partial x} \right] + \frac{\partial}{\partial x} \left[EI \frac{\partial^2 \eta(l,t)}{\partial x^2} \right] \delta \eta(l,t) \right.$$

$$\left. -\int_0^l \frac{\partial^2}{\partial x^2} \left[EI \frac{\partial^2 \eta(x,t)}{\partial x^2} \right] \delta \eta(x,t) \, dx \right\} \quad (9\text{-}71)$$

Thus, we have converted the variation from a variation on (a portion of) the lagrangian to a variation on the generalized coordinate. Note that the first term on the right-hand side of

Eq. (9-71) cannot be evaluated further. This first term, which has a variation on the slope evaluated at $x = l$, cannot be converted further because it is a spacewise derivative that is no longer subjected to a spacewise integral and furthermore that is not subjected to any geometric admissibility requirement. In the context of our integrations by parts, it is complete. Also, the second term on the right-hand side of Eq. (9-71) cannot be evaluated further since it too is not subjected to any geometric admissibility requirement. As we shall see below, both the first and second terms on the right-hand side of Eq. (9-71) will participate in the enunciation of the natural boundary conditions. Thus, in converting term ⑩, both of the geometric admissibility requirements given in Eq. (9-61) have been imposed in order to ensure that only geometrically admissible variations remain in this term.

Now, we are prepared to collect all of the converted terms into the variational indicator in Eq. (9-59). In doing this, it is convenient to aggregate terms containing alike variations on the generalized coordinate. Accordingly, all terms containing $\delta\eta(x, t)$ should be grouped together, all terms containing $\delta\eta(l, t)$ should be grouped together, and all terms containing $\delta\left[\dfrac{\partial\eta(l, t)}{\partial x}\right]$ should be grouped together. Thus, the variational indicator becomes

$$
\begin{aligned}
\text{V.I.} = \int_{t_1}^{t_2} \Bigg[\int_0^l & \left\{ -\rho A\left[\frac{\partial^2\eta(x, t)}{\partial t^2}\right] - \frac{\partial^2}{\partial x^2}\left[EI\frac{\partial^2\eta(x, t)}{\partial x^2}\right] + f(x, t) \right\} \delta\eta(x, t)\, dx \\
& + \left\{ -M\left[\frac{\partial^2\eta(l, t)}{\partial t^2}\right] - k\eta(l, t) + \frac{\partial}{\partial x}\left[EI\frac{\partial^2\eta(l, t)}{\partial x^2}\right] + F(t) \right\} \delta\eta(l, t) \\
& + \left\{ -EI\frac{\partial^2\eta(l, t)}{\partial x^2} \right\} \delta\left[\frac{\partial\eta(l, t)}{\partial x}\right] \Bigg]\, dt \quad (9\text{-}72)
\end{aligned}
$$

Since all the variations in terms ⑦ through ⑩ are geometrically admissible, the variational indicator in Eq. (9-72) has been converted into a geometrically admissible form. Hamilton's principle requires that the variational indicator in Eq. (9-72) vanishes for arbitrary admissible variations of the generalized coordinate. Thus, the necessary conditions that Eq. (9-72) vanishes for arbitrary values of $\delta\eta(x, t)$ throughout $0 < x < l$, for arbitrary values of $\delta\eta(l, t)$ at $x = l$, and for arbitrary values of $\delta[\partial\eta(l, t)/\partial x]$ at $x = l$ are that each of the terms in flower brackets in Eq. (9-72) vanishes. Thus,

$$
\rho A\frac{\partial^2\eta}{\partial t^2} = -\frac{\partial^2}{\partial x^2}\left(EI\frac{\partial^2\eta}{\partial x^2}\right) + f(x, t) \qquad 0 < x < l \qquad (9\text{-}73)
$$

which is the governing partial differential equation (PDE) for the flexural motion of the beam. Furthermore,

$$
M\frac{\partial^2\eta}{\partial t^2} + k\eta = \frac{\partial}{\partial x}\left(EI\frac{\partial^2\eta}{\partial x^2}\right) + F(t) \qquad x = l \qquad (9\text{-}74)
$$

and

$$
EI\frac{\partial^2\eta}{\partial x^2} = 0 \qquad x = l \qquad (9\text{-}75)
$$

which are the *natural boundary conditions*. (Recall that the equally important geometric boundary conditions were imposed immediately following Eqs. (9-68) and (9-70).) Note that the natural boundary conditions are statements of the dynamic requirements on the shear force (Eq. (9-74)) and bending moment (Eq. (9-75)) at the right-hand end of the beam.

Equation (9-74) establishes the relation between the applied force $F(t)$, the force due to the mass M times its acceleration, the force due to the spring's extension, and the shear force $\dfrac{EI\,\partial^3\eta}{\partial x^3}$ in the beam, all at $x = l$. Similarly, because $\dfrac{EI\,\partial^2\eta}{\partial x^2}$ is the bending moment in the beam, Eq. (9-75) indicates that the bending moment in the beam is zero at $x = l$.

Not only has Hamilton's principle provided the partial differential equation governing the dynamic behavior of the beam continuum, it has also furnished the natural boundary conditions. In many systems without simple boundary conditions, it may be rather difficult to ensure the proper signs on the various terms that are present in boundary conditions such as represented in Eqs. (9-74) and (9-75); Hamilton's principle automatically gives these complete dynamic relationships. Finally, in order to solve Eq. (9-73) subjected to Eqs. (9-60) and to Eqs. (9-74) and (9-75) in a particular problem, it is necessary to prescribe the system's *initial conditions* and the external forces $f(x, t)$ and $F(t)$. For the Bernoulli-Euler beam continuum in this example, these initial conditions would be represented by the complete specification of

$$\eta(x, 0) \qquad \text{and} \qquad \frac{\partial \eta(x, 0)}{\partial t} \qquad 0 < x < l \tag{9-76}$$

which are the transverse displacement and transverse velocity throughout the beam at time $t = 0$.

9-2.5   Summaries

In the preceding subsections, we have derived the equations of motion for four system models containing one-dimensional continuous elements: rods, shafts, electric transmission lines, and beams. We have seen that the application of Hamilton's principle to these models requires the definition of the variational indicator and the specification of the geometric boundary conditions while furnishing both (1) the governing partial differential equation for the continuum and (2) the natural boundary conditions at the end(s) of the continuum. The governing equation of motion for the continuum represents the inherent dynamic characteristics of the continuum, while the geometric and natural boundary conditions are determined by the specific end terminations.

The governing equations of motion for various one-dimensional continua are summarized in Table 9-3. In addition to the elements considered in the previous subsections, the governing equation of motion for a taut string—subjected to a very large tension P and undergoing transverse displacement $\eta(x, t)$—is also given in Table 9-3, without derivation.[7] Such an equation can be derived via the same approach as in the previous subsections, making use of the strain energy function for strings derived in Example K–4.

The equations of motion in Table 9-3 exhibit two distinct forms and so may be classified into two categories: one containing Bernoulli-Euler beams, and another containing all the other elements. This classification is based on the *order* of the spatial derivative on the right-hand side of the equation of motion. That is, the right-hand side of the equation of motion for a Bernoulli-Euler beam contains a fourth-order spacewise derivative while the equations of motion for all the other elements contain a second-order spacewise derivative. We shall designate those models resulting in equations of motion having a fourth-order

---

[7]See Problem 9-25.

TABLE 9-3 Summary of Equations of Motion for Various One-Dimensional Continuous Elements of Extent l, Without Internal Mechanical Damping or Electrical Resistance

Element	Generalized coordinate	Equation of motion
String	$\eta(x,t)$: transverse displacement	$\rho A \dfrac{\partial^2 \eta}{\partial t^2} = P \dfrac{\partial^2 \eta}{\partial x^2} + f(x,t)$
Rod	$\xi(x,t)$: longitudinal displacement	$\rho A \dfrac{\partial^2 \xi}{\partial t^2} = \dfrac{\partial}{\partial x}\left(EA \dfrac{\partial \xi}{\partial x}\right) + f(x,t)$
Shaft	$\varphi(x,t)$: angular displacement	$\rho J \dfrac{\partial^2 \varphi}{\partial t^2} = \dfrac{\partial}{\partial x}\left(GJ \dfrac{\partial \varphi}{\partial x}\right) + \tau(x,t)$
Electric Transmission Line	$q(x,t)$: electrical charge	$L \dfrac{\partial^2 q}{\partial t^2} = \dfrac{\partial}{\partial x}\left(\dfrac{1}{C} \dfrac{\partial q}{\partial x}\right)$
Electric Transmission Line	$\lambda(x,t)$: flux linkage	$C \dfrac{\partial^2 \lambda}{\partial t^2} = \dfrac{\partial}{\partial x}\left(\dfrac{1}{L} \dfrac{\partial \lambda}{\partial x}\right)$
Bernoulli-Euler Beam	$\eta(x,t)$: transverse displacement	$\rho A \dfrac{\partial^2 \eta}{\partial t^2} = -\dfrac{\partial^2}{\partial x^2}\left(EI \dfrac{\partial^2 \eta}{\partial x^2}\right) + f(x,t)$

spatial derivative as fourth-order systems and those having a second-order spatial derivative as second-order systems.

The benefit of this classification is that we can find a generic equation of motion for each category and thereby study the characteristics of systems via their generic equations of motion without specifying the underlying physical system. First, we note that the generic equation of motion for all second-order systems may be written as

$$p_1 \frac{\partial^2 z(x,t)}{\partial t^2} = \frac{\partial}{\partial x}\left(p_2 \frac{\partial z(x,t)}{\partial x}\right) + g(x,t) \tag{9-77}$$

where $z(x,t)$ is the generic generalized coordinate that is specified for each element in Table 9-3; p_1 and p_2 are coefficients that depend on the element parameters and can be easily identified from the equations of motion listed in the third column of Table 9-3; and $g(x,t)$ is the generic distributed generalized force along the length of the element, which can also be easily identified from the equations of motion listed in the third column of Table 9-3. Similarly, the generic equation of motion for the fourth-order element may be written as

$$p_1 \frac{\partial^2 z(x,t)}{\partial t^2} = -\frac{\partial^2}{\partial x^2}\left(p_2 \frac{\partial^2 z(x,t)}{\partial x^2}\right) + g(x,t) \tag{9-78}$$

where $z(x,t)$ is the generic generalized coordinate ($\eta(x,t)$ for Bernoulli-Euler beams as indicated in Table 9-3); p_1 and p_2 can be identified from the equation of motion given in Table 9-3 (that is, $p_1 = \rho A$ and $p_2 = EI$ for Bernoulli-Euler beams); and the generic distributed generalized force $g(x,t)$ is the force distributed along the length of the element, which can also be easily identified from the equation of motion in the third column of Table 9-3 (that is, $g(x,t)$ for Bernoulli-Euler beams is the transverse force distributed along the beam). For all

TABLE 9-4 Generic Equation of Motion for Several One-Dimensional Continua

Generic equation of motion	Specific element	Specific generalized coordinate	Parameters		
			p_1	p_2	c_q^2
Second-order: $$p_1 \frac{\partial^2 z}{\partial t^2} = \frac{\partial}{\partial x}\left(p_2 \frac{\partial z}{\partial x}\right) + g(x,t)$$ or $$\frac{\partial^2 z}{\partial t^2} = c_q^2 \frac{\partial^2 z}{\partial x^2} + \frac{1}{p_1} g(x,t)$$	String	$\eta(x,t)$	ρA	P	$\dfrac{P}{\rho A}$
	Rod	$\xi(x,t)$	ρA	EA	$\dfrac{E}{\rho}$
	Shaft	$\varphi(x,t)$	ρJ	GJ	$\dfrac{G}{\rho}$
	Electric transmission line	$q(x,t)$	L	$\dfrac{1}{C}$	$\dfrac{1}{LC}$
Fourth-order: $$p_1 \frac{\partial^2 z}{\partial t^2} = -\frac{\partial^2}{\partial x^2}\left(p_2 \frac{\partial^2 z}{\partial x^2}\right) + g(x,t)$$ or $$\frac{\partial^2 z}{\partial t^2} = -c_q^2 \frac{\partial^4 z}{\partial x^4} + \frac{1}{p_1} g(x,t)$$	Bernoulli-Euler beam	$\eta(x,t)$	ρA	EI	$\dfrac{EI}{\rho A}$

elements considered in Table 9-3, the generalized coordinate $z(x,t)$ and the parameters p_1 and p_2 in the generic equations of motion—Eqs. (9-77) and (9-78)—are listed in Table 9-4.

The generic equations of motion become simpler if we make the additional assumption that the parameters p_1 and p_2 are constant throughout the extent of the one-dimensional continua, meaning that the continua are uniform throughout their lengths. This assumption of uniform parameters is valid in many engineering analyses. Under this assumption, Eqs. (9-77) and (9-78) become, respectively,

$$p_1 \frac{\partial^2 z(x,t)}{\partial t^2} = p_2 \frac{\partial^2 z(x,t)}{\partial x^2} + g(x,t) \qquad (9\text{-}79)$$

and

$$p_1 \frac{\partial^2 z(x,t)}{\partial t^2} = -p_2 \frac{\partial^4 z(x,t)}{\partial x^4} + g(x,t) \qquad (9\text{-}80)$$

By defining a new parameter[8] $c_q^2 \equiv \dfrac{p_2}{p_1}$, Eqs. (9-79) and (9-80) reduce further to

$$\frac{\partial^2 z(x,t)}{\partial t^2} = c_q^2 \frac{\partial^2 z(x,t)}{\partial x^2} + \frac{1}{p_1} g(x,t) \qquad (9\text{-}81)$$

and

$$\frac{\partial^2 z(x,t)}{\partial t^2} = -c_q^2 \frac{\partial^4 z(x,t)}{\partial x^4} + \frac{1}{p_1} g(x,t) \qquad (9\text{-}82)$$

[8]In wave mechanics, c_q for *second-order* systems is often called the *phase velocity* of the medium, meaning the velocity of propagation of disturbances of the medium.

TABLE 9-5 Boundary Conditions for (Longitudinal) Vibration of Rod for Various End Conditions

End condition		Boundary condition at $x = 0$ (left)	Boundary condition at $x = \ell$ (right)
Clamped (displacement $= 0$)		$\xi(0, t) = 0$	$\xi(\ell, t) = 0$
Free (force$=0$)		$\dfrac{\partial \xi}{\partial x} = 0$	$\dfrac{\partial \xi}{\partial x} = 0$
Spring k		$EA\dfrac{\partial \xi}{\partial x} = k\xi$	$EA\dfrac{\partial \xi}{\partial x} = -k\xi$
Mass m		$EA\dfrac{\partial \xi}{\partial x} = m\dfrac{\partial^2 \xi}{\partial t^2}$	$EA\dfrac{\partial \xi}{\partial x} = -m\dfrac{\partial^2 \xi}{\partial t^2}$
Dashpot c		$EA\dfrac{\partial \xi}{\partial x} = c\dfrac{\partial \xi}{\partial t}$	$EA\dfrac{\partial \xi}{\partial x} = -c\dfrac{\partial \xi}{\partial t}$

The new parameter c_q in Eqs. (9-81) and (9-82) is also listed in Table 9-4 for each element. In the remainder of this chapter, we shall continue to assume uniformity of the parameters p_1 and p_2 throughout the lengths of the continua. That is, we shall consider the generic equations of motion—Eqs. (9-81) and (9-82)—for second- and fourth-order systems where c_q is specified in Table 9-4 for each element.

While the governing equations of motion for one-dimensional continua represent the inherent dynamic characteristics of the extended element's medium, as previously indicated,

TABLE 9-6 Boundary Conditions for (Twisting) Vibration of Shaft for Various End Conditions

End condition		Boundary condition at $x = 0$ (left)	Boundary condition at $x = \ell$ (right)
Clamped (angular displacement $= 0$)		$\varphi(0, t) = 0$	$\varphi(\ell, t) = 0$
Free (torque $= 0$)		$\dfrac{\partial \varphi}{\partial x} = 0$	$\dfrac{\partial \varphi}{\partial x} = 0$
Torsional spring k_t		$GJ\dfrac{\partial \varphi}{\partial x} = k_t\varphi$	$GJ\dfrac{\partial \varphi}{\partial x} = -k_t\varphi$
Rotating inertia I_1 about shaft axis		$GJ\dfrac{\partial \varphi}{\partial x} = I_1\dfrac{\partial^2 \varphi}{\partial t^2}$	$GJ\dfrac{\partial \varphi}{\partial x} = -I_1\dfrac{\partial^2 \varphi}{\partial t^2}$
Torsional dashpot c_t		$GJ\dfrac{\partial \varphi}{\partial x} = c_t\dfrac{\partial \varphi}{\partial t}$	$GJ\dfrac{\partial \varphi}{\partial x} = -c_t\dfrac{\partial \varphi}{\partial t}$

TABLE 9-7 Boundary Conditions for Electric Transmission Line for Various End Conditions when Charge Formulation is Used[†]

End condition	Boundary condition at $x = 0$ (left)	Boundary condition at $x = \ell$ (right)
Open circuit (charge = 0)	$q = 0$	$q = 0$
Short circuit (voltage drop = 0)	$\dfrac{\partial q}{\partial x} = 0$	$\dfrac{\partial q}{\partial x} = 0$
Capacitor C_1	$\dfrac{1}{C(x)}\dfrac{\partial q}{\partial x} = \dfrac{q}{C_1}$	$\dfrac{1}{C(x)}\dfrac{\partial q}{\partial x} = -\dfrac{q}{C_1}$
Inductor L_1	$\dfrac{1}{C(x)}\dfrac{\partial q}{\partial x} = L_1\dfrac{\partial^2 q}{\partial t^2}$	$\dfrac{1}{C(x)}\dfrac{\partial q}{\partial x} = -L_1\dfrac{\partial^2 q}{\partial t^2}$
Resistor R	$\dfrac{1}{C(x)}\dfrac{\partial q}{\partial x} = R\dfrac{\partial q}{\partial t}$	$\dfrac{1}{C(x)}\dfrac{\partial q}{\partial x} = -R\dfrac{\partial q}{\partial t}$

[†]This table is based on the differential model of an electric transmission line having distributed series inductance and distributed shunt capacitance, as shown in Figure 9-5.

TABLE 9-8 Boundary Conditions for (Flexural) Vibration of Bernoulli-Euler Beam for Various End Conditions

End condition		Boundary condition at $x = 0$ (left)	Boundary condition at $x = \ell$ (right)
Clamped (displacement and slope = 0)		$\eta = 0$ $\dfrac{\partial \eta}{\partial x} = 0$	$\eta = 0$ $\dfrac{\partial \eta}{\partial x} = 0$
Free (bending moment and shear force = 0)		$EI\dfrac{\partial^2 \eta}{\partial x^2} = 0$ $\dfrac{\partial}{\partial x}\left(EI\dfrac{\partial^2 \eta}{\partial x^2}\right) = 0$	$EI\dfrac{\partial^2 \eta}{\partial x^2} = 0$ $\dfrac{\partial}{\partial x}\left(EI\dfrac{\partial^2 \eta}{\partial x^2}\right) = 0$
Simply Supported (displacement and bending moment = 0)		$\eta(0, t) = 0$ $EI\dfrac{\partial^2 \eta}{\partial x^2} = 0$	$\eta(\ell, t) = 0$ $EI\dfrac{\partial^2 \eta}{\partial x^2} = 0$
Sliding-Rotationally Restrained (slope and shear force = 0)		$\dfrac{\partial \eta}{\partial x} = 0$ $\dfrac{\partial}{\partial x}\left(EI\dfrac{\partial^2 \eta}{\partial x^2}\right) = 0$	$\dfrac{\partial \eta}{\partial x} = 0$ $\dfrac{\partial}{\partial x}\left(EI\dfrac{\partial^2 \eta}{\partial x^2}\right) = 0$
Mass m and moment of inertia I_1 orthogonal to beam axis		$EI\dfrac{\partial^2 \eta}{\partial x^2} = I_1\dfrac{\partial^3 \eta}{\partial x \partial t^2}$ $\dfrac{\partial}{\partial x}\left(EI\dfrac{\partial^2 \eta}{\partial x^2}\right) = -m\dfrac{\partial^2 \eta}{\partial t^2}$	$EI\dfrac{\partial^2 \eta}{\partial x^2} = -I_1\dfrac{\partial^3 \eta}{\partial x \partial t^2}$ $\dfrac{\partial}{\partial x}\left(EI\dfrac{\partial^2 \eta}{\partial x^2}\right) = m\dfrac{\partial^2 \eta}{\partial t^2}$
Dashpot c and spring k (bending moment = 0)		$\dfrac{\partial}{\partial x}\left(EI\dfrac{\partial^2 \eta}{\partial x^2}\right) = -k\eta - c\dfrac{\partial \eta}{\partial t}$ $EI\dfrac{\partial^2 \eta}{\partial x^2} = 0$	$\dfrac{\partial}{\partial x}\left(EI\dfrac{\partial^2 \eta}{\partial x^2}\right) = k\eta + c\dfrac{\partial \eta}{\partial t}$ $EI\dfrac{\partial^2 \eta}{\partial x^2} = 0$

the natural boundary conditions as well as the geometric boundary conditions depend also on the specific end terminations in each system model. Without proof, we summarize the boundary conditions that result from various end conditions (including those examined in the previous subsections) in Tables 9-5 through 9-8.

9-3 NATURAL MODES OF VIBRATION

In this section, we seek solutions to the governing partial differential equations listed in Table 9-3 or Table 9-4, for system models containing no (nonconservative) generalized forces and having several boundary conditions among those listed in Tables 9-5 through 9-8. We consider only models in which the continua are uniform throughout their lengths. Therefore,

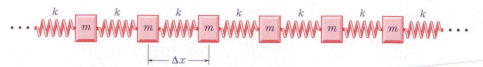

FIGURE 9-7   A discrete model approximating a rod undergoing longitudinal motion.

we seek the solutions to the generic equations of motion—Eqs. (9-81) and (9-82)—with $g(x, t) = 0$; namely,

$$\frac{\partial^2 z(x, t)}{\partial t^2} = c_q^2 \frac{\partial^2 z(x, t)}{\partial x^2} \tag{9-83}$$

and

$$\frac{\partial^2 z(x, t)}{\partial t^2} = -c_q^2 \frac{\partial^4 z(x, t)}{\partial x^4} \tag{9-84}$$

where c_q is specified in Table 9-4 for each one-dimensional continuum, and for several of the elementary boundary conditions contained in Tables 9-5 through 9-8.

Recall from the vibration of lumped-parameter systems in Chapter 8, systems can vibrate in such a manner that all discrete generalized coordinates oscillate at the same frequency (that is, at one of the natural frequencies) while maintaining a fixed amplitude ratio between them (that is, the corresponding mode shape). Heuristically, continuous systems may be considered as the limiting cases of the corresponding discrete models in which the elements are made increasingly finer as their numbers go to infinity. For example, a uniform continuous rod undergoing longitudinal motion may be modeled as sketched in Figure 9-7, where each of the elements approximating a small length Δx of the continuous rod is represented by a mass m and a spring of spring constant k. As the element size Δx is made increasingly smaller (thus, the number of elements becomes increasingly larger), the mass m and spring constant k must change accordingly. In the limit of this process, as the element size approaches zero and the number of elements approaches infinity, the discretized model becomes equivalent to the continuous system upon which it was based.

Because of this equivalence of continuous systems with limiting-case discrete systems, it is natural to expect that continuous systems also exhibit features similar to the natural modes of vibration for lumped-parameter systems. That is, the natural vibration of continuous systems is expected to be a motion in which all particles in the system oscillate at the same frequency. Furthermore, since continuous systems are equivalent to limiting-case lumped-parameter systems having an infinite number of elements (and, thus, an infinite number of degrees of freedom), their number of natural frequencies is expected to be infinite. Also, since the generalized coordinates for continuous systems are continuous functions of the spatial variable, the mode shapes are expected to be given in the form of continuous functions of the spatial variable.

In this section, we investigate the free vibration of continuous systems and obtain the natural frequencies and mode shapes as solutions to the generic homogeneous equations of motion, Eqs. (9-83) and (9-84), for various boundary conditions.

9-3.1   Method of Separation of Variables

One of the simplest techniques for solving the governing partial differential equations given by Eqs. (9-83) and (9-84) is the *method of separation of variables*. In this method it is assumed that the solution can be expressed as the product of two functions: one is a function of the spatial variable x only and the other is a function of the time variable t only. Therefore, the solutions to the generic equations of motion, Eqs. (9-83) and (9-84), are *assumed* to be

$$z(x, t) = X(x)T(t) \tag{9-85}$$

where $X(x)$ is a function of x only, and $T(t)$ is a function of t only.

First, we consider second-order systems whose generic homogeneous equation of motion is given by Eq. (9-83). Substitution of Eq. (9-85) into Eq. (9-83) gives

$$X(x)\frac{d^2T}{dt^2} = c_q^2 T(t)\frac{d^2X}{dx^2} \tag{9-86}$$

Dividing both sides of Eq. (9-86) by the product $X(x)T(t)$ yields

$$\frac{1}{T(t)}\frac{d^2T}{dt^2} = c_q^2 \frac{1}{X(x)}\frac{d^2X}{dx^2} \tag{9-87}$$

Note that the left-hand side of Eq. (9-87) is a function of t only while the right-hand side is a function of x only. Therefore, in order for Eq. (9-87) to hold for all values of t and for all values of x, $0 < x < l$, both sides of Eq. (9-87) must be equal to a constant. Denoting the constant by α, which has units of (time)$^{-2}$, and equating both sides of Eq. (9-87) to α yield the equations

$$\frac{d^2T}{dt^2} - \alpha T(t) = 0 \tag{9-88}$$

$$\frac{d^2X}{dx^2} - \frac{\alpha}{c_q^2}X(x) = 0 \tag{9-89}$$

Once the solutions $T(t)$ and $X(x)$ to Eqs. (9-88) and (9-89) have been found, the solution $z(x, t)$ to Eq. (9-83) can be expressed directly via Eq. (9-85). In this sense, we have transformed one partial differential equation, Eq. (9-83), into two ordinary differential equations, Eqs. (9-88) and (9-89). The solutions $T(t)$ and $X(x)$ to Eqs. (9-88) and (9-89), respectively, will be found in the following subsections.

Second, we consider fourth-order systems whose generic homogeneous equation of motion is given by Eq. (9-84). Substitution of Eq. (9-85) into Eq. (9-84) gives

$$X(x)\frac{d^2T}{dt^2} = -c_q^2 T(t)\frac{d^4X}{dx^4} \tag{9-90}$$

Dividing both sides of Eq. (9-90) by the product $X(x)T(t)$ yields

$$\frac{1}{T(t)}\frac{d^2T}{dt^2} = -c_q^2 \frac{1}{X(x)}\frac{d^4X}{dx^4} \tag{9-91}$$

For the reasons cited immediately following Eq. (9-87), equating both sides of Eq. (9-91) to a constant α having units of (time)$^{-2}$ yields

$$\frac{d^2T}{dt^2} - \alpha T(t) = 0 \tag{9-92}$$

and

$$\frac{d^4X}{dx^4} + \frac{\alpha}{c_q^2}X(x) = 0 \tag{9-93}$$

Again, we have transformed one partial differential equation, Eq. (9-84), into two ordinary differential equations, Eqs. (9-92) and (9-93). The solutions $T(t)$ and $X(x)$ to Eqs. (9-92) and (9-93), respectively, will also be found in the following subsections.

Note that the equation for $T(t)$ is the same for both the second-order and fourth-order systems, as represented by Eqs. (9-88) and (9-92). (The equations for $X(x)$ are different for the two types of systems.) Thus, the timewise behavior of the second-order and fourth-order systems is identical. The solution $T(t)$ which satisfies this common ordinary differential equation, will be found next in Subsection 9-3.2, without specifying the order of the system under consideration.

9-3.2   Time Response

We seek the solution $T(t)$ to Eq. (9-88), or equivalently Eq. (9-92). As suggested by the theory of ordinary differential equations, we *assume* the solution to have the form

$$T(t) = c'e^{\gamma t} \tag{9-94}$$

where c' and γ are unknown constants. Substitution of Eq. (9-94) into Eq. (9-88) and cancellation of the common term $e^{\gamma t}$ require

$$\gamma^2 = \alpha \tag{9-95}$$

or

$$\gamma = \pm \sqrt{\alpha} \tag{9-96}$$

Therefore, by substitution of Eq. (9-96) into Eq. (9-94), the solution becomes

$$T(t) = a'e^{\sqrt{\alpha}t} + b'e^{-\sqrt{\alpha}t} \tag{9-97}$$

where, because there are two solutions in Eq. (9-96), a' and b' are two unknown coefficients. Note that if α is positive, the first exponential term in Eq. (9-97) grows exponentially to infinity as t increases without limit, which is physically unacceptable. Therefore, α must be negative (or zero as considered below), and so we may write

$$\alpha = -\omega^2 \tag{9-98}$$

where ω is a real number having units of $(\text{time})^{-1}$. Experience has shown it to be convenient to assign it units of rad/s. Then, Eq. (9-97) becomes

$$\begin{aligned} T(t) &= a'e^{i\omega t} + b'e^{-i\omega t} \\ &= a'\cos \omega t + a'i \sin \omega t + b' \cos \omega t - b'i \sin \omega t \\ &= (a' + b') \cos \omega t + (a' - b')i \sin \omega t \\ &= a \cos \omega t + b \sin \omega t \end{aligned} \tag{9-99}$$

where $i \equiv \sqrt{-1}$ and Euler's formula,[9] $e^{\pm i\omega t} = \cos \omega t \pm i \sin \omega t$, has been used and where a, b, and ω are yet unknown.[10]

---

[9]See Eq. (H-12) in Appendix H.

[10]Note that since the function $T(t)$ is a real function, the coefficients a and b in Eq. (9-99) must also be real, from which it may be deduced that the coefficients a' and b' in Eq. (9-99) must be complex conjugates. Such details occur repeatedly; see, for example, Eqs. (8-100) through (8-102) for a specific case.

A special case that should be considered separately is the case when α in Eq. (9-88) is zero. In this case, Eq. (9-88) reduces to

$$\frac{d^2T}{dt^2} = 0 \tag{9-100}$$

whose general solution by direct integration is

$$T(t) = a + bt \tag{9-101}$$

where a and b are yet unknown. (Clearly a and b in Eq. (9-99) and a and b in Eq. (9-101) are not the same. By using the same constants in these two solutions—which never appear simultaneously for the same frequency—we shall be able to express the complete solutions in an otherwise more consistent format.)

In summary, we have found the time response function $T(t)$, which represents the time dependence of the solutions to the generic homogeneous equations of motion. In finding $T(t)$, we found via physical considerations that the constant α to which each side of Eqs. (9-87) and (9-91) was equated must be negative. Therefore, we replaced α by $-\omega^2$ in accordance with Eq. (9-98). Then, the time response function was found to be

$$T(t) = a\cos\omega t + b\sin\omega t \qquad \omega \neq 0 \tag{9-102}$$

And, in the special case when $\alpha = 0$ (or equivalently $\omega = 0$), the time response function was found to be

$$T(t) = a + bt \qquad \omega = 0 \tag{9-103}$$

Therefore, when $\omega \neq 0$, the solution $z(x, t)$ can be found by substituting Eq. (9-102) into Eq. (9-85), which gives

$$z(x, t) = (a\cos\omega t + b\sin\omega t)X(x) \tag{9-104}$$

Equation (9-104) exhibits the expected timewise behavior as it describes a motion in which all particles throughout the continuum $(0 < x < l)$ oscillate at the same frequency ω. Therefore, the frequency ω is called the *natural frequency* (ω^2 is called the *eigenvalue*) and the spatial function $X(x)$, which is not yet known, is called the *natural mode shape* (or, *eigenfunction*) corresponding to the frequency ω, in analogy with the natural modes of vibration of two- (or higher) degrees-of-freedom lumped-parameter systems. Furthermore, when $\omega = 0$, substitution of Eq. (9-103) into Eq. (9-85) yields the solution

$$z(x, t) = (a + bt)X(x) \tag{9-105}$$

Equation (9-105) describes a nonoscillatory motion that is analogous to the rigid-body mode of semidefinite lumped-parameter systems and, therefore, $X(x)$ in this case is called the *rigid-body mode eigenfunction* corresponding to the zero natural frequency, $\omega = 0$. As we saw in Chapter 8 for lumped-parameter systems, here too the rigid-body mode of a continuous system may or may not exist, depending on the boundary conditions. This will be shown later in this chapter.

In the following two subsections, we shall find the eigenfunction $X(x)$, which represents the spatial dependence of the solutions to each of the generic equations of motion. As indicated in Subsection 9-3.1, unlike the time response function $T(t)$, the eigenfunction $X(x)$ is governed by different expressions for different orders of the equations of motion; that is, the governing equations for $X(x)$ for second- and fourth-order systems are given by Eqs. (9-89) and (9-93), respectively. Therefore, the eigenfunctions for second- and fourth-order systems will be found separately in Subsections 9-3.3 and 9-3.4, respectively.

9-3.3 Eigenfunctions for Second-Order Systems

In this subsection, we seek the eigenfunction $X(x)$ that satisfies Eq. (9-89). In Subsection 9-3.2 we found that α in Eq. (9-89) must be negative (or zero in a special case); so combining Eqs. (9-89) and (9-98) gives

$$\frac{d^2X}{dx^2} + \frac{\omega^2}{c_q^2}X = 0 \tag{9-106}$$

The solution $X(x)$ to Eq. (9-106) is *assumed* to be

$$X(x) = e^{\gamma x} \tag{9-107}$$

Substitution of Eq. (9-107) into Eq. (9-106) and cancellation of the common term $e^{\gamma x}$ yield

$$\gamma = \pm \frac{\omega}{c_q}i \tag{9-108}$$

Therefore, the general solution for $X(x)$ becomes

$$X(x) = c \cos \frac{\omega}{c_q}x + d \sin \frac{\omega}{c_q}x \tag{9-109}$$

where Euler's formula, $e^{i\theta} = \cos \theta + i \sin \theta$, has been used and where c, d, and ω are yet unknown.

The special case of $\omega = 0$ needs to be considered separately, as in Subsection 9-3.2. When $\omega = 0$, Eq. (9-106) reduces to

$$\frac{d^2X}{dx^2} = 0 \tag{9-110}$$

whose general solution by direct integration is

$$X(x) = c + dx \tag{9-111}$$

where c and d are yet unknown.

Therefore, the solution $z(x, t)$ to the generic equation of motion, Eq. (9-83), for second-order systems is given by either

$$z(x, t) = (a \cos \omega t + b \sin \omega t)\left(c \cos \frac{\omega}{c_q}x + d \sin \frac{\omega}{c_q}x\right) \qquad \omega \neq 0 \tag{9-112}$$

or

$$z(x, t) = (a + bt)(c + dx) \qquad \omega = 0 \tag{9-113}$$

As we shall see in the following analyses, the unknowns ω, c, and d in Eqs. (9-112) and (9-113) are evaluated in accordance with the boundary conditions of the system. Also, the unknowns a and b in Eqs. (9-112) and (9-113) are evaluated in accordance with the initial conditions imposed on the system, as we shall see in Section 9-4. In determining the unknowns ω, c, and d in Eqs. (9-112) and (9-113), we consider the various boundary conditions listed in Tables 9-5 through 9-7.

9-3.3.1 *Examples with Simple Boundary Conditions* First of all, the simpler cases—those listed in the first two rows of Tables 9-5 through 9-7—are considered. These simpler cases require that either the generalized coordinate or its spacewise derivative is zero at

each end of the continuum. For second-order systems, we shall designate these boundary conditions as "simple" boundary conditions. There are four such cases:

1. Clamped–Clamped
   $z(0,t) = z(l,t) = 0$ for all time t that require $X(0) = X(l) = 0$

2. Clamped–Free
   $z(0,t) = z'(l,t) = 0$ for all time t that require $X(0) = X'(l) = 0$

3. Free–Clamped
   $z'(0,t) = z(l,t) = 0$ for all time t that require $X'(0) = X(l) = 0$

4. Free–Free
   $z'(0,t) = z'(l,t) = 0$ for all time t that require $X'(0) = X'(l) = 0$

where $z' \equiv \dfrac{\partial z}{\partial x}$ and $X' \equiv \dfrac{dX}{dx}$, and where we have used Eq. (9-85) to obtain the conditions on $X(x)$ and its spatial derivative from the conditions on $z(x,t)$ and its spatial derivative. Note that the boundary conditions listed above have been designated by such terms as "clamped" and "free," which are appropriate only for the mechanical systems as listed in Tables 9-5 and 9-6. We adopt this terminology because in the remainder of this chapter we shall be concerned primarily with mechanical systems, even though the same solutions may be applied to the electric transmission line. We now find the eigenfunction $X(x)$ for each of the four cases listed immediately above.

CASE 1: *Clamped–Clamped* $(X(0) = X(l) = 0)$

First, we consider the case when $\omega \neq 0$. In which case, the condition $X(0) = 0$ requires that $c = 0$ in Eq. (9-109). (That is, from Eq. (9-109), $X(0) = 0 = c\cos(0) + d\sin(0) = c$, since $\cos(0) = 1$ and $\sin(0) = 0$.) Therefore, Eq. (9-109) becomes

$$X(x) = d\sin\frac{\omega}{c_q}x \qquad (9\text{-}114)$$

The condition $X(l) = 0$, when combined with Eq. (9-114), yields

$$\sin\frac{\omega}{c_q}l = 0 \qquad (9\text{-}115)$$

since d cannot be zero for a nontrivial solution. Equation (9-115) imposes a restriction on ω, namely,

$$\frac{\omega}{c_q}l = n\pi \qquad n = 1,2,3,\ldots \qquad (9\text{-}116)$$

since for any arbitrary function, say ξ, $\sin(\xi) = 0$ only when ξ is some integral multiple of π. In Eq. (9-116) we exclude the value $n = 0$ because it corresponds to $\omega = 0$; here we are considering the case when $\omega \neq 0$. If we denote the frequency ω in Eq. (9-116) for each value of $n = 1,2,3,\ldots$ by ω_n, Eq. (9-116) yields

$$\omega_n = \frac{n\pi c_q}{l} \qquad n = 1,2,3,\ldots \qquad (9\text{-}117)$$

Equation (9-117) gives the natural frequencies of the system in Case 1, and it demonstrates that continuous systems have an infinite number of natural frequencies (as we might have expected). By combining Eqs. (9-114) and (9-117), the eigenfunction $X_n(x)$ corresponding to the nth natural frequency ω_n is given by

$$X_n(x) = d_n\sin\frac{n\pi x}{l} \qquad n = 1,2,3,\ldots \qquad (9\text{-}118)$$

which is the mode shape for each $n = 1,2,3,\ldots$. The combination of ω_j and $X_j(x)$ comprises what is known as the jth *natural mode of vibration* or simply the jth *mode*. Note that the eigenfunctions given by Eq. (9-118) contain the unknown coefficients d_n. That is, the boundary conditions $(X(0) = X(l) = 0)$ are sufficient to determine the eigenfunctions, but only to within a multiplicative constant. (Indeed, we are revisiting the concepts and the terminology of Subsection 8-4.1.)

In order to assess whether the boundary conditions in Case 1 accommodate a rigid-body mode, we consider the case when $\omega = 0$. In this case, we need to consider the eigenfunction in the form of Eq. (9-111). Substitution of

$X(0) = 0$ into Eq. (9-111) gives $c = 0$ and substitution of $X(l) = 0$ into Eq. (9-111) gives $c + dl = 0$, leading to $c = d = 0$, since l cannot be zero. Therefore, when $\omega = 0$, the eigenfunction solution becomes $X(x) = c + dx = 0$, a trivial solution that holds no interest for us. This means that second-order systems subjected to the boundary conditions of Case 1 exhibit no rigid-body mode. This fact, which has been demonstrated mathematically, should seem obvious to the reader.

CASE 2: *Clamped–Free* $(X(0) = X'(l) = 0)$

First, we consider the case when $\omega \neq 0$. In which case, the condition $X(0) = 0$ requires that $c = 0$ in Eq. (9-109). Therefore, Eq. (9-109) becomes

$$X(x) = d \sin \frac{\omega}{c_q} x \qquad (9\text{-}119)$$

The condition $X'(l) = 0$, when combined with Eq. (9-119), yields

$$\cos \frac{\omega}{c_q} l = 0 \qquad (9\text{-}120)$$

since d cannot be zero for a nontrivial solution. Equation (9-120) imposes a restriction on ω, namely,

$$\frac{\omega}{c_q} l = \left(n - \frac{1}{2}\right)\pi \qquad n = 1, 2, 3, \ldots \qquad (9\text{-}121)$$

If we denote the frequency ω in Eq. (9-121) for each value of $n = 1, 2, 3, \ldots$ by ω_n, Eq. (9-121) yields

$$\omega_n = \left(n - \frac{1}{2}\right)\frac{\pi c_q}{l} \qquad n = 1, 2, 3, \ldots \qquad (9\text{-}122)$$

Equation (9-122) gives an infinite number of natural frequencies for the system in Case 2. By combining Eqs. (9-119) and (9-122), the eigenfunction $X_n(x)$ corresponding to the nth natural frequency ω_n is given by

$$X_n(x) = d_n \sin\left(n - \frac{1}{2}\right)\frac{\pi x}{l} \qquad n = 1, 2, 3, \ldots \qquad (9\text{-}123)$$

which is the mode shape for each $n = 1, 2, 3, \ldots$. The combination of ω_j and $X_j(x)$ comprises what is known as the jth *natural mode of vibration* or simply the jth *mode*.

In order to assess whether the boundary conditions in Case 2 accommodate a rigid-body mode, we consider the case when $\omega = 0$. In this case, we need to consider the eigenfunction in the form of Eq. (9-111). Substitution of $X(0) = 0$ into Eq. (9-111) gives $c = 0$ and substitution of $X'(l) = 0$ into Eq. (9-111) gives $d = 0$. Therefore, when $\omega = 0$, the eigenfunction solution becomes $X(x) = c + dx = 0$, a trivial solution that holds no interest for us. This means that second-order systems subjected to the boundary conditions of Case 2 exhibit no rigid-body mode; again, a mathematically demonstrated fact that should seem obvious to the reader.

CASE 3: *Free–Clamped* $(X'(0) = X(l) = 0)$

First, we consider the case when $\omega \neq 0$. In which case, the condition $X'(0) = 0$ requires that $d = 0$ in Eq. (9-109). Therefore, Eq. (9-109) becomes

$$X(x) = c \cos \frac{\omega}{c_q} x \qquad (9\text{-}124)$$

The condition $X(l) = 0$, when combined with Eq. (9-124), yields

$$\cos \frac{\omega}{c_q} l = 0 \qquad (9\text{-}125)$$

Equation (9-125) imposes a restriction on ω, namely,

$$\frac{\omega}{c_q} l = \left(n - \frac{1}{2}\right)\pi \qquad n = 1, 2, 3, \ldots \qquad (9\text{-}126)$$

If we denote the frequency ω in Eq. (9-126) for each value of $n = 1, 2, 3, \ldots$ by ω_n, Eq. (9-126) yields

$$\omega_n = \left(n - \frac{1}{2}\right)\frac{\pi c_q}{l} \qquad n = 1, 2, 3, \ldots \qquad (9\text{-}127)$$

Equation (9-127) gives an infinite number of natural frequencies for the system in Case 3. By combining Eqs. (9-124) and (9-127), the eigenfunction $X_n(x)$ corresponding to the nth natural frequency ω_n is given by

$$X_n(x) = c_n \cos\left(n - \frac{1}{2}\right)\frac{\pi x}{l} \qquad n = 1, 2, 3, \ldots \qquad (9\text{-}128)$$

which is the mode shape for each $n = 1, 2, 3, \ldots$. The combination of ω_j and $X_j(x)$ comprises what is known as the jth *natural mode of vibration* or simply the jth *mode*.

In order to assess whether the boundary conditions in Case 3 accommodate a rigid-body mode, we consider the case when $\omega = 0$. In this case, we need to consider the eigenfunction in the form of Eq. (9-111). Substitution of $X'(0) = 0$ into Eq. (9-111) gives $d = 0$ and substitution of $X(l) = 0$ into Eq. (9-111) gives $c + dl = 0$, leading to $c = d = 0$, since l cannot be zero. Therefore, when $\omega = 0$, the eigenfunction solution becomes $X(x) = c + dx = 0$, a trivial solution that holds no interest for us. This means that second-order systems subjected to the boundary conditions of Case 3 exhibit no rigid-body mode.

We may have obtained the natural frequencies in Eq. (9-127) and the eigenfunctions in Eq. (9-128) by noting that the boundary conditions in Case 3 are simply the mirror image of those in Case 2, about the midpoint $x = \dfrac{l}{2}$. That is, if we had taken the generalized coordinate for Case 2 such that its origin was at the right-hand end of the system and it had increased toward the left instead of toward the right,

we should have obtained the result for Case 3. Actually, in Eq. (9-123), replacing x by $(l - x)$ yields

$$X_n(x) = d_n \sin\left\{\left(n - \frac{1}{2}\right)\frac{\pi(l-x)}{l}\right\}$$

$$= d_n \sin\left\{\left(n - \frac{1}{2}\right)\left(\pi - \frac{\pi x}{l}\right)\right\}$$

$$= d_n \sin\left\{n\pi - \frac{\pi}{2} - \left(n - \frac{1}{2}\right)\frac{\pi x}{l}\right\}$$

$$= (-1)^{n+1} d_n \cos\left\{\left(n - \frac{1}{2}\right)\frac{\pi x}{l}\right\} \quad (9\text{-}129)$$

which is equal to Eq. (9-128) with understanding that the arbitrary multiplicative coefficient in Eq. (9-129) can be considered to be c_n in Eq. (9-128). The natural frequencies are identical for both Cases 2 and 3 as indicated by Eqs. (9-122) and (9-127). This is because natural frequencies are inherent characteristics of the system that do not depend on the choice of the generalized coordinate.

CASE 4: *Free–Free ($X'(0) = X'(l) = 0$)*

First, we consider the case when $\omega \neq 0$. In which case, the condition $X'(0) = 0$ requires that $d = 0$ in Eq. (9-109). Therefore, Eq. (9-109) becomes

$$X(x) = c \cos\frac{\omega}{c_q}x \quad (9\text{-}130)$$

The condition $X'(l) = 0$, when combined with Eq. (9-130), yields

$$\sin\frac{\omega}{c_q}l = 0 \quad (9\text{-}131)$$

Equation (9-131) imposes a restriction on ω, namely,

$$\frac{\omega}{c_q}l = n\pi \qquad n = 1, 2, 3, \ldots \quad (9\text{-}132)$$

in which we exclude the value $n = 0$ because it corresponds to $\omega = 0$; here we are considering the case $\omega \neq 0$. If we denote the frequency ω in Eq. (9-132) for each value of $n = 1, 2, 3, \ldots$ by ω_n, Eq. (9-132) yields

$$\omega_n = \frac{n\pi c_q}{l} \qquad n = 1, 2, 3, \ldots \quad (9\text{-}133)$$

Equation (9-133) gives an infinite number of natural frequencies for the system in Case 4. By combining Eqs. (9-130) and (9-133), the eigenfunction $X_n(x)$ corresponding to the nth natural frequency ω_n is given by

$$X_n(x) = c_n \cos\frac{n\pi x}{l} \qquad n = 1, 2, 3, \ldots \quad (9\text{-}134)$$

which is the mode shape for each $n = 1, 2, 3, \ldots$. The combination of ω_j and $X_j(x)$ comprises the jth *natural mode of vibration* or simply the jth *mode*.

In order to assess whether the boundary conditions in Case 4 accommodate a rigid-body mode, we consider the case when $\omega = 0$. In this case, we need to consider the eigenfunction in the form of Eq. (9-111). Substitution of $X'(0) = 0$ into Eq. (9-111) gives $d = 0$ and substitution of $X'(l) = 0$ into Eq. (9-111) gives $d = 0$ again, leaving c undetermined. That is, when $\omega = 0$, the eigenfunction is a nonzero arbitrary constant. Therefore, second-order systems subjected to the boundary conditions of Case 4 possess a rigid-body mode. The eigenfunction of the rigid-body mode is conventionally denoted by $X_0(x)$, and the mode is often called the *zeroth mode*. This rigid-body mode eigenfunction $X_0(x) = c_0$ (that is, a constant throughout the extent of the system) represents a uniform (undeformed) motion of all particles in the system, which is a translation or rotation in a mechanical sense. Note that the existence of a rigid-body mode depends on the boundary conditions; a rigid-body mode exists for Case 4 while none exists for the other cases considered for second-order systems. (This case may be compared conceptually with the semidefinite system in Example 8-13.)

TABLE 9-9 Natural Frequencies and Eigenfunctions of Second-Order System for Various Simple Boundary Conditions

Boundary conditions	Rigid-body mode ($\omega_0 = 0$)	Nonrigid-body modes (c_n, d_n: arbitrary constants, $n = 1, 2, 3, \ldots$)				
	$X_0(x)$	ω_n	$X_n(x)$	$X_1(x)$	$X_2(x)$	$X_3(x)$
Clamped–Clamped	0	$\dfrac{n\pi c_q}{\ell}$	$d_n \sin \dfrac{\omega_n}{c_q} x$			
Clamped–Free	0	$\left(n - \dfrac{1}{2}\right)\dfrac{\pi c_q}{\ell}$	$d_n \sin \dfrac{\omega_n}{c_q} x$			
Free–Clamped	0	$\left(n - \dfrac{1}{2}\right)\dfrac{\pi c_q}{\ell}$	$c_n \cos \dfrac{\omega_n}{c_q} x$			
Free–Free	c_0 (constant)	$n\dfrac{\pi c_q}{\ell}$	$c_n \cos \dfrac{\omega_n}{c_q} x$			

In summary, the equations for the natural frequencies and mode shapes for different boundary conditions are listed in Table 9-9. Also, the first three deformable mode shapes in each case are illustrated in Table 9-9. (The longitudinal extent of the system is sketched horizontally and the generalized coordinate is sketched vertically.) Note that the mode shapes indicate that particles at certain locations do not move during the modal motion. These stationary points are called *nodes*. As can be seen, the number of nodes increases with increasing mode number.

9-3.3.2 *Orthogonality* Now, we shall derive an interesting and useful property of the eigenfunctions of second-order systems subjected to the "simple" boundary conditions summarized in Table 9-9. This property is that the eigenfunctions are *orthogonal* to each other in the sense that they satisfy the relation[11]

$$\int_0^l X_n X_m \, dx = 0 \qquad n \neq m \tag{9-135}$$

where X_n and X_m are the nth and m^{th} eigenfunctions, respectively. In order to prove Eq. (9-135), note that all eigenfunctions $X_n(x)$ satisfy Eq. (9-106) along with the corresponding natural frequency ω_n. Therefore, we can write Eq. (9-106) for two different modes (say, the mth and n^{th} modes) as

---

[11]This orthogonality relation is not general in the sense that it holds only when the continuum is uniform *and* the boundary conditions are such that Eq. (9-143) vanishes, as in all cases of simple boundary conditions listed in Table 9-9 *plus* the cases having stiffness element(s) at the end(s) (see Problem 9-9). More general orthogonality relations are required for more general cases when the continuum is not uniform or when the end(s) is (are) terminated by inertial loads (see Problem 9-10). However, we shall consider only this simplest version of orthogonality, merely to provide a tool that significantly simplifies calculations in free vibration analyses, as will be shown in Section 9-4.

$$\frac{d^2X_m}{dx^2} + \frac{\omega_m^2}{c_q^2}X_m = 0 \tag{9-136}$$

$$\frac{d^2X_n}{dx^2} + \frac{\omega_n^2}{c_q^2}X_n = 0 \tag{9-137}$$

Now multiply Eq. (9-136) by X_n and multiply Eq. (9-137) by X_m, which give

$$X_n\frac{d^2X_m}{dx^2} + \frac{\omega_m^2}{c_q^2}X_nX_m = 0 \tag{9-138}$$

$$X_m\frac{d^2X_n}{dx^2} + \frac{\omega_n^2}{c_q^2}X_mX_n = 0 \tag{9-139}$$

Subtracting Eq. (9-139) from Eq. (9-138) yields

$$X_n\frac{d^2X_m}{dx^2} - X_m\frac{d^2X_n}{dx^2} = \frac{\omega_n^2 - \omega_m^2}{c_q^2}X_nX_m \tag{9-140}$$

Integrate both sides of Eq. (9-140) with respect to x from 0 to l. The first term on the left-hand side of Eq. (9-140) can be integrated by parts via Eq. (9-13), yielding

$$\int_0^l X_n\frac{d^2X_m}{dx^2}\,dx = X_n\frac{dX_m}{dx}\Big|_0^l - \int_0^l \frac{dX_m}{dx}\frac{dX_n}{dx}\,dx \tag{9-141}$$

Similarly, the second term on the left-hand side of Eq. (9-140) can be integrated to yield

$$\int_0^l X_m\frac{d^2X_n}{dx^2}\,dx = X_m\frac{dX_n}{dx}\Big|_0^l - \int_0^l \frac{dX_n}{dx}\frac{dX_m}{dx}\,dx \tag{9-142}$$

Therefore, subtracting Eq. (9-142) from Eq. (9-141) reduces the integral on the left-hand side of Eq. (9-140) to

$$\int_0^l \left(X_n\frac{d^2X_m}{dx^2} - X_m\frac{d^2X_n}{dx^2}\right)dx$$

$$= X_n\frac{dX_m}{dx}\Big|_0^l - X_m\frac{dX_n}{dx}\Big|_0^l$$

$$= X_n(l)X_m'(l) - X_n(0)X_m'(0) - X_m(l)X_n'(l) + X_m(0)X_n'(0) \tag{9-143}$$

where $X_j' \equiv \dfrac{dX_j}{dx}$. Equating Eq. (9-143) to the integral on the right-hand side of Eq. (9-140) gives

$$X_n(l)X_m'(l) - X_n(0)X_m'(0) - X_m(l)X_n'(l) + X_m(0)X_n'(0) = \frac{\omega_n^2 - \omega_m^2}{c_q^2}\int_0^l X_nX_m\,dx \tag{9-144}$$

Since all boundary conditions considered above (Cases 1 through 4) require at each end that either the eigenfunction itself or its first derivative must be zero, the left-hand side of Eq. (9-144) vanishes for any of the boundary conditions in Table 9-9. Therefore, Eq. (9-144) reduces to

$$0 = \frac{\omega_n^2 - \omega_m^2}{c_q^2}\int_0^l X_nX_m\,dx \tag{9-145}$$

Because $\omega_n^2 - \omega_m^2 \neq 0$ in accordance with our statement that $n \neq m$, Eq. (9-145) implies that the eigenfunctions for two different modes must satisfy the expression

$$\int_0^l X_n X_m \, dx = 0 \qquad n \neq m \tag{9-146}$$

which is the *orthogonality relation* in Eq. (9-135).

9-3.3.3   An Example with Less Simple Boundary Conditions

All boundary conditions considered so far are of the "simple" type, in which either the generalized coordinate or its spacewise derivative is zero. For other end conditions, the analysis becomes more complicated and the resulting equations for the natural frequencies may not have closed-form solutions. For example and for illustrative purposes only, consider the natural frequencies and mode shapes for the system sketched in Figure 9-1. The boundary conditions for this system are given by Eqs. (9-27) and (9-28). In considering the modes of vibration, we must use the complete solution forms given by Eqs. (9-112) and (9-113). The reason for this is because the boundary conditions, Eqs. (9-27) and (9-28), involve time derivatives as well as spatial derivatives, whereas all boundary conditions in Cases 1 through 4 above involve only spatial derivatives, requiring consideration of only the spatial eigenfunctions of the solutions.

First, we consider the case when $\omega \neq 0$. Substitution of Eq. (9-112) into Eq. (9-27) at $x = 0$ yields

$$M_1[-\omega^2 T(t)]c + kT(t)c = EAT(t)\frac{\omega}{c_q}d \tag{9-147}$$

where $T(t) \equiv a\cos\omega t + b\sin\omega t$ and $c_q \equiv \sqrt{\dfrac{E}{\rho}}$. Since Eq. (9-147) must hold for all values of t, dividing both sides of Eq. (9-147) by $T(t)$ and rearranging terms give

$$(k - \omega^2 M_1)c - \frac{EA\omega}{c_q}d = 0 \tag{9-148}$$

Now, we consider the boundary condition given by Eq. (9-28) at $x = l$. In order to find the natural modes, we omit the damping term on the left-hand side of Eq. (9-28) and the applied force $F(t)$ on the right-hand side of Eq. (9-28). Therefore, the boundary condition at $x = l$ reduces to

$$M_2\frac{\partial^2 \xi}{\partial t^2} + EA\frac{\partial \xi}{\partial x} = 0 \tag{9-149}$$

Substituting Eq. (9-112) into Eq. (9-149) at $x = l$ gives

$$-M_2\omega^2 T(t)\left(c\cos\frac{\omega}{c_q}l + d\sin\frac{\omega}{c_q}l\right) + EAT(t)\frac{\omega}{c_q}\left(-c\sin\frac{\omega}{c_q}l + d\cos\frac{\omega}{c_q}l\right) = 0 \tag{9-150}$$

where $T(t) \equiv a\cos\omega t + b\sin\omega t$. Since Eq. (9-150) must hold for all values of t, dividing both sides of Eq. (9-150) by $T(t)$ and rearranging terms give

$$\left(-\omega^2 M_2 \cos\frac{\omega}{c_q}l - \frac{EA\omega}{c_q}\sin\frac{\omega}{c_q}l\right)c + \left(-\omega^2 M_2 \sin\frac{\omega}{c_q}l + \frac{EA\omega}{c_q}\cos\frac{\omega}{c_q}l\right)d = 0 \tag{9-151}$$

Equations (9-148) and (9-151) constitute a set of two linear equations in the two unknowns c and d. Rewriting Eqs. (9-148) and (9-151) in matrix form yields

$$\begin{bmatrix} (k - \omega^2 M_1) & -\dfrac{EA\omega}{c_q} \\ \left(-\omega^2 M_2 \cos \dfrac{\omega}{c_q}l - \dfrac{EA\omega}{c_q}\sin\dfrac{\omega}{c_q}l\right) & \left(-\omega^2 M_2 \sin\dfrac{\omega}{c_q}l + \dfrac{EA\omega}{c_q}\cos\dfrac{\omega}{c_q}l\right) \end{bmatrix} \begin{Bmatrix} c \\ d \end{Bmatrix} = 0 \quad (9\text{-}152)$$

The condition for nontrivial solutions for c and d to exist is that the determinant of the coefficient matrix in Eq. (9-152) must vanish. Therefore,

$$(k - \omega^2 M_1)\left(-\omega^2 M_2 \sin\frac{\omega}{c_q}l + \frac{EA\omega}{c_q}\cos\frac{\omega}{c_q}l\right)$$

$$+ \frac{EA\omega}{c_q}\left(-\omega^2 M_2 \cos\frac{\omega}{c_q}l - \frac{EA\omega}{c_q}\sin\frac{\omega}{c_q}l\right) = 0 \quad (9\text{-}153)$$

or, upon rearrangement,

$$\left\{(k - \omega^2 M_1)\frac{EA\omega}{c_q} - \frac{EAM_2\omega^3}{c_q}\right\} = \left\{(k - \omega^2 M_1)\omega^2 M_2 + \frac{E^2 A^2 \omega^2}{c_q^2}\right\}\tan\frac{\omega l}{c_q} \quad (9\text{-}154)$$

which is the equation for the natural frequencies of the system sketched in Figure 9-1. Note that, if it exists, a closed-form solution to Eq. (9-154) for ω cannot be easily found. In any event, because the tangent function is periodic, Eq. (9-154) contains an infinite number of solutions for ω. Once those ω's have been found, say, by numerical methods, they may be arranged in ascending order of their magnitudes and designated as ω_n ($n = 1, 2, 3, \ldots$) in the same order.

Note that c and d (or more precisely, their ratio) in Eqs. (9-148) and (9-151) have different values for different ω_n's. Therefore, by designating c and d for each ω_n as c_n and d_n, respectively, their ratio can be found from either Eq. (9-148) or Eq. (9-151) as

$$\frac{c_n}{d_n} = \frac{EA\omega_n}{c_q(k - \omega_n^2 M_1)} = \frac{-\omega_n^2 M_2 \sin\dfrac{\omega_n}{c_q}l + \dfrac{EA\omega_n}{c_q}\cos\dfrac{\omega_n}{c_q}l}{\omega_n^2 M_2 \cos\dfrac{\omega_n}{c_q}l + \dfrac{EA\omega_n}{c_q}\sin\dfrac{\omega_n}{c_q}l} \quad (9\text{-}155)$$

Therefore, from Eq. (9-109) or Eq. (9-112), the nth eigenfunction corresponding to ω_n is given by

$$X_n(x) = d_n\left(\frac{c_n}{d_n}\cos\frac{\omega_n}{c_q}x + \sin\frac{\omega_n}{c_q}x\right) \quad (9\text{-}156)$$

where the ratio $\dfrac{c_n}{d_n}$ is a known factor once ω_n is known, as provided by Eq. (9-155). We note that the boundary conditions in Figure 9-1 do not result in the elimination of either the sine or the cosine terms in Eq. (9-156), whereas the boundary conditions in Cases 1 through 4 in Table 9-9 did. This is an additional reason why the analysis of the system in Figure 9-1 is more complicated. Also, we note again that as Eq. (9-156) shows, the boundary conditions determine the eigenfunctions only to within a multiplicative constant, d_n in this case.

In order to complete the analysis of this system, we consider the possibility of a rigid-body mode. For this case, we suppose $\omega = 0$ and use the solution in the form of Eq. (9-113). Note that the second derivative of Eq. (9-113) in a timewise sense is identically zero; that is, $\dfrac{\partial^2 \xi}{\partial t^2} \equiv 0$ in Eqs. (9-27) and (9-28). Therefore, the boundary condition represented by Eq. (9-27) at $x = 0$, when combined with Eq. (9-113), yields

$$kT_0(t)c = EAT_0(t)d \tag{9-157}$$

where $T_0(t) \equiv a + bt$. Since Eq. (9-157) must hold for all t, dividing it by $T_0(t)$ yields

$$kc = EAd \tag{9-158}$$

Substituting Eq. (9-113) into Eq. (9-149) gives

$$EAT_0(t)d = 0 \tag{9-159}$$

where $\dfrac{\partial^2 \xi}{\partial t^2} \equiv 0$ has been used and where $T_0(t) \equiv a + bt$. Since Eq. (9-159) must hold for all t, it requires that $d = 0$, which in turn requires that $c = 0$ according to Eq. (9-158). Therefore, the solution for the case when $\omega = 0$ turns out to be the trivial solution, zero, indicating that the system shown in Figure 9-1 has no rigid-body mode.

9-3.4   Eigenfunctions for Fourth-Order Systems

In this subsection, we seek the eigenfunction $X(x)$ that satisfies Eq. (9-93). In Subsection 9-3.2, we found that α in Eq. (9-93) must be negative (or zero in a special case); so combining Eqs. (9-93) and (9-98) gives

$$\frac{d^4 X}{dx^4} - \frac{\omega^2}{c_q^2} X = 0 \tag{9-160}$$

The solution $X(x)$ to Eq. (9-160) is *assumed* to be

$$X(x) = e^{\gamma x} \tag{9-161}$$

Substitution of Eq. (9-161) into Eq. (9-160) and cancellation of the common term $e^{\gamma x}$ yield four solutions for γ:[12]

$$\gamma = \pm \sqrt{\frac{\omega}{c_q}}, \pm \sqrt{\frac{\omega}{c_q}}\, i \tag{9-162}$$

where $i = \sqrt{-1}$. It is convenient to define a new factor β such that

$$\beta \equiv \sqrt{\frac{\omega}{c_q}} \tag{9-163}$$

which reduces Eq. (9-162) to

$$\gamma = \pm\beta, \pm\beta i \tag{9-164}$$

Therefore, combining Eqs. (9-161) and (9-164) gives the general solution for $X(x)$ as

$$\begin{aligned} X(x) &= c'e^{i\beta x} + d'e^{-i\beta x} + e'e^{\beta x} + f'e^{-\beta x} \\ &= c\cos\beta x + d\sin\beta x + e\cosh\beta x + f\sinh\beta x \end{aligned} \tag{9-165}$$

where the general identities, $e^{i\theta} = \cos\theta + i\sin\theta$, $\cosh\theta = \dfrac{(e^{\theta} + e^{-\theta})}{2}$ and $\sinh\theta = \dfrac{(e^{\theta} - e^{-\theta})}{2}$, have been used and where β, c, d, e and f are yet unknown.[13]

---

[12]See Problem 9-22.

[13]For convenience, we shall use the letter e for the base of the system of natural logarithms as well as for an undetermined constant. This should cause no confusion, however, since e as the natural logarithm base *always* appears with an exponent and e as an undetermined constant *never* appears with an exponent.

As in Subsections 9-3.2 and 9-3.3, the case when $\omega = 0$ in Eq. (9-160) requires special treatment. When $\omega = 0$, Eq. (9-160) reduces to

$$\frac{d^4X}{dx^4} = 0 \tag{9-166}$$

whose general solution by direct integration is

$$X(x) = c + dx + ex^2 + fx^3 \tag{9-167}$$

where $c, d, e,$ and f are yet unknown.

Therefore, the solution $z(x, t)$ to the generic equation of motion, Eq. (9-84), for fourth-order systems is given by either

$$z(x, t) = (a \cos \omega t + b \sin \omega t)(c \cos \beta x + d \sin \beta x + e \cosh \beta x + f \sinh \beta x) \qquad \omega \neq 0 \tag{9-168}$$

or

$$z(x, t) = (a + bt)(c + dx + ex^2 + fx^3) \qquad \omega = 0 \tag{9-169}$$

As we shall see in the following analyses, the unknowns ω (or β), $c, d, e,$ and f in Eqs. (9-168) and (9-169) are evaluated in accordance with the boundary conditions of the system. Also, the unknowns a and b in Eqs. (9-168) and (9-169) are evaluated in accordance with the initial conditions imposed on the system, as we shall see in Section 9-4. In determining the unknowns ω (or β), c, d, e and f, we consider several of the various boundary conditions listed in Table 9-8.

9-3.4.1 *Examples with Simple Boundary Conditions*   First of all, the simpler cases—those listed in the first four rows of Table 9-8, are considered. These simpler cases require that two of the following quantities must vanish at each end of the continuum: the generalized coordinate itself or the first, second, or third spacewise derivative of the generalized coordinate. For fourth-order systems, we shall designate these boundary conditions as "simple" boundary conditions. It is convenient to have, at hand, expressions for the spatial derivatives of the general solutions given by Eqs. (9-165) and (9-167). For the case when $\omega \neq 0$, Eq. (9-165) yields

$$X'(x) = -c\beta \sin \beta x + d\beta \cos \beta x + e\beta \sinh \beta x + f\beta \cosh \beta x \tag{9-170}$$

$$X''(x) = -c\beta^2 \cos \beta x - d\beta^2 \sin \beta x + e\beta^2 \cosh \beta x + f\beta^2 \sinh \beta x \tag{9-171}$$

$$X'''(x) = c\beta^3 \sin \beta x - d\beta^3 \cos \beta x + e\beta^3 \sinh \beta x + f\beta^3 \cosh \beta x \tag{9-172}$$

And, for the case when $\omega = 0$, Eq. (9-167) yields

$$X'(x) = d + 2ex + 3fx^2 \tag{9-173}$$

$$X''(x) = 2e + 6fx \tag{9-174}$$

$$X'''(x) = 6f \tag{9-175}$$

There are ten possible combinations of the simple boundary conditions. Along with the associated requirements on $z(x, t)$ and $X(x)$, and their spatial derivatives, we list these ten combinations:

1. *Clamped–Clamped*
   $z(0, t) = z'(0, t) = z(l, t) = z'(l, t) = 0$ for all time t that require $X(0) = X'(0) = X(l) = X'(l) = 0$

2. *Clamped–Simply supported*
   $z(0, t) = z'(0, t) = z(l, t) = z''(l, t) = 0$ for all time t that require $X(0) = X'(0) = X(l) = X''(l) = 0$

3. *Clamped–Sliding-rotationally restrained*
   $z(0, t) = z'(0, t) = z'(l, t) = z'''(l, t) = 0$ for all time t that require $X(0) = X'(0) = X'(l) = X'''(l) = 0$

4. *Clamped–Free*
   $z(0, t) = z'(0, t) = z''(l, t) = z'''(l, t) = 0$ for all time t that require $X(0) = X'(0) = X''(l) = X'''(l) = 0$

5. *Simply supported–Simply supported*
   $z(0, t) = z''(0, t) = z(l, t) = z''(l, t) = 0$ for all time t that require $X(0) = X''(0) = X(l) = X''(l) = 0$

6. *Simply supported–Sliding-rotationally restrained*
   $z(0, t) = z''(0, t) = z'(l, t) = z'''(l, t) = 0$ for all time t that require $X(0) = X''(0) = X'(l) = X'''(l) = 0$

7. *Simply supported–Free*
   $z(0, t) = z''(0, t) = z''(l, t) = z'''(l, t) = 0$ for all time t that require $X(0) = X''(0) = X''(l) = X'''(l) = 0$

8. *Sliding-rotationally restrained–Sliding-rotationally restrained*
   $z'(0, t) = z'''(0, t) = z'(l, t) = z'''(l, t) = 0$ for all time t that require $X'(0) = X'''(0) = X'(l) = X'''(l) = 0$

9. *Sliding-rotationally restrained–Free*
   $z'(0, t) = z'''(0, t) = z''(l, t) = z'''(l, t) = 0$ for all time t that require $X'(0) = X'''(0) = X''(l) = X'''(l) = 0$

10. *Free–Free*
    $z''(0, t) = z'''(0, t) = z''(l, t) = z'''(l, t) = 0$ for all time t that require $X''(0) = X'''(0) = X''(l) = X'''(l) = 0$

where $z' \equiv \dfrac{\partial z}{\partial x}$, $z'' \equiv \dfrac{\partial^2 z}{\partial x^2}$, $z''' \equiv \dfrac{\partial^3 z}{\partial x^3}$, $X' \equiv \dfrac{dX}{dx}$, $X'' \equiv \dfrac{d^2 X}{dx^2}$, and $X''' \equiv \dfrac{d^3 X}{dx^3}$; and, where we have used Eq. (9-85) to obtain the conditions on $X(x)$ from the conditions on $z(x, t)$. Note that we do not consider the boundary conditions that are mirror images about $x = \dfrac{l}{2}$ of the above cases because the natural frequencies and eigenfunctions for those cases can be obtained from those for the above cases by the relationship found between Cases 2 and 3 in Subsection 9-3.3. (Refer to Eq. (9-129).) We shall find the natural frequencies and eigenfunctions only for several typical cases and provide summarily the natural frequencies and eigenfunctions for all other cases without derivation. In particular, we shall derive the eigenfunctions for Cases 1, 7, 9, and 10; Cases 7, 9, and 10 being selected because they possess rigid-body modes.

CASE 1: *Clamped–Clamped ($X(0) = X'(0) = X(l) = X'(l) = 0$)*

First, we consider the case when $\omega \neq 0$ or, equivalently, $\beta \neq 0$. In which case, the conditions $X(0) = 0$ and $X'(0) = 0$ require from Eqs. (9-165) and (9-170), respectively, that

$$c + e = 0 \qquad (9\text{-}176)$$

$$d\beta + f\beta = 0 \qquad (9\text{-}177)$$

Also, the conditions $X(l) = 0$ and $X'(l) = 0$ require from Eqs. (9-165) and (9-170), respectively, that

$$c \cos \beta l + d \sin \beta l + e \cosh \beta l + f \sinh \beta l = 0 \qquad (9\text{-}178)$$

$$-c\beta \sin \beta l + d\beta \cos \beta l + e\beta \sinh \beta l + f\beta \cosh \beta l = 0 \qquad (9\text{-}179)$$

Equations (9-176) through (9-179) constitute a set of simultaneous linear equations for the unknown coefficients c, d, e and f. Equations (9-176) and (9-177) yield, respectively, that

$$e = -c \qquad (9\text{-}180)$$

$$f = -d \qquad (9\text{-}181)$$

where the condition $\beta \neq 0$ has been used in dividing both sides of Eq. (9-177) by β. Eliminating e and f from Eqs. (9-178) and (9-179) via Eqs. (9-180) and (9-181) yields

$$\begin{bmatrix} (\cos \beta l - \cosh \beta l) & (\sin \beta l - \sinh \beta l) \\ -(\sin \beta l + \sinh \beta l) & (\cos \beta l - \cosh \beta l) \end{bmatrix} \begin{Bmatrix} c \\ d \end{Bmatrix} = 0 \qquad (9\text{-}182)$$

where the condition $\beta \neq 0$ has been used in dividing both sides of Eq. (9-179) by β. In order for nonzero c and d to exist, the determinant of the coefficient matrix in Eq. (9-182) must vanish, giving

$$(\cos \beta l - \cosh \beta l)^2 + (\sin \beta l - \sinh \beta l)(\sin \beta l + \sinh \beta l)$$

$$= 0 \qquad (9\text{-}183)$$

or, upon rearrangement,

$$\cos \beta l \cosh \beta l = 1 \qquad (9\text{-}184)$$

where the relations $\cos^2 \theta + \sin^2 \theta = 1$ and $\cosh^2 \theta -$ $\sinh^2 \theta = 1$ have been used in obtaining Eq. (9-184) from Eq. (9-183).

There is an infinite number of solutions to Eq. (9-184) for β. These solutions may be arranged in ascending order of their magnitudes and designated as β_n ($n = 1, 2, 3, \ldots$) in the same order. If we define $\omega_n = c_q \beta_n^2$ using Eq. (9-163), these ω_n's are the natural frequencies of the system. By substituting β_n into Eq. (9-182), the ratio of c_n to d_n corresponding to β_n can be found as

$$\frac{c_n}{d_n} = -\frac{\sin \beta_n l - \sinh \beta_n l}{\cos \beta_n l - \cosh \beta_n l} = \frac{\cos \beta_n l - \cosh \beta_n l}{\sin \beta_n l + \sinh \beta_n l} \qquad (9\text{-}185)$$

By combining Eqs. (9-165), (9-180), and (9-181), the eigenfunction $X_n(x)$ corresponding to the nth natural frequency ω_n is given by

$$X_n(x)$$
$$= c_n(\cos \beta_n x - \cosh \beta_n x) + d_n(\sin \beta_n x - \sinh \beta_n x)$$
$$= d_n \left\{ \frac{c_n}{d_n}(\cos \beta_n x - \cosh \beta_n x) + (\sin \beta_n x - \sinh \beta_n x) \right\} \qquad (9\text{-}186)$$

where the ratio c_n/d_n is determined by Eq. (9-185) for each β_n found from Eq. (9-184). Therefore, Eq. (9-186) gives the eigenfunctions ($n = 1, 2, 3, \ldots$) for fourth-order systems subjected to the boundary conditions of Case 1, to within a multiplicative constant d_n.

In order to complete the analysis for Case 1, consider the instance where $\omega = \beta = 0$. In this case, the spatial function solution is given by the third-order polynomial in Eq. (9-167) and its derivatives are given by Eqs. (9-173) through (9-175). Therefore, the boundary conditions $X(0) = 0$ and $X'(0) = 0$, when combined with Eqs. (9-167) and (9-173), respectively, yield $c = 0$ and $d = 0$. Similarly, combining the boundary conditions $X(l) = 0$ and $X'(l) = 0$ with Eqs. (9-167) and (9-173), respectively, gives $c + dl + el^2 + fl^3 = 0$ and $d + 2el + 3fl^2 = 0$, which yield $e = f = 0$ because $c = d = 0$. Therefore, when $\omega = \beta = 0$, we get a trivial solution (zero), which means fourth-order systems subjected to the boundary conditions of Case 1 possess no rigid-body mode.

CASE 7: *Simply supported–Free ($X(0) = X''(0) = X''(l) = X'''(l) = 0$)*

First, we consider the case when $\omega \neq 0$ or, equivalently, $\beta \neq 0$. In which case, the conditions $X(0) = 0$ and

$X''(0) = 0$ require from Eqs. (9-165) and (9-171), respectively, that

$$c + e = 0 \qquad (9\text{-}187)$$

$$-c\beta^2 + e\beta^2 = 0 \qquad (9\text{-}188)$$

Also, the conditions $X''(l) = 0$ and $X'''(l) = 0$ require from Eqs. (9-171) and (9-172), respectively, that

$$-c\beta^2 \cos\beta l - d\beta^2 \sin\beta l + e\beta^2 \cosh\beta l + f\beta^2 \sinh\beta l = 0 \qquad (9\text{-}189)$$

$$c\beta^3 \sin\beta l - d\beta^3 \cos\beta l + e\beta^3 \sinh\beta l + f\beta^3 \cosh\beta l = 0 \qquad (9\text{-}190)$$

Because $\beta \neq 0$, Eqs. (9-187) and (9-188) require that

$$c = e = 0 \qquad (9\text{-}191)$$

Eliminating terms containing c and e from Eqs. (9-189) and (9-190) yields

$$\begin{bmatrix} -\sin\beta l & \sinh\beta l \\ -\cos\beta l & \cosh\beta l \end{bmatrix} \begin{Bmatrix} d \\ f \end{Bmatrix} = 0 \qquad (9\text{-}192)$$

where the condition $\beta \neq 0$ has been used in dividing both sides of Eqs. (9-189) and (9-190) by β^2 and β^3, respectively. In order for nonzero d and f to exist, the determinant of the coefficient matrix in Eq. (9-192) must vanish, giving

$$-\sin\beta l \cosh\beta l + \cos\beta l \sinh\beta l = 0 \qquad (9\text{-}193)$$

or, upon rearrangement and dividing through by $\cos\beta l \cosh\beta l$,

$$\tan\beta l = \tanh\beta l \qquad (9\text{-}194)$$

There is an infinite number of solutions to Eq. (9-194) for β. These solutions may be arranged in ascending order of their magnitudes and designated as β_n ($n = 1, 2, 3, \ldots$) in the same order. If we define $\omega_n = c_q\beta_n^2$ using Eq. (9-163), these ω_n's are the natural frequencies of the system. By substituting β_n into Eq. (9-192), the ratio of d_n to f_n corresponding to β_n can be found as

$$\frac{d_n}{f_n} = \frac{\sinh\beta_n l}{\sin\beta_n l} = \frac{\cosh\beta_n l}{\cos\beta_n l} \qquad (9\text{-}195)$$

By combining Eqs. (9-165) and (9-191), the eigenfunction $X_n(x)$ corresponding to the nth natural frequency ω_n is given by

$$X_n(x) = d_n \sin\beta_n x + f_n \sinh\beta_n x$$
$$= f_n\left(\frac{d_n}{f_n}\sin\beta_n x + \sinh\beta_n x\right) \qquad (9\text{-}196)$$

where the ratio $\dfrac{d_n}{f_n}$ is determined by Eq. (9-195) for each β_n found from Eq. (9-194). Therefore, Eq. (9-196) gives the eigenfunctions ($n = 1, 2, 3, \ldots$) for fourth-order systems subjected to the boundary conditions of Case 7, to within a multiplicative constant f_n.

Now, in order to assess whether the boundary conditions of Case 7 accommodate a rigid-body mode, consider the instance where $\omega = \beta = 0$. In this case, the spatial function solution is given by the third-order polynomial in Eq. (9-167) and its derivatives are given by Eqs. (9-173) through (9-175). Therefore, the boundary conditions $X(0) = 0$ and $X''(0) = 0$, when combined with Eqs. (9-167) and (9-174), respectively, yield $c = 0$ and $e = 0$. Similarly, combining the boundary conditions $X''(l) = 0$ and $X'''(l) = 0$ with Eqs. (9-174) and (9-175), respectively, gives $2e + 6fl = 0$ and $6f = 0$, which yield $e = f = 0$. Therefore, the boundary conditions impose no restriction on the unknown constant d. Thus, substituting $c = e = f = 0$ into Eq. (9-167) yields

$$X_0(x) = d_0 x \qquad (9\text{-}197)$$

where d_0 is arbitrary and the subscript 0 indicates that this eigenfunction is for the zeroth mode (or, equivalently, the rigid-body mode). We note that the zeroth eigenfunction given by Eq. (9-197) represents a rotation of the beam about the simply supported point at $x = 0$.

In summary, fourth-order systems subjected to the boundary conditions of Case 7 possess an infinite number of modes, consisting of a rigid-body mode given by Eq. (9-197) plus an infinite number of flexible modes given by Eq. (9-196) for $n = 1, 2, 3, \ldots$.

CASE 9: *Sliding-rotationally restrained–Free* ($X'(0) = X'''(0) = X''(l) = X'''(l) = 0$)

First, we consider the case when $\omega \neq 0$ or, equivalently, $\beta \neq 0$. In which case, the conditions $X'(0) = 0$ and $X'''(0) = 0$ require from Eqs. (9-170) and (9-172), respectively, that

$$d\beta + f\beta = 0 \qquad (9\text{-}198)$$
$$-d\beta^3 + f\beta^3 = 0 \qquad (9\text{-}199)$$

Also, the conditions $X''(l) = 0$ and $X'''(l) = 0$ require from Eqs. (9-171) and (9-172), respectively, that

$$-c\beta^2 \cos\beta l - d\beta^2 \sin\beta l + e\beta^2 \cosh\beta l + f\beta^2 \sinh\beta l = 0 \qquad (9\text{-}200)$$

$$c\beta^3 \sin\beta l - d\beta^3 \cos\beta l + e\beta^3 \sinh\beta l + f\beta^3 \cosh\beta l = 0 \qquad (9\text{-}201)$$

Because $\beta \neq 0$, Eqs. (9-198) and (9-199) require that

$$d = f = 0 \qquad (9\text{-}202)$$

Eliminating terms containing d and f from Eqs. (9-200) and (9-201) yields

$$\begin{bmatrix} -\cos\beta l & \cosh\beta l \\ \sin\beta l & \sinh\beta l \end{bmatrix}\begin{Bmatrix} c \\ e \end{Bmatrix} = 0 \qquad (9\text{-}203)$$

where the condition $\beta \neq 0$ has been used in dividing both sides of Eqs. (9-200) and (9-201) by β^2 and β^3, respectively. In order for nonzero c and e to exist, the determinant of the coefficient matrix in Eq. (9-203) must vanish, giving

$$-\cos\beta l \sinh\beta l - \sin\beta l \cosh\beta l = 0 \qquad (9\text{-}204)$$

or, upon dividing through by $-\cos\beta l \cosh\beta l$,

$$\tanh\beta l + \tan\beta l = 0 \qquad (9\text{-}205)$$

There is an infinite number of solutions to Eq. (9-205) for β. These solutions may be arranged in ascending order of their magnitudes and designated as β_n ($n = 1, 2, 3, \ldots$) in the same order. If we define $\omega_n = c_q\beta_n^2$ using Eq. (9-163), these ω_n's are the natural frequencies of the system. By substituting β_n into Eq. (9-203), the ratio of c_n to e_n corresponding to β_n can be found as

$$\frac{c_n}{e_n} = \frac{\cosh\beta_n l}{\cos\beta_n l} = -\frac{\sinh\beta_n l}{\sin\beta_n l} \qquad (9\text{-}206)$$

By combining Eqs. (9-165) and (9-202), the eigenfunction $X_n(x)$ corresponding to the nth natural frequency ω_n is given by

$$X_n(x) = c_n \cos\beta_n x + e_n \cosh\beta_n x$$
$$= e_n\left(\frac{c_n}{e_n}\cos\beta_n x + \cosh\beta_n x\right) \qquad (9\text{-}207)$$

where the ratio $\frac{c_n}{e_n}$ is determined by Eq. (9-206) for each β_n found from Eq. (9-205). Therefore, Eq. (9-207) gives the eigenfunctions ($n = 1, 2, 3, \ldots$) for fourth-order systems subjected to the boundary conditions of Case 9, to within a multiplicative constant e_n.

Now, in order to assess whether the boundary conditions of Case 7 accommodate a rigid-body mode, consider the instance where $\omega = \beta = 0$. In this case, the spatial function solution is given by the third-order polynomial in Eq. (9-167) and its derivatives are given by Eqs. (9-173) through (9-175). Therefore, the boundary conditions $X'(0) = 0$ and $X'''(0) = 0$, when combined with Eqs. (9-173) and (9-175), respectively, yield $d = 0$ and $f = 0$. Similarly, combining the boundary conditions $X''(l) = 0$ and $X'''(l) = 0$ with Eqs. (9-174) and (9-175), respectively, gives $2e + 6fl = 0$ and $6f = 0$, which yield $e = f = 0$. Therefore, the boundary conditions impose no restriction on the unknown constant c. Thus, substituting $d = e = f = 0$ into Eq. (9-167) yields

$$X_0(x) = c_0 \qquad (9\text{-}208)$$

where c_0 is an arbitrary constant and the subscript 0 indicates that this eigenfunction is for the zeroth mode (or, equivalently, the rigid-body mode). We note that the zeroth eigenfunction given by Eq. (9-208) represents a lateral translation of the beam as a rigid body.

In summary, fourth-order systems subjected to the boundary conditions of Case 9 possess an infinite number of modes, consisting of a rigid-body mode given by Eq. (9-208) plus an infinite number of flexible modes given by Eq. (9-207) for $n = 1, 2, 3, \ldots$.

CASE 10: *Free–Free* ($X''(0) = X'''(0) = X''(l) = X'''(l) = 0$)

First, we consider the case when $\omega \neq 0$ or, equivalently, $\beta \neq 0$. In which case, the conditions $X''(0) = 0$ and $X'''(0) = 0$ require from Eqs. (9-171) and (9-172), respectively, that

$$-c\beta^2 + e\beta^2 = 0 \qquad (9\text{-}209)$$
$$-d\beta^3 + f\beta^3 = 0 \qquad (9\text{-}210)$$

Also, the conditions $X''(l) = 0$ and $X'''(l) = 0$ require from Eqs. (9-171) and (9-172), respectively, that

$$-c\beta^2\cos\beta l - d\beta^2\sin\beta l + e\beta^2\cosh\beta l + f\beta^2\sinh\beta l = 0 \qquad (9\text{-}211)$$

$$c\beta^3\sin\beta l - d\beta^3\cos\beta l + e\beta^3\sinh\beta l + f\beta^3\cosh\beta l = 0 \qquad (9\text{-}212)$$

Because $\beta \neq 0$, Eqs. (9-209) and (9-210) yield, respectively, that

$$e = c \qquad (9\text{-}213)$$
$$f = d \qquad (9\text{-}214)$$

Eliminating e and f from Eqs. (9-211) and (9-212) via Eqs. (9-213) and (9-214) yields

$$\begin{bmatrix} (\cos\beta l - \cosh\beta l) & (\sin\beta l - \sinh\beta l) \\ (\sin\beta l + \sinh\beta l) & -(\cos\beta l - \cosh\beta l) \end{bmatrix}\begin{Bmatrix} c \\ d \end{Bmatrix} = 0 \qquad (9\text{-}215)$$

where the condition $\beta \neq 0$ has been used in dividing both sides of Eqs. (9-211) and (9-212) by β^2 and β^3, respectively.

In order for nonzero c and d to exist, the determinant of the coefficient matrix in Eq. (9-215) must vanish, giving

$$-(\cos \beta l - \cosh \beta l)^2 - (\sin \beta l - \sinh \beta l)(\sin \beta l + \sinh \beta l)$$

$$= 0 \qquad (9\text{-}216)$$

or, upon rearrangement,

$$\cos \beta l \cosh \beta l = 1 \qquad (9\text{-}217)$$

where the relations $\cos^2 \theta + \sin^2 \theta = 1$ and $\cosh^2 \theta - \sinh^2 \theta = 1$ have been used in the rearrangement.

There is an infinite number of solutions to Eq. (9-217) for β. These solutions may be arranged in ascending order of their magnitudes and designated as β_n ($n = 1, 2, 3, \ldots$) in the same order. If we define $\omega_n = c_q \beta_n^2$ using Eq. (9-163), these ω_n's are the natural frequencies of the system. By substituting β_n into Eq. (9-215), the ratio of c_n to d_n corresponding to β_n can be found as

$$\frac{c_n}{d_n} = -\frac{\sin \beta_n l - \sinh \beta_n l}{\cos \beta_n l - \cosh \beta_n l} = \frac{\cos \beta_n l - \cosh \beta_n l}{\sin \beta_n l + \sinh \beta_n l}$$

$$(9\text{-}218)$$

By combining Eqs. (9-165), (9-213), and (9-214), the eigenfunction $X_n(x)$ corresponding to the nth natural frequency ω_n is given by

$$X_n(x)$$

$$= c_n(\cos \beta_n x + \cosh \beta_n x) + d_n(\sin \beta_n x + \sinh \beta_n x)$$

$$= d_n \left\{ \frac{c_n}{d_n}(\cos \beta_n x + \cosh \beta_n x) + (\sin \beta_n x + \sinh \beta_n x) \right\}$$

$$(9\text{-}219)$$

where the ratio $\dfrac{c_n}{d_n}$ is determined by Eq. (9-218) for each β_n found from Eq. (9-217). Therefore, Eq. (9-219) gives the eigenfunctions ($n = 1, 2, 3, \ldots$) for fourth-order systems subjected to the boundary conditions of Case 10, to within a multiplicative constant d_n.

In order to complete the analysis for Case 10, consider the instance where $\omega = \beta = 0$. In this case, the spatial function solution is given by the third-order polynomial in Eq. (9-167) and its derivatives are given by Eqs. (9-173) through (9-175). Therefore, the boundary conditions $X''(0) = 0$ and $X'''(0) = 0$, when combined with Eqs. (9-174) and (9-175), respectively, yield $e = 0$ and $f = 0$. Similarly, combining the boundary conditions $X''(l) = 0$ and $X'''(l) = 0$ with Eqs. (9-174) and (9-175), respectively, gives $2e + 6fl = 0$ and $6f = 0$, which yield $e = f = 0$ again. We note that the boundary conditions at $x = l$ provide no additional information, leaving c and d arbitrary. Therefore, substituting $e = f = 0$ into Eq. (9-167) yields

$$X_0(x) = c_0 + d_0 x \qquad (9\text{-}220)$$

where c_0 and d_0 are arbitrary constants and the subscript 0 indicates that this eigenfunction is for the zeroth mode (or, equivalently, the rigid-body mode). We note that the zeroth eigenfunction given by Eq. (9-220) represents a combination of translation (when $d_0 = 0$) of the beam and rotation (when $c_0 = 0$) of the beam about $x = 0$.

In summary, fourth-order systems subjected to the boundary conditions of Case 10 possess an infinite number of modes, consisting of a rigid-body mode given by Eq. (9-220) and an infinite number of flexible modes given by Eq. (9-219) for $n = 1, 2, 3, \ldots$.

In conclusion, the equations for the natural frequencies and mode shapes for all ten sets of "simple" boundary conditions listed above are summarized in Table 9-10a. In the second column of Table 9-10a, the equations for the rigid-body mode eigenfunctions $X_0(x)$ are shown only for the four cases (in the bottom four rows) that accommodate rigid-body modes. The third through fifth columns of Table 9-10a are concerned with the deformable (or nonrigid-body) modes. That is, the third column of Table 9-10a lists the equations for

$$\beta_n l \equiv \left(\sqrt{\frac{\omega_n}{c_q}} \right) l$$ for the ten "simple" boundary conditions; the fourth column gives the general form of the nth eigenfunction $X_n(x)$, where we note that the $X_n(x)$'s possess either one or two unknown coefficients. So, the third column is related directly to the *natural frequencies* for each of the specified boundary conditions, and the fourth column is related directly to the *mode shapes* for each set of the specified boundary conditions. Note the general correspondence here with two-degree-of-freedom systems in Chapter 8, where natural frequencies and mode shapes were calculated, and the more direct correspondence with Table 9-9 in this chapter.

TABLE 9-10a Equations for Natural Frequencies and Eigenfunctions of Fourth-Order System (Bernoulli-Euler Beam) for Various Simple Boundary Conditions

Boundary conditions	Rigid-body mode ($\omega_0 = 0$) $X_0(x)$	Equation for $\beta_n\ell \equiv \sqrt{\omega_n/c_q}\,\ell$	Nonrigid-body modes ($n = 1, 2, 3, \ldots$)	
			$X_n(x)$	Ratio of coefficients
Clamped–Clamped	—	$\cos\beta_n\ell\cosh\beta_n\ell = 1$	$c_n(\cos\beta_n x - \cosh\beta_n x)$ $+\,d_n(\sin\beta_n x - \sinh\beta_n x)$	$\frac{c_n}{d_n} = -\frac{\sin\beta_n\ell - \sinh\beta_n\ell}{\cos\beta_n\ell - \cosh\beta_n\ell} = \frac{\cos\beta_n\ell - \cosh\beta_n\ell}{\sin\beta_n\ell + \sinh\beta_n\ell}$
Clamped–Simply supported	—	$\tan\beta_n\ell = \tanh\beta_n\ell$	$c_n(\cos\beta_n x - \cosh\beta_n x)$ $+\,d_n(\sin\beta_n x - \sinh\beta_n x)$	$\frac{c_n}{d_n} = -\frac{\sin\beta_n\ell - \sinh\beta_n\ell}{\cos\beta_n\ell - \cosh\beta_n\ell} =$ $-\frac{\sin\beta_n\ell + \sinh\beta_n\ell}{\cos\beta_n\ell + \cosh\beta_n\ell}$
Clamped–Sliding-rotationally restrained	—	$\tan\beta_n\ell + \tanh\beta_n\ell = 0$	$c_n(\cos\beta_n x - \cosh\beta_n x)$ $+\,d_n(\sin\beta_n x - \sinh\beta_n x)$	$\frac{c_n}{d_n} = \frac{\cos\beta_n\ell + \cosh\beta_n\ell}{\sin\beta_n\ell + \sinh\beta_n\ell} = \frac{\cos\beta_n\ell + \cosh\beta_n\ell}{\sin\beta_n\ell - \sinh\beta_n\ell}$
Clamped–Free	—	$\cos\beta_n\ell\cosh\beta_n\ell = -1$	$c_n(\cos\beta_n x - \cosh\beta_n x)$ $+\,d_n(\sin\beta_n x - \sinh\beta_n x)$	$\frac{c_n}{d_n} = -\frac{\sin\beta_n\ell + \sinh\beta_n\ell}{\cos\beta_n\ell + \cosh\beta_n\ell} = \frac{\cos\beta_n\ell + \cosh\beta_n\ell}{\sin\beta_n\ell - \sinh\beta_n\ell}$
Simply supported–Simply supported	—	$\sin\beta_n\ell = 0$	$d_n\sin\beta_n x$	d_n: arbitrary
Simply supported–Sliding-rotationally restrained	—	$\cos\beta_n\ell = 0$	$d_n\sin\beta_n x$	d_n: arbitrary
Simply supported–Free	$d_0 x$ (d_0: arbitrary)	$\tan\beta_n\ell = \tanh\beta_n\ell$	$d_n\sin\beta_n x + f_n\sinh\beta_n x$	$\frac{d_n}{f_n} = \frac{\sinh\beta_n\ell}{\sin\beta_n\ell} = \frac{\cosh\beta_n\ell}{\cos\beta_n\ell}$
Sliding-rotationally restrained–Sliding-rotationally restrained	c_0 (c_0: arbitrary)	$\sin\beta_n\ell = 0$	$c_n\cos\beta_n x$	c_n: arbitrary
Sliding-rotationally restrained–Free	c_0 (c_0: arbitrary)	$\tan\beta_n\ell + \tanh\beta_n\ell = 0$	$c_n\cos\beta_n x + e_n\cosh\beta_n x$	$\frac{c_n}{e_n} = \frac{\cosh\beta_n\ell}{\cos\beta_n\ell} = -\frac{\sinh\beta_n\ell}{\sin\beta_n\ell}$
Free–Free	$c_0 + d_0 x$ (c_0, d_0: arbitrary)	$\cos\beta_n\ell\cosh\beta_n\ell = 1$	$c_n(\cos\beta_n x + \cosh\beta_n x)$ $+\,d_n(\sin\beta_n x + \sinh\beta_n x)$	$\frac{c_n}{d_n} = -\frac{\sin\beta_n\ell - \sinh\beta_n\ell}{\cos\beta_n\ell - \cosh\beta_n\ell} = \frac{\cos\beta_n\ell - \cosh\beta_n\ell}{\sin\beta_n\ell + \sinh\beta_n\ell}$

TABLE 9-10b Several Lowest Mode Natural Frequencies and Eigenfunctions of Fourth-Order System (Bernoulli-Euler Beam) for Various Simple Boundary Conditions

Boundary conditions	Rigid-body mode $X_0(x)(\omega_0 = 0)$	Nonrigid-body modes ($n = 1, 2, 3$) $\beta_1\ell, \beta_2\ell, \beta_3\ell$	$X_1(x)$	$X_2(x)$	$X_3(x)$
Clamped–Clamped		4.730 7.853 10.996			
Clamped–Simply supported		3.927 7.069 10.210			
Clamped–Sliding-rotationally restrained		2.365 5.498 8.639			
Clamped–Free		1.875 4.694 7.855			
Simply supported–Simply supported		3.142 6.283 9.425			
Simply supported–Sliding-rotationally restrained		1.571 4.712 7.854			
Simply supported–Free		3.927 7.069 10.210			
Sliding-rotationally restrained–Sliding-rotationally restrained		3.142 6.283 9.425			
Sliding-rotationally restrained–Free		2.365 5.498 8.639			
Free–Free		4.730 7.853 10.996			

Recall that the eigenfunctions can be determined only to within a multiplicative constant. Therefore, as indicated in the fifth column of Table 9-10a, in the cases where the eigenfunction possesses only one unknown coefficient, the coefficient is arbitrary; and in the cases where the eigenfunction possesses two unknown coefficients, a relation between those two coefficients is given, leaving only one of the two coefficients unknown. For example, for the clamped–clamped boundary conditions as derived in Eqs. (9-176) through (9-186), the eigenfunction $X_n(x)$ is given in the fourth column as

$$X_n(x) = c_n(\cos\beta_n x - \cosh\beta_n x) + d_n(\sin\beta_n x - \sinh\beta_n x) \qquad (9\text{-}221)$$

which is the same as Eq. (9-186). Equation (9-221) can be rewritten as

$$X_n(x) = d_n \left\{ \frac{c_n}{d_n} (\cos \beta_n x - \cosh \beta_n x) + (\sin \beta_n x - \sinh \beta_n x) \right\} \qquad (9\text{-}222)$$

Note that the ratio of the coefficients $\dfrac{c_n}{d_n}$ in Eq. (9-222) is given in the fifth column of the first row of Table 9-10*a* as

$$\frac{c_n}{d_n} = -\frac{\sin \beta_n l - \sinh \beta_n l}{\cos \beta_n l - \cosh \beta_n l} = \frac{\cos \beta_n l - \cosh \beta_n l}{\sin \beta_n l + \sinh \beta_n l} \qquad (9\text{-}223)$$

which follows from Eq. (9-185). For each value of n ($n = 1, 2, 3, \ldots$), the ratio $\dfrac{c_n}{d_n}$ is a known quantity once the value of $\beta_n l$ has been found as the root of the equation, in the third column, given as

$$\cos \beta_n l \cosh \beta_n l = 1 \qquad n = 1, 2, 3, \ldots \qquad (9\text{-}224)$$

Therefore, the eigenfunction $X_n(x)$ in Eq. (9-222) is determined to within a multiplicative constant d_n, once $\beta_n l$ has been found from Eq. (9-224) and $\dfrac{c_n}{d_n}$ has been found from Eq. (9-223) using the thus found value for $\beta_n l$. The results obtained for each of the boundary conditions via this procedure are shown in Table 9-10*b*.

In Table 9-10*b*, the rigid-body mode eigenfunctions $X_0(x)$'s are illustrated in the second column, for those boundary conditions that accommodate rigid-body modes. The third column of Table 9-10*b* lists the first three roots $\beta_n l$ ($n = 1, 2, 3$) to the equations given in the third column of Table 9-10*a*. The corresponding natural frequencies ω_n's can be calculated from the values for $\beta_n l$ via the definition of β_n; that is, $\omega_n \equiv c_q \beta_n^2 = \dfrac{c_q (\beta_n l)^2}{l^2}$. The deformable mode eigenfunctions $X_n(x)$'s ($n = 1, 2, 3$) corresponding to the first three nonzero natural frequencies ω_n's ($n = 1, 2, 3$) are illustrated in the fourth through sixth columns of Table 9-10*b*. In plotting these mode shapes, the extent of the system l has been set equal to unity. Since the mode shapes are determined only to within a multiplicative constant, the mode shapes shown in Table 9-10*b* are plotted with arbitrary amplitudes. Note that the mode shapes indicate that particles at certain locations do not move during the modal motion. These stationary points are called *nodes*. As can be seen, the number of nodes increases with increasing mode number.

9-3.4.2   *Orthogonality*   Now, we shall derive an interesting and useful property of the eigenfunctions of fourth-order systems subjected to the "simple" boundary conditions considered in the ten cases above. This property is that the eigenfunctions are *orthogonal* to each other in the sense that they satisfy the relation[14]

$$\int_0^l X_n X_m \, dx = 0 \qquad n \neq m \qquad (9\text{-}225)$$

---

[14]This orthogonality relation is not general in the sense that it holds only when the continuum is uniform *and* the boundary conditions are such that Eq. (9-233) vanishes, as in all cases of simple boundary conditions listed in Table 9-10 *plus* the cases having stiffness element(s) at the end(s). More general orthogonality relations are required for more general cases when the continuum is not uniform or when the end(s) is (are) terminated by inertial loads. However, we shall consider only this simplest version of orthogonality, merely to provide a tool that significantly simplifies calculations in free vibration analyses, as will be shown in Section 9-4.

where X_n and X_m are the nth and m^{th} eigenfunctions, respectively. In order to prove Eq. (9-225), note that all eigenfunctions $X_n(x)$ satisfy Eq. (9-160) along with the corresponding natural frequency ω_n. Therefore, we can write Eq. (9-160) for two different modes (say, the mth and n^{th} modes) as

$$\frac{d^4 X_m}{dx^4} - \frac{\omega_m^2}{c_q^2} X_m = 0 \tag{9-226}$$

$$\frac{d^4 X_n}{dx^4} - \frac{\omega_n^2}{c_q^2} X_n = 0 \tag{9-227}$$

Now multiply Eq. (9-226) by X_n and multiply Eq. (9-227) by X_m, which give

$$X_n \frac{d^4 X_m}{dx^4} - \frac{\omega_m^2}{c_q^2} X_n X_m = 0 \tag{9-228}$$

$$X_m \frac{d^4 X_n}{dx^4} - \frac{\omega_n^2}{c_q^2} X_m X_n = 0 \tag{9-229}$$

Subtracting Eq. (9-229) from Eq. (9-228) yields

$$X_n \frac{d^4 X_m}{dx^4} - X_m \frac{d^4 X_n}{dx^4} = \frac{\omega_m^2 - \omega_n^2}{c_q^2} X_n X_m \tag{9-230}$$

Integrate both sides of Eq. (9-230) with respect to x from 0 to l. The first term on the left-hand side of Eq. (9-230) can be integrated by parts via two successive applications of Eq. (9-13). Therefore,

$$\int_0^l X_n \frac{d^4 X_m}{dx^4}\, dx$$

$$= X_n \frac{d^3 X_m}{dx^3}\bigg|_0^l - \int_0^l \frac{dX_n}{dx} \frac{d^3 X_m}{dx^3}\, dx$$

$$= X_n \frac{d^3 X_m}{dx^3}\bigg|_0^l - \frac{dX_n}{dx} \frac{d^2 X_m}{dx^2}\bigg|_0^l + \int_0^l \frac{d^2 X_n}{dx^2} \frac{d^2 X_m}{dx^2}\, dx \tag{9-231}$$

Similarly, the second term on the left-hand side of Eq. (9-230) can be integrated to yield

$$\int_0^l X_m \frac{d^4 X_n}{dx^4}\, dx$$

$$= X_m \frac{d^3 X_n}{dx^3}\bigg|_0^l - \frac{dX_m}{dx} \frac{d^2 X_n}{dx^2}\bigg|_0^l + \int_0^l \frac{d^2 X_m}{dx^2} \frac{d^2 X_n}{dx^2}\, dx \tag{9-232}$$

Subtracting Eq. (9-232) from Eq. (9-231) reduces the integral on the left-hand side of Eq. (9-230) to

$$\int_0^l \left(X_n \frac{d^4 X_m}{dx^4} - X_m \frac{d^4 X_n}{dx^4} \right) dx$$

$$= X_n \frac{d^3 X_m}{dx^3}\bigg|_0^l - \frac{dX_n}{dx} \frac{d^2 X_m}{dx^2}\bigg|_0^l - X_m \frac{d^3 X_n}{dx^3}\bigg|_0^l + \frac{dX_m}{dx} \frac{d^2 X_n}{dx^2}\bigg|_0^l$$

$$= X_n(l)X_m'''(l) - X_n(0)X_m'''(0) - X_n'(l)X_m''(l) + X_n'(0)X_m''(0)$$

$$- X_m(l)X_n'''(l) + X_m(0)X_n'''(0) + X_m'(l)X_n''(l) - X_m'(0)X_n''(0) \tag{9-233}$$

where $X_i' \equiv \dfrac{dX_i}{dx}$, $X_i'' \equiv \dfrac{d^2X_i}{dx^2}$, and $X_i''' \equiv \dfrac{d^3X_i}{dx^3}$. Equating Eq. (9-233) to the integral on the right-hand side of Eq. (9-230) gives

$$X_n(l)X_m'''(l) - X_n(0)X_m'''(0) - X_n'(l)X_m''(l) + X_n'(0)X_m''(0)$$

$$- X_m(l)X_n'''(l) + X_m(0)X_n'''(0) + X_m'(l)X_n''(l) - X_m'(0)X_n''(0)$$

$$= \frac{\omega_n^2 - \omega_m^2}{c_q^2} \int_0^l X_n X_m \, dx \tag{9-234}$$

Note that all boundary conditions considered above (Cases 1 through 10 in Table 9-10) require at each end of the continuum that either the eigenfunction itself or its third derivative be zero *and* that either the first or second derivative of the eigenfunction be zero. Therefore, for the boundary conditions of Cases 1 through 10, all terms on the left-hand side of Eq. (9-234) vanish. Thus, Eq. (9-234) reduces to

$$0 = \frac{\omega_m^2 - \omega_n^2}{c_q^2} \int_0^l X_n X_m \, dx \tag{9-235}$$

Because $\omega_n^2 \neq \omega_m^2$ in accordance with our statement that $n \neq m$, Eq. (9-235) implies that the eigenfunctions for two different modes must satisfy the expression

$$\int_0^l X_n X_m \, dx = 0 \qquad n \neq m \tag{9-236}$$

which is the *orthogonality relation* in Eq. (9-225).

9-3.4.3 *An Example with Less Simple Boundary Conditions*

All boundary conditions considered so far in this subsection have been of the "simple" type, defined to be those boundary conditions for which two of the four quantities—the generalized coordinate itself and its first, second, and third spacewise derivatives—are zero at each end. For other end conditions, the analysis becomes more complicated. For example and for illustrative purposes only, we consider the natural frequencies and mode shapes for the system sketched in Figure 9-6.

The boundary conditions for this system consist of the *geometric boundary conditions* at $x = 0$, given by Eq. (9-60) or Table 9-8 as

$$\eta = 0 \qquad x = 0 \tag{9-237}$$

$$\frac{\partial \eta}{\partial x} = 0 \qquad x = 0 \tag{9-238}$$

and the *natural boundary conditions* at $x = l$, given by Eqs. (9-74) and (9-75) as

$$M\frac{\partial^2 \eta}{\partial t^2} + k\eta = EI\frac{\partial^3 \eta}{\partial x^3} \qquad x = l \tag{9-239}$$

$$EI\frac{\partial^2 \eta}{\partial x^2} = 0 \qquad x = l \tag{9-240}$$

where in writing Eq. (9-239) we have set $F(t) = 0$ and we have assumed the beam is uniform; that is, EI is constant throughout the length of the beam. Note that Eqs. (9-237) through (9-240) must hold for all time t. Also, note that Eqs. (9-237), (9-238), and (9-240) contain only spatial derivatives, while Eq. (9-239) contains a time derivative as well as a spatial derivative. Therefore, substituting Eq. (9-85) into Eqs. (9-237), (9-238), and (9-240) where $z(x, t)$ corresponds to $\eta(x, t)$ here, and requiring the resulting equations to hold for all time t yield

$$X(0) = X'(0) = X''(l) = 0 \tag{9-241}$$

We seek the natural frequencies and eigenfunctions for the system sketched in Figure 9-6 subjected to the boundary conditions now given by Eqs. (9-239) and (9-241). First, we consider the case when $\omega \neq 0$. Then, substitution of Eqs. (9-165), (9-170), and (9-171) into Eq. (9-241) yields

$$c + e = 0 \tag{9-242}$$
$$d\beta + f\beta = 0 \tag{9-243}$$
$$-c\beta^2 \cos \beta l - d\beta^2 \sin \beta l + e\beta^2 \cosh \beta l + f\beta^2 \sinh \beta l = 0 \tag{9-244}$$

Note that we must use the complete solution form given by Eq. (9-168) for the boundary condition in Eq. (9-239) because of the time derivative contained in it. Substitution of Eq. (9-168) into Eq. (9-239) yields

$$(-M\omega^2 + k)(c \cos \beta l + d \sin \beta l + e \cosh \beta l + f \sinh \beta l)$$
$$= EI(c\beta^3 \sin \beta l - d\beta^3 \cos \beta l + e\beta^3 \sinh \beta l + f\beta^3 \cosh \beta l) \tag{9-245}$$

where $T(t) = a \cos \omega t + b \sin \omega t$ has been deleted from both sides because Eq. (9-245) must hold for all values of t. Equations (9-242) through (9-245) constitute a set of simultaneous linear equations in the four unknowns c, d, e, and f. Combining Eqs. (9-242) and (9-243) gives $e = -c$ and $f = -d$, which can be used to eliminate e and f from Eqs. (9-244) and (9-245). Thus, Eqs. (9-244) and (9-245) in terms of c and d become

$$\begin{bmatrix} (\cos \beta l + \cosh \beta l) & (\sin \beta l + \sinh \beta l) \\ \left\{ \begin{matrix} (k - \omega^2 M)(\cos \beta l - \cosh \beta l) \\ -EI\beta^3(\sin \beta l - \sinh \beta l) \end{matrix} \right\} & \left\{ \begin{matrix} (k - \omega^2 M)(\sin \beta l - \sinh \beta l) \\ +EI\beta^3(\cos \beta l + \cosh \beta l) \end{matrix} \right\} \end{bmatrix} \begin{Bmatrix} c \\ d \end{Bmatrix} = 0 \tag{9-246}$$

where the condition $\beta \neq 0$ has been used in dividing both sides of Eq. (9-244) by $-\beta^2$.

The condition for nontrivial solutions for c and d to exist is that the determinant of the coefficient matrix in Eq. (9-246) must vanish. Therefore,

$$\left\{ (k - \omega^2 M)(\sin \beta l - \sinh \beta l) + EI\beta^3(\cos \beta l + \cosh \beta l) \right\} \left\{ \cos \beta l + \cosh \beta l \right\}$$
$$- \left\{ (k - \omega^2 M)(\cos \beta l - \cosh \beta l) - EI\beta^3(\sin \beta l - \sinh \beta l) \right\} \left\{ \sin \beta l + \sinh \beta l \right\} = 0$$
$$\tag{9-247}$$

or, upon rearrangement,

$$(k - \omega^2 M)(\sin \beta l \cosh \beta l - \cos \beta l \sinh \beta l) + EI\beta^3(1 + \cos \beta l \cosh \beta l) = 0 \tag{9-248}$$

where the identities $\cos^2 \theta + \sin^2 \theta = 1$ and $\cosh^2 \theta - \sinh^2 \theta = 1$ have been used in obtaining the second term of Eq. (9-248). Note that, if they exist, closed-form solutions to Eq. (9-248) for β cannot be easily found. However, due to the periodicity of the various harmonic and hyperbolic functions, Eq. (9-248) contains an infinite number of solutions for β (or equivalently ω; recall $\omega = c_q \beta^2$). Once those ω's have been found, say, by numerical methods, they may be arranged in ascending order of their magnitudes and designated as ω_n ($n = 1, 2, 3, \ldots$) in the same order. These ω_n's are the natural frequencies of the system in Figure 9-6.

Note that c and d (or more precisely, their ratio because c and d cannot be found explicitly) in Eq. (9-246) have different values for different ω_n's. Therefore, by designating c and d for each ω_n by c_n and d_n, respectively, their ratio can be found from Eq. (9-246) as

$$
\frac{c_n}{d_n} = -\frac{\sin \beta_n l + \sinh \beta_n l}{\cos \beta_n l + \cosh \beta_n l}
$$

$$
= -\frac{(k - \omega_n^2 M)(\sin \beta_n l - \sinh \beta_n l) + EI\beta_n^3(\cos \beta_n l + \cosh \beta_n l)}{(k - \omega_n^2 M)(\cos \beta_n l - \cosh \beta_n l) - EI\beta_n^3(\sin \beta_n l - \sinh \beta_n l)} \tag{9-249}
$$

Recalling from Eqs. (9-242) and (9-243) that $e_n = -c_n$ and $f_n = -d_n$ and using these expressions to eliminate e_n and f_n in Eq. (9-165) yield the nth eigenfunction $X_n(x)$ corresponding to ω_n as

$$
X_n(x) = c_n(\cos \beta_n x - \cosh \beta_n x) + d_n(\sin \beta_n x - \sinh \beta_n x)
$$

$$
= d_n \left\{ \frac{c_n}{d_n}(\cos \beta_n x - \cosh \beta_n x) + (\sin \beta_n x - \sinh \beta_n x) \right\} \tag{9-250}
$$

where the ratio $\dfrac{c_n}{d_n}$ is a known factor once ω_n is known, as given by Eq. (9-249). We note again that Eq. (9-250) shows that boundary conditions determine the eigenfunctions only to within a multiplicative constant, d_n in this case.

In order to complete the analysis for this system, we consider the possibility of a rigid-body mode. To this end, we suppose $\omega = 0$ and use the solution in the form of Eqs. (9-167) and (9-169). Substitution of Eqs. (9-167), (9-173), and (9-174) into Eq. (9-241) requires that

$$
c = d = 0 \tag{9-251}
$$

$$
2e + 6lf = 0 \tag{9-252}
$$

As in the case for nonzero ω, we consider the solution form given by Eq. (9-169) for the boundary condition in Eq. (9-239). However, we note that the second derivative of Eq. (9-169) in a timewise sense is identically zero: that is, $\dfrac{\partial^2 \eta}{\partial t^2} = 0$. Therefore, the mass term in Eq. (9-239) vanishes. Thus, substituting Eqs. (9-167) and (9-175) into Eq. (9-239), without the mass term, yields

$$
k(el^2 + fl^3) = 6EIf \tag{9-253}
$$

where Eq. (9-251) has been used. Eliminating e in Eq. (9-253) via Eq. (9-252) and rearranging terms give

$$
(2kl^3 + 6EI)f = 0 \tag{9-254}
$$

which requires $f = 0$, and which in turn requires $e = 0$ via Eq. (9-252). Thus, c through f are all zero, which makes the assumed solution, Eq. (9-169), identically zero. Therefore, the solution for the case of $\omega = 0$ is a trivial solution of zero, and so the system in Figure 9-6 possesses no rigid-body mode.

9-3.5 General Solutions for Free Undamped Vibration

In this section, we have studied the free undamped vibration of two types of continuous systems governed by their generic equations of motion: Eq. (9-83) for second-order systems and Eq. (9-84) for fourth-order systems. We have obtained the *natural frequencies* or the algebraic expressions for obtaining the natural frequencies; and we have obtained the corresponding *eigenfunctions* (mode shapes) for various boundary conditions. We have also found that continuous systems possess an infinite number of *natural modes*. Recall that the combination of a natural frequency and its corresponding eigenfunction (mode shape) is called a natural mode. Tables 9-9 and 9-10 summarize the natural frequencies and the eigenfunctions for second-order and fourth-order systems, respectively, subjected to various simple boundary conditions. The complete modal motion for each mode is given by Eqs. (9-112) and (9-113) for second-order systems and by Eqs. (9-168) and (9-169) for fourth-order systems. That is, the complete modal motions are given by the product of a time response function and an eigenfunction. Since each of these modal motions satisfies the corresponding governing partial differential equation (Eq. (9-83) or Eq. (9-84), depending upon the order of system), the general solution for the natural vibration is given by the superposition of all modal motions. The emphatic correspondence between this section and Subsection 8-4.1 should not elude the reader.

First, for second-order systems, the general solution for natural vibration (that is, the general solution to Eq. (9-83)) is given by combining Eqs. (9-112) and (9-113) as

$$z(x, t) = p_0 T_0(t) X_0(x) + \sum_{n=1}^{\infty} T_n(t) X_n(x) \qquad (9\text{-}255)$$

where p_0 is a factor that is either one or zero depending on whether a rigid-body mode (that is, the zeroth mode) exists for the given boundary conditions: $p_0 = 1$ if the zeroth mode exists and $p_0 = 0$ if the zeroth mode does not exist. The most general forms for $T_0(t)$, $X_0(x)$, $T_n(t)$, and $X_n(x)$ in Eq. (9-255) are recalled and are given by

$$T_0(t) \equiv a_0 + b_0 t \qquad (9\text{-}256)$$

$$X_0(x) \equiv c_0 + d_0 x \qquad (9\text{-}257)$$

$$T_n(t) \equiv a_n \cos \omega_n t + b_n \sin \omega_n t \qquad (9\text{-}258)$$

$$X_n(x) \equiv c_n \cos \frac{\omega_n}{c_q} x + d_n \sin \frac{\omega_n}{c_q} x \qquad (9\text{-}259)$$

where ω_n is the nth natural frequency, and c_q is given in Table 9-4 for several one-dimensional continua. Note that the *specific* eigenfunctions $X_0(x)$ and $X_n(x)$ are given in Table 9-9 for each of the "simple" boundary conditions.

From Table 9-9, it may be observed that only the case of free–free boundary conditions accommodates a rigid-body mode, in which instance the rigid-body mode eigenfunction is given by $X_0(x) = c_0$ (that is, $d_0 = 0$ in Eq. (9-257)). Therefore, for the free–free boundary conditions, p_0 in Eq. (9-255) is unity; whereas for all other "simple" boundary conditions, p_0 in Eq. (9-255) is zero, leaving only the summation term of the deformable (or nonrigid-body) modes. Also, Table 9-9 gives the deformable mode eigenfunctions $X_n(x)$ for each of the "simple" boundary conditions, where it may be observed that for all "simple" boundary conditions either c_n or d_n in Eq. (9-259) is zero. In conclusion, the free undamped response of a second-order system subjected to one of the "simple" boundary conditions can be expressed in the form of Eq. (9-255) either (1) by substituting $p_0 = 0$ and the corresponding $X_n(x)$ found in Table 9-9 into Eq. (9-255) when the boundary conditions do not accommodate a rigid-body mode or (2) by substituting $p_0 = 1$ and the corresponding $X_0(x)$ and $X_n(x)$ found in Table 9-9 into Eq. (9-255) when the boundary conditions accommodate a rigid-body mode.

Second, for fourth-order systems, the general solution for natural vibration (that is, the general solution to Eq. (9-84)) is given by combining Eqs. (9-168) and (9-169) as

$$z(x,t) = p_0 T_0(t) X_0(x) + \sum_{n=1}^{\infty} T_n(t) X_n(x) \tag{9-260}$$

where p_0 is a factor that is either one or zero depending on whether a rigid-body mode (that is, the zeroth mode) exists for the given boundary conditions: $p_0 = 1$ if the zeroth mode exists and $p_0 = 0$ if the zeroth mode does not exist. The most general forms for $T_0(t)$, $X_0(x)$, $T_n(t)$, and $X_n(x)$ in Eq. (9-260) are recalled and are given by

$$T_0(t) \equiv a_0 + b_0 t \tag{9-261}$$

$$X_0(x) \equiv c_0 + d_0 x + e_0 x^2 + f_0 x^3 \tag{9-262}$$

$$T_n(t) \equiv a_n \cos \omega_n t + b_n \sin \omega_n t \tag{9-263}$$

$$X_n(x) \equiv c_n \cos \beta_n x + d_n \sin \beta_n x + e_n \cosh \beta_n x + f_n \sinh \beta_n x \tag{9-264}$$

where β_n is related to the nth natural frequency ω_n in accordance with $\beta_n \equiv \sqrt{\dfrac{\omega_n}{c_q}}$, c_q being defined as $c_q = \sqrt{\dfrac{EI}{\rho A}}$ in Table 9-4 for Bernoulli-Euler beams. Note that the *specific* eigenfunctions $X_0(x)$ and $X_n(x)$ are given in Table 9-10a for each of the "simple" boundary conditions.

From Table 9-10a, it may be observed that only four cases of the "simple" boundary conditions (simply supported–free, sliding-rotationally restrained–sliding-rotationally restrained, sliding-rotationally restrained–free, and free–free) accommodate rigid-body modes, in which instances the rigid-body mode eigenfunctions are given in the second column of Table 9-10a. Therefore, for these four cases, p_0 in Eq. (9-260) is unity; whereas for all other "simple" boundary conditions, p_0 in Eq. (9-260) is zero, leaving only the summation term of the deformable (or nonrigid-body) modes. Also, Table 9-10a gives the deformable mode eigenfunctions $X_n(x)$ for each of the "simple" boundary conditions, where it may be observed that for all "simple" boundary conditions at least two of the coefficients c_n through f_n in Eq. (9-264) are zero. In conclusion, the free undamped response of a fourth-order system subjected to one of the "simple" boundary conditions can be expressed in the form of Eq. (9-260) either (1) by substituting $p_0 = 0$ and the corresponding $X_n(x)$ found in Table 9-10a into Eq. (9-260) when the boundary conditions do not accommodate a rigid-body mode or (2) by substituting $p_0 = 1$ and the corresponding $X_0(x)$ and $X_n(x)$ found in Table 9-10a into Eq. (9-260) when the boundary conditions accommodate a rigid-body mode.

Also, recall that while all spatial functions in Eqs. (9-255) and (9-260) have been determined for various boundary conditions, they have been found only to within a multiplicative constant, as summarized in Tables 9-9 and 9-10a. Remember, these eigenfunctions are the spatial representations of the system's inherent configurations during free undamped vibration, and are *not* the response to any specified initial conditions. In Chapter 8, we called these eigenfunctions mode shapes or eigenvectors. Thus, the multiplicative constant reflects the fact that a specific eigenfunction represents only relative displacements between any two locations during free vibration at the corresponding natural frequency, and not absolute displacements. Indeed, in the absence of specific initial conditions, it would be meaningless to talk about an absolute or specific free vibration displacement. The constants associated with these eigenfunctions will be collapsed into the timewise constants a_0, b_0, a_n, and b_n, and then the constants a_0, b_0, a_n, and b_n in Eqs. (9-256), (9-258), (9-261), and (9-263) will be determined in accordance with the initial conditions imposed on the system, as will now be shown.

9-4 RESPONSE TO INITIAL CONDITIONS

If a continuous system is initially in an equilibrium configuration, the system would remain so unless disturbed by an external agent. The external agent may simply change the system's *initial conditions,* no longer acting beyond an initial time that we can conveniently set as zero. Or, the external agent may persist, acting during a portion or all of the subsequent motion of the system. In this section, we consider the subsequent motion of several one-dimensional continua due to imposed initial conditions. In defining each example, the problem statement will represent an idealization that will allow us to formulate a specific *boundary value problem.*

9-4.1 An Example: Release of Compressed Rod

We consider a uniform rod having cross-sectional area A, initial length l_0, density ρ, modulus of elasticity E, and coefficient of thermal expansion α_t. The left-hand end of the rod is fixed (that is, clamped) in a rigid immovable wall and the right-hand end is in contact with, but not fixed in, another rigid immovable wall, as sketched in Figure 9-8. Also, the rod is initially strain-free throughout its length. Now, suppose the temperature of the entire rod is increased by ΔT and remains constant. We seek the subsequent motion of the rod after the right-hand wall is suddenly removed at time $t = 0$.

First of all, we need to define clearly the generalized coordinate for the system. One convenient choice is the longitudinal displacement $\xi(x, t)$ of a section whose *stress-free equilibrium position at the elevated temperature* is x relative to the left-hand wall. Therefore, the range of x in $\xi(x, t)$ is not 0 to l_0, but 0 to l where l is given by

$$l = (1 + \alpha_t \Delta T)l_0 \qquad (9\text{-}265)$$

which, from the technical mechanics of solids, is the equilibrium length of the rod at the elevated temperature, without restraint by the right-hand wall.

After the right-hand wall is removed, the end conditions of the system become clamped at $x = 0$ and free at $x = l$. Examining Table 9-9 for the rod subjected to clamped–free boundary conditions reveals that there exists no rigid-body mode and that the eigenfunction for the deformable modes is given by

$$X_n(x) = d_n \sin \frac{\omega_n}{c_q} x \qquad (9\text{-}266)$$

where d_n is an arbitrary multiplicative constant; $c_q \equiv \sqrt{\dfrac{E}{\rho}}$ from Table 9-4; and ω_n is the nth natural frequency and is given in Table 9-9 as

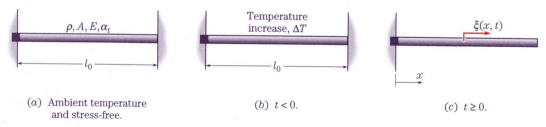

(a) Ambient temperature and stress-free.   (b) $t < 0$.   (c) $t \geq 0$.

FIGURE 9-8   Response of rod uniformly compressed due to temperature increase.

$$\omega_n = \left(n - \frac{1}{2}\right)\frac{\pi c_q}{l} \tag{9-267}$$

Therefore, substituting $p_0 = 0$ (due to the absence of a rigid-body mode) and Eq. (9-266) into Eq. (9-255) gives the free undamped response of the rod in Figure 9-8 as

$$\xi(x, t) = \sum_{n=1}^{\infty}(a_n \cos \omega_n t + b_n \sin \omega_n t)d_n \sin \frac{\omega_n}{c_q}x \tag{9-268}$$

where the time function $T_n(t)$ defined in Eq. (9-258) has been substituted into Eq. (9-255). Because the arbitrary multiplicative constant d_n in Eq. (9-268) is multiplied by other unknown constants a_n and b_n, the products of two unevaluated constants such as $a_n d_n$ and $b_n d_n$ can be designated by single unevaluated constants such as a_n and b_n. Therefore, Eq. (9-268) can be rewritten as

$$\xi(x, t) = \sum_{n=1}^{\infty}(a_n \cos \omega_n t + b_n \sin \omega_n t) \sin \frac{\omega_n}{c_q}x \tag{9-269}$$

In order to determine the unknown coefficients a_n and b_n in Eq. (9-269), the initial conditions for the system immediately upon the removal of the right-hand wall must be specified. From the technical mechanics of solids, the initial longitudinal thermal strain of the rod, clamped at both ends, is given by

$$\epsilon_0 = -\alpha_t \Delta T \tag{9-270}$$

where the negative sign signifies that the strain is compressive. Because the strain ϵ_0 given by Eq. (9-270) is uniform throughout the length of the rod, in terms of the generalized coordinate the initial displacement in the rod is given by

$$\xi(x, 0) = \epsilon_0 x = -\alpha_t \Delta T x \tag{9-271}$$

Further, the initial velocity of the rod is zero throughout its length, or when expressed in terms of the generalized coordinate the initial velocity is

$$\dot{\xi}(x, 0) = 0 \tag{9-272}$$

As indicated earlier, these initial conditions can be used to find the unknown coefficients a_n and b_n in Eq. (9-269) for all values of n. Substituting Eq. (9-269) for $t = 0$ into Eq. (9-271) and substituting the time derivative of Eq. (9-269) for $t = 0$ into Eq. (9-272) give, respectively,

$$\sum_{n=1}^{\infty} a_n \sin \frac{\omega_n}{c_q}x = -\alpha_t \Delta T x \tag{9-273}$$

$$\sum_{n=1}^{\infty} b_n \omega_n \sin \frac{\omega_n}{c_q}x = 0 \tag{9-274}$$

Noting that Eq. (9-273) contains exclusively a_n and Eq. (9-274) contains exclusively b_n, the coefficient a_n can be evaluated from Eq. (9-273) and the coefficient b_n can be evaluated from Eq. (9-274).

In order to evaluate a_n from Eq. (9-273), multiply both sides of Eq. (9-273) by the mth eigenfunction $X_m(x) = \sin \frac{\omega_m}{c_q}x$ (where m is an arbitrary integer) and integrate the resulting expression with respect to x from 0 to l, which give

$$\sum_{n=1}^{\infty} a_n \int_0^l \sin\left\{\left(n - \frac{1}{2}\right)\frac{\pi}{l}x\right\} \sin\left\{\left(m - \frac{1}{2}\right)\frac{\pi}{l}x\right\} dx = -\int_0^l \alpha_t \Delta T x \sin\left\{\left(m - \frac{1}{2}\right)\frac{\pi}{l}x\right\} dx$$

(9-275)

where Eq. (9-267) has been used for ω_n and ω_m and where l is the equilibrium length of the rod at the elevated temperature as given by Eq. (9-265). We note that the integrand on the left-hand side of Eq. (9-275) is the product of two eigenfunctions. Therefore, utilizing the orthogonality relation given by Eq. (9-135), all terms for $n \neq m$ vanish, so we need to compute only the term for $n = m$. By elementary integration, we find that

$$\int_0^l \sin^2\left\{\left(m - \frac{1}{2}\right)\frac{\pi}{l}x\right\} dx = \frac{l}{2}$$

(9-276)

Therefore, the left-hand side of Eq. (9-275) becomes

$$\sum_{n=1}^{\infty} a_n \int_0^l \sin\left\{\left(n - \frac{1}{2}\right)\frac{\pi}{l}x\right\} \sin\left\{\left(m - \frac{1}{2}\right)\frac{\pi}{l}x\right\} dx = a_m \frac{l}{2}$$

(9-277)

The right-hand side of Eq. (9-275) can be evaluated via integration by parts. That is, identifying corresponding terms in Eq. (9-13) and the right-hand side of Eq. (9-275) gives[15]

$$-\int_0^l \alpha_t \Delta T x \sin\left\{\left(m - \frac{1}{2}\right)\frac{\pi}{l}x\right\} dx$$

$$= -\alpha_t \Delta T \left[\frac{-xl}{\left(m - \frac{1}{2}\right)\pi}\cos\left\{\left(m - \frac{1}{2}\right)\frac{\pi}{l}x\right\}\Bigg|_0^l + \frac{l}{\left(m - \frac{1}{2}\right)\pi}\int_0^l \cos\left\{\left(m - \frac{1}{2}\right)\frac{\pi}{l}x\right\} dx\right]$$

(9-278)

where the first term in the square brackets is zero both at $x = 0$ and at $x = l$. Performing integration of the second term in the square brackets of Eq. (9-278) yields

$$-\int_0^l \alpha_t \Delta T x \sin\left\{\left(m - \frac{1}{2}\right)\frac{\pi}{l}x\right\} dx$$

$$= -\alpha_t \Delta T \left\{\frac{l}{\left(m - \frac{1}{2}\right)\pi}\right\}^2 \sin\left\{\left(m - \frac{1}{2}\right)\frac{\pi}{l}x\right\}\Bigg|_0^l$$

$$= \alpha_t \Delta T \left\{\frac{l}{\left(m - \frac{1}{2}\right)\pi}\right\}^2 (-1)^m$$

(9-279)

---

[15]In the integration by parts, the terms in Eq. (9-13) may be identified by

$$u = x \qquad dv = \sin\left\{\left(m - \frac{1}{2}\right)\frac{\pi}{l}x\right\} dx \qquad du = dx \qquad v = \frac{-l}{\left(m - \frac{1}{2}\right)\pi}\cos\left\{\left(m - \frac{1}{2}\right)\frac{\pi}{l}x\right\}$$

Using Eqs. (9-277) and (9-279), Eq. (9-275) reduces to

$$a_m \frac{l}{2} = \alpha_t \Delta T \left\{ \frac{l}{\left(m - \frac{1}{2}\right)\pi} \right\}^2 (-1)^m \tag{9-280}$$

or, upon rearrangement,

$$a_m = \frac{2}{l} \alpha_t \Delta T (-1)^m \left\{ \frac{2l}{(2m-1)\pi} \right\}^2$$

$$= \alpha_t \Delta T (-1)^m \frac{8l}{(2m-1)^2 \pi^2} \tag{9-281}$$

Since Eq. (9-281) holds for all values of $m = 1, 2, 3, \ldots$, we have found the unknown coefficient a_n in Eq. (9-269).

Now, in order to evaluate b_n from Eq. (9-274), multiply both sides of Eq. (9-274) by $X_m(x) = \sin \frac{\omega_m}{c_q} x$ and integrate the resulting expression with respect to x from 0 to l, which yield

$$b_m \omega_m \frac{l}{2} = 0 \tag{9-282}$$

where the orthogonality relations in Eq. (9-135) and Eq. (9-276) have been used. Since Eq. (9-282) holds for all values of $m = 1, 2, 3, \ldots$, the coefficient b_n in Eq. (9-269) is zero for all values of $n = 1, 2, 3, \ldots$.

Therefore, substitution of Eq. (9-281) and $b_n = 0$ (via Eq. (9-282)) into Eq. (9-269) and use of Eq. (9-267) yield

$$\xi(x, t) = \sum_{n=1}^{\infty} \alpha_t \Delta T (-1)^n \frac{8l}{(2n-1)^2 \pi^2} \cos\left\{ \left(n - \frac{1}{2}\right) \frac{\pi c_q}{l} t \right\} \sin\left\{ \left(n - \frac{1}{2}\right) \frac{\pi}{l} x \right\} \tag{9-283}$$

which is the solution for the response of the rod, subjected to the initial conditions given by Eqs. (9-271) and (9-272).

Figure 9-9 is a three-dimensional representation of the displacement $\xi(x, t)$ along the rod's length l given by Eq. (9-283), for various times t. And Figure 9-10 shows the displacement plots in x-ξ space at the indicated times, accompanied by the corresponding velocity[16] plots in the right-hand column. The primary purpose in presenting both displacements and velocities simultaneously in Figure 9-10 is merely to aid the reader in understanding how the various sections of the rod move with time. As shown in Figures 9-9 and 9-10, the initial negative displacement given by Eq. (9-271) is relieved by the removal of the right-hand wall at $x = l$, at $t = 0$. During the interval $0 < t < \frac{2l}{c_q}$, the displacement becomes increasingly positive throughout the rod. At $t = \frac{2l}{c_q}$, the tension is identically the negative of the initial compression. During the interval $\frac{2l}{c_q} < t < \frac{3l}{c_q}$, the tension decreases until,

[16]See Problem 9-31.

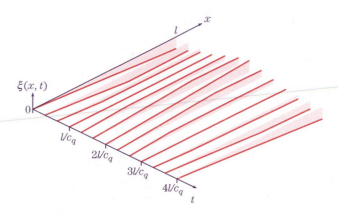

FIGURE 9-9 Three-dimensional representation of longitudinal displacement $\xi(x, t)$ of rod subjected to initial compression.

at $t = \dfrac{3l}{c_q}$, the rod reaches its equilibrium length l. During the interval $\dfrac{3l}{c_q} < t < \dfrac{4l}{c_q}$, the compression increases until, at $t = \dfrac{4l}{c_q}$, the rod reaches its initial compression as given in Eq. (9-271). Beyond $t = \dfrac{4l}{c_q}$, the sequence of changes repeats itself. Although the velocity plots are included primarily to enhance the understanding of the displacement evolution, several observations regarding the velocity plots can be made. In particular, the velocity distribution is zero throughout the rod when the rod's displacement distribution is maximum: see Figures 9-10a, e, and i. Analogously, the displacement distribution is zero throughout the rod when the rod's velocity distribution is maximum: see Figures 9-10c and g. These are simple manifestations of the free vibration of conservative systems; that is, systems for which there are no nonconservative forces and no time-varying constraints.

This periodic change of displacement can be explained by the propagation of the *disturbance*, due to the removal of the right-hand wall, along the rod and its *reflection* at the ends of the rod. In particular, the duration $\dfrac{l}{c_q}$ is the time required for a disturbance to travel the distance l, thus making c_q the speed at which the disturbance propagates. Also, note from Figures 9-9 and 9-10 that the displacement is continuous along the length of the rod at all times, but its spatial slope is discontinuous at the location where the *disturbance front* (the kink in $\xi(x, t)$) exists.

Figure 9-11 displays the convergence behavior of the series solution in Eq. (9-283) at a time $t = \dfrac{l}{3c_q}$ by showing the individual terms of the series in the left-hand column and the partial sums, up through the individual term, in the right-hand column for various values of n. Qualitatively, it can be noted from Figure 9-11 that a fairly accurate solution has been obtained by summing the first twenty terms.

In some engineering analyses, there is substantial interest in the stresses in the system as well as in the displacements because, for example, it is necessary in designing a system to ensure that the stress does not exceed a safe fraction of the yield strength of the material. For this purpose, the axial stress $\sigma_x(x, t)$ along the rod caused by the displacement given in Eq. (9-283) may be obtained (see Appendix K) as

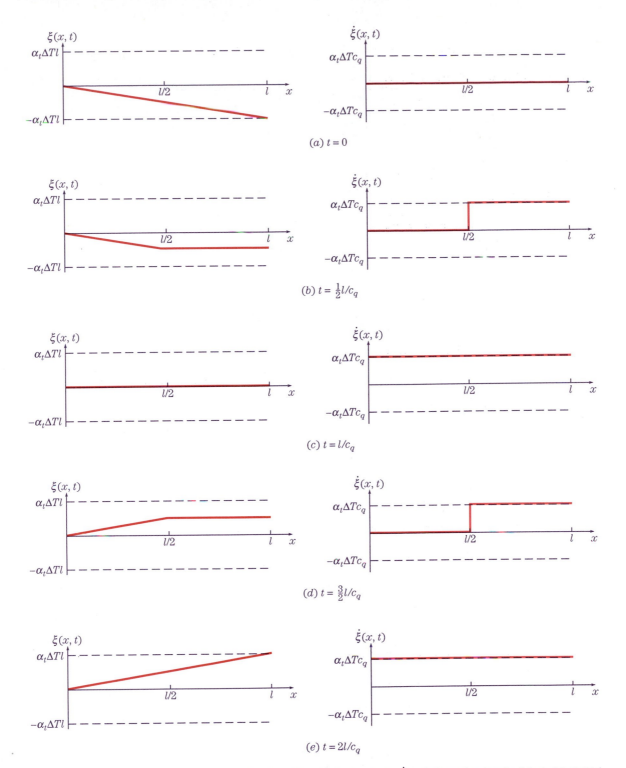

FIGURE 9-10a Longtudinal displacement $\xi(x,t)$ and longitudinal velocity $\dot{\xi}(x,t)$ throughout rod subjected to initial compression.

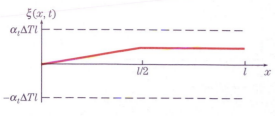
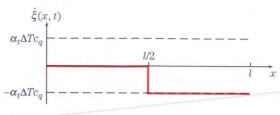

(f) $t = \frac{5}{2}l/c_q$

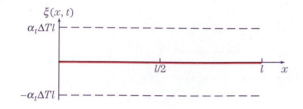
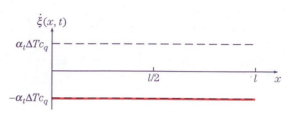

(g) $t = 3l/c_q$

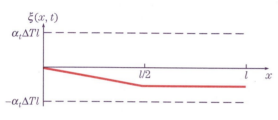
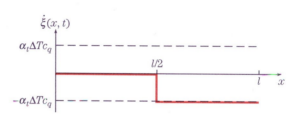

(h) $t = \frac{7}{2}l/c_q$

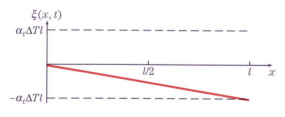
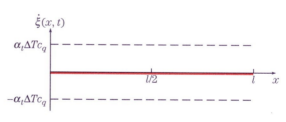

(i) $t = 4l/c_q$

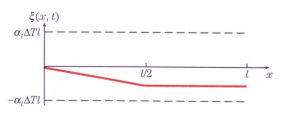
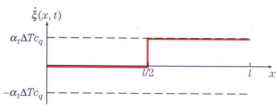

(j) $t = \frac{9}{2}l/c_q$

FIGURE 9-10b

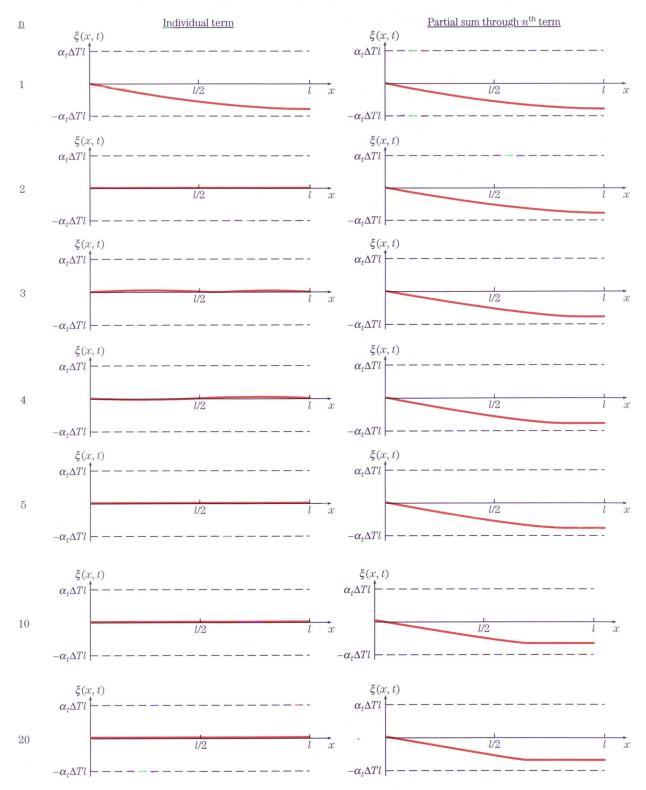

FIGURE 9-11 Individual terms and partial sums of displacement series solution in Eq. (9-283) at $t = \dfrac{l}{3c_q}$, for various values of n.

$$\sigma_x(x,t) = E\epsilon_x = E\frac{\partial \xi}{\partial x}$$

$$= E\alpha_t \Delta T \sum_{n=1}^{\infty} (-1)^n \frac{8l}{(2n-1)^2\pi^2} \cos\left\{\left(n-\frac{1}{2}\right)\frac{\pi c_q}{l}t\right\}\left(n-\frac{1}{2}\right)\frac{\pi}{l}\cos\left\{\left(n-\frac{1}{2}\right)\frac{\pi}{l}x\right\}$$

$$= E\alpha_t \Delta T \sum_{n=1}^{\infty} (-1)^n \frac{4}{(2n-1)\pi} \cos\left\{\left(n-\frac{1}{2}\right)\frac{\pi c_q}{l}t\right\}\cos\left\{\left(n-\frac{1}{2}\right)\frac{\pi}{l}x\right\} \qquad (9\text{-}284)$$

In analogy with Figures 9-9 and 9-10, Figures 9-12 and 9-13 show the behavior of the axial stress $\sigma_x(x,t)$ given by Eq. (9-284). As in the case of the displacement, the change in stress with time repeats itself with the time period of $\dfrac{4l}{c_q}$. Note from Figures 9-12 and 9-13 that the stress at $x = l$ is always zero for all positive times because of the free boundary condition at $x = l$. By comparing Figures 9-9 and 9-12, or Figures 9-10 and 9-13, note that the stress $\sigma_x(x,t)$ is given by the spatial derivative of the displacement $\xi(x,t)$ and therefore, the stress itself is discontinuous at the location of the disturbance front where the displacement has been observed in Figures 9-9 and 9-10 to exhibit a discontinuity in its spatial slope. This discontinuous stress at the disturbance front elucidates the nature of the disturbance caused by the removal of the right-hand wall; that is, because the removal of the right-hand wall renders the end at $x = l$ (initially under compressive stress at $t = 0$) "free" (zero stress), the disturbance due to the removal of the right-hand wall is simply a positive-step increase in the stress, annulling the initial compressive stress.

In analogy with Figure 9-11, Figure 9-14 shows the individual terms and the partial sums of the stress series solution in Eq. (9-284) at $t = \dfrac{l}{3c_q}$, for various values of n. Note that due to the discontinuities in the stress, the convergence of the series solution in Eq. (9-284) is, by comparison with the displacement convergence, slow and the summation of even two hundred terms may not give a solution of acceptable accuracy for very strident requirements.

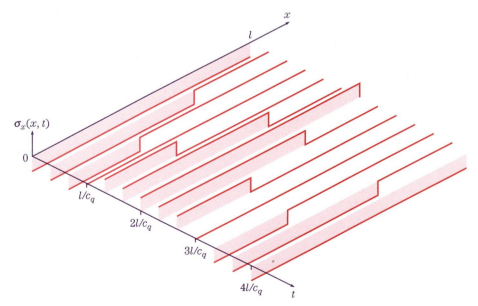

FIGURE 9-12 Three-dimensional representation of axial stress $\sigma_x(x,t)$ of rod subjected to initial compression.

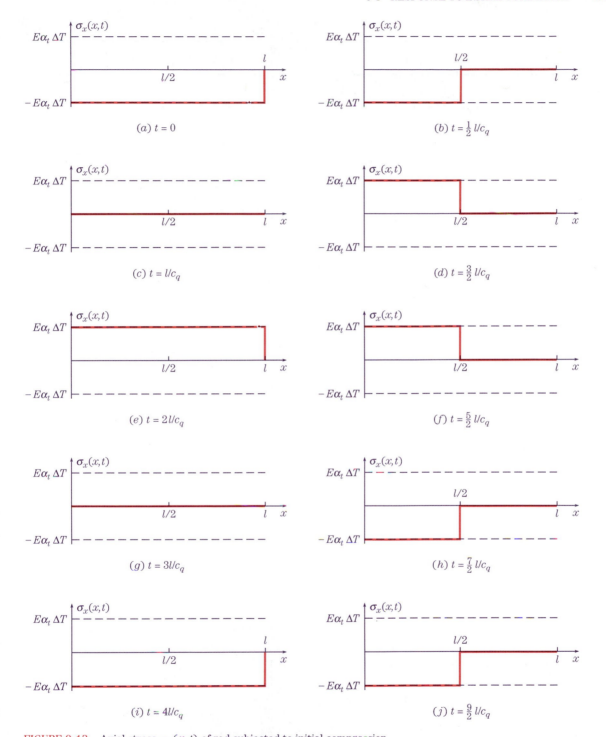

FIGURE 9-13 Axial stress $\sigma_x(x, t)$ of rod subjected to initial compression.

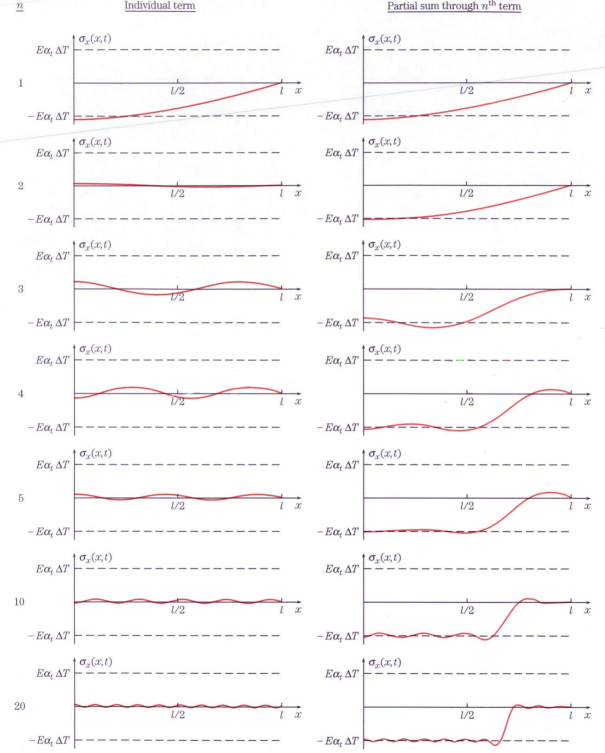

FIGURE 9-14a Individual terms and partial sums of stress series solution in Eq. (9-284) at $t = \dfrac{l}{3c_q}$, for various values of n.

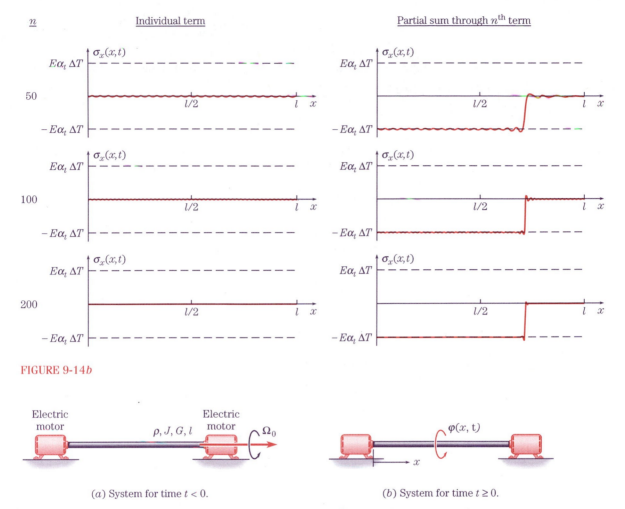

FIGURE 9-14b

(a) System for time $t < 0$.   (b) System for time $t \geq 0$.

FIGURE 9-15   Rotary system before and after the shaft is suddenly stopped following steady rotation.

9-4.2   An Example: Shaft Stopped after Rotation

We consider a uniform shaft having density ρ, polar moment of the cross-sectional area about the axis of the shaft J, shear modulus G, and equilibrium length l. Each end of the shaft is fixed (or clamped) in and driven by an electric motor at a constant angular velocity Ω_0; thus, the entire shaft is strain-free during this steady rotation, as sketched in Figure 9-15a. Suppose, at $t = 0$, both motors are suddenly stopped due to a power outage and remain rigid thereafter. We seek the subsequent motion of the shaft.

We define the generalized coordinate for this system to be the angular displacement $\varphi(x, t)$ measured from the undeformed configuration of the shaft. After the motors are stopped, the end conditions of the shaft are modeled as clamped–clamped. Examining Table 9-9 for the shaft subjected to clamped–clamped boundary conditions reveals that there exists no rigid-body mode and that the eigenfunction for the deformable modes is given by

$$X_n(x) = d_n \sin \frac{\omega_n}{c_q} x \tag{9-285}$$

where d_n is an arbitrary multiplicative constant; $c_q \equiv \sqrt{\dfrac{G}{\rho}}$ from Table 9-4; and ω_n is the nth natural frequency and is given in Table 9-9 as

$$\omega_n = \frac{n \pi c_q}{l} \tag{9-286}$$

Therefore, substituting $p_0 = 0$ (due to the absence of a rigid-body mode) and Eq. (9-285) into Eq. (9-255) gives the free undamped response of the shaft in Figure 9-15 as

$$\varphi(x, t) = \sum_{n=1}^{\infty} (a_n \cos \omega_n t + b_n \sin \omega_n t) d_n \sin \frac{\omega_n}{c_q} x \tag{9-287}$$

where the time function $T_n(t)$ defined in Eq. (9-258) has been substituted into Eq. (9-255). As indicated in the discussion between Eqs. (9-268) and (9-269), because the arbitrary multiplicative constant d_n in Eq. (9-287) is multiplied by other unknown constants a_n and b_n, the products of two unevaluated constants such as $a_n d_n$ and $b_n d_n$ can be designated by single unevaluated constants such as a_n and b_n. Therefore, Eq. (9-287) can be rewritten as

$$\varphi(x, t) = \sum_{n=1}^{\infty} (a_n \cos \omega_n t + b_n \sin \omega_n t) \sin \frac{\omega_n}{c_q} x \tag{9-288}$$

Next, as in Subsection 9-4.1, we establish the initial conditions for the system at $t = 0$, immediately after the motors are stopped. The initial displacement along the shaft is zero because the entire shaft has been free of strain during the steady rotation up to the instant when the motors are stopped. That is,

$$\varphi(x, 0) = 0 \tag{9-289}$$

The initial velocity of all sections in the shaft is Ω_0; so

$$\dot{\varphi}(x, 0) = \Omega_0 \tag{9-290}$$

As indicated earlier, these initial conditions can be used to find the unknown coefficients a_n and b_n in Eq. (9-288) for all values of n. Substituting Eq. (9-288) for $t = 0$ into Eq. (9-289) and substituting the time derivative of Eq. (9-288) for $t = 0$ into Eq. (9-290) give, respectively,

$$\sum_{n=1}^{\infty} a_n \sin \frac{\omega_n}{c_q} x = 0 \tag{9-291}$$

$$\sum_{n=1}^{\infty} b_n \omega_n \sin \frac{\omega_n}{c_q} x = \Omega_0 \tag{9-292}$$

Noting that Eq. (9-291) contains exclusively a_n and Eq. (9-292) contains exclusively b_n, the coefficient a_n can be evaluated from Eq. (9-291) and the coefficient b_n can be evaluated from Eq. (9-292).

In order to evaluate a_n from Eq. (9-291), multiply both sides of Eq. (9-291) by the mth eigenfunction $X_m(x) = \sin \dfrac{\omega_m}{c_q} x$ and integrate the resulting expression with respect to x from 0 to l, which yield

$$\sum_{n=1}^{\infty} a_n \int_0^l \sin \frac{\omega_n}{c_q} x \sin \frac{\omega_m}{c_q} x \, dx = 0 \tag{9-293}$$

Use of the orthogonality relation in Eq. (9-135) reduces the summation on the left-hand side of Eq. (9-293) to a single term for $n = m$:

$$\sum_{n=1}^{\infty} a_n \int_0^l \sin \frac{\omega_n}{c_q} x \sin \frac{\omega_m}{c_q} x \, dx = a_m \int_0^l \sin^2 \frac{\omega_m}{c_q} x \, dx \tag{9-294}$$

or, upon elementary integration,

$$a_m \int_0^l \sin^2 \frac{m\pi x}{l} \, dx = a_m \int_0^l \frac{1}{2}\left(1 - \cos \frac{2m\pi x}{l}\right) dx$$

$$= a_m \left(\frac{x}{2}\Big|_0^l - \frac{l}{2m\pi} \sin \frac{2m\pi x}{l}\Big|_0^l\right) = a_m \frac{l}{2} \tag{9-295}$$

where Eq. (9-286) has been used for ω_m. Therefore, Eq. (9-293) becomes

$$a_m \frac{l}{2} = 0 \tag{9-296}$$

Since Eq. (9-296) holds for all values of $m = 1, 2, 3, \ldots$, the coefficient a_n in Eq. (9-288) is zero for all values of $n = 1, 2, 3, \ldots$.

Now, in order to evaluate the other coefficient b_n from Eq. (9-292), multiply both sides of Eq. (9-292) by $X_m(x) = \sin \frac{\omega_m}{c_q} x$ and integrate the resulting expression with respect to x from 0 to l, which give

$$\sum_{n=1}^{\infty} b_n \frac{n\pi c_q}{l} \int_0^l \sin \frac{n\pi x}{l} \sin \frac{m\pi x}{l} \, dx = \int_0^l \Omega_0 \sin \frac{m\pi x}{l} \, dx \tag{9-297}$$

where Eq. (9-286) has been used for ω_n and ω_m. The right-hand side of Eq. (9-297) can be computed by elementary integration. That is,

$$\int_0^l \Omega_0 \sin \frac{m\pi x}{l} \, dx = -\frac{\Omega_0 l}{m\pi} \cos \frac{m\pi x}{l}\Big|_0^l = \frac{\Omega_0 l}{m\pi} \left\{ (-1)^{m+1} + 1 \right\} \tag{9-298}$$

Now, we compute the left-hand side of Eq. (9-297). Using Eq. (9-295) and the orthogonality relation in Eq. (9-135) reduces the left-hand side of Eq. (9-297) to

$$\sum_{n=1}^{\infty} b_n \frac{n\pi c_q}{l} \int_0^l \sin \frac{n\pi x}{l} \sin \frac{m\pi x}{l} \, dx = b_m \frac{m\pi c_q}{l} \int_0^l \sin^2 \frac{m\pi x}{l} \, dx$$

$$= b_m \frac{m\pi c_q}{l} \frac{l}{2}$$

$$= b_m \frac{m\pi c_q}{2} \tag{9-299}$$

Therefore, substituting Eqs. (9-298) and (9-299) into the right-hand side and left-hand side, respectively, of Eq. (9-297) gives

$$b_m \frac{m\pi c_q}{2} = \frac{\Omega_0 l}{m\pi} \left\{ (-1)^{m+1} + 1 \right\} \tag{9-300}$$

or, upon rearrangement,

$$b_m = \frac{2\Omega_0 l}{m^2 \pi^2 c_q} \left\{ (-1)^{m+1} + 1 \right\} \tag{9-301}$$

Since Eq. (9-301) holds for all values of $m = 1, 2, 3, \ldots$, we have found the unknown coefficient b_n in Eq. (9-288).

Therefore, substituting $a_n = 0$ (via Eq. (9-296)) and Eq. (9-301) into Eq. (9-288) and using Eq. (9-286) yield

$$\varphi(x,t) = \sum_{n=1}^{\infty} \frac{2\Omega_0 l}{n^2 \pi^2 c_q} \left\{ (-1)^{n+1} + 1 \right\} \sin \frac{n\pi c_q t}{l} \sin \frac{n\pi x}{l} \tag{9-302}$$

which is the solution for the response of the shaft subjected to the initial conditions given by Eqs. (9-289) and (9-290).

9-4.3   An Example: Sliding–Free Beam Initially Bent

We consider a uniform Bernoulli-Euler beam having density ρ, cross-sectional area A, modulus of elasticity E, second moment of the cross-sectional area about the neutral axis I, and equilibrium length l. The end conditions of the beam are, as sketched in Figure 9-16a, sliding-rotationally restrained and free. Suppose the left-hand end (the sliding-rotationally restrained end) is held fixed and a bending moment M_0 is applied to the right-hand end (the free end), resulting in the upward deflection and rotation of the beam tip, as sketched in Figure 9-16b. Now, suppose at $t = 0$ we suddenly release the left-hand end and simultaneously remove the moment M_0 from the right-hand end. We seek the ensuing motion of the beam for $t \geq 0$.

We denote the generalized coordinate as $\eta(x,t)$ to be defined as the transverse displacement of the section whose position is x along the beam's neutral axis, measured from its initial undeformed configuration, which is equivalent to the horizontal line passing through the left-hand end before the system is released.

As in Subsections 9-4.1 and 9-4.2, the general form of the free undamped solution is first established. Note that after $t = 0$, the end conditions of the system are sliding-rotationally restrained–free. Examining Table 9-10a for the beam subjected to sliding-rotationally restrained–free boundary conditions reveals that there exists a rigid-body mode whose eigenfunction is

$$X_0(x) = c_0 \tag{9-303}$$

and the eigenfunction for the deformable modes is given by

$$X_n(x) = c_n \cos \beta_n x + e_n \cosh \beta_n x \tag{9-304}$$

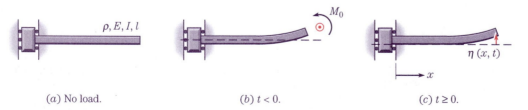

(a) No load.   (b) $t < 0$.   (c) $t \geq 0$.

FIGURE 9-16   Response of sliding-rotationally restrained–free beam initially bent due to moment M_0.

The coefficients c_n and e_n are related to each other by the expression

$$\frac{c_n}{e_n} = \frac{\cosh \beta_n l}{\cos \beta_n l} = \frac{-\sinh \beta_n l}{\sin \beta_n l} \tag{9-305}$$

where $\beta_n \equiv \sqrt{\dfrac{\omega_n}{c_q}}$; ω_n being the nth natural frequency; $c_q = \sqrt{\dfrac{EI}{\rho A}}$ from Table 9-4; and β_n satisfies the equation

$$\tan \beta_n l + \tanh \beta_n l = 0 \tag{9-306}$$

Again, Eqs. (9-303) through (9-306) are all retrieved from Table 9-10a.

As illustrated via Eqs. (9-221) through (9-224) in Subdivision 9-3.4.1, either of the two coefficients c_n and e_n in Eq. (9-304) can be eliminated by using Eq. (9-305), resulting in only one multiplicative constant in the eigenfunction. That is, for example, in order to eliminate c_n in Eq. (9-304), substitution of the first expression in Eq. (9-305) into Eq. (9-304) gives

$$\begin{aligned}
X_n(x) &= e_n \frac{\cosh \beta_n l}{\cos \beta_n l} \cos \beta_n x + e_n \cosh \beta_n x \\
&= \frac{e_n}{\cos \beta_n l} (\cosh \beta_n l \cos \beta_n x + \cos \beta_n l \cosh \beta_n x) \\
&= e_n'(\cosh \beta_n l \cos \beta_n x + \cos \beta_n l \cosh \beta_n x) \tag{9-307}
\end{aligned}$$

where $e_n' \equiv \dfrac{e_n}{\cos \beta_n l}$ is an unknown coefficient that may be used in lieu of the otherwise unknown e_n.

Substituting $p_0 = 1$ (due to the presence of a rigid-body mode, unlike the examples in Subsections 9-4.1 and 9-4.2), and Eqs. (9-303) and (9-307) into Eq. (9-260) yields

$$\eta(x, t) = (a_0 + b_0 t)c_0$$
$$+ \sum_{n=1}^{\infty} (a_n \cos \omega_n t + b_n \sin \omega_n t)e_n'(\cosh \beta_n l \cos \beta_n x + \cos \beta_n l \cosh \beta_n x) \tag{9-308}$$

where the time functions $T_0(t)$ and $T_n(t)$ defined by Eqs. (9-261) and (9-263) have been substituted into Eq. (9-260). Note in Eq. (9-308) that all the eigenfunctions have a surplus unknown multiplicative constant, namely c_0 or e_n'; thus the product of two unevaluated constants such as $a_0 c_0$, $b_0 c_0$, $a_n e_n'$, or $b_n e_n'$ can be designated as a single corresponding unevaluated constant such as a_0, b_0, a_n, or b_n, respectively. Therefore, Eq. (9-308) may be rewritten as

$$\eta(x, t) = (a_0 + b_0 t)$$
$$+ \sum_{n=1}^{\infty} (a_n \cos \omega_n t + b_n \sin \omega_n t)(\cosh \beta_n l \cos \beta_n x + \cos \beta_n l \cosh \beta_n x) \tag{9-309}$$

where β_n satisfies Eq. (9-306).

Next, as in Subsections 9-4.1 and 9-4.2, we establish the initial conditions for the system immediately after the system is released. While the left-hand end of the beam is held fixed, the beam behaves as a cantilever beam. From the technical mechanics of solids,[17] the initial displacement along the beam is given by

---

[17] See, for example, S. H. Crandall, N. C. Dahl, and T. J. Lardner, *An Introduction to the Mechanics of Solids*, McGraw-Hill, New York, 1978, 2nd ed., Table 8-1 on page 531.

$$\eta(x, 0) = \frac{M_0 x^2}{2EI} \tag{9-310}$$

Further, the initial velocity of the beam is zero throughout its length; so

$$\dot{\eta}(x, 0) = 0 \tag{9-311}$$

As indicated earlier, these initial conditions can be used to find the unknown coefficients a_0, b_0, a_n, and b_n in Eq. (9-309). Substituting Eq. (9-309) for $t = 0$ into Eq. (9-310) and substituting the time derivative of Eq. (9-309) for $t = 0$ into Eq. (9-311) give, respectively,

$$a_0 + \sum_{n=1}^{\infty} a_n(\cosh \beta_n l \cos \beta_n x + \cos \beta_n l \cosh \beta_n x) = \frac{M_0 x^2}{2EI} \tag{9-312}$$

$$b_0 + \sum_{n=1}^{\infty} b_n \omega_n(\cosh \beta_n l \cos \beta_n x + \cos \beta_n l \cosh \beta_n x) = 0 \tag{9-313}$$

Note that Eq. (9-312) contains exclusively the set of coefficients a_0 and a_n whereas Eq. (9-313) contains exclusively the set of coefficients b_0 and b_n. Therefore, the coefficients a_0 and a_n can be evaluated from Eq. (9-312) and the coefficients b_0 and b_n can be evaluated from Eq. (9-313).

Since the beam in Figure 9-16c possesses a rigid-body mode and its eigenfunction $X_n(x)$ has two terms, unlike the rod system and the shaft system in Subsections 9-4.1 and 9-4.2, respectively, each of which has a one-term eigenfunction, the calculation required in the evaluation of the coefficients a_0, a_n, b_0, and b_n in Eqs. (9-312) and (9-313) is relatively lengthy. Nevertheless, the procedure is precisely analogous to that used in both Subsections 9-4.1 and 9-4.2. That is, in order to evaluate a_0 from Eq. (9-312), Eq. (9-312) is integrated with respect to x from 0 to l. And, in order to evaluate a_n from Eq. (9-312), Eq. (9-312) is multiplied by the mth eigenfunction $X_m(x)$ and then is integrated with respect to x from 0 to l. Similarly, b_0 and b_n are evaluated from Eq. (9-313).

First, in order to evaluate a_0, integrate both sides of Eq. (9-312) with respect to x from 0 to l, which yields

$$a_0 l + \sum_{n=1}^{\infty} \frac{a_n}{\beta_n} (\cosh \beta_n l \sin \beta_n l + \cos \beta_n l \sinh \beta_n l) = \frac{M_0 l^3}{6EI} \tag{9-314}$$

where the following elementary integrals have been used:

$$\int_0^l \cos \beta_n x \, dx = \frac{1}{\beta_n} \sin \beta_n x \Big|_0^l = \frac{1}{\beta_n} \sin \beta_n l \tag{9-315}$$

$$\int_0^l \cosh \beta_n x \, dx = \frac{1}{\beta_n} \sinh \beta_n x \Big|_0^l = \frac{1}{\beta_n} \sinh \beta_n l \tag{9-316}$$

$$\int_0^l x^2 \, dx = \frac{1}{3} x^3 \Big|_0^l = \frac{l^3}{3} \tag{9-317}$$

We recall that β_n must satisfy Eq. (9-306) for all values of $n = 1, 2, 3, \ldots$. By comparing Eq. (9-306) with the quantity in parentheses on the left-hand side of Eq. (9-314), we note that multiplying Eq. (9-306) by $\cos \beta_n l \cosh \beta_n l$ gives

$$\cos \beta_n l \sinh \beta_n l + \sin \beta_n l \cosh \beta_n l = 0 \qquad (9\text{-}318)$$

which is identical to the quantity in parentheses on the left-hand side of Eq. (9-314). Thus, Eq. (9-318) is an alternative form of Eq. (9-306), which will be used repeatedly in subsequent calculations. In particular, the summation term on the left-hand side of Eq. (9-314) vanishes, giving

$$a_0 = \frac{M_0 l^2}{6EI} \qquad (9\text{-}319)$$

Note here that the direct integration of Eq. (9-312) with respect to x is equivalent to multiplying both sides of Eq. (9-312) by the zeroth eigenfunction and integrating with respect to x because the zeroth eigenfunction is a constant, $X_0(x) = c_0$, as noted in Eq. (9-303). If this alternative procedure is performed, Eq. (9-312) becomes

$$\int_0^l a_0 c_0 \, dx + \sum_{n=0}^{\infty} \int_0^l c_0 a_n (\cosh \beta_n l \cos \beta_n x + \cos \beta_n l \cosh \beta_n x) \, dx = \int_0^l c_0 \frac{M_0 x^2}{2EI} \, dx \qquad (9\text{-}320)$$

where the commutativity of summation and integration has been used for the second term on the left-hand side of Eq. (9-320). The summation term in Eq. (9-320) can be evaluated without executing any integration, by recalling the orthogonality relation in Eq. (9-225) between two different modes:

$$\int_0^l X_n X_0 = 0 \qquad n \neq 0 \qquad (9\text{-}321)$$

which is a special form of Eq. (9-225) for $m = 0$. Therefore, the summation term in Eq. (9-320) vanishes due to Eq. (9-321) and the direct integration of the other terms yields

$$a_0 c_0 l = c_0 \frac{M_0 l^3}{6EI} \qquad (9\text{-}322)$$

which gives the same solution for a_0 as Eq. (9-319).

Now, to find the coefficients a_n for $n \geq 1$, multiply both sides of Eq. (9-312) by the mth eigenfunction $X_m = \cosh \beta_m l \cos \beta_m x + \cos \beta_m l \cosh \beta_m x$ and integrate the resulting equation with respect to x from 0 to l. Then, Eq. (9-312) becomes

$$\int_0^l a_0 (\cosh \beta_m l \cos \beta_m x + \cos \beta_m l \cosh \beta_m x) \, dx$$

$$+ \sum_{n=1}^{\infty} a_n \int_0^l (\cosh \beta_n l \cos \beta_n x + \cos \beta_n l \cosh \beta_n x) \times$$

$$(\cosh \beta_m l \cos \beta_m x + \cos \beta_m l \cosh \beta_m x) \, dx$$

$$= \int_0^l \frac{M_0 x^2}{2EI} (\cosh \beta_m l \cos \beta_m x + \cos \beta_m l \cosh \beta_m x) \, dx \qquad (9\text{-}323)$$

where the commutativity of summation and integration has been used for the summation term in Eq. (9-323). The first term of Eq. (9-323) is zero for the reason discussed following Eq. (9-317) or due to Eq. (9-321). The direct integration of the terms in the summation term of Eq. (9-323) is straightforward but laborious. Much effort can be saved by recalling the orthogonality relation, Eq. (9-225), between two different modes:

$$\int_0^l X_n X_m \, dx = 0 \qquad n \neq m \qquad (9\text{-}324)$$

Consequently, all terms in the summation of Eq. (9-323) vanish except the term for which $n = m$, which must be integrated directly. Therefore, the left-hand side of Eq. (9-323) reduces to

$$a_m \int_0^l (\cosh \beta_m l \cos \beta_m x + \cos \beta_m l \cosh \beta_m x)^2 \, dx$$

$$= a_m \int_0^l (\cosh^2 \beta_m l \cos^2 \beta_m x + 2 \cosh \beta_m l \cos \beta_m l \cos \beta_m x \cosh \beta_m x$$

$$+ \cos^2 \beta_m l \cosh^2 \beta_m x) \, dx \qquad (9\text{-}325)$$

The terms in Eq. (9-325) may now be integrated one by one. The first term can be integrated, without rewriting the constant $a_m \cosh^2 \beta_m l$, as

$$\int_0^l \cos^2 \beta_m x \, dx = \frac{1}{2} \int_0^l (1 + \cos 2\beta_m x) \, dx = \frac{1}{2} \left(l + \frac{1}{2\beta_m} \sin 2\beta_m l \right) \qquad (9\text{-}326)$$

where the identity $\cos^2 \theta = \dfrac{(1 + \cos 2\theta)}{2}$ has been used in the first equality. The third term in Eq. (9-325) can be integrated, without rewriting the constant $a_m \cos^2 \beta_m l$, as

$$\int_0^l \cosh^2 \beta_m x \, dx = \frac{1}{2} \int_0^l (1 + \cosh 2\beta_m x) \, dx = \frac{1}{2} \left(l + \frac{1}{2\beta_m} \sinh 2\beta_m l \right) \qquad (9\text{-}327)$$

where the identity $\cosh^2 \theta = \dfrac{(1 + \cosh 2\theta)}{2}$ has been used in the first equality.

The integration of the second term in Eq. (9-325) is slightly more complicated: Integration by parts will be applied twice. Without rewriting the constant $2a_m \cosh \beta_m l \cos \beta_m l$ and temporarily denoting the integral by I_0, use of Eq. (9-13) twice sequentially gives[18]

$$I_0 \equiv \int_0^l \cos \beta_m x \cosh \beta_m x \, dx$$

$$= \frac{1}{\beta_m} \cos \beta_m l \sinh \beta_m l + \int_0^l \sin \beta_m x \sinh \beta_m x \, dx$$

$$= \frac{1}{\beta_m} \cos \beta_m l \sinh \beta_m l + \frac{1}{\beta_m} \sin \beta_m l \cosh \beta_m l - \int_0^l \cos \beta_m x \cosh \beta_m x \, dx$$

$$= \frac{1}{\beta_m} \cos \beta_m l \sinh \beta_m l + \frac{1}{\beta_m} \sin \beta_m l \cosh \beta_m l - I_0 \qquad (9\text{-}328)$$

---

[18]In the first integration by parts, the terms in Eq. (9-13) may be identified by

$$u = \cos \beta_m x \qquad dv = \cosh \beta_m x \, dx \qquad du = -\beta_m \sin \beta_m x \, dx \qquad v = \frac{1}{\beta_m} \sinh \beta_m x$$

and in the second integration by parts, the terms in Eq. (9-13) may be identified by

$$u = \sin \beta_m x \qquad dv = \sinh \beta_m x \, dx \qquad du = \beta_m \cos \beta_m x \, dx \qquad v = \frac{1}{\beta_m} \cosh \beta_m x.$$

where, as indicated, the two consecutive integrations by parts reproduced the integral I_0 being computed. Solving Eq. (9-328) for I_0 gives

$$I_0 \equiv \int_0^l \cos \beta_m x \cosh \beta_m x \, dx = \frac{\cos \beta_m l \sinh \beta_m l + \sin \beta_m l \cosh \beta_m l}{2\beta_m} = 0 \quad (9\text{-}329)$$

where Eq. (9-318) has been used to establish the final result. Substituting Eqs. (9-326), (9-327), and (9-329) into Eq. (9-325), multiplied by the appropriate constants, which were set aside prior to integration, gives

$$a_m \int_0^l (\cosh \beta_m l \cos \beta_m x + \cos \beta_m l \cosh \beta_m x)^2 \, dx$$

$$= \frac{a_m}{2} \left\{ \cosh^2 \beta_m l \left(l + \frac{1}{2\beta_m} \sin 2\beta_m l \right) + \cos^2 \beta_m l \left(l + \frac{1}{2\beta_m} \sinh 2\beta_m l \right) \right\}$$

$$= \frac{a_m}{2} \left\{ l(\cos^2 \beta_m l + \cosh^2 \beta_m l) \right.$$

$$\left. + \frac{1}{2\beta_m} \left(\cosh^2 \beta_m l \sin 2\beta_m l + \cos^2 \beta_m l \sinh 2\beta_m l \right) \right\}$$

$$= \frac{a_m}{2} \left\{ l(\cos^2 \beta_m l + \cosh^2 \beta_m l) \right.$$

$$\left. + \frac{1}{\beta_m} \left(\cosh^2 \beta_m l \cos \beta_m l \sin \beta_m l + \cos^2 \beta_m l \cosh \beta_m l \sinh \beta_m l \right) \right\}$$

$$= \frac{a_m}{2} \left\{ l(\cos^2 \beta_m l + \cosh^2 \beta_m l) \right.$$

$$\left. + \frac{1}{\beta_m} \cos \beta_m l \cosh \beta_m l [\cosh \beta_m l \sin \beta_m l + \cos \beta_m l \sinh \beta_m l] \right\} \quad (9\text{-}330)$$

where the identities, $\sin 2\theta = 2 \sin \theta \cos \theta$ and $\sinh 2\theta = 2 \sinh \theta \cosh \theta$, have been used. Note that the quantity in square brackets of Eq. (9-330) is zero because of Eq. (9-318). Therefore, Eq. (9-330) reduces to

$$a_m \int_0^l (\cosh \beta_m l \cos \beta_m x + \cos \beta_m l \cosh \beta_m x)^2 \, dx = \frac{a_m l}{2} (\cos^2 \beta_m l + \cosh^2 \beta_m l)$$

$$(9\text{-}331)$$

which is the result of the computation for the summation term on the left-hand side of Eq. (9-323).

Now, we consider the right-hand side of Eq. (9-323). Again, we use integration by parts twice as given by Eq. (9-13). The first integration by parts gives[19]

---

[19]In the integration by parts for the first term, the terms in Eq. (9-13) may be identified by

$$u = x^2 \qquad dv = \cos \beta_m x \, dx \qquad du = 2x \, dx \qquad v = \frac{1}{\beta_m} \sin \beta_m x$$

and in the integration by parts for the second term, the terms in Eq. (9-13) may be identified by

$$u = x^2 \qquad dv = \cosh \beta_m x \, dx \qquad du = 2x \, dx \qquad v = \frac{1}{\beta_m} \sinh \beta_m x.$$

$$\int_0^l \frac{M_0 x^2}{2EI} (\cosh \beta_m l \cos \beta_m x + \cos \beta_m l \cosh \beta_m x) \, dx$$

$$= \frac{M_0}{2EI} \cosh \beta_m l \int_0^l x^2 \cos \beta_m x \, dx + \frac{M_0}{2EI} \cos \beta_m l \int_0^l x^2 \cosh \beta_m x \, dx$$

$$= \frac{M_0}{2EI} \left\{ \underline{\frac{\cosh \beta_m l}{\beta_m}} \left(l^2 \sin \beta_m l - \int_0^l 2x \sin \beta_m x \, dx \right) \right.$$

$$\left. + \underline{\frac{\cos \beta_m l}{\beta_m}} \left(l^2 \sinh \beta_m l - \int_0^l 2x \sinh \beta_m x \, dx \right) \right\} \quad \text{(9-332)}$$

where the terms, except the integrals, sum to zero due to Eq. (9-318) (that is, the underlined terms). Computing the integrals of Eq. (9-332) via integration by parts yields[20]

$$\int_0^l \frac{M_0 x^2}{2EI} (\cosh \beta_m l \cos \beta_m x + \cos \beta_m l \cosh \beta_m x) \, dx$$

$$= -\frac{M_0}{EI\beta_m^2} \left\{ \cosh \beta_m l \left(-l \cos \beta_m l + \int_0^l \cos \beta_m x \, dx \right) \right.$$

$$\left. + \cos \beta_m l \left(l \cosh \beta_m l - \int_0^l \cosh \beta_m x \, dx \right) \right\}$$

$$= -\frac{M_0}{EI\beta_m^2} \left\{ \cosh \beta_m l \int_0^l \cos \beta_m x \, dx - \cos \beta_m l \int_0^l \cosh \beta_m x \, dx \right\}$$

$$= -\frac{M_0}{EI\beta_m^3} (\cosh \beta_m l \sin \beta_m l - \cos \beta_m l \sinh \beta_m l) \quad \text{(9-333)}$$

which is the result of the computation for the right-hand side of Eq. (9-323). Therefore, substituting Eqs. (9-331) and (9-333) into Eq. (9-323) and solving for a_m give

$$a_m = -\frac{2M_0}{EIl\beta_m^3} \left(\frac{\cosh \beta_m l \sin \beta_m l - \cos \beta_m l \sinh \beta_m l}{\cos^2 \beta_m l + \cosh^2 \beta_m l} \right) \quad \text{(9-334)}$$

Therefore, in summary, the coefficients a_0 and a_n have been found from Eq. (9-312) via the multiplication of Eq. (9-312) by orthogonal eigenfunctions and subsequent spacewise integration. These results for a_0 and a_n are given in Eqs. (9-319) and (9-334), respectively, since the subscript m in Eq. (9-334) is arbitrary and may be set to n.

Now, the set of unknown coefficients b_0 and b_n may be evaluated from Eq. (9-313). Integrating both sides of Eq. (9-313) (or, equivalently, multiplying both sides by $X_0 = c_0$ and integrating) with respect to x from 0 to l gives $b_0 l = 0$ (or, $c_0 b_0 l = 0$ from Eq. (9-308)) because the integral of the summation term in Eq. (9-313) vanishes due to the discussion immediately following Eq. (9-317) (or, due to Eq. (9-321)). Therefore, $b_0 = 0$. In order to

---

[20]In the integration by parts for the first term, the terms in Eq. (9-13) may be identified by

$$u = 2x \qquad dv = \sin \beta_m x \, dx \qquad du = 2 \, dx \qquad v = -\frac{1}{\beta_m} \cos \beta_m x$$

and in the integration by parts for the second term, the terms in Eq. (9-13) may be identified by

$$u = 2x \qquad dv = \sinh \beta_m x \, dx \qquad du = 2 \, dx \qquad v = \frac{1}{\beta_m} \cosh \beta_m x.$$

find the coefficients b_n for $n \geq 1$, multiply both sides of Eq. (9-313) by the mth eigenfunction X_m and integrate the resulting equation with respect to x from 0 to l. Then, Eq. (9-313) becomes

$$\sum_{n=1}^{\infty} b_n \omega_n \int_0^l X_n X_m dx = 0 \qquad (9\text{-}335)$$

where $b_0 = 0$ has been used. Due to the orthogonality relation in Eq. (9-324), the summation term in Eq. (9-335) reduces to the single term for which $n = m$. That is, Eq. (9-335) reduces to

$$b_m \omega_m \int_0^l X_m^2 dx = 0 \qquad (9\text{-}336)$$

Since neither ω_m nor the integral in Eq. (9-336) is zero, $b_m = 0$ for $m = 1, 2, 3, \ldots$.

Therefore, substitution of $b_0 = b_n = 0$ and Eqs. (9-319) and (9-334) into Eq. (9-309) gives the solution to the problem as

$$\eta(x, t) = \frac{M_0 l^2}{6EI} + \sum_{n=1}^{\infty} a_n \cos \omega_n t (\cosh \beta_n l \cos \beta_n x + \cos \beta_n l \cosh \beta_n x) \qquad (9\text{-}337)$$

where a_n is given by Eq. (9-334) as

$$a_n = -\frac{2M_0}{EIl\beta_n^3} \left(\frac{\cosh \beta_n l \sin \beta_n l - \cos \beta_n l \sinh \beta_n l}{\cos^2 \beta_n l + \cosh^2 \beta_n l} \right) \qquad (9\text{-}338)$$

Due to the lengthy derivation of the solution given by Eqs. (9-337) and (9-338), it may be useful to check the validity of Eq. (9-337). One easy way to do so is to verify the conservation of linear momentum of the beam in the transverse direction because there is no external

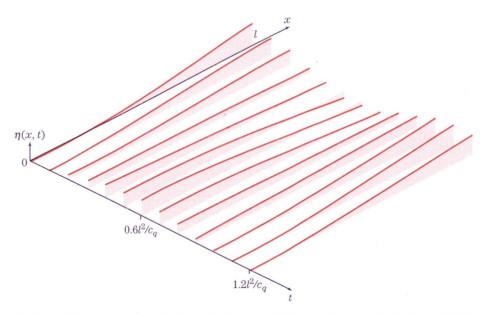

FIGURE 9-17 Three-dimensional representation of transverse displacement $\eta(x, t)$ of beam subjected to initial bending.

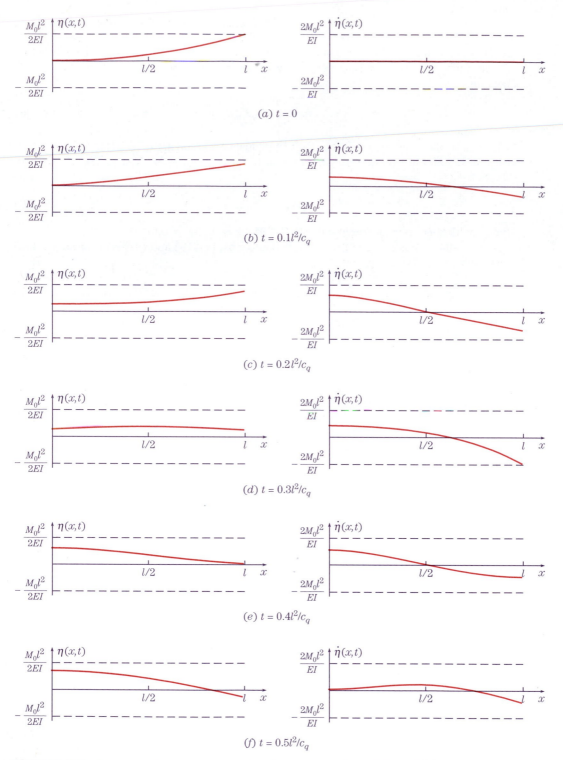

FIGURE 9-18a Transverse displacement $\eta(x,t)$ and transverse velocity $\dot{\eta}(x,t)$ throughout beam subjected to initial bending.

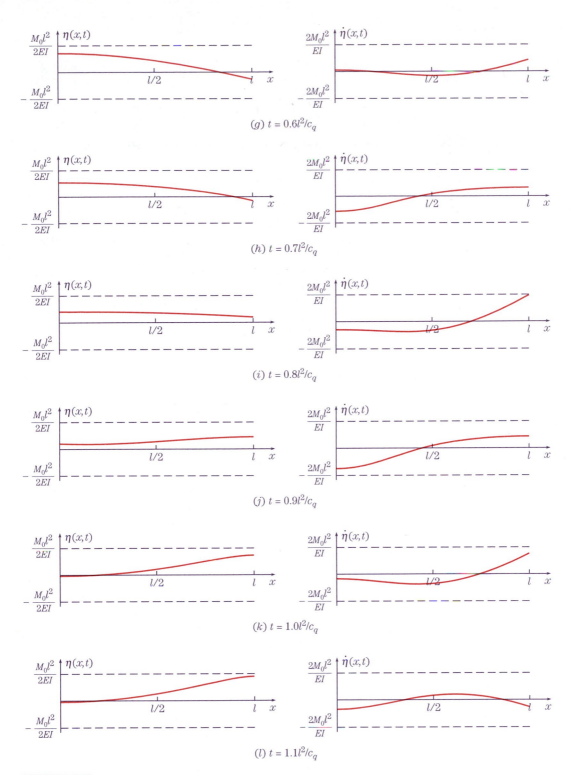

(g) $t = 0.6l^2/c_q$

(h) $t = 0.7l^2/c_q$

(i) $t = 0.8l^2/c_q$

(j) $t = 0.9l^2/c_q$

(k) $t = 1.0l^2/c_q$

(l) $t = 1.1l^2/c_q$

FIGURE 9-18b

transverse force exerted on the (conservative) system. That is, the solution given by Eqs. (9-337) and (9-338) must satisfy, for all positive time,[21]

$$\int_0^l \rho A \frac{\partial \eta}{\partial t} \, dx = 0 \qquad (9\text{-}339)$$

where $\frac{\partial \eta(x, t)}{\partial t}$ is the transverse velocity of the differential element of the beam at location x, having mass $\rho A \, dx$.

Figure 9-17 is a three-dimensional representation of the transverse displacement $\eta(x, t)$ along the beam's length l given by Eq. (9-337), for various times t. Figure 9-18 shows the same displacement plots in x-η space at the indicated times, accompanied by the corresponding velocity[22] plots in the right-hand column. As in Figure 9-10, the velocity plots are presented along with the displacement plots primarily to enhance the reader's understanding of the displacement evolution. Careful examination of Figures 9-17 and 9-18 will reveal that changes in the spatial configuration of the system are not periodic, at least for the time period shown, in contrast with the periodic changes in the spatial configuration of the rod shown in Figures 9-9 and 9-10. This is due to the fact that the propagation speed of disturbances in Bernoulli-Euler beams depends on the frequency of the oscillation—a characteristic called *dispersion*, a topic beyond the scope of our interest. Therefore, the evolution of displacement in Figures 9-17 and 9-18 cannot be explained in elementary terms of propagation and reflection of a single disturbance as was the case in Figures 9-9 and 9-10. Although the velocity plots are presented primarily to enhance the understanding of how the displacements evolve, it is interesting to note that in each velocity plot, the area above the x-axis must be equal to the area below the x-axis. This is in accordance with the conservation of linear momentum, as expressed in Eq. (9-339).

9-5 RESPONSE TO HARMONIC EXCITATIONS

In Section 9-4, we considered the motion of several one-dimensional continuous systems due to nonzero initial conditions. Motion in continuous systems can also be generated by external excitations that persist during some or all of the subsequent motion of the system. The most common excitations include *specified* or *prescribed motion* of the boundaries, applied forces distributed throughout the extent of the systems, and applied localized forces at the boundaries. We shall consider the forced response of several one-dimensional continua due to each of these types of external excitations.

Our interest in this section will be restricted to the cases where the time dependence of the excitations is harmonic. Furthermore, we also assume that the systems have no damping. In defining each example, the problem statement will represent an idealization that will allow us to formulate a specific boundary value problem.

9-5.1 An Example: Specified Harmonic Motion of Boundary

To demonstrate the method for finding the forced response of continuous systems due to *harmonic motion* of a boundary, we consider the system sketched in Figure 9-19a. The system contains a uniform shaft having density ρ, polar moment of the cross-sectional area

[21]See Problem 9-37.
[22]See Problem 9-38.

about the axis of the shaft J, shear modulus G, and equilibrium length l. The left-hand end of the shaft is clamped, and the right-hand end of the shaft is attached to an electric motor, which imposes a time-harmonic *specified motion* to the right-hand end. The generalized coordinate for the shaft is $\varphi(x, t)$, which is the angular displacement of the section whose position is x, where x is referenced with respect to the clamped end. Observing that the specified harmonic motion of the motor is $\varphi_0 \sin \Omega t$, the boundary condition of the shaft at $x = l$ may be stated as

$$\varphi(l, t) = \varphi_0 \sin \Omega t \tag{9-340}$$

where φ_0 is the amplitude of the specified or prescribed harmonic motion of the motor. Because the left-hand end of the shaft is fixed, the boundary condition at $x = 0$ is given by

$$\varphi(0, t) = 0 \tag{9-341}$$

Because the excitation is harmonic, the forced response will also be harmonic and at the same frequency Ω, as shown in Subsections 8-2.4, 8-3.5, and 8-4.3 in Chapter 8. This is a property of linear systems, not nonlinear systems. Therefore, we *assume* a solution of the form

$$\varphi(x, t) = \Phi(x) \sin \Omega t \tag{9-342}$$

where $\Phi(x)$ is the unknown spatial distribution of the generalized coordinate $\varphi(x, t)$. Note that the equation of motion for this forced shaft is given by the generic equation of motion, Eq. (9-81), with $c_q = \sqrt{\dfrac{G}{\rho}}$, $z(x, t) = \varphi(x, t)$, and $g(x, t) = 0$.

In order to find the functional form of $\Phi(x)$, Eq. (9-342) is substituted into Eq. (9-81). Then

$$-\Omega^2 \Phi(x) \sin \Omega t = c_q^2 \frac{d^2\Phi}{dx^2} \sin \Omega t \tag{9-343}$$

which, after cancellation of the $\sin \Omega t$ term, yields

$$\frac{d^2\Phi}{dx^2} + \frac{\Omega^2}{c_q^2}\Phi = 0 \tag{9-344}$$

Equation (9-344) has the same form as Eq. (9-106) with $X = \Phi$ and $\omega = \Omega$. Thus, the solution to Eq. (9-344) is given by an equation analogous to Eq. (9-109) such that

$$\Phi(x) = C_1 \cos \frac{\Omega}{c_q}x + C_2 \sin \frac{\Omega}{c_q}x \tag{9-345}$$

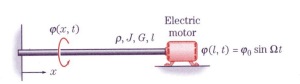

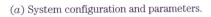

(a) System configuration and parameters. (b) Cross-section of shaft.

FIGURE 9-19 Shaft system subjected to specified harmonic motion at $x = l$.

where C_1 and C_2 are unknown coefficients. Combining Eqs. (9-342) and (9-345) gives

$$\varphi(x, t) = \left(C_1 \cos \frac{\Omega}{c_q} x + C_2 \sin \frac{\Omega}{c_q} x \right) \sin \Omega t \qquad (9\text{-}346)$$

The unknown coefficients C_1 and C_2 are evaluated in accordance with the boundary conditions; there are two coefficients and two boundary conditions. Substituting Eq. (9-346) into Eqs. (9-340) and (9-341) gives, respectively,

$$\left(C_1 \cos \frac{\Omega}{c_q} l + C_2 \sin \frac{\Omega}{c_q} l \right) \sin \Omega t = \varphi_0 \sin \Omega t \qquad (9\text{-}347)$$

and

$$C_1 \sin \Omega t = 0 \qquad (9\text{-}348)$$

Because Eqs. (9-347) and (9-348) must hold for all time t, both equations can be divided by $\sin \Omega t$. Then, Eq. (9-348) requires

$$C_1 = 0 \qquad (9\text{-}349)$$

and Eq. (9-347) requires

$$C_2 = \frac{\varphi_0}{\sin \dfrac{\Omega l}{c_q}} \qquad (9\text{-}350)$$

where Eq. (9-349) has been used in obtaining C_2 from Eq. (9-347). Substituting Eqs. (9-349) and (9-350) into Eq. (9-346) gives

$$\varphi(x, t) = \frac{\varphi_0}{\sin \dfrac{\Omega l}{c_q}} \sin \frac{\Omega x}{c_q} \sin \Omega t \qquad (9\text{-}351)$$

which is the forced response of the shaft subjected to the specified harmonic motion at its right-hand end.

In addition to the solution in Eq. (9-351), the stresses in the shaft due to the motion may be of considerable interest in the design of the shaft. The shear stress $\tau_{x\theta}$ (see Figure 9-19b for polar coordinates (r, θ)) due to the rotational motion $\varphi(x, t)$ given by Eq. (9-351) can be obtained via the technical mechanics of solids (see Appendix K) as

$$\tau_{x\theta}(r, x, t) = Gr \frac{\partial \varphi}{\partial x} = Gr \frac{\varphi_0}{\sin \dfrac{\Omega l}{c_q}} \frac{\Omega}{c_q} \cos \frac{\Omega x}{c_q} \sin \Omega t$$

$$= \frac{G \varphi_0 \Omega}{c_q \sin \dfrac{\Omega l}{c_q}} r \cos \frac{\Omega x}{c_q} \sin \Omega t \qquad (9\text{-}352)$$

where r is the radial distance from the axis of the shaft.

9-5.2 An Example: Distributed Harmonic Force

To demonstrate the method for finding the forced response of continuous systems due to *harmonic force distributions* throughout their extent, we consider the system sketched in Figure 9-20. The system contains a uniform rod having density ρ, cross-sectional area A, equilibrium length l, and modulus of elasticity E. Suppose both ends of the rod are fixed within rigid walls, and there is a longitudinal force distribution $f(x, t)$ along the rod.

The equation of motion for this system can be found from Eq. (9-26) or Eq. (9-81) as

$$\rho A \frac{\partial^2 \xi}{\partial t^2} = EA \frac{\partial^2 \xi}{\partial x^2} + f(x, t) \tag{9-353}$$

where the generalized coordinate $\xi(x, t)$ is the longitudinal displacement of the section whose equilibrium position is x, where x is referenced with respect to the left-hand wall. In general, the solution to Eq. (9-353) can be found in closed form only when $f(x, t)$ is separable into spatial and temporal functions *and* the temporal function is harmonic. Therefore, we suppose the harmonic force distribution function is of the form

$$f(x, t) = f_0 \cos \Omega t \tag{9-354}$$

where f_0 is the constant amplitude of the force. Substituting Eq. (9-354) into Eq. (9-353) gives

$$\rho A \frac{\partial^2 \xi}{\partial t^2} = EA \frac{\partial^2 \xi}{\partial x^2} + f_0 \cos \Omega t \tag{9-355}$$

Because the excitation is harmonic, the forced response will also be harmonic and at the same frequency Ω, as shown in Subsections 8-2.4, 8-3.5, and 8-4.3 in Chapter 8. This is a property of linear systems, not nonlinear systems. Therefore, we *assume* a solution of the form

$$\xi(x, t) = X(x) \cos \Omega t \tag{9-356}$$

where $X(x)$ is the unknown spatial distribution of the generalized coordinate $\xi(x, t)$. In order to find the functional form of $X(x)$, we substitute Eq. (9-356) into Eq. (9-355). Then,

$$\left(\rho A \Omega^2 X(x) + EA \frac{d^2 X}{dx^2} \right) \cos \Omega t = -f_0 \cos \Omega t \tag{9-357}$$

Because Eq. (9-357) holds for all time t, both sides of Eq. (9-357) may be divided by $\cos \Omega t$, yielding

$$\frac{d^2 X}{dx^2} + \frac{\Omega^2}{c_q^2} X = -\frac{f_0}{EA} \tag{9-358}$$

where $c_q = \sqrt{\dfrac{EA}{\rho A}} = \sqrt{\dfrac{E}{\rho}}$. Note that Eq. (9-358) has the same form as Eq. (9-106) with $\omega = \Omega$ except that Eq. (9-358) has a nonzero right-hand side.

The elementary theory of ordinary differential equations gives the solution to Eq. (9-358) as the sum of a homogeneous solution and a particular solution. Due to the near

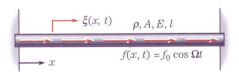

$\xi(x, t)$ ρ, A, E, l

$f(x, t) = f_0 \cos \Omega t$

x

FIGURE 9-20   Rod system subjected to distributed harmonic force.

identity of the left-hand sides of Eqs. (9-106) and (9-358), the homogeneous solution to Eq. (9-358) is essentially the same as Eq. (9-109), or

$$X_h(x) = C_1 \cos \frac{\Omega}{c_q} x + C_2 \sin \frac{\Omega}{c_q} x \tag{9-359}$$

where C_1 and C_2 are unknown coefficients and the subscript h on $X_h(x)$ denotes the "homogeneous" solution. The particular solution due to the nonzero and constant right-hand side of Eq. (9-358) is a constant. That is, the particular solution is

$$X_p(x) = C_3 \tag{9-360}$$

where C_3 is another unknown coefficient and the subscript p on $X_p(x)$ denotes the "particular" solution. Substituting Eq. (9-360) into Eq. (9-358) yields

$$\frac{\Omega^2}{c_q^2} C_3 = -\frac{f_0}{EA} \tag{9-361}$$

or,

$$X_p(x) = C_3 = -\frac{f_0 c_q^2}{\Omega^2 EA} = -\frac{f_0}{\rho A \Omega^2} \tag{9-362}$$

where $c_q^2 = \dfrac{(EA)}{(\rho A)}$ has been used in Eq. (9-362). Summation of Eqs. (9-359) and (9-362) gives the solution to Eq. (9-358) as

$$X(x) = C_1 \cos \frac{\Omega}{c_q} x + C_2 \sin \frac{\Omega}{c_q} x - \frac{f_0}{\rho A \Omega^2} \tag{9-363}$$

Also, substitution of Eq. (9-363) into Eq. (9-356) gives the solution to Eq. (9-355) as

$$\xi(x, t) = \left(C_1 \cos \frac{\Omega}{c_q} x + C_2 \sin \frac{\Omega}{c_q} x - \frac{f_0}{\rho A \Omega^2} \right) \cos \Omega t. \tag{9-364}$$

In order to find C_1 and C_2 in Eq. (9-364), we must establish the boundary conditions and use them in Eq. (9-364). Since both ends of the rod are clamped, the boundary conditions are given by

$$\xi(0, t) = 0 \tag{9-365}$$

$$\xi(l, t) = 0 \tag{9-366}$$

There are two unknown coefficients and two boundary conditions. Substituting Eq. (9-364) into Eqs. (9-365) and (9-366) gives, respectively,

$$\left(C_1 - \frac{f_0}{\rho A \Omega^2} \right) \cos \Omega t = 0 \tag{9-367}$$

and

$$\left(C_1 \cos \frac{\Omega}{c_q} l + C_2 \sin \frac{\Omega}{c_q} l - \frac{f_0}{\rho A \Omega^2} \right) \cos \Omega t = 0 \tag{9-368}$$

Because Eqs. (9-367) and (9-368) must hold for all time t, both equations can be divided by $\cos \Omega t$. Then, Eq. (9-367) requires

$$C_1 = \frac{f_0}{\rho A \Omega^2} \tag{9-369}$$

and Eq. (9-368) requires

$$C_2 = \frac{\dfrac{f_0}{\rho A \Omega^2} - C_1 \cos \dfrac{\Omega}{c_q} l}{\sin \dfrac{\Omega}{c_q} l}$$

$$= \frac{f_0}{\rho A \Omega^2} \frac{1 - \cos \dfrac{\Omega}{c_q} l}{\sin \dfrac{\Omega}{c_q} l} \tag{9-370}$$

where Eq. (9-369) has been used in evaluating C_2. Substituting Eqs. (9-369) and (9-370) into Eq. (9-364) gives

$$\xi(x, t) = \frac{f_0}{\rho A \Omega^2} \left\{ \cos \frac{\Omega}{c_q} x + \frac{1 - \cos \dfrac{\Omega}{c_q} l}{\sin \dfrac{\Omega}{c_q} l} \sin \frac{\Omega}{c_q} x - 1 \right\} \cos \Omega t \tag{9-371}$$

which is the forced response of the rod subjected to the constant-amplitude harmonic force distribution given in Eq. (9-354).

Figure 9-21 is a three-dimensional representation of the longitudinal displacement $\xi(x, t)$ along the rod's length l given by Eq. (9-371), for various times t. And Figure 9-22 shows the same displacement plots in x-ξ space at the indicated times, accompanied by the corresponding velocity[23] plots in the right-hand column. Again, as in Figures 9-10 and 9-18, the velocity plots are presented along with the displacement plots primarily to enhance the reader's understanding of the displacement evolution. It may be noted from Figures 9-21 and 9-22 that the change in the spatial configuration of the rod is periodic, repeating itself with the time period of $\frac{2\pi}{\Omega}$, Ω being the forcing frequency. It should be emphasized here that the absolute values of time t indicated in both Figures 9-21 and 9-22 are not meaningful because the solution in Eq. (9-371) represents the forced response found without consideration of initial conditions. There is no beginning or ending to either the loading or the response. Therefore, the time $t = 0$ in Figures 9-21 and 9-22 can be thought of as an arbitrary time, say $t = t_0$, and all the other times can be thought of as the time relative to t_0; that is, for example, $t = \frac{\pi}{\Omega}$ can be thought of as $t = t_0 + \frac{\pi}{\Omega}$.

9-5.3 An Example: Harmonic Force on Boundary

To demonstrate the method for finding the forced response of continuous systems due to *harmonic forces* on a boundary, we consider the system sketched in Figure 9-23. The system contains a uniform Bernoulli-Euler beam having density ρ, cross-sectional area A, modulus of elasticity E, second moment of the cross-sectional area about the neutral axis I, and equilibrium length l. The left-hand end of the beam ($x = 0$) is clamped or built into a rigid immovable wall. The right-hand end of the beam ($x = l$) is restrained vertically by a linear

[23]See Problem 9-40.

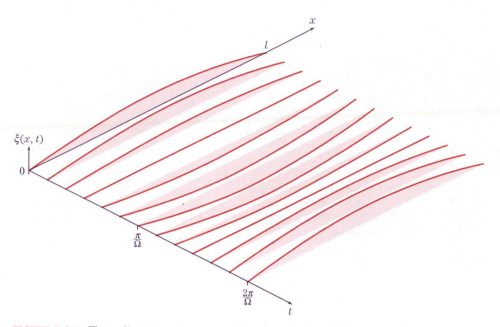

FIGURE 9-21 Three-dimensional representation of longitudinal displacement $\xi(x,t)$ of rod subjected to distributed harmonic excitation.

spring having a spring constant k and is also attached to a mass M. We note that the system in Figure 9-23 is the same as the system sketched in Figure 9-6; so the generalized coordinate is $\eta(x,t)$, which is the transverse displacement of the section whose position is x, where x is referenced with respect to the left-hand wall.

Now, we suppose the mass M at $x = l$ represents an electric motor, which in operation exerts a vertical force due to its unbalanced rotor. Further, if we denote the unbalance force of the motor by $F_0 \sin \Omega t$ (F_0 being a constant), one of the *natural boundary conditions* at $x = l$ is given by

$$M\frac{\partial^2 \eta}{\partial t^2} + k\eta = EI\frac{\partial^3 \eta}{\partial x^3} + F_0 \sin \Omega t \qquad x = l \qquad (9\text{-}372)$$

which is obtained from Eq. (9-74) by setting $F(t) = F_0 \sin \Omega t$. The other *natural boundary condition* at $x = l$ is given by Eq. (9-75), namely,

$$EI\frac{\partial^2 \eta}{\partial x^2} = 0 \qquad x = l \qquad (9\text{-}373)$$

The boundary conditions at $x = 0$ are the *geometric boundary conditions* due to the clamped left-hand end, and are given by

$$\eta = 0 \qquad x = 0 \qquad (9\text{-}374)$$

$$\frac{\partial \eta}{\partial x} = 0 \qquad x = 0 \qquad (9\text{-}375)$$

as presented in Eqs. (9-60).

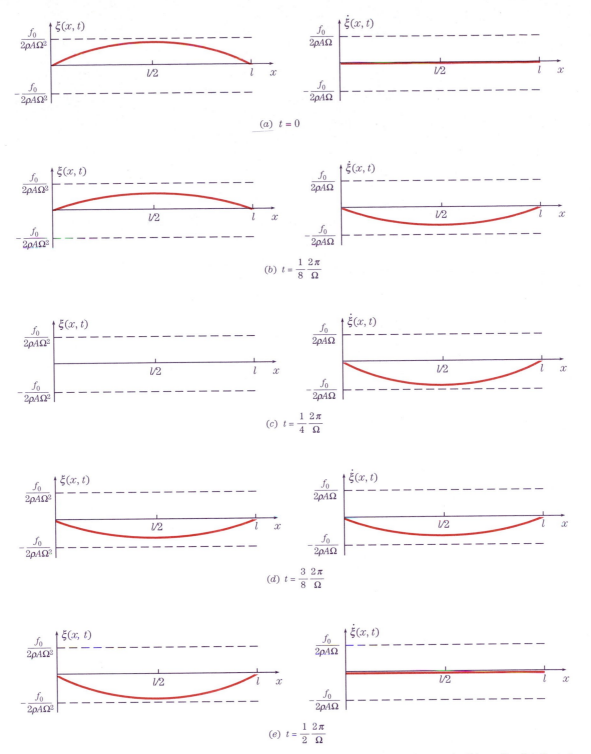

FIGURE 9-22a Longitudinal displacement $\xi(x, t)$ and longitudinal velocity throughout rod subjected to distributed harmonic excitation.

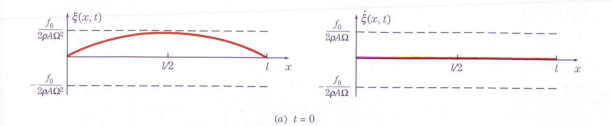

(a) $t = 0$

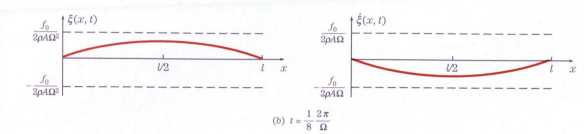

(b) $t = \dfrac{1}{8}\dfrac{2\pi}{\Omega}$

(c) $t = \dfrac{1}{4}\dfrac{2\pi}{\Omega}$

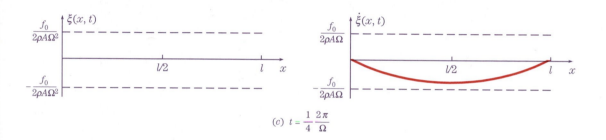

(d) $t = \dfrac{3}{8}\dfrac{2\pi}{\Omega}$

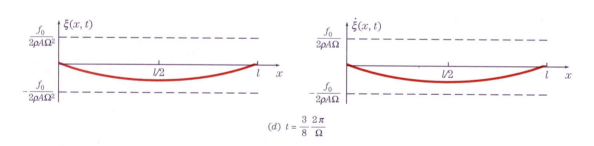

(e) $t = \dfrac{1}{2}\dfrac{2\pi}{\Omega}$

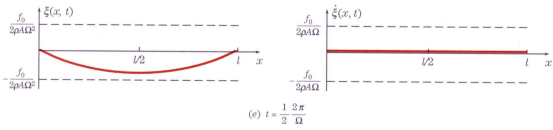

FIGURE 9-22b

668

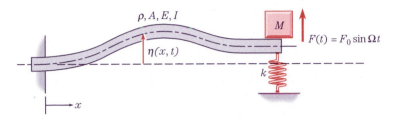

FIGURE 9-23 Beam system subjected to harmonic force on boundary.

Because the excitation is harmonic, the forced response will also be harmonic and at the same frequency Ω, as shown in Subsections 8-2.4, 8-3.5, and 8-4.3 in Chapter 8. This is a property of linear systems, not nonlinear systems. Therefore, we *assume* a solution of the form

$$\eta(x,t) = X(x) \sin \Omega t \qquad (9\text{-}376)$$

where $X(x)$ is the unknown spatial distribution of the generalized coordinate $\eta(x,t)$. Note that the equation of motion for this forced beam is given by the generic equation of motion, Eq. (9-82), with $c_q = \sqrt{\dfrac{EI}{\rho A}}$, $z(x,t) = \eta(x,t)$, and $g(x,t) = 0$. (Also, refer to Tables 9-3 and 9-4.) In order to find the functional form of $X(x)$, we substitute Eq. (9-376) into Eq. (9-82). Thus

$$-\Omega^2 X(x) \sin \Omega t = -c_q^2 \frac{d^4 X}{dx^4} \sin \Omega t \qquad (9\text{-}377)$$

Since Eq. (9-377) must hold for all time t, the common $\sin \Omega t$ term can be canceled, yielding

$$\frac{d^4 X}{dx^4} - \frac{\Omega^2}{c_q^2} X = 0 \qquad (9\text{-}378)$$

Equation (9-378) has the same form as Eq. (9-160). Thus, the solution to Eq. (9-378) is given by an equation analogous to Eq. (9-165) such that

$$X(x) = C_1 \cos Bx + C_2 \sin Bx + C_3 \cosh Bx + C_4 \sinh Bx \qquad (9\text{-}379)$$

where the C_j's are the unknown coefficients and $B \equiv \sqrt{\dfrac{\Omega}{c_q}}$. Combining Eqs. (9-376) and (9-379) gives

$$\eta(x,t) = (C_1 \cos Bx + C_2 \sin Bx + C_3 \cosh Bx + C_4 \sinh Bx) \sin \Omega t \qquad (9\text{-}380)$$

The unknown coefficients C_1 through C_4 must be evaluated in accordance with the boundary conditions; there are four coefficients and four boundary conditions. Substituting Eq. (9-380) into Eq. (9-372) and rearranging the resulting expression give

$$C_1 \big\{ (k - \Omega^2 M) \cos Bl - EIB^3 \sin Bl \big\} + C_2 \big\{ (k - \Omega^2 M) \sin Bl + EIB^3 \cos Bl \big\}$$
$$+ C_3 \big\{ (k - \Omega^2 M) \cosh Bl - EIB^3 \sinh Bl \big\} + C_4 \big\{ (k - \Omega^2 M) \sinh Bl - EIB^3 \cosh Bl \big\}$$

$$= F_0 \qquad (9\text{-}381)$$

where the common time function $\sin \Omega t$ has been canceled because Eq. (9-381) must hold for all time t. Similarly, substituting Eq. (9-380) into Eq. (9-373) gives

$$-C_1 \cos Bl - C_2 \sin Bl + C_3 \cosh Bl + C_4 \sinh Bl = 0 \qquad (9\text{-}382)$$

where the common $B^2 \sin \Omega t$ has been canceled from Eq. (9-382). Also, substituting Eq. (9-380) into Eqs. (9-374) and (9-375) yields, respectively,

$$C_1 + C_3 = 0 \qquad (9\text{-}383)$$

$$C_2 + C_4 = 0 \qquad (9\text{-}384)$$

Equations (9-381) through (9-384) can be solved for C_1, C_2, C_3, and C_4. First, Eqs. (9-383) and (9-384) yield

$$C_3 = -C_1 \qquad (9\text{-}385)$$

$$C_4 = -C_2 \qquad (9\text{-}386)$$

By using Eqs. (9-385) and (9-386), C_3 and C_4 can be eliminated from Eqs. (9-381) and (9-382). That is, Eqs. (9-381) and (9-382) reduce, respectively, to

$$C_1 \left\{ (k - \Omega^2 M)(\cos Bl - \cosh Bl) - EIB^3(\sin Bl - \sinh Bl) \right\}$$
$$+ C_2 \left\{ (k - \Omega^2 M)(\sin Bl - \sinh Bl) + EIB^3(\cos Bl + \cosh Bl) \right\} = F_0 \qquad (9\text{-}387)$$

and

$$C_1(\cos Bl + \cosh Bl) + C_2(\sin Bl + \sinh Bl) = 0 \qquad (9\text{-}388)$$

Equation (9-388), upon rearrangement, yields

$$C_2 = -\frac{\cos Bl + \cosh Bl}{\sin Bl + \sinh Bl} C_1 \qquad (9\text{-}389)$$

Substituting Eq. (9-389) into Eq. (9-387) gives

$$C_1 \left[(k - \Omega^2 M)(\cos Bl - \cosh Bl) - EIB^3(\sin Bl - \sinh Bl) \right.$$
$$- \frac{\cos Bl + \cosh Bl}{\sin Bl + \sinh Bl} \left\{ (k - \Omega^2 M)(\sin Bl - \sinh Bl) + EIB^3(\cos Bl + \cosh Bl) \right\} \right]$$
$$= F_0 \qquad (9\text{-}390)$$

Equation (9-390) can be solved to obtain C_1 as

$$C_1 = \frac{F_0(\sin Bl + \sinh Bl)}{\Delta} \qquad (9\text{-}391)$$

where

$$\Delta \equiv 2(k - \Omega^2 M)(\cos Bl \sinh Bl - \cosh Bl \sin Bl) - 2EIB^3(1 + \cos Bl \cosh Bl) \qquad (9\text{-}392)$$

Substituting Eq. (9-391) into Eq. (9-389) gives

$$C_2 = \frac{-F_0(\cos Bl + \cosh Bl)}{\Delta} \tag{9-393}$$

The solution to the problem can be found by substituting Eqs. (9-391) and (9-393), along with Eqs. (9-385) and (9-386), into Eq. (9-380), thus yielding

$$\eta(x, t) = \frac{F_0}{\Delta} \sin \Omega t \left\{ (\sin Bl + \sinh Bl)(\cos Bx - \cosh Bx) \right.$$
$$\left. - (\cos Bl + \cosh Bl)(\sin Bx - \sinh Bx) \right\} \tag{9-394}$$

TABLE 9-11 Procedure to Obtain Modes of Vibration and Free Undamped Solution via Separation of Variables

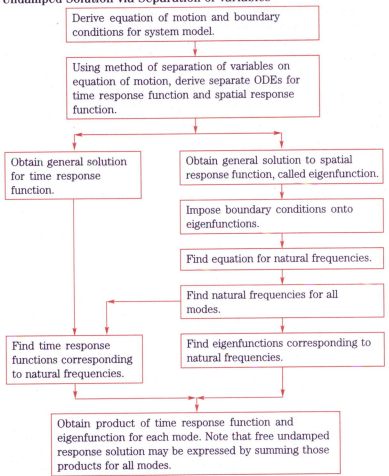

where Δ is defined by Eq. (9-392). Therefore, Eq. (9-394) gives the forced response of the beam subjected to the harmonic force at its right-hand end.

9-6 SUMMARIES

In this chapter on the dynamics of one-dimensional continua, we have accomplished four significant goals:

1. Derivation of equations of motion.
2. Derivation of the modes of vibration.
3. Calculation of the system response to initial conditions for a few cases.
4. Calculation of the system response to harmonic excitations for a few cases.

More could have been done; yet for an introduction much has been accomplished. Extension of this material might include more complicated structures (membranes, plates, shells, or combinations of these elements); more complicated boundary conditions; and more complicated dynamic loading.

As aids for both the understanding and the use of the topics considered, summary tables are provided for analyses involving modes of vibration (Table 9-11), response to initial conditions (Table 9-12), and response to harmonic excitations (Table 9-13).

TABLE 9-12   Procedure to Obtain Response to Initial Conditions

TABLE 9-13 Procedure to Obtain Response to Harmonic Excitation

Derive equation of motion and boundary conditions for system model, including motion or force harmonic excitation.

↓

Assume product solution as unknown spatial function multiplied by time-harmonic function, having same frequency as excitation.

↓

Substitute assumed solution into equation of motion, and obtain ODE for spatial function.

↓

Find general solution for spatial function, having unknown coefficients.

↓

Impose boundary conditions onto general solution of spatial function, and determine unknown coefficients.

↓

Obtain complete solution as product of time harmonic function and spatial function with determined coefficients.

Problems for Chapter 9

Category II: Intermediate

Equation of motion for lumped-parameter systems via Hamilton's principle: Problem 9-1

Problem 9-1: Derive the equation of motion for the lumped-parameter system sketched in Figure P9-1 using Hamilton's principle, not Lagrange's equation.

Longitudinal vibration of rods: Problems 9-2 through 9-13

In Problems 9-2 through 9-8, do the following:

(a) By simply using Table 9-5, state the appropriate boundary conditions at each end of the rod and identify them as "geometric" or "natural" boundary conditions.

(b) Find the equation for the natural frequencies of the system.

Assume that the rods described in Problems 9-2 through 9-8 are all uniform with density ρ, cross-sectional area A, modulus of elasticity E, and equilibrium length l; and the systems are all constrained to move only along the rod's longitudinal direction.

Problem 9-2: The rod sketched in Figure P9-2 is clamped at one end and is attached to a mass M at the other end.

Problem 9-3: The rod sketched in Figure P9-3 is free at one end and is connected to a rigid wall via a spring of spring constant k at the other end.

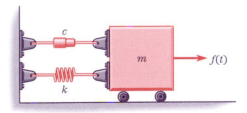

FIGURE P9-1

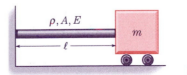

FIGURE P9-2

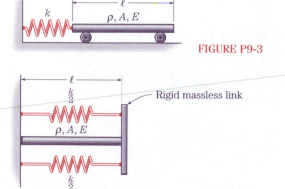

FIGURE P9-3

FIGURE P9-4

Problem 9-4: The rod sketched in Figure P9-4 is in parallel with two springs, each of spring constant $\frac{k}{2}$; and the three elements are connected to a rigid wall at one end and a rigid massless vertical link at the other end.

Problem 9-5: The rod sketched in Figure P9-5 is attached to masses M_1 and M_2 at its ends.

Problem 9-6: The rod sketched in Figure P9-6 is attached to springs of spring constants k_1 and k_2 at its ends.

Problem 9-7: The rod sketched in Figure P9-7 is attached to a mass M at one end and to a spring of spring constant k at the other end.

Problem 9-8: One end of the rod sketched in Figure P9-8 is attached to a mass M that is connected to a rigid wall via a spring of spring constant k_1 and a dashpot of dashpot constant c. The other end of the rod is attached to a second spring of spring constant k_2.

Problem 9-9: Consider the system in Problem 9-4.

(a) Find the eigenfunctions.

(b) Show that the resulting eigenfunctions in part (a) are orthogonal to each other in the

sense that they satisfy Eq. (9-135), by direct evaluation of the integral in Eq. (9-153).

(c) Do the task in part (b) by using the boundary conditions found in Problem 9-4 to show that Eq. (9-143) vanishes and so Eqs. (9-145) and (9-146) hold.

Problem 9-10: Consider the system in Problem 9-2.

(a) Find the eigenfunctions.

(b) Show that the resulting eigenfunctions in part (a) do *not* satisfy Eq. (9-135), by direct evaluation of the integral in Eq. (9-153).

(c) Do the task in part (b) by using the boundary conditions found in Problem 9-2 to show that Eq. (9-143) does *not* vanish and so Eqs. (9-145) and (9-146) do *not* hold.

Problem 9-11: Consider free longitudinal vibration of the oil drill sketched in Figure P9-11. The oil drill is modeled as a uniform cylindrical rod of density $\rho = 7000$ kg/m^3, diameter $d = 0.2$ m, modulus of elasticity $E = 2.0 \times 10^{11}$ N/m^2, and equilibrium length $l = 1$ km. It is assumed that the upper end of the drill is fixed. Find the lowest natural frequency and the corresponding mode shape in each of the following two cases of lower end condition. Ignore gravity.

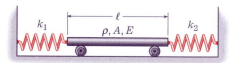

FIGURE P9-6

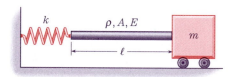

FIGURE P9-7

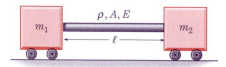

FIGURE P9-5

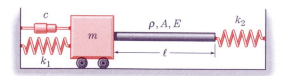

FIGURE P9-8

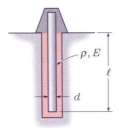

FIGURE P9-11

(a) When the drill bit at the lower end is stuck in the ground.

(b) When the drill bit at the lower end is free.

Problem 9-12: As sketched in Figure P9-12, a uniform rod of density ρ_1, cross-sectional area A_1, modulus of elasticity E_1 and equilibrium length l_1 is clamped at one end; and at the other end, it is welded to another uniform rod of corresponding properties ρ_2, A_2, E_2 and l_2. The longitudinal vibration of the composite rod system is considered.

(a) Find the boundary conditions and identify them as "geometric" or "natural" boundary conditions.

(b) Find the equation for the natural frequencies of the system.

Problem 9-13: The bar sketched in Figure P9-13 has variable area $A(x)$ and mass per unit length $m(x)$ along its length. At its end $x = L$, it is acted upon by an exciting force $F(t)$ and a dissipative element

that is modeled as viscous damping having an equivalent dashpot constant c. Derive the equation(s) of motion and clearly indicate all geometric and natural boundary conditions.

Torsional vibration of shafts: Problems 9-14 through 9-21

Problem 9-14: The uniform circular cylindrical shaft sketched in Figure 9-3 is modeled as a one-dimensional continuum of density ρ, polar moment of the cross-sectional area about the axis of the shaft J. Show that the moment of inertia per unit length of the uniform shaft about its axis I_p is given as $I_p = \rho J$.

In Problems 9-15 through 9-18 do the following:

(a) By simply using Table 9-6, state the appropriate boundary condition at each end of the shaft and identify them as "geometric" or "natural" boundary conditions.

(b) Find the equation for the natural frequencies of the system.

Assume that the shafts described in Problems 9-15 through 9-18 are all uniform with density ρ, polar moment of the cross-sectional area about the axis of the shaft J, shear modulus G and length l; and that the systems are all constrained to rotate only about the shaft's axis.

Problem 9-15: The shaft sketched in Figure P9-15 is attached to rotating inertias I_1 and I_2 at its ends.

Problem 9-16: The shaft sketched in Figure P9-16 is attached to a rotational spring of spring constant k_t at one end and it is free at the other end.

Problem 9-17: A torsional system with a rigid flywheel at one end and a torsional spring at the other is sketched in Figure P9-17. The shaft is uniform throughout its length L, with density ρ, shear modulus G, and polar moment of the cross-sectional area about its axis J. The rotary inertia of the flywheel about its axis is I_0 and the torsional spring has a spring constant of k_t.

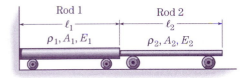

FIGURE P9-12

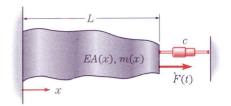

FIGURE P9-13

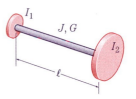

FIGURE P9-15

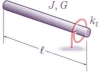

FIGURE P9-16

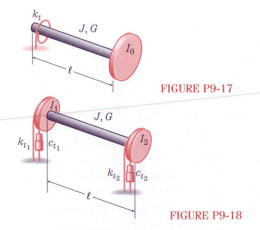

FIGURE P9-17

FIGURE P9-18

Problem 9-18: The shaft sketched in Figure P9-18 is attached to a combination of a rotating inertia I_1, torsional spring of spring constant k_{t1}, and torsional dashpot of dashpot constant c_{t1} at one end; and at the other end, it is attached to a combination of rotating inertia I_2, torsional spring of spring constant k_{t2}, and torsional dashpot of dashpot constant c_{t2}.

Problem 9-19: As sketched in Figure P9-19, a uniform shaft of density ρ_1, polar moment of the cross-sectional area about the axis of the shaft J_1, shear modulus G_1, and length l_1 is clamped at one end; and at the other end, it is welded to another uniform shaft of corresponding properties ρ_2, J_2, G_2 and l_2. The torsional vibration of the composite shaft system is considered.

(a) Find the boundary conditions and identify them as "geometric" or "natural" boundary conditions.

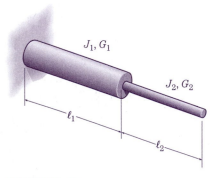

FIGURE P9-19

(b) Find the equation for the natural frequencies of the system.

Problem 9-20: As sketched in Figure P9-20, a solid shaft of density ρ_1, polar moment of the cross-sectional area about the axis of the shaft J_1, shear modulus G_1, and length l is assembled at the center of a cylindrical tube. The tube is of same length l, but different density ρ_2, polar moment of the cross-sectional area about the axis of the tube J_2, and shear modulus G_2. The left-hand ends of both the shaft and tube are clamped into a rigid wall and their right-hand ends are welded to a rigid disk of rotating inertia I_0. The torsional vibration of this assembly about the common axis of the shaft and tube is considered.

(a) Find the boundary conditions and identify them as "geometric" or "natural" boundary conditions.

(b) Find the equation for the natural frequencies of the system.

Problem 9-21: As sketched in Figure P9-21, two shafts (Shaft 1 and Shaft 2) having densities ρ_1 and ρ_2, polar moments of the cross-sectional area J_1 and

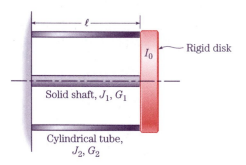

FIGURE P9-20

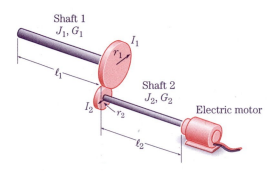

FIGURE P9-21

J_2, shear moduli G_1 and G_2, and lengths l_1 and l_2, respectively, are coupled by two cylindrical gears of rotating inertias I_1 and I_2 and radii r_1 and r_2. The left-hand end of Shaft 1 is free whereas the right-hand end of Shaft 2 is clamped in a stationary electrical motor. Assume that the shafts are supported by frictionless bearings (not sketched) that are fixed to ground and thus the shafts can rotate only about their respective axes.

(a) Find the boundary conditions and identify them as "geometric" or "natural" boundary conditions.

(b) Find the equation for the natural frequencies of the system.

Flexural vibration of beams: Problems 9-22 through 9-24

Problem 9-22: Show that the solution to the fourth-order ordinary differential equation (Eq. (9-160))

$$\frac{d^4 X}{dx^4} - \frac{\omega^2}{c_q^2} X = 0$$

is in the form (Eq. (9-161)) of

$$X(x) = e^{\gamma x}$$

where γ is found to be (Eq. (9-162))

$$\gamma = \pm \sqrt{\frac{\omega}{c_q}}, \ \pm \sqrt{\frac{\omega}{c_q}} i$$

where $i = \sqrt{-1}$.

Problem 9-23: The uniform beam sketched in Figure P9-23 is modeled as a Bernoulli-Euler beam having density ρ, modulus of elasticity E, cross-sectional area A, second moment of the cross-sectional area about its neutral axis I, and length l. The left-hand end of the beam is simply supported and the right-hand end is attached to a body of mass m and moment of inertia I_0 orthogonal to the beam's axis, which is sup-

ported by a spring of spring constant k. Assume that no gravity acts upon the system.

(a) By simply using Table 9-8, state the boundary conditions at each end of the beam and identify them as "geometric" or "natural" boundary conditions.

(b) Find the equation for the natural frequencies of the system.

Problem 9-24: The system sketched in Figure P9-24 consists of a Bernoulli-Euler beam with one clamped end and a thin disk of mass M and radius R connected to its other end. The system parameters are as sketched in Figure P9-24. The lateral flexural vibration of the system is considered.

(a) By simply using Table 9-8, state the boundary conditions at both ends of the beam and identify them as "geometric" or "natural" boundary conditions.

(b) Find the equation for the natural frequencies of the system.

Category III: More Difficult

Equations of motion, boundary conditions, natural frequencies: Problems 9-25 through 9-30

Problem 9-25: A uniform string of mass ρA per unit length is stretched to a length l under a large tension P and is fastened between two fixed supports as indicated in Figure P9-25. The string is also subjected to a distributed transverse force $f(x, t)$ per unit length. Consider the transverse motion of the string. Derive

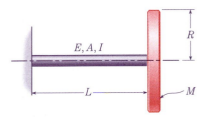

FIGURE P9-24

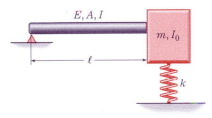

FIGURE P9-23

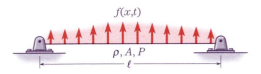

FIGURE P9-25

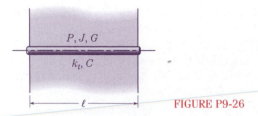

FIGURE P9-26

the equation of motion and clearly indicate all geo-metric and natural boundary conditions.

(*Hint*: For the potential energy function of the string, see Example L-4 in Appendix L.)

Problem 9-26: Consider the uniform shaft sketched in Figure P9-26 having density ρ, polar moment of the cross-sectional area about its axis J, shear modulus G, and length l. The shaft is supported throughout its length by a continuous elastic foundation that exerts a restoring torque $k_t\varphi$ per unit length when the angular displacement is φ and a retarding torque $C\dfrac{\partial\varphi}{\partial t}$ per unit length when the angular velocity is $\dfrac{\partial\varphi}{\partial t}$. Small torsional motion of the shaft is considered.

    (*a*) Derive the equation of motion and clearly indicate all geometric and natural boundary conditions.

    (*b*) Estimate the natural frequency of the rigid-body mode of the shaft. (*Note*: The rigid-body mode is when the shaft does not deform but rotates as a rigid body.)

Problem 9-27: Figure P9-27a depicts a simply sup-ported uniform Bernoulli-Euler beam having mass m, modulus of elasticity E, second moment of the cross-sectional area about its neutral axis I, and length L. Figure P9-27b depicts another simply supported uniform Bernoulli-Euler beam that is massless and otherwise identical to the beam sketched in Figure P9-27a. The massless beam sketched in Figure P9-27b is loaded with a mass M at its midpoint. Find the relationship between m and M for which the natu-ral frequency of the system sketched in Figure P9-27b

coincides with the lowest natural frequency of the system sketched in Figure P9-27a. It is known that the deflection at the midspan of a simply supported Bernoulli-Euler beam subjected to a transverse force P applied at its midspan is $\delta = \dfrac{PL^3}{48EI}$.

Problem 9-28: A uniform Bernoulli-Euler beam sketched in Figure P9-28 has density ρ, cross-sectional area A, modulus of elasticity E, second-moment of the cross-sectional area about its neutral axis I, and length l. The beam is embedded in a con-tinuous concrete foundation that exerts a restoring elastic force $K_0\eta$ per unit length and a retarding force $C_0\dfrac{\partial\eta}{\partial t}$ per unit length when the transverse displace-ment and velocity of the beam are η and $\dfrac{\partial\eta}{\partial t}$, respec-tively. The left-hand end of the beam is attached to a massless follower by a ball joint and the follower is supported by a spring of spring constant k and a dashpot of dashpot constant c, as shown. Also, the right-hand end of the beam is welded to a cart of mass M that is constrained to move only horizontally and is acted upon by a constant axial compressive force P_0. Transverse motion of the beam is considered. De-rive the equation of motion and clearly indicate all geometric and natural boundary conditions.

    (*Hint*: The potential energy function of the beam due to the constant compressive axial force P_0 is neg-ative; otherwise, it is the same as the potential energy

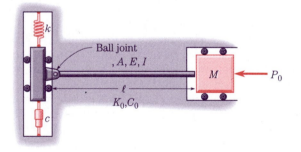

FIGURE P9-28

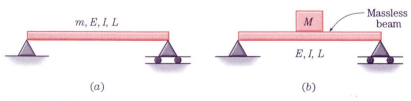

FIGURE P9-27

for a string under a constant tension, which may be found in Example L-4 in Appendix L.)

Problem 9-29: A model for studying the earthquake response of a building having a massive foundation is sketched in Figure P9-29. The foundation is assumed to be of mass M. The soil is modeled as having some stiffness characterized by a spring of spring constant k and some damping characterized by a dashpot of dashpot constant c.

The building of height h is assumed to be a uniform *shear beam*; that is, a beam that deforms in shear only. The building is assumed to have a mass density ρ, shear modulus G and cross-sectional area A. The earthquake excitation is assumed to impose a prescribed or specified displacement $x_0(t)$ as shown. Small transverse motion of the building $\eta(y, t)$, defined with respect to an inertial reference frame, is investigated. Assume that no gravity acts. Derive the equation of motion and clearly indicate all geometric and natural boundary conditions.

(*Hint*: For a shear beam as shown, the potential energy function is a function of shear strain only (see Table 9-1) and is given by

$$V = \int_0^h \frac{1}{2} \kappa G A \left(\frac{\partial \eta}{\partial y} \right)^2 dy$$

where κ is a constant shear coefficient, which, from advanced beam theory, depends on the shape of the beam's cross-section and an averaging process.)

Problem 9-30: A large but thin circular flywheel having mass M and principal centroidal moments of inertia I_1, I_1, and I_3 is fastened to the end of a revolving bar. (See Figure P9-30.) Consider the bar to be a Bernoulli-Euler beam constructed of a compos-

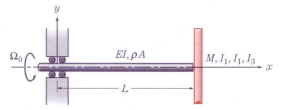

FIGURE P9-30

ite material that bends but does not undergo twisting deformation. The beam has a flexural rigidity EI, mass per unit length ρA, and length L. Due to a slight imbalance, the shaft bends such that its neutral axis no longer is coincident with the axis of rotation (x axis), but lies in the y-x plane where the y axis is rotating about the x axis at a constant rate Ω_0. Assume that the mass of the beam is very nearly concentrated along its neutral axis and do not assume that Ω_0 is small. Derive the equation of motion and clearly indicate all geometric and natural boundary conditions.

Response to initial conditions: Problems 9-31 through 9-39

Problem 9-31: Find the expression for the velocity of the rod described in Subsection 9-4.1, from the displacement solution in Eq. (9-283).

Problem 9-32: Figure P9-32 depicts a uniform rod of density ρ, cross-sectional area A, modulus of elasticity E and equilibrium length l. The left-hand end of the rod is clamped and the right-hand end is free.

(*a*) Using Table 9-9, write the general form of the free response for the longitudinal vibration of the rod.

(*b*) Find the dynamic response of the rod when it is subjected to an initial displacement $\xi(x, 0) = A_0 \sin\left(\frac{\pi x}{2l}\right)$ and zero initial velocity throughout its length.

(*c*) During the free response in part (*b*), find the maximum tensile stress in the rod.

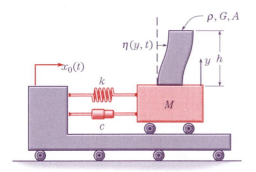

FIGURE P9-29

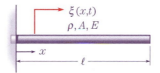

FIGURE P9-32

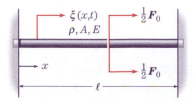

FIGURE P9-33

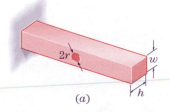

(a)

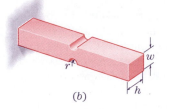

(b)

FIGURE P9-35

Problem 9-33: A uniform rod of density ρ, cross-sectional area A, modulus of elasticity E, and equilibrium length l is clamped at both ends. Static equilibrium is established after a concentrated constant axial force F_0 is applied at the rod's midpoint, as sketched in Figure P9-33. Find the response of the rod after the force F_0 is suddenly removed.

Problem 9-34: A uniform rod of density ρ, cross-sectional area A, modulus of elasticity E, and equilibrium length l falls freely upright and gets stuck into a tapered hole in the ground, as sketched in Figure P9-34. Find the response of the rod immediately after the rod's lower end hits the bottom of the hole and remains stuck to the rigid ground. Assume that the rod's lower end is located h above the bottom of the hole prior to its free fall and that the hole's depth is small enough to ignore the effects of the friction between the hole's wall and the rod. Also, it may be assumed that the rod falls as a rigid body until it contacts the ground and that the rod's deformation due to its own weight is negligible compared with its deformation due to the collision. Note that gravity acts on the system.

Problem 9-35: Consider the free vibration of the rod sketched in Figure 9-8 and studied in Subsection 9-4.1. Suppose the rod is made of steel whose modulus of elasticity, coefficient of thermal expansion, and tensile strength are known to be $E = 2.0 \times 10^{11} \ \text{N/m}^2$, $\alpha_t = 1.1 \times 10^{-5} \text{°C}^{-1}$, and $\sigma_Y = 4.2 \times 10^8 \ \text{N/m}^2$, respectively.

(a) Find the maximum allowable temperature increase ΔT without causing the material to yield.

(b) Redo part (a) above, when the rod is rectangular and it has a through-thickness circular hole at its midheight, as sketched in Figure P9-35a. Assume the ratio of the hole's diameter to the rod's height is $\frac{2r}{w} = 0.1$ and that the free vibration solution given in Subsection 9-4.1 is not affected by the hole.

(c) Redo part (a) above, when the rod is rectangular and it has two symmetrical semicircular edge notches, as sketched in Figure P9-35b. Assume the ratio of the notch's radius to the rod's width is $\frac{r}{w} = 0.1$ and that the free vibration solution given in Subsection 9-4.1 is not affected by the notches.

(*Hint*: In parts (b) and (c) above, find the stress concentration factors from the

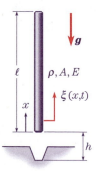

FIGURE P9-34

corresponding plots in elementary mechanics books.)

Problem 9-36: A torsional assembly consists of a uniform shaft of density ρ, shear modulus G, length l, and polar moment of the cross-sectional area about its axis J (see Figure P9-36). Also, a massless cylindrical pulley is welded to each end of the shaft. Suppose the shaft has been twisted by rotating one of the pulleys by φ_0 with respect to the other pulley. Find the subsequent torsional vibration of the shaft when the system, including this twisted pulley, is released from rest on a frictionless surface.

Problem 9-37: Show that the conservation of linear momentum of the freely vibrating beam in Subsection 9-4.3 (in its transverse direction) can be expressed as Eq. (9-339). Further, show that the transverse linear momentum of the beam is indeed conserved during its free vibration by proving that the free vibration solution given by Eqs. (9-337) and (9-338) satisfies Eq. (9-339).

Problem 9-38: Find the expression for the velocity of the beam in Subsection 9-4.3, from the displacement solution in Eqs. (9-337) and (9-338).

Problem 9-39: Consider a simply supported Bernoulli-Euler beam as sketched in Figure P9-39, which has density ρ, modulus of elasticity E, cross-sectional area A, second moment of the cross-sectional area about its neutral axis I, and length l. Suppose at time $t = 0$, the beam is struck by a hammer, imparting a localized initial velocity v_0 to a portion of the beam Δl located as shown in Figure P9-39. Find the response of the beam to this input.

Response to harmonic excitation: Problems 9-40 through 9-44

Problem 9-40: Find the expression for the velocity of the rod in Subsection 9-5.2, from the displacement solution in Eq. (9-371).

Problem 9-41: Consider the clamped–clamped rod subjected to a distributed harmonic force, as studied in Subsection 9-5.2.

(a) Derive the rod's dynamic stress response associated with the displacement response found in Eq. (9-371).

(b) Find the lowest natural frequency Ω_1 for the dynamic stress response in part (a); and for the range of forcing frequency $0 < \Omega < \Omega_1$, find the maximum axial stress in the rod.

(c) Find the static displacement and the corresponding static stress for the case of static loading $f(x,t) = f_0$, via the techniques of elementary mechanics of solids.

(d) Find the maximum axial stress in the rod for the static case in part (c). Compare the result with the results of part (b) for the following particular values of forcing frequency, $\Omega = \dfrac{\Omega_1}{4}, \dfrac{\Omega_1}{2}$ and $\dfrac{3\Omega_1}{4}$.

(e) Suppose the rod is made of a material whose yield strength is σ_Y. Find the maximum allowable amplitude f_0 of the applied force distribution without causing the material to yield, for the static loading case and the three dynamic loading cases of $\Omega = \dfrac{\Omega_1}{4}, \dfrac{\Omega_1}{2}$ and $\dfrac{3\Omega_1}{4}$.

Problem 9-42: A machine is mounted on a rigid platform that is supported by a uniform column of density ρ, modulus of elasticity E, cross-sectional area A, and equilibrium length l, as sketched in Figure P9-42. It is observed that the platform undergoes a harmonic vertical motion, $y_1(t) = Y_0 \cos \Omega t$. Find the harmonic longitudinal vibration response of the column due to the harmonic displacement of the platform. Ignore the effects of gravity.

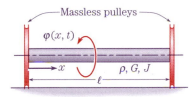

FIGURE P9-36

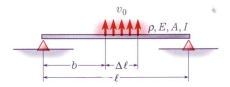

FIGURE P9-39

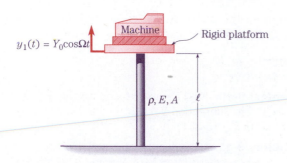

$y_1(t) = Y_0 \cos \Omega t$

Machine   Rigid platform

ρ, E, A l

FIGURE P9-42

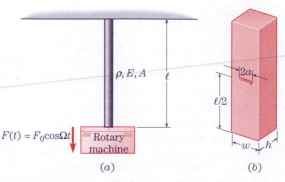

ρ, E, A l

$F(t) = F_0 \cos \Omega t$ Rotary machine

$|2a|$

$l/2$

w h

(a)   (b)

FIGURE P9-44

Problem 9-43: To analyze the response of a tall building to an earthquake, the building is modeled as a uniform Bernoulli-Euler beam having density ρ, modulus of elasticity E, cross-sectional area A, second moment of the cross-sectional area about its neutral axis I, and length l (see Figure P9-43). The building is assumed to be clamped in the ground at its bottom and the earthquake is modeled as the harmonic ground motion $y_1(t) = Y_0 \cos \Omega t$, as indicated in Figure P9-43. Find the harmonic response of the building to this harmonic seismic loading.

Problem 9-44: A small rotary machine of mass M is suspended by a uniform rod of density ρ, modulus of elasticity E, cross-sectional area A, and equilibrium length l, as sketched in Figure P9-44a. Due to a mass unbalance in the rotary machine, the machine exerts a harmonic vertical force $F(t) = F_0 \cos \Omega t$ on the lower end of the rod. Ignore gravity.

(a) Determine the harmonic longitudinal displacement response of the rod to the harmonic loading $F(t) = F_0 \cos \Omega t$.

(b) Find the axial stress associated with the displacement response in part (a).

(c) Find the static displacement and the static stress from the results of parts (a) and (b), for $F(t) = F_0 \cos \Omega t$ as $\Omega \to 0$.

(d) Suppose the rod is rectangular with dimensions as sketched in Figure P9-44b and there is a crack of length $2a$ ($a \ll w$) in the middle of the width of the rod. When the rod is under plane strain and under a static loading ($\Omega \to 0$), evaluate the stress intensity factor K_I using the expression

$$K_I = \sigma_0 \sqrt{\pi a}$$

where σ_0 is the uniform longitudinal stress measured far from the crack. Also, when the rod is made of a material whose critical stress intensity factor is known as K_{Ic}, find the critical size of the initial crack.

(e) Consider the rectangular rod in part (d), which is now subjected to the dynamic loading $F(t) = F_0 \cos \Omega t$. In general, under cyclic loading, the fatigue crack growth becomes a significant issue that should be considered in design. Assume the loading frequency Ω is very small such that the quasi-static condition holds. For many structural materials subjected to such a quasi-static cyclic loading, the fatigue crack growth rate is governed approximately by

$$\frac{da}{dN} = B(\Delta K_I)^p$$

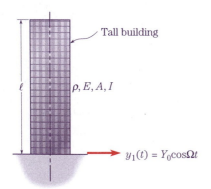

Tall building

l ρ, E, A, I

$y_1(t) = Y_0 \cos \Omega t$

FIGURE P9-43

where N is the number of cycles of loading; B is a constant; ΔK_I is the difference between the maximum and minimum values of K_I, corresponding to the maximum and minimum stress in a cycle of loading; and p is a constant ranging from 2 to 7 for most structural materials (specifically, $p = 2.5$ for the particular steel under consideration).

If the rod sketched in Figure P9-44b is made of the steel under consideration and its crack growth rate is to be reduced by half, how should the dimensions w and h be changed for the given F_0 and a?

(f) By integrating the crack growth rate equation given in part (e) and using the defini-

tion of K_I in part (d), the fatigue life can be found as

$$(\Delta\sigma)^p N_f = \text{Constant}$$

where $\Delta\sigma = |(\sigma_0)_{max} - (\sigma_0)_{min}|$ and N_f is the cycles-to-failure.

If the cycles-to-failure of the rod sketched in Figure P9-44b is to be increased by a factor of two, how should the dimensions w and h be changed for the given F_0? Assume the material of the rod is a steel for which $p = 2.5$. Compare the result with the result of part (e).

BIBLIOGRAPHY

The principal references and background documents used in writing this textbook are listed alphabetically within the eight indicated sections.

1 Historical

ALONZO, GUALBERTO Z., *Descriptive Guidebook to Chichén Itzá,* Libros, Revistas y Folletos de Yucatán, Merida, Yucatá, Mexico, February 1994.

Ancient Cities of the Indus, Gregory L. Possehl, ed., Vikas Publishing House Pvt Ltd., New Delhi, India, 1979.

ARNOLD, DIETER, *Building in Egypt. Pharaonic Stone Masonry,* Oxford University Press, New York, 1991.

BAINES, JOHN AND MALEK, JAROMIR, *Atlas of Ancient Egypt,* Facts on File, New York, 1980.

BENJAMIN, PARK, *A History of Electricity,* Arno Press, New York, 1975.

BERNAL, MARTIN, *Black Athena, Vol. I,* Rutgers University Press, New Brunswick, NJ, 1987.

BERNAL, MARTIN, *Black Athena, Vol. II,* Rutgers University Press, New Brunswick, NJ, 1991.

BERNAL, MARTIN, "Animadversions on the Origins of Western Science," *ISIS,* 83 (4), December 1992.

BOYER, CARL B., *A History of Mathematics,* Wiley, New York, 1968.

BREASTED, JAMES H., *A History of Egypt from Earliest Times to the Persian Conquest,* 2nd ed., Hodden & Stoughton, London, 1912.

BUCCIARELLI, LOUIS L. AND DWORSKY, NANCY, *Sophie Germain,* D. Reidel Publishing, Dordrecht, Holland.

CAJORI, FLORIAN, *A History of Mathematics,* Macmillan, New York, 1931.

CLARK, SOMERS AND ENGELBACH, R., *Ancient Egyptian Construction and Architecture,* Dover Publications, New York, 1990.

Correspondence, *The New Republic,* pp. 6–7, March 2, 1992 and pp. 4–5, March 9, 1992.

COWELL, ALAN, "After 350 Years, Vatican Says Galileo Was Right: It Moves," *New York Times,* October 31, 1992, New York, p. 1.

DIBNER, BERN, *Oersted and the Discovery of Electromagnetism,* Burndy Library, Norwalk, CT, 1961.

Dictionary of Scientific Biography: Sixteen Volumes, Charles C. Gillispie, ed., Charles Scribner's Sons, New York, 1970–1980.

DIOP, CHEIKH ANTA, *Civilization or Barbarism,* Lawrence Hill Books, Brooklyn, NY, 1991.

DUGAS, RENÉ, *A History of Mechanics,* translated by J.R. Maddox, Éditions du Griffon, Neuchatel, Switzerland, 1955.

Egypt Revisited, Ivan Van Sertima, ed., Transaction Publishers, New Brunswick, NJ, 1989.

FARADAY, MICHAEL, *Experimental Researches in Electricity,* Dover Publications, New York, 1965.

FLETCHER, SIR BANISTER, *A History of Architecture,* John Musgrove, ed., Butterworths, London, 1987.

GILLINGS, RICHARD J., *Mathematics in the Time of the Pharaohs,* Dover Publications, New York, 1982.

HALL, RICHARD S., *About Mathematics,* Prentice-Hall, Englewood Cliffs, NJ, 1973.

HAYNES, JOYCE L., *Nubia—Ancient Kingdoms of Africa,* Museum of Fine Arts, Boston, 1992.

HEILBRON, J. L., *Electricity in the 17th and 18th Centuries,* University of California Press, Berkeley, 1979.

A History of Architecture, John Musgrove, ed., Butterworths, London, 1987.

JAMES, GEORGE G. M., *Stolen Legacy,* United Brothers Communications Systems, Newport News, VA, 1989.

JONES, TOM B., *From the Tigris to the Tiber,* Dorsey Press, Homewood, IL, 1969.

KING, HENRY C., *The History of the Telescope,* Sky Publishing, Cambridge, MA, 1955.

KLINE, MORRIS, *Mathematics—The Loss of Certainty,* Oxford University Press, Oxford, England, 1980.

KNAPP, A. BERNARD, *The History and Culture of Ancient Western Asia and Egypt,* Wadsworth Publishing, Belmont, CA, 1988.

KRAMER, SAMUEL NOAH, *The Sumerians—Their History, Culture, and Character,* University of Chicago Press, Chicago, 1963.

"Lessons of the Galileo Case," *Origins,* 22 (22), November 12, 1992.

LEFKOWITZ, MARY, "Not Out of Africa," *The New Republic,* pp. 29–36, February 10, 1992.

LUMPKIN, BEATRICE, "The Egyptians and Pythagorean Triples," *Historia Mathematica,* 7, 1980, pp. 186–187.

MACAULAY, DAVID, *Pyramid,* Houghton Mifflin, Boston, 1975.

Mathematics–Dictionaries, Kluwer Academic Publishers, Netherlands, 1987.

MAXWELL, JAMES C., *A Treatise on Electricity and Magnetism,* Clarendon Press, Oxford, England, 1904.

MCCLAIN, ERNEST G., *The Myth of Invariance,* Nicolas Hays, Ltd., New York, 1976.

MELLAART, JAMES, "Egyptian and Near Eastern Chronology: A dilemma?" *Antiquity,* LIII (207), March 1979, pp. 6–19.

MICHALOWSKI, KAZIMIERZ AND DZIEWANOWSKI, ANDRZEJ, *Karnak,* Pall Mall Press, London, 1970.

MYEROWITZ LEVINE, MOLLY, "Review Article—The Use and Abuse of Black Athena," *American Historical Review,* April 1992, pp. 440–460.

NEUGEBAUER, OTTO, *Astronomy and History—Selected Essays,* Springer-Verlag, New York, 1983, pp. 211–213.

NEUGEBAUER, OTTO, *The Exact Sciences in Antiquity,* Brown University Press, Providence, RI, 1957.

The New Encyclopædia Britannica, 30 Volumes, 15th ed., Encyclopædia Britannica, Chicago, 1984.

NEWTON, ISAAC, *Philosphiae Naturalis Principia Mathematica,* 1687, translated by A. Motte, revised by Florian Cajori, University of California Press, Berkeley, 1934.

PAGE, S. G., *Mathematics: A Second Start,* Halsted Press, New York, 1986.

"A Papal Address on the Church and Science," *Origins,* 13 (3), June 2, 1983.

PETRIE, W. M. FLINDERS, *A History of Egypt,* Methuen, London, 1899.

PLEDGE, H. T., *Science Since 1500,* Philosophical Library, New York, 1947.

POUNDER, ROBERT L., "Review Article—Black Athena 2: History without Rules," *American Historical Review,* April 1992, pp. 461–464.

RICHARDSON, MOSES, *Fundamentals of Mathematics,* 3rd ed., Macmillan, New York, 1966.

ROBINS, GAY AND SHUTE, CHARLES, *The Rhind Mathematical Papyrus,* Dover Publications, New York, 1987.

Science and Technology in Ancient India, Abdur Rahman, Ashok Jain, and Sushil K. Mukherjee, general advisors, Firma KLM Privte, Ltd., Calcutta, India, 1986.

SCOTT, J. F., *A History of Mathematics,* Taylor & Francis Ltd., London, and Barnes & Noble, New York, 1969.

TIMOSHENKO, STEPHEN P., *History of Strength of Materials,* McGraw-Hill, New York, 1953.

TOMPKINS, PETER, *Secrets of the Great Pyramid,* Harper & Row, New York, 1971.

TOMPKINS, PETER, *The Magic of Obelisks,* Harper & Row, New York, 1981.

TRUESDELL, CLIFFORD A., III, *An Idiot's Fugitive Essays on Science,* Springer-Verlag, New York, 1984.

TRUESDELL, CLIFFORD A., III, *Essays in the History of Mechanics,* Springer-Verlag, New York, 1968.

WALKER, C. B. F., *Cuneiform,* University of California Press, Berkeley, 1987.

WALLIS BUDGE, E. A., *The Eqyptian Book of the Dead,* Dover Publications, New York, 1967.

WALLIS BUDGE, E. A., *The Rosetta Stone,* Dover Publications, New York, 1989.

YOKE, HO PENG, *Li, Qi and Shu: An Introduction to Science and Civilization in China,* University of Washington Press, Seattle, 1987.

2 Astronomy

AGELL, GEORGE O., MORRISON, DAVID, AND WOLFF, SIDNEY C., *Realm of the Universe,* 4th ed., Saunders College Publishing, Philadelphia, 1988.

MOTZ, LLOYD AND DUVEEN, ANNETA, *Essentials of Astronomy,* Columbia University Press, New York, 1977.

PASACHOFF, JAY M. AND KUTNER, MARC L., *University Astronomy,* W.B. Saunders, Philadelphia, 1978.

ZEILIK, MICHAEL AND GAUSTAD, JOHN, *Astronomy: The Cosmic Perspective,* 2nd ed., Wiley, New York, 1990.

ZEILIK, MICHAEL AND SMITH, ELSKE V. P., *Introductory Astronomy and Astrophysics,* 2nd ed., Saunders College Publishing, Philadelphia, 1987.

3 Design, Systems, and Modeling

BERG, CHARLES A., "On Teaching Design: Identifying the Subject," *International Journal of Mechanical Engineering Education,* 20 (4), October 1992, pp. 235–240.

CLOSE, CHARLES M. AND FREDERICK, DEAN K., *Modeling and Analysis of Dynamic Systems,* 2nd ed., Houghton Mifflin, Boston, 1993.

COCHIN, IRA AND PLASS, HAROLD J., JR., *Analysis and Design of Dynamic Systems,* 2nd ed., Harper & Row, New York, 1990.

EDWARDS, DILWYN AND HAMSON, MICHAEL, *Guide to Mathematical Modelling,* CRC Press, Boca Rotan, FL, 1990.

KECMAN, V., *State–Space Models of Lumped and Distributed Systems,* Springer-Verlag Berlin, Heidelberg, Germany, 1988.

MANN, ROBERT W., *The Killian Award Lecture: 1983–1984,* MIT.

MISCHKE, CHARLES R., *Mathematical Model Building,* 2nd ed., Iowa State University Press, Ames, IA, 1980.

SHEARER, J. LOWEN, MURPHY, ARTHUR T., AND RICHARDSON, HERBERT H., *Introduction to System Dynamics,* Addison-Wesley, Reading, MA, 1967.

SUH, NAM P., *The Principles of Design,* Oxford University Press, New York, 1990.

WHITE, HARRY J. AND TAUBER, SELMO, *Systems Analysis,* W.B. Saunders, Philadelphia, 1969.

4 Elementary Dynamics

DEN HARTOG, JACOB P., *Mechanics,* McGraw-Hill, New York, 1948.

HALLIDAY, DAVID AND RESNICK, ROBERT, *Physics, Part II,* Wiley, New York, 1962.

HALLIDAY, DAVID, RESNICK, ROBERT AND WALKER, JEARL, *Fundamentals of Physics,* 4th ed., Wiley, New York, 1993.

HIBBELER, RUSSELL C., *Engineering Mechanics– Dynamics,* 6th ed., Macmillan, New York, 1992.

INGARD, UNO AND KRAUSHAAR, WILLIAM L., *Introduction to Mechanics, Matter, and Waves,* Addison-Wesley, Reading, MA, 1960.

JONG, I. C. AND ROGERS, B. G., *Engineering Mechanics: Dynamics,* Saunders College Publishing, Philadelphia, 1991.

KLEPPNER, DANIEL AND KOLENKOW, ROBERT J., *An Introduction to Mechanics,* McGraw-Hill, New York, 1973.

MERIAM, JAMES L. AND KRAIGE, L. GLENN, *Engineering Mechanics: Volume 2—Dynamics,* 3rd ed., Wiley, New York, 1992.

O'DWYER, JOHN J., *College Physics,* 2nd ed., Wadsworth, Belmont, CA, 1984.

OLENIC, RICHARD P., APOSTOL, TOM M., AND GOODSTEIN, DAVID L., *The Mechanical Universe,* Cambridge University Press, New York, 1985.

RILEY, WILLIAM F. AND STURGES, LEROY D., *Engineering Mechanics: Dynamics,* Wiley, New York, 1993.

SANDOR, BELA I., *Engineering Mechanics: Dynamics,* 2nd ed., Prentice-Hall, Englewood Cliffs, NJ, 1987.

SHAMES, IRVING H., *Engineering Mechanics: Volume II Dynamics,* 3rd ed., Prentice-Hall, Englewood Cliffs, NJ, 1980.

5 Intermediate/Advanced Dynamics

CRANDALL, S. H. ET AL., *Dynamics of Mechanical and Electromechanical Systems,* McGraw-Hill, New York, 1968.

FOWLER, GRANT R., *Analytical Mechanics,* 3rd ed., Holt, Rinehart and Winston, New York, 1977.

GOLDSTEIN, HERBERT, *Classical Mechanics,* Addison-Wesley, Reading, MA, 1950.

GREENWOOD, DONALD T., *Principles of Dynamics*, 2nd ed., Prentice-Hall, Englewood Cliffs, NJ, 1988.

HAUG, EDWARD J., *Intermediate Dynamics*, Prentice-Hall, Englewood Cliffs, NY, 1992.

KANE, THOMAS R. AND LEVINSON, DAVID A., *Dynamics: Theory and Applications*, McGraw-Hill, New York, 1985.

LANCZOS, CORNELIUS, *The Variational Principles of Mechanics*, 4th ed., University of Toronto Press, Toronto, Canada, 1970.

MARION, JERRY B., *Classical Dynamics of Particles and Systems*, 2nd ed., Academic Press, New York, 1970.

MEIROVITCH, LEONARD, *Methods of Analytical Dynamics*, McGraw-Hill, New York, 1970.

PARS, L. A., *A Treatise on Analytical Dynamics*, Ox Bow Press, Woodbridge, CT, 1979.

ROSENBERG, REINHART M., *Analytical Dynamics of Discrete Systems*, Plenum Press, New York, 1977.

SLOTINE, JEAN-JACQUES E. AND LI, WEIPING, *Applied Nonlinear Control*, Prentice-Hall, Englewood Cliffs, NJ, 1991.

TRUESDELL, CLIFFORD A., III AND TOUPIN, RICHARD A., "The Classical Field Theories," *Encyclopedia of Physics*, III (1), S. Flugge, ed., Springer, Berlin, 1960.

YIM, HYUNJUNE, "Momentum Principles for Continuum Models," Composite Materials and Nondestructive Evaluation Laboratory, MIT, 1995.

6 Hamilton's Law of Varying Action and Hamilton's Principle

BAILEY, C.D., "A New Look at Hamilton's Principle," *Foundations of Physics*, 5 (3), 1975, pp. 433–451.

BAILEY, C. D., "Application of Hamilton's Law of Varying Action," *AIAA Journal*, 3 (9), 1975, pp. 1539–1540.

HAMILTON, W. R., "On a General Method in Dynamics," *Philosophical Transactions of Royal Society*, 1834, pp. 247–308.

HAMILTON, W. R., "Second Essay on a General Method in Dynamics," *Philosophical Transactions of Royal Society*, 1835, pp. 95–144.

PAPASTAVRIDIS, J. G., "The Variational Principle of Mechanics, and a Reply to C.D. Bailey," *Journal of Sound and Vibration*, 118 (2), 1987, pp. 378–393.

YIN, SHUMEI, "Hamilton's Law of Varying Action: Its Interpretation and Application in Dynamics Problems of Lumped-Parameter Systems," S.M. Thesis, MIT, June 1991.

7 Electrical and Electromechanical Systems

MAGID, LEONID M., *Electromagnetic Fields, Energy, and Waves*, Krieger Publishing, Malabar, FL, 1981.

SEELY, SAMUEL, *Introduction to Electromagnetic Fields*, McGraw-Hill, New York, 1958.

SENTURIA, STEPHEN D. AND WEDLOCK, BRUCE D., *Electronic Circuits and Applications*, Wiley, New York, 1975.

WOODSON, HERBERT H. AND MELCHER, JAMES R., *Electromechanical Dynamics Part I: Discrete Systems*, Wiley, New York, 1968.

8 Vibration

ANDERSON, ROGER A., *Fundamentals of Vibration*, Macmillan, New York, 1967.

BLAND, DAVID R., *Wave Theory and Applications*, Clarendon Press, Oxford, England, 1988.

DIMAROGONAS, ANDREW D. AND HADDAD, SAM, *Vibration for Engineers*, Prentice-Hall, Englewood Cliffs, NJ, 1992.

HUTTON, DAVID V., *Applied Mechanical Vibrations*, McGraw-Hill, New York, 1981.

MEIROVITCH, LEONARD, *Elements of Vibration Analysis*, McGraw-Hill, New York, 1975.

PAIN, HERBERT J., *The Physics of Vibrations and Waves*, 2nd ed., Wiley, London, 1976.

RAO, SINGIVESU R., *Mechanical Vibration, Volumes I and II*, Springer-Verlag, New York, 1991.

THOMSON, WILLIAM T., *Theory of Vibration with Applications*, 3rd ed., Prentice-Hall, Englewood Cliffs, NJ, 1988.

TSE, FRANCIS S., MORSE, IVAN E., AND HINKLE, ROLLAND T., *Mechanical Vibrations—Theory and Applications*, 2nd ed., Allyn and Bacon, Boston, 1978.

FINITE ROTATION

In this appendix, we investigate some of the characteristics of rotation of a rigid body, with a focus on finite rotation. In general, the motion of a rigid body may be described by the combination of a translation of a point and a rotation about an axis through that point. Here, without loss of generality, we may fix the origin of a body-coordinate frame at that point so that the remaining motions consist of rotations of this frame. In other words, the motions we shall discuss are rotations of a rigid body about axes through a fixed point.

A-1   CHANGE IN POSITION VECTOR DUE TO FINITE ROTATION

Here we consider the change in a position vector of fixed magnitude, due to a finite rotation $\boldsymbol{\theta} = \theta\boldsymbol{n}$, where θ is the magnitude of the rotation about an axis having the unit vector $\boldsymbol{n}$. We denote an arbitrary point of interest as P, which is located by the position vector $\boldsymbol{r}$, before the rotation, as sketched in Figure A-1. After the rotation, point P has moved to P' and $\boldsymbol{r}$ has moved to $\boldsymbol{r}'$. So, after the rotation, the position vector $\boldsymbol{r}'$ can be expressed as

$$\boldsymbol{r}' = \boldsymbol{r} + \Delta\boldsymbol{r} \tag{A-1}$$

The purpose of this section is to derive an expression for $\boldsymbol{r}'$ in terms of $\boldsymbol{r}$, $\boldsymbol{n}$ and θ, which will be found to be very useful throughout this appendix.

We denote ON as the projection of the position vector $\boldsymbol{r}$ along the axis of rotation. Then, the rotation of the position vector $\boldsymbol{r}$ about the $\boldsymbol{n}$ direction can be visualized by considering the rotation of the right triangle $\triangle ONP$ about one of its two perpendicular sides, ON. After rotating by the amount θ, $\triangle ONP$ has moved to a location designated by $\triangle ONP'$. Because the rotation $\boldsymbol{\theta}$ does not change the magnitude of the position vector locating points P' and P,

$$|\boldsymbol{r}'| = |\boldsymbol{r}| = r \tag{A-2}$$

Also, since both NP and NP' are normal to ON, the plane formed by NP and NP', which we shall call plane NPP', is normal to ON, the axis of rotation.

In plane NPP', point P moves in a circular trajectory centered at N. Projections of $\boldsymbol{r}$ and $\boldsymbol{r}'$ onto plane NPP' are denoted by $\boldsymbol{e}$ and $\boldsymbol{e}'$, respectively. We define the angle between $\boldsymbol{r}$ and $\boldsymbol{n}$ as ϕ, which is also the angle between $\boldsymbol{r}'$ and $\boldsymbol{n}$. Then, since $\triangle ONP$ and $\triangle ONP'$ are a pair of congruent right triangles,

$$|\boldsymbol{e}'| = r'\sin\phi = r\sin\phi = |\boldsymbol{e}| \tag{A-3}$$

where Eq. (A-2) has been used.

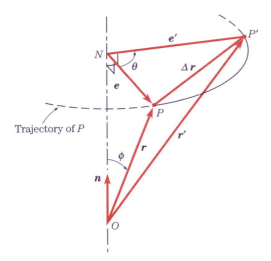

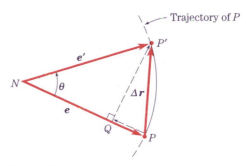

FIGURE A-2   Plane normal to rotating axis, illustrating geometrical relations.

FIGURE A-1   Motion of point P due to θ rotation of r about n.

In addition, as sketched in Fig. A-2, the vector Δr may be decomposed into a vector along e, PQ, and a vector normal to e, QP'. The vector QP' is normal to the triangle $\triangle ONP$ containing the vectors n and r. Recall from vector algebra that the cross-product $n \times r$ is normal to the plane containing n and r, and its direction is determined by the right-hand rule. Using the right-hand rule, we note that QP' and $n \times r$ have the same direction.

We now calculate the magnitude of vector QP'. Since in plane NPP', QP' is normal to e, we note from Fig. A-2 that

$$|QP'| = |e'| \sin \theta = r \sin \phi \sin \theta \tag{A-4}$$

where Eq. (A-3) has been used.

So, we have found that the direction of vector QP' is the same as that of $n \times r$, and the magnitude of QP' is given by Eq. (A-4). We can combine these to express the vector QP' as its magnitude times the unit vector in its direction; that is,

$$QP' = |QP'| \frac{n \times r}{|n \times r|} \tag{A-5}$$

where the elementary idea that a unit vector along any direction is simply any vector in that direction ($n \times r$) divided by its own magnitude $|n \times r|$. Recalling from vector algebra that

$$|n \times r| = |r||n| \sin \phi \tag{A-6}$$

and noting that n is a unit vector; that is, $|n| = 1$, we observe that

$$|n \times r| = r \sin \phi \tag{A-7}$$

where Eq. (A-2) has been used. Substitution of Eqs. (A-4) and (A-7) into Eq. (A-5) gives

$$QP' = r \sin \phi \sin \theta \frac{n \times r}{r \sin \phi}$$

$$= \sin \theta (n \times r) \tag{A-8}$$

A similar procedure can be used to determine an expression for the vector $\boldsymbol{PQ}$. The vector $\boldsymbol{PQ}$ is normal to $\boldsymbol{n} \times \boldsymbol{r}$, and it is also normal to $\boldsymbol{n}$ because it lies in the plane NPP'. Therefore, $\boldsymbol{PQ}$ is along the line of $\boldsymbol{n} \times (\boldsymbol{n} \times \boldsymbol{r})$. Again, by use of the right-hand rule for vector cross-products, we find that $\boldsymbol{n} \times (\boldsymbol{n} \times \boldsymbol{r})$ has the same direction as $\boldsymbol{PQ}$. Furthermore, in plane NPP',

$$
\begin{aligned}
|\boldsymbol{PQ}| = |\boldsymbol{e}| - |\boldsymbol{NQ}| &= |\boldsymbol{e}| - |\boldsymbol{e}'| \cos \theta \\
&= r \sin \phi - r \sin \phi \cos \theta \\
&= r \sin \phi (1 - \cos \theta)
\end{aligned}
\tag{A-9}
$$

where Eq. (A-3) has been used. Also, since $\boldsymbol{n}$ is normal to $\boldsymbol{n} \times \boldsymbol{r}$, by definition,

$$
|\boldsymbol{n} \times (\boldsymbol{n} \times \boldsymbol{r})| = |\boldsymbol{n}||\boldsymbol{n} \times \boldsymbol{r}| = r \sin \phi
\tag{A-10}
$$

where Eq. (A-7) has been used.

Combining Eqs. (A-9) and (A-10) in a manner analogous to Eq. (A-5) gives

$$
\begin{aligned}
\boldsymbol{PQ} = |\boldsymbol{PQ}| \frac{\boldsymbol{n} \times (\boldsymbol{n} \times \boldsymbol{r})}{|\boldsymbol{n} \times (\boldsymbol{n} \times \boldsymbol{r})|} \\
= r \sin \phi (1 - \cos \theta) \frac{\boldsymbol{n} \times (\boldsymbol{n} \times \boldsymbol{r})}{r \sin \phi} \\
= (1 - \cos \theta)[\boldsymbol{n} \times (\boldsymbol{n} \times \boldsymbol{r})]
\end{aligned}
\tag{A-11}
$$

Thus, from Fig. A-2, the expression for $\Delta \boldsymbol{r}$ is

$$
\begin{aligned}
\Delta \boldsymbol{r} &= \boldsymbol{PQ} + \boldsymbol{QP'} \\
&= (1 - \cos \theta)[\boldsymbol{n} \times (\boldsymbol{n} \times \boldsymbol{r})] + \sin \theta (\boldsymbol{n} \times \boldsymbol{r})
\end{aligned}
\tag{A-12}
$$

where Eqs. (A-8) and (A-11) have been used. Or, upon substitution of Eq. (A-12) into (A-1), the position vector locating point P' after the $\boldsymbol{\theta}$ rotation of $\boldsymbol{r}$ may be expressed as

$$
\boldsymbol{r}' = \boldsymbol{r} + (1 - \cos \theta)[\boldsymbol{n} \times (\boldsymbol{n} \times \boldsymbol{r})] + \sin \theta (\boldsymbol{n} \times \boldsymbol{r})
\tag{A-13}
$$

A-2   FINITE ROTATIONS ARE NOT VECTORS

Recall that vectors can be characterized by the following three attributes:

1. A vector has magnitude.
2. A vector has direction.
3. Vectors obey the commutative law of addition; that is, $\boldsymbol{A} + \boldsymbol{B} = \boldsymbol{B} + \boldsymbol{A}$, where $\boldsymbol{A}$ and $\boldsymbol{B}$ are arbitrary dimensionally homogeneous[1] vectors.

Finite rotations do not obey this third attribute.

---

[1]The addition of physical quantities may be true only if all terms are of the same kind. Whereas the expression

$$
12 \text{ golf clubs} + 3 \text{ golf balls} = 15 \text{ weeks of bliss}
\tag{A-a}
$$

may be desirable, it is not true. On the other hand, the expression

$$
2 \text{ apples} + 3 \text{ apples} = 5 \text{ apples}
\tag{A-b}
$$

is true. Equation (A-b) is *dimensionally homogeneous;* Eq. (A-a) is not.

Consider the case in which two consecutive rotations $\boldsymbol{\theta}_1 = \theta_1 \boldsymbol{n}_1$ and $\boldsymbol{\theta}_2 = \theta_2 \boldsymbol{n}_2$, the first of magnitude θ_1 about an axis having a unit vector $\boldsymbol{n}_1$ and the second of magnitude θ_2 about another axis having a unit vector $\boldsymbol{n}_2$, are applied to a position vector locating point P and having an initial value designated as $\boldsymbol{r}$.

We denote the position vectors locating point P after the first and second rotations as $\boldsymbol{r}'$ and $\boldsymbol{r}''$, respectively. Then, according to Fig. A-1 and Eq. (A-13), the position vector after the first rotation is

$$\boldsymbol{r}' = \boldsymbol{r} + (1 - \cos\theta_1)[\boldsymbol{n}_1 \times (\boldsymbol{n}_1 \times \boldsymbol{r})] + \sin\theta_1(\boldsymbol{n}_1 \times \boldsymbol{r}) \qquad \text{(A-14)}$$

Next, applying Eq. A-13 again, we note that for this second rotation, $\boldsymbol{r}$ in Eq. (A-13) should be replaced by $\boldsymbol{r}'$. Then, the position vector locating point P after the second rotation is

$$\boldsymbol{r}'' = \boldsymbol{r}' + (1 - \cos\theta_2)[\boldsymbol{n}_2 \times (\boldsymbol{n}_2 \times \boldsymbol{r}')] + \sin\theta_2(\boldsymbol{n}_2 \times \boldsymbol{r}') \qquad \text{(A-15)}$$

Substitution of Eq. (A-14) into Eq. (A-15) gives

$$\begin{aligned}
\boldsymbol{r}'' = {}& \boldsymbol{r} + (1 - \cos\theta_1)[\boldsymbol{n}_1 \times (\boldsymbol{n}_1 \times \boldsymbol{r})] + \sin\theta_1(\boldsymbol{n}_1 \times \boldsymbol{r}) \\
& + (1 - \cos\theta_2)[\boldsymbol{n}_2 \times (\boldsymbol{n}_2 \times \boldsymbol{r})] \\
& + (1 - \cos\theta_2)(1 - \cos\theta_1)\big[\boldsymbol{n}_2 \times \{\boldsymbol{n}_2 \times [\boldsymbol{n}_1 \times (\boldsymbol{n}_1 \times \boldsymbol{r})]\}\big] \\
& + (1 - \cos\theta_2)\sin\theta_1\{\boldsymbol{n}_2 \times [\boldsymbol{n}_2 \times (\boldsymbol{n}_1 \times \boldsymbol{r})]\} \\
& + \sin\theta_2(\boldsymbol{n}_2 \times \boldsymbol{r}) + \sin\theta_2(1 - \cos\theta_1)\{\boldsymbol{n}_2 \times [\boldsymbol{n}_1 \times (\boldsymbol{n}_1 \times \boldsymbol{r})]\} \\
& + \sin\theta_2\sin\theta_1[\boldsymbol{n}_2 \times (\boldsymbol{n}_1 \times \boldsymbol{r})]
\end{aligned} \qquad \text{(A-16)}$$

If the sequence of the rotations is reversed, that is, $\boldsymbol{\theta}_2$ is applied first and then $\boldsymbol{\theta}_1$ is applied, and if we denote the position vectors after the first and second rotations as $\widetilde{\boldsymbol{r}}'$ and $\widetilde{\boldsymbol{r}}''$, respectively, then the application of Eq. (A-13) in precisely the manner as above for $\widetilde{\boldsymbol{r}}'$ and $\widetilde{\boldsymbol{r}}''$ yields

$$\widetilde{\boldsymbol{r}}' = \boldsymbol{r} + (1 - \cos\theta_2)[\boldsymbol{n}_2 \times (\boldsymbol{n}_2 \times \boldsymbol{r})] + \sin\theta_2(\boldsymbol{n}_2 \times \boldsymbol{r}) \qquad \text{(A-17)}$$

and

$$\widetilde{\boldsymbol{r}}'' = \widetilde{\boldsymbol{r}}' + (1 - \cos\theta_1)[\boldsymbol{n}_1 \times (\boldsymbol{n}_1 \times \widetilde{\boldsymbol{r}}')] + \sin\theta_1(\boldsymbol{n}_1 \times \widetilde{\boldsymbol{r}}') \qquad \text{(A-18)}$$

Or, substitution of Eq. (A-17) into Eq. (A-18) gives

$$\begin{aligned}
\widetilde{\boldsymbol{r}}'' = {}& \boldsymbol{r} + (1 - \cos\theta_2)[\boldsymbol{n}_2 \times (\boldsymbol{n}_2 \times \boldsymbol{r})] + \sin\theta_2(\boldsymbol{n}_2 \times \boldsymbol{r}) \\
& + (1 - \cos\theta_1)[\boldsymbol{n}_1 \times (\boldsymbol{n}_1 \times \boldsymbol{r})] \\
& + (1 - \cos\theta_1)(1 - \cos\theta_2)\big[\boldsymbol{n}_1 \times \{\boldsymbol{n}_1 \times [\boldsymbol{n}_2 \times (\boldsymbol{n}_2 \times \boldsymbol{r})]\}\big] \\
& + (1 - \cos\theta_1)\sin\theta_2\{\boldsymbol{n}_1 \times [\boldsymbol{n}_1 \times (\boldsymbol{n}_2 \times \boldsymbol{r})]\} \\
& + \sin\theta_1(\boldsymbol{n}_1 \times \boldsymbol{r}) + \sin\theta_1(1 - \cos\theta_2)\{\boldsymbol{n}_1 \times [\boldsymbol{n}_2 \times (\boldsymbol{n}_2 \times \boldsymbol{r})]\} \\
& + \sin\theta_1\sin\theta_2[\boldsymbol{n}_1 \times (\boldsymbol{n}_2 \times \boldsymbol{r})]
\end{aligned} \qquad \text{(A-19)}$$

where Eqs. (A-17) and (A-19) are completely analogous to Eqs. (A-14) and (A-16), respectively.

Although each of the finite rotations has been assigned a direction and a magnitude, in order that these finite rotations may be added like vectors, the parallelogram law implies that the ending position vector locating point P must be unique, irrespective of the sequence in which the rotations are applied. For the present analysis, this provision is equivalent to requiring that

$$\tilde{r}'' = r'' \tag{A-20}$$

Comparison of Eq. (A-16) with Eq. (A-19) readily reveals that, in general, Eq. (A-20) is *not* true. A simple and well-known example of this inequality is where a change in the sequence of two orthogonal 90° rotations leads to different end configurations.

A-3   DO ROTATIONS EVER BEHAVE AS VECTORS?

While finite rotations are not vectors, *in general*, two important instances for which rotations do behave as vectors may be noted.

A-3.1   Infinitesimal Rotations Are Vectors

For infinitesimal rotations designated as $\boldsymbol{\theta}_1$ and $\boldsymbol{\theta}_2$, terms higher than first order in trigonometric expressions of θ_1 and θ_2 may be neglected since in the limit as θ goes to zero,

$$\sin \theta \approx \theta \qquad \text{and} \qquad 1 - \cos \theta \approx 0$$

Hence, for infinitesimal rotations, Eqs. (A-16) and (A-19) become

$$r'' \approx r + \theta_1 n_1 \times r + \theta_2 n_2 \times r \tag{A-21}$$

and

$$\tilde{r}'' \approx r + \theta_2 n_2 \times r + \theta_1 n_1 \times r \tag{A-22}$$

Or

$$r'' \approx \tilde{r}'' \approx r + (\theta_1 n_1 + \theta_1 n_2) \times r \tag{A-23}$$

Thus, the parallelogram law, which is explicitly expressed in Eq (A-20), is satisfied for infinitesimal rotations. In other words, infinitesimal rotations are vectors.

Since infinitesimal rotations are vectors, it is a direct consequence that *angular velocity and angular acceleration are also vectors*. This is because the angular velocity is defined as the limit of an infinitesimal rotation divided by an infinitesimal *scalar* time interval, resulting in a vector quantity; and the angular acceleration is the time derivative of a vector, and thus must retain the vector character of the angular velocity.

A-3.2   Consecutive Finite Rotations about a Common Axis Are Vectors

Assume $\boldsymbol{n}_2 = \boldsymbol{n}_1$. Then recalling from vector algebra the relation,

$$a \times [b \times (c \times d)] = (b \cdot d)(a \times c) - (b \cdot c)(a \times d) \tag{A-24}$$

leads to

$$n_1 \times [n_1 \times (n_1 \times r)] = -n_1 \times r \tag{A-25}$$

and

$$n_1 \times \{n_1 \times [n_1 \times (n_1 \times r)]\} = -n_1 \times (n_1 \times r) \tag{A-26}$$

Now, since $\boldsymbol{n}_1 = \boldsymbol{n}_2$, Eq. (A-16) may be rewritten if $\boldsymbol{n}_1$ is substituted for $\boldsymbol{n}_2$ throughout; and if Eqs. (A-25) and (A-26) are used to simplify the resulting expression, the equation corresponding to Eq. (A-16) becomes

$$\begin{aligned}
\boldsymbol{r}'' = {}&\boldsymbol{r} + (1 - \cos\theta_1)[\boldsymbol{n}_1 \times (\boldsymbol{n}_1 \times \boldsymbol{r})] + \sin\theta_1(\boldsymbol{n}_1 \times \boldsymbol{r})\\
&+ (1 - \cos\theta_2)[\boldsymbol{n}_1 \times (\boldsymbol{n}_1 \times \boldsymbol{r})] - (1 - \cos\theta_1)(1 - \cos\theta_2)[\boldsymbol{n}_1 \times (\boldsymbol{n}_1 \times \boldsymbol{r})]\\
&- (1 - \cos\theta_2)\sin\theta_1(\boldsymbol{n}_1 \times \boldsymbol{r})\\
&+ \sin\theta_2(\boldsymbol{n}_1 \times \boldsymbol{r}) - \sin\theta_2(1 - \cos\theta_1)(\boldsymbol{n}_1 \times \boldsymbol{r}) + \sin\theta_2\sin\theta_1[\boldsymbol{n}_1 \times (\boldsymbol{n}_1 \times \boldsymbol{r})]\\
= {}&\boldsymbol{r} + (1 - \cos\theta_1\cos\theta_2 + \sin\theta_1\sin\theta_2)[\boldsymbol{n}_1 \times (\boldsymbol{n}_1 \times \boldsymbol{r})]\\
&+ (\sin\theta_1\cos\theta_2 + \cos\theta_1\sin\theta_2)(\boldsymbol{n}_1 \times \boldsymbol{r})\\
= {}&\boldsymbol{r} + [1 - \cos(\theta_1 + \theta_2)][\boldsymbol{n}_1 \times (\boldsymbol{n}_1 \times \boldsymbol{r})] + \sin(\theta_1 + \theta_2)(\boldsymbol{n}_1 \times \boldsymbol{r})
\end{aligned} \tag{A-27}$$

where

$$\sin(\theta_1 + \theta_2) = \sin\theta_1\cos\theta_2 + \cos\theta_1\sin\theta_2$$

and

$$\cos(\theta_1 + \theta_2) = \cos\theta_1\cos\theta_2 - \sin\theta_1\sin\theta_2$$

have been used. Further, by following the same procedure that resulted in Eq. (A-27), Eq. (A-19) can be simplified also to the identical expression; thus giving

$$\widetilde{\boldsymbol{r}}'' = \boldsymbol{r}'' \tag{A-28}$$

So, when consecutive rotations are about a common axis, finite rotations behave like vectors, and thus Eq. (A-20) is satisfied.

This special case is more intuitively understood if viewed from above the plane normal to the rotating axis. As discussed in Section A-1, in this plane, point P moves in a circular trajectory centered at the point at which the rotating axis passes through the plane. The final location of point P is determined by the total angle through which NP is rotated. As sketched in Figs. A-3a and A-3b, irrespective of the sequence of the rotations, the total angle is a scalar summation of the individual rotations, where the identical scalar summation in both cases is the result of commutative contributions.

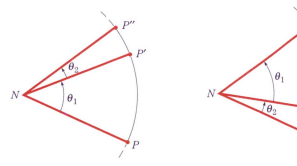

(a) θ_1 is applied, then θ_2 is applied.   (b) θ_2 is applied, then θ_1 is applied

FIGURE A-3   Geometrical relations in plane normal to axis of rotation show motions of point P caused by two successive rotations about common axis, but in different sequence, result in identical final configuration.

GENERAL KINEMATIC ANALYSIS

In this appendix, we find the velocity and acceleration of a point P with respect to the fixed reference frame $OXYZ$, when the kinematics of P are given with respect to an arbitrary set of intermediate frames.

Figure B-1 shows a point P, which is defined in the nth intermediate frame $o_n x_n y_n z_n$ where n is an arbitrary positive integer. The angular velocity of frame $o_i x_i y_i z_i$ is $\boldsymbol{\omega}_i$, $i = 1, 2, 3, \ldots, n$. We shall denote the desired velocity as $\boldsymbol{v}_{P(OXYZ)}$ and the desired acceleration as $\boldsymbol{a}_{P(OXYZ)}$.

The Two Cases

We shall consider two cases. In case 1, the angular velocity of each intermediate frame is defined with respect to the fixed reference frame; that is, $\boldsymbol{\omega}_1, \boldsymbol{\omega}_2, \ldots, \boldsymbol{\omega}_n$ are all defined with respect to $OXYZ$. In case 2, the angular velocity of each intermediate frame is defined with respect to the immediately preceding frame. That is, $\boldsymbol{\omega}_n$ is defined with respect to $o_{n-1} x_{n-1} y_{n-1} z_{n-1}, \ldots, \boldsymbol{\omega}_2$ is defined with respect to $o_1 x_1 y_1 z_1$, and $\boldsymbol{\omega}_1$ is defined with respect to $OXYZ$. Thus, the kinematic analyses presented here are general only within the context of these two cases. Nevertheless, these two cases can be combined to obtain the kinematics for an arbitrary combination of references for the angular velocities.

B-1   ALL ANGULAR VELOCITIES DEFINED WITH RESPECT TO FIXED REFERENCE FRAME (CASE 1)

We shall assume that all the angular velocities of the intermediate frames are defined with respect to the fixed reference frame $OXYZ$, and initially we shall derive the expression for the velocity of point P with respect to $OXYZ$. In order to find the velocity of point P, we shall use the result for a single intermediate frame. Recall that for a point P defined in a single intermediate frame $oxyz$,

$$\boldsymbol{v}_{P(OXYZ)} = \frac{d\boldsymbol{R}_0}{dt} + \boldsymbol{v}_{\text{rel}} + \boldsymbol{\omega} \times \boldsymbol{r} \tag{B-1}$$

where $\boldsymbol{v}_{P(OXYZ)}$ is the velocity of point P with respect to $OXYZ$; $\dfrac{d\boldsymbol{R}_0}{dt}$ is the time rate of change of the position vector $\boldsymbol{R}_0$ that locates o with respect to O; $\boldsymbol{v}_{\text{rel}}$ is the relative velocity of point P with respect to the intermediate frame $oxyz$; $\boldsymbol{\omega}$ is the angular velocity of $oxyz$ with respect to $OXYZ$; and $\boldsymbol{r}$ is the position vector that locates P with respect to $oxyz$.

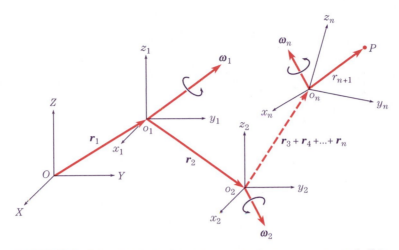

FIGURE B-1   Point P is defined in intermediate frame $o_n x_n y_n z_n$; and all intermediate frames, $o_1 x_1 y_1 z_1$ through $o_n x_n y_n z_n$, rotate at angular velocities $\boldsymbol{\omega}_1$ through $\boldsymbol{\omega}_n$, respectively. The references for $\boldsymbol{\omega}_1$ through $\boldsymbol{\omega}_n$ remain to be defined.

We now consider point P, which is defined in $o_n x_n y_n z_n$, and proceed to find its velocity with respect to $OXYZ$. Using Eq. (B-1), we make a series of parameter substitutions. First, as sketched in Figure B-2,

$$\frac{d\boldsymbol{R}_0}{dt} = \boldsymbol{v}_{o_n(OXYZ)} \tag{B-2}$$

where $\boldsymbol{v}_{o_n(OXYZ)}$ is the velocity of point o_n with respect to $OXYZ$, which we do not presently know. The position of point P defined in $o_n x_n y_n z_n$ is given by the relative position vector as

$$\boldsymbol{r} = \boldsymbol{r}_{n+1} \tag{B-3}$$

and consequently, the relative velocity of point P defined in $o_n x_n y_n z_n$ is the scalar time derivative of $\boldsymbol{r}_{n+1}$.[1] Thus,

$$\boldsymbol{v}_{\text{rel}} = \dot{\boldsymbol{r}}_{n+1} \tag{B-4}$$

In this phase of the analysis, we are considering only the $OXYZ$ and $o_n x_n y_n z_n$ frames. Note that $o_n x_n y_n z_n$ rotates with respect to $OXYZ$ with an angular velocity $\boldsymbol{\omega}_n$. So,

$$\boldsymbol{\omega} = \boldsymbol{\omega}_n \tag{B-5}$$

Substituting Eqs. (B-2) through (B-5) into Eq. (B-1) yields

$$\boldsymbol{v}_{P(OXYZ)} = \boldsymbol{v}_{o_n(OXYZ)} + \dot{\boldsymbol{r}}_{n+1} + \boldsymbol{\omega}_n \times \boldsymbol{r}_{n+1} \tag{B-6}$$

The only unknown quantity on the right-hand side of Eq. (B-6) is $\boldsymbol{v}_{o_n(OXYZ)}$.

---

[1]Great care is taken here to distinguish between $\dfrac{d}{dt} = \left[\left(\dfrac{\partial}{\partial t}\right)_{\text{rel}} + \boldsymbol{\omega} \times\right]$ and $\left(\dfrac{\partial}{\partial t}\right)$ only. Here, $\left(\dfrac{\partial}{\partial t}\right)_{\text{rel}}$ will be denoted by (˙) to indicate time differentiation of the scalar components (coefficients) of the vector in the reference frame in which the vector is defined. Thus, we may simply state that the notation ($\dot{\Box}$) represents the time derivative of the quantity $\Box$, as observed in the reference frame in which $\Box$ is defined. The result of such a time differentiation is often called a *scalar time derivative*.

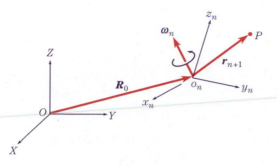

FIGURE B-2 Motion of point P, defined in $o_n x_n y_n z_n$, with respect to $OXYZ$.

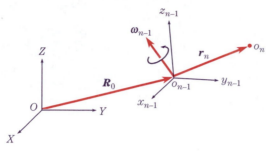

FIGURE B-3 Motion of point o_n, defined in $o_{n-1} x_{n-1} y_{n-1} z_{n-1}$, with respect to $OXYZ$.

Next, we obtain an expression for the velocity of o_n, with respect to $OXYZ$, by use of Eq. (B-1) and Figure B-3. The intermediate frame to consider in this step is $o_{n-1} x_{n-1} y_{n-1} z_{n-1}$ where

$$\boldsymbol{r} = \boldsymbol{r}_n \tag{B-7}$$

$$\boldsymbol{v}_{\text{rel}} = \dot{\boldsymbol{r}}_n \tag{B-8}$$

$$\boldsymbol{\omega} = \boldsymbol{\omega}_{n-1} \tag{B-9}$$

in direct analogy with Eqs. (B-3), (B-4), and (B-5). Substitution of Eqs. (B-7), (B-8), and (B-9) into Eq. (B-1) gives the velocity of point o_n with respect to $OXYZ$ as

$$\boldsymbol{v}_{o_n(OXYZ)} = \boldsymbol{v}_{o_{n-1}(OXYZ)} + \dot{\boldsymbol{r}}_n + \boldsymbol{\omega}_{n-1} \times \boldsymbol{r}_n \tag{B-10}$$

where it has been assumed that $\dfrac{d\boldsymbol{R}_0}{dt} = \boldsymbol{v}_{o_{n-1}(OXYZ)}$ in analogy with Eq. (B-2).

In view of Eq. (B-10), the velocity of the first intermediate frame origin, that is o_1, can be inferred as

$$\boldsymbol{v}_{o_1(OXYZ)} = \boldsymbol{v}_{O(OXYZ)} + \dot{\boldsymbol{r}}_1 + \boldsymbol{\omega}_0 \times \boldsymbol{r}_1 = \dot{\boldsymbol{r}}_1 \tag{B-11}$$

since, by definition, $\boldsymbol{\omega}_0 = 0$ because the reference frame $OXYZ$ does not rotate and $\boldsymbol{v}_{O(OXYZ)} = 0$ because the origin of $OXYZ$ is fixed. In view of Eqs. (B-6), (B-10), and (B-11), we can write

$$\boldsymbol{v}_{P(OXYZ)} = \boldsymbol{v}_{o_n(OXYZ)} + \dot{\boldsymbol{r}}_{n+1} + \boldsymbol{\omega}_n \times \boldsymbol{r}_{n+1}$$

$$\boldsymbol{v}_{o_n(OXYZ)} = \boldsymbol{v}_{o_{n-1}(OXYZ)} + \dot{\boldsymbol{r}}_n + \boldsymbol{\omega}_{n-1} \times \boldsymbol{r}_n$$

$$\boldsymbol{v}_{o_{n-1}(OXYZ)} = \boldsymbol{v}_{o_{n-2}(OXYZ)} + \dot{\boldsymbol{r}}_{n-1} + \boldsymbol{\omega}_{n-2} \times \boldsymbol{r}_{n-1}$$

$$\vdots$$

$$\boldsymbol{v}_{o_2(OXYZ)} = \boldsymbol{v}_{o_1(OXYZ)} + \dot{\boldsymbol{r}}_2 + \boldsymbol{\omega}_1 \times \boldsymbol{r}_2$$

$$\boldsymbol{v}_{o_1(OXYZ)} = \dot{\boldsymbol{r}}_1$$

Summing all of these expressions gives

$$v_{P(OXYZ)} = \sum_{i=0}^{n} \dot{r}_{i+1} + \sum_{i=1}^{n} (\omega_i \times r_{i+1}) \qquad \text{(B-12)}$$

where, in accordance with Eq. (B-11), it has been shown that $\dfrac{dr_1}{dt}$ reduces to $\dot{r}_1$.

In Eq. (B-12), i is a dummy integer variable and n is the total number of intermediate frames. The first summation in Eq. (B-12) is the contribution of the scalar time derivatives of the relative position vectors, whereas the second summation is the contribution of the directional changes of the relative position vectors. Clearly, the $i = 0$ term in Eq. (B-12) accounts for the operation on r_1, corresponding to the first term on the right-hand side of Eq. (B-1).

The same procedure outlined above for the velocity of P can be followed to derive an equation for $a_{P(OXYZ)}$, the acceleration of point P with respect to the fixed reference frame $OXYZ$. Recall that for a point P defined in a single intermediate frame, the acceleration of P with respect to $OXYZ$ is

$$a_{P(OXYZ)} = \frac{d^2 R_0}{dt^2} + a_{\text{rel}} + 2\omega \times v_{\text{rel}} + \dot{\omega} \times r + \omega \times (\omega \times r). \qquad \text{(B-13)}$$

Analogous to the velocity calculations conducted above, first we choose to identify the substitutional parameters for each term in Eq. (B-13) corresponding to Figure B-2. These substitutional parameters are

$$\frac{d^2 R_0}{dt^2} = a_{o_n(OXYZ)} \qquad \text{(B-14)}$$

$$r = r_{n+1} \qquad \text{(B-15)}$$

$$v_{\text{rel}} = \dot{r}_{n+1} \qquad \text{(B-16)}$$

$$a_{\text{rel}} = \ddot{r}_{n+1} \qquad \text{(B-17)}$$

$$\omega = \omega_n \qquad \text{and} \qquad \dot{\omega} = \dot{\omega}_n \qquad \text{(B-18)}$$

Substituting Eqs. (B-14) through (B-18) into Eq. (B-13) yields

$$a_{P(OXYZ)} = a_{o_n(OXYZ)} + \ddot{r}_{n+1} + 2\omega_n \times \dot{r}_{n+1} + \dot{\omega}_n \times r_{n+1} + \omega_n \times (\omega_n \times r_{n+1}) \qquad \text{(B-19)}$$

As in the derivation for the velocity, we shift toward $OXYZ$ to find the acceleration of $a_{o_n(OXYZ)}$ in terms of the appropriate kinematic vectors.

We isolate the motion of o_n defined in $o_{n-1}x_{n-1}y_{n-1}z_{n-1}$ (see Figure B-3). By use of Eq. (B-13) and Figure B-3, for $a_{o_n(OXYZ)}$ we identify

$$\frac{d^2 R_0}{dt^2} = a_{o_{n-1}(OXYZ)} \qquad \text{(B-20)}$$

$$r = r_n \qquad \text{(B-21)}$$

$$v_{\text{rel}} = \dot{r}_n \qquad \text{(B-22)}$$

$$a_{\text{rel}} = \ddot{r}_n \qquad \text{(B-23)}$$

$$\omega = \omega_{n-1} \qquad \text{and} \qquad \dot{\omega} = \dot{\omega}_{n-1} \qquad \text{(B-24)}$$

Substitution of Eqs. (B-20) through (B-24) into Eq. (B-13) gives the acceleration of point o_n with respect to $OXYZ$ as

$$\boldsymbol{a}_{o_n(OXYZ)} = \boldsymbol{a}_{o_{n-1}(OXYZ)} + \ddot{\boldsymbol{r}}_n + 2\boldsymbol{\omega}_{n-1} \times \dot{\boldsymbol{r}}_n + \dot{\boldsymbol{\omega}}_{n-1} \times \boldsymbol{r}_n + \boldsymbol{\omega}_{n-1} \times (\boldsymbol{\omega}_{n-1} \times \boldsymbol{r}_n) \quad \text{(B-25)}$$

Furthermore, analogous to Eq. (B-11) and by use of Eq. (B-13), the acceleration of point o_1 with respect to $OXYZ$ reduces to

$$\boldsymbol{a}_{o_1(OXYZ)} = \ddot{\boldsymbol{r}}_1 \quad \text{(B-26)}$$

because, by definition, the fixed reference frame $OXYZ$ neither rotates nor translates. In view of Eqs. (B-19), (B-25), and (B-26), we may compile the following set of expressions:

$$\boldsymbol{a}_{P(OXYZ)} = \boldsymbol{a}_{o_n(OXYZ)} + \ddot{\boldsymbol{r}}_{n+1} + 2\boldsymbol{\omega}_n \times \dot{\boldsymbol{r}}_{n+1} + \dot{\boldsymbol{\omega}}_n \times \boldsymbol{r}_{n+1} + \boldsymbol{\omega}_n \times (\boldsymbol{\omega}_n \times \boldsymbol{r}_{n+1})$$

$$\boldsymbol{a}_{o_n(OXYZ)} = \boldsymbol{a}_{o_{n-1}(OXYZ)} + \ddot{\boldsymbol{r}}_n + 2\boldsymbol{\omega}_{n-1} \times \dot{\boldsymbol{r}}_n + \dot{\boldsymbol{\omega}}_{n-1} \times \boldsymbol{r}_n + \boldsymbol{\omega}_{n-1} \times (\boldsymbol{\omega}_{n-1} \times \boldsymbol{r}_n)$$

$$\boldsymbol{a}_{o_{n-1}(OXYZ)} = \boldsymbol{a}_{o_{n-2}(OXYZ)} + \ddot{\boldsymbol{r}}_{n-1} + 2\boldsymbol{\omega}_{n-2} \times \boldsymbol{r}_{n-1} + \dot{\boldsymbol{\omega}}_{n-2} \times \boldsymbol{r}_{n-1}$$
$$+ \boldsymbol{\omega}_{n-2} \times (\boldsymbol{\omega}_{n-2} \times \boldsymbol{r}_{n-1})$$

$$\vdots$$

$$\boldsymbol{a}_{o_2(OXYZ)} = \boldsymbol{a}_{o_1(OXYZ)} + \ddot{\boldsymbol{r}}_2 + 2\boldsymbol{\omega}_1 \times \boldsymbol{r}_2 + \dot{\boldsymbol{\omega}}_1 \times \boldsymbol{r}_2 + \boldsymbol{\omega}_1 \times (\boldsymbol{\omega}_1 \times \boldsymbol{r}_2)$$

$$\boldsymbol{a}_{o_1(OXYZ)} = \ddot{\boldsymbol{r}}_1$$

where ($\dot{\Box}$) and ($\ddot{\Box}$) signify scalar time derivatives, in (or as "seen" from) the frame in which the differentiated vector is defined. Summing all of these expressions gives

$$\boxed{\boldsymbol{a}_{P(OXYZ)} = \sum_{i=0}^{n} \ddot{\boldsymbol{r}}_{i+1} + \sum_{i=1}^{n} [2\boldsymbol{\omega}_i \times \dot{\boldsymbol{r}}_{i+1} + \dot{\boldsymbol{\omega}}_i \times \boldsymbol{r}_{i+1} + \boldsymbol{\omega}_i \times (\boldsymbol{\omega}_i \times \boldsymbol{r}_{i+1})]} \quad \text{(B-27)}$$

As in Eq. (B-12), in Eq. (B-27) i is a dummy integer variable and n is the total number of intermediate frames. The first summation in Eq. (B-27) is the contribution of the scalar time derivatives of the relative position vectors, where $i = 0$ accounts for acceleration corresponding to the first term on the right-hand side of Eq. (B-13) and $i = 1, \ldots, n$ in the first term on the right-hand side of Eq. (B-27) accounts for accelerations corresponding to the second term on the right-hand side of Eq. (B-13).

B-2   EACH ANGULAR VELOCITY DEFINED WITH RESPECT TO IMMEDIATELY PRECEDING FRAME (CASE 2)

In this case, we shall assume that the angular velocity of each intermediate frame is defined with respect to its immediately preceding frame. Thus, referring to Figure B-1, $\boldsymbol{\omega}_i$, the angular velocity of $o_i x_i y_i z_i$, is defined with respect to the $o_{i-1} x_{i-1} y_{i-1} z_{i-1}$ frame. As before, we want to derive general expressions for the velocity and acceleration of point P with respect to $OXYZ$.

We begin by considering the motion of point P, defined in $o_n x_n y_n z_n$, with respect to $o_{n-1} x_{n-1} y_{n-1} z_{n-1}$, as depicted in Figure B-4. We note that $\boldsymbol{\omega}_n$, the angular velocity of frame

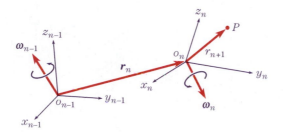

FIGURE B-4   Motion of point P, defined in $o_n x_n y_n z_n$, with respect to $o_{n-1} x_{n-1} y_{n-1} z_{n-1}$.

$o_n x_n y_n z_n$, is defined with respect to frame $o_{n-1} x_{n-1} y_{n-1} z_{n-1}$. (Contrast Figures B-2 and B-4.) First we derive the velocity of point P with respect to $OXYZ$.

Initially, we derive the equation for the velocity of point P, defined in $o_n x_n y_n z_n$, with respect to $o_{n-1} x_{n-1} y_{n-1} z_{n-1}$. In view of Figure B-4, for this initial partial result, the parameters to be substituted into Eq. (B-1) are[2]

$$\frac{d\boldsymbol{R}_0}{dt} = \dot{\boldsymbol{r}}_n \tag{B-28}$$

$$\boldsymbol{r} = \boldsymbol{r}_{n+1} \tag{B-29}$$

$$\boldsymbol{v}_{\text{rel}} = \dot{\boldsymbol{r}}_{n+1} \tag{B-30}$$

$$\boldsymbol{\omega} = \boldsymbol{\omega}_n \tag{B-31}$$

Note that in order to use Eq. (B-1), it is not necessary to require that the frame $o_{n-1} x_{n-1} y_{n-1} z_{n-1}$ is fixed; only that the frame $o_n x_n y_n z_n$ rotates at $\boldsymbol{\omega}_n$ with respect to the frame $o_{n-1} x_{n-1} y_{n-1} z_{n-1}$. Substituting Eqs. (B-28) through (B-31) into Eq. (B-1) gives

$$\boldsymbol{v}_{P(o_{n-1} x_{n-1} y_{n-1} z_{n-1})} = \dot{\boldsymbol{r}}_n + \dot{\boldsymbol{r}}_{n+1} + \boldsymbol{\omega}_n \times \boldsymbol{r}_{n+1} \tag{B-32}$$

Next, we derive the equation for the velocity of point P, defined in $o_{n-1} x_{n-1} y_{n-1} z_{n-1}$, with respect to $o_{n-2} x_{n-2} y_{n-2} z_{n-2}$. In view of Figure B-5, for this next partial result, the parameters to be substituted into Eq. (B-1) are

$$\frac{d\boldsymbol{R}_0}{dt} = \dot{\boldsymbol{r}}_{n-1} \tag{B-33}$$

$$\boldsymbol{v}_{\text{rel}} = \boldsymbol{v}_{P(o_{n-1} x_{n-1} y_{n-1} z_{n-1})} \tag{B-34}$$

$$\boldsymbol{\omega} = \boldsymbol{\omega}_{n-1} \tag{B-35}$$

$$\boldsymbol{r} = \boldsymbol{r}_n + \boldsymbol{r}_{n+1} \tag{B-36}$$

Notice that in order to define the kinematic variables for point P in frame $o_{n-1} x_{n-1} y_{n-1} z_{n-1}$, the position vector $\boldsymbol{r}$ must be as given by Eq. (B-36). Also, the appropriate expression for $\boldsymbol{v}_{\text{rel}}$ in Eq. (B-34) has been found in Eq. (B-32). Substituting Eqs. (B-33) through (B-36) into Eq. (B-1) yields

$$\boldsymbol{v}_{P(o_{n-2} x_{n-2} y_{n-2} z_{n-2})} = \dot{\boldsymbol{r}}_{n-1} + \boldsymbol{v}_{P(o_{n-1} x_{n-1} y_{n-1} z_{n-1})} + \boldsymbol{\omega}_{n-1} \times (\boldsymbol{r}_n + \boldsymbol{r}_{n+1}) \tag{B-37}$$

---

[2]As before, $(\dot{\square})$ denotes the scalar time derivative of $\square$ and signifies the change of the scalar components of the vector $\square$ in the frame in which it is defined.

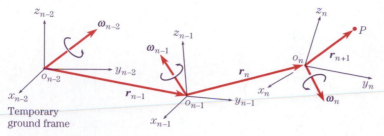

FIGURE B-5   Motion of point P, defined in $o_{n-1}x_{n-1}y_{n-1}z_{n-1}$, with respect to $o_{n-2}x_{n-2}y_{n-2}z_{n-2}$.

Substituting the right-hand side of Eq. (B-32) into Eq. (B-37) yields

$$\boldsymbol{v}_{P(o_{n-2}x_{n-2}y_{n-2}z_{n-2})} = \dot{\boldsymbol{r}}_{n-1} + \dot{\boldsymbol{r}}_n + \dot{\boldsymbol{r}}_{n+1} + \boldsymbol{\omega}_n \times \boldsymbol{r}_{n+1} + \boldsymbol{\omega}_{n-1} \times (\boldsymbol{r}_n + \boldsymbol{r}_{n+1}) \quad \text{(B-38)}$$

Similarly, for the velocity of point P, defined in $o_{n-2}x_{n-2}y_{n-2}z_{n-2}$, with respect to $o_{n-3}x_{n-3}y_{n-3}z_{n-3}$, we consider the motion of point P as sketched in Figure B-6. Then

$$\frac{d\boldsymbol{R}_0}{dt} = \dot{\boldsymbol{r}}_{n-2} \quad \text{(B-39)}$$

$$\boldsymbol{v}_{\text{rel}} = \boldsymbol{v}_{P(o_{n-2}x_{n-2}y_{n-2}z_{n-2})} \quad \text{(B-40)}$$

$$\boldsymbol{\omega} = \boldsymbol{\omega}_{n-2} \quad \text{(B-41)}$$

$$\boldsymbol{r} = \boldsymbol{r}_{n-1} + \boldsymbol{r}_n + \boldsymbol{r}_{n+1} \quad \text{(B-42)}$$

Substituting Eqs. (B-39) through (B-42) into Eq. (B-1) yields

$$\boldsymbol{v}_{P(o_{n-3}x_{n-3}y_{n-3}z_{n-3})} = \dot{\boldsymbol{r}}_{n-2} + \boldsymbol{v}_{P(o_{n-2}x_{n-2}y_{n-2}z_{n-2})} + \boldsymbol{\omega}_{n-2} \times (\boldsymbol{r}_{n-1} + \boldsymbol{r}_n + \boldsymbol{r}_{n+1}) \quad \text{(B-43)}$$

Also, substituting the right-hand side of Eq. (B-38) into Eq. (B-43), in order to eliminate $\boldsymbol{v}_{P(o_{n-2}x_{n-2}y_{n-2}z_{n-2})}$ from Eq. (B-43), yields

$$\boldsymbol{v}_{P(o_{n-3}x_{n-3}y_{n-3}z_{n-3})} = \dot{\boldsymbol{r}}_{n-2} + \dot{\boldsymbol{r}}_{n-1} + \dot{\boldsymbol{r}}_n + \dot{\boldsymbol{r}}_{n+1} + \boldsymbol{\omega}_n \times \boldsymbol{r}_{n+1}$$
$$+ \boldsymbol{\omega}_{n-1} \times (\boldsymbol{r}_n + \boldsymbol{r}_{n+1}) + \boldsymbol{\omega}_{n-2} \times (\boldsymbol{r}_{n-1} + \boldsymbol{r}_n + \boldsymbol{r}_{n+1}) \quad \text{(B-44)}$$

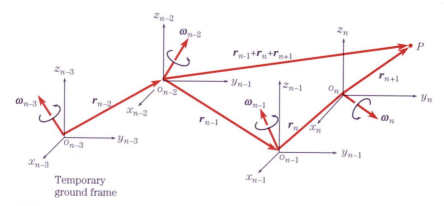

FIGURE B-6   Motion of point P, defined in $o_{n-2}x_{n-2}y_{n-2}z_{n-2}$, with respect to $o_{n-3}x_{n-3}y_{n-3}z_{n-3}$.

Inspection of Eq. (B-44) suggests that as we move toward the fixed reference frame $OXYZ$, the velocity of point P, with respect to $OXYZ$, becomes a structured sum of the time derivatives of the r_j, $j = 1, 2, 3, \ldots, n + 1$. These scalar time derivatives are the first four terms in Eq. (B-44), and the remaining terms are the directional derivatives. Thus, in view of Eqs. (B-32), (B-38), and (B-44), and by mathematical induction, the velocity of point P, defined in $o_n x_n y_n z_n$, with respect to $OXYZ$ may be written as

$$v_{P(OXYZ)} = \sum_{j=1}^{n+1} \dot{r}_j + \sum_{j=0}^{n-1} \omega_{n-j} \times \left(\sum_{k=0}^{j} r_{n-k+1} \right) \tag{B-45}$$

where n is the total number of intermediate frames, and k and j are dummy integer variables.

Finally, we derive the equation for the acceleration of point P, with respect to $OXYZ$, for the case where the angular velocity of each intermediate frame is defined with respect to the immediately preceding frame. We begin this last task by deriving the equation for the acceleration of point P, defined in $o_n x_n y_n z_n$, with respect to $o_{n-1} x_{n-1} y_{n-1} z_{n-1}$. In view of Figure B-4, we must define the parameters to be substituted into Eq. (B-13). Note that Eqs. (B-28) through (B-31) remain valid for this analysis; and the additional parameters are

$$\frac{d^2 R_0}{dt^2} = \ddot{r}_n \tag{B-46}$$

$$a_{\text{rel}} = \ddot{r}_{n+1} \tag{B-47}$$

$$\dot{\omega} = \dot{\omega}_n \tag{B-48}$$

Substituting Eqs. (B-28) through (B-31) and Eqs. (B-46) through (B-48) into Eq. (B-13) yields

$$a_{P(o_{n-1} x_{n-1} y_{n-1} z_{n-1})} = \ddot{r}_n + \ddot{r}_{n+1} + 2\omega_n \times \dot{r}_{n+1} + \dot{\omega}_n \times r_{n+1} + \omega_n \times (\omega_n \times r_{n+1}). \tag{B-49}$$

Next, consider the motion of point P, defined in $o_{n-1} x_{n-1} y_{n-1} z_{n-1}$, with respect to $o_{n-2} x_{n-2} y_{n-2} z_{n-2}$. In view of Figure B-5, we must define the parameters to be substituted into Eq. (B-13). Note that Eqs. (B-33) through (B-36) remain valid for this analysis; and the additional parameters are

$$\frac{d^2 R_0}{dt^2} = \ddot{r}_{n-1} \tag{B-50}$$

$$a_{\text{rel}} = a_{P(o_{n-1} x_{n-1} y_{n-1} z_{n-1})} \tag{B-51}$$

$$\dot{\omega} = \dot{\omega}_{n-1} \tag{B-52}$$

Substituting Eqs. (B-33) through (B-36) and Eqs. (B-50) through (B-52) into Eq. (B-13) yields

$$a_{P(o_{n-2} x_{n-2} y_{n-2} z_{n-2})} = \ddot{r}_{n-1} + a_{P(o_{n-1} x_{n-1} y_{n-1} z_{n-1})} + 2\omega_{n-1} \times v_{P(o_{n-1} x_{n-1} y_{n-1} z_{n-1})}$$
$$+ \dot{\omega}_{n-1} \times (r_n + r_{n+1}) + \omega_{n-1} \times [\omega_{n-1} \times (r_n + r_{n+1})] \tag{B-53}$$

Further, substitution of the right-hand sides of Eqs. (B-32) and (B-49) into Eq. (B-53) gives

$$\boldsymbol{a}_{P(o_{n-2}x_{n-2}y_{n-2}z_{n-2})} = \ddot{\boldsymbol{r}}_{n-1} + \ddot{\boldsymbol{r}}_n + \ddot{\boldsymbol{r}}_{n+1} + 2\boldsymbol{\omega}_n \times \dot{\boldsymbol{r}}_{n+1} + \dot{\boldsymbol{\omega}}_n \times \boldsymbol{r}_{n+1}$$
$$+ \boldsymbol{\omega}_n \times (\boldsymbol{\omega}_n \times \boldsymbol{r}_{n+1}) + 2\boldsymbol{\omega}_{n-1} \times \dot{\boldsymbol{r}}_n + 2\boldsymbol{\omega}_{n-1} \times \dot{\boldsymbol{r}}_{n+1}$$
$$+ 2\boldsymbol{\omega}_{n-1} \times (\boldsymbol{\omega}_n \times \boldsymbol{r}_{n+1}) + \dot{\boldsymbol{\omega}}_{n-1} \times (\boldsymbol{r}_n + \boldsymbol{r}_{n+1})$$
$$+ \boldsymbol{\omega}_{n-1} \times [\boldsymbol{\omega}_{n-1} \times (\boldsymbol{r}_n + \boldsymbol{r}_{n+1})] \tag{B-54}$$

Rearranging the terms in Eq. (B-54) gives

$$\boldsymbol{a}_{P(o_{n-2}x_{n-2}y_{n-2}z_{n-2})} = \ddot{\boldsymbol{r}}_{n-1} + \ddot{\boldsymbol{r}}_n + \ddot{\boldsymbol{r}}_{n+1} + 2\boldsymbol{\omega}_n \times \dot{\boldsymbol{r}}_{n+1} + 2\boldsymbol{\omega}_{n-1} \times (\dot{\boldsymbol{r}}_n + \dot{\boldsymbol{r}}_{n+1})$$
$$+ \dot{\boldsymbol{\omega}}_n \times \boldsymbol{r}_{n+1} + \dot{\boldsymbol{\omega}}_{n-1} \times (\boldsymbol{r}_n + \boldsymbol{r}_{n+1}) + \boldsymbol{\omega}_n \times (\boldsymbol{\omega}_n \times \boldsymbol{r}_{n+1})$$
$$+ \boldsymbol{\omega}_{n-1} \times [\boldsymbol{\omega}_{n-1} \times (\boldsymbol{r}_n + \boldsymbol{r}_{n+1})] + 2\boldsymbol{\omega}_{n-1} \times (\boldsymbol{\omega}_n \times \boldsymbol{r}_{n+1})$$
$$\tag{B-55}$$

Similarly, for the acceleration of point P, defined in $o_{n-2}x_{n-2}y_{n-2}z_{n-2}$, with respect to $o_{n-3}x_{n-3}y_{n-3}z_{n-3}$, we must consider the motion of point P as illustrated in Figure B-6. Note that Eqs. (B-39) through (B-42) remain valid for this analysis; and the additional parameters are

$$\frac{d^2\boldsymbol{R}_0}{dt^2} = \ddot{\boldsymbol{r}}_{n-2} \tag{B-56}$$

$$\boldsymbol{a}_{\text{rel}} = \boldsymbol{a}_{P(o_{n-2}x_{n-2}y_{n-2}z_{n-2})} \tag{B-57}$$

$$\dot{\boldsymbol{\omega}} = \dot{\boldsymbol{\omega}}_{n-2} \tag{B-58}$$

Substituting Eqs. (B-39) through (B-42) and Eqs. (B-56) through (B-58) into Eq. (B-13) yields

$$\boldsymbol{a}_{P(o_{n-3}x_{n-3}y_{n-3}z_{n-3})} = \ddot{\boldsymbol{r}}_{n-2} + \boldsymbol{a}_{P(o_{n-2}x_{n-2}y_{n-2}z_{n-2})}$$
$$+ 2\boldsymbol{\omega}_{n-2} \times \boldsymbol{v}_{P(o_{n-2}x_{n-2}y_{n-2}z_{n-2})}$$
$$+ \dot{\boldsymbol{\omega}}_{n-2} \times (\boldsymbol{r}_{n-1} + \boldsymbol{r}_n + \boldsymbol{r}_{n+1})$$
$$+ \boldsymbol{\omega}_{n-2} \times [\boldsymbol{\omega}_{n-2} \times (\boldsymbol{r}_{n-1} + \boldsymbol{r}_n + \boldsymbol{r}_{n+1})] \tag{B-59}$$

Further, substitution of the right-hand sides of Eqs. (B-38) and (B-55) into Eq. (B-59) gives

$$\boldsymbol{a}_{P(o_{n-3}x_{n-3}y_{n-3}z_{n-3})} = \ddot{\boldsymbol{r}}_{n-2} + \ddot{\boldsymbol{r}}_{n-1} + \ddot{\boldsymbol{r}}_n + \ddot{\boldsymbol{r}}_{n+1}$$
$$+ 2\boldsymbol{\omega}_n \times \dot{\boldsymbol{r}}_{n+1} + 2\boldsymbol{\omega}_{n-1} \times (\dot{\boldsymbol{r}}_n + \dot{\boldsymbol{r}}_{n+1})$$
$$+ \dot{\boldsymbol{\omega}}_n \times \boldsymbol{r}_{n+1} + \dot{\boldsymbol{\omega}}_{n-1} \times (\boldsymbol{r}_n + \boldsymbol{r}_{n+1})$$
$$+ \boldsymbol{\omega}_n \times (\boldsymbol{\omega}_n \times \boldsymbol{r}_{n+1}) + \boldsymbol{\omega}_{n-1} \times [\boldsymbol{\omega}_{n-1} \times (\boldsymbol{r}_n + \boldsymbol{r}_{n+1})]$$
$$+ 2\boldsymbol{\omega}_{n-1} \times (\boldsymbol{\omega}_n \times \boldsymbol{r}_{n+1}) + 2\boldsymbol{\omega}_{n-2} \times (\dot{\boldsymbol{r}}_{n-1} + \dot{\boldsymbol{r}}_n + \dot{\boldsymbol{r}}_{n+1})$$
$$+ 2\boldsymbol{\omega}_{n-2} \times (\boldsymbol{\omega}_n \times \boldsymbol{r}_{n+1}) + 2\boldsymbol{\omega}_{n-2} \times [\boldsymbol{\omega}_{n-1} \times (\boldsymbol{r}_n + \boldsymbol{r}_{n+1})]$$
$$+ \dot{\boldsymbol{\omega}}_{n-2} \times (\boldsymbol{r}_{n-1} + \boldsymbol{r}_n + \boldsymbol{r}_{n+1})$$
$$+ \boldsymbol{\omega}_{n-2} \times [\boldsymbol{\omega}_{n-2} \times (\boldsymbol{r}_{n-1} + \boldsymbol{r}_n + \boldsymbol{r}_{n+1})] \tag{B-60}$$

Rearranging the terms in Eq. (B-60) gives

$$\boldsymbol{a}_{P(o_{n-3}x_{n-3}y_{n-3}z_{n-3})} = \ddot{\boldsymbol{r}}_{n-2} + \ddot{\boldsymbol{r}}_{n-1} + \ddot{\boldsymbol{r}}_n + \ddot{\boldsymbol{r}}_{n+1} + 2\boldsymbol{\omega}_n \times \dot{\boldsymbol{r}}_{n+1}$$
$$+ 2\boldsymbol{\omega}_{n-1} \times (\dot{\boldsymbol{r}}_n + \dot{\boldsymbol{r}}_{n+1}) + 2\boldsymbol{\omega}_{n-2} \times (\dot{\boldsymbol{r}}_{n-1} + \dot{\boldsymbol{r}}_n + \dot{\boldsymbol{r}}_{n+1})$$
$$+ 2\boldsymbol{\omega}_{n-1} \times (\boldsymbol{\omega}_n \times \boldsymbol{r}_{n+1}) + 2\boldsymbol{\omega}_{n-2} \times (\boldsymbol{\omega}_n \times \boldsymbol{r}_{n+1})$$
$$+ 2\boldsymbol{\omega}_{n-2} \times [\boldsymbol{\omega}_{n-1} \times (\boldsymbol{r}_n + \boldsymbol{r}_{n+1})]$$
$$+ \dot{\boldsymbol{\omega}}_n \times \boldsymbol{r}_{n+1} + \dot{\boldsymbol{\omega}}_{n-1} \times (\boldsymbol{r}_n + \boldsymbol{r}_{n+1}) + \dot{\boldsymbol{\omega}}_{n-2} \times (\boldsymbol{r}_{n-1} + \boldsymbol{r}_n + \boldsymbol{r}_{n+1})$$
$$+ \boldsymbol{\omega}_n \times (\boldsymbol{\omega}_n \times \boldsymbol{r}_{n+1}) + \boldsymbol{\omega}_{n-1} \times [\boldsymbol{\omega}_{n-1} \times (\boldsymbol{r}_n + \boldsymbol{r}_{n+1})]$$
$$+ \boldsymbol{\omega}_{n-2} \times [\boldsymbol{\omega}_{n-2} \times (\boldsymbol{r}_{n-1} + \boldsymbol{r}_n + \boldsymbol{r}_{n+1})] \tag{B-61}$$

In consideration of Eq. (B-61), perhaps the most difficult summation to generalize via induction is the combination of the third and fourth rows. In clarifying this summation, it may be beneficial to pursue this process of acceleration derivations one step further to $\boldsymbol{a}_{P(o_{n-4}x_{n-4}y_{n-4}z_{n-4})}$. If such a derivation is accomplished, then the terms corresponding to the third and fourth rows of Eq. (B-61), where the acceleration is $\boldsymbol{a}_{P(o_{n-4}x_{n-4}y_{n-4}z_{n-4})}$, may be grouped in the following manner:

$$2\boldsymbol{\omega}_{n-1} \times (\boldsymbol{\omega}_n \times \boldsymbol{r}_{n+1}) \qquad\qquad\qquad\qquad \left.\vphantom{\int}\right\} \quad \text{Ⓐ}$$
$$+ 2\boldsymbol{\omega}_{n-2} \times (\boldsymbol{\omega}_n \times \boldsymbol{r}_{n+1}) \qquad\qquad\qquad \left.\vphantom{\int\int}\right\} \quad \text{Ⓑ}$$
$$+ 2\boldsymbol{\omega}_{n-2} \times [\boldsymbol{\omega}_{n-1} \times (\boldsymbol{r}_n + \boldsymbol{r}_{n+1})]$$
$$+ 2\boldsymbol{\omega}_{n-3} \times (\boldsymbol{\omega}_n \times \boldsymbol{r}_{n+1})$$
$$+ 2\boldsymbol{\omega}_{n-3} \times [\boldsymbol{\omega}_{n-1} \times (\boldsymbol{r}_n + \boldsymbol{r}_{n+1})] \qquad\qquad \left.\vphantom{\int\int\int}\right\} \quad \text{Ⓒ}$$
$$+ 2\boldsymbol{\omega}_{n-3} \times [\boldsymbol{\omega}_{n-2} \times (\boldsymbol{r}_{n-1} + \boldsymbol{r}_n + \boldsymbol{r}_{n+1})] \tag{B-62}$$

Notice that the groupings labeled Ⓐ, Ⓑ, and Ⓒ are three different classes of terms that can be condensed into a *single* summation. Below, in Eq. (B-63), we call this summation term the "pseudocentripetal acceleration." Therefore, one choice of summation formulation for the acceleration of point P, defined in $o_n x_n y_n z_n$, with respect to $OXYZ$ becomes

$$\boldsymbol{a}_{P(OXYZ)} = \sum_{j=1}^{n+1} \ddot{\boldsymbol{r}}_j$$

(which we call the intermediate acceleration contribution)

$$+ 2\sum_{j=0}^{n-1} \boldsymbol{\omega}_{n-j} \times \left(\sum_{k=0}^{j} \dot{\boldsymbol{r}}_{n-k+1} \right)$$

(which we call the Coriolis acceleration contribution)

$$+ 2\sum_{i=1}^{n-1} \boldsymbol{\omega}_{n-i} \times \left[\sum_{j=1}^{i} \boldsymbol{\omega}_{n-j+1} \times \left(\sum_{k=0}^{j-1} \boldsymbol{r}_{n-k+1} \right) \right]$$

(which we call the pseudocentripetal acceleration contribution)

$$+ \sum_{j=0}^{n-1} \dot{\boldsymbol{\omega}}_{n-j} \times \left(\sum_{k=0}^{j} \boldsymbol{r}_{n-k+1} \right)$$

(which we call the Euler acceleration contribution)

$$+ \sum_{j=0}^{n-1} \boldsymbol{\omega}_{n-j} \times \left[\boldsymbol{\omega}_{n-j} \times \left(\sum_{k=0}^{j} \boldsymbol{r}_{n-k+1} \right) \right] \tag{B-63}$$

(which we call the centripetal acceleration contribution)

where n is the total number of intermediate reference frames, and k and j are dummy integer variables.

If, for example, $n = 2$ (that is, there are two intermediate reference frames), Eq. (B-63) gives

$$
\begin{aligned}
\boldsymbol{a}_{P(OXYZ)} &= \sum_{j=1}^{3} \ddot{\boldsymbol{r}}_j + 2\sum_{j=0}^{1} \boldsymbol{\omega}_{2-j} \times \left(\sum_{k=0}^{j} \dot{\boldsymbol{r}}_{2-k+1}\right) \\
&\quad + 2\sum_{i=1}^{1} \boldsymbol{\omega}_{2-i} \times \left[\sum_{j=1}^{i} \boldsymbol{\omega}_{2-j+1} \times \left(\sum_{k=0}^{j-1} \boldsymbol{r}_{2-k+1}\right)\right] \\
&\quad + \sum_{j=0}^{1} \dot{\boldsymbol{\omega}}_{2-j} \times \left(\sum_{k=0}^{j} \boldsymbol{r}_{2-k+1}\right) \\
&\quad + \sum_{j=0}^{1} \boldsymbol{\omega}_{2-j} \times \left[\boldsymbol{\omega}_{2-j} \times \left(\sum_{k=0}^{j} \boldsymbol{\omega}_{2-k+1}\right)\right] \\
&= \ddot{\boldsymbol{r}}_1 + \ddot{\boldsymbol{r}}_2 + \ddot{\boldsymbol{r}}_3 + 2\boldsymbol{\omega}_2 \times \dot{\boldsymbol{r}}_3 + 2\boldsymbol{\omega}_1 \times (\dot{\boldsymbol{r}}_3 + \dot{\boldsymbol{r}}_2) \\
&\quad + 2\boldsymbol{\omega}_1 \times (\boldsymbol{\omega}_2 \times \boldsymbol{r}_3) + \dot{\boldsymbol{\omega}}_2 \times \boldsymbol{r}_3 + \dot{\boldsymbol{\omega}}_1 \times (\boldsymbol{r}_3 + \boldsymbol{r}_2) \\
&\quad + \boldsymbol{\omega}_2 \times (\boldsymbol{\omega}_2 \times \boldsymbol{r}_3) + \boldsymbol{\omega}_1 \times [\boldsymbol{\omega}_1 \times (\boldsymbol{r}_3 + \boldsymbol{r}_2)]
\end{aligned}
\tag{B-64}
$$

which upon rearranging the terms becomes

$$
\begin{aligned}
\boldsymbol{a}_{P(OXYZ)} &= \ddot{\boldsymbol{r}}_1 + \ddot{\boldsymbol{r}}_2 + \ddot{\boldsymbol{r}}_3 + 2\boldsymbol{\omega}_1 \times \dot{\boldsymbol{r}}_2 + 2(\boldsymbol{\omega}_1 + \boldsymbol{\omega}_2) \times \dot{\boldsymbol{r}}_3 \\
&\quad + \dot{\boldsymbol{\omega}}_1 \times \boldsymbol{r}_2 + (\dot{\boldsymbol{\omega}}_1 + \dot{\boldsymbol{\omega}}_2) \times \boldsymbol{r}_3 + \boldsymbol{\omega}_2 \times (\boldsymbol{\omega}_2 \times \boldsymbol{r}_3) \\
&\quad + 2\boldsymbol{\omega}_1 \times (\boldsymbol{\omega}_2 \times \boldsymbol{r}_3) + \boldsymbol{\omega}_1 \times [\boldsymbol{\omega}_1 \times (\boldsymbol{r}_2 + \boldsymbol{r}_3)]
\end{aligned}
\tag{B-65}
$$

You now know more—or less—than you think you do; neo–Dr. Spock.

MOMENTUM PRINCIPLES FOR SYSTEMS OF PARTICLES

In this appendix, the two momentum principles—the *linear momentum principle* and the *angular momentum principle*—are asserted (not derived) with respect to an inertial reference frame for the motions of two models of mass-bearing systems, that is, a single particle and a system of particles. It is shown that in the case of a system of particles, the two momentum principles are independent of each other; whereas in the case of a single particle, they are not independent. Also, it is shown that certain conditions must be satisfied by the internal interactions between the constituents in the case of a system of particles. Also, in the case of a system of particles, convenient and general forms of the two momentum principles are derived.

Finally, the angular momentum principle is derived in terms of the variables defined with respect to a noninertial reference frame.

C-1   ASSERTED MOMENTUM PRINCIPLES

In an inertial reference frame, the governing equations of motion for a mass-bearing system—a particle or a system of particles—may be derived from two generally *independent* momentum principles: the linear momentum principle and the angular momentum principle.

The linear momentum principle prescribes the relationship between the resultant or net external force and the linear momentum of the mass-bearing system. That is, the resultant force acting on the system (or any subset thereof in the case of a system of particles) is equal to the time rate of change of its linear momentum with respect to an inertial reference frame:

$$F = \frac{dP}{dt} \tag{C-1}$$

where F denotes the resultant force vector and P denotes the linear momentum vector defined with respect to an inertial reference frame.

The angular momentum principle similarly provides the relationship between the resultant or net external torque and the time rate of change of the angular momentum, which is defined with respect to an inertial reference frame and about the origin of that inertial reference frame. That is, the resultant torque acting on the system (or any subset thereof in the case of a system of particles), about the origin of an inertial reference frame, is equal to the time rate of change of its angular momentum:

$$\boldsymbol{\tau}_o = \frac{d\boldsymbol{H}_o}{dt} \tag{C-2}$$

where $\boldsymbol{\tau}_o$ denotes the resultant torque vector about the origin O of an inertial reference frame and $\boldsymbol{H}_o$ denotes the angular momentum vector of the system about the origin of that inertial reference frame.

It should be noted that the two momentum principles—of linear momentum and of angular momentum—are generally independent of each other, and that, in general, neither of these may be derived exclusively from the other.

Finally, we emphasize once again that the two momentum principles given by Eqs. (C-1) and (C-2) hold for a particle, a system of particles, and any subset of the latter. In the following sections, Eqs. (C-1) and (C-2) are applied to each of these mass-bearing systems in order to study the momentum principles for each case in detail. It is to be understood that throughout this discussion, we are concerned with velocities of bodies, which are small in comparison with the velocity of light.

C-2 PRINCIPLES FOR SINGLE PARTICLE

The simplest case to which Eqs. (C-1) and (C-2) may be applied is the case of a single particle as depicted in Figure C-1. The constitutive relations for the motion of a single particle are *defined* for linear momentum as

$$\boldsymbol{p} = m\boldsymbol{v} \tag{C-3}$$

and, for angular momentum about the origin of the inertial reference frame, as

$$\boldsymbol{h}_o = \boldsymbol{r} \times m\boldsymbol{v} \tag{C-4}$$

where $\boldsymbol{r}$ and $\boldsymbol{v}$ are the position and velocity vectors of the particle, respectively, with respect to the inertial reference frame $OXYZ$, and m denotes the constant mass of the particle.

The momentum principles for a single particle may be found by substituting Eqs. (C-3) and (C-4) into Eqs. (C-1) and (C-2), respectively, where $\boldsymbol{H}_o$ is replaced by $\boldsymbol{h}_o$ and $\boldsymbol{F}$ is replaced by $\boldsymbol{f}$, the resultant force on the particle. Thus,

$$\boldsymbol{f} = \frac{d}{dt}(m\boldsymbol{v})$$

$$= m\frac{d\boldsymbol{v}}{dt} \tag{C-5}$$

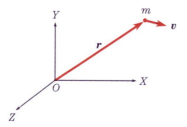

FIGURE C-1   Motion of single particle.

and

$$\tau_o = \frac{d}{dt}(\boldsymbol{r} \times m\boldsymbol{v})$$

$$= \frac{d\boldsymbol{r}}{dt} \times m\boldsymbol{v} + \boldsymbol{r} \times \frac{d}{dt}(m\boldsymbol{v})$$

$$= \boldsymbol{v} \times m\boldsymbol{v} + \boldsymbol{r} \times m\frac{d\boldsymbol{v}}{dt}$$

$$= \boldsymbol{r} \times m\frac{d\boldsymbol{v}}{dt} \tag{C-6}$$

The first term on the right-hand side of the line preceding Eq. (C-6) is zero since $\boldsymbol{v} \times \boldsymbol{v} = 0$. Also, the kinematic relationship $\boldsymbol{v} = \dfrac{d\boldsymbol{r}}{dt}$ was used in deriving Eq. (C-6). Note that τ_o in Eq. (C-6) is simply the moment of the force $\boldsymbol{f}$ about the fixed point O, since, by definition, a particle cannot otherwise sustain a torque. Therefore, the left-hand side of Eq. (C-6) may be rewritten as

$$\tau_o = \boldsymbol{r} \times \boldsymbol{f} \tag{C-7}$$

Using Eq. (C-7) in Eq. (C-6) gives

$$\boldsymbol{r} \times \boldsymbol{f} = \boldsymbol{r} \times m\frac{d\boldsymbol{v}}{dt} \tag{C-8}$$

or

$$\boldsymbol{r} \times \left(\boldsymbol{f} - m\frac{d\boldsymbol{v}}{dt}\right) = 0 \tag{C-9}$$

Equation (C-9) is satisfied if the quantity in the parentheses vanishes, or equivalently, if Eq. (C-5) holds. Conversely, if Eq. (C-9) holds, Eq. (C-5) is also satisfied because $\boldsymbol{r}$ is generally nonzero and because $\boldsymbol{r}$ is generally not in the same direction as the vector quantity in the parentheses in Eq. (C-9). That is, the linear momentum principle is a necessary and sufficient condition for the validity of the angular momentum principle, and (with the exception of the isolated cases when $\boldsymbol{r} = 0$ or when $\boldsymbol{r}$ and $\boldsymbol{f}$ are collinear) vice versa. Alternatively, for a particle, Eq. (C-6) is an equivalent and sometimes more convenient form of Eq. (C-5).

In summary, it may be stated that the only momentum principle required for the formulation of the equations of motion of a single newtonian particle is the linear momentum principle given by Eq. (C-5): *The resultant force acting on a particle is proportional in magnitude to and is in the same direction as the time rate of change of the particle's velocity with respect to an inertial reference frame, where the proportionality constant is the particle's mass.*

C-3   PRINCIPLES FOR SYSTEM OF PARTICLES

Next in level of difficulty is the case of a system of N particles as shown in Figure C-2. In this section, the momentum principles for a system of particles are *asserted* as Eqs. (C-1) and (C-2), and then expressed in terms of the momentum principles for each constituent particle. Next, the momentum principles for a system of particles are derived by summing

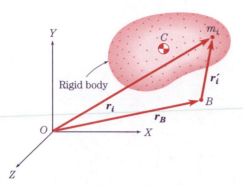

FIGURE C-2   Motion of system of particles.

the principles for an individual particle to obtain the corresponding principles for a system of particles. Then, the conditions on the nature of the mutual internal forces between the constituent particles are found by comparing the *asserted* and the *derived* momentum principles. Finally, a useful form of the linear momentum principle is derived in terms of the motion of the centroid; and a more general form of the angular momentum principle is derived about an arbitrary point.

C-3.1   Asserted System Momentum Principles

In this subsection, the momentum principles for a system of particles are *asserted* as Eqs. (C-1) and (C-2). Now we shall express these asserted principles in terms of variables of the constituent particles.

For a system of particles, from Eq. (C-3), the linear momentum is given by

$$P = \sum_{i=1}^{N} p_i = \sum_{i=1}^{N} m_i v_i \tag{C-10}$$

where m_i, p_i, and v_i are the mass, linear momentum, and inertial velocity of the ith particle, respectively; and the angular momentum about the fixed point O, with respect to the inertial reference frame, is expressed, using Eq. (C-4), as

$$H_o = \sum_{i=1}^{N} h_{oi} = \sum_{i=1}^{N} r_i \times m_i v_i \tag{C-11}$$

where h_{oi} is the angular momentum of the ith particle about the fixed point O and r_i is the position vector of the ith particle measured from the fixed point O. By substituting Eqs. (C-10) and (C-11) into Eqs. (C-1) and (C-2), the *asserted* momentum principles for a system of particles may be obtained as

$$F = \frac{d}{dt} \sum_{i=1}^{N} m_i v_i$$

$$= \sum_{i=1}^{N} m_i \frac{dv_i}{dt} \tag{C-12}$$

and

$$\boldsymbol{\tau}_o = \frac{d}{dt} \sum_{i=1}^{N} (\boldsymbol{r}_i \times m_i \boldsymbol{v}_i)$$

$$= \sum_{i=1}^{N} \left(\frac{d\boldsymbol{r}_i}{dt} \times m_i \boldsymbol{v}_i \right) + \sum_{i=1}^{N} \left(\boldsymbol{r}_i \times m_i \frac{d\boldsymbol{v}_i}{dt} \right)$$

$$= \sum_{i=1}^{N} (\boldsymbol{v}_i \times m_i \boldsymbol{v}_i) + \sum_{i=1}^{N} \left(\boldsymbol{r}_i \times m_i \frac{d\boldsymbol{v}_i}{dt} \right)$$

$$= \sum_{i=1}^{N} \left(\boldsymbol{r}_i \times m_i \frac{d\boldsymbol{v}_i}{dt} \right) \tag{C-13}$$

Equations (C-12) and (C-13) are sometimes called Euler's first law and Euler's second law, respectively, although we shall not adopt these names.

Note that $\boldsymbol{F}$ and $\boldsymbol{\tau}_o$ in Eqs. (C-1) and (C-2) (and, thus, also those in Eqs. (C-12) and (C-13)) are the net *external* (meaning external to the system of particles) force and torque, respectively, acting on the system of particles. Therefore,

$$\boldsymbol{F} = \sum_{i=1}^{N} \boldsymbol{f}_i \tag{C-14}$$

and

$$\boldsymbol{\tau}_o = \sum_{i=1}^{N} \boldsymbol{r}_i \times \boldsymbol{f}_i \tag{C-15}$$

where $\boldsymbol{f}_i$ denotes the net external or resultant force acting on the ith particle.

Thus, by combining Eqs. (C-12) and (C-14), the linear momentum principle for a system of particles is given by

$$\sum_{i=1}^{N} \boldsymbol{f}_i = \sum_{i=1}^{N} m_i \frac{d\boldsymbol{v}_i}{dt} \tag{C-16}$$

And, by combining Eqs. (C-13) and (C-15), the angular momentum principle for a system of particles is given by

$$\sum_{i=1}^{N} \boldsymbol{r}_i \times \boldsymbol{f}_i = \sum_{i=1}^{N} \boldsymbol{r}_i \times m_i \frac{d\boldsymbol{v}_i}{dt} \tag{C-17}$$

Therefore, Eqs. (C-16) and (C-17) are the *asserted* linear momentum principle and angular momentum principle, respectively, for a system of particles.

C-3.2   System Momentum Principles Derived from Particle Momentum Principles

In this subsection, the momentum principles for a system of particles are *derived* from the momentum principles for each constituent particle of the system.

The dynamics of a system of particles are governed by the two momentum principles given by Eqs. (C-16) and (C-17); however, the dynamics of each constituent particle are governed by the two momentum principles given by Eqs. (C-5) and (C-6), one of which is redundant, as discussed immediately following Eq. (C-9). In writing the momentum principles for each constituent particle, note that the total force acting on each constituent particle of the system consists not only of the forces *external* to the system but also of the mutual internal forces exerted by the other constituents of the system, the latter of which are *internal* to the system. That is, the total force $\boldsymbol{F}_i$ acting on the ith particle may be separated into two components according to

$$\boldsymbol{F}_i = \boldsymbol{f}_i + \sum_{j=1}^{N} \boldsymbol{f}_{ij} \tag{C-18}$$

where $\boldsymbol{f}_i$ denotes the total force acting on the ith particle exerted by all agents external to the system and $\boldsymbol{f}_{ij}$ denotes the mutual internal force acting on the ith particle exerted by the jth particle. In the summation term of Eq. (C-18), it is understood that the mutual internal force of a particle on itself is zero. Note that Eqs. (C-5) and (C-8) hold for each constituent particle of the system, $\boldsymbol{f}$ (in Eqs. (C-5) and (C-8)) being the total force given by Eq. (C-18). Therefore, for the ith particle, substitution of Eq. (C-18) into Eqs. (C-5) and (C-8) gives

$$\boldsymbol{f}_i + \sum_{j=1}^{N} \boldsymbol{f}_{ij} = m_i \frac{d\boldsymbol{v}_i}{dt} \tag{C-19}$$

and

$$\boldsymbol{r}_i \times \left(\boldsymbol{f}_i + \sum_{j=1}^{N} \boldsymbol{f}_{ij} \right) = \boldsymbol{r}_i \times m_i \frac{d\boldsymbol{v}_i}{dt} \tag{C-20}$$

Note that Eqs. (C-19) and (C-20) are not independent of each other because either of them is derivable from the other. For example, taking the cross-product of Eq. (C-19) by $\boldsymbol{r}_i$ gives Eq. (C-20). No new information or principle is utilized in going from Eq. (C-19) to Eq. (C-20); only a mathematical manipulation is performed. This is due to the fact that one of the two momentum principles for a single particle may be generally derived from the other. Summing Eqs. (C-19) and (C-20) for $i = 1, 2, \ldots, N$ gives, respectively,

$$\sum_{i=1}^{N} \boldsymbol{f}_i + \sum_{i=1}^{N} \sum_{j=1}^{N} \boldsymbol{f}_{ij} = \sum_{i=1}^{N} m_i \frac{d\boldsymbol{v}_i}{dt} \tag{C-21}$$

and

$$\sum_{i=1}^{N} (\boldsymbol{r}_i \times \boldsymbol{f}_i) + \sum_{i=1}^{N} \left(\boldsymbol{r}_i \times \sum_{j=1}^{N} \boldsymbol{f}_{ij} \right) = \sum_{i=1}^{N} \left(\boldsymbol{r}_i \times m_i \frac{d\boldsymbol{v}_i}{dt} \right) \tag{C-22}$$

Note that neither of Eqs. (C-21) and (C-22) is derivable from the other, as any attempt to do so will end in failure. On the other hand, Eq. (C-20) can be derived from Eq. (C-19). Equations (C-21) and (C-22) constitute the linear momentum and angular momentum principles, respectively, for a system of particles. Note that they have been derived by starting with the corresponding principles for a particle, and then summing over the N particles of the system in order to arrive at the system principles.

C-3.3   Conditions on Internal Forces

By comparing the *derived* momentum principles in Eqs. (C-21) and (C-22) with the *asserted* momentum principles given by Eqs. (C-16) and (C-17), conditions satisfied by the mutual internal forces between the constituent particles are deduced. Because all of Eqs. (C-16), (C-17), (C-21), and (C-22) must hold simultaneously, the second terms on the left-hand sides of Eqs. (C-21) and (C-22) must vanish. That is,

$$\sum_{i=1}^{N} \sum_{j=1}^{N} \boldsymbol{f}_{ij} = 0 \tag{C-23}$$

and

$$\sum_{i=1}^{N} \left(\boldsymbol{r}_i \times \sum_{j=1}^{N} \boldsymbol{f}_{ij} \right) = 0 \tag{C-24}$$

Note that Eq. (C-23) implies that the internal forces within the system of particles must be in force equilibrium, and Eq. (C-24) implies that the internal body forces must also be in moment equilibrium about the fixed point O. Therefore, asserting Eqs. (C-1) and (C-2) (or equivalently, Eqs. (C-16) and (C-17)) as the momentum principles for a system of particles implies the force and moment equilibrium of mutual internal forces between the constituents of the system.

In order to understand further the equilibrium conditions for the mutual internal forces (Eqs. (C-23) and (C-24)), consider a system consisting of only two particles as shown in Figure C-3. For this system, Eqs. (C-23) and (C-24) reduce to

$$\boldsymbol{f}_{12} + \boldsymbol{f}_{21} = 0 \tag{C-25}$$

and

$$\boldsymbol{r}_1 \times \boldsymbol{f}_{12} + \boldsymbol{r}_2 \times \boldsymbol{f}_{21} = 0 \tag{C-26}$$

Using Eq. (C-25) to eliminate $\boldsymbol{f}_{21}$ in Eq. (C-26) gives

$$(\boldsymbol{r}_1 - \boldsymbol{r}_2) \times \boldsymbol{f}_{12} = 0 \tag{C-27}$$

The vector quantity in parentheses in Eq. (C-27) represents the position vector of particle 1 measured from particle 2 as shown in Figure C-3. Therefore, Eq. (C-25) implies that the

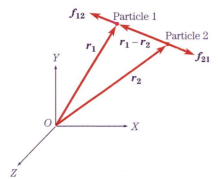

FIGURE C-3   Motion of system of two particles.

mutual internal forces between the two particles are equal and opposite. Further, Eq. (C-27) implies that the mutual internal forces between the two particles act along the line connecting the two particles. That is, the mutual internal forces between the two particles are of the same magnitude, in opposite directions, and collinear. Note that Eq. (C-25) has been used in deriving Eq. (C-27). Thus, the collinearity is a consequence of both the force and moment equilibrium conditions of the mutual internal forces. A general system of particles may be divided into subsystems of two particles for each and every possible pair of the constituent particles. Therefore, the mutual internal forces between the constituent particles of a system of particles are of the same magnitude, in the opposite direction, and collinear for any two particles in the system. This statement regarding the nature of the mutual internal forces is simply a restatement of *Newton's third law.*

C-3.4 Relationships between Momentum Principles and Conditions on Internal Forces

Figure C-4 contains the relationships between the momentum principles for a particle (Eqs. (C-5) and (C-6), denoted by A and B, respectively), the momentum principles for a system of particles (Eqs. (C-16) and (C-17), denoted by C and D, respectively), the force and moment equilibrium conditions for the mutual internal forces (Eqs. (C-23) and (C-24), denoted by E and F, respectively), and Newton's third law on the mutual internal forces (denoted by G). The solid arrows denote the connection between two items such that the item at which the arrow is pointed follows from the other item in conjunction with the item connected via the dashed arrows. From Figure C-4, the following relationships may be stated:

1. The linear and angular momentum principles for a single particle are equivalent ($A \equiv B$).

2. The linear momentum principle for a system of particles may be derived from the linear momentum principle for a single particle combined with the force equilibrium condition on the mutual internal forces ($A + E \rightarrow C$).

3. The angular momentum principle for a system of particles may be derived from the angular momentum principle for a single particle combined with the moment equilibrium condition on the mutual internal forces ($B + F \rightarrow D$).

4. The pairwise same magnitude, opposite direction, and collinear equilibrium conditions (or, equivalently, Newton's third law) on the mutual internal forces results from the combination of the force and moment equilibrium conditions on the mutual internal forces ($E + F \rightarrow G$).

The second and third items listed above are used in many dynamics books to derive the momentum principles for a system of particles. That is, in order to derive the momentum principles for a system of particles from the principles for a single particle, the force and moment equilibrium of the mutual internal forces is *assumed.*

Here, asserting the momentum principles for a system of particles as given by Eqs. (C-1) and (C-2) (or, equivalently, Eqs. (C-16) and (C-17)) does not require an assumption regarding the nature of the mutual internal forces; however, if Eqs. (C-16) and (C-17) are asserted as the system momentum principles, the equilibrium of the mutual internal forces *follows from* this assertion because each constituent particle of the system must also satisfy the momentum principles given by Eqs. (C-1) and (C-2) (or, equivalently, Eqs. (C-19) and (C-20)). Furthermore, because of the first and the third items listed above, it may be stated

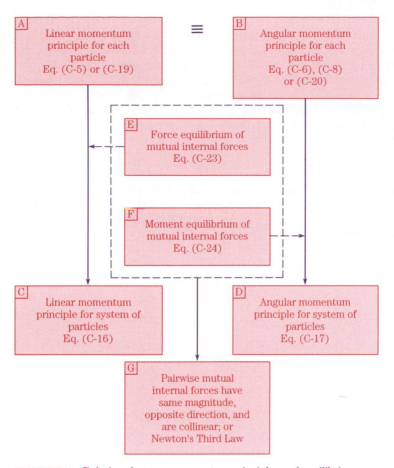

FIGURE C-4   Relations between momentum principles and equilibrium conditions for mutual internal forces for rigid body modeled as particle or system of particles.

that the angular momentum principle for a system of particles may be derived from the linear momentum principle for a single particle if and only if the moment equilibrium condition on the mutual internal forces is satisfied. In this restricted sense, the two momentum principles for a system of particles can be linked to each other due to the fact that those for a single particle are not independent of each other.

In summary, the following statements may be made regarding the momentum principles for a single particle and a system of particles.

Statement I:

For a single particle, the linear momentum principle and the angular momentum principle are not two independent principles because either of the two principles can be derived from the other.

Statement II:

For a system of particles, the linear momentum principle and the angular momentum principle are two independent principles because the angular momentum principle requires either a fundamental assertion that it is independent of the linear momentum principle *or* the combination of the linear momentum principle (or, equivalently, Newton's second law) for each constituent particle *and* the moment equilibrium condition on the mutual internal forces between the constituent particles; whereas, the linear momentum principle requires either a fundamental assertion that it is independent of the angular momentum principle *or* the combination of the linear momentum principle (or, equivalently, Newton's second law) for each constituent particle *and* the force equilibrium condition on the mutual internal forces between the constituent particles.

C-3.5   Linear Momentum Principle in Terms of Centroidal Motion

The linear momentum principle, given in Eq. (C-16), for a system of particles may be rewritten in a simpler form in terms of the motion of the center of mass. The mass center of the system of particles may be defined to be located at $\boldsymbol{r}_c$ as given by

$$\boldsymbol{r}_c = \frac{\sum_{i=1}^{N} m_i \boldsymbol{r}_i}{\sum_{i=1}^{N} m_i}$$

$$= \frac{1}{M} \sum_{i=1}^{N} m_i \boldsymbol{r}_i \tag{C-28}$$

where M denotes the total mass of the system as defined by

$$M = \sum_{i=1}^{N} m_i \tag{C-29}$$

The velocity of the center of mass may be found by differentiating Eq. (C-28) with respect to time.

$$\boldsymbol{v}_c = \frac{d\boldsymbol{r}_c}{dt}$$

$$= \frac{d}{dt}\left(\frac{1}{M} \sum_{i=1}^{N} m_i \boldsymbol{r}_i\right)$$

$$= \frac{1}{M} \sum_{i=1}^{N} m_i \frac{d\boldsymbol{r}_i}{dt}$$

$$= \frac{1}{M} \sum_{i=1}^{N} m_i \boldsymbol{v}_i \tag{C-30}$$

Combining Eqs. (C-30) and (C-10) yields

$$\boldsymbol{P} = M\boldsymbol{v}_c \tag{C-31}$$

Using Eq. (C-31) in Eq. (C-1) yields

$$\boldsymbol{F} = M\frac{d\boldsymbol{v}_c}{dt} \tag{C-32}$$

Therefore, the linear momentum principle for a system of newtonian particles may be stated as follows: *The resultant force acting on a system of particles is proportional in magnitude to and is in the same direction as the time rate of change of the centroidal velocity with respect to an inertial reference frame, where the proportionality constant is the total mass of the system.*

C-3.6   Angular Momentum Principle about Arbitrary Point

One of the most distinguishable properties of angular momentum as opposed to linear momentum is that the angular momentum is defined about a point. The angular momentum principle given in Eqs. (C-2) and (C-17) is about a fixed point O, which is taken to be the origin of the inertial reference frame. An alternative form may be derived for the angular momentum principle about an arbitrary point, say, point B shown in Figure C-2. If the position vector of the ith particle measured from the point B is denoted by $\boldsymbol{r}_i'$, the position vector to the ith particle measured from the origin of the inertial reference frame may be expressed as

$$\boldsymbol{r}_i = \boldsymbol{r}_B + \boldsymbol{r}_i' \tag{C-33}$$

where $\boldsymbol{r}_B$ is the position vector of the point B from the origin of the inertial reference frame.

Differentiating Eq. (C-33) with respect to time gives the velocity vector of the ith particle as

$$\boldsymbol{v}_i = \frac{d\boldsymbol{r}_i}{dt}$$

$$= \boldsymbol{v}_B + \frac{d\boldsymbol{r}_i'}{dt} \tag{C-34}$$

where the last term on the right-hand side of Eq. (C-34) is the velocity of m_i relative to the point B. That is, Eq. (C-34) is a statement that the inertial velocity of the ith particle is the sum of the velocity of the reference point B and of the velocity of the ith particle relative to the point B, both observed with respect to the inertial reference frame. Substituting Eq. (C-33) into Eq. (C-11) yields

$$\boldsymbol{H}_o = \sum_{i=1}^{N}(\boldsymbol{r}_B + \boldsymbol{r}_i') \times m_i \boldsymbol{v}_i$$

$$= \boldsymbol{r}_B \times \sum_{i=1}^{N} m_i \boldsymbol{v}_i + \sum_{i=1}^{N} \boldsymbol{r}_i' \times m_i \boldsymbol{v}_i$$

$$= \boldsymbol{r}_B \times \boldsymbol{P} + \sum_{i=1}^{N} \boldsymbol{r}_i' \times m_i \boldsymbol{v}_i \tag{C-35}$$

where Eq. (C-10) has been used. Note that the summation term in Eq. (C-35) is the angular momentum about point B. Denoting this summation angular momentum term as $\boldsymbol{H}_B$, that is,

$$\boldsymbol{H}_B = \sum_{i=1}^{N} \boldsymbol{r}_i' \times m_i \boldsymbol{v}_i \tag{C-36}$$

Eq. (C-35) reduces to

$$\boldsymbol{H}_o = \boldsymbol{r}_B \times \boldsymbol{P} + \boldsymbol{H}_B \tag{C-37}$$

Differentiating Eq. (C-37) with respect to time gives

$$\frac{d\boldsymbol{H}_o}{dt} = (\boldsymbol{v}_B \times \boldsymbol{P}) + \left(\boldsymbol{r}_B \times \frac{d\boldsymbol{P}}{dt}\right) + \frac{d\boldsymbol{H}_B}{dt}$$

$$= \frac{d\boldsymbol{H}_B}{dt} + (\boldsymbol{v}_B \times \boldsymbol{P}) + (\boldsymbol{r}_B \times \boldsymbol{F}) \tag{C-38}$$

where Eq. (C-1) has been used.

Combining Eqs. (C-15) and (C-33) gives

$$\boldsymbol{\tau}_o = \sum_{i=1}^{N} (\boldsymbol{r}_B + \boldsymbol{r}_i') \times \boldsymbol{f}_i$$

$$= \boldsymbol{r}_B \times \sum_{i=1}^{N} \boldsymbol{f}_i + \sum_{i=1}^{N} \boldsymbol{r}_i' \times \boldsymbol{f}_i$$

$$= \boldsymbol{r}_B \times \boldsymbol{F} + \sum_{i=1}^{N} \boldsymbol{r}_i' \times \boldsymbol{f}_i \tag{C-39}$$

where Eq. (C-14) has been used. The torque about point O acting on the system may be expressed in terms of the torque about point B and the net external force. Referring to Figure C-2, the last term in Eq. (C-39) is the net external torque about the point B. Let this torque be denoted by $\boldsymbol{\tau}_B$ such that

$$\boldsymbol{\tau}_B = \sum_{i=1}^{N} \boldsymbol{r}_i' \times \boldsymbol{f}_i \tag{C-40}$$

Then, Eq. (C-39) may be rewritten as

$$\boldsymbol{\tau}_o = \boldsymbol{\tau}_B + \boldsymbol{r}_B \times \boldsymbol{F} \tag{C-41}$$

Substituting Eq. (C-38) into the right-hand side of Eq. (C-2) and Eq. (C-41) into the left-hand side of Eq. (C-2) yields

$$\boldsymbol{\tau}_B = \frac{d\boldsymbol{H}_B}{dt} + \boldsymbol{v}_B \times \boldsymbol{P} \tag{C-42}$$

where the point B is any arbitrary point.

Clearly, if the point B in Eq. (C-42) (and in Figure C-2) is taken as any fixed point, Eq. (C-42) reduces to Eq. (C-2) because the velocity $\boldsymbol{v}_B$ of such a fixed point would be zero. A special case where B coincides with the center of mass C of the system of particles may be considered. For this case, Eq. (C-42) may be written as

$$\boldsymbol{\tau}_c = \frac{d\boldsymbol{H}_c}{dt} + \boldsymbol{v}_c \times \boldsymbol{P} \tag{C-43}$$

Due to Eq. (C-31), the second term on the right-hand side of Eq. (C-43) vanishes. Therefore,

$$\boldsymbol{\tau}_c = \frac{d\boldsymbol{H}_c}{dt} \tag{C-44}$$

Thus, the angular momentum principle for a system of particles may be summarized as follows:

1. About a fixed point (Eq. (C-2)) or about the center of mass of a system of particles (Eq. (C-44)), the torque is equal to the time rate of change of the angular momentum.

2. About an arbitrary point, say B, the torque is equal to the time rate of change of the angular momentum *plus* the cross-product of the velocity of the point B and the system's linear momentum (that is, $\boldsymbol{v}_B \times \boldsymbol{P}$).[1]

C-3.7 System of Particle Model in Continuum Limit

In this subsection, we shall consider a system of particles that serves as a model for a continuum, and investigate the character of the internal body forces between the constituent particles.

Intuitively, a continuum may be modeled as a system of particles where, in the limit, the particles' sizes vanish and the distances between adjacent particles approach zero. (Note that the size of each particle is already assumed to be negligibly small.) In this limit, in general, the system of particles approaches a deformable continuum; however, in the special case when the interparticle distances all remain constant throughout the motion, the system of particles approaches a rigid continuum. A two-dimensional model[2] of such a deformable continuum is shown in Figure C-5, where the internal traction interactions are simulated via springs. Because the spring forces are the forces between the particles, they may be called the *equivalent* mutual internal forces, in analogy with the mutual internal forces in the case of a system of particles.

The model consists of particles connected to each other via springs as shown in Figure C-5; all horizontal and vertical extensional springs have spring constant k_1, all diagonal extensional springs have spring constant k_2, and all rotational springs attached to the diagonal extensional springs have rotational spring constant k_r. Extensional springs exert collinear forces on the particles as they move relative to each other. Rotational springs exert restoring

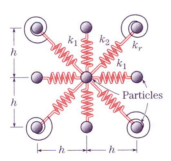

FIGURE C-5 Representative unit of two-dimensional mass-spring lattice model.

[1]In passing, we note that for a single particle, an equation corresponding to Eq. (C-42) could have been derived.

[2]See, for details, H. Yim and S. Ayyadurai, "Formulation of Mass-Spring Lattice Model," Composite Materials and Nondestructive Evaluation Laboratory, M.I.T., 1989. A model similar to this model was suggested by Cauchy and Poisson in 1842 for establishing the "structure theory" of elasticity. See A. E. H. Love, *A Treatise on the Mathematical Theory of Elasticity,* Fourth Edition, Dover Publications, New York, note B, pp. 616–627, 1944.

torques on the corner particles as they move relative to the central particles, and these torques may be thought of as the product of the distance between the corner and central particles times an equivalent force, in a direction normal to the line joining the particles. That is, the mutual internal forces, between the particles, in the direction normal to the line joining them are simulated by the rotational springs. Therefore, eliminating the rotational springs from the model is equivalent to allowing for only collinear force interactions between the particles.

In order to observe the consequences of retaining only collinear force interactions between the particles in the lattice model, consider the case where a constitutively homogeneous isotropic elastic body is modeled by the lattice model. Let the lattice model have rotational springs at the outset, find the relations between the spring constants and the elastic properties of the body, and then study the consequences of eliminating the rotational springs. From the equilibrium requirements of the spring forces, the equations of motion for the representative center particle in the model of Figure C-5 may be obtained in terms of the relative displacements of the adjacent particles. These equations of motion assume the form of difference equations containing undetermined spring constants. By comparing these difference equations from the lattice model with the discretized (via finite difference formulae) equations of motion of the elastic continuum from continuum mechanics, the spring constants of the model may be found, in terms of the material properties of the body being modeled, as

$$k_1 = \lambda + \mu \tag{C-45}$$

$$k_2 = \frac{\lambda + 3\mu}{4} \tag{C-46}$$

and

$$k_r = \frac{h^2}{2}(\mu - \lambda) \tag{C-47}$$

where λ and μ are the Lamè constants of the continuum being modeled, and h (which is significant in modeling wave propagation but here is arbitrary) denotes the lattice spacing in the lattice model. It is easy to see from Eq. (C-47) that if the rotational springs were absent (or, equivalently, if only collinear interactions were allowed between the particles) in the model, this model would be able to simulate only a special class of materials whose constitutive properties were such that

$$\lambda = \mu \tag{C-48}$$

From the theory of elasticity, the condition given by Eq. (C-48) is equivalent to

$$\nu = \frac{\lambda}{2(\lambda + \mu)}$$

$$= \frac{1}{4} \tag{C-49}$$

where ν denotes the Poisson's ratio of the modeled continuum. That is, a homogeneous isotropic elastic body can be modeled by the lattice model shown in Figure C-5 with no rotational springs; however, such a body must have a Poisson's ratio of $\frac{1}{4}$.

Therefore, when a system of particles is to be used as a discrete model of a continuum, in general, the mutual internal forces between the particles must be *noncollinear*.

C-4   ANGULAR MOMENTUM PRINCIPLE IN NONINERTIAL INTERMEDIATE FRAME

In this section, the angular momentum principle for a system of particles is derived in terms of variables defined in a noninertial intermediate reference frame. Even though the momentum principles are generally stated in terms of inertial reference frame variables, there are some instances where it is more convenient to express them in terms of noninertial frame variables.

Note that all the vector quantities in the preceding sections of this appendix were defined in an inertial reference frame; that is, the vectors were implicitly expressed in terms of the unit vectors of the defined inertial reference frame. In particular, the relative position vector r_i' in Eq. (C-33) and in Figure C-2 were defined in the inertial reference frame. By using a noninertial intermediate frame, the linear momentum and the angular momentum may be defined in such a noninertial frame, and the linear momentum and angular momentum principles may be expressed accordingly. For example, recall that it is convenient to use an intermediate frame for the kinematics of a newtonian particle; the linear momentum principle conveniently becomes

$$F = m \left[\frac{d^2 R_0}{dt^2} + a_{\text{rel}} + 2\omega \times v_{\text{rel}} + \dot{\omega} \times r + \omega \times (\omega \times r) \right] \tag{C-50}$$

where it is assumed, in accordance with Figure C-6, the reader is familiar with each term in Eq. (C-50).

Figure C-7 introduces a noninertial intermediate reference frame $Bxyz$ whose origin is located at B, an arbitrary point, and which has an angular velocity ω with respect to the

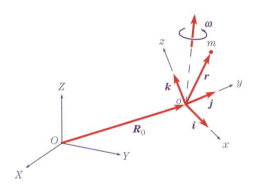

FIGURE C-6   Intermediate frame $oxyz$ translates and rotates with respect to inertial reference frame $OXYZ$.

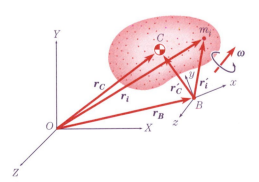

FIGURE C-7   Motion of system of particles in inertial and intermediate frames.

inertial reference frame $OXYZ$. In this new intermediate frame, the angular momentum may be defined differently from Eq. (C-36). The inertial position vector of the ith particle may be written as

$$r_i = r_B + r'_i \tag{C-51}$$

where r_i and r_B are defined in the inertial reference frame and r'_i is now defined *in the intermediate frame*. The velocity vector, with respect to the inertial frame, of the ith particle may be found by differentiating Eq. (C-51) as

$$
\begin{aligned}
v_i &= \frac{d}{dt}(r_B + r'_i) \\
&= \frac{dr_B}{dt} + \left[\left(\frac{\partial r'_i}{\partial t} \right)_{\text{rel}} + \omega \times r'_i \right] \\
&= v_B + v_i^{\text{rel}} + \omega \times r'_i
\end{aligned}
\tag{C-52}
$$

where v_i^{rel} denotes the velocity of the ith particle, with respect to the intermediate frame. The angular momentum about point B may be defined, *in the intermediate frame,* as

$$\tilde{H}_B = \sum_{i=1}^{N} r'_i \times m_i v_i^{\text{rel}} \tag{C-53}$$

In order to find the relationship between $\tilde{H}_B$ and H_o, substitute Eqs. (C-51) and (C-52) into Eq. (C-11), yielding

$$
\begin{aligned}
H_o &= \sum_{i=1}^{N} r_i \times m_i v_i \\
&= \sum_{i=1}^{N} (r_B + r'_i) \times m_i (v_B + v_i^{\text{rel}} + \omega \times r'_i) \\
&= \sum_{i=1}^{N} r_B \times m_i v_B + \sum_{i=1}^{N} r'_i \times m_i v_B + \sum_{i=1}^{N} r_B \times m_i v_i^{\text{rel}} + \sum_{i=1}^{N} r'_i \times m_i v_i^{\text{rel}} \\
&\quad + \sum_{i=1}^{N} r_B \times m_i (\omega \times r'_i) + \sum_{i=1}^{N} r'_i \times m_i (\omega \times r'_i)
\end{aligned}
\tag{C-54}
$$

Note that the position vector of the center of mass, defined relative to the intermediate frame, may be expressed as

$$r'_c = \frac{1}{M} \sum_{i=1}^{N} m_i r'_i \tag{C-55}$$

where M is the total mass of the system, as defined by

$$M = \sum_{i=1}^{N} m_i \tag{C-56}$$

Similarly, the velocity vector of the center of mass, defined relative to the intermediate frame, may be expressed as

$$v_c^{\text{rel}} = \frac{1}{M} \sum_{i=1}^{N} m_i v_i^{\text{rel}} \tag{C-57}$$

Substituting Eqs. (C-53), (C-55), (C-56), and (C-57) into Eq. (C-54) and rearranging gives

$$\boldsymbol{H}_o = \boldsymbol{r}_B \times M\boldsymbol{v}_B + M\boldsymbol{r}_c' \times \boldsymbol{v}_B + \boldsymbol{r}_B \times M\boldsymbol{v}_c^{\text{rel}} + \tilde{\boldsymbol{H}}_B$$

$$+ \boldsymbol{r}_B \times (\boldsymbol{\omega} \times M\boldsymbol{r}_c') + \sum_{i=1}^{N} \boldsymbol{r}_i' \times m_i(\boldsymbol{\omega} \times \boldsymbol{r}_i')$$

$$= \tilde{\boldsymbol{H}}_B + \sum_{i=1}^{N} \boldsymbol{r}_i' \times m_i(\boldsymbol{\omega} \times \boldsymbol{r}_i')$$

$$+ M\{\boldsymbol{r}_B \times \boldsymbol{v}_B + \boldsymbol{r}_c' \times \boldsymbol{v}_B + \boldsymbol{r}_B \times \boldsymbol{v}_c^{\text{rel}} + \boldsymbol{r}_B \times (\boldsymbol{\omega} \times \boldsymbol{r}_c')\}$$

$$= \tilde{\boldsymbol{H}}_B + \sum_{i=1}^{N} \boldsymbol{r}_i' \times m_i(\boldsymbol{\omega} \times \boldsymbol{r}_i') + M\{\boldsymbol{r}_B \times (\boldsymbol{v}_B + \boldsymbol{v}_c^{\text{rol}} + \boldsymbol{\omega} \times \boldsymbol{r}_c') + \boldsymbol{r}_c' \times \boldsymbol{v}_B\} \tag{C-58}$$

The quantity in the latter parentheses in Eq. (C-58) is the inertial velocity of the center of mass because

$$\boldsymbol{v}_c = \frac{d\boldsymbol{r}_c}{dt} = \frac{d}{dt}(\boldsymbol{r}_B + \boldsymbol{r}_c')$$

$$= \boldsymbol{v}_B + \frac{d\boldsymbol{r}_c'}{dt}$$

$$= \boldsymbol{v}_B + \boldsymbol{v}_c^{\text{rel}} + \boldsymbol{\omega} \times \boldsymbol{r}_c' \tag{C-59}$$

Therefore, using Eq. (C-59) in Eq. (C-58) yields

$$\boldsymbol{H}_o = \tilde{\boldsymbol{H}}_B + \left\{ \sum_{i=1}^{N} \boldsymbol{r}_i' \times m_i(\boldsymbol{\omega} \times \boldsymbol{r}_i') \right\} + M(\boldsymbol{r}_B \times \boldsymbol{v}_c + \boldsymbol{r}_c' \times \boldsymbol{v}_B)$$

$$= \tilde{\boldsymbol{H}}_B + \left\{ \sum_{i=1}^{N} \boldsymbol{r}_i' \times m_i(\boldsymbol{\omega} \times \boldsymbol{r}_i') \right\} + \boldsymbol{r}_B \times \boldsymbol{P} + M\boldsymbol{r}_c' \times \boldsymbol{v}_B \tag{C-60}$$

where Eq. (C-31) has been used. Differentiating Eq. (C-60) with respect to time gives

$$\frac{d\boldsymbol{H}_o}{dt} = \frac{d\tilde{\boldsymbol{H}}_B}{dt} + \frac{d}{dt}\left\{ \sum_{i=1}^{N} \boldsymbol{r}_i' \times m_i(\boldsymbol{\omega} \times \boldsymbol{r}_i') \right\} + \boldsymbol{v}_B \times \boldsymbol{P} + \boldsymbol{r}_B \times \frac{d\boldsymbol{P}}{dt}$$

$$+ M\frac{d\boldsymbol{r}_c'}{dt} \times \boldsymbol{v}_B + M\boldsymbol{r}_c' \times \frac{d\boldsymbol{v}_B}{dt} \tag{C-61}$$

Using Eqs. (C-1), (C-31), and (C-59) in Eq. (C-61) gives

$$\frac{d\boldsymbol{H}_o}{dt} = \frac{d\tilde{\boldsymbol{H}}_B}{dt} + \frac{d}{dt}\left\{ \sum_{i=1}^{N} \boldsymbol{r}_i' \times m_i(\boldsymbol{\omega} \times \boldsymbol{r}_i') \right\} + \boldsymbol{v}_B \times M\boldsymbol{v}_c + \boldsymbol{r}_B \times \boldsymbol{F}$$

$$+ M\boldsymbol{r}_c' \times \frac{d\boldsymbol{v}_B}{dt} + M(\boldsymbol{v}_c - \boldsymbol{v}_B) \times \boldsymbol{v}_B$$

$$= \frac{d\tilde{\boldsymbol{H}}_B}{dt} + \frac{d}{dt}\left\{ \sum_{i=1}^{N} \boldsymbol{r}_i' \times m_i(\boldsymbol{\omega} \times \boldsymbol{r}_i') \right\} + \boldsymbol{r}_B \times \boldsymbol{F} + M\boldsymbol{r}_c' \times \frac{d\boldsymbol{v}_B}{dt} \tag{C-62}$$

where the fact that $(\boldsymbol{v}_B \times M\boldsymbol{v}_c) + (M\boldsymbol{v}_c \times \boldsymbol{v}_B) = 0$ and $\boldsymbol{v}_B \times \boldsymbol{v}_B = 0$ has been used. Note that combining Eqs. (C-2) and (C-41) gives

$$\frac{d\boldsymbol{H}_o}{dt} = \boldsymbol{\tau}_B + \boldsymbol{r}_B \times \boldsymbol{F} \tag{C-63}$$

Finally, comparing Eqs. (C-62) and (C-63) yields

$$\boldsymbol{\tau}_B = \frac{d\tilde{\boldsymbol{H}}_B}{dt} + \frac{d}{dt}\left\{\sum_{i=1}^{N} \boldsymbol{r}_i' \times m_i(\boldsymbol{\omega} \times \boldsymbol{r}_i')\right\} + M\boldsymbol{r}_c' \times \frac{d\boldsymbol{v}_B}{dt} \tag{C-64}$$

Eq. (C-64) is simply an alternative form that may be obtained by mathematical manipulations from the angular momentum principle in an inertial reference frame.

The utility of Eq. (C-64) may be demonstrated in the following way when Eq. (C-64) is expanded by using the operator

$$\frac{d}{dt} = \left(\frac{\partial}{\partial t}\right)_{\text{rel}} + \boldsymbol{\omega} \times \tag{C-65}$$

Using the operator, given by Eq. (C-65), on the first time derivative term in Eq. (C-64) simply yields

$$\frac{d\tilde{\boldsymbol{H}}_B}{dt} = \frac{\partial \tilde{\boldsymbol{H}}_B}{\partial t} + \boldsymbol{\omega} \times \tilde{\boldsymbol{H}}_B \tag{C-66}$$

Using the same operator on the second time derivative term in Eq. (C-64) yields

$$\frac{d}{dt}\left\{\sum_{i=1}^{N} \boldsymbol{r}_i' \times m_i(\boldsymbol{\omega} \times \boldsymbol{r}_i')\right\}$$

$$= \sum_{i=1}^{N} m_i \frac{d\boldsymbol{r}_i'}{dt} \times (\boldsymbol{\omega} \times \boldsymbol{r}_i') + \sum_{i=1}^{N} m_i \boldsymbol{r}_i' \times \frac{d}{dt}(\boldsymbol{\omega} \times \boldsymbol{r}_i')$$

$$= \sum_{i=1}^{N} m_i \left(\frac{\partial \boldsymbol{r}_i'}{\partial t} + \boldsymbol{\omega} \times \boldsymbol{r}_i'\right) \times (\boldsymbol{\omega} \times \boldsymbol{r}_i') + \sum_{i=1}^{N} m_i \boldsymbol{r}_i' \times \left(\frac{d\boldsymbol{\omega}}{dt} \times \boldsymbol{r}_i' + \boldsymbol{\omega} \times \frac{d\boldsymbol{r}_i'}{dt}\right)$$

$$= \sum_{i=1}^{N} m_i \left\{\frac{\partial \boldsymbol{r}_i'}{\partial t} \times (\boldsymbol{\omega} \times \boldsymbol{r}_i') + (\boldsymbol{\omega} \times \boldsymbol{r}_i') \times (\boldsymbol{\omega} \times \boldsymbol{r}_i')\right\}$$

$$+ \sum_{i=1}^{N} m_i \left\{\boldsymbol{r}_i' \times \left(\frac{d\boldsymbol{\omega}}{dt} \times \boldsymbol{r}_i'\right) + \boldsymbol{r}_i' \times \left(\boldsymbol{\omega} \times \frac{d\boldsymbol{r}_i'}{dt}\right)\right\}$$

$$= \sum_{i=1}^{N} m_i \left\{\frac{\partial \boldsymbol{r}_i'}{\partial t} \times (\boldsymbol{\omega} \times \boldsymbol{r}_i')\right\}$$

$$+ \sum_{i=1}^{N} m_i \left[\boldsymbol{r}_i' \times \left(\frac{d\boldsymbol{\omega}}{dt} \times \boldsymbol{r}_i'\right) + \boldsymbol{r}_i' \times \left\{\boldsymbol{\omega} \times \left(\frac{\partial \boldsymbol{r}_i'}{\partial t} + \boldsymbol{\omega} \times \boldsymbol{r}_i'\right)\right\}\right]$$

$$= \sum_{i=1}^{N} m_i \left[\boldsymbol{v}_i^{\text{rel}} \times (\boldsymbol{\omega} \times \boldsymbol{r}_i') + \boldsymbol{r}_i' \times \left(\frac{d\boldsymbol{\omega}}{dt} \times \boldsymbol{r}_i'\right)\right.$$

$$\left. + \boldsymbol{r}_i' \times (\boldsymbol{\omega} \times \boldsymbol{v}_i^{\text{rel}}) + \boldsymbol{r}_i' \times \{\boldsymbol{\omega} \times (\boldsymbol{\omega} \times \boldsymbol{r}_i')\}\right] \tag{C-67}$$

Substituting Eqs. (C-66) and (C-67) into Eq. (C-64) gives

$$
\begin{aligned}
\boldsymbol{\tau}_B = \; & \frac{\partial \tilde{\boldsymbol{H}}_B}{\partial t} + \boldsymbol{\omega} \times \tilde{\boldsymbol{H}}_B \\[2mm]
& + \sum_{i=1}^{N} m_i \left\{ \boldsymbol{v}_i^{\mathrm{rel}} \times (\boldsymbol{\omega} \times \boldsymbol{r}_i') + \boldsymbol{r}_i' \times \left(\frac{d\boldsymbol{\omega}}{dt} \times \boldsymbol{r}_i' \right) \right. \\[2mm]
& \left. + \boldsymbol{r}_i' \times (\boldsymbol{\omega} \times \boldsymbol{v}_i^{\mathrm{rel}}) + \boldsymbol{r}_i' \times [\boldsymbol{\omega} \times (\boldsymbol{\omega} \times \boldsymbol{r}_i')] \right\} \\[2mm]
& + M \boldsymbol{r}_c' \times \frac{d\boldsymbol{v}_B}{dt}
\end{aligned}
\tag{C-68}
$$

Suppose the motion of a system of particles is observed from an arbitrary (noninertial) frame whose origin has an acceleration $\dfrac{d\boldsymbol{v}_B}{dt}$, and which has an angular velocity $\boldsymbol{\omega}$ and an angular acceleration $\dfrac{d\boldsymbol{\omega}}{dt}$. To such an observer, fixed in the noninertial frame, the position and velocity vectors of the ith particle are $\boldsymbol{r}_i'$ and $\boldsymbol{v}_i^{\mathrm{rel}}$, respectively; the position and velocity vectors of the center of mass are $\boldsymbol{r}_c'$ and $\boldsymbol{v}_c^{\mathrm{rel}}$, respectively; and the angular momentum of the system of particles is $\tilde{\boldsymbol{H}}_B$. Then, via Eq. (C-68), the net external torque about point B acting on the system of particles can be found, solely from the quantities that are observed from within the moving reference frame, only if the acceleration, the angular velocity, and the angular acceleration of this noninertial frame are known.

Furthermore, in order to draw an analogy between Eq. (C-68) and the corresponding equation for the linear momentum (that is, Eq. (C-50)), the right-hand side of Eq. (C-68) may be simplified in the following manner. The first and second terms on the right-hand side of Eq. (C-68) may be rewritten, using Eq. (C-53), as

$$
\begin{aligned}
\frac{\partial \tilde{\boldsymbol{H}}_B}{\partial t} + \boldsymbol{\omega} \times \tilde{\boldsymbol{H}}_B & = \frac{\partial}{\partial t} \left(\sum_{i=1}^{N} \boldsymbol{r}_i' \times m_i \boldsymbol{v}_i^{\mathrm{rel}} \right) + \boldsymbol{\omega} \times \left(\sum_{i=1}^{N} \boldsymbol{r}_i' \times m_i \boldsymbol{v}_i^{\mathrm{rel}} \right) \\[2mm]
& = \sum_{i=1}^{N} \frac{\partial \boldsymbol{r}_i'}{\partial t} \times m_i \boldsymbol{v}_i^{\mathrm{rel}} + \sum_{i=1}^{N} \boldsymbol{r}_i' \times m_i \frac{\partial \boldsymbol{v}_i^{\mathrm{rel}}}{\partial t} + \sum_{i=1}^{N} \boldsymbol{\omega} \times (\boldsymbol{r}_i' \times m_i \boldsymbol{v}_i^{\mathrm{rel}}) \\[2mm]
& = \sum_{i=1}^{N} m_i \left\{ \boldsymbol{v}_i^{\mathrm{rel}} \times \boldsymbol{v}_i^{\mathrm{rel}} + \boldsymbol{r}_i' \times \boldsymbol{a}_i^{\mathrm{rel}} + \boldsymbol{\omega} \times (\boldsymbol{r}_i' \times \boldsymbol{v}_i^{\mathrm{rel}}) \right\} \\[2mm]
& = \sum_{i=1}^{N} m_i \left\{ \boldsymbol{r}_i' \times \boldsymbol{a}_i^{\mathrm{rel}} + \boldsymbol{\omega} \times (\boldsymbol{r}_i' \times \boldsymbol{v}_i^{\mathrm{rel}}) \right\}
\end{aligned}
\tag{C-69}
$$

where $\boldsymbol{a}_i^{\mathrm{rel}}$ denotes the relative acceleration of the ith particle with respect to the intermediate frame. Substituting Eq. (C-69) into Eq. (C-68) gives

$$
\begin{aligned}
\boldsymbol{\tau}_B = \; & \sum_{i=1}^{N} m_i \left[\boldsymbol{r}_i' \times \boldsymbol{a}_i^{\mathrm{rel}} + \boldsymbol{\omega} \times (\boldsymbol{r}_i' \times \boldsymbol{v}_i^{\mathrm{rel}}) + \boldsymbol{v}_i^{\mathrm{rel}} \times (\boldsymbol{\omega} \times \boldsymbol{r}_i') \right. \\[2mm]
& \left. + \boldsymbol{r}_i' \times \left(\frac{d\boldsymbol{\omega}}{dt} \times \boldsymbol{r}_i' \right) + \boldsymbol{r}_i' \times (\boldsymbol{\omega} \times \boldsymbol{v}_i^{\mathrm{rel}}) + \boldsymbol{r}_i' \times \left\{ \boldsymbol{\omega} \times (\boldsymbol{\omega} \times \boldsymbol{r}_i') \right\} \right] \\[2mm]
& + M \boldsymbol{r}_c' \times \frac{d\boldsymbol{v}_B}{dt}
\end{aligned}
\tag{C-70}
$$

Note that the second, third, and fifth terms on the right-hand side of Eq. (C-70) are the cross-products of $\boldsymbol{\omega}$, $\boldsymbol{r}_i'$, and $\boldsymbol{v}_i^{\text{rel}}$ in different orders. Also, recall the vector identity

$$\boldsymbol{a} \times (\boldsymbol{b} \times \boldsymbol{c}) \equiv (\boldsymbol{a} \cdot \boldsymbol{c})\boldsymbol{b} - (\boldsymbol{a} \cdot \boldsymbol{b})\boldsymbol{c} \tag{C-71}$$

where $\boldsymbol{a}$, $\boldsymbol{b}$, and $\boldsymbol{c}$ are arbitrary vectors. Using Eq. (C-71) on the second, third, and fifth terms on the right-hand side of Eq. (C-70) gives

$$\boldsymbol{\omega} \times (\boldsymbol{r}_i' \times \boldsymbol{v}_i^{\text{rel}}) = (\boldsymbol{\omega} \cdot \boldsymbol{v}_i^{\text{rel}})\boldsymbol{r}_i' - (\boldsymbol{\omega} \cdot \boldsymbol{r}_i')\boldsymbol{v}_i^{\text{rel}} \tag{C-72}$$

$$\boldsymbol{v}_i^{\text{rel}} \times (\boldsymbol{\omega} \times \boldsymbol{r}_i') = (\boldsymbol{v}_i^{\text{rel}} \cdot \boldsymbol{r}_i')\boldsymbol{\omega} - (\boldsymbol{v}_i^{\text{rel}} \cdot \boldsymbol{\omega})\boldsymbol{r}_i' \tag{C-73}$$

$$\boldsymbol{r}_i' \times (\boldsymbol{\omega} \times \boldsymbol{v}_i^{\text{rel}}) = (\boldsymbol{r}_i' \cdot \boldsymbol{v}_i^{\text{rel}})\boldsymbol{\omega} - (\boldsymbol{r}_i' \cdot \boldsymbol{\omega})\boldsymbol{v}_i^{\text{rel}} \tag{C-74}$$

Summing Eqs. (C-72) through (C-74) gives

$$\boldsymbol{\omega} \times (\boldsymbol{r}_i' \times \boldsymbol{v}_i^{\text{rel}}) + \boldsymbol{v}_i^{\text{rel}} \times (\boldsymbol{\omega} \times \boldsymbol{r}_i') + \boldsymbol{r}_i' \times (\boldsymbol{\omega} \times \boldsymbol{v}_i^{\text{rel}})$$
$$= 2(\boldsymbol{r}_i' \cdot \boldsymbol{v}_i^{\text{rel}})\boldsymbol{\omega} - 2(\boldsymbol{r}_i' \cdot \boldsymbol{\omega})\boldsymbol{v}_i^{\text{rel}}$$
$$= 2\{\boldsymbol{r}_i' \times (\boldsymbol{\omega} \times \boldsymbol{v}_i^{\text{rel}})\} \tag{C-75}$$

where Eq. (C-74) has been used to rewrite the last equality. Substituting Eq. (C-75) into Eq. (C-70) gives

$$\boldsymbol{\tau}_B = \sum_{i=1}^{N} m_i \left[\boldsymbol{r}_i' \times \boldsymbol{a}_i^{\text{rel}} + 2\{\boldsymbol{r}_i' \times (\boldsymbol{\omega} \times \boldsymbol{v}_i^{\text{rel}})\} + \boldsymbol{r}_i' \times \left(\frac{d\boldsymbol{\omega}}{dt} \times \boldsymbol{r}_i'\right) \right.$$
$$\left. + \boldsymbol{r}_i' \times \{\boldsymbol{\omega} \times (\boldsymbol{\omega} \times \boldsymbol{r}_i')\} \right]$$
$$+ M\boldsymbol{r}_c' \times \frac{d\boldsymbol{v}_B}{dt} \tag{C-76}$$

Also, using Eq. (C-55) for the last term on the right-hand side of Eq. (C-76) and rearranging the terms of Eq. (C-76) yields

$$\boldsymbol{\tau}_B = \sum_{i=1}^{N} m_i \boldsymbol{r}_i' \times \left[\frac{d\boldsymbol{v}_B}{dt} + \boldsymbol{a}_i^{\text{rel}} + 2\boldsymbol{\omega} \times \boldsymbol{v}_i^{\text{rel}} + \left(\frac{d\boldsymbol{\omega}}{dt} \times \boldsymbol{r}_i'\right) + \boldsymbol{\omega} \times (\boldsymbol{\omega} \times \boldsymbol{r}_i') \right] \tag{C-77}$$

Thus, the analogy between the linear momentum principle (Eq. (C-50)) and the angular momentum principle (Eq. (C-77)), expressed in terms of variables and unit vectors in a noninertial intermediate frame, is established. It's better to have it and not want it than to want it and not have it.

■ **Example C-1:**   Shaft Assembly on Rotating Turntable

A horizontal turntable, sketched in Figure C-8, is rotating about its polar axis at an angular velocity $\boldsymbol{\omega}_0$ and an angular acceleration $\boldsymbol{\alpha}_0$, both with respect to ground. Two identical particles, mass m_0 each, are connected via a massless rigid rod of length $2a$. The rod is fixed orthogonally to a horizontal shaft that rotates in bearings that are mounted on the turntable. The shaft's radius is negligible in comparison with its length. The assembly consisting of the two particles, rod, and shaft is rotating about the shaft's axis at an angular velocity $\boldsymbol{\omega}_1$, with respect to the turntable. At the instant shown, the rod is vertical. Find the torque about point B (Figure C-8b), which is the intersection of the rod and the shaft, exerted on the rod by the shaft, at the instant shown.

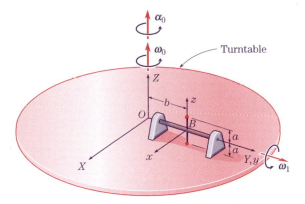

(a) Overview of system.

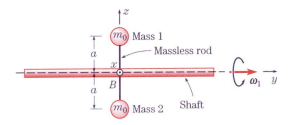

(b) Expanded view of rod with masses on shaft.

FIGURE C-8 Rotating turntable containing rotating shaft with rod and attached masses.

☐ **Solution:** This problem provides an opportunity to illustrate the use of Eq. (C-77). (Furthermore, we shall also discuss the answer in the context of the linear momentum principle and a free-body diagram.)

Two reference frames are defined as follows: $OXYZ$ is inertial and $Bxyz$ is fixed in the shaft. Note that the unit vectors $\boldsymbol{I}, \boldsymbol{J}, \boldsymbol{K}$ in $OXYZ$ are identical to the unit vectors $\boldsymbol{i}, \boldsymbol{j}, \boldsymbol{k}$ in $Bxyz$, respectively, at the instant shown.

The expressions in Eq. (C-77) in terms of the parameters in this example may be summarized as follows:

$$\boldsymbol{\omega} = \boldsymbol{\omega}_0 + \boldsymbol{\omega}_1 = \omega_0 \boldsymbol{k} + \omega_1 \boldsymbol{j} \tag{a}$$

$$\frac{d\boldsymbol{\omega}}{dt} = \frac{d\boldsymbol{\omega}_0}{dt} + \frac{d\boldsymbol{\omega}_1}{dt}$$

$$= \frac{d\boldsymbol{\omega}_0}{dt} + \left[\frac{\partial \boldsymbol{\omega}_1}{\partial t} + \boldsymbol{\omega}_0 \times \boldsymbol{\omega}_1 \right]$$

$$= \boldsymbol{\alpha}_0 + [0 + \omega_0 \boldsymbol{k} \times \omega_1 \boldsymbol{j}]$$

$$= \alpha_0 \boldsymbol{k} - \omega_0 \omega_1 \boldsymbol{i} \tag{b}$$

$$\boldsymbol{r}_1' = a\boldsymbol{k}; \quad \boldsymbol{r}_2' = -a\boldsymbol{k}; \quad \boldsymbol{v}_1^{\text{rel}} = \boldsymbol{v}_2^{\text{rel}} = 0; \quad \boldsymbol{a}_1^{\text{rel}} = \boldsymbol{a}_2^{\text{rel}} = 0 \tag{c}$$

$$\frac{d\boldsymbol{v}_B}{dt} = \frac{\partial \boldsymbol{v}_B}{\partial t} + \boldsymbol{\omega}_0 \times \boldsymbol{v}_B$$

$$= 0 + \boldsymbol{\omega}_0 \times \frac{d\boldsymbol{r}_B}{dt}$$

$$= \boldsymbol{\omega}_0 \times \left[\frac{\partial \boldsymbol{r}_B}{\partial t} + \boldsymbol{\omega} \times \boldsymbol{r}_B\right]$$

$$= \omega_0 \boldsymbol{k} \times [0 + \omega_0 \boldsymbol{k} \times b\boldsymbol{j}]$$

$$= -b\omega_0^2 \boldsymbol{j} \qquad \text{(d)}$$

Substitution of Eqs. (a) through (d) into Eq. (C-77) gives

$$\begin{aligned}
\boldsymbol{\tau}_B &= m_0 a\boldsymbol{k} \times \left[-b\omega_0^2 \boldsymbol{j} + 0 + 0 + (\alpha_0 \boldsymbol{k} - \omega_0 \omega_1 \boldsymbol{i}) \times a\boldsymbol{k}\right. \\
&\quad \left.+ (\omega_0 \boldsymbol{k} + \omega_1 \boldsymbol{j}) \times \{(\omega_0 \boldsymbol{k} + \omega_1 \boldsymbol{j}) \times a\boldsymbol{k}\}\right] \\
&\quad + m_0(-a\boldsymbol{k}) \times \left[-b\omega_0^2 \boldsymbol{j} + 0 + 0 + (\alpha_0 \boldsymbol{k} - \omega_0 \omega_1 \boldsymbol{i}) \times (-a\boldsymbol{k})\right. \\
&\quad \left.+ (\omega_0 \boldsymbol{k} + \omega_1 \boldsymbol{j}) \times \{(\omega_0 \boldsymbol{k} + \omega_1 \boldsymbol{j}) \times (-a\boldsymbol{k})\}\right] \\
&= m_0 a\boldsymbol{k} \times \left(-b\omega_0^2 \boldsymbol{j} + a\omega_0 \omega_1 \boldsymbol{j} + a\omega_0 \omega_1 \boldsymbol{j} - a\omega_1^2 \boldsymbol{k}\right) \\
&\quad + m_0(-a\boldsymbol{k}) \times \left(-b\omega_0^2 \boldsymbol{j} - a\omega_0 \omega_1 \boldsymbol{j} - a\omega_0 \omega_1 \boldsymbol{j} + a\omega_1^2 \boldsymbol{k}\right) \\
&= m_0 a\boldsymbol{k} \times \left(-b\omega_0^2 \boldsymbol{j} + 2a\omega_0 \omega_1 \boldsymbol{j} - a\omega_1^2 \boldsymbol{k}\right) - m_0 a\boldsymbol{k} \times \left(-b\omega_0^2 \boldsymbol{j} - 2a\omega_0 \omega_1 \boldsymbol{j} + a\omega_1^2 \boldsymbol{k}\right) \qquad \text{(e)} \\
&= -4m_0 a^2 \omega_0 \omega_1 \boldsymbol{i} \qquad \text{(f)}
\end{aligned}$$

Thus, $\boldsymbol{\tau}_B$ is the torque on the rod applied by the shaft.

Discussion

Although the answer we sought has been found in Eq. (f), it is instructive to examine an alternative means of finding $\boldsymbol{\tau}_B$. Given Eqs. (a) through (d) and Eq. (C-50), the forces on mass 1 and mass 2 exerted by the massless rigid rod are given by

$$\boldsymbol{F}_1 = m_0(-b\omega_0^2 \boldsymbol{j} + 2a\omega_0 \omega_1 \boldsymbol{j} - a\omega_1^2 \boldsymbol{k}) \qquad \text{(g)}$$

and

$$\boldsymbol{F}_2 = m_0(-b\omega_0^2 \boldsymbol{j} - 2a\omega_0 \omega_1 \boldsymbol{j} + a\omega_1^2 \boldsymbol{k}) \qquad \text{(h)}$$

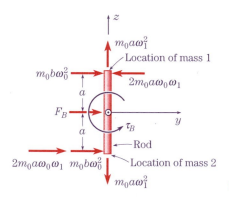

FIGURE C-9   Free-body diagram of massless rigid rod.

where $\boldsymbol{F}_1$ and $\boldsymbol{F}_2$ are the forces on mass 1 and mass 2, respectively. Note that the quantities in the first and second sets of parentheses in Eq. (e) are the inertial accelerations of mass 1 and mass 2, respectively, as found in Eqs. (g) and (h).

The free-body diagram of the massless rigid rod is shown in Figure C-9. In accordance with Newton's third law, the forces at the ends of the rod in Figure C-9 are the opposite of those given in Eqs. (g) and (h). Also, an unknown force $\boldsymbol{F}_B$ and an unknown torque $\boldsymbol{\tau}_B$ are exerted on the rod by the shaft. Because the rod is an extended massless body, both the sum of the forces and the sum of the torques must vanish; otherwise, its linear acceleration or its angular acceleration would be infinite. Thus, from Figure C-9,

$$\sum \boldsymbol{F} = 0 : \quad \boldsymbol{F}_B = -2m_0 b \omega_0^2 \boldsymbol{j} \tag{i}$$

$$\sum \boldsymbol{\tau} = 0 : \quad \boldsymbol{\tau}_B = -4m_0 a^2 \omega_0 \omega_1 \boldsymbol{i} \tag{j}$$

As expected, Eq. (j) is the same as Eq. (f), providing a confirmation of our use of Eq. (C-77).

ELEMENTARY RESULTS OF THE CALCULUS OF VARIATIONS

In this appendix we summarize some of the elementary results from the calculus of variations. Such common phrases as the "path of least resistance" when encountering some political or social circumstances or "optimize the rate of return" when considering investment opportunities or "get the best value" when making a purchase suggest the extent to which we wish to extremize (find minima or maxima) considerations in our daily lives. The calculus of variations is a mathematical technique for comparing neighboring functions as a means of selecting the one that either maximizes or minimizes a certain integral. The purpose of this appendix is simply to tabulate the rules for mathematical operations involving variation, as summarized in Table D–1.

D-1 INTRODUCTION

Consider a set of problems, say, Set I:

1. Find the dimensions of the rectangle having the largest area, which can be inscribed in a circle of radius 5 meters.

As in Figure D-1, we inscribe an arbitrary rectangle $JKLM$ and let KL equal x. Then, LM is $\sqrt{100 - x^2}$, giving the area of the rectangle as

$$A(x) = x\sqrt{100 - x^2}$$

From elementary calculus, we recall that the maximum value of $A(x)$ can be found according to

$$\frac{dA(x)}{dx} = \frac{100 - 2x^2}{\sqrt{100 - x^2}} = 0$$

Thus,

$$x = 5\sqrt{2} = 7.07$$

which leads to the desired answer.

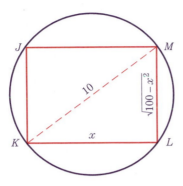

FIGURE D-1 Circle containing rectangle $JKLM$ whose area is to be maximized.

2. Select for study any of the vast number of examples in elementary calculus under the title of "maximum and minimum values of a function." Such examples will be intrinsically the same as the preceding problem.

Two observations should be made. First, in evaluating $\dfrac{dA}{dx}$ only the positive sign of the radical was taken because from the character of the problem, the negative sign has no significance. Second, although we could investigate analytically the issue of whether the answer is a maximum or a minimum of the problem, we shall note simply that the character of the problem ensures that the answer is a maximum.

Consider another set of problems, say, Set II:

1. Find the function that gives the minimum distance between two points in a (flat) plane.
2. Find the function that gives the maximum area that can be enclosed by a line of a specified length in a (flat) plane.

The answers to these two problems are (1) a straight line and (2) a circle. Neither of these answers needs to be taken as an intuitive axiom, however, as each can be proved. Furthermore, although these two answers are "intuitively" correct, slightly more difficult problems of this kind are neither intuitive nor easily answered.

The character of the problems in Set I and Set II is different, and it is this distinguishing difference that gave rise to the calculus of variations. Set I consists of problems from (ordinary) differential calculus in which one seeks extrema (maxima or minima) of a function: The answer is one or more numbers (or variables). Set II consists of problems from the calculus of variations where one seeks extrema of a *functional*, where a functional is a function of functions. For example, imagine the collection of all the functions lying in a flat plane between two specified points. This entire collection of functions may be considered to be the functional. Then, the calculus of variations provides the mathematical techniques that enable us to extract the straight line from this collection, subject to the specification of "minimum distance between the two points." So, in Set II the answer is a function, such as a "straight line" or a "circle." In geometric terminology, the calculus of variations may be considered as a technique for finding a path (straight line), area (circle), or region (hypersurface) that produces the desired extrema, subject to the specified constraints.

From ancient times through Newton and Leibniz, problems of the character of the calculus of variations had been considered. (For example, that the sphere is the minimum surface of a given volume had been "known" for millennia but had to await the calculus of variations for its scientific proof.) John Bernoulli is often regarded as being the first person

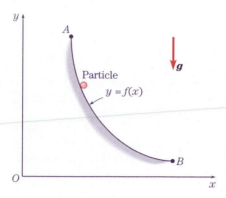

FIGURE D-2   Curve between two points A and B in analysis of brachistochrone.

who formally proposed and solved such a problem (ca. 1696). The problem, illustrated in Fig. D-2, is to find the frictionless curve (that is, the function), joining two fixed points A and B, for which a particle, subjected only to gravity, will descend in the shortest time. From the Greek words *brachus* (short) and *chronon* (time), such a curve is called the *brachistochrone*. The brachistochrone was found to be a *cycloid*,[1] lying in the vertical plane and connecting points A and B. A closer reading of history suggests that this solution, which was published by John Bernoulli, is probably more properly attributable to his brother James as well as to himself.

Euler and Lagrange were the major contributors to the development of the calculus of variations from about 1744 and 1755, respectively. The name "calculus of variations" is variously attributed to both Euler and Lagrange; the symbol δ to denote *admissible change*, versus d, which represents *actual change*, was introduced by Lagrange; and of the two of them, Lagrange is often credited with the greater contributions to the subject. However, Euler and Lagrange by no means left the subject complete. Like many of the disciplines of mathematics, science, and engineering, the calculus of variations was developed further by many mathematicians of the nineteenth century.

The calculus of variations has become a broad and somewhat complicated discipline within mathematics. For our purposes, however, its results can be characterized quite simply. First, although the meanings of δ and d are conceptually different, *they behave the same mathematically*. Thus, no intrinsically new mathematical operations must be learned. Second, for our limited goal of deriving equations of motion, the application of the calculus of variations often reduces to the elementary technique of *integration by parts*.

D-2   SUMMARY OF ELEMENTARY RESULTS

If the *admissible motion* of a system, such as a particle, is defined by its position or generalized coordinate $\xi(t)$ as sketched in Fig. D-3, an arbitrary infinitesimal geometrically compatible change of this motion is defined as $\delta\xi(t)$, an *admissible variation*. The *varied motion*, as shown by the dashed line, is $\xi(t) + \delta\xi(t)$. Note that the variation $\delta\xi(t')$ occurs at

---

[1]The reader may recall that the trajectory of a point on the periphery of a planar rolling nonslipping disk on a flat surface is a cycloid.

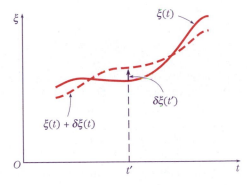

FIGURE D-3 Two trajectories and the variation $\delta\xi(t)$ between them.

an instant in time t'; in this manner, the variations are said to be *contemporaneous*. All variations considered here will be contemporaneous.

Functions such as energies and coenergies can be formulated in terms of ξ and its derivatives. A variation in a function $f(\xi)$ is defined by

$$\delta f(\xi) = f(\xi + \delta\xi) - f(\xi) \tag{D-1}$$

When the function represents the velocity $\dfrac{d\xi}{dt}$, the variation based on Eq. (D-1) is

$$\delta\left(\frac{d\xi}{dt}\right) = \frac{d(\xi + \delta\xi)}{dt} - \frac{d(\xi)}{dt}$$

$$= \frac{d\xi}{dt} + \frac{d}{dt}(\delta\xi) - \frac{d\xi}{dt}$$

$$= \frac{d}{dt}(\delta\xi) \tag{D-2}$$

Equation (D-2) expresses the important result that the variation operation and time differentiation are commutative; that is, their order is *interchangeable*.

If the generalized coordinate ξ depends on both time and space, as in the case of a generalized coordinate in a continuous medium, then $\xi = \xi(x, t)$, and the variation of $\dfrac{\partial\xi}{\partial x}$ based on Eq. (D-1) is

$$\delta\left(\frac{\partial\xi}{\partial x}\right) = \frac{\partial(\xi + \delta\xi)}{\partial x} - \frac{\partial\xi}{\partial x}$$

$$= \frac{\partial\xi}{\partial x} + \frac{\partial}{\partial x}(\delta\xi) - \frac{\partial\xi}{\partial x}$$

$$= \frac{\partial}{\partial x}(\delta\xi) \tag{D-3}$$

Equation (D-3) is a statement that the variation operation and space differentiation are also commutative.

When an integral $\int_A^B f(\xi)\,dt$ is considered, its variation based on Eq. (D-1) is

$$\delta\int_A^B f(\xi)\,dt = \int_A^B f(\xi + \delta\xi)\,dt - \int_A^B f(\xi)\,dt = \int_A^B [f(\xi + \delta\xi) - f(\xi)]\,dt \tag{D-4}$$

Substitution of Eq. (D-1) into Eq. (D-4) gives

$$\delta \int_A^B f(\xi)\, dt = \int_A^B \delta f(\xi)\, dt \tag{D-5}$$

Equation (D-5) indicates that the variation operation and time integration are commutative. Similarly, it can be shown that the variation operation and space integration are also commutative, as in

$$\delta \int_A^B f(\xi)\, dx = \int_A^B \delta f(\xi)\, dx \tag{D-6}$$

Recalling that the admissible variations are infinitesimal changes, Eq. (D-1) may be usefully recast. The variation of a function can also be obtained from a Taylor series expansion of the function. The expansion of the varied function $f(\xi + \delta\xi)$ about ξ is

$$f(\xi + \delta\xi) = f(\xi) + \frac{df(\xi)}{d\xi}\delta\xi + \frac{d^2 f(\xi)}{d\xi^2}\frac{(\delta\xi)^2}{2} + \cdots \tag{D-7}$$

Substitution of Eq. (D-7) into Eq. (D-1) and retention of only terms up to the first order in $\delta\xi$ give

$$\delta f(\xi) = \frac{df(\xi)}{d\xi}\delta\xi \tag{D-8}$$

Similarly, it can be shown that

$$\delta f(\dot{\xi}) = \frac{df(\dot{\xi})}{d\dot{\xi}}\delta\dot{\xi} \tag{D-9}$$

and

$$\delta f(\xi') = \frac{df(\xi')}{d\xi'}\delta\xi' \tag{D-10}$$

where for the time derivative

$$\dot{\xi} = \frac{d\xi}{dt} \tag{D-11}$$

has been used and for the space derivative

$$\xi' = \frac{\partial\xi}{\partial x} \tag{D-12}$$

has been used. When an integral $\int_0^\xi f(\theta)\, d\theta$ is considered, its variation based on Eq. (D-8) is

$$\delta \int_0^\xi f(\theta)\, d\theta = \frac{d}{d\xi}\left[\int_0^\xi f(\theta)\, d\theta\right]\delta\xi$$

$$= f(\xi)\,\delta\xi \tag{D-13}$$

where Leibniz's rule from elementary calculus has been used to obtain the final term on the right-hand side of Eq. (D-13).

TABLE D-1 Summary of Rules of Variation

$$\delta(\xi_1 + \xi_2) = \delta\xi_1 + \delta\xi_2 \tag{D-17}$$

$$\delta(\xi_1\xi_2) = \xi_2\,\delta\xi_1 + \xi_1\,\delta\xi_2 \tag{D-18}$$

$$\delta(\xi^n) = n\xi^{n-1}\delta\xi \tag{D-19}$$

$$\delta\left(\frac{\xi_1}{\xi_2}\right) = \frac{\xi_2\,\delta\xi_1 - \xi_1\,\delta\xi_2}{\xi_2^2} \tag{D-20}$$

$$\delta\left(\frac{d\xi}{dt}\right) = \frac{d}{dt}(\delta\xi) \tag{D-21}$$

$$\delta\left(\frac{\partial\xi}{\partial t}\right) = \frac{\partial}{\partial t}(\delta\xi) \tag{D-22}$$

$$\delta f(\xi) = \frac{df(\xi)}{d\xi}\delta\xi \tag{D-23}$$

$$\delta f(\dot{\xi}) = \frac{df(\dot{\xi})}{d\dot{\xi}}\delta\dot{\xi}, \quad \text{where} \quad \dot{\xi} = \frac{d\xi}{dt} \tag{D-24}$$

$$\delta f(\xi') = \frac{df(\xi')}{d\xi'}\delta\xi', \quad \text{where} \quad \xi' = \frac{d\xi}{dt} \tag{D-25}$$

$$\delta f(\xi, \dot{\xi}) = \frac{\partial f(\xi, \dot{\xi})}{\partial\xi}\delta\xi + \frac{\partial f(\xi, \dot{\xi})}{\partial\dot{\xi}}\delta\dot{\xi} \tag{D-26}$$

$$\delta[f(\xi_1, \xi_2, \cdots, \xi_n)] = \sum_{j=1}^{n}\frac{\partial f}{\partial\xi_j}\delta\xi_j \tag{D-27}$$

$$\delta\int_A^B f(\xi)\,dt = \int_A^B \delta f(\xi)\,dt \tag{D-28}$$

$$\delta\int_A^B f(\xi)\,dx = \int_A^B \delta f(\xi)\,dx \tag{D-29}$$

$$\delta\int_0^\xi f(\theta)\,d\theta = f(\xi)\,\delta\xi \tag{D-30}$$

If the function f depends on more than one variable such as $f(\xi, \dot{\xi})$, its variation can be derived from the Taylor series for multiple variables[2] to give

$$\delta f(\xi, \dot{\xi}) = \frac{\partial f(\xi, \dot{\xi})}{\partial\xi}\delta\xi + \frac{\partial f(\xi, \dot{\xi})}{\partial\dot{\xi}}\delta\dot{\xi} \tag{D-14}$$

Similarly, for a function $f(\xi_1, \xi_2, \ldots, \xi_n)$, its variation is

$$\delta f(\xi_1, \xi_2, \ldots, \xi_n) = \sum_{j=1}^{n}\frac{\partial f}{\partial\xi_j}\delta\xi_j \tag{D-15}$$

Note the similarity between the variation of a function and the total differential of a function. For the function $f(\xi_1, \xi_2, \ldots, \xi_n)$, the total differential is

[2]For example, see F. B. Hildebrand, *Advanced Calculus for Applications*, 2nd ed., Prentice-Hall, Englewood Cliffs, NJ, 1976, p. 355.

$$df(\xi_1, \xi_2, \ldots, \xi_n) = \sum_{j=1}^{n} \frac{\partial f}{\partial \xi_j} d\xi_j \tag{D-16}$$

as may be verified from most elementary calculus textbooks.

A summary of useful equations is shown in Table D-1. Equations (D-17), (D-18), and (D-20) can be easily derived from Eq. (D-26), as can Eq. (D-19) from Eq. (D-23). For example, an exchange of ξ_1 and ξ_2 for ξ and $\dot\xi$, respectively, in Eq. (D-26) will lead directly to the results in Eqs. (D-17), (D-18) or (D-20). Also, Eq. (D-27) is simply a generalization of Eq. (D-26). Finally, note that the rules governing the variation of sums, products, exponents, and ratios are identical to the corresponding rules for ordinary differentials.

D-3   EULER EQUATION: NECESSARY CONDITION FOR A VARIATIONAL INDICATOR TO VANISH

Consider a variational indicator of the form

$$V.I. = \delta \int_{t_1}^{t_2} F(\xi, \dot\xi, t)\, dt \tag{D-31}$$

where, in a particular application, the integral in Eq. (D-31) is a known *functional* of its arguments and $\xi(t)$ is the unknown motion, which is to be determined in accordance with the requirement that Eq. (D-31) vanishes for arbitrary admissible variations of the variable ξ; that is, for arbitrary $\delta\xi$. Furthermore, and solely for convenience, $\delta\xi$ is required to vanish at the integration end points t_1 and t_2, as sketched in Fig. D-4.

The vanishing of the variations $\delta\xi$ at t_1 and t_2 is a widely adopted convention in the calculus of variations. As indicated in the preceding paragraph, it is a convention adopted purely for simplicity and, in general, does not alter the sought-after goal. Hamilton's principle is an example of this type. On the other hand, it is not a universal convention as the sought-after goal may depend on evaluations made at t_1 and t_2. Hamilton's law of varying action is an example of this type. The goals of this appendix are conveniently and fully met when $\delta\xi$ vanishes at t_1 and t_2.

Application of Eq. (D-28) in Eq. (D-31) gives

$$V.I. = \int_{t_1}^{t_2} \delta F(\xi, \dot\xi, t)\, dt \tag{D-32}$$

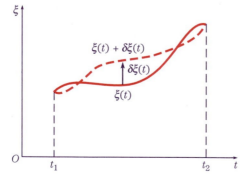

FIGURE D-4   Two trajectories and the variation $\delta\xi(t)$, which is arbitrary within $t_1 < t < t_2$ but restricted to vanish at t_1 and t_2.

Then application of Eq. (D-26) in Eq. (D-32) gives

$$V.I. = \int_{t_1}^{t_2} \left(\frac{\partial F}{\partial \xi} \delta\xi + \frac{\partial F}{\partial \dot{\xi}} \delta\dot{\xi} \right) dt \tag{D-33}$$

where the explicit dependence on t in Eq. (D-32) does not appear in Eq. (D-33) because we have restricted our considerations to contemporaneous variations.

The next task is to make the arbitrary variation $\delta\xi$ a common factor. Application of Eq. (D-21) to the second term on the right-hand side of Eq. (D-33) gives

$$\int_{t_1}^{t_2} \frac{\partial F}{\partial \dot{\xi}} \delta\dot{\xi} dt = \int_{t_1}^{t_2} \frac{\partial F}{\partial \dot{\xi}} \frac{d}{dt} (\delta\xi) \, dt \tag{D-34}$$

The standard formula for integration by parts is often written as

$$\int u \, dv = uv - \int v \, du \tag{D-35}$$

where the constant of integration has been arbitrarily set to zero. Let

$$u = \frac{\partial F}{\partial \dot{\xi}} \quad \text{and} \quad dv = \frac{d}{dt} (\delta\xi) \, dt \tag{D-36}$$

which give

$$du = \frac{d}{dt} \left(\frac{\partial F}{\partial \dot{\xi}} \right) dt \quad \text{and} \quad v = \delta\xi \tag{D-37}$$

Substitution of Eqs. (D-36) and (D-37) into Eq. (D-35) gives the equivalent form of Eq. (D-34) as

$$\int_{t_1}^{t_2} \frac{\partial F}{\partial \dot{\xi}} \delta\dot{\xi} dt = \frac{\partial F}{\partial \dot{\xi}} \delta\xi \Big|_{t_1}^{t_2} - \int_{t_1}^{t_2} \delta\xi \frac{d}{dt} \left(\frac{\partial F}{\partial \dot{\xi}} \right) dt \tag{D-38}$$

Substitution of Eq. (D-38) into Eq. (D-33) gives

$$V.I. = \int_{t_1}^{t_2} \left[\frac{\partial F}{\partial \xi} - \frac{d}{dt} \left(\frac{\partial F}{\partial \dot{\xi}} \right) \right] \delta\xi dt + \frac{\partial F}{\partial \dot{\xi}} \delta\xi \Big|_{t_1}^{t_2} \tag{D-39}$$

Recall that the variations $\delta\xi$ are required to vanish at the end points at t_1 and t_2 as sketched in Fig. D-4. This restriction on $\delta\xi$ simplifies Eq. (D-39) into

$$V.I. = \int_{t_1}^{t_2} \left[\frac{\partial F}{\partial \xi} - \frac{d}{dt} \left(\frac{\partial F}{\partial \dot{\xi}} \right) \right] \delta\xi dt \tag{D-40}$$

In order for the variational indicator to vanish for arbitrary $\delta\xi$, the terms in the square bracket must vanish altogether in the interval t_1 to t_2. Thus, we conclude that a *necessary* condition for Eq. (D-39) to vanish is

$$\frac{\partial F}{\partial \xi} - \frac{d}{dt} \left(\frac{\partial F}{\partial \dot{\xi}} \right) = 0 \quad t_1 < t < t_2 \tag{D-41}$$

Equation (D-41) is generally known as the Euler equation, first discovered by him in 1744. Later, Lagrange derived it more rigorously after having worked on it during the period

1762–1770. In mathematical treatises on the calculus of variations, it is variously called the Euler equation, the Lagrange equation, and the Euler-Lagrange equation. As a result that follows from Hamilton's principle, we shall refer to equations of the form of Eq. (D-41) as Lagrange's equation. As a general result of the calculus of variations, however, we shall simply call it the Euler equation.

The argument leading from Eq. (D-40) to the final result in Eq. (D-41) is sometimes called the Fundamental Lemma of the calculus of variations. For convenience, let the square bracket in Eq. (D-40) be denoted by $\Lambda(t)$. The argument of the Fundamental Lemma goes as follows: Suppose $\Lambda(t)$ does not vanish over some interval $[t_3$ to $t_4]$ where $t_1 < t_3 < t_4 < t_2$. Then for some (possibly very small) arbitrary region within the interval $[t_3$ to $t_4]$, the sign of $\Lambda(t)$ must be constant over this region. Let the arbitrary $\delta\xi(t)$ be zero over the intervals $[t_1$ to $t_3]$ and $[t_4$ to $t_2]$, and be nonzero and have the same sign as $\Lambda(t)$ over the interval $[t_3$ to $t_4]$. This situation is illustrated in Fig. D-5. Then, because both $\Lambda(t)$ and $\delta\xi(t)$ are nonzero and have the same sign over $[t_3$ to $t_4]$, the variational indicator in Eq. (D-40) must be greater than zero. It follows, therefore, that the assumption of nonzero $\Lambda(t)$ within the interval $[t_1$ to $t_2]$ violates the requirement of a vanishing $V.I.$ in Eq. (D-40). Thus, it is necessary that $\Lambda(t)$ vanishes throughout the interval $[t_1$ to $t_2]$ for arbitrary variations; that is, Eq. (D-41) must hold. Such proofs are said to follow from a *reductio ad absurdum* argument.

In dynamics, Hamilton's principle for holonomic systems is a problem of the calculus of variations. The idea that Maupertuis (incompletely) expressed is that as systems move from one state to another, Nature is efficient and minimizes a certain quantity that he called the "action." Along the way toward its maturity, action was defined variously including the space integral of the momentum $\left(\int m\boldsymbol{v} \cdot d\boldsymbol{R}\right)$ and the time integral of the *vis viva* $\left(\int mv^2 \, dt\right)$, both by Lagrange in 1760. Ultimately, these notions evolved into Hamilton's principle where, for systems subjected to no nonconservative forces, the time integral of the lagrangian $\left(\int \mathcal{L}(\xi, \dot{\xi}, t) \, dt\right)$ is to be *stationary*. (In this context, a stationary value of a function simply means that the slope of the function is zero at the specific value of the dependent variation, irrespective of whether it is a maximum or a minimum.) The stationarity of the hamiltonian action is $V.I. = \delta \int \mathcal{L}(\xi, \dot{\xi}, t) \, dt = 0$, which leads directly to Lagrange's equations.

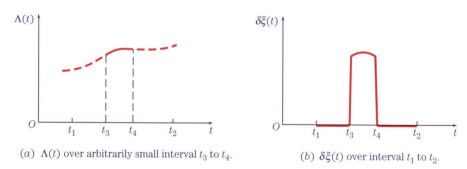

(a) $\Lambda(t)$ over arbitrarily small interval t_3 to t_4.

(b) $\delta\xi(t)$ over interval t_1 to t_2.

FIGURE D-5   Values of $\Lambda(t)$ and $\delta\xi(t)$ that contradict Eq. (D-41).

SOME FORMULATIONS OF THE PRINCIPLES OF HAMILTON

In this appendix, several forms of Hamilton's principle will be shown to be related to other laws or principles with which the reader is likely already familiar. Here, we limit our analyses to mechanical and electromechanical formulations; electrical formulations are derivative simply from the electromechanical formulations. We shall begin with familiar principles and then show their relationship to Hamilton's principle. We emphasize the view that we have adopted: We are *not* deriving Hamilton's principle; Hamilton's principle is a fundamental law of nature which, based upon its generality, is arguably more fundamental than the principles with which we begin.

Also, an important and useful relation between the total energy of the system and the work done on the system by all the nonconservative forces is derived from Lagrange's equations, which themselves are derived from Hamilton's principle.

E-1 MECHANICAL FORMULATIONS

We consider an arbitrarily complex holonomic mechanical system consisting of N particles m_i ($i = 1, 2, \ldots, N$) where the ith particle is acted upon by the ith resultant force $\boldsymbol{f}_i$ ($i = 1, 2, \ldots, N$). It is noted that the forces $\boldsymbol{f}_i$ represent all the various forces, including those that are conservative (such as elastic spring forces) as well as those that are nonconservative (such as dashpot forces). Below, we shall treat the contributions of these two types of forces separately. As sketched in Figure E-1, the ith particle can be located with respect to an inertial reference frame by an associated position vector $\boldsymbol{R}_i$ ($i = 1, 2, \ldots, N$). Since the system is holonomic, we may let the system be characterized by n complete and independent generalized coordinates ξ_j ($j = 1, 2, \ldots, n$) and the n associated admissible variations $\delta \xi_j$. The geometric requirements on the motions are therefore satisfied. Furthermore, the position vector of the ith particle can be expressed in terms of the generalized coordinates (and time if there are time-varying constraints) as

$$\boldsymbol{R}_i = \boldsymbol{R}_i(\xi_1, \xi_2, \ldots, \xi_n, t) \tag{E-1}$$

Newton's second law for each particle states that

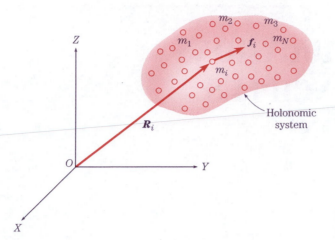

FIGURE E-1 Holonomic system consisting of N particles.

$$\boldsymbol{f}_i = \frac{d\boldsymbol{p}_i}{dt} \qquad i = 1, 2, \ldots, N \tag{E-2}$$

where $\boldsymbol{p}_i$ is the linear momentum of the ith particle. Equation (E-2) can be rewritten as

$$\boldsymbol{f}_i - \frac{d\boldsymbol{p}_i}{dt} = 0 \qquad i = 1, 2, \ldots, N \tag{E-3}$$

Equations (E-2) or (E-3) satisfy the dynamic requirements on the forces. We form a scalar *variational indicator* for the entire system by taking the vector dot product of Eq. (E-3) with $\delta\boldsymbol{R}_i$ which we write as

$$(\text{V.I.})_0 = \sum_{i=1}^{N} \left(\boldsymbol{f}_i - \frac{d\boldsymbol{p}_i}{dt} \right) \cdot \delta\boldsymbol{R}_i$$

$$= \sum_{i=1}^{N} \boldsymbol{f}_i \cdot \delta\boldsymbol{R}_i - \sum_{i=1}^{N} \frac{d\boldsymbol{p}_i}{dt} \cdot \delta\boldsymbol{R}_i \tag{E-4}$$

The variational indicator $(\text{V.I.})_0$ is a scalar by virtue of the vector dot product in Eq. (E-4), and it must vanish for arbitrary variations $\delta\boldsymbol{R}_i$ if each equation in Eq. (E-3) is satisfied.

Considering the first term on the right-hand side of Eq. (E-4) gives

$$\sum_{i=1}^{N} \boldsymbol{f}_i \cdot \delta\boldsymbol{R}_i = -\delta V + \sum_{j=1}^{n} \Xi_j\, \delta\xi_j \tag{E-5}$$

where we have separated the contributions due to conservative forces, given by $-\delta V$, from the contributions due to nonconservative forces, given by $\sum_{j=1}^{n} \Xi_j\, \delta\xi_j$. V is the potential energy function of the conservative forces and Ξ_j represents the generalized forces corresponding to the nonconservative forces. If there is ambiguity regarding whether a particular force is conservative or nonconservative, that force can always be treated in Eq. (E-5) as a nonconservative force.

In calculating the generalized forces Ξ_j, an abbreviated form of Eq. (E-5) in which the $-\delta V$ term is omitted is often used. In such a use of Eq. (E-5), the summation on the left-hand side of the equation is effectively a summation over all the nonconservative forces in the system.

The second term on the right-hand side of Eq. (E-4) can be usefully explored by recalling the product differentiation rule as follows:

$$\sum_{i=1}^{N} \frac{d}{dt}(\boldsymbol{p}_i \cdot \delta\boldsymbol{R}_i) = \sum_{i=1}^{N} \frac{d\boldsymbol{p}_i}{dt} \cdot \delta\boldsymbol{R}_i + \sum_{i=1}^{N} \boldsymbol{p}_i \cdot \frac{d\delta\boldsymbol{R}_i}{dt} \tag{E-6}$$

Rearranging the terms in Eq. (E-6) gives

$$-\sum_{i=1}^{N} \frac{d\boldsymbol{p}_i}{dt} \cdot \delta\boldsymbol{R}_i = -\sum_{i=1}^{N} \frac{d}{dt}(\boldsymbol{p}_i \cdot \delta\boldsymbol{R}_i) + \sum_{i=1}^{N} \boldsymbol{p}_i \cdot \frac{d\delta\boldsymbol{R}_i}{dt} \tag{E-7}$$

Further, the second term on the right-hand side of Eq. (E-7) can be expressed as

$$\sum_{i=1}^{N} \boldsymbol{p}_i \cdot \frac{d\delta\boldsymbol{R}_i}{dt} = \sum_{i=1}^{N} \boldsymbol{p}_i \cdot \delta\left(\frac{d\boldsymbol{R}_i}{dt}\right) = \sum_{i=1}^{N} \boldsymbol{p}_i \cdot \delta\boldsymbol{v}_i = \delta T^* \tag{E-8}$$

since we note that

$$\delta(T_i^*) = \delta\left(\frac{1}{2}m\boldsymbol{v}_i \cdot \boldsymbol{v}_i\right) = \left(\frac{1}{2}m\boldsymbol{v}_i \cdot \delta\boldsymbol{v}_i\right) + \left(\frac{1}{2}m\delta\boldsymbol{v}_i \cdot \boldsymbol{v}_i\right)$$

or

$$\delta(T_i^*) = m\boldsymbol{v}_i \cdot \delta\boldsymbol{v}_i = \boldsymbol{p}_i \cdot \delta\boldsymbol{v}_i \tag{E-9}$$

where T^* is the kinetic coenergy of the system. Substitution of Eq. (E-8) into Eq. (E-7) gives

$$-\sum_{i=1}^{N} \frac{d\boldsymbol{p}_i}{dt} \cdot \delta\boldsymbol{R}_i = -\sum_{i=1}^{N} \frac{d}{dt}(\boldsymbol{p}_i \cdot \delta\boldsymbol{R}_i) + \delta T^* \tag{E-10}$$

where the left-hand side of Eq. (E-10) is observed to be the same as the second term on the right-hand side of Eq. (E-4).

Substitution of Eq. (E-5) and Eq. (E-10) into Eq. (E-4) gives

$$(\text{V.I.})_0 = \delta T^* - \delta V + \sum_{j=1}^{n} \Xi_j\,\delta\xi_j - \sum_{i=1}^{N} \frac{d}{dt}(\boldsymbol{p}_i \cdot \delta\boldsymbol{R}_i) \tag{E-11}$$

The variational indicator $(\text{V.I.})_0$ in Eq. (E-11) must vanish for arbitrary admissible variations at each instant in time in order to ensure that Newton's second law is satisfied. Further, it is important to note that the *geometric requirements* have been introduced via Eq. (E-1); the *force–dynamic* requirements have been introduced via Eq. (E-2); and the *constitutive requirements* have been introduced via the state functions T^* and V and the generalized forces Ξ_j. Thus, Eq. (E-11) clearly incorporates the three fundamental requirements for the formulation of the equations of motion for mechanical systems.

In order to obtain any of the various forms of Hamilton's principle(s), we integrate Eq. (E-11) with respect to time over an interval from $t = t_1$ to $t = t_2$, where t_1 is less than t_2. (The meanings of the times t_1 and t_2 vary depending on the particular principle or law under consideration; their significance will be discussed below.) The resulting variational indicator obtained by integration is

$$\text{V.I.}' = \int_{t_1}^{t_2} \left[\delta(T^* - V) + \sum_{j=1}^{n} \Xi_j \, \delta\xi_j - \sum_{i=1}^{N} \frac{d}{dt}(\boldsymbol{p}_i \cdot \delta\boldsymbol{R}_i) \right] dt \qquad \text{(E-12)}$$

which may be interpreted as a global variational indicator that is valid for all times from t_1 to t_2.

E-1.1   Hamilton's Law of Varying Action

The last term in Eq. (E-12) is a total differential on which the integration can be performed, enabling Eq. (E-12) to be rewritten as

$$\text{V.I.}' = \int_{t_1}^{t_2} \left[\delta(T^* - V) + \sum_{j=1}^{n} \Xi_j \, \delta\xi_j \right] dt - \sum_{i=1}^{N} \boldsymbol{p}_i \cdot \delta\boldsymbol{R}_i \Big|_{t_1}^{t_2} \qquad \text{(E-13)}$$

Eq. (E-13) is a less common presentation of the *general form of Hamilton's law of varying action.* Before proceeding we shall consider the integrated term in order to cast it into a more commonly written form.

Without engaging in an extended discussion, we recall that the following expressions can be obtained in a straightforward manner.[1]

$$\boldsymbol{p}_i = \frac{\partial T^*}{\partial \boldsymbol{v}_i} = \frac{\partial T^*}{\partial \dot{\boldsymbol{R}}_i} \qquad \text{(E-14)}$$

$$\delta\boldsymbol{R}_i = \sum_{j=1}^{n} \frac{\partial \boldsymbol{R}_i}{\partial \xi_j} \, \delta\xi_j \qquad \text{(E-15)}$$

and

$$\frac{\partial \boldsymbol{R}_i}{\partial \xi_j} = \frac{\partial \dot{\boldsymbol{R}}_i}{\partial \dot{\xi}_j} \qquad \text{(E-16)}$$

where $\frac{d}{dt}(\square) = (\dot{\square})$ in Eqs. (E-14) and (E-16). Then, reformulating the integrated term in Eq. (E-13) can be accomplished as follows:

$$-\sum_{i=1}^{N} (\boldsymbol{p}_i \cdot \delta\boldsymbol{R}_i) = -\sum_{i=1}^{N} \left(\frac{\partial T^*}{\partial \dot{\boldsymbol{R}}_i} \cdot \sum_{j=1}^{n} \frac{\partial \boldsymbol{R}_i}{\partial \xi_j} \, \delta\xi_j \right) \qquad \text{(E-17)}$$

where Eqs. (E-14) and (E-15) have been used in Eq. (E-17) to express $\boldsymbol{p}_i$ and $\delta\boldsymbol{R}_i$, respectively. Now, substituting Eq. (E-16) into Eq. (E-17) yields

$$-\sum_{i=1}^{N} (\boldsymbol{p}_i \cdot \delta\boldsymbol{R}_i) = -\sum_{j=1}^{n} \left[\sum_{i=1}^{N} \frac{\partial T^*}{\partial \dot{\boldsymbol{R}}_i} \cdot \frac{\partial \dot{\boldsymbol{R}}_i}{\partial \dot{\xi}_j} \right] \delta\xi_j$$

$$= -\sum_{j=1}^{n} \left[\frac{\partial T^*}{\partial \dot{\xi}_j} \right] \delta\xi_j \qquad \text{(E-18)}$$

where the summation over i has been performed.

---

[1]Each of these results is derived in detail elsewhere in this book.

Substitution of Eq.(E-18) into Eq.(E-13) gives

$$\text{V.I.}' = \int_{t_1}^{t_2} \left[\delta(T^* - V) + \sum_{j=1}^{n} \Xi_j \, \delta\xi_j \right] dt - \sum_{j=1}^{n} \left[\frac{\partial T^*}{\partial \dot{\xi}_j} \, \delta\xi_j \right]_{t_1}^{t_2} \tag{E-19}$$

where Eq.(E-19) is the more common presentation of the *general form of Hamilton's law of varying action.* Equation (E-19) will be discussed in Subsection E-1.4, where the meanings of t_1 and t_2 will also be explained.

E-1.2   Hamilton's Principle

If we proceed from the beginning of the analysis in Section E-1 and arrive at Eq. (E-12), we may immediately deduce *Hamilton's principle* by adopting the widely adopted convention that all admissible variations vanish at $t = t_1$ and $t = t_2$, as indicated in Figure E-2. In this case, the last term in Eq. (E-12) vanishes (refer to the last term in Eq. (E-13), which is its equivalent), resulting in

$$\text{V.I.} = \int_{t_1}^{t_2} \left[\delta(T^* - V) + \sum_{j=1}^{n} \Xi_j \, \delta\xi_j \right] dt \tag{E-20}$$

Equation (E-20) is *Hamilton's principle* which, in words, we express as follows:

An admissible motion of the system between specified configurations at t_1 and t_2 is a natural motion if, and only if, the variational indicator V.I. *vanishes for arbitrary admissible variations.*

In general, we shall call Eq.(E-20) *Hamilton's principle;* and occasionally we shall refer to it as the *classical Hamilton's principle* in order to distinguish it from the corresponding forms of the principle for electrical and electromechanical systems. The terminology and the statement of these principles vary among authors of dynamics literature. For example, only when $\Xi_j = 0$ is Eq. (E-20) called Hamilton's principle by some authors. In such cases, when $\Xi_j \neq 0$, many of those authors refer to Eq. (E-20) as the *extended form of Hamilton's principle.*

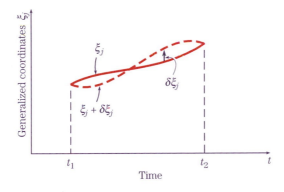

FIGURE E-2   An admissible trajectory and a varied admissible trajectory, where variations vanish at t_1 and t_2.

E-1.3   Lagrange's Equations

By a direct application of the calculus of variations to either of the variational indicators, that is, either Eq. (E-13) (or its equivalent Eq. (E-19)) or Eq. (E-20), we can obtain the necessary conditions for each of these variational indicators to vanish for arbitrary admissible variations $\delta\xi_j$. We shall use Eq. (E-20), and in the discussion below describe the minor differences between this choice and the other choice, namely, Eq. (E-13).

The *lagrangian function,* which in general we shall simply call the *lagrangian,* is defined as

$$\mathcal{L}(\xi_j, \dot{\xi}_j, t) \equiv T^*(\xi_j, \dot{\xi}_j, t) - V(\xi_j) \tag{E-21}$$

Because V is a function of ξ_j only while T^* may be a function of ξ_j, $\dot{\xi}_j$, and t, where time t enters explicitly if there are time-varying constraints, the lagrangian is also a function of these variables. The variation on the lagrangian is

$$\delta\mathcal{L} = \sum_{j=1}^{n} \left(\frac{\partial\mathcal{L}}{\partial\dot{\xi}_j} \delta\dot{\xi}_j + \frac{\partial\mathcal{L}}{\partial\xi_j} \delta\xi_j \right) \tag{E-22}$$

Note that although the lagrangian is a function of ξ_j, $\dot{\xi}_j$, and t, δt does not appear in expressions such as Eq. (E-22) because all variations are considered to be contemporaneous; that is, they occur at an instant in time. Substitution of Eq. (E-22) into Eq. (E-20) gives

$$\text{V.I.} = \int_{t_1}^{t_2} \sum_{j=1}^{n} \left[\frac{\partial\mathcal{L}}{\partial\dot{\xi}_j} \delta\dot{\xi}_j + \frac{\partial\mathcal{L}}{\partial\xi_j} \delta\xi_j + \Xi_j \, \delta\xi_j \right] dt \tag{E-23}$$

In order to apply Hamilton's principle, we require that the V.I. vanishes for arbitrary admissible variations on the generalized coordinates. While the second and third terms in Eq. (E-23) contain an arbitrary variation on the generalized coordinates, the first term is an arbitrary variation on the generalized velocities. The conversion of the first term, from a variation on the generalized velocities to a variation on the generalized coordinates, can be accomplished by using integration by parts. Let's consider the first term only.

$$\int_{t_1}^{t_2} \sum_{j=1}^{n} \frac{\partial\mathcal{L}}{\partial\dot{\xi}_j} \delta\dot{\xi}_j \, dt = \int_{t_1}^{t_2} \sum_{j=1}^{n} \frac{\partial\mathcal{L}}{\partial\dot{\xi}_j} \delta\left(\frac{d\xi_j}{dt} \right) dt = \int_{t_1}^{t_2} \sum_{j=1}^{n} \frac{\partial\mathcal{L}}{\partial\dot{\xi}_j} \frac{d\delta\xi_j}{dt} \, dt \tag{E-24}$$

where we have used the fact that differentiation and variation are commutative. A common expression for integration by parts is

$$\int u \, dv = uv - \int v \, du \tag{E-25}$$

where we have omitted any constants of integration. If we *imagine* the cancellation of the two dt's in Eq. (E-24), which is a conceptually useful maneuver but which does not really occur, it is straightforward to analogize from Eqs. (E-24) and (E-25) the following equivalents:

$$\frac{\partial\mathcal{L}}{\partial\dot{\xi}_j} \equiv u \qquad d(\delta\xi_j) \equiv dv \qquad d\left(\frac{\partial\mathcal{L}}{\partial\dot{\xi}_j} \right) \equiv du \qquad \delta\xi_j \equiv v \tag{E-26}$$

Then, use of these analogies assists in the integration by parts of Eq. (E-24) according to

$$\int_{t_1}^{t_2} \sum_{j=1}^{n} \frac{\partial\mathcal{L}}{\partial\dot{\xi}_j} \delta\xi_j \, dt = \sum_{j=1}^{n} \left[\frac{\partial\mathcal{L}}{\partial\dot{\xi}_j} \delta\xi_j \Big|_{t_1}^{t_2} - \int_{t_1}^{t_2} \frac{d}{dt}\left(\frac{\partial\mathcal{L}}{\partial\dot{\xi}_j} \right) \delta\xi_j \, dt \right] \tag{E-27}$$

where the two dt's that were never cancelled appear in their proper places in the last term in Eq. (E-27). In accordance with Hamilton's principle and Figure E-2,

$$\sum_{j=1}^{n} \frac{\partial \mathcal{L}}{\partial \dot{\xi}_j} \delta \xi_j \bigg|_{t_1}^{t_2} = 0 \tag{E-28}$$

since the variations vanish at t_1 and t_2.

Thus, substitution of Eqs. (E-27) and (E-28) into Eq. (E-23) gives

$$\text{V.I.} = -\int_{t_1}^{t_2} \sum_{j=1}^{n} \left[\frac{d}{dt} \left(\frac{\partial \mathcal{L}}{\partial \dot{\xi}_j} \right) - \frac{\partial \mathcal{L}}{\partial \xi_j} - \Xi_j \right] \delta \xi_j \, dt \tag{E-29}$$

The necessary condition for Eq. (E-29) to vanish for arbitrary $\delta \xi_j$ is that the n equations,

$$\frac{d}{dt} \left(\frac{\partial \mathcal{L}}{\partial \dot{\xi}_j} \right) - \frac{\partial \mathcal{L}}{\partial \xi_j} = \Xi_j \qquad j = 1, 2, \ldots, n \tag{E-30}$$

must be satisfied at each instant within the interval from t_1 to t_2. Equations (E-30) are *Lagrange's equations,* and for many holonomic systems they provide the easiest route to obtaining the equations of motion for lumped-parameter systems. The derivation of the equations of motion for continuous-parameter systems is most directly obtained by use of Hamilton's principle as given in Eq. (E-20).

E-1.4   Discussion

Two of the most obvious questions resulting from the above analyses are "What are the differences between *Hamilton's law of varying action* and *Hamilton's principle?*" and "What are the meanings of the times t_1 and t_2?" Indeed, these two questions are intimately related.

Hamilton's law of varying action (HLVA) is more general than Hamilton's principle (HP) in that the latter can be shown to follow from the former, but not vice versa. As presented in Eq. (E-19), HLVA can be used to integrate directly the formulation of the system model, that is, without ever specifically deriving the equations of motion. In this use, the time t_1 represents some time at which the state of the system is somehow known. For example, this known state might be represented as the initial generalized coordinates and the initial generalized velocities at t_1. The time t_2 represents some final time at which the state of the system is desired to be known. If a numerical integration procedure is required, then the interval $t' \equiv t_2 - t_1$, which is theoretically arbitrary, should be kept suitably small, such that at each time step the old t_2 becomes the new t_1 as the integration procedure is conducted for a series of time intervals. Thus, HLVA may be used to integrate the system model directly to obtain the dynamic response over the interval $t' \equiv t_2 - t_1$ when the system's initial conditions are specified at t_1; again, all such calculations are conducted *without* ever deriving the equations of motion.

Hamilton's principle cannot be used to integrate directly the formulation of the system model, as described above for HLVA. The primary use of Hamilton's principle is in deriving the *equations of motion* for continuous-parameter or lumped-parameter models of systems. For continuous models, this is accomplished by applying Eq. (E-20) directly. For lumped-parameter models, this is accomplished either by applying Eq. (E-20) directly or by applying the generally more convenient Lagrange's equations given as Eqs. (E-30). In these applications of HP, the times t_1 and t_2 are arbitrary except that the period of interest in the analysis is contained *within* the arbitrary interval, t_1 to t_2. Thus, no physical interpretation can or need be associated with t_1 and t_2, as they do not appear any place as variables or parameters in the equations of motion.

In the matter of equation formulation, HLVA may be used also for this purpose. In this regard, note that Eq. (E-28) can be rewritten slightly. Recall that of the two terms T^* and V in the lagrangian, only T^* is a function of the generalized velocities $\dot{\xi}_j$. So, without loss of information, Eq. (E-28) may be rewritten as

$$\sum_{j=1}^{n} \frac{\partial T^*}{\partial \dot{\xi}_j} \delta\xi_j \bigg|_{t_1}^{t_2} = 0 \qquad (E\text{-}31)$$

But this term, which appears in the use of HP to derive Lagrange's equations and whose vanishing is a consequence of requiring that $\delta\xi_j = 0$ at t_1 and t_2, is precisely the integrated term in Eq. (E-19) of HLVA, but with the opposite sign. In other words, the requirement that $\delta\xi_j = 0$ at t_1 and t_2 in the statement of HP is simply an artifact that allows the dropping of the integrated term in Eq. (E-19) of HLVA to obtain Eq. (E-20) of HP. And, although it may be convenient to drop this term in HP, it is unnecessary to do so because it would have been canceled anyhow by the integrated term in Eq. (E-19)! Therefore, if the goal is to integrate the variational indicator directly to compute the system response, HLVA and HP are very different; and this difference is expressed in the differences of the meanings of t_1 and t_2 in each case. Whereas, if the goal is simply to derive the equations of motion for a dynamic system, HLVA and HP are equivalent.

Both HLVA and HP can be pursued in the context of electrical and electromechanical systems. Doing this is straightforward and directly analogous to the above presentation for mechanical systems. This will be accomplished in Sections E-2 and E-3 below, where we shall illustrate the connections between these variational principles and the corresponding more familiar momentum principle of Newton for mechanical systems (again) and the rules of Kirchhoff for electrical networks.

Finally, we cannot overemphasize the fact that the above mathematical manipulations do *not* represent derivations of HLVA or HP. We adopt the philosophical view that HLVA and HP *cannot* be derived—they are fundamental principles about the intrinsic behavior of the physical universe. The above mathematical manipulations simply represent a demonstration of the fact that for mechanical systems, HLVA and HP are strongly related to Newton's momentum principle (or second law). To suggest (as is often done) that HP is equivalent to Newton's second law is wrong. For example, Eqs. (E-19) and (E-20) apply equally well to the dynamics of extended (rigid or flexible) bodies that may be subjected to torques (or moments) as well as forces. Newton's second law is incomplete in this regard. Furthermore, the application of HLVA and HP to electrical and electromechanical systems is obviously out of the realm of newtonian mechanics. The fact that we can take either a train or airplane from Boston to New York may lead us to conclude that they are equivalent means of transportation, a conclusion that we should be wise to reject if Boston to London were our goal.

E-2   HAMILTON'S PRINCIPLE FOR ELECTROMECHANICAL SYSTEMS USING A DISPLACEMENT–CHARGE FORMULATION

In formulating the equations of motion for an electromechanical system, both the admissibility and dynamic requirements for the mechanical portion of the system as well as the admissibility and dynamic requirements for the electrical portion of the system must be satisfied. Further, the constitutive relations for each ideal transducer in the system will produce coupling between one or more of the mechanical variables and one or more of the electrical variables. In ideal conservative transducers, this coupling will be expressed in either an energy or coenergy function that will be defined as part of the lagrangian.

When displacement and charge variables are used, the mechanical admissibility requirements are represented by the geometric requirements including the relations between the displacements and velocities, and the electrical admissibility requirements are represented by Kirchhoff's current rule (KCR) including the relations between the charges and currents.

The dynamic requirements embody Newton's momentum principle as represented by the requirements between the forces and the time rates of change of the momenta, and Kirchhoff's voltage rule (KVR), including the relations between the flux linkages and voltages.

In analogy with Eq. (E-4), we may write

$$(\text{V.I.})_0 = \sum_{h=1}^{N} \left(\boldsymbol{f}_h - \frac{d\boldsymbol{p}_h}{dt} \right) \cdot \delta \boldsymbol{R}_h + \sum_{i=1}^{M} \left(e_i - \frac{d\lambda_i}{dt} \right) \delta q_i \qquad (\text{E-32})$$

where N is the number of particles and M is the number of circuit elements. Also, e_i, λ_i and q_i are the voltage, flux linkage, and charge associated with the ith circuit element.

The first summation on the right-hand side of Eq. (E-32) represents contributions due to the "mechanical side" of the electromechanical system. Recalling Eqs. (E-5) and (E-10) for this first summation leads directly to

$$\sum_{h=1}^{N} \left(\boldsymbol{f}_h - \frac{d\boldsymbol{p}_h}{dt} \right) \cdot \delta \boldsymbol{R}_h = \delta T^* - \delta V + \sum_{j=1}^{n_m} \Xi_j \, \delta \xi_j - \sum_{h=1}^{N} \frac{d}{dt} (\boldsymbol{p}_h \cdot \delta \boldsymbol{R}_h) \qquad (\text{E-33})$$

where n_m is the number of generalized geometric coordinates in a complete and independent set for the *mechanical* portion of the system.

The second summation on the right-hand side of Eq. (E-32) represents contributions due to the "electrical side" of the electromechanical system. Considering the first term within the second summation leads to

$$\sum_{i=1}^{M} e_i \, \delta q_i = -\delta W_e + \sum_{k=1}^{n_e} \mathcal{E}_k \, \delta q_k \qquad (\text{E-34})$$

where we have separated contributions due to conservative electrical elements, given by $-\delta W_e$, and contributions due to nonconservative electrical elements, given by $\sum_{k=1}^{n_e} \mathcal{E}_k \, \delta q_k$. W_e is the electrical energy function for the system, $\mathcal{E}_k$ represents the generalized voltages corresponding to the nonconservative electrical elements in the system, and n_e is the number of generalized charge coordinates in a complete and independent set for the *electrical* portion of the system. The negative sign on the electrical energy function in Eq. (E-34) is due to the fact that when the voltage across the ith conservative circuit element does work on the circuit, the electrical energy in that element decreases. In calculating the generalized voltages $\mathcal{E}_k$, an abbreviated form of Eq. (E-34) in which the $-\delta W_e$ term is omitted is often used. In this use of Eq. (E-34), the summation on the left-hand side of the equation is effectively a summation over all the nonconservative elements in the network.

Considering the second term within the second summation on the right-hand side of Eq. (E-32) and recalling the product differentiation rule give

$$\sum_{i=1}^{M} \frac{d}{dt} (\lambda_i \, \delta q_i) = \sum_{i=1}^{M} \frac{d\lambda_i}{dt} \delta q_i + \sum_{i=1}^{M} \lambda_i \frac{d(\delta q_i)}{dt} \qquad (\text{E-35})$$

Rearranging terms in Eq. (E-35) gives

$$-\sum_{i=1}^{M} \frac{d\lambda_i}{dt} \delta q_i = -\sum_{i=1}^{M} \frac{d}{dt}(\lambda_i \, \delta q_i) + \sum_{i=1}^{M} \lambda_i \frac{d(\delta q_i)}{dt} \tag{E-36}$$

Further, the second term on the right-hand side of Eq. (E-36) can be expressed as

$$\sum_{i=1}^{M} \lambda_i \frac{d(\delta q_i)}{dt} = \sum_{i=1}^{M} \lambda_i \, \delta\left(\frac{dq_i}{dt}\right) = \sum_{i=1}^{M} \lambda_i \, \delta i_i = \delta W_m^* \tag{E-37}$$

where W_m^* is the magnetic coenergy function for the system. Substitution of Eq. (E-37) into Eq. (E-36) gives

$$-\sum_{i=1}^{M} \frac{d\lambda_i}{dt} \delta q_i = -\sum_{i=1}^{M} \frac{d}{dt}(\lambda_i \, \delta q_i) + \delta W_m^* \tag{E-38}$$

where the left-hand side of Eq. (E-38) is observed to be the same as the second term within the second summation on the right-hand side of Eq. (E-32). So, substitution of Eqs. (E-33), (E-34), and (E-38) into Eq. (E-32) gives

$$(V.I.)_0 = \delta T^* - \delta V + \delta W_m^* - \delta W_e + \sum_{j=1}^{n_m} \Xi_j \, \delta\xi_j + \sum_{k=1}^{n_e} \mathcal{E}_k \, \delta q_k$$
$$- \sum_{h=1}^{N} \frac{d}{dt}(\boldsymbol{p}_h \cdot \delta \boldsymbol{R}_h) - \sum_{i=1}^{M} \frac{d}{dt}(\lambda_i \, \delta q_i) \tag{E-39}$$

From Eqs. (E-34) and (E-37) it should be clear and interesting how the charge variable formulation leads naturally to W_e and W_m^*, as opposed to the alternative functions W_e^* and W_m. Such an energy-coenergy distinction is very important if the constitutive relations for any of the elements are nonlinear.

In order to obtain Hamilton's principle for electromechanical systems using a displacement–charge formulation, Eq. (E-39) is integrated with respect to time over an interval from $t = t_1$ to $t = t_2$, where t_1 is less than t_2. The resulting variational indicator is

$$(V.I.)' = \int_{t_1}^{t_2} \left[\delta(T^* - V + W_m^* - W_e) + \sum_{j=1}^{n_m} \Xi_j \, \delta\xi_j + \sum_{k=1}^{n_e} \mathcal{E}_k \, \delta q_k \right] dt$$
$$- \sum_{h=1}^{N} \boldsymbol{p}_h \cdot \delta \boldsymbol{R}_h \Big|_{t_1}^{t_2} - \sum_{i=1}^{M} \lambda_i \, \delta q_i \Big|_{t_1}^{t_2} \tag{E-40}$$

which may be interpreted as a form of *Hamilton's law of varying action* for electromechanical systems using a displacement–charge formulation. By adopting the widely accepted convention that all admissible variations vanish at $t = t_1$ and $t = t_2$, the last two terms on the right-hand side of Eq. (E-40) vanish, resulting in

$$V.I. = \int_{t_1}^{t_2} \left[\delta(T^* - V + W_m^* - W_e) + \sum_{j=1}^{n_m} \Xi_j \, \delta\xi_j + \sum_{k=1}^{n_e} \mathcal{E}_k \, \delta q_k \right] dt \tag{E-41}$$

Equation (E-41) is Hamilton's principle for electromechanical systems using a displacement–charge formulation, which is expressed as follows:

An admissible motion of the system between specified configurations at t_1 and t_2 is a natural motion if, and only if, the variational indicator V.I. vanishes for arbitrary admissible variations.

Clearly, by dropping all the mechanical terms in Eq. (E-41), Hamilton's principle for electrical networks using a charge formulation becomes

$$\text{V.I.} = \int_{t_1}^{t_2} \left[\delta(W_m^* - W_e) + \sum_{k=1}^{n_e} \mathcal{E}_k\, \delta q_k \right] dt \qquad \text{(E-42)}$$

Furthermore, Lagrange's equations for electromechanical systems using a charge formulation may be obtained by operating on Eq. (E-41) using the same procedures of Subsection E-1.3. And Lagrange's equations for electrical networks using a displacement–charge formulation may be obtained identically from Eq. (E-42).

E-3   HAMILTON'S PRINCIPLE FOR ELECTROMECHANICAL SYSTEMS USING A DISPLACEMENT–FLUX LINKAGE FORMULATION

In formulating the equations of motion for an electromechanical system, both the admissibility and dynamic requirements for the mechanical portion of the system as well as the admissibility and dynamic requirements for the electrical portion of the system must be satisfied. Further, the constitutive relations for each ideal transducer in the system will produce coupling between one or more of the mechanical variables and one or more of the electrical variables. In ideal conservative transducers, this coupling will be expressed in either an energy or coenergy function that will be defined as part of the lagrangian.

When displacement and flux linkage variables are used, the mechanical admissibility requirements are represented by the geometric requirements including the relations between the displacements and velocities, and the electrical admissibility requirements are represented by Kirchhoff's voltage rule (KVR) including the relations between the flux linkages and voltages.

The dynamic requirements embody Newton's momentum principle as represented by the requirements between the forces and the time rates of change of the momenta, and Kirchhoff's current rule (KCR), including the relations between the charges and currents.

In analogy with Eqs. (E-4) and (E-32), we may write

$$(\text{V.I.})_0 = \sum_{h=1}^{N} \left(\boldsymbol{f}_h - \frac{d\boldsymbol{p}_h}{dt} \right) \cdot \delta\boldsymbol{R}_h + \sum_{i=1}^{M'} \left(i_i - \frac{dq_i}{dt} \right) \delta\lambda_i \qquad \text{(E-43)}$$

where N is the number of particles and M' is the number of circuit elements. Also, i_i, q_i and λ_i are the current, charge, and flux linkage associated with the ith circuit element.

The first summation on the right-hand side of Eq. (E-43) represents contributions due to the "mechanical side" of the electromechanical system. Recalling Eqs. (E-5) and (E-10) for this first summation leads directly to

$$\sum_{h=1}^{N} \left(\boldsymbol{f}_h - \frac{d\boldsymbol{p}_h}{dt} \right) \cdot \delta\boldsymbol{R}_h = \delta T^* - \delta V + \sum_{j=1}^{n_m} \Xi_j\, \delta\xi_j - \sum_{h=1}^{N} \frac{d}{dt}(\boldsymbol{p}_h \cdot \delta\boldsymbol{R}_h) \qquad \text{(E-44)}$$

where n_m is the number of generalized geometric coordinates in a complete and independent set for the *mechanical* portion of the system.

The second summation on the right-hand side of Eq. (E-43) represents contributions due to the "electrical side" of the electromechanical system. Considering the first term within the second summation leads to

$$\sum_{i=1}^{M'} i_i \, \delta\lambda_i = -\delta W_m + \sum_{k=1}^{n'_e} \mathcal{I}_k \, \delta\lambda_k \tag{E-45}$$

where we have separated contributions due to conservative electrical elements, given by $-\delta W_m$, and contributions due to nonconservative electrical elements, given by $\sum_{k=1}^{n'_e} \mathcal{I}_k \delta\lambda_k$. W_m is the magnetic energy function for the system, $\mathcal{I}_k$ represents the generalized currents corresponding to the nonconservative electrical elements in the system, and n'_e is the number of generalized flux linkage coordinates in a complete and independent set for the *electrical* portion of the system. The negative sign on the magnetic energy function in Eq. (E-45) is due to the fact that when the current in the ith conservative circuit element does work on the circuit, the magnetic energy in that element decreases. In calculating the generalized current $\mathcal{I}_k$, an abbreviated form of Eq. (E-45) in which the $-\delta W_m$ term is omitted is often used. In this use of Eq. (E-45), the summation on the left-hand side of the equation is effectively a summation over all the nonconservative elements in the network.

Considering the second term within the second summation on the right-hand side of Eq. (E-43) and recalling the product differentiation rule give

$$\sum_{i=1}^{M'} \frac{d}{dt}(q_i \, \delta\lambda_i) = \sum_{i=1}^{M'} \frac{dq_i}{dt} \, \delta\lambda_i + \sum_{i=1}^{M'} q_i \frac{d(\delta\lambda_i)}{dt} \tag{E-46}$$

Rearranging terms in Eq. (E-46) gives

$$-\sum_{i=1}^{M'} \frac{dq_i}{dt} \, \delta\lambda_i = -\sum_{i=1}^{M'} \frac{d}{dt}(q_i \, \delta\lambda_i) + \sum_{i=1}^{M'} q_i \frac{d(\delta\lambda_i)}{dt} \tag{E-47}$$

Further, the second term on the right-hand side of Eq. (E-47) can be expressed as

$$\sum_{i=1}^{M'} q_i \frac{d(\delta\lambda_i)}{dt} = \sum_{i=1}^{M'} q_i \, \delta\left(\frac{d\lambda_i}{dt}\right) = \sum_{i=1}^{M'} q_i \, \delta e_i = \delta W_e^* \tag{E-48}$$

where W_e^* is the electrical coenergy function for the system. Substitution of Eq. (E-48) into Eq. (E-47) gives

$$-\sum_{i=1}^{M'} \frac{dq_i}{dt} \, \delta\lambda_i = -\sum_{i=1}^{M'} \frac{d}{dt}(q_i \, \delta\lambda_i) + \delta W_e^* \tag{E-49}$$

where the left-hand side of Eq. (E-49) is observed to be the same as the second term within the second summation on the right-hand side of Eq. (E-43). So, substitution of Eqs. (E-44), (E-45) and (E-49) into Eq. (E-43) gives

$$(\text{V.I.})_0 = \delta T^* - \delta V + \delta W_e^* - \delta W_m + \sum_{j=1}^{n_m} \Xi_j \, \delta\xi_j + \sum_{k=1}^{n'_e} \mathcal{I}_k \, \delta\lambda_k$$

$$- \sum_{h=1}^{N} \frac{d}{dt}(\boldsymbol{p}_h \cdot \delta\boldsymbol{R}_h) - \sum_{i=1}^{M'} \frac{d}{dt}(q_i \, \delta\lambda_i) \tag{E-50}$$

From Eqs. (E-45) and (E-48) it should be clear and interesting how the flux linkage variable formulation leads naturally to W_m and W_e^*, as opposed to the alternative functions W_m^* and

W_e. Such an energy-coenergy distinction is very important if the constitutive relations for any of the elements are nonlinear.

In order to obtain Hamilton's principle for electromechanical systems using a displacement–flux linkage formulation, Eq. (E-50) is integrated with respect to time over an interval from $t = t_1$ to $t = t_2$, where t_1 is less than t_2. The resulting variational indicator is

$$(\text{V.I.})' = \int_{t_1}^{t_2} \left[\delta(T^* - V + W_e^* - W_m) + \sum_{j=1}^{n_m} \Xi_j \, \delta\xi_j + \sum_{k=1}^{n_e'} \mathcal{I}_k \, \delta\lambda_k \right] dt$$

$$- \sum_{h=1}^{N} \boldsymbol{p}_h \cdot \delta\boldsymbol{R}_h \Big|_{t_1}^{t_2} - \sum_{i=1}^{M'} q_i \, \delta\lambda_i \Big|_{t_1}^{t_2} \tag{E-51}$$

which may be interpreted as a form of *Hamilton's law of varying action* for electromechanical systems using a displacement–flux linkage formulation. By adopting the widely accepted convention that all admissible variations vanish at $t = t_1$ and $t = t_2$, the last two terms on the right-hand side of Eq. (E-51) vanish, resulting in

$$\text{V.I.} = \int_{t_1}^{t_2} \left[\delta(T^* - V + W_e^* - W_m) + \sum_{j=1}^{n_m} \Xi_j \, \delta\xi_j + \sum_{k=1}^{n_e'} \mathcal{I}_k \, \delta\lambda_k \right] dt \tag{E-52}$$

Equation (E-52) is Hamilton's principle for electromechanical systems using a displacement–flux linkage formulation, which is expressed as follows:

An admissible motion of the system between specified configurations at t_1 and t_2 is a natural motion if, and only if, the variational indicator V.I. vanishes for arbitrary admissible variations.

Clearly, by dropping all the mechanical terms in Eq. (E-52), Hamilton's principle for electrical networks using a flux linkage formulation becomes

$$\text{V.I.} = \int_{t_1}^{t_2} \left[\delta(W_e^* - W_m) + \sum_{k=1}^{n_e'} \mathcal{I}_k \, \delta\lambda_k \right] dt \tag{E-53}$$

Furthermore, Lagrange's equations for electromechanical systems using a displacement–flux linkage formulation may be obtained by operating on Eq. (E-52) using the same procedures of Subsection E-1.3. And Lagrange's equations for electrical networks using a flux linkage formulation may be obtained identically from Eq. (E-53).

E-4   WORK–ENERGY RELATION DERIVED FROM LAGRANGE'S EQUATIONS

It can be shown that when a system is free of time-varying constraints, the work done on the system by all the nonconservative forces is equal to the increase in the total energy of the system. For exclusively mechanical systems, the work–energy relation can be derived most easily via the direct approach (that is, via momentum principles).

In this section, we shall derive the *work–energy relation* from Lagrange's equations. One significant advantage of this approach over the direct approach follows from the universality of Lagrange's equations; that is, Lagrange's equations for mechanical, electrical and electromechanical systems are all of the same form, but with different variables. Therefore, the derivation of the work–energy relation directly from Lagrange's equations is valid for all these systems and thus it is clear that the work–energy relation is equally applicable to mechanical, electrical, *and* electromechanical systems. In this section, we shall derive the work–energy relation from Lagrange's equations for mechanical systems; however, as just indicated, we could equally well select as our initiation point Lagrange's equations for electrical or electromechanical systems.

Lagrange's equations for mechanical systems are given by Eqs. (E-30). Multiplying both sides of Eqs. (E-30) by $\dot{\xi}_j$ gives

$$\dot{\xi}_j \frac{d}{dt}\left(\frac{\partial \mathcal{L}}{\partial \dot{\xi}_j}\right) - \dot{\xi}_j \frac{\partial \mathcal{L}}{\partial \xi_j} = \Xi_j \dot{\xi}_j \qquad j = 1, 2, \ldots, n \tag{E-54}$$

Using the chain rule for product differentiation, $\dfrac{d(AB)}{dt} = A\left(\dfrac{dB}{dt}\right) + B\left(\dfrac{dA}{dt}\right)$ for arbitrary functions $A(t)$ and $B(t)$, the first term on the left-hand side of Eqs. (E-54) can be rewritten, yielding

$$\frac{d}{dt}\left(\dot{\xi}_j \frac{\partial \mathcal{L}}{\partial \dot{\xi}_j}\right) - \ddot{\xi}_j \frac{\partial \mathcal{L}}{\partial \dot{\xi}_j} - \dot{\xi}_j \frac{\partial \mathcal{L}}{\partial \xi_j} = \Xi_j \dot{\xi}_j \qquad j = 1, 2, \ldots, n \tag{E-55}$$

Summing Eqs. (E-55) over $j = 1, 2, \ldots, n$ gives the single equation

$$\frac{d}{dt}\left(\sum_{j=1}^{n} \dot{\xi}_j \frac{\partial \mathcal{L}}{\partial \dot{\xi}_j}\right) - \sum_{j=1}^{n}\left(\ddot{\xi}_j \frac{\partial \mathcal{L}}{\partial \dot{\xi}_j} + \dot{\xi}_j \frac{\partial \mathcal{L}}{\partial \xi_j}\right) = \sum_{j=1}^{n} \Xi_j \dot{\xi}_j \tag{E-56}$$

Recall from the discussion immediately following Eq. (E-21) that the lagrangian for mechanical systems subject to *no* time-varying constraints is a function of only the generalized coordinates and the generalized velocities; that is, in such cases the lagrangian can be defined as

$$\mathcal{L}(\xi_j, \dot{\xi}_j) = T^*(\xi_j, \dot{\xi}_j) - V(\xi_j) \tag{E-57}$$

Differentiating Eq. (E-57) with respect to time gives

$$\frac{d\mathcal{L}}{dt} = \sum_{j=1}^{n}\left(\dot{\xi}_j \frac{\partial \mathcal{L}}{\partial \xi_j} + \ddot{\xi}_j \frac{\partial \mathcal{L}}{\partial \dot{\xi}_j}\right) \tag{E-58}$$

which is identical to the second summation term on the left-hand side of Eq. (E-56). Therefore, Eq. (E-56) can be rewritten as

$$\frac{d}{dt}\left(\sum_{j=1}^{n} \dot{\xi}_j \frac{\partial \mathcal{L}}{\partial \dot{\xi}_j} - \mathcal{L}\right) = \sum_{j=1}^{n} \Xi_j \dot{\xi}_j \tag{E-59}$$

Note from Eq. (E-57) that

$$\frac{\partial \mathcal{L}}{\partial \dot{\xi}_j} = \frac{\partial T^*}{\partial \dot{\xi}_j} \tag{E-60}$$

because the potential energy function V is not a function of the generalized velocities. Substitution of Eq. (E-60) into Eq. (E-59) gives

$$\frac{d}{dt}\left(\sum_{j=1}^{n} \dot{\xi}_j \frac{\partial T^*}{\partial \dot{\xi}_j} - \mathcal{L}\right) = \sum_{j=1}^{n} \Xi_j \dot{\xi}_j \tag{E-61}$$

Note from Eq. (E-18) that equating the two expressions in the two pairs of square brackets there gives

$$\frac{\partial T^*}{\partial \dot{\xi}_j} = \sum_{i=1}^{n} \frac{\partial T^*}{\partial \dot{R}_i} \cdot \frac{\partial \dot{R}_i}{\partial \dot{\xi}_j} \tag{E-62}$$

Using Eqs. (E-14) and (E-16) to rewrite the two vector terms on the right-hand side of Eq. (E-62) yields

$$\frac{\partial T^*}{\partial \dot{\xi}_j} = \sum_{i=1}^{N} \boldsymbol{p}_i \cdot \frac{\partial \boldsymbol{R}_i}{\partial \xi_j} \tag{E-63}$$

Using Eq. (E-63), the summation term on the left-hand side of Eq. (E-61) can be rewritten as

$$\sum_{j=1}^{n} \dot{\xi}_j \frac{\partial T^*}{\partial \dot{\xi}_j} = \sum_{j=1}^{n} \sum_{i=1}^{N} \boldsymbol{p}_i \cdot \frac{\partial \boldsymbol{R}_i}{\partial \xi_j} \dot{\xi}_j$$

$$= \sum_{i=1}^{N} \boldsymbol{p}_i \cdot \left(\sum_{j=1}^{n} \frac{\partial \boldsymbol{R}_i}{\xi_j} \dot{\xi}_j \right) \tag{E-64}$$

where the order of the summations over i and j has been reversed. From Eq. (E-1) it can be shown that for a system having no time-varying constraints, the velocity vector for the ith particle can be written as

$$\boldsymbol{v}_i = \frac{d\boldsymbol{R}_i}{dt} = \sum_{j=1}^{n} \frac{\partial \boldsymbol{R}_i}{\partial \xi_j} \frac{d\xi_j}{dt} = \sum_{j=1}^{n} \frac{\partial \boldsymbol{R}_i}{\partial \xi_j} \dot{\xi}_j \tag{E-65}$$

Substituting Eq. (E-65) into Eq. (E-64) gives

$$\sum_{j=1}^{n} \dot{\xi}_j \frac{\partial T^*}{\partial \dot{\xi}_j} = \sum_{i=1}^{N} \boldsymbol{p}_i \cdot \boldsymbol{v}_i = T + T^* \tag{E-66}$$

where the relation between the kinetic energy and coenergy functions, $T = \sum_{i=1}^{N} (\boldsymbol{p}_i \cdot \boldsymbol{v}_i) - T^*$, has been used.

Now, substitution of Eq. (E-66) into Eq. (E-61) gives

$$\frac{d}{dt}(T + T^* - \mathcal{L}) = \sum_{j=1}^{n} \Xi_j \dot{\xi}_j \tag{E-67}$$

And, substitution of of Eq. (E-57) into Eq. (E-67) yields

$$\frac{d}{dt}(T + V) = \sum_{j=1}^{n} \Xi_j \dot{\xi}_j \tag{E-68}$$

The right-hand side of Eq. (E-68) can be rewritten in terms of the physical nonconservative forces $\boldsymbol{f}_i^{nc}$ and the physical velocities $\boldsymbol{v}_i$ associated with the ith particle. Recall that the generalized forces Ξ_j are frequently formulated from a nonconservative work increment expression; namely,

$$\delta \mathcal{W}^{nc} = \sum_{i=1}^{N} \boldsymbol{f}_i^{nc} \cdot \delta \boldsymbol{R}_i = \sum_{j=1}^{n} \Xi_j \, \delta \xi_j \tag{E-69}$$

Using Eq. (E-15) to rewrite $\delta \boldsymbol{R}_i$ in Eq. (E-69) yields

$$\delta \mathcal{W}^{nc} = \sum_{i=1}^{N} \boldsymbol{f}_i^{nc} \cdot \sum_{j=1}^{n} \frac{\partial \boldsymbol{R}_i}{\partial \xi_j} \delta \xi_j = \sum_{j=1}^{n} \left(\sum_{i=1}^{N} \boldsymbol{f}_i^{nc} \cdot \frac{\partial \boldsymbol{R}_i}{\partial \xi_j} \right) \delta \xi_j \tag{E-70}$$

$$= \sum_{j=1}^{n} \Xi_j \, \delta \xi_j \tag{E-71}$$

where in Eq. (E-70) the order of the i-summation and j-summation has been reversed. Comparing Eqs. (E-70) and (E-71) yields

$$\Xi_j = \sum_{i=1}^{N} \boldsymbol{f}_i^{nc} \cdot \frac{\partial \boldsymbol{R}_i}{\partial \xi_j} \tag{E-72}$$

Substituting Eq. (E-72) into the right-hand side of Eq. (E-68) gives

$$\sum_{j=1}^{n} \Xi_j \dot{\xi}_j = \sum_{j=1}^{n} \sum_{i=1}^{N} \boldsymbol{f}_i^{nc} \cdot \frac{\partial \boldsymbol{R}_i}{\partial \xi_j} \dot{\xi}_j = \sum_{i=1}^{N} \boldsymbol{f}_i^{nc} \cdot \sum_{j=1}^{n} \frac{\partial \boldsymbol{R}_i}{\partial \xi_j} \dot{\xi}_j = \sum_{i=1}^{N} \boldsymbol{f}_i^{nc} \cdot \boldsymbol{v}_i \tag{E-73}$$

where again the order of the i-summation and j-summation has been reversed, and Eq. (E-65) has been used. Therefore, substituting Eq. (E-73) into the right-hand side of Eq. (E-68) gives

$$\frac{d}{dt}(T + V) = \sum_{i=1}^{N} \boldsymbol{f}_i^{nc} \cdot \boldsymbol{v}_i \tag{E-74}$$

The sum of the system's kinetic energy T and the system's potential energy V is often called the *total energy* of the system. That is, the total energy of the mechanical system, to which we have restricted this derivation, is defined by

$$\hat{E}(\xi_j(t), \dot{\xi}_j(t)) \equiv T(\xi_j, \dot{\xi}_j) + V(\xi_j) = E(t) \tag{E-75}$$

where $\hat{E}$ is a function of the generalized coordinates and the generalized velocities, which themselves are functions of time. Hence, the total energy is a function of time, which we denote as $E(t)$ in Eq. (E-75). Thus, using this function $E(t)$, Eq. (E-74) can be rewritten as

$$\frac{dE}{dt} = \sum_{i=1}^{N} \boldsymbol{f}_i^{nc} \cdot \boldsymbol{v}_i \tag{E-76}$$

Note that the right-hand side of Eq. (E-76) is the time rate of the energy input (or it is simply the power input) into the system by all nonconservative forces. For example, if an applied force does work on the system, the associated power input is positive whereas the power input by dashpot forces is always negative. Therefore, Eq. (E-76) states that the time rate of change of the system's total energy is equal to the power input to the system by all nonconservative forces. Denoting the nonconservative power input as $\dfrac{d\mathcal{W}^{nc}}{dt}$, Eq. (E-76) can be rewritten as

$$\frac{dE}{dt} = \frac{d\mathcal{W}^{nc}}{dt} \tag{E-77}$$

where $\mathcal{W}^{nc}$ is the work done by all nonconservative forces and is defined as

$$\mathcal{W}^{nc} = \sum_{i=1}^{N} \int_{\boldsymbol{R}_{i0}}^{\boldsymbol{R}_i} \boldsymbol{f}_i^{nc} \cdot d\boldsymbol{R}_i = \sum_{i=1}^{N} \int_{0}^{t} \boldsymbol{f}_i^{nc} \cdot \boldsymbol{v}_i \, dt \tag{E-78}$$

where $\boldsymbol{R}_{i0}$ is the initial position vector of the ith particle and where the definition $\boldsymbol{v}_i \equiv \dfrac{d\boldsymbol{R}_i}{dt}$ has been used.

Eq. (E-77) can be integrated to give

$$E(t) = \mathcal{W}^{\mathrm{nc}}(t) + E_0 \tag{E-79}$$

where E_0 is the constant of integration, which can be evaluated by the initial conditions. Thus, Eq. (E-79) states that the total energy of the system at time t is equal to the initial total energy plus the nonconservative work done on the system up to the time t.

LAGRANGE'S FORM OF D'ALEMBERT'S PRINCIPLE

In this appendix, we examine a frequently discussed principle of dynamics: *Lagrange's form of d'Alembert's principle.* It is an important principle both historically and conceptually to the entire discipline of variational dynamics; there is extremely little disagreement on this point. There are professional fissures regarding this principle, however, and they arise concerning the issue of whether it is a valuable *intermediate* result that leads to more efficient formulations for the derivation of equations of motion or a valuable *terminal* result that itself is a useful analytical device for the derivation of equations of motion. We reside distinctly among those who believe the former.

F-1   FUNDAMENTAL CONCEPTS AND DERIVATIONS

Consider a holonomic mechanical system of N particles m_i $(i = 1, 2, \ldots, N)$ as indicated in Figure F-1. Let the system be described by a complete and independent set of n generalized coordinates $\xi_1, \ldots, \xi_n$ and let $\delta\xi_1, \ldots, \delta\xi_n$ represent their associated admissible variations. Note that the admissible variations $\delta\xi_1, \ldots, \delta\xi_n$ are independent since the system is holonomic. As always, the generalized coordinates may be inertial or noninertial.

If the position vector $\boldsymbol{R}_i$ of the particle m_i is expressed as $\boldsymbol{R}_i = \boldsymbol{R}_i(\xi_1, \ldots, \xi_n, t)$, then the inertial velocity $\boldsymbol{v}_i$ of this particle is

$$\boldsymbol{v}_i = \frac{d\boldsymbol{R}_i}{dt} = \sum_{j=1}^{n} \frac{\partial \boldsymbol{R}_i}{\partial \xi_j} \frac{d\xi_j}{dt} + \frac{\partial \boldsymbol{R}_i}{\partial t} \tag{F-1}$$

Equation (F-1) may be rewritten in a slightly more conventional form as

$$\boldsymbol{v}_i = \dot{\boldsymbol{R}}_i = \sum_{j=1}^{n} \frac{\partial \boldsymbol{R}_i}{\partial \xi_j} \dot{\xi}_j + \frac{\partial \boldsymbol{R}_i}{\partial t} \tag{F-2}$$

where the $\dot{\xi}_j$ are typically called the generalized velocities. Differentiating Eq. (F-2) with respect to $\dot{\xi}_j$ gives

$$\frac{\partial \dot{\boldsymbol{R}}_i}{\partial \dot{\xi}_j} = \frac{\partial \boldsymbol{R}_i}{\partial \xi_j} \tag{F-3}$$

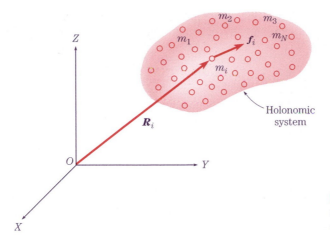

FIGURE F-1 System of N particles subject to holonomic constraints that are not shown.

The variation of the position vector $\boldsymbol{R}_i$ can be expressed in terms of the variations of the generalized coordinates as

$$\delta \boldsymbol{R}_i = \sum_{j=1}^{n} \frac{\partial \boldsymbol{R}_i}{\partial \xi_j} \delta \xi_j \tag{F-4}$$

where Eq. (F-4) is a standard result from the calculus of variations. Recall that, by definition, a term corresponding to the last term in Eq. (F-1) does not appear in Eq. (F-4) because the admissible variations are considered to be the examinations of neighboring geometrically admissible contemporaneous (that is, at the same instant) configurations.

We form the scalar variational indicator

Newton's second law for each particle can be written as

$$\boldsymbol{f}_i - \frac{d\boldsymbol{p}_i}{dt} = 0 \qquad i = 1, 2, \ldots, N \tag{F-5}$$

where $\boldsymbol{f}_i$ is the resultant force acting on the particle m_i and $\boldsymbol{p}_i$ is the linear momentum of that particle. Historically, Eq. (F-5) has been often called *d'Alembert's principle;* but, as written, it is difficult to justify calling Eq. (F-5) anything other than Newton's second law or the linear momentum principle.

We form the scalar variational indicator

$$\sum_{i=1}^{n} \left(\boldsymbol{f}_i - \frac{d\boldsymbol{p}_i}{dt} \right) \cdot \delta \boldsymbol{R}_i = 0 \tag{F-6}$$

which must vanish for arbitrary variations if each equation of Eq. (F-5) is satisfied.

The resultant force $\boldsymbol{f}_i$ may be considered to be expressible as

$$\boldsymbol{f}_i = \boldsymbol{f}_i' + \boldsymbol{f}_i'' \tag{F-7}$$

where $\boldsymbol{f}_i'$ is the total applied or impressed force on m_i and $\boldsymbol{f}_i''$ is the total constraint reaction force on m_i. By definition,

$$\sum_{i=1}^{N} \boldsymbol{f}_i'' \cdot \delta \boldsymbol{R}_i = 0 \tag{F-8}$$

since the constraint forces do no work during an admissible variation of the system. Equation (F-8) is a significant consequence of forming the scalar variational indicator. Therefore,

$$\sum_{i=1}^{N} \left(\boldsymbol{f}_i' - \frac{d\boldsymbol{p}_i}{dt} \right) \cdot \delta\boldsymbol{R}_i = 0 \tag{F-9}$$

Equation (F-9) is called by many names: the *fundamental equation* by Pars [1979, p. 28], *d'Alembert's principle* by Lanczos [1970, p. 90], Goldstein [1950, p. 16], and Greenwood [1988, p. 446], the *generalized principle of d'Alembert* by Meirovitch [1970, p. 65], and *Lagrange's form of d'Alembert's principle* by Rosenberg [1977, p. 126]; although we can find no evidence that d'Alembert ever wrote such an equation. We prefer to call Eq. (F-9) "Lagrange's form of d'Alembert's principle," believing that such a choice most accurately reflects the historical record, although we attach no great significance to this choice.

Equation (F-9) can be expressed in terms of the n generalized coordinates. Substitution of Eq. (F-4) into Eq. (F-9) gives

$$\sum_{j=1}^{n} \left[\sum_{i=1}^{N} \left(\boldsymbol{f}_i' - \frac{d\boldsymbol{p}_i}{dt} \right) \cdot \frac{\partial\boldsymbol{R}_i}{\partial\xi_j} \right] \delta\xi_j = 0 \tag{F-10}$$

Because the system is holonomic, the arbitrary admissible variations $\delta\xi_j$ are independent; thus Eq. (F-10) gives

$$\sum_{i=1}^{N} \left(\boldsymbol{f}_i' - \frac{d\boldsymbol{p}_i}{dt} \right) \cdot \frac{\partial\boldsymbol{R}_i}{\partial\xi_j} = 0 \qquad j = 1, 2, \ldots, n \tag{F-11}$$

or

$$Q_j + Q_j^* = 0 \qquad (j = 1, 2, \ldots, n) \tag{F-12}$$

where

$$Q_j \equiv \sum_{i=1}^{N} \boldsymbol{f}_i' \cdot \frac{\partial\boldsymbol{R}_i}{\partial\xi_j} \tag{F-13}$$

and

$$Q_j^* \equiv \sum_{i=1}^{N} \left(-\frac{d\boldsymbol{p}_i}{dt} \right) \cdot \frac{\partial\boldsymbol{R}_i}{\partial\xi_j} \tag{F-14}$$

Equations (F-11) or (F-12) are called *d'Alembert's principle in terms of generalized coordinates* by Greenwood [1988, p. 447] and they are called *Kane's equations* by Kane and Levinson [1985, p. 159].

The fact that various authors assign different names to several of the terms in Eqs. (F-11) and (F-12) does not change anything, except their names, of course. In particular, the term $\dfrac{\partial\dot{\boldsymbol{R}}_i}{\partial\dot{\xi}_j}$ is given the name *partial velocity* by Kane and Levinson [1985], and appears in lieu of $\dfrac{\partial\boldsymbol{R}}{\partial\xi_j}$ in equations which correspond to Eqs. (F-13) and (F-14) above. Yet, as we see from Eq. (F-3), the equality of these two terms is purely a kinematic identity. So, the use of $\dfrac{\partial\dot{\boldsymbol{R}}_i}{\partial\dot{\xi}_j}$ in place of its undifferentiated equivalent is simply an equivalent substitution. Furthermore, we note that by differentiating Eq. (F-3) once more to obtain the kinematically equivalent term $\dfrac{\partial \ddot{\boldsymbol{R}}_i}{\partial \ddot{\xi}_j}$, Eqs. (F-12) become the so-called Gibbs-Appell equations (due mostly

to Appell) where $\dfrac{\partial \dot{\boldsymbol{R}}_i}{\partial \dot{\xi}_j}$ simply replaces $\dfrac{\partial \boldsymbol{R}_i}{\partial \xi_j}$ in Eqs. (F-13) and (F-14). To illustrate the application of Eqs. (F-12), three examples are given below.

Equations of motion obtained via the method expressed by Eq. (F-11) or the equivalent combination of Eqs. (F-12), (F-13), and (F-14) possess the unattractive feature of requiring an analytic expression for the accelerations. As we understand from our study of kinematics, accelerations are significantly more difficult to derive than velocities. On the other hand, derivations via this technique have the advantage over the strictly newtonian direct approach that workless constraint forces do not have to be considered.

F-2   EXAMPLES

In this section, we present three examples of the derivation of the equations of motion via *Lagrange's form of d'Alembert's principle.* The examples are of increasing difficulty although they are not difficult examples, especially if the derivations were accomplished via Lagrange's equations.

■ **Example F-1:**   Bead on Rigid Rod

The bead of mass m in Figure F-2, which is modeled as a particle, slides along the rigid rod without friction. A known force $\boldsymbol{f}(t)$ acts on the bead at a constant angle θ, with respect to the horizontal. Derive the equation(s) of motion via Lagrange's form of d'Alembert's principle.

❑ **Solution:**   In order to utilize Eq. (F-12), first of all, we need to identify the generalized coordinate(s) of the system, and then derive the expressions for the position, velocity, and linear momentum in terms of the generalized coordinate(s).

1. *Generalized coordinate(s):*   Without an extended discussion, we simply note that there is only one generalized coordinate for the system, which can be expressed as

$$\xi_j : x \tag{a}$$

where, as indicated in Figure F-2, x is the position of the bead relative to the left-hand wall.

It is also noted that $N = 1$ and $n = 1$ for the system, where N and n are the number of particles and the number of generalized coordinates, respectively.

2. *System kinematics in terms of generalized coordinate:*   If we denote the unit vectors along the inertial X and Y axes as $\boldsymbol{I}$ and $\boldsymbol{J}$, respectively, then from Figure F-2, we can find the position vector of the bead as

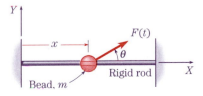

FIGURE F-2   Bead, modelled as particle, on rigid rod, subjected to force at constant angle θ.

$$\boldsymbol{R}_1 = x\boldsymbol{I} \tag{b}$$

and its partial derivative with respect to the generalized coordinate x as

$$\frac{\partial \boldsymbol{R}_1}{\partial \xi_1} = \frac{\partial \boldsymbol{R}_1}{\partial x} = \boldsymbol{I} \tag{c}$$

Then, according to Eq. (F-2), the velocity of the bead is

$$\boldsymbol{v}_1 = \frac{\partial \boldsymbol{R}_1}{\partial x}\dot{x} = \dot{x}\boldsymbol{I} \tag{d}$$

and the linear momentum of the bead, by definition, is

$$\boldsymbol{p}_1 = m\boldsymbol{v}_1 = m\dot{x}\boldsymbol{I} \tag{e}$$

3. *Applied forces:* There is only one applied or impressed force acting on the particle, expressed as

$$\boldsymbol{f}_1' = f(t)\cos\theta\boldsymbol{I} + f(t)\sin\theta\boldsymbol{J} \tag{f}$$

4. *Equation of motion:* By substituting Eqs. (c), (e), and (f) into Eqs. (F-13) and (F-14), the equation of motion can be derived as follows:

$$
\begin{aligned}
Q_1 &= \sum_{i=1}^{1}\boldsymbol{f}_i' \cdot \frac{\partial \boldsymbol{R}_i}{\partial \xi_1} = \boldsymbol{f}_1' \cdot \frac{\partial \boldsymbol{R}_1}{\partial x} \\
&= [f(t)\cos\theta\boldsymbol{I} + f(t)\sin\theta\boldsymbol{J}] \cdot \boldsymbol{I} \\
&= f(t)\cos\theta
\end{aligned}
\tag{g}
$$

$$
\begin{aligned}
Q_1^* &= \sum_{i=1}^{1}\left(-\frac{d\boldsymbol{p}_i}{dt}\right) \cdot \frac{\partial \boldsymbol{R}_i}{\partial \xi_1} = -\frac{d\boldsymbol{p}_1}{dt} \cdot \frac{\partial \boldsymbol{R}_1}{\partial x} \\
&= -\frac{d}{dt}(m\dot{x})\boldsymbol{I} \cdot \boldsymbol{I} \\
&= -m\ddot{x}
\end{aligned}
\tag{h}
$$

Thus, the equation of motion for the bead, according to Eq. (F-12), is

$$Q_1 + Q_1^* = 0 \tag{i}$$

or

$$f(t)\cos\theta - m\ddot{x} = 0 \tag{j}$$

■ **Example F-2:** Simple Plane Pendulum

Derive the equation(s) of motion via Lagrange's form of d'Alembert's principle for the simple plane pendulum in Figure F-3.

❏ **Solution:**

1. *Generalized coordinate(s):* Without an extended discussion, we simply note that the system has only one generalized coordinate, which can be expressed formally as

$$\xi_j : \theta \tag{a}$$

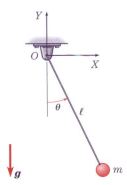

FIGURE F-3 Simple plane pendulum.

where θ is the counterclockwise angular displacement of the pendulum from its downward-hanging equilibrium position.

It is also noted that $N = 1$ and $n = 1$ for this system, where N and n are the number of particles and the number of generalized coordinates, respectively.

2. *System kinematics in terms of generalized coordinate:* In addition to the chosen generalized coordinate, we select an inertial reference frame $OXYZ$, which is fixed at the joint of the pendulum O as sketched in Figure F-3, and denote the unit vectors along the X and Y axes as $\boldsymbol{I}$ and $\boldsymbol{J}$, respectively. Then, the position vector of the mass m of the pendulum is

$$\boldsymbol{R}_1 = \ell(\sin\theta\boldsymbol{I} - \cos\theta\boldsymbol{J}) \tag{b}$$

and its partial derivative with respect to the generalized coordinate θ is

$$\frac{\partial\boldsymbol{R}_1}{\partial\xi_1} = \frac{\partial\boldsymbol{R}_1}{\partial\theta} = \ell(\cos\theta\boldsymbol{I} + \sin\theta\boldsymbol{J}) \tag{c}$$

According to Eq. (F-2), the velocity of m is

$$\boldsymbol{v}_1 = \frac{d\boldsymbol{R}_1}{dt} = \frac{\partial\boldsymbol{R}_1}{\partial\theta}\dot{\theta} = \ell\dot{\theta}(\cos\theta\boldsymbol{I} + \sin\theta\boldsymbol{J}) \tag{d}$$

and the time rate of change of the linear momentum of the mass of the pendulum is

$$\begin{aligned}
\frac{d\boldsymbol{p}_1}{dt} &= \frac{d}{dt}(m\boldsymbol{v}_1) \\
&= m\ell\frac{d}{dt}\big[\dot{\theta}(\cos\theta\boldsymbol{I} + \sin\theta\boldsymbol{J})\big] \\
&= m\ell\big[\ddot{\theta}(\cos\theta\boldsymbol{I} + \sin\theta\boldsymbol{J}) + \dot{\theta}^2(-\sin\theta\boldsymbol{I} + \cos\theta\boldsymbol{J})\big] \\
&= m\ell\big[(\ddot{\theta}\cos\theta - \dot{\theta}^2\sin\theta)\boldsymbol{I} + (\ddot{\theta}\sin\theta + \dot{\theta}^2\cos\theta)\boldsymbol{J}\big]
\end{aligned} \tag{e}$$

3. *Applied forces:* There is only one applied or impressed force, the gravitational force, acting on the mass, expressed as

$$\boldsymbol{f}_1 = -mg\boldsymbol{J} \tag{f}$$

Note that the force in the rod is a workless constraint force.

4. *Equation of motion:* By substituting Eqs. (c), (e), and (f) into Eqs. (F-13) and (F-14), the equation of motion can be derived as follows:

$$Q_1 = \sum_{i=1}^{1} \boldsymbol{f}_i' \cdot \frac{\partial \boldsymbol{R}_i}{\partial \xi_1} = \boldsymbol{f}_1' \cdot \frac{\partial \boldsymbol{R}_1}{\partial \theta}$$

$$= -mg\boldsymbol{J} \cdot \ell(\cos\theta \boldsymbol{I} + \sin\theta \boldsymbol{J})$$

$$= -mg\ell\sin\theta \tag{g}$$

$$Q_1^* = \sum_{i=1}^{1} \left(-\frac{d\boldsymbol{p}_i}{dt}\right) \cdot \frac{\partial \boldsymbol{R}_i}{\partial \xi_1} = -\frac{d\boldsymbol{p}_1}{dt} \cdot \frac{\partial \boldsymbol{R}_1}{\partial \theta}$$

$$= -m\ell[(\ddot\theta\cos\theta - \dot\theta^2\sin\theta)\boldsymbol{I} + (\ddot\theta\sin\theta + \dot\theta^2\cos\theta)\boldsymbol{J}]$$
$$\ell(\cos\theta \boldsymbol{I} + \sin\theta \boldsymbol{J})$$

$$= -m\ell^2[\ddot\theta\cos^2\theta - \dot\theta^2\sin\theta\cos\theta + \ddot\theta\sin^2\theta + \dot\theta^2\sin\theta\cos\theta]$$

$$= -m\ell^2\ddot\theta(\sin^2\theta + \cos^2\theta) = -m\ell^2\ddot\theta \tag{h}$$

So, the equation of motion for the simple plane pendulum, according to Eq. (F-12), is

$$Q_1 + Q_1^* = 0 \tag{i}$$

or

$$-mg\ell\sin\theta - m\ell\ddot\theta = 0 \tag{j}$$

■ **Example F-3:** Extensible Plane Pendulum

A mass m is attached to one end of a massless spring having spring constant k and undeformed length L. The other end of the spring is hinged at a fixed point O in such a way that the spring can rotate in the plane of the sketch about point O without friction, as sketched in Figure F-4. Derive the equation(s) of motion via Lagrange's form of d'Alembert's principle.

❑ **Solution:**

1. *Generalized coordinate(s):* Our choice for the generalized coordinates for this system is

$$\xi_j : \xi_1 = r, \quad \xi_2 = \theta \tag{a}$$

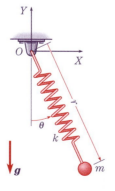

FIGURE F-4 Extensible plane pendulum.

where r is the radial position of the mass relative to the hinge O and θ is the counterclockwise angular displacement of the pendulum from its downward-hanging equilibrium position.

It is also noted that there is one particle, that is, $N = 1$; and there are two generalized coordinates, that is, $n = 2$.

2. *System kinematics in terms of generalized coordinates:*   To facilitate the analysis, we define an inertial reference frame $OXYZ$ as sketched in Figure F-4, and denote the unit vectors along the X and Y axes as I and J, respectively. Then, the position vector of the mass m is

$$\boldsymbol{R}_1 = r(\sin\theta\boldsymbol{I} - \cos\theta\boldsymbol{J}) \tag{b}$$

and the partial derivatives with respect to the generalized coordinates are

$$\frac{\partial \boldsymbol{R}_1}{\partial \xi_1} = \frac{\partial \boldsymbol{R}_1}{\partial r} = \sin\theta\boldsymbol{I} - \cos\theta\boldsymbol{J} \tag{c}$$

$$\frac{\partial \boldsymbol{R}_1}{\partial \xi_2} = \frac{\partial \boldsymbol{R}_1}{\partial \theta} = r(\cos\theta\boldsymbol{I} + \sin\theta\boldsymbol{J}) \tag{d}$$

According to Eq. (F-2), the velocity of the mass is

$$\boldsymbol{v}_1 = \frac{d\boldsymbol{R}_1}{dt} = \frac{\partial \boldsymbol{R}_1}{\partial r}\dot{r} + \frac{\partial \boldsymbol{R}_1}{\partial \theta}\dot{\theta}$$
$$= (\dot{r}\sin\theta + r\dot{\theta}\cos\theta)\boldsymbol{I} + (-\dot{r}\cos\theta + r\dot{\theta}\sin\theta)\boldsymbol{J} \tag{e}$$

and the time rate of change of the linear momentum of the mass is

$$\frac{d\boldsymbol{p}_1}{dt} = m\frac{d\boldsymbol{v}_1}{dt}$$
$$= m\big(\ddot{r}\sin\theta + 2\dot{r}\dot{\theta}\cos\theta + r\ddot{\theta}\cos\theta - r\dot{\theta}^2\sin\theta\big)\boldsymbol{I}$$
$$+ m\big(-\ddot{r}\cos\theta + 2\dot{r}\dot{\theta}\sin\theta + r\ddot{\theta}\sin\theta + r\dot{\theta}^2\cos\theta\big)\boldsymbol{J} \tag{f}$$

3. *Applied forces:*   There are two applied or impressed forces acting on the particle. The first is the gravitational force, which is always in the negative $\boldsymbol{J}$ direction. The second is the force applied by the spring along the radial direction. Noting that the elongation of the spring is $r - L$, we find the total applied or impressed force on the mass as

$$\boldsymbol{f}_1' = -mg\boldsymbol{J} + k(r - L)(-\sin\theta\boldsymbol{I} + \cos\theta\boldsymbol{J})$$
$$= -k(r - L)\sin\theta\boldsymbol{I} - [mg - k(r - L)\cos\theta]\boldsymbol{J} \tag{g}$$

4. *Equations of motion:*   By substituting Eqs. (c), (d), (f), and (g) into Eqs. (F-13) and (F-14) and by noting that $n = 2$, the equations of motion can be derived as follows:

$$Q_1 = \sum_{i=1}^{1}\boldsymbol{f}_i' \cdot \frac{\partial \boldsymbol{R}_i}{\partial \xi_1} = \boldsymbol{f}_1' \cdot \frac{\partial \boldsymbol{R}_1}{\partial r}$$
$$= \{-k(r - L)\sin\theta\boldsymbol{I} - [mg - k(r - L)\cos\theta]\boldsymbol{J}\} \cdot (\sin\theta\boldsymbol{I} - \cos\theta\boldsymbol{J})$$
$$= -k(r - L)\sin^2\theta + mg\cos\theta - k(r - L)\cos^2\theta$$
$$= mg\cos\theta - k(r - L) \tag{h}$$

$$Q_2 = \sum_{i=1}^{1} \boldsymbol{f}_i' \cdot \frac{\partial \boldsymbol{R}_i}{\partial \xi_2} = \boldsymbol{f}_1' \cdot \frac{\partial \boldsymbol{R}_1}{\partial \theta}$$

$$= \{-k(r - L)\sin\theta\boldsymbol{I} - [mg - k(r - L)\cos\theta]\boldsymbol{J}\} \cdot r(\cos\theta\boldsymbol{I} + \sin\theta\boldsymbol{J})$$

$$= -kr(r - L)\sin\theta\cos\theta - mgr\sin\theta + kr(r - L)\sin\theta\cos\theta$$

$$= -mgr\sin\theta \tag{i}$$

$$Q_1^* = \sum_{i=1}^{1} \left(-\frac{d\boldsymbol{p}_i}{dt}\right) \cdot \frac{\partial \boldsymbol{R}_i}{\partial \xi_1} = -\frac{d\boldsymbol{p}_1}{dt} \cdot \frac{\partial \boldsymbol{R}_1}{\partial r}$$

$$= -m[(\ddot{r}\sin\theta + 2\dot{r}\dot{\theta}\cos\theta + r\ddot{\theta}\cos\theta - r\dot{\theta}^2\sin\theta)\boldsymbol{I}$$

$$+ (-\ddot{r}\cos\theta + 2\dot{r}\dot{\theta}\sin\theta + r\ddot{\theta}\sin\theta + r\dot{\theta}^2\cos\theta)\boldsymbol{J}] \cdot (\sin\theta\boldsymbol{I} - \cos\theta\boldsymbol{J})$$

$$= -m[\ddot{r}\sin^2\theta + 2\dot{r}\dot{\theta}\sin\theta\cos\theta + r\ddot{\theta}\sin\theta\cos\theta - r\dot{\theta}^2\sin^2\theta$$

$$+ \ddot{r}\cos^2\theta - 2\dot{r}\dot{\theta}\sin\theta\cos\theta - r\ddot{\theta}\sin\theta\cos\theta - r\dot{\theta}^2\cos^2\theta]$$

$$= -m(\ddot{r} - r\dot{\theta}^2) \tag{j}$$

$$Q_2^* = \sum_{i=1}^{1} \left(-\frac{d\boldsymbol{p}_i}{dt}\right) \cdot \frac{\partial \boldsymbol{R}_i}{\partial \xi_2} = -\frac{d\boldsymbol{p}_1}{dt} \cdot \frac{\partial \boldsymbol{R}_1}{\partial \theta}$$

$$= -m[(\ddot{r}\sin\theta + 2\dot{r}\dot{\theta}\cos\theta + r\ddot{\theta}\cos\theta - r\dot{\theta}^2\sin\theta)\boldsymbol{I}$$

$$+ (-\ddot{r}\cos\theta + 2\dot{r}\dot{\theta}\sin\theta + r\ddot{\theta}\sin\theta + r\dot{\theta}^2\cos\theta)\boldsymbol{J}] \cdot r(\cos\theta\boldsymbol{I} + \sin\theta\boldsymbol{J})$$

$$= -mr[\ddot{r}\sin\theta\cos\theta + 2\dot{r}\dot{\theta}\cos^2\theta + r\dot{\theta}^2\sin\theta\cos\theta - r\ddot{\theta}\cos^2\theta$$

$$- \ddot{r}\sin\theta\cos\theta + 2\dot{r}\dot{\theta}\sin^2\theta + r\ddot{\theta}\sin^2\theta + r\dot{\theta}^2\sin\theta\cos\theta]$$

$$= -mr(2\dot{r}\dot{\theta} + r\ddot{\theta}) \tag{k}$$

Thus,

$$Q_1 + Q_1^* = 0 = mg\cos\theta - k(r - L) - m(\ddot{r} - r\dot{\theta}^2) \tag{l}$$

and

$$Q_2 + Q_2^* = 0 = -mgr\sin\theta - mr(2\dot{r}\dot{\theta} + r\ddot{\theta}) \tag{m}$$

Therefore, the equations of motion for the extensible plane pendulum are

$$mg\cos\theta - k(r - L) - m(\ddot{r} - r\dot{\theta}^2) = 0$$
$$-mgr\sin\theta - mr(2\dot{r}\dot{\theta} + r\ddot{\theta}) = 0 \tag{n}$$

Readers may decide for themselves whether they prefer to use Lagrange's form of d'Alembert's principle or Lagrange's equations to derive the equations of motion of lumped-parameter models.

A BRIEF REVIEW OF ELECTROMAGNETIC (EM) THEORY AND APPROXIMATIONS

In this appendix, Maxwell's equations, the quantitative phenomenological descriptions of electromagnetism, are briefly reviewed. The primary goal of this appendix is to deduce from the complete form of Maxwell's equations the corresponding equations for quasistatics, quasistatics being the theoretical basis for modern circuit theory; the equations for statics are deduced as a byproduct along the way. In particular, the quantitative criteria for the validity of the quasistatic approximation are derived where it is shown that the electric and magnetic fields decouple substantially. Furthermore, it is shown that in quasistatics, the character of capacitive elements is dominated by the electric fields and the character of inductive elements is dominated by the magnetic fields. Finally, the so-called "Kirchhoff's laws" are deduced from quasistatics and are, therefore, not laws in the fundamental sense, but *rules*.

G-1 MAXWELL'S EQUATIONS: COMPLETE FORM

The fundamental laws of electromagnetism can be summarized by a set of expressions known as Maxwell's equations. In this section, the integral and differential forms of Maxwell's equations are presented. It is assumed that the reader is already familiar with the terms *electric field intensity* $\boldsymbol{E}$, *magnetic flux density* $\boldsymbol{B}$, *charge density* ρ_e and *current density* $\boldsymbol{J}$; all of which are defined in most first-year college textbooks on electromagnetism.

G-1.1 Integral Form

The integral forms of Maxwell's equations along with the experimental bases for these laws are presented in Table G-1.

In Table G-1, $d\boldsymbol{\ell}, d\boldsymbol{S}$ and dV are the differentials of the contour C, surface S, and volume V, respectively. Also, the parameters ϵ_0 and μ_0 are the permittivity constant and permeability constant, respectively. Both ϵ_0 and μ_0 are defined for a vacuum and have the measured values

$$\epsilon_0 = 8.854 \times 10^{-12} \frac{\text{coulomb}}{\text{volt-meter}} \tag{G-1}$$

763

TABLE G-1 Complete Form of Maxwell's Equations in Integral Form[†]

Equation	Integral Form	Experimental Basis
Gauss's law for electricity	$\epsilon_0 \int_S \boldsymbol{E} \cdot d\boldsymbol{S} = \int_V \rho_e \, dV$	Like charges repel and unlike charges attract, as the inverse square of their separation.
Gauss's law for magnetism	$\int_S \boldsymbol{B} \cdot d\boldsymbol{S} = 0$	Single magnetic poles do not exist.
Ampere's law (as extended by Maxwell)	$\oint_C \boldsymbol{B} \cdot d\boldsymbol{\ell} = \mu_0 \epsilon_0 \dfrac{d}{dt} \int_S \boldsymbol{E} \cdot d\boldsymbol{S}$ $+ \mu_0 \int_S \boldsymbol{J} \cdot d\boldsymbol{S}^{\ddagger}$	A magnetic field is produced by (i) a changing electric field and (ii) a current.
Faraday's law of induction	$\oint_C \boldsymbol{E} \cdot d\boldsymbol{\ell} = -\dfrac{d}{dt} \int_S \boldsymbol{B} \cdot d\boldsymbol{S}$	An electric field is produced by a changing magnetic field.

[†]Written on the assumption of a vacuum; that is, no dielectric or magnetizable material is present.

[‡]Electromagnetic waves at the speed of light were predicted as a consequence.

and

$$\mu_0 = 4\pi \times 10^{-7} \frac{\text{volt-s}^2}{\text{coulomb-meter}} \tag{G-2}$$

Note that Ampere's law, in the third row of Table G-1, has two terms on the right-hand side. The first term on the right-hand side was added by Maxwell. This additional term, which is known as Maxwell's *displacement current correction* to Ampere's law, was one of the keys to coupling electric and magnetic fields mathematically, and was also among the seminal ideas for the prediction of the existence of electromagnetic waves.

TABLE G-2 Complete Form of Maxwell's Equations in Differential Form[†]

Equation	Differential Form
Gauss's law for electricity	$\nabla \cdot \epsilon_0 \boldsymbol{E} = \rho_e$
Gauss's law for magnetism	$\nabla \cdot \boldsymbol{B} = 0$
Ampere's law (as extended by Maxwell)	$\nabla \times \boldsymbol{B} = \mu_0 \epsilon_0 \dfrac{\partial \boldsymbol{E}}{\partial t} + \mu_0 \boldsymbol{J}$
Faraday's law of induction	$\nabla \times \boldsymbol{E} = -\dfrac{\partial \boldsymbol{B}}{\partial t}$

[†]Written on the assumption of a vacuum; that is, no dielectric or magnetizable material is present.

G-1.2   Differential Form

The differential forms of Maxwell's equations are presented in Table G-2. Since these equations are equivalent to those in Table G-1, they also apply to free space (vacuum).

G-2   MAXWELL'S EQUATIONS: ELECTROSTATICS AND MAGNETOSTATICS

The static forms of Maxwell's equations are derived from the complete forms by imposing the conditions that there are no time-varying E and B fields. Mathematically, these conditions are represented by the requirements that

$$\frac{\partial E}{\partial t} = 0 \tag{G-3}$$

$$\frac{\partial B}{\partial t} = 0 \tag{G-4}$$

By substituting Eqs. (G-3) and (G-4) into the complete forms of Maxwell's equations, the static forms of Maxwell's equations are derived. In Table G-3, the static forms of the resulting equations are summarized in both integral and differential forms.

The most significant feature of the equations in Table G-3 is that each equation is a function of either the electric field E or the magnetic field B; no equation contains *both* E and B. This observation reveals the uncoupling of the electric and magnetic fields.

So, when the E and B fields are, at most, constant in accordance with Eqs. (G-3) and (G-4), electricity and magnetism can be treated as independent subjects called *electrostatics* and *magnetostatics,* thus quantitatively illustrating why the disciplines of electrostatics and magnetostatics developed independently until the early nineteenth century. In Table G-3, there are two equations that are a function of the E field: Gauss's law for electricity and Faraday's law of induction. These two laws in their static forms are the basic equations of

TABLE G-3   Maxwell's Equations for Electrostatics and Magnetostatics[†]

Equation	Integral Form	Differential Form
Gauss's law for electricity	$\epsilon_0 \int_S E \cdot dS = \int_V \rho_e \, dV$	$\nabla \cdot \epsilon_0 E = \rho_e$
Gauss's law for magnetism	$\int_S B \cdot dS = 0$	$\nabla \cdot B = 0$
Ampere's law (as extended by Maxwell)	$\oint_C B \cdot d\ell = \mu_0 \int_S J \cdot dS$	$\nabla \times B = \mu_0 J$
Faraday's law of induction	$\oint_C E \cdot d\ell = 0$	$\nabla \times E = 0$

[†]Written on the assumption of a vacuum; that is, no dielectric or magnetizable material is present.

electrostatics. Similarly, there are two equations that are a function of the B field: Gauss's law for magnetism and Ampere's law. These two laws in their static forms are the fundamental equations of magnetostatics.

G-3 MAXWELL'S EQUATIONS: ELECTROQUASISTATICS AND MAGNETOQUASISTATICS

In this section, we consider the important case of quasistatics. Quasistatics reside between electrostatics and magnetostatics (no time changes) on the one hand, and on the other hand electrodynamics (a general term incorporating both fully dynamic electric and magnetic effects where the complete coupling of E and B exhibit wave behavior). We shall develop the quantitative criteria for the appropriate modeling of quasistatics.

The primary significance of quasistatics is that it establishes the foundations of modern circuit theory. As we shall show, quasistatics demonstrates that the character of circuit capacitive elements is dominated by electric fields and the character of circuit inductive elements is dominated by magnetic fields.

In the derivation of the quasistatics equations, the approach is to examine the effect on the complete forms of Maxwell's equations of considering the dominance of either the electric field or the magnetic field. The first case, where the electric field effects dominate the magnetic field effects, will yield the laws of electroquasistatics. The second case, where the magnetic field effects dominate the electric field effects, will yield the laws of magnetoquasistatics. In this way, the quantitative criteria under which these laws are applicable will also be derived.

G-3.1 Electroquasistatics

In the derivation of the equations of electroquasistatics, the main strategy is to consider an arbitrary system where the electric field effects are so dominant that the magnetic field can be neglected. In the process of imposing this dominance of the electric field over the magnetic field, *a priori* conditions will be discovered that must hold in order for electroquasistatics to be applicable. This assumption is equivalent to restricting consideration to systems or elements having *high impedance* or, equivalently, that behave as an *open circuit*. (In the next subsection, we adopt the opposite assumption, leading to systems or elements having *low impedance* or, equivalently, that behave as a *short circuit*.)

The mathematical approach in this derivation is via a dimensional argument. In this dimensional argument, L is taken as a characteristic length of the system, and τ is taken as a characteristic time such as the period of the electromagnetic process. This dimensional argument is imposed first on Faraday's law of induction and then on Ampere's law.

Recall Faraday's law from Table G-2 as

$$\nabla \times \boldsymbol{E} = -\frac{\partial \boldsymbol{B}}{\partial t} \tag{G-5}$$

When the dimensional parameters are assigned appropriately to the length and time elements of Eq. (G-5), we get

$$\frac{E}{L} \sim \frac{B}{\tau} \tag{G-6}$$

In order for the magnetic field effects to be negligible compared with the electric field effects, Eq. (G-6) requires that

$$\frac{E}{L} \gg \frac{B}{\tau} \tag{G-7}$$

Now, a similar dimensional argument will be imposed on Ampere's law. Recall Ampere's law from Table G-2 as

$$\nabla \times \boldsymbol{B} = \mu_0 \epsilon_0 \frac{\partial \boldsymbol{E}}{\partial t} + \mu_0 \boldsymbol{J} \tag{G-8}$$

In a high-impedance system, such as two conducting surfaces separated by an open space (for example, a capacitor), the current density term $(\mu_0 \boldsymbol{J})$ will be negligibly small compared with the displacement current correction term because no (or little) charge can be transported across the open space. Equivalently, it may be stated that it is the field-induced current rather than the flow of charge across the gap of a capacitor that is primarily responsible for the current through a capacitor.

So, neglecting the current density term in Eq. (G-8) and applying the dimensional argument on the remaining expression gives

$$\frac{B}{L} \sim \mu_0 \epsilon_0 \frac{E}{\tau} \tag{G-9}$$

Recall that the speed of electromagnetic propagation c (that is, the speed of light) can be computed according to

$$c^2 = \frac{1}{\mu_0 \epsilon_0} \tag{G-10}$$

Substitution of an expression for B from Eq. (G-9) into Eq. (G-7) and use of Eq. (G-10) to eliminate $\mu_0 \epsilon_0$ give

$$\frac{E}{L} \gg \frac{LE}{c^2 \tau^2} \tag{G-11}$$

which simplifies to

$$\frac{L}{c\tau} \ll 1 \tag{G-12}$$

Equation (G-12) is an important result; it delivers a quantitative criterion for the appropriate application of electroquasistatics. Thus, the characteristic time or period of the electromagnetic process must be sufficiently long to ensure the satisfaction of Eq. (G-12).

There are several alternative expressions for Eq. (G-12); two will be derived. Recall that the period τ of a harmonic process can be written alternatively as either

$$\tau = \frac{\lambda}{c} \tag{G-13}$$

or

$$\tau = \frac{1}{f} \tag{G-14}$$

where λ and f are the electromagnetic wavelength and frequency in hertz, respectively. Substitution of Eq. (G-13) into Eq. (G-12) gives

$$\frac{L}{\lambda} \ll 1 \tag{G-15}$$

and substitution of Eq. (G-14) into Eq. (G-12) gives

$$\frac{Lf}{c} \ll 1 \tag{G-16}$$

Equations (G-15) and (G-16) are alternative expressions for Eq. (G-12); any one of them may be used to assess the electroquasistatic approximation. For example, suppose the capacitive element has a characteristic length of 1 cm, then recalling that $c \approx 3 \times 10^{10}$ cm/s and using Eq. (G-16) lead to

$$f \ll 30 \times 10^9 \, \text{Hz} \tag{G-17}$$

Thus, if the frequency of the voltage across a capacitor having a 1 cm characteristic length is much less than tens of gigahertz, the electroquasistatic approximation is valid.

The *a priori* conditions that must be satisfied to ensure the validity of electroquasistatics are summarized in the Table G-4. Also, the differential and integral forms of Maxwell's equations for electroquasistatics are summarized in Tables G-5 and G-6, respectively. In various references, the reader will find alternative presentations of the magnetoquasistatics expression of Gauss's law for electricity and both the magnetoquasistatics and electroquasistatics expressions of Gauss's law for magnetism. Nevertheless, for the purposes of this appendix, the presentations in Tables G-5 and G-6 are easier to deduce and are adequate.

G-3.2   Magnetoquasistatics

In the derivations of the equations of magnetoquasistatics, the strategy is to consider an arbitrary system where the magnetic field effects are so dominant that the electric field can be neglected. In the process of imposing this dominance of the magnetic field over the electric field, *a priori* conditions will be discovered that must hold in order for magnetoquasistatics to be applicable.

As in the case of electroquasistatics, the approach here is also via a dimensional argument. Again, L is taken as a characteristic length of the system, and τ is taken as a characteristic time such as the period of the electromagnetic process. These dimensional arguments are imposed on both Faraday's law of induction (by use of Eq. (G-6)) and Ampere's law.

TABLE G-4   Conditions for Electroquasistatics and Magnetoquasistatics

	Electroquasistatics	Magnetoquasistatics
Circuit conditions	(1) Open Circuit ($\boldsymbol{J}$ effect is small compared with $\boldsymbol{E}$ effect)	(1) Short Circuit ($\boldsymbol{J}$ effect is large compared with $\boldsymbol{E}$ effect)
Temporal conditions	(2) Slow time variation, $\dfrac{L}{\lambda} \ll 1$ or $\dfrac{Lf}{c} \ll 1$	(2) Slow time variation, $\dfrac{L}{\lambda} \ll 1$ or $\dfrac{Lf}{c} \ll 1$

TABLE G-5 Maxwell's Equations in Differential Form for Electroquasistatics and Magnetoquasistatics[†]

Equation	Electroquasistatics	Magnetoquasistatics
Gauss's law for electricity	$\nabla \cdot \epsilon_0 \boldsymbol{E} = \rho_e$	$\nabla \cdot \epsilon_0 \boldsymbol{E} = \rho_e$
Gauss's law for magnetism	$\nabla \cdot \boldsymbol{B} = 0$	$\nabla \cdot \boldsymbol{B} = 0$
Ampere's law (as extended by Maxwell)	$\nabla \times \boldsymbol{B} = \mu_0 \epsilon_0 \dfrac{\partial \boldsymbol{E}}{\partial t} + \mu_0 \boldsymbol{J}$	$\nabla \times \boldsymbol{B} = \mu_0 \boldsymbol{J}$
Faraday's law of induction	$\nabla \times \boldsymbol{E} = 0$	$\nabla \times \boldsymbol{E} = -\dfrac{\partial \boldsymbol{B}}{\partial t}$

[†]Written on the assumption of a vacuum; that is, no dielectric or magnetizable material is present.

TABLE G-6 Maxwell's Equations in Integral Form for Electroquasistatics and Magnetoquasistatics[†]

Equation	Electroquasistatics	Magnetoquasistatics
Gausss' law for electricity	$\epsilon_0 \displaystyle\int_S \boldsymbol{E} \cdot d\boldsymbol{S} = \int_V \rho_e \, dV$	$\epsilon_0 \displaystyle\int_S \boldsymbol{E} \cdot d\boldsymbol{S} = \int_V \rho_e \, dV$
Gauss's law for magnetism	$\displaystyle\int_S \boldsymbol{B} \cdot d\boldsymbol{S} = 0$	$\displaystyle\int_S \boldsymbol{B} \cdot d\boldsymbol{S} = 0$
Ampere's law (as extended by Maxwell)	$\displaystyle\oint_C \boldsymbol{B} \cdot d\boldsymbol{\ell} = \mu_0 \epsilon_0 \dfrac{d}{dt} \int_S \boldsymbol{E} \cdot d\boldsymbol{S}$ $+ \mu_0 \displaystyle\int_S \boldsymbol{J} \cdot d\boldsymbol{S}$	$\displaystyle\oint_C \boldsymbol{B} \cdot d\boldsymbol{\ell} = \mu_0 \int_S \boldsymbol{J} \cdot d\boldsymbol{S}$
Faraday's law of induction	$\displaystyle\oint_C \boldsymbol{E} \cdot d\boldsymbol{\ell} = 0$	$\displaystyle\oint_C \boldsymbol{E} \cdot d\boldsymbol{\ell} = -\dfrac{d}{dt} \int_S \boldsymbol{B} \cdot d\boldsymbol{S}$

[†]Written on the assumption of a vacuum; that is, no dielectric or magnetizable material is present.

Recall Ampere's law from Table G-2 as

$$\nabla \times \boldsymbol{B} = \mu_0 \epsilon_0 \frac{\partial \boldsymbol{E}}{\partial t} + \mu_0 \boldsymbol{J} \tag{G-18}$$

In order to apply the dimensional argument, it is necessary for the $\boldsymbol{B}$ field to be much more dominant than the $\boldsymbol{E}$ field; this implies that the entire displacement current term (that is, the $\boldsymbol{E}$ field term) must be negligible in Eq. (G-18). This is possible only if it is much smaller than both the current density term (that is, the $\boldsymbol{J}$ term) and the magnetic field term (that is, the $\boldsymbol{B}$ term). Alternatively, it means that both the $\boldsymbol{J}$ and $\boldsymbol{B}$ terms need to be much larger than the displacement current term. In order for the $\boldsymbol{J}$ term to be large, the system under consideration must be of *low impedance* or behave as a *short circuit.* For instance, in a system consisting of a conducting wire loop (for example, an inductor), the $\boldsymbol{J}$ effects will be large compared with the $\boldsymbol{E}$ effects.

Now, using the dimensional argument on Ampere's law and requiring that the $\boldsymbol{B}$ field term be much larger than the $\boldsymbol{E}$ field term yield the inequality

$$\frac{B}{L} \gg \mu_0 \epsilon_0 \frac{E}{\tau} \tag{G-19}$$

Solving Eq. (G-6) for E and substituting the resulting expression into Eq. (G-19) to eliminate E give

$$\frac{B}{L} \gg \mu_0 \epsilon_0 \frac{BL}{\tau^2} \tag{G-20}$$

Use of Eq. (G-10) in Eq. (G-20) to eliminate $\mu_0 \epsilon_0$ gives

$$\frac{L}{c\tau} \ll 1 \tag{G-21}$$

which is identical to Eq. (G-12). Thus, the quantitative *a priori* condition that must be satisfied to ensure the validity of magnetoquasistatics is the same as that for electroquasistatics. Furthermore, Eq. (G-21) can be represented alternatively by Eqs. (G-15) and (G-16) by the identical procedures leading to those two expressions. Tables G-4, G-5, and G-6 also contain the summaries for magnetoquasistatics corresponding to those for electroquasistatics. Although the forms of the equations in Tables G-5 and G-6 are not best suited for subsequent calculation, they will serve our goals in this appendix.

The equations in Table G-5 are the differential equations of electroquasistatics and magnetoquasistatics. In electroquasistatics, the change from the complete form of Maxwell's equations is in Faraday's law. Here, the term $\dfrac{-\partial \boldsymbol{B}}{\partial t}$ has been removed from the complete form of Faraday's law. In magnetoquasistatics, the change is in Ampere's law. Here, the displacement current term has been removed.

In quasistatics there is a decoupling of the electric and magnetic fields, similar to statics; however, this decoupling is partial but, unlike statics, it maintains a time variation of the $\boldsymbol{E}$ and $\boldsymbol{B}$ fields. The middle column of Table G-5 illustrates that electroquasistatics is dependent only on time variations in the $\boldsymbol{E}$ field. Similarly, the right-hand column of Table G-5 illustrates that magnetoquasistatics is dependent only on time variations in the $\boldsymbol{B}$ field. In the next section, this decoupling will be shown even more clearly by illustrating how energy in electroquasistatics and magnetoquasistatics is stored via the $\boldsymbol{E}$ and $\boldsymbol{B}$ fields, respectively.

G-4   ENERGY STORAGE IN ELECTROQUASISTATICS AND MAGNETOQUASISTATICS

In this section, energy storage is considered in electroquasistatics and magnetoquasistatics. The results of this analysis illustrate further the consequences of the decoupling of the $\boldsymbol{E}$ and $\boldsymbol{B}$ fields in quasistatics. In this discussion, it will be shown that the electromagnetic energy is stored in the electric field for systems characterized by the electroquasistatics approximation while the electromagnetic energy is stored in the magnetic field for systems characterized by the magnetoquasistatics approximation.

In investigating the energy storage, the energy balance for a volume V bounded by the surface S, as sketched in Figure G-1, is considered. This energy balance can be described as

Rate of energy flow into system through surface S (E_{in}) $=$ Energy accumulation rate in system volume V (E_a)
$+$ Energy dissipation rate in system volume V (E_d)

Mathematically, this is expressed as

$$-\int_S \boldsymbol{P} \cdot d\boldsymbol{S} = \frac{\partial}{\partial t} \int_V w\, dV + \int_V p_d\, dV \tag{G-22}$$

where $\boldsymbol{P}$ is the rate of energy flow per unit area through the surface $\boldsymbol{S}$ into the volume V; w is the energy density within V and p_d is the rate of dissipation of the energy density within V.

Because the direction of $d\boldsymbol{S}$ is defined as positive outward from the enclosed volume, the negative sign on the left-hand side of Eq. (G-22) ensures that the integral is positive for energy flow into the volume. So, the left-hand side of Eq. (G-22) is the rate of energy flow through the surface S *into* the volume V. The first term on the right-hand side of Eq. (G-22) is the rate of energy accumulation within the volume V. The second term on the right-hand side of Eq. (G-22) is the rate of energy dissipation within the volume V.

The electroquasistatic and magnetoquasistatic equations will be manipulated into the form given in Eq. (G-22) to investigate the energy balance of an electromagnetic system. The goal of this mathematical manipulation is to obtain expressions for the energy storage per unit volume in V (that is, w, the energy density within V) for both electroquasistatic and magnetoquasistatic fields.

G-4.1   Energy Storage in Electroquasistatics

In electroquasistatics, Ampere's law is

$$\nabla \times \boldsymbol{B} = \mu_0 \epsilon_0 \frac{\partial \boldsymbol{E}}{\partial t} + \mu_0 \boldsymbol{J} \tag{G-23}$$

and Faraday's law is

$$\nabla \times \boldsymbol{E} = 0 \tag{G-24}$$

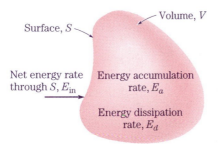

Surface, S

Volume, V

Net energy rate through S, E_{in}

Energy accumulation rate, E_a

Energy dissipation rate, E_d

FIGURE G-1   Energy balance in volume V bounded by closed surface S.

Dot multiplying Eqs. (G-23) and (G-24) by $\dfrac{E}{\mu_0}$ and $\dfrac{B}{\mu_0}$, respectively, and then subtracting the resulting second equation from the resulting first equation give

$$\frac{E}{\mu_0} \cdot (\nabla \times B) - \frac{B}{\mu_0} \cdot (\nabla \times E) = \epsilon_0 E \cdot \frac{\partial E}{\partial t} + E \cdot J \tag{G-25}$$

The left-hand side of Eq. (G-25) can be simplified, by inspection, by using the general vector identity

$$C \cdot (\nabla \times A) - A \cdot (\nabla \times C) = \nabla \cdot (A \times C) \tag{G-26}$$

where C and A are arbitrary vectors. (Note that in Eqs. (G-25) and (G-26), the electromagnetic vectors B and E correspond to the arbitrary vectors A and C, respectively.) Also, rewriting the right-hand side of Eq. (G-25) gives

$$\nabla \cdot \left(\frac{1}{\mu_0} B \times E \right) = \frac{\partial}{\partial t} \left(\frac{1}{2} \epsilon_0 E \cdot E \right) + E \cdot J \tag{G-27}$$

where the left-hand side of Eq. (G-27) corresponds to the left-hand side of Eq. (G-25), having been rewritten in accordance with the right-hand side of Eq. (G-26).

Equation (G-27) can now be rewritten into the form of Eq. (G-22) by the following series of steps: (1) recognize that $B \times E = -E \times B$, (2) integrate the entire equation over the volume V, and (3) apply the divergence theorem to the left-hand side of the resulting expression by recalling that for an arbitrary vector A

$$\int_S A \cdot dS = \int_V \nabla \cdot A \, dV \tag{G-28}$$

where, in the calculation performed here, the vector A is replaced by the term $\dfrac{1}{\mu_0} E \times B$ to arrive at

$$-\int_S \left(\frac{1}{\mu_0} E \times B \right) \cdot dS = \frac{\partial}{\partial t} \int_V \frac{1}{2} \epsilon_0 E \cdot E \, dV + \int_V E \cdot J \, dV \tag{G-29}$$

Equation (G-29) is in the form of the energy balance expression of Eq. (G-22). By comparing Eqs. (G-22) and (G-29), we can identify from the left-hand side of each equation that

$$P = \frac{1}{\mu_0} E \times B \tag{G-30}$$

The vector P, by convention, is called the *Poynting vector* and is defined as shown in Eq. (G-30). This vector represents the rate of electromagnetic energy flow per unit area, or the rate of energy flux. From the first term on the right-hand side of Eq. (G-29), we can identify

$$w = \frac{1}{2} \epsilon_0 E \cdot E \tag{G-31}$$

as the electromagnetic energy density in V. Note that this energy accumulation is accomplished via the E field. This is an important result. It reveals that *in an element or system characterized by the electroquasistatic approximation (for example, an ideal capacitor), the energy is stored in the electric field.* The second term on the right-hand side of Eq.

(G-29) simply represents the energy dissipation rate, as commonly presented in elementary electromagnetic theory.

G-4.2 Energy Storage in Magnetoquasistatics

In magnetoquasistatics, Ampere's law is

$$\nabla \times \boldsymbol{B} = \mu_0 \boldsymbol{J} \tag{G-32}$$

and Faraday's law is

$$\nabla \times \boldsymbol{E} = -\frac{\partial \boldsymbol{B}}{\partial t} \tag{G-33}$$

Dot multiplying Eqs. (G-32) and (G-33) by $\dfrac{\boldsymbol{E}}{\mu_0}$ and $\dfrac{\boldsymbol{B}}{\mu_0}$, respectively, and then subtracting the resulting second equation from the resulting first equation give

$$\frac{\boldsymbol{E}}{\mu_0} \cdot (\nabla \times \boldsymbol{B}) - \frac{\boldsymbol{B}}{\mu_0} \cdot (\nabla \times \boldsymbol{E}) = \frac{\boldsymbol{B}}{\mu_0} \cdot \frac{\partial \boldsymbol{B}}{\partial t} + \boldsymbol{E} \cdot \boldsymbol{J} \tag{G-34}$$

The left-hand side of Eq. (G-34) can be simplified, by inspection, by using the general vector identity in Eq. (G-26) and the right-hand side of Eq. (G-34) can be rewritten slightly, leading to

$$\nabla \cdot \left(\frac{1}{\mu_0} \boldsymbol{E} \times \boldsymbol{B} \right) = \frac{\partial}{\partial t} \left(\frac{1}{2} \frac{1}{\mu_0} \boldsymbol{B} \cdot \boldsymbol{B} \right) + \boldsymbol{E} \cdot \boldsymbol{J} \tag{G-35}$$

Note that in Eqs. (G-34) and (G-26), the electromagnetic vectors $\boldsymbol{B}$ and $\boldsymbol{E}$ correspond to the arbitrary vectors $\boldsymbol{A}$ and $\boldsymbol{C}$, respectively.

Equation (G-35) can now be rewritten into the form of Eq. (G-22) by the following series of steps: (1) recognize that $\boldsymbol{B} \times \boldsymbol{E} = -\boldsymbol{E} \times \boldsymbol{B}$, (2) integrate the entire equation over the volume V, and (3) apply the divergence theorem in Eq. (G-28) to the left-hand side of the resulting expression to arrive at

$$-\int_S \left(\frac{1}{\mu_0} \boldsymbol{E} \times \boldsymbol{B} \right) \cdot d\boldsymbol{S} = \frac{\partial}{\partial t} \int_V \frac{1}{2} \frac{1}{\mu_0} \boldsymbol{B} \cdot \boldsymbol{B} \, dV + \int_V \boldsymbol{E} \cdot \boldsymbol{J} \, dV \tag{G-36}$$

Equation (G-36) is in the form of the energy balance expression in Eq. (G-22). The left-hand side of Eq. (G-36) contains the Poynting vector $\boldsymbol{P}$ and the second term on the right-hand side is the energy dissipation rate. The first term on the right-hand side of Eq. (G-36) can be identified as defining

$$w = \frac{1}{2} \frac{1}{\mu_0} \boldsymbol{B} \cdot \boldsymbol{B} \tag{G-37}$$

which is the electromagnetic energy density in V. Note that here the energy accumulation is accomplished via the $\boldsymbol{B}$ field. Thus, it is concluded that *in an element or system characterized by the magnetoquasistatic approximation (for example, an ideal inductor), the energy is stored in the magnetic field.*

To summarize, in this section we have shown that the electromagnetic energy is stored in the electric field for systems characterized by the electroquasistatic approximation while the electromagnetic energy is stored in the magnetic field for systems characterized by the magnetoquasistatic approximation. Since the electroquasistatic approximation requires high-impedance or capacitive devices, analytical expressions for their energy storage are well approximated by the appropriate electrical energy function W_e or the corresponding electrical coenergy function W_e^*. Similarly, since the magnetoquasistatic approximation requires low-impedance or inductive devices, analytical expressions for their energy storage are well approximated by the appropriate magnetic energy function W_m or the corresponding magnetic coenergy function W_m^*.

G-5   KIRCHHOFF'S "LAWS"

The fundamental concepts of modern circuit theory can be derived as a consequence of quasistatics. The formulation of the equations of motion for an electrical network via circuit theory requires the satisfaction of three steps.

1. Kirchhoff's current "law";
2. Kirchhoff's voltage "law"; and
3. Constitutive relations for the individual elements.

In the previous section, we showed that quasistatics establishes the bases for the characteristic constitutive behavior of capacitive and inductive elements. In this section, we briefly review the derivation of Kirchhoff's current and voltage "laws."

One of the major consequences of quasistatics that is utilized in circuit theory is that the electric fields and the magnetic fields are highly localized in the capacitive and inductive elements, respectively, and negligible elsewhere. Further, in circuit theory, Kirchhoff's current and voltage "laws" represent statements about conditions either at the nodes or between one node and another node throughout the network. By definition via quasistatics, nodes in electric circuits are points at which no fields exist; fields exist only within the capacitive and inductive elements.

G-5.1   Kirchhoff's Current "Law"

Kirchhoff's current "law" can be derived from the quasistatic form of Ampere's law. Consider any node in an electrical network, such as sketched in Figure G-2, where, in accordance with quasistatics, the E and B fields are negligible. Thus, for an arbitrary surface S enclosing the node, but no other circuit elements, Ampere's law reduces to

$$\int_S \boldsymbol{J} \cdot \boldsymbol{S} = 0 \tag{G-38}$$

By definition of current density $\boldsymbol{J}$,

$$i_t = \int_S \boldsymbol{J} \cdot d\boldsymbol{S} \tag{G-39}$$

where i_t is the total or net current passing out through the surface $\boldsymbol{S}$. Thus, in accordance with Eqs. (G-38) and (G-39)

$$i_t = 0 \tag{G-40}$$

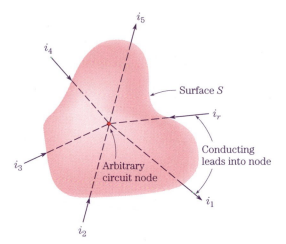

i_5

i_4

Surface S

i_r

Conducting
leads into node

i_3

Arbitrary
circuit node

i_1

FIGURE G-2 Arbitrary circuit node, showing currents passing along leads through enclosing surface S.

i_2

or more conventionally, for a total of r electrical leads passing through the surface

$$\sum_{k=1}^{r} i_k = 0 \tag{G-41}$$

Either of Eqs. (G-40) or (G-41) is commonly called Kirchhoff's current "law."

G-5.2 Kirchhoff's Voltage "Law"

Kirchhoff's voltage "law" can be derived from the quasistatic form of Faraday's law (see
Table G-6). In order to eliminate the $\boldsymbol{B}$ field from the magnetoquasistatics form of Faraday's
law, we recall that the $\boldsymbol{B}$ field is dominant only *within* inductive elements, being negligible
elsewhere, *and* we select a hypothetical path around the network as illustrated in Figure
G-3.

By definition, the voltage at point a with respect to point b is

$$e = \int_{a}^{b} \boldsymbol{E} \cdot d\boldsymbol{\ell} \tag{G-42}$$

where a and b are points along the integration path. Further, recall from quasistatics that
for a path such as that shown in Figure G-3, Faraday's law reduces to

$$\oint_{C} \boldsymbol{E} \cdot d\boldsymbol{\ell} = 0 \tag{G-43}$$

Equations (G-42) and (G-43) require that the sum of the voltage drops around a closed
contour C must be zero, or

$$\sum_{k=1}^{n} e_k = 0 \tag{G-44}$$

where e_k is the voltage drop across the kth element of a total of n elements in the closed
contour or loop. Equation (G-44) is called Kirchhoff's voltage "law."

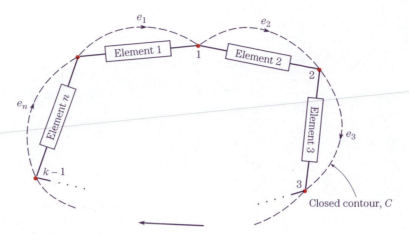

FIGURE G-3   Hypothetical integration path around network, illustrating voltage drops along closed path.

G-5.3   Summary

The fundamental concepts of modern circuit theory, in particular, the so-called Kirchhoff's laws, are a consequence of quasistatics. Thus, since "Kirchhoff's laws" are derivable from Maxwell's equations, they are not *laws* in the fundamental sense but are theorems or corollaries or rules. As such, we shall refer to them as *Kirchhoff's rules*. Furthermore, because of the fact that Kirchhoff's rules are the consequence of the quasistatic approximation plus the assumption of a negligible time rate of change of the ***B*** field (see Faraday's law in Table G-5), Kirchhoff's rules too are approximations.

COMPLEX NUMBERS AND SOME USEFUL FORMULAS OF COMPLEX VARIABLES AND TRIGONOMETRY

In this appendix we review a few useful elementary formulas from the theory of complex variables. Probably the most significant obstacle at this elementary level is the unfortunate terminology, namely, "imaginary" and "complex" numbers, which is largely historically accidental. In the introduction of this appendix, we focus on the issues of perspective and terminology concerning numbers, in general. Then, with no attempt toward a thorough presentation—the reader has likely encountered more than all the complex variables discussed here—we present a few elementary results that are useful in vibration analysis.

H-1 INTRODUCTION

Every number system represents an invention of humankind to solve some useful or potentially useful physical problem. Most often, the physical problem arises first, followed by the number system to address it; when the number system precedes the physical application, the study of the number system is often categorized as *pure mathematics*. The first number system involved the counting numbers 1, 2, 3, ..., now called the *positive integers*. This system was used initially as a system of unitary digits (1 = "1", 2 = "11", 3 = "111", ...) to keep tabs on one's children, mate(s), or possessions. This was an effective system of numbers, serving almost all of the purposes for which it was invented. It was also a closed system with respect to addition and multiplication, meaning that such mathematical operations could be performed without producing other numbers that were outside the system. For example, the addition of any two numbers within the positive system always results in another positive integer. There was no zero in this system; it wasn't needed.

Without attempting to review the history of number systems, we note that early in humankind's history, equations of the form

$$x + 5 = 2 \tag{H-1}$$

were encountered. Such encounters were extraordinarily disturbing, and when first confronted were evaded or rejected as being unreasonable, sometimes with considerable ingenuity. (Solutions to Eq. (H-1) were called "nonnumbers," which later from the Latin became "negative numbers.") Evidence for this disavowal need go no further than the fact that it took *thousands* of years for humankind to fully accept the notion of a negative number as the solution to Eq. (H-1). Thus, the number system that had begun with the positive integers evolved through the invention of some fractions, the invention of zero, and ultimately the inclusion of negative numbers, historically in that order. Further, the encounter with the so-called Pythagorean theorem for the unit square (Figure H-1), required confrontation with the equation

$$x^2 = 2 \tag{H-2}$$

and such numbers as $\sqrt{2}$, resulting in the invention of the system of *irrational numbers*. Given Eq. (H-2), could the system of complex numbers, required in considering equations of the form

$$x^2 + 2 = 0 \tag{H-3}$$

be far behind?

As with all the previous systems of numbers, solutions to expressions of the form of Eq. (H-3) have a fascinating and variegated history. The great Italian algebraists of the sixteenth century considered such solutions as *fictitious* (a name, as far as we can tell, due to Hieronimo Cardano (1501–1576)); the name *imaginary number* appears to have been introduced by René Descartes (1596–1650); Leibniz and many others considered such solutions absurd. It was the steady progress by Euler during the period 1731–1747 that ultimately led to a resolution, though not immediate acceptance, of such solutions. As with the early use of the now-common notation of π for $3.141\ldots$ and the first use of e for $2.718\ldots$, Euler is credited with the introduction of the notation of i for the quantity $\sqrt{-1}$. Nevertheless, it was not until 1801, when the eminent mathematician Carl Friedrich Gauss (1777–1855) adopted the use of i for $\sqrt{-1}$, that i became secure in mathematical notation.

In the course of solving the ordinary differential equation of motion for systems as simple as a mass spring model or a capacitor resistor model, we encounter an algebraic expression of the form of Eq. (H-3), which has no immediately obvious solution among the set of real numbers. In this confrontation we have, at least, two distinct choices: (1) Surrender to the suggestion that such equations are not meaningful despite their very physical origin, or (2) enlarge the set of numbers that we are willing to consider in order to obtain a solution to improve our understanding of the dynamics of some obviously useful physical models. Despite considerable difficulty and numerous historical detours along the lines of the first choice, the history of mathematics inexorably evolved along the path of the second choice.

Concerning our interest in solving ordinary differential equations, the following ideas are at the very heart of the use of complex numbers. The Taylor series expansions of three

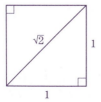

FIGURE H-1   Unit square displaying diagonal of irrational length.

useful simple functions, with which the reader is likely to feel a great deal of comfort, are as follows:

$$\sin z = z - \frac{z^3}{3!} + \frac{z^5}{5!} - \cdots \tag{H-4}$$

$$\cos z = 1 - \frac{z^2}{2!} + \frac{z^4}{4!} - \cdots \tag{H-5}$$

$$e^z = 1 + z + \frac{z^2}{2!} + \frac{z^3}{3!} + \cdots \tag{H-6}$$

each of which converges for all z. Even if we knew nothing about the generally assumed form of the solutions to ordinary differential equations, staring at and thinking about and staring at and thinking about these three series might suggest some possible relationship between them. In particular, suppose there existed a mathematical operator, say γ, that periodically changed sign and that such a change in sign was the consequence of multiplying the operator by itself. (After all, negative numbers, such as $-a$, which took centuries to accept as real numbers, possess just such a sign periodicity: $(-a)^2$ is positive, $(-a)^3$ is negative, etc.) Hence, we could write

$$e^{\gamma z} = 1 + \gamma z + \frac{(\gamma z)^2}{2!} + \frac{(\gamma z)^3}{3!} + \frac{(\gamma z)^4}{4!} + \frac{(\gamma z)^5}{5!} + \cdots \tag{H-7}$$

So, if this operator γ has the single convenient property that

$$\gamma^2 = -1 \tag{H-8}$$

then, as a result of collecting alternate terms, Eq. (H-7) may be rewritten as

$$e^{\gamma z} = \left(1 - \frac{z^2}{2!} + \frac{z^4}{4!} - \cdots\right) + \gamma\left(z - \frac{z^3}{3!} + \frac{z^5}{5!} - \cdots\right) \tag{H-9}$$

Equation (H-9) is precisely what is sought because we recognize the first and second sets of parentheses as $\cos z$ and $\sin z$, respectively. So, Eq. (H-9) may be written as

$$e^{\gamma z} = \cos z + \gamma \sin z \tag{H-10}$$

where γ is defined in Eq. (H-8).

While it cannot be stated that the identical arguments in the preceding paragraph were ever made, the fact that several mathematicians—perhaps most notably John Bernoulli—were somewhat familiar with Eq. (H-10) before Euler clearly articulated it in 1747 suggests that such arguments could have been made. Nevertheless, it appears that of the great mathematicians of the mid-eighteenth century and before, only Euler was sufficiently confident to make the bold step that led to the extension of the number system to include complex numbers.

There has been no physics involved in this discussion, only a simple mathematical idea expressed in Eq. (H-8) that led to Eq. (H-10). The particular value of the idea is that it provides a (potentially) useful relationship between Eqs. (H-4), (H-5), and (H-6). Among other things, we arrive at the issue of what to call the operator. It appears that the name *friend* or *buddy* would be quite appropriate, since it establishes a special relationship between very important mathematical ideas; that is, between sines, cosines, and exponentials. After all, we know that to name something is to own it. However, it was found that others who had encountered Eq. (H-8) earlier, though not as indicated above but perhaps as $\gamma = \sqrt{-1}$, were greatly disturbed by their encounter. After all, the square root of a negative number *is* a disconcerting idea. Thus, γ was called an *imaginary* number. Nevertheless, despite all the concern and discomfort such a name has generated for the past three centuries, the name *imaginary number* has survived. So even though in our desire to communicate with our colleagues we may use the phrase "imaginary number," in our hearts we know that $\gamma^2 = -1$ is simply a convenient mathematical operator, or even a *friend*. Furthermore, having relinquished the name of γ^2 (at least, in public), we might as well relinquish the symbol also and use the established notation, namely,

$$\left(\gamma^2 =\right)i^2 = -1 \tag{H-11}$$

Therefore, use of Eq. (H-11) in Eq.(H-10) gives

$$e^{iz} = \cos z + i \sin z \tag{H-12}$$

which is called *Euler's formula*.

H-2   ELEMENTARY ALGEBRAIC OPERATIONS OF COMPLEX NUMBERS

The four elementary algebraic operations (addition, subtraction, multiplication, and division) of complex numbers are now defined. Consider two arbitrary complex numbers z_1 and z_2, which may be expressed as

$$z_1 = x_1 + iy_1 \tag{H-13}$$

$$z_2 = x_2 + iy_2 \tag{H-14}$$

The complex numbers z_1 and z_2 are equal only when both x_1 equals x_2 *and* y_1 equals y_2. The addition of z_1 and z_2 is defined by

$$\begin{aligned} z_1 + z_2 &= (x_1 + iy_1) + (x_2 + iy_2) \\ &= (x_1 + x_2) + i(y_1 + y_2) \end{aligned} \tag{H-15}$$

The subtraction of z_2 from z_1 is similarly defined by

$$\begin{aligned} z_1 - z_2 &= (x_1 + iy_1) - (x_2 + iy_2) \\ &= (x_1 - x_2) + i(y_1 - y_2) \end{aligned} \tag{H-16}$$

The multiplication of z_1 by z_2 is defined by

$$\begin{aligned} z_1 \cdot z_2 &= (x_1 + iy_1) \cdot (x_2 + iy_2) \\ &= x_1 x_2 + i x_1 y_2 + i y_1 x_2 + i^2 y_1 y_2 \\ &= x_1 x_2 + i(x_1 y_2 + x_2 y_1) - y_1 y_2 \\ &= (x_1 x_2 - y_1 y_2) + i(x_1 y_2 + x_2 y_1) \end{aligned} \tag{H-17}$$

where $i^2 = -1$ and Eqs. (H-15) and (H-16) have been used. And, the division of z_1 by z_2 is defined by

$$\frac{z_1}{z_2} = \frac{x_1 + iy_1}{x_2 + iy_2}$$

$$= \frac{(x_1 + iy_1)(x_2 - iy_2)}{(x_2 + iy_2)(x_2 - iy_2)}$$

$$= \frac{x_1 x_2 - i x_1 y_2 + i y_1 x_2 - i^2 y_1 y_2}{x_2^2 - i x_2 y_2 + i y_2 x_2 - i^2 y_2^2}$$

$$= \frac{(x_1 x_2 + y_1 y_2) + i(x_2 y_1 - x_1 y_2)}{x_2^2 + y_2^2}$$

$$= \frac{x_1 x_2 + y_1 y_2}{x_2^2 + y_2^2} + i \frac{x_2 y_1 - x_1 y_2}{x_2^2 + y_2^2} \tag{H-18}$$

where $i^2 = -1$ and Eqs. (H-15) and (H-16) have been used. Thus, the ratio of two complex numbers can be expressed as a single complex number.

H-3   COMPLEX CONJUGATES

Consider an arbitrary complex number expressed by

$$z = x + iy \tag{H-19}$$

Then it is convenient to define another complex number that is associated with z such that

$$z^* = x - iy \tag{H-20}$$

where z^* is called the *complex conjugate* of z. That is, the complex conjugate of a complex number has the same real part as the original complex number, and the same imaginary part as the original complex number but with the opposite sign. Therefore, a complex number and its complex conjugate have the same magnitude, given by $\sqrt{x^2 + y^2}$. It is a widely used convention to denote the complex conjugate of a complex number by a superscript asterisk on the original complex number, as may be seen in Eqs. (H-19) and (H-20).

One of the interesting and useful consequences of defining the complex conjugate is that the product of a complex number and its complex conjugate gives the square of their identical magnitude. That is,

$$z \cdot z^* = (x + iy) \cdot (x - iy)$$

$$= (x^2 + y^2) + i(-xy + xy)$$

$$= x^2 + y^2$$

$$= |z|^2 \tag{H-21}$$

where Eq. (H-17) has been used. Also, note that in expressing the division of two complex numbers as a single complex number, as in Eq. (H-18), the numerator and the denominator of the original ratio should be multiplied by the complex conjugate of the denominator.

H-4   A USEFUL FORMULA OF COMPLEX VARIABLES

Using Euler's formula, the following formula can be derived:

$$x + iy = Ze^{i\phi} \tag{H-22}$$

where x and y are arbitrary real numbers and where the *magnitude* (or modulus) of the complex number $(x + iy)$ is defined by

$$Z = \sqrt{x^2 + y^2} \tag{H-23}$$

and an angle is defined by

$$\phi = \tan^{-1}\left(\frac{y}{x}\right) \tag{H-24}$$

In order to derive Eq. (H-22), we rewrite the left-hand side of Eq. (H-22), by multiplying and dividing by $\sqrt{x^2 + y^2}$, as

$$x + iy = \sqrt{x^2 + y^2}\left\{\frac{x}{\sqrt{x^2 + y^2}} + \frac{iy}{\sqrt{x^2 + y^2}}\right\} \tag{H-25}$$

Now, simply as a matter of convenience, we define an angle ϕ such that

$$\frac{x}{\sqrt{x^2 + y^2}} = \cos\phi \tag{H-26}$$

$$\frac{y}{\sqrt{x^2 + y^2}} = \sin\phi \tag{H-27}$$

which yields

$$\tan\phi = \frac{y}{x} \tag{H-28}$$

where we note that the angle ϕ defined by Eq. (H-28) is equal to the angle ϕ in Eq. (H-24). Equations (H-26), (H-27), and (H-28) are easily visualized with the use of the polar coordinate representation sketched in Figure H-2. Equation (H-28) has been obtained by dividing Eq. (H-27) by Eq. (H-26), yielding a single equation for ϕ. However, as we shall see shortly, Eq. (H-28) does not completely represent the combination of Eqs. (H-26) and (H-27). Indeed, if Eq. (H-28) (or, equivalently, Eq. (H-24)) is to be used for evaluating ϕ, we need to carefully define the range of the angle ϕ. Otherwise, there exists some ambiguity in

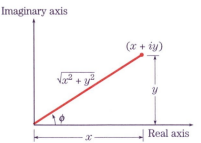

Imaginary axis

FIGURE H-2   Polar coordinate representation of complex number $x + iy$.

evaluating ϕ from Eq. (H-28) only. This will be discussed in Example H-1 after the derivation of Eq. (H-22) is completed.

Substituting Eqs. (H-26) and (H-27) into Eq. (H-25) gives

$$x + iy = \sqrt{x^2 + y^2}\,(\cos\phi + i\sin\phi) \qquad \text{(H-29)}$$

which, when combined with Euler's formula in Eq. (H-12), yields

$$x + iy = \sqrt{x^2 + y^2}\,e^{i\phi} = Ze^{i\phi} \qquad \text{(H-30)}$$

where Z and ϕ are defined by Eqs. (H-23) and (H-24), respectively. Therefore, Eq. (H-22) has been derived using Euler's formula.

As discussed above, the single equation Eq. (H-24) (or, equivalently Eq. (H-28)) is an alternative form of the combination of Eqs. (H-26) and (H-27). In the following example, we shall explore the uniqueness of the solution for such an inverse tangent equation, as opposed to an "isolated" inverse tangent equation, "isolated" meaning the inverse tangent equation stands alone and does not represent such a combination of the cosine and sine equations as in the first case.

■ **Example H-1:**   Uniqueness of Solution to Simple Trigonometric Equations

Given arbitrary values of x and y, determine whether the solution for ϕ is unique in the following two cases:

1.
$$\cos\phi = \frac{x}{\sqrt{x^2 + y^2}} \qquad \text{(a)}$$

and

$$\sin\phi = \frac{y}{\sqrt{x^2 + y^2}} \qquad \text{(b)}$$

which can be reduced to

$$\tan\phi = \frac{y}{x} \qquad \text{(c)}$$

2.
$$\tan\phi = \frac{y}{x} \qquad \text{(d)}$$

□ **Solution:**   Before we proceed, note that all trigonometric functions are periodic with a periodicity of 2π for both the cosine and sine functions and π for the tangent function. Thus, inverse trigonometric functions are multivalued; that is, for example, there is an infinite number of values for the angle θ that satisfy the relation $\theta = \sin^{-1} a$ for a given a, $-1 \le a \le 1$. Therefore, in order to obtain a finite number of solutions, it is necessary to designate a range for the angle. A commonly considered range for the angle is the interval between $-\pi$ and π.

Note here that part (1) shows a case where a tangent equation represents an alternative form of the combination of a sine equation and a cosine equation as in Eqs. (H-26) through (H-28); part (2) shows an "isolated" case, as defined above.

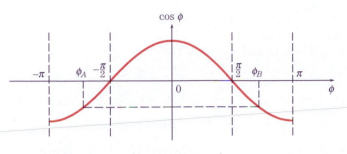

(a) $\cos\phi$, $-\pi \leq \phi \leq \pi$.

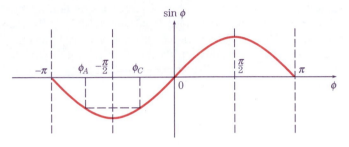

(b) $\sin\phi$, $-\pi \leq \phi \leq \pi$.

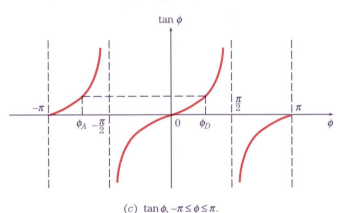

(c) $\tan\phi$, $-\pi \leq \phi \leq \pi$.

FIGURE H-3 Trigonometric functions in the range between $-\pi$ and π, illustrating their inverse functions being multivalued.

☐ **Part (1):**

As will be shown, in this part the solution for ϕ is determined uniquely within the range of $-\pi \leq \phi \leq \pi$. First, in the range of $-\pi \leq \phi \leq \pi$, there are two solutions for Eq. (a) for all values of x and y. For example, consider the case where x and y are both negative; then, $\cos\phi$ is negative and the two solutions of Eq. (a) are ϕ_A and ϕ_B, as indicated in Figure H-3a. Second, in the range of $-\pi \leq \phi \leq \pi$, there are also two solutions for Eq. (b) for all values of x and y. For the case where x and y are both negative, $\sin\phi$ is negative and the two solutions of Eq. (b) are indicated as ϕ_A and ϕ_C in Figure H-3b. Therefore, the simultaneous solution of both Eqs. (a) and (b) is uniquely determined as ϕ_A.

Now, as an alternative requirement on ϕ to the simultaneous satisfaction of Eqs. (a) and (b), consider Eq. (c). There are two solutions of Eq. (c) within the range of $-\pi \leq \phi \leq \pi$. For example, when x and y are both negative (thus, $\frac{y}{x} > 0$), the two solutions of Eq. (d) are indicated as ϕ_A and ϕ_D in Figure H-3c. However, recall that the simultaneous solution to both Eqs. (a) and (b) for negative x and y was uniquely found to be ϕ_A. Therefore, in order for Eq. (c) to be equivalent to the combination of Eqs. (a) and (b), a scheme must be devised to select the correct solution of ϕ_A or ϕ_D that also satisfies both Eqs. (a) and (b).

Such a scheme can be devised by designating the appropriate range of the angle ϕ in Eq. (c) in accordance with the various combinations of the signs of x and y. For example, in the case of Figure H-3, Eq. (c) along with a specified range of $-\pi < \phi < -\dfrac{\pi}{2}$ is equivalent to the combination of Eqs. (a) and (b). Such appropriate ranges of angle ϕ are summarized in Table H-1 for all possible combinations of signs and zero values of x and y.

Part (2):   As indicated in part (1), within the range of $-\pi \le \phi \le \pi$, there are always two solutions that satisfy Eq. (d) when x and y are arbitrary. For example, in the case of Figure H-3, where x and y are both negative, we found two solutions of Eq. (d), that is, ϕ_A and ϕ_D as indicated in Figure H-3c. Due to the period of π for the tangent function, two such solutions will always be π apart. Since Eq. (d) is the only equation that must be satisfied by ϕ, both ϕ_A and ϕ_D are acceptable solutions. Therefore, when Eq. (d) is an "isolated" equation, its solution is not unique, as there are always two solutions within the range of $-\pi \le \phi \le \pi$.

In part (1) of the preceding example, we have shown that it is necessary to designate an appropriate range for ϕ in Eq. (H-28) and that, in general, this appropriate range of ϕ may be determined unambiguously via Table H-1. Note that x and y in Eqs. (H-22) through (H-28) and in Table H-1 were subject to no physical restrictions; the considerations were purely mathematical. However, in certain cases where x and y are physical quantities (and, thus may be subject to restrictions in terms of their signs), the use of Table H-1 may not be required since the appropriate range of ϕ might be determinable via physical limitations. The following example contains two such cases, the physical meanings of which will be made clear in Examples H-3 and H-4.

TABLE H-1   Appropriate Range of Angle ϕ Defined by Eq. (d) as Alternative to Combination of Eqs. (a) and (b)

Sign of x	Sign of y	Range of ϕ
$x < 0$	$y = 0$	$\phi = -\pi$ or π
$x < 0$	$y < 0$	$-\pi < \phi < -\dfrac{\pi}{2}$
$x = 0$	$y < 0$	$\phi = -\dfrac{\pi}{2}$
$x > 0$	$y < 0$	$-\dfrac{\pi}{2} < \phi < 0$
$x > 0$	$y = 0$	$\phi = 0$
$x > 0$	$y > 0$	$0 < \phi < \dfrac{\pi}{2}$
$x = 0$	$y > 0$	$\phi = \dfrac{\pi}{2}$
$x < 0$	$y > 0$	$\dfrac{\pi}{2} < \phi < \pi$

■ **Example H-2:** Examples of Use of Eqs. (H-22) through (H-24) without Use of Table H-1

1. Use Eq. (H-22) to rewrite the complex function

$$X = \frac{f_0(1 - i\tau\Omega)}{1 + \tau^2\Omega^2} \tag{a}$$

as

$$X = \frac{f_0}{\sqrt{1 + \tau^2\Omega^2}}e^{-i\phi} \tag{b}$$

where the angle ϕ is given by

$$\phi = \tan^{-1}\tau\Omega \qquad 0 \le \phi < \frac{\pi}{2} \tag{c}$$

Assume that $f_0 \ne 0$, $\tau > 0$, and $\Omega \ge 0$.

2. Use Eq. (H-22) to rewrite the complex function

$$X_1 = \frac{f_0\{(k - m\Omega^2) - ic\Omega\}}{(k - m\Omega^2)^2 + c^2\Omega^2} \tag{d}$$

as

$$X_1 = \frac{f_0}{\sqrt{(k - m\Omega^2)^2 + c^2\Omega^2}}e^{-i\phi} \tag{e}$$

where the angle ϕ is given by

$$\phi = \tan^{-1}\left(\frac{c\Omega}{k - m\Omega^2}\right) \qquad 0 \le \phi < \pi \tag{f}$$

Assume that $f_0 \ne 0$; $m, k > 0$; and $c, \Omega \ge 0$.

❏ **Solution:**

❏ **Part (1):** Using Eqs. (H-22) through (H-24), the expression in the parentheses of Eq. (a) can be rewritten as

$$1 - i\tau\Omega = \sqrt{1 + \tau^2\Omega^2}e^{i\phi'} \tag{g}$$

where

$$\phi' = \tan^{-1}(-\tau\Omega) \\ = -\tan^{-1}\tau\Omega \tag{h}$$

where the identity $\tan^{-1}(-A) = -\tan^{-1}A$ for arbitrary A, due to the odd character of the tangent function, has been used.

Before the range of the angle ϕ' in Eq. (h) is considered, in order to reduce Eqs. (g) and (h) into the forms in Eqs. (b) and (c), define another angle ϕ such that

$$\phi \equiv -\phi' \tag{i}$$

Substitution of Eq. (i) into Eqs. (g) and (h) gives

$$1 - i\tau\Omega = \sqrt{1 + \tau^2\Omega^2}e^{-i\phi} \tag{j}$$

$$\phi = \tan^{-1}\tau\Omega \tag{k}$$

Now, we consider the range of the angle ϕ defined in Eq. (k). To this end, Table H-1 may be used; however, it is not required because the range may follow naturally from the restrictions imposed on τ and Ω, namely, $\tau > 0$ and $\Omega \geq 0$. That is, when $\Omega = 0$, $\phi = \tan^{-1} 0 = 0$; and as Ω increases from 0 toward ∞, ϕ increases from 0 toward $\frac{\pi}{2}$. Therefore, the range of ϕ is naturally found to be $0 \leq \phi < \frac{\pi}{2}$.

Therefore, substituting Eq. (j) into Eq. (a) yields Eq. (b) where ϕ is defined by Eq. (k) and its range is $0 \leq \phi < \frac{\pi}{2}$, as indicated in Eq. (c).

☐ **Part (2):**   Using Eqs. (H-22) through (H-24), the expression in the flower brackets of Eq. (d) can be rewritten as

$$(k - m\Omega^2) - ic\Omega = \sqrt{(k - m\Omega^2)^2 + c^2\Omega^2}\, e^{i\phi'} \tag{l}$$

where

$$\phi' = \tan^{-1}\left(\frac{-c\Omega}{k - m\Omega^2}\right)$$

$$= -\tan^{-1}\left(\frac{c\Omega}{k - m\Omega^2}\right) \tag{m}$$

where the identity $\tan^{-1}(-A) = -\tan^{-1} A$ for arbitrary A, due to the odd character of the tangent function, has been used.

Before the range of the angle ϕ' in Eq. (m) is considered, in order to reduce Eqs. (l) and (m) into the forms in Eqs. (e) and (f), define another angle ϕ such that

$$\phi \equiv -\phi' \tag{n}$$

Substitution of Eq. (n) into Eqs. (l) and (m) gives

$$(k - m\Omega^2) - ic\Omega = \sqrt{(k - m\Omega^2)^2 + c^2\Omega^2}\, e^{-i\phi} \tag{o}$$

$$\phi = \tan^{-1}\left(\frac{c\Omega}{k - m\Omega^2}\right) \tag{p}$$

Now, we consider the range of the angle ϕ defined in Eq. (p). To this end, Table H-1 may be used; however, it is not required because the range may follow naturally from the restrictions imposed on m, k, c, and Ω, namely, $m, k > 0$ and $c, \Omega \geq 0$. That is, when $\Omega = 0$, $\phi = \tan^{-1} 0 = 0$; as Ω increases up to $\Omega = \sqrt{\frac{k}{m}}$, ϕ increases up to $\frac{\pi}{2}$; and as Ω increases further and beyond $\sqrt{\frac{k}{m}}$, ϕ increases further up toward π. Therefore, the range of ϕ is naturally found to be $0 \leq \phi < \pi$.

Therefore, substituting Eq. (o) into Eq. (d) yields Eq. (e), where ϕ is defined by Eq. (p) and its range is $0 \leq \phi < \pi$, as indicated in Eq. (f).

H-5   USE OF COMPLEX VARIABLES IN HARMONIC RESPONSE ANALYSES

When the response of linear dynamic systems to harmonic excitations is sought, complex variable expressions can be used to find the solutions for both sinusoidal and cosinusoidal excitations simultaneously and efficiently.

First, note that the governing equations of all linear dynamic systems can be expressed in terms of a *linear* differential operator as

$$L[\nu] = f(t) \tag{H-31}$$

where $f(t)$ is the forcing function, $\nu(t)$ is the response variable, which is a function of time, and the operator $L[\ ]$ is a linear differential operator such that

$$L[\nu_1 + \nu_2] = L[\nu_1] + L[\nu_2] \tag{H-32}$$

where ν_1 and ν_2 are arbitrary functions of time, and such that

$$L[a\nu] = aL[\nu] \tag{H-33}$$

where a is an arbitrary constant, real, or complex. For example, the linear differential operator for many single-degree-of-freedom first-order systems is

$$L = \left(\tau\frac{d}{dt} + 1\right) \tag{H-34}$$

and the linear differential operator for many single-degree-of-freedom second-order systems is

$$L = \left(m\frac{d^2}{dt^2} + c\frac{d}{dt} + k\right) \tag{H-35}$$

Furthermore, the equations of motion for many two-degree-of-freedom systems can also be written in the form of Eq. (H-31); in which case there are four differential operators L_{pq}'s, $p, q = 1, 2$. That is, the equations of motion for many two-degree-of-freedom systems can be written as

$$L_{11}[x_1] + L_{12}[x_2] = f_1(t) \tag{H-36}$$

$$L_{21}[x_1] + L_{22}[x_2] = f_2(t) \tag{H-37}$$

where $x_1(t)$ and $x_2(t)$ are the generalized coordinates that are functions of time and where

$$L_{11} = \left(m_1\frac{d^2}{dt^2} + c_{11}\frac{d}{dt} + k_{11}\right) \tag{H-38}$$

$$L_{12} = \left(c_{12}\frac{d}{dt} + k_{12}\right) \tag{H-39}$$

$$L_{21} = \left(c_{21}\frac{d}{dt} + k_{21}\right) \tag{H-40}$$

$$L_{22} = \left(m_2\frac{d^2}{dt^2} + c_{22}\frac{d}{dt} + k_{22}\right) \tag{H-41}$$

For simplicity, the focus of this section will be on single-degree-of-freedom systems. Nevertheless, the same technique can be applied to any linear dynamic system of higher order and higher degrees of freedom.

For the system response to harmonic excitations, let $f(t)$ in Eq. (H-31) equal $Ae^{i\Omega t}$, giving

$$L[\nu] = Ae^{i\Omega t} \tag{H-42}$$

where A is a real constant, representing the constant amplitude of the harmonic forcing function having angular frequency Ω. In accordance with the theory of ordinary differential equations, the complete solution to Eq. (H-42) is given by the sum of the particular solution and the homogeneous solution. First, consider the particular solution. Generally, the particular solution $\nu_p(t)$ found via the theory of ordinary differential equations will be a complex function. If the complex particular solution $\nu_p(t)$ is written in terms of its real and imaginary parts, it becomes

$$\nu_p(t) = \nu_{pr}(t) + i\nu_{pi}(t) \tag{H-43}$$

where $\nu_{pr}(t)$ and $\nu_{pi}(t)$ are both real functions of time and are respectively called the real and imaginary parts of the complex function $\nu_p(t)$.

Substituting Eq. (H-43) into the left-hand side of Eq. (H-42) and expanding the right-hand side of Eq. (H-42) via Euler's formula in Eq. (H-12) yield

$$L[\nu_{pr} + i\nu_{pi}] = A\cos\Omega t + iA\sin\Omega t \tag{H-44}$$

Furthermore, the left-hand side of Eq. (H-44) can be rewritten via Eqs. (H-32) and (H-33), leading to

$$L[\nu_{pr}] + iL[\nu_{pi}] = A\cos\Omega t + iA\sin\Omega t \tag{H-45}$$

Since the real parts on each side of Eq. (H-45) must be equal and since the imaginary parts on each side must also be equal, Eq. (H-45) yields

$$L[\nu_{pr}] = A\cos\Omega t \tag{H-46}$$

$$L[\nu_{pi}] = A\sin\Omega t \tag{H-47}$$

Therefore, the real and imaginary parts of the complex particular solution to Eq. (H-42) have been shown to provide the particular responses to the cosinusoidal forcing function $f(t) = A\cos\Omega t$ and the sinusoidal forcing function $f(t) = A\sin\Omega t$, respectively. That is, expressing the harmonic forcing function in terms of a complex function provides a mechanism for finding the particular harmonic responses to both sinusoidal and cosinusoidal forcing functions simultaneously, and thus somewhat efficiently.

Via the principle of superposition, the particular solution for an excitation consisting of both cosinusoidal and sinusoidal functions can be found by superposing the real and imaginary parts of the complex particular solution. For example, consider the case where the (real) harmonic excitation for $f(t)$ on the right-hand side of Eq. (H-31) is specified as

$$f(t) = f_1\cos\Omega t + f_2\sin\Omega t \tag{H-48}$$

where the constant coefficients f_1 and f_2 are generally unequal. Then, in view of Eqs. (H-46) and (H-47) and via the principle of superposition, the (real) particular solution for the harmonic excitation specified by Eq. (H-48) may be found as the sum of the real part of the complex particular solution to Eq. (H-42) evaluated when $A = f_1$ and the imaginary part of the complex particular solution to Eq. (H-42) evaluated when $A = f_2$. That is, the particular solution for the excitation in Eq. (H-48) may be given as $\nu_{pr}(t)\big|_{A=f_1} + \nu_{pi}(t)\big|_{A=f_2}$, where Eqs. (H-46) and (H-47) are respectively used. Thus, note that the particular solution is, as it must be, a real function.

Now, consider the homogeneous solution $\nu_h(t)$, which is a real function that satisfies Eq. (H-31) when $f(t) = 0$ and which we assume has been found; that is,

$$L[\nu_h] = 0 \tag{H-49}$$

Since the particular solution above has been found as a real function by taking the real and/or imaginary parts of the complex particular solution and the homogeneous solution has also been evaluated as a real function, the sum of these two real functions gives a real complete solution. Then, and only then, the unknown coefficients of the homogeneous solution can be evaluated by imposing the (real) physical initial conditions onto the complete solution. Thus, the real complete solution can be found.

Now, we consider a special case where we can utilize the *complex* complete solution obtained as the sum of the complex particular solution and the homogeneous solution. Suppose one wants to obtain the complete solutions for two *totally separate* problems but where an identical system is subject to a cosinusoidal excitation and a sinusoidal excitation, respectively, each of the *same amplitude* and each subject to the *same initial conditions*. In this special case, we can obtain the complete solutions for these two problems simultaneously by establishing a complex complete solution as the sum of the complex particular solution and real homogeneous solution and evaluating the coefficients of the homogeneous solution in accordance with the initial conditions. That is, in this approach, the complete solution is established without taking the real and/or imaginary parts of the complex particular solution, but using the complex particular solution itself.

The cautionary comment regarding this approach is that in the process of determining the constants of the homogeneous solution, one must use *complex initial conditions*, which are obtained by multiplying the real physical initial conditions by the complex quantity $(1 + i)$. This is because the two problems of cosinusoidal and sinusoidal excitations are being solved simultaneously and the real and imaginary parts of the thus-obtained complex complete solution must both satisfy the physical initial conditions. That is, for example, in the case of single-degree-of-freedom first-order systems, if the actual initial condition on the real complete solution is given as $v(0) = v_0$, then we must use a modified complex initial condition for the complex complete solution as $v(0) = v_0(1 + i) = v_0 + iv_0$.

Once the coefficients of the complex complete solution are evaluated, the real and imaginary parts of this complex solution give the complete solutions for the two totally separate problems having cosinusoidal and sinusoidal excitations, respectively, and the same initial conditions. The cases of single-degree-of-freedom first-order and second-order systems are considered in the following examples. The use of complex initial conditions in these examples constitutes a somewhat unconventional application of the complex variables approach, and thus represents an aside.

■ **Example H-3:**   Complete Solutions of Single-Degree-of-Freedom First-Order Systems Subject to Cosinusoidal and Sinusoidal Excitations

Find the two complete solutions for the single-degree-of-freedom first order systems, governed by

$$\tau \frac{dv}{dt} + v(t) = f(t) \tag{a}$$

when $f(t)$ in the two separate cases is given by

1.
$$f(t) = \begin{cases} 0 & t < 0 \\ f_0 \cos \Omega t & t \geq 0 \end{cases} \tag{b}$$

2.
$$f(t) = \begin{cases} 0 & t < 0 \\ f_0 \sin \Omega t & t \geq 0 \end{cases} \tag{c}$$

Furthermore, in each case, the initial condition is given by

$$v(0) = v_0 \tag{d}$$

◻ **Solution:**   The particular solutions for the two problems can be found simultaneously by using complex functions. That is, we replace the problem in Eqs. (a) through (c) by

$$\tau \frac{dv}{dt} + v(t) = f(t) \tag{e}$$

where

$$f(t) = \begin{cases} 0 & t < 0 \\ f_0 e^{i\Omega t} & t \geq 0 \end{cases} \tag{f}$$

Furthermore, as we shall see below, Eq. (d) will also be replaced.

Then, in accordance with the theory of ordinary differential equations, the complex particular solution to Eqs. (e) and (f) can be assumed to be

$$v_p(t) = X e^{i\Omega t} \tag{g}$$

where, by substitution of Eq. (g) into Eq. (e), the complex amplitude X can be found to be

$$X = \frac{f_0(1 - i\tau\Omega)}{1 + \tau^2\Omega^2} \tag{h}$$

Recall from part (1) of Example H-2 that the complex amplitude X in Eq. (h) can be rewritten as

$$X = \frac{f_0}{\sqrt{1 + \tau^2\Omega^2}} e^{-i\phi} \tag{i}$$

where the angle ϕ is given by

$$\phi = \tan^{-1}\tau\Omega \qquad 0 \leq \phi < \frac{\pi}{2} \tag{j}$$

Thus, substituting Eq. (i) into Eq. (g) yields the complex particular solution as

$$v_p(t) = \frac{f_0}{\sqrt{1 + \tau^2\Omega^2}} e^{i(\Omega t - \phi)} \tag{k}$$

Also, in accordance with the theory of ordinary differential equations, the general homogeneous solution to Eq. (e) with $f(t) = 0$ may be found as

$$v_h(t) = C e^{-\frac{t}{\tau}} \tag{l}$$

where C is an unknown coefficient.

Therefore, the complex complete solution can be found by summing Eqs. (k) and (l), which yields

$$v(t) = C e^{-\frac{t}{\tau}} + \frac{f_0}{\sqrt{1 + \tau^2\Omega^2}} e^{i(\Omega t - \phi)} \tag{m}$$

The only unknown coefficient C in Eq. (m) can be found by imposing the initial condition onto Eq. (m). In doing so, we need to replace the initial condition in Eq. (d) by

$$\nu(0) = \nu_0(1 + i) \tag{n}$$

because we are essentially solving the two problems, each having cosinusoidal or sinusoidal excitation, simultaneously, and thus we must impose the same initial condition onto each case. Therefore, setting $t = 0$ in Eq. (m) and equating the resulting expression with $\nu(0)$ in accordance with Eq. (n) give

$$\nu_0(1 + i) = C + \frac{f_0}{\sqrt{1 + \tau^2\Omega^2}}e^{-i\phi} \tag{o}$$

Because of the definition for ϕ in Eq. (j), the exponential function in Eq. (o) can be rewritten as

$$e^{-i\phi} = \cos\phi - i\sin\phi$$
$$= \frac{1}{\sqrt{1 + \tau^2\Omega^2}} - i\frac{\tau\Omega}{\sqrt{1 + \tau^2\Omega^2}} \tag{p}$$

where in the first equality of Eq. (p) we have used Euler's formula from Eq. (H-12); and in the second equality of Eq. (p), in accordance with Eq. (j), we have imagined a right triangle having an orthogonal opposite side of $\tau\Omega$ and adjacent side of 1, ϕ being the included angle. (This is analogous to Figure H-2.) Replacing the exponential function on the right-hand side of Eq. (o) by Eq. (p) leads to

$$C = \nu_0(1 + i) - \frac{f_0}{\sqrt{1 + \tau^2\Omega^2}}e^{-i\phi}$$
$$= \nu_0(1 + i) - \frac{f_0}{\sqrt{1 + \tau^2\Omega^2}}\frac{1 - i\tau\Omega}{\sqrt{1 + \tau^2\Omega^2}}$$
$$= \nu_0(1 + i) + \frac{-f_0 + i\tau\Omega f_0}{1 + \tau^2\Omega^2} \tag{q}$$

Therefore, substitution of Eq. (q) into Eq. (m) and use of Euler's formula give the complex complete solution as

$$\nu(t) = \left\{\nu_0(1 + i) + \frac{-f_0 + i\tau\Omega f_0}{1 + \tau^2\Omega^2}\right\}e^{-\frac{t}{\tau}} + \frac{f_0}{\sqrt{1 + \tau^2\Omega^2}}e^{i(\Omega t - \phi)}$$
$$= \left\{\left(\nu_0 - \frac{f_0}{1 + \tau^2\Omega^2}\right)e^{-\frac{t}{\tau}} + \frac{f_0}{\sqrt{1 + \tau^2\Omega^2}}\cos(\Omega t - \phi)\right\}$$
$$+ i\left\{\left(\nu_0 + \frac{\tau\Omega f_0}{1 + \tau^2\Omega^2}\right)e^{-\frac{t}{\tau}} + \frac{f_0}{\sqrt{1 + \tau^2\Omega^2}}\sin(\Omega t - \phi)\right\} \tag{r}$$

where ϕ is given by Eq. (j).

Since the real and imaginary parts of the complex complete solution in Eq. (r) both satisfy the initial condition in Eq. (d), they give the complete solutions for the system in part (1) and the system in part (2), respectively. That is, the complete solution for part (1) is

$$\nu(t) = \left\{\nu_0 - \frac{f_0}{1 + \tau^2\Omega^2}\right\}e^{-\frac{t}{\tau}} + \frac{f_0}{\sqrt{1 + \tau^2\Omega^2}}\cos(\Omega t - \phi) \tag{s}$$

And, the complete solution for part (2) is

$$v(t) = \left\{ v_0 + \frac{\tau\Omega f_0}{1 + \tau^2\Omega^2} \right\} e^{-\frac{t}{\tau}} + \frac{f_0}{\sqrt{1 + \tau^2\Omega^2}} \sin(\Omega t - \phi) \tag{t}$$

■ **Example H-4:** Complete Solutions of Single-Degree-of-Freedom Second-Order Systems Subject to Cosinusoidal and Sinusoidal Excitations

Find the two complete solutions for the single-degree-of-freedom second-order systems, governed by

$$m\frac{d^2x}{dt^2} + c\frac{dx}{dt} + kx(t) = f(t) \tag{a}$$

when $f(t)$ in the two separate cases is given by

1.
$$f(t) = \begin{cases} 0 & t < 0 \\ f_0 \cos \Omega t & t \geq 0 \end{cases} \tag{b}$$

2.
$$f(t) = \begin{cases} 0 & t < 0 \\ f_0 \sin \Omega t & t \geq 0 \end{cases} \tag{c}$$

Furthermore, in each case, the initial conditions are given by

$$x(0) = x_0 \qquad \dot{x}(0) = v_0 \tag{d}$$

□ **Solution:** The particular solutions for the two problems can be found simultaneously by using complex functions. That is, we replace the problem in Eqs. (a) through (c) by

$$m\frac{d^2x}{dt^2} + c\frac{dx}{dt} + kx(t) = f(t) \tag{e}$$

where

$$f(t) = \begin{cases} 0 & t < 0 \\ f_0 e^{i\Omega t} & t \geq 0 \end{cases} \tag{f}$$

Furthermore, as we shall see below, Eq. (d) will also be replaced.

Then, in accordance with the theory of ordinary differential equations, the complex particular solution to Eqs. (e) and (f) can be assumed to be

$$x_p(t) = X_1 e^{i\Omega t} \tag{g}$$

where, by substitution of Eq. (g) into Eq. (e), the complex amplitude X_1 can be found to be

$$X_1 = \frac{f_0\{(k - m\Omega^2) - ic\Omega\}}{(k - m\Omega^2)^2 + c^2\Omega^2} \tag{h}$$

Recall from part (2) of Example H-2 that the complex amplitude X_1 in Eq. (h) can be rewritten as

$$X_1 = \frac{f_0}{\sqrt{(k - m\Omega^2)^2 + c^2\Omega^2}} e^{-i\phi} \tag{i}$$

where the angle ϕ is given by

$$\phi = \tan^{-1}\left(\frac{c\Omega}{k - m\Omega^2}\right) \qquad 0 \le \phi < \pi \tag{j}$$

Thus, substituting Eqs. (i) and (j) into Eq. (g) yields the the complex particular solution as

$$x_p(t) = \frac{f_0}{\sqrt{(k - m\Omega^2)^2 + c^2\Omega^2}} e^{i(\Omega t - \phi)} \tag{k}$$

Also, in accordance with the theory of ordinary differential equations, the (real) general homogeneous solution $\nu_h(t)$ for the system governed by Eq. (e) may be found. When the eigenvalues of the system are not equal (which is the case when $c \ne c_c = 2\sqrt{mk}$),

$$x_h(t) = C_1 e^{\lambda_1 t} + C_2 e^{\lambda_2 t} \qquad \lambda_1 \ne \lambda_2 \tag{l}$$

where C_1 and C_2 are unknown coefficients and λ_1 and λ_2 are the eigenvalues of the system. When the eigenvalues of the system are equal (which is the case when $c = c_c = 2\sqrt{mk}$, often called the critically damped case),

$$x_h(t) = C_3 e^{\lambda t} + C_4 t e^{\lambda t} \qquad \lambda_1 = \lambda_2 \tag{m}$$

Therefore, when the system is not critically damped, the complex complete solution can be found by summing Eqs. (k) and (l); that is,

$$x(t) = C_1 e^{\lambda_1 t} + C_2 e^{\lambda_2 t} + X_1 e^{i\Omega t} \qquad \lambda_1 \ne \lambda_2 \tag{n}$$

where X_1 is given by Eq. (i). And, when the system is critically damped, the complex complete solution can be found by summing Eqs. (k) and (m); that is,

$$x(t) = C_3 e^{\lambda t} + C_4 t e^{\lambda t} + X_1 e^{i\Omega t} \qquad \lambda_1 = \lambda_2 \tag{o}$$

Now that the complex complete solutions have been found, we are prepared to find the unknown coefficients C_1 and C_2 in Eq. (n) or C_3 and C_4 in Eq. (o), by imposing the initial conditions. The initial conditions in Eqs. (d) need to be replaced by

$$x(0) = x_0(1 + i); \qquad \dot{x}(0) = v_0(1 + i) \tag{p}$$

because we are essentially solving the two problems, each having cosinusoidal or sinusoidal excitation, simultaneously, and thus we must impose the same initial conditions onto each of the two cases embedded within Eq. (n) and onto each of the two cases embedded within Eq. (o).

First, when the system is not critically damped, combining Eqs. (p) with Eq. (n) for $x(0)$ and with the time derivative of Eq. (n) for $\dot{x}(0)$, respectively, gives

$$x_0 + ix_0 = C_1 + C_2 + X_1 \tag{q}$$

$$v_0 + iv_0 = C_1\lambda_1 + C_2\lambda_2 + i\Omega X_1 \tag{r}$$

And, solving for C_1 and C_2 gives

$$C_1 = \frac{v_0(1 + i) - \lambda_2 x_0(1 + i) - (i\Omega - \lambda_2)X_1}{\lambda_1 - \lambda_2} \tag{s}$$

$$C_2 = \frac{-v_0(1 + i) + \lambda_1 x_0(1 + i) + (i\Omega - \lambda_1)X_1}{\lambda_1 - \lambda_2} \tag{t}$$

Second, in the critically damped case, combining Eqs. (p) with Eq. (o) for $x(0)$ and with the time derivative of Eq. (o) for $\dot{x}(0)$, respectively, gives

$$x_0 + ix_0 = C_3 + X_1 \tag{u}$$

$$v_0 + iv_0 = C_3\lambda + C_4 + i\Omega X_1 \tag{v}$$

And, solving for C_3 and C_4 gives

$$C_3 = x_0(1 + i) - X_1 \tag{w}$$

$$C_4 = v_0(1 + i) - \lambda x_0(1 + i) + (\lambda - i\Omega)X_1 \tag{x}$$

Therefore, the complex complete solution is given by Eqs. (n), (s), and (t) when $c \neq c_c$, or by Eqs. (o), (w), and (x) when $c = c_c$.

We shall not write out the complex complete solutions here. Nevertheless, for each system, since the real and imaginary parts of the complex complete solution both satisfy the initial conditions in Eq. (d), these real and imaginary solutions would give the complete solutions for the problem in part (1) and the problem in part (2), respectively. These complete solutions will be evaluated in detail in the next example for the undamped case.

■ **Example H-5:** Complete Response of Single-Degree-of-Freedom Second-Order Undamped Systems Subject to Cosinusoidal and Sinusoidal Excitations

Find the *complete* responses for the problems involving two loading functions in Example H-4 when the system is undamped.

❑ **Solution:** In the undamped case (that is, $c = 0$), Eq. (j) in Example H-4 yields

$$\phi = 0 \tag{a}$$

and, therefore, Eq. (i) in Example H-4 reduces to

$$X_1 = \frac{f_0}{k - m\Omega^2} \tag{b}$$

Also, in the undamped case, the eigenvalues of the system may be found as $\lambda_{1,2} = \pm i\omega_n$, ω_n being the natural frequency of the system. Substituting these eigenvalues into Eq. (n) in Example H-4 gives the complete solution, with unknown coefficients, as

$$x(t) = C_1 e^{i\omega_n t} + C_2 e^{-i\omega_n t} + X_1 e^{i\Omega t}$$

$$= (C_1 + C_2)\cos \omega_n t + (C_1 - C_2)i \sin \omega_n t + X_1 e^{i\Omega t} \tag{c}$$

where Euler's formula has been used.

The coefficients C_1 and C_2 can be found by substituting the eigenvalues $\lambda_{1,2} = \pm i\omega_n$ into Eqs. (s) and (t) in Example H-4:

$$C_1 = \frac{v_0(1 + i) + i\omega_n x_0(1 + i) - (i\Omega + i\omega_n)X_1}{2i\omega_n} \tag{d}$$

$$C_2 = \frac{-v_0(1 + i) + i\omega_n x_0(1 + i) + (i\Omega - i\omega_n)X_1}{2i\omega_n} \tag{e}$$

Or, from Eqs. (d) and (e), we find

$$C_1 + C_2 = x_0(1 + i) - X_1 \tag{f}$$

$$i(C_1 - C_2) = \frac{v_0}{\omega_n}(1 + i) - i\frac{\Omega}{\omega_n}X_1 \tag{g}$$

where X_1 is given by Eq. (b). Substituting Eqs. (f) and (g) into Eq. (c) and using Euler's formula yield

$$x(t) = (x_0 + ix_0 - X_1)\cos \omega_n t + \left(\frac{v_0}{\omega_n} + i\frac{v_0}{\omega_n} - i\frac{\Omega}{\omega_n}X_1\right)\sin \omega_n t$$

$$+ X_1(\cos \Omega t + i\sin \Omega t)$$

$$= \left\{ (x_0 - X_1)\cos \omega_n t + \frac{v_0}{\omega_n}\sin \omega_n t + X_1 \cos \Omega t \right\}$$

$$+ i\left\{ x_0 \cos \omega_n t + \left(\frac{v_0}{\omega_n} - \frac{\Omega}{\omega_n}X_1\right)\sin \omega_n t + X_1 \sin \Omega t \right\} \tag{h}$$

which is the complex complete solution.

Since the real and imaginary parts of Eq. (h) both satisfy the initial conditions in Eq. (d) of Example H-4, they give the complete solutions for part (1) and part (2) in Example H-4, when the system is undamped. Thus, the complete solution for part (1) in Example H-4 with $c = 0$ is given by the real part of Eq. (h); that is,

$$x(t) = \left(x_0 - \frac{f_0}{k - m\Omega^2}\right)\cos \omega_n t + \frac{v_0}{\omega_n}\sin \omega_n t + \frac{f_0}{k - m\Omega^2}\cos \Omega t \tag{i}$$

where Eq. (b) has been used to write out X_1. Similarly, the complete solution for part (2) in Example H-4 with $c = 0$ is given by the imaginary part of Eq. (h); that is,

$$x(t) = x_0 \cos \omega_n t + \left(\frac{v_0}{\omega_n} - \frac{\Omega}{\omega_n}\frac{f_0}{k - m\Omega^2}\right)\sin \omega_n t + \frac{f_0}{k - m\Omega^2}\sin \Omega t \tag{j}$$

Therefore, the complex forcing function in Eq. (f) in Example H-4 combined with the complex initial conditions in Eqs. (p) in Example H-4 provided a mechanism for the simultaneous evaluation of the complete responses of the system to both sinusoidal and cosinusoidal forcing functions, having the identical initial conditions.

In calculating C_1 and C_2, the reader may have wondered which of the eigenvalues $+i\omega_n$ or $-i\omega_n$ should have been interpreted as λ_1 and λ_2 in Eqs. (s) and (t) in Example H-4. Does it even matter? The answer is no. The reader is encouraged to show why not.

H-6   USEFUL FORMULAS OF TRIGONOMETRY

The following expression in various forms is frequently encountered in vibration analyses:

$$X = A\sin \theta + B\cos \theta \tag{H-50}$$

where A, B, and θ are variable or constant. The following alternative expressions are often preferred to Eq. (H-50):

$$X = C\sin(\theta + \phi_1) \tag{H-51}$$

or

$$X = C\cos(\theta - \phi_2) \tag{H-52}$$

where

$$C = \sqrt{A^2 + B^2} \tag{H-53}$$

$$\phi_1 = \tan^{-1}\left(\frac{B}{A}\right) \tag{H-54}$$

$$\phi_2 = \tan^{-1}\left(\frac{A}{B}\right) \tag{H-55}$$

We shall derive Eqs. (H-51) and (H-52), with Eq. (H-50) as our starting point.
First, we rewrite Eq. (H-50), by definition, as

$$X = \sqrt{A^2 + B^2}\left\{\frac{A}{\sqrt{A^2 + B^2}}\sin\theta + \frac{B}{\sqrt{A^2 + B^2}}\cos\theta\right\}$$

$$= C\left\{\frac{A}{C}\sin\theta + \frac{B}{C}\cos\theta\right\} \tag{H-56}$$

where Eq. (H-53) has been used to simplify the expression. Now, simply as a matter of
convenience, we define an angle ϕ_1 such that

$$\cos\phi_1 = \frac{A}{C} \tag{H-57}$$

$$\sin\phi_1 = \frac{B}{C} \tag{H-58}$$

where we note that ϕ_1 in Eqs. (H-57) and (H-58) also satisfies Eq. (H-54). Refer to Figure H-4. Nevertheless, we shall see shortly that even though the ϕ_1 that satisfies both Eqs.
(H-57) and (H-58) also satisfies Eq. (H-54), the ϕ_1 that satisfies Eq. (H-54) does not in general satisfy both Eqs. (H-57) and (H-58). By using Eqs. (H-57) and (H-58), Eq. (H-56) can
be rewritten as

$$X = C(\cos\phi_1\sin\theta + \sin\phi_1\cos\theta) \tag{H-59}$$

By recalling the trigonometric identity

$$\sin(a + b) = \cos b\sin a + \sin b\cos a \tag{H-60}$$

Eq. (H-60) may be used to write Eq. (H-59) as

$$X = C\sin(\theta + \phi_1) \tag{H-61}$$

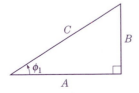

FIGURE H-4   Arbitrarily defined angle ϕ_1 for illustrating relationships
between parameters $A, B,$ and C.

which is identical to Eq. (H-51), where C and ϕ_1 are defined by Eqs. (H-53) and (H-54), respectively.

Now, in order to derive Eq. (H-52), we define another angle ϕ_2 such that

$$\sin \phi_2 = \frac{A}{C} \tag{H-62}$$

$$\cos \phi_2 = \frac{B}{C} \tag{H-63}$$

where we note that ϕ_2 in Eqs. (H-62) and (H-63) also satisfies Eq. (H-55). Refer to Figure H-5. Nevertheless, we shall see shortly that even though the ϕ_2 that satisfies both Eqs. (H-62) and (H-63) also satisfies Eq. (H-55), the ϕ_2 that satisfies Eq. (H-55) does not in general satisfy both Eqs. (H-62) and (H-63). By using Eqs. (H-62) and (H-63), Eq. (H-56) can be rewritten as

$$X = C(\sin \phi_2 \sin \theta + \cos \phi_2 \cos \theta) \tag{H-64}$$

By recalling the trigonometric identity

$$\cos(a - b) = \sin a \sin b + \cos a \cos b \tag{H-65}$$

Eq. (H-65) can be used to write Eq. (H-64) as

$$X = C \cos(\theta - \phi_2) \tag{H-66}$$

which is identical to Eq. (H-52), where C and ϕ_2 are defined by Eqs. (H-53) and (H-55), respectively.

A cautionary comment is made here regarding the use of Eqs. (H-54) and (H-55). The discussion in part (1) of Example H-1 in Section H-4 applies here due to the analogy of Eqs. (H-54), (H-57), and (H-58) with Eqs. (c), (a), and (b) in Example H-1, respectively, and due to the analogy of Eqs. (H-55), (H-62), and (H-63) with Eqs. (c), (b), and (a) in Example H-1, respectively. That is, Eq. (H-54) is not completely equivalent to the combination of Eqs. (H-57) and (H-58) unless an appropriate range for the angle ϕ_1 is defined. As in Table H-1, the appropriate ranges for the angle ϕ_1 in Eq. (H-54) are summarized in Table H-2, depending on the signs of A and B. Similarly, Eq. (H-55) is not completely equivalent to the combination of Eqs. (H-62) and (H-63) unless an appropriate range for the angle ϕ_2 is similarly defined, as shown in Table H-3. Even though Tables H-2 and H-3 must be used in general for the evaluation of the angle from Eqs. (H-54) and (H-55), in many instances the appropriate ranges can be derived straightforwardly from physical considerations, and so the use of Tables H-2 and H-3 is not required. The reader is referred to Example H-2.

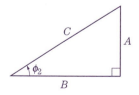

FIGURE H-5   Arbitrarily defined angle ϕ_2 for illustrating relationships between parameters $A, B,$ and C.

TABLE H-2 Appropriate Range of Angle ϕ_1 Defined by Eq. (H-54) as Alternative to Combination of Eqs. (H-57) and (H-58)

Sign of A	Sign of B	Range of ϕ_1
$A < 0$	$B = 0$	$\phi_1 = -\pi$ or π
$A < 0$	$B < 0$	$-\pi < \phi_1 < -\dfrac{\pi}{2}$
$A = 0$	$B < 0$	$\phi_1 = -\dfrac{\pi}{2}$
$A > 0$	$B < 0$	$-\dfrac{\pi}{2} < \phi_1 < 0$
$A > 0$	$B = 0$	$\phi_1 = 0$
$A > 0$	$B > 0$	$0 < \phi_1 < \dfrac{\pi}{2}$
$A = 0$	$B > 0$	$\phi_1 = \dfrac{\pi}{2}$
$A < 0$	$B > 0$	$\dfrac{\pi}{2} < \phi_1 < \pi$

TABLE H-3 Appropriate Range of Angle ϕ_2 Defined by Eq. (H-55) as Alternative to Combination of Eqs. (H-62) and (H-63)

Sign of A	Sign of B	Range of ϕ_2
$B < 0$	$A = 0$	$\phi_2 = -\pi$ or π
$B < 0$	$A < 0$	$-\pi < \phi_2 < -\dfrac{\pi}{2}$
$B = 0$	$A < 0$	$\phi_2 = -\dfrac{\pi}{2}$
$B > 0$	$A < 0$	$-\dfrac{\pi}{2} < \phi_2 < 0$
$B > 0$	$A = 0$	$\phi_2 = 0$
$B > 0$	$A > 0$	$0 < \phi_2 < \dfrac{\pi}{2}$
$B = 0$	$A > 0$	$\phi_2 = \dfrac{\pi}{2}$
$B < 0$	$A > 0$	$\dfrac{\pi}{2} < \phi_2 < \pi$

In summary, the expressions in Eqs. (H-50), (H-51), and (H-52) are all equivalent, if their defined parameters satisfy Eqs. (H-53), (H-54), and (H-55), where the angles ϕ_1 and ϕ_2 in Eqs. (H-54) and (H-55), respectively, are in the appropriate ranges either designated by Tables H-2 and H-3, respectively, or derived from physical considerations.

TEMPORAL FUNCTION FOR SYNCHRONOUS MOTION OF TWO-DEGREE-OF-FREEDOM SYSTEMS

In this appendix, we find the common temporal function shared by the two generalized coordinates of two-degree-of-freedom conservative systems executing the same temporal motion. In many vibration analyses, a particular motion in which all the generalized coordinates of the system execute the same temporal motion is of special interest. This type of motion is often called *synchronous motion* of the system.

I-1 FREE UNDAMPED EQUATIONS OF MOTION

The free undamped equations of motion for two-degree-of-freedom conservative systems are given by

$$m_1 \ddot{x}_1 + k_{11} x_1 + k_{12} x_2 = 0 \qquad (I\text{-}1)$$

$$m_2 \ddot{x}_1 + k_{21} x_1 + k_{22} x_2 = 0 \qquad (I\text{-}2)$$

where $x_1(t)$ and $x_2(t)$ are the generalized coordinates of the system, and the m's and k's are the given system parameters; and $k_{21} = k_{12}$.

I-2 SYNCHRONOUS MOTION

Since synchronous motion is of interest, the desired solutions of Eqs. (I-1) and (I-3) are assumed to take the form

$$x_1(t) = u_1 g(t) \qquad (I\text{-}3)$$

$$x_2(t) = u_2 g(t) \qquad (I\text{-}4)$$

where u_1 and u_2 are real constants, representing the generalized coordinate amplitudes during synchronous motion, and $g(t)$ is the common temporal function. The goal is to determine the function $g(t)$.

Substituting the assumed solutions, Eqs. (I-3) and (I-4), into the equations of motion, Eqs. (I-1) and (I-2), yields

$$m_1 u_1 \ddot{g}(t) + (k_{11}u_1 + k_{12}u_2)g(t) = 0 \tag{I-5}$$

$$m_2 u_2 \ddot{g}(t) + (k_{12}u_1 + k_{22}u_2)g(t) = 0 \tag{I-6}$$

In order for Eqs.(I-5) and (I-6) to have a solution,

$$-\frac{\ddot{g}(t)}{g(t)} = \frac{k_{11}u_1 + k_{12}u_2}{m_1 u_1} = \frac{k_{12}u_1 + k_{22}u_2}{m_2 u_2} = \lambda \tag{I-7}$$

where λ is a real constant because m_1, m_2, k_{11}, k_{12}, k_{22}, u_1, and u_2 are all real and constant. After straightforwardly rearranging Eqs. (I-5) and (I-6) to obtain the far left-hand term and the two middle terms in Eq. (I-7), the argument for obtaining a solution goes as follows: The far left-hand term in Eq. (I-7) is a function of time t only, and the two middle terms consist of all real constants. Thus, the two middle terms are constant, are equal to each other, and may be set equal to a simple constant, λ. As t changes, λ does not change; so the far left-hand term of Eq. (I-7) does not change as it is equal to λ, which is sometimes called the *separation constant*. Thus, from Eq. (I-7), synchronous motion is described by the solution of

$$\ddot{g}(t) + \lambda g(t) = 0 \tag{I-8}$$

and

$$(k_{11} - \lambda m_1)u_1 + k_{12}u_2 = 0 \tag{I-9}$$

$$k_{12}u_1 + (k_{22} - \lambda m_2)u_2 = 0 \tag{I-10}$$

That is, the common temporal function $g(t)$ must satisfy the ordinary differential equation given by Eqs. (I-8) and the amplitudes u_1 and u_2 must satisfy the expressions given by Eqs. (I-9) and (I-10). This is a form of *separation of variables*. Equations, (I-9) and (I-10) govern the spatial behavior of synchronous motion, and are considered elsewhere in this textbook; Eqs. (I-8) governs the temporal behavior of synchronous motion, and is considered in this appendix.

I-3   GENERAL TEMPORAL SOLUTION

According to the theory of ordinary differential equations, the solution of Eq. (I-8) is *assumed* to have the exponential form

$$g(t) = Ae^{st} \tag{I-11}$$

Substitution of Eq. (I-11) into Eq. (I-8) and cancellation of the common e^{st} from the resulting equation give

$$s^2 + \lambda = 0 \tag{I-12}$$

which has two roots

$$\left.\begin{array}{c} s_1 \\ s_2 \end{array}\right\} = \pm\sqrt{-\lambda} \tag{I-13}$$

Since both s_1 and s_2 are the roots of Eq. (I-12), the general solution to Eq. (I-8) can be found by substituting Eq. (I-13) into Eq. (I-11) and combining the resulting two terms. That is, the general solution of Eq. (I-8) is

$$g(t) = A_1 e^{\sqrt{-\lambda} t} + A_2 e^{-\sqrt{-\lambda} t} \tag{I-14}$$

where A_1 and A_2 are unknown coefficients.

Note that if λ is negative, the first term on the right-hand side of Eq. (I-14) goes to infinity as time t goes to infinity. This fact is not consistent with a conservative system described by Eqs. (I-1) and (I-2). Thus, λ must be nonnegative. The case when λ equals zero is somewhat special and will be considered in the next section. The case when λ is positive will now be considered.

Letting $\lambda = \omega^2$, where ω is real and positive, Eq. (I-13) gives

$$\left. \begin{array}{r} s_1 \\ s_2 \end{array} \right\} = \pm i\omega \tag{I-15}$$

Therefore, Eq. (I-14) becomes

$$g(t) = A_1 e^{i\omega t} + A_2 e^{-i\omega t} \tag{I-16}$$

Recalling Euler's formula, $e^{\pm i\omega t} = \cos \omega t \pm i \sin \omega t$, Eq. (I-16) becomes

$$g(t) = (A_1 + A_2) \cos \omega t + i(A_1 - A_2) \sin \omega t \tag{I-17}$$

The displacements in Eqs. (I-3) and (I-4) must be real. Further, because the constants u_1 and u_2 in Eqs. (I-3) and (I-4) are real, $g(t)$ in Eqs. (I-3) and (I-4) must also be real. Thus, Eq. (I-17) may be rewritten as

$$g(t) = C_1 \cos \omega t + C_2 \sin \omega t \tag{I-18}$$

where C_1 and C_2 are real unknown constants.

Rather than simply proceeding from Eq. (I-18), it may be useful to explore the transition from Eq. (I-17) to Eq. (I-18); after all, the bothersome i was somehow eliminated. Essentially, the transition can be easily understood by observing that the function $g(t)$ in Eq. (I-17) will be real if, and only if, A_1 and A_2 are complex conjugates.

For example, by comparing Eqs. (I-17) and (I-18), we note that

$$C_1 = A_1 + A_2 \qquad C_2 = i(A_1 - A_2) \tag{I-19}$$

By solving the expressions in Eq. (I-19) for A_1 and A_2, we find

$$A_1 = \frac{1}{2}(C_1 - iC_2) \qquad A_2 = \frac{1}{2}(C_1 + iC_2) \tag{I-20}$$

Since C_1 and C_2 are real constants as imposed by Eq. (I-18), A_1 and A_2 are complex conjugates as hypothesized, and thus the coefficients of $\cos \omega t$ and $\sin \omega t$ in Eq. (I-17) are indeed real.

The complex conjugates A_1 and A_2 are never evaluated because there is no reason to do so; they are simply *transitional mathematical* results. The real constants C_1 and C_2 become a part of the solution and will be evaluated, directly or indirectly, based upon the subsequent imposition of initial conditions on the system model represented by Eqs. (I-1) and (I-2). So, the use of complex numbers simply provided a shorthand round trip from real variables

(Eq. (I-8)) through complex variables (Eq. (I-15) through Eq. (I-17)) back to real variables (Eq.(I-18)), exclusively for mathematical convenience.

I-4   SPECIAL (SEMIDEFINITE) TEMPORAL SOLUTION

The special case when $\lambda = 0$ is now considered. When $\lambda = 0$, Eq. (I-8) becomes

$$\ddot{g}(t) = 0 \tag{I-21}$$

which, by direct integration, has the general solution

$$g(t) = D_1 + D_2 t \tag{I-22}$$

where D_1 and D_2 are unknown coefficients. The temporal function $g(t)$ given by Eq. (I-22) goes to infinity as time goes to infinity. This temporal behavior, not acceptable in most cases, *is* acceptable when $\lambda = 0$, which occurs in a special class of systems without inertial restraints and which are capable of moving indefinitely. This special kind of system is called a *semidefinite* system.

I-5   GENERALIZATION TO SYSTEMS HAVING MORE DEGREES OF FREEDOM

Even though the temporal function $g(t)$ has been derived only for two-degree-of-freedom systems, the same procedure can be executed to find the temporal function for higher-degree-of-freedom systems. Essentially, only the number of generalized coordinates will increase, resulting in more terms in Eqs. (I-1) and (I-2) and more equations of motion such as Eqs. (I-1) and (I-2). Assuming the solutions for all generalized coordinates have the same form as those in Eq. (I-3) and Eq. (I-4), the same equation as Eq. (I-7) will be obtained, but with more expressions in the middle terms. Therefore, the governing equation for $g(t)$ for higher-degree-of-freedom systems will be identical to Eq. (I-8) and thus, the solution $g(t)$ for systems having more than two degrees of freedom will be the same as found here for two-degree-of-freedom systems.

STABILITY ANALYSES
OF NONLINEAR SYSTEMS

In this appendix two methods for stability analysis of nonlinear systems are presented. The first section provides a formal, yet brief, presentation of a method called the *state-space stability analysis method* for finding the equilibrium configurations of a nonlinear system and investigating the stability of each equilibrium configuration. Even though the emphasis in the first section is the stability of a *linearized system,* the local stability of the original nonlinear system is also addressed. In the second section, a method that can be used for studying the stability of the original nonlinear system is presented; however, this method, which uses the potential energy function of the system, is applicable only to conservative systems.

J-1   STATE-SPACE STABILITY FORMULATION

In this section, we explore briefly a formal presentation of a method for finding the equilibrium configurations of a nonlinear system and investigating the stability of each equilibrium configuration. This method is commonly used in control theory, and it will be presented without derivations, in the context of an inverted pendulum system for illustration. The purpose of this presentation is not only to provide a systematic discussion of stability analysis but also to provide a prelude to the notation commonly used in modern control theory.

J-1.1   State-Space Representation of Equations of Motion

The equations of motion for a linear or nonlinear *autonomous* lumped-parameter system can be written in the form of

$$\dot{x}_1 = f_1(x_1, x_2, x_3, \ldots, x_n)$$
$$\dot{x}_2 = f_2(x_1, x_2, x_3, \ldots, x_n)$$
$$\dot{x}_3 = f_3(x_1, x_2, x_3, \ldots, x_n) \qquad \text{(J-1)}$$
$$\vdots$$
$$\dot{x}_n = f_n(x_1, x_2, x_3, \ldots, x_n)$$

or

$$\dot{\boldsymbol{x}} = \boldsymbol{f}(\boldsymbol{x}) \qquad \text{(J-2)}$$

where the bold-faced letters in Eq. (J-2) and in all subsequent equations are vectors or matrices. Autonomous systems are defined as those whose equations of motion do not possess the time variable t explicitly. For instance, linear time-invariant systems are autonomous whereas linear time-variant systems are nonautonomous. In an autonomous system, neither the forces nor the constraints may vary explicitly with time. On account of our interests in stability as opposed to general system response, we shall be interested in Eq. (J-2) in a form in which all the time-varying force excitation and time-varying motion excitation vanish. Equation (J-2) implies that for each component of the vector $\dot{\boldsymbol{x}}$, $\dot{x}_i = f_i(x_1, x_2, \ldots, x_n)$ for $i = 1, 2, \ldots, n$, where n is the dimension of the vectors $\boldsymbol{x}$ and $\boldsymbol{f}$. The vector $\boldsymbol{x}$ is called the *state vector*, and its components x_i are called the *state variables*; and $\boldsymbol{f}$ is a nonlinear vector function.

The state variables of a lumped-parameter dynamic system are a minimal set of variables such that once their values at a certain time are known, along with the external input to the system, the system behavior for all subsequent time can be completely determined. Equation (J-2) is often called the *state-space representation* of the equations of motion. A tremendous amount of software has been developed for analyzing equations of motion in the form of Eq. (J-2).

Note that just as for generalized coordinates, the choice of the state variables for a given system is *not* unique; however, the number of the state variables (or, equivalently, the dimension of the state vector for a given system) is unique. A convenient set of state variables often consists of the generalized coordinates (say, ξ_j) and the generalized velocities (say, $\dot{\xi}_j$). That is, when the generalized coordinates of a system are ξ_j ($j = 1, 2, \ldots, p$), a convenient state vector for the system is

$$\boldsymbol{x} = \begin{Bmatrix} \xi_1 \\ \xi_2 \\ \vdots \\ \xi_p \\ \dot{\xi}_1 \\ \dot{\xi}_2 \\ \vdots \\ \dot{\xi}_p \end{Bmatrix} \tag{J-3}$$

where p is the number of generalized coordinates. Therefore, the dimension of a state vector of a system having p generalized coordinates is $2p \times 1$. In which case, $2p = n$.

As an example, we consider the inverted pendulum sketched in Figure J-1a consisting of a massless rigid rod of length L and a mass m at the tip of the rod. The mass is attached to one end of a linear spring of spring constant k, the other end of which is connected to a massless follower that in turn can move vertically without friction. Note that the massless follower maintains the spring in a horizontal orientation. Note also that gravity acts. When the generalized coordinate is chosen as the angle θ of the rod measured from the vertical inertial reference as shown, the equation of motion for the system is given by

$$mL^2\ddot{\theta} + kL^2 \sin\theta \cos\theta - mgL \sin\theta = 0 \tag{J-4}$$

which is a nonlinear equation in θ, due to the sine and cosine terms. Note that the nonlinear equation of motion holds for all values of θ between $\pm\dfrac{\pi}{2}$ radians, which is the allowable

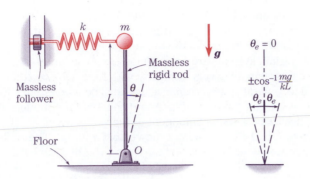

(a) Inverted pendulum.   (b) Equilibrium positions.

FIGURE J-1   Inverted pendulum and equilibrium positions.

range of θ, given the geometric constraint of the floor at the pivot. Also, it is assumed[1] that mg may not equal kL; otherwise, the parameters m, k, and L can have arbitrary values.

Since the generalized coordinate for the pendulum in Figure J-1 has been chosen as the angle θ, a convenient set of state variables for the system can be chosen as $x_1 = \theta$ and $x_2 = \dot{\theta}$. So, corresponding to Eq. (J-3), we may write

$$\boldsymbol{x} = \begin{Bmatrix} \theta \\ \dot{\theta} \end{Bmatrix} = \begin{Bmatrix} x_1 \\ x_2 \end{Bmatrix} \tag{J-5}$$

Then, the equation of motion given by Eq. (J-4) can be rewritten in the state-space representation as

$$\dot{x}_1 = f_1 = \dot{\theta} = x_2 \tag{J-6}$$

$$\begin{aligned} \dot{x}_2 = f_2 &= \ddot{\theta} \\ &= -\frac{k}{m} \sin\theta \cos\theta + \frac{g}{L} \sin\theta \\ &= -\frac{k}{m} \sin x_1 \cos x_1 + \frac{g}{L} \sin x_1 \end{aligned} \tag{J-7}$$

Note that Eqs. (J-6) and (J-7) are in the form of Eq. (J-1) where $\dot{x}_i$ are on the left-hand sides and x_i are on the right-hand sides.

J-1.2   Equilibrium States

The equilibrium configuration(s) of a system, when its equations of motion are written in the form of Eq. (J-2), can be found by simply setting $\dot{\boldsymbol{x}} = \boldsymbol{0}$ in Eq. (J-2). Note that for a mechanical system the requirement of $\dot{\boldsymbol{x}} = \boldsymbol{0}$ implies zero velocity and zero acceleration in view of the state vector of Eq. (J-3). Therefore, the state vector at equilibrium (denoted by $\boldsymbol{x}_e$) can be found by solving the vector equation

$$\boldsymbol{f}(\boldsymbol{x}_e) = \boldsymbol{0} \tag{J-8}$$

---

[1]The case of $mg = kL$ will be considered at the end of Section J-2.

which follows directly from Eq. (J-2) and the requirement $\dot{\boldsymbol{x}} = \boldsymbol{0}$. The state vector $\boldsymbol{x}_e$ satisfying Eq. (J-8) represents a state called the *equilibrium state*. For ease of interpretation, Eq. (J-8) can be written out in terms of the components of the vector function $\boldsymbol{f}$ and the vector $\boldsymbol{x}_e$ as

$$f_1(x_{1e}, x_{2e}, x_{3e}, \ldots, x_{ne}) = 0$$
$$f_2(x_{1e}, x_{2e}, x_{3e}, \ldots, x_{ne}) = 0$$
$$f_3(x_{1e}, x_{2e}, x_{3e}, \ldots, x_{ne}) = 0 \tag{J-9}$$
$$\vdots$$
$$f_n(x_{1e}, x_{2e}, x_{3e}, \ldots, x_{ne}) = 0$$

where x_{ie} ($i = 1, 2, 3, \ldots, n$) is the equilibrium value of the state variable x_i.

For the inverted pendulum system in Figure J-1, the equilibrium states can be found by letting $f_1 = 0$ and $f_2 = 0$ in Eqs. (J-6) and (J-7). That is,

$$\dot{x}_1 = f_1 = x_{2e} = 0 \tag{J-10}$$
$$\dot{x}_2 = f_2 = -\frac{k}{m}\sin x_{1e}\cos x_{1e} + \frac{g}{L}\sin x_{1e}$$
$$= \sin x_{1e}\left(\frac{g}{L} - \frac{k}{m}\cos x_{1e}\right) = 0 \tag{J-11}$$

Therefore, the equilibrium values x_{1e} for the state variable x_1 can be found by solving Eq. (J-11) as

$$x_{1e} = 0; \qquad x_{1e} = \cos^{-1}(mg/kL); \qquad x_{1e} = -\cos^{-1}\left(\frac{mg}{kL}\right) \tag{J-12}$$

Note that Eq. (J-10) requires that $x_{2e} = 0$, which is by definition equivalent to $\dot{\theta}_e = 0$ (that is, zero velocity), which must hold for all equilibrium states. Therefore, combining each of the values of x_{1e} in Eq. (J-12) with $x_{2e} = 0$ gives the equilibrium states. For example, the first equilibrium state is specified by $x_{1e} = 0$ via the first of Eq. (J-12) and $x_{2e} = 0$ via Eq. (J-10), yielding an equilibrium state vector

$$\boldsymbol{x}_e = \begin{Bmatrix} 0 \\ 0 \end{Bmatrix} \tag{J-13}$$

Similarly, the other two equilibrium state vectors $\boldsymbol{x}_e$ are given by

$$\boldsymbol{x}_e = \begin{Bmatrix} \cos^{-1}\left(\dfrac{mg}{kL}\right) \\ 0 \end{Bmatrix} \tag{J-14}$$

and

$$\boldsymbol{x}_e = \begin{Bmatrix} -\cos^{-1}\left(\dfrac{mg}{kL}\right) \\ 0 \end{Bmatrix} \tag{J-15}$$

Thus, there are three equilibrium states and these are shown in Figure J-1*b*. We note that the two equilibrium states in Eqs. (J-14) and (J-15) exist only when $mg < kL$ because the cosine function cannot exceed unity, and the possibility of $mg = kL$ was excluded in the description of the example.

J-1.3   Linearization about Equilibrium States

As a first step for examining the stability of a nonlinear system about its equilibrium states, the state-space equations in the form of Eq. (J-2) should be linearized in the vicinity of the equilibrium states.

The linearization about an equilibrium state can be performed by considering perturbations—small changes—$\epsilon_i(t)$ $(i = 1, 2, \ldots, n)$ of the state variables from their equilibrium values x_{ie} $(i = 1, 2, \ldots, n)$. That is, by letting

$$x_1(t) = x_{1e} + \epsilon_1(t)$$
$$x_2(t) = x_{2e} + \epsilon_2(t)$$
$$x_3(t) = x_{3e} + \epsilon_3(t) \tag{J-16}$$
$$\vdots$$
$$x_n(t) = x_{ne} + \epsilon_n(t)$$

where the ϵ_i's are sufficiently small that $\epsilon_i^2 \ll \epsilon_i$ where $i = 1, 2, \ldots, n$. Equations (J-16) can be written in vector form as

$$\boldsymbol{x}(t) = \boldsymbol{x}_e + \boldsymbol{\epsilon}(t) \tag{J-17}$$

where $\boldsymbol{\epsilon}(t)$ is an n-dimensional vector whose components are $\epsilon_i(t)$ $(i = 1, 2, \ldots, n)$.

Substituting Eq. (J-17) into Eq. (J-2) yields

$$\dot{\boldsymbol{x}}_e + \dot{\boldsymbol{\epsilon}} = \boldsymbol{f}(\boldsymbol{x}_e + \boldsymbol{\epsilon}) \tag{J-18}$$

The first term on the left-hand side of Eq. (J-18) is zero because of the requirement $\dot{\boldsymbol{x}}_e = \boldsymbol{0}$ for equilibrium. Therefore, Eq. (J-18) becomes

$$\dot{\boldsymbol{\epsilon}} = \boldsymbol{f}(\boldsymbol{x}_e + \boldsymbol{\epsilon}) \tag{J-19}$$

The ith component of the vector relation in Eq. (J-19) can be written as

$$\dot{\epsilon}_i = f_i(x_{1e} + \epsilon_1, x_{2e} + \epsilon_2, \ldots, x_{ne} + \epsilon_n) \qquad i = 1, 2, \ldots, n \tag{J-20}$$

The multivariable function f_i having n variables can be approximated by a Taylor series for small ϵ_i's. That is,

$$\dot{\epsilon}_i = f_i(x_{1e}, x_{2e}, \ldots, x_{ne}) + \left.\frac{\partial f_i}{\partial x_1}\right|_e \epsilon_1 + \left.\frac{\partial f_i}{\partial x_2}\right|_e \epsilon_2 + \cdots + \left.\frac{\partial f_i}{\partial x_n}\right|_e \epsilon_n$$
$$+ \text{ (higher order terms)} \qquad i = 1, 2, \ldots, n \tag{J-21}$$

where $|_e$ designates the evaluation of the quantity at equilibrium. Note that the first term on the right-hand side of Eq. (J-21) is zero because of Eq. (J-8), which represents the condition satisfied by the state variables at equilibrium. Using this cancellation of the $f_i|_e$ term and omitting the higher-order terms in Eq. (J-21) lead to a linear equation

$$\dot{\epsilon}_i = \left.\frac{\partial f_i}{\partial x_1}\right|_e \epsilon_1 + \left.\frac{\partial f_i}{\partial x_2}\right|_e \epsilon_2 + \cdots + \left.\frac{\partial f_i}{\partial x_n}\right|_e \epsilon_n \qquad i = 1, 2, \ldots, n \tag{J-22}$$

The linearized equations for ϵ_i where $i = 1, 2, \ldots, n$ can be assembled into a matrix equation for the vector $\boldsymbol{\epsilon}$ as

$$
\dot{\boldsymbol{\epsilon}} =
\begin{bmatrix}
\dfrac{\partial f_1}{\partial x_1}\bigg|_e & \dfrac{\partial f_1}{\partial x_2}\bigg|_e & \cdots & \dfrac{\partial f_1}{\partial x_n}\bigg|_e \\[2ex]
\dfrac{\partial f_2}{\partial x_1}\bigg|_e & \dfrac{\partial f_2}{\partial x_2}\bigg|_e & \cdots & \dfrac{\partial f_2}{\partial x_n}\bigg|_e \\[2ex]
\vdots & \vdots & \ddots & \vdots \\[2ex]
\dfrac{\partial f_n}{\partial x_1}\bigg|_e & \dfrac{\partial f_n}{\partial x_2}\bigg|_e & \cdots & \dfrac{\partial f_n}{\partial x_n}\bigg|_e
\end{bmatrix}
\begin{Bmatrix}
\epsilon_1(t) \\[1ex]
\epsilon_2(t) \\[1ex]
\vdots \\[1ex]
\epsilon_n(t)
\end{Bmatrix}
$$

$$
\equiv \boldsymbol{A}\boldsymbol{\epsilon}(t) \tag{J-23}
$$

where $\boldsymbol{A}$ is an $n \times n$ matrix whose elements, the A_{ij}'s (the element in the ith row and the jth column) are given by

$$
A_{ij} = \frac{\partial f_i}{\partial x_j}\bigg|_e \qquad i, j = 1, 2, \ldots, n \tag{J-24}
$$

For the inverted pendulum, the elements of $\boldsymbol{A}$ can be found, by substituting Eqs. (J-6) and (J-7) into Eq. (J-24), as

$$
A_{11} = \frac{\partial f_1}{\partial x_1}\bigg|_e = 0 \tag{J-25}
$$

$$
A_{12} = \frac{\partial f_1}{\partial x_2}\bigg|_e = 1 \tag{J-26}
$$

$$
A_{21} = \frac{\partial f_2}{\partial x_1}\bigg|_e = \frac{k}{m}\left(\sin^2 x_{1e} - \cos^2 x_{1e}\right) + \frac{g}{L}\cos x_{1e} \tag{J-27}
$$

$$
A_{22} = \frac{\partial f_2}{\partial x_2}\bigg|_e = 0 \tag{J-28}
$$

which constitute the matrix $\boldsymbol{A}$ for the linearized matrix equation of motion in the form of Eq. (J-23).

J-1.4 Concept and Types of Stability

In order to better understand the general concept of stability, a convenient operator called the *norm* is introduced. The following presentation is not intended to be mathematically rigorous, but is devised to assist in the explanation of several fundamental ideas relating to the concept of stability.

For a general n-dimensional vector, the (Euclidian) norm is conventionally denoted by the symbol $\|\cdots\|$. The norm of an n-dimensional vector $\boldsymbol{y}$ is defined as

$$
\|\boldsymbol{y}\| \equiv \sqrt{y_1^2 + y_2^2 + \cdots + y_n^2} \tag{J-29}
$$

The concept of the norm may be easily understood by considering the case of one, two, or three dimensions. In these cases, vectors can be considered to represent a position in the corresponding spaces; that is, as shown in Figure J-2a, a one-dimensional vector $\boldsymbol{y}$ (which is called a scalar as a degenerate case of a vector) represents a position along an axis that has an origin O. Similarly, as shown in Figures J-2b and c, respectively, two-dimensional and three-dimensional vectors $\boldsymbol{y}$ represent positions in the two and three-dimensional spaces.

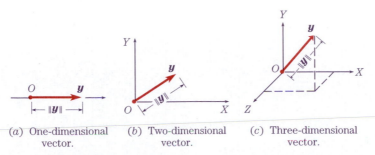

(a) One-dimensional vector.   (b) Two-dimensional vector.   (c) Three-dimensional vector.

FIGURE J-2   Graphical representation of vector and its norm in one, two, and three dimensions.

Higher dimensional vectors could be imagined to represent a position in the corresponding hyperspace. Therefore, it may be noted from the definition in Eq. (J-29) that for one, two, and three-dimensional vectors, the norm is simply the distance from the origin of the corresponding space to a point represented by the vector. Similarly, the norm of a general n-dimensional vector represents the "distance" from the origin of the n-dimensional space to the point represented by the vector.

In general, the stability of both linear and nonlinear systems can be defined in accordance with the time evolution of the norm of the difference between the state vector $x(t)$ and the equilibrium state x_e. That is, stability of a system about an equilibrium state is determined by the time evolution of the norm as given by

$$\|x(t) - x_e\| \equiv \sqrt{(x_1(t) - x_{1e})^2 + (x_2(t) - x_{2e})^2 + \cdots + (x_n(t) - x_{ne})^2} \quad \text{(J-30)}$$

Many different definitions of stability have been devised; the following are among the most commonly used.

1. An equilibrium state x_e is said to be *stable* if, for any positive R, there exists some positive r, such that if $\|x(0) - x_e\| < r$ at $t = 0$, then $\|x(t) - x_e\| < R$ for all subsequent time, $t \geq 0$. If an equilibrium state is not stable, then it is said to be *unstable*.

2. An equilibrium state x_e is said to be *asymptotically stable,* if it is stable *and* if there exists some positive r such that $\|x(0) - x_e\| < r$ at $t = 0$ implies that $\|x(t) - x_e\|$ goes to zero as time t goes to infinity; that is, $x(t)$ goes to the equilibrium state x_e as time t goes to infinity.

3. An equilibrium state x_e is said to be *marginally stable* if it is stable but not asymptotically stable.

Note that stable equilibrium as elucidated in definition 1 is further categorized, in definitions 2 and 3, into two classes: asymptotical and marginal.

In Figure J-3, these concepts of stability are illustrated in *state space,* which is a hyperspace whose dimension is the same as that of the state vector, and each point in the state space represents a particular value of the state vector. From the discussion of the norm of a vector, recall that the inequality $\|x - x_e\| < R$ (R being a constant) implies geometrically that x is inside of a hypersphere, in the state space, having its center at x_e and a radius R. This geometrical interpretation may be seen more easily, for example, in a two-dimensional case, where $\|x - x_e\| < R$ means that the state vector x is inside a circle, in the two-dimensional

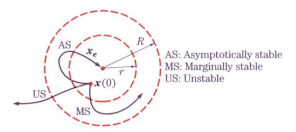

AS: Asymptotically stable
MS: Marginally stable
US: Unstable

FIGURE J-3 Schematic of asymptotical stability, marginal stability, and instability in state space.

state space, having its center at $\boldsymbol{x}_e$ and a radius R. (Although it is not necessary that $\boldsymbol{x}_e$ be located at the center of the state space, this is generally done for convenience.)

In Figure J-3, three different time evolutions of the state vector $\boldsymbol{x}(t)$ are shown in the n-dimensional state space when the initial value of the state vector is $\boldsymbol{x}(0)$ as indicated. The curves labeled "AS" (asymptotically stable), "MS" (marginally stable), and "US" (unstable) in Figure J-3 show the time evolutions of the state vector $\boldsymbol{x}(t)$ when the initial state $\boldsymbol{x}(0)$ is in the neighborhood of an equilibrium state $\boldsymbol{x}_e$ (or graphically within a hypersphere having a small radius r with its center at $\boldsymbol{x}_e$) and when the equilibrium state $\boldsymbol{x}_e$ is asymptotically stable, marginally stable, and unstable, respectively. That is, if the equilibrium state $\boldsymbol{x}_e$ is asymptotically stable, the state vector $\boldsymbol{x}(t)$ ultimately approaches the equilibrium state $\boldsymbol{x}_e$ and remains at $\boldsymbol{x}_e$ as time elapses, as indicated by the curve labeled "AS". If the equilibrium state $\boldsymbol{x}_e$ is marginally stable, the state vector $\boldsymbol{x}(t)$ remains within the hypersphere having a finite radius R for all subsequent times but does not ultimately approach $\boldsymbol{x}_e$, as indicated by the curve labeled "MS". Also, if the equilibrium state $\boldsymbol{x}_e$ is unstable, the state vector $\boldsymbol{x}(t)$ departs from the equilibrium state as time elapses, beyond the finite radius R, as indicated by the curve labeled "US".

J-1.5   Stability of Linearized Systems

The stability of a linearized system may be examined from the linearized equations of motion, which are valid only in the vicinity of the equilibrium state. It should be emphasized here that since this analysis is based on the linearized equations, the stability to be explored here is related *to some extent*[2] to the local stability of the original nonlinear system.

There exists a well-developed systematic technique for the stability analysis of linearized systems, which we now explore. The linearized system is given by Eq. (J-23) where the components of the matrix $\boldsymbol{A}$ may be obtained by Eq. (J-24). The solution to Eq. (J-23) may be *assumed* to be of the form[3]

$$\boldsymbol{\epsilon} = \mathbf{w}e^{\lambda t} \tag{J-31}$$

where $\mathbf{w}$ is an unknown n-dimensional vector and λ is a constant. Substituting Eq. (J-31) into the linearized equation, Eq. (J-23), and canceling the common $e^{\lambda t}$ term give

$$\lambda \mathbf{w} = \boldsymbol{A}\mathbf{w} \tag{J-32}$$

[2] As will be shown later, the stability analysis for linearized systems may *not* provide decisive information about the local stability of the original nonlinear system.

[3] Note that this form of solution is consistent with the solution form assumed in the linear vibration analyses. This solution form is a major result from the theory of ordinary differential equations.

which may be rewritten as

$$[\lambda I - A]\mathbf{w} = 0 \tag{J-33}$$

where I is the $n \times n$ identity matrix. In order for Eq. (J-33) to have nontrivial solutions for $\mathbf{w}$, the determinant of its coefficient matrix must be zero. That is,

$$\det[\lambda I - A] = 0 \tag{J-34}$$

When the matrices I and A are $n \times n$ matrices, Eq. (J-34) leads to an nth order polynomial equation such as

$$a_n \lambda^n + a_{n-1} \lambda^{n-1} + a_{n-2} \lambda^{n-2} + \cdots + a_2 \lambda^2 + a_1 \lambda + a_0 = 0 \tag{J-35}$$

which is known as the *characteristic equation,* which generally has n roots, say λ_k for $k = 1, 2, \ldots, n$.

Using Eq. (J-31), the general solution to Eq. (J-23) can be written as

$$\boldsymbol{\epsilon}(t) = \mathbf{w}_1 e^{\lambda_1 t} + \mathbf{w}_2 e^{\lambda_2 t} + \cdots + \mathbf{w}_n e^{\lambda_n t} \tag{J-36}$$

where the $\mathbf{w}_k$'s ($k = 1, 2, \ldots, n$) are the n-dimensional *eigenvectors* associated with the λ_k's ($k = 1, 2, \ldots, n$), respectively. The roots of Eq. (J-35) are generally complex numbers; therefore, they can be generally written as

$$\lambda_k = p_k + i q_k \qquad k = 1, 2, \ldots, n \tag{J-37}$$

where the p_k and q_k are the real and imaginary parts, respectively, of the complex number λ_k and $i = \sqrt{-1}$. Therefore, using Eq. (J-37), the exponential function of the kth term on the right-hand side of Eq. (J-36) can be written as

$$\begin{aligned} e^{\lambda_k t} &= e^{(p_k + i q_k)t} \\ &= e^{p_k t} e^{i q_k t} \\ &= e^{p_k t}(\cos q_k t + i \sin q_k t) \end{aligned} \tag{J-38}$$

where Euler's formula, $e^{i q_k t} = \cos q_k t + i \sin q_k t$, has been used. Note that the imaginary part in Eq. (J-38) has no meaning in terms of the physical behavior of the system, but it is simply the result of a mathematical treatment, and it must vanish in subsequent analysis. Therefore, taking only the real part of Eq. (J-38) gives

$$e^{\lambda_k t} = e^{p_k t} \cos q_k t \tag{J-39}$$

Since the cosine term in Eq. (J-39) simply oscillates between ± 1, the magnitude of the exponential term $e^{\lambda_k t}$ largely depends on the sign of the exponent p_k on the right-hand side of Eq. (J-39). That is, if p_k is positive, the exponential function $e^{\lambda_k t}$ increases indefinitely as time increases; if p_k is negative, it decreases monotonically and asymptotically approaches zero; and if p_k is zero, it oscillates harmonically.

Since the general solution $\boldsymbol{\epsilon}(t)$ in Eq. (J-36) of the linearized equation is a sum of terms like that in Eq. (J-39), the following criteria may be established for the stability of the linearized system about each equilibrium state $\boldsymbol{x}_e$:

1. If at least one of the real parts of the roots of the characteristic equation is positive, then the equilibrium state of the linearized system is unstable; otherwise, the linearized system is stable.

2. If all the real parts of the roots of the characteristic equation are negative, then the equilibrium state of the linearized system is asymptotically stable.

3. If at least one of the real parts of the roots of the characteristic equation is zero and all the rest of the real parts of the roots are negative, then the equilibrium state of the linearized system is marginally stable.

Here, it should be recalled that the stability depends on the specific equilibrium state because the **A** matrix in the linearized equation, Eq. (J-23), is evaluated at a specific equilibrium, and the characteristic equation depends on this specific **A** matrix.

As an example, the characteristic equation for the inverted pendulum may be found by substituting the **A** matrix whose components are given by Eqs. (J-25) through (J-28) into Eq. (J-34). Then, the characteristic equation becomes

$$
\det\left\{\lambda\begin{bmatrix}1&0\\0&1\end{bmatrix}-\begin{bmatrix}0&1\\\left\{\dfrac{k}{m}(\sin^2 x_{1e}-\cos^2 x_{1e})+\dfrac{g}{L}\cos x_{1e}\right\}&0\end{bmatrix}\right\}
$$

$$
=\det\begin{bmatrix}\lambda&-1\\-\left\{\dfrac{k}{m}(\sin^2 x_{1e}-\cos^2 x_{1e})+\dfrac{g}{L}\cos x_{1e}\right\}&\lambda\end{bmatrix}
$$

$$
=\lambda^2+\left[\dfrac{k}{m}(2\cos^2 x_{1e}-1)-\dfrac{g}{L}\cos x_{1e}\right]
$$

$$
=0 \tag{J-40}
$$

or,

$$
\lambda^2=\dfrac{g}{L}\cos x_{1e}-\dfrac{k}{m}(2\cos^2 x_{1e}-1) \tag{J-41}
$$

The three equilibrium states given by Eqs. (J-13) through (J-15) are considered. Substitution of the first equilibrium state in Eq. (J-13) (that is, $x_{1e}=0$ and $x_{2e}=0$) into Eq. (J-41) yields

$$
\lambda^2=\dfrac{g}{L}-\dfrac{k}{m} \tag{J-42}
$$

When $mg>kL$, one of the two roots of Eq. (J-42) has a positive real value; thus, in this case, the first equilibrium state is unstable. When $mg<kL$, the roots λ's of Eq. (J-42) are purely imaginary having opposite signs; thus, in this case, this equilibrium state is marginally stable because both roots have zero real parts.

Now, the second and third equilibrium states given in Eqs. (J-14) and (J-15), respectively, are considered. Since $\cos x_{1e}=\left(\dfrac{mg}{kL}\right)$ for both of these two equilibrium states, substitution of $\cos x_{1e}=\left(\dfrac{mg}{kL}\right)$ into Eq. (J-41) yields

$$
\lambda^2=\dfrac{g}{L}\dfrac{mg}{kL}-\dfrac{k}{m}\left(2\dfrac{m^2g^2}{k^2L^2}-1\right)
$$

$$
=\dfrac{mg^2}{kL^2}-2\dfrac{mg^2}{kL^2}+\dfrac{k}{m}
$$

$$
=\dfrac{k}{m}\left\{1-\left(\dfrac{mg}{kL}\right)^2\right\} \tag{J-43}
$$

which is valid for both equilibrium states in Eqs. (J-14) and (J-15), and which is always positive because these equilibrium states exist only when $(mg) < (kL)$. Therefore, one of the two roots of Eq. (J-43) is positive real; thus the two equilibrium states in Eqs. (J-14) and (J-15) are both unstable.

J-1.6   Local Stability of Nonlinear Systems

So far, only the stability of linearized systems has been discussed. As indicated earlier, this stability is only somewhat related to the local stability of the original nonlinear system because the linearized equations of motion hold only in the vicinity of the equilibrium states.

A question may arise: Once the linearized system has turned out to be stable about an equilibrium, is the original nonlinear system also *locally*[4] stable about that equilibrium? This question may be rephrased as follows: Is the linearized equation about an equilibrium always fully representative of the local behavior of the original nonlinear system? This set of questions may be answered by *Lyapunov's indirect method** (sometimes called *Lyapunov's*

---

*This technique is named in honor of the Russian mathematician and mechanician Aleksandr M. Lyapunov (1857–1918) for his seminal research in nonlinear systems. The proof of Lyapunov's indirect method is based upon nonlinear stability analysis and will not be given here.

---

linearization method), which provides the following relationships between the stability of the linearized system and the local stability of the original nonlinear system:

1.   If the linearized system is asymptotically stable, then the equilibrium state of the original nonlinear system is locally asymptotically stable.

2.   If the linearized system is unstable, then the equilibrium state of the original nonlinear system is unstable.

3.   If the linearized system is marginally stable, then no conclusion can be made about the stability of the original nonlinear system.

That is, the local stability of the original nonlinear system can be found from the stability analysis of the linearized system, *except* when the linearization analysis reveals that the equilibrium state is marginally stable. The invalidity of the linearization stability analysis in the case of marginal stability may be seen in some special cases.[5] In this special case of

---

[4]The local stability of a nonlinear system concerns the behavior of the system when its state is perturbed only in the vicinity of the equilibrium state. Since a state vector generally consists of both displacement and velocity, the perturbed displacement must be near the equilibrium position *and* the perturbed velocity must be small near the equilibrium position.

[5]For example, the linearized stability analysis of two systems whose equations of motion are $m\ddot{y} \pm c\dot{y}^3 + ky = 0$ will conclude marginal stability for both systems. However, a nonlinear stability analysis of both systems will reveal that $m\ddot{y} + c\dot{y}^3 + ky = 0$ represents an asymptotically stable system whereas $m\ddot{y} - c\dot{y}^3 + ky = 0$ represents an unstable system. This means that via linearization, the important information about stability characteristics may be lost.

marginal stability, a full nonlinear analysis must be performed in order to assess the nonlinear system stability in the vicinity of an equilibrium.

Recall that in the example of the inverted pendulum, the linearized system turned out to be marginally stable about the equilibrium state $x_{1e} = 0$ and $x_{2e} = 0$ in Eq. (J-13) (see the discussion following Eq. (J-42)), and unstable about $x_{1e} = \pm \cos^{-1}\left(\dfrac{mg}{kL}\right)$ and $x_{2e} = 0$ in Eqs. (J-14) and (J-15), respectively (see the discussion following Eq. (J-43)). Therefore, the local stability of the nonlinear system about the equilibrium state in Eq. (J-13) is unknown whereas the original nonlinear system can be concluded to be unstable about the equilibrium states in Eqs. (J-14) and (J-15).

J-1.7 Nonlinear Stability Analyses

As indicated above, when the linearized system turns out to be marginally stable about an equilibrium state, the local stability of the original nonlinear system is not known by the stability analysis of the linearized system. Therefore, in this case, the stability of the nonlinear system must be explored via a full nonlinear stability analysis using the original nonlinear equations of motion.

Such a full nonlinear analysis is also needed in the following case. It is often desired to examine the *global* stability of a nonlinear system, which concerns the stability of the system in the case when the perturbation of the state vector is not restricted to the vicinity of the equilibrium state. In this case, linearization cannot be justified and thus the stability analysis of linearized systems does not provide valid information. Thus, in general, for global stability analysis, the complete nonlinear equations of motion must be examined for the stability of the equilibrium state.

Techniques for the full nonlinear analysis of stability are explored under the titles of *perturbation methods, graphical methods* (such as the *phase-plane method*), and *Lyapunov's direct method.* All of these methods are beyond the scope of this textbook and thus will not be detailed here. However, there is a simple stability analysis method that can be applied to *conservative* systems and it will be presented in the next section.

J-1.8 Summary of State-Space Stability Analysis

The state-space method for finding the equilibrium positions of nonlinear systems and their stability has been presented. Table J-1 summarizes the procedure. First, derive the equations of motion and establish the state-space equations from the equations of motion. Second, from the state-space equations, find the equilibrium states. Third, the nonlinear state-space equations are linearized about each equilibrium state and the characteristic equation is established from the linearized state-space equation. Fourth, solve the characteristic equation, and based on the real parts of the roots, determine the stability of the linearized system about each equilibrium state. Fifth, according to the relationships provided by Lyapunov's linearization method, determine the local stability of the nonlinear system from the stability of the linearized system except when the linearized system is marginally stable; in which case, a nonlinear stability analysis such as Lyapunov's direct method should be used. Furthermore, if the global stability of the nonlinear system is desired, a nonlinear analysis must also be performed. As a demonstration, the procedure summarized in Table J-1 has been followed in this section for the inverted pendulum in Figure J-1.

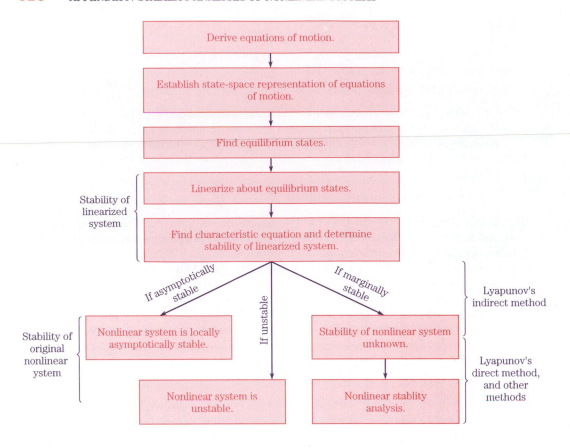

J-2   NONLINEAR STABILITY ANALYSIS FOR CONSERVATIVE SYSTEMS

It is noteworthy that there exists a nonlinear stability analysis method that can be easily applied directly to nonlinear *conservative* systems to find equilibrium positions and their stability without linearizations or solutions to equations of motion. The procedure in this method is briefly presented below where, for simplicity, the degree of freedom of the system is restricted to one.

1. **Potential Energy Function:** Find the potential energy function $V(x)$ of the system, where x is the only generalized coordinate.

2. **Equilibrium Positions:** The equilibrium positions are the positions where the potential energy is *stationary* (that is, maximum or minimum).
    This condition is given mathematically by

    $$\frac{dV}{dx} = 0 \tag{J-44}$$

3. **Local Stability:** The nonlinear system is *locally stable*[6] about an equilibrium if the potential energy is *locally* minimum at the equilibrium.

[6]Stability of conservative systems always means *marginal stability* because conservative systems do not possess damping and thus do not exhibit asymptotical stability.

The potential energy $V(x)$ is locally minimum at the equilibrium when

$$\left.\frac{d^2V}{dx^2}\right|_e > 0 \tag{J-45}$$

or when

$$\left.\frac{d^2V}{dx^2}\right|_e = 0 \tag{J-46}$$

and the first nonzero higher derivative is of even-order and is positive at the equilibrium.[7]

The nonlinear system is *unstable* about an equilibrium if the potential energy is neither *locally* minimum at the equilibrium nor constant about the equilibrium.

Therefore, the nonlinear system is *unstable* when

$$\left.\frac{d^2V}{dx^2}\right|_e < 0 \tag{J-47}$$

or when

$$\left.\frac{d^2V}{dx^2}\right|_e = 0 \tag{J-48}$$

and the first nonzero higher derivative is of an odd order;[8]
or when

$$\left.\frac{d^2V}{dx^2}\right|_e = 0 \tag{J-49}$$

and the first nonzero higher derivative is of even-order and is negative.[9]

The nonlinear system is *locally neutrally stable* about an equilibrium if the potential energy is *locally* constant about the equilibrium.[10]

That is, the nonlinear system is locally neutrally stable when

$$\left.\frac{d^nV}{dx^n}\right|_e = 0 \qquad \text{for all integers } n \geq 1 \tag{J-50}$$

The term "locally" has been emphasized because the criteria in Step 3 here are for the local stability of the nonlinear system; that is, the criteria are concerned only with the local behavior of the potential energy. Therefore, our concern in this section is restricted to the motion of the system when the initial displacement of the system configuration from the equilibrium is sufficiently small *and* the initial velocity is also sufficiently small; in essence,

---

[7]For example, consider the case where $V(x) = x^4$. The only equilibrium position is $x_e = 0$ due to the condition $\frac{dV}{dx} = 0$. In this case, $\left.\frac{d^2V}{dx^2}\right|_e = \left.\frac{d^3V}{dx^3}\right|_e = 0$ and $\left.\frac{d^4V}{dx^4}\right|_e = 24$. Since the first nonzero derivative is of the fourth order (an even order) and is positive, the potential energy is minimum at the equilibrium. This can be easily seen by plotting $V(x)$ versus x.

[8]Consider the case of $V(x) = x^3$.

[9]Consider the case of $V(x) = -x^4$.

[10]Neutral stability means that when the system is displaced from an equilibrium and released with zero velocity, the system stays there at rest.

the phrase "sufficiently small" limits consideration to the immediate vicinity of the point for which the equilibrium has been found.

The criterion in Step 2 for equilibrium positions is often called the *principle of stationary potential energy,* and the criteria in Step 3 for stability of equilibrium positions are often called the *principle of minimum potential energy.* Because the procedure listed above can be used for finding both the equilibrium positions and their stability, this technique provides a complete (local) stability analysis for nonlinear conservative systems.

The above criteria for equilibrium positions and their local stability can be easily appreciated by considering a simple system. Such a simple system is shown in Figure J-4, which shows a particle of mass m, subjected to gravity, moving along a frictionless surface having hills and valleys. Note that the system is conservative and the potential energy is simply due to the gravitational field. Therefore, the potential energy of the system is given by

$$V = mgz \tag{J-51}$$

where z is the vertical height of the particle measured from an arbitrarily chosen datum, in this case, the abscissa. Note that the vertical height z of the particle cannot be an appropriate choice for the generalized coordinate because the value of z does not completely specify the system configuration (in this case, the location of the particle); that is, for a given value of z, there may be several (different) locations of the particle having that value of z. A good choice for the generalized coordinate is the horizontal location of the particle, say x. Mathematically, z in Eq. (J-51) is a function of x and therefore the potential energy for the system may be written as

$$V(x) = mgz(x) \tag{J-52}$$

Because of the proportionality of the potential function to the vertical height z in Eq. (J-52), the surface along which the particle moves can also be regarded as the plot of the potential energy function, as indicated in Figure J-4.

According to the criterion in Step 2, the equilibrium positions correspond to the positions where the potential energy has zero slope, as given by Eq. (J-44). Thus, the positions A, B, C, and the finite region labeled D in Figure J-4 are all equilibrium positions, where C is an inflection point. A physical explanation for the criterion in Step 2 may be obtained by considering the fact that for (static: $\dot{x} = \ddot{x} = 0$) equilibrium of a conservative system, the net force exerted on the system is zero. In Figure J-4, at the positions A, B, C, and within the region D, the particle is subjected to zero net force, resulting from the cancellation of

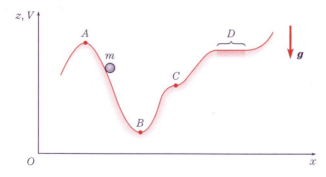

FIGURE J-4   Schematic of conservative system, illustrating various types of stability characteristics.

the gravitational force by the reaction force normal to the surface. Therefore, the particle will remain at rest at those equilibrium positions when subjected to no disturbance.

In order to explore the stability of these equilibrium positions, the criteria in Step 3 are used. The equilibrium position A is unstable because the potential energy is maximum there; that is, Eq. (J-47) is satisfied at A. The physical reason for this is that when the particle is displaced slightly from this equilibrium position and then released, the net force on the particle is nonzero and acts tangentially along the surface in a direction away from the equilibrium position.[11] Therefore, the particle tends to move away from the equilibrium position, indicating instability.

On the other hand, according to the criteria in Step 3, the equilibrium position B is stable because the potential energy is locally minimum there; that is, Eq. (J-45) is satisfied at B. By the same physical reasoning cited in the immediately preceding paragraph for the unstable equilibrium position A, when the particle is displaced slightly from the equilibrium position B and then released, the net force on the particle is nonzero and acts tangentially along the surface in the direction toward the equilibrium position. Therefore, the particle tends to move back to the equilibrium position, and thus the equilibrium position B is stable.

According to the criteria in Step 3, the equilibrium position C is unstable because the potential energy is not minimum there; indeed, although it is not obvious from Figure J-4, calculations show that, because C is an inflection point, this is an example where Eq. (J-48) is satisfied *and* the first nonzero higher derivative is of an odd order. By the same physical reasoning used above, when the particle is displaced slightly to the *right* of the equilibrium position C and then released, the net force on the particle is nonzero and acts tangentially along the surface in the direction toward the equilibrium position. However, the particle will pass through the equilibrium position because of the absence of dissipation. And, when the particle is displaced slightly to the *left* of the equilibrium position C and then released, the net force on the particle is nonzero and acts tangentially along the surface in the direction away from the equilibrium position. Therefore, the particle tends to depart from the equilibrium position C under a perturbation, indicating instability.

The finite equilibrium region D is neutral because the potential energy is constant throughout the region D; that is, Eq. (J-50) is satisfied in the region D. Physically, when the particle is displaced from any point inside the region D to a neighboring point inside the region D and then released, the particle is subjected to no net force and thus remains there at rest.

As an example of a conservative system, the inverted pendulum in Figure J-1 is reconsidered. The potential energy for the system can be shown to be

$$V(\theta) = \frac{k}{2}(L\sin\theta)^2 - mgL(1 - \cos\theta) \qquad \text{(J-53)}$$

whose first and second derivatives with respect to the generalized coordinate θ are given by

$$\frac{dV}{d\theta} = kL^2\sin\theta\cos\theta - mgL\sin\theta \qquad \text{(J-54)}$$

---

[11]The net force is the vector sum of the gravitational force and the surface reaction force. At this displaced location, the component of the gravitational force in the direction normal to the surface is canceled by the surface reaction force and therefore the net force is downward along the surface.

$$\frac{d^2V}{d\theta^2} = kL^2(\cos^2\theta - \sin^2\theta) - mgL\cos\theta$$

$$= kL^2(2\cos^2\theta - 1) - mgL\cos\theta \tag{J-55}$$

Therefore, the equilibrium positions may be found by setting Eq. (J-54) to zero: that is,

$$kL^2\sin\theta\cos\theta - mgL\sin\theta = L\sin\theta(kL\cos\theta - mg) = 0 \tag{J-56}$$

which leads to the three equilibrium positions given by

$$\theta_e = 0; \qquad \theta_e = \cos^{-1}\frac{mg}{kL}; \qquad \theta_e = -\cos^{-1}\frac{mg}{kL} \tag{J-57}$$

where the second and third equilibrium positions exist only when $kL > mg$ because the cosine function cannot exceed unity and the possibility of $kL = mg$ has been excluded in the description of Figure J-1. The equilibrium positions in Eqs. (J-57) are consistent with those in Eqs. (J-13) through (J-15).

Substitution of the first of Eqs. (J-57) into Eq. (J-55) gives

$$\left.\frac{d^2V}{d\theta^2}\right|_e = L(kL - mg) \tag{J-58}$$

which is positive when $kL > mg$ and negative when $kL < mg$. Therefore, according to Step 3 above, about the equilibrium position $\theta_e = 0$, the nonlinear system is stable when $kL > mg$ (because in this case Eq. (J-45) is satisfied) and is unstable when $kL < mg$ (because in this case Eq. (J-47) is satisfied). Recall from Subsection J-1.5 that when $kL > mg$, the stability of the *linearized system* about the equilibrium position $\theta_e = 0$ was found to be marginally stable.[12] Also, recall from Subsection J-1.6 that according to the statements of Lyapunov's indirect method, no conclusion could be made regarding the local stability of the *nonlinear* system about this equilibrium position, $\theta_e = 0$. This undetermined local stability of the nonlinear system has now been found via the nonlinear analysis, utilizing the principle of minimum potential energy.

For both the second and third equilibrium positions given in Eqs. (J-57), Eq. (J-55) reduces to

$$\left.\frac{d^2V}{d\theta^2}\right|_e = kL^2\left\{2\left(\frac{mg}{kL}\right)^2 - 1\right\} - mgL\frac{mg}{kL}$$

$$= 2\frac{m^2g^2}{k} - kL^2 - \frac{m^2g^2}{k}$$

$$= \frac{1}{k}(m^2g^2 - k^2L^2) \tag{J-59}$$

which is always negative because these two equilibrium positions exist only when $mg < kL$. Therefore, the nonlinear system is unstable about the two equilibrium positions $\theta_e = \cos^{-1}\frac{mg}{kL}$ and $\theta_e = -\cos^{-1}\frac{mg}{kL}$.

Finally, recall that in all analyses so far, we have assumed that $mg \neq kL$, as indicated in the problem statement following Eq. (J-4). In summary, when $mg \neq kL$, the only (locally)

---

[12]See the discussion following Eq. (J-42).

stable equilibrium position of the nonlinear system of the inverted pendulum in Figure J-1 is $\theta_e = 0$. In concluding, we shall consider the case where $mg = kL$.

First, note that when $mg = kL$, there is a single solution to Eq. (J-56); that is,

$$\theta_e = 0 \tag{J-60}$$

which may also have been obtained by substituting $mg = kL$ into each of Eqs. (J-57). In order to explore the local stability about this single equilibrium position, substitution of Eq. (J-60) into Eq. (J-55) gives

$$\left.\frac{d^2V}{d\theta^2}\right|_e = kL^2 - mgL = kL^2 - kL^2 = 0 \tag{J-61}$$

which may also have been obtained by substituting $mg = kL$ into Eq. (J-59). Since the second derivative of the potential energy function vanishes at the equilibrium position $\theta_e = 0$, higher derivatives of the potential energy function must be considered, in accordance with the criteria for stability in Step 3. Therefore, the third derivative of the potential energy function may be derived from Eq. (J-55) as

$$\frac{d^3V}{d\theta^3} = -4kL^2 \cos\theta \sin\theta + mgL \sin\theta \tag{J-62}$$

which at $\theta_e = 0$ gives

$$\left.\frac{d^3V}{d\theta^3}\right|_e = -4kL^2 \cos 0 \sin 0 + mgL \sin 0 = 0 \tag{J-63}$$

Since the third derivative of the potential energy function also vanishes at the equilibrium position $\theta_e = 0$, the fourth derivative must be considered. The fourth derivative of the potential energy function is derived from Eq. (J-62) as

$$\frac{d^4V}{d\theta^4} = -4kL^2(\cos^2\theta - \sin^2\theta) + mgL \cos\theta \tag{J-64}$$

which at $\theta_e = 0$ yields

$$\left.\frac{d^4V}{d\theta^4}\right|_e = -4kL^2 + mgL = -4kL^2 + kL^2 = -3kL^2 < 0 \tag{J-65}$$

Since at the equilibrium position $\theta_e = 0$, the second derivative of the potential energy function is zero in Eq. (J-61) *and* the first nonzero higher derivative is of the fourth order (an even order) and is negative in Eq. (J-65), Eq. (J-49) and the immediately following criterion in Step 3 are satisfied. Therefore, when $mg = kL$, there is a single equilibrium position $\theta_e = 0$ and the nonlinear system is unstable about that equilibrium position.

STRAIN ENERGY FUNCTIONS

In this appendix, we explore the concepts and some specific examples of strain energy functions for several deformable continuous structural members. In the formulation of the equations of motion via the variational approach, contributions to the potential energy function due to the elastic deformation of these structural members are represented by their strain energy functions. We shall restrict our considerations to one-dimensional models of structural members, namely, rods, shafts, beams, and strings; however, these concepts can be extended to any elastic structural member, including those modelled as membranes, plates, and shells, as well as three-dimensional continua.

K-1 CONCEPT

When a body is deformed by external forces and torques, work is done by those forces and torques. In all of the discussions that follow, we assume that the forces and torques are applied so slowly that inertia effects due to them may be neglected and that the elements of the body are in equilibrium at every instant. Furthermore, if the body is elastic and if during the deformation process a negligibly small amount of energy is lost as heat, the energy absorbed by the body due to external work can be recovered when the external forces and torques are removed. When expressed in terms of the strain state in the body, this energy is a state function and is called the *strain energy function*. For simplicity, the following discussions are limited to the idealizations of homogeneous isotropic linear elasticity, an idealization that is broadly practical.

K-2 STRAIN ENERGY DENSITY FUNCTION

The uniaxial tension in a linearly elastic rod is represented in Figure K-1a, where u is the rod's elongation due to the force F_x. By definition, the work done on the body by F_x is

$$\text{Work} = \int_0^u F_x \, du \tag{K-1}$$

which, as sketched in Figure K-1b, is equal to the area under the force–elongation curve up to the point corresponding to the particular elongation u. Note that it is the assumed linearity of the elastic material of the rod that produces the linear force–elongation behavior displayed in Figure K-1b. This area also corresponds to the stored elastic energy in the rod, based upon our assumption of negligible energy dissipation.

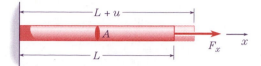

(a) Rod undergoing uniaxial tension.

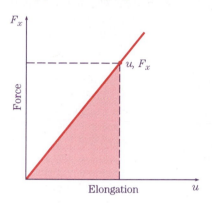

(b) Linearly elastic force–elongation diagram.

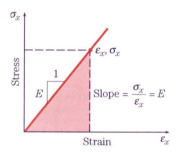

(c) Linearly elastic stress–strain diagram.

FIGURE K-1   Uniaxial tension of linearly elastic rod.

We shall be interested in evaluating the elastic energy in structural members where the strain varies within the body. In such cases, the total elastic energy, as represented by the cross-hatched area in Figure K-1b for the force–elongation curve, is not adequate because for each operating point in Figure K-1b, the stress and strain are constant throughout the body. A more useful function is the strain energy per unit volume, which is defined as the *strain energy density function*. For the rod of cross-sectional area A and undeformed length L sketched in Figure K-1, the strain energy density function, or simply the strain energy density V_0, may be written as

$$V_0 = \frac{\text{Work}}{\text{Volume}} = \int_0^u \frac{F_x \, du}{AL} = \int_0^{\epsilon_x} \sigma_x \, d\epsilon_x \qquad \text{(K-2)}$$

where we have used the elementary relations $\sigma_x = \dfrac{F_x}{A}$ and $d\epsilon_x = \dfrac{du}{L}$, where σ_x and ϵ_x are the uniaxial stress and strain, respectively. Common units of strain energy density are inch-pound per cubic inch (in-lb/in^3) and newton-meter per cubic meter (N-m/m^3). As indicated in Figure K-1c, because the stress and strain are linearly related, Eq.(K-2) integrates to

$$V_0 = \int_0^{\epsilon_x} \sigma_x d\epsilon_x = \int_0^{\epsilon_x} E\epsilon_x \, d\epsilon_x = \frac{1}{2}E\epsilon_x^2 = \frac{1}{2}(E\epsilon_x)\epsilon_x = \frac{1}{2}\sigma_x\epsilon_x \qquad \text{(K-3)}$$

where E is the *modulus of elasticity*.

Eq.(K-3), which has been written for a linearly elastic rod subjected to uniaxial stress, may be extended to include all stress components. As shown in many elementary textbooks on the mechanics of solids, for a general three-dimensional state of stress in a linearly elastic body, the strain energy density is

$$V_0 = \frac{1}{2}\left[\sigma_x\epsilon_x + \sigma_y\epsilon_y + \sigma_z\epsilon_z + \tau_{xy}\gamma_{xy} + \tau_{xz}\gamma_{xz} + \tau_{yz}\gamma_{yz}\right] \qquad \text{(K-4)}$$

where the widely adopted engineering notation for normal stresses (σ_i), normal strains (ϵ_i), shear stresses (τ_{ij}), and shear strains (γ_{ij}) is used.

In analogy with concepts of energy-coenergy functions, it is possible to define a coenergy density function. Such a coenergy density function is a very useful concept in nonlinear elasticity and could be developed by writing the analogous expression for Eq.(K-3) in terms of stress instead of strain. For linear elasticity such complementary functions are equal, but for nonlinear elasticity the strain energy and coenergy density functions are not equal. We do not need the coenergy density function in our discussions here, and therefore we shall not pursue it further.

K-3   STRAIN ENERGY FUNCTION

The *strain energy function* for an entire linearly elastic body may be obtained by integrating the strain energy density function over the volume of the body as

$$V = \int\int\int_{\text{Vol}} V_0 \, dx \, dy \, dz \qquad \text{(K-5)}$$

where "Vol" denotes volume. In the "technical mechanics of solids," with an appropriate choice of coordinate axes (to be defined), Eq.(K-5) effectively reduces to

$$V = \frac{1}{2}\int\int\int_{\text{Vol}} [\sigma_x\epsilon_x + \tau_{xy}\gamma_{xy}] \, dx \, dy \, dz \qquad \text{(K-6)}$$

where, even though the coordinate axes will be defined later, Eq.(K-6) is written on the basis of the x coordinate being along the longitudinal axis of the member. By the phrase "technical mechanics of solids," we restrict our consideration to slender structural members typically called rods (which undergo uniaxial deformation due to longitudinal forces), shafts (which undergo twisting due to torsional moments), and beams (which undergo transverse deflections due to bending moments). By an appropriate choice of coordinate axes as illustrated in the examples below, Eq.(K-6) can be used to calculate the strain energy functions for rods, shafts, and beams.

Finally, because for a linearly elastic slender structural member the constitutive relations effectively reduce to $\sigma_x = E\epsilon_x$ and $\tau_{xy} = G\gamma_{xy}$ where G is the *shear modulus*, Eq.(K-6) can be rewritten as

$$V = \int \int \int_{\text{Vol}} \left[\frac{E\epsilon_x^2}{2} \right] dx\, dy\, dz + \int \int \int_{\text{Vol}} \left[\frac{G\gamma_{xy}^2}{2} \right] dx\, dy\, dz \qquad \text{(K-7)}$$

K-4 EXAMPLES

In this section, we derive the strain energy functions for four of the most widely used structural models: rods, circular shafts, Bernoulli-Euler beams, and taut strings.

■ **Example K-1:** Strain Energy Function for Rods

Find the strain energy function for a rod subjected to uniaxial longitudinal loading. The equilibrium length L, modulus of elasticity E, and cross-sectional area A define the parameters of the rod.

▢ **Solution:**

An arbitrary element of the rod is sketched in Figure K-2 where the generalized coordinate is denoted by $\xi(x, t)$. $\xi(x, t)$ is the longitudinal displacement of the section whose equilibrium position is x. In the technical mechanics of solids, the strain energy function is given by Eq.(K-7). For the coordinate axes sketched in Figure K-2 and because $\gamma_{xy} = 0$ for a rod subjected to uniaxial forces, Eq.(K-7) reduces to

$$V = \int \int \int_{\text{Vol}} \left[\frac{E\epsilon_x^2}{2} \right] dx\, dy\, dz \qquad \text{(a)}$$

If we note that the area A, the modulus of elasticity E, and the strain ϵ_x may be functions of x, but do not vary at a given value of x, then we may write

$$A = \int_y \int_z dy\, dz \qquad \text{(b)}$$

which can be extracted from the volume integral in Eq.(a). Further, the axial strain may be expressed in terms of the generalized coordinate as

$$\epsilon_x = \frac{\partial \xi(x, t)}{\partial x} \qquad \text{(c)}$$

Substitution of Eqs.(b) and (c) into Eq.(a) gives

$$V = \int_0^L \frac{1}{2} EA \left[\frac{\partial \xi(x, t)}{\partial x} \right]^2 dx \qquad \text{(d)}$$

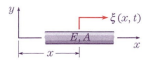

FIGURE K-2   Element of rod located at x, showing generalized coordinate $\xi(x, t)$.

which is the strain energy function for a linearly elastic rod, where we have introduced the limits on the integral on the assumption that the x coordinate ranges from 0 to L.

■ **Example K-2:**   **Strain Energy Function for Circular Shafts**

Find the strain energy function for a solid circular shaft. The length L, shear modulus G, and outer radius r_0 define the parameters of the shaft.

❑ **Solution:**

An arbitrary element of the shaft is sketched in Figure K-3a, where the generalized coordinate is denoted by $\varphi(x, t)$. $\varphi(x, t)$ is the angular displacement of the cross-section whose position is x. In the technical mechanics of solids, the strain energy function is given by Eq.(K-7). For the coordinate axes sketched in Figure K-3 and because $\epsilon_x = 0$ for a shaft subjected to a torsional moment, Eq.(K-7) reduces to

$$V = \int \int \int_{\text{Vol}} \left[\frac{G\gamma_{xy}^2}{2} \right] dx \, dy \, dz \tag{a}$$

or

$$V = \int \int \int_{\text{Vol}} \left[\frac{G\gamma_{x\theta}^2}{2} \right] r \, d\theta \, dr \, dx \tag{b}$$

where Eq.(b) represents an equivalent but more convenient formulation due to the circular cylindrical geometry of the shaft, and where r and θ are the polar coordinates at a cross-section of the shaft as sketched in Figure K-3b.

Recall from the technical mechanics of solids that for a solid circular cylindrical shaft, the shear strain may be written as

$$\gamma_{x\theta} = \left(\frac{r}{r_0} \right) \gamma_{\text{max}} \tag{c}$$

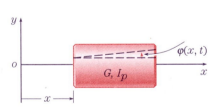

(a) Longitudinal section.

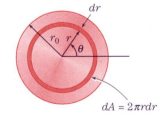

(b) Cross-section at arbitrarty x, indicating polar integration.

FIGURE K-3   Element of shaft located at x, showing generalized coordinate $\varphi(x, t)$.

where

$$\gamma_{\max} = r_0 \frac{\partial \varphi(x, t)}{\partial x} \tag{d}$$

where Eq.(c) indicates the geometrically linear character of the shear strain with radius and Eq.(d) gives the value of the maximum shear strain that occurs at the outermost radius of the shaft. Combining Eqs.(c) and (d) gives

$$\gamma_{x\theta} = r \frac{\partial \varphi(x, t)}{\partial x} \tag{e}$$

Substitution of Eq.(e) into Eq.(b) gives

$$V = \int_0^L \int_0^{r_0} \int_0^{2\pi} \frac{G}{2} \left[\frac{\partial \varphi(x, t)}{\partial x} \right]^2 r^3 \, d\theta \, dr \, dx \tag{f}$$

or

$$V = \int_0^L \frac{G \pi r_0^4}{4} \left[\frac{\partial \varphi(x, t)}{\partial x} \right]^2 dx \tag{g}$$

which is the strain energy function for a linearly elastic solid circular cylindrical shaft of outer radius r_0.

Equation (g), which is the specific answer that was sought, can be written in a slightly more general form. Recall that J_p, the polar moment of the cross-sectional area about the axis of the shaft, is defined by

$$J_p = \int_{\text{Area}} r^2 \, dA \tag{h}$$

Then for the shaft in Figure K-3b

$$J_p = \int_0^{r_0} r^2 (2\pi r \, dr) = \int_0^{r_0} 2\pi r^3 dr = \frac{\pi r_0^4}{2} \tag{i}$$

Substitution of Eq.(i) into Eq.(g) leads to the slightly more general result

$$V = \int_0^L \frac{1}{2} G J_p \left[\frac{\partial \varphi(x, t)}{\partial x} \right]^2 dx \tag{j}$$

which is valid for circular cylindrical shafts with or without a concentric hole.

■ **Example K-3:**   Strain Energy Function for Bernoulli-Euler Beams

There are several one-dimensional continuum models for elastic beams. The most widely used model is the Bernoulli-Euler beam, in which all the strain energy is assumed to be due exclusively to flexural deformation; thus, strain energy due to shear deformation is ignored.

Find the strain energy function for a Bernoulli-Euler beam subjected to bending. The length L, modulus of elasticity E, and the second moment of the cross-sectional area about the neutral axis I define the parameters of the beam.

☐ **Solution:**

An arbitrary element of the beam is sketched in Figure K-4, where the generalized coordinate is denoted by $\eta(x, t)$. $\eta(x, t)$ is the transverse displacement of the section whose position is x. In the technical mechanics of solids, the strain energy function is given by Eq.(K-7). For the coordinate axes sketched in Figure K-4 and because $\gamma_{xy} = 0$ for a beam undergoing flexural deformation only, sometimes referred to as pure bending, Eq.(K-7) reduces to

$$V = \int\int\int_{\text{Vol}} \left[\frac{E\epsilon_x^2}{2}\right] dx\, dy\, dz \tag{a}$$

Recall from the technical mechanics of solids that for a Bernoulli-Euler beam

$$\epsilon_x = -\frac{M_b y}{EI} \tag{b}$$

and

$$\frac{1}{\tilde{\rho}} = \frac{\dfrac{\partial^2 \eta}{\partial x^2}}{\left[1 + \left(\dfrac{\partial \eta}{\partial x}\right)^2\right]^{\frac{3}{2}}} = \frac{M_b}{EI} \tag{c}$$

where M_b is the bending moment and $\tilde{\rho}$ is the radius of curvature of the beam in the plane of Figure K-4. The minus sign in Eq.(b) is due to the sign convention that results in a negative longitudinal strain above the neutral axis for positive bending moments. That is, when the bending moment causes the beam to "smile," the beam experiences compression above the neutral axis and tension below the neutral axis. It is commonly assumed that the beam's slope throughout its length is very small compared with unity; thus Eq.(c) may be rewritten as

$$\frac{M_b}{EI} \approx \frac{\partial^2 \eta(x, t)}{\partial x^2} \tag{d}$$

since the denominator in the middle portion of Eq.(c) is approximately unity. Within the validity of Eq.(d), combining Eqs.(b) and (d) gives

$$\epsilon_x = -y\frac{\partial^2 \eta(x, t)}{\partial x^2} \tag{e}$$

Substitution of Eq.(e) into Eq.(a) gives

$$V = \int\int\int_{\text{Vol}} \frac{E}{2} \left[\frac{\partial^2 \eta(x, t)}{\partial x^2}\right]^2 y^2 \, dx\, dy\, dz \tag{f}$$

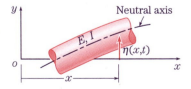

FIGURE K-4   Element of Bernoulli-Euler beam located at x, showing generalized coordinate $\eta(x, t)$.

Because neither E nor $\dfrac{\partial^2 \eta}{\partial x^2}$ is a function of y or z, Eq. (f) can be rewritten as

$$V = \int_0^L \frac{E}{2} \left[\frac{\partial^2 \eta(x,t)}{\partial x^2} \right]^2 \left[\int_y \int_z y^2 \, dy \, dz \right] dx \tag{g}$$

where the term in the second set of square brackets in Eq. (g) simply represents the definition of I. Therefore,

$$V = \int_0^L \frac{EI}{2} \left[\frac{\partial^2 \eta(x,t)}{\partial x^2} \right]^2 dx \tag{h}$$

which is the strain energy function for a linearly elastic Bernoulli-Euler beam.

■ **Example K-4:** Strain Energy Function for Taut String

Although it is not a structure within the technical mechanics of solids, a tightly stretched flexible string is a useful one-dimensional continuum model in some engineering applications. For example, taut cables having negligible flexural rigidity may be usefully modeled in this way. Here we want to find the strain energy function for such a taut linearly elastic flexible string, when it undergoes *small* transverse deflections. The length L and the *very large* tensile force P in the transversely undeflected string define the parameters in the problem. It is assumed that because P is very large, changes in P are negligible during the small transverse deflection of the string. Also, the unstretched length of the string is denoted by L_0. So, under the very large tension P but in its transversely undeflected configuration, the string has already undergone an elongation of $(L - L_0)$.

❑ **Solution:**

An arbitrary element of the string is sketched in Figure K-5, where the generalized coordinate is denoted by $\eta(x,t)$. $\eta(x,t)$ is the transverse displacement of the section whose position is x. The strain energy in the string is due to the energy that is stored in the deflected string as it elongates during its transverse motion $\eta(x,t)$. Note that this strain energy is in addition to the stored energy in the string in its transversely undeflected configuration. Thus, the strain energy to be calculated is due to the work done by the tension in the string as it deflects transversely.

If we consider the element dx of the string sketched in Figure K-5, the elongated length ds of the element dx due to the transverse motion $\eta(x,t)$ is given by

$$ds = \sqrt{dx^2 + d\eta^2} \tag{a}$$

Then, the increase in the total length of the string may be written as

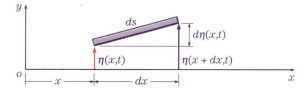

FIGURE K-5 Element of taut string located at x, showing generalized coordinate $\eta(x,t)$.

$$\Delta L = \int_0^L ds - \int_0^L dx = \int_0^L \left[\sqrt{dx^2 + d\eta^2} - dx \right] \qquad \text{(b)}$$

where Eq.(a) has been used, or

$$\Delta L = \int_0^L \left[\sqrt{1 + \left(\frac{\partial \eta}{\partial x}\right)^2} - 1 \right] dx \qquad \text{(c)}$$

Recall from the binomial series* that

$$(1 \pm x)^n = 1 \pm nx + \frac{n(n-1)}{2!}x^2 \pm \frac{n(n-1)(n-2)}{3!}x^3$$
$$+ \text{(higher-order terms)} \qquad x^2 < 1 \qquad \text{(d)}$$

---

*The binomial series was discovered by I. Newton in 1665 but characteristically was not announced by Newton until 1676 nor published until 1685. Furthermore, like so many mathematical results, there is evidence that centuries earlier the Indians, Arabs, and Chinese all had the rudiments of the binomial series in their possession but, like several Europeans in the late sixteenth and early seventeenth centuries, never succeeded in the critical steps required for enunciation. Even Greek scholars in the fifth century B.C. and the Mesopotamians before them were familiar with the two-term approximation to the binomial series. Nevertheless, to have a crab on the line is not to net it.

Note that for $x^2 \ll 1$, Eq.(d) can be reduced to

$$(1 \pm x)^n \approx 1 \pm nx \qquad \text{(e)}$$

(Verify this approximation on your calculator.) Now if our interest is restricted to small transverse deflections for which $(\partial \eta / \partial x)^2 \ll 1$, Eq.(e) can be used to rewrite the square-root term in Eq.(c) as

$$\left[1 + \left(\frac{\partial \eta}{\partial x}\right)^2 \right]^{1/2} \approx 1 + \frac{1}{2}\left(\frac{\partial \eta}{\partial x}\right)^2 \qquad \text{(f)}$$

Thus, Eq.(f) is a quantitative consequence of the phrase "*small* transverse deflections" in the problem statement! Substitution of Eq.(f) into Eq.(c) gives

$$\Delta L = \int_0^L \frac{1}{2}\left(\frac{\partial \eta}{\partial x}\right)^2 dx \qquad \text{(g)}$$

Thus, we may approximate the potential energy of the deflected string by

$$V = P\,\Delta L = P \int_0^L \frac{1}{2}\left(\frac{\partial \eta}{\partial x}\right)^2 dx \qquad \text{(h)}$$

where P is assumed to remain constant. Clearly, this is the additional potential energy that is due to the transverse displacement $\eta(x, t)$.

Equation (h) can be interpreted by considering Figure K-6, which shows the string tension as a function of the string elongation. In Figure K-6, the unstretched length of the string is L_0. During the extension of the string from L_0 to its transversely undeflected equilibrium length L, the tension is linearly increased from zero to the *very large* value P. Therefore,

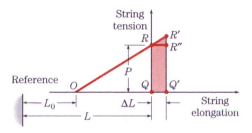

FIGURE   K-6   String tension–string elongation sketch.

the work done by the tension as the string is stretched from L_0 to L is represented by the triangular area ORQ. The focus of this strain energy analysis, however, is on the increment of stored energy when the string undergoes the additional elongation ΔL due to its transverse deflection $\eta(x, t)$. This increment of stored energy is represented by the trapezoidal area $RR'Q'Q$. Because P in Eq.(h) is assumed to be very large and constant, however, the energy given by Eq.(h) is the rectangular area $RR''Q'Q$, which is an approximation to the trapezoidal area $RR'Q'Q$. Thus, neglecting the small triangular area $RR'R''$ within the trapezoid $RR'Q'Q$ is the quantitative consequence of the phrase *very large* and constant tensile force P in the problem statement!

ANSWERS TO MOST OF THE ODD-NUMBERED PROBLEMS

CHAPTER 3

3-1 1.289 s.

3-7 223.14 ft.

3-9 $d \geq 0.6V_0 + 2.77$ ft where V_0 in ft/sec.

For $V_0 = 50$ mi/hr, $d \geq 46.77$ ft; for $V_0 = 60$ mi/hr,

$d \geq 55.57$ ft; for $V_0 = 70$ mi/hr, $d \geq 64.37$ ft.

3-11 $\overline{v} = 0$, $\overline{\omega} = 0.0628$ rad/s.

3-13 2 rotations.

3-15 23 hr 56 min 4.1 sec;   366.25 sidereal days.

3-17 $\omega_2\omega_0 i - \omega_1\omega_0 j$.

3-19 $\boldsymbol{a}_A = -\dfrac{v^2}{R}\boldsymbol{j}$; $\boldsymbol{a}_B = \dfrac{v^2}{R}\boldsymbol{j}$.

3-21 $\boldsymbol{v}_① = 20\boldsymbol{i}$ m/s, $\boldsymbol{a}_① = 2\boldsymbol{i} - 20\boldsymbol{j}$ m/s^2;

$\boldsymbol{v}_② = -20\boldsymbol{j}$ m/s, $\boldsymbol{a}_② = -20\boldsymbol{i} - 2\boldsymbol{j}$ m/s^2;

$\boldsymbol{v}_③ = -20\boldsymbol{i}$ m/s, $\boldsymbol{a}_③ = -2\boldsymbol{i} + 20\boldsymbol{j}$ m/s^2.

3-23 $\boldsymbol{v}_A = -\ell\dot\theta\sin\theta\boldsymbol{J}$, $\boldsymbol{a}_A = -\ell\left(\ddot\theta\sin\theta + \dot\theta^2\cos\theta\right)\boldsymbol{J}$;

$\boldsymbol{v}_B = \ell\dot\theta\cos\theta\boldsymbol{I}$, $\boldsymbol{a}_B = \ell\left(\ddot\theta\cos\theta - \dot\theta^2\sin\theta\right)\boldsymbol{I}$;

$\boldsymbol{v}_C = \dfrac{\ell}{2}\dot\theta\left(\cos\dot\theta\boldsymbol{I} - \sin\theta\boldsymbol{J}\right)$;

$\boldsymbol{a}_C = \dfrac{\ell}{2}\left(\ddot\theta\cos\theta - \dot\theta^2\sin\theta\right)\boldsymbol{I} - \dfrac{\ell}{2}\left(\ddot\theta\sin\theta + \dot\theta^2\cos\theta\right)\boldsymbol{J}$.

3-25 (a) 84.3 s.   (b) 113.6 s.   (c) 75.6 s.   (d) 72.0 s.

3-27 $\boldsymbol{v} = -\omega h\boldsymbol{i} + (v_0 + \omega d)\boldsymbol{j}$;

$\boldsymbol{a} = -\left(\omega^2 d + h\dot\omega + 2\omega v_0\right)\boldsymbol{i} + \left(a_0 + \dot\omega d - \omega^2 h\right)\boldsymbol{j}$.

3-29 $\boldsymbol{v} = \dot r\boldsymbol{i} - \dot\theta r\boldsymbol{k}$; $\boldsymbol{a} = \left(\ddot r - r\dot\theta^2\right)\boldsymbol{i} - \left(r\ddot\theta + 2\dot r\dot\theta\right)\boldsymbol{k}$.

3-31 $\boldsymbol{v} = [v_0 - (\omega_1 + \omega_2)\ell_0\sin\beta + \omega_1\ell\cos\gamma]\boldsymbol{i}$

$+ [(\omega_1 + \omega_2)\ell_0\cos\beta + \omega_1\ell\sin\gamma]\boldsymbol{j}$;

$\boldsymbol{a} = [a_0 - \dot\omega_2\ell_0\sin\beta - (\omega_1 + \omega_2)^2\ell_0\cos\beta - \omega_1^2\ell\sin\gamma]\boldsymbol{i}$

$+ [\dot\omega_2\ell_0\cos\beta - (\omega_1 + \omega_2)^2\ell_0\sin\beta + \omega_1^2\ell\cos\gamma]\boldsymbol{j}$.

3-33 $\boldsymbol{v}_P = (\omega_1 R - \omega_2 r)\boldsymbol{j}$; $\boldsymbol{a}_P = -\left(\omega_1^2 R + \omega_2^2 r\right)\boldsymbol{i}$.

3-35 $\boldsymbol{v}_A = 6.7\boldsymbol{i} + 0.8\boldsymbol{j} + 0.9\boldsymbol{k}$ m/s;

$\boldsymbol{a}_A = -0.8\boldsymbol{i} + 4.0\boldsymbol{j}$ m/s^2.

3-37 $\boldsymbol{v}_P = 2.059\boldsymbol{i}$ m/s; $\boldsymbol{a}_P = 4.923\boldsymbol{i} + 3.611\boldsymbol{j}$ m/s^2.

3-39 $\boldsymbol{v} = \boldsymbol{v}_d + (\boldsymbol{\omega}_1 + \boldsymbol{\omega}_2) \times \boldsymbol{s}$;

$\boldsymbol{a} = \boldsymbol{a}_d + (\dot{\boldsymbol{\omega}}_1 + \dot{\boldsymbol{\omega}}_2) \times \boldsymbol{s} + (\boldsymbol{\omega}_2 + 2\boldsymbol{\omega}_1) \times (\boldsymbol{\omega}_2 \times \boldsymbol{s}) + \boldsymbol{\omega}_1 \times (\boldsymbol{\omega}_1 \times \boldsymbol{s})$.

3-41 $\boldsymbol{v}_A = (-v_0\sin\theta + b\omega_2\cos\theta)\boldsymbol{i} + (v_0\cos\theta + b\omega_2\sin\theta)\boldsymbol{j}$

$- b\omega_1\sin\theta\boldsymbol{k}$;

$\boldsymbol{a}_A = -(a_0\sin\theta + 2v_0\omega_2\cos\theta + b\omega_2^2\sin\theta - b\omega_1^2\sin\theta)\boldsymbol{i}$

$+ (a_0\cos\theta - 2v_0\omega_2\sin\theta + b\omega_2^2\cos\theta)\boldsymbol{j}$

$+ 2\omega_1(v_0\sin\theta - b\omega_2\cos\theta)\boldsymbol{k}$.

3-43 $\boldsymbol{v}_P = -(\omega_1 + \omega_2)(a\sin\phi + b\sin\theta)\boldsymbol{I}$

$+ (\omega_1 + \omega_2)(a\cos\phi + b\cos\theta)\boldsymbol{J}$

$- \omega_3(a\cos\phi + b\cos\theta)\boldsymbol{K}$;

$\boldsymbol{a}_P = -\{[(\omega_1 + \omega_2)^2 + \omega_3^2](a\cos\phi + b\cos\theta)$

$+ (\dot\omega_1 + \dot\omega_2)(a\sin\phi + b\sin\theta)\}\boldsymbol{I}$

$+ [-(\omega_1 + \omega_2)^2(a\sin\phi + b\sin\theta)$

$+ (\dot\omega_1 + \dot\omega_2)(a\cos\phi + b\cos\theta)]\boldsymbol{J}$

$+ [-\dot\omega_3(a\cos\phi + b\cos\theta)$

$+ \omega_3(2\omega_1 + \omega_2)(a\sin\phi + b\sin\theta)]\boldsymbol{K}$.

3-45 $\boldsymbol{v}_P = 69.13\boldsymbol{j} - 0.4189\boldsymbol{k}$ m/s;

$\boldsymbol{a}_P = -2.475\boldsymbol{i} - 0.0220\boldsymbol{j} - 0.0393\boldsymbol{k}$ m/s^2.

3-47 $\boldsymbol{v}_C = (v_0\cos\theta - L_2\omega_2\sin\theta)\boldsymbol{I} + [(v_0\sin\theta - L_2\omega_3)$

$+ \omega_2(L_1\cos\theta + L_2)]\boldsymbol{J} - \omega_1(L_1\cos\theta + L_2)\boldsymbol{K}$;

$\boldsymbol{a}_C = -\{[\omega_1^2 + (\omega_2 + \omega_3)^2]L_2 + (\omega_1^2 + \omega_2^2)L_1\cos\theta$

$+ 2\omega_2 v_0\sin\theta + \dot\omega_2 L_1\sin\theta\}\boldsymbol{I}$

$+ [\omega_2(2v_0\cos\theta - \omega_2 L_1\sin\theta)$

$+ \dot\omega_2(L_1\cos\theta + L_2)]\boldsymbol{J}$

$- \omega_1(2v_0\cos\theta - \omega_2 L_1\sin\theta)\boldsymbol{K}$.

3-49 $\boldsymbol{v} = (v_0 + \omega_2 L)\boldsymbol{j} + \omega_1 L\boldsymbol{k}$; $\boldsymbol{a} = [(\omega_1^2 + \omega_2^2)L - a_0]\boldsymbol{i}$.

3-51 $\boldsymbol{v}_{ED} = 1.975\boldsymbol{i} - 3.573\boldsymbol{j} + 1.624\boldsymbol{k}$ m/s;

$\boldsymbol{a}_{ED} = -5.198\boldsymbol{i} - 3.814\boldsymbol{j} + 14.02\boldsymbol{k}$ m/s^2.

3-53 See eqns. (3–33) and (3–34).

3-55 $\boldsymbol{v}_B = -\omega_1 r\boldsymbol{j}$; $\boldsymbol{a}_B = 2\omega_1\omega_2 r\boldsymbol{i} - \omega_1^2 r\boldsymbol{k}$.

3-57 $\boldsymbol{v}_C = 0.3214\boldsymbol{I} + 0.3830\boldsymbol{J} - 1.202\boldsymbol{K}$ m/s;

$\boldsymbol{a}_C = 0.1853\boldsymbol{I} + 0.7346\boldsymbol{J} - 0.8819\boldsymbol{K}$ m/s^2.

3-59 $\boldsymbol{v}_B = 2.4749\boldsymbol{I} + 3.1820\boldsymbol{J} - 19.1421\boldsymbol{K}$ m/s;

$\boldsymbol{a}_B = -95.0947\boldsymbol{I} + 1.2728\boldsymbol{J} - 24.7487\boldsymbol{K}$ m/s^2.

CHAPTER 4

4-1 2268 N.

4-3 $\overline{\boldsymbol{F}} = 1474\boldsymbol{i} + 2532\boldsymbol{j}$ N.

4-5 The vine will not break.

4-7 $2160\,\text{m}$.

4-9   (a) $0.1771\,\text{m}$.   (b) $1.9\,\text{m}$.

4-11 (a) $7.38\,\text{m/s}$.   (b) $150\,\text{m/s}$.

4-13 (a) $1.10\,\text{N}$.   (b) $2.10\,\text{N}$.

4-15 (a) $7.18\,\text{m/s}$.   (b) $1202\,\text{N}$.

4-17 (a) $10.99\,\text{N}$.   (b) $0.144\,\text{m/s}^2$.   (c) $1.52\,\text{m/s}^2$.

4-19 (b) $4540\,\text{ft}$. The highway turns left.

4-21 (a) $16m_0 r \left(\dfrac{\omega_0}{\pi d}\right)^2$.   (b) B.   (c) No.

(d) Newton's first law.

4-23 $\boldsymbol{F}_① = 180\boldsymbol{i} - 2682.9\boldsymbol{j}\,\text{N};$ $\boldsymbol{F}_② = -1800\boldsymbol{i}$
$- 1062.9\boldsymbol{j}\,\text{N};$

$\boldsymbol{F}_③ = -180\boldsymbol{i} + 917.1\boldsymbol{j}\,\text{N}.$

4-25 $\boldsymbol{F}_L = -m(a_0 + \omega_0^2\ell)\boldsymbol{i} + m(-2\omega_0 v_0 + \dot\omega_0\ell + g)\boldsymbol{j};$
$\boldsymbol{F}_R = m(a_0 + \omega_0^2\ell)\boldsymbol{i} + m(2\omega_0 v_0 - \dot\omega_0\ell + g)\boldsymbol{j}.$

4-27 $\boldsymbol{F} = -m[(\omega_1 - \omega_2)^2 r + \omega_1^2 R]\boldsymbol{i} + mg\boldsymbol{j}.$

4-29 $\boldsymbol{F} = 0.6660\boldsymbol{i} - 0.3704\boldsymbol{j} - 0.01248\boldsymbol{k}\,\text{lb}.$

4-31 $\boldsymbol{F} = 98.2\boldsymbol{I} + 214\boldsymbol{J} - 142\boldsymbol{K}\,\text{N}.$

4-33 $\boldsymbol{F} = -m\big\{[(\omega_1 + \omega_2)^2 + \omega_3^2](a\cos\phi + b\cos\theta)$
$\qquad + (\dot\omega_1 + \dot\omega_2)(a\sin\phi + b\sin\theta)\big\}\,\boldsymbol{I}$
$\qquad + m[-(\omega_1 + \omega_2)^2(a\sin\phi + b\sin\theta)$
$\qquad + (\dot\omega_1 + \dot\omega_2)(a\cos\phi + b\cos\theta) + g]\,\boldsymbol{J}$
$\qquad + m[-\dot\omega_3(a\cos\phi + b\cos\theta)$
$\qquad + \omega_3(2\omega_1 + \omega_2)(a\sin\phi + b\sin\theta)]\,\boldsymbol{K}.$

4-35 $\boldsymbol{F} = -198\boldsymbol{i} + 767.2\boldsymbol{j} - 3.144\boldsymbol{k}\,\text{N}.$

4-37 $\boldsymbol{F} = m[(\omega_1^2 + \omega_2^2)L - a_0]\boldsymbol{i} + mg\boldsymbol{j}.$

4-39 $-195\,\text{N}.$

4-41 $\dfrac{m\Omega^2 r}{A}.$

4-43 Axial force: $\boldsymbol{T} = -4.79\boldsymbol{j}\,\text{lb};$
Shear force: $\boldsymbol{S} = 1.58\boldsymbol{i} - 0.25\boldsymbol{k}\,\text{lb}.$

4-45 If $\cos\theta_0 \neq 0$, $y = -\dfrac{g}{2v_0^2 \cos^2\theta_0}(x - x_0)^2 + \dfrac{v_0^2 \sin^2\theta_0}{2g}$
$+ y_0;$ if $\cos\theta_0 = 0$, $x = x_0$.

4-47 $3.27\,\text{ft/sec}.$

4-49 $\dfrac{1}{6}\ell.$

4-53 (a) $5.76°$.   (b) $137.7\,\text{mi/hr}$ in a dry day; $103.9\,\text{mi/hr}$
in a rainy day; $67.4\,\text{mi/hr}$ in icy conditions.

4-55 $(m_1 + m_2)\ddot{x} + kx = m_2 g\sin\theta.$

4-57 $c\dot{x}_1 - c\dot{x}_2 + kx_1 = 0,$
$m\ddot{x}_2 - c\dot{x}_1 + c\dot{x}_2 = F(t).$

4-59 $m_1\ddot{x}_1 + (c_1 + c_2)\dot{x}_1 - c_2\dot{x}_2 + (k_1 + k_2)x_1 - k_2 x_2 = F(t),$
$m_2\ddot{x}_2 - c_2\dot{x}_1 + (c_2 + c_3)\dot{x}_2 - k_2 x_1 + (k_2 + k_3)x_2 = 0.$

4-61 $\theta = \cos^{-1}\left(\dfrac{2}{3} - \dfrac{v_0^2}{3gR}\right);$ $2gR < v_0^2 < 5gR.$

4-63 $(m_1 r_1^2 + m_2 r_2^2)\ddot{x}_1 + (k_1 r_1^2 + k_2 r_2^2)x_1 = 0.$

4-65 $m\ddot{x} + c\dot{x} + \left(2k\sin\theta_0 - \dfrac{mg\cos^3\theta_0}{a\sin\theta_0}\right)x = 0.$

4-67 (a) $0.443\,\text{rad/s}.$ (b) $\boldsymbol{F} = -44.3\boldsymbol{i} + 39.3\boldsymbol{j}\,\text{N}.$

4-69 $x(t) = \dfrac{\mu_k \ell}{1 + \mu_k} + \left(\ell_0 - \dfrac{\mu_k \ell}{1 + \mu_k}\right)\cosh\sqrt{\dfrac{(1 + \mu_k)g}{\ell}}\,t.$

CHAPTER 5

5-1   (a) $0.294\,\text{J}$.   (b) $-1.8\,\text{J}$.   (c) $3.47\,\text{m/s}$.
(d) $0.230\,\text{m}$.

5-3 $825.2\,\text{ft}$.

5-5   The block comes to rest midway between points B
and C.

5-7   (a) $34670\,\text{lb}$.   (b) Yes.

5-13 (a) 2.   (b) Nonholonomic.

5-15 (a) 3.   (b) Nonholonomic.

5-17 (a) 3.   (b) Nonholonomic.

5-19 (a) $\mathcal{L} = \dfrac{1}{2}m\dot\eta^2 - \dfrac{1}{2}k\eta^2 + mg\eta,$
$m\ddot\eta + k\eta = mg;$
(b) $\mathcal{L} = \dfrac{1}{2}m\dot\xi^2 - \dfrac{1}{2}k(\xi - e_0)^2 + mg\xi,$ $m\ddot\xi + k\xi = 0.$

5-21 $m\ddot{x} + c\dot{x} + f(t) = 0.$

5-23 $\left[m_1 + \left(\dfrac{r_2}{r_1}\right)^2 m_2\right]\ddot{x} + \left[k_1 + \left(\dfrac{r_2}{r_1}\right)^2 k_2\right]x = \left(\dfrac{r_2}{r_1}m_2 - m_1\right)g.$

5-25 $m\ell^2\ddot\theta + (k_1 + k_2)h^2 \sin\theta\cos\theta - mg\ell\sin\theta = 0.$

5-27 $m\ell^2\ddot\theta - mg\ell\sin\theta + \dfrac{1}{2}kb^2 \sin\theta\cos\theta = 0.$

5-29 $7m\ell^2\ddot\theta + 2c\ell^2\dot\theta\cos^2\theta + 2k\ell^2 \sin\theta\cos\theta - 3mg\ell\sin\theta = 0.$

5-31 $m(3\ell^2 + 2b^2)\ddot\theta + c(\ell - a)^2 \cos^2\theta\dot\theta + k(\ell - a)^2 \sin\theta\cos\theta$
$- 3mg\ell\sin\theta = 0.$

5-33 $md^2\ddot\theta + ca^2\dot\theta + (k_1 a^2 + k_2 b^2)\theta - mgd = 0.$

5-35 $m\ddot{x} + \dfrac{c\left(\sqrt{3}\ell + x\right)^2}{4\ell^2 - \left(\sqrt{3}\ell + x\right)^2}\dot{x} + mg$
$\qquad - k\left[\sqrt{4\ell^2 - \left(\sqrt{3}\ell + x\right)^2} - \ell\right]\dfrac{\sqrt{3}\ell + x}{\sqrt{4\ell^2 - \left(\sqrt{3}\ell + x\right)^2}}$
$\qquad = 0.$

5-37 $m\ddot{x} + c\dot{x}$
$\qquad + k\left[\dfrac{a}{\cos\theta_0} - \sqrt{a^2 + (a\tan\theta_0 - x)^2}\right]$
$\qquad \times \dfrac{x - a\tan\theta_0}{\sqrt{a^2 + (a\tan\theta_0 - x)^2}} = 0.$

5-39 $c\dot{x} + kx = c\dot{x}_0(t).$

5-41 $m\ddot{x} + c\dot{x} + kx = ca\Omega\cos\Omega t + ka\sin\Omega t.$

5-43 $m_2\ell^2\ddot\theta_2 + ca^2\dot\theta_2 + [k(a + b)^2 + m_2 g\ell]\theta_2 = ca^2\dot\theta_1(t)$
$\qquad + k(a + b)^2\theta_1(t) - F(t)\ell$

5-45 $m\ddot{x} - m\Omega^2 \cos^2\theta x - mg\sin\theta + k(x - \ell_0) = 0.$

5-47 $\begin{cases} m\ddot{x}_1 + c(\dot{x}_1 - \dot{x}_2) = 0, \\ c(\dot{x}_2 - \dot{x}_1) = F(t). \end{cases}$ Or, $m\ddot{x}_1 = F(t).$

5-49 $c\dot{x}_1 - c\dot{x}_2 + kx_1 = 0.$
$m\ddot{x}_2 - c\dot{x}_1 + c\dot{x}_2 = F(t).$

5-51 $m_1\ddot{x}_1 + k_1x_1 - k_1x_2 = 0,$
$m_2\ddot{x}_2 - k_1x_1 + (k_1 + k_2)x_2 = 0.$

5-53 $m_1\ddot{x}_1 + (c_1 + c_2)\dot{x}_1 - c_2\dot{x}_2 + (k_1 + k_2)x_1 - k_2x_2 = 0,$
$m_2\ddot{x}_2 - c_2\dot{x}_1 + (c_2 + c_3)\dot{x}_2 - k_2x_1 + (k_2 + k_3)x_2 = F(t).$

5-55 $\left(m_1 + \dfrac{1}{4}m_3\right)\ddot{x}_1 + \dfrac{1}{4}m_3\ddot{x}_2 + \dfrac{1}{4}kx_1 + \dfrac{1}{4}kx_2$

$= F(t) - \left(m_1 - \dfrac{1}{2}m_3\right)g,$

$\dfrac{1}{4}m_3\ddot{x}_1 + \left(m_2 + \dfrac{1}{4}m_3\right)\ddot{x}_2 + c\dot{x}_2 + \dfrac{1}{4}kx_1 + \dfrac{1}{4}kx_2 = \dfrac{1}{2}m_3g.$

5-57 $m\ddot{x}_1 + \dfrac{3}{2}kx_1 - \dfrac{1}{2}kx_2 = -m\ddot{x}_0(t),$
$m\ddot{x}_2 - kx_1 + kx_2 = -m\ddot{x}_0(t).$

5-59 $(m_1L^2 + m_2x^2)\ddot{\theta} + 2m_2x\dot{x}\dot{\theta} + m_1gL\sin\theta$
$\qquad + m_2gx\sin\theta = 0,$
$m_2\ddot{x} - m_2r\dot{\theta}^2 + k(x - \ell_0) - m_2g\cos\theta = 0.$

5-61 $\ddot{\theta} - \sin\theta\cos\theta\,\dot{\phi}^2 + \dfrac{g}{\ell}\sin\theta = 0,$

$m\ell^2\sin^2\theta\,\ddot{\phi} + 2m\ell^2\sin\theta\cos\theta\,\dot{\theta}\dot{\phi} = T(t).$

5-63 $m_1\ell_1^2\ddot{\theta}_1 + m_1g\ell_1\theta_1 + k\ell_1(\ell_1\theta_1 - \ell_2\theta_2) = 0,$
$m_2\ell_2^2\ddot{\theta}_2 + m_2g\ell_2\theta_2 + k\ell_2(\ell_2\theta_2 - \ell_1\theta_1) = 0.$

5-65 $m\ell\ddot{\theta} + k\ell\theta + kx = 0,$

$2m\ddot{x} - \dfrac{1}{2}k\ell\theta + 3kx = 0.$

5-67 $m\ddot{r} - mr\dot{\theta}^2 = F(t),$
$r\ddot{\theta} + 2\dot{r}\dot{\theta} = 0.$

5-69 $\ddot{y} + \ell\ddot{\theta}\sin\theta + \ell\cos\theta\,\dot{\theta}^2 + g = 0,$
$m\ell^2\ddot{\theta} + m\ell\sin\theta\,\ddot{y} + k_t\theta + mg\ell\sin\theta = 0.$

5-71 $2m\ddot{x} + m\ell\ddot{\theta} + 2kx = 0,$
$\ell\ddot{\theta} + \ddot{x} + g\theta = 0.$

5-73 $(m_0 + m_1 + m_2 + m_3)\ddot{x} + (m_3b - m_2a)(\ddot{\theta}\cos\theta - \dot{\theta}^2\sin\theta)$
$\qquad + (c_0 + c_2)\dot{x} + c_2b\dot{\theta}\cos\theta + (k_0 + k_2)x$
$\qquad - k_2a\sin\theta = F(t), (m_2a^2 + m_3b^2)\ddot{\theta}$
$\qquad + (m_3b - m_2a)\ddot{x}\cos\theta + b^2(c_1\sin^2\theta + c_2\cos^2\theta)\dot{\theta}$
$\qquad + c_2b\cos\theta\dot{x} - k_2a\cos\theta x + k_t\theta$
$\qquad + a^2(k_2 - k_1)\sin\theta\cos\theta$
$\qquad - (m_2ga - m_3gb - k_1a^2)\sin\theta = 0.$

5-75 $m\ddot{x} + \dfrac{24}{29}kx = 0.$

5-77 $m_1\ddot{x}_1 + c_1\dot{x}_1 + \left(k_1 + \dfrac{1}{2}k_0\right)x_1 - \dfrac{1}{2}k_0x_2 = 0,$

$m_2\ddot{x}_2 + c_2\dot{x}_2 + \dfrac{1}{2}k_0x_1 + \left(k_2 + \dfrac{3}{2}k_0\right)x_2 = 0.$

5-79 $m_0\ell^2\ddot{\theta} + c\ell^2\dot{\theta} - c\ell\dot{x}_1 + k\ell^2\theta - kx_2 + mg\ell\theta = 0,$
$m_1\ddot{x}_1 - c\ell\dot{\theta} + c\dot{x}_1 = 0,$
$m_2\ddot{x}_2 - k\ell\theta + kx_2 = F(t).$

5-81 $(m_1 + m_2 + m_3)\ddot{x}_1 + m_2\ddot{x}_2 + m_3\ddot{x}_3 = F(t),$
$m_2\ddot{x}_1 + m_2\ddot{x}_2 + c\dot{x}_2 + (k_1 + k_2)x_2 - k_1x_3 = 0,$
$m_3\ddot{x}_1 + m_3\ddot{x}_3 + c\dot{x}_3 - k_1x_2 + k_1x_3 = 0.$

5-83 $(m_0 + m_1 + m_2 + M)\ddot{x} = 0,$
$(m_0 + m_1)\ddot{y}_1 + (2k_1 + k_2)y_1 - k_2y_2 = 0,$
$m_2\ddot{y}_2 - k_2y_1 + k_2y_2 = 0.$

5-85 $M\ddot{x}_1 + (c_1 + c_2)\dot{x}_1 - c_2\dot{x}_2 + k_1x_1 - k_1x_2 = F(t),$
$m_1\ddot{x}_2 - c_2\dot{x}_1 + c_2\dot{x}_2 - k_1x_1 + (k_1 + k_2)x_2 - k_2x_3 = 0,$
$m_2\ddot{x}_3 + c_3\dot{x}_3 - k_2x_2 + k_2x_3 = 0.$

CHAPTER 6

6-1 36.8 m.

6-3 $(a)\ \dfrac{1}{2}H.$ $(b)\ \dfrac{1}{2}H.$ $(d)\ x = \dfrac{MH}{m}\left(\sqrt{1 + \dfrac{m}{M}} - 1\right).$

6-5 $\omega_0 = -\dfrac{m}{M + m}\dfrac{v_0}{R}.$

6-7 (a) 47.7 cm. (b) 0.205 kg · m².

6-9 $\dfrac{1}{2}MR^2.$

6-11 $[I]_o = M\begin{bmatrix} \dfrac{b^2 + 3c^3}{12} & -\dfrac{ac}{4} & 0 \\[2mm] -\dfrac{ac}{4} & \dfrac{3a^2 + b^2}{12} & 0 \\[2mm] 0 & 0 & \dfrac{a^2 + c^2}{3} \end{bmatrix}.$

6-13 46.4 slug · ft².

6-15 $(a)\ y_c = -\dfrac{1}{20}R.$ $(b)\ I_o = \dfrac{237}{480}MR^2.$

6-17 $[I]_O = [I]_o + 3m_0\begin{bmatrix} a_2^2 + a_3^2 & -a_1a_2 & -a_1a_3 \\ -a_1a_2 & a_1^2 + a_3^2 & -a_2a_3 \\ -a_1a_3 & -a_2a_3 & a_1^2 + a_2^2 \end{bmatrix}$

$+3m_0\begin{bmatrix} 2\left(a_2r + \dfrac{1}{3}a_3l\right) & -a_1r & -\dfrac{1}{3}a_1l \\[2mm] -a_1r & \dfrac{2}{3}a_3l & -\left(a_3r + \dfrac{1}{3}a_2l\right) \\[2mm] -\dfrac{1}{3}a_1l & -\left(a_3r + \dfrac{1}{3}a_2l\right) & 2a_2r \end{bmatrix}.$

6-19 1440 slug · ft²;   195.8 slug.

6-21 $(a)\ \begin{bmatrix} I_{xx} & -I_{xy} & -I_{xz} \\ -I_{xy} & I_{yy} & I_{yz} \\ -I_{xz} & I_{yz} & I_{zz} \end{bmatrix}.$ $(b)\ \begin{bmatrix} I_{xx} & I_{xz} & -I_{xy} \\ I_{xz} & I_{zz} & -I_{yz} \\ -I_{xy} & -I_{yz} & I_{yy} \end{bmatrix}.$

$(c)\ \begin{bmatrix} I_{xx} & -I_{xy} & I_{xz} \\ -I_{xy} & I_{yy} & -I_{yz} \\ I_{xz} & -I_{yz} & I_{zz} \end{bmatrix}.$ $(d)\ \begin{bmatrix} I_{zz} & -I_{yz} & -I_{xz} \\ -I_{yz} & I_{yy} & I_{xy} \\ -I_{xz} & I_{xy} & I_{xx} \end{bmatrix}.$

$(e)\ \begin{bmatrix} I_{xx} & I_{xy} & -I_{xz} \\ I_{xy} & I_{yy} & -I_{yz} \\ -I_{xz} & -I_{yz} & I_{zz} \end{bmatrix}.$ $(f)\ \begin{bmatrix} I_{yy} & -I_{xy} & I_{yz} \\ -I_{xy} & I_{xx} & -I_{xz} \\ I_{yz} & -I_{xz} & I_{zz} \end{bmatrix}.$

6-23 The z-axis is a principal axis. The other two principal axes lie in the x–y plane and form an angle of 26.57° with respect to the x- and y-axes, respectively;
$I_1 = 0.2896 \text{ kg} \cdot \text{m}^2$,　　$I_2 = 0.008324 \text{ kg} \cdot \text{m}^2$,
$I_3 = 0.3750 \text{ kg} \cdot \text{m}^2$.

6-25 (a) Yes, $h = \sqrt{3}R$.　　(b) No.　　(c) Yes, $h = 2R$.

6-29 $\dfrac{25}{52}Mg$.

6-31 Wheel B;　$\dfrac{\mu_s g L}{2L - \mu_s d}$.

6-33 2.94 m/s^2;　0.3.

6-35 9.84 m.

6-37 $I_c \ddot{\theta} + Wh\theta = 0$.

6-39 (a) $T = \rho t w r_0^2 \omega_0^2$;　$\sigma_{\theta\theta} = \rho r_0^2 \omega_0^2$.
(b) $(\omega_0)_{\max} = \dfrac{1}{r_0}\sqrt{\dfrac{\sigma_{\text{frac}}}{\rho}}$.

6-43 $T^* = \dfrac{1}{2}\left[I_c + M\left(r^2 - 2rd\cos\theta + d^2\right)\right]\dot{\theta}^2$.

6-45 $\dfrac{1}{2}mR^2\ddot{\theta} + kr^2\theta = 0$.

6-47 $\ddot{\theta} + 17.58\sin\theta = 0$　where θ is in radians.

6-49 $I\ddot{\theta} + k_t\theta = 0$.

6-51 $\ddot{r} - r\dot{\theta}^2 + g\sin\theta = 0$,
$\left(I_o + \dfrac{1}{12}mL^2 + mr^2\right)\ddot{\theta} + 2mrr\dot{\theta} + mgr\cos\theta = 0$.

6-53 $\dfrac{3}{2}m\ddot{x} + m\ddot{y}\cos\theta_0 + 4kx - mg\sin\theta_0 = 0$,
$(m + m_0)\ddot{y} + m\ddot{x}\cos\theta_0 = 0$.

6-55 $\dfrac{mr^2}{2}\ddot{\theta} + \dfrac{3GJ}{L}\theta = 0$.

6-59 $\dfrac{3}{2}r\ddot{\theta}_1 + r\ddot{\theta}_2 - g = 0$,
$\dfrac{3}{2}r\ddot{\theta}_2 + r\ddot{\theta}_1 - g = 0$.

6-61 $m\ddot{\theta} + c\dot{\theta} + 4k\theta = \dfrac{6}{L}f(t)$.

6-63 $I_1\ddot{\theta}_1 + (c_1 + c_2)\dot{\theta}_1 - c_2\dot{\theta}_2 + (k_1 + k_2)\theta_1 - k_2\theta_2 = T_1(t)$,
$I_2\ddot{\theta}_2 - c_2\dot{\theta}_1 + (c_2 + c_3)\dot{\theta}_2 - c_3\dot{\theta}_3 - k_2\theta_1 + (k_2 + k_3)\theta_2 - k_3\theta_3 = T_2(t)$,
$I_3\ddot{\theta}_3 - c_3\dot{\theta}_2 + (c_3 + c_4)\dot{\theta}_3 - k_3\theta_2 + (k_3 + k_4)\theta_3 = T_3(t)$.

6-65 (a) $y_c = \dfrac{r\sin\alpha}{\alpha} = \dfrac{2r^2}{l_0}\sin\dfrac{l_0}{2r}$.　　(b) Never.
(c) $V = Mg\dfrac{r\sin\alpha}{\alpha} = Mg\dfrac{2r^2}{l_0}\sin\dfrac{l_0}{2r}$.

6-67 1.814 rad/s.

6-69 3.86 cm.

6-71 85.34 rad/s.

6-73 $\omega = \sqrt{\dfrac{10g(R + r)(1 - \cos\theta)}{7r^2}}$;
$v_c = \sqrt{\dfrac{10g(R + r)(1 - \cos\theta)}{7}}$.

6-75 17.56 rad/s.

6-77 $\sqrt{\dfrac{2\pi aF}{I_c}}$.

6-81 $\left[\dfrac{1}{12}ab(4a^2 + b^2) - \pi R^2\left(R^2 + 2d^2 + \dfrac{1}{2}a^2\right)\right]\ddot{\theta}$
$+ \dfrac{1}{2}(ab - 2\pi R^2)ag\sin\theta = 0$.

6-83 $\left[\dfrac{1}{3}m_1l_1^2 + m_2\left(l_1^2 + \dfrac{1}{12}l_2^2\right) + m_3\left(2l_1^2 + \dfrac{1}{2}l_2^2 + \dfrac{1}{6}l_3^2\right)\right]\ddot{\theta}$
$+ \left[ka^2 + \dfrac{1}{2}m_1gl_1 + (m_2 + 2m_3)gl_1\right]\theta = 0$.

6-85 $\left(\dfrac{3}{2}m_1 + 2m_2 + 4m_0\right)\ddot{x}_1 + (m_2 + 4m_0)\ddot{x}_2 + 4c\dot{x}_1 + 4c\dot{x}_2$
$+ kx_1 = 2m_0g$,
$(m_2 + 4m_0)\ddot{x}_1 + \left(\dfrac{3}{2}m_2 + 4m_0\right)\ddot{x}_2 + 4c\dot{x}_1 + 4c\dot{x}_2 + k_2x_2$
$= (2m_0 - m_2)g$.

6-89 (a) $m\ddot{x}_1 + 2\dfrac{AE}{L}x_1 - \dfrac{AE}{L}x_2 = 0$,
$m\ddot{x}_2 - \dfrac{AE}{L}x_1 + 2\dfrac{AE}{L}x_2 = 0$.
(b) $\dfrac{mr^2}{2}\ddot{\theta}_1 + 2\dfrac{GJ}{L}\theta_1 - \dfrac{GJ}{L}\theta_2 = 0$,
$\dfrac{mr^2}{2}\ddot{\theta}_2 - \dfrac{GJ}{L}\theta_1 + 2\dfrac{GJ}{L}\theta_2 = 0$.

6-91 $\left(\dfrac{1}{2}m_1R_1^2 + \dfrac{1}{2}m_2R_2^2 + m_0R_1^2\right)\ddot{\theta} + cR_2^2\dot{\theta} + kR_2^2\theta = m_0gR_1$.

6-93 $\left(\dfrac{1}{3}m_1l_1^2 + \dfrac{1}{2}m_2r_2^2\right)\ddot{\theta} + k_t\theta + (k_1b^2 + k_2a^2)\sin\theta\cos\theta$
$= 0$.

6-95 (a) $\left(\dfrac{1}{12}l^2 + \dfrac{1}{3}h^2 + R^2\theta^2\right)\ddot{\theta} + R^2\dot{\theta}^2\theta$
$+ g\left(-\dfrac{1}{2}h\sin\theta + R\theta\cos\theta\right) = 0$.
(b) $\left(\dfrac{1}{12}l^2 + \dfrac{1}{3}h^2\right)\ddot{\theta} + \left(R - \dfrac{1}{2}h\right)g\theta = 0$.

6-97 (a) $\dfrac{3}{2}(R - r)\ddot{\theta} + g\sin\theta = 0$.
(b) $\ddot{\theta} + \dfrac{2}{3}\dfrac{g}{R - r}\theta = 0$.

6-99 (a) $\left(\dfrac{111}{12} - 2\cos\theta\right)\ddot{\theta} + 2\sin\theta\dot{\theta}^2 + \dfrac{g}{a}\sin\theta = 0$.
(b) $\dfrac{87}{12}\ddot{\theta} + \dfrac{g}{a}\theta = 0$.

6-101 (a) $m\ddot{x} = F$,
$\left[I_c + m(R - h)^2\sin^2\theta\right]\ddot{\theta} + m(R - h)^2\sin\theta\cos\theta\dot{\theta}^2$
$+ mg(R - h)\sin\theta$
$= -F\left[b\sin\theta + (h - a)\cos\theta\right]$.
(b) $(m + M)\ddot{x}' = F$,
$\left[I_c + I_p + \dfrac{mM}{m + M}(H - h)^2 + (m + M)\right.$
$\left. \times \left(R - \dfrac{mh + MH}{m + M}\right)^2\sin^2\theta\right]\ddot{\theta}$

$$+(m+M)\left(R-\frac{mh+MH}{m+M}\right)^2\sin\theta\cos\theta\,\dot\theta^2$$

$$+(m+M)g\left(R-\frac{mh+MH}{m+M}\right)\sin\theta$$

$$=-F\left[b\sin\theta+\left(\frac{mh+MH}{m+M}-a\right)\cos\theta\right].$$

6-103 $(M+m)\ddot{x}+m(R-r)\ddot\theta\cos\theta-m(R-r)\dot\theta^2\sin\theta$

$\qquad+c\dot{x}-c\dot{y}+kx-ky=F(t),$

$\qquad M_0\ddot{y}+c\dot{y}-c\dot{x}+ky-kx=0,$

$\qquad\dfrac{3}{2}(R-r)\ddot\theta+\ddot{x}\cos\theta+g\sin\theta=0.$

6-105 $(m_0+m_1+m)\ddot{y}+(c_1+c_2)\dot{y}-\dfrac{1}{2}l_0(c_1-c_2)\dot\theta$

$\qquad+(k_1+k_2)y-\dfrac{1}{2}l_0(k_1-k_2)\theta=0,$

$\qquad\left[\dfrac{1}{12}m_0l_0^2+\dfrac{1}{3}m_1l_1^2+m(l_1+r)^2+\dfrac{2}{5}mr^2\right]\ddot\theta$

$\qquad+\dfrac{1}{4}l_0^2(c_1+c_2)\dot\theta-\dfrac{1}{2}l_0(c_1-c_2)\dot{y}$

$\qquad+\left[\dfrac{1}{4}l_0^2(k_1+k_2)-\dfrac{1}{2}m_1gl_1-mg(l_1+r)\right]\theta$

$\qquad-\dfrac{1}{2}l_0(k_1-k_2)y=0.$

6-107 $\left[\dfrac{1}{2}m_1l+m_2\left(l+\dfrac{1}{2}a\right)\right](\ddot{x}_0\cos\theta-\dot{x}_0\dot\theta\sin\theta)$

$\qquad+\left[\dfrac{1}{3}m_1l^2+m_2\left(l+\dfrac{1}{2}a\right)^2+\dfrac{1}{12}m_2(a^2+b^2)\right]\ddot\theta$

$\qquad+\left[\dfrac{1}{2}m_1gl+m_2g\left(l+\dfrac{1}{2}a\right)\right]\sin\theta=0.$

6-109 $\left(m_1+\dfrac{1}{2}m_2\right)b^2\ddot\theta+cb^2\dot\theta+2kb^2\theta=af(t).$

6-111 $\left[m_1+m_2+\dfrac{1}{2}(m_3+m_4)\right]\ddot{x}+c\dot{x}+\left(k+\dfrac{k_t}{R^2}\right)x$

$\qquad=(m_1+m_2)g.$

6-113 $\left[I_1+m_1a^2+\dfrac{1}{3}m_3b^2+\left(I_2+m_2d^2+\dfrac{1}{3}m_4e^2\right)\dfrac{r^2}{r_1^2}\right]\ddot\theta$

$\qquad+ce^2\dfrac{r^2}{r_1^2}\dot\theta$

$\qquad\left[k_1b^2+k_{t1}-m_1ga+\dfrac{1}{2}m_3gb+\left(k_2e^2+m_2gd+k_{t2}\right)\dfrac{r^2}{r_1^2}\right]$

$\qquad\times\theta=bF(t).$

6-115 $m\ddot{r}+mr_0\ddot\phi\sin(\theta\ phi)-mr_0\dot\phi^2\cos(\theta-\phi)$

$\qquad-mr\dot\theta^2+k(r-l_0)-mg\cos\theta=0,$

$\qquad r\ddot\theta+2\dot{r}\dot\theta+r_0\ddot\phi\cos(\theta-\phi)$

$\qquad+r_0\dot\phi^2\sin(\theta-\phi)+g\sin\theta=0,$

$\qquad\dfrac{4}{3}r_0\ddot\phi+(r\ddot\theta+2\dot{r}\dot\theta)\cos(\theta-\phi)$

$\qquad+(\ddot{r}-r\dot\theta^2)\sin(\theta-\phi)+g\sin\phi=0.$

6-117 $\left[\dfrac{1}{2}(m_0+m_1)R^2-I_1-m_1\left(\dfrac{1}{2}b-r+a\right)^2\right.$

$\qquad\left.+ma^2+\dfrac{1}{2}mr^2\right]\ddot\theta$

$$mr\left(a+\dfrac{1}{2}r\right)\ddot\phi+2mr^2\phi\dot\phi\dot\theta+mr^2\phi^2\ddot\theta$$

$$+\dfrac{2GJ}{L}\theta=\tau(t),$$

$$\dfrac{2}{3}mr^2\ddot\phi+mr\left(a+\dfrac{1}{2}r\right)\ddot\theta+cr^2\dot\phi-mr^2\phi\dot\theta^2$$

$$+kr^2\phi=0.$$

6-119 $\left[\dfrac{1}{3}ml^2+I+I_0+(M+m_0)\left(l+\dfrac{1}{2}b\right)^2\right]\ddot\theta$

$\qquad+m_0\left(l+\dfrac{1}{2}b\right)\ddot{x}$

$\qquad+m_0x^2\ddot\theta+2m_0x\dot{x}\dot\theta+\left[\dfrac{1}{2}ml+(M+m_0)\right.$

$\qquad\left.\times\left(l+\dfrac{1}{2}b\right)\right]g\sin\theta+mgx\cos\theta=0,$

$\qquad m_0\ddot{x}+m_0\left(l+\dfrac{1}{2}b\right)\ddot\theta-m_0x\dot\theta^2+m_0g\sin\theta$

$\qquad+2kx=0.$

CHAPTER 7

7-1 (a) $\Phi_B=BA.$ $\qquad(b)$ $e=\dfrac{d\Phi_B}{dt}.$

$\qquad(c)$ $e=\dfrac{d(N\Phi_B)}{dt}.$ $\qquad(d)$ $\lambda=N\Phi_B.$

7-3 (a) $\quad(R_1+R_2)\dot{q}=E_2-E_1.$

$\qquad(b)$ $\quad R_2\dot{q}_1=-E_1+E_2+E_3,\quad R_1\dot{q}_2=-E_2.$

$\qquad(c)$ $\quad(R_1+R_2)\dot{q}_1-(R_1+R_2)\dot{q}_2=E_0,$

$\qquad\qquad-(R_1+R_2+R_3)\dot{q}_1+(R_1+R_2+R_3+R_5)\dot{q}_2$

$\qquad\qquad-R_3\dot{q}_3=0,$

$\qquad\qquad-R_3\dot{q}_2+(R_3+R_4)\dot{q}_3=0.$

7-5 (a) $\quad(C_1+C_2)\ddot\lambda_1=C_1\ddot{E}_0(t).$

$\qquad(b)$ $\quad\left(\dfrac{1}{R_1}+\dfrac{1}{R_2}\right)\dot\lambda_1-\dfrac{1}{R_2}\dot\lambda_2=\dfrac{E_0}{R_1},$

$\qquad\qquad-\dfrac{1}{R_2}\dot\lambda_1+\left(\dfrac{1}{R_2}+\dfrac{1}{R_5}\right)\dot\lambda_2-\dfrac{1}{R_5}\dot\lambda_3=0,$

$\qquad\qquad-\dfrac{1}{R_5}\dot\lambda_2+\left(\dfrac{1}{R_3}+\dfrac{1}{R_4}+\dfrac{1}{R_5}\right)\dot\lambda_3=\left(\dfrac{1}{R_3}+\dfrac{1}{R_4}\right)E_0.$

$\qquad(c)$ $\quad\dfrac{\dot\lambda_1}{R}+\dfrac{\dot\lambda_1}{L}=I_0.$

7-7 $RC\ddot\lambda_2+\dot\lambda_2+\dfrac{R}{L}\lambda_2=e_s(t).$

7-9 $C_1\ddot\lambda_2-C_1\ddot\lambda_3+\left(\dfrac{1}{R_s}+\dfrac{1}{R_1}\right)\dot\lambda_2-\dfrac{1}{R_1}\dot\lambda_3+\dfrac{1}{L}\lambda_2=\dfrac{e_s}{R_s},$

$\qquad-C_1\ddot\lambda_2+(C_1+C_2)\ddot\lambda_3-\dfrac{1}{R_1}\dot\lambda_2$

$\qquad+\left(\dfrac{1}{R_1}+\dfrac{1}{R_2}\right)\dot\lambda_3=0.$

7-11 $(L_1+L_2)\ddot{q}_1-L_2\ddot{q}_2+R_1\dot{q}_1-R_1\dot{q}_2+\dfrac{1}{C_1}q_1$

$\qquad-\dfrac{1}{C_1}q_4=E(t),$

$$-L_2\ddot{q}_1 + L_2\ddot{q}_2 - R_1\dot{q}_1 + (R_1 + R_3)\dot{q}_2 + \frac{1}{C_2}q_2$$
$$= -R_3 I(t),$$
$$L_3\ddot{q}_4 + R_2\dot{q}_4 - \frac{1}{C_1}q_1 + \frac{1}{C_1}q_4 = 0.$$

7-13 $(L_1 + L_2)\ddot{q}_1 - L_2\ddot{q}_2 + R_1\dot{q}_1 - R_1\dot{q}_4 + \frac{1}{C_1}q_1$
$$- \frac{1}{C_1}q_2 = E(t),$$
$$- L_2\ddot{q}_1 + L_2\ddot{q}_2 + (R_2 + R_3)\dot{q}_2 - \frac{1}{C_1}q_1$$
$$+ \left(\frac{1}{C_1} + \frac{1}{C_2}\right)q_2 = -R_3 I(t) - \frac{1}{C_3}q_0,$$
$$L_3\ddot{q}_4 - R_1\dot{q}_1 + R_1\dot{q}_4 + \frac{1}{C_1}q_4 = 0,$$
where $\dot{q}_0 = I(t)$.

7-15 $(L_1 + L_2)\ddot{q}_1 - L_2\ddot{q}_2 + \frac{1}{C_1}q_1 - \frac{1}{C_1}q_4 = E_1(t),$
$$- L_2\ddot{q}_1 + L_2\ddot{q}_2 + R_1\dot{q}_2 - R_1\dot{q}_3 - R_1\dot{q}_4 + \frac{1}{C_2}q_2 = 0,$$
$$L_3\ddot{q}_3 + L_3\ddot{q}_4 - R_1\dot{q}_2 + (R_1 + R_2)\dot{q}_3 + (R_1 + R_2)\dot{q}_4$$
$$= E_2(t), L_3\ddot{q}_3$$
$$+ L_3\ddot{q}_4 - R_1\dot{q}_2 + (R_1 + R_2)\dot{q}_3 + (R_1 + R_2$$
$$+ R_3)\dot{q}_4 - \frac{1}{C_1}q_1 + \left(\frac{1}{C_1} + \frac{1}{C_3}\right)q_4 = 0.$$

7-17 (a) $L\ddot{q}_1 + \frac{1}{C}q_1 - \frac{1}{C}q_2 = E(t),$
$$L\ddot{q}_2 - \frac{1}{C}q_1 + \frac{2}{C}q_2 - \frac{1}{C}q_3 = 0,$$
$$L\ddot{q}_3 - \frac{1}{C}q_2 + \frac{2}{C}q_3 - \frac{1}{C}q_4 = 0,$$
$$L\ddot{q}_4 - \frac{1}{C}q_3 + \frac{2}{C}q_4 = 0.$$
(b) $C\ddot{\lambda}_1 + \frac{2}{L}\lambda_1 - \frac{1}{L}\lambda_2 = \frac{\lambda_0}{L},$
$$C\ddot{\lambda}_2 - \frac{1}{L}\lambda_1 + \frac{2}{L}\lambda_2 - \frac{1}{L}\lambda_3 = 0,$$
$$C\ddot{\lambda}_3 - \frac{1}{L}\lambda_2 + \frac{2}{L}\lambda_3 - \frac{1}{L}\lambda_4 = 0,$$
$$C\ddot{\lambda}_4 - \frac{1}{L}\lambda_3 + \frac{1}{L}\lambda_4 = 0, \quad \text{where } \dot{\lambda}_0 = E(t).$$

7-19 (a) $C_1\ddot{\lambda}_1 + \left(\frac{1}{R_1} + \frac{1}{R_2} + \frac{1}{R_3}\right)\dot{\lambda}_1 = \frac{e_s}{R_1} + \frac{e_o}{R_3},$
$$\frac{1}{R_2}\dot{\lambda}_1 = -C_2\dot{e}_o.$$
(b) $R_1\dot{q}_1 + \frac{1}{C_1}q_1 - \frac{1}{C_1}q_3 = e_s,$
$$(R_2 + R_3)\dot{q}_2 + R_2\dot{q}_3 + \frac{1}{C_2}q_2 - \frac{1}{C_2}q_3 = 0,$$
$$- R_2\dot{q}_2 + R_2\dot{q}_3 - \frac{1}{C_1}q_1 - \frac{1}{C_2}q_2$$
$$+ \left(\frac{1}{C_1} + \frac{1}{C_2}\right)q_3 = 0.$$

7-21 $\lambda_3 = -\frac{L}{R_1}e_s(t),$
$$\frac{1}{R_2}\dot{\lambda}_3 + \frac{1}{L}\lambda_3 = \frac{e_o(t)}{R_2};$$
$$e_0(t) = -\frac{L}{R_1}\dot{e}_s(t) - \frac{R_2}{R_1}e_s(t).$$

7-25 $C\ddot{\lambda}_2 + \left(\frac{1}{R_s} + \frac{1}{R}\right)\dot{\lambda}_2 + \frac{x_0}{L_0(x_0 + x)}\lambda_2 = \frac{E(t)}{R_s}.$

7-27 $(m + M)\ddot{x} - m(x + c_0)\dot{\theta}^2 + \frac{BE^2(t)}{2(x + a_0)^2} = -Mg,$
$$\frac{d}{dt}\left[m(x + c_0)^2\dot{\theta}\right] = 0$$
where c_0 is a constant.

7-29 $m_2\ddot{x}_2 + (c_1 + c_2)\dot{x}_2 - c_2\dot{x}_3 + (k_1 + k_2)x_2 - k_3x_3$
$$= c_1\dot{x}_0(t) + k_1x_0(t),$$
$$m_3\ddot{x}_3 - c_2\dot{x}_2 + c_2\dot{x}_3 - k_2x_2 + k_2x_3 + \frac{1}{2C_0d_0}q_1^2 = 0,$$
$$L\ddot{q}_1 + R_1\dot{q}_1 + \left(\frac{d_0 - x}{C_0d_0} - \frac{1}{C_1}\right)q_1 - \frac{1}{C_1}q_2 = E(t),$$
$$C_1R_2\dot{q}_2 - q_1 + q_2 = 0.$$

7-31 (a) $f(x, i) = -\frac{1}{4}Ai^4 + kx.$
(b) $f(x, \lambda) = kx - \frac{1}{4}\frac{\lambda^{4/3}}{A^{1/3}(x + \ell)^{4/3}}.$

CHAPTER 8

8-3 (a) 8.27 s.   (b) Yes.

8-11 (a) 22.15 rad/s.   (b) 0.05 m.

8-13 0.216 m.

8-15 (b) $\tilde{\nu}(t) = (\nu_0 - \nu_f)e^{-t/\tau},$ $t \geq 0.$
(c) $\nu(t) = (\nu_0 - \nu_f)e^{-t/\tau} + \nu_f,$ $t \geq 0.$

8-17 The solution is identical to that in eqn. (8-37).

8-21 (a) $x_p(t) = a_0,$ $t \geq 0.$
(b) $x_p(t) = a_0\left(t - \frac{c}{k}\right),$ $t \geq 0.$
(c) $x_p(t) = a_0\left(t - \frac{c}{k}\right) + a_0,$ $t \geq 0.$
(d) Yes.

8-23 $v(t) = \frac{m_2 g}{c}\left(1 - e^{-t/\tau}\right)$ where $\tau = \frac{2(m_1 + m_2) + m_0}{2c}.$

8-25 $q(t) = (q_0 - CE_0)e^{-\frac{t}{RC}} + CE_0,$ $t \geq 0.$

8-27 $C_0 = \sqrt{x_0^2 + \left(\frac{v_0}{\omega_n}\right)^2};$ $\psi_1 = \tan^{-1}\left(\frac{x_0\omega_n}{v_0}\right);$
$$\psi_2 = \tan^{-1}\left(\frac{v_0}{x_0\omega_n}\right).$$

8-29 $C = \sqrt{x_0^2 + \left(\frac{v_0 + \zeta\omega_n x_0}{\omega_d}\right)^2};$
$$\psi_1 = \tan^{-1}\left(\frac{x_0\omega_d}{v_0 + \zeta\omega_n x_0}\right);$$

$$\psi_2 = \tan^{-1}\left(\frac{v_0 + \zeta\omega_n x_0}{x_0\omega_d}\right).$$

8-31   122 steps.

8-33   6.21 cm.

8-35   7.85 rad/s.

8-37   (a) $\dfrac{Mgl^3}{51EI}.$ (b) $\sqrt{\dfrac{51EI}{Ml^3}}.$

8-39   (a) $\theta(t) = -\dfrac{m_0 b}{l^2}\sqrt{\dfrac{6gh}{mk}}\sin\sqrt{\dfrac{3k}{m}}t.$

   (b) $b \le \dfrac{f_{max}l}{m_0 b}\sqrt{\dfrac{m}{6ghk}}.$

8-41   (a) $L - t_0 < l_0.$ (b) $l_1 = l_0 - \dfrac{k_2}{k_1 + k_2}$

   $\times (2l_0 + t_0 - L).$ (c) $\sqrt{\dfrac{k_1 + k_2}{m}}.$

8-43   (a) $\omega_n = \sqrt{\dfrac{2kR_2^2}{(2m_0 + m_1)R_1^2 + m_2 R_2^2}}.$

   (b) $\theta(t) = e^{-\zeta\omega_n t}\dfrac{m_0 gR_1}{kR_2^2}\left[\cos\left(\sqrt{1 - \zeta^2}\,\omega_n t\right)\right.$

   $\left. + \dfrac{\zeta}{\sqrt{1 - \zeta^2}}\sin\left(\sqrt{1 - \zeta^2}\,\omega_n t\right)\right],$

   where $\zeta = \dfrac{cR_2}{\sqrt{2k[(2m_0 + m_1)R_1^2 + m_2 R_2^2]}}.$

8-45 $\omega_n = \sqrt{\dfrac{12ka^2 + 6(m_1 + 2m_2 + 4m_3)gl_1}{4m_1 l_1^2 + m_2(12l_1^2 + l_2^2) + 2m_3(12l_1^2 + 3l_2^2 + l_3^2)}}.$

8-47 $\omega_n = \sqrt{\dfrac{6(k_t + k_1 b^2 + k_2 a^2)}{2m_1 l_1^2 + 3m_2 r_2^2}}.$

8-49   (a) $x(t) = te^{-\omega_n t}\sqrt{2gh}.$ (b) $h \le 3.695\dfrac{f_{max}^2}{mgk}.$

8-51   (a) 8.0 rad/s.   (b) $\theta(t) = 5.0\cos 8.0t$ deg.

8-55 $\omega_1 = 1.095\sqrt{\dfrac{k}{m}},\quad \omega_2 = 1.732\sqrt{\dfrac{k}{m}};$

   $\{u^{(1)}\} = \left\{\begin{array}{c} 1 \\ 2.5 \end{array}\right\},\quad \{u^{(2)}\} = \left\{\begin{array}{c} 1 \\ -2.0 \end{array}\right\}.$

8-57 $\omega_1 = 0.848\sqrt{\dfrac{k}{m}},\quad \omega_2 = 1.67\sqrt{\dfrac{k}{m}};$

   $\{u^{(1)}\} = \left\{\begin{array}{c} 1 \\ 0.640 \end{array}\right\},\quad \{u^{(2)}\} = \left\{\begin{array}{c} 1 \\ -0.390 \end{array}\right\}.$

8-59 $\omega_1 = 0.835\sqrt{\dfrac{k}{m}},\quad \omega_2 = 2.07\sqrt{\dfrac{k}{m}};$

   $\{u^{(1)}\} = \left\{\begin{array}{c} 1 \\ -\dfrac{0.434}{d} \end{array}\right\},\quad \{u^{(2)}\} = \left\{\begin{array}{c} 1 \\ \dfrac{0.768}{d} \end{array}\right\}.$

8-61 $\omega_1 = 0.618\sqrt{\dfrac{k}{m}},\quad \omega_2 = 1.62\sqrt{\dfrac{k}{m}};$

   $\{u^{(1)}\} = \left\{\begin{array}{c} 1 \\ \dfrac{3.24}{l} \end{array}\right\},\quad \{u^{(2)}\} = \left\{\begin{array}{c} 1 \\ -\dfrac{1.24}{l} \end{array}\right\}.$

8-63 $\omega_1 = 0.61803\sqrt{\dfrac{GJ}{I_0 l}},\quad \omega_2 = 1.61803\sqrt{\dfrac{GJ}{I_0 l}};$

   $\{u^{(1)}\} = \left\{\begin{array}{c} 1 \\ 1.61803 \end{array}\right\},\quad \{u^{(2)}\} = \left\{\begin{array}{c} 1 \\ -0.61803 \end{array}\right\}.$

8-65 $\omega_1 = \sqrt{\dfrac{k}{5m}},\quad \omega_2 = \sqrt{\dfrac{k}{m}};\quad \{u^{(1)}\} = \left\{\begin{array}{c} 1 \\ 1 \end{array}\right\},$

   $\{u^{(2)}\} = \left\{\begin{array}{c} 1 \\ -1 \end{array}\right\}.$

8-67   (a) $\omega_1 = 0,\quad \omega_2 = \sqrt{\dfrac{3k}{m_1}};\quad \{u^{(1)}\} = \left\{\begin{array}{c} 1 \\ 1 \end{array}\right\},$

   $\{u^{(2)}\} = \left\{\begin{array}{c} 1 \\ -2 \end{array}\right\}.$

   (b) $\{x(t)\} = \left\{\begin{array}{c} x_1(t) \\ x_2(t) \end{array}\right\}$

   $= \left\{\begin{array}{c} \dfrac{v_0}{3}t - \dfrac{v_0}{3}\sqrt{\dfrac{m_1}{3k}}\sin\sqrt{\dfrac{3k}{m_1}}t \\[2mm] \dfrac{v_0}{3}t + \dfrac{2v_0}{3}\sqrt{\dfrac{m_1}{3k}}\sin\sqrt{\dfrac{3k}{m_1}}t \end{array}\right\}.$

   (c) $f_l = -kv_0\sqrt{\dfrac{m_1}{3k}}\sin\sqrt{\dfrac{3k}{m_1}}t.$

   (d) $(v_0)_{max} = f_{max}\sqrt{\dfrac{3}{m_1 k}}.$

8-69 $\left\{\begin{array}{c} x_1(t) \\ x_2(t) \end{array}\right\} = 2.36v_0\sqrt{\dfrac{m_0}{k_0}}\left\{\begin{array}{c} 1 \\ 0.562 \end{array}\right\}\sin\left(0.468\sqrt{\dfrac{k_0}{m_0}}t\right)$

   $- 0.0704v_0\sqrt{\dfrac{m_0}{k_0}}\left\{\begin{array}{c} 1 \\ -3.56 \end{array}\right\}\sin\left(1.51\sqrt{\dfrac{k_0}{m_0}}t\right).$

8-71   (a) $\omega_1 = 0.222\sqrt{\dfrac{k}{m_0}},\quad \omega_2 = 1.29\sqrt{\dfrac{k}{m_0}};$

   $\{u^{(1)}\} = \left\{\begin{array}{c} 1 \\ 0.0206 \end{array}\right\},\quad \{u^{(2)}\} = \left\{\begin{array}{c} 1 \\ -32.4 \end{array}\right\}.$

   (b) $\left\{\begin{array}{c} x_1(t) \\ x_2(t) \end{array}\right\} = 4.50v_0\sqrt{\dfrac{m_0}{k_1}}\left\{\begin{array}{c} 1 \\ 0.0206 \end{array}\right\}\sin\left(2.22\sqrt{\dfrac{k_1}{m_0}}t\right)$

   $+ 0.000493v_0\sqrt{\dfrac{m_0}{k_1}}\left\{\begin{array}{c} 1 \\ -32.4 \end{array}\right\}\sin\left(1.29\sqrt{\dfrac{k_1}{m_0}}t\right).$

   (c) $v_0 \le \dfrac{F_Y}{4.41\sqrt{m_0 k_1}}.$

8-75   (a) $\dot{\boldsymbol{x}} = \begin{bmatrix} 0 & 1 \\ 0 & -\dfrac{c_t}{I} \end{bmatrix}\boldsymbol{x} + \left\{\begin{array}{c} 0 \\ -\dfrac{1}{I} \end{array}\right\}T(t).$

   (b) $\dot{\boldsymbol{x}} = \begin{bmatrix} 0 & 1 & 0 & 0 \\ -\dfrac{m_1 g + kl_1}{m_1 l_1} & 0 & \dfrac{kl_2}{m_1 l_1} & 0 \\ 0 & 0 & 0 & 1 \\ \dfrac{kl_1}{m_2 l_2} & 0 & -\dfrac{m_2 g + kl_2}{m_2 l_2} & 0 \end{bmatrix}\boldsymbol{x}.$

8-77   (a) $h > 0.$ (b) $\sqrt{\dfrac{Wh}{I_c}}.$

8-79   (a) $\theta_e = 0, \pi.$

   (b)   At $\theta_e = 0,$

   $(m_1 a^2 + m_2 b^2)\ddot{\epsilon} + (-m_1 a + m_2 b)g\epsilon = 0;$

At $\theta_e = \pi$, $(m_1 a^2 + m_2 b^2)\ddot{\epsilon}$
$-(-m_1 a + m_2 b)g\epsilon = 0$.

(c) The equilibrium position $\theta_e = 0$ is stable only when $m_2 b > m_1 a$, in which case,

$\tilde{\omega}_n = \sqrt{\dfrac{-m_1 a + m_2 b}{m_1 a^2 + m_2 b^2}\, g}$; the equilibrium position $\theta_e = \pi$ is stable only when $m_1 a > m_2 b$, in which case, $\tilde{\omega}_n = \sqrt{\dfrac{m_1 a - m_2 b}{m_1 a^2 + m_2 b^2}\, g}$.

8-81 (a) $\omega_n = 2\sqrt{\dfrac{k}{m}}$.   (c) $\theta(t) = \sqrt{\dfrac{m}{3k}}\, e^{-\sqrt{\frac{k}{m}}t}\sin\sqrt{\dfrac{3k}{m}}\,t$.

8-83 (a) 8.245 N.   (b) No.   (c) 95.54 rpm.

8-85 (a) $\sqrt{\dfrac{2EI}{mL^3}}$.   (b) $\sqrt{\dfrac{2EI}{mL^3}}$.   (c) $\sqrt{\dfrac{2AE}{3mL}}$.

     (d) $\sqrt{\dfrac{2GJ}{mr^2 L}}$.

8-87 (a) $x(t) = \dfrac{a\cos(\Omega t - \phi) + b\sin(\Omega t - \phi)}{A}$

$+ e^{-\zeta\omega_n t}\left\{\left(x_0 - \dfrac{a\cos\phi - b\sin\phi}{A}\right)\cos\omega_d t\right.$

$+ \dfrac{1}{\omega_d}\left[v_0 + \zeta\omega_n x_0\right.$

$\left.\left. - \dfrac{a(\Omega\sin\phi + \zeta\omega_n\cos\phi) - b(\zeta\omega_n\sin\phi - \Omega\cos\phi)}{A}\right]\sin\omega_d t\right\}$,

where $A = \sqrt{(k - m\Omega^2)^2 + (c\Omega)^2}$,

$\phi = \tan^{-1}\dfrac{c\Omega}{k - m\Omega^2}$.

(b) $x(t) = \dfrac{a\cos(\Omega t - \phi) + b\sin(\Omega t - \phi)}{A}$

$+ e^{-\omega_n t}\left\{\left(x_0 - \dfrac{a\cos\phi - b\sin\phi}{A}\right)\right.$

$+ \left[v_0 + \omega_n x_0\right.$

$\left.\left. - \dfrac{a(\Omega\sin\phi + \omega_n\cos\phi) - b(\omega_n\sin\phi - \Omega\cos\phi)}{A}\right]t\right\}$.

(c) $x(t) = \dfrac{a\cos(\Omega t - \phi) + b\sin(\Omega t - \phi)}{A}$

$+ e^{-\zeta\omega_n t}\left\{\left(x_0 - \dfrac{a\cos\phi - b\sin\phi}{A}\right)\right.$

$\times\cosh\left(\sqrt{\zeta^2 - 1}\,\omega_n t\right) + \dfrac{1}{\omega_n\sqrt{\zeta^2 - 1}}\left[v_0 + \zeta\omega_n x_0\right.$

$\left. - \dfrac{a(\Omega\sin\phi + \zeta\omega_n\cos\phi) - b(\zeta\omega_n\sin\phi - \Omega\cos\phi)}{A}\right]$

$\left.\times\sinh\left(\sqrt{\zeta^2 - 1}\,\omega_n t\right)\right\}$.

8-89 (a) $\{u^{(1)}\} = \left\{\begin{array}{c} 1 \\ -\dfrac{0.3}{l} \end{array}\right\}$, $\{u^{(2)}\} = \left\{\begin{array}{c} 1 \\ \dfrac{3.3}{l} \end{array}\right\}$.

(b) Any initial condition that satisfies $r_1 x_{10} - x_{20} = 0$ and $r_1 v_{10} - v_{20} = 0$.

8-91 (a) $\omega_1 = \sqrt{\dfrac{g}{l}}$, $\omega_2 = \sqrt{\dfrac{mgl + 2k_t}{ml^2}}$;

$\{u^{(1)}\} = \left\{\begin{array}{c} 1 \\ 1 \end{array}\right\}$, $\{u^{(2)}\} = \left\{\begin{array}{c} 1 \\ -1 \end{array}\right\}$.

(b) 52.4 s.

8-93 (a) $\omega_1 = \sqrt{\dfrac{k}{5m}}$, $\omega_2 = \sqrt{\dfrac{k}{2m}}$;

$\{u^{(1)}\} = \left\{\begin{array}{c} 1 \\ 1 \\ \dfrac{1}{2a} \end{array}\right\}$, $\{u^{(2)}\} = \left\{\begin{array}{c} 1 \\ -1 \\ -\dfrac{1}{a} \end{array}\right\}$.

(b) $x(t) = \dfrac{2}{3}a\theta_0\left[\cos\left(\sqrt{\dfrac{k}{5m}}\,t\right) - \cos\left(\sqrt{\dfrac{k}{2m}}\,t\right)\right]$,

$\theta(t) = \dfrac{\theta_0}{3}\left[\cos\left(\sqrt{\dfrac{k}{5m}}\,t\right) + 2\cos\left(\sqrt{\dfrac{k}{2m}}\,t\right)\right]$.

8-95 $mI_c\Omega^4 - \left(\dfrac{mkL^2}{4} + kI_c + mk_t - \dfrac{m^2 gL}{2}\right)\Omega^2$

$+ k\left(k_t - \dfrac{mgL}{2}\right) = 0$.

8-97 (a) $y_c(t) = A\sin\Omega t$, $\theta(t) = B\sin\Omega t$,

where $A = \dfrac{F_0 ml^2}{3\Delta}\left(\dfrac{7}{8}\dfrac{k}{m} - 2\Omega^2\right)$,

$B = \dfrac{F_0 ml}{\Delta}\left(\dfrac{k}{m} - \Omega^2\right)$,

$\Delta = \dfrac{m^2 l^2}{8}\left(16\Omega^4 - 15\dfrac{k}{m}\Omega^2 + 4\dfrac{k^2}{m^2}\right)$.

(b) $\Omega = \sqrt{\dfrac{k}{m}}$.   (c) $\Omega = \sqrt{\dfrac{7k}{16m}}$.

(d) $\Omega = \sqrt{\dfrac{15 \pm \sqrt{97}}{32}}\sqrt{\dfrac{k}{m}}$.

8-99 (a) $\theta_{e_1} = 0$, $\theta_{e_2} = \cos^{-1}\dfrac{g}{l\Omega_0^2}$ $\left(\dfrac{g}{l} < \Omega_0^2\right)$.

(b) The equilibrium position $\theta_{e_1} = 0$ is stable when $\dfrac{g}{l} > \Omega_0^2$, unstable when $\dfrac{g}{l} < \Omega_0^2$; the equilibrium position θ_{e_2} is stable.

(c) $\theta_{e_1} : \tilde{\omega}_n = \sqrt{\dfrac{g}{l} - \Omega_0^2}$;

$\theta_{e_2} : \tilde{\omega}_n = \sqrt{\Omega_0^2 - \left(\dfrac{g}{l\Omega_0}\right)^2}$.

8-103 (b) $m\ddot{x} + \dfrac{L_0}{2a}\dfrac{1}{\left(1 + \dfrac{x}{a}\right)^2}[I + G(x - x_A)]^2 = mg$.

(c) $m\ddot{\epsilon} + \dfrac{L_0 Ia}{(a + x_A)^2}\left(G - \dfrac{I}{a + x_A}\right)\epsilon = 0$.

(d) $G > \dfrac{I}{a + x_A}$.

(e) $\tilde{\omega}_n^2 = \dfrac{L_0 Ia}{m(a + x_A)^2}\left(G - \dfrac{I}{a + x_A}\right)$.

CHAPTER 9

9-1 $m\ddot{x} + c\dot{x} + kx = f(t).$

9-3 (a) $EA\dfrac{\partial \xi}{\partial x} = k\xi$ at $x = 0,$ $\quad \dfrac{\partial \xi}{\partial x} = 0$ at $x = l.$

(b) $\tan\dfrac{\omega l}{c_q} = \dfrac{kc_q}{EA\omega}$ $\quad (\omega \neq 0)$ where $c_q = \sqrt{\dfrac{E}{\rho}}.$

9-5 (a) $EA\dfrac{\partial \xi}{\partial x} = M_1\dfrac{\partial^2 \xi}{\partial t^2}$ at $x = 0,$

$EA\dfrac{\partial \xi}{\partial x} = -M_2\dfrac{\partial^2 \xi}{\partial t^2}$ at $x = l.$

(b) $\tan\dfrac{\omega l}{c_q} = \dfrac{(M_1 + M_2)EAc_q\omega}{M_1M_2c_q^2\omega^2 - E^2A^2}$ $\quad (\omega \neq 0)$

where $c_q = \sqrt{\dfrac{E}{\rho}}.$

9-7 (a) $EA\dfrac{\partial \xi}{\partial x} = k\xi$ at $x = 0,$

$EA\dfrac{\partial \xi}{\partial x} = -M\dfrac{\partial^2 \xi}{\partial t^2}$ at $x = l.$

(b) $\tan\dfrac{\omega l}{c_q} = \dfrac{c_q EA(k - M\omega^2)}{\omega(kMc_q^2 + E^2A^2)}$ $\quad (\omega \neq 0)$

where $c_q = \sqrt{\dfrac{E}{\rho}}.$

9-9 (a) $X_j(x) = d_j \sin\dfrac{\omega_j}{c_q}x$ where ω_j's satisfy $\tan\dfrac{\omega_j l}{c_q}$

$\quad = -\dfrac{EA\omega_j}{kc_q}$ for $j = 1, 2, \ldots.$

9-11 (a) $\omega_1 = 16.8\,\text{rad/s};$ $\quad X_1(x) = d_1 \sin 0.00314x.$

(b) $\omega_1 = 8.40\,\text{rad/s};$ $\quad X_1(x) = d_1 \sin 0.00157x.$

9-13 $m(x)\dfrac{\partial^2 \xi}{\partial t^2} = \dfrac{\partial}{\partial x}\left(EA(x)\dfrac{\partial \xi}{\partial x}\right)$ $\quad 0 < x < l;$

Natural boundary condition: $EA\dfrac{\partial \xi}{\partial x} + c\dfrac{\partial \xi}{\partial t} = F(t)$
at $x = l,$

Geometric boundary condition: $\xi(0, t) = 0$
at $x = 0.$

9-15 (a) $GJ\dfrac{\partial \phi}{\partial x} = I_1\dfrac{\partial^2 \phi}{\partial t^2}$ at $x = 0$ (Natural),

$GJ\dfrac{\partial \phi}{\partial x} = -I_2\dfrac{\partial^2 \phi}{\partial t^2}$ at $x = l$ (Natural).

(b) $\tan\dfrac{\omega l}{c_q} = \dfrac{c_q\omega GJ(I_1 + I_2)}{c_q^2 I_1 I_2 \omega^2 - G^2J^2}$ $\quad (\omega \neq 0)$ where

$c_q = \sqrt{\dfrac{G}{\rho}}.$

9-17 (a) $GJ\dfrac{\partial \phi}{\partial x} = k_t\phi$ at $x = 0$ (Natural),

$GJ\dfrac{\partial \phi}{\partial x} = -I_0\dfrac{\partial^2 \phi}{\partial t^2}$ at $x = l$ (Natural).

(b) $\tan\dfrac{\omega l}{c_q} = \dfrac{c_q GJ(k_t - I_0\omega^2)}{k_t c_q^2 I_0\omega + G^2J^2\omega}$ $\quad (\omega \neq 0)$ where

$c_q = \sqrt{\dfrac{G}{\rho}}.$

9-19 (a) $\phi_1(0, t) = 0$ at $x_1 = 0$ (Geometric),

$\dfrac{\partial \phi_2}{\partial x_2} = 0$ at $x_2 = l_2$ (Natural);

$\phi_1(l_1, t) = \phi_2(0, t)$ at $x_1 = l_1$ and $x_2 = 0$ (Geometric),

$G_1 J_1\dfrac{\partial \phi_1}{\partial x_1} = G_2 J_2\dfrac{\partial \phi_2}{\partial x_2}$ at $x_1 = l_1$ and $x_2 = 0$ (Natural).

(b) $\tan\dfrac{\omega l_1}{c_{q1}}\tan\dfrac{\omega l_2}{c_{q2}} = \dfrac{c_{q2} G_1 J_1}{c_{q1} G_2 J_2}$ $\quad (\omega \neq 0)$ where

$c_{q1} = \sqrt{\dfrac{G_1}{\rho_1}}$ and $c_{q2} = \sqrt{\dfrac{G_2}{\rho_2}}.$

9-21 (a) $\dfrac{\partial \phi_1}{\partial x_1} = 0$ at $x_1 = 0$ (Natural),

$\phi_2(l_2, t) = 0$ at $x_2 = l_2$ (Geometric),

$r_1\phi_1(l_1, t) = -r_2\phi_2(0, t)$ at $x_1 = l_1$ and $x_2 = 0$ (Geometric),

$r_2 I_1\dfrac{\partial^2 \phi_1}{\partial t^2} - r_1 I_2\dfrac{\partial^2 \phi_2}{\partial t^2} = -r_2 G_1 J_1\dfrac{\partial \phi_1}{\partial x_1}$

$\quad - r_1 G_2 J_2\dfrac{\partial \phi_2}{\partial x_2}$

at $x_1 = l_1$ and $x_2 = 0$ (Natural).

(b) $\omega(r_1^2 I_2 + r_2^2 I_1)\tan\dfrac{\omega l_2}{c_{q2}} + \dfrac{r_2^2 G_1 J_1}{c_{q1}}\tan\dfrac{\omega l_1}{c_{q1}}\tan\dfrac{\omega l_2}{c_{q2}}$

$\quad = \dfrac{r_1^2 G_2 J_2}{c_{q2}}$ $\quad (\omega \neq 0)$

where $c_{q1} = \sqrt{\dfrac{G_1}{\rho_1}}$ and $c_{q2} = \sqrt{\dfrac{G_2}{\rho_2}}.$

9-23 (a) $\eta(0, t) = 0$ at $x = 0$ (Geometric),

$EI\dfrac{\partial^2 \eta}{\partial x^2} = 0$ at $x = 0$ (Natural),

$EI\dfrac{\partial^3 \eta}{\partial x^3} = m\dfrac{\partial^2 \eta}{\partial t^2} + k\eta$ at $x = l$ (Natural),

$EI\dfrac{\partial^2 \eta}{\partial x^2} = -I_0\dfrac{\partial^3 \eta}{\partial x\partial t^2}$ at $x = l$ (Natural).

(b) $[-EI\beta^3\cos\beta l + (m\omega^2 - k)\sin\beta l]$
$\times(EI\beta\sinh\beta l - I_0\omega^2\cosh\beta l)$
$+ [EI\beta^3\cosh\beta l + (m\omega^2 - k)\sinh\beta l]$
$\times(EI\beta\sin\beta l + I_0\omega^2\cos\beta l) = 0.$

where $\beta = \sqrt{\dfrac{\omega}{c_q}},\ c_q = \sqrt{\dfrac{EI}{\rho A}},\ \omega \neq 0$ and $\beta \neq 0.$

9-25 $\rho A\dfrac{\partial^2 \eta}{\partial t^2} = P\dfrac{\partial^2 \eta}{\partial x^2} + f(x, t)$ $\quad 0 < x < l;$

Boundary conditions: $\eta(0, t) = 0$ (Geometric) and
$\eta(l, t) = 0$ (Geometric).

9-27 $\dfrac{M}{m} = \dfrac{48}{\pi^4}.$

9-29 $\rho A\dfrac{\partial^2 \eta}{\partial t^2} = \kappa GA\dfrac{\partial^2 \eta}{\partial y^2}$ $\quad 0 < y < h;$

$\kappa GA\dfrac{\partial \eta}{\partial y} = 0$ at $y = h$ (Natural),

$M\dfrac{\partial^2 \eta}{\partial t^2} + c\left[\dfrac{\partial \eta}{\partial t} - \dot{x}_0(t)\right] + k[\eta - x_0(t)] = \kappa GA\dfrac{\partial \eta}{\partial y}$

at $y = 0$ (Natural).

9-31 $\dot{\xi}(x,t) = \sum_{n=1}^{\infty} \alpha_t \Delta T (-1)^{n+1} \dfrac{4c_q}{(2n-1)\pi}$

$\times \sin\left[\left(n - \dfrac{1}{2}\right)\dfrac{\pi c_q t}{l}\right] \sin\left[\left(n - \dfrac{1}{2}\right)\dfrac{\pi x}{l}\right].$

9-33 $\xi(x,t) = \sum_{n=1}^{\infty} \dfrac{F_0 l}{n\pi AE}\left(\dfrac{2}{n\pi}\sin\dfrac{n\pi}{2} - \cos n\pi\right)$

$\times \cos\dfrac{n\pi c_q t}{l}\sin\dfrac{n\pi x}{l}, \text{ where } c_q = \sqrt{\dfrac{E}{\rho}}.$

9-35 (a) 191 °C.   (b) 62.5 °C.   (c) 62.3 °C.

9-39 $\eta(x,t) = \sum_{n=1}^{\infty} \dfrac{-2v_0}{n\pi\omega_n}\left(\cos\dfrac{n\pi(b+\Delta l)}{l} - \cos\dfrac{n\pi b}{l}\right)$

$\times \sin\omega_n t \sin\dfrac{n\pi x}{l},$

where $\omega_n = \left(\dfrac{n\pi}{l}\right)^2\sqrt{\dfrac{EI}{\rho A}}.$

9-41 (a) $\sigma(x,t) = \dfrac{Ef_0}{\rho A\Omega c_q}$

$\times \left[-\sin\dfrac{\Omega x}{c_q} + \dfrac{1 - \cos\dfrac{\Omega l}{c_q}}{\sin\dfrac{\Omega l}{c_q}}\cos\dfrac{\Omega x}{c_q}\right]\cos\Omega t.$

(b) $\Omega_1 = \dfrac{\pi c_q}{l};$

$\sigma_{max} = \dfrac{Ef_0}{\rho A\Omega c_q}\dfrac{1}{\sin\dfrac{\Omega l}{c_q}}\left(1 - \cos\dfrac{\Omega l}{c_q}\right).$

(c) $\xi(x) = \dfrac{f_0}{2EA}x(l - x).$

(d) Static case: $\sigma_{max} = \dfrac{f_0 l}{2A};$

For $\Omega = \dfrac{1}{4}\Omega_1,$ $\sigma_{max} = 0.527\dfrac{f_0 l}{A};$

For $\Omega = \dfrac{1}{2}\Omega_1,$ $\sigma_{max} = 0.637\dfrac{f_0 l}{A};$

For $\Omega = \dfrac{3}{4}\Omega_1,$ $\sigma_{max} = 1.025\dfrac{f_0 l}{A};$

(e) For $\Omega = \dfrac{1}{4}\Omega_1,$ $f_0 \le 1.90\dfrac{A\sigma_Y}{l};$

For $\Omega = \dfrac{1}{2}\Omega_1,$ $f_0 \le 1.57\dfrac{A\sigma_Y}{l};$

For $\Omega = \dfrac{3}{4}\Omega_1,$ $f_0 \le 0.976\dfrac{A\sigma_Y}{l}.$

9-43 $\eta(x,t) = \dfrac{Y_0}{\Delta}[(\cosh Bl\cos Bl - \sinh Bl\sin Bl + 1)$

$\times (\cos Bx - \cosh Bx)$

$+ (\sinh Bl\cos Bl + \cosh Bl\sin Bl)(\sin Bx - \sinh Bx)$

$+ 2(1 + \cos Bl\cosh Bl)\cosh Bx]\cos\Omega t,$

where $\Delta = 2(1 + \cos Bl\cosh Bl)$ and $B = \sqrt{\Omega\sqrt{\dfrac{\rho A}{EI}}}.$

LIST OF TABLES

INDEX

Lumped-Parameter Offering of Variational Electricity

	Charge Formulation	Flux Linkage Formulation
• *Generalized Coordinates*	$q_k \quad k = 1, 2, \ldots, n_e$	$\lambda_k \quad k = 1, 2, \ldots, n'_e$
• *Lagrangian*	$\mathcal{L}(q_k, \dot{q}_k) = W^*_m - W_e$	$\mathcal{L}(\lambda_k, \dot{\lambda}_k) = W^*_e - W_m$

• *State Functions*

Linear: $\quad W^*_m = \dfrac{1}{2} L \dot{q}^2$ $\qquad$ Linear: $\quad W^*_e = \dfrac{1}{2} C \dot{\lambda}^2$

$\qquad\qquad\qquad W_e = \dfrac{q^2}{2C}$ $\qquad\qquad\qquad\qquad W_m = \dfrac{\lambda^2}{2L}$

Nonlinear: $\quad W^*_m = \displaystyle\int_0^{\dot{q}} \lambda \, d\dot{q}$ $\qquad$ Nonlinear: $\quad W^*_e = \displaystyle\int_0^{\dot{\lambda}} q \, d\dot{\lambda}$

$\qquad\qquad\qquad W_e = \displaystyle\int_0^{q} \dot{\lambda} \, dq$ $\qquad\qquad\qquad\qquad W_m = \displaystyle\int_0^{\lambda} \dot{q} \, d\lambda$

• *Work Expressions*

$$\delta \mathcal{W}^{\text{nc}} = \sum_{j=1}^{M} e_j^{\text{nc}} \, \delta q_j = \sum_{k=1}^{n_e} \mathcal{E}_k \, \delta q_k \qquad \delta \mathcal{W}^{\text{nc}} = \sum_{j=1}^{M'} i_j^{\text{nc}} \, \delta \lambda_j = \sum_{k=1}^{n'_e} \mathcal{I}_k \, \delta \lambda_k$$

Resistor (Linear)	$-R\dot{q}\,\delta q$	$-R^{-1}\dot{\lambda}\,\delta\lambda$
Voltage Source	$E(t)\,\delta q$	No Work Expression: KVR Admissibility
Current Source	No Work Expression: KCR Admissibility	$I(t)\,\delta\lambda$
Operational Amplifier	Akin to Voltage Source Plus Idealized Characteristics	Akin to Voltage Source Plus Idealized Characteristics

• *Equations of Motion*

$$\frac{d}{dt}\left(\frac{\partial \mathcal{L}}{\partial \dot{q}_k}\right) - \frac{\partial \mathcal{L}}{\partial q_k} = \mathcal{E}_k \qquad\qquad \frac{d}{dt}\left(\frac{\partial \mathcal{L}}{\partial \dot{\lambda}_k}\right) - \frac{\partial \mathcal{L}}{\partial \lambda_k} = \mathcal{I}_k$$

$$k = 1, 2, \ldots, n_e \qquad\qquad\qquad k = 1, 2, \ldots, n'_e$$

• *Linear Constitutive Relations*

Capacitor: $q = C\dot{\lambda}$ $\qquad$ Inductor: $\lambda = L\dot{q}$ $\qquad\qquad$ Resistor: $\dot{\lambda} = R\dot{q}$

• *Linear Op-Amp*

Voltage-Transfer Characteristic: $e_0 = A(e_+ - e_-)$

Summary of Responses for Single-Degree-of-Freedom First-Order System
$\tau \nu + \nu = f(t)$ Subjected to Various Excitations and Initial Condition $\nu(0) = \nu_0$

Forcing function. $f(t), t \geq 0$	Particular solution, $\nu_p(t)$	Homogeneous solution, $\nu_h(t)$
0 (No input)	—	$\nu_0 e^{-t/\tau}$
ν_f (constant) (Step input)	ν_f	$(\nu_0 - \dot{\nu}_f)e^{-t/\tau}$
at (a: constant) (Ramp input)	$at - \tau a$	$(\nu_0 + \tau a)e^{-t/\tau}$
$a\cos\Omega t + b\sin\Omega t$ (a, b, Ω: constant) (Harmonic input)	$\dfrac{a\cos(\Omega t - \phi) + b\sin(\Omega t - \phi)}{\sqrt{1 + \tau^2\Omega^2}}$ where $\phi = \tan^{-1}\tau\Omega$, $0 \leq \phi < \dfrac{\pi}{2}$	$\left\{\nu_0 - \dfrac{a\cos\phi - b\sin\phi}{\sqrt{1 + \tau^2\Omega^2}}\right\}e^{-t/\tau}$ where $\phi = \tan^{-1}\tau\Omega$, $0 \leq \phi < \dfrac{\pi}{2}$